Lecture Notes in Computer Science 12349

T0205247

More information about this series at http://www.springer.com/series/7412

Andrea Vedaldi · Horst Bischof ·
Thomas Brox · Jan-Michael Frahm (Eds.)

Computer Vision – ECCV 2020

16th European Conference
Glasgow, UK, August 23–28, 2020
Proceedings, Part IV

Springer

Editors
Andrea Vedaldi 🆔
University of Oxford
Oxford, UK

Thomas Brox 🆔
University of Freiburg
Freiburg im Breisgau, Germany

Horst Bischof 🆔
Graz University of Technology
Graz, Austria

Jan-Michael Frahm
University of North Carolina at Chapel Hill
Chapel Hill, NC, USA

ISSN 0302-9743   ISSN 1611-3349   (electronic)
Lecture Notes in Computer Science
ISBN 978-3-030-58547-1   ISBN 978-3-030-58548-8   (eBook)
https://doi.org/10.1007/978-3-030-58548-8

LNCS Sublibrary: SL6 – Image Processing, Computer Vision, Pattern Recognition, and Graphics

This Springer imprint is published by the registered company Springer Nature Switzerland AG
The registered company address is: Gewerbestrasse 11, 6330 Cham, Switzerland

Foreword

Hosting the European Conference on Computer Vision (ECCV 2020) was certainly an exciting journey. From the 2016 plan to hold it at the Edinburgh International Conference Centre (hosting 1,800 delegates) to the 2018 plan to hold it at Glasgow's Scottish Exhibition Centre (up to 6,000 delegates), we finally ended with moving online because of the COVID-19 outbreak. While possibly having fewer delegates than expected because of the online format, ECCV 2020 still had over 3,100 registered participants.

Although online, the conference delivered most of the activities expected at a face-to-face conference: peer-reviewed papers, industrial exhibitors, demonstrations, and messaging between delegates. In addition to the main technical sessions, the conference included a strong program of satellite events with 16 tutorials and 44 workshops.

Furthermore, the online conference format enabled new conference features. Every paper had an associated teaser video and a longer full presentation video. Along with the papers and slides from the videos, all these materials were available the week before the conference. This allowed delegates to become familiar with the paper content and be ready for the live interaction with the authors during the conference week. The live event consisted of brief presentations by the oral and spotlight authors and industrial sponsors. Question and answer sessions for all papers were timed to occur twice so delegates from around the world had convenient access to the authors.

As with ECCV 2018, authors' draft versions of the papers appeared online with open access, now on both the Computer Vision Foundation (CVF) and the European Computer Vision Association (ECVA) websites. An archival publication arrangement was put in place with the cooperation of Springer. SpringerLink hosts the final version of the papers with further improvements, such as activating reference links and supplementary materials. These two approaches benefit all potential readers: a version available freely for all researchers, and an authoritative and citable version with additional benefits for SpringerLink subscribers. We thank Alfred Hofmann and Aliaksandr Birukou from Springer for helping to negotiate this agreement, which we expect will continue for future versions of ECCV.

August 2020

Vittorio Ferrari
Bob Fisher
Cordelia Schmid
Emanuele Trucco

Preface

Welcome to the proceedings of the European Conference on Computer Vision (ECCV 2020). This is a unique edition of ECCV in many ways. Due to the COVID-19 pandemic, this is the first time the conference was held online, in a virtual format. This was also the first time the conference relied exclusively on the Open Review platform to manage the review process. Despite these challenges ECCV is thriving. The conference received 5,150 valid paper submissions, of which 1,360 were accepted for publication (27%) and, of those, 160 were presented as spotlights (3%) and 104 as orals (2%). This amounts to more than twice the number of submissions to ECCV 2018 (2,439). Furthermore, CVPR, the largest conference on computer vision, received 5,850 submissions this year, meaning that ECCV is now 87% the size of CVPR in terms of submissions. By comparison, in 2018 the size of ECCV was only 73% of CVPR.

The review model was similar to previous editions of ECCV; in particular, it was double blind in the sense that the authors did not know the name of the reviewers and vice versa. Furthermore, each conference submission was held confidentially, and was only publicly revealed if and once accepted for publication. Each paper received at least three reviews, totalling more than 15,000 reviews. Handling the review process at this scale was a significant challenge. In order to ensure that each submission received as fair and high-quality reviews as possible, we recruited 2,830 reviewers (a 130% increase with reference to 2018) and 207 area chairs (a 60% increase). The area chairs were selected based on their technical expertise and reputation, largely among people that served as area chair in previous top computer vision and machine learning conferences (ECCV, ICCV, CVPR, NeurIPS, etc.). Reviewers were similarly invited from previous conferences. We also encouraged experienced area chairs to suggest additional chairs and reviewers in the initial phase of recruiting.

Despite doubling the number of submissions, the reviewer load was slightly reduced from 2018, from a maximum of 8 papers down to 7 (with some reviewers offering to handle 6 papers plus an emergency review). The area chair load increased slightly, from 18 papers on average to 22 papers on average.

Conflicts of interest between authors, area chairs, and reviewers were handled largely automatically by the Open Review platform via their curated list of user profiles. Many authors submitting to ECCV already had a profile in Open Review. We set a paper registration deadline one week before the paper submission deadline in order to encourage all missing authors to register and create their Open Review profiles well on time (in practice, we allowed authors to create/change papers arbitrarily until the submission deadline). Except for minor issues with users creating duplicate profiles, this allowed us to easily and quickly identify institutional conflicts, and avoid them, while matching papers to area chairs and reviewers.

Papers were matched to area chairs based on: an affinity score computed by the Open Review platform, which is based on paper titles and abstracts, and an affinity score computed by the Toronto Paper Matching System (TPMS), which is based on the paper's full text, the area chair bids for individual papers, load balancing, and conflict avoidance. Open Review provides the program chairs a convenient web interface to experiment with different configurations of the matching algorithm. The chosen configuration resulted in about 50% of the assigned papers to be highly ranked by the area chair bids, and 50% to be ranked in the middle, with very few low bids assigned.

Assignments to reviewers were similar, with two differences. First, there was a maximum of 7 papers assigned to each reviewer. Second, area chairs recommended up to seven reviewers per paper, providing another highly-weighed term to the affinity scores used for matching.

The assignment of papers to area chairs was smooth. However, it was more difficult to find suitable reviewers for all papers. Having a ratio of 5.6 papers per reviewer with a maximum load of 7 (due to emergency reviewer commitment), which did not allow for much wiggle room in order to also satisfy conflict and expertise constraints. We received some complaints from reviewers who did not feel qualified to review specific papers and we reassigned them wherever possible. However, the large scale of the conference, the many constraints, and the fact that a large fraction of such complaints arrived very late in the review process made this process very difficult and not all complaints could be addressed.

Reviewers had six weeks to complete their assignments. Possibly due to COVID-19 or the fact that the NeurIPS deadline was moved closer to the review deadline, a record 30% of the reviews were still missing after the deadline. By comparison, ECCV 2018 experienced only 10% missing reviews at this stage of the process. In the subsequent week, area chairs chased the missing reviews intensely, found replacement reviewers in their own team, and managed to reach 10% missing reviews. Eventually, we could provide almost all reviews (more than 99.9%) with a delay of only a couple of days on the initial schedule by a significant use of emergency reviews. If this trend is confirmed, it might be a major challenge to run a smooth review process in future editions of ECCV. The community must reconsider prioritization of the time spent on paper writing (the number of submissions increased a lot despite COVID-19) and time spent on paper reviewing (the number of reviews delivered in time decreased a lot presumably due to COVID-19 or NeurIPS deadline). With this imbalance the peer-review system that ensures the quality of our top conferences may break soon.

Reviewers submitted their reviews independently. In the reviews, they had the opportunity to ask questions to the authors to be addressed in the rebuttal. However, reviewers were told not to request any significant new experiment. Using the Open Review interface, authors could provide an answer to each individual review, but were also allowed to cross-reference reviews and responses in their answers. Rather than PDF files, we allowed the use of formatted text for the rebuttal. The rebuttal and initial reviews were then made visible to all reviewers and the primary area chair for a given paper. The area chair encouraged and moderated the reviewer discussion. During the discussions, reviewers were invited to reach a consensus and possibly adjust their ratings as a result of the discussion and of the evidence in the rebuttal.

After the discussion period ended, most reviewers entered a final rating and recommendation, although in many cases this did not differ from their initial recommendation. Based on the updated reviews and discussion, the primary area chair then made a preliminary decision to accept or reject the paper and wrote a justification for it (meta-review). Except for cases where the outcome of this process was absolutely clear (as indicated by the three reviewers and primary area chairs all recommending clear rejection), the decision was then examined and potentially challenged by a secondary area chair. This led to further discussion and overturning a small number of preliminary decisions. Needless to say, there was no in-person area chair meeting, which would have been impossible due to COVID-19.

Area chairs were invited to observe the consensus of the reviewers whenever possible and use extreme caution in overturning a clear consensus to accept or reject a paper. If an area chair still decided to do so, she/he was asked to clearly justify it in the meta-review and to explicitly obtain the agreement of the secondary area chair. In practice, very few papers were rejected after being confidently accepted by the reviewers.

This was the first time Open Review was used as the main platform to run ECCV. In 2018, the program chairs used CMT3 for the user-facing interface and Open Review internally, for matching and conflict resolution. Since it is clearly preferable to only use a single platform, this year we switched to using Open Review in full. The experience was largely positive. The platform is highly-configurable, scalable, and open source. Being written in Python, it is easy to write scripts to extract data programmatically. The paper matching and conflict resolution algorithms and interfaces are top-notch, also due to the excellent author profiles in the platform. Naturally, there were a few kinks along the way due to the fact that the ECCV Open Review configuration was created from scratch for this event and it differs in substantial ways from many other Open Review conferences. However, the Open Review development and support team did a fantastic job in helping us to get the configuration right and to address issues in a timely manner as they unavoidably occurred. We cannot thank them enough for the tremendous effort they put into this project.

Finally, we would like to thank everyone involved in making ECCV 2020 possible in these very strange and difficult times. This starts with our authors, followed by the area chairs and reviewers, who ran the review process at an unprecedented scale. The whole Open Review team (and in particular Melisa Bok, Mohit Unyal, Carlos Mondragon Chapa, and Celeste Martinez Gomez) worked incredibly hard for the entire duration of the process. We would also like to thank René Vidal for contributing to the adoption of Open Review. Our thanks also go to Laurent Charling for TPMS and to the program chairs of ICML, ICLR, and NeurIPS for cross checking double submissions. We thank the website chair, Giovanni Farinella, and the CPI team (in particular Ashley Cook, Miriam Verdon, Nicola McGrane, and Sharon Kerr) for promptly adding material to the website as needed in the various phases of the process. Finally, we thank the publication chairs, Albert Ali Salah, Hamdi Dibeklioglu, Metehan Doyran, Henry Howard-Jenkins, Victor Prisacariu, Siyu Tang, and Gul Varol, who managed to compile these substantial proceedings in an exceedingly compressed schedule. We express our thanks to the ECVA team, in particular Kristina Scherbaum for allowing open access of the proceedings. We thank Alfred Hofmann from Springer who again serve as the publisher. Finally, we thank the other chairs of ECCV 2020, including in particular the general chairs for very useful feedback with the handling of the program.

August 2020

Andrea Vedaldi
Horst Bischof
Thomas Brox
Jan-Michael Frahm

Organization

General Chairs

Vittorio Ferrari	Google Research, Switzerland
Bob Fisher	University of Edinburgh, UK
Cordelia Schmid	Google and Inria, France
Emanuele Trucco	University of Dundee, UK

Program Chairs

Andrea Vedaldi	University of Oxford, UK
Horst Bischof	Graz University of Technology, Austria
Thomas Brox	University of Freiburg, Germany
Jan-Michael Frahm	University of North Carolina, USA

Industrial Liaison Chairs

Jim Ashe	University of Edinburgh, UK
Helmut Grabner	Zurich University of Applied Sciences, Switzerland
Diane Larlus	NAVER LABS Europe, France
Cristian Novotny	University of Edinburgh, UK

Local Arrangement Chairs

Yvan Petillot	Heriot-Watt University, UK
Paul Siebert	University of Glasgow, UK

Academic Demonstration Chair

Thomas Mensink	Google Research and University of Amsterdam, The Netherlands

Poster Chair

Stephen Mckenna	University of Dundee, UK

Technology Chair

Gerardo Aragon Camarasa	University of Glasgow, UK

Tutorial Chairs

Carlo Colombo   University of Florence, Italy
Sotirios Tsaftaris   University of Edinburgh, UK

Publication Chairs

Albert Ali Salah   Utrecht University, The Netherlands
Hamdi Dibeklioglu   Bilkent University, Turkey
Metehan Doyran   Utrecht University, The Netherlands
Henry Howard-Jenkins   University of Oxford, UK
Victor Adrian Prisacariu   University of Oxford, UK
Siyu Tang   ETH Zurich, Switzerland
Gul Varol   University of Oxford, UK

Website Chair

Giovanni Maria Farinella   University of Catania, Italy

Workshops Chairs

Adrien Bartoli   University of Clermont Auvergne, France
Andrea Fusiello   University of Udine, Italy

Area Chairs

Lourdes Agapito   University College London, UK
Zeynep Akata   University of Tübingen, Germany
Karteek Alahari   Inria, France
Antonis Argyros   University of Crete, Greece
Hossein Azizpour   KTH Royal Institute of Technology, Sweden
Joao P. Barreto   Universidade de Coimbra, Portugal
Alexander C. Berg   University of North Carolina at Chapel Hill, USA
Matthew B. Blaschko   KU Leuven, Belgium
Lubomir D. Bourdev   WaveOne, Inc., USA
Edmond Boyer   Inria, France
Yuri Boykov   University of Waterloo, Canada
Gabriel Brostow   University College London, UK
Michael S. Brown   National University of Singapore, Singapore
Jianfei Cai   Monash University, Australia
Barbara Caputo   Politecnico di Torino, Italy
Ayan Chakrabarti   Washington University, St. Louis, USA
Tat-Jen Cham   Nanyang Technological University, Singapore
Manmohan Chandraker   University of California, San Diego, USA
Rama Chellappa   Johns Hopkins University, USA
Liang-Chieh Chen   Google, USA

Yung-Yu Chuang	National Taiwan University, Taiwan
Ondrej Chum	Czech Technical University in Prague, Czech Republic
Brian Clipp	Kitware, USA
John Collomosse	University of Surrey and Adobe Research, UK
Jason J. Corso	University of Michigan, USA
David J. Crandall	Indiana University, USA
Daniel Cremers	University of California, Los Angeles, USA
Fabio Cuzzolin	Oxford Brookes University, UK
Jifeng Dai	SenseTime, SAR China
Kostas Daniilidis	University of Pennsylvania, USA
Andrew Davison	Imperial College London, UK
Alessio Del Bue	Fondazione Istituto Italiano di Tecnologia, Italy
Jia Deng	Princeton University, USA
Alexey Dosovitskiy	Google, Germany
Matthijs Douze	Facebook, France
Enrique Dunn	Stevens Institute of Technology, USA
Irfan Essa	Georgia Institute of Technology and Google, USA
Giovanni Maria Farinella	University of Catania, Italy
Ryan Farrell	Brigham Young University, USA
Paolo Favaro	University of Bern, Switzerland
Rogerio Feris	International Business Machines, USA
Cornelia Fermuller	University of Maryland, College Park, USA
David J. Fleet	Vector Institute, Canada
Friedrich Fraundorfer	DLR, Austria
Mario Fritz	CISPA Helmholtz Center for Information Security, Germany
Pascal Fua	EPFL (Swiss Federal Institute of Technology Lausanne), Switzerland
Yasutaka Furukawa	Simon Fraser University, Canada
Li Fuxin	Oregon State University, USA
Efstratios Gavves	University of Amsterdam, The Netherlands
Peter Vincent Gehler	Amazon, USA
Theo Gevers	University of Amsterdam, The Netherlands
Ross Girshick	Facebook AI Research, USA
Boqing Gong	Google, USA
Stephen Gould	Australian National University, Australia
Jinwei Gu	SenseTime Research, USA
Abhinav Gupta	Facebook, USA
Bohyung Han	Seoul National University, South Korea
Bharath Hariharan	Cornell University, USA
Tal Hassner	Facebook AI Research, USA
Xuming He	Australian National University, Australia
Joao F. Henriques	University of Oxford, UK
Adrian Hilton	University of Surrey, UK
Minh Hoai	Stony Brooks, State University of New York, USA
Derek Hoiem	University of Illinois Urbana-Champaign, USA

Timothy Hospedales	University of Edinburgh and Samsung, UK
Gang Hua	Wormpex AI Research, USA
Slobodan Ilic	Siemens AG, Germany
Hiroshi Ishikawa	Waseda University, Japan
Jiaya Jia	The Chinese University of Hong Kong, SAR China
Hailin Jin	Adobe Research, USA
Justin Johnson	University of Michigan, USA
Frederic Jurie	University of Caen Normandie, France
Fredrik Kahl	Chalmers University, Sweden
Sing Bing Kang	Zillow, USA
Gunhee Kim	Seoul National University, South Korea
Junmo Kim	Korea Advanced Institute of Science and Technology, South Korea
Tae-Kyun Kim	Imperial College London, UK
Ron Kimmel	Technion-Israel Institute of Technology, Israel
Alexander Kirillov	Facebook AI Research, USA
Kris Kitani	Carnegie Mellon University, USA
Iasonas Kokkinos	Ariel AI, UK
Vladlen Koltun	Intel Labs, USA
Nikos Komodakis	Ecole des Ponts ParisTech, France
Piotr Koniusz	Australian National University, Australia
M. Pawan Kumar	University of Oxford, UK
Kyros Kutulakos	University of Toronto, Canada
Christoph Lampert	IST Austria, Austria
Ivan Laptev	Inria, France
Diane Larlus	NAVER LABS Europe, France
Laura Leal-Taixe	Technical University Munich, Germany
Honglak Lee	Google and University of Michigan, USA
Joon-Young Lee	Adobe Research, USA
Kyoung Mu Lee	Seoul National University, South Korea
Seungyong Lee	POSTECH, South Korea
Yong Jae Lee	University of California, Davis, USA
Bastian Leibe	RWTH Aachen University, Germany
Victor Lempitsky	Samsung, Russia
Ales Leonardis	University of Birmingham, UK
Marius Leordeanu	Institute of Mathematics of the Romanian Academy, Romania
Vincent Lepetit	ENPC ParisTech, France
Hongdong Li	The Australian National University, Australia
Xi Li	Zhejiang University, China
Yin Li	University of Wisconsin-Madison, USA
Zicheng Liao	Zhejiang University, China
Jongwoo Lim	Hanyang University, South Korea
Stephen Lin	Microsoft Research Asia, China
Yen-Yu Lin	National Chiao Tung University, Taiwan, China
Zhe Lin	Adobe Research, USA

Haibin Ling	Stony Brooks, State University of New York, USA
Jiaying Liu	Peking University, China
Ming-Yu Liu	NVIDIA, USA
Si Liu	Beihang University, China
Xiaoming Liu	Michigan State University, USA
Huchuan Lu	Dalian University of Technology, China
Simon Lucey	Carnegie Mellon University, USA
Jiebo Luo	University of Rochester, USA
Julien Mairal	Inria, France
Michael Maire	University of Chicago, USA
Subhransu Maji	University of Massachusetts, Amherst, USA
Yasushi Makihara	Osaka University, Japan
Jiri Matas	Czech Technical University in Prague, Czech Republic
Yasuyuki Matsushita	Osaka University, Japan
Philippos Mordohai	Stevens Institute of Technology, USA
Vittorio Murino	University of Verona, Italy
Naila Murray	NAVER LABS Europe, France
Hajime Nagahara	Osaka University, Japan
P. J. Narayanan	International Institute of Information Technology (IIIT), Hyderabad, India
Nassir Navab	Technical University of Munich, Germany
Natalia Neverova	Facebook AI Research, France
Matthias Niessner	Technical University of Munich, Germany
Jean-Marc Odobez	Idiap Research Institute and Swiss Federal Institute of Technology Lausanne, Switzerland
Francesca Odone	Università di Genova, Italy
Takeshi Oishi	The University of Tokyo, Tokyo Institute of Technology, Japan
Vicente Ordonez	University of Virginia, USA
Manohar Paluri	Facebook AI Research, USA
Maja Pantic	Imperial College London, UK
In Kyu Park	Inha University, South Korea
Ioannis Patras	Queen Mary University of London, UK
Patrick Perez	Valeo, France
Bryan A. Plummer	Boston University, USA
Thomas Pock	Graz University of Technology, Austria
Marc Pollefeys	ETH Zurich and Microsoft MR & AI Zurich Lab, Switzerland
Jean Ponce	Inria, France
Gerard Pons-Moll	MPII, Saarland Informatics Campus, Germany
Jordi Pont-Tuset	Google, Switzerland
James Matthew Rehg	Georgia Institute of Technology, USA
Ian Reid	University of Adelaide, Australia
Olaf Ronneberger	DeepMind London, UK
Stefan Roth	TU Darmstadt, Germany
Bryan Russell	Adobe Research, USA

Kwang Moo Yi   University of Victoria, Canada
Zhaozheng Yin   Stony Brook, State University of New York, USA
Chang D. Yoo   Korea Advanced Institute of Science and Technology,
                              South Korea
Shaodi You   University of Amsterdam, The Netherlands
Jingyi Yu   ShanghaiTech University, China
Stella Yu   University of California, Berkeley, and ICSI, USA
Stefanos Zafeiriou   Imperial College London, UK
Hongbin Zha   Peking University, China
Tianzhu Zhang   University of Science and Technology of China, China
Liang Zheng   Australian National University, Australia
Todd E. Zickler   Harvard University, USA
Andrew Zisserman   University of Oxford, UK

Technical Program Committee

Sathyanarayanan
  N. Aakur
Wael Abd Almgaeed
Abdelrahman
  Abdelhamed
Abdullah Abuolaim
Supreeth Achar
Hanno Ackermann
Ehsan Adeli
Triantafyllos Afouras
Sameer Agarwal
Aishwarya Agrawal
Harsh Agrawal
Pulkit Agrawal
Antonio Agudo
Eirikur Agustsson
Karim Ahmed
Byeongjoo Ahn
Unaiza Ahsan
Thalaiyasingam Ajanthan
Kenan E. Ak
Emre Akbas
Naveed Akhtar
Derya Akkaynak
Yagiz Aksoy
Ziad Al-Halah
Xavier Alameda-Pineda
Jean-Baptiste Alayrac

Samuel Albanie
Shadi Albarqouni
Cenek Albl
Hassan Abu Alhaija
Daniel Aliaga
Mohammad
  S. Aliakbarian
Rahaf Aljundi
Thiemo Alldieck
Jon Almazan
Jose M. Alvarez
Senjian An
Saket Anand
Codruta Ancuti
Cosmin Ancuti
Peter Anderson
Juan Andrade-Cetto
Alexander Andreopoulos
Misha Andriluka
Dragomir Anguelov
Rushil Anirudh
Michel Antunes
Oisin Mac Aodha
Srikar Appalaraju
Relja Arandjelovic
Nikita Araslanov
Andre Araujo
Helder Araujo

Pablo Arbelaez
Shervin Ardeshir
Sercan O. Arik
Anil Armagan
Anurag Arnab
Chetan Arora
Federica Arrigoni
Mathieu Aubry
Shai Avidan
Angelica I. Aviles-Rivero
Yannis Avrithis
Ismail Ben Ayed
Shekoofeh Azizi
Ioan Andrei Bârsan
Artem Babenko
Deepak Babu Sam
Seung-Hwan Baek
Seungryul Baek
Andrew D. Bagdanov
Shai Bagon
Yuval Bahat
Junjie Bai
Song Bai
Xiang Bai
Yalong Bai
Yancheng Bai
Peter Bajcsy
Slawomir Bak

Mahsa Baktashmotlagh
Kavita Bala
Yogesh Balaji
Guha Balakrishnan
V. N. Balasubramanian
Federico Baldassarre
Vassileios Balntas
Shurjo Banerjee
Aayush Bansal
Ankan Bansal
Jianmin Bao
Linchao Bao
Wenbo Bao
Yingze Bao
Akash Bapat
Md Jawadul Hasan Bappy
Fabien Baradel
Lorenzo Baraldi
Daniel Barath
Adrian Barbu
Kobus Barnard
Nick Barnes
Francisco Barranco
Jonathan T. Barron
Arslan Basharat
Chaim Baskin
Anil S. Baslamisli
Jorge Batista
Kayhan Batmanghelich
Konstantinos Batsos
David Bau
Luis Baumela
Christoph Baur
Eduardo
  Bayro-Corrochano
Paul Beardsley
Jan Bednavr'ik
Oscar Beijbom
Philippe Bekaert
Esube Bekele
Vasileios Belagiannis
Ohad Ben-Shahar
Abhijit Bendale
Róger Bermúdez-Chacón
Maxim Berman
Jesus Bermudez-cameo

Florian Bernard
Stefano Berretti
Marcelo Bertalmio
Gedas Bertasius
Cigdem Beyan
Lucas Beyer
Vijayakumar Bhagavatula
Arjun Nitin Bhagoji
Apratim Bhattacharyya
Binod Bhattarai
Sai Bi
Jia-Wang Bian
Simone Bianco
Adel Bibi
Tolga Birdal
Tom Bishop
Soma Biswas
Mårten Björkman
Volker Blanz
Vishnu Boddeti
Navaneeth Bodla
Simion-Vlad Bogolin
Xavier Boix
Piotr Bojanowski
Timo Bolkart
Guido Borghi
Larbi Boubchir
Guillaume Bourmaud
Adrien Bousseau
Thierry Bouwmans
Richard Bowden
Hakan Boyraz
Mathieu Brédif
Samarth Brahmbhatt
Steve Branson
Nikolas Brasch
Biagio Brattoli
Ernesto Brau
Toby P. Breckon
Francois Bremond
Jesus Briales
Sofia Broomé
Marcus A. Brubaker
Luc Brun
Silvia Bucci
Shyamal Buch

Pradeep Buddharaju
Uta Buechler
Mai Bui
Tu Bui
Adrian Bulat
Giedrius T. Burachas
Elena Burceanu
Xavier P. Burgos-Artizzu
Kaylee Burns
Andrei Bursuc
Benjamin Busam
Wonmin Byeon
Zoya Bylinskii
Sergi Caelles
Jianrui Cai
Minjie Cai
Yujun Cai
Zhaowei Cai
Zhipeng Cai
Juan C. Caicedo
Simone Calderara
Necati Cihan Camgoz
Dylan Campbell
Octavia Camps
Jiale Cao
Kaidi Cao
Liangliang Cao
Xiangyong Cao
Xiaochun Cao
Yang Cao
Yu Cao
Yue Cao
Zhangjie Cao
Luca Carlone
Mathilde Caron
Dan Casas
Thomas J. Cashman
Umberto Castellani
Lluis Castrejon
Jacopo Cavazza
Fabio Cermelli
Hakan Cevikalp
Menglei Chai
Ishani Chakraborty
Rudrasis Chakraborty
Antoni B. Chan

Kwok-Ping Chan
Siddhartha Chandra
Sharat Chandran
Arjun Chandrasekaran
Angel X. Chang
Che-Han Chang
Hong Chang
Hyun Sung Chang
Hyung Jin Chang
Jianlong Chang
Ju Yong Chang
Ming-Ching Chang
Simyung Chang
Xiaojun Chang
Yu-Wei Chao
Devendra S. Chaplot
Arslan Chaudhry
Rizwan A. Chaudhry
Can Chen
Chang Chen
Chao Chen
Chen Chen
Chu-Song Chen
Dapeng Chen
Dong Chen
Dongdong Chen
Guanying Chen
Hongge Chen
Hsin-yi Chen
Huaijin Chen
Hwann-Tzong Chen
Jianbo Chen
Jianhui Chen
Jiansheng Chen
Jiaxin Chen
Jie Chen
Jun-Cheng Chen
Kan Chen
Kevin Chen
Lin Chen
Long Chen
Min-Hung Chen
Qifeng Chen
Shi Chen
Shixing Chen
Tianshui Chen

Weifeng Chen
Weikai Chen
Xi Chen
Xiaohan Chen
Xiaozhi Chen
Xilin Chen
Xingyu Chen
Xinlei Chen
Xinyun Chen
Yi-Ting Chen
Yilun Chen
Ying-Cong Chen
Yinpeng Chen
Yiran Chen
Yu Chen
Yu-Sheng Chen
Yuhua Chen
Yun-Chun Chen
Yunpeng Chen
Yuntao Chen
Zhuoyuan Chen
Zitian Chen
Anchieh Cheng
Bowen Cheng
Erkang Cheng
Gong Cheng
Guangliang Cheng
Jingchun Cheng
Jun Cheng
Li Cheng
Ming-Ming Cheng
Yu Cheng
Ziang Cheng
Anoop Cherian
Dmitry Chetverikov
Ngai-man Cheung
William Cheung
Ajad Chhatkuli
Naoki Chiba
Benjamin Chidester
Han-pang Chiu
Mang Tik Chiu
Wei-Chen Chiu
Donghyeon Cho
Hojin Cho
Minsu Cho

Nam Ik Cho
Tim Cho
Tae Eun Choe
Chiho Choi
Edward Choi
Inchang Choi
Jinsoo Choi
Jonghyun Choi
Jongwon Choi
Yukyung Choi
Hisham Cholakkal
Eunji Chong
Jaegul Choo
Christopher Choy
Hang Chu
Peng Chu
Wen-Sheng Chu
Albert Chung
Joon Son Chung
Hai Ci
Safa Cicek
Ramazan G. Cinbis
Arridhana Ciptadi
Javier Civera
James J. Clark
Ronald Clark
Felipe Codevilla
Michael Cogswell
Andrea Cohen
Maxwell D. Collins
Carlo Colombo
Yang Cong
Adria R. Continente
Marcella Cornia
John Richard Corring
Darren Cosker
Dragos Costea
Garrison W. Cottrell
Florent Couzinie-Devy
Marco Cristani
Ioana Croitoru
James L. Crowley
Jiequan Cui
Zhaopeng Cui
Ross Cutler
Antonio D'Innocente

Rozenn Dahyot
Bo Dai
Dengxin Dai
Hang Dai
Longquan Dai
Shuyang Dai
Xiyang Dai
Yuchao Dai
Adrian V. Dalca
Dima Damen
Bharath B. Damodaran
Kristin Dana
Martin Danelljan
Zheng Dang
Zachary Alan Daniels
Donald G. Dansereau
Abhishek Das
Samyak Datta
Achal Dave
Titas De
Rodrigo de Bem
Teo de Campos
Raoul de Charette
Shalini De Mello
Joseph DeGol
Herve Delingette
Haowen Deng
Jiankang Deng
Weijian Deng
Zhiwei Deng
Joachim Denzler
Konstantinos G. Derpanis
Aditya Deshpande
Frederic Devernay
Somdip Dey
Arturo Deza
Abhinav Dhall
Helisa Dhamo
Vikas Dhiman
Fillipe Dias Moreira
  de Souza
Ali Diba
Ferran Diego
Guiguang Ding
Henghui Ding
Jian Ding

Mingyu Ding
Xinghao Ding
Zhengming Ding
Robert DiPietro
Cosimo Distante
Ajay Divakaran
Mandar Dixit
Abdelaziz Djelouah
Thanh-Toan Do
Jose Dolz
Bo Dong
Chao Dong
Jiangxin Dong
Weiming Dong
Weisheng Dong
Xingping Dong
Xuanyi Dong
Yinpeng Dong
Gianfranco Doretto
Hazel Doughty
Hassen Drira
Bertram Drost
Dawei Du
Ye Duan
Yueqi Duan
Abhimanyu Dubey
Anastasia Dubrovina
Stefan Duffner
Chi Nhan Duong
Thibaut Durand
Zoran Duric
Iulia Duta
Debidatta Dwibedi
Benjamin Eckart
Marc Eder
Marzieh Edraki
Alexei A. Efros
Kiana Ehsani
Hazm Kemal Ekenel
James H. Elder
Mohamed Elgharib
Shireen Elhabian
Ehsan Elhamifar
Mohamed Elhoseiny
Ian Endres
N. Benjamin Erichson

Jan Ernst
Sergio Escalera
Francisco Escolano
Victor Escorcia
Carlos Esteves
Francisco J. Estrada
Bin Fan
Chenyou Fan
Deng-Ping Fan
Haoqi Fan
Hehe Fan
Heng Fan
Kai Fan
Lijie Fan
Linxi Fan
Quanfu Fan
Shaojing Fan
Xiaochuan Fan
Xin Fan
Yuchen Fan
Sean Fanello
Hao-Shu Fang
Haoyang Fang
Kuan Fang
Yi Fang
Yuming Fang
Azade Farshad
Alireza Fathi
Raanan Fattal
Joao Fayad
Xiaohan Fei
Christoph Feichtenhofer
Michael Felsberg
Chen Feng
Jiashi Feng
Junyi Feng
Mengyang Feng
Qianli Feng
Zhenhua Feng
Michele Fenzi
Andras Ferencz
Martin Fergie
Basura Fernando
Ethan Fetaya
Michael Firman
John W. Fisher

Matthew Fisher
Boris Flach
Corneliu Florea
Wolfgang Foerstner
David Fofi
Gian Luca Foresti
Per-Erik Forssen
David Fouhey
Katerina Fragkiadaki
Victor Fragoso
Jean-Sébastien Franco
Ohad Fried
Iuri Frosio
Cheng-Yang Fu
Huazhu Fu
Jianlong Fu
Jingjing Fu
Xueyang Fu
Yanwei Fu
Ying Fu
Yun Fu
Olac Fuentes
Kent Fujiwara
Takuya Funatomi
Christopher Funk
Thomas Funkhouser
Antonino Furnari
Ryo Furukawa
Erik Gärtner
Raghudeep Gadde
Matheus Gadelha
Vandit Gajjar
Trevor Gale
Juergen Gall
Mathias Gallardo
Guillermo Gallego
Orazio Gallo
Chuang Gan
Zhe Gan
Madan Ravi Ganesh
Aditya Ganeshan
Siddha Ganju
Bin-Bin Gao
Changxin Gao
Feng Gao
Hongchang Gao

Jin Gao
Jiyang Gao
Junbin Gao
Katelyn Gao
Lin Gao
Mingfei Gao
Ruiqi Gao
Ruohan Gao
Shenghua Gao
Yuan Gao
Yue Gao
Noa Garcia
Alberto Garcia-Garcia
Guillermo
 Garcia-Hernando
Jacob R. Gardner
Animesh Garg
Kshitiz Garg
Rahul Garg
Ravi Garg
Philip N. Garner
Kirill Gavrilyuk
Paul Gay
Shiming Ge
Weifeng Ge
Baris Gecer
Xin Geng
Kyle Genova
Stamatios Georgoulis
Bernard Ghanem
Michael Gharbi
Kamran Ghasedi
Golnaz Ghiasi
Arnab Ghosh
Partha Ghosh
Silvio Giancola
Andrew Gilbert
Rohit Girdhar
Xavier Giro-i-Nieto
Thomas Gittings
Ioannis Gkioulekas
Clement Godard
Vaibhava Goel
Bastian Goldluecke
Lluis Gomez
Nuno Gonçalves

Dong Gong
Ke Gong
Mingming Gong
Abel Gonzalez-Garcia
Ariel Gordon
Daniel Gordon
Paulo Gotardo
Venu Madhav Govindu
Ankit Goyal
Priya Goyal
Raghav Goyal
Benjamin Graham
Douglas Gray
Brent A. Griffin
Etienne Grossmann
David Gu
Jiayuan Gu
Jiuxiang Gu
Lin Gu
Qiao Gu
Shuhang Gu
Jose J. Guerrero
Paul Guerrero
Jie Gui
Jean-Yves Guillemaut
Riza Alp Guler
Erhan Gundogdu
Fatma Guney
Guodong Guo
Kaiwen Guo
Qi Guo
Sheng Guo
Shi Guo
Tiantong Guo
Xiaojie Guo
Yijie Guo
Yiluan Guo
Yuanfang Guo
Yulan Guo
Agrim Gupta
Ankush Gupta
Mohit Gupta
Saurabh Gupta
Tanmay Gupta
Danna Gurari
Abner Guzman-Rivera

JunYoung Gwak
Michael Gygli
Jung-Woo Ha
Simon Hadfield
Isma Hadji
Bjoern Haefner
Taeyoung Hahn
Levente Hajder
Peter Hall
Emanuela Haller
Stefan Haller
Bumsub Ham
Abdullah Hamdi
Dongyoon Han
Hu Han
Jungong Han
Junwei Han
Kai Han
Tian Han
Xiaoguang Han
Xintong Han
Yahong Han
Ankur Handa
Zekun Hao
Albert Haque
Tatsuya Harada
Mehrtash Harandi
Adam W. Harley
Mahmudul Hasan
Atsushi Hashimoto
Ali Hatamizadeh
Munawar Hayat
Dongliang He
Jingrui He
Junfeng He
Kaiming He
Kun He
Lei He
Pan He
Ran He
Shengfeng He
Tong He
Weipeng He
Xuming He
Yang He
Yihui He

Zhihai He
Chinmay Hegde
Janne Heikkila
Mattias P. Heinrich
Stéphane Herbin
Alexander Hermans
Luis Herranz
John R. Hershey
Aaron Hertzmann
Roei Herzig
Anders Heyden
Steven Hickson
Otmar Hilliges
Tomas Hodan
Judy Hoffman
Michael Hofmann
Yannick Hold-Geoffroy
Namdar Homayounfar
Sina Honari
Richang Hong
Seunghoon Hong
Xiaopeng Hong
Yi Hong
Hidekata Hontani
Anthony Hoogs
Yedid Hoshen
Mir Rayat Imtiaz Hossain
Junhui Hou
Le Hou
Lu Hou
Tingbo Hou
Wei-Lin Hsiao
Cheng-Chun Hsu
Gee-Sern Jison Hsu
Kuang-jui Hsu
Changbo Hu
Di Hu
Guosheng Hu
Han Hu
Hao Hu
Hexiang Hu
Hou-Ning Hu
Jie Hu
Junlin Hu
Nan Hu
Ping Hu

Ronghang Hu
Xiaowei Hu
Yinlin Hu
Yuan-Ting Hu
Zhe Hu
Binh-Son Hua
Yang Hua
Bingyao Huang
Di Huang
Dong Huang
Fay Huang
Haibin Huang
Haozhi Huang
Heng Huang
Huaibo Huang
Jia-Bin Huang
Jing Huang
Jingwei Huang
Kaizhu Huang
Lei Huang
Qiangui Huang
Qiaoying Huang
Qingqiu Huang
Qixing Huang
Shaoli Huang
Sheng Huang
Siyuan Huang
Weilin Huang
Wenbing Huang
Xiangru Huang
Xun Huang
Yan Huang
Yifei Huang
Yue Huang
Zhiwu Huang
Zilong Huang
Minyoung Huh
Zhuo Hui
Matthias B. Hullin
Martin Humenberger
Wei-Chih Hung
Zhouyuan Huo
Junhwa Hur
Noureldien Hussein
Jyh-Jing Hwang
Seong Jae Hwang

Sung Ju Hwang
Ichiro Ide
Ivo Ihrke
Daiki Ikami
Satoshi Ikehata
Nazli Ikizler-Cinbis
Sunghoon Im
Yani Ioannou
Radu Tudor Ionescu
Umar Iqbal
Go Irie
Ahmet Iscen
Md Amirul Islam
Vamsi Ithapu
Nathan Jacobs
Arpit Jain
Himalaya Jain
Suyog Jain
Stuart James
Won-Dong Jang
Yunseok Jang
Ronnachai Jaroensri
Dinesh Jayaraman
Sadeep Jayasumana
Suren Jayasuriya
Herve Jegou
Simon Jenni
Hae-Gon Jeon
Yunho Jeon
Koteswar R. Jerripothula
Hueihan Jhuang
I-hong Jhuo
Dinghuang Ji
Hui Ji
Jingwei Ji
Pan Ji
Yanli Ji
Baoxiong Jia
Kui Jia
Xu Jia
Chiyu Max Jiang
Haiyong Jiang
Hao Jiang
Huaizu Jiang
Huajie Jiang
Ke Jiang

Lai Jiang
Li Jiang
Lu Jiang
Ming Jiang
Peng Jiang
Shuqiang Jiang
Wei Jiang
Xudong Jiang
Zhuolin Jiang
Jianbo Jiao
Zequn Jie
Dakai Jin
Kyong Hwan Jin
Lianwen Jin
SouYoung Jin
Xiaojie Jin
Xin Jin
Nebojsa Jojic
Alexis Joly
Michael Jeffrey Jones
Hanbyul Joo
Jungseock Joo
Kyungdon Joo
Ajjen Joshi
Shantanu H. Joshi
Da-Cheng Juan
Marco Körner
Kevin Köser
Asim Kadav
Christine Kaeser-Chen
Kushal Kafle
Dagmar Kainmueller
Ioannis A. Kakadiaris
Zdenek Kalal
Nima Kalantari
Yannis Kalantidis
Mahdi M. Kalayeh
Anmol Kalia
Sinan Kalkan
Vicky Kalogeiton
Ashwin Kalyan
Joni-kristian Kamarainen
Gerda Kamberova
Chandra Kambhamettu
Martin Kampel
Meina Kan

Christopher Kanan
Kenichi Kanatani
Angjoo Kanazawa
Atsushi Kanehira
Takuhiro Kaneko
Asako Kanezaki
Bingyi Kang
Di Kang
Sunghun Kang
Zhao Kang
Vadim Kantorov
Abhishek Kar
Amlan Kar
Theofanis Karaletsos
Leonid Karlinsky
Kevin Karsch
Angelos Katharopoulos
Isinsu Katircioglu
Hiroharu Kato
Zoltan Kato
Dotan Kaufman
Jan Kautz
Rei Kawakami
Qiuhong Ke
Wadim Kehl
Petr Kellnhofer
Aniruddha Kembhavi
Cem Keskin
Margret Keuper
Daniel Keysers
Ashkan Khakzar
Fahad Khan
Naeemullah Khan
Salman Khan
Siddhesh Khandelwal
Rawal Khirodkar
Anna Khoreva
Tejas Khot
Parmeshwar Khurd
Hadi Kiapour
Joe Kileel
Chanho Kim
Dahun Kim
Edward Kim
Eunwoo Kim
Han-ul Kim

Hansung Kim
Heewon Kim
Hyo Jin Kim
Hyunwoo J. Kim
Jinkyu Kim
Jiwon Kim
Jongmin Kim
Junsik Kim
Junyeong Kim
Min H. Kim
Namil Kim
Pyojin Kim
Seon Joo Kim
Seong Tae Kim
Seungryong Kim
Sungwoong Kim
Tae Hyun Kim
Vladimir Kim
Won Hwa Kim
Yonghyun Kim
Benjamin Kimia
Akisato Kimura
Pieter-Jan Kindermans
Zsolt Kira
Itaru Kitahara
Hedvig Kjellstrom
Jan Knopp
Takumi Kobayashi
Erich Kobler
Parker Koch
Reinhard Koch
Elyor Kodirov
Amir Kolaman
Nicholas Kolkin
Dimitrios Kollias
Stefanos Kollias
Soheil Kolouri
Adams Wai-Kin Kong
Naejin Kong
Shu Kong
Tao Kong
Yu Kong
Yoshinori Konishi
Daniil Kononenko
Theodora Kontogianni
Simon Korman

Adam Kortylewski
Jana Kosecka
Jean Kossaifi
Satwik Kottur
Rigas Kouskouridas
Adriana Kovashka
Rama Kovvuri
Adarsh Kowdle
Jedrzej Kozerawski
Mateusz Kozinski
Philipp Kraehenbuehl
Gregory Kramida
Josip Krapac
Dmitry Kravchenko
Ranjay Krishna
Pavel Krsek
Alexander Krull
Jakob Kruse
Hiroyuki Kubo
Hilde Kuehne
Jason Kuen
Andreas Kuhn
Arjan Kuijper
Zuzana Kukelova
Ajay Kumar
Amit Kumar
Avinash Kumar
Suryansh Kumar
Vijay Kumar
Kaustav Kundu
Weicheng Kuo
Nojun Kwak
Suha Kwak
Junseok Kwon
Nikolaos Kyriazis
Zorah Lähner
Ankit Laddha
Florent Lafarge
Jean Lahoud
Kevin Lai
Shang-Hong Lai
Wei-Sheng Lai
Yu-Kun Lai
Iro Laina
Antony Lam
John Wheatley Lambert

Xiangyuan lan
Xu Lan
Charis Lanaras
Georg Langs
Oswald Lanz
Dong Lao
Yizhen Lao
Agata Lapedriza
Gustav Larsson
Viktor Larsson
Katrin Lasinger
Christoph Lassner
Longin Jan Latecki
Stéphane Lathuilière
Rynson Lau
Hei Law
Justin Lazarow
Svetlana Lazebnik
Hieu Le
Huu Le
Ngan Hoang Le
Trung-Nghia Le
Vuong Le
Colin Lea
Erik Learned-Miller
Chen-Yu Lee
Gim Hee Lee
Hsin-Ying Lee
Hyungtae Lee
Jae-Han Lee
Jimmy Addison Lee
Joonseok Lee
Kibok Lee
Kuang-Huei Lee
Kwonjoon Lee
Minsik Lee
Sang-chul Lee
Seungkyu Lee
Soochan Lee
Stefan Lee
Taehee Lee
Andreas Lehrmann
Jie Lei
Peng Lei
Matthew Joseph Leotta
Wee Kheng Leow

Gil Levi
Evgeny Levinkov
Aviad Levis
Jose Lezama
Ang Li
Bin Li
Bing Li
Boyi Li
Changsheng Li
Chao Li
Chen Li
Cheng Li
Chenglong Li
Chi Li
Chun-Guang Li
Chun-Liang Li
Chunyuan Li
Dong Li
Guanbin Li
Hao Li
Haoxiang Li
Hongsheng Li
Hongyang Li
Houqiang Li
Huibin Li
Jia Li
Jianan Li
Jianguo Li
Junnan Li
Junxuan Li
Kai Li
Ke Li
Kejie Li
Kunpeng Li
Lerenhan Li
Li Erran Li
Mengtian Li
Mu Li
Peihua Li
Peiyi Li
Ping Li
Qi Li
Qing Li
Ruiyu Li
Ruoteng Li
Shaozi Li

Sheng Li
Shiwei Li
Shuang Li
Siyang Li
Stan Z. Li
Tianye Li
Wei Li
Weixin Li
Wen Li
Wenbo Li
Xiaomeng Li
Xin Li
Xiu Li
Xuelong Li
Xueting Li
Yan Li
Yandong Li
Yanghao Li
Yehao Li
Yi Li
Yijun Li
Yikang LI
Yining Li
Yongjie Li
Yu Li
Yu-Jhe Li
Yunpeng Li
Yunsheng Li
Yunzhu Li
Zhe Li
Zhen Li
Zhengqi Li
Zhenyang Li
Zhuwen Li
Dongze Lian
Xiaochen Lian
Zhouhui Lian
Chen Liang
Jie Liang
Ming Liang
Paul Pu Liang
Pengpeng Liang
Shu Liang
Wei Liang
Jing Liao
Minghui Liao

Renjie Liao
Shengcai Liao
Shuai Liao
Yiyi Liao
Ser-Nam Lim
Chen-Hsuan Lin
Chung-Ching Lin
Dahua Lin
Ji Lin
Kevin Lin
Tianwei Lin
Tsung-Yi Lin
Tsung-Yu Lin
Wei-An Lin
Weiyao Lin
Yen-Chen Lin
Yuewei Lin
David B. Lindell
Drew Linsley
Krzysztof Lis
Roee Litman
Jim Little
An-An Liu
Bo Liu
Buyu Liu
Chao Liu
Chen Liu
Cheng-lin Liu
Chenxi Liu
Dong Liu
Feng Liu
Guilin Liu
Haomiao Liu
Heshan Liu
Hong Liu
Ji Liu
Jingen Liu
Jun Liu
Lanlan Liu
Li Liu
Liu Liu
Mengyuan Liu
Miaomiao Liu
Nian Liu
Ping Liu
Risheng Liu

Sheng Liu
Shu Liu
Shuaicheng Liu
Sifei Liu
Siqi Liu
Siying Liu
Songtao Liu
Ting Liu
Tongliang Liu
Tyng-Luh Liu
Wanquan Liu
Wei Liu
Weiyang Liu
Weizhe Liu
Wenyu Liu
Wu Liu
Xialei Liu
Xianglong Liu
Xiaodong Liu
Xiaofeng Liu
Xihui Liu
Xingyu Liu
Xinwang Liu
Xuanqing Liu
Xuebo Liu
Yang Liu
Yaojie Liu
Yebin Liu
Yen-Cheng Liu
Yiming Liu
Yu Liu
Yu-Shen Liu
Yufan Liu
Yun Liu
Zheng Liu
Zhijian Liu
Zhuang Liu
Zichuan Liu
Ziwei Liu
Zongyi Liu
Stephan Liwicki
Liliana Lo Presti
Chengjiang Long
Fuchen Long
Mingsheng Long
Xiang Long

Yang Long
Charles T. Loop
Antonio Lopez
Roberto J. Lopez-Sastre
Javier Lorenzo-Navarro
Manolis Lourakis
Boyu Lu
Canyi Lu
Feng Lu
Guoyu Lu
Hongtao Lu
Jiajun Lu
Jiasen Lu
Jiwen Lu
Kaiyue Lu
Le Lu
Shao-Ping Lu
Shijian Lu
Xiankai Lu
Xin Lu
Yao Lu
Yiping Lu
Yongxi Lu
Yongyi Lu
Zhiwu Lu
Fujun Luan
Benjamin E. Lundell
Hao Luo
Jian-Hao Luo
Ruotian Luo
Weixin Luo
Wenhan Luo
Wenjie Luo
Yan Luo
Zelun Luo
Zixin Luo
Khoa Luu
Zhaoyang Lv
Pengyuan Lyu
Thomas Möllenhoff
Matthias Müller
Bingpeng Ma
Chih-Yao Ma
Chongyang Ma
Huimin Ma
Jiayi Ma

K. T. Ma
Ke Ma
Lin Ma
Liqian Ma
Shugao Ma
Wei-Chiu Ma
Xiaojian Ma
Xingjun Ma
Zhanyu Ma
Zheng Ma
Radek Jakob Mackowiak
Ludovic Magerand
Shweta Mahajan
Siddharth Mahendran
Long Mai
Ameesh Makadia
Oscar Mendez Maldonado
Mateusz Malinowski
Yury Malkov
Arun Mallya
Dipu Manandhar
Massimiliano Mancini
Fabian Manhardt
Kevis-kokitsi Maninis
Varun Manjunatha
Junhua Mao
Xudong Mao
Alina Marcu
Edgar Margffoy-Tuay
Dmitrii Marin
Manuel J. Marin-Jimenez
Kenneth Marino
Niki Martinel
Julieta Martinez
Jonathan Masci
Tomohiro Mashita
Iacopo Masi
David Masip
Daniela Massiceti
Stefan Mathe
Yusuke Matsui
Tetsu Matsukawa
Iain A. Matthews
Kevin James Matzen
Bruce Allen Maxwell
Stephen Maybank

Helmut Mayer
Amir Mazaheri
David McAllester
Steven McDonagh
Stephen J. Mckenna
Roey Mechrez
Prakhar Mehrotra
Christopher Mei
Xue Mei
Paulo R. S. Mendonca
Lili Meng
Zibo Meng
Thomas Mensink
Bjoern Menze
Michele Merler
Kourosh Meshgi
Pascal Mettes
Christopher Metzler
Liang Mi
Qiguang Miao
Xin Miao
Tomer Michaeli
Frank Michel
Antoine Miech
Krystian Mikolajczyk
Peyman Milanfar
Ben Mildenhall
Gregor Miller
Fausto Milletari
Dongbo Min
Kyle Min
Pedro Miraldo
Dmytro Mishkin
Anand Mishra
Ashish Mishra
Ishan Misra
Niluthpol C. Mithun
Kaushik Mitra
Niloy Mitra
Anton Mitrokhin
Ikuhisa Mitsugami
Anurag Mittal
Kaichun Mo
Zhipeng Mo
Davide Modolo
Michael Moeller

Pritish Mohapatra
Pavlo Molchanov
Davide Moltisanti
Pascal Monasse
Mathew Monfort
Aron Monszpart
Sean Moran
Vlad I. Morariu
Francesc Moreno-Noguer
Pietro Morerio
Stylianos Moschoglou
Yael Moses
Roozbeh Mottaghi
Pierre Moulon
Arsalan Mousavian
Yadong Mu
Yasuhiro Mukaigawa
Lopamudra Mukherjee
Yusuke Mukuta
Ravi Teja Mullapudi
Mario Enrique Munich
Zachary Murez
Ana C. Murillo
J. Krishna Murthy
Damien Muselet
Armin Mustafa
Siva Karthik Mustikovela
Carlo Dal Mutto
Moin Nabi
Varun K. Nagaraja
Tushar Nagarajan
Arsha Nagrani
Seungjun Nah
Nikhil Naik
Yoshikatsu Nakajima
Yuta Nakashima
Atsushi Nakazawa
Seonghyeon Nam
Vinay P. Namboodiri
Medhini Narasimhan
Srinivasa Narasimhan
Sanath Narayan
Erickson Rangel
    Nascimento
Jacinto Nascimento
Tayyab Naseer

Lakshmanan Nataraj
Neda Nategh
Nelson Isao Nauata
Fernando Navarro
Shah Nawaz
Lukas Neumann
Ram Nevatia
Alejandro Newell
Shawn Newsam
Joe Yue-Hei Ng
Trung Thanh Ngo
Duc Thanh Nguyen
Lam M. Nguyen
Phuc Xuan Nguyen
Thuong Nguyen Canh
Mihalis Nicolaou
Andrei Liviu Nicolicioiu
Xuecheng Nie
Michael Niemeyer
Simon Niklaus
Christophoros Nikou
David Nilsson
Jifeng Ning
Yuval Nirkin
Li Niu
Yuzhen Niu
Zhenxing Niu
Shohei Nobuhara
Nicoletta Noceti
Hyeonwoo Noh
Junhyug Noh
Mehdi Noroozi
Sotiris Nousias
Valsamis Ntouskos
Matthew O'Toole
Peter Ochs
Ferda Ofli
Seong Joon Oh
Seoung Wug Oh
Iason Oikonomidis
Utkarsh Ojha
Takahiro Okabe
Takayuki Okatani
Fumio Okura
Aude Oliva
Kyle Olszewski

Björn Ommer
Mohamed Omran
Elisabeta Oneata
Michael Opitz
Jose Oramas
Tribhuvanesh Orekondy
Shaul Oron
Sergio Orts-Escolano
Ivan Oseledets
Aljosa Osep
Magnus Oskarsson
Anton Osokin
Martin R. Oswald
Wanli Ouyang
Andrew Owens
Mete Ozay
Mustafa Ozuysal
Eduardo Pérez-Pellitero
Gautam Pai
Dipan Kumar Pal
P. H. Pamplona Savarese
Jinshan Pan
Junting Pan
Xingang Pan
Yingwei Pan
Yannis Panagakis
Rameswar Panda
Guan Pang
Jiahao Pang
Jiangmiao Pang
Tianyu Pang
Sharath Pankanti
Nicolas Papadakis
Dim Papadopoulos
George Papandreou
Toufiq Parag
Shaifali Parashar
Sarah Parisot
Eunhyeok Park
Hyun Soo Park
Jaesik Park
Min-Gyu Park
Taesung Park
Alvaro Parra
C. Alejandro Parraga
Despoina Paschalidou

Nikolaos Passalis
Vishal Patel
Viorica Patraucean
Badri Narayana Patro
Danda Pani Paudel
Sujoy Paul
Georgios Pavlakos
Ioannis Pavlidis
Vladimir Pavlovic
Nick Pears
Kim Steenstrup Pedersen
Selen Pehlivan
Shmuel Peleg
Chao Peng
Houwen Peng
Wen-Hsiao Peng
Xi Peng
Xiaojiang Peng
Xingchao Peng
Yuxin Peng
Federico Perazzi
Juan Camilo Perez
Vishwanath Peri
Federico Pernici
Luca Del Pero
Florent Perronnin
Stavros Petridis
Henning Petzka
Patrick Peursum
Michael Pfeiffer
Hanspeter Pfister
Roman Pflugfelder
Minh Tri Pham
Yongri Piao
David Picard
Tomasz Pieciak
A. J. Piergiovanni
Andrea Pilzer
Pedro O. Pinheiro
Silvia Laura Pintea
Lerrel Pinto
Axel Pinz
Robinson Piramuthu
Fiora Pirri
Leonid Pishchulin
Francesco Pittaluga

Daniel Pizarro
Tobias Plötz
Mirco Planamente
Matteo Poggi
Moacir A. Ponti
Parita Pooj
Fatih Porikli
Horst Possegger
Omid Poursaeed
Ameya Prabhu
Viraj Uday Prabhu
Dilip Prasad
Brian L. Price
True Price
Maria Priisalu
Veronique Prinet
Victor Adrian Prisacariu
Jan Prokaj
Sergey Prokudin
Nicolas Pugeault
Xavier Puig
Albert Pumarola
Pulak Purkait
Senthil Purushwalkam
Charles R. Qi
Hang Qi
Haozhi Qi
Lu Qi
Mengshi Qi
Siyuan Qi
Xiaojuan Qi
Yuankai Qi
Shengju Qian
Xuelin Qian
Siyuan Qiao
Yu Qiao
Jie Qin
Qiang Qiu
Weichao Qiu
Zhaofan Qiu
Kha Gia Quach
Yuhui Quan
Yvain Queau
Julian Quiroga
Faisal Qureshi
Mahdi Rad

Filip Radenovic
Petia Radeva
Venkatesh
  B. Radhakrishnan
Ilija Radosavovic
Noha Radwan
Rahul Raguram
Tanzila Rahman
Amit Raj
Ajit Rajwade
Kandan Ramakrishnan
Santhosh
  K. Ramakrishnan
Srikumar Ramalingam
Ravi Ramamoorthi
Vasili Ramanishka
Ramprasaath R. Selvaraju
Francois Rameau
Visvanathan Ramesh
Santu Rana
Rene Ranftl
Anand Rangarajan
Anurag Ranjan
Viresh Ranjan
Yongming Rao
Carolina Raposo
Vivek Rathod
Sathya N. Ravi
Avinash Ravichandran
Tammy Riklin Raviv
Daniel Rebain
Sylvestre-Alvise Rebuffi
N. Dinesh Reddy
Timo Rehfeld
Paolo Remagnino
Konstantinos Rematas
Edoardo Remelli
Dongwei Ren
Haibing Ren
Jian Ren
Jimmy Ren
Mengye Ren
Weihong Ren
Wenqi Ren
Zhile Ren
Zhongzheng Ren

Zhou Ren
Vijay Rengarajan
Md A. Reza
Farzaneh Rezaeianaran
Hamed R. Tavakoli
Nicholas Rhinehart
Helge Rhodin
Elisa Ricci
Alexander Richard
Eitan Richardson
Elad Richardson
Christian Richardt
Stephan Richter
Gernot Riegler
Daniel Ritchie
Tobias Ritschel
Samuel Rivera
Yong Man Ro
Richard Roberts
Joseph Robinson
Ignacio Rocco
Mrigank Rochan
Emanuele Rodolà
Mikel D. Rodriguez
Giorgio Roffo
Grégory Rogez
Gemma Roig
Javier Romero
Xuejian Rong
Yu Rong
Amir Rosenfeld
Bodo Rosenhahn
Guy Rosman
Arun Ross
Paolo Rota
Peter M. Roth
Anastasios Roussos
Anirban Roy
Sebastien Roy
Aruni RoyChowdhury
Artem Rozantsev
Ognjen Rudovic
Daniel Rueckert
Adria Ruiz
Javier Ruiz-del-solar
Christian Rupprecht

Chris Russell
Dan Ruta
Jongbin Ryu
Ömer Sümer
Alexandre Sablayrolles
Faraz Saeedan
Ryusuke Sagawa
Christos Sagonas
Tonmoy Saikia
Hideo Saito
Kuniaki Saito
Shunsuke Saito
Shunta Saito
Ken Sakurada
Joaquin Salas
Fatemeh Sadat Saleh
Mahdi Saleh
Pouya Samangouei
Leo Sampaio
  Ferraz Ribeiro
Artsiom Olegovich
  Sanakoyeu
Enrique Sanchez
Patsorn Sangkloy
Anush Sankaran
Aswin Sankaranarayanan
Swami Sankaranarayanan
Rodrigo Santa Cruz
Amartya Sanyal
Archana Sapkota
Nikolaos Sarafianos
Jun Sato
Shin'ichi Satoh
Hosnieh Sattar
Arman Savran
Manolis Savva
Alexander Sax
Hanno Scharr
Simone Schaub-Meyer
Konrad Schindler
Dmitrij Schlesinger
Uwe Schmidt
Dirk Schnieders
Björn Schuller
Samuel Schulter
Idan Schwartz

William Robson Schwartz
Alex Schwing
Sinisa Segvic
Lorenzo Seidenari
Pradeep Sen
Ozan Sener
Soumyadip Sengupta
Arda Senocak
Mojtaba Seyedhosseini
Shishir Shah
Shital Shah
Sohil Atul Shah
Tamar Rott Shaham
Huasong Shan
Qi Shan
Shiguang Shan
Jing Shao
Roman Shapovalov
Gaurav Sharma
Vivek Sharma
Viktoriia Sharmanska
Dongyu She
Sumit Shekhar
Evan Shelhamer
Chengyao Shen
Chunhua Shen
Falong Shen
Jie Shen
Li Shen
Liyue Shen
Shuhan Shen
Tianwei Shen
Wei Shen
William B. Shen
Yantao Shen
Ying Shen
Yiru Shen
Yujun Shen
Yuming Shen
Zhiqiang Shen
Ziyi Shen
Lu Sheng
Yu Sheng
Rakshith Shetty
Baoguang Shi
Guangming Shi

Hailin Shi
Miaojing Shi
Yemin Shi
Zhenmei Shi
Zhiyuan Shi
Kevin Jonathan Shih
Shiliang Shiliang
Hyunjung Shim
Atsushi Shimada
Nobutaka Shimada
Daeyun Shin
Young Min Shin
Koichi Shinoda
Konstantin Shmelkov
Michael Zheng Shou
Abhinav Shrivastava
Tianmin Shu
Zhixin Shu
Hong-Han Shuai
Pushkar Shukla
Christian Siagian
Mennatullah M. Siam
Kaleem Siddiqi
Karan Sikka
Jae-Young Sim
Christian Simon
Martin Simonovsky
Dheeraj Singaraju
Bharat Singh
Gurkirt Singh
Krishna Kumar Singh
Maneesh Kumar Singh
Richa Singh
Saurabh Singh
Suriya Singh
Vikas Singh
Sudipta N. Sinha
Vincent Sitzmann
Josef Sivic
Gregory Slabaugh
Miroslava Slavcheva
Ron Slossberg
Brandon Smith
Kevin Smith
Vladimir Smutny
Noah Snavely

Roger
   D. Soberanis-Mukul
Kihyuk Sohn
Francesco Solera
Eric Sommerlade
Sanghyun Son
Byung Cheol Song
Chunfeng Song
Dongjin Song
Jiaming Song
Jie Song
Jifei Song
Jingkuan Song
Mingli Song
Shiyu Song
Shuran Song
Xiao Song
Yafei Song
Yale Song
Yang Song
Yi-Zhe Song
Yibing Song
Humberto Sossa
Cesar de Souza
Adrian Spurr
Srinath Sridhar
Suraj Srinivas
Pratul P. Srinivasan
Anuj Srivastava
Tania Stathaki
Christopher Stauffer
Simon Stent
Rainer Stiefelhagen
Pierre Stock
Julian Straub
Jonathan C. Stroud
Joerg Stueckler
Jan Stuehmer
David Stutz
Chi Su
Hang Su
Jong-Chyi Su
Shuochen Su
Yu-Chuan Su
Ramanathan Subramanian
Yusuke Sugano

Masanori Suganuma
Yumin Suh
Mohammed Suhail
Yao Sui
Heung-Il Suk
Josephine Sullivan
Baochen Sun
Chen Sun
Chong Sun
Deqing Sun
Jin Sun
Liang Sun
Lin Sun
Qianru Sun
Shao-Hua Sun
Shuyang Sun
Weiwei Sun
Wenxiu Sun
Xiaoshuai Sun
Xiaoxiao Sun
Xingyuan Sun
Yifan Sun
Zhun Sun
Sabine Susstrunk
David Suter
Supasorn Suwajanakorn
Tomas Svoboda
Eran Swears
Paul Swoboda
Attila Szabo
Richard Szeliski
Duy-Nguyen Ta
Andrea Tagliasacchi
Yuichi Taguchi
Ying Tai
Keita Takahashi
Kouske Takahashi
Jun Takamatsu
Hugues Talbot
Toru Tamaki
Chaowei Tan
Fuwen Tan
Mingkui Tan
Mingxing Tan
Qingyang Tan
Robby T. Tan

Xiaoyang Tan
Kenichiro Tanaka
Masayuki Tanaka
Chang Tang
Chengzhou Tang
Danhang Tang
Ming Tang
Peng Tang
Qingming Tang
Wei Tang
Xu Tang
Yansong Tang
Youbao Tang
Yuxing Tang
Zhiqiang Tang
Tatsunori Taniai
Junli Tao
Xin Tao
Makarand Tapaswi
Jean-Philippe Tarel
Lyne Tchapmi
Zachary Teed
Bugra Tekin
Damien Teney
Ayush Tewari
Christian Theobalt
Christopher Thomas
Diego Thomas
Jim Thomas
Rajat Mani Thomas
Xinmei Tian
Yapeng Tian
Yingli Tian
Yonglong Tian
Zhi Tian
Zhuotao Tian
Kinh Tieu
Joseph Tighe
Massimo Tistarelli
Matthew Toews
Carl Toft
Pavel Tokmakov
Federico Tombari
Chetan Tonde
Yan Tong
Alessio Tonioni

Andrea Torsello
Fabio Tosi
Du Tran
Luan Tran
Ngoc-Trung Tran
Quan Hung Tran
Truyen Tran
Rudolph Triebel
Martin Trimmel
Shashank Tripathi
Subarna Tripathi
Leonardo Trujillo
Eduard Trulls
Tomasz Trzcinski
Sam Tsai
Yi-Hsuan Tsai
Hung-Yu Tseng
Stavros Tsogkas
Aggeliki Tsoli
Devis Tuia
Shubham Tulsiani
Sergey Tulyakov
Frederick Tung
Tony Tung
Daniyar Turmukhambetov
Ambrish Tyagi
Radim Tylecek
Christos Tzelepis
Georgios Tzimiropoulos
Dimitrios Tzionas
Seiichi Uchida
Norimichi Ukita
Dmitry Ulyanov
Martin Urschler
Yoshitaka Ushiku
Ben Usman
Alexander Vakhitov
Julien P. C. Valentin
Jack Valmadre
Ernest Valveny
Joost van de Weijer
Jan van Gemert
Koen Van Leemput
Gul Varol
Sebastiano Vascon
M. Alex O. Vasilescu

Subeesh Vasu
Mayank Vatsa
David Vazquez
Javier Vazquez-Corral
Ashok Veeraraghavan
Erik Velasco-Salido
Raviteja Vemulapalli
Jonathan Ventura
Manisha Verma
Roberto Vezzani
Ruben Villegas
Minh Vo
MinhDuc Vo
Nam Vo
Michele Volpi
Riccardo Volpi
Carl Vondrick
Konstantinos Vougioukas
Tuan-Hung Vu
Sven Wachsmuth
Neal Wadhwa
Catherine Wah
Jacob C. Walker
Thomas S. A. Wallis
Chengde Wan
Jun Wan
Liang Wan
Renjie Wan
Baoyuan Wang
Boyu Wang
Cheng Wang
Chu Wang
Chuan Wang
Chunyu Wang
Dequan Wang
Di Wang
Dilin Wang
Dong Wang
Fang Wang
Guanzhi Wang
Guoyin Wang
Hanzi Wang
Hao Wang
He Wang
Heng Wang
Hongcheng Wang

Hongxing Wang
Hua Wang
Jian Wang
Jingbo Wang
Jinglu Wang
Jingya Wang
Jinjun Wang
Jinqiao Wang
Jue Wang
Ke Wang
Keze Wang
Le Wang
Lei Wang
Lezi Wang
Li Wang
Liang Wang
Lijun Wang
Limin Wang
Linwei Wang
Lizhi Wang
Mengjiao Wang
Mingzhe Wang
Minsi Wang
Naiyan Wang
Nannan Wang
Ning Wang
Oliver Wang
Pei Wang
Peng Wang
Pichao Wang
Qi Wang
Qian Wang
Qiaosong Wang
Qifei Wang
Qilong Wang
Qing Wang
Qingzhong Wang
Quan Wang
Rui Wang
Ruiping Wang
Ruixing Wang
Shangfei Wang
Shenlong Wang
Shiyao Wang
Shuhui Wang
Song Wang

Tao Wang
Tianlu Wang
Tiantian Wang
Ting-chun Wang
Tingwu Wang
Wei Wang
Weiyue Wang
Wenguan Wang
Wenlin Wang
Wenqi Wang
Xiang Wang
Xiaobo Wang
Xiaofang Wang
Xiaoling Wang
Xiaolong Wang
Xiaosong Wang
Xiaoyu Wang
Xin Eric Wang
Xinchao Wang
Xinggang Wang
Xintao Wang
Yali Wang
Yan Wang
Yang Wang
Yangang Wang
Yaxing Wang
Yi Wang
Yida Wang
Yilin Wang
Yiming Wang
Yisen Wang
Yongtao Wang
Yu-Xiong Wang
Yue Wang
Yujiang Wang
Yunbo Wang
Yunhe Wang
Zengmao Wang
Zhangyang Wang
Zhaowen Wang
Zhe Wang
Zhecan Wang
Zheng Wang
Zhixiang Wang
Zilei Wang
Jianqiao Wangni

Anne S. Wannenwetsch
Jan Dirk Wegner
Scott Wehrwein
Donglai Wei
Kaixuan Wei
Longhui Wei
Pengxu Wei
Ping Wei
Qi Wei
Shih-En Wei
Xing Wei
Yunchao Wei
Zijun Wei
Jerod Weinman
Michael Weinmann
Philippe Weinzaepfel
Yair Weiss
Bihan Wen
Longyin Wen
Wei Wen
Junwu Weng
Tsui-Wei Weng
Xinshuo Weng
Eric Wengrowski
Tomas Werner
Gordon Wetzstein
Tobias Weyand
Patrick Wieschollek
Maggie Wigness
Erik Wijmans
Richard Wildes
Olivia Wiles
Chris Williams
Williem Williem
Kyle Wilson
Calden Wloka
Nicolai Wojke
Christian Wolf
Yongkang Wong
Sanghyun Woo
Scott Workman
Baoyuan Wu
Bichen Wu
Chao-Yuan Wu
Huikai Wu
Jiajun Wu

Jialin Wu
Jiaxiang Wu
Jiqing Wu
Jonathan Wu
Lifang Wu
Qi Wu
Qiang Wu
Ruizheng Wu
Shangzhe Wu
Shun-Cheng Wu
Tianfu Wu
Wayne Wu
Wenxuan Wu
Xiao Wu
Xiaohe Wu
Xinxiao Wu
Yang Wu
Yi Wu
Yiming Wu
Ying Nian Wu
Yue Wu
Zheng Wu
Zhenyu Wu
Zhirong Wu
Zuxuan Wu
Stefanie Wuhrer
Jonas Wulff
Changqun Xia
Fangting Xia
Fei Xia
Gui-Song Xia
Lu Xia
Xide Xia
Yin Xia
Yingce Xia
Yongqin Xian
Lei Xiang
Shiming Xiang
Bin Xiao
Fanyi Xiao
Guobao Xiao
Huaxin Xiao
Taihong Xiao
Tete Xiao
Tong Xiao
Wang Xiao

Yang Xiao
Cihang Xie
Guosen Xie
Jianwen Xie
Lingxi Xie
Sirui Xie
Weidi Xie
Wenxuan Xie
Xiaohua Xie
Fuyong Xing
Jun Xing
Junliang Xing
Bo Xiong
Peixi Xiong
Yu Xiong
Yuanjun Xiong
Zhiwei Xiong
Chang Xu
Chenliang Xu
Dan Xu
Danfei Xu
Hang Xu
Hongteng Xu
Huijuan Xu
Jingwei Xu
Jun Xu
Kai Xu
Mengmeng Xu
Mingze Xu
Qianqian Xu
Ran Xu
Weijian Xu
Xiangyu Xu
Xiaogang Xu
Xing Xu
Xun Xu
Yanyu Xu
Yichao Xu
Yong Xu
Yongchao Xu
Yuanlu Xu
Zenglin Xu
Zheng Xu
Chuhui Xue
Jia Xue
Nan Xue

Tianfan Xue
Xiangyang Xue
Abhay Yadav
Yasushi Yagi
I. Zeki Yalniz
Kota Yamaguchi
Toshihiko Yamasaki
Takayoshi Yamashita
Junchi Yan
Ke Yan
Qingan Yan
Sijie Yan
Xinchen Yan
Yan Yan
Yichao Yan
Zhicheng Yan
Keiji Yanai
Bin Yang
Ceyuan Yang
Dawei Yang
Dong Yang
Fan Yang
Guandao Yang
Guorun Yang
Haichuan Yang
Hao Yang
Jianwei Yang
Jiaolong Yang
Jie Yang
Jing Yang
Kaiyu Yang
Linjie Yang
Meng Yang
Michael Ying Yang
Nan Yang
Shuai Yang
Shuo Yang
Tianyu Yang
Tien-Ju Yang
Tsun-Yi Yang
Wei Yang
Wenhan Yang
Xiao Yang
Xiaodong Yang
Xin Yang
Yan Yang

Yanchao Yang
Yee Hong Yang
Yezhou Yang
Zhenheng Yang
Anbang Yao
Angela Yao
Cong Yao
Jian Yao
Li Yao
Ting Yao
Yao Yao
Zhewei Yao
Chengxi Ye
Jianbo Ye
Keren Ye
Linwei Ye
Mang Ye
Mao Ye
Qi Ye
Qixiang Ye
Mei-Chen Yeh
Raymond Yeh
Yu-Ying Yeh
Sai-Kit Yeung
Serena Yeung
Kwang Moo Yi
Li Yi
Renjiao Yi
Alper Yilmaz
Junho Yim
Lijun Yin
Weidong Yin
Xi Yin
Zhichao Yin
Tatsuya Yokota
Ryo Yonetani
Donggeun Yoo
Jae Shin Yoon
Ju Hong Yoon
Sung-eui Yoon
Laurent Younes
Changqian Yu
Fisher Yu
Gang Yu
Jiahui Yu
Kaicheng Yu

Ke Yu
Lequan Yu
Ning Yu
Qian Yu
Ronald Yu
Ruichi Yu
Shoou-I Yu
Tao Yu
Tianshu Yu
Xiang Yu
Xin Yu
Xiyu Yu
Youngjae Yu
Yu Yu
Zhiding Yu
Chunfeng Yuan
Ganzhao Yuan
Jinwei Yuan
Lu Yuan
Quan Yuan
Shanxin Yuan
Tongtong Yuan
Wenjia Yuan
Ye Yuan
Yuan Yuan
Yuhui Yuan
Huanjing Yue
Xiangyu Yue
Ersin Yumer
Sergey Zagoruyko
Egor Zakharov
Amir Zamir
Andrei Zanfir
Mihai Zanfir
Pablo Zegers
Bernhard Zeisl
John S. Zelek
Niclas Zeller
Huayi Zeng
Jiabei Zeng
Wenjun Zeng
Yu Zeng
Xiaohua Zhai
Fangneng Zhan
Huangying Zhan
Kun Zhan

Xiaohang Zhan
Baochang Zhang
Bowen Zhang
Cecilia Zhang
Changqing Zhang
Chao Zhang
Chengquan Zhang
Chi Zhang
Chongyang Zhang
Dingwen Zhang
Dong Zhang
Feihu Zhang
Hang Zhang
Hanwang Zhang
Hao Zhang
He Zhang
Hongguang Zhang
Hua Zhang
Ji Zhang
Jianguo Zhang
Jianming Zhang
Jiawei Zhang
Jie Zhang
Jing Zhang
Juyong Zhang
Kai Zhang
Kaipeng Zhang
Ke Zhang
Le Zhang
Lei Zhang
Li Zhang
Lihe Zhang
Linguang Zhang
Lu Zhang
Mi Zhang
Mingda Zhang
Peng Zhang
Pingping Zhang
Qian Zhang
Qilin Zhang
Quanshi Zhang
Richard Zhang
Rui Zhang
Runze Zhang
Shengping Zhang
Shifeng Zhang

Shuai Zhang
Songyang Zhang
Tao Zhang
Ting Zhang
Tong Zhang
Wayne Zhang
Wei Zhang
Weizhong Zhang
Wenwei Zhang
Xiangyu Zhang
Xiaolin Zhang
Xiaopeng Zhang
Xiaoqin Zhang
Xiuming Zhang
Ya Zhang
Yang Zhang
Yimin Zhang
Yinda Zhang
Ying Zhang
Yongfei Zhang
Yu Zhang
Yulun Zhang
Yunhua Zhang
Yuting Zhang
Zhanpeng Zhang
Zhao Zhang
Zhaoxiang Zhang
Zhen Zhang
Zheng Zhang
Zhifei Zhang
Zhijin Zhang
Zhishuai Zhang
Ziming Zhang
Bo Zhao
Chen Zhao
Fang Zhao
Haiyu Zhao
Han Zhao
Hang Zhao
Hengshuang Zhao
Jian Zhao
Kai Zhao
Liang Zhao
Long Zhao
Qian Zhao
Qibin Zhao

Qijun Zhao
Rui Zhao
Shenglin Zhao
Sicheng Zhao
Tianyi Zhao
Wenda Zhao
Xiangyun Zhao
Xin Zhao
Yang Zhao
Yue Zhao
Zhichen Zhao
Zijing Zhao
Xiantong Zhen
Chuanxia Zheng
Feng Zheng
Haiyong Zheng
Jia Zheng
Kang Zheng
Shuai Kyle Zheng
Wei-Shi Zheng
Yinqiang Zheng
Zerong Zheng
Zhedong Zheng
Zilong Zheng
Bineng Zhong
Fangwei Zhong
Guangyu Zhong
Yiran Zhong
Yujie Zhong
Zhun Zhong
Chunluan Zhou
Huiyu Zhou
Jiahuan Zhou
Jun Zhou
Lei Zhou
Luowei Zhou
Luping Zhou
Mo Zhou
Ning Zhou
Pan Zhou
Peng Zhou
Qianyi Zhou
S. Kevin Zhou
Sanping Zhou
Wengang Zhou
Xingyi Zhou

Yanzhao Zhou
Yi Zhou
Yin Zhou
Yipin Zhou
Yuyin Zhou
Zihan Zhou
Alex Zihao Zhu
Chenchen Zhu
Feng Zhu
Guangming Zhu
Ji Zhu
Jun-Yan Zhu
Lei Zhu
Linchao Zhu
Rui Zhu
Shizhan Zhu
Tyler Lixuan Zhu

Wei Zhu
Xiangyu Zhu
Xinge Zhu
Xizhou Zhu
Yanjun Zhu
Yi Zhu
Yixin Zhu
Yizhe Zhu
Yousong Zhu
Zhe Zhu
Zhen Zhu
Zheng Zhu
Zhenyao Zhu
Zhihui Zhu
Zhuotun Zhu
Bingbing Zhuang
Wei Zhuo

Christian Zimmermann
Karel Zimmermann
Larry Zitnick
Mohammadreza
  Zolfaghari
Maria Zontak
Daniel Zoran
Changqing Zou
Chuhang Zou
Danping Zou
Qi Zou
Yang Zou
Yuliang Zou
Georgios Zoumpourlis
Wangmeng Zuo
Xinxin Zuo

Additional Reviewers

Victoria Fernandez
  Abrevaya
Maya Aghaei
Allam Allam
Christine
  Allen-Blanchette
Nicolas Aziere
Assia Benbihi
Neha Bhargava
Bharat Lal Bhatnagar
Joanna Bitton
Judy Borowski
Amine Bourki
Romain Brégier
Tali Brayer
Sebastian Bujwid
Andrea Burns
Yun-Hao Cao
Yuning Chai
Xiaojun Chang
Bo Chen
Shuo Chen
Zhixiang Chen
Junsuk Choe
Hung-Kuo Chu

Jonathan P. Crall
Kenan Dai
Lucas Deecke
Karan Desai
Prithviraj Dhar
Jing Dong
Wei Dong
Turan Kaan Elgin
Francis Engelmann
Erik Englesson
Fartash Faghri
Zicong Fan
Yang Fu
Risheek Garrepalli
Yifan Ge
Marco Godi
Helmut Grabner
Shuxuan Guo
Jianfeng He
Zhezhi He
Samitha Herath
Chih-Hui Ho
Yicong Hong
Vincent Tao Hu
Julio Hurtado

Jaedong Hwang
Andrey Ignatov
Muhammad
  Abdullah Jamal
Saumya Jetley
Meiguang Jin
Jeff Johnson
Minsoo Kang
Saeed Khorram
Mohammad Rami Koujan
Nilesh Kulkarni
Sudhakar Kumawat
Abdelhak Lemkhenter
Alexander Levine
Jiachen Li
Jing Li
Jun Li
Yi Li
Liang Liao
Ruochen Liao
Tzu-Heng Lin
Phillip Lippe
Bao-di Liu
Bo Liu
Fangchen Liu

Hanxiao Liu
Hongyu Liu
Huidong Liu
Miao Liu
Xinxin Liu
Yongfei Liu
Yu-Lun Liu
Amir Livne
Tiange Luo
Wei Ma
Xiaoxuan Ma
Ioannis Marras
Georg Martius
Effrosyni Mavroudi
Tim Meinhardt
Givi Meishvili
Meng Meng
Zihang Meng
Zhongqi Miao
Gyeongsik Moon
Khoi Nguyen
Yung-Kyun Noh
Antonio Norelli
Jaeyoo Park
Alexander Pashevich
Mandela Patrick
Mary Phuong
Bingqiao Qian
Yu Qiao
Zhen Qiao
Sai Saketh Rambhatla
Aniket Roy
Amelie Royer
Parikshit Vishwas
  Sakurikar
Mark Sandler
Mert Bülent Sarıyıldız
Tanner Schmidt
Anshul B. Shah

Ketul Shah
Rajvi Shah
Hengcan Shi
Xiangxi Shi
Yujiao Shi
William A. P. Smith
Guoxian Song
Robin Strudel
Abby Stylianou
Xinwei Sun
Reuben Tan
Qingyi Tao
Kedar S. Tatwawadi
Anh Tuan Tran
Son Dinh Tran
Eleni Triantafillou
Aristeidis Tsitiridis
Md Zasim Uddin
Andrea Vedaldi
Evangelos Ververas
Vidit Vidit
Paul Voigtlaender
Bo Wan
Huanyu Wang
Huiyu Wang
Junqiu Wang
Pengxiao Wang
Tai Wang
Xinyao Wang
Tomoki Watanabe
Mark Weber
Xi Wei
Botong Wu
James Wu
Jiamin Wu
Rujie Wu
Yu Wu
Rongchang Xie
Wei Xiong

Yunyang Xiong
An Xu
Chi Xu
Yinghao Xu
Fei Xue
Tingyun Yan
Zike Yan
Chao Yang
Heran Yang
Ren Yang
Wenfei Yang
Xu Yang
Rajeev Yasarla
Shaokai Ye
Yufei Ye
Kun Yi
Haichao Yu
Hanchao Yu
Ruixuan Yu
Liangzhe Yuan
Chen-Lin Zhang
Fandong Zhang
Tianyi Zhang
Yang Zhang
Yiyi Zhang
Yongshun Zhang
Yu Zhang
Zhiwei Zhang
Jiaojiao Zhao
Yipu Zhao
Xingjian Zhen
Haizhong Zheng
Tiancheng Zhi
Chengju Zhou
Hao Zhou
Hao Zhu
Alexander Zimin

Contents – Part IV

Making an Invisibility Cloak: Real World Adversarial Attacks on Object Detectors

Zuxuan Wu[1,2]($\boxtimes$), Ser-Nam Lim[2], Larry S. Davis[1], and Tom Goldstein[1,2]

[1] University of Maryland, College Park, USA
zxwu@cs.umd.edu
[2] Facebook AI, New York, USA

Abstract. We present a systematic study of the transferability of adversarial attacks on state-of-the-art object detection frameworks. Using standard detection datasets, we train patterns that suppress the objectness scores produced by a range of commonly used detectors, and ensembles of detectors. Through extensive experiments, we benchmark the effectiveness of adversarially trained patches under both white-box and black-box settings, and quantify transferability of attacks between datasets, object classes, and detector models. Finally, we present a detailed study of physical world attacks using printed posters and wearable clothes, and rigorously quantify the performance of such attacks with different metrics.

1 Introduction

Adversarial examples are security vulnerabilities of machine learning systems in which an attacker makes small or unnoticeable perturbations to system inputs with the goal of manipulating system outputs. These attacks are most effective in the digital world, where attackers can directly manipulate image pixels. However, many studies assume a white box threat model, in which the attacker knows the dataset, architecture, and model parameters used by the victim. In addition, most attacks have real security implications only when they cross into the physical realm.

In a "physical" attack, the adversary modifies a real-world *object*, rather than a digital image, so that it confuses systems that observe it. These objects must maintain their adversarial effects when observed with different cameras, resolutions, lighting conditions, distances, and angles.

While a range of physical attacks have been proposed in the literature, these attacks are frequently confined to digital simulations, or are demonstrated against simple classifiers rather than object detectors. However, in most realistic situations the attacker has only black or grey box knowledge, and their attack

---

Electronic supplementary material The online version of this chapter (https://doi.org/10.1007/978-3-030-58548-8_1) contains supplementary material, which is available to authorized users.

© Springer Nature Switzerland AG 2020
A. Vedaldi et al. (Eds.): ECCV 2020, LNCS 12349, pp. 1–17, 2020.
https://doi.org/10.1007/978-3-030-58548-8_1

Fig. 1. In this demonstration, the YOLOv2 detector is evaded using a pattern trained on the COCO dataset with a carefully constructed objective.

must *transfer* from the digital world into the physical world, from the attacker model to the victim model, or from models trained on one dataset to another.

In this paper, we study the transferability of attacks on object detectors across different architectures, classes, and datasets, with the ultimate goal of generalizing digital attacks to the real-world. Our study has the following goals:

– We focus on industrial-strength detectors under both black-box and white-box settings. Unlike *classifiers*, which output one feature vector per image, object *detectors* output a map of vectors, one for each prior (*i.e.*, candidate bounding box), centered at each output pixel. Since any of these priors can detect an object, attacks must simultaneously manipulate hundreds or thousands of priors operating at different positions, scales, and aspect ratios.
– In the digital setting, we systematically quantify how well attacks on detectors transfer between models, classes and datasets.
– We break down the incremental process of getting attacks out of a digital simulation and into the real world. We explore how real-world nuisance variables cause major differences between the digital and physical performance of attacks, and present experiments for quantifying and identifying the sources of these differences.
– We *quantify* the success rate of attacks under various conditions, and measure how algorithm and model choices impact success rates. We rigorously study how attacks degrade classifiers using standard metrics (average precision) that best describe the strength of detectors, and also more interpretable success/failure metrics.
– We push physical attacks to their limits with wearable adversarial clothing (See Fig. 1) and quantify the success rate of our attacks under complex fabric distortions.

2   Related Work

Attacks on Object Detection and Semantic Segmentation. While there is a plethora of work on attacking image classifiers [11,23,25], less work has been done on more complex vision tasks like object detection and semantic segmentation. Metzen *et al.* demonstrate that nearly imperceptible adversarial perturbations can fool segmentation models to produce incorrect outputs [24]. Arnab *et al.* also show that segmentation models are vulnerable to attacks [1], and claim that adversarial perturbations fail to transfer across network architectures. Xie *et al.* introduce Dense Adversary Generation (DAG), a method that produces incorrect predictions for pixels in segmentation models or proposals in object detection frameworks [34]. Wei *et al.* further extend the attack from images to videos [33]. In contrast to [33,34], which Attacks the classifier stage of object detectors, Li *et al.* attack region proposal networks by decreasing the confidence scores of positive proposals [19]. DPatch causes misclassification of detectors, by placing a patch that does not overlap with the objects of interest [22]. Li *et al.* add imperceptible patches to the background to fool object detectors [18]. Note that all of these studies focus on digital (as opposed to physical) attacks with a specific detector, without studying the transferability of attacks. In this paper, we systematically evaluate a wide range of popular detectors in both the digital and physical world, and benchmark how attacks transfer in different settings.

Physical Attacks in the Real World. Kurakin *et al.* took photos of adversarial images with a camera and input them to a pretrained classifier [16]; they demonstrate that a large fraction of images are misclassified. Eykholt *et al.* consider physical attacks on stop sign classifiers using images cropped from video frames [9]. They successfully fool classifiers using both norm bounded perturbations, and also sparse perturbations using carefully placed stickers. Stop signs attacks on object detectors are considered in [5,8]. Lu *et al.* showed that the perturbed sign images from [9] can be reliably recognized by popular detectors like Faster-RCNN [27] and Yolov2 [26], and showed that detectors are much more robust to attacks than classifiers. Note that fooling stop sign detectors differs from fooling person detectors because stop sign perturbations can cover the whole object, whereas our person attacks leave the face, hands, and legs uncovered.

Zeng *et al.* use rendering tools to perform attacks in 3D environments [37]. Sitawarin *et al.* [30] propose large out-of-distribution perturbations, producing toxic signs to deceive autonomous vehicles. Athalye *et al.* introduce expectation over transformations (EoT) to generate physically robust adversarial samples, and they produce 3D physical adversarial objects that can attack classifiers in different conditions. Sharif *et al.* explore adversarial eyeglass frames that fool face classifiers [28]. Brown *et al.* placed adversarial patches [3] on raw images, forcing classifiers to output incorrect predictions. Komkov *et al.* generate stickers attached to hats to attack face classifiers [15]. Huang *et al.* craft attacks by simulations to cause misclassification of detectors [14]. Thys *et al.* produce printed adversarial patches [31] that deceive person detectors instantiated by Yolov2 [26]. This proof-of-concept study was the first to consider physical attacks on detectors, although it was restricted to the white-box setting. Furthermore the authors did not quantify the performance, or address issues like robustness to distance/distortions and detectors beyond Yolov2, and the transferability of their attacks are limited. Xu *et al.* learn TPS transformations to generate T-shirts [36]. To the best of our knowledge, no paper has conclusively demonstrated the transferability of attacks on object detectors, or quantified the reliability of transfer attacks.

2.1   Object Detector Basics

We briefly review the inner workings of object detectors [4,17,20,29], most of which can be described as two-stage frameworks (*e.g.*, Fast(er) RCNN [10,27], Mask RCNN [12], *etc.*) or one-stage pipelines (*e.g.*, YOLOv2 [26], SSD [21], *etc.*).

Two-stage Detectors. These detectors use a region proposal network (RPN) to identify potential bounding boxes (Stage I), and then classify the contents of these bounding boxes (Stage II). An RPN passes an image through a *backbone* network to produce a stack of 2D feature maps with resolution $W' \times H'$ (or a feature pyramid containing features at different resolutions). The RPN considers k "priors", or candidate bounding boxes with a range of aspect ratios and sizes, centered on every output pixel. For each of the $W' \times H' \times k$ priors, the RPN produces an "objectness score", and also the offset to the center coordinates and dimensions of the prior to best fit the closest object. Finally, proposals with high objectness scores are sent to a Stage-II network for classification.

One-stage Detectors. These networks generate object proposals and at the same time predict their class labels. Similar as RPNs, these networks typically transform an image into a $W' \times H'$ feature map, and each pixel on the output contains the locations of a set of default bounding boxes, their class prediction scores, as well as objectness scores.

Why are Detectors Hard to Fool? A detector usually produces hundreds or thousands of priors that overlap with an object. Usually, non-maximum supression (NMS) is used to select the bounding box with highest confidence, and reject overlapping boxes of lower confidence so that an object is only detected once. Suppose an adversarial attack evades detection by one prior. In this case, the NMS will simply select a different prior to represent the object. For an object to be completely erased from an image, the attack must simultaneously fool the ensemble of all priors that overlap with the object—a much harder task than fooling the output of a single classifier.

Detection Metrics. In the field of object detection, the standard performance metric is average precision (AP) per class, which balances the trade-off between precision and recall. In contrast, success rates (using a fixed detection threshold) are more often used when evaluating physical attacks, due to their interpretability. However, manually selected thresholds are required to compute success rates.

To mitigate this issue, we report both AP (averaging over all confidence thresholds) and success rates.

3 Approach

Our goal is to generate an adversarial pattern that, when placed over an object either digitally or physically, makes that object invisible to detectors. Further, we expect the pattern to be (1) universal (image-agnostic)—the pattern must be effective against a range of objects and within different scenes; (2) transferable—it breaks a variety of detectors with different backbone networks; (3) dataset agnostic—it should fool detectors trained on disparate datasets; (4) robust to viewing conditions—it can withstand field-of-view changes when observed from different perspectives and distances; (5) realizable—patterns should remain adversarial when printed over real-world 3D objects.

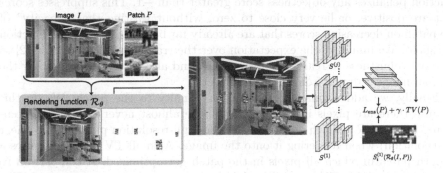

Fig. 2. An overview of the framework. Given a patch and an image, the rendering function uses translations and scalings, plus random augmentations, to overlay the patch onto detected persons. The patch is then updated to minimize the objectness scores produced by a detector while maintaining its smoothness.

3.1 Creating a Universal Adversarial Patch

Our strategy is to "train" a patch using a large set of images containing people. On each iteration, we draw a random batch of images, and pass them through an object detector to obtain bounding boxes containing people. We then place a randomly transformed patch over each detected person, and update the patch pixels to minimize the objectness scores in the output feature map (See Fig. 2).

More formally, we consider a patch $P \in \mathbb{R}^{w \times h \times 3}$ and a randomized rendering function $\mathcal{R}_\theta$. The rendering function takes a patch P and image I, and renders a rescaled copy of P over every detected person in the image I. In addition to scaling and translating the patch to place it into each bounding box, the rendering function also applies an augmentation transform parameterized by the (random) vector θ.

These transforms are a composition of brightness, contrast, rotation, translation, and sheering transforms that help make patches robust to variations caused by lighting and viewing angle that occur in the real world. We also consider more complex thin-plate-spline (TPS) transforms to simulate the random "crumpling" of fabrics.

A detector network takes a patched image $\mathcal{R}_\theta(I, P)$ as its input, and outputs a vector of objectness scores, $\mathcal{S}(\mathcal{R}_\theta(I, P))$ one for each prior. These scores rank general objectness for a two-stage detector, and the strength of the "person" class for a one-stage detectors. A positive score is taken to mean that an object/person overlaps with the corresponding prior, while a negative score denotes the absence of a person. To minimize objectness scores, we use the objectness loss function

$$L_{\mathrm{obj}}(P) = \mathbb{E}_{\theta, I} \sum_i \max\{\mathcal{S}_i(\mathcal{R}_\theta(I, P)) + 1, 0\}^2. \tag{1}$$

Here, i indexes the priors produced by the detector's score mapping. The loss function penalizes any objectness score greater than -1. This suppresses scores that are positive, or lie very close to zero, without wasting the "capacity" of the patch on decreasing scores that are already far below the standard detection threshold. We minimize the expectation over the transform vector θ as in [2] to promote robustness to real-world distortions, and also the expectation over the random image I drawn from the training set.

Finally, we add a small total-variation penalty to the patch. We do this because there are pixels in the patch that are almost never used by the rendering function $\mathcal{R}_\theta$, which down-samples the high-resolution patch (using linear interpolation) when rendering it onto the image. A small TV penalty ensures a smooth patch in which all pixels in the patch get optimized. A comparison to results without this TV penalty is shown in the supplementary material. The final optimization problem we solve is

$$\underset{P}{\mathrm{minimize}}\, L_{\mathrm{obj}}(P) + \gamma \cdot TV(P), \tag{2}$$

where γ was chosen to be small enough to prevent outlier pixels without visibly distorting the patch.

Ensemble Training. To help adversarial patterns generalize to detectors that were not used for training (*i.e.*, to create a black-box attack), we also consider training patches that fool an ensemble of detectors. In this case we replace the objectness loss (1) with the ensemble loss

$$L_{\mathrm{ens}}(P) = \mathbb{E}_{\theta, I} \sum_{i,j} \max\{\mathcal{S}_i^{(j)}(\mathcal{R}_\theta(I, P)) + 1, 0\}^2, \tag{3}$$

where $\mathcal{S}^{(j)}$ denotes the jth detector in an ensemble.

4   Crafting Attacks in the Digital World

Datasets and Metrics. We craft attack patches using the COCO dataset,[1] which contains a total of 123,000 images. After removing images from the dataset that do not contain people, we then chose a random subset of 10,000 images for training. We compute average precision (AP) for the category of interest to measure the effectiveness of patches.

Object Detectors Attacked. We experiment with both one-stage detectors, *i.e.*, YOLOv2 and YOLOv3, and two-stage detectors, *i.e.*, R50-C4 and R50-FPN, both of which are based on Faster RCNN with a ResNet-50 [13] backbone. R50-C4 and R50-FPN use different features for region proposal—R50-C4 uses single-resolution features, while R50-FPN uses a multi-scale feature pyramid. For all these detectors, we adopt standard models pre-trained on COCO, in addition to our own models retrained from scratch (models denoted with "-r") to test for attack transferability across network weights. Finally, we consider patches crafted using three different ensembles of detectors—ENS2: YOLOv2 + R50-FPN, ENS2-r: YOLOv2 + R50-FPN-r, and ENS3-r: YOLOv2 + YOLOv3 + R50-FPN-r.

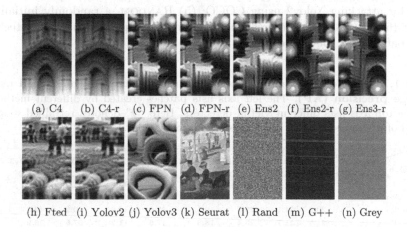

(a) C4   (b) C4-r   (c) FPN   (d) FPN-r   (e) Ens2   (f) Ens2-r   (g) Ens3-r

(h) Fted   (i) Yolov2   (j) Yolov3   (k) Seurat   (l) Rand   (m) G++   (n) Grey

Fig. 3. Adversarial patches, and comparisons with control patches. Here, (a)-(d) are based on R50, and G++ denotes Grey++.

Implementation Details. We use PyTorch for implementation, and we start with a random uniform patch of size $3 \times 250 \times 150$ (the patch is dynamically re-sized by $\mathcal{R}_\theta$ during the forward pass). We use the size since the aspect ratio is similar to that of a body trunk and it has sufficient capacity to reduce

---

[1] We focus on the COCO dataset for its wide diversity of scenes, although we consider the effect of the dataset later.

objectness scores. We use the Adam optimizer with a learning rate 10^{-3}, and decay the rate every 100 epochs until 400 is reached. For YOLOv2/v3, images are resized to 640×640 for both training and testing. For Faster RCNN detectors, the shortest side of images is 250^2 for training, and 800 for testing.

4.1 Evaluation of Digital Attacks

We begin by evaluating patches in digital simulated settings: we consider white-box attacks (detector weights are used for patch learning) and black-box attacks (patch is crafted on a surrogate model and tested on a victim model with different parameters).

Effectiveness of Learned Patches for White-Box Attack. We optimize patches using the aforementioned detectors, and denote the learned patch with the corresponding model it is trained on. We further compare with the following alternative patches: (1) FTED, a learned YOLOV2 patch that is further fine-tuned on a R50-FPN model; (2) SEURAT, a crop from the famous paining "A Sunday Afternoon on the Island of La Grande Jatte" by Georges Seurat, which is visually similar to the top-performing YOLOV2 patch with objects like persons, umbrellas *etc.* (see Fig. 3k); (3) GREY, a grey patch; (4) GREY++, the most powerful RGB value for attacking Yolov2 using COCO; (5) RANDOM, a randomly initialized patch; (6) CLEAN, which corresponds to the oracle performance of detectors when patches are not applied.

Table 1. Impact of different patches on various detectors, measured using average precision (AP). The left axis lists patches created by different methods, and the top axis lists different victim detectors. Here, "r" denotes retrained weights instead of pretrained weights downloaded from model zoos.

Patch \ Victim	R50-C4	R50-C4-r	R50-FPN	R50-FPN-r	YOLOv2	YOLOv2-r	YOLOv3	YOLOv3-r
R50-C4	24.5	24.5	31.4	31.4	37.9	42.6	57.6	48.3
R50-C4-r	25.4	23.9	30.6	30.2	37.7	42.1	57.5	47.4
R50-FPN	20.9	21.1	23.5	19.6	22.6	12.9	40.2	40.3
R50-FPN-r	21.5	21.7	25.4	18.8	17.6	11.2	37.5	36.9
YOLOV2	21.1	19	21.5	21.4	10.7	7.5	18.1	25.7
YOLOV3	28.3	28.9	31.5	27.2	20	15.9	17.8	36.1
FTED	25.6	23.9	24.2	24.4	18.9	16.4	31.6	28.2
ENS2	20	20.3	23.2	19.3	17.5	11.3	39	38.8
ENS2-r	19.7	20.2	23.3	16.8	14.9	9.7	36.3	34.1
ENS3-r	21.1	21.4	24.2	17.4	13.4	9.0	29.8	33.6
SEURAT	47.9	52	51.6	52.5	43.4	39.5	62.6	57.1
RANDOM	53	58.2	59.8	59.7	52	52.5	70	63.5
GREY	45.9	49.6	50	50.8	48	47.1	65.6	57.5
GREY++	46.5	49.8	51.4	52.7	48.5	49.4	64.8	58.6
CLEAN	78.7	78.7	82.2	82.1	63.6	62.7	81.6	74.5

[2] We found that using a lower resolution produced more effective attacks.

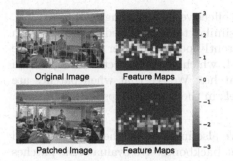

Original Image   Feature Maps

Patched Image   Feature Maps

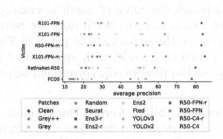

Patches		Random		Ens2		R50-FPN-r
Clean		Seurat		Fted		R50-FPN
Grey++		Ens3-r		YOLOv3		R50-C4-r
Grey		Ens2-r		YOLOv2		R50-C4

Fig. 4. Images and their feature maps, w. and w/o patches, using YOLOv2. Each pixel in the feature map represents an objectness score.

Fig. 5. Performance of different patches, when tested on detectors with different backbones.

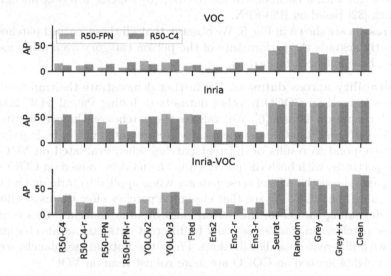

Fig. 6. Results of different patches, trained on COCO, tested on the person category of different datasets. Top two panels: COCO patches tested on VOC and Inria, respectively, using backbones learned on COCO; The bottom panel: COCO patches tested on Inria with backbones trained on VOC.

Adversarial and control patches are shown in Fig. 3, and results are summarized in Table 1. We observe that all adversarially learned patches are highly effective in digital simulations, where the AP of all detectors degrades by at least 29%, going as low as 7.5% AP when the YOLOv2 patch is tested on the retrained YOLOv2 model (YOLOv2-r). All patches transferred well to the corresponding retrained models. In addition, the ensemble patches perform better compared to Faster RCNN patches but are worse than YOLO patches. It is also interesting to see that YOLO patches can be effectively transferred to Faster RCNN models, while Faster RCNN patches are not very effective at attacking YOLO models. Although the SEURAT patch is visually similar to the learned YOLOv2 patch (cf. Fig. 3i and Fig. 3k), it does not consistently perform better than GREY. We visualize the impact of the patch in Fig. 4, which shows objectness maps from the YOLOv2 model with and without patches. We see that when patches are applied to persons, the corresponding pixels in the feature maps are suppressed.

Transferability Across Backbones. We also investigate whether the learned patches transfer to detectors with different backbones. We evaluate the patches on the following detectors: (1) R101-FPN, which uses ResNet-101 with FPN as its backbone; (2) X101-FPN, replaces the R101-FPN with ResNeXt-101 [35]; (3) R50-FPN-M, a Mask RCNN model [12] based on R50-FPN; (4) X101-FPN-M, a Mask RCNN model based on X101-FPN; (5) RETINANET-R50, a RetinaNet [20] with a backbone of ResNet-50; (6) FCOS, a recent anchor-free framework [32] based on R50-FPN.

The results are shown in Fig. 5. We observe that all these learned patches can significantly degrade the performance of the person category even using models that they have not been trained on.

Transferability across datasets. We further demonstrate the transferability of patches learned on COCO to other datasets including Pascal VOC 2007 [7] and the Inria person dataset [6]. We evaluate the patches on the person category using R50-FPN and R50-C4, and the results are presented in Fig. 6. The top two panels correspond to results of different patches when evaluated on VOC and Inria, respectively, with both the patches and the models trained on COCO; the bottom panel shows the APs of these patches when applied to Inria images using models trained on VOC. We see that ensemble patches offer the most effective attacks, degrading the AP of the person class by large margins. This confirms that these patches can transfer not only to different datasets but also backbones trained with different data distributions. From the bottom two panels, we can see that weights learned on COCO are more robust than on VOC.

Table 2. Transferability of patches across classes from VOC, measured with average precision (AP).

Patch \ Class	aero	bike	bird	boat	bottle	bus	car	cat	chair	cow	table	dog	horse	mbike	person	plant	sheep	sofa	train	tv
PERSON	2.0	14.6	1.0	1.8	2.7	13.5	10.7	2.3	0.1	2.4	6.4	2.3	8.3	12.3	5.5	0.3	2.2	1.3	3.8	12.4
HORSE	5.0	31.9	4.7	4.1	2.5	26.4	17.6	10.6	2.3	26.0	24.7	9.5	27.9	26.6	16.0	7.6	12.4	13.4	13.2	35.3
BUS	3.1	30.6	8.5	4.4	1.9	18.4	15.6	7.8	2.7	25.7	39.8	5.3	20.8	20.7	16.0	8.9	12.3	9.5	9.3	29.5
GREY	3.0	19.0	6.4	14.6	8.5	26.9	19.6	9.9	9.8	28.6	24.4	7.4	22.7	15.9	35.8	6.1	18.7	8.7	11.4	61.8
CLEAN	77.5	82.2	76.3	63.6	64.5	82.9	86.5	83.0	57.2	83.3	66.2	84.9	84.5	81.4	83.3	48.0	76.7	70.1	80.1	75.4

Transferability Scross Classes. We find that patches effectively suppress multiple classes, even when they are trained to suppress only one. In addition to the "person" patch, we also train patches on "bus" and "horse," and then evaluate these patches on all 20 categories in VOC.[3] Table 2 summarizes the results. We can see that the "person" patch transfers to almost all categories, possibly because they co-occur with most classes. We also compare with the GREY patch to rule out the possibility that the performance drops are due to occlusion.

5 Physical World Attacks

We now discuss the results of physical world attacks with printed posters. In addition to the standard average precision,[4] we also quantify the performance of attacks with "success rates," which we define as (1) a SUCCESS attack: when there is no bounding box predicted for the person with adversarial patterns; (2) a PARTIAL SUCCESS attack: when there is a bounding box covering less than 50% of a person; (3) a FAILURE attack: when the person is successfully detected.

Examples of detections in each category are shown in Fig. 7. To compute these scores, we use cutoff zero for YOLOv2, and we tune the threshold of other detectors to achieve the best F-1 score on the COCO minival set.

5.1 Printed Posters

We printed posters and took photos at 15 different locations using 10 different patches. At each location, we took four photos for each printed patch corresponding to two distances from the camera and two heights where the patch is held. We also took photos without printed posters as controls (CONTROL). In

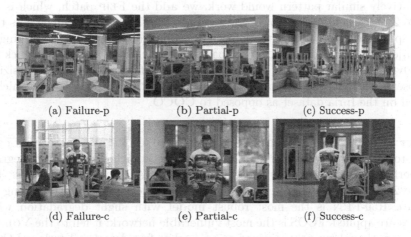

(a) Failure-p   (b) Partial-p   (c) Success-p

(d) Failure-c   (e) Partial-c   (f) Success-c

Fig. 7. Examples of attack failure, partial success, and full success, using posters (top) and shirts (bottom).

---

[3] We observe similar trends on COCO.

[4] We only consider the person with adversarial patterns to calculate AP by eliminating boxes without any overlapping with the GT box.

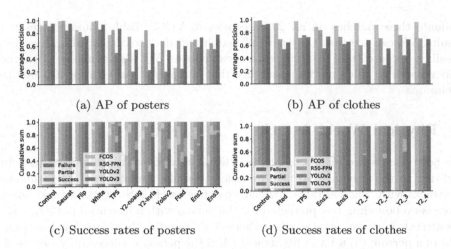

(a) AP of posters　　　　　　　(b) AP of clothes

(c) Success rates of posters　　(d) Success rates of clothes

Fig. 8. AP and success rates for physical attacks. Top: AP of different printed posters (left) and clothes (right). Lower is better. Bottom: success rates of different printed posters (left) and clothes (right). Y2 denotes YOLOv2.

total, we collected 630 photos (see the top row of Fig. 7 for examples). We use four patches that perform well digitally (*i.e.*, YOLOv2, ENS2, ENS3, FTED), and three baseline patches (SEURAT patch, FLIP patch, WHITE).

To better understand the impact of the training process on attack power, we also consider several variants of the YOLOv2 patch (the best digital performer). To assess whether the learned patterns are "truly" adversarial, or whether any qualitatively similar pattern would work, we add the FLIP patch, which is the YOLOv2 patch held upside-down. We compare to a TPS patch, which uses thin plate spline transformations to potentially enhance robustness to warping in real objects. We consider a YOLOv2-noaug patch, which is trained to attack the YOLOv2 model without any augmentations/transformations beyond resizing. To observe the effect of the dataset, we add the YOLOv2-Inria patch, which is trained on the Inria dataset as opposed to COCO.

Poster Results. Figure 8a and 8c summarize the results. We can see that compared to baseline patches, Adversarially optimized patches successfully degrade the performance of detectors measured by both AP and success rates. The YOLOv2 patch achieves the best performance measured by AP among all patches. R50-FPN is the most robust model with slight degradation when patches are applied. FCOS is the most vulnerable network; it fell to the YOLOv2 patch even though we never trained on an anchor-free detector, let alone FCOS. This may be because anchor-free models predict the "center-ness" of pixels for bounding boxes, and the performance drops when center pixels of persons are occluded by printed posters. Interestingly though, simply using baseline patches for occlusion fails to deceive FCOS.

Fig. 9. Paper dolls are made by dressing up printed images with paper patches. We use dolls to observe the effects of camera distortions, and "scrumpled" patches to test against physical deformations that are not easily simulated.

Beyond the choice of detector, several other training factors impact performance. Surprisingly, the TPS patch is worse than YOLOV2, and we believe this results from the fact that adding such complicated transformation makes optimization more difficult during training. It is also surprising to see that the YOLOV2-Inria patch offers impressive success rates on YOLOV2, but it does not transfer as well to other detectors. Not surprisingly, the YOLOV2 patch outperforms the YOLOV2-noaug in terms of AP, however these gains shrink when measured in terms of success rates.

We included the FLIP patch to evaluate whether patches are generic, *i.e.*, any texture with similar shapes and scales would defeat the detector, or whether they are "truly adversarial." The poor performance of the FLIP patch seems to indicate that the learned patches are exploiting specialized behaviors learned by the detector, rather than a generic weakness of the detector model. From the left column of Fig. 8 and Table 1, we see that performance in digital simulations correlates well with physical world performance. However, we observe that patches lose effectiveness when transferring from the digital world into the physical world, demonstrating that physical world attacks are more challenging.

5.2   Paper Dolls

We found that a useful technique for crafting physical attacks was to make "paper dolls"—printouts of images that we could dress up with different patches at different scales. This facilitates quick experiments with physical world effects and camera distortions without the time and expense of fabricating textiles. We use paper dolls to gain insights into why physical attacks are not as effective as digital attacks. The reasons might be three-fold: (1) Pixelation at the detector and compression algorithms incur subtle changes; (2) the rendering artifacts around patch borders assist digital attacks; (3) there exists differences in appearance and texture between large-format digital patches and the digital patches.

In our paper doll study, we print out patches and photos separately. We then overlay patches onto objects and photograph them. We used the first 20 images from the COCO minival set. We use the same patches from the poster experiment, we also compare with "scrumpled" versions of YOLOV2, *i.e.*, YOLOV2-s1 and YOLOV2-s2, to test for robustness to physical deformation, where "-s1" and "-s2" denote the level of crumpling ("s1" < "s2", see Fig. 9).

We compute success rates of different patches when tested with YOLOv2 and present the results in Fig. 10. Comparing across Fig. 10 and the left side of Fig. 8, we see that paper dolls perform only slightly better than large-format posters. The performance drop of paper dolls compared to digital simulations, combined with the high fidelity of the paper printouts, leads us to believe that the dominant factor in the performance difference between digital and physical attacks can be attributed to the imaging process, like camera post-processing, pixelation, and compression.

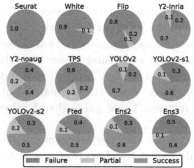

Fig. 10. Effectiveness of different patches on paper dolls.

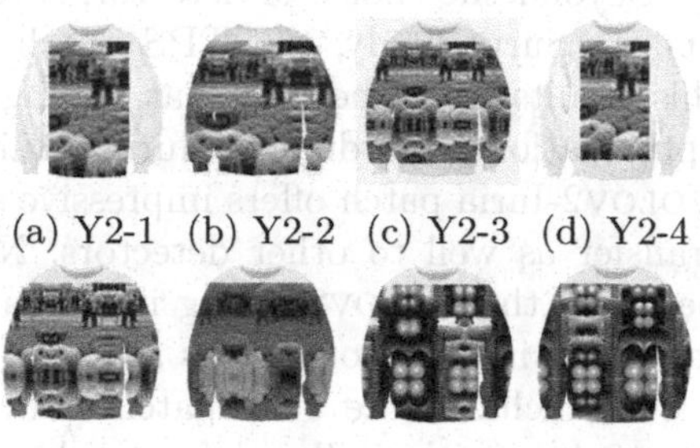

(a) Y2-1 (b) Y2-2 (c) Y2-3 (d) Y2-4

(e) FTED (f) TPS (g) E2-r (h) E3-r

Fig. 11. Adversarial shirts tested in Sect. 6. Y2 denotes YOLOV2.

6   Wearable Adversarial Examples

Printed posters provide a controlled setting under which to test the real-world transferability of adversarial attacks. However the success of printed posters does not inform us about whether attacks can survive the complex deformations and textures of real objects.

To experiment with complex real-world transfer, we printed adversarial patterns on shirts using various strategies. We consider four versions of the YOLOV2 patch representing two different scalings of the patch, both with and without boundary reflections to cover the entire shirt (see Fig. 11). We also consider the TPS patch to see if complex data augmentation can help the attack survive fabric deformations. Finally, we include the FTED, ENS2, ENS3 patches to see if these more complex crafting methods facilitate transfer. We collected photos of a person wearing these shirts at ten different locations. For each location and shirt, we took 4 photos with two orientations (front and back) and two distances from the camera. We also took control photos where the person was not wearing an attack. We collected 360 photos in total (see Supple. for a gallery of samples).

We tested the collected images under the same settings as the poster study, and measure the performance of the patches using both AP and success rates. The results are shown in Fig. 8b and Fig. 8d. We can see that these wearable attacks significantly degrade the performance of detectors. This effect is most pronounced when measured in AP because, when persons are detected, they tend to generate multiple fragmented boxes. It is also interesting to see that FCOS, which is vulnerable to printed posters, is quite robust with wearable attacks, possibly because shirts more closely resemble the clothing that appears in the training set. When measured in success rates, sweatshirts with YOLOV2 patterns achieve ~50% success rates, yet they do not transfer well to other detectors. Among all YOLOV2 shirts, smaller patterns (*i.e.*, YOLOV2-2) perform worse as compared to larger patterns. We also found that tiling/reflecting a patch to cover the whole shirt did not negatively impact performance, even though the patch was not designed for this use. Finally, we found that augmenting attacks with non-rigid TPS transforms did not improve transferability, and in fact was detrimental. This seems to be a result of the difficulty of training a patch with such transformations, as the patch also under-performs other patches digitally.

7   Conclusion

It is widely believed that fooling detectors is a much harder task than fooling classifiers; the ensembling effect of thousands of distinct priors, combined with complex texture, lighting, and measurement distortions in the real world, makes detectors naturally robust. Despite these complexities, the experiments conducted here show that digital attacks can indeed transfer between models, classes, datasets, and also into the real world, although with less reliability than attacks on simple classifiers.

Acknowledgement. Thanks to Ross Girshick for helping us improve our experiments. This work is partially supported by Facebook AI.

References

1. Arnab, A., Miksik, O., Torr, P.H.: On the robustness of semantic segmentation models to adversarial attacks. In: CVPR (2018)
2. Athalye, A., Engstrom, L., Ilyas, A., Kwok, K.: Synthesizing robust adversarial examples. In: ICML (2018)
3. Brown, T.B., Mané, D., Roy, A., Abadi, M., Gilmer, J.: Adversarial patch. arXiv preprint arXiv:1712.09665 (2017)
4. Cai, Z., Vasconcelos, N.: Cascade R-CNN: high quality object detection and instance segmentation. arXiv preprint arXiv:1906.09756 (2019)
5. Chen, S.-T., Cornelius, C., Martin, J., Chau, D.H.P.: ShapeShifter: robust physical adversarial attack on faster R-CNN object detector. In: Berlingerio, M., Bonchi, F., Gärtner, T., Hurley, N., Ifrim, G. (eds.) ECML PKDD 2018. LNCS (LNAI), vol. 11051, pp. 52–68. Springer, Cham (2019). https://doi.org/10.1007/978-3-030-10925-7_4

6. Dalal, N., Triggs, B.: Histograms of oriented gradients for human detection. In: CVPR (2005)
7. Everingham, M., Eslami, S.A., Van Gool, L., Williams, C.K., Winn, J., Zisserman, A.: The pascal visual object classes challenge: a retrospective. IJCV **111**, 98–136 (2015). https://doi.org/10.1007/s11263-014-0733-5
8. Eykholt, K., et al.: Physical adversarial examples for object detectors. In: WOOT (2018)
9. Eykholt, K., et al.: Robust physical-world attacks on deep learning models. In: CVPR (2018)
10. Girshick, R.: Fast r-cnn. In: ICCV (2015)
11. Goodfellow, I.J., Shlens, J., Szegedy, C.: Explaining and harnessing adversarial examples. In: ICLR (2015)
12. He, K., Gkioxari, G., Dollár, P., Girshick, R.: Mask r-cnn. In: ICCV (2017)
13. He, K., Zhang, X., Ren, S., Sun, J.: Deep residual learning for image recognition. In: CVPR (2016)
14. Huang, L., et al.: Universal physical camouflage attacks on object detectors. In: CVPR (2020)
15. Komkov, S., Petiushko, A.: Advhat: Real-world adversarial attack on arcface face id system. arXiv preprint arXiv:1908.08705 (2019)
16. Kurakin, A., Goodfellow, I., Bengio, S.: Adversarial examples in the physical world. In: ICLR Workshop (2017)
17. Li, Y., Chen, Y., Wang, N., Zhang, Z.: Scale-aware trident networks for object detection. In: ICCV (2019)
18. Li, Y., Bian, X., Chang, M.C., Lyu, S.: Exploring the vulnerability of single shot module in object detectors via imperceptible background patches. In: BMVC (2019)
19. Li, Y., Tian, D., Chang, M., Bian, X., Lyu, S.: Robust adversarial perturbation on deep proposal-based models. In: BMVC (2018)
20. Lin, T.Y., Goyal, P., Girshick, R., He, K., Dollár, P.: Focal loss for dense object detection. In: ICCV (2017)
21. Liu, W., et al.: SSD: single shot multibox detector. In: Leibe, B., Matas, J., Sebe, N., Welling, M. (eds.) ECCV 2016. LNCS, vol. 9905, pp. 21–37. Springer, Cham (2016). https://doi.org/10.1007/978-3-319-46448-0_2
22. Liu, X., Yang, H., Liu, Z., Song, L., Li, H., Chen, Y.: Dpatch: An adversarial patch attack on object detectors. arXiv preprint arXiv:1806.02299 (2018)
23. Madry, A., Makelov, A., Schmidt, L., Tsipras, D., Vladu, A.: Towards deep learning models resistant to adversarial attacks. arXiv preprint arXiv:1706.06083 (2017)
24. Metzen, J.H., Kumar, M.C., Brox, T., Fischer, V.: Universal adversarial perturbations against semantic image segmentation. In: ICCV (2017)
25. Moosavi-Dezfooli, S.M., Fawzi, A., Fawzi, O., Frossard, P.: Universal adversarial perturbations. In: CVPR (2017)
26. Redmon, J., Farhadi, A.: Yolo9000: better, faster, stronger. In: CVPR (2017)
27. Ren, S., He, K., Girshick, R., Sun, J.: Faster r-cnn: towards real-time object detection with region proposal networks. In: NIPS (2015)
28. Sharif, M., Bhagavatula, S., Bauer, L., Reiter, M.K.: Accessorize to a crime: real and stealthy attacks on state-of-the-art face recognition. In: ACM CCS (2016)
29. Singh, B., Najibi, M., Davis, L.S.: Sniper: efficient multi-scale training. In: NeurIPS (2018)
30. Sitawarin, C., Bhagoji, A.N., Mosenia, A., Chiang, M., Mittal, P.: Darts: deceiving autonomous cars with toxic signs. arXiv preprint arXiv:1802.06430 (2018)

31. Thys, S., Van Ranst, W., Goedemé, T.: Fooling automated surveillance cameras: adversarial patches to attack person detection. In: CVPR Workshop (2019)
32. Tian, Z., Shen, C., Chen, H., He, T.: FCOS: fully convolutional one-stage object detection. In: ICCV (2019)
33. Wei, X., Liang, S., Chen, N., Cao, X.: Transferable adversarial attacks for image and video object detection. In: IJCAI (2019)
34. Xie, C., Wang, J., Zhang, Z., Zhou, Y., Xie, L., Yuille, A.: Adversarial examples for semantic segmentation and object detection. In: ICCV (2017)
35. Xie, S., Girshick, R., Dollár, P., Tu, Z., He, K.: Aggregated residual transformations for deep neural networks. In: CVPR (2017)
36. Xu, K., et al.: Adversarial t-shirt! evading person detectors in a physical world. In: ECCV (2020)
37. Zeng, X., et al.: Adversarial attacks beyond the image space. In: CVPR (2019)

TuiGAN: Learning Versatile Image-to-Image Translation with Two Unpaired Images

Jianxin Lin[1], Yingxue Pang[1], Yingce Xia[2], Zhibo Chen[1(✉)], and Jiebo Luo[3]

[1] CAS Key Laboratory of Technology in Geo-spatial Information Processing and Application System, University of Science and Technology of China, Hefei, China
{linjx,pangyx}@mail.ustc.edu.cn,chenzhibo@ustc.edu.cn
[2] Microsoft Research Asia, Beijing, China
yingce.xia@microsoft.com
[3] University of Rochester, Rochester, USA
jluo@cs.rochester.edu

Abstract. An unsupervised image-to-image translation (UI2I) task deals with learning a mapping between two domains without paired images. While existing UI2I methods usually require numerous unpaired images from different domains for training, there are many scenarios where training data is quite limited. In this paper, we argue that even if each domain contains a single image, UI2I can still be achieved. To this end, we propose TuiGAN, a generative model that is trained on only two unpaired images and amounts to one-shot unsupervised learning. With TuiGAN, an image is translated in a coarse-to-fine manner where the generated image is gradually refined from global structures to local details. We conduct extensive experiments to verify that our versatile method can outperform strong baselines on a wide variety of UI2I tasks. Moreover, TuiGAN is capable of achieving comparable performance with the state-of-the-art UI2I models trained with sufficient data.

Keywords: Image-to-Image Translation · Generative adversarial network · One-shot unsupervised learning

1 Introduction

Unsupervised image-to-image translation (UI2I) tasks aim to map images from a source domain to a target domain with the main source content preserved and the target style transferred, while no paired data is available to train the models. Recent UI2I methods have achieved remarkable successes [3,22,26,38].

J. Lin and Y. Pang—The first two authors contributed equally to this work.

Electronic supplementary material The online version of this chapter (https://doi.org/10.1007/978-3-030-58548-8_2) contains supplementary material, which is available to authorized users.

© Springer Nature Switzerland AG 2020
A. Vedaldi et al. (Eds.): ECCV 2020, LNCS 12349, pp. 18–35, 2020.
https://doi.org/10.1007/978-3-030-58548-8_2

Among them, conditional UI2I gets much attention, where two images are given: an image from the source domain used to provide the main content, and the other one from the target domain used to specify which style the main content should be converted to. To achieve UI2I, typically one needs to collect numerous unpaired images from both the source and target domains.

However, we often come across cases for which there might not be enough unpaired data to train the image translator. An extreme case resembles one-shot unsupervised learning, where only one image in the source domain and one image in the target domain are given but unpaired. Such a scenario has a wide range of real-world applications, e.g., taking a photo and then converting it to a specific style of a given picture, or replacing objects in an image with target objects for image manipulation. In this paper, we take the first step towards this direction and study UI2I given only two unpaired images.

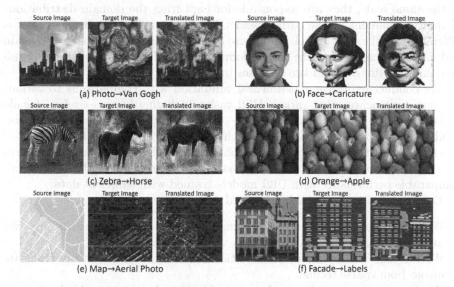

Fig. 1. Several results of our proposed method on various tasks ranging from image style transfer (Figures (a), (b)) to object transformation (Figures (c), (d)) and appearance transformation (Figures (e), (f)). In each sub-figure, the three pictures from left to right refer to the source image (providing the main content), target image (providing the style and high-level semantic information), and translated image.

Note that the above problem subsumes the conventional image style transfer task. Both problems require one source image and one target image, which serve as the content and style images, respectively. In image style transfer, the features used to describe the styles (such as the Gram matrix of pre-trained deep features [7]) of the translated image and the style image should match (e.g., Fig. 1(a)). In our generalized problem, not only the style but the higher-level semantic information should also match. As shown in Fig. 1(c), on the zebra-to-horse translation, not only the background style (e.g., prairie) is transferred, but the high-level semantics (i.e., the profile of the zebra) is also changed.

Achieving UI2I requires the models to effectively capture the variations of domain distributions between two domains, which is the biggest challenge for our problem since there are only two images available. To realize such one-shot translation, we propose a new conditional generative adversarial network, TuiGAN, which is able to transfer the domain distribution of input image to the target domain by progressively translating image from coarse to fine. The progressive translation enables the model to extract the underlying relationship between two images by continuously varying the receptive fields at different scales. Specifically, we use two pyramids of generators and discriminators to refines the generated result progressively from global structures to local details. For each pair of generators at the same scale, they are responsible for producing images that look like the target domain ones. For each pair of discriminators at the same scale, they are responsible for capturing the domain distributions of the two domains at the current scale. The "one-shot" term in our paper is different from the ones in [1,4], which use a single image from the source domain and a set of images from the target domain for UI2I. In contrast, we only use two unpaired images from two domains in our work.

We conduct extensive experimental validation with comparisons to various baseline approaches using various UI2I tasks, including horse ↔ zebra, facade ↔ labels, aerial maps ↔ maps, apple ↔ orange, and so on. The experimental results show that the versatile approach effectively addresses the problem of one-shot image translation. We show that our model can not only outperform existing UI2I models in the one-shot scenario, but more remarkably, also achieve comparable performance with UI2I models trained with sufficient data.

Our contributions can be summarized as follows:

- We propose a TuiGAN to realize image-to-image translation with only two unpaired images.
- We leverage two pyramids of conditional GANs to progressively translate image from coarse to fine.
- We demonstrate that the a wide range of UI2I tasks can be tackled using our versatile model.

2   Related Works

2.1   Image-to-Image Translation

The earliest concept of image-to-image translation (I2I) may be raised in [11] which supports a wide variety of "image filter" effects. Rosales et al. [31] propose to infer correspondences between a source image and another target image using Bayesian framework. With the development of deep neural networks, the advent of Generative Adversarial Networks (GAN) [8] really inspires many works in I2I. Isola et al. [15] propose a conditional GAN called "pix2pix" model for a wide range of supervised I2I tasks. However, paired data may be difficult or even impossible to obtain in many cases. DiscoGAN [20], CycleGAN [38] and Dual-GAN [35] are proposed to tackle the unsupervised image-to-image translation (UI2I) problem by constraining two cross-domain translation models to maintain cycle-consistency. Liu et al. [27] propose a FUNIT model for few-shot UI2I. However, FUNIT requires not only a large amount of training data and computation resources to infer unseen domains, but also the training data and unseen domains to share similar attributes. Our work does not require any pre-training and specific form of data. Related to our work, Benaim et al. [1] and Cohen et al. [4] propose to solve the one-shot cross-domain translation problem, which aims to learn an unidirectional mapping function given a single image from the source domain and a set of images from the target domain. Moreover, their methods cannot translate images in the opposite direction as they claim that one seen sample in the target domain is difficult for capturing domain distribution. However, in this work, we focus on solving UI2I given only two unpaired image from two domains and realizing I2I in both directions.

2.2 Image Style Transfer

Image style transfer can be traced back to Hertzmann et al.'s work [10]. More recent approaches use neural networks to learn the style statistics. Gatys et al. [7] first model image style transfer by minimizing the Gram matrix of pre-trained deep features. Luan et al. [28] further propose to realize photorealistic style transfer which can preserve the photorealism of the content image. To avoid inconsistent stylizations in semantically uniform regions, Li et al. [24] introduce a two-step framework in which both steps have a closed-form solution. However, it is difficult for these models to transfer higher-level semantic structures, such as object transformation. We demonstrate that our model can outperform Li et al. [24] in various UI2I tasks.

2.3 Single Image Generative Models

Single image generative models aim to capture the internal distribution of an image. Conditional GAN based models have been proposed for texture expansion [37] and image retargeting [33]. InGAN [33] is trained with a single natural input and learns its internal patch-distribution by an image-specific GAN. Unconditional GAN based models also have been proposed for texture synthesis [2,16,23] and image manipulation [32]. In particular, SinGAN [32] employs an unconditional pyramidal generative model to learn the patch distribution based on images of different scales. However, these single image generative models usually take one image into consideration and do not capture the relationship between two images. In contrast, our model aims to capture the distribution variations between two unpaired images. In this way, our model can transfer an image from a source distribution to a target distribution while maintaining its internal content consistency.

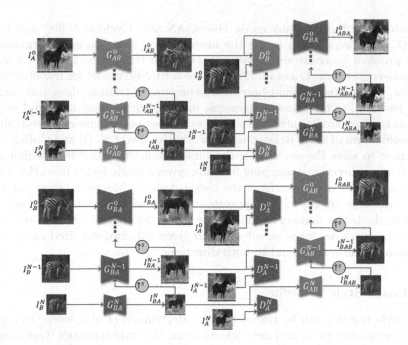

Fig. 2. TuiGAN network architecture: TuiGAN consists of two symmetric pyramids of generators (G_{AB}^ns and G_{BA}^ns) and discriminators (D_B^ns and D_A^ns), $0 \leq n \leq N$. At each scale, the generators take the downsampled source image and previously translated image to generate the new translated image. The discriminators learn the domain distribution by progressively narrowing the receptive fields. The whole framework is learned in a scale-to-scale fashion and the final result is obtained at the finest scale.

3 Method

Given two images $I_A \in A$ and $I_B \in B$, where A and B are two image domains, our goal is to convert I_A to $I_{AB} \in B$ and I_B to $I_{BA} \in A$ without any other data accessible. Since we have only two unpaired images, the translated result (e.g., I_{AB}) should inherit the domain-invariant features of the source image (e.g., I_A) and replace the domain-specific features with the ones of the target image (e.g., I_B) [13,22,25]. To realize such image translation, we need to obtain a pair of mapping functions $G_{AB} : A \mapsto B$ and $G_{BA} : B \mapsto A$, such that

$$I_{AB} = G_{AB}(I_A), \quad I_{BA} = G_{BA}(I_B). \tag{1}$$

Our formulation aims to learn the internal domain distribution variation between I_A and I_B. Considering that the training data is quite limited, G_{AB} and G_{BA} are implemented as two multi-scale conditional GANs that progressively translate images from coarse to fine. In this way, the training data can be fully leveraged at different resolution scales. We downsample I_A and I_B to N different scales, and then obtain $\mathcal{I}_A = \{I_A^n | n = 0, 1, \cdots, N\}$ and $\mathcal{I}_B = \{I_B^n | n = 0, 1, \cdots, N\}$, where I_A^n and I_B^n are downsampled from I_A and I_B, respectively, by a scale factor $(1/s)^n$ ($s \in \mathbb{R}$).

In previous literature, multi-scale architectures have been explored for unconditional image generation with multiple training images [5,12,18,19], conditional image generation with multiple paired training images [34] and image generation with a single training image [32]. In this paper, we leverage the benefit of multi-scale architecture for one-shot unsupervised learning, in which only two unpaired images are used to learn UI2I.

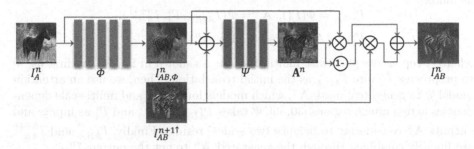

Fig. 3. Architecture of the generator G_{AB}^n, which achieves the $I_A^n \rightarrow I_{AB}^n$ translation. There are two modules, Φ and Ψ. The input I_A^n is first transformerd via Φ to obtain $I_{AB,\Phi}^n$. Then, the transformed $I_{AB,\Phi}^n$, original input I_A^n and the output of previous scale $I_{AB}^{n+1\uparrow}$ are fused by model Ψ to generated a mask $\mathbf{A}^n$. Finally, $I_{AB,\Phi}^n$ and $I_{AB}^{n+1\uparrow}$ are linearly combined through $\mathbf{A}^n$ to obtain the final output.

3.1   Network Architecture

The network architecture of the proposed TuiGAN is shown in Fig. 2. The entire framework consists of two symmetric translation models: G_{AB} for $I_A \rightarrow I_{AB}$ (the top part in Fig. 2) and G_{BA} for $I_B \rightarrow I_{BA}$ (the bottom part in Fig. 2). G_{AB} and G_{BA} are made up of a series of generators, $\{G_{AB}^n\}_{n=0}^N$ and $\{G_{BA}^n\}_{n=0}^N$, which can achieve image translation at the corresponding scales. At each image scale, we also need discriminators D_A^n and D_B^n ($n \in \{0, 1, \cdots, N\}$), which is used to verify whether the input image is a natural one in the corresponding domain.

Progressive Translation. The translation starts from images with the lowest resolution and gradually moves to the higher resolutions. G_{AB}^N and G_{BA}^N first map I_A^N and I_B^N to the corresponding target domains:

$$I_{AB}^N = G_{AB}^N(I_A^N); \;\; I_{BA}^N = G_{BA}^N(I_B^N). \tag{2}$$

For images with scales $n < N$, the generator G_{AB}^n has two inputs, I_A^n and the previously generated I_{AB}^{n+1}. Similarly, G_{BA}^n takes I_B^n and I_{BA}^{n+1} as inputs. Mathematically,

$$I_{AB}^n = G_{AB}^n(I_A^n, I_{AB}^{n+1\uparrow}), \;\; I_{BA}^n = G_{BA}^n(I_B^n, I_{BA}^{n+1\uparrow}), \tag{3}$$

where ↑ means to use bicubic upsampling to resize image by a scale factor s. Leveraging I_{AB}^{n+1}, G_{AB}^n could refine the previous output with more details, and I_{AB}^{n+1} also provides the global structure of the target image for current resolution. Eqn.(3) is iteratively applied until the eventual output I_{AB}^0 and I_{BA}^0 are obtained.

Scale-aware Generator. The network architecture of G_{AB}^n is shown in Fig. 3. Note that G_{AB}^n and G_{BA}^n shares the same architecture but have different weights. G_{AB}^n consists of two fully convolutional networks. Mathematically, G_{AB}^n works as follows:

$$I_{AB,\Phi}^n = \Phi(I_A^n), \quad \mathbf{A}^n = \Psi(I_{AB,\Phi}^n, I_A^n, I_{AB}^{n+1\uparrow}),$$
$$I_{AB}^n = \mathbf{A}^n \otimes I_{AB,\Phi}^n + (1 - \mathbf{A}^n) \otimes I_{AB}^{n+1\uparrow}, \tag{4}$$

where $\otimes$ represents pixel-wise multiplication. As shown in Eq. (4), we first use Φ to preprocess I_A^n into $I_{AB,\Phi}^n$ as the initial translation. Then, we use an attention model Ψ to generate a mask $\mathbf{A}^n$, which models long term and multi-scale dependencies across image regions [30,36]. Ψ takes $I_{AB,\Phi}^n$, $I_{AB}^{n+1\uparrow}$ and I_A^n as inputs and outputs $\mathbf{A}^n$ considering to balance two scales' results. Finally, $I_{AB,\Phi}^n$ and $I_{AB}^{n+1\uparrow}$ are linearly combined through the generated $\mathbf{A}^n$ to get the output I_{AB}^n.

Similarly, the translation $I_B \rightarrow I_{BA}$ at n-th scale is implemented as follows:

$$I_{BA,\Phi}^n = \Phi(I_B^n); \quad \mathbf{A}^n = \Psi(I_{BA,\Phi}^n, I_B^n, I_{BA}^{n+1\uparrow}),$$
$$I_{BA}^n = \mathbf{A}^n \otimes I_{BA,\Phi}^n + (1 - \mathbf{A}^n) \otimes I_{BA}^{n+1\uparrow}. \tag{5}$$

In this way, the generator focuses on regions of the image that are responsible of synthesizing details in current scale and keeps the previously learned global structure untouched in the previous scale. As shown in Fig. 3, the previous generator has generated global structure of a zebra in $I_{AB}^{n+1\uparrow}$, but still fails to generate stripe details. In the n-th scale, the current generator generates an attention map to add stripe details on the zebra and produces better result I_{AB}^n.

3.2 Loss Functions

Our model is progressively trained from low resolution to high resolution. Each scale keeps fixed after training. For any $n \in \{0, 1, \cdots, N\}$, the overall loss function of the n-th scale is defined as follows:

$$\mathcal{L}_{ALL}^n = \mathcal{L}_{ADV}^n + \lambda_{CYC}\mathcal{L}_{CYC}^n + \lambda_{IDT}\mathcal{L}_{IDT}^n + \lambda_{TV}\mathcal{L}_{TV}^n, \tag{6}$$

where $\mathcal{L}_{ADV}^n$, $\mathcal{L}_{CYC}^n$, $\mathcal{L}_{IDT}^n$, $\mathcal{L}_{TV}^n$ refer to adversarial loss, cycle-consistency loss, identity loss and total variation loss respectively, and $\lambda_{CYC}, \lambda_{IDT}, \lambda_{TV}$ are hyperparameters to balance the tradeoff among each loss term. At each scale, the generators aim to minimize $\mathcal{L}_{ALL}^n$ while the discriminators is trained to maximize $\mathcal{L}_{ALL}^n$. We will introduce details of these loss functions.

Adversarial Loss. The adversarial loss builds upon that fact that the discriminator tries to distinguish real images from synthetic images and generator tries to fool the discriminator by generating realistic images. At each scale n, there are two discriminators D_A^n and D_B^n, which take an image as input and output the probability that the input is a natural image in the corresponding domain. We choose WGAN-GP [9] as adversarial loss which can effectively improve the stability of adversarial training by weight clipping and gradient penalty:

$$\mathcal{L}_{\text{ADV}}^n = D_B^n(I_B^n) - D_B^n(G_{AB}^n(I_A^n)) + D_A^n(I_A^n) - D_A^n(G_{BA}^n(I_B^n))$$
$$- \lambda_{\text{PEN}}(\|\nabla_{\hat{I}_B^n} D_B^n(\hat{I}_B^n)\|_2 - 1)^2 - \lambda_{\text{PEN}}(\|\nabla_{\hat{I}_A^n} D_A^n(\hat{I}_A^n)\|_2 - 1)^2, \tag{7}$$

where $\hat{I}_B^n = \alpha I_B^n + (1 - \alpha)I_{AB}^n$, $\hat{I}_A^n = \alpha I_A^n + (1 - \alpha)I_{BA}^n$ with $\alpha \sim U(0,1)$, λ_{PEN} is the penalty coefficient.

Cycle-Consistency Loss. One of the training problems of conditional GAN is mode collapse, i.e., a generator produces an especially plausible output whatever the input is. We utilize cycle-consistency loss [38] to constrain the model to retain the inherent properties of input image after translation: $\forall n \in \{0, 1, \cdots, N\}$,

$$\mathcal{L}_{\text{CYC}}^n = \|I_A^n - I_{ABA}^n\|_1 + \|I_B^n - I_{BAB}^n\|_1, \quad \text{where}$$
$$I_{ABA}^n = G_{BA}^n(I_{AB}^n, I_{ABA}^{n+1\uparrow}), \ I_{BAB}^n = G_{AB}^n(I_{BA}^n, I_{BAB}^{n+1\uparrow}), \quad \text{if } n < N; \tag{8}$$
$$I_{ABA}^N = G_{BA}^N(I_{AB}^N), \ I_{BAB}^N = G_{AB}^N(I_{BA}^N), \quad \text{if } n = N.$$

Identity Loss. We noticed that relying on the two losses mentioned above for one-shot image translation could easily lead to color [38] and texture misaligned results. To tackle the problem, we introduce the identity loss at each scale, which is denoted as L_{IDT}^n. Mathematically,

$$\mathcal{L}_{\text{IDT}}^n = \|I_A^n - I_{AA}^n\|_1 + \|I_B^n - I_{BB}^n\|_1, \quad \text{where}$$
$$I_{AA}^n = G_{BA}^n(I_A^n, I_{AA}^{n+1\uparrow}), \ I_{BB}^n = G_{AB}^n(I_B^n, I_{BB}^{n+1\uparrow}), \quad \text{if } n < N; \tag{9}$$
$$I_{AA}^N = G_{BA}^N(I_A^N), \quad I_{BB}^N = G_{AB}^N(I_B^N), \quad \text{if } n = N.$$

We found that identity loss can effectively preserve the consistency of color and texture tone between the input and the output images as shown in Sect. 4.4.

Total Variation Loss. To avoid noisy and overly pixelated, following [29], we introduce total variation (TV) loss to help in removing rough texture of the generated image and get more spatial continuous and smoother result. It encourages images to consist of several patches by calculating the differences of neighboring pixel values in the image. Let $x[i, j]$ denote the pixel located in the i-th row and j-th column of image x. The TV loss at the n-th scale is defined as follows:

$$\mathcal{L}_{TV}^n = L_{tv}(I_{AB}^n) + L_{tv}(I_{BA}^n),$$

$$L_{tv}(x) = \sum_{i,j} \sqrt{(x[i,j+1] - x[i,j])^2 + (x[i+1,j] - x[i,j])^2}, \ x \in \{I_{AB}^n, I_{BA}^n\}.$$

(10)

3.3   Implementation Details

Network Architecture. As mentioned before, all generators share the same architecture and they are all fully convolutional networks. In detail, Φ is constructed by 5 blocks of the form 3×3 Conv-BatchNorm-LeakyReLU [14] with stride 1. Ψ is constructed by 4 blocks of the form 3×3 Conv-BatchNorm-LeakyReLU. For each discriminator, we use the Markovian discriminator (PatchGANs) [15] which has the same 11×11 patch-size as Φ to keep the same receptive field as generator.

Training Settings. We train our networks using Adam [21] with initial learning rate 0.0005, and we decay the learning rate after every 1600 iterations. We set our scale factor $s = 4/3$ and train 4000 iterations for each scale. The number of scale N is set to 4. For all experiments, we set weight parameters $\lambda_{CYC} = 1$, $\lambda_{IDT} = 1$, $\lambda_{TV} = 0.1$ and $\lambda_{PEN} = 0.1$. Our model requires 3–4 Hrs on a single 2080-Ti GPU with the images of 250×250 size.

4   Experiments

We conduct experiments on several tasks of unsupervised image-to-image translation, including the general UI2I tasks[1], image style transfer, animal face translation and paint-to-image translation, to verify our versatile TuiGAN. To construct datasets of one-shot image translation, given a specific task (like horse↔zebra translation [38]), we randomly sample an image from the source domain and the other one from the target domain, respectively, and train models on the selected data.

4.1   Baselines

We compare TuiGAN with two types of baselines. The first type leverages the full training data without subsampling. We choose CycleGAN [38] and DRIT [22] algorithms for image synthesis. The second type leverages partial data, even one or two images only. We choose the following baselines:

(1) OST [1], where one image from the source domain and a set of images in the target domain are given;

---

[1] In this paper, we refer to general UI2I as tasks where there are multiple images in the source and target domains, i.e., the translation tasks studied in [38].

(2) SinGAN [32], which is a pyramidal unconditional generative model trained on only one image from the target domain, and injects an image from the source domain to the trained model for image translation.

(3) PhotoWCT [24], which can be considered as a special kind of image-to-image translation model, where a content photo is transferred to the reference photo's style while remaining photorealistic.

(4) FUNIT [27], which targets few-shot UI2I and requires lots of data for pre-training. We test the one-shot translation of FUNIT.

(5) ArtStyle [6], which is a classical art style transfer model.

For all the above baselines, we use their official released code to produce the results.

4.2   Evaluation Metrics

(1) **Single Image Fréchet Inception Distance (SIFID)** [32]: SIFID captures the difference of internal distributions between two images, which is implemented by computing the Fréchet Inception Distance (FID) between deep features of two images. A lower SIFID score indicates that the style of two images is more similar. We compute SIFID between translated image and corresponding target image.

(2) **Perceptual Distance (PD)** [17]: PD computes the perceptual distance between images. A lower PD score indicates that the content of two images is more similar. We compute PD between translated image and corresponding source image.

(3) **User Preference (UP)**: We conduct user preference studies for performance evaluation since the qualitative assessment is highly subjective.

4.3   Results

General UI2I Tasks. Following [38], we first conduct general experiments on Facade↔Labels, Apple↔Orange, Horse↔Zebra and Map↔Aerial Photo translation tasks to verify the effectiveness of our algorithm. The visual results of our proposed TuiGAN and the baselines are shown in Fig. 4.

Overall, the images generated by TuiGAN exhibit better translation quality than OST, SinGAN, PhotoWCT and FUNIT. While both SinGAN and PhotoWCT change global colors of the source image, they fail to transfer the high-level semantic structures as our model (e.g., in Facade↔Labels and Horse↔Zebra). Although OST is trained with the full training set of the target domain and transfers high-level semantic structures in some cases, the generated results contain many noticeable artifacts, e.g., the irregular noises on apples and oranges. Compared with CycleGAN and DRIT trained on full datasets, TuiGAN achieves comparable results to them. There are some cases that TuiGAN produces better results than these two models in Labels→Facade, Zebra→Horse

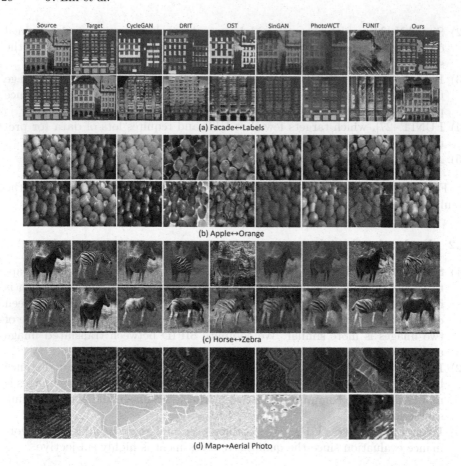

Source Target CycleGAN DRIT OST SinGAN PhotoWCT FUNIT Ours

(a) Facade↔Labels

(b) Apple↔Orange

(c) Horse↔Zebra

(d) Map↔Aerial Photo

Fig. 4. Results of general UI2I tasks using CycleGAN (trained with full training dataset), DRIT (trained with the full training dataset), OST (trained with 1 sample in the source domain and full data in the target domain), SinGAN (trained with one target image), PhotoWCT (trained with two unpaired images), FUNIT (pre-trained) and our TuiGAN (trained with two unpaired images).

tasks, which further verifies that our model can actually capture domain distributions with only two unpaired images.

The results of average SIFID, PD and UP are reported in Table 1. For user preference study, we randomly select 8 unpaired images, and generate 8 translated images for each general UI2I task. In total, we collect 32 translated images for each subject to evaluate. We display the source image, target image and two translated images from our model and another baseline method respectively on a webpage in random order. We ask each subject to select the better translated image at each page. We finally collect the feedback from 18 subjects of total 576 votes and 96 votes for each comparison. We compute the percentage from a method is selected as the User Preference (UP) score.

Table 1. Average SIFID, PD and UP across different general UI2I tasks.

Metrics	CycleGAN	DRIT	OST	SinGAN	PhotoWCT	FUNIT	Ours
SIFID ($\times 10^{-2}$)	0.091	0.142	0.123	0.384	717.622	1510.494	0.080
PD	5.56	8.24	10.26	7.55	3.27	7.55	7.28
UP	61.45%	52.08%	26.04%	6.25%	25.00%	2.08%	–

We can see that TuiGAN obtains the best SIFID score among all the base-lines, which shows that our model successfully captures the distributions of images in the target domain. In addition, our model achieves the third place in PD score after CycleGAN and PhotoWCT. From the visual results, we can see that PhotoWCT can only change global colors of the source image, which is the reason why it achieves the best PD score. As for user study, we can see that most of the users prefer the translation results generated by TuiGAN than OST, SinGAN, PhotoWCT and FUNIT. Compared with DRIT trained on full data, our model also achieves similar votes from subjects.

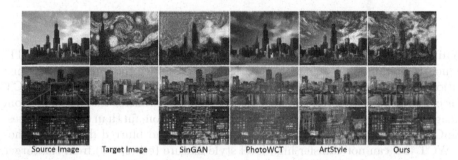

Source Image   Target Image   SinGAN   PhotoWCT   ArtStyle   Ours

Fig. 5. Results of image style transfer. The first row represents the results of art style transfer, and the second row is the results of photorealistic style transfer. We amplify the green boxes in photorealistic style transfer results at the third row to show more details. (Color figure online)

Image Style Transfer. We demonstrate the effectiveness of our TuiGAN on image style transfer: art style transfer, which is to convert image to the target artistic style with specific strokes or textures, and photorealistic style transfer, which is to obtain stylized photo that remains photorealistic. Results are shown in Fig. 5. As can be seen in the first row of Fig. 5, TuiGAN retains the architec-tural contour and generates stylized result with vivid strokes, which just looks like Van Gogh's painting. Instead, SinGAN fails to generate clear stylized image, and PhotoWCT [24] only changes the colors of real photo without capturing the salient painting patterns. In the second row, we transfer the night image to photorealistic day image with the key semantic information retained. Although

SinGAN and ArtStyle produce realistic style, they fail to the maintain detailed edges and structures. The result of PhotoWCT is also not as clean as ours. Overall, our model achieves competitive performance on both types of image style transfer, while other methods usually can only target on a specific task but fail in another one.

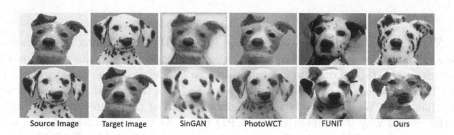

Fig. 6. Results of animal face translation. Our model can accurately transfer the fur colors, while FUNIT, a model pre-trained on animal face dataset, does not work as well as our model.

Animal Face Translation. To compare with the few-shot model FUNIT, which is pretained on animal face dataset, we conduct the animal face translation experiments as shown in Fig. 6. We also include SinGAN and PhotoWCT for comparison. As we can see, our model can better transfer the fur colors from image in the target domain to the that of the source domain than other baselines: SinGAN [32] generates results with faint artifacts and blurred dog shape; PhotoWCT [24] can not transfer high-level style feature (e.g. spots) from the target image although it preserves the content well; and FUNIT generates results that are not consistent with the target dog's appearance.

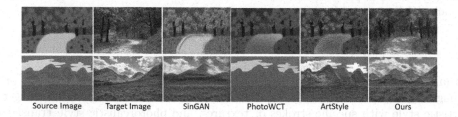

Fig. 7. Results of painting-to-image translation. TuiGAN can translate more specific style patterns of the target image (e.g., leaves on the road in the first row) and maintain more accurate content of the source images (e.g., mountains and clouds in the second row).

Painting-to-Image Translation. This task focuses to generate photo-realistic image with more details based on a roughly related clipart as described in Sin-GAN [32]. We use the two samples provided by SinGAN for comparison. The results are shown in Fig. 7. Although two testing images share similar elements (e.g., trees and road), their styles are extremely different. Therefore, PhotoWCT and ArtStyle fail to transfer the target style in two translation cases. SinGAN also fails to generate specific details, such as leaves on the road in the first row of Fig. 7, and maintain accurate content, such as mountains and clouds in the second row of Fig. 7. Instead, our method preserves the crucial components of input and generates rich local details in two cases.

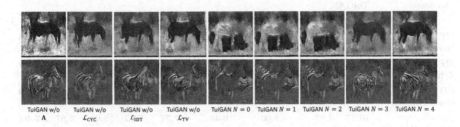

TuiGAN w/o A   TuiGAN w/o $\mathcal{L}_{CYC}$ TuiGAN w/o $\mathcal{L}_{IDT}$ TuiGAN w/o $\mathcal{L}_{TV}$ TuiGAN $N=0$ TuiGAN $N=1$ TuiGAN $N=2$ TuiGAN $N=3$ TuiGAN $N=4$

Fig. 8. Visual results of ablation study.

4.4   Ablation Study

To investigate the influences of different training losses, generator architecture and multi-scale structure, we conduct several ablation studies based on Horse↔Zebra task. Specifically,

(1) Fixing $N = 4$, we remove the cycle-consistent loss (TuiGAN w/o L_{CYC}), identity loss (TuiGAN w/o L_{IDT}), total variation loss (TuiGAN w/o L_{TV}) and compare the differences.
(2) We range N from 0 to 4 to see the effect of different scales. When $N = 0$, our model can be roughly viewed as the CycleGAN [38] that is trained with two unpaired images.
(3) We remove the attention model Ψ in the generators, and combine $I_{AB,\Phi}^{n}$ and $I_{AB}^{n+1\uparrow}$ by simply addition (briefly denoted as TuiGAN w/o **A**).

The qualitative results are shown in Fig. 8. Without L_{IDT}, the generated results suffers from inaccurate color and texture (e.g., green color on the transferred zebra). Without attention mechanism or L_{CYC}, our model can not guarantee the completeness of the object shape (e.g., missed legs in the transferred horse). Without L_{TV}, our model produces images with artifacts (e.g., colour spots around the horse). The results from $N = 0$ to $N = 3$ either have poor global content information contained (e.g. the horse layout) or have obvious artifacts (e.g. the zebra stripes). Our full model (TuiGAN $N = 4$) could capture the salient content of the source image and transfer remarkable style patterns of the target image.

We compute the quantitative ablations by assessing SIFID and PD scores of different variants of TuiGAN. As shown in Table 2, our full model still obtains the lowest SIFID score and PD score, which indicates that our TuiGAN could generate more realistic and stylized outputs while preserving the content unchanged.

Table 2. Quantitative comparisons between different variants of TuiGAN in terms of SIFID and PD scores. The best scores are in bold.

Metrics	TuiGAN								
	w/o **A**	w/o $\mathcal{L}_{CYC}$	w/o $\mathcal{L}_{IDT}$	w/o $\mathcal{L}_{TV}$	$N=0$	$N=1$	$N=2$	$N=3$	$N=4$
SIFID ($\times 10^{-4}$) Horse→Zebra	1.08	3.29	2.43	2.41	2.26	2.32	2.31	2.38	**1.03**
SIFID ($\times 10^{-4}$) Zebra→Horse	2.09	5.61	5.54	10.85	3.75	3.86	3.77	6.30	**1.79**
PD Horse→Zebra	8.00	6.98	8.24	6.90	6.40	6.82	6.76	6.25	**6.16**
PD Zebra→Horse	10.77	7.92	8.00	6.48	7.77	7.92	8.68	6.87	**5.91**

5 Conclusion

In this paper, we propose TuiGAN, a versatile conditional generative model that is trained on only two unpaired image, for image-to-image translation. Our model is designed in a coarse-to-fine manner, in which two pyramids of conditional GANs refine the result progressively from global structures to local details. In addition, a scale-aware generator is introduced to better combine two scales' results. We validate the capability of TuiGAN on a wide variety of unsupervised image-to-image translation tasks by comparing with several strong baselines. Ablation studies also demonstrate that the losses and network scales are reasonably designed. Our work represents a further step toward the possibility of unsupervised learning with extremely limited data.

Acknowledgements. This work was supported in part by NSFC under Grant U1908209, 61632001 and the National Key Research and Development Program of China 2018AAA0101400. This work was also supported in part by NSF award IIS-1704337.

References

1. Benaim, S., Wolf, L.: One-shot unsupervised cross domain translation. In: Proceedings of the 32nd International Conference on Neural Information Processing Systems, pp. 2108–2118. Curran Associates Inc. (2018)

2. Bergmann, U., Jetchev, N., Vollgraf, R.: Learning texture manifolds with the periodic spatial gan. In: Proceedings of the 34th International Conference on Machine Learning, vol. 70, pp. 469–477. JMLR. org (2017)

3. Choi, Y., Uh, Y., Yoo, J., Ha, J.W.: Stargan v2: Diverse image synthesis for multiple domains. arXiv preprint arXiv:1912.01865 (2019)

4. Cohen, T., Wolf, L.: Bidirectional one-shot unsupervised domain mapping. In: Proceedings of the IEEE International Conference on Computer Vision, pp. 1784–1792 (2019)

5. Denton, E.L., Chintala, S., Fergus, R., et al.: Deep generative image models using a laplacian pyramid of adversarial networks. In: Advances in Neural Information Processing Systems, pp. 1486–1494 (2015)

6. Gatys, L.A., Ecker, A.S., Bethge, M.: A neural algorithm of artistic style. Nat. Commun. (2015)

7. Gatys, L.A., Ecker, A.S., Bethge, M.: Image style transfer using convolutional neural networks. In: Proceedings of the IEEE Conference on Computer Vision and Pattern Recognition, pp. 2414–2423 (2016)

8. Goodfellow, I., et al.: Generative adversarial nets. In: Advances in Neural Information Processing Systems, pp. 2672–2680 (2014)

9. Gulrajani, I., Ahmed, F., Arjovsky, M., Dumoulin, V., Courville, A.C.: Improved training of wasserstein gans. In: Advances in Neural Information Processing Systems, pp. 5767–5777 (2017)

10. Hertzmann, A.: Painterly rendering with curved brush strokes of multiple sizes. In: Proceedings of the 25th Annual Conference on Computer Graphics and Interactive Techniques, pp. 453–460 (1998)

11. Hertzmann, A., Jacobs, C.E., Oliver, N., Curless, B., Salesin, D.H.: Image analogies. In: Proceedings of the 28th Annual Conference on Computer Graphics and Interactive Techniques, pp. 327–340 (2001)

12. Huang, X., Li, Y., Poursaeed, O., Hopcroft, J., Belongie, S.: Stacked generative adversarial networks. In: Proceedings of the IEEE Conference on Computer Vision and Pattern Recognition, pp. 5077–5086 (2017)

13. Huang, X., Liu, M.Y., Belongie, S., Kautz, J.: Multimodal unsupervised image-to-image translation. In: ECCV (2018)

14. Ioffe, S., Szegedy, C.: Batch normalization: accelerating deep network training by reducing internal covariate shift. In: International Conference on Machine Learning, pp. 448–456 (2015)

15. Isola, P., Zhu, J.Y., Zhou, T., Efros, A.A.: Image-to-image translation with conditional adversarial networks. In: Proceedings of the IEEE Conference on Computer Vision and Pattern Recognition, pp. 1125–1134 (2017)

16. Jetchev, N., Bergmann, U., Vollgraf, R.: Texture synthesis with spatial generative adversarial networks. arXiv preprint arXiv:1611.08207 (2016)

17. Johnson, J., Alahi, A., Fei-Fei, L.: Perceptual losses for real-time style transfer and super-resolution. In: Leibe, B., Matas, J., Sebe, N., Welling, M. (eds.) ECCV 2016. LNCS, vol. 9906, pp. 694–711. Springer, Cham (2016). https://doi.org/10.1007/978-3-319-46475-6_43

18. Karras, T., Aila, T., Laine, S., Lehtinen, J.: Progressive growing of gans for improved quality, stability, and variation. arXiv preprint arXiv:1710.10196 (2017)

19. Karras, T., Laine, S., Aila, T.: A style-based generator architecture for generative adversarial networks. In: Proceedings of the IEEE Conference on Computer Vision and Pattern Recognition, pp. 4401–4410 (2019)

20. Kim, T., Cha, M., Kim, H., Lee, J.K., Kim, J.: Learning to discover cross-domain relations with generative adversarial networks. In: Proceedings of the 34th International Conference on Machine Learning, pp. 1857–1865 (2017)
21. Kingma, D., Ba, J.: Adam: A method for stochastic optimization. arXiv preprint arXiv:1412.6980 (2014)
22. Lee, H.Y., Tseng, H.Y., Huang, J.B., Singh, M.K., Yang, M.H.: Diverse image-to-image translation via disentangled representations. In: European Conference on Computer Vision (2018)
23. Li, C., Wand, M.: Precomputed real-time texture synthesis with markovian generative adversarial networks. In: Leibe, B., Matas, J., Sebe, N., Welling, M. (eds.) ECCV 2016. LNCS, vol. 9907, pp. 702–716. Springer, Cham (2016). https://doi.org/10.1007/978-3-319-46487-9_43
24. Li, Y., Liu, M.Y., Li, X., Yang, M.H., Kautz, J.: A closed-form solution to photorealistic image stylization. In: Proceedings of the European Conference on Computer Vision (ECCV), pp. 453–468 (2018)
25. Lin, J., Xia, Y., Qin, T., Chen, Z., Liu, T.Y.: Conditional image-to-image translation. In: The IEEE Conference on Computer Vision and Pattern Recognition (CVPR), July 2018, pp. 5524–5532 (2018)
26. Liu, M.Y., Breuel, T., Kautz, J.: Unsupervised image-to-image translation networks. In: Advances in Neural Information Processing Systems, pp. 700–708 (2017)
27. Liu, M.Y., et al.: Few-shot unsupervised image-to-image translation. In: Proceedings of the IEEE International Conference on Computer Vision, pp. 10551–10560 (2019)
28. Luan, F., Paris, S., Shechtman, E., Bala, K.: Deep photo style transfer. In: Proceedings of the IEEE Conference on Computer Vision and Pattern Recognition, pp. 4990–4998 (2017)
29. Mahendran, A., Vedaldi, A.: Understanding deep image representations by inverting them. In: Proceedings of the IEEE Conference on Computer Vision and Pattern Recognition, pp. 5188–5196 (2015)
30. Pumarola, A., Agudo, A., Martinez, A.M., Sanfeliu, A., Moreno-Noguer, F.: Ganimation: anatomically-aware facial animation from a single image. In: Proceedings of the European Conference on Computer Vision (ECCV), pp. 818–833 (2018)
31. Rosales, R., Achan, K., Frey, B.J.: Unsupervised image translation. In: ICCV, pp. 472–478 (2003)
32. Shaham, T.R., Dekel, T., Michaeli, T.: Singan: learning a generative model from a single natural image. In: Proceedings of the IEEE International Conference on Computer Vision, pp. 4570–4580 (2019)
33. Shocher, A., Bagon, S., Isola, P., Irani, M.: Ingan: capturing and remapping the"DNA" of a natural image. arXiv preprint arXiv:1812.00231 (2018)
34. Wang, T.C., Liu, M.Y., Zhu, J.Y., Tao, A., Kautz, J., Catanzaro, B.: High-resolution image synthesis and semantic manipulation with conditional gans. In: Proceedings of the IEEE Conference on Computer Vision and Pattern Recognition, pp. 8798–8807 (2018)
35. Yi, Z., Zhang, H., Tan, P., Gong, M.: Dualgan: unsupervised dual learning for image-to-image translation. In: Proceedings of the IEEE International Conference on Computer Vision, pp. 2849–2857 (2017)
36. Zhang, H., Goodfellow, I., Metaxas, D., Odena, A.: Self-attention generative adversarial networks. In: International Conference on Machine Learning, pp. 7354–7363 (2019)

37. Zhou, Y., Zhu, Z., Bai, X., Lischinski, D., Cohen-Or, D., Huang, H.: Non-stationary texture synthesis by adversarial expansion. ACM Trans. Graph. (TOG) **37**(4), 1–13 (2018)

38. Zhu, J.Y., Park, T., Isola, P., Efros, A.A.: Unpaired image-to-image translation using cycle-consistent adversarial networks. In: The IEEE International Conference on Computer Vision (ICCV) (2017)

Semi-Siamese Training for Shallow Face Learning

Hang Du[1,2], Hailin Shi[2], Yuchi Liu[2], Jun Wang[2], Zhen Lei[3], Dan Zeng[1(✉)], and Tao Mei[2]

[1] Shanghai University, Shanghai, China
{duhang,dzeng}@shu.edu.cn
[2] JD AI Research, Beijing, China
{shihailin,wangjun492,tmei}@jd.com, u6009551@anu.edu.au
[3] NLPR, Institute of Automation, Chinese Academy of Sciences, Beijing, China
zlei@nlpr.ia.ac.cn

Abstract. Most existing public face datasets, such as MS-Celeb-1M and VGGFace2, provide abundant information in both breadth (large number of IDs) and depth (sufficient number of samples) for training. However, in many real-world scenarios of face recognition, the training dataset is limited in depth, *i.e.* only two face images are available for each ID. *We define this situation as Shallow Face Learning, and find it problematic with existing training methods.* Unlike deep face data, the shallow face data lacks intra-class diversity. As such, it can lead to collapse of feature dimension and consequently the learned network can easily suffer from degeneration and over-fitting in the collapsed dimension. In this paper, we aim to address the problem by introducing a novel training method named Semi-Siamese Training (SST). A pair of Semi-Siamese networks constitute the forward propagation structure, and the training loss is computed with an updating gallery queue, conducting effective optimization on shallow training data. Our method is developed without extra-dependency, thus can be flexibly integrated with the existing loss functions and network architectures. Extensive experiments on various benchmarks of face recognition show the proposed method significantly improves the training, not only in shallow face learning, but also for conventional deep face data.

Keywords: Face recognition · Shallow face learning

1 Introduction

Face Recognition (FR) has made remarkable advance and has been widely applied in the last few years. It can be attributed to three aspects, including con-

H. Du and H. Shi—Equal contribution. This work was performed at JD AI Research.

Electronic supplementary material The online version of this chapter (https://doi.org/10.1007/978-3-030-58548-8_3) contains supplementary material, which is available to authorized users.

© Springer Nature Switzerland AG 2020
A. Vedaldi et al. (Eds.): ECCV 2020, LNCS 12349, pp. 36–53, 2020.
https://doi.org/10.1007/978-3-030-58548-8_3

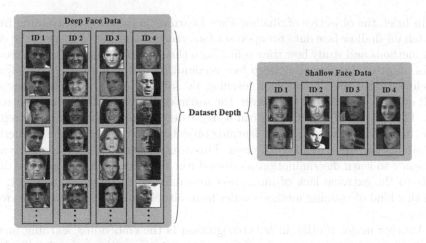

Fig. 1. Deep face data and shallow face data comparison in terms of data depth. Usually, only two images are available for each ID in shallow face data.

volution neural networks (CNNs) [15,16,26,31], loss functions [23,28,29,36,37, 44] and large-scale training datasets [1,12,18,40]. In recent years, the commonly used public training datasets, such as CASIA-WebFace [40], MS-Celeb-1M [12] and VGGFace2 [1] *etc.*, provide abundant information in not only breadth (large number of IDs), but also depth (dozens of face images for each ID). In this paper, we call this type of dataset as deep face data (Fig. 1). Unfortunately, such deep face data is not available in many real-world scenarios. Usually, the training encounters the problem of "shallow face data" in which only two face images are available for each ID (generally a registration photo and a spot photo, so-called "gallery" and "probe"). As a result, it lacks intra-class diversity, which prevents the network from effective optimization and leads to the collapse of feature dimension. In such situation, we find the existing training methods suffer from either the model degeneration or the over-fitting issue.

In this paper, we regard the training on shallow face data as a particular task, named **Shallow Face Learning** (SFL). SFL is similar to the existing problem of Low-shot Learning (LSL) [9] in face recognition, but they have two significant differences. First, LSL performs close-set recognition [3,11,34,38], while SFL includes open-set recognition task in which test IDs are excluded from training IDs. Second, LSL requires pretraining in the source domain (with deep data) before finetuning to the target domain [3,41,47], however, the pretraining is not always a good choice for practical development of face recognition *w.r.t* the following reasons: (1) the network architecture is fixed once the pretraining is done, thus it is inconvenient to change the architecture in the finetuning; (2) deploying new architectures needs restarting from the pretraining, while the pretraining is often time-consuming; (3) there exists domain gap between pretraining data and finetuning data, so the finetuning still suffers from the shallow data problem. Therefore, SFL argues to directly train from scratch on shallow face data.

In brief, the objective of Shallow Face Learning is the effective training from scratch on shallow face data for open-set face recognition. We retrospect the current methods and study how they suffer from the shallow data problem. In recent years, most of the prevailing deep face recognition methods [7,20,21,32,33] are developed from the classification learning by softmax or its variants. They are built on a fully connected (FC) layer, the softmax function and the cross-entropy loss. The weights of the FC layer can be regarded as the prototypes which represent the center of each class. The learning objective is to maximize the prediction probability on the ground-truth class. This routine shows great capability and efficiency to learn discrimination on deep data. However, since the shallow data leads to the extreme lack of intra-class information, as shown in Sect. 3.1, we find this kind of training methods suffer from either model degeneration or over-fitting.

Another major routine in face recognition is the embedding learning methods [6,13,24,27,28], which can learn face representation without the classification layer. For example, Contrastive loss [28] and Triplet loss [24] calculate pair-wise Euclidean distance and optimize the model over the sample relation. Generally, the embedding learning performs better than the classification learning when data becomes shallow. The potential reason is that the embedding learning employs feature comparison between samples, instead of classifying them to the specific classes whose prototypes include large amount of parameters.

However, the performance and efficiency of the embedding learning routine depends on the number of sample pairs matched batch-wisely, which is limited by the GPU memory and hard sampling strategy. In this paper, we desire to draw the advantage of embedding learning for achieving successful classification learning on shallow data. If we address the issues of model degeneration and over-fitting, the training can greatly benefit from the capability and efficiency of the classification learning. A straightforward solution comes up from the plain combination of the two routines, which employs sample features as the prototypes to initialize the FC weights, and runs classification learning with them. The similar modification on softmax has been suggested by the previous methods [47]. Specifically, for each ID of the shallow data, one photo is employed as the initial prototype, and the other photo is employed as training sample. However, such prototype initialization brings still limited improvement when training on shallow data (e.g. DP-softmax in Sect. 4.3). To explain this result, we assume that the prototype becomes too similar to its intra-class training sample, which leads to the extreme small gradient and impedes the optimization.

To overcome this issue, we propose to improve the training method from the perspective of enlarging intra-class diversity. Taking Contrastive or Triplet loss as an example, the features are extracted by the backbone. The backbone can be regarded as a pair (or a triplet) of Siamese networks, since the parameters are fully shared between the networks. We find the crucial technique for the solution is to enforce the backbone being **Semi-Siamese**, which means the two networks have close (but not identical) parameters. One of the networks extracts the feature from gallery as the prototype, and the other network extracts the feature from probe as the training sample, for each ID in the training. The intra-class diversity between the features is guaranteed by the difference between the networks. There are many ways to constrain the two networks to have slight difference. For example, one can add a network constraint between their parameters during SGD (stochastic gradient descent) updating; or SGD updating for one, and moving-average updating for the other (like momentum proposed by [14]). We conduct extensive experiments and find all of them contribute to the shallow face learning effectively. Furthermore, we incorporate the Semi-Siamese backbone with an updating feature-based prototype queue (*i.e.* the gallery queue), and achieve significant improvement on shallow face learning. We name this training scheme as Semi-Siamese Training, which can be integrated with any existing loss functions and network architectures. As shown in Sect. 4.3, whatever loss function, a large improvement can be obtained by using the proposed method for shallow face learning.

Moreover, we conduct two extra experiments to demonstrate more advantage of SST in a wide range. (1) Although SST is proposed for the shallow data problem, an experiment on conventional deep data shows that leading performance can still be obtained by using SST. (2) Another experiment for verifying the effectiveness of SST for real-world scenario, with pretrain-finetune setting, also shows that SST outperforms the conventional training.

In summary, the paper includes the following contributions:

- We formally depict a critical problem of face recognition, *i.e.* Shallow Face Learning, from which the training of face recognition suffers severely. This problem exists in many real-world scenarios but has been overlooked before.
- We study the Shallow Face Learning problem with thorough experiments, and find the lack of intra-class diversity impedes the optimization and leads to the collapse of the feature space. In such situation, the model suffers from degeneration and over-fitting in the training.
- We propose Semi-Siamese Training (SST) method to address the issues in Shallow Face Learning. SST is able to perform with flexible combination with the existing loss functions and network architectures.
- We conduct comprehensive experiments to show the significant improvement by SST on Shallow Face Learning. Besides, the extra experiments show SST also prevails in both conventional deep data and pretrain-finetune task.

2   Related Work

2.1   Deep Face Recognition

There are two major schemes in the deep face recognition. On one hand, the classification based methods is developed from softmax loss and its variants. SphereFace [21] introduces the angular margin to enlarge gaps between classes. CosFace [33] and AM-softmax [32] propose an additive margin to the positive logit. ArcFace [7] employs an additive angular margin inside the cosine and gives a more clear geometric interpretation. On the other hand, the feature embedding methods, such as Contrastive loss [6,13,28] and Triplet loss [24] calculate pairwise Euclidean distance and optimize the network over the relation between samples pairs or triplets. N-pairs loss [27] optimizes positive and negative pairs following a local softmax formulation each mini-batch. Beyond the two schemes, Zhu *et al.* [47] proposes a classification-verification-classification training strategy and DP-softmax loss to progressively enhance the performance on ID versus spot face recognition task.

2.2 Low-Shot Face Recognition

Low-shot Learning (LSL) in face recognition aims at close-set ID recognition by few face samples. Choe *et al.* [5] use data augmentation and generation methods to enlarge the training dataset. Cheng *et al.* [3] propose an enforced softmax that contains optimal dropout, selective attenuation, L_2 normalization and model-level optimization. Wu *et al.* [38] develop the hybrid classifiers by using a CNN and a nearest neighbor model. Guo *et al.* [11] propose to align the norms of the weight vectors of the one-shot classes and the normal classes. Yin *et al.* [41] augment feature space of low-shot classes by transferring the principal components from normal to low-shot classes. The above methods focus on the MS-Celeb-1M Low-shot Learning benchmark [11], which has relatively sufficient samples for each ID in a base set and only one sample for each ID in a novel set, and the target is to recognize faces from both the base and novel set. However, as discussed in the previous section, the differences between Shallow Face Learning and LSL have two aspects. First, the LSL methods aim at close-set classification, for example, in the MS-Celeb-1M Low-shot Learning benchmark, the test IDs are included in the training set; but Shallow Face Learning includes open-set recognition where the test samples belong to unseen classes. Second, unlike the LSL generally employing transfer learning from source dataset (pretraining) to target low-shot dataset (finetuning), Shallow Face Learning argues to train from scratch on target shallow dataset.

2.3 Self-supervised Learning

The recent self-supervised methods [8,14,39,48] have achieved exciting progress in visual representation learning. Exemplar CNN [8] introduces the surrogate class concept for the first time, which adopts a parametric paradigm during training and test. Memory Bank [39] formulates the instance-level discrimination as a metric learning problem, where the similarity between instances are calculated from the features in a non-parametric way. MoCo [14] proposes a dynamic dictionary with a queue and a momentum-updating encoder, which can build a large and consistent dictionary on-the-fly that facilitates the contrastive unsupervised learning. These methods regard each training sample as an instance-level class. Although they employ the data augmentation for each sample, the instance-level classes still lack the intra-class diversity, which is similar to the Shallow Face Learning problem. Inspired by the effectiveness of the self-supervised learning methods, we tackle the issues in Shallow Face Learning with similar techniques, such as the moving-average updating for the Semi-Siamese backbone, and the prototype queue for the supervised loss. Nonetheless, SST is quite different with the self-supervised methods. For example, the gallery queue of SST is built based on the gallery samples rather than the sample augmentation technique; SST aims to deal with Shallow Face Learning which is a specific task in supervised learning. From the perspective of learning against the lack of intra-class diversity, our method generalize the advantages of the self-supervised scheme to the supervised scheme on shallow data.

3 The Proposed Approach

3.1 Shallow Face Learning Problem

Shallow face learning is a practical problem in real-world face recognition scenario. For example, in the authentication application, the face data usually contains a registration photo (gallery) and a spot photo (probe) for each ID. The ID number could be large, but the shallow depth leads to extreme lack of intra-class information. Here, we study how the current classification-based methods suffer from this problem, and what the consequence is brought by the shallow data.

Most of the current prevailing methods are developed from softmax or its variants, which includes a FC layer, the softmax function, and the cross-entropy loss. The output of the FC layer is the inner product $w_j^T x_i$ of the i-th sample feature x_i and j-th class weight w_j. When the feature and weight are normalized by their L_2 norm, the inner product equals to the cosine similarity $w_j^T x_i = \cos(\theta_{i,j})$. Without loss of generality, we take the conventional softmax as an example, and the loss function (omitting the bias term) can be formulated by

$$\mathcal{L} = -\frac{1}{N} \sum_{i=1}^{N} \log \frac{e^{s\cos(\theta_{i,y})}}{e^{s\cos(\theta_{i,y})} + \sum_{j=1,j\neq y}^{n} e^{s\cos(\theta_{i,j})}}, \tag{1}$$

where N is the batch size, n is the class number, s is the scaling parameter, and y is the ground truth label of the i-th sample. The learning objective is maximizing the intra-class pair similarity $w_y^T x_i$ and minimizing the inter-class pairs $w_j^T x_i$ to achieve compact features for intra-class and separate for inter-class. The term inside the logarithm is the prediction probability on the ground truth class $P_y = \frac{e^{s\cos(\theta_{i,y})}}{e^{s\cos(\theta_{i,y})}+\sum_{j=1,j\neq y}^{n} e^{s\cos(\theta_{i,j})}}$, which can be written as $\frac{P_y}{1-P_y} = \frac{e^{s\cos(\theta_{i,y})}}{\sum_{j=1,j\neq y}^{n} e^{s\cos(\theta_{i,j})}}$. This equation implies that the optimal solution of the prototype w_y satisfies two conditions,

$$\begin{cases} w_y = \frac{1}{n_y}\sum_{i=1}^{n_y} x_i, & (i) \\ w_j^T x_i|_{j\neq y} = 0, & (ii) \end{cases} \tag{2}$$

where n_y is the sample number in this class. The Condition (i) means, ideally, the optimal prototype w_y will be the class center which equals to the average of the features in this classes. Meanwhile, the Condition (ii) pushes the prototype w_y to the risk of collapse to zeros in many dimensions. When n_y is large enough (deep

Table 1. The performance(%) on training data and LFW test.

Data	Softmax		A-softmax		AM-softmax		Arc-softmax	
	Training	Test	Training	Test	Training	Test	Training	Test
Deep	99.83	99.10	99.96	99.38	99.42	99.32	98.74	99.40
Shallow	99.40 ↓	92.64 ↓	99.42 ↓	94.67 ↓	99.98 ↑	92.75 ↓	99.99 ↑	94.32 ↓

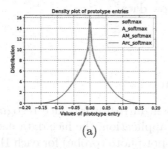
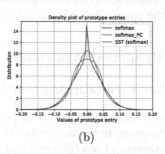

(a)   (b)

Fig. 2. (a) The distribution of the entry values of prototype w_y by the conventional training. The loss functions are softmax, A-softmax, AM-softmax and Arc-softmax. (b) The red curve is the distribution of the entry values of prototype x_y by SST. The green curve is that by prototype constraint. The loss function is softmax. Best viewed in color. (Color figure online)

data), x_i's have large diversity, so keeping the prototype $w_y = \frac{1}{n_y}\sum_{i=1}^{n_y} x_i$ away from collapse. While in shallow data ($n_y = 2$), the prototype w_y is determined by only two samples in a class, *i.e.* the gallery x_g and probe x_p. As a result, the three vectors w_y, x_y and x_p will rapidly become very close ($w_y \approx x_g \approx x_p$), and this class will achieve very small loss value. Considering the network is trained batch-wisely by SGD, in every iteration the network is well-fitted on a small number of classes and badly-fitted on the other classes, thus the total loss value will be oscillating and the training will be harmed (as shown in Fig. 5a dot curves). Moreover, since all the classes gradually lose the intra-class diversity in features space $x_g \approx x_p$, the prototype w_y is pushed to zeros in most dimensions by Condition (ii), and unable to span a discriminative feature space (Fig. 2).

To explore the consequence brought by the shallow data problem, we conduct an experiment on both deep data and shallow data with the loss functions of softmax, A-softmax [21], AM-softmax [32] and Arc-softmax [7]. The deep data is MS1M-v1c [30] (cleaned version of MS-Celeb-1M [12]). Shallow data is a subset of MS1M-v1c, with two face images selected randomly per ID from the deep data. Table 1 shows not only the test accuracy on LFW [17] but also the accuracy on the training data. We can find that the softmax and A-softmax get lower performance both in training and test when training data becomes from deep to shallow, while the AM-softmax and Arc-softmax get higher in training but lower in test. Therefore, we argue that the softmax and A-softmax suffer from the model degeneration issue, while the AM-softmax and Arc-softmax suffer from the over-fitting issue. To further support this argument, we inspect the value of each entry in the prototype w_y, and compute the distribution with Parzen window. The distribution is displayed in Fig. 2a, with the horizontal axis represents the entry values, and the vertical axis represents the density. We can find that most entries of the prototypes degrade to zeros, which means the feature space collapses in most dimensions. In such reduced-dimension space, the models could be easily degenerated or over-fitted.

3.2   Semi-Siamese Training

From the above analysis, we can see, when the data becomes shallow, the current methods are damaged by the model degeneration and over-fitting issues, and the essential reason consists in feature space collapse. To cope with this problem, there are two directions for us to proceed: (1) to make w_y and x_i updating correctly, and (2) to keep the entries of w_y away from zeros.

In the first direction, the major issue is the network is prevented from effective optimization. We retrospect the Condition (i) in Eq. 2 for Shallow Face Learning in which only two face images are available for each ID. We denote them by I_g (gallery) and I_p (probe) and their features $x_g = \phi(I_g)$ and $x_p = \phi(I_p)$, where ϕ is the Siamese backbone. According to Condition (i), $w_y = \frac{1}{2}(x_g + x_p)$. Due to the lack of intra-class diversity, the gallery and probe often have close features, and thus $w_y = \frac{1}{2}(x_g + x_p) \approx x_g \approx x_p$. As studied in the previous subsection, this situation will lead to loss value oscillation, preventing the network from effective optimization. The basic idea to deal with the problem is to keep x_g some distance from x_p, i.e. $x_g = x_p + \epsilon, \forall \epsilon > 0$. To maintain the distance between x_g and x_p, we propose to make the Siamese backbone ϕ being Semi-Siamese. Specifically, a gallery-set network ϕ_g gets input of gallery, and a probe-set network ϕ_p gets input of probe. ϕ_g and ϕ_p have the same architecture but non-identical parameters, $\phi_g = \phi_p + \epsilon'$, so the features prevent being attracted to each other $\phi_g(I_g) = \phi_p(I_p) + \epsilon$. There are certain choices to implement the Semi-Siamese networks. For example, one can add a network constraint $\|\phi_g - \phi_p\| < \epsilon'$ in the training loss, such as $\mathcal{L}_{\text{total}} = \mathcal{L} + \lambda * \|\phi_g - \phi_p\|$, and the non-negative parameter λ is used to balance the network constraint in the training loss. Another choice, as suggested by MoCo [14], aims to update the gallery-set network in the momentum way,

$$\phi_g = m \cdot \phi_g + (1 - m) \cdot \phi_p, \tag{3}$$

where m is the weight of moving-average, and the probe-set network ϕ_p updates with SGD w.r.t. the training loss. Both λ and m are the instantiation of ϵ' which keeps ϕ_g and ϕ_p similar. We compare different implementations for the Semi-Siamese networks, and find the moving-average style gives significant improvement in the experiments. Owing to the intra-class diversity maintaining, the training loss decreases steadily without oscillation (solid curves in Fig. 5a).

In the second direction, a straightforward idea is to add a prototype constraint in the training loss to enlarge the entries of prototype, such like $\mathcal{L} + \beta(\alpha - \|w_y\|)$ with parameters α and β. However, we find this technique

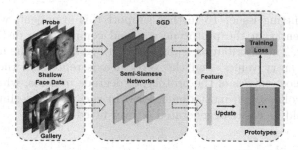

Fig. 3. The overview of Semi-Siamese Training (SST). SST includes a pair of Semi-Siamese networks, which have a probe-set network (the top dark blue network) to embed the probe features, and a gallery-set network (the bottom pale blue network) to update prototypes by gallery features. SST employs the probe features and the feature-based prototypes to compute the training losses which can be any existing loss such as softmax, Arc-softmax, Contrastive *etc*. Finally, the probe-set network is optimized via SGD *w.r.t.* the training loss, and the gallery-set network is updated by the moving-average. Best viewed in color. (Color figure online)

enlarges the entries in most dimension indiscriminately (Fig. 2b the green distribution), and results in decrease (Table 2). Instead of manipulating w_y, we argue to replace w_y by the gallery feature x_g as the prototype. Thus, the prototype totally depends on the output of the backbone, avoiding the zero issue of the parameters (entries) of w_y. The red distribution in Fig. 2b shows the feature-based prototype avoids the issue of collapse while keeping more discriminative components compared with the prototype constraint. Removing w_y also alleviates the over-fitting risk of heavy parameters. The entire prototype set updates by maintaining a gallery queue. Certain self-learning methods [14,39] have studied this technique and its further advantages, such as better generalization when encountering unseen test IDs.

In summary, our Semi-Siamese Training method is developed to address the Shallow Face Learning problem along the two directions. The forward propagation backbone is constituted by a pair of Semi-Siamese networks, each of which is in charge of feature encoding for gallery and probe, respectively; the training loss is computed with an updating gallery queue, so the networks are optimized effectively on the shallow data. This training scheme can be integrated with any form of existing loss function (no matter classification loss or embedding loss) and network architectures (Fig. 3).

4   Experiments

This section is structured as follows. Section 4.1 introduces the datasets and experimental settings. Section 4.2 includes the ablation study on SST. Section 4.3 demonstrates the significant improvement by SST on Shallow Face Learning with various loss functions. Section 4.4 shows the convergence of SST with various backbones. Section 4.5 shows SST can also achieve leading performance on deep face data. Section 4.6 studies SST also outperforms conventional training for the pretrain-finetune task.

4.1 Datasets and Experimental Settings

Training Data. To prove the reproducibility[1], we employ the public datasets for training. To construct shallow data, two images are randomly selected for each ID from the MS1M-v1c [30] dataset. Thus, the shallow data includes 72,778 IDs and 145,556 images. For deep data, we use the full MS1M-v1c which has 44 images per ID in average. Besides, we utilize a real-world surveillance face recognition benchmark QMUL-SurvFace [4] for the experiment of pretrain-finetune.

Test Data. For a thorough evaluation, we adopt LFW [17], BLUFR [19], AgeDB-30 [22], CFP-FP [25], CALFW [46], CPLFW [45], MegaFace [18] and QMUL-SurvFace [4] datasets. AgeDB-30 and CALFW focus on large age gap face verification. CFP-FP and CPLFW aim at cross-pose variants face verification. BLUFR is dedicated for the evaluation with focus at low false accept rates (FAR), and we report the verification rate at the lowest FAR (1e-5) on BLUFR. MegaFace also evaluates the performance of large-scale face recognition with the millions of distractors. QMUL-SurvFace test set aims at real-world surveillance face recognition and has a large domain gap compared to above benchmarks.

Prepossessing. All face images are detected by the FaceBoxes [42]. Then, we align and crop faces to 144×144 RGB images by five facial landmarks [10].

CNN Architecture. To balance the performance and the time cost, we use the MobileFaceNet [2] in the ablation study and the experiments with various loss functions. Besides, we employ Attention-56 [31] in the deep data and pretrain-finetune experiments. The output is a 512-dimension feature. In addition, we also employ extra backbones including VGG-16 [26], SE-ResNet-18 [16], ResNet-50 and -101 [15] to prove the convergence of SST with various architectures.

Training and Evaluation. Four NVIDIA Tesla P40 GPUs are employed for training. The batch size is 256 and the learning rate begins with 0.05. In the shallow data experiments, the learning rate is divided by 10 at the 36k, 54k iterations and the training process is finished at 64k iterations. For the deep data, we divide the learning rate at the 72k, 96k, 112k iterations and finish at 120k iterations. For pretrain-finetune experiments, the learning rate starts from 0.001 and is divided by 10 at the 6k, 9k iterations and finished at 10k iterations. The size of the gallery queue depends on the number of classes in training datasets, so we empirically set it as 16,384 for shallow and deep data, and 2,560 for QMUL-SurvFace. In the evaluation stage, we extract the last

---

[1] Our code will be available at https://github.com/JDAI-CV/faceX-Zoo.

layer output from the probe-set network as the face representation. The cosine similarity is utilized as the similarity metric. For strict and precise evaluation, all the overlapping IDs between training and test datasets are removed according to the list [35].

Loss Function. SST can be flexible integrated with the existing training loss functions. Both classification and embedding learning loss functions are considered as the baseline, and compared with the integration with SST. The classification loss functions include A-softmax [21], AM-softmax [32], Arc-softmax [7], AdaCos [43], MV-softmax [36], DP-softmax [47] and Center loss [37]. The embedding learning methods include Contrastive [28], Triplet [24] and N-pairs [27].

Table 2. Ablation study. Performance (%) on LFW, AgeDB-30, CFP-FP, CALFW, CPLFW and BLUFR.

	LFW	AgeDB	CFP	CALFW	CPLFW	BLUFR
softmax						
Org.	92.64	73.96	70.80	73.05	62.64	27.05
A	91.36	71.85	69.00	72.14	61.35	24.87
B	93.43	76.00	71.46	74.65	62.68	30.65
C	96.62	82.63	79.10	80.18	67.55	52.05
D	98.32	88.77	84.81	86.63	74.80	69.93
SST	**98.77**	**91.60**	**88.63**	**89.82**	78.43	**77.58**
A-softmax						
Org.	94.67	77.88	72.90	75.85	64.00	37.16
A	93.76	76.79	71.35	74.56	62.80	35.18
B	94.62	78.08	74.03	76.35	63.87	38.35
C	96.32	82.28	81.30	81.05	68.77	57.13
D	97.52	85.83	81.87	83.88	71.03	60.79
SST	**98.98**	**91.88**	**89.54**	**89.73**	**77.68**	**80.65**

	LFW	AgeDB	CFP	CALFW	CPLFW	BLUFR
AM-softmax						
Org.	92.75	75.30	68.74	76.63	63.63	33.23
A	92.35	74.12	68.08	74.89	62.76	32.12
B	93.25	76.16	69.17	77.78	63.88	36.59
C	98.02	86.37	85.17	85.72	72.83	62.07
D	98.30	88.18	87.31	87.93	76.27	75.46
SST	**98.97**	**92.25**	**88.97**	**90.23**	**79.45**	**84.95**
Arc-softmax						
Org.	94.32	77.80	71.25	78.15	65.45	40.34
A	93.60	77.35	70.59	77.78	64.28	40.08
B	94.48	78.42	72.15	78.65	65.78	42.50
C	98.20	85.28	81.50	83.50	71.32	60.67
D	98.08	88.68	84.54	86.92	74.40	68.84
SST	**98.95**	**91.73**	**88.59**	**89.85**	**79.60**	**82.74**

4.2 Ablation Study

We analyze each technique in SST, and compare them with the other choices mentioned in the previous section, such as the network constraint ($\|\phi_g - \phi_p\| < \epsilon'$) and the prototype constraint ($\beta(\alpha - \|w_y\|)$). Table 2 compares their performance with four basic loss functions (softmax, A-Softmax, AM-Softmax and Arc-softmax). In this table, "Org." denotes the plain training, "A" denotes the prototype constraint, "B" denotes the network constraint, "C" denotes the gallery queue, "D" denotes the combination of "B" and "C", "SST" denotes the ultimate scheme of Semi-Siamese Training which includes the moving-average updating Semi-Siamese networks and the training scheme with gallery queue. From Table 2, we can conclude: (1) the naive prototype constraint "A" leads to decrease in most terms, which means enlarging w_y in every dimension indiscriminatively does not help on Shallow Face Learning; (2) the network constraint "B" and the gallery queue "C" results in progressive increase, and the combination of them "D" obtains further improvement; (3) finally, SST employs moving-average updating and gallery queue, and achieves the best results by all terms. The comparison indicates SST well addresses the problem in Shallow Face Learning, and obtains significantly improvements in test accuracy.

4.3   SST with Various Loss Functions

First, we train the network on the shallow data with various loss functions and test it on BLUFR at FAR=1e-5 (the blue bars in Fig. 4). The loss functions include classification and embedding ones such as softmax, A-softmax, AM-softmax, Arc-softmax, AdaCos, MV-softmax, DP-softmax, Center loss, Contrastive, Triplet and N-pairs. Then, we train the same network with the same loss functions on the shallow data, but with SST scheme. As shown in Fig. 4, SST can be flexibly integrated with every loss function, and obtains large increase for Shallow Face Learning (the orange bars). Moreover, we employ hard example mining strategies when training on MV-softmax and embedding losses. The results prove SST can also work well with the hard example mining strategies.

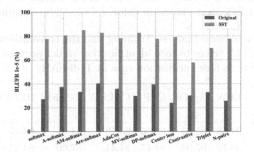

Fig. 4. After being integrated with SST, every loss function obtains large increase on Shallow Face Learning. The blue bars correspond the conventional training on shallow data, the orange bars correspond to SST on shallow data. The test results are the verification rates at FAR=1e-5 on BLUFR. Best viewed in color. (Color figure online)

4.4   SST with Various Network Architectures

To demonstrate the stable convergence in the training, we employ SST to train different CNN architectures, including MobileFaceNet, VGG-16, SE-ResNet-18, Attention-56, ResNet-50 and -101. As shown in Fig. 5b, the loss curves of conventional training (the dot curves) suffer from oscillation. But every loss curve of SST (the solid curves) decreases steadily, indicating the convergence of each network along with the training of SST. Besides, the digits in the legend of Fig. 5b indicates the test result of each network on BLUFR. For conventional training, the test accuracy decreases with the deeper network architectures, showing that the larger model size exacerbates the model degeneration and over-fitting. In contrast, as the network becomes heavy, the test accuracy of SST increases, showing that SST makes increasing contribution with more complicated architectures.

4.5   SST on Deep Data Learning

The previous experiments show SST has well tackled the problems in Shallow Face Learning and obtained significant improvement in test accuracy. To further

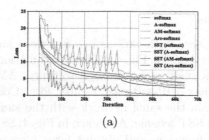 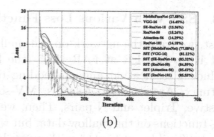

(a)   (b)

Fig. 5. (a) The loss values of conventional training (the dot curves) and SST (the solid curves) with different training loss functions along training iteration. By maintaining the intra-class diversity, SST can prevent the oscillation and achieve steady convergence. (b) The loss values of conventional training and SST with various network architectures along training iteration. The digits in the legend are the test accuracy of each network on BLUFR. Best viewed in color. (Color figure online)

Table 3. Comparison of Semi-Siamese Training and the conventional training on deep data. In MegaFace, "Id." refers to face identification rank1 accuracy with 1M distractors, and "Veri." refers to face verification rate at 1e-6 FAR.

Method	LFW	AgeDB	CFP	CALFW	CPLFW	BLUFR	MegaFace	
							Id	Veri
Softmax	99.58	95.33	92.66	93.18	84.47	93.15	89.89	92.00
AM-softmax	99.70	97.03	94.17	94.41	87.00	94.25	95.67	96.35
Arc-softmax	99.73	97.18	94.37	**95.25**	87.05	**95.29**	96.10	96.81
SST(softmax)	99.67	96.37	94.96	94.18	85.82	94.56	91.01	93.23
SST(AM-softmax)	99.75	**97.20**	95.10	94.62	**88.35**	94.84	**96.27**	**96.96**
SST(Arc-softmax)	**99.77**	97.12	**95.96**	94.78	87.15	94.76	95.63	96.50

explore the advantage of SST for wider application, we adopt SST scheme on the deep data (full version of MS1M-v1c), and make comparison with the conventional training. Table 3 shows the performance on LFW, AgeDB-30, CFP-FP, CALFW, CPLFW and BLUFR and MegaFace. SST gains the leading accuracy in most of the test sets, and also the competitive results on CALFW and BLUFR. SST (softmax) achieves at least one percent improvement on AgeDB-30, CFP-FP, CALFW and CPLFW which include the hard cases of large face pose or large age gap. Notably, SST reduces large amount of FC parameters by which the classification loss is computed for the conventional training. One can refer to the supplementary material for more results on deep data.

4.6 Pretrain and Finetune

In real-world face recognition, there is a large domain gap between the public training datasets and the captured face images. The public training datasets, such as MS-Celeb-1M and VGGFace2, are well-posed face images collected from internet. But the real-world applications are usually quite different. To cope

Table 4. QMUL-SurvFace evaluation. "TPR(%)@FAR" includes the true positive verification rate at varying FARs, and "TPIR20(%)@FPIR" includes rank-20 true positive identification rate at varying false positive identification rates.

Method	TPR(%)@FAR				TPIR20(%)@FPIR			
	0.3	0.1	0.01	0.001	0.3	0.2	0.1	0.01
Softmax	73.09	52.29	26.07	12.54	8.09	6.25	3.98	1.13
AM-softmax [32]	69.59	47.67	23.90	13.24	9.07	7.14	4.65	1.34
Arc-softmax [7]	68.14	48.65	24.12	11.34	8.77	6.88	4.79	1.36
DP-softmax [47]	76.32	55.85	25.32	11.64	7.50	5.38	3.38	0.95
Contrastive [28]	84.48	67.99	31.87	5.31	9.16	6.91	4.44	0.10
Triplet [24]	85.59	69.61	33.76	7.20	10.14	7.70	4.75	0.37
N-pairs [27]	87.26	67.04	29.67	12.07	10.75	8.09	4.87	0.41
SST(softmax)	81.08	63.41	34.20	19.03	11.24	8.49	5.28	1.21
SST(AM-softmax)	86.49	69.41	36.21	18.51	12.22	9.51	5.85	1.63
SST(Arc-softmax)	87.00	68.21	35.72	**22.18**	**12.38**	**9.71**	**6.61**	**1.72**
SST(DP-softmax)	87.69	69.69	**36.32**	14.83	10.20	7.83	5.14	1.08
SST(Contrastive)	87.54	69.91	32.15	9.58	9.87	7.38	4.76	0.78
SST(Triplet)	**90.65**	**73.35**	33.85	12.48	11.09	8.14	5.27	0.92
SST(N-pairs)	89.31	71.26	32.34	15.96	11.30	9.13	5.68	1.22

with this issue, the typical routine is to pretrain a network on the public training datasets and fine-tune it on real-world face data. Although SST is dedicated to the training from scratch on shallow data, we are still interested in employing SST to deal with the challenge in finetuning task. So, we conduct an extra experiment with pretraining on MS1M-v1c and finetuning on QMUL-SurvFace in this subsection. The network is first pretrained with softmax on MS1M-v1c. We randomly select two samples for each ID from the QMUL-SurvFace to construct the shallow data. The network is then finetuned on QMUL-SurvFace shallow data with/without SST. The evaluation is performed on the QMUL-SurvFace test set. From Table 4, we can find that, no matter for classification learning or embedding learning, SST boosts the performance significantly in both verification and identification, compared with the conventional training.

5 Conclusions

In this paper, we first study a critical problem in real-world face recognition, *i.e.* Shallow Face Learning, which has been overlooked before. We analyze how the existing training methods suffer from Shallow Face Learning. The core issues consist in the training difficulty and feature space collapse, which leads to the model degeneration and over-fitting. Then, we propose a novel training method, namely Semi-Siamese Training (SST), to address challenges in Shallow Face Learning.

Specifically, SST employs the Semi-Siamese networks and constructs the gallery queue with gallery features to overcome the issues. SST can perform with flexible integration with the existing training loss functions and network architectures. Experiments on shallow data show SST significantly improves the conventional training. Besides, extra experiments further explore the advantage of SST in a wide range, such as deep data training and pretrain-finetune development.

Acknowledgement. This work was supported in part by the National Key Research & Development Program (No. 2020YFC2003901), Chinese National Natural Science Foundation Projects #61872367, and #61572307, and Beijing Academy of Artificial Intelligence (BAAI).

References

1. Cao, Q., Shen, L., Xie, W., Parkhi, O.M., Zisserman, A.: Vggface2: a dataset for recognising faces across pose and age. In: 2018 13th IEEE International Conference on Automatic Face & Gesture Recognition (FG 2018), pp. 67–74. IEEE (2018)
2. Chen, S., Liu, Y., Gao, X., Han, Z.: MobileFaceNets: efficient CNNs for accurate real-time face verification on mobile devices. In: Zhou, J., et al. (eds.) CCBR 2018. LNCS, vol. 10996, pp. 428–438. Springer, Cham (2018). https://doi.org/10.1007/978-3-319-97909-0_46
3. Cheng, Y., et al.: Know you at one glance: a compact vector representation for low-shot learning. In: Proceedings of the IEEE International Conference on Computer Vision Workshops, pp. 1924–1932 (2017)
4. Cheng, Z., Zhu, X., Gong, S.: Surveillance face recognition challenge. arXiv preprint arXiv:1804.09691 (2018)
5. Choe, J., Park, S., Kim, K., Hyun Park, J., Kim, D., Shim, H.: Face generation for low-shot learning using generative adversarial networks. In: Proceedings of the IEEE International Conference on Computer Vision Workshops, pp. 1940–1948 (2017)
6. Chopra, S., Hadsell, R., LeCun, Y.: Learning a similarity metric discriminatively, with application to face verification. In: IEEE Computer Society Conference on Computer Vision and Pattern Recognition, vol. 1, pp. 539–546 (2005)
7. Deng, J., Guo, J., Xue, N., Zafeiriou, S.: Arcface: additive angular margin loss for deep face recognition. In: Proceedings of the IEEE Conference on Computer Vision and Pattern Recognition, pp. 4690–4699 (2019)
8. Dosovitskiy, A., Springenberg, J.T., Riedmiller, M., Brox, T.: Discriminative unsupervised feature learning with convolutional neural networks. In: Advances in Neural Information Processing Systems, pp. 766–774 (2014)
9. Fei-Fei, L., Fergus, R., Perona, P.: One-shot learning of object categories. IEEE Trans. Pattern Anal. Mach. Intelligence **28**(4), 594–611 (2006)
10. Feng, Z.H., Kittler, J., Awais, M., Huber, P., Wu, X.J.: Wing loss for robust facial landmark localisation with convolutional neural networks. In: Proceedings of the IEEE Conference on Computer Vision and Pattern Recognition, pp. 2235–2245 (2018)
11. Guo, Y., Zhang, L.: One-shot face recognition by promoting underrepresented classes. arXiv preprint arXiv:1707.05574 (2017)

12. Guo, Y., Zhang, L., Hu, Y., He, X., Gao, J.: MS-Celeb-1M: a dataset and benchmark for large-scale face recognition. In: Leibe, B., Matas, J., Sebe, N., Welling, M. (eds.) ECCV 2016. LNCS, vol. 9907, pp. 87–102. Springer, Cham (2016). https://doi.org/10.1007/978-3-319-46487-9_6

13. Hadsell, R., Chopra, S., LeCun, Y.: Dimensionality reduction by learning an invariant mapping. In: IEEE Computer Society Conference on Computer Vision and Pattern Recognition, vol. 2, pp. 1735–1742 (2006)

14. He, K., Fan, H., Wu, Y., Xie, S., Girshick, R.: Momentum contrast for unsupervised visual representation learning. In: Proceedings of the IEEE Conference on Computer Vision and Pattern Recognition (2020)

15. He, K., Zhang, X., Ren, S., Sun, J.: Deep residual learning for image recognition. In: Proceedings of the IEEE Conference on Computer Vision and Pattern Recognition, pp. 770–778 (2016)

16. Hu, J., Shen, L., Sun, G.: Squeeze-and-excitation networks. In: Proceedings of the IEEE Conference on Computer Vision and Pattern Recognition, pp. 7132–7141 (2018)

17. Huang, G.B., Ramesh, M., Berg, T., Learned-Miller, E.: Labeled faces in the wild: a database for studying face recognition in unconstrained environments. Technical Report 07-49, University of Massachusetts, Amherst (2007)

18. Kemelmacher-Shlizerman, I., Seitz, S.M., Miller, D., Brossard, E.: The megaface benchmark: 1 million faces for recognition at scale. In: Proceedings of the IEEE Conference on Computer Vision and Pattern Recognition, pp. 4873–4882 (2016)

19. Liao, S., Lei, Z., Yi, D., Li, S.Z.: A benchmark study of large-scale unconstrained face recognition. In: IEEE International Joint Conference on Biometrics, pp. 1–8 (2014)

20. Liu, H., Zhu, X., Lei, Z., Li, S.Z.: Adaptiveface: adaptive margin and sampling for face recognition. In: Proceedings of the IEEE Conference on Computer Vision and Pattern Recognition, pp. 11947–11956 (2019)

21. Liu, W., Wen, Y., Yu, Z., Li, M., Raj, B., Song, L.: Sphereface: deep hypersphere embedding for face recognition. In: Proceedings of the IEEE Conference on Computer Vision and Pattern Recognition, pp. 212–220 (2017)

22. Moschoglou, S., Papaioannou, A., Sagonas, C., Deng, J., Kotsia, I., Zafeiriou, S.: Agedb: the first manually collected, in-the-wild age database. In: Proceedings of the IEEE Conference on Computer Vision and Pattern Recognition Workshops, pp. 51–59 (2017)

23. Ranjan, R., Castillo, C.D., Chellappa, R.: L2-constrained softmax loss for discriminative face verification. arXiv preprint arXiv:1703.09507 (2017)

24. Schroff, F., Kalenichenko, D., Philbin, J.: Facenet: A unified embedding for face recognition and clustering. In: Proceedings of the IEEE Conference on Computer Vision and Pattern Recognition, pp. 815–823 (2015)

25. Sengupta, S., Chen, J.C., Castillo, C., Patel, V.M., Chellappa, R., Jacobs, D.W.: Frontal to profile face verification in the wild. In: 2016 IEEE Winter Conference on Applications of Computer Vision, pp. 1–9. IEEE (2016)

26. Simonyan, K., Zisserman, A.: Very deep convolutional networks for large-scale image recognition. CoRR abs/1409.1556 (2015)

27. Sohn, K.: Improved deep metric learning with multi-class n-pair loss objective. In: Advances in Neural Information Processing Systems, pp. 1857–1865 (2016)

28. Sun, Y., Chen, Y., Wang, X., Tang, X.: Deep learning face representation by joint identification-verification. In: Advances in Neural Information Processing Systems, pp. 1988–1996 (2014)

29. Taigman, Y., Yang, M., Ranzato, M., Wolf, L.: Deepface: closing the gap to human-level performance in face verification. In: Proceedings of the IEEE Conference on Computer Vision and Pattern Recognition, pp. 1701–1708 (2014)
30. trillionpairs.org: http://trillionpairs.deepglint.com/overview
31. Wang, F., et al.: Residual attention network for image classification. In: Proceedings of the IEEE Conference on Computer Vision and Pattern Recognition, pp. 3156–3164 (2017)
32. Wang, H., et al.: Cosface: large margin cosine loss for deep face recognition. In: Proceedings of the IEEE Conference on Computer Vision and Pattern Recognition, pp. 5265–5274 (2018)
33. Wang, L., Li, Y., Wang, S.: Feature learning for one-shot face recognition. In: 2018 25th IEEE International Conference on Image Processing, pp. 2386–2390 (2018)
34. Wang, L., Li, Y., Wang, S.: Feature learning for one-shot face recognition. In: 2018 25th IEEE International Conference on Image Processing, pp. 2386–2390 (2018)
35. Wang, X., Wang, S., Wang, J., Shi, H., Mei, T.: Co-mining: deep face recognition with noisy labels. In: Proceedings of the IEEE International Conference on Computer Vision, pp. 9358–9367 (2019)
36. Wang, X., Zhang, S., Wang, S., Fu, T., Shi, H., Mei, T.: Mis-classified vector guided softmax loss for face recognition. In: Proceedings of the AAAI Conference on Artificial Intelligence (2020)
37. Wen, Y., Zhang, K., Li, Z., Qiao, Yu.: A discriminative feature learning approach for deep face recognition. In: Leibe, B., Matas, J., Sebe, N., Welling, M. (eds.) ECCV 2016. LNCS, vol. 9911, pp. 499–515. Springer, Cham (2016). https://doi.org/10.1007/978-3-319-46478-7_31
38. Wu, Y., Liu, H., Fu, Y.: Low-shot face recognition with hybrid classifiers. In: Proceedings of the IEEE International Conference on Computer Vision Workshops, pp. 1933–1939 (2017)
39. Wu, Z., Xiong, Y., Yu, S.X., Lin, D.: Unsupervised feature learning via non-parametric instance discrimination. In: Proceedings of the IEEE Conference on Computer Vision and Pattern Recognition, pp. 3733–3742 (2018)
40. Yi, D., Lei, Z., Liao, S., Li, S.Z.: Learning face representation from scratch. arXiv preprint arXiv:1411.7923 (2014)
41. Yin, X., Yu, X., Sohn, K., Liu, X., Chandraker, M.: Feature transfer learning for face recognition with under-represented data. In: Proceedings of the IEEE Conference on Computer Vision and Pattern Recognition, pp. 5704–5713 (2019)
42. Zhang, S., Zhu, X., Lei, Z., Shi, H., Wang, X., Li, S.Z.: Faceboxes: a cpu real-time face detector with high accuracy. In: 2017 IEEE International Joint Conference on Biometrics, pp. 1–9. IEEE (2017)
43. Zhang, X., Zhao, R., Qiao, Y., Wang, X., Li, H.: Adacos: adaptively scaling cosine logits for effectively learning deep face representations. In: Proceedings of the IEEE Conference on Computer Vision and Pattern Recognition, pp. 10823–10832 (2019)
44. Zhao, K., Xu, J., Cheng, M.M.: Regularface: deep face recognition via exclusive regularization. In: Proceedings of the IEEE Conference on Computer Vision and Pattern Recognition, pp. 1136–1144 (2019)
45. Zheng, T., Deng, W.: Cross-pose lfw: A database for studying crosspose face recognition in unconstrained environments. Beijing University of Posts and Telecommunications, Technical Report, pp. 18–01 (2018)
46. Zheng, T., Deng, W., Hu, J.: Cross-age lfw: A database for studying cross-age face recognition in unconstrained environments. arXiv preprint arXiv:1708.08197 (2017) 10

47. Zhu, X., et al.: Large-scale bisample learning on id versus spot face recognition. Int. J. Comput. Vis. **127**(6–7), 684–700 (2019)
48. Zhuang, C., Zhai, A.L., Yamins, D.: Local aggregation for unsupervised learning of visual embeddings. In: Proceedings of the IEEE International Conference on Computer Vision, pp. 6002–6012 (2019)

GAN Slimming: All-in-One GAN Compression by a Unified Optimization Framework

Haotao Wang[1], Shupeng Gui[2], Haichuan Yang[2], Ji Liu[3],
and Zhangyang Wang[1(✉)]

[1] University of Texas at Austin, Austin, TX 78712, USA
{htwang,atlaswang}@utexas.edu
[2] University of Rochester, Rochester, NY 14627, USA
{sgui2,hyang36}@ur.rochester.edu
[3] AI Platform, Ytech Seattle AI Lab, FeDA Lab, Kwai Inc., Seattle, WA 98004, USA
ji.liu.uwisc@gmail.com

Abstract. Generative adversarial networks (GANs) have gained increasing popularity in various computer vision applications, and recently start to be deployed to resource-constrained mobile devices. Similar to other deep models, state-of-the-art GANs suffer from high parameter complexities. That has recently motivated the exploration of compressing GANs (usually generators). Compared to the vast literature and prevailing success in compressing deep classifiers, the study of GAN compression remains in its infancy, so far leveraging individual compression techniques instead of more sophisticated combinations. We observe that due to the notorious instability of training GANs, heuristically stacking different compression techniques will result in unsatisfactory results. To this end, we propose the first unified optimization framework combining multiple compression means for GAN compression, dubbed **GAN Slimming** (GS). GS seamlessly integrates three mainstream compression techniques: model distillation, channel pruning and quantization, together with the GAN minimax objective, into one unified optimization form, that can be efficiently optimized from end to end. Without bells and whistles, GS largely outperforms existing options in compressing image-to-image translation GANs. Specifically, we apply GS to compress CartoonGAN, a state-of-the-art style transfer network, by up to **47**× times, with minimal visual quality degradation. Codes and pre-trained models can be found at https://github.com/TAMU-VITA/GAN-Slimming.

1 Introduction

Generative adversarial networks (GANs) [12], especially, image-to-image translation GANs, have been successfully applied to image synthesis [30], style

Electronic supplementary material The online version of this chapter (https://doi.org/10.1007/978-3-030-58548-8_4) contains supplementary material, which is available to authorized users.

© Springer Nature Switzerland AG 2020
A. Vedaldi et al. (Eds.): ECCV 2020, LNCS 12349, pp. 54–73, 2020.
https://doi.org/10.1007/978-3-030-58548-8_4

	Original result	Our result (32bit)	Our result (8bit)
Source image	56.46 GFLOPs	1.34 GFLOPs	1.20 GFLOPs
	42.34 MB	0.80 MB	0.18 MB

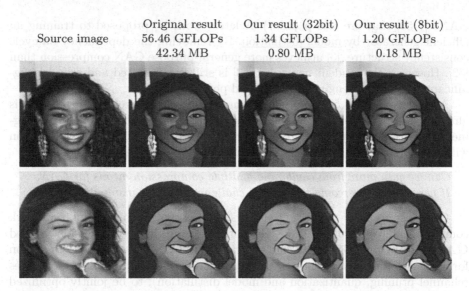

Fig. 1. Representative visual examples by GAN Slimming on CartoonGAN [7].

transfer [7], image editing and enhancement [26,34,35], to name just a few. Due to the growing usage, there has been an increasing demand to deploy them on resource-constrained devices [38]. For example, many filter-based image editing applications now desire to run image-to-image translation GANs locally. However, GANs, just like most other deep learning models, bear explosive parameter amounts and computational complexities. For example, in order to process a 256×256 image, a state-of-the-art style transfer network, CartoonGAN [7], would cost over 56 GFLOPs. Launching such models on mobile devices requires considerable memory and computation costs, which would be infeasible for most devices, or at least degrades user experience due to the significant latency (Fig. 1).

Existing deep model compression methods mainly focus on image classification or segmentation tasks, and were not directly applicable on GAN compression tasks due to notorious instability of GAN minimax training. For example, [52] shows that generators compressed by state-of-the-art classifier compression methods [23,39,42] all suffer great performance decay compared with the original generator. Combining (either heuristically cascading or jointly training) multiple different compression techniques, such as channel pruning, model distillation, quantization and weight sharing, has been shown to outperform separately using single compression techniques alone in traditional classification tasks [43,47,57]. In comparison, current methods [3,52] have so far only tried to apply one single technique to compressing GANs. [52] proposed the first dedicated GAN compression algorithm: an evolutionary method based channel pruning algorithm. However, the method is specifically designed for CycleGAN and non-straightforward to extend to GANs without cycle consistency structure (e.g., encoder-decoder

GANs [7,50] that are also popular). A latest work [3] proposed to training an efficient generator by <u>model distillation</u>. By removing the dependency on cycle consistency structure, [3] achieves more general-purpose GAN compression than [52]. However, the student network in [3] is still hand-crafted and relies on significant architecture engineering for good performance.

As discussed in [3,52], applying a single compression technique to GANs is already challenging due to their notorious training instability. As one may imagine, integrating multiple compression techniques together for GAN compression will only further amplify the instability, putting an open question:

> *Can we gain more from combining multiple compression means for GANs?*
> *If yes, how to overcome the arising challenge of GAN instability?*

Our answer is by presenting the first end-to-end optimization framework combining multiple compression means for general GAN compression, named **GAN Slimming (GS)**. The core contribution of GS is a unified optimization form, that seamlessly integrates three popular model compression techniques (channel pruning, quantization and model distillation), to be jointly optimized in a minimax optimization framework. GS pioneers to advance GAN compression into jointly leveraging multiple model compression methods, and demonstrate the feasibility and promise of doing so, despite the GAN instability.

Experiments demonstrate that GS overwhelms state-of-the-art GAN compression options that rely on single compression means. For example, we compress the heavily-parameterized CartoonGAN by up to 47×, achieving nearly real-time cartoon style transfer on mobile devices, with minimal visual quality loss. Moreover, we have included a detailed ablation study for a deeper understanding of GS. Specifically, we demonstrate that *naively stacking different compression techniques cannot achieve satisfactory GAN compression, sometimes even hurting catastrophically*, therefore testifying the necessity of our unified optimization. We also verify the effectiveness of incorporating the minimax objective into this specific problem.

2   Related Works

2.1   Deep Model Compression

Many model compression methods, including knowledge distillation [47], pruning [18,20] and quantization [18], have been investigated to compress large deep learning models, primarily classifiers [2,4,6,17,51,53,58,60,62]. Structured pruning [61], such as channel pruning [24,39,45,61,63], result in hardware-friendly compressed models and thus are widely adopted in real-world applications. The authors of [61] enforced structured sparsity constraint on each layer's kernel weights, aided by group Lasso [31] to solve the optimization. [39] added ℓ_1 constraint on the learnable scale parameters of batch normalization layers in order to encourage channel sparsity, and used subgradient descent to optimize the ℓ_1 loss. Similarly, [24] also utilized sparsity constraint on channel-wise scale parameters, solved with an accelerated proximal gradient method [46].

Quantization, as another popular compression means, reduces the bit width of the element-level numerical representations of weights and activations. Earlier works [18,37,66] presented to quantize all layer-wise weights and activations to the same low bit width, *e.g.*, from 32 bits to 8 bits or less. The model could even consist of only binary weights in the extreme case [8,48]. Note that, introducing quantization into network weights or activations will result in notable difficulty for propagating gradients [25]. Straight-through estimator (STE) [8] is a successful tool to solve this problem by a proxy gradient for back propagation.

Knowledge distillation was first developed in [22] to transfer the knowledge in an ensemble of models to a single model, using a soft target distribution produced by the former models. It was later on widely used to obtain a smaller network (student model), by fitting the "soft labels" (probabilistic outputs) generated from a trained larger network (teacher model). [1] used distillation to train a more efficient and accurate predictor. [41] unified distillation and privileged information into one generalized distillation framework to learn better representations. [5,59] used generative adversarial training for model distillation.

Combination of Multiple Compression Techniques. For compressing a deep classifier, [55,64] proposed to jointly train (unstructured) pruning and quantization together. [54,57] adopted knowledge distillation to fine-tune a pruned student network, by utilizing the original dense network as teacher, which essentially followed a two-step cascade pipeline. Similarly, [43,47] used full-precision networks as teachers to distill low-precision student networks. [14] showed jointly training pruning and quantization can obtain compact classifiers with state-of-the-art trade-off between model efficiency and adversarial robustness.

Up to our best knowledge, all above methods cascade or unify two compression techniques, besides that they investigate compressing deep classifiers only. In comparison, our proposed framework jointly optimize three methods in one unified form[1], that is innovative even for general model compression. It is further adapted for the special GAN scenario, by incorporating the minimax loss.

2.2 GAN Compression

GANs have been successful on many image generation and translation tasks [12,13,16,29,44], yet their training remains notoriously unstable. Numerous techniques were developed to stabilize the GAN training, *e.g.*, spectral normalization [44], gradient penalty [15] and progressive training [29]. As discussed in [3,52], the training difficulty causes extra challenges for compressing GANs, and failed many traditional pruning methods for classifiers such as [23,39,42].

The authors of [52] proposed the first dedicated GAN compression method: a co-evolution algorithm based channel pruning method for CycleGAN. Albeit successfully demonstrated on the style transfer application, their method faces

---

[1] A concurrent work [65] jointly optimized pruning, decomposition, and quantization, into one unified framework for reducing the memory storage/access.

several limitations. First, their co-evolution algorithm relies on the cycle consistency loss to simultaneously compress generators of both directions. It is hence non-straightforward to extend to image-to-image GANs without cycle consistent loss (*e.g.*, encoder-decoder GANs [7,50]). Second, in order to avoid the instability in GAN training, the authors model GAN compression as a "dense prediction" process by fixing the original discriminator instead of jointly updating it with the generator in a minimax optimization framework. This surrogate leads to degraded performance of the compressed generator, since the fixed discriminator may not suit the changed (compressed) generator capacity. These limitations hurdle both its broader application scope and performance.

The latest concurrent work [3] explored model distillation: to guide the student to effectively inherit knowledge from the teacher, the authors proposed to jointly distill generator and discriminator in a minimax two-player game. [3] improved over [52] by removing the above two mentioned hurdles. However, as we observe from experiments (and also confirmed with their authors), the success of [3] hinges notably on the appropriate design of student network architectures. Our method could be considered as another important step over [3], that "learns" the student architecture jointly with the distillation, via pruning and quantization, as to be explained by the end of Sect. 3.1.

3   The GAN Slimming Framework

Considering a dense full-precision generator G_0 which converts the images from one domain $\mathcal{X}$ to another $\mathcal{Y}$, our aim is to obtain a more efficient generator G from G_0, such that their generated images $\{G_0(x), x \in \mathcal{X}\}$ and $\{G(x), x \in \mathcal{X}\}$ have similar style transfer qualities. In this section, we first outline the unified optimization form of our GS framework combining model distillation, channel pruning and quantization (Sect. 3.1). We then show how to solve each part of the optimization problem respectively (Sect. 3.2), and eventually present the overall algorithm (Sect. 3.3).

3.1   The Unified Optimization Form

We start formulating our GS objective from the traditional minimax optimization problem in GAN:

$$\min_{G} \max_{D} L_{GAN}, \text{ where } L_{GAN} = \mathbb{E}_{y \in \mathcal{Y}}[\log(D(y))] + \mathbb{E}_{x \in \mathcal{X}}[\log(1 - D(G(x)))], \quad (1)$$

where D is the discriminator jointly trained with efficient generator G by minimax optimization. Since G is the functional part to be deployed on mobile devices and D can be discarded after training, we do not need to compress D. Inspired by the success of model distillation in previous works [3,22], we add a model distillation loss term L_{dist} to enforce the small generator G to mimic the behaviour of original large generator G_0, where $d(\cdot, \cdot)$ is some distance metric:

$$L_{dist} = \mathbb{E}_{x \in \mathcal{X}}[d(G(x), G_0(x))], \quad (2)$$

The remaining key question is: how to properly define the architecture of G? Previous methods [3,22] first hand-crafted the smaller student model's architecture and then performed distillation. However, it is well known that the choice of the student network structure will affect the final performance notably too, in addition to the teacher model's strength.

Unlike existing distillation methods [3], we propose to *jointly infer* the G architecture together with the distillation process. Specifically, we assume that G can be "slimmed" from G_0, through two popular compression operations: channel pruning and quantization. For channel pruning, we follow [39] to apply L_1 norm on the trainable scale parameters γ in the normalization layers to encourage channel sparsity: $L_{cp} = \|\gamma\|_1$. Denoting all other trainable weights in G as W, we could incorporate the channel pruning via such sparsity constraint into the distillation loss in Eq. (2) as below:

$$L_{dist}(W,\gamma) + \rho L_{cp}(\gamma) = \mathbb{E}_{x\in\mathcal{X}}[\mathrm{d}(G(x;W,\gamma),G_0(x))] + \rho\|\gamma\|_1, \qquad (3)$$

where ρ is the trade-off parameter controlling the network sparsity level. Further, to integrate quantization, we propose to quantize both activations and weights,[2] using two quantizers $q_a(\cdot)$ and $q_w(\cdot)$, respectively, to enable the potential flexibility for hybrid quantization [56]. While it is completely feasible to adopt learnable quantization intervals [28], we adopt uniform quantizers with pre-defined bit-width for $q_a(\cdot)$ and $q_w(\cdot)$, respectively, for the sake of simplicity (including hardware implementation ease). The quantized weights can be expressed as $q_w(W)$, while we use G_q to denote generators equipped with activation quantization $q_a(\cdot)$ for notation compactness. Eventually, the final objective combining model distillation, channel pruning and quantization has the following form:

$$
\begin{aligned}
L(W,\gamma,\theta) = \ & L_{GAN}(W,\gamma,\theta) + \beta L_{dist}(W,\gamma) + \rho L_{cp}(\gamma) \\
= \ & \mathbb{E}_{y\in\mathcal{Y}}\left[\log(D(y;\theta))\right] + \mathbb{E}_{x\in\mathcal{X}}\left[\log(1 - D(G_q(x;q_w(W),\gamma);\theta))\right] \\
& + \mathbb{E}_{x\in\mathcal{X}}\left[\beta\mathrm{d}(G_q(x;q_w(W),\gamma),G_0(x))\right] \\
& + \rho\|\gamma\|_1,
\end{aligned}
$$

$$(4)$$

where θ represents the parameters in D. The blue parts represent the distillation component, green represents channel pruning red represents quantization. The above Eq. (4) is the target objective of GS, which is to be solved in a minimax optimization framework:

$$\min_{W,\gamma}\max_{\theta}\ L(W,\gamma,\theta) \qquad (5)$$

Connection to AutoML Compression. Our framework could be alternatively interpreted as performing a special neural architecture search (NAS) [11,19] to obtain the student model, where the student's architecture needs be "morphable" from the teacher's through only pruning and quantization operations.

---

[2] We only quantize W, while always leaving γ unquantized.

Interestingly, two concurrent works [9,36] have successfully applied NAS to search efficient generator architectures, and both achieved very promising performance too. We notice that a notable portion of the performance gains shall be attributed to the carefully designed search spaces, as well as computationally intensive search algorithms. In comparison, our framework is based on an end-to-end optimization formulation, that (1) has explainable and well-understood behaviors; (2) is lighter and more stable to solve; and (3) is also free of the NAS algorithm's typical engineering overhead (such as defining the search space and tuning search algorithms). Since our method directly shrinks the original dense model via pruning and quantization only, it cannot introduce any new operator not existing in the original model. That inspires us to combine the two streams of compression ideas (optimization-based versus NAS-based), as future work.

3.2   End-to-End Optimization

The difficulties of optimizing (5) can be summarized in three-folds. First, the minimax optimization problem itself is unstable. Second, updating W involves non-differentiable quantization operations. Third, updating γ involves a sparse loss term that is also non-differentiable. Below we discuss how to optimize them.

Updating W. The sub-problem for updating W in Eq. (5) is:

$$\min_{W} L_W(W),$$
$$\text{where } L_W(W) = \mathbb{E}_{x \in \mathcal{X}}[\log(1 - D(G_q(x; q_w(W), \gamma); \theta)) \qquad (6)$$
$$+ \beta d(G_q(x; q_w(W), \gamma), G_0(x))],$$

To solve (6) with gradient-based methods, we need to calculate $\nabla_W L_W$, which is difficult due to the non-differentiable $q_a(\cdot)$ and $q_w(\cdot)$. We now define the concrete form of $q_a(\cdot)$, $q_w(\cdot)$ and then demonstrate how to back propagate through them in order to calculate $\nabla_W L_W$. Since both $q_a(\cdot)$ and $q_w(\cdot)$ are elementwise operations, we only discuss how they work on scalars. We use a and w to denote a scalar element in the activation and convolution kernel tensors respectively.

When quantizing activations, we first clamp activations into range $[0, p]$ to bound the values, and then use $s_a = p/2^m$ as a scale factor to convert the floating point number to m bits integers: round(min(max(0, a), p)/s_a). Thus the activation quantization operator is as follows:

$$q_a(a) = \text{round}(\min(\max(0, a), p)/s_a) \cdot s_a. \qquad (7)$$

For weights quantization, we keep the range of the original weights and use symmetric coding for positive and negative ranges to quantize weights to n bits. Specifically, the scale factor $s_w = \|w\|_\infty/2^{(n-1)}$, leading to the quantization operator for weights:

$$q_w(w) = \text{round}(w/s_w) \cdot s_w. \qquad (8)$$

Since both quantization operators are non-differentiable, we use a proxy as the "pseudo" gradient in the backward pass, known as the straight through estimator (STE). For the activation quantization, we use

$$\frac{\partial q_a(a)}{\partial a} = \begin{cases} 1 & \text{if } 0 \le a \le p; \\ 0 & \text{otherwise.} \end{cases} \tag{9}$$

Similarly for the weight quantization, the pseudo gradient is set to

$$\frac{\partial q_w(w)}{\partial w} = 1. \tag{10}$$

Now that we have defined the derivatives of $q_a(\cdot)$ and $q_w(\cdot)$, we can calculate $\nabla_W L_W$ through back propagation and update W using the Adam optimizer [32].

Updating γ. The sub-problem for updating γ in Eq. (4) is a sparse optimization problem with a non-conventional fidelity term:

$$\min_{\gamma} L_\gamma(\gamma) + \rho\|\gamma\|_1,$$
$$\text{where } L_\gamma(\gamma) = \mathbb{E}_{x \in \mathcal{X}}[\log(1 - D(G_q(x; q_w(W), \gamma); \theta)) \tag{11}$$
$$+ \beta d(G_q(x; q_w(W), \gamma), G_0(x))],$$

We use the proximal gradient to update γ as follows:

$$g_\gamma^{(t)} \leftarrow \nabla_\gamma L_\gamma(\gamma)\Big|_{\gamma=\gamma^{(t)}} \tag{12}$$

$$\gamma^{(t+1)} \leftarrow \text{prox}_{\rho\eta^{(t)}}(\gamma^{(t)} - \eta^{(t)} g_\gamma^{(t)}) \tag{13}$$

where $\gamma^{(t)}$ and $\eta^{(t)}$ are the values of γ and learning rate at step t, respectively. The proximal function $\text{prox}_\lambda(\cdot)$ for the ℓ_1 constraint is the soft threshold function:

$$\text{prox}_\lambda(x) = \text{sgn}(x) \odot \max(|x| - \lambda\mathbf{1}, \mathbf{0}) \tag{14}$$

where $\odot$ is element-wise product, $\text{sgn}(\cdot)$ and $\max(\cdot, \cdot)$ are element-wise sign and maximum functions respectively.

Updating θ. The sub-problem of updating θ is the inner maximization problem in Eq. (5), which we solve by the gradient ascent method:

$$\max_{\theta} L_\theta(\theta),$$
$$\text{where } L_\theta(\theta) = \mathbb{E}_{y \in \mathcal{Y}}[\log(D(y; \theta))] + \mathbb{E}_{x \in \mathcal{X}}[\log(1 - D(G_q(x); \theta))], \tag{15}$$

We iteratively update D (parameterized by θ) and G (parameterized by W and γ) following [67].

3.3   Algorithm Implementation

Equipped with the above gradient computation, the last missing piece in solving problem (4) is to choose d. Note that, most previous distillation works are for classification-type models with softmax outputs (soft labels), and therefore adopt KL divergence. For GAN compression, the goal of distillation shall minimize the discrepancy between two sets of generated images. To this end, we adopt the perceptual loss [27] as our choice of d. It has shown to effectively measure not only low-level visual cue, but also high-level semantic differences between images, and has been popularly adopted to regularizing GAN-based image generation.

Finally, Algorithm 1 summarizes our GS algorithm with end-to-end optimization. By default, we quantize both activation and kernel weights uniformly to 8-bit (*i.e.*, $m = n = 8$) and set activation clamping threshold p to 4. We use Adam ($\beta_1 = 0.9$, $\beta_2 = 0.5$, following [67]) to update W and θ, and SGD to update γ. We also use two groups of learning rates $\alpha^{(t)}$ and $\eta^{(t)}$ for updating $\{W, \theta\}$ and γ respectively. $\alpha^{(t)}$ starts to be decayed linearly to zero, from the $T/2$-th iteration, while $\eta^{(t)}$ is decayed using a cosine annealing scheduler.

Algorithm 1: GAN Slimming (GS)

Input: $\mathcal{X}, \mathcal{Y}, \beta, \rho, T,$
$\quad \{\alpha^{(t)}\}_{t=1}^{T}, \{\eta^{(t)}\}_{t=1}^{T}.$

Output: W, γ

1 Random initialization: $W^{(1)}, \gamma^{(1)}, \theta^{(1)}$
2 **for** $t \leftarrow 1$ **to** T **do**
3 $\quad$ Get a batch of data from $\mathcal{X}$ and $\mathcal{Y}$;
4 $\quad W^{(t+1)} \leftarrow W^{(t)} - \alpha^{(t)} \nabla_W L_W$;
5 $\quad \gamma^{(t+1)} \leftarrow \text{prox}_{\rho\eta^{(t)}}(\gamma^{(t)} - \eta^{(t)} \nabla_\gamma L_\gamma)$;
6 $\quad \theta^{(t+1)} \leftarrow \theta^{(t)} + \alpha^{(t)} \nabla_\theta L_\theta$;
7 **end**
8 $W \leftarrow q_w(W^{T+1})$
9 $\gamma \leftarrow \gamma^{T+1}$

---

4   Experiments

4.1   Unpaired Image Translation with CycleGAN

Image translation and stylization is currently an important motivating application to deploy GANs on mobile devices. In this section, we compare GS with the only two published GAN compression methods CEC [52] and GD [3] on horse2zebra [67] and summer2winter [67] datasets. Following [52], we use model size and FLOPs to measure the efficiency of generator and use FID [21] between source style test set transfer results and target style test set to quantitatively measure the effectiveness of style transfer. We used the same implementation of FID as [52] for fair comparison. The metric statistic of original CycleGAN is summarized in Table 1. We denote the original dense model as G_0 and an arbitrary compressed generator as G. Following [52], we further define the following three metrics to evaluate efficiency-quality trade-off of different compression methods:

$$r_c = \frac{\text{ModelSize}_{G_0}}{\text{ModelSize}_G}, \quad r_s = \frac{\text{FLOPs}_{G_0}}{\text{FLOPs}_G}, \quad r_f = \frac{\text{FID}_{G_0}}{\text{FID}_G}.$$

Larger r_c and r_s indicate more model compactness and efficiency and larger r_f indicates better style transfer quality.

Quantitative comparison results on four different tasks are shown in Table 2. GS-32 outperforms both CEC and GD on all four tasks, in terms that it achieves better FID (larger r_f) with less computational budgets (larger r_c and r_s). For example, on horse-to-zebra task, GS-32 has much better FID than both CEC and GD, while achieving more model compactness. Combined with quantization, our method can further boost the model efficiency (much larger r_c) with minimal loss of performance (similar r_f). For example, on horse-to-zebra task, GS-8 achieves 4× larger r_c compared with GS-32 with negligible FID drop. On winter-to-summer task, GS-8 compress CycleGAN by 31× and achieve even slightly better FID. The visual comparison results are collectively displayed in Fig. 2. We compare the transfer results of four images reported in [52] for fair comparison. As we can see, the visual quality of GS is better than or at least comparable to those of CEC and GD.

Table 1. Statistics of the original CycleGAN model: FLOPs, model size and FID on different tasks.

GFLOPs	Memory (MB)	FID			
		horse-to-zebra	zebra-to-horse	summer-to-winter	winter-to-summer
52.90	43.51	74.04	148.81	79.12	73.31

Table 2. Compassion with the state-of-the-art GAN compression methods [52] and [3] on CycleGAN compression. The best metric is shown in bold and the second best is underlined.

Task	Metric	Method			
		CEC [52]	GD [3]	GS-32	GS-8
horse-to-zebra	r_s	4.23	3.91	<u>4.66</u>	**4.81**
	r_c	4.27	4.00	<u>5.05</u>	**21.75**
	r_f	0.77	0.76	**0.86**	<u>0.84</u>
zebra-to-horse	r_s	4.35	3.91	<u>4.39</u>	**4.40**
	r_c	4.34	4.00	<u>4.81</u>	**21.00**
	r_f	0.94	0.99	<u>1.24</u>	**1.25**
summer-to-winter	r_s	5.14	3.91	<u>6.21</u>	**7.18**
	r_c	5.44	4.00	<u>6.77</u>	**38.10**
	r_f	1.01	1.08	**1.13**	<u>1.12</u>
winter-to-summer	r_s	5.17	3.91	<u>6.01</u>	**6.36**
	r_c	5.70	4.00	<u>6.17</u>	**31.22**
	r_f	0.93	0.97	<u>0.98</u>	**1.01**

Source image	Original [67]	CEC [52]	GD [3]	GS-32	GS-8
	52.90 G	12.51 G	13.51 G	11.34 G	10.99 G
	43.51 MB	10.19 MB	10.86 MB	8.61 MB	2.15 MB

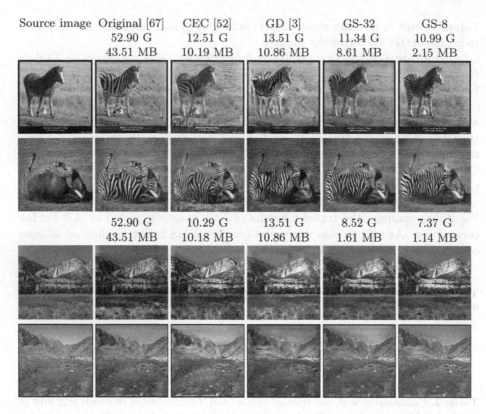

	52.90 G	10.29 G	13.51 G	8.52 G	7.37 G
	43.51 MB	10.18 MB	10.86 MB	1.61 MB	1.14 MB

Fig. 2. CycleGAN compression results. Top two rows: horse-to-zebra task. Bottom two rows: summer-to-winter task. Six columns from left to right: source image, style transfer results by original CycleGAN, CEC, GD, GS-32 and GS-8 respectively. FLOPs (in G) and model size (in MB) of each method on each task are annotated above the images.

4.2 Ablation Study

In order to show the superiority of our unified optimization framework over single compression methods and their naive combinations, we conduct thorough ablation studies by comparing the following methods:

- Distillation (*i.e.*, GD [3]): Use model distillation alone to train a slim student generator.[3]
- Channel pruning (CP): Directly use channel pruning during GAN minimax training process. This is implemented by adding L_{cp} to L_{GAN}. After channel pruning, we finetune the sub-network by minimax optimizing Eq. (1).
- Cascade: Distillation + CP (D+CP): Use channel pruning to further compress on the student network obtained by model distillation. Then finetune the sub-network by minimax optimizing Eq. (1).

---

[3] Following [3], we use student networks with 1/2 channels of the original generator.

- Cascade: CP + Distillation (CP+D): First do channel pruning on the original network, then use distillation to finetune the pruned network. This method is shown to outperform using channel pruning alone on classification tasks [57].
- GS-32: Jointly optimizing channel pruning and distillation.
- Cascade: GS-32 + quantization (postQ): First use GS-32 to compress the original network, then use 8 bit quantization as post processing and also do quantization-aware finetune on the quantized model by solving problem (1).
- GS-8: Jointly optimizing channel pruning, distillation and quantization.
- GS-8 (MSE): Replace the perceptual loss in GS-8 by MSE loss.
- GAN compression with fixed discriminator (i.e., CEC [52]): Co-evolution based channel pruning. Modeling GAN compression as dense prediction process instead of minimax problem by fixing the discriminator (both network structure and parameter values) during compression process.

Numerical and visualization results on horse2zebra dataset are shown in Fig. 3 and Fig. 4 respectively. As we can see, our unified optimization method achieves superior trade-off between style transfer quality and model efficiency compared with single compression techniques used separately (e.g., CP, GD) and their naive combinations (e.g., CP+D, D+CP, postQ), showing the effectiveness of our unified optimization framework. For example, directly injecting channel sparsity in GAN minimax optimization (CP) greatly increases the training instability and achieves degraded image generation quality as shown in Fig. 4. This aligns with the conclusions in [52] that model compression methods developed for classifiers are not directly applicable on GAN compression tasks. Using model distillation to finetune channel pruned models (CP+D) can indeed improve image generation quality compared with CP, however the generation quality is still much more inferior to our methods at similar compression ratio, as shown in Fig. 3 and Fig. 4. Compared with GD, which uses a hand-crafted student network, GS-32 achieves much better FID with even considerably larger compression ratio, showing the effectiveness of jointly searching slim student network structures by channel pruning and training the student network with model distillation. In contrast, directly using channel pruning to further compress the student generator trained by GD (D+CP) will catastrophically hurt the image translation performance. Doing post quantization and quantization-aware finetune (postQ) on GS-32 models also suffers great degradation in image translation quality compared with GS-8, showing the necessity to jointly train quantization with channel pruning and model distillation in our unified optimization framework. Replacing perceptual loss with MSE loss as d in Eq. (2) fails to generate satisfying target images, since MSE loss cannot effectively capture the high-level semantic differences between images. Last but not least, GS-32 largely outperforms CEC, verifying the effectiveness of incorporating minimax objective into GAN compression problem.

4.3   Real-World Application: CartoonGAN

Finally, we apply GS to a recently proposed style transfer network Cartoon-GAN, which transforms photos to cartoon images, in order to deploy the model on mobile devices. CartoonGAN has its heavily parameterized generator (56.46 GFLOPs on 256×256 images) publicly available.[4] Since CartoonGAN has a feed-forward encoder-decoder structure, without using cycle consistent loss, CEC [52] is not directly applicable to compress it. So we only compare GS with the other published state-of-the-art method GD [3] on this task. Experiments are conducted on the CelebA dataset [40]. Following [3], we use a student generator with 1/6 channels of the teacher generator for GD, which achieves similar (but less) compression ratio compared with GS.

The visual results of cartoon style transfer, together with model statistics (FLOPs and model sizes), are shown in Fig. 5.[5] All FLOPs are calculated for input images with shape 256 × 256. At large compression ratio, the style transfer results of GD have obvious visual artifacts (*e.g.*, abnormal white spots). In contrast, GS-32 can remarkably compress the original generator by around 42× (in terms of FLOPs) with minimal degradation in the visual quality. GS-8 can further improve the FLOPs compression ratio to 47× with almost identical visual quality. These results again show the superiority of our student generator jointly learned by channel pruning, quantization and distillation, over the hand-crafted student generator used in GD. Part of the proposed GS framework is integrated into some style transfer products in Kwai Inc.'s Apps.

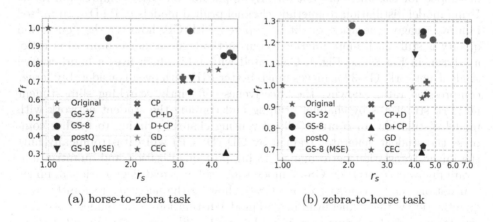

(a) horse-to-zebra task   (b) zebra-to-horse task

Fig. 3. Numerical results of ablation studies on horse2zebra dataset.

---

[4] Available at https://github.com/maciej3031/comixify.
[5] Following [10], we use color matching as the post-processing on all compared methods, for better visual display quality.

Original [67]   GD [3]   D+CP   CP
Source image 52.90 GFLOPs 13.51 GFLOPs 11.71 GFLOPs 16.95 GFLOPs
43.51 MB   10.86 MB   8.92 MB   11.07 MB

CP+D   postQ   GS-32   GS-8   GS-8 (MSE)
16.95 GFLOPs 15.90 GFLOPs 11.34 GFLOPs 10.99 GFLOPs 15.66 GFLOPs
11.07 MB   3.13 MB   8.61 MB   2.00 MB   3.22 MB

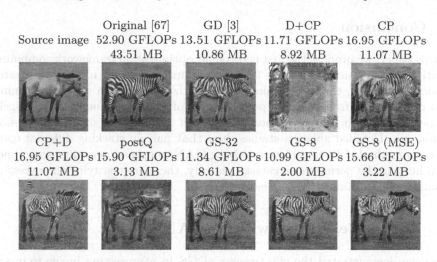

Fig. 4. Visualization results of ablation studies on horse2zebra dataset. FLOPs and model size of each method are annotated above the images.

Photo images Original CartoonGAN   GD   GS-32   GS-8
56.46 G   1.41 G   1.34 G   1.20 G
42.34 MB   1.04 MB   0.80 MB   0.18 MB

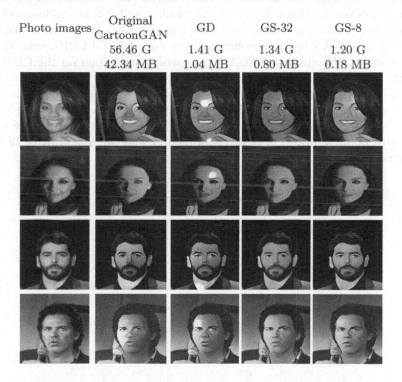

Fig. 5. CartoonGAN compression results. From left to right columns: original photo images, cartoon images generated by original CartoonGAN, GD [3], GS-32 and GS-8 compressed models, respectively. Corresponding FLOPs (in G) and model size (in MB) are annotated on top of each column.

5   Conclusion

In this paper, we propose the first end-to-end optimization framework combining multiple compression techniques for GAN compression. Our method integrates model distillation, channel pruning and quantization, within one unified mini-max optimization framework. Experimental results show that our method largely outperforms existing GAN compression options which utilize single compression techniques. Detailed ablation studies show that naively stacking different compression methods fails to achieve satisfying GAN compression results, sometimes even hurting the performance catastrophically, therefore testifying the necessity of our unified optimization framework.

A   Image Generation with SNGAN

We have demonstrated the effectiveness of GS in compressing image-to-image GANs (e.g., CycleGAN [67], StyleGAN [50]) in the main text. Here we show GS is also generally applicable to noise-to-image GANs (e.g., SNGAN [44]). SNGAN with the ResNet backbone is one of the most popular noise-to-image GANs, with state-of-the-art performance on a few datasets such as CIFAR10 [33]. The generator in SNGAN has 7 convolution layers with 1.57 GFLOPs, with 32×32 image outputs. We evaluate SNGAN generator compression on the CIFAR-10 dataset. Inception Score (IS) [49] is used to measure image generation and style transfer quality. We use latency (FLOPs) and model size to evaluate the network efficiency. Quantitative and visualization results are shown in Table 3 and Fig. 6 respectively. GS is able to compress SNGAN by up to 8× (in terms of model size), with minimum drop in both visual quality and the quantitative IS value of generated images.

Table 3. SNGAN compression results.

Method	MFLOPs	Model Size (MB)	IS
Original	1602.75	16.28	8.27
GS-32	1108.78	12.88	8.01
	509.39	8.32	7.65
GS-8	1115.11	3.24	8.14
	510.33	2.01	7.62

Original SNGAN   GS-8   GS-8

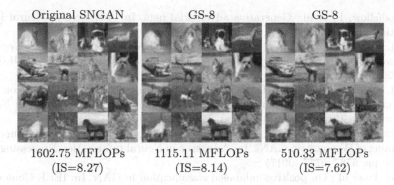

1602.75 MFLOPs   1115.11 MFLOPs   510.33 MFLOPs
(IS=8.27)   (IS=8.14)   (IS=7.62)

Fig. 6. CIFAR-10 images generation by SNGAN (original and compressed). Leftmost column: images generated by original SNGAN. The rest columns: images generated by GS-8 compressed SNGAN, with different compression ratios. Images are randomly selected instead of cherry-picked.

References

1. Bulò, S.R., Porzi, L., Kontschieder, P.: Dropout distillation. In: International Conference on Machine Learning, pp. 99–107 (2016)
2. Chen, H., et al.: Frequency domain compact 3D convolutional neural networks. In: IEEE Conference on Computer Vision and Pattern Recognition, pp. 1641–1650 (2020)
3. Chen, H., et al.: Distilling portable generative adversarial networks for image translation. In: AAAI Conference on Artificial Intelligence (2020)
4. Chen, H., Wang, Y., Xu, C., Xu, C., Tao, D.: Learning student networks via feature embedding. IEEE Trans. Neural Netw. Learn. Syst. (2020)
5. Chen, H., et al.: Data-free learning of student networks. In: IEEE International Conference on Computer Vision, pp. 3514–3522 (2019)
6. Chen, H., et al.: AdderNet: do we really need multiplications in deep learning? In: IEEE Conference on Computer Vision and Pattern Recognition, pp. 1468–1477 (2020)
7. Chen, Y., Lai, Y.K., Liu, Y.J.: CartoonGAN: generative adversarial networks for photo cartoonization. In: IEEE Conference on Computer Vision and Pattern Recognition, pp. 9465–9474 (2018)
8. Courbariaux, M., Bengio, Y., David, J.P.: BinaryConnect: training deep neural networks with binary weights during propagations. In: Advances in Neural Information Processing Systems, pp. 3123–3131 (2015)
9. Fu, Y., Chen, W., Wang, H., Li, H., Lin, Y., Wang, Z.: AutoGAN-Distiller: searching to compress generative adversarial networks. In: International Conference on Machine Learning (2020)
10. Gatys, L.A., Bethge, M., Hertzmann, A., Shechtman, E.: Preserving color in neural artistic style transfer. arXiv preprint arXiv:1606.05897 (2016)
11. Gong, X., Chang, S., Jiang, Y., Wang, Z.: AutoGAN: neural architecture search for generative adversarial networks. In: IEEE International Conference on Computer Vision, pp. 3224–3234 (2019)

12. Goodfellow, I., et al.: Generative adversarial nets. In: Advances in Neural Information Processing Systems, pp. 2672–2680 (2014)
13. Gui, J., Sun, Z., Wen, Y., Tao, D., Ye, J.: A review on generative adversarial networks: algorithms, theory, and applications. arXiv preprint arXiv:2001.06937 (2020)
14. Gui, S., Wang, H., Yang, H., Yu, C., Wang, Z., Liu, J.: Model compression with adversarial robustness: a unified optimization framework. In: Advances in Neural Information Processing Systems, pp. 1283–1294 (2019)
15. Gulrajani, I., Ahmed, F., Arjovsky, M., Dumoulin, V., Courville, A.C.: Improved training of Wasserstein GANs. In: Advances in Neural Information Processing Systems, pp. 5767–5777 (2017)
16. Guo, T., et al.: On positive-unlabeled classification in GAN. In: IEEE Conference on Computer Vision and Pattern Recognition, pp. 8385–8393 (2020)
17. Han, K., Wang, Y., Tian, Q., Guo, J., Xu, C., Xu, C.: GhostNet: more features from cheap operations. In: IEEE Conference on Computer Vision and Pattern Recognition, pp. 1580–1589 (2020)
18. Han, S., Mao, H., Dally, W.J.: Deep compression: compressing deep neural networks with pruning, trained quantization and Huffman coding. arXiv preprint arXiv:1510.00149 (2015)
19. He, Y., Lin, J., Liu, Z., Wang, H., Li, L.-J., Han, S.: AMC: AutoML for model compression and acceleration on mobile devices. In: Ferrari, V., Hebert, M., Sminchisescu, C., Weiss, Y. (eds.) ECCV 2018. LNCS, vol. 11211, pp. 815–832. Springer, Cham (2018). https://doi.org/10.1007/978-3-030-01234-2_48
20. He, Y., Zhang, X., Sun, J.: Channel pruning for accelerating very deep neural networks. In: IEEE International Conference on Computer Vision, pp. 1389–1397 (2017)
21. Heusel, M., Ramsauer, H., Unterthiner, T., Nessler, B., Hochreiter, S.: GANs trained by a two time-scale update rule converge to a local Nash equilibrium. In: Advances in Neural Information Processing Systems, pp. 6626–6637 (2017)
22. Hinton, G., Vinyals, O., Dean, J.: Distilling the knowledge in a neural network. arXiv preprint arXiv:1503.02531 (2015)
23. Hu, H., Peng, R., Tai, Y.W., Tang, C.K.: Network trimming: a data-driven neuron pruning approach towards efficient deep architectures. arXiv preprint arXiv:1607.03250 (2016)
24. Huang, Z., Wang, N.: Data-driven sparse structure selection for deep neural networks. In: Ferrari, V., Hebert, M., Sminchisescu, C., Weiss, Y. (eds.) ECCV 2018. LNCS, vol. 11220, pp. 317–334. Springer, Cham (2018). https://doi.org/10.1007/978-3-030-01270-0_19
25. Hubara, I., Courbariaux, M., Soudry, D., El-Yaniv, R., Bengio, Y.: Quantized neural networks: training neural networks with low precision weights and activations. J. Mach. Learn. Res. 18(1), 6869–6898 (2017)
26. Jiang, Y., et al.: EnlightenGAN: deep light enhancement without paired supervision. arXiv preprint arXiv:1906.06972 (2019)
27. Johnson, J., Alahi, A., Fei-Fei, L.: Perceptual losses for real-time style transfer and super-resolution. In: Leibe, B., Matas, J., Sebe, N., Welling, M. (eds.) ECCV 2016. LNCS, vol. 9906, pp. 694–711. Springer, Cham (2016). https://doi.org/10.1007/978-3-319-46475-6_43
28. Jung, S., et al.: Learning to quantize deep networks by optimizing quantization intervals with task loss. In: IEEE Conference on Computer Vision and Pattern Recognition, pp. 4350–4359 (2019)

29. Karras, T., Aila, T., Laine, S., Lehtinen, J.: Progressive growing of GANs for improved quality, stability, and variation. In: International Conference on Learning Representations (2018)
30. Karras, T., Laine, S., Aila, T.: A style-based generator architecture for generative adversarial networks. In: IEEE Conference on Computer Vision and Pattern Recognition, pp. 4401–4410 (2019)
31. Kim, S., Xing, E.P.: Tree-guided group Lasso for multi-task regression with structured sparsity. Ann. Appl. Stat. **6**(3), 1095–1117 (2012)
32. Kingma, D.P., Ba, J.: Adam: a method for stochastic optimization. arXiv preprint arXiv:1412.6980 (2014)
33. Krizhevsky, A.: Learning multiple layers of features from tiny images. Master's thesis, University of Toronto (2009)
34. Kupyn, O., Martyniuk, T., Wu, J., Wang, Z.: DeblurGAN-v2: deblurring (orders-of-magnitude) faster and better. In: IEEE International Conference on Computer Vision, pp. 8878–8887 (2019)
35. Ledig, C., et al.: Photo-realistic single image super-resolution using a generative adversarial network. In: IEEE Conference on Computer Vision and Pattern Recognition, pp. 4681–4690 (2017)
36. Li, M., Lin, J., Ding, Y., Liu, Z., Zhu, J.Y., Han, S.: GAN compression: efficient architectures for interactive conditional GANs. In: IEEE Conference on Computer Vision and Pattern Recognition, pp. 5284–5294 (2020)
37. Lin, J., Rao, Y., Lu, J., Zhou, J.: Runtime neural pruning. In: Advances in Neural Information Processing Systems, pp. 2181–2191 (2017)
38. Liu, S., Du, J., Nan, K., Wang, A., Lin, Y., et al.: AdaDeep: a usage-driven, automated deep model compression framework for enabling ubiquitous intelligent mobiles. arXiv preprint arXiv:2006.04432 (2020)
39. Liu, Z., Li, J., Shen, Z., Huang, G., Yan, S., Zhang, C.: Learning efficient convolutional networks through network slimming. In: IEEE International Conference on Computer Vision, pp. 2736–2744 (2017)
40. Liu, Z., Luo, P., Wang, X., Tang, X.: Deep learning face attributes in the wild. In: IEEE International Conference on Computer Vision, pp. 3730–3738 (2015)
41. Lopez-Paz, D., Bottou, L., Schölkopf, B., Vapnik, V.: Unifying distillation and privileged information. arXiv preprint arXiv:1511.03643 (2015)
42. Luo, J.H., Wu, J., Lin, W.: ThiNet: a filter level pruning method for deep neural network compression. In: IEEE International Conference on Computer Vision, pp. 5058–5066 (2017)
43. Mishra, A., Marr, D.: Apprentice: using knowledge distillation techniques to improve low-precision network accuracy. arXiv preprint arXiv:1711.05852 (2017)
44. Miyato, T., Kataoka, T., Koyama, M., Yoshida, Y.: Spectral normalization for generative adversarial networks. In: International Conference on Learning Representations (2018)
45. Molchanov, P., Mallya, A., Tyree, S., Frosio, I., Kautz, J.: Importance estimation for neural network pruning. In: IEEE Conference on Computer Vision and Pattern Recognition, pp. 11264–11272 (2019)
46. Parikh, N., Boyd, S., et al.: Proximal algorithms. Found. Trends Optim. **1**(3), 127–239 (2014)
47. Polino, A., Pascanu, R., Alistarh, D.: Model compression via distillation and quantization. arXiv preprint arXiv:1802.05668 (2018)

48. Rastegari, M., Ordonez, V., Redmon, J., Farhadi, A.: XNOR-Net: ImageNet classification using binary convolutional neural networks. In: Leibe, B., Matas, J., Sebe, N., Welling, M. (eds.) ECCV 2016. LNCS, vol. 9908, pp. 525–542. Springer, Cham (2016). https://doi.org/10.1007/978-3-319-46493-0_32

49. Salimans, T., Goodfellow, I., Zaremba, W., Cheung, V., Radford, A., Chen, X.: Improved techniques for training GANs. In: Advances in Neural Information Processing Systems, pp. 2234–2242 (2016)

50. Sanakoyeu, A., Kotovenko, D., Lang, S., Ommer, B.: A style-aware content loss for real-time HD style transfer. In: Ferrari, V., Hebert, M., Sminchisescu, C., Weiss, Y. (eds.) ECCV 2018. LNCS, vol. 11212, pp. 715–731. Springer, Cham (2018). https://doi.org/10.1007/978-3-030-01237-3_43

51. Shen, M., Han, K., Xu, C., Wang, Y.: Searching for accurate binary neural architectures. In: IEEE International Conference on Computer Vision Workshops (2019)

52. Shu, H., et al.: Co-evolutionary compression for unpaired image translation. In: IEEE International Conference on Computer Vision, pp. 3235–3244 (2019)

53. Singh, P., Verma, V.K., Rai, P., Namboodiri, V.: Leveraging filter correlations for deep model compression. In: IEEE Winter Conference on Applications of Computer Vision, pp. 835–844 (2020)

54. Theis, L., Korshunova, I., Tejani, A., Huszár, F.: Faster gaze prediction with dense networks and Fisher pruning. arXiv preprint arXiv:1801.05787 (2018)

55. Tung, F., Mori, G.: CLIP-Q: deep network compression learning by in-parallel pruning-quantization. In: IEEE Conference on Computer Vision and Pattern Recognition, pp. 7873–7882 (2018)

56. Wang, K., Liu, Z., Lin, Y., Lin, J., Han, S.: HAQ: hardware-aware automated quantization with mixed precision. In: IEEE Conference on Computer Vision and Pattern Recognition, pp. 8612–8620 (2019)

57. Wang, M., Zhang, Q., Yang, J., Cui, X., Lin, W.: Graph-adaptive pruning for efficient inference of convolutional neural networks. arXiv preprint arXiv:1811.08589 (2018)

58. Wang, Y., Xu, C., Chunjing, X., Xu, C., Tao, D.: Learning versatile filters for efficient convolutional neural networks. In: Advances in Neural Information Processing Systems, pp. 1608–1618 (2018)

59. Wang, Y., Xu, C., Xu, C., Tao, D.: Adversarial learning of portable student networks. In: AAAI Conference on Artificial Intelligence, pp. 4260–4267 (2018)

60. Wang, Y., Xu, C., Xu, C., Tao, D.: Packing convolutional neural networks in the frequency domain. IEEE Trans. Pattern Anal. Mach. Intell. 41(10), 2495–2510 (2018)

61. Wen, W., Wu, C., Wang, Y., Chen, Y., Li, H.: Learning structured sparsity in deep neural networks. In: Advances in Neural Information Processing Systems, pp. 2074–2082 (2016)

62. Wu, J., Wang, Y., Wu, Z., Wang, Z., Veeraraghavan, A., Lin, Y.: Deep-k-means: re-training and parameter sharing with harder cluster assignments for compressing deep convolutions. In: International Conference on Machine Learning, pp. 5363–5372 (2018)

63. Yang, H., Zhu, Y., Liu, J.: ECC: platform-independent energy-constrained deep neural network compression via a bilinear regression model. In: IEEE Conference on Computer Vision and Pattern Recognition, pp. 11206–11215 (2019)

64. Yang, S., Wang, Z., Wang, Z., Xu, N., Liu, J., Guo, Z.: Controllable artistic text style transfer via shape-matching GAN. In: IEEE International Conference on Computer Vision, pp. 4442–4451 (2019)

65. Zhao, Y., et al.: SmartExchange: trading higher-cost memory storage/access for lower-cost computation. arXiv preprint arXiv:2005.03403 (2020)
66. Zhu, C., Han, S., Mao, H., Dally, W.J.: Trained ternary quantization. arXiv preprint arXiv:1612.01064 (2016)
67. Zhu, J.Y., Park, T., Isola, P., Efros, A.A.: Unpaired image-to-image translation using cycle-consistent adversarial networks. In: IEEE International Conference on Computer Vision, pp. 2223–2232 (2017)

Human Interaction Learning on 3D Skeleton Point Clouds for Video Violence Recognition

Yukun Su[1,2], Guosheng Lin[3(✉)], Jinhui Zhu[1,2], and Qingyao Wu[1,2(✉)]

[1] School of Software Engineering, South China University of Technology, Guangzhou, China
suyukun666@gmail.com, {csjhzhu,qyw}@scut.edu.cn
[2] Key Laboratory of Big Data and Intelligent Robot, Ministry of Education, Beijing, China
[3] Nanyang Technological University, Singapore, Singapore
gslin@ntu.edu.sg

Abstract. This paper introduces a new method for recognizing violent behavior by learning contextual relationships between related people from human skeleton points. Unlike previous work, we first formulate 3D skeleton point clouds from human skeleton sequences extracted from videos and then perform interaction learning on these 3D skeleton point clouds. A novel **S**keleton **P**oints **I**nteraction **L**earning (SPIL) module, is proposed to model the interactions between skeleton points. Specifically, by constructing a specific weight distribution strategy between local regional points, SPIL aims to selectively focus on the most relevant parts of them based on their features and spatial-temporal position information. In order to capture diverse types of relation information, a multi-head mechanism is designed to aggregate different features from independent heads to jointly handle different types of relationships between points. Experimental results show that our model outperforms the existing networks and achieves new state-of-the-art performance on video violence datasets.

1 Introduction

Generally, the concept of video-based violence recognition is defined as detecting violent behaviors in video data, which is of vital importance in some video surveillance scenarios like railway stations, prisons or psychiatric centers. Consider some sample frames from public datasets as shown in Fig. 1(a). When we humans see the sequences of images, we can easily recognize those violent actions through the human body's torso movements such as "kick", "beat", "push", etc.

However, current deep learning approaches fail to capture these ingredients precisely in a multi-dynamic and complex multi-people scene. For instance, the approaches based on two-stream ConvNets [26,31] are learning to classify actions based on individual video frames or local motion vectors. However, such local

© Springer Nature Switzerland AG 2020
A. Vedaldi et al. (Eds.): ECCV 2020, LNCS 12349, pp. 74–90, 2020.
https://doi.org/10.1007/978-3-030-58548-8_5

(a)   (b)

Fig. 1. (a) Sample frames from the Hockey-Fight [20] dataset (first row), the Crowd Violence [14] dataset (second row) and the RWF-2000 Violence [4] dataset (third row). In each row, the left two columns are non-violent scenes while the right three columns are violent scenes. (b) Skeleton point clouds for a certain video.

motions that are captured by optical flow [2] sometimes fail to satisfy the dynamics modeling of shape change in multiple motion states. To tackle this limitation, recent Recurrent Neural Networks [9,36] and 3D Convolutions [3,28,29] works have also focused on modeling long term temporal information. However, all these frameworks focus on the features extracted from the whole scenes, leading to the interference by irrelevant information in the scenarios, and fail to capture region-based relationships. Meanwhile, the existing vision-based methods are mainly based on hand-crafted features such as statistic features between motion regions, leading to poor adaptability to another dataset. In violence recognition, extracting such appearance features and dynamics information of objects suffer from a number of complexities. Therefore, the above methods are often not very effective.

The movements of people in the video are reflected in human skeletal point sequences, which can be converted into 3D point clouds. We can then perform feature extraction on this data. Our experiments show that existing 3D point clouds methods [24,32,34] can readily be applied to the violence recognition task. However, current methods, while excellent at extracting pertinent features on ordinary point clouds, they lack the ability to focus on relevant points and their interactions.

Based on these observations, we introduce a novel approach to perform video violence recognition via a human skeleton point convolutional reasoning framework. We first represent the input video as the cluster of 3D point clouds data as shown in Fig. 1(b) through extracting the human skeleton sequences pose coordinates from each frames in the video. In order to better observe the characteristics of the skeleton points and determine whether there is violence, specifically, (i) **Skeleton Points Interaction Learning (SPIL) module**: the weight distribution among regional points with high coupling degree or strong semantic correlation is relatively high. Based on this, we can capture the appearance features and spatial-temporal structured position relation of skeleton points uniformly avoiding feature contamination between objects, and model how the state of the same object changes and the dependencies between different objects in frames. (ii) **Multi-head mechanism**: the single head is responsible for processing the action information for the skeleton points without interference from the scene. Multiple heads attend to fuse features from different independent heads to capture different types of information among points parallelly, which can enhance the robustness of the network.

In summary, we highlight the major contributions of this paper in three folds:

– We formulate the video violence recognition task as 3D skeleton point clouds recognition problem, and we propose an effective interaction learning method on skeleton point clouds for video recognition.
– We propose a novel SPIL module, which can be learned on the human skeleton points to capture both feature and position relation information simultaneously. And the multi-head mechanism allows SPIL to capture different types of points interactions to improve robustness.
– Different from the previous methods, we use the skeleton point clouds technique on recognition in violent videos, and our approach significantly outperforms state-of-the-art methods by a large margin.

2 Related Work

Video Classification with Deep Learning: Most recent works on video classification are based on deep learning. Initial approaches explored methods to combine temporal information based on pooling or temporal convolution [16,36]. To jointly explore spatial and temporal information of videos, 3D convolutional networks have been widely used. Tran et al. [28] trained 3D ConvNets on the large-scale video datasets, where they experimentally found that a $3 \times 3 \times 3$ convolutional kernel can learn both appearance and motion features. In a later work, Hara et al. [13] studied the use of a Resnet architecture with 3D convolutions and Xie et al. [35] exploited aggregated residual transformations to show the improvements. Two-stream networks [3,31] have also been attracting high attention, they took the input of a single RGB frame (captures appearance information) and a stack of optical flow frames (captures motion information). An alternative way to model the temporal relation between frames is by using recurrent networks [5,18]. However, these above approaches encountered bottlenecks in feature extraction when faced with more complex scenes and more irregular dynamic features in video violence recognition tasks. They fail to fully capture the comprehensive information in the entire video and are difficult to focus on distinguishing violent behavior in multiple characters and action features.

3D Point Clouds: To adapt the 3D points coordinates data for convolution, one straightforward approach is to voxelize it in a 3D grid structure [11,27]. OctNet [25] explored the sparsity of voxel data and alleviated this problem. However, since voxels are the discrete representations of space, this method still requires high-resolution grids with large memory consumption as a trade-off to keep a level of representation quality. Because the body keypoints themselves are a very sparse spatial structure, the application of the 3D voxel method to the skeleton points will lead to insufficient data characterization and make it difficult to train the model. In this trend, PointNet [23] first discusses the irregular format and permutation invariance of point sets, and presents a network that directly consumes point clouds. PointNet++ [24] extends PointNet by further considering not only the global information but also the local details with a farthest-sampling-layer and a grouping-layer. Deep learning in graph [1] is a modern term for a set of emerging technologies that attempt to address non-Euclidean structured data (e.g., 3D point clouds, social networks or genetic networks) by deep neural networks. Graph CNNs [6,19] show advantages of graph representation in many tasks for non-Euclidean data, as it can naturally deal with these irregular structures. [39] builds a graph CNN architecture to capture the local structure and classify point clouds, which also proves that deep geometric learning has enormous potential for unordered point clouds analysis. Nonetheless, these works ignore the different importance of each point's contribution, especially in skeleton points. Even though some works [12,30] suggest the use of attention, they seem to be of little use in the processing of specific spatial-temporal relation in the skeleton point clouds.

To this end, different from these works, our approach encodes dependencies between objects with both feature and position relations, which focus on specific human dynamic action expressions ignoring action-independent information. This framework provides a significant boost over the state-of-the-art.

3 Proposed Method

Our goal task is to represent the video as human skeleton point clouds of objects and perform reasoning for video violence recognition. To this end, we propose the multi-head Skeleton Points Interaction Learning (SPIL) module to deal with the interrelationships between points. In this section, we will give detailed descriptions of our approach. Section 3.1 presents an overview of our framework. Section 3.2 introduces the detail of the SPIL module. Section 3.3 shows the multi-head mechanism and Sect. 3.4 describes how to train and infer this method for skeleton point clouds.

3.1 Framework

We propose to tackle this problem with an architecture, as illustrated in Fig. 2. Our approach takes raw video clips as input streams. First, we follow the human pose extraction strategy used in [10], which can detect body points from each frames in the video. The coordinate (x, y, z) of each keypoint represents the position of the current point in each frame, where z represents the tth frame. Then we collect all the skeleton points sequences and transform the dynamic representation of the people in the video into a point clouds structure.

Subsequently, a SPIL abstraction level module takes an $N \times (3 + C)$ matrix as input that is from N centroid points which are sampled following the scheme

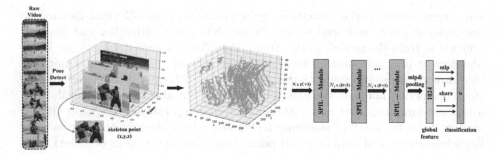

Fig. 2. Model Pipeline Overview. Our model uses the pose detection method to extract skeleton coordinates from each frames of the video. These human skeleton points are provided as point clouds inputs to the SPIL modules which perform information propagation based on assigning different weights to different skeleton points. Finally, a global feature is extracted to perform classification.

of [24] with 3-dim coordinates and C-dim points feature, and it outputs an $N_1 \times (3 + D)$ matrix of N_1 subsampled points with 3-dim coordinates and new D-dim feature vectors representing local features. To obtain various relation information between points, we design a multi-head mechanism in SPIL.

With this representation, we apply several SPIL modules to sample and extract the skeleton point clouds, and finally, classify the global feature through a fully connected layer to complete the video recognition.

3.2   Skeleton Points Interaction Learning Module

To compute the interaction weights between points, the standard general approach is determined by the K neighbors. However, not all nearby points have an effect on the current point. For example, if there are multiple characters in a scene, irrelevant skeleton points can sometimes be confusing and the learned feature characterizes all of its neighbors indistinguishably. To address this problem, in our SPIL module, as shown in Fig. 3 intuitively, the points interaction weights on different skeleton points are distributed based on the relationships between points. We learn to mask or weaken part of the convolution weights according to the neighbors' feature attributes. In this way, the network can focus on the skeleton points for prediction.

Consider a point set $\{p_1, p_2, ..., p_K\} \in \mathbb{R}^3$ according to a centroid point's K neighbors, where the local region is grouped within a radius. We set the radius to $(r \times T_{frame})$ that guarantees local region to cover more inter- and intra-frames points across space and time. Particularly, the pair-wise interaction weights between points can be mathematically considered to use W to represent, where the weight W_{ij} indicates the connection of point j to point i. In the case of $W \in \mathbb{R}^{K \times K}$, we define the set of points as $\mathcal{S} = \left\{ (p_i^f, p_i^l) | i = 1, ..., K \right\}$. Among them, $p_i^f \in \mathbb{R}^C$ is point i's C-dim feature, here we take points' confidence as initial feature, and $p_i^l = (l_i^x, l_i^y, l_i^z) \in \mathbb{R}^3$ is the 3-dim position coordinates.

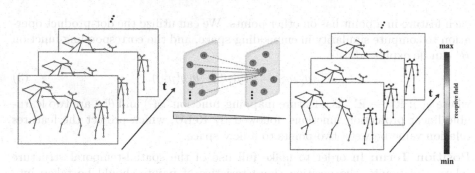

Fig. 3. Illustration of a single SPIL module on a subgraph of the human skeleton point clouds. Input points with a constant scalar feature (in grey) are convolved through a human skeleton points interaction filter. The output is a dynamically attentional weighted combination of the neighbor's points. The weights on irrelevant points (the dotted arrows) are masked so that the convolution kernel can focus on the correlate points for prediction. (α, β, γ of different colors denote parts of different human joint points respectively, such as left hand, right elbow, left ankle). (Color figure online)

For the distribution of the weights W, unlike the traditional point operation, in order to obtain distinguishable representation ability to capture the correlation between different skeletal points, it is necessary to consider both feature similarity and position characteristic relation. To this end, we separately explore the feature and position information and then perform high-level modeling of them to dynamically adapt to the structure of the objects. Concretely, the interaction weight of each neighboring point is computed as follows:

$$W_{ij} = \Phi(R^F(p_i^f, p_j^f), R^L(p_i^l, p_j^l)), \tag{1}$$

where $R^F(p_i^f, p_j^f)$ implies the feature relation between points and $R^L(p_i^l, p_j^l)$ denotes the position relation. Φ function plays the role of a combination of feature information and position information.

In this work, we follow [33] and use the differentiable architecture but different in building R^F and R^L functions detailedly to compute points interaction value, which can be formulated as:

$$W_{ij} = \frac{R^L(p_i^l, p_j^l) \ exp(R^F(p_i^f, p_j^f))}{\sum_{j=1}^{K} R^L(p_i^l, p_j^l) \ exp(R^F(p_i^f, p_j^f))}, \tag{2}$$

where the points interaction weights are normalized across all the neighbors of a point i to handle the size-varying neighbors across different points and spatial scales.

Feature Term: Intuitively, the points of different features in the local area exert various influences to enhance the expressive power of each point. Our feature modulator solves this problem by adaptively learning the amount of influence each feature in a point has on other points. We can utilize the dot-product operation to compute similarity in embedding space, and the corresponding function R^F can be expressed as:

$$R^F(p_i^f, p_j^f) = \phi(g(p_i^f))^\mathsf{T} \theta(g(p_j^f)), \tag{3}$$

where $g(\cdot)$: $\mathbb{R}^C \to \mathbb{R}^{C'}$ is a feature mapping function. $\phi(\cdot)$ and $\theta(\cdot)$ are two learnable linear projection functions, followed by ReLU, which project the features relation value between two points to a new space.

Position Term: In order to make full use of the spatial-temporal structure relation of points, the position characteristics of points should be taken into account. In our work, we consider the following three choices:

(1) Eu-distance in space: Considering the Euclidean distance between the points, the relatively distant points contribute less to the connection of the current point than the local points. With this in mind, we directly calculate the distance information and act on the points. The R^P is formed as:

$$R^L(p_i^l, p_j^l) = -ln(\sigma(\mathcal{D}(p_i^l, p_j^l))), \tag{4}$$

where $\sigma(\cdot)$ is a sigmoid activation function and the output range is controlled between [0, 1] to fed to the ln function. $\mathcal{D}$ is a function to calculate the distance between points.

(2) Eu-distance Spanning: Alternatively, we can first encode the relations between two points to a high-dimensional representation based on the position distance. Then the difference between two terms encourages to span the relations to a new subspace. Specifically, the position relation value is computed as:

$$R^L(p_i^l, p_j^l) = \frac{\psi(M_1(p_i^l) - M_2(p_j^l))}{\mathcal{D}(p_i^l, p_j^l)}, \tag{5}$$

where $M_1(\cdot)$ and $M_2(\cdot)$ are two multilayer perceptrons functions and $\psi(\cdot)$ is a linear projection function followed by ReLU that generates the embedded feature into a scalar.

(3) Eu-distance Masking: In addition, a more intuitive approach is to ignore some distant points and retain the characteristic contribution of the relative local points. For this purpose, we set a threshold to ignore the contribution of points and the function can be defined as:

$$R^L(p_i^l, p_j^l) = \begin{cases} 0, & if \ \underset{(l_i^z = l_j^z)}{\mathcal{D}}(p_i^l, p_j^l) > d, \\ \psi(M_1(p_i^l) || M_2(p_j^l)), & else. \end{cases} \tag{6}$$

Note that the radius ($r \times T_{frame}$) ensures that the local region cover points spatially and temporally. However, spatially, we try to mask out some weak correlation points within the same frames. The implication is that we preserve the globality in time and the locality in space. || is the concatenation operation and the embedded feature between two points is transformed into a scalar by a learnable linear function, followed by a ReLU activation. d acts as a distance threshold which is a hyper-parameter.

3.3 Multi-head Mechanism

Although a single head SPIL module can perform interaction feature extraction on skeleton points, because the connection between the joint points of the human body is ever-changing, each points may have different types of features. For example, a joint point has an information effect on its own posture and also a dynamic information effect on the interaction between human bodies at the same time, we call it a point with different types of features. Specifically, as shown in Fig. 4, for a certain skeletal point such as the elbow joint, the first head may be sensitive to the information of the human elbow joint's own posture, such as judging whether it is a "punch" posture; while the second head is more concerned with the connection of elbow joint motion information between people to extract dynamic features. In the same way, the remaining (H-2) heads extract different features for other types that may be related.

For this reason, the designed multi-head mechanism allows the SPIL module to work in parallel to capture diverse types of relation points. Every weights W_ι is computed in the same way according to Eq. 2, where $\iota \in H$ is the number of heads. It should be noted that independent headers do not share weights during the calculation. By using the multi-head mechanism, the model can make more robust relational reasoning upon the points.

3.4 Skeleton Point Convolution

To perform reasoning on the skeleton points, unlike the standard 2D or 3D convolutions that run on a local regular grid. For a target point, the outputs are updated features of each object points from all its neighbors. We can represent one layer of convolutions as:

$$X^{(l+1)} = WX^{(l)}\mathcal{Z}^{(l)}, \tag{7}$$

where $W \in \mathbb{R}^{K \times K}$ represents the interaction weights we have introduced in Sect. 3.2. $X^{(l)} \in \mathbb{R}^{K \times C'}$ is the input feature projected by $g(\cdot)$ mapping function of a centroid grouping skeleton point set. $\mathcal{Z}^{(l)} \in \mathbb{R}^{C' \times d}$ is the layer-specific learnable weight matrix. After each layer of convolutions, we adopt non-linear functions for activating before the feature $X^{(l+1)}$ is forwarded to the next layer. We stack the convolution operation into two layers in our work.

To combine multi-head weights, in this work, we employ the concatenation fusion function. We can extend Eq. 7 as:

$$X^{(l+1)} = \overset{H}{\underset{\iota}{\|}} (W_\iota X^{(l)} \mathcal{Z}_\iota^{(l)}, dim = 1), \tag{8}$$

where W_ι indicates different types of weights and the different $\mathcal{Z}_\iota$ are not shared. $\|(\cdot)$ function aggregates and fuses the output information of all H heads. Namely, all $K \times d$-dim features will concatenate together to form a new feature $\in \mathbb{R}^{K \times D}$, where $D = \sum_1^H d$. Thus the points can be increased in dimensionality to obtain

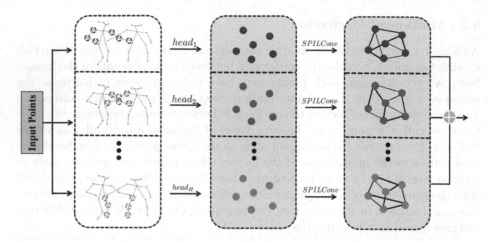

Fig. 4. Illustration of the multi-head SPIL. A single head encodes the skeleton point clouds from the input independently, and multiple different headers are responsible for processing different types of information from the points and eventually aggregate them together. Each node denotes a skeleton joint point and each edge is a scalar weight, which is computed according to two points' features and their relative position.

more characteristic information. Afterward, for each subsampled centroid grouping points, $Maxpooling$ is applied for fusing K local region points that are updated by Eq. 8.

Finally, as illustrated in Fig. 2. The output relational features are forwarded to an average pooling layer, which calculates the mean of all the proposal features and leads to a 1×1024 dimensions representation. Then it is fed to the classifier to generate predictions. Y_W denotes the labels and $\hat{Y}_W$ are predictions, with standard cross-entropy loss, the final loss function is formed as:

$$\mathcal{L} = \mathcal{L}(Y_W, \hat{Y}_W). \tag{9}$$

4   Experiments

In this section, we evaluate our method on public violent video datasets. The results show that our method outperforms representative baseline methods and achieves state-of-the-art performance.

Dataset Details: We train and evaluate our model on four datasets (Hockey-Fight dataset [20], Crowd Violence dataset [14], Movies-Fight dataset [21] and the RWF-200 Violence dataset [4]) for video violence recognition. To the best of our knowledge, RWF-2000 is the largest dataset which consists of 2,000 video clips captured by surveillance cameras in real-world scenes. Each video file is a 5-s video clip with 30 fps. Half of the videos contain violent behaviors, while others belong to non-violent actions. All videos in this dataset are captured by surveillance cameras in the real world, none of them are modified by multimedia technologies. Unlike ordinary action video datasets, where there are only one or two people, these violent video datasets usually have a lot of characters in each video and they are not fixed. The background information is complicated and the dynamic information of each character is also more changeable.

Implementation Details: Our implementation is based on PyTorch deep learning framework. Under the setting of batch-size 8 and 2048 sampled points, dropout [15] is applied to the last global fully connected layer with a ratio of 0.4. We train our model from scratch for 200 epochs using SGD with momentum and a learning rate of 10-3 on 4 Nvidia 2080Ti GPUs. Empirically, we set the layer number of our SPIL modules to 3 and $T_{frame} = 5$. Augmentation is used during the training with randomly jittering and rotating ($\pm 10\%$). For the baseline of our skeleton point convolution network, we give a vanilla model without using SPIL and multi-head mechanism. In other words, we simply use the feature information of the points and single header, ignoring the position information, and then use the multi-layer perceptrons to perform operations. We run all experiments four times with different random seeds and report mean accuracies. For simplicity, we define the method using Eq. 4 as **SPIL-Space**, and **SPIL-Span** represents using Eq. 5 while **SPIL-Mask** implies Eq. 6's strategy.

4.1   Ablation Study

To show the impact of the SPIL module and the hyper-parameter in our network, we conduct several ablation studies on the RWF-2000 dataset. First, we compare the baseline model with the SPIL convolution of the three discussed solutions we designed. Among them, we all use a single head method for a fair comparison. At the same time, we set the default value as 0.02 in the hyper-parameter of the threshold d in Eq. 6.

Table 1. Exploration of different skeleton points interaction learning strategies.

Method	Baseline	SPIL-Space	SPIL-Span	SPIL-Mask
Accuracy (%)	84.3	86.4	88.0	**88.7**

Table 2. Exploration of number of multi-heads.

Num-head	1	2	4	8	16	32	64
SPIL-Mask (%)	88.7	88.9	89.0	**89.2**	89.1	88.9	89.0
Baseline (%)	84.3	84.6	**84.8**	84.7	84.7	84.6	84.6

As shown in Table 1, In the case of point convolution operation without using SPIL to extract both feature and position information of points, our baseline has an accuracy of only **84.3%**. After using our proposed SPIL module, base on a single head, all methods outperform the based model, demonstrating the effectiveness of modeling interaction weights between points. And the **SPIL-Mask** yields the best accuracy with **88.7%** than the other two ways. We conjecture that the function of the mask can filter out some redundant information and make the information more stable. In the rest of the paper, we choose **SPIL-Mask** to represent our main method.

We also reveal the effectiveness of building a multi-head mechanism to capture diverse types of related information. As depicted in Table 2, we compare the performance of using different numbers of heads in baseline and our model, which the results indicate that multi-head mechanism can improve both methods in different degrees. When head number is set to 8, **SPIL-Mask** is able to further boost accuracy from **88.7%** to **89.2%**. Meanwhile, too many heads will lead to redundant learning features and will increase computational costs.

Table 3. Exploration of different distance threshold d in Eq. 6.

Eu-Distance	$d_1 = 0.01$	$d_2 = 0.02$	$d_3 = 0.04$	$d_4 = 0.08$
Accuracy (%)	88.7	89.2	**89.3**	89.0

Furthermore, we also implement experiments to reveal the effect of the hyperparameter threshold of d on network performance. As shown in Table 3, We see that when $d = 0.04$ achieves the best performance approaching **89.3%**. We conjecture that a too-small threshold will cause network information to be lost, and a too-large threshold will cause a negative effect of excess network information. Thus, we adopt head number $H = 8$, $d = 0.04$ in the following experiments.

Table 4. Comparison with state-of-the-art on the RWF-2000 dataset.

Method	Core operator	Accuracy (%)
TSN [31]	Two-stream	81.5
I3D [3]	Two-stream	83.4
3D-ResNet101 [13]	3D convolution	82.6
ECO [40]	3D convolution + RGB	83.7
Representation flow [22]	Flow + Flow	85.3
Flow gated network [4]	3D convolution + flow	87.3
PointNet++ [24]	Multiscale point MLP	78.2
PointConv [34]	Dynamic filter	76.8
DGCNN [32]	Graph convolution	80.6
Ours	**SPIL convolution**	**89.3**

4.2 Comparison with the State of the Art

We compare our approach with state-of-the-art approaches. Here, we adopt several models that perform well in traditional video action recognition tasks, and apply them to the violent video recognition tasks in this paper (they all are pre-trained on Kinetics [17]). At the same time, we also compared the excellent models in processing 3D point clouds tasks with our method.

As shown in Table 4 on RWF-200 dataset, above the solid line it can be seen that although these methods reach the state-of-the-art level in general video action recognition tasks, their best results can only reach **87.3%** in violent video recognition which is 2% lower than our proposed method with accuracy in **89.3%**. This shows that traditional action video recognition methods lack the ability to extract the dynamic characteristics of people in the violence videos and the long-term correlation performance in each frame. At the same time, in violent videos, due to the diversity of characters and the variety of scenes, as a result, these methods rely on information such as optical flow information, global scene characteristics are invalidated, resulting in low recognition results. When we use human skeleton points to identify action characteristics, previous point clouds methods are not targeted and are not sensitive to the skeleton points, therefore, the recognition results are not accurate enough. By learning different skeleton points interactions through our method, we can reach the leading level.

Table 5. Comparison with state-of-the-art on the Hockey-Fight and Crowed Violence dataset.

Method	Hockey-Fight (%)	Method	Crowd (%)
3D CNN [8]	91.0	3D CNN [8]	-
MoWLD + BoW [38]	91.9	MoIWLD + KDE [37]	93.1
TSN [31]	91.5	TSN [31]	81.5
I3D [3]	93.4	I3D [3]	83.4
ECO [40]	94.0	ECO [40]	84.7
Representation flow [22]	92.5	Representation Flow [22]	85.9
Flow Gated Network [4]	**98.0**	Flow Gated Network [4]	88.8
PointNet++ [24]	89.7	PointNet++ [24]	89.2
PointConv [34]	88.6	PointConv [34]	88.9
DGCNN [32]	90.2	DGCNN [32]	87.4
Ours	96.8	**Ours**	**94.5**

We further evaluate the proposed model on the Hockey-Fight and Crowd Violence dataset. As shown in Table 5, specifically, in terms of accuracy, the average level on the crowd dataset is lower than the hockey dataset. This is because the crowd dataset has more people and its background information is relatively complicated. Therefore, some previous work is difficult to have a high degree of recognition. Our SPIL module is not affected by complex scenes. It obtains features by extracting the skeleton point clouds information of the people, which can achieve a high accuracy rate. This outstanding performance shows the effectiveness and generality of the proposed SPIL for capturing the related points information in multiple people scene.

Table 6. Comparison with state-of-the-art on the Movies-Fight dataset.

Method	Accuracy (%)
Extreme acceleration [7]	85.4
TSN [31]	94.2
I3D [3]	95.8
3D-ResNet101 [13]	96.9
ECO [40]	96.3
Representation Flow [22]	97.3
PointNet++ [24]	89.2
PointConv [34]	91.3
DGCNN [32]	92.6
Ours	**98.5**

Finally, we validate our model on the Movies-Fight dataset. As can be seen in Table 6, our method can outperform the existing methods. Under different datasets and different environmental scenarios, our method does not rely on other prior knowledge and will not overdone in one dataset and causes inadaptability in other scenarios.

4.3   Failure Case

To further analyze the network effect, we study the misclassification results of the above four datasets. We found that some similar to violent actions are misclassified as violence. As shown in Fig. 5, some movements in the picture that have physical contact, models consider they are fighting or some other violence. Therefore, we collect all these easily confused samples from 4 datasets evenly and formed them into a new small dataset for classification.

Fig. 5. From left to right: "hockey ball scrambling", "hug", "dance", "high five" and "wave".

Figure 6 shows the confusion matrix comparison. In those actions that are very similar to violence but are not actual acts of violence, compared with the traditional video action recognition technique, our method can have a large degree of discrimination to classify these actions. The above results illustrate that the proposed network is an effective method for video violence recognition. Essentially paying more attention to the information about the skeletal modality movements of the characters will help to distinguish the behaviors of the characters.

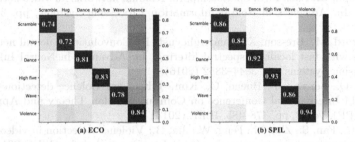

Fig. 6. Confusion matrix comparison on the confused sample. (a) ECO [40] method. (b) Our SPIL method.

5 Conclusion

In this paper, we propose a novel and effective approach for video violence recognition. To the best of our knowledge, we are the first one to solve this task by using a 3D point clouds technique to extract action feature information of human skeleton points. By introducing the Skeleton Points Interaction Learning (SPIL) module, our model is able to assign different weights according to different skeleton points to obtain the motion characteristics of different people. Furthermore, we also design a multi-head mechanism to process different types of information in parallel and eventually aggregate them together. The experiment results on four violent video datasets are promising, demonstrating that our proposed network outperforms the existing state-of-the-art violent video recognition approaches.

Acknowledgment. This work was supported by NSFC 61876208, Key-Area Research and Development Program of Guangdong 2018B010108002, National Research Foundation Singapore under its AI Singapore Programme (AISG-RP-2018-003) and the MOE Tier-1 research grants: RG28/18 (S) and RG22/19 (S).

References

1. Bronstein, M.M., Bruna, J., LeCun, Y., Szlam, A., Vandergheynst, P.: Geometric deep learning: going beyond Euclidean data. IEEE Signal Process. Mag. **34**(4), 18–42 (2017)

2. Brox, T., Bruhn, A., Papenberg, N., Weickert, J.: High accuracy optical flow estimation based on a theory for warping. In: Pajdla, T., Matas, J. (eds.) ECCV 2004. LNCS, vol. 3024, pp. 25–36. Springer, Heidelberg (2004). https://doi.org/10.1007/978-3-540-24673-2_3

3. Carreira, J., Zisserman, A.: Quo vadis, action recognition? A new model and the kinetics dataset. In: Proceedings of the IEEE Conference on Computer Vision and Pattern Recognition, pp. 6299–6308 (2017)

4. Cheng, M., Cai, K., Li, M.: RWF-2000: an open large scale video database for violence detection. arXiv preprint arXiv:1911.05913 (2019)

5. Christoph, R., Pinz, F.A.: Spatiotemporal residual networks for video action recognition. In: Advances in Neural Information Processing Systems, pp. 3468–3476 (2016)

6. Defferrard, M., Bresson, X., Vandergheynst, P.: Convolutional neural networks on graphs with fast localized spectral filtering. In: Advances in Neural Information Processing Systems, pp. 3844–3852 (2016)

7. Deniz, O., Serrano, I., Bueno, G., Kim, T.K.: Fast violence detection in video. In: 2014 International Conference on Computer Vision Theory and Applications (VISAPP), vol. 2, pp. 478–485. IEEE (2014)

8. Ding, C., Fan, S., Zhu, M., Feng, W., Jia, B.: Violence detection in video by using 3D convolutional neural networks. In: Bebis, G., et al. (eds.) ISVC 2014. LNCS, vol. 8888, pp. 551–558. Springer, Cham (2014). https://doi.org/10.1007/978-3-319-14364-4_53

9. Donahue, J., et al.: Long-term recurrent convolutional networks for visual recognition and description. In: Proceedings of the IEEE Conference on Computer Vision and Pattern Recognition, pp. 2625–2634 (2015)

10. Fang, H.S., Xie, S., Tai, Y.W., Lu, C.: RMPE: regional multi-person pose estimation. In: Proceedings of the IEEE International Conference on Computer Vision, pp. 2334–2343 (2017)

11. Garcia-Garcia, A., Gomez-Donoso, F., Garcia-Rodriguez, J., Orts-Escolano, S., Cazorla, M., Azorin-Lopez, J.: PointNet: a 3D convolutional neural network for real-time object class recognition. In: 2016 International Joint Conference on Neural Networks (IJCNN), pp. 1578–1584. IEEE (2016)

12. Gehring, J., Auli, M., Grangier, D., Dauphin, Y.N.: A convolutional encoder model for neural machine translation. arXiv preprint arXiv:1611.02344 (2016)

13. Hara, K., Kataoka, H., Satoh, Y.: Can spatiotemporal 3D CNNs retrace the history of 2D CNNs and ImageNet? In: Proceedings of the IEEE Conference on Computer Vision and Pattern Recognition, pp. 6546–6555 (2018)

14. Hassner, T., Itcher, Y., Kliper-Gross, O.: Violent flows: real-time detection of violent crowd behavior. In: 2012 IEEE Computer Society Conference on Computer Vision and Pattern Recognition Workshops, pp. 1–6. IEEE (2012)

15. Hinton, G.E., Srivastava, N., Krizhevsky, A., Sutskever, I., Salakhutdinov, R.R.: Improving neural networks by preventing co-adaptation of feature detectors. arXiv preprint arXiv:1207.0580 (2012)

16. Karpathy, A., Toderici, G., Shetty, S., Leung, T., Sukthankar, R., Fei-Fei, L.: Large-scale video classification with convolutional neural networks. In: Proceedings of the IEEE Conference on Computer Vision and Pattern Recognition, pp. 1725–1732 (2014)

17. Kay, W., et al.: The kinetics human action video dataset. arXiv preprint arXiv:1705.06950 (2017)

18. Lev, G., Sadeh, G., Klein, B., Wolf, L.: RNN Fisher vectors for action recognition and image annotation. In: Leibe, B., Matas, J., Sebe, N., Welling, M. (eds.) ECCV 2016. LNCS, vol. 9910, pp. 833–850. Springer, Cham (2016). https://doi.org/10. 1007/978-3-319-46466-4_50
19. Niepert, M., Ahmed, M., Kutzkov, K.: Learning convolutional neural networks for graphs. In: International Conference on Machine Learning, pp. 2014–2023 (2016)
20. Nievas, E.B., Suarez, O.D., Garcia, G.B., Sukthankar, R.: Hockey fight detection dataset. In: Computer Analysis of Images and Patterns, pp. 332–339. Springer (2011)
21. Nievas, E.B., Suarez, O.D., Garcia, G.B., Sukthankar, R.: Movies fight detection dataset. In: Computer Analysis of Images and Patterns, pp. 332–339. Springer (2011)
22. Piergiovanni, A., Ryoo, M.S.: Representation flow for action recognition. In: Proceedings of the IEEE Conference on Computer Vision and Pattern Recognition, pp. 9945–9953 (2019)
23. Qi, C.R., Su, H., Mo, K., Guibas, L.J.: PointNet: deep learning on point sets for 3D classification and segmentation. In: Proceedings of the IEEE Conference on Computer Vision and Pattern Recognition, pp. 652–660 (2017)
24. Qi, C.R., Yi, L., Su, H., Guibas, L.J.: PointNet++: deep hierarchical feature learning on point sets in a metric space. In: Advances in Neural Information Processing Systems, pp. 5099–5108 (2017)
25. Riegler, G., Osman Ulusoy, A., Geiger, A.: OctNet: learning deep 3D representations at high resolutions. In: Proceedings of the IEEE Conference on Computer Vision and Pattern Recognition, pp. 3577–3586 (2017)
26. Simonyan, K., Zisserman, A.: Two-stream convolutional networks for action recognition in videos. In: Advances in Neural Information Processing Systems, pp. 568–576 (2014)
27. Song, S., Yu, F., Zeng, A., Chang, A.X., Savva, M., Funkhouser, T.: Semantic scene completion from a single depth image. In: Proceedings of the IEEE Conference on Computer Vision and Pattern Recognition, pp. 1746–1754 (2017)
28. Tran, D., Bourdev, L., Fergus, R., Torresani, L., Paluri, M.: Learning spatiotemporal features with 3D convolutional networks. In: Proceedings of the IEEE International Conference on Computer Vision, pp. 4489–4497 (2015)
29. Tran, D., Wang, H., Torresani, L., Ray, J., LeCun, Y., Paluri, M.: A closer look at spatiotemporal convolutions for action recognition. In: Proceedings of the IEEE Conference on Computer Vision and Pattern Recognition, pp. 6450–6459 (2018)
30. Veličković, P., Cucurull, G., Casanova, A., Romero, A., Lio, P., Bengio, Y.: Graph attention networks. arXiv preprint arXiv:1710.10903 (2017)
31. Wang, L., et al.: Temporal segment networks: towards good practices for deep action recognition. In: Leibe, B., Matas, J., Sebe, N., Welling, M. (eds.) ECCV 2016. LNCS, vol. 9912, pp. 20–36. Springer, Cham (2016). https://doi.org/10.1007/ 978-3-319-46484-8_2
32. Wang, Y., Sun, Y., Liu, Z., Sarma, S.E., Bronstein, M.M., Solomon, J.M.: Dynamic graph CNN for learning on point clouds. ACM Trans. Graph. (TOG) 38(5), 1–12 (2019)
33. Wu, J., Wang, L., Wang, L., Guo, J., Wu, G.: Learning actor relation graphs for group activity recognition. In: Proceedings of the IEEE Conference on Computer Vision and Pattern Recognition, pp. 9964–9974 (2019)
34. Wu, W., Qi, Z., Fuxin, L.: PointConv: deep convolutional networks on 3D point clouds. In: Proceedings of the IEEE Conference on Computer Vision and Pattern Recognition, pp. 9621–9630 (2019)

35. Xie, S., Girshick, R., Dollár, P., Tu, Z., He, K.: Aggregated residual transformations for deep neural networks. In: Proceedings of the IEEE Conference on Computer Vision and Pattern Recognition, pp. 1492–1500 (2017)
36. Yue-Hei Ng, J., Hausknecht, M., Vijayanarasimhan, S., Vinyals, O., Monga, R., Toderici, G.: Beyond short snippets: deep networks for video classification. In: Proceedings of the IEEE Conference on Computer Vision and Pattern Recognition, pp. 4694–4702 (2015)
37. Zhang, T., Jia, W., He, X., Yang, J.: Discriminative dictionary learning with motion weber local descriptor for violence detection. IEEE Trans. Circuits Syst. Video Technol. **27**(3), 696–709 (2016)
38. Zhang, T., Jia, W., Yang, B., Yang, J., He, X., Zheng, Z.: MoWLD: a robust motion image descriptor for violence detection. Multimed. Tools Appl. **76**(1), 1419–1438 (2015). https://doi.org/10.1007/s11042-015-3133-0
39. Zhang, Y., Rabbat, M.: A graph-CNN for 3D point cloud classification. In: 2018 IEEE International Conference on Acoustics, Speech and Signal Processing (ICASSP), pp. 6279–6283. IEEE (2018)
40. Zolfaghari, M., Singh, K., Brox, T.: ECO: efficient convolutional network for online video understanding. In: Ferrari, V., Hebert, M., Sminchisescu, C., Weiss, Y. (eds.) ECCV 2018. LNCS, vol. 11206, pp. 713–730. Springer, Cham (2018). https://doi.org/10.1007/978-3-030-01216-8_43

Binarized Neural Network for Single Image Super Resolution

Jingwei Xin[1], Nannan Wang[2(✉)], Xinrui Jiang[2], Jie Li[1], Heng Huang[3],
and Xinbo Gao[1]

[1] State Key Laboratory of Integrated Services Networks, School of Electronic
Engineering, Xidian University, Xi'an 710071, China
jwxintt@gmail.com, {leejie,xbgao}@mail.xidian.edu.cn
[2] State Key Laboratory of Integrated Services Networks, School of
Telecommunications Engineering, Xidian University, Xi'an 710071, China
nnwang@xidian.edu.cn, xrjiang@stu.xidian.edu.cn
[3] School of Electrical and Computer Engineering, University of Pittsburgh,
Pittsburgh, PA 15261, USA
heng.huang@pitt.edu

Abstract. Lighter model and faster inference are the focus of current
single image super-resolution (SISR) research. However, existing methods are still hard to be applied in real-world applications due to the
heavy computation requirement. Model quantization is an effective way
to significantly reduce model size and computation time. In this work, we
investigate the binary neural network-based SISR problem and propose a
novel model binarization method. Specially, we design a bit-accumulation
mechanism (BAM) to approximate the full-precision convolution with a
value accumulation scheme, which can gradually refine the precision of
quantization along the direction of model inference. In addition, we further construct an efficient model structure based on the BAM for lower
computational complexity and parameters. Extensive experiments show
the proposed model outperforms the state-of-the-art binarization methods by large margins on 4 benchmark datasets, specially by average more
than 0.7 dB in terms of Peak Signal-to-Noise Ratio on Set5 dataset.

Keywords: Single image super-resolution · Model quantization ·
Binary neural network · Bit-accumulation mechanism

1 Introduction

Single image super-resolution (SISR) aims to recover a high-resolution (HR)
version from a low-resolution (LR) input image, which has been widely used in
many fields, such as medical imaging [21], satellite imaging [25], security and
surveillance [35] and so on. As a classical low-level problem, SISR is still an
active yet challenging research topic in the field of computer vision due to its
ill-poseness nature and high practical values.

© Springer Nature Switzerland AG 2020
A. Vedaldi et al. (Eds.): ECCV 2020, LNCS 12349, pp. 91–107, 2020.
https://doi.org/10.1007/978-3-030-58548-8_6

SRResNet	SRResNet-BNN	SRResNet-ABC-Net	SRResNet-BAM(Ours)
(W-32 A-32)	(W-1 A-1)	(W-1 A-1)	(W-1 A-1)

Fig. 1. Results of binarized networks based on SRResNet.

Recently, convolutional neural network-based (CNN-based) super-resolution methods have been demonstrated state-of-the-art performance by learning a mapping from LR to HR image patches. Dong et al. [6] proposed the SRCNN model with only three convolution layers, which is the first deep learning method in super-resolution community. From then, researchers had carried out its study with different perspectives and obtained plentiful achievements, and various model structure and learning strategies are used in SR networks (*e.g.*, residual learning [10], recursive learning [11,23,24], skip connection [13,27], channel attention [32] and so on). These CNN-based methods can often achieve satisfactory results, but the increasing model size and the computational complexity severely restrict their applications in the real world. Recently, there appear some lightweight approaches to reduce the computational complexity [1,8]. However, it is still a huge burden for mobile devices with limited computing resources.

As a way to significantly reduce model size and computation time, binarized neural network (BNN) can replace the floating point operations with the bitcounting operations, and has shown excellent performance on many semantic-level tasks such as image classification and recognition. However, quantifying the weights and activations will lead to serious information loss in the process of network inference, which is unacceptable for super-resolution tasks because of its highly dependence on the accuracy of pixel values. Ma et al. [17] tried to apply the binarization to the residual blocks for image super resolution, and improve the performance by learning a gain term. However, this work only investigate the binary weights and full-precision activations model, the convolution calculation is not the bitcounting operation. Then the model's inference speed cannot be simplified enough.

In this paper, we introduce an efficient and accurate CNN-based SISR model by binarizing the weights and even the intermediate activations. Our binarization approach aims to find the best approximations of the convolutions using binary operations, and perform image super resolution task on the devices with limited computing resources. As shown in Fig. 1, our method achieves better visual SR results compared with state-of-the-art methods, and could even be comparable to the performance of the full-precision convolutional model.

Overall, our contributions are mainly threefold: (1) To the best of our knowledge, this is the first work to introduce the binary neural network (both the weights and activations are binary values) to the field of image super resolution, in which the convolutions can be estimated by the bitcounting operations. As a result, our model could obtain about 58× speed up (the model size is also about 32× lighter) than that of an equivalent network with single-precision weight values. Its inference can be done very efficiently on CPU. (2) A bit accumulation mechanism is proposed to approximate the full-precision convolution with an iterative scheme of cascaded binary weights and activations. What's more, it implements highly accurate one-bit estimation of filter and activation only relying on existing basic models and without introducing any other additional inference modules. (3) We construct an architecture of binary super-resolution network (BSRN) for highly accurate image SR problem. Experimental results show that the proposed BSRN can achieve the better SR performance with a lighter network structure and fewer operands.

2 Related Work

2.1 Single Image Super Resolution

Recently numerous deep learning based methods have been explored and shown dramatic improvements on the SISR tasks. Deep SISR network is first introduced by SRCNN [6] which is an end-to-end model with only three convolution layers. Considering the effectiveness of deep learning and the natural sparsity of images [30], Wang et al. [28] proposed a Sparse Coding Network (SCN) to make full use of the natural sparsity of images. Later on, Kim et al. [10] proposed a 20-layers network VDSR, which demonstrates significant improvement by increasing the network depth. After this, many others followed up with this strategy for network design. Tong et al. [27] adopted dense blocks to construct a 69-layers network SRDenseNet. Extended from it, Lim et al. [15] developed a more in-depth and broader residual network known as EDSR, which exhibits comparable performance for SR task. Zhang et al. [33] introduced a residual dense network RDN, which combines residuals and dense blocks to achieve higher image reconstruction performance with higher feature extraction capability. To overcome the gradient vanishing problem, residual channel attention network is adopted in RCAN [32], which proposes long and short skip connections in the residual structure to obtain deep residual network.

Aiming to achieve better performance with less parameters, recursive learning have been employed in SISR. Kim et al. proposed a Deeply-Recursive Convolutional Network(DRCN) [11] for SISR task. Then, Tai et al. also proposed Deep Recursive Residual Network(DRRN) [23], which introduces a very deep model (52 layers) with residual learning and recursive module. The authors further proposed Memory Network (MemNet) [24], which could adaptively combine the multi-scale features by the memory blocks. Another research direction which is time-saving network designing. For instance, the deconvolution layer has been proposed in FSRCNN [7] and sub-pixel layer has been introduced in ESPCN [20], which are the better up-sampling operator for accelerating super-resolution network. Ahn et al. [1] introduced a cascading residual network CARN combining the efficient recursive scheme and multiple residual connections. In view of the above-mentioned methods which is heavily dependent on practical experience, He et al. [8] adopted an ordinary differential equation (ODE)-inspired design scheme to single image super resolution.

2.2 Quantitative Model

In the development of CNN, a great amount of efforts have been explored for model compression, which can speed-up the inference process of deep networks. Recently, published strategies for reducing precision (number of bits used for numerical representation) have achieved significant progress in computer vision tasks. Among the existing methods, Soudry et al. [22] introduced a variational Bayesian approach to calculate the posterior distribution of the weights (the weights are constrained to +1 or −1). Courbariaux et al. [4] proposed a Bina-ryConnect method, which binarizes network weights during the forward and updates the full-precision weights during the backward propagations. Extended from the BinaryConnect, Hubara et al. [5] proposed a network named BinaryNet, weights and activations in BinaryNet are both binarized. Rastegari et al. [19] proposed a similar model called XNOR-Net. XNOR-Net includes major steps from the original BNN, but adds a gain term to compensate for lost information during binarization. Zhou et al. [34] tried to generalize quantization and take advantage of bitwise operations for fixed point data with widths of various sizes, and proposed the DoReFa-Net method. Lin et al. [16] proposed ABC-Net to reconcile the accuracy gap between BNNs and full-precision networks. Little effort has been spent on model quantization for the image SR task. We design a binarization strategy for the SR task to make the large SR networks being configured on mobile device.

3   Proposed Approach

In this section, we present our proposed BNN-based SISR approach. After presenting the motivation of our approach in Sect. 3.1, we give details on our quantification of weights and activations in Sect. 3.2 and Sect. 3.3 respectively. The proposed binary SR framework is presented in Sect. 3.4.

3.1   Motivation

The process in existing deep learning-based SISR methods could be divided into three stages: feature extraction, nonlinear mapping and image reconstruction.

Let x denote the LR input image and y as the final recovered HR image, the function of these models can be formulated as follows,

$$y = \mathcal{R}(\mathcal{M}(\mathcal{E}(x))), \tag{1}$$

where $\mathcal{E}, \mathcal{M}$ and $\mathcal{R}$ are the three aforementioned stages. Generally, $\mathcal{E}$ and $\mathcal{R}$ are composed of only one convolutional layer to realize the transformation from image to deep features and its inverse transformation. $\mathcal{M}$ realizes the mapping process from low-precision features to high-precision features through multiple cascading convolutional layers. The structure of $\mathcal{M}$ will directly determine the model's performance, parameters and computational complexity. Therefore, replacing the $\mathcal{M}$'s full-precision convolution with binary convolution can greatly reduce the model's consumption of computing and storage resources.

In BNN model, the full-precision convolution $W*A$ is estimated by the binary convolution $W^B \oplus A^B$. Where W^B and A^B are the binary weights and activations, $\oplus$ indicates the bitcounting operations. To find an optimal estimation, a straightforward way is to solve the following optimization problem:

$$J(W^B, A^B) = \left\| W * A - W^B \oplus A^B \right\|,$$
$$W^{B^*}, A^{B^*} = \underset{W^B, A^B}{arg\min} J(W^B, A^B), \tag{2}$$

The key to solve this optimization problem lies in the generation of W^B and A^B. In the existing quantitative models, two gain terms α, β are usually introduced to compensate for the lost information during binarization ($W^B = \alpha|W|_{\text{Bin}}$, $A^B = \beta|A|_{\text{Bin}}$). Then the optimal solution (gain terms) can be obtained by calculating or learning from the weights and activations before binarization. However the improvement of convolution precision by gain term is limited, and the back propagation of network gradient is inefficient to update it. In this work, we propose a novel quantization method of filter weights and activations based on bit accumulation mechanism, which achieves better performance for the image SR task.

3.2 Quantization of Weights

The start point of our method is refining the precision of binary filters along the direction of model inference gradually. The processing of model inference is shown in Fig. 2, we first accumulate the weight of the each past layer, then use the current weights to offset the accumulated weights in a positive or negative direction, lead to the accumulated weight can be quantified in a more accurate state (-1 or $+1$). Considering that different convolutional layers have different preferences for image features (color, edge, texture, etc.). We also set a group of combination coefficients α ($\alpha = [\alpha_1, \alpha_2, ..., \alpha_n]$) for the accumulation process.

$$W_n{}^B = Sign(BN(\alpha_1 W_1 + \alpha_2 W_2 + ... + \alpha_n W_n)) * E(|W_n|), \tag{3}$$

where W_n denotes the n_{th} full-precision weights, and $W_n{}^B$ denotes the actual convolution weights of the n_{th} binary filter after updating. $E(|W_n|)$ is the mean

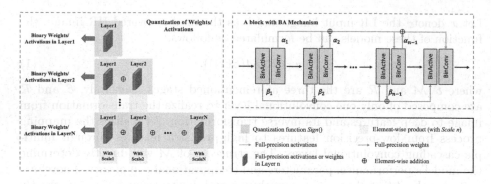

Fig. 2. Bit Accumulation Mechanism (BAM).

of absolute value of each output channel of weights. The utilization of $E(|W_n|)$ could increase the value range of weights and is beneficial to estimate the high-precision binary weights. $BN()$ is the batch normalization operation. $Sign()$ is the symbolic function which transform full-precision values to $+1$ or -1:

$$x^B = Sign(x) \begin{cases} +1, & \text{if} \quad x \geq 0, \\ -1, & \text{otherwise}, \end{cases} \quad (4)$$

An optimal estimation could be find by solving the following optimization problem:

$$\underset{W_n^B}{arg\min} \, J(W_n^B) = \left\| W_n - W_n^B \right\|^2. \quad (5)$$

For W_n^B, full-precision weights W_n is given, and the gain term $E(|W_n|)$ could be determined from W_n^B. The optimization error directly depends on the choice of the combination coefficients α. Therefore, the Eq. (5) can be rewritten as the following optimization problem:

$$\alpha^* = \underset{W_n^B}{arg\min} \, J(W_n^B) \quad (6)$$

Straight-Through Estimator (STE) is defined for Bernoulli sampling with probability $p \in [0,1]$, which could be thought of as an operator that has arbitrary forward and backward operations:

$$Forward : q \sim Bernoulli(p).$$
$$Backward : \frac{\partial c}{\partial p} = \frac{\partial c}{\partial q}. \quad (7)$$

Here we adopt the STE method for back-propagate through W^B. Assume c as the cost function, A_o^B as the output tensor of a convolution respectively, the gradient at n_{th} filter in back propagation can be computed as follows:

$$\frac{\partial c}{\partial W_n} = \frac{\partial c}{\partial A_o} \frac{\partial A_o}{\partial W_n^B} \frac{\partial W_n^B}{\partial W_n} \overset{STE}{\approx} \frac{\partial c}{\partial A_o} \frac{\partial A_o}{\partial W_n^B} = \frac{\partial c}{\partial W_n^B}. \quad (8)$$

Algorithm 1. Training a N-layers CNN-block with BAM:

Input: Full-precision activations A_1 and weights W from the N filters ($W = [W_1, W_2, ..., W_N]$), activation combination coefficients $\beta_1, \beta_2, ..., \beta_N$, cost function c and current learning rate η^t.

Output: updated weight W^{t+1}, updated activation combination coefficients $\beta_1, \beta_2, ..., \beta_N$, and updated learning rate η^{t+1}.

1: **for** $n = 1$ to N **do**
2:    $A_n^B \leftarrow$ calculate the equation (9)
3:    $\alpha \leftarrow$ solve the equation (6)
4:    $W_n^B \leftarrow$ calculate the equation (3)
5:    $A_{n+1} = BinaryConv(W_n^B, A_n^B)$
6: **end for**
7: $g^w \leftarrow$ **BinaryBackward**$(\frac{\partial c}{\partial A_o}, W^B)$, computed using the equation (8)
8: **for** $n = N$ to 1 **do**
9:    $g_n^w \leftarrow$ **BackwardGradient**(g^w, W_n)
10:    $\beta_n =$ **UpdateParameters**$(\beta_n^{scale}, g_{\beta_n^{scale}}, \eta^t)$
11:    $W_n^{t+1} =$ **UpdateParameters**(W_n^t, g_n^w, η^t)
12: **end for**
13: $\eta^{t+1} =$ **UpdateLearningrate**$(\eta^t; t)$

Bear in mind that during training, the full-precision weights are reserved and updated at every epoch, while in test-time only binary weights are used in convolution. The training details of our model are summarized in Algorithm 1.

3.3 Quantization of Activations

The operation of floating-point convolution could be implemented without multiplications when weights are binarized, but the computation is still much more than the bitcounting operation. Next we detail our approach to getting binary activations that are input to convolutions, which is of critical importance in replacing floating-point convolutions by bitcounting-based convolutions.

The proposed bit accumulation mechanism is not only applicable to the quantification of weights, but also applicable to the quantization of activations ass shown in Fig. 2. Here we also make linear combination of the past multi-layer's activations to approximate the current layer's activations.

$$A_n^B \approx Sign(BN(\beta_1 A_1 + \beta_2 A_2 + ... + \beta_n A_n)), \quad (9)$$

Considering that the process of image super resolution depends heavily on the details of the activation, the rough quantification ($Sign()$) of the activation can lead to a sharp decline in model performance. Instead of setting the combination weights of the activations to a single scale, we use a two-dimensional array to represent these weights.

$$\beta_n \leftarrow Update(\beta_n^{scale} E(A_n), g_{\beta_n^{scale}}, \eta) \quad (10)$$

where β_n^{scale} is a scale factor, and only β_n^{scale} will be updated during the training process. $g_{\beta_n^{scale}}$ and η are the gradient and learning rate. $E(A_n)$ is the activation statistics along the channel dimension, which could reflect the spatial information of the input image.

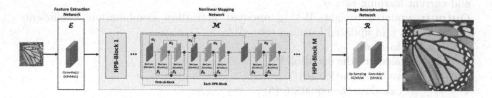

Fig. 3. Network architecture of our proposed BSRN.

3.4 Binary Super Resolution Network

As described in Sect. 3.1, our binary super resolution network (BSRN) also consists of three subnetworks: a feature extraction network $\mathcal{E}$, a nonlinear mapping network $\mathcal{M}$ and an image reconstruction network $\mathcal{R}$. $\mathcal{E}$ contains only one convolutional layer and $\mathcal{R}$ contain only one convolutional layer, while $\mathcal{R}$ contains an additional upsampling layer. Consistent with existing binarization methods, all convolution operations in $\mathcal{E}$ and $\mathcal{R}$ are full-precision.

Particularly, we construct a high-precision binarization (HPB) block structure to form our nonlinear mapping network $\mathcal{M}$, which could achieve the approximation of high precision convolution by accumulation the multiple binary convolution. The details are shown in Fig. 3. HPB block is also composed of several cascaded local binarization (LB) block, and each LB-block consists of one 3×3 convolution layer and two 1×1 convolution layers. The novelty of HPB block lies in two ranges of bit accumulation. Firstly, the short-range accumulation exists in each LB-block. The filters and activations of intermediary layers are accumulated into the higher layers, and finally converge on the last 1×1 convolution layer. Then, the long-range accumulation occurs in the first convolutional layer of each LB-block, and only filters are accumulated in the same way as the short-range accumulation.

4 Experiments

4.1 Datasets

DIV2K is a high quality image dataset which consists of 800 training images, 100 validation images and 100 testing images and it is widely used for super-resolution in recent years. Following [1,3,8,14], we use 800 training images from DIV2K dataset [26] as training set. To illustrate the performance of our proposed methods, we conduct test experiments on four standard datasets, Set5 [2], Set14 [31], BSD100 [18] and Urban100 [9] which includes 5, 14, 100 and 100 images respectively.

Three upscaling factors are evaluated, including ×2, ×3 and ×4. The input LR image is generated by bicubic down sampling the HR image with ×2, ×3 and ×4 scale respectively. The size of input patches are 48 × 48, and the output patch size is 96 × 96, 144 × 144 and 192 × 192 for ×2, ×3 and ×4 upscaling factor respectively.

4.2 Implementations

Data augmentation is performed on aforementioned 800 training images, which are randomly rotated by 90°, 180°, 270° and flipped horizontally. Batch size is set to 16. Adam is utilized to optimize the network. The momentum parameter is set to 0.5, weight decay is set to 2×10^{-4}, and the initial learning rate is set to 1×10^{-4} and will be divided a half every 200 epochs. All of our models are implemented under the PyTorch environment with Python 3.6 on Ubuntu 16.04 system with a 12G NVIDIA Titan Xp GPU.

For assessing the quality of SR results, we employ two objective image quality assessment metrics: Peak Signal to Noise Ratio (PSNR) and structural similarity (SSIM) [29]. All metrics are performed on the Y-channel (YCbCr color space) of center-cropped, removaling of a s-pixel wide strip from each ×s upscaling image border.

4.3 Evaluation

We choose two simple and practical super-resolution networks to evaluate model quantization methods: VDSR [10] and SRResNet [13]. VDSR and SRResNet can be regarded as the most typical methods of convolution-cascade model and block-cascade model respectively. Experimental evaluation on these two networks can more intuitively reflect the performance of model quantization. The performance evaluation of the proposed bit accumulation mechanism (BAM) is carry out on these two SR network, the compared methods including BNN [5] DoReFa-Net [34] and ABC-Net [16].

Evaluation on VDSR: Considering that the operation of our bit accumulation method needs to work in multiple convolutional layers, we divide the middle 18 convolutional layers of VDSR into 6 blocks on average. Each block contains three convolutional layers. The weight and activation accumulation process in each convolution block is consistent with Sects. 3.2 and 3.3. Other methods quantify each convolutional layer at the middle of VDSR separately. One should be noted is that ABC-Net simulates a full-precision convolution through multiple one-bit convolutions. Here we set the number of one-bit convolution to 3, that is, the model size of VDSR_ABC is nearly three times larger than other methods.

Table 1 shows the quantitative comparisons of the performances over the benchmark datasets. For binarization of weight and activation, there is no compensation process for quantization error in VDSR_BNN and VDSR_DoReFa methods. The information carried in the activations is limited, and the image's

Table 1. Quantitative evaluation of VDSR-based state-of-the-art model quantization methods.

Methods	Scale	Set5		Set14		B100		Urban100	
		PSRN	SSIM	PSRN	SSIM	PSRN	SSIM	PSRN	SSIM
VDSR	×2	37.53	0.959	33.05	0.913	31.90	0.896	30.77	0.914
Bicubic	×2	33.66	0.930	30.24	0.869	29.56	0.843	26.88	0.840
VDSR_BNN	×2	34.43	0.936	30.94	0.882	30.05	0.856	27.54	0.860
VDSR_DoReFa	×2	34.70	0.933	31.22	0.876	30.25	0.849	28.25	0.865
VDSR_ABC	×2	35.35	0.939	31.71	0.886	30.68	0.861	28.77	0.878
VDSR_BAM	×2	**36.60**	**0.953**	**32.41**	**0.905**	**31.32**	**0.886**	**29.43**	**0.895**
VDSR	×3	33.66	0.921	29.77	0.831	28.82	0.798	27.14	0.828
Bicubic	×3	30.39	0.868	27.55	0.774	27.21	0.739	24.46	0.735
VDSR_BNN	×3	31.01	0.874	28.15	0.791	27.57	0.755	25.01	0.758
VDSR_DoReFa	×3	31.79	0.895	28.68	0.806	27.98	0.766	25.53	0.782
VDSR_ABC	×3	32.01	0.898	28.86	0.808	28.08	0.770	25.80	0.787
VDSR_BAM	×3	**32.52**	**0.907**	**29.17**	**0.819**	**28.29**	**0.782**	**26.07**	**0.799**
VDSR	×4	31.35	0.884	28.01	0.767	27.29	0.725	25.18	0.752
Bicubic	×4	28.42	0.810	26.00	0.703	25.96	0.668	23.14	0.658
VDSR_BNN	×4	29.02	0.827	26.55	0.724	26.29	0.685	23.55	0.685
VDSR_DoReFa	×4	29.39	0.837	26.79	0.728	26.45	0.689	23.81	0.696
VDSR_ABC	×4	29.59	0.841	29.63	0.730	26.51	0.687	23.96	0.699
VDSR_BAM	×4	**30.31**	**0.860**	**27.46**	**0.749**	**26.83**	**0.706**	**24.38**	**0.720**

high frequency details are difficult to predict. These two methods achieve inferior performance than VDSR_ABC and our proposed VDSR_BAM. VDSR_ABC approximates the full-precision convolution by linear combination of multiple binary convolutions. Its performance is significantly higher than VDSR_DoReFa. However, the single convolution process in VDSR_ABC is not significantly improved compared with that of VDSR_DoReFa. Besides, the model parameters and computational operands of VDSR_ABC are 3 times (in this work) higher than other methods. Benefit from BAM's ability to retain and compensate for lost information during the quantization process, our VDSR_BAM exceeds all previous methods on four benchmark datasets.

Evaluation on SRResNet: SRResNet is a modular (residual-block) network structure. Then our BAM can be directly applied to each module. The setup of other model quantization methods is consistent with evaluation on VDSR. Each residual block in SRResNet_ABC contains six one-bit convolutions. The comparison results are shown in Table 2.

Compared with VDSR-based models, the performance of model quantization methods with SRResNet has been significantly improved. This is not only

Table 2. Quantitative evaluation of SRResNet-based state-of-the-art model quantization methods.

Methods	Scale	Set5		Set14		B100		Urban100	
		PSRN	SSIM	PSRN	SSIM	PSRN	SSIM	PSRN	SSIM
SRResNet	×2	37.76	0.958	33.27	0.914	31.95	0.895	31.28	0.919
Bicubic	×2	33.66	0.930	30.24	0.869	29.56	0.843	26.88	0.840
SRResNet_BNN	×2	35.21	0.942	31.55	0.896	30.64	0.876	28.01	0.869
SRResNet_DoReFa	×2	36.09	0.950	32.09	0.902	31.02	0.882	28.87	0.880
SRResNet_ABC	×2	36.34	0.952	32.28	0.903	31.16	0.884	29.29	0.891
SRResNet_BAM	×2	**37.21**	**0.956**	**32.74**	**0.910**	**31.60**	**0.891**	**30.20**	**0.906**
SRResNet	×3	34.07	0.922	30.04	0.835	28.91	0.798	27.50	0.837
Bicubic	×3	30.39	0.868	27.55	0.774	27.21	0.739	24.46	0.735
SRResNet_BNN	×3	31.18	0.877	28.29	0.799	27.73	0.765	25.03	0.758
SRResNet_DoReFa	×3	32.44	0.903	28.99	0.811	28.21	0.778	25.84	0.783
SRResNet_ABC	×3	32.69	0.908	29.24	0.820	28.35	0.782	26.12	0.797
SRResNet_BAM	×3	**33.33**	**0.915**	**29.63**	**0.827**	**28.61**	**0.790**	**26.69**	**0.816**
SRResNet	×4	31.76	0.888	28.25	0.773	27.38	0.727	25.54	0.767
Bicubic	×4	28.42	0.810	26.00	0.703	25.96	0.668	23.14	0.658
SRResNet_BNN	×4	29.33	0.826	26.72	0.728	26.45	0.692	23.68	0.683
SRResNet_DoReFa	×4	30.38	0.862	27.48	0.754	26.87	0.708	24.45	0.720
SRResNet_ABC	×4	30.78	0.868	27.71	0.756	27.00	0.713	24.54	0.729
SRResNet_BAM	×4	**31.24**	**0.878**	**27.97**	**0.765**	**27.15**	**0.719**	**24.95**	**0.745**

attributed to the stronger learning ability of nonlinear mapping of the residual block, but also to its highly gain brought by the increase of model parameters. Especially, the performance of SRResNet_DoReFa improved significantly. The main reason is that the updating and generating modes of weights and activations in VDSR is monotonous, especially under the interference of binaryzation function $Sign()$. The difficulty of weights updating in gradient back propagation is greatly increased. While the gain of the residuals to the gradient back propagation is beyond doubt, and its skip connection also can effectively enrich the information of the activations in forward inference processing. In some cases, *e.g.* x3 on B100, x4 on Set14 and x4 on B100, the gap between our SRResNet_BAM and the full-precision SRResNet is shrunk to no more than 0.3dB.

Qualitative Evaluation: As demonstrated in Fig. 4, we present a subjective comparison with state-of-the-art model quantization methods based on VDSR. We enlarge the texture on the general real-world images to compare the subjective visual effects of different SR methods. It is obvious that the compared SR methods fails to extract realistic details from LR inputs and they are prone to

Fig. 4. VDSR-based subjective quality assessment for ×4 upscaling on the structured image: *img*001 from Urban100.

produce a blurry texture. Our model could reveal the most accurate and realistic details and generates the correct direction of texture.

Figure 5 presents the results of different quantization SR methods based on SRResNet. Parts including lines or holes in the buildings are magnified for more obvious comparison. It is observed that after 4 minification and magnification by bicubic, the direction of the line and the outline of the hole is hard to be distinguished. Most methods can effectively recover the lines in the regions close to the shooting point. However, most of them can do nothing for the regions far from the shooting point. Benefit from the retention of information in the BAM quantization process, our proposed SRResNet_BAM method could recover more accurate details for these regions.

4.4 Model Analysis

In this section, we first evaluated the performance of our proposed BSRN model. Then we investigate the effects of different quantization approaches. Finally, we compare our method with existing quantized convolution-based super resolution method [17].

Evaluation on Proposed BSRN Framework: The purpose of our work is to reduce model parameters and improve the reasoning speed so that it can be applied to mobile devices such as mobile phones. We restrict our study to the binary networks with a low number of parameters and do not further investigate potentially beneficial extensions such as width and depth of network [32] or different loss functions [12,13].

In our experiment, we set the number of HPB-Block (see Fig. 3) as 20. Each HPB-Block contains 2 LB-Blocks. The number of feature channels is set

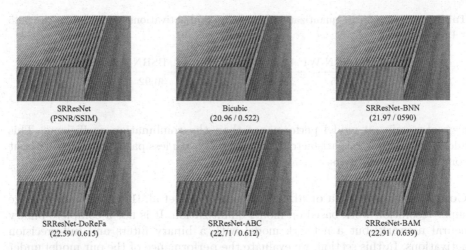

Fig. 5. SRResNet-based subjective quality assessment for ×4 upscaling on the structured image: $img045$ from Urban100.

Table 3. Comparison between VDSR_BAM and SRResNet_BAM and BSRN on ×4. MAC is the number of multiply-accumulate operations. We assume that the generated SR image is 720P (1280 × 720).

Methods	Paras	MAC	Set5		Set14		B100		Urban100	
			PSRN	SSIM	PSRN	SSIM	PSRN	SSIM	PSRN	SSIM
VDSR_BAM	$668K$	$616.9G$	30.31	0.860	27.46	0.749	26.83	0.706	24.38	0.720
SRResNet_BAM	$1547K$	$127.9G$	31.24	0.878	27.97	0.765	27.15	0.719	24.95	0.745
BSRN	$1216K$	$85G$	**31.35**	**0.880**	**28.04**	**0.768**	**27.18**	**0.720**	**25.11**	**0.749**

to 48. Then, based on our proposed BAM quantization method, BSRN is compared with the other two super-resolution networks (VDSR and SRResNet). The results are shown in Table 3. It is obvious that our model can achieve superior performance with lower model parameters and calculation operands.

Ablation Study on Quantization of Weights and Activations: Table 4 presents the ablation study on the effect of our BAM quantization method. In this table, BSRN-W1 is BSRN without linear combination coefficients in the process of weight quantization, that is, $\alpha_1 = \alpha_2 = ... = \alpha_n = 1$. BSRN-W2 refers to the process of weight quantization of BSRN without bit accumulation operation, i.e. $W_n{}^B = Sign(BN(\alpha W_n)) * E(|W_n|)$. Corresponding to BSRN-W1 and BSRN-W2, BSRN-A1 and BSRN-A2 quantize the activations according to the same way.

From the results, we can see that the processing mode of activation has a great impact on the performance of the model, which also indicates that for the image super-resolution task, the ability to activate information can directly determine the performance of the model. In addition, the bit accumulation operation has

Table 4. Effects of the quantization of weights and activations measured on the Set5 ×4 dataset.

Models	BSRN-W1	BSRN-W2	BSRN-A1	BSRN-A2	BSRN
PSNR	31.25	31.14	31.22	30.92	31.35

a greater gain on model performance than the combination coefficients. This allows us to achieve a satisfactory performance with less parameters and without combination coefficients.

Comparison with Ma et al. [17]: Recently, Ma et al. [17] proposed an image super-resolution work based on model quantization. It is not a complete binary neural network, but a network model with a binary filters and full-precision activations. In this section, we evaluate the performance of the our model under the binary filters and with full-precision activations. The results are shown in Table 5. The superiority of our performance can be clearly seen from the results.

Table 5. Quantitative evaluation with the work of Ma et al. on SRResNet.

Methods	Scale	Set5		Set14		Urban100	
		PSRN	SSIM	PSRN	SSIM	PSRN	SSIM
SRResNet_Ma et al.	×2	35.66	0.946	31.56	0.897	28.76	0.882
SRResNet_BAM	×2	**37.51**	**0.956**	**33.03**	**0.912**	**30.79**	**0.915**
SRResNet_Ma et al.	×4	30.34	0.864	27.16	0.756	24.48	0.728
SRResNet_BAM	×4	**31.57**	**0.883**	**28.16**	**0.769**	24.30	**0.755**

In general, the experiments not only illustrate the effectiveness of design binarization method but suggest the reasonability of bit-accumulation perspective.

5   Conclusions

In this paper, we proposed a BNN-based model for SISR, in which a novel binarization method named bit-accumulation mechanism and a lightweight network structure are designed to approximate the full-precision CNN. The evaluation and analysis in this paper indicates that the presented method can gradually refine the precision of quantization along the direction of model inference and significantly improve the model performance. Extensive experiments compared with the state-of-the-art methods demonstrated the superiority of the our proposed binarization method. We believe that this bit-accumulation mechanism could be more widely applicable in practice. Moreover, it is readily to be used in other machine vision problems, especially image reconstruction related tasks, such as image/video deblurring, compression artifact removal, image/video restoration and so on.

Acknowledgement. This work was supported in part by the National Key Research and Development Program of China under Grant 2018AAA0103202, in part by the National Natural Science Foundation of China under Grant 61922066, Grant 61876142, Grant 61671339, Grant 61772402, Grant 62036007, and Grant U1605252, in part by the National High-Level Talents Special Support Program of China under Grant CS31117200001, in part by the Fundamental Research Funds for the Central Universities under Grant JB190117, in part by the Xidian University Intellifusion Joint Innovation Laboratory of Artificial Intelligence, and in part by the Innovation Fund of Xidian University.

References

1. Ahn, N., Kang, B., Sohn, K.-A.: Fast, accurate, and lightweight super-resolution with cascading residual network. In: Ferrari, V., Hebert, M., Sminchisescu, C., Weiss, Y. (eds.) ECCV 2018. LNCS, vol. 11214, pp. 256–272. Springer, Cham (2018). https://doi.org/10.1007/978-3-030-01249-6_16
2. Bevilacqua, M., Roumy, A., Guillemot, C., Alberi-Morel, M.L.: Low-complexity single-image super-resolution based on nonnegative neighbor embedding. BMVA Press (2012)
3. Choi, J.S., Kim, M.: A deep convolutional neural network with selection units for super-resolution. In: Proceedings of the IEEE Conference on Computer Vision and Pattern Recognition Workshops, pp. 154–160 (2017)
4. Courbariaux, M., Bengio, Y., David, J.P.: BinaryConnect: training deep neural networks with binary weights during propagations. In: Advances in Neural Information Processing Systems, pp. 3123–3131 (2015)
5. Courbariaux, M., Hubara, I., Soudry, D., El-Yaniv, R., Bengio, Y.: Binarized neural networks: training deep neural networks with weights and activations constrained to + 1 or −1. arXiv preprint arXiv:1602.02830 (2016)
6. Dong, C., Loy, C.C., He, K., Tang, X.: Image super-resolution using deep convolutional networks. IEEE Trans. Pattern Anal. Mach. Intell. **38**(2), 295–307 (2015)
7. Dong, C., Loy, C.C., Tang, X.: Accelerating the super-resolution convolutional neural network. In: Leibe, B., Matas, J., Sebe, N., Welling, M. (eds.) ECCV 2016. LNCS, vol. 9906, pp. 391–407. Springer, Cham (2016). https://doi.org/10.1007/978-3-319-46475-6_25
8. He, X., Mo, Z., Wang, P., Liu, Y., Yang, M., Cheng, J.: ODE-inspired network design for single image super-resolution. In: 2019 IEEE Conference on Computer Vision and Pattern Recognition, July 2019
9. Huang, J.B., Singh, A., Ahuja, N.: Single image super-resolution from transformed self-exemplars. In: Proceedings of the IEEE Conference on Computer Vision and Pattern Recognition, pp. 5197–5206 (2015)
10. Jiwon, K., Jung Kwon, L., Kyoung Mu, L.: Accurate image super-resolution using very deep convolutional networks. In: Proceedings of the IEEE Conference on Computer Vision and Pattern Recognition, pp. 1646–1654 (2016)
11. Kim, J., Kwon Lee, J., Mu Lee, K.: Deeply-recursive convolutional network for image super-resolution. In: Proceedings of the IEEE Conference on Computer Vision and Pattern Recognition, pp. 1637–1645 (2016)
12. Lai, W.S., Huang, J.B., Ahuja, N., Yang, M.H.: Deep Laplacian pyramid networks for fast and accurate super-resolution. In: Proceedings of the IEEE Conference on Computer Vision and Pattern Recognition, pp. 624–632 (2017)

13. Ledig, C., et al.: Photo-realistic single image super-resolution using a generative adversarial network. In: Proceedings of the IEEE Conference on Computer Vision and Pattern Recognition, pp. 4681–4690 (2017)
14. Li, J., Fang, F., Mei, K., Zhang, G.: Multi-scale residual network for image super-resolution. In: Ferrari, V., Hebert, M., Sminchisescu, C., Weiss, Y. (eds.) ECCV 2018. LNCS, vol. 11212, pp. 527–542. Springer, Cham (2018). https://doi.org/10.1007/978-3-030-01237-3_32
15. Lim, B., Son, S., Kim, H., Nah, S., Mu Lee, K.: Enhanced deep residual networks for single image super-resolution. In: Proceedings of the IEEE Conference on Computer Vision and Pattern Recognition Workshops, pp. 136–144 (2017)
16. Lin, X., Zhao, C., Pan, W.: Towards accurate binary convolutional neural network. In: Advances in Neural Information Processing Systems, pp. 345–353 (2017)
17. Ma, Y., Xiong, H., Hu, Z., Ma, L.: Efficient super resolution using binarized neural network. In: Proceedings of the IEEE Conference on Computer Vision and Pattern Recognition Workshops, p. 0 (2019)
18. Martin, D., Fowlkes, C., Tal, D., Malik, J., et al.: A database of human segmented natural images and its application to evaluating segmentation algorithms and measuring ecological statistics. In: ICCV, Vancouver (2001)
19. Rastegari, M., Ordonez, V., Redmon, J., Farhadi, A.: XNOR-Net: ImageNet classification using binary convolutional neural networks. In: Leibe, B., Matas, J., Sebe, N., Welling, M. (eds.) ECCV 2016. LNCS, vol. 9908, pp. 525–542. Springer, Cham (2016). https://doi.org/10.1007/978-3-319-46493-0_32
20. Shi, W., et al.: Real-time single image and video super-resolution using an efficient sub-pixel convolutional neural network. In: Proceedings of the IEEE Conference on Computer Vision and Pattern Recognition, pp. 1874–1883 (2016)
21. Shi, W., et al.: Cardiac image super-resolution with global correspondence using multi-atlas PatchMatch. In: Mori, K., Sakuma, I., Sato, Y., Barillot, C., Navab, N. (eds.) MICCAI 2013. LNCS, vol. 8151, pp. 9–16. Springer, Heidelberg (2013). https://doi.org/10.1007/978-3-642-40760-4_2
22. Soudry, D., Hubara, I., Meir, R.: Expectation backpropagation: parameter-free training of multilayer neural networks with continuous or discrete weights. In: Advances in Neural Information Processing Systems, pp. 963–971 (2014)
23. Tai, Y., Yang, J., Liu, X.: Image super-resolution via deep recursive residual network. In: Proceedings of the IEEE Conference on Computer Vision and Pattern Recognition, pp. 3147–3155 (2017)
24. Tai, Y., Yang, J., Liu, X., Xu, C.: MemNet: a persistent memory network for image restoration. In: Proceedings of the IEEE International Conference on Computer Vision, pp. 4539–4547 (2017)
25. Thornton, M.W., Atkinson, P.M., Holland, D.: Sub-pixel mapping of rural land cover objects from fine spatial resolution satellite sensor imagery using super-resolution pixel-swapping. Int. J. Remote Sens. 27(3), 473–491 (2006)
26. Timofte, R., Agustsson, E., Van Gool, L., Yang, M.H., Zhang, L.: NTIRE 2017 challenge on single image super-resolution: methods and results. In: Proceedings of the IEEE Conference on Computer Vision and Pattern Recognition Workshops, pp. 114–125 (2017)
27. Tong, T., Li, G., Liu, X., Gao, Q.: Image super-resolution using dense skip connections. In: Proceedings of the IEEE International Conference on Computer Vision, pp. 4799–4807 (2017)
28. Wang, Z., Liu, D., Yang, J., Han, W., Huang, T.: Deep networks for image super-resolution with sparse prior. In: Proceedings of the IEEE International Conference on Computer Vision, pp. 370–378 (2015)

29. Wang, Z., Bovik, A.C., Sheikh, H.R., Simoncelli, E.P., et al.: Image quality assessment: from error visibility to structural similarity. IEEE Trans. Image Process. **13**(4), 600–612 (2004)
30. Yang, J., Wright, J., Huang, T.S., Ma, Y.: Image super-resolution via sparse representation. IEEE Trans. Image Process. **19**(11), 2861–2873 (2010)
31. Zeyde, R., Elad, M., Protter, M.: On single image scale-up using sparse-representations. In: Boissonnat, J.-D., et al. (eds.) Curves and Surfaces 2010. LNCS, vol. 6920, pp. 711–730. Springer, Heidelberg (2012). https://doi.org/10.1007/978-3-642-27413-8_47
32. Zhang, Y., Li, K., Li, K., Wang, L., Zhong, B., Fu, Y.: Image super-resolution using very deep residual channel attention networks. In: Ferrari, V., Hebert, M., Sminchisescu, C., Weiss, Y. (eds.) ECCV 2018. LNCS, vol. 11211, pp. 294–310. Springer, Cham (2018). https://doi.org/10.1007/978-3-030-01234-2_18
33. Zhang, Y., Tian, Y., Kong, Y., Zhong, B., Fu, Y.: Residual dense network for image super-resolution. In: Proceedings of the IEEE Conference on Computer Vision and Pattern Recognition, pp. 2472–2481 (2018)
34. Zhou, S., Wu, Y., Ni, Z., Zhou, X., Wen, H., Zou, Y.: DoReFa-Net: training low bitwidth convolutional neural networks with low bitwidth gradients. arXiv preprint arXiv:1606.06160 (2016)
35. Zou, W.W., Yuen, P.C.: Very low resolution face recognition problem. IEEE Trans. Image Process. **21**(1), 327–340 (2011)

Axial-DeepLab: Stand-Alone Axial-Attention for Panoptic Segmentation

Huiyu Wang[1]([envelope]), Yukun Zhu[2], Bradley Green[2], Hartwig Adam[3], Alan Yuille[1], and Liang-Chieh Chen[3]

[1] Johns Hopkins University, Baltimore, USA
hwang157@jhu.edu
[2] Google Research, Seattle, USA
[3] Google Research, Los Angeles, USA

Abstract. Convolution exploits locality for efficiency at a cost of missing long range context. Self-attention has been adopted to augment CNNs with non-local interactions. Recent works prove it possible to stack self-attention layers to obtain a fully attentional network by restricting the attention to a local region. In this paper, we attempt to remove this constraint by factorizing 2D self-attention into two 1D self-attentions. This reduces computation complexity and allows performing attention within a larger or even global region. In companion, we also propose a position-sensitive self-attention design. Combining both yields our position-sensitive axial-attention layer, a novel building block that one could stack to form axial-attention models for image classification and dense prediction. We demonstrate the effectiveness of our model on four large-scale datasets. In particular, our model outperforms all existing stand-alone self-attention models on ImageNet. Our Axial-DeepLab improves 2.8% PQ over *bottom-up* state-of-the-art on COCO test-dev. This previous state-of-the-art is attained by our small variant that is 3.8× parameter-efficient and 27× computation-efficient. Axial-DeepLab also achieves state-of-the-art results on Mapillary Vistas and Cityscapes.

Keywords: Bottom-up panoptic segmentation · Self-attention

1 Introduction

Convolution is a core building block in computer vision. Early algorithms employ convolutional filters to blur images, extract edges, or detect features. It has been heavily exploited in modern neural networks [46,47] due to its efficiency and

---

H. Wang—Work done while an intern at Google.

Electronic supplementary material The online version of this chapter (https://doi.org/10.1007/978-3-030-58548-8_7) contains supplementary material, which is available to authorized users.

© Springer Nature Switzerland AG 2020
A. Vedaldi et al. (Eds.): ECCV 2020, LNCS 12349, pp. 108–126, 2020.
https://doi.org/10.1007/978-3-030-58548-8_7

generalization ability, in comparison to fully connected models [2]. The success of convolution mainly comes from two properties: translation equivariance, and locality. Translation equivariance, although not exact [93], aligns well with the nature of imaging and thus generalizes the model to different positions or to images of different sizes. Locality, on the other hand, reduces parameter counts and M-Adds. However, it makes modeling long range relations challenging.

A rich set of literature has discussed approaches to modeling long range interactions in convolutional neural networks (CNNs). Some employ atrous convolutions [12,33,64,74], larger kernel [67], or image pyramids [82,94], either designed by hand or searched by algorithms [11,57,99]. Another line of works adopts attention mechanisms. Attention shows its ability of modeling long range interactions in language modeling [80,85], speech recognition [10,21], and neural captioning [88]. Attention has since been extended to vision, giving significant boosts to image classification [6], object detection [36], semantic segmentation [39], video classification [84], and adversarial defense [86]. These works enrich CNNs with non-local or long-range attention modules.

Recently, stacking attention layers as stand-alone models without any spatial convolution has been proposed [37,65] and shown promising results. However, naive attention is computationally expensive, especially on large inputs. Applying local constraints to attention, proposed by [37,65], reduces the cost and enables building fully attentional models. However, local constraints limit model receptive field, which is crucial to tasks such as segmentation, especially on high-resolution inputs. In this work, we propose to adopt axial-attention [32,39], which not only allows efficient computation, but recovers the large receptive field in stand-alone attention models. The core idea is to factorize 2D attention into two 1D attentions along height- and width-axis sequentially. Its efficiency enables us to attend over large regions and build models to learn long range or even global interactions. Additionally, most previous attention modules do not utilize positional information, which degrades attention's ability in modeling position-dependent interactions, like shapes or objects at multiple scales. Recent works [6,37,65] introduce positional terms to attention, but in a context-agnostic way. In this paper, we augment the positional terms to be context-dependent, making our attention position-sensitive, with marginal costs.

We show the effectiveness of our axial-attention models on ImageNet [70] for classification, and on three datasets (COCO [56], Mapillary Vistas [62], and Cityscapes [22]) for panoptic segmentation [45], instance segmentation, and semantic segmentation. In particular, on ImageNet, we build an Axial-ResNet by replacing the 3 × 3 convolution in all residual blocks [31] with our position-sensitive axial-attention layer, and we further make it fully attentional [65] by adopting axial-attention layers in the 'stem'. As a result, our Axial-ResNet attains state-of-the-art results among stand-alone attention models on ImageNet. For segmentation tasks, we convert Axial-ResNet to Axial-DeepLab by replacing the backbones in Panoptic-DeepLab [18]. On COCO [56], our Axial-DeepLab outperforms the current *bottom-up* state-of-the-art, Panoptic-DeepLab [19], by 2.8% PQ on test-dev set. We also show state-of-the-art segmentation results on Mapillary Vistas [62], and Cityscapes [22].

To summarize, our contributions are four-fold:

- The proposed method is the first attempt to build stand-alone attention models with large or global receptive field.
- We propose position-sensitive attention layer that makes better use of positional information without adding much computational cost.
- We show that axial attention works well, not only as a stand-alone model on image classification, but also as a backbone on panoptic segmentation, instance segmentation, and segmantic segmentation.
- Our Axial-DeepLab improves significantly over bottom-up state-of-the-art on COCO, achieving comparable performance of two-stage methods. We also surpass previous state-of-the-art methods on Mapillary Vistas and Cityscapes.

2   Related Work

Top-Down Panoptic Segmentation: Most state-of-the-art panoptic segmentation models employ a two-stage approach where object proposals are firstly generated followed by sequential processing of each proposal. We refer to such approaches as top-down or proposal-based methods. Mask R-CNN [30] is commonly deployed in the pipeline for instance segmentation, paired with a lightweight stuff segmentation branch. For example, Panoptic FPN [44] incorporates a semantic segmentation head to Mask R-CNN [30], while Porzi et al. [68] append a light-weight DeepLab-inspired module [13] to the multi-scale features from FPN [55]. Additionally, some extra modules are designed to resolve the overlapping instance predictions by Mask R-CNN. TASCNet [49] and AUNet [52] propose a module to guide the fusion between 'thing' and 'stuff' predictions, while Liu et al. [61] adopt a Spatial Ranking module. UPSNet [87] develops an efficient parameter-free panoptic head for fusing 'thing' and 'stuff', which is further explored by Li et al. [50] for end-to-end training of panoptic segmentation models. AdaptIS [77] uses point proposals to generate instance masks.

Bottom-up Panoptic Segmentation: In contrast to top-down approaches, bottom-up or proposal-free methods for panoptic segmentation typically start with the semantic segmentation prediction followed by grouping 'thing' pixels into clusters to obtain instance segmentation. DeeperLab [89] predicts bounding box four corners and object centers for class-agnostic instance segmentation. SSAP [28] exploits the pixel-pair affinity pyramid [60] enabled by an efficient graph partition method [43]. BBFNet [7] obtains instance segmentation results by Watershed transform [4,81] and Hough-voting [5,48]. Recently, Panoptic-DeepLab [19], a simple, fast, and strong approach for bottom-up panoptic segmentation, employs a class-agnostic instance segmentation branch involving a simple instance center regression [42,63,79], coupled with DeepLab semantic segmentation outputs [12,14,15]. Panoptic-DeepLab has achieved state-of-the-art results on several benchmarks, and our method builds on top of it.

Self-attention: Attention, introduced by [3] for the encoder-decoder in a neural sequence-to-sequence model, is developed to capture correspondence of tokens between two sequences. In contrast, self-attention is defined as applying attention to a single context instead of across multiple modalities. Its ability to directly encode long-range interactions and its parallelizability, has led to state-of-the-art performance for various tasks [24,25,38,53,66,72,80]. Recently, self-attention has been applied to computer vision, by augmenting CNNs with non-local or long-range modules. Non-local neural networks [84] show that self-attention is an instantiation of non-local means [9] and achieve gains on many vision tasks such as video classification and object detection. Additionally, [6,17] show improvements on image classification by combining features from self-attention and convolution. State-of-the-art results on video action recognition tasks [17] are also achieved in this way. On semantic segmentation, self-attention is developed as a context aggregation module that captures multi-scale context [26,39,95,98]. Efficient attention methods are proposed to reduce its complexity [39,53,73]. Additionally, CNNs augmented with non-local means [9] are shown to be more robust to adversarial attacks [86]. Besides discriminative tasks, self-attention is also applied to generative modeling of images [8,32,91]. Recently, [37,65] show that self-attention layers alone could be stacked to form a fully attentional model by restricting the receptive field of self-attention to a *local* square region. Encouraging results are shown on both image classification and object detection. In this work, we follow this direction of research and propose a stand-alone self-attention model with large or global receptive field, making self-attention models *non-local* again. Our models are evaluated on bottom-up panoptic segmentation and show significant improvements.

3   Method

We begin by formally introducing our position-sensitive self-attention mechanism. Then, we discuss how it is applied to axial-attention and how we build stand-alone Axial-ResNet and Axial-DeepLab with axial-attention layers.

3.1   Position-Sensitive Self-attention

Self-attention: Self-attention mechanism is usually applied to vision models as an add-on to augment CNNs outputs [39,84,91]. Given an input feature map $x \in \mathbb{R}^{h \times w \times d_{in}}$ with height h, width w, and channels d_{in}, the output at position $o = (i, j)$, $y_o \in \mathbb{R}^{d_{out}}$, is computed by pooling over the projected input as:

$$y_o = \sum_{p \in \mathcal{N}} \text{softmax}_p(q_o^T k_p) v_p \tag{1}$$

where $\mathcal{N}$ is the whole location lattice, and queries $q_o = W_Q x_o$, keys $k_o = W_K x_o$, values $v_o = W_V x_o$ are all linear projections of the input $x_o \ \forall o \in \mathcal{N}$. $W_Q, W_K \in \mathbb{R}^{d_q \times d_{in}}$ and $W_V \in \mathbb{R}^{d_{out} \times d_{in}}$ are all learnable matrices. The softmax$_p$ denotes a softmax function applied to all possible $p = (a, b)$ positions, which in this case is also the whole 2D lattice.

This mechanism pools values v_p globally based on affinities $x_o^T W_Q^T W_K x_p$, allowing us to capture related but non-local context in the whole feature map, as opposed to convolution which only captures local relations.

However, self-attention is extremely expensive to compute ($\mathcal{O}(h^2 w^2)$) when the spatial dimension of the input is large, restricting its use to only high levels of a CNN (*i.e.*, downsampled feature maps) or small images. Another drawback is that the global pooling does not exploit positional information, which is critical to capture spatial structures or shapes in vision tasks.

These two issues are mitigated in [65] by adding local constraints and positional encodings to self-attention. For each location o, a local $m \times m$ square region is extracted to serve as a memory bank for computing the output y_o. This significantly reduces its computation to $\mathcal{O}(hwm^2)$, allowing self-attention modules to be deployed as stand-alone layers to form a fully self-attentional neural network. Additionally, a learned relative positional encoding term is incorporated into the affinities, yielding a dynamic prior of where to look at in the receptive field (*i.e.*, the local $m \times m$ square region). Formally, [65] proposes

$$y_o = \sum_{p \in \mathcal{N}_{m \times m}(o)} \text{softmax}_p(q_o^T k_p + q_o^T r_{p-o}) v_p \tag{2}$$

where $\mathcal{N}_{m \times m}(o)$ is the local $m \times m$ square region centered around location $o = (i, j)$, and the learnable vector $r_{p-o} \in \mathbb{R}^{d_q}$ is the added relative positional encoding. The inner product $q_o^T r_{p-o}$ measures the compatibility from location $p = (a, b)$ to location $o = (i, j)$. We do not consider absolute positional encoding $q_o^T r_p$, because they do not generalize well compared to the relative counterpart [65]. In the following paragraphs, we drop the term relative for conciseness.

In practice, d_q and d_{out} are much smaller than d_{in}, and one could extend single-head attention in Eq. (2) to multi-head attention to capture a mixture of affinities. In particular, multi-head attention is computed by applying N single-head attentions in parallel on x_o (with different $W_Q^n, W_K^n, W_V^n, \forall n \in \{1, 2, \ldots, N\}$ for the n-th head), and then obtaining the final output z_o by concatenating the results from each head, *i.e.*, $z_o = \text{concat}_n(y_o^n)$. Note that positional encodings are often shared across heads, so that they introduce marginal extra parameters.

Position-Sensitivity: We notice that previous positional bias only depends on the query pixel x_o, not the key pixel x_p. However, the keys x_p could also have information about which location to attend to. We therefore add a key-dependent positional bias term $k_p^T r_{p-o}^k$, besides the query-dependent bias $q_o^T r_{p-o}^q$.

Similarly, the values v_p do not contain any positional information in Eq. (2). In the case of large receptive fields or memory banks, it is unlikely that y_o contains the precise location from which v_p comes. Thus, previous models have to trade-off between using smaller receptive fields (*i.e.*, small $m \times m$ regions) and throwing away precise spatial structures. In this work, we enable the output y_o to retrieve relative positions r_{p-o}^v, besides the content v_p, based on query-key affinities $q_o^T k_p$. Formally,

$$y_o = \sum_{p \in \mathcal{N}_{m \times m}(o)} \text{softmax}_p(q_o^T k_p + q_o^T r_{p-o}^q + k_p^T r_{p-o}^k)(v_p + r_{p-o}^v) \tag{3}$$

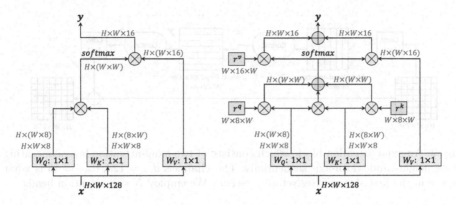

Fig. 1. A non-local block (left) *vs.* our position-sensitive axial-attention applied along the width-axis (right). "⊗" denotes matrix multiplication, and "⊕" denotes element-wise sum. The softmax is performed on the last axis. Blue boxes denote 1×1 convolutions, and red boxes denote relative positional encoding. The channels $d_{in} = 128$, $d_q = 8$, and $d_{out} = 16$ is what we use in the first stage of ResNet after 'stem' (Color figure online)

where the learnable $r^k_{p-o} \in \mathbb{R}^{d_q}$ is the positional encoding for keys, and $r^v_{p-o} \in \mathbb{R}^{d_{out}}$ is for values. Both vectors do not introduce many parameters, since they are shared across attention heads in a layer, and the number of local pixels $|\mathcal{N}_{m \times m}(o)|$ is usually small.

We call this design *position-sensitive* self-attention (Fig. 1), which captures long range interactions with precise positional information at a reasonable computation overhead, as verified in our experiments.

3.2 Axial-Attention

The local constraint, proposed by the stand-alone self-attention models [65], significantly reduces the computational costs in vision tasks and enables building fully self-attentional model. However, such constraint sacrifices the global connection, making attention's receptive field no larger than a depthwise convolution with the same kernel size. Additionally, the local self-attention, performed in local square regions, still has complexity quadratic to the region length, introducing another hyper-parameter to trade-off between performance and computation complexity. In this work, we propose to adopt axial-attention [32,39] in stand-alone self-attention, ensuring both global connection and efficient computation. Specifically, we first define an axial-attention layer on the *width*-axis of an image as simply a one dimensional *position-sensitive* self-attention, and use the similar definition for the *height*-axis. To be concrete, the axial-attention layer along the width-axis is defined as follows.

$$y_o = \sum_{p \in \mathcal{N}_{1 \times m}(o)} \text{softmax}_p(q_o^T k_p + q_o^T r^q_{p-o} + k_p^T r^k_{p-o})(v_p + r^v_{p-o}) \tag{4}$$

114   H. Wang et al.

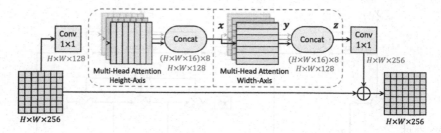

Fig. 2. An axial-attention block, which consists of two axial-attention layers operating along height- and width-axis sequentially. The channels $d_{in} = 128$, $d_{out} = 16$ is what we use in the first stage of ResNet after 'stem'. We employ $N = 8$ attention heads

One axial-attention layer propagates information along one particular axis. To capture global information, we employ two axial-attention layers consecutively for the height-axis and width-axis, respectively. Both of the axial-attention layers adopt the multi-head attention mechanism, as described above.

Axial-attention reduces the complexity to $\mathcal{O}(hwm)$. This enables global receptive field, which is achieved by setting the span m directly to the whole input features. Optionally, one could also use a fixed m value, in order to reduce memory footprint on huge feature maps.

Axial-ResNet: To transform a ResNet [31] to an *Axial*-ResNet, we replace the 3×3 convolution in the residual bottleneck block by two multi-head axial-attention layers (one for height-axis and the other for width-axis). Optional striding is performed on each axis after the corresponding axial-attention layer. The two 1×1 convolutions are kept to shuffle the features. This forms our (residual) axial-attention block, as illustrated in Fig. 2, which is stacked multiple times to obtain Axial-ResNets. Note that we do not use a 1×1 convolution in-between the two axial-attention layers, since matrix multiplications (W_Q, W_K, W_V) follow immediately. Additionally, the stem (*i.e.*, the first strided 7×7 convolution and 3×3 max-pooling) in the original ResNet is kept, resulting in a *conv-stem* model where convolution is used in the first layer and attention layers are used everywhere else. In *conv-stem* models, we set the span m to the whole input from the first block, where the feature map is 56×56.

In our experiments, we also build a full axial-attention model, called Full Axial-ResNet, which further applies axial-attention to the stem. Instead of designing a special spatially-varying attention stem [65], we simply stack three axial-attention bottleneck blocks. In addition, we adopt local constraints (*i.e.*, a local $m \times m$ square region as in [65]) in the first few blocks of Full Axial-ResNets, in order to reduce computational cost.

Axial-DeepLab: To further convert Axial-ResNet to Axial-DeepLab for segmentation tasks, we make several changes as discussed below.

Firstly, to extract dense feature maps, DeepLab [12] changes the stride and atrous rates of the last one or two stages in ResNet [31]. Similarly, we remove the stride of the last stage but we do not implement the 'atrous' attention module, since our axial-attention already captures global information for the whole input. In this work, we extract feature maps with output stride (*i.e.*, the ratio of input resolution to the final backbone feature resolution) 16. We do not pursue output stride 8, since it is computationally expensive.

Secondly, we do not adopt the atrous spatial pyramid pooling module (ASPP) [13,14], since our axial-attention block could also efficiently encode the multi-scale or global information. We show in the experiments that our Axial-DeepLab without ASPP outperforms Panoptic-DeepLab [19] with and without ASPP.

Lastly, following Panoptic-DeepLab [19], we adopt exactly the same stem [78] of three convolutions, dual decoders, and prediction heads. The heads produce semantic segmentation and class-agnostic instance segmentation, and they are merged by majority voting [89] to form the final panoptic segmentation.

In cases where the inputs are extremely large (*e.g.*, 2177×2177) and memory is constrained, we resort to a large span $m = 65$ in all our axial-attention blocks. Note that we do not consider the axial span as a hyper-parameter because it is already sufficient to cover long range or even global context on several datasets, and setting a smaller span does not significantly reduce M-Adds.

4   Experimental Results

We conduct experiments on four large-scale datasets. We first report results with our Axial-ResNet on ImageNet [70]. We then convert the ImageNet pretrained Axial-ResNet to Axial-DeepLab, and report results on COCO [56], Mapillary Vistas [62], and Cityscapes [22] for panoptic segmentation, evaluated by panoptic quality (PQ) [45]. We also report average precision (AP) for instance segmentation, and mean IoU for semantic segmentation on Mapillary Vistas and Cityscapes. Our models are trained using TensorFlow [1] on 128 TPU cores for ImageNet and 32 cores for panoptic segmentation.

Training Protocol: On ImageNet, we adopt the same training protocol as [65] for a fair comparison, except that we use batch size 512 for Full Axial-ResNets and 1024 for all other models, with learning rates scaled accordingly [29].

For panoptic segmentation, we strictly follow Panoptic-DeepLab [19], except using a linear warm up Radam [58] Lookahead [92] optimizer (with the same learning rate 0.001). All our results on panoptic segmentation use this setting. We note this change does not improve the results, but smooths our training curves. Panoptic-DeepLab yields similar result in this setting.

4.1   ImageNet

For ImageNet, we build Axial-ResNet-L from ResNet-50 [31]. In detail, we set $d_{in} = 128$, $d_{out} = 2d_q = 16$ for the first stage after the 'stem'. We double them when spatial resolution is reduced by a factor of 2 [76]. Additionally, we multiply all the channels [34,35,71] by 0.5, 0.75, and 2, resulting in Axial-ResNet-{S, M, XL}, respectively. Finally, *Stand-Alone* Axial-ResNets are further generated by replacing the 'stem' with three axial-attention blocks where the

Table 1. ImageNet validation set results. **BN:** Use batch normalizations in attention layers. **PS:** Our position-sensitive self-attention. **Full:** Stand-alone self-attention models without spatial convolutions

Method	BN	PS	Full	Params	M-Adds	Top-1
Conv-Stem methods						
ResNet-50 [31,65]				25.6M	4.1B	76.9
Conv-Stem + Attention [65]				18.0M	3.5B	77.4
Conv-Stem + Attention	✓			18.0M	3.5B	77.7
Conv-Stem + PS-Attention	✓	✓		18.0M	3.7B	78.1
Conv-Stem + Axial-Attention	✓	✓		12.4M	2.8B	77.5
Fully self-attentional methods						
LR-Net-50 [37]			✓	23.3M	4.3B	77.3
Full Attention [65]			✓	18.0M	3.6B	77.6
Full Axial-Attention	✓	✓	✓	**12.5M**	**3.3B**	**78.1**

first block has stride 2. Due to the computational cost introduced by the early layers, we set the axial span $m = 15$ in all blocks of Stand-Alone Axial-ResNets. We always use $N = 8$ heads [65]. In order to avoid careful initialization of $W_Q, W_K, W_V, r^q, r^k, r^v$, we use batch normalizations [40] in all attention layers.

Table 1 summarizes our ImageNet results. The baselines ResNet-50 [31] (done by [65]) and Conv-Stem + Attention [65] are also listed. In the conv-stem setting, adding BN to attention layers of [65] slightly improves the performance by 0.3%. Our proposed position-sensitive self-attention (Conv-Stem + PS-Attention) further improves the performance by 0.4% at the cost of extra marginal computation. Our Conv-Stem + Axial-Attention performs on par with Conv-Stem + Attention [65] while being more parameter- and computation-efficient. When comparing with other full self-attention models, our Full Axial-Attention outperforms Full Attention [65] by 0.5%, while being 1.44× more parameter-efficient and 1.09× more computation-efficient.

Following [65], we experiment with different network widths (i.e., Axial-ResNets-{S,M,L,XL}), exploring the trade-off between accuracy, model parameters, and computational cost (in terms of M-Adds). As shown in Fig. 3, our proposed Conv-Stem + PS-Attention and Conv-Stem + Axial-Attention already outperforms ResNet-50 [31,65] and attention models [65] (both Conv-Stem + Attention, and Full Attention) at all settings. Our Full Axial-Attention further attains the best accuracy-parameter and accuracy-complexity trade-offs.

4.2 COCO

The ImageNet pretrained Axial-ResNet model variants (with different channels) are then converted to Axial-DeepLab model variant for panoptic segmentation tasks. We first demonstrate the effectiveness of our Axial-DeepLab on the chal-

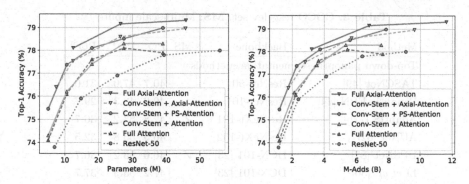

Fig. 3. Comparing parameters and M-Adds against accuracy on ImageNet classification. Our position-sensitive self-attention (Conv-Stem + PS-Attention) and axial-attention (Conv-Stem + Axial-Attention) consistently outperform ResNet-50 [31,65] and attention models [65] (both Conv-Stem + Attention, and Full Attention), across a range of network widths (*i.e.*, different channels). Our Full Axial-Attention works the best in terms of both parameters and M-Adds

Table 2. COCO val set. **MS:** Multi-scale inputs

Method	Backbone	MS	Params	M-Adds	PQ	PQTh	PQSt
DeeperLab [89]	Xception-71				33.8	-	-
SSAP [28]	ResNet-101	✓			36.5	-	-
Panoptic-DeepLab [19]	Xception-71		46.7M	274.0B	39.7	43.9	33.2
Panoptic-DeepLab [19]	Xception-71	✓	46.7M	3081.4B	41.2	44.9	35.7
Axial-DeepLab-S	Axial-ResNet-S		12.1M	110.4B	41.8	46.1	35.2
Axial-DeepLab-M	Axial-ResNet-M		25.9M	209.9B	42.9	47.6	35.8
Axial-DeepLab-L	Axial-ResNet-L		44.9M	343.9B	43.4	48.5	35.6
Axial-DeepLab-L	Axial-ResNet-L	✓	44.9M	3867.7B	43.9	48.6	36.8

lenging COCO dataset [56], which contains objects with various scales (from less than 32×32 to larger than 96×96).

Val Set: In Table 2, we report our validation set results and compare with other bottom-up panoptic segmentation methods, since our method also belongs to the bottom-up family. As shown in the table, our *single-scale* Axial-DeepLab-S outperforms DeeperLab [89] by 8% PQ, *multi-scale* SSAP [28] by 5.3% PQ, and *single-scale* Panoptic-DeepLab by 2.1% PQ. Interestingly, our *single-scale* Axial-DeepLab-S also outperforms *multi-scale* Panoptic-DeepLab by 0.6% PQ while being **3.8×** parameter-efficient and **27×** computation-efficient (in M-Adds). Increasing the backbone capacity (via large channels) continuously improves the performance. Specifically, our *multi-scale* Axial-DeepLab-L attains 43.9% PQ, outperforming Panoptic-DeepLab [19] by 2.7% PQ.

Test-dev Set: As shown in Table 3, our Axial-DeepLab variants show consistent improvements with larger backbones. Our *multi-scale* Axial-DeepLab-L

Table 3. COCO test-dev set. **MS:** Multi-scale inputs

Method	Backbone	MS	PQ	PQTh	PQSt
Top-down panoptic segmentation methods					
TASCNet [49]	ResNet-50		40.7	47.0	31.0
Panoptic-FPN [44]	ResNet-101		40.9	48.3	29.7
AdaptIS [77]	ResNeXt-101	✓	42.8	53.2	36.7
AUNet [52]	ResNeXt-152		46.5	55.8	32.5
UPSNet [87]	DCN-101 [23]	✓	46.6	53.2	36.7
Li et al. [50]	DCN-101 [23]		47.2	53.5	37.7
SpatialFlow [16]	DCN-101 [23]	✓	47.3	53.5	37.9
SOGNet [90]	DCN-101 [23]	✓	47.8	-	-
Bottom-up panoptic segmentation methods					
DeeperLab [89]	Xception-71		34.3	37.5	29.6
SSAP [28]	ResNet-101	✓	36.9	40.1	32.0
Panoptic-DeepLab [19]	Xception-71	✓	41.4	45.1	35.9
Axial-DeepLab-S	Axial-ResNet-S		42.2	46.5	35.7
Axial-DeepLab-M	Axial-ResNet-M		43.2	48.1	35.9
Axial-DeepLab-L	Axial-ResNet-L		43.6	48.9	35.6
Axial-DeepLab-L	Axial-ResNet-L	✓	44.2	49.2	36.8

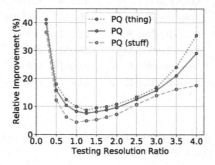

Fig. 4. Scale stress test on COCO val set. Axial-DeepLab gains the most when tested on extreme resolutions. On the x-axis, ratio 4.0 means inference with resolution 4097×4097

attains the performance of 44.2% PQ, outperforming DeeperLab [89] by 9.9% PQ, SSAP [28] by 7.3% PQ, and Panoptic-DeepLab [19] by 2.8% PQ, setting a new state-of-the-art among bottom-up approaches. We also list several top-performing methods adopting the top-down approaches in the table for reference.

Scale Stress Test: In order to verify that our model learns long range interactions, we perform a scale stress test besides standard testing. In the stress test, we train Panoptic-DeepLab (X-71) and our Axial-DeepLab-L with the standard setting, but test them on out-of-distribution resolutions (i.e., resize the input

Table 4. Mapillary Vistas validation set. **MS:** Multi-scale inputs

Method	MS	Params	M-Adds	PQ	PQTh	PQSt	AP	mIoU
Top-down panoptic segmentation methods								
TASCNet [49]				32.6	31.1	34.4	18.5	-
TASCNet [49]	✓			34.3	34.8	33.6	20.4	-
AdaptIS [77]				35.9	31.5	41.9	-	-
Seamless [68]				37.7	33.8	42.9	16.4	50.4
Bottom-up panoptic segmentation methods								
DeeperLab [89]				32.0	-	-	-	55.3
Panoptic-DeepLab (Xception-71 [20,69]) [19]		46.7M	1.24T	37.7	30.4	47.4	14.9	55.4
Panoptic-DeepLab (Xception-71 [20,69]) [19]	✓	46.7M	31.35T	40.3	33.5	49.3	17.2	56.8
Panoptic-DeepLab (HRNet-W48 [83]) [19]	✓	71.7M	58.47T	39.3	-	-	17.2	55.4
Panoptic-DeepLab (Auto-XL++ [57]) [19]	✓	72.2M	60.55T	40.3	-	-	16.9	57.6
Axial-DeepLab-L		44.9M	1.55T	40.1	32.7	49.8	16.7	57.6
Axial-DeepLab-L	✓	44.9M	39.35T	41.1	33.4	51.3	17.2	58.4

to different resolutions). Figure 4 summarizes our relative improvements over
Panoptic-DeepLab on PQ, PQ (thing) and PQ (stuff). When tested on huge
images, Axial-DeepLab shows large gain (30%), demonstrating that it encodes
long range relations better than convolutions. Besides, Axial-DeepLab improves
40% on small images, showing that axial-attention is more robust to scale variations.

4.3   Mapillary Vistas

We evaluate our Axial-DeepLab on the large-scale Mapillary Vistas dataset [62].
We only report validation set results, since the test server is not available.

Val Set: As shown in Table 4, our Axial-DeepLab-L outperforms all the state-
of-the-art methods in both single-scale and multi-scale cases. Our *single-scale*
Axial-DeepLab-L performs 2.4% PQ better than the previous best *single-scale*
Panoptic-DeepLab (X-71) [19]. In multi-scale setting, our lightweight Axial-
DeepLab-L performs better than Panoptic-DeepLab (Auto-DeepLab-XL++),
not only on panoptic segmentation (0.8% PQ) and instance segmentation (0.3%
AP), but also on semantic segmentation (0.8% mIoU), the task that Auto-
DeepLab [57] was searched for. Additionally, to the best of our knowledge, our
Axial-DeepLab-L attains the best *single-model* semantic segmentation result.

4.4   Cityscapes

Val Set: In Table 5(a), we report our Cityscapes validation set results. Without using extra data (*i.e.*, only Cityscapes fine annotation), our Axial-DeepLab

Table 5. Cityscapes val set and test set. **MS:** Multi-scale inputs. **C:** Cityscapes coarse annotation. **V:** Cityscapes video. **MV:** Mapillary Vistas

(a) Cityscapes validation set

Method	Extra Data	MS	PQ	AP	mIoU
AdaptIS [77]		✓	62.0	36.3	79.2
SSAP [28]		✓	61.1	37.3	-
Panoptic-DeepLab [19]			63.0	35.3	80.5
Panoptic-DeepLab [19]		✓	64.1	38.5	**81.5**
Axial-DeepLab-L			63.9	35.8	81.0
Axial-DeepLab-L		✓	64.7	37.9	**81.5**
Axial-DeepLab-XL			64.4	36.7	80.6
Axial-DeepLab-XL		✓	**65.1**	**39.0**	81.1
SpatialFlow [16]	COCO	✓	62.5	-	-
Seamless [68]	MV		65.0	-	80.7
Panoptic-DeepLab [19]	MV		65.3	38.8	82.5
Panoptic-DeepLab [19]	MV	✓	67.0	42.5	83.1
Axial-DeepLab-L	MV		66.5	40.2	83.2
Axial-DeepLab-L	MV	✓	67.7	42.9	83.8
Axial-DeepLab-XL	MV		67.8	41.9	84.2
Axial-DeepLab-XL	MV	✓	**68.5**	**44.2**	**84.6**

(b) Cityscapes test set

Method	Extra Data	PQ	AP	mIoU
GFF-Net [51]		-	-	82.3
Zhu et al. [97]	C, V, MV	-	-	83.5
AdaptIS [77]		-	32.5	-
UPSNet [87]	COCO	-	33.0	-
PANet [59]	COCO	-	36.4	-
PolyTransform [54]	COCO	-	40.1	
SSAP [28]		58.9	32.7	-
Li et al. [50]		61.0	-	-
Panoptic-DeepLab [19]		62.3	34.6	79.4
TASCNet [49]	COCO	60.7	-	-
Seamless [68]	MV	62.6	-	-
Li et al. [50]	COCO	63.3	-	-
Panoptic-DeepLab [19]	MV	65.5	39.0	84.2
Axial-DeepLab-L		62.7	33.3	79.5
Axial-DeepLab-XL		62.8	34.0	79.9
Axial-DeepLab-L	MV	65.6	38.1	83.1
Axial-DeepLab-XL	MV	**66.6**	39.6	84.1

achieves 65.1% PQ, which is 1% better than the current best bottom-up Panoptic-DeepLab [19] and 3.1% better than proposal-based AdaptIS [77]. When using extra data (*e.g.*, Mapillary Vistas [62]), our *multi-scale* Axial-DeepLab-XL attains 68.5% PQ, 1.5% better than Panoptic-DeepLab [19] and 3.5% better than Seamless [68]. Our instance segmentation and semantic segmentation results are respectively 1.7% and 1.5% better than Panoptic-DeepLab [19].

Test Set: Table 5(b) shows our test set results. Without extra data, Axial-DeepLab-XL attains 62.8% PQ, setting a new state-of-the-art result. Our model further achieves 66.6% PQ, 39.6% AP, and 84.1% mIoU with Mapillary Vistas pretraining. Note that Panoptic-DeepLab [19] adopts the trick of output stride 8 during inference on test set, making their M-Adds comparable to our XL models.

4.5   Ablation Studies

We perform ablation studies on Cityscapes validation set.

Importance of Position-Sensitivity and Axial-Attention: In Table 1, we experiment with attention models on ImageNet. In this ablation study, we transfer them to Cityscapes segmentation tasks. As shown in Table 6, all variants outperform ResNet-50 [31]. Position-sensitive attention performs better than previous self-attention [65], which aligns with ImageNet results in Table 1. However,

Table 6. Ablating self-attention variants on Cityscapes val set. **ASPP**: Atrous spatial pyramid pooling. **PS**: Our position-sensitive self-attention

Backbone	ASPP	PS	Params	M-Adds	PQ	AP	mIoU
ResNet-50 [31] (our impl.)			24.8M	374.8B	58.1	30.0	73.3
ResNet-50 [31] (our impl.)	✓		30.0M	390.0B	59.8	32.6	77.8
Attention [65] (our impl.)			17.3M	317.7B	58.7	31.9	75.8
Attention [65] (our impl.)	✓		22.5M	332.9B	60.9	30.0	78.2
PS-Attention		✓	17.3M	326.7B	59.9	32.2	76.3
PS-Attention	✓	✓	22.5M	341.9B	**61.5**	**33.1**	**79.1**
Axial-DeepLab-S		✓	**12.1M**	**220.8B**	**62.6**	**34.9**	**80.5**
Axial-DeepLab-M		✓	25.9M	419.6B	63.1	35.6	80.3
Axial-DeepLab-L		✓	44.9M	687.4B	63.9	35.8	81.0
Axial-DeepLab-XL		✓	173.0M	2446.8B	64.4	36.7	80.6

Table 7. Varying axial-attention span on Cityscapes val set

Backbone	Span	Params	M-Adds	PQ	AP	mIoU
ResNet-101	-	43.8M	530.0B	59.9	31.9	74.6
Axial-ResNet-L	5 × 5	44.9M	617.4B	59.1	31.3	74.5
Axial-ResNet-L	9 × 9	44.9M	622.1B	61.2	31.1	77.6
Axial-ResNet-L	17 × 17	44.9M	631.5B	62.8	34.0	79.5
Axial-ResNet-L	33 × 33	44.9M	650.2B	63.8	35.9	80.2
Axial-ResNet-L	65 × 65	44.9M	687.4B	**64.2**	**36.3**	**80.6**

employing axial-attention, which is on-par with position-sensitive attention on ImageNet, gives more than 1% boosts on all three segmentation tasks (in PQ, AP, and mIoU), without ASPP, and with fewer parameters and M-Adds, suggesting that the ability to encode long range context of axial-attention significantly improves the performance on segmentation tasks with large input images.

Importance of Axial-Attention Span: In Table 7, we vary the span m (*i.e.*, spatial extent of local regions in an axial block), without ASPP. We observe that a larger span consistently improves the performance at marginal costs.

5   Conclusion and Discussion

In this work, we have shown the effectiveness of proposed position-sensitive axial-attention on image classification and segmentation tasks. On ImageNet, our Axial-ResNet, formed by stacking axial-attention blocks, achieves state-of-the-art results among stand-alone self-attention models. We further convert Axial-ResNet to Axial-DeepLab for bottom-up segmentation tasks, and also show state-of-the-art performance on several benchmarks, including COCO, Mapillary Vistas, and Cityscapes. We hope our promising results could establish that axial-attention is an effective building block for modern computer vision models.

Our method bears a similarity to decoupled convolution [41], which factorizes a depthwise convolution [20,35,75] to a column convolution and a row convolution. This operation could also theoretically achieve a large receptive field, but its convolutional template matching nature limits the capacity of modeling multi-scale interactions. Another related method is deformable convolution [23,27,96], where each point attends to a few points dynamically on an image. However, deformable convolution does not make use of key-dependent positional bias or content-based relation. In addition, axial-attention propagates information densely, and more efficiently along the height- and width-axis sequentially.

Although our axial-attention model saves M-Adds, it runs slower than convolutional counterparts, as also observed by [65]. This is due to the lack of specialized kernels on various accelerators for the time being. This might well be improved if the community considers axial-attention as a plausible direction.

Acknowledgments. We thank Niki Parmar for discussion and support; Ashish Vaswani, Xuhui Jia, Raviteja Vemulapalli, Zhuoran Shen for their insightful comments and suggestions; Maxwell Collins and Blake Hechtman for technical support. This work is supported by Google Faculty Research Award and NSF 1763705.

References

1. Abadi, M., et al.: Tensorflow: a system for large-scale machine learning. In: Proceedings of the 12th USENIX Conference on Operating Systems Design and Implementation (2016)
2. Ackley, D.H., Hinton, G.E., Sejnowski, T.J.: A learning algorithm for boltzmann machines. Cogn. Sci. **9**(1), 147–169 (1985)
3. Bahdanau, D., Cho, K., Bengio, Y.: Neural machine translation by jointly learning to align and translate. arXiv:1409.0473 (2014)
4. Bai, M., Urtasun, R.: Deep watershed transform for instance segmentation. In: CVPR (2017)
5. Ballard, D.H.: Generalizing the hough transform to detect arbitrary shapes. Pattern Recogn. **3**, 111–122 (1981)
6. Bello, I., Zoph, B., Vaswani, A., Shlens, J., Le, Q.V.: Attention augmented convolutional networks. In: ICCV (2019)
7. Bonde, U., Alcantarilla, P.F., Leutenegger, S.: Towards bounding-box free panoptic segmentation. arXiv:2002.07705 (2020)
8. Brock, A., Donahue, J., Simonyan, K.: Large scale GAN training for high fidelity natural image synthesis. In: ICLR (2019)
9. Buades, A., Coll, B., Morel, J.M.: A non-local algorithm for image denoising. In: CVPR (2005)
10. Chan, W., Jaitly, N., Le, Q., Vinyals, O.: Listen, attend and spell: a neural network for large vocabulary conversational speech recognition. In: ICASSP (2016)
11. Chen, L.C., et al.: Searching for efficient multi-scale architectures for dense image prediction. In: NeurIPS (2018)

12. Chen, L.C., Papandreou, G., Kokkinos, I., Murphy, K., Yuille, A.L.: Semantic image segmentation with deep convolutional nets and fully connected CRFs. In: ICLR (2015)
13. Chen, L.C., Papandreou, G., Kokkinos, I., Murphy, K., Yuille, A.L.: DeepLab: semantic image segmentation with deep convolutional nets, atrous convolution, and fully connected CRFs. IEEE TPAMI (2017)
14. Chen, L.C., Papandreou, G., Schroff, F., Adam, H.: Rethinking atrous convolution for semantic image segmentation. arXiv:1706.05587 (2017)
15. Chen, L.-C., Zhu, Y., Papandreou, G., Schroff, F., Adam, H.: Encoder-decoder with atrous separable convolution for semantic image segmentation. In: Ferrari, V., Hebert, M., Sminchisescu, C., Weiss, Y. (eds.) ECCV 2018. LNCS, vol. 11211, pp. 833–851. Springer, Cham (2018). https://doi.org/10.1007/978-3-030-01234-2_49
16. Chen, Q., Cheng, A., He, X., Wang, P., Cheng, J.: SpatialFlow: bridging all tasks for panoptic segmentation. arXiv:1910.08787 (2019)
17. Chen, Y., Kalantidis, Y., Li, J., Yan, S., Feng, J.: Aˆ 2-nets: double attention networks. In: NeurIPS (2018)
18. Cheng, B., et al.: Panoptic-deeplab. In: ICCV COCO + Mapillary Joint Recognition Challenge Workshop (2019)
19. Cheng, B., et al.: Panoptic-deeplab: a simple, strong, and fast baseline for bottom-up panoptic segmentation. In: CVPR (2020)
20. Chollet, F.: Xception: deep learning with depthwise separable convolutions. In: CVPR (2017)
21. Chorowski, J.K., Bahdanau, D., Serdyuk, D., Cho, K., Bengio, Y.: Attention-based models for speech recognition. In: NeurIPS (2015)
22. Cordts, M., et al.: The cityscapes dataset for semantic urban scene understanding. In: CVPR (2016)
23. Dai, J., et al.: Deformable convolutional networks. In: ICCV (2017)
24. Dai, Z., Yang, Z., Yang, Y., Carbonell, J.G., Le, Q., Salakhutdinov, R.: Transformer-XL: Attentive language models beyond a fixed-length context. In: ACL (2019)
25. Devlin, J., Chang, M.W., Lee, K., Toutanova, K.: Bert: Pre-training of deep bidirectional transformers for language understanding. arXiv:1810.04805 (2018)
26. Fu, J., et al.: Dual attention network for scene segmentation. In: CVPR (2019)
27. Gao, H., Zhu, X., Lin, S., Dai, J.: Deformable kernels: adapting effective receptive fields for object deformation. arXiv:1910.02940 (2019)
28. Gao, N., et al.: SSAP: single-shot instance segmentation with affinity pyramid. In: ICCV (2019)
29. Goyal, P., et al.: Accurate, large minibatch SGD: training imagenet in 1 hour. arXiv:1706.02677 (2017)
30. He, K., Gkioxari, G., Dollár, P., Girshick, R.: Mask R-CNN. In: ICCV (2017)
31. He, K., Zhang, X., Ren, S., Sun, J.: Deep residual learning for image recognition. In: CVPR (2016)
32. Ho, J., Kalchbrenner, N., Weissenborn, D., Salimans, T.: Axial attention in multi-dimensional transformers. arXiv:1912.12180 (2019)
33. Holschneider, M., Kronland-Martinet, R., Morlet, J., Tchamitchian, P.: A real-time algorithm for signal analysis with the help of the wavelet transform. In: Combes, J.M., Grossmann, A., Tchamitchian, P. (eds.) Wavelets, pp. 286–297. Springer, Heidelberg (1990). https://doi.org/10.1007/978-3-642-75988-8_28
34. Howard, A., et al.: Searching for mobilenetv3. In: ICCV (2019)
35. Howard, A.G., et al.: MobileNets: efficient convolutional neural networks for mobile vision applications. arXiv:1704.04861 (2017)

36. Hu, H., Gu, J., Zhang, Z., Dai, J., Wei, Y.: Relation networks for object detection. In: CVPR (2018)
37. Hu, H., Zhang, Z., Xie, Z., Lin, S.: Local relation networks for image recognition. In: ICCV (2019)
38. Huang, C.A., et al.: Music transformer: Generating music with long-term structure. In: ICLR (2019)
39. Huang, Z., Wang, X., Huang, L., Huang, C., Wei, Y., Liu, W.: CCNet: criss-cross attention for semantic segmentation. In: ICCV (2019)
40. Ioffe, S., Szegedy, C.: Batch normalization: accelerating deep network training by reducing internal covariate shift. In: ICML (2015)
41. Jaderberg, M., Vedaldi, A., Zisserman, A.: Speeding up convolutional neural networks with low rank expansions. In: BMVC (2014)
42. Kendall, A., Gal, Y., Cipolla, R.: Multi-task learning using uncertainty to weigh losses for scene geometry and semantics. In: CVPR (2018)
43. Keuper, M., Levinkov, E., Bonneel, N., Lavoué, G., Brox, T., Andres, B.: Efficient decomposition of image and mesh graphs by lifted multicuts. In: ICCV (2015)
44. Kirillov, A., Girshick, R., He, K., Dollár, P.: Panoptic feature pyramid networks. In: CVPR (2019)
45. Kirillov, A., He, K., Girshick, R., Rother, C., Dollár, P.: Panoptic segmentation. In: CVPR (2019)
46. Krizhevsky, A., Sutskever, I., Hinton, G.E.: Imagenet classification with deep convolutional neural networks. In: NeurIPS (2012)
47. LeCun, Y., Bottou, L., Bengio, Y., Haffner, P.: Gradient-based learning applied to document recognition. Proc. IEEE **86**(11), 2278–2324 (1998)
48. Leibe, B., Leonardis, A., Schiele, B.: Combined object categorization and segmentation with an implicit shape model. In: Workshop on Statistical Learning in Computer Vision, ECCV (2004)
49. Li, J., Raventos, A., Bhargava, A., Tagawa, T., Gaidon, A.: Learning to fuse things and stuff. arXiv:1812.01192 (2018)
50. Li, Q., Qi, X., Torr, P.H.: Unifying training and inference for panoptic segmentation. arXiv:2001.04982 (2020)
51. Li, X., Zhao, H., Han, L., Tong, Y., Yang, K.: GFF: gated fully fusion for semantic segmentation. arXiv:1904.01803 (2019)
52. Li, Y., Chen, X., Zhu, Z., Xie, L., Huang, G., Du, D., Wang, X.: Attention-guided unified network for panoptic segmentation. In: CVPR (2019)
53. Li, Y., et al.: Neural architecture search for lightweight non-local networks. In: CVPR (2020)
54. Liang, J., Homayounfar, N., Ma, W.C., Xiong, Y., Hu, R., Urtasun, R.: PolyTransform: deep polygon transformer for instance segmentation. arXiv:1912.02801 (2019)
55. Lin, T.Y., Dollár, P., Girshick, R., He, K., Hariharan, B., Belongie, S.: Feature pyramid networks for object detection. In: CVPR (2017)
56. Lin, T.-Y., et al.: Microsoft COCO: common objects in context. In: Fleet, D., Pajdla, T., Schiele, B., Tuytelaars, T. (eds.) ECCV 2014. LNCS, vol. 8693, pp. 740–755. Springer, Cham (2014). https://doi.org/10.1007/978-3-319-10602-1_48
57. Liu, C., et al.: Auto-deeplab: Hierarchical neural architecture search for semantic image segmentation. In: CVPR (2019)
58. Liu, L., et al.: On the variance of the adaptive learning rate and beyond. In: ICLR (2020)
59. Liu, S., Qi, L., Qin, H., Shi, J., Jia, J.: Path aggregation network for instance segmentation. In: CVPR (2018)

60. Liu, Y., et al.: Affinity derivation and graph merge for instance segmentation. In: Ferrari, V., Hebert, M., Sminchisescu, C., Weiss, Y. (eds.) ECCV 2018. LNCS, vol. 11207, pp. 708–724. Springer, Cham (2018). https://doi.org/10.1007/978-3-030-01219-9_42

61. Liu1, H., et al.: An end-to-end network for panoptic segmentation. In: CVPR (2019)

62. Neuhold, G., Ollmann, T., Rota Bulo, S., Kontschieder, P.: The mapillary vistas dataset for semantic understanding of street scenes. In: ICCV (2017)

63. Neven, D., Brabandere, B.D., Proesmans, M., Gool, L.V.: Instance segmentation by jointly optimizing spatial embeddings and clustering bandwidth. In: CVPR (2019)

64. Papandreou, G., Kokkinos, I., Savalle, P.A.: Modeling local and global deformations in deep learning: epitomic convolution, multiple instance learning, and sliding window detection. In: CVPR (2015)

65. Parmar, N., Ramachandran, P., Vaswani, A., Bello, I., Levskaya, A., Shlens, J.: Stand-alone self-attention in vision models. In: NeurIPS (2019)

66. Parmar, N., et al.: Image transformer. In: ICML (2018)

67. Peng, C., Zhang, X., Yu, G., Luo, G., Sun, J.: Large kernel matters-improve semantic segmentation by global convolutional network. In: CVPR (2017)

68. Porzi, L., Bulò, S.R., Colovic, A., Kontschieder, P.: Seamless scene segmentation. In: CVPR (2019)

69. Qi, H., et al.: Deformable convolutional networks - COCO detection and segmentation challenge 2017 entry. In: ICCV COCO Challenge Workshop (2017)

70. Russakovsky, O., et al.: Imagenet large scale visual recognition challenge. IJCV 115, 211–252 (2015)

71. Sandler, M., Howard, A., Zhu, M., Zhmoginov, A., Chen, L.C.: MobileNetV2: inverted residuals and linear bottlenecks. In: CVPR (2018)

72. Shaw, P., Uszkoreit, J., Vaswani, A.: Self-attention with relative position representations. In: NAACL (2018)

73. Shen, Z., Zhang, M., Zhao, H., Yi, S., Li, H.: Efficient attention: attention with linear complexities. arXiv:1812.01243 (2018)

74. Shensa, M.J.: The discrete wavelet transform: wedding the a trous and mallat algorithms. IEEE Trans. Signal Process. 40(10), 2464–2482 (1992)

75. Sifre, L.: Rigid-motion scattering for image classification. Ph.D. thesis (2014)

76. Simonyan, K., Zisserman, A.: Very deep convolutional networks for large-scale image recognition. arXiv:1409.1556 (2014)

77. Sofiiuk, K., Barinova, O., Konushin, A.: AdaptiS: adaptive instance selection network. In: ICCV (2019)

78. Szegedy, C., Vanhoucke, V., Ioffe, S., Shlens, J., Wojna, Z.: Rethinking the inception architecture for computer vision. In: CVPR (2016)

79. Uhrig, J., Rehder, E., Fröhlich, B., Franke, U., Brox, T.: Box2pix: single-shot instance segmentation by assigning pixels to object boxes. In: IEEE Intelligent Vehicles Symposium (IV) (2018)

80. Vaswani, A., et al.: Attention is all you need. In: NeurIPS (2017)

81. Vincent, L., Soille, P.: Watersheds in digital spaces: an efficient algorithm based on immersion simulations. IEEE TPAMI (1991)

82. Wang, H., Kembhavi, A., Farhadi, A., Yuille, A.L., Rastegari, M.: Elastic: improving CNNs with dynamic scaling policies. In: CVPR (2019)

83. Wang, J., et al.: Deep high-resolution representation learning for visual recognition. arXiv:1908.07919 (2019)

84. Wang, X., Girshick, R., Gupta, A., He, K.: Non-local neural networks. In: CVPR (2018)
85. Wu, Y., et al.: Google's neural machine translation system: bridging the gap between human and machine translation. arXiv:1609.08144 (2016)
86. Xie, C., Wu, Y., Maaten, L.v.d., Yuille, A.L., He, K.: Feature denoising for improving adversarial robustness. In: CVPR (2019)
87. Xiong, Y., et al.: UPSNet: a unified panoptic segmentation network. In: CVPR (2019)
88. Xu, K., et al.: Show, attend and tell: Neural image caption generation with visual attention. In: ICML (2015)
89. Yang, T.J., et al.: DeeperLab: single-shot image parser. arXiv:1902.05093 (2019)
90. Yang, Y., Li, H., Li, X., Zhao, Q., Wu, J., Lin, Z.: SOGNet: scene overlap graph network for panoptic segmentation. arXiv:1911.07527 (2019)
91. Zhang, H., Goodfellow, I., Metaxas, D., Odena, A.: Self-attention generative adversarial networks. arXiv:1805.08318 (2018)
92. Zhang, M., Lucas, J., Ba, J., Hinton, G.E.: Lookahead optimizer: k steps forward, 1 step back. In: NeurIPS (2019)
93. Zhang, R.: Making convolutional networks shift-invariant again. In: ICML (2019)
94. Zhao, H., Shi, J., Qi, X., Wang, X., Jia, J.: Pyramid scene parsing network. In: CVPR (2017)
95. Zhu, X., Cheng, D., Zhang, Z., Lin, S., Dai, J.: An empirical study of spatial attention mechanisms in deep networks. In: ICCV, pp. 6688–6697 (2019)
96. Zhu, X., Hu, H., Lin, S., Dai, J.: Deformable ConvNets v2: more deformable, better results. In: CVPR (2019)
97. Zhu, Y., et al.: Improving semantic segmentation via video propagation and label relaxation. In: CVPR (2019)
98. Zhu, Z., Xu, M., Bai, S., Huang, T., Bai, X.: Asymmetric non-local neural networks for semantic segmentation. In: CVPR (2019)
99. Zoph, B., Le, Q.V.: Neural architecture search with reinforcement learning. In: ICLR (2017)

Adaptive Computationally Efficient Network for Monocular 3D Hand Pose Estimation

Zhipeng Fan[1], Jun Liu[2]([⊠]), and Yao Wang[1]

[1] Tandon School of Engineering, New York University, Brooklyn, NY, USA
{zf606,yw523}@nyu.edu
[2] Information Systems Technology and Design Pillar, Singapore University
of Technology and Design, Singapore, Singapore
jun_liu@sutd.edu.sg

Abstract. 3D hand pose estimation is an important task for a wide range of real-world applications. Existing works in this domain mainly focus on designing advanced algorithms to achieve high pose estimation accuracy. However, besides accuracy, the computation efficiency that affects the computation speed and power consumption is also crucial for real-world applications. In this paper, we investigate the problem of reducing the overall computation cost yet maintaining the high accuracy for 3D hand pose estimation from video sequences. A novel model, called Adaptive Computationally Efficient (ACE) network, is proposed, which takes advantage of a Gaussian kernel based Gate Module to dynamically switch the computation between a light model and a heavy network for feature extraction. Our model employs the light model to compute efficient features for most of the frames and invokes the heavy model only when necessary. Combined with the temporal context, the proposed model accurately estimates the 3D hand pose. We evaluate our model on two publicly available datasets, and achieve state-of-the-art performance at 22% of the computation cost compared to traditional temporal models.

Keywords: 3D hand pose estimation · Computation efficiency · Dynamic adaption · Gaussian gate

1 Introduction

Understanding human hand poses is a long lasting problem in computer vision community, due to the great amount of potential applications in action recognition, AR/VR [28], robotics and human computer interactions (HCI) [11]. The problem of inferring 3D configurations of human hands from images and videos

---

Electronic supplementary material The online version of this chapter (https://doi.org/10.1007/978-3-030-58548-8_8) contains supplementary material, which is available to authorized users.

A. Vedaldi et al. (Eds.): ECCV 2020, LNCS 12349, pp. 127–144, 2020.
https://doi.org/10.1007/978-3-030-58548-8_8

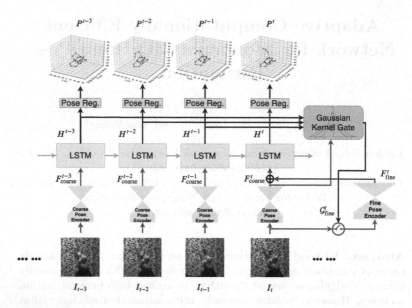

Fig. 1. Illustration of our Adaptive Computationally Efficient (ACE) network. In most of the time, the LSTM takes features from the coarse pose encoder and refines the predicted pose. Occasionally, when the pose varies a lot or severely occluded, the Gaussian Kernel Gate opts to compute fine features with the computationally heavy model to inject more accurate features to the LSTM.

is inherently challenging because of the frequent self-occlusion and the large variance of hand poses. A large body of existing works address the problem of hand pose estimation from depth data [7,37], as it reduces ambiguities in the depth dimension and makes it easier to acquire the 3D poses of the corresponding hand. However, depth cameras, such as Kinect, are not always available and are prone to measurement errors if deployed in outdoor settings. Therefore, in this work we address the problem of 3D hand pose estimation with a monocular RGB commercial camera.

Recent successes in 3D hand pose estimation [2,3,22,26,46] mainly focus on employing the same computation framework for all video frames, without considering the redundancy that exists across adjacent frames and the variation of the pose estimation difficulties over frames. The moving speed and occlusion status of the human hands vary when performing different actions, which inspires us to design a new scheme to dynamically allocate the computation resources based on the ambiguity determined by the current input frame and temporal context status. This kind of adaption mechanism is useful for both online and offline applications. For offline pose estimation from videos, being able to use a simpler computation module in most of the frames saves the amount of the resource usage and reduces the total inference time for the entire video. For online pose estimation applications (e.g. HCI and robots), multiple tasks often run concurrently under a total computation resource constraint, thus the saved resources at most of the frames could be released for other important tasks at those time steps, which meanwhile also reduces the amount of energy consumed by the pose estimation task.

Motivated by our idea of reducing computation consumption, and given the fact that the information among video frames could be redundant and the pose estimation difficulty varies over frames, we propose a novel Adaptive Computationally Efficient (ACE) network using a recurrent 3D hand pose estimator with adaptive input. In our method, we design two base pose encoders based on the hourglass(HG) [27] architecture with different computational costs. A Long Short-Term Memory (LSTM) [14] model was introduced to refine the predicted pose and features from the single-frame base pose encoder, by considering the temporal consistency. We propose a new **Gaussian Gate Module** to automatically determine whether the low complexity coarse encoder output alone is sufficient for the LSTM, or the high complexity fine encoder is needed. The fine encoder is only invoked when necessary and its output is combined with the output of the coarse encoder to generate the input for the LSTM. The proposed network architecture is illustrated in Fig. 1. To facilitate the training of our switch module, which is naturally a discrete operation, an effective Gumbel-SoftMax strategy, as an approximation of sampling from discrete distributions, is introduced.

To summarize, a novel end-to-end ACE network is proposed for 3D hand pose estimation from monocular video. It dynamically switches between using coarse v.s. fine features at each time step, which eliminates the computational cost of the fine encoder when the prediction from the coarse encoder is deemed sufficient. We evaluate our network on two broadly used datasets, First-Person Hand Action (FPHA) and Stereo Tracking Benchmark (STB), and obtain state-of-the-art pose estimation accuracy, while greatly reducing the overall computation cost (around 78% on STB dataset), compared to baseline models that constantly use the fine encoder for all time steps.

2 Related Work

Most of the existing works focus on the accuracy of 3D hand pose estimation without explicitly considering the important computation cost issue. We will briefly review the recent works in both the 3D hand pose estimation domain as well as the recent endeavor in designing computationally efficient architectures for image and video understanding.

3D Hand Pose Estimation. 3D hand pose estimation is a long-standing problem in computer vision domain, and various methods have been proposed. We restrict ourselves to the more recent deep learning based approaches since they are more related to our work.

A large body of the works on hand pose estimation operate on the depth input, which greatly reduces the depth ambiguity of the task. Deephand proposes a ConvNet model with an additional matrix completion algorithm to retrieve the actual poses [34]. Volumetric representation was adopted to better encode the depth image recently [7,8]. The volumetric representation is projected to multiple views and then processed by several 2D ConvNets followed by fusion in [7]. Rather than tedious projections to multiple views, a 3D ConvNet is directly introduced to infer the 3D position from the volumetric representations [8]. This line of work is further summarized in [9], in which the completeness of the 3D hand surface is leveraged as additional supervision. Rather than volumetric representations, the skeleton annotation could be represented as dense pixel-wise labels [37]. The predicted dense estimations are then converted back to 3D coordinates with a vote casting mechanism. Recently, self-supervised methods are also explored on a mixture of synthetic and unlabelled dataset by exploring the approximate depth and the kinematic feasibility as the weak supervision [36].

Rather than performing pose estimation on depth data, we lay more focus on the works with RGB inputs, which are often less restricted in real-world applications. Zimmermann and Brox proposed a multi-stage network, which performs hand segmentation, localization, 2D and 3D pose estimations one by one [46]. Similar to the depth based method, depth regularization was employed to enable weakly supervised learning [2]. Instead of regressing the joint positions independently, kinematic model could be naturally integrated into the model to yield anatomically plausible results [26]. A latent 2.5D representation is introduced in [16], where the ConvNet also learns the implicit depth map of the entire palm. Numerous graphic models are also proposed to better handle the joint relationships [3,22]. Spatial dependencies and temporal consistencies could be modeled explicitly with graph neural net [3] and could further boost the quality of estimated features [22] from hourglass models [27]. Another line of works reconstruct the shape and the pose of hands at the same time [1,10,25,42,45], in which either a hand mesh model [25,33] or a generative GNN [45] is leveraged to map the low-dimensional hand pose & shape manifold to the full 3D meshes.

Despite all the success in accurate hand pose estimation, we argue that the efficiency problem is also of vital importance, especially for AR/VR [28] and mobile devices [11], where resources are often limited. To harvest the redundancy present in the consecutive frames, we propose an adaptive dynamic gate to efficiently switch between an efficient light pose estimator and a computationally heavy pose estimator for 3D hand pose estimation from sequences of frames.

Computationally Efficient Architectures. Recent progresses have shown that the computation efficiency of neural net models could be improved in various ways. Neural network pruning was first realized using second-order derivative [13,19] and then evolved into pruning weights with relatively small magnitude [12]. Different from the pruning technique operated on fully trained models [12,13,19], recent developments reveal that pruning while training often results in better performance. This was achieved by enforcing additional loss ($L1$ norm [23], Group LASSO [39] or $L0$ norm approximations [24].) during training. Other innovative ideas include specially designed architectures for high-efficiency computing [15,44] and network quantization [4,5,20,31].

In videos, consecutive frames are often quite similar and strongly co-dependent, which leave lots of space for efficiency optimization. Recently, various works have been developed to improve the computation efficiency for video classification [18,29,38,40]. Leveraging the fact that most of the computational expansive layers (w/o activation) are linear and sparse feature updates are more efficient, a recurrent residual model was introduced [29] to incur minimum amount of feature updates between consecutive frames. Hierarchical coarse-to-fine architectures are also introduced for more efficient video inference [40]. Recently, RL frameworks are adopted to learn an efficient sampling agent to filter out salient parts/frames from videos for fast recognition [18,41].

In this work, we address the problem of dense hand pose estimation from video sequences, where we need to derive corresponding poses for each individual frame. We take advantage of the fact that at most of the time, when the motion of the hand is not extreme or the hand pose is not severely occluded, the 3D hand pose could be safely derived from the temporal context. We thus propose a novel Gaussian Kernel-based Adaptive Dynamic Gate module that explicitly measures the necessity to compute fine features with a costly model, which significantly reduces the total amount of computation in general. Our scheme is also orthogonal to many of the aforementioned methods, such as the pruning methods, which leaves the potential to further boost the efficiency.

3   Method

3.1   Overview

Given a sequence of video frames $\{I^t\}_{t=1}^T$, our task is to infer the 3D pose $\mathbf{P}^t = \{P_k^t\}_{k=1}^K$ of the hand at each frame t, where K denotes the number of hand joints, and P_k^t denotes the 3D position of the joint k at frame t.

The overall pipeline of our proposed ACE network is illustrated in Fig. 1. In our method, at each time step, both a less accurate yet computationally light model and an accurate but computationally heavy model can be selected as the pose encoder for the RGB input. The features from either models could be fed into a LSTM to refine the inferred features and the estimated pose based on the temporal coherence. To reduce the computation cost, inspired by the idea that temporal context can provide sufficient information when the motion of the target hand is slow or the pose is less challenging, we propose a novel Gaussian Kernel-based gate module as the key component of our ACE network, which compares the temporal context information provided by the LSTM model with the coarse features computed by the light encoder to assess the necessity of extracting fine features with the heavier encoder for the current time step. Below we introduce each component in more detail.

3.2   Single Frame Hand Pose Estimator

We first introduce two base pose encoders: coarse pose encoder and fine pose encoder, which have significantly different computation profiles for a single

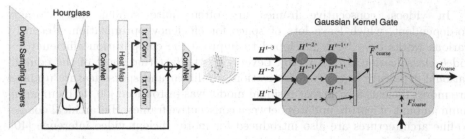

Architecture of pose encoder   Schema of the Gaussian Kernel Gate.

Fig. 2. Graphical illustration of the Single frame pose encoder and the Gaussian Kernel Based Gate.

frame. Both models are constructed with the state-of-the-art hourglass (HG) network [27]. Furthermore, as illustrated in Fig. 2a, we augment it to directly regress the hand joint coordinates $\mathbf{P}^t$ via a ConvNet from the heat map of joint probabilities $\mathbf{H}^t = \{H^t_k\}^K_{k=1}$, output feature map from HG as well as feature maps from early downsampling layers. The complexity of the models is adjusted by changing the number of convolutional layers and the size of the inputs of the hourglass module. We denote the light-weight coarse pose encoder model as $\mathbf{M}_{\text{Coarse-Enc}}$, and the heavy model as $\mathbf{M}_{\text{Fine-Enc}}$. These encoders extract pose related features, $\mathbf{F}^t_{\text{coarse}}$ and $\mathbf{F}^t_{\text{fine}}$, based on the input frame I^t, as follows:

$$\mathbf{F}^t_{\text{coarse}} = \mathbf{M}_{\text{Coarse-Enc}}(I^t) \tag{1}$$

$$\mathbf{F}^t_{\text{fine}} = \mathbf{M}_{\text{Fine-Enc}}(I^t) \tag{2}$$

Note that in our final ACE network with the gate mechanism, we compute the coarse features ($\mathbf{F}^t_{\text{coarse}}$) for each frame, while the fine features ($\mathbf{F}^t_{\text{fine}}$) are computed for a fraction of time only, thus reducing the overall computation cost.

3.3   Pose Refinement Recurrent Model

In pose estimation from videos, a natural idea is to exploit the temporal context information for more smooth and accurate estimations, i.e., instead of solely relying on the information of the current frame, historical context can also be incorporated to reduce the ambiguities in pose estimation [21,30,32,35]. Thus we introduce a LSTM model to refine the estimations from the hourglass modules. The LSTM module, denoted as $\mathbf{M}_{\text{LSTM}}$, takes the sequential features from pose encoder as inputs, and refine these input features using the temporal context.

More formally, at the t-th time step, the LSTM takes the pose-related features from the current frame as the input, and infer the 3D pose ($\mathbf{P}^t$) for the current step based on the hidden state, as follows:

$$h^t, c^t = \mathbf{M}_{\text{LSTM}}(\mathbf{F}^t_{\text{frame}}, (h^{t-1}, c^{t-1})) \tag{3}$$

$$\mathbf{P}^t = W^\top_{\text{pose}} h^t + b_{\text{pose}} \tag{4}$$

where h^t and c^t are the hidden state and cell state of the LSTM module respectively. W_{pose} and b_{pose} are the parameters of the output linear layer for regressing the final 3D hand joint coordinates. Here we denote the features from the single frame pose estimator as $\mathbf{F}^t_{\text{frame}}$, which is controlled by our adaptive dynamic gate model (introduced next) and could be either the coarse features $\mathbf{F}^t_{\text{coarse}}$ or the weighted combination of the coarse features $\mathbf{F}^t_{\text{coarse}}$ and fine features $\mathbf{F}^t_{\text{Fine}}$.

3.4 Adaptive Dynamic Gate Model

Recall that when humans perform activities with their hands, the motion speed and the self-occlusion status of the hands vary across different activities and different frames. In some of the actions like "high five", the palm is often less occluded and the pose pattern is relatively static and simple, while in some other actions, like "open soda can" and "handshake", the human hand is often under severe occlusions and presents rich and delicate movements of the fingers.

This inspires us to rely more on the temporal context information (and only use a brief glimpse over the current frame with the coarse pose encoder) for pose inference when the pose pattern is simple, stable and could be safely derived from the temporal context. However, if the temporal context is not consistent with the current frame information, this means either the current frame could be challenging for pose inference (i.e. pose inaccurately estimated by coarse pose encoder but temporal context is reliable) or significantly differs from previous frames due to large motions (i.e. temporal context becomes unstable), and thus the network needs to take a more careful examination for the current frame by using the fine pose encoder. Therefore, we propose an adaptive dynamic gate model in our ACE framework to dynamically determine the granularity of the features needed for pose estimation with our LSTM model.

Assuming the motion of the hand is smooth, the first and second-order statistics of the hand's status over different frames provide useful context information for estimating the evolution of the hand pose over time. Accordingly, we compute the first-order difference $(h^{t'})$ and second-order difference $(h^{t''})$ over the history hidden states of the LSTM to estimate the motion status information of the hand pose as:

$$h^{t'} = h^t - h^{t-1} \tag{5}$$

$$h^{t''} = (h^t - h^{t-1}) - (h^{t-1} - h^{t-2}) \tag{6}$$

At the time step t, we feed the hidden state of the previous frame(h^{t-1}), as well as its first and second-order information $(h^{t-1'}$ and $h^{t-1''})$ as the history context information, to our gate module, which then estimates the pose feature information of current frame (t) with a sub-network, as follows:

$$\widetilde{\mathbf{F}}^t = W_g^\top [h^{t-1}, h^{t-1'}, h^{t-1''}] + b_g \tag{7}$$

We then measure the similarity of the predicted pose feature information $(\widetilde{\mathbf{F}}^t)$ that is completely estimated from the temporal context of previous frames, with the pose features ($\mathbf{F}^t_{\text{coarse}}$) that are extracted with the coarse pose encoder solely based on current frame I^t, via a Gaussian Kernel with a fixed spread ω as follows:

$$\mathcal{G}^t_{\text{coarse}} = \left[\exp\left(-\frac{(\widetilde{\mathbf{F}}^t - \mathbf{F}^t_{\text{coarse}})^2}{\omega^2} \right) \right]_{\text{Mean}} \tag{8}$$

This Gaussian Kernel based gate outputs mean value ($\mathcal{G}^t_{coarse}$) between 0 and 1, which provides an explicit measurement of the consistency and similarity between $\widetilde{\mathbf{F}}^t$ and $\mathbf{F}^t_{\text{coarse}}$, implying the pose estimation difficulty of current frame, i.e., higher $\mathcal{G}^t_{coarse}$ value indicates simple pose and stable movements of the hand.

If the hand pose status at this step changes a lot and pose feature becomes unpredictable from the temporal context, or the pose at current frame becomes challenging, leading to the pose features ($\mathbf{F}^t_{\text{coarse}}$) extracted by the coarse pose encoder not reliable and therefore inconsistent with the temporal context, the discrepancy between $\widetilde{\mathbf{F}}^t$ and $\mathbf{F}^t_{\text{coarse}}$ grows larger, and thus our Gaussian gate will output a relatively small value close to 0.

With an estimation of the difficulty of current frame, we then decide if we need to employ the more powerful fine pose encoder to carefully examine the input frame of current time step. Specifically, we can use the $\mathcal{G}^t_{\text{coarse}}$ from our Gaussian gate as the confidence score of staying with the coarse pose encoder for current time step, and naturally $\mathcal{G}^t_{\text{fine}} = 1 - \mathcal{G}^t_{\text{coarse}}$ becomes the score that we need to use the more powerful fine pose encoder.

A straight-forward switching mechanism would be to directly follow the one with a larger confidence score, i.e., if $\mathcal{G}^t_{\text{fine}} > \mathcal{G}^t_{\text{coarse}}$, we need to involve the fine pose encoder for the current frame. This switching operation is however a discrete operation that is not differentiable. To facilitate the network training, following the recent work on reparameterization for the categorical distribution [17], we reparameterize the Bernoulli distribution with the Gumbel-Softmax trick, which introduces a simple yet efficient way to draw samples z from a categorical distribution parameterized by the unnormalized probability π. Specifically, we can approximately sample from π_i following:

$$z_i = \underset{i \in \mathcal{M}}{\text{argmax}} \, [g_i + \log \pi_i] \qquad \mathcal{M} = \{\text{coarse, fine}\} \tag{9}$$

where at each time step t, we set $\pi^t_i = -\log(1 - \mathcal{G}^t_i)$, which is the unnormalized version of the predict probability $\mathcal{G}^t_i$ in Bernoulli distribution $\mathcal{G}^t_i \in \{\mathcal{G}^t_{\text{coarse}}, \mathcal{G}^t_{\text{fine}}\}$. g_i is the Gumbel noise. Here we draw samples from the Gumbel distribution following $g_i = -\log(-\log(u_i))$, where u_i is the i.i.d. samples drawn from Uniform(0,1). We further relax the non-differentiable operation argmax with softmax to facilitate back propagation. The final sampled probability is obtained with:

$$z^t_i = \frac{\exp\left((g_i + \log \pi^t_i)/\tau \right)}{\sum_j \exp\left((g_j + \log \pi^t_j)/\tau \right)} \qquad \text{for } i,j \in \{\text{coarse, fine}\} \tag{10}$$

where τ is the hyper-parameter of temperature, which controls the discreteness of the sampling mechanism. When $\tau \to \infty$, the sample approximates the uniform sampling, and when $\tau \to 0$, it yields the argmax operation while allows the gradient to be back-propagated.

During training, we obtain the confidence scores of using rough glimpse for the input frame via the coarse pose encoder or using careful derived features with the fine encoder, via the Gumbel-SoftMax trick following Eq. (10), and then we combine the coarse features $\mathbf{F}^t_{\text{coarse}}$ and fine features $\mathbf{F}^t_{\text{fine}}$ as:

$$\mathbf{F}^t_{\text{weighted}} = z^t_{\text{coarse}}\mathbf{F}^t_{\text{coarse}} + z^t_{\text{fine}}\mathbf{F}^t_{\text{fine}} \tag{11}$$

During evaluation, we omit the sampling process and directly use the coarse features when $\mathcal{G}^t_{\text{fine}} \leq \lambda$, and use the weighted average features when $\mathcal{G}^t_{\text{fine}} > \lambda$ with weight $\mathcal{G}^t_{\text{fine}}$ and $\mathcal{G}^t_{\text{coarse}}$. In general, λ is set to 0.5, which essentially follows the larger probability. This threshold λ could also be tweaked to balance between accuracy and efficiency during inference.

3.5 Training Strategy and Losses

We employ a two-step training strategy, in which we separately train the single-frame coarse pose encoder and fine pose encoder first, and then fine-tune them during the training of the LSTM pose refinement module and the adaptive gate module. To train the single frame pose encoder, we use the combination of 2D heat map regression loss and 3D coordinate regression loss:

$$\mathcal{L}_{\text{single}} = \frac{1}{K}\sum_{k=1}^{K}(\widetilde{H}^k - H^k)^2 + \beta \cdot \text{Smooth L1}(\widetilde{\mathbf{P}}, \mathbf{P}) \tag{12}$$

where H^k corresponds to the 2D heat map of joint k and $\mathbf{P}$ is the 3D joint coordinates. We use the mean squared loss for the heat map and Smooth L1 loss for the 3D coordinates, which has a squared term when the absolute element-wise difference is below 1 (otherwise it is essentially a L1 term).

The single frame pose estimator is then fine-tuned when training the pose refinement LSTM and the gate module. To prevent the gate module from constantly using the fine features, we set an expected activation frequency (γ_g) for the gate, and optimize the mean square error between mean probability of using fine encoder and the expected activation frequency. Specifically, we define the loss mathematically given the expected activate rate, γ_g as:

$$\mathcal{L}_{\text{whole}} = \sum_{d \in \mathcal{S}} \text{Smooth L1}(\widetilde{\mathbf{P}}_d, \mathbf{P}_d) + \delta \cdot \mathbb{E}_{z^t \sim \text{Bernoulli}(\mathcal{G}^t|\theta_g)}(\frac{1}{T}\sum_{t=1}^{T} z^t_{\text{fine}} - \gamma_g)^2 \tag{13}$$

where $\mathcal{S} = \{\text{coarse}, \text{fine}, \text{LSTM}\}$ and z^t_{fine} is the sample probability based on the prediction $\mathcal{G}^t$ given by the adaptive dynamic gate model. θ_g denotes the parameter of the gate and δ balances the accuracy and the efficiency.

4   Experiments

4.1   Datasets and Metrics

We evaluate our ACE network on two publicly available datasets, namely the Stereo Tracking Benchmark (STB) [43] dataset and the First-Person Hand Action (FPHA) [6] dataset.

Stereo Tracking Benchmark (STB) provides 2D and 3D pose annotations of the 21 hand keypoints for 12 stereo video sequences. Each sequence consists of 1500 RGB frames for both the left camera and right camera. In total, this dataset consists of 18000 frames and the resolution of the frame is 640×480. Within the dataset, 6 set of different backgrounds are captured, with each background appears in two video sequences. Following the setting of [46], we separate the dataset into a training set of 10 videos (15000 frames) and a evaluation set of 2 video sequences (3000 frames).

First Person Hand Action (FPHA) contains video sequences for 45 different daily actions from 6 different subjects in egocentric views. In total, FPHA contains more than 100k frames with a resolution of 1920×1080. The ground truth is provided via a mo-cap system and derived with inverse kinematics. Similar to the STB dataset, 21 keypoints on the human hand are annotated. Object interaction with 26 different objects is involved, which introduces additional challenges to hand pose estimation. We follow the official split of the dataset.

Metrics. We report the Percentage of Correct Keypoints (PCKs) under 20 mm and the Area Under Curve (AUC) of PCKs under the error thresholds from 20 mm to 50 mm for STB dataset following [46], and from 0 mm to 50 mm for FPHA dataset. We report average GFLOPs[1] per frame for speed comparison, which does not rely on the hardware configurations and thus provides more objective evaluations.

4.2   Implementation Details

Although the proposed ACE module is theoretically compatible with different pose encoder architectures, we mainly evaluate it with the hourglass (HG) architecture [27] as it is widely used and works well in many existing works [22,45]. Compared to the FPHA dataset, STB is less challenging as no hand-object interaction is involved. Therefore, different HG architectures are employed for different datasets. For the STB dataset, the coarse pose encoder contains one hourglass module with 32 feature channels, while for the fine pose encoder, we employ 64 channels. In addition to the different configurations of the module, the input images to the coarse and fine modules are set to 64×64 and 256×256 respectively, which greatly reduce the amount of computation. For the more challenging FPHA dataset, we keep the configurations of the fine pose encoder as STB, while for the coarse pose encoder, we double the size of input to 128×128. Please see the supplementary materials for more details of the pose encoder.

---

[1] Computed based on the public toolbox: PyTorch-OpCounter.

Table 1. Results of various models (vanilla single frame coarse/fine models and their variants considering temporal dynamics) for 3D hand pose estimation. Our adaptive model uses much less computation with minor accuracy drops.

Method	STB			FPHA		
	3D PCK20	AUC(20-50)	GFLOPs	3D PCK20	AUC(0-50)	GFLOPs
Coarse-HG	85.1%	0.946	0.28	72.6%	0.674	1.10
Fine-HG	96.3%	0.994	6.96	79.7%	0.714	6.96
Vanilla-LSTM-Coarse-HG	92.1%	0.973	0.28	78.9%	0.707	1.10
Vanilla-LSTM-Fine-HG	98.7%	0.997	6.96	83.9%	0.740	6.96
Vanilla-LSTM-Mix-HG	98.7%	0.997	7.24	83.1%	0.734	8.06
Adaptive-LSTM-Mix-HG	97.9%	0.996	1.56	82.9%	0.731	1.37

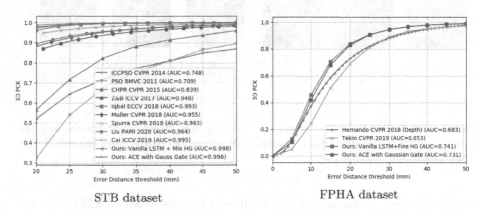

STB dataset   FPHA dataset

Fig. 3. Quantiatitative evaluations. We achieve state-of-the-art performance on STB, and outperform the existing methods on FPHA by a large margin.

For the LSTM refinement module, we use one layer of LSTM with hidden state dimension of 256. The hidden states and its order statistics are first mapped to a fixed dimension of 256 and then concatenated as the input to our adaptive Gaussian gate. During training, we set $\gamma_g = 0.05$ for STB and $\gamma_g = 0.01$ for FPHA and $\omega = 0.1$.

4.3   Main Results

We conduct extensive experiments to show the advantages of our proposed ACE framework for hand pose estimation from videos. We compare the accuracy and computation efficiency among different models and further visualize the prediction results of our model. To facilitate the understanding of the gate behaviour, we also present the frames selected for fine feature computation.

Quantitative Comparison. We present the comparison among our adaptive dynamic gate model and various baselines in Table 1, where Coarse-HG/fine-HG indicates that the baseline pose encoder (hourglass structure) is employed to

Table 2. Comparison of the computation cost with state-of-the arts on STB. Our method achieves higher AUC yet consumes significantly less computation.

Method	3D PCK20	AUC	GFLOPs
Z& B [46]	0.870	0.948	78.2
Liu et al. [22]	0.895	0.964	16.0
HAMR [45]	0.982	0.995	8.0
Cai et al. [3]	0.973	0.995	6.2
Ours	0.979	0.996	1.6

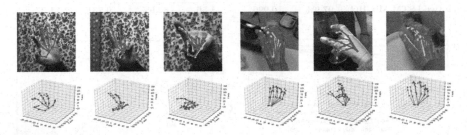

Fig. 4. Visualization of pose estimation. The top row shows input frames and the bottom row visualizes the predicted poses (red) and ground-truth poses (green). (Color figure online)

predict 3D joint coordinates frame by frame. For the Vanilla-LSTM variants, we take features from either coarse pose encoder, fine pose encoder, or average features from coarse and fine pose encoders, and then feed them into an ordinary LSTM model without gate module. The detailed results are in Table. 1.

As shown in Table 1, our adaptive model obtains comparable performance to our designed baseline model "Vanilla-LSTM-Fine-HG" that constantly takes the fine features for pose estimation with less than 1/4 computation cost by computing the fine features only on selected frames. Besides, our proposed method obtains state-of-the-art performance on both benchmarks, which is presented in Fig. 3a and 3b, where we plot the area under the curve (AUC) on the percentage of the correct key points (PCK) with various thresholds.

In addition to the comparison in terms of the accuracy, we further evaluate the speed of our model compared to the existing art. The detailed comparison are illustrated in Table 2. As FPHA dataset is relatively new and fewer works report their performance, we mainly conduct the evaluation on the STB dataset.

Visualization. To verify our model works well in terms of accurately deriving poses from the RGB images. We visualize a few predictions by our network in Fig. 4. Our model is capable of inferring precise poses from RGB input images even under severe occlusion and challenging lightning conditions.

We further look into the mechanism of the Gaussian kernel based gate. We visualize a few test sequences as in Fig. 5. In (a), the fine pose encoder activates

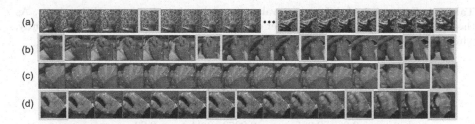

Fig. 5. Visualization of frames selected (marked with yellow boxes) to adopt the fine pose encoder. The fine encoder activates sparsely when the pose is straight-forward while is frequently used when the pose becomes challenging (left part v.s. right part of (a)). When the hand pose status becomes less stable (see rightmost part of (c) and (d)) or occlusions become more severe (see rightmost part of (b)), our model tends to use the fine encoder more frequently. The frequency of invoking fine pose encoder is much lower when the poses are relatively stable. (Color figure online)

Table 3. Evaluation of using different gate architectures on STB dataset. P_{fine} denotes the frequency of using the fine pose encoder. Our Gaussian kernel gate achieves highest accuracy yet at lowest computation cost.

Gate	γ_g	3D PCK20	AUC	GFLOPs	P_{fine}
Neural Gate	0.1	0.981	0.995	2.54	0.32
Neural Temporal Gate	0.1	0.977	0.996	2.20	0.43
Gaussian Kernel Gate	0.1	0.983	0.997	2.09	0.26

less often for the straightforward poses while more densely used for the challenging poses close to the end of the sequence. For (b), the gate tends to invoke fine pose encoder more often when occlusion presents (1st half v.s. 2nd half), while in (c) and (d), when large motion presents (see the rightmost blurry frames from both sequences), the gate chooses to examine the frame more closely with the fine pose encoder. Those observations are in par with our motivations that to only invoke the computationally heavy pose encoders when necessary.

4.4 Ablation Study

We first study on the design choice of the Gaussian kernel based adaptive gate. Instead of explicitly parameterize the difference with Gaussian function, one straight forward way would be to directly predict the probability via a linear module. The linear module takes the hidden state, 1st and 2nd order statistics and coarse feature as the input and yields the probability of introducing the fine module. This model is referred as **Neural Gate**. Going one step further, although the coarse pose encoder is relatively light, we could still obtain performance gains by avoiding it and derive probability solely based on the temporal context. Therefore, we also evaluate the model that make decisions based on

Table 4. Evaluation of different γ_g values for network training on STB dataset. As the expected usage of fine pose encoder drops, the computation cost falls significantly, while the accuracy decreases marginally

γ_g	3D PCK20	AUC	GFLOPs	P_{fine}
1	0.987	0.9978	6.96	1
0.3	0.984	0.9972	3.30	0.43
0.2	0.985	0.9973	2.34	0.29
0.1	0.983	0.9970	2.09	0.26
0.05	0.979	0.9962	1.56	0.18
0.01	0.977	0.9956	1.37	0.15
0.001	0.955	0.9897	1.43	0.16

Table 5. Evaluation of different λ ($\gamma_g = 0.1$) during testing on the STB dataset. For the same trained model, with higher λ, fine encoder is used less often, i.e., we can configure λ to balance the trade-off between the efficiency and accuracy.

λ	3D PCK20	AUC	GFLOPs	P_{fine}
0.1	0.987	0.9977	7.01	0.97
0.3	0.986	0.9976	3.31	0.43
0.5	0.983	0.9970	2.09	0.26
0.7	0.943	0.9894	0.88	0.08
0.9	0.505	0.8277	0.30	0

the temporal context only, which is referred as **Neural Temporal Gate**. The detailed results are in Table 3.

As shown in Table 3, different gates offer similar performance while the Gaussian Kernel is slightly more accurate and more efficient. We further investigate the impact of a few hyper parameters on the overall performance. Specifically, we look into the γ_g in Table 4 and λ in Table 5, which could be tweaked to adjust the rate of computing fine features before and after training.

When varying γ_g from 0.3 to 0.01, the accuracy of the models does not vary much while the frequency of using fine features drops from 0.43 to 0.15, which suggests the large amount redundancy in consecutive frames are exploited by the ACE model. While for λ, with a larger threshold, we greatly reduce the frequency of using fine encoders at the cost of accuracy. λ could be adjusted during inference to balance the trade off between efficiency and accuracy.

5   Conclusion

We present the ACE framework, an adaptive dynamic model for efficient hand pose estimation from monocular videos. At the core of the ACE model is the Gaussian kernel based gate, which determines whether to carefully examine the current frame using a computationally heavy pose encoder based on a quick glimpse of the current frame with a light pose encoder and the temporal context. We further introduce the Gumbel-SoftMax trick to enable the learning of the discrete decision gate. As a result, we obtain state of the art performance on 2 widely used datasets, STB and FPHA, while with less than 1/4 of the computation compared to the baseline models. The proposed ACE model is general and could be built upon any single frame pose encoder, which indicates the efficiency could be further improved by harvesting more efficient structures as single frame pose encoder.

Acknowledgements. This work is partially supported by the National Institutes of Health under Grant R01CA214085 as well as SUTD Projects PIE-SGP-Al-2020-02 and SRG-ISTD-2020-153.

References

1. Boukhayma, A., de Bem, R., Torr, P.H.: 3D hand shape and pose from images in the wild. In: Proceedings of the IEEE Conference on Computer Vision and Pattern Recognition, pp. 10843–10852 (2019)
2. Cai, Y., Ge, L., Cai, J., Yuan, J.: Weakly-supervised 3D hand pose estimation from monocular RGB images. In: Ferrari, V., Hebert, M., Sminchisescu, C., Weiss, Y. (eds.) ECCV 2018. LNCS, vol. 11210, pp. 678–694. Springer, Cham (2018). https://doi.org/10.1007/978-3-030-01231-1_41
3. Cai, Y., et al.: Exploiting spatial-temporal relationships for 3d pose estimation via graph convolutional networks. In: Proceedings of the IEEE International Conference on Computer Vision, pp. 2272–2281 (2019)
4. Courbariaux, M., Bengio, Y., David, J.P.: Binaryconnect: training deep neural networks with binary weights during propagations. In: Advances in Neural Information Processing Systems, pp. 3123–3131 (2015)
5. Courbariaux, M., Hubara, I., Soudry, D., El-Yaniv, R., Bengio, Y.: Binarized neural networks: Training deep neural networks with weights and activations constrained to+ 1 or −1. arXiv preprint arXiv:1602.02830 (2016)
6. Garcia-Hernando, G., Yuan, S., Baek, S., Kim, T.K.: First-person hand action benchmark with RGB-D videos and 3D hand pose annotations. In: Proceedings of the IEEE Conference on Computer Vision and Pattern Recognition, pp. 409–419 (2018)
7. Ge, L., Liang, H., Yuan, J., Thalmann, D.: Robust 3D hand pose estimation in single depth images: from single-view CNN to multi-view CNNs. In: Proceedings of the IEEE Conference on Computer Vision and Pattern Recognition, pp. 3593–3601 (2016)
8. Ge, L., Liang, H., Yuan, J., Thalmann, D.: 3D convolutional neural networks for efficient and robust hand pose estimation from single depth images. In: Proceedings of the IEEE Conference on Computer Vision and Pattern Recognition, pp. 1991–2000 (2017)

9. Ge, L., Liang, H., Yuan, J., Thalmann, D.: Real-time 3D hand pose estimation with 3D convolutional neural networks. IEEE Trans. Pattern Anal. Mach. Intell. **41**(4), 956–970 (2018)
10. Ge, L., et al.: 3D hand shape and pose estimation from a single RGB image. In: Proceedings of the IEEE Conference on Computer Vision and Pattern Recognition, pp. 10833–10842 (2019)
11. Gouidis, F., Panteleris, P., Oikonomidis, I., Argyros, A.: Accurate hand keypoint localization on mobile devices. In: 2019 16th International Conference on Machine Vision Applications (MVA), pp. 1–6. IEEE (2019)
12. Han, S., Pool, J., Tran, J., Dally, W.: Learning both weights and connections for efficient neural network. In: Advances in Neural Information Processing Systems, pp. 1135–1143 (2015)
13. Hassibi, B., Stork, D.G.: Second order derivatives for network pruning: optimal brain surgeon. In: Advances in Neural Information Processing Systems, pp. 164–171 (1993)
14. Hochreiter, S., Schmidhuber, J.: Long short-term memory. Neural Comput. **9**(8), 1735–1780 (1997)
15. Howard, A.G., et al.: MobileNets: efficient convolutional neural networks for mobile vision applications. arXiv preprint arXiv:1704.04861 (2017)
16. Iqbal, U., Molchanov, P., Breuel, T., Gall, J., Kautz, J.: Hand pose estimation via latent 2.5D heatmap regression. In: Ferrari, V., Hebert, M., Sminchisescu, C., Weiss, Y. (eds.) ECCV 2018. LNCS, vol. 11215, pp. 125–143. Springer, Cham (2018). https://doi.org/10.1007/978-3-030-01252-6_8
17. Jang, E., Gu, S., Poole, B.: Categorical reparameterization with gumbel-softmax. arXiv preprint arXiv:1611.01144 (2016)
18. Korbar, B., Tran, D., Torresani, L.: SCSampler: sampling salient clips from video for efficient action recognition. In: Proceedings of the IEEE International Conference on Computer Vision, pp. 6232–6242 (2019)
19. LeCun, Y., Denker, J.S., Solla, S.A.: Optimal brain damage. In: Advances in Neural Information Processing Systems, pp. 598–605 (1990)
20. Li, Z., Ni, B., Zhang, W., Yang, X., Gao, W.: Performance guaranteed network acceleration via high-order residual quantization. In: Proceedings of the IEEE International Conference on Computer Vision, pp. 2584–2592 (2017)
21. Lin, M., Lin, L., Liang, X., Wang, K., Cheng, H.: Recurrent 3D pose sequence machines. In: Proceedings of the IEEE Conference on Computer Vision and Pattern Recognition, pp. 810–819 (2017)
22. Liu, J., et al.: Feature boosting network for 3D pose estimation. IEEE Trans. Pattern Anal. Mach. Intell. **42**(2), 494–501 (2020)
23. Liu, Z., Li, J., Shen, Z., Huang, G., Yan, S., Zhang, C.: Learning efficient convolutional networks through network slimming. In: Proceedings of the IEEE International Conference on Computer Vision, pp. 2736–2744 (2017)
24. Louizos, C., Welling, M., Kingma, D.P.: Learning sparse neural networks through l_0 regularization. arXiv preprint arXiv:1712.01312 (2017)
25. Malik, J., Elhayek, A., Nunnari, F., Varanasi, K., Tamaddon, K., Heloir, A., Stricker, D.: DeepHPS: end-to-end estimation of 3d hand pose and shape by learning from synthetic depth. In: 2018 International Conference on 3D Vision (3DV), pp. 110–119. IEEE (2018)
26. Mueller, F., et al.: Ganerated hands for real-time 3d hand tracking from monocular RGB. In: Proceedings of the IEEE Conference on Computer Vision and Pattern Recognition, pp. 49–59 (2018)

27. Newell, A., Yang, K., Deng, J.: Stacked hourglass networks for human pose estimation. In: Leibe, B., Matas, J., Sebe, N., Welling, M. (eds.) ECCV 2016. LNCS, vol. 9912, pp. 483–499. Springer, Cham (2016). https://doi.org/10.1007/978-3-319-46484-8_29

28. Oculus: Hand tracking SDK for oculus quest available with v12 release. https://developer.oculus.com/blog/hand-tracking-sdk-for-oculus-quest-available

29. Pan, B., Lin, W., Fang, X., Huang, C., Zhou, B., Lu, C.: Recurrent residual module for fast inference in videos. In: Proceedings of the IEEE Conference on Computer Vision and Pattern Recognition, pp. 1536–1545 (2018)

30. Pavllo, D., Feichtenhofer, C., Grangier, D., Auli, M.: 3D human pose estimation in video with temporal convolutions and semi-supervised training. In: Proceedings of the IEEE Conference on Computer Vision and Pattern Recognition, pp. 7753–7762 (2019)

31. Rastegari, M., Ordonez, V., Redmon, J., Farhadi, A.: XNOR-Net: ImageNet classification using binary convolutional neural networks. In: Leibe, B., Matas, J., Sebe, N., Welling, M. (eds.) ECCV 2016. LNCS, vol. 9908, pp. 525–542. Springer, Cham (2016). https://doi.org/10.1007/978-3-319-46493-0_32

32. Hossain, M.R.I., Little, J.J.: Exploiting temporal information for 3D human pose estimation. In: Ferrari, V., Hebert, M., Sminchisescu, C., Weiss, Y. (eds.) ECCV 2018. LNCS, vol. 11214, pp. 69–86. Springer, Cham (2018). https://doi.org/10.1007/978-3-030-01249-6_5

33. Romero, J., Tzionas, D., Black, M.J.: Embodied hands: modeling and capturing hands and bodies together. ACM Trans. Graph. (ToG) 36(6), 245 (2017)

34. Sinha, A., Choi, C., Ramani, K.: Deephand: robust hand pose estimation by completing a matrix imputed with deep features. In: Proceedings of the IEEE Conference on Computer Vision and Pattern Recognition, pp. 4150–4158 (2016)

35. Tekin, B., Rozantsev, A., Lepetit, V., Fua, P.: Direct prediction of 3D body poses from motion compensated sequences. In: Proceedings of the IEEE Conference on Computer Vision and Pattern Recognition, pp. 991–1000 (2016)

36. Wan, C., Probst, T., Gool, L.V., Yao, A.: Self-supervised 3D hand pose estimation through training by fitting. In: Proceedings of the IEEE Conference on Computer Vision and Pattern Recognition, pp. 10853–10862 (2019)

37. Wan, C., Probst, T., Van Gool, L., Yao, A.: Dense 3D regression for hand pose estimation. In: Proceedings of the IEEE Conference on Computer Vision and Pattern Recognition, pp. 5147–5156 (2018)

38. Wang, F., Wang, G., Huang, Y., Chu, H.: Sast: learning semantic action-aware spatial-temporal features for efficient action recognition. IEEE Access 7, 164876–164886 (2019)

39. Wen, W., Wu, C., Wang, Y., Chen, Y., Li, H.: Learning structured sparsity in deep neural networks. In: Advances in Neural Information Processing Systems, pp. 2074–2082 (2016)

40. Wu, Z., Xiong, C., Jiang, Y.G., Davis, L.S.: LiteEval: a coarse-to-fine framework for resource efficient video recognition. In: Advances in Neural Information Processing Systems, pp. 7778–7787 (2019)

41. Wu, Z., Xiong, C., Ma, C.Y., Socher, R., Davis, L.S.: AdaFrame: adaptive frame selection for fast video recognition. In: Proceedings of the IEEE Conference on Computer Vision and Pattern Recognition, pp. 1278–1287 (2019)

42. Xiang, D., Joo, H., Sheikh, Y.: Monocular total capture: Posing face, body, and hands in the wild. In: Proceedings of the IEEE Conference on Computer Vision and Pattern Recognition, pp. 10965–10974 (2019)

43. Zhang, J., Jiao, J., Chen, M., Qu, L., Xu, X., Yang, Q.: 3D hand pose tracking and estimation using stereo matching. arXiv preprint arXiv:1610.07214 (2016)
44. Zhang, X., Zhou, X., Lin, M., Sun, J.: Shufflenet: an extremely efficient convolutional neural network for mobile devices. In: Proceedings of the IEEE Conference on Computer Vision and Pattern Recognition, pp. 6848–6856 (2018)
45. Zhang, X., Li, Q., Mo, H., Zhang, W., Zheng, W.: End-to-end hand mesh recovery from a monocular RGB image. In: Proceedings of the IEEE International Conference on Computer Vision, pp. 2354–2364 (2019)
46. Zimmermann, C., Brox, T.: Learning to estimate 3D hand pose from single RGB images. In: Proceedings of the IEEE International Conference on Computer Vision, pp. 4903–4911 (2017)

Chained-Tracker: Chaining Paired Attentive Regression Results for End-to-End Joint Multiple-Object Detection and Tracking

Jinlong Peng[1], Changan Wang[1], Fangbin Wan[2], Yang Wu[3(✉)], Yabiao Wang[1],
Ying Tai[1], Chengjie Wang[1], Jilin Li[1], Feiyue Huang[1], and Yanwei Fu[2]

[1] Tencent Youtu Lab, Shanghai, China
{jeromepeng,changanwang,caseywang,yingtai,jasoncjwang,
jerolinli,garyhuang}@tencent.com
[2] Fudan University, Shanghai, China
{fbwan18,yanweifu}@fudan.edu.cn
[3] Nara Institute of Science and Technology, Ikoma, Japan
wuyang0321@gmail.com, yangwu@rsc.naist.jp

Abstract. Existing Multiple-Object Tracking (MOT) methods either follow the tracking-by-detection paradigm to conduct object detection, feature extraction and data association separately, or have two of the three subtasks integrated to form a partially end-to-end solution. Going beyond these sub-optimal frameworks, we propose a simple online model named Chained-Tracker (CTracker), which naturally integrates all the three subtasks into an end-to-end solution (the first as far as we know). It chains paired bounding boxes regression results estimated from overlapping nodes, of which each node covers two adjacent frames. The paired regression is made attentive by object-attention (brought by a detection module) and identity-attention (ensured by an ID verification module). The two major novelties: chained structure and paired attentive regression, make CTracker simple, fast and effective, setting new MOTA records on MOT16 and MOT17 challenge datasets (67.6 and 66.6, respectively), without relying on any extra training data. The source code of CTracker can be found at: github.com/pjl1995/CTracker.

Keyword: Multiple-object Tracking, Chained-Tracker, End-to-end solution, Joint detection and tracking

J. Peng and C. Wang—Equal contribution.

Electronic supplementary material The online version of this chapter (https:// doi.org/10.1007/978-3-030-58548-8_9) contains supplementary material, which is available to authorized users.

© Springer Nature Switzerland AG 2020
A. Vedaldi et al. (Eds.): ECCV 2020, LNCS 12349, pp. 145–161, 2020.
https://doi.org/10.1007/978-3-030-58548-8_9

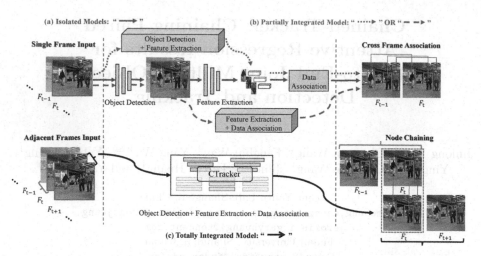

Fig. 1. Comparison of our CTracker (Bottom) with other typical MOT methods (Top), which are either isolated models or partially integrated models. Our CTracker significantly differs from other methods in two aspects: 1) It is a totally end-to-end model using adjacent frame pair as input and generating the box pair representing the same target. 2) We convert the challenging cross-frame association problem into pair-wise object detection problem.

1   Introduction

Video-based scene understanding and human behavior analysis are important high-level tasks in computer vision with many valuable real applications. They rely on many other tasks, within which *Multiple-Object Tracking (MOT)* is a significant one. However, MOT remains challenging due to occlusions, object trajectory overlap, challenging background, *etc.*, especially for crowded scenes.

Despite the great efforts and encouraging progress in the past years, there are two major problems of existing MOT solutions. One is that most methods are based on the tracking-by-detection paradigm [1], which is plausible but sub-optimal due to the infeasibility of global (end-to-end) optimization. It usually contains three sequential subtasks: object detection, feature extraction and data association. However, splitting the whole task into isolated subtasks may lead to local optima and more computation cost than end-to-end solutions. Moreover, data association heavily relies on the quality of object detection, which by itself is hard to generate reliable and stable results across frames as it discards the temporal relationships of adjacent frames.

The other problem is that recent MOT methods get more and more complex as they try to gain better performances. *Re-identification* and *attention* are two major points found to be helpful for improving the performance of MOT. Re-identification (or ID verification) is used to extract more robust features for data association. Attention helps the model to be more focused, avoiding the distraction by irrelevant yet confusing information (e.g. the complex background).

Despite their effectiveness, the involvement of them in existing solutions greatly increases the model complexity and computational cost.

In order to solve the above problems, we propose a novel online tracking method named *Chained-Tracker* (CTracker), which unifies object detection, feature extraction and data association into a single end-to-end model. As can be seen in Fig. 1, our novel CTracker model is cleaner and simpler than the classical tracking-by-detection or partially end-to-end MOT methods. It takes adjacent frame pairs as input to perform joint detection and tracking in a single regression model that simultaneously regress the paired bounding boxes for the targets that appear in both of the two adjacent frames.

Furthermore, we introduce a joint attention module using predicted confidence maps to further improve the performance of our CTracker. It guides the paired boxes regression branch to focus on informative spatial regions with two other branches. One is the object classification branch, which predicts the confidence scores for the first box in the detected box pairs, and such scores are used to guide the regression branch to focus on the foreground regions. The other one is the ID verification branch whose prediction facilitates the regression branch to focus on regions corresponding to the same target. Finally, the bounding box pairs are filtered according to the classification confidence. Then, the generated box pairs belonging to the adjacent frame pairs could be associated using simple methods like IoU (Intersection over Union) matching [2] according to their boxes in the common frame. In this way, the tracking process could be achieved by chaining all the adjacent frame pairs (*i.e.* chain nodes) sequentially.

Benefiting from the end-to-end optimization of joint detection and tracking network, our model shows significant superiority over strong competitors while remaining simple. With the temporal information of the combined features from adjacent frames, the detector becomes more robust, which in turn makes data association easier, and finally results in better tracking performance.

The contribution of this paper can be summarized into the following aspects:

1. We propose an end-to-end online Multiple-Object Tracking model, to optimize object detection, feature extraction and data association simultaneously. Our proposed CTracker is the first method that converts the challenging data association problem to a pair-wise object detection problem.
2. We design a joint attention module to highlight informative regions for box pair regression and the performance of our CTracker is further improved.
3. Our online CTracker achieves state-of-the-art performance on the tracking result list with private detection of MOT16 and MOT17.

2   Related Work

2.1   Detection-Based MOT Methods

Yu *et al.* [3] proposed the POI algorithm, which conducted a high-performance detector based on Faster R-CNN [4] by adding several extra pedestrian detection datasets. Chen *et al.* [5] incorporated an enhanced detection model by simultaneously modeling the detection-scene relation and detection-detection relation, called EDMT. Furthermore, Henschel *et al.* [6] added a head detection model to support MOT in addition to original pedestrian detection, which also needed extra training data and annotations. Bergmann *et al.* [7] proposed the Tracktor by exploiting the bounding box regression to predict the position of the pedestrian in the next frame, which was equal to modifying the detection box. However, the detection model and the tracking model in these detection-based methods are completely **independent**, which is complex and time-consuming. While our CTracker algorithm only needs one **integrated** model to perform detection and tracking, which is simple and efficient.

2.2 Partially End-to-End MOT Methods

Lu *et al.* [8] proposed RetinaTrack, which combined detection and feature extraction in the network and used greedy bipartite matching for data association. Sun *et al.* [9] harnessed the power of deep learning for data association in tracking by jointly modeling object appearances and their affinities between different frames. Similarly, Chu *et al.* [10] designed the FAMNet to jointly optimize the feature extraction, affinity estimation and multi-dimensional assignment. Li *et al.* [11] proposed TrackNet by using frame tubes as input to do joint detection and tracking, however the links among tubes are not modeled which limits the trajectory lengths. Moreover, the model is designed and tested only for rigid object (vehicle) tracking, leaving its generalization ability questionable. Despite their differences, all these methods are just **partially** end-to-end MOT methods, because they just integrated some parts of the whole model, *i.e.* [8] combined the detection and feature extraction module in a network, [9,10] combined the feature extraction and data association module. Differently, our CTracker is a **totally** end-to-end joint detection and tracking methods, unifying the object detection, feature extraction and data association in a single model.

2.3 Attention-Assistant MOT Methods

Chu *et al.* [12] introduced a Spatial-Temporal Attention Mechanism (STAM) to handle the tracking drift caused by the occlusion and interaction among targets. Similarly, Zhu *et al.* [13] proposed a Dual Matching Attention Networks (DMAN) with both spatial and temporal attention mechanisms to perform the tracklet data association. Gao *et al.* [14] also utilized an attention-based appearance model to solve the inter-object occlusion. All these attention-assistant MOT methods used a complex attention model to optimize data association in the **local** bounding box level. While our CTracker can improve both the detection and tracking performance through the simple object-attention and identity-attention in the **global** image level, which is more efficient.

3   Methodology

3.1   Problem Settings

Given an image sequence $\{F_t\}_{t=1}^{N}$ with totally N frames, Multiple-Object Tracking task aims to output all the bounding boxes $\{\mathcal{G}_t\}_{t=1}^{N}$ and identity labels $\{\mathcal{Y}_t^{GT}\}_{t=1}^{N}$ for all the objects of interest in all the frames where they appear. $F_t \in \mathbb{R}^{c \times w \times h}$ indicates the t-th frame, $\mathcal{G}_t \subset \mathbb{R}^4$ represents the ground-truth bounding boxes of the K_t number of targets in t-th frame and $\mathcal{Y}_t^{GT} \subset \mathbb{Z}$ denotes their identities. Most of the recent MOT algorithms divide the MOT task into three components, which are object detection, feature extraction and data association. However, many researches and experiments demonstrate that the association's effectiveness relies heavily on the performance of detection. Therefore, in order to better utilize their correlation, in this paper, we propose a novel Chained-Tracker (abbr. CTracker), which uses a single network to simultaneously achieve object detection, feature extraction and data association. We introduce the pipeline of our CTracker in the Subsect. 3.2. The details of the network and loss design are described separately in the Subsects. 3.3 and 3.4.

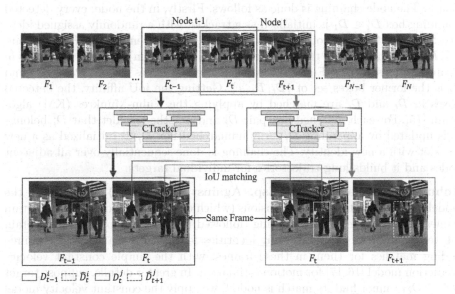

Fig. 2. Illustration of the node chaining. After generating bounding box pairs $\{\mathcal{D}_{t-1}, \hat{\mathcal{D}}_t\}$ by CTracker for two arbitrary adjacent nodes (F_{t-1}, F_t) and (F_t, F_{t+1}), we chain these two nodes by doing IoU matching on the shared common frame. Such a chaining is done sequentially over all adjacent nodes to generate long trajectories for the whole video sequence. More detailed can be found in the main text.

3.2   Chained-Tracker Pipeline

Framework. Different from other MOT models that only takes *a single frame* as input, our CTracker model requires *two adjacent frames* as input, which is called a chain node. The first chain node is (F_1, F_2) and the last (*i.e.,* the N-th) is (F_N, F_{N+1}). Note that F_N is the last frame, so we just take the copy version of F_N as F_{N+1}. Given the node (F_{t-1}, F_t) as input, CTracker can generate bounding box pairs $\{(D^i_{t-1}, \hat{D}^i_t)\}^{n_{t-1}}_{i=1}$ of the same targets appearing in both frames, where n_{t-1} is the total pair number, $D^i_{t-1} \in \mathcal{D}_{t-1} \subset \mathbb{R}^4$ and $\hat{D}^i_t \in \mathcal{D}_t \subset \mathbb{R}^4$ denote the two bounding boxes of the same target. Similarly, we can also get the box pairs $\{(D^j_t, \hat{D}^j_{t+1})\}^{n_t}_{j=1}$ in the next node (F_t, F_{t+1}). As can be seen in Fig. 2, assume that $\hat{D}^i_t$ and D^j_t represent detected boxes of the same target located in the common frame of the adjacent nodes, there shall be only slight difference between the two boxes. We can further use an extremely simple matching strategy (as detailed below) to chain the two boxes, instead of using complicated appearance features as in canonical MOT methods. By chaining nodes sequentially over the given sequence, we can obtain long trajectories of all the detected targets.

Node Chaining. We use $\{\mathcal{D}_{t-1}, \hat{\mathcal{D}}_t\}$ to represent $\{(D^i_{t-1}, \hat{D}^i_t)\}^{n_{t-1}}_{i=1}$ for convenience. The node chaining is done as follows. Firstly, in the node, every detected bounding box $D^i_1 \in \mathcal{D}_1$ is initialized as a tracklet with a randomly assigned identity. Secondly, for any another node t, we chain the adjacent nodes (F_{t-1}, F_t) and (F_t, F_{t+1}) by calculating the IoU (Intersection over Union) between the boxes in $\hat{\mathcal{D}}_t$ and $\mathcal{D}_t$ as shown in Fig. 2, where $\hat{\mathcal{D}}_t$ is the last boxes set of $\{\mathcal{D}_{t-1}, \hat{\mathcal{D}}_t\}$ and $\mathcal{D}_t$ is the former boxes set of $\{\mathcal{D}_t, \hat{\mathcal{D}}_{t+1}\}$. Getting the IoU affinity, the detected boxes in $\hat{\mathcal{D}}_t$ and $\mathcal{D}_t$ are matched by applying the Kuhn-Munkres (KM) algorithm [15]. For each matched box pair $\hat{D}^i_t$ and D^j_t, the tracklet that $\hat{D}^i_t$ belongs to is updated by appending D^j_t. Any unmatched box D^k_t is initialized as a new tracklet with a new identity. The chaining is done sequentially over all adjacent nodes and it builds long trajectories for individual targets.

Robustness Enhancement (Esp. Against Occlusions). To enhance the model's robustness to serious occlusions (which can make detection fail in certain frames) and short-term disappearing (followed by quick reappearing), we retain the terminated tracklets and their identities for up to σ frames and continue finding matches for them in these frames, with the simple constant velocity prediction model [16,17] for motion estimation. In greater details, suppose target $(D^l_{t-1}, \hat{D}^l_t)$ cannot find its match is node t, we apply the constant velocity model to predict its bounding box $P^l_{t+\tau}$ in frame $t + \tau$ ($1 <= \tau <= \sigma$) according to D^l_{t-1} (not the less reliable $\hat{D}^l_t$). When we chain node $t + \tau - 1$ and node $t + \tau$ with $\{\mathcal{D}_{t+\tau-1}, \hat{\mathcal{D}}_{t+\tau}\}$ and $\{\mathcal{D}_{t+\tau}, \hat{\mathcal{D}}_{t+\tau+1}\}$, the current set of all the predicted bounding boxes of retained targets denoted by $\mathcal{P}_{t+\tau}$, is appended to $\hat{\mathcal{D}}_{t+\tau}$ for matching with $\mathcal{D}_{t+\tau}$. If $P^i_{t+\tau}$ gets a match, its tracklet will be extended by linking to the new bounding boxes.

Effectiveness and Limitations. Our model is effective for handling the cases when targets appear or disappear (*i.e.,* enter or leave camera view), which are quite common for MOT. When a target is not in frame $t-1$ but appears in frame t, it is likely that no bounding box pair for it gets generated in the chain node (F_{t-1}, F_t). However, as long as this target continues to appear in frame $t+1$, it will be detected in the next chain node (F_t, F_{t+1}) and get a new tracklet and identity there. Similarly, if a target is in the frame $t-1$ but disappears from frame t, it will not be detected in node (F_t, F_{t+1}), resulting the termination of its tracklet in node $t-1$ or even $t-2$. Note that the chaining operation itself cannot be fully parameterized and therefore it cannot be optimized together with the regressions. Since the regression model (as detailed below) does the major work and there is no need to get feedback for it from the chaining operation, we still use the "end-to-end" property to describe CTracker. A pure end-to-end trainable model requires a differentiable replacement to the current IoU matching based chaining strategy.

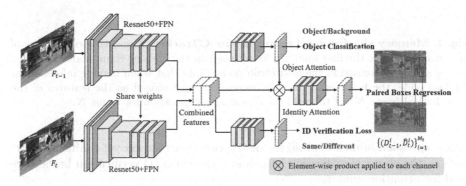

Fig. 3. Network architecture of CTracker. Given two adjacent frames, we firstly use two backbone branches with tied weights to extract the features for each frame separately. Then we concatenate features of the two frames on channel level and the combined features are used to predict the paired boxes. To highlight local informative regions for paired boxes regression, the combined features are multiplied with the attention maps from the object classification branch and the ID verification branch.

3.3 Network Architecture

Overview. Our proposed CTracker network uses two adjacent frames as input and regresses the bounding box pair of the same target. To do this, we adopt ResNet-50 [18] as the backbone to extract high-level semantic features. It then integrates Feature Pyramid Networks (FPN) to generate multi-scale feature representation for subsequent prediction. In order to associate targets in adjacent frames, the scale-level feature maps from individual frames are firstly concatenated together, and then fed into the prediction network to regress bounding box pairs. As can be seen in Fig. 3, the paired boxes regression branch generates a box pair for each target, and the object classification branch predicts a score for each pair indicating the confidence of being foreground. To help the paired

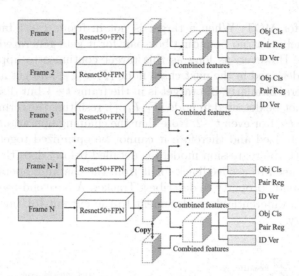

Fig. 4. Memory sharing mechanism in our CTracker. The extracted features of each frame (except the first one) are firstly used in the current chain node, and then can be saved and reused in the next chain node. Note that when making inference for the last node, the features of the last frame N is also reused as the features of the hypothetical frame $N + 1$ to avoid the repeated computation for frame N.

boxes regression branch to avoid the distraction by irrelevant yet confusing information, the object classification branch and the extra ID verification branch are used for attention guidance.

Paired Boxes Regression. Inspired by predicting the offsets relative to pre-defined (default) anchor boxes in object detection, we propose Chained-Anchors for the paired boxes regression branch to regress two boxes simultaneously. As a novel natural derivative of the anchors used in most object detection methods, Chained-Anchors are densely arranged on a spatial grid, each of them allows predicting two bounding boxes of the same object instance in two adjacent frames. In order to handle the large scale variation in real scenes, the K-means clustering as used in [19] is conducted on all ground-truth bounding boxes in the dataset for getting the scales of chained-anchors. And each cluster is assigned to the corresponding level of FPN for later scale specific predictions. The detected bounding box pairs are firstly post-processed with soft-NMS [20] according to the IoU of the first box in each pair, and then filtered based on the confidence scores from the classification branch. Finally, the remaining box pairs are chained into the whole tracking trajectories using the method described in Sect. 3.2. To keep our model simple, both the paired boxes regression branch and the classification branch only stack four consecutive 3×3 Conv layers interleaved with ReLU activations before the final convolution layer.

Joint Attention Module. We design an attention mechanism based component called Joint Attention Module (JAM) to highlight local informative regions in the combined features before the regression branch. As shown from the right of Fig. 3, the ID verification branch is introduced to get confidence scores, indicating whether the two boxes in the detected pair belong to the same target. Then both the predicted confidence map of ID verification branch and object classification branch are used as attention maps. Note that the guidance from the two branches is complementary, the confidence maps from the classification branch focuses on foreground regions while the prediction from the ID verification branch is used to highlight the features of the same target.

Feature Reuse. Since the input of the network contains two adjacent frames, the common frame of two adjacent nodes has to be used twice in the tracking process. To avoid the nearly double cost of computation and memory in inference, we propose a Memory Sharing Mechanism (MSM) to temporarily save the extracted features of the current frame and reuse them until the next node is processed, as shown in Fig. 4. Besides, in order to make inference for the last node, we make a copy of frame N as the hypothetical frame $N + 1$. To further avoid the repeated computation for the frame $N + 1$, we also apply the trick of feature resue to frame N, and the feature of frame N is copied as the feature of the hypothetical frame $N + 1$. We demonstrate that the proposed MSM can reduce almost half of the overall computation and time cost.

3.4   Label Assignment and Loss Design

For an arbitrary chain node (F_t, F_{t+1}), let $A_t^i = (x_a^{t,i}, y_a^{t,i}, w_a^{t,i}, h_a^{t,i})$ denote its i-th chained-anchor (where $x_a^{t,i}$ and $y_a^{t,i}$ are the box center coordinates; $w_a^{t,i}$ and $h_a^{t,i}$ are the width and height, respectively), we adopt a ground-truth bounding box matching strategy similar to that of SSD [21]. We use a matrix M to denote the result of such a matching. If G_t^j is the corresponding ground-truth bounding box in F_t for A_t^i, which is judged by the IoU ratio (higher than a threshold T_p), then we have $M_{ij} = 1$. If the IoU ratio is lower than another smaller threshold T_n, then $M_{ij} = 0$. Based on M, we can assign the ground-truth label c_{cls}^i to CTracker's classification branch for A_t^i as:

$$c_{\text{cls}}^i = \begin{cases} 1, & \text{if } \Sigma_{j=1}^{K_t} M_{ij} = 1, \\ 0, & \text{if } \Sigma_{j=1}^{K_t} M_{ij} = 0, \end{cases} \tag{1}$$

where K_t is the total number of ground-truth bounding boxes for frame F_t.

With A_t^i, suppose the predicted pair of bounding boxes are $(D_t^i, \hat{D}_{t+1}^i)$ and the corresponding ground-truth bounding boxes are (G_t^j, G_{t+1}^k) when they exist, the ID verification branch of CTracker shall get its ground-truth label as:

$$c_{\text{id}}^i = \begin{cases} 1, & \text{if } c_{\text{cls}}^i = 1 \text{ and } \mathcal{I}[G_t^j] = \mathcal{I}[G_{t+1}^k], \\ 0, & \text{otherwise,} \end{cases} \tag{2}$$

where $\mathcal{I}[\cdot]$ represents the identity of the target in the bounding box.

We follow Faster R-CNN [4] to regress offsets of $(D_t^i, \hat{D}_{t+1}^i)$ w.r.t. A_t^i, where $D_t^i = (x_d^{t,i}, y_d^{t,i}, w_d^{t,i}, h_d^{t,i})$. Let $(\Delta_d^{t,i}, \Delta_d^{t+1,i})$ denote these offsets and

$(\Delta_g^{t,j}, \Delta_g^{t+1,k})$ be the offsets for the ground-truths, we list the details of $\Delta_d^{t,i} = (\Delta_{d,x}^{t,i}, \Delta_{d,y}^{t,i}, \Delta_{d,w}^{t,i}, \Delta_{d,h}^{t,i})$ as an example (the others are similar):

$$\Delta_{d,x}^{t,i} = (x_d^{t,i} - x_a^{t,i})/w_a^{t,i}, \quad \Delta_{d,y}^{t,i} = (y_d^{t,i} - y_a^{t,i})/h_a^{t,i},$$
$$\Delta_{d,w}^{t,i} = \log(w_d^{t,i}/w_a^{t,i}), \quad \Delta_{d,h}^{t,i} = \log(h_d^{t,i}/h_a^{t,i}). \tag{3}$$

The loss for the paired boxes regression branch is defined as follows:

$$L_{reg}(\Delta_d^{t,i}, \Delta_{\hat{d}}^{t+1,i}, \Delta_g^{t,j}, \Delta_g^{t+1,k})$$
$$= \sum_{l \in \{x,y,w,h\}} \left[\text{smooth}_{L_1}(\Delta_{d,l}^{t,i} - \Delta_{g,l}^{t,j}) + \text{smooth}_{L_1}(\Delta_{\hat{d},l}^{t+1,i} - \Delta_{g,l}^{t+1,k}) \right]/8, \tag{4}$$

where smooth_{L_1} is the smooth L_1 loss.

The total loss of CTracker is

$$L_{all} = \sum_{t,i} \left[L_{reg}(\Delta_d^{t,i}, \Delta_{\hat{d}}^{t+1,i}, \Delta_g^{t,j}, \Delta_g^{t+1,k}) + \alpha \mathcal{F}(p_{cls}^i, c_{cls}^i) + \beta \mathcal{F}(p_{id}^i, c_{id}^i) \right], \tag{5}$$

where $\mathcal{F}(p_{cls}^i, c_{cls}^i)$ and $\mathcal{F}(p_{id}^i, c_{id}^i)$ are the focal losses [22] for the classification branch and the ID verification branch (for mitigating the sample imbalance problem), respectively, with p_{cls}^i and p_{id}^i denoting their predictions (confidence scores); α and β are the weighting factors.

4   Experiment

4.1   Datasets and Evaluation Metrics

We conduct the experiments on two public datasets: MOT16 [23] and MOT17. Which contain the same image sequences including 7 training sequences and 7 test sequences. However, MOT16 and MOT17 contain different detection input, and different ground-truth labels (bounding boxes and identities), which would influence the training of CTracker. In public detection, MOT16 includes DPM [24] detector while MOT17 includes DPM, Faster R-CNN [4] and SDP [25] detectors. For a fair comparison with other methods, we trained two models separately using the training data from MOT16 and MOT17, and separately applied the two models on the MOT16 test set and MOT17 test set.

In the MOTChallenge benchmark, tracking performance is measured by the widely used CLEAR MOT Metrics [26], including Multiple-Object Tracking Accuracy (MOTA), Multiple-Object Tracking Precision (MOTP), the total number of False Negatives (FN), False Positives (FP), Identity Switches (IDS), and the percentage of Mostly Tracked Trajectories (MT), Mostly Lost Trajectories (ML). ID F1 Score (IDF1) is also used to measure the trajectory identity accuracy. Among these metrics, MOTA is the primary metric to measure the overall detection and tracking performance. In addition, we use Tracker Speed in Frames Per Seconds (Hz) to measure the tracking speed of all methods.

4.2   Implementation Details

All the experiments are done with PyTorch. During training, the ground-truth boxes with a visible score above 0.1 are selected for training. To avoid overfitting, we use several data augmentation strategies such as photometric distortions, random flip and random crop. The same augmentation operation is guaranteed to apply for each image in the same training pair. Then the augmented image pair are resized or padded to the half of their original images' shorter side. We also add a novel data augmentation strategy in the temporal dimension to form chain nodes: instead of always choosing two adjacent frames, we sample two frames close to each other with a random temporal gap (1 to 3 frames).

As a speed-accuracy trade-off, we use the Resnet50 [18] network as the backbone in all the following experiments. All trainable weights except the BN parameters in Resnet50 are trained end-to-end using the Adam optimizer. We initialize the parameters for all the newly added convolutional layers with the Kaiming initialization method in [27] and set the initial learning rate to $5 \times e^{-5}$. The model training process takes 100 epochs with the batch size of 8 (4 training pairs). The weighting factors α and β in the loss function are both set to 1. In the anchor matching stage, we use 0.5 for the positive threshold and 0.4 for the negative threshold. For paired boxes post-processing, we use a threshold of 0.7 for the soft-nms, and then further filter remaining pairs with the confidence threshold of 0.4. In the chaining stage, the IoU matching threshold is 0.5, and the retention threshold of σ is 10.

Table 1. Ablation study on MOT17 test dataset.

Method	MOTA↑	IDF1↑	MOTP↑	MT↑	ML↓	FP↓	FN↓	IDS↓
Baseline	64.4	51.6	78.2	28.5%	28.0%	**16089**	178704	6336
Baseline+ObjAtten	66.0	55.7	**78.8**	31.3%	24.5%	17724	168522	5595
Baseline+ObjAtten+IDVer	65.6	55.2	78.3	**32.6%**	24.7%	25815	162489	5769
Baseline+JointAtten	**66.6**	**57.4**	78.2	32.2%	**24.2%**	22284	**160491**	**5529**

4.3   Ablation Study

Performance Analysis. We compare the following models on MOT17 dataset to show the effectiveness of CTracker's parts:

(1) *Baseline*. It only covers the classification branch and the paired boxes regression branch, without guidance from any attention map. This is the simplest implementation of our CTracker.

(2) *Baseline+ObjAtten*. In addition to the Baseline, the predicted confidence map of the object classification branch is used as an attention map, which is multiplied to the combined features before the paired boxes regression branch.

Table 2. Time cost analysis of CTracker.

Methods	Time cost (ms)			
	Backbone	Prediction	Chaining	Total
CTracker-Det	80.27	38.78	-	119.05
CTracker w/o MSM	154.53	66.93	2.10	223.56
CTracker	80.29	65.71	2.10	148.10

(3) *Baseline+ObjAtten+IDVer*. Except for the object classification branch with attention map and the paired boxes regression branch, we add the ID verification branch but do not use it as attention guidance.

(4) *Baseline+JointAtten (CTracker)*. This is the full version of our approach.

Results presented in Table 1 show that:

(1) *Baseline+ObjAtten* performs significantly better than *Baseline*, which proves the effectiveness of the object attention operation. By applying the object classification branch as the attention map of the paired boxes regression branch, we can get more accurate bounding boxes. There is a significant improvement of MOTA, which increases from 64.4 to 66.0 and MOTP also increases from 78.2 to 78.8. The more accurate bounding boxes also result in better performance of data association, with IDF1 increasing from 51.6 to 55.7.

(2) *Baseline+ObjAtten+IDVer* performs slightly worse than *Baseline+ObjAtten*. Simply adding the independent ID verification branch is weak due to the lack of bounding boxes information. Reliable identification needs good bounding boxes.

(3) *Baseline+JointAtten* further outperforms *Baseline+ObjAtten*, indicating that the ID attention operation is also beneficial. By adding the ID verification branch and using it as another guidance of the paired boxes regression branch, the association of the regressed bounding boxes is more accurate. Though MOTA is only improved by 0.6, the IDF1 is improved by 1.7, and IDF1 can better reflect the accuracy of data association more clearly. On the other hand, by adding the ID attention, the model pays more attention to the data association and sacrifices slightly of the regression bounding box precision, thus the MOTP is decreased from 78.8 to 78.2. Qualitative results of CTracker are illustrated in Fig. 5.

Time Cost Analysis. We analyze the inference speed for each module in CTracker, displayed in Table 2. The time cost is measured for 1080×1920 images using single Tesla P40 and cuDNN v7 with Intel Xeon E5-2699v4@2.20GHz. In Table 2, CTracker-Det only predicts boxes for a single frame, which is the initial detection network of CTracker. Since nearly 70% of the forward time is spent on the backbone network, our original CTracker costs about double-time

MOT17-03 #Frame 100 | MOT17-03 #Frame 150 | MOT17-03 #Frame 200

MOT17-07 #Frame 250 | MOT17-07 #Frame 275 | MOT17-07 #Frame 300

Fig. 5. Qualitative results of our CTracker on MOT17 test dataset. MOT17-03 sequence is captured by a static camera and MOT17-07 sequence is captured by a moving camera. The detected bounding boxes and the tracking trajectory with the same identity are displayed by the same color.

to perform joint detection and tracking compared with the initial detection network, the time increasing from 119.05 ms to 223.56 ms. With the help of the proposed Memory Sharing Mechanism (MSM) in Sect. 3.3, we achieve a faster joint detection and tracking model with only 29.05 ms extra cost compared with the detection network. There is just a small increase of time from 119.05 ms to 148.10 ms. To some extent, 29.05 ms per frame means the tracking module runs at 34.4 FPS, demonstrating the efficiency of our online approach.

4.4   Benchmark Evaluation

We compare our CTracker approach with other MOT methods on both MOT16 and MOT17 test datasets. For comparison, we trained our model separately using the MOT16 training data and MOT17 training data. Tables 3 and 4 compare the tracking results of all the methods separately on MOT16 and MOT17 test dataset. From Tables 3 and 4 we can find that:

(1) In the private detection part of both MOT16 and MOT17, our CTracker significantly outperforms existing online MOT methods in terms of MOTA. In MOT16, the MOTA of our approach is only 0.6 lower than the best offline method KDNT [3], while it is 1.5 higher than its online version POI [3]. In addition, KDNT and POI use many extra training data, including ETHZ pedestrian dataset [38], Caltech pedestrian dataset [39] and their own collected surveillance dataset [3]. While we only use the training data of MOT16. MOTA is the primary metric reflecting the overall detection and tracking performance, which proves the effectiveness of our approach.
(2) In the public detection part, Tracktor [7] performs the best in terms of MOTA. To have a comparison with Tracktor using the same detection result,

Table 3. Comparisons of tracking results on MOT16 test dataset.

Public detection										
Process	Method	MOTA↑	IDF1↑	MOTP↑	MT↑	ML↓	FP↓	FN↓	IDS↓	Hz↑
Offline	MHT-bLSTM [28]	42.1	47.8	75.9	14.9%	44.4%	11637	93172	753	**1.8**
	Quad-CNN [29]	44.1	38.3	76.4	14.6%	44.9%	**6388**	94775	745	**1.8**
	EDMT [5]	45.3	47.9	75.9	17.0%	**39.9%**	11122	87890	639	**1.8**
	LMP [30]	**48.8**	**51.3**	**79.0**	**18.2%**	40.1%	6654	**86245**	481	0.5
Online	CDA-DDAL [31]	43.9	45.1	74.7	10.7%	44.4%	6450	95175	676	-
	STAM [12]	46.0	50.0	74.9	14.6%	43.6%	6895	91117	**473**	-
	DMAN [13]	46.1	**54.8**	73.8	17.4%	42.7%	7909	89874	532	-
	MOTDT [32]	47.6	50.9	74.8	15.2%	38.3%	9253	85431	792	**20.6**
	Tracktor [7]	**54.4**	52.5	**78.2**	**19.0%**	**36.9%**	**3280**	**79149**	682	-
Private detection										
Process	Method	MOTA↑	IDF1↑	MOTP↑	MT↑	ML↓	FP↓	FN↓	IDS↓	Hz↑
Offline	NOMT [33]	62.2	**62.6**	**79.6**	32.5%	31.1%	**5119**	63352	**406**	11.5
	MCMOT-HDM [34]	62.4	51.6	78.3	31.5%	24.2%	9855	57257	1394	**34.9**
	KDNT [3]	**68.2**	60.0	79.4	**41.0%**	**19.0%**	11479	**45605**	933	0.7
Online	EAMTT [35]	52.5	53.3	78.8	19.0%	34.9%	**4407**	81223	910	12.0
	DeepSORT [16]	61.4	62.2	79.1	32.8%	**18.2%**	12852	56668	**781**	20.0
	CNNMTT [36]	65.2	62.2	78.4	32.4%	21.3%	6578	55896	946	11.2
	POI [3]	66.1	**65.1**	**79.5**	**34.0%**	20.8%	5061	55914	805	9.9
	CTracker (Ours)	**67.6**	57.2	78.4	32.9%	23.1%	8934	**48305**	1897	**34.4**

Table 4. Comparisons of tracking results on MOT17 test dataset.

Public detection										
Process	Method	MOTA↑	IDF1↑	MOTP↑	MT↑	ML↓	FP↓	FN↓	IDS↓	Hz↑
Offline	MHT-bLSTM [28]	47.5	51.9	**77.5**	18.2%	41.7%	25981	268042	2069	**1.8**
	EDMT [5]	50.0	51.3	77.3	**21.6%**	36.3%	32279	**247297**	2264	**1.8**
	JCC [37]	51.2	**54.5**	75.9	20.9%	37.0%	25937	247822	**1802**	-
	FWT [6]	**51.3**	47.6	77.0	21.4%	**35.2%**	**24101**	247921	2648	-
Online	DMAN [13]	48.2	**55.7**	75.9	19.3%	38.3%	26218	263608	2194	-
	MOTDT [32]	50.9	52.7	76.6	17.5%	**35.7%**	24069	250768	2474	**20.6**
	Tracktor [7]	**53.5**	52.3	**78.0**	**19.5%**	36.6%	**12201**	**248047**	2072	-
Private Detection										
Process	Method	MOTA↑	IDF1↑	MOTP↑	MT↑	ML↓	FP↓	FN↓	IDS↓	Hz↑
Online	Tracktor+CTdet [7]	54.4	56.1	78.1	25.7%	29.8%	44109	210774	2574	-
	DeepSORT [16]	60.3	**61.2**	**79.1**	31.5%	**20.3%**	36111	185301	**2442**	20.0
	CTracker (Ours)	**66.6**	57.4	78.2	**32.2%**	24.2%	**22284**	**160491**	5529	**34.4**

we reproduce Tracktor using its code. Tracktor+CTdet in Table 4 is the tracking result of Tracktor using the detection result of our CTracker. Compared with the results of public detection, the MOTA of Tracktor+CTdet increases from 53.5 to 54.4 and IDF1 increases from 52.3 to 56.1, which indicates that the performance of our detection is better than the public detection. Besides, our CTracker outperforms Tracktor+CTdet in terms of all the metrics except IDS, which further proves the superior tracking performance of our CTracker.

(3) On the other hand, to keep the simplicity and efficiency of our CTracker, we abandon using the patch-level ReID features of the detected boxes like other

MOT methods to enhance cross-frame data association. Thus, the IDF1 and IDS of our CTracker approach are lower than several methods. We conduct an extra experiment by adding features, introduced in the supplementary. To further prove the efficiency of our approach, we compare the time cost of CTracker with other state-of-the-art MOT methods on the MOT16 and MOT17 benchmark, as shown in the Hz column of Tables 3 and 4. From Tables 3 and 4 we can find that CTracker achieves the best tracking speed among all online MOT methods, although the fastest offline method runs at a similar tracking speed as our CTracker, but has a much lower MOTA than our CTracker, demonstrating the effectiveness and efficiency of our approach.

5   Conclusion

We designed a novel joint multiple-object detection and tracking framework named Chained-Tracker in this paper, which is the first totally end-to-end solution as far as we are aware. Different from existing methods, we use two adjacent frames as the input of our network, which is called a chain node. The network regresses a pair of bounding boxes for the same target in the two adjacent frames, guided by a simple yet novel joint attention module: an interplay of detection-driven object attention and ID verification-injected identity attention. Using the simple IoU information, two adjacent and overlapping nodes can be chained by their boxes in the common frame. The tracking trajectories can be generated by alternately applying the paired boxes regression and node chaining. Extensive experiments on widely used MOT benchmarks demonstrate the superiority of our approach in terms of both effectiveness and efficiency.

Acknowledgement. This work was supported by a MSRA Collaborative Research 2019 Grant.

References

1. Breitenstein, M.D., Reichlin, F., Leibe, B., Koller-Meier, E., Gool, L.V.: Robust tracking-by-detection using a detector confidence particle filter. In: ICCV (2009)
2. Bochinski, E., Eiselein, V., Sikora, T.: High-speed tracking-by-detection without using image information. In: AVSS (2017)
3. Yu, F., Li, W., Li, Q., Liu, Y., Shi, X., Yan, J.: POI: multiple object tracking with high performance detection and appearance feature. In: Hua, G., Jégou, H. (eds.) ECCV 2016. LNCS, vol. 9914, pp. 36–42. Springer, Cham (2016). https://doi.org/10.1007/978-3-319-48881-3_3
4. Ren, S., He, K., Girshick, R., Sun, J.: Faster R-CNN: towards real-time object detection with region proposal networks. In: NIPS (2015)
5. Chen, J., Sheng, H., Zhang, Y., Xiong, Z.: Enhancing detection model for multiple hypothesis tracking. In: CVPRW (2017)
6. Henschel, R., Leal-Taixé, L., Cremers, D., Rosenhahn, B.: Fusion of head and full-body detectors for multi-object tracking. In: CVPRW (2018)

7. Bergmann, P., Meinhardt, T., Leal-Taixe, L.: Tracking without bells and whistles. In: ICCV (2019)
8. Lu, Z., Rathod, V., Votel, R., Huang, J.: Retinatrack: online single stage joint detection and tracking. In: CVPR (2020)
9. Sun, S., Akhtar, N., Song, H., Mian, A.S., Shah, M.: Deep affinity network for multiple object tracking. TPAMI (2019)
10. Chu, P., Ling, H.: FAMNet: joint learning of feature, affinity and multi-dimensional assignment for online multiple object tracking. In: ICCV (2019)
11. Li, C., Dobler, G., Feng, X., Wang, Y.: Tracknet: simultaneous object detection and tracking and its application in traffic video analysis. arXiv preprint arXiv:1902.01466 (2019)
12. Chu, Q., Ouyang, W., Li, H., Wang, X., Liu, B., Yu, N.: Online multi-object tracking using CNN-based single object tracker with spatial-temporal attention mechanism. In: ICCV (2017)
13. Zhu, J., Yang, H., Liu, N., Kim, M., Zhang, W., Yang, M.-H.: Online multi-object tracking with dual matching attention networks. In: Ferrari, V., Hebert, M., Sminchisescu, C., Weiss, Y. (eds.) ECCV 2018. LNCS, vol. 11209, pp. 379–396. Springer, Cham (2018). https://doi.org/10.1007/978-3-030-01228-1_23
14. Gao, X., Jiang, T.: OSMO: online specific models for occlusion in multiple object tracking under surveillance scene. In: ACMMM (2018)
15. Kuhn, H.W.: The hungarian method for the assignment problem. NRL **2**, 83–97 (1955)
16. Wojke, N., Bewley, A., Paulus, D.: Simple online and realtime tracking with a deep association metric. In: ICIP (2017)
17. Peng, J., et al.: TPM: multiple object tracking with tracklet-plane matching. PR (2020)
18. He, K., Zhang, X., Ren, S., Sun, J.: Deep residual learning for image recognition. In: CVPR (2016)
19. Redmon, J., Farhadi, A.: Yolo9000: better, faster, stronger. In: CVPR (2017)
20. Bodla, N., Singh, B., Chellappa, R., Davis, L.S.: Soft-NMS - improving object detection with one line of code. In: ICCV (2017)
21. Liu, W., et al.: SSD: single shot MultiBox detector. In: Leibe, B., Matas, J., Sebe, N., Welling, M. (eds.) ECCV 2016. LNCS, vol. 9905, pp. 21–37. Springer, Cham (2016). https://doi.org/10.1007/978-3-319-46448-0_2
22. Lin, T.Y., Goyal, P., Girshick, R., He, K., Dollár, P.: Focal loss for dense object detection. In: CVPR (2017)
23. Milan, A., Leal-Taixé, L., Reid, I., Roth, S., Schindler, K.: Mot16: a benchmark for multi-object tracking. arXiv preprint arXiv:1603.00831 (2016)
24. Felzenszwalb, P.F., Girshick, R.B., McAllester, D., Ramanan, D.: Object detection with discriminatively trained part-based models. TPAMI (2010)
25. Yang, F., Choi, W., Lin, Y.: Exploit all the layers: fast and accurate CNN object detector with scale dependent pooling and cascaded rejection classifiers. In: CVPR (2016)
26. Bernardin, K., Stiefelhagen, R.: Evaluating multiple object tracking performance: the clear mot metrics. JIVP (2008)
27. He, K., Zhang, X., Ren, S., Sun, J.: Delving deep into rectifiers: surpassing human-level performance on imagenet classification. In: ICCV (2015)
28. Kim, C., Li, F., Rehg, J.M.: Multi-object tracking with neural gating using bilinear LSTM. In: Ferrari, V., Hebert, M., Sminchisescu, C., Weiss, Y. (eds.) ECCV 2018. LNCS, vol. 11212, pp. 208–224. Springer, Cham (2018). https://doi.org/10.1007/978-3-030-01237-3_13

29. Son, J., Baek, M., Cho, M., Han, B.: Multi-object tracking with quadruplet convolutional neural networks. In: CVPR (2017)
30. Tang, S., Andriluka, M., Andres, B., Schiele, B.: Multiple people tracking by lifted multicut and person re-identification. In: CVPR (2017)
31. Bae, S.H., Yoon, K.J.: Confidence-based data association and discriminative deep appearance learning for robust online multi-object tracking. TPAMI (2018)
32. Chen, L., Ai, H., Zhuang, Z., Shang, C.: Real-time multiple people tracking with deeply learned candidate selection and person re-identification. In: ICME (2018)
33. Choi, W.: Near-online multi-target tracking with aggregated local flow descriptor. In: ICCV (2015)
34. Lee, B., Erdenee, E., Jin, S., Nam, M.Y., Jung, Y.G., Rhee, P.K.: Multi-class multi-object tracking using changing point detection. In: Hua, G., Jégou, H. (eds.) ECCV 2016. LNCS, vol. 9914, pp. 68–83. Springer, Cham (2016). https://doi.org/10.1007/978-3-319-48881-3_6
35. Sanchez-Matilla, R., Poiesi, F., Cavallaro, A.: Online multi-target tracking with strong and weak detections. In: Hua, G., Jégou, H. (eds.) ECCV 2016. LNCS, vol. 9914, pp. 84–99. Springer, Cham (2016). https://doi.org/10.1007/978-3-319-48881-3_7
36. Mahmoudi, N., Ahadi, S.M., Rahmati, M.: Multi-target tracking using CNN-based features: CNNMTT. MTAP (2019)
37. Keuper, M., Tang, S., Andres, B., Brox, T., Schiele, B.: Motion segmentation & multiple object tracking by correlation co-clustering. TPAMI (2018)
38. Ess, A., Leibe, B., Schindler, K., Van Gool, L.: A mobile vision system for robust multi-person tracking. In: CVPR (2008)
39. Dollár, P., Wojek, C., Schiele, B., Perona, P.: Pedestrian detection: a benchmark. In: CVPR (2009)

Distribution-Balanced Loss for Multi-label Classification in Long-Tailed Datasets

Tong Wu[1]([✉])[iD], Qingqiu Huang[2][iD], Ziwei Liu[2][iD], Yu Wang[1][iD],
and Dahua Lin[2][iD]

[1] Tsinghua University, Beijing, China
wutong16.thu@gmail.com, yu-wang@mail.tsinghua.edu.cn
[2] The Chinese University of Hong Kong, Hong Kong, China
{hq016,dhlin}@ie.cuhk.edu.hk, zwliu.hust@gmail.com

Abstract. We present a new loss function called Distribution-Balanced Loss for the multi-label recognition problems that exhibit long-tailed class distributions. Compared to conventional single-label classification problem, multi-label recognition problems are often more challenging due to two significant issues, namely the co-occurrence of labels and the dominance of negative labels (when treated as multiple binary classification problems). The Distribution-Balanced Loss tackles these issues through two key modifications to the standard binary cross-entropy loss: 1) a new way to re-balance the weights that takes into account the impact caused by label co-occurrence, and 2) a negative tolerant regularization to mitigate the over-suppression of negative labels. Experiments on both Pascal VOC and COCO show that the models trained with this new loss function achieve significant performance gains over existing methods. Code and models are available at: https://github.com/wutong16/DistributionBalancedLoss.

Keywords: Multi-label classification · Long-tailed data · Distribution-balanced loss

1 Introduction

Along with the wide adoption of deep learning, recent years have seen great progress in visual recognition, especially the remarkable breakthroughs in classification tasks. However, mainstream benchmarks are often constructed under two common conditions: 1) all classes have comparable numbers of instances and 2) each instance belongs to a unique class. While providing a clean setting for various studies, this conventional setting conceals a number of complexities

Electronic supplementary material The online version of this chapter (https://doi.org/10.1007/978-3-030-58548-8_10) contains supplementary material, which is available to authorized users.

A. Vedaldi et al. (Eds.): ECCV 2020, LNCS 12349, pp. 162–178, 2020.
https://doi.org/10.1007/978-3-030-58548-8_10

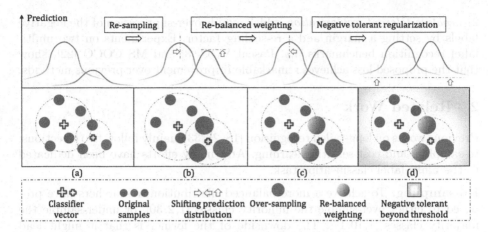

Fig. 1. Our *Distribution-Balanced Loss* performs re-balanced weighting along with re-sampling that takes label co-ocurrence into consideration, and it leverages negative-tolerant regularization to avoid over-suppression of the negative labels caused by the dominance of negative classes in *binary cross entropy (BCE)*

that often arise in real-world applications [14,16,25,26]. In contrast, the distribution of different object categories typically exhibit a long tail in practical contexts while individual images can generally be associated with multiple semantic labels [15,23,24,35]. Previous works [1,12,17] have repeatedly shown that such issues can cause substantial performance drop if not appropriately handled.

A widely adopted approach to multi-label problem is to use binary cross-entropy [7] in the place of the softmax loss, and use class-specific re-weighting to balance the contributions of different classes, e.g. setting the class weights to be inversely proportional to the class sizes. Such simple methods often result in limited improvement, as they fail to take into account the impacts of two important issues, namely *label co-occurrence* and *the dominance of negative labels*.

First, label co-occurrence is very common in natural images. For example, an image that contains unusual concepts, *e.g.* "tigers" and "leopards", is likely to be also associated with more common labels, *e.g.* "trees" and "river". Therefore, re-sampling such images may not necessarily result in a more balanced distribution of classes. Second, each image is usually associated with a very small fraction of all the classes in the list. Consequently, given an image, most classes are negative. However, the *binary cross entropy (BCE)* loss is designed to be symmetric, where positive and negative classes are treated uniformly. This symmetric formulation in conjunction with the dominant portion of negative classes would lead to over-suppression of the negative side, thus introducing significant bias to the classification boundaries. In response to issues above, we propose a new loss function, called *Distribution-Balanced Loss*. This loss function consists of two key modifications to the standard BCE loss: 1) *re-balanced weighting*, which adjust the weights in a way that closes the gap between expected sampling times and actual sampling times, with label co-occurrence taken into account; and

2) *negative-tolerant regularization*, which avoids over-suppression of the negative labels by setting a margin and a re-scaling factor. Experiments on two multi-label recognition benchmarks, *i.e.* Pascal VOC [8] and MS COCO [22], show that the proposed loss achieves remarkable improvement over previous methods.

2   Related Work

Previous works on long-tailed recognition [18,26,33] mainly follow two directions: re-sampling and cost-sensitive learning. And many efforts have been dedicated to the multi-label classification task.

Re-sampling. To achieve a more balanced distribution, researchers have proposed to either over-sample the minority classes [1,2,30], or under-sample the majority classes [1,10,17]. The downside of the former is that it might lead to over-fitting on minority classes with duplicated samples, while the latter might weaken feature learning capacity due to omitting a number of valuable instances. While previous works mainly focus on single label datasets, we extend re-sampling to the multi-label scenario.

Cost-Sensitive Learning. Assigning different costs to different training samples is proved to be an effective strategy dealing with imbalanced data. Typically, researchers apply class-level re-weighting by the proportional inverse of class frequency [13,33], or the square root for smoothing. Recently, Cui *et al* [5] proposed to re-weight by the inverse of effective number of samples, and Cao *et al* [3] emphasized larger margin for rare classes. Further, various works adopted sample-level control of cost based on individual properties, *e.g.* example difficulty [21], estimated Bayesian uncertainty [19], gradient direction [29]. Our method applies re-weighting based on class frequency and individual ground truth labels and modifies the loss gradient with a regularization as well for a better optimization.

Multi-label Classification. Earlier solutions for multi-label recognition include decomposing it into independent binary classification tasks [31], and k-nearest neighbor named ML-kNN [36], etc. Recently, many approaches attempted to take label relationships into consideration to better regularize the embedding space. CNN-RNN [32] utilized the RNNs combined with CNN to learn a joint image-label embedding, and Wang *et al* [34] took advantages of a spatial transformer layer and long short-term memory (LSTM) units to capture contextual dependencies. There's also a popular trend to model label correlation with graph structure [4,20]. Our method is based on the widely used binary cross-entropy loss [7] and gains improvement by combining it with re-sampling and re-weighting.

3   Distribution-Balanced Loss

The problem we want to exploit here is how to train a model effectively when training samples follow a long-tailed distribution. Suppose the dataset we use is

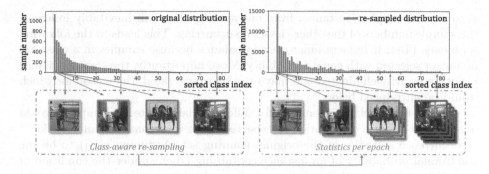

Fig. 2. Visualization of the class-aware re-sampling procedure, and the sample number distribution before (left) and after (right) re-sampling. The distribution may not necessarily be balanced due to label co-occurrence, and the unexpected sampling by associated labels introduces inner-class imbalance

$\mathcal{D} = \{(\mathbf{x}^1, \mathbf{y}^1), \cdots, (\mathbf{x}^N, \mathbf{y}^N)\}$, where N is the number of training samples and $(\mathbf{x}^k, \mathbf{y}^k), k \in \{1, ..., N\}$ is a sample-label pair. Let's denote the number of classes as C, then we have $\mathbf{y}^k = [y_1^k, \cdots, y_C^k] \in \{0,1\}^C$. Let $n_i = \sum_{k=1}^N y_i^k$ denote the number of training examples that contain class i. Please note that $N \leq \sum_{i=0}^C n_i$ since a single example can be counted several times for each class it contains.

As we mentioned before, our distribution-balanced loss consists of two components, namely re-balanced weighting and negative-tolerant regularization. In Sect. 3.1, we would introduce the reason why we need a re-balanced weight in long-tailed multi-label classification and the mathematical derivation of the optimal value of this weight. In Sect. 3.2, we would demonstrate the over-suppression for negative samples brought by *sigmoid* and how to overcome the problem with our negative-tolerant regularization. Finally, these two components can be integrated as a unified loss function, *i.e.* distribution-balanced loss, for end-to-end training, which would be shown in Sect. 3.3.

3.1 Re-balanced Weighting After Re-sampling

The most common sampling rule is to select each example from training set with equal probability, and the probability of a sampled example containing class i would be $p_i = n_i/N$. To alleviate the discrepancy of imbalanced sampling probability among classes caused by data distribution, many re-sampling strategies are proposed. One popular strategy is known as class-aware sampling [30]. It first uniformly samples a class from the whole C classes, and then samples an example from the selected class randomly. This process runs iteratively in each training epoch. Let N_e denote the times for each class to be visited by the iterator in one epoch, which is usually set as $N_e = \max(n_1, \cdots, n_C)$. In cases of extreme imbalance, N_e can be set smaller to control the data scale in one epoch.

However, in the multi-label scenario, an example usually contains several ground-truth labels, making the selection for classes no longer independent. That is to say, re-sampling instances from one specific class will inevitably influence the sample numbers of the other classes co-occurring. This leads to the following problems. First, it induces inner-class imbalance because samples in a class are no longer selected with equal probability. More importantly, the class imbalance is not necessarily eliminated and may even be exaggerated, the reason for which would be introduced below.

In fact, the numbers of samples for different classes after re-sampling would not follow a uniform distribution as expected. Here we estimate them using label co-occurrence statistics of the original training set. Assuming $p(i|j)$ to be the conditional probability of an instance containing label i under the condition of containing label j, so that $p(i|j) = n_{i \cap j}/n_j$, where $n_{i \cap j}$ denotes the number of examples that contain both label i and label j. Therefore, when we randomly choose a class and sample an instance from it, the probability that it contains label i would be shown as Eq. 1.

$$\hat{p}_i = \frac{1}{C} \sum_{j=0}^{C} p(i|j) = \frac{1}{C} \sum_{j=0}^{C} \frac{n_{i \cap j}}{n_j} \tag{1}$$

The class distribution after re-sampling is show in Fig. 2, and the theoretical estimation matches our statistics of data sampled in one epoch during real training procedure. According to the distribution, we proposed a re-balanced weighting strategy to overcome the extra imbalance caused by re-sampling. First, without taking label co-occurrence into consideration, for each instance k and class i with $y_i^k = 1$, the expectation of class-level sampling frequency can be calculated as $P_i^C(x^k)$ in Eq. 2. Then given an instance x^k and its corresponding label y^k, it is supposed to be repeatedly sampled by each positive class i it contains, thus the expectation of instance-level sampling frequency can be estimated as $P^I(x^k)$ in Eq. 2. Correspondingly, we define a re-balancing weight, namely r_i^k, to close the gap between expected sampling times and actual sampling times, as shown in Eq. 3.

$$P_i^C(x^k) = \frac{1}{C}\frac{1}{n_i}, \quad P^I(x^k) = \frac{1}{C} \sum_{y_i^k=1} \frac{1}{n_i} \tag{2}$$

$$r_i^k = \frac{P_i^C(x^k)}{P^I(x^k)} \tag{3}$$

$$\hat{r} = \alpha + \frac{1}{1 + exp(-\beta \times (r - \mu))} \tag{4}$$

However, the weight elements are sometimes towards zero and may increase the difficulty of optimization. To make the optimization process stable, we further designed a smoothing function to map r into a proper range of values, which is demonstrated in Eq. 4. Here α is an overall lift in weight, while β and μ controls the shape of the mapping function, which rapidly increases near 0 and goes flat near 1.

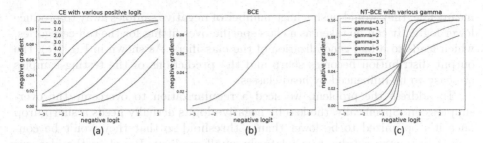

Fig. 3. Visualization of gradient to a negative logit. (a) The gradient for CE loss can be relatively small with a high positive logit; (b) for BCE loss it's only effected by the negative logit itself which results in continuous suppression; (c) NTR encourages a sharp decrease when the logit is lower than a threshold and slowers down the optimization

Finally, the loss function, which we name as Re-balanced-BCE, becomes Eq. 5, where z^k denotes the output of the classifier.

$$\mathcal{L}_{R-BCE}(x^k, y^k) = \frac{1}{C} \sum_{i=0}^{C} \left[y_i^k log(1 + e^{-z_i^k}) + (1 - y_i^k)log(1 + e^{z_i^k}) \right] \times \hat{r}_i^k \quad (5)$$

What's worth noting is that $\hat{r}_i^k$ is applicable to both positive and negative labels although it was originally deduced from the sampling procedure regarding only the positive ones, in order to keep a consistency at class-level.

3.2 Negative-Tolerant Regularization

As mentioned above, *binary cross entropy (BCE) loss*, which is widely used for multi-label classification, sometimes suffers from over-suppression for negative labels because of the dominance of negative classes. To be more specific, BCE considers the recognition task as a series of binary classification tasks, calculating independent class-wise probability with *sigmoid* function. In contrast, *cross entropy (CE) loss*, which is popular in single-label classification, utilizes *softmax* to emphasize mutual exclusion. Unlike *softmax* where the optimization step would be rather small once the logit for positive class is much higher than those of negative classes, *sigmoid* treats them independently and encourages the logits of both positive and negative classes to be away from zero in the same gradient declining manner. The difference between them can be observed by their gradients shown in Eq. 6 and visualized in Fig. 3(a)(b).

$$\begin{cases} \dfrac{\partial \mathcal{L}_{CE}(z_j, y)}{\partial(z_j)} = \dfrac{e^{z_j}}{\sum_{i=0}^{C} e^{z_i}}, y_j = 0 \\ \dfrac{\partial \mathcal{L}_{BCE}(z_j, y)}{\partial(z_j)} = \dfrac{1}{C} \dfrac{e^{z_j}}{1 + e^{z_j}}, y_j = 0 \end{cases} \quad (6)$$

A straightforward consequence is that the classifiers for the tail classes would over-fit to a limited number of positive samples in the feature space, and meanwhile, they would push a huge number of negative samples away to produce lower logits. It can be taken as a class-specific over-fitting for the tail categories, which leads to a bad generalization of the classifiers. As shown in Fig. 1(c), the output distribution becomes sharp and the predictions of the testing samples are easy to be influenced by head classes.

To address the problem, we need a regularization to overcome the over-suppression. Specifically, the loss by negative logits actually needs a sharp drop once it's optimized to be lower than a threshold so that they won't be continuously suppressed due to a relatively small gradient. Based on the idea, we propose a negative-tolerant regularization (NTR) by first using a non-zero bias initializaiton to act as the thresholds, and then applying a linear scaling to the negative logits before their calculation in the standard BCE, together with a regularization parameter to constrain the gradient between 0 and 1. The Negative-Tolerant-BCE thus becomes Eq. 7.

$$\mathcal{L}_{NT-BCE}(x^k, y^k) = \frac{1}{C}\sum_{i=0}^{C} y_i^k log(1 + e^{z_i^k - \nu_i}) + \frac{1}{\lambda}(1 - y_i^k)log(1 + e^{-\lambda(z_i^k - \nu_i)})$$

$$(7)$$

λ is the scale factor that effects the loss gradient as shown in Fig. 3(c), controlling how "tolerant" we are to z_i, and ν is a class-specific bias. The design for ν is supposed to take intrinsic model bias into consideration. Concretely, a network trained with imbalanced data is likely to give passive predictions on those tail classes on average, the thresholds for them should correspondingly be lower, assuring that they won't be too easily achieved. It shares a similar idea with [3] that a larger margin is needed for rare classes. Assuming that we use a fully-connect layer as the classifier, the intrinsic bias of the model can be estimated by minimizing the loss function at the very beginning of training, where the classifiers are randomly initialized, and the dot-product distance between classifier vectors and instance features are at an average of zero. For a regular BCE loss, considering the bias b_i as the only variable, and assuming the class prior to be $p_i = n_i/N_0$, we can deduce an approximation of averaged loss by class i:

$$L_i = p_i \log(1 + e^{-b_i}) + (1 - p_i)\log(1 + e^{b_i}) \tag{8}$$

$$\hat{b}_i = -\log(\frac{1}{p_i} - 1), \quad \nu_i = -\kappa \hat{b}_i \tag{9}$$

We minimize Eq. 8 at $\hat{b}_i$, and use κ as a scale factor to get ν_i, which is further applied to Eq. 7.

3.3 Distribution-Balanced Loss

So far, R-BCE performs a re-balanced weighting strategy and the weight vector is fixed given an instance, while NT-BCE conducts regularization to the classifier outputs and affects the training by modifying the loss gradient. They can be

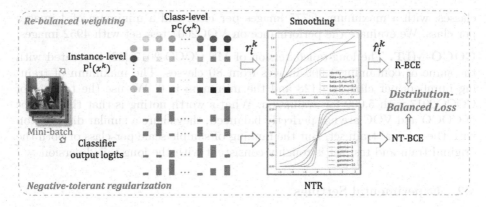

Fig. 4. Pipeline of the training procedure. Given a mini-batch of instances out of class-aware sampling, the calculation of re-balanced weight is shown in the upper stream, while NTR is shown in the lower. The two techniques are combined to our final distribution-balanced loss

naturally integrated as a unified loss function for end-to-end training, as shown in Fig. 4, and we finally get our distribution-balanced loss as Eq. 10.

$$\mathcal{L}_{DB}(x^k, y^k) = \frac{1}{C} \sum_{i=0}^{C} \hat{r}_i^k \left[y_i^k log(1 + e^{z_i^k - \nu_i}) + \frac{1}{\lambda}(1 - y_i^k)log(1 + e^{-\lambda(z_i^k - \nu_i)}) \right]$$

(10)

DB loss helps to smooth the distribution of the classifier outputs especially for those tail classes. It achieves superior performance in multi-label datasets with a long-tailed distribution, as will be validated in Sect. 4.

4   Experiments

4.1   Datasets

The proposed *Distribution Balanced Loss* is analyzed on artificially created long-tailed versions of two multi-label image recognition benchmarks, named VOC-MLT and COCO-MLT, respectively. They're subsets sampled from the original datasets by *pareto distribution pdf* $(x) = \alpha \frac{x_{min}^{\alpha}}{x^{\alpha+1}}$ following [26]. α controls how fast data scale decays. Regarding the interaction of sampling among classes, we construct the datasets in a head-to-tail manner so that we can strictly constrain the scale of tail classes: we first rank all the classes by $\hat{p}_i$ calculated with original data, and for each class i from head to tail, we *add* or *eliminate* instances that contain class i from the subset by referring to the expected distribution. Details on the construction of VOC-MLT and COCO-MLT can be found in the appendix.

VOC-MLT. We construct the long-tailed version of VOC [8] from its 2012 train-val set, with the power parameter $\alpha = 6$. It contains 1,142 images from 20 classes, with a maximum of 775 images per class and a minimum of 4 images per class. We evaluate the performance on VOC2007 test set with 4952 images.

COCO-MLT. The long-tailed version of MS COCO-2017 [22] is created with the same α, containing 1,909 images from 80 classes. The maximum of training numbers per class is 1,128 and the minimum is 6. We use the test set of COCO2017 with 5,000 for evaluation. What's worth noting is that the test set of COCO and VOC are not perfectly balanced, they share a similar distribution with the original train set. But the ranking of sample scale per-class of both the original train and test set is roughly consistent with the long-tailed version.

4.2 Experimental Settings

Evaluation Metrics. Following [26], we split the classes into three groups by the number of their training examples: *head* classes each contains over 100 samples, *medium* classes each has between 20 and 100 samples, and *tail* classes with under 20 samples each. We evaluate mean average precision(mAP) for all the classes, and we also report mAP for each subset to observe how the techniques effect on them.

Comparing Methods. We compare our methods with several state-of-the-art techniques dealing with multi-label classification or long-tailed recognition. We also report the results of their effective combinations for fair comparison. The standard binary cross entropy loss with sigmoid function is used or modified by all the methods. The compared methods include: (1) Empirical risk minimization: The plain model with all the examples having the same weight and sampling probability. (2) Re-weighting (RW): we perform a smooth version of re-weighting to be inversely proportional to the square root of class frequency, and we normalize the weights to be between 0 and 1 in a mini-batch. (2) Re-sampling (RS) [30]: we use class-aware re-sampling without extra tricks as a baseline, and we also evaluate the combination of RS and other techniques in comparison. (3) ML-GCN [4]: a recently proposed method by for multi-label classification with *graph convolutional network* (GCN). (4) Focal loss [21]: we use $\gamma = 2$ with a balance parameter of 2 for focal loss. (5) Class-balanced loss (CB) [5]: a class-wise re-weighting guided by the effective number of each class $E_n = (1 - \beta^n)/(1 - \beta)$. (6) Label-distribution-aware margin loss (LDAM) [3]: a recently proposed margin-loss which is proved to be effective for softmax classifier.

Implementation Details. We adopt Resnet50 [11] pretrianed on ImageNet [6] as backbone feature extractor, followed by global average pooling and a 2048×256 fully connection(FC) layer to obtain image-level features. The final classifier is a $256 \times C$ FC layer which outputs the logits. The input images are organized with a batch size of 32, randomly cropped and resized to 224×224 together with standard data augmentation. We use SGD with momentum of 0.9 and weight decay of 1×10^{-4} as our optimizer, and we also use linear warm-up learning rate schedule [9] for the first 500 iterations with a ratio of $\frac{1}{3}$. Training not

Table 1. Experimental results of mAP by our methods and other comparing approaches on VOC-MLT and COCO-MLT. We evaluate the results on the whole class set and the three subsets, respectively

Datasets	VOC-MLT				COCO-MLT			
Methods	Total	Head	Medium	Tail	Total	Head	Medium	Tail
ERM	70.86	68.91	80.20	65.31	41.27	48.48	49.06	24.25
RW	74.70	67.58	82.81	73.96	42.27	48.62	45.80	32.02
Focal Loss [21]	73.88	69.41	81.43	71.56	49.46	49.80	54.77	42.14
RS [30]	75.38	70.95	82.94	73.05	46.97	47.58	50.55	41.7
RS-Focal	76.45	72.05	83.42	74.52	51.14	48.90	54.79	48.30
ML-GCN [4]	68.92	70.14	76.41	62.39	44.24	44.04	48.36	38.96
LDAM [3]	70.73	68.73	80.38	69.09	40.53	48.77	48.38	22.92
CB-Focal [5]	75.24	70.30	83.53	72.74	49.06	47.91	53.01	44.85
R-BCE	76.34	71.40	82.76	75.22	49.43	48.77	53.00	45.33
R-BCE-Focal	77.39	72.44	83.16	76.77	52.75	50.20	56.52	50.02
DB	78.65	73.16	84.11	78.66	52.53	50.25	56.33	49.54
DB-Focal	**78.94**	**73.22**	**84.18**	**79.30**	**53.55**	**51.13**	**57.05**	**51.06**

combined with re-sampling is trained for 80 epochs with an initial learning rate of 0.02, which decays by a factor of 10 after 55 and 70 epochs, respectively. Re-sampling enhanced methods are trained for 8 epochs with the same learning rate initialization decaying step, and it decays after 5 and 7 epochs, respectively. We use the class-aware re-sampling [30] and the times that the iterator visits each class in one epoch is set as $\frac{N_{max}}{4}$ where N_{max} denotes the maximum number of training samples per class. Once $N_e < N_{max}$, it actually controls how we over-sample the tail classes and under-sample the head classes with no tail ones co-occurring. The experiments are implemented in PyTorch.

4.3 Benchmarking Results

VOC-MLT. VOC-MLT contains 6, 6, and 8 classes for the head, medium, and tail classes, respectively. We adjusted the hyper-parameters in other methods so that they currently work best in our dataset. For our method, we choose $\alpha = 0.1$, $\beta = 10$, and $\mu = 0.3$ for the smoothing function during re-balanced weighting. And we set $\lambda = 5, \kappa = 0.05$ for NTR. The experimental results compared with other traditional and state-of-the-art approaches can be seen in Table 1. What's worth noting is that unlike COCO-MLT whose tail classes always co-occur with head classes, the tail classes for VOC usually appear as single-label, which lower the complexity and difficulty for the classification task on them, bringing higher performance for the tail that even outperforms the head. This character also alleviate the defect of inner-class imbalance cause by head-tail connection and notice that after regular re-sampling, all subsets including the head have an

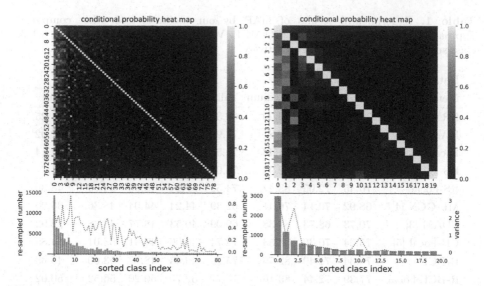

Fig. 5. In the matrix heatmap, the element in the ith column, j^{th} row represents the conditional probability of class existence, $p(i|j)$, which is usually higher for head and medium classes. The histograms below show the data distribution after re-sampling, confirming that the imbalance is not eliminated by re-sampling. The sampling frequency for each instance is different, and the line chart shows the variance within each class. The high variance at the head indicates inner-class imbalance

improvement, especially the tail classes. Comparing with the best baseline, re-sampling trained with focal loss (RS-Focal), our re-balanced weighting strategy gains further improvement by about 1.0% in total mAP, with 0.4% and 2.2% for head and tail classes and drop 0.3% for medium classes. Our final DB-Loss further achieves remarkable improvements by 1.5% compared with R-BCE, and by 0.8%, 1.0% and 2.6% for the three subsets, respectively. It can be seen that NTR is especially beneficial for the tail classes.

COCO-MLT. The whole 80 classes of COCO-MLT are split into 22, 33, and 25 classes for the head, medium, and tail classes, respectively. We choose $\alpha = 0.1$, $\beta = 10$, and $\mu = 0.2$ for the smoothing function during re-balanced weighting. And we set $\lambda = 2, \kappa = 0.05$ for NTR. The experimental results compared with other traditional and state-of-the-art approaches can be seen in Table 1. COCO-MLT has a heavy head-tail connection, *i.e.* some tail classes has a 100% probability of oc-curring with certain head classes, as shown in Fig. 5. A direct result of class-aware re-sampling is a sharp rise of mAP to the tail classes, while the performance for head classes drops by about 0.9%. With re-balanced weighting, the negative effect on the head classes is fixed and mAP for head, medium, and tail classes all have an improvement, by 2.3%, 1.1% and 2.1%, respectively. With focal loss combined with either BCE trained re-sampling or R-BCE trained re-sampling, we see an extra improvement. Replacing R-BCE with DB-Loss would further bring an average improvement of about 0.8%, and brings up mAP for tail classes by about 1.1%.

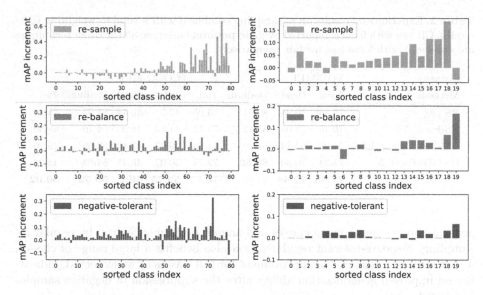

Fig. 6. Corresponding to the four training strategies *i.e.* ERM, regular class-aware re-sampling, re-sampling with R-BCE and the final DB-Loss, we show the per-class mAP increments between each two steps to evaluate our pipeline piecemeal, presented from left to right. Result on COCO-MLT is in the left and VOC-MLT is in the right

4.4   Ablation Study

Visualization of the Imbalance Caused by Re-sampling. We visualize the conditional probability matrix that reveals label co-occurrence relationship in Fig. 5. As can be seen that the most frequently appearing classes usually have the highest co-existing probability on the condition on of other classes. This makes them repeatedly sampled, and the imbalance is not eliminated after re-sampling. We also roughly estimate the inner-class imbalance by the variance of normalized sampling times: for each class and each of the instances containing it, we calculate its expected sampling times. We normalize them within a class to a mean value of 1, and the variance of the normalized sampling times is calculated, as shown in Fig. 5. Variance can roughly represent the extent of imbalance in sampling. Classes with high variance gain little or negative increment in mAP despite heavy sampling on them. A more precise and complicated cooperation of sampling variance and data scale is out of the scope of this paper, which can be reserved as future work.

Step-Wise Evaluation of Our Framework. We perform a step-wise evaluation on the test set by showing mAP increment per-class to have a better understanding of how re-balanced weighting and negative-tolerant regularization work on different parts of the dataset distribution. As shown in Fig. 6 and mentioned above, regular re-sampling is not friendly to head and medium classes, with little or negative increment. While using re-balanced weighting has a

Table 2. Experimental results on re-sampling combined with several re-weighting techniques. CB loss with focal is reported by [5] to perform better, so all the other techniques are enhanced with focal loss for fair comparison

Datasets	VOC-MLT				COCO-MLT			
Methods	Total	Head	Medium	Tail	Total	Head	Medium	Tail
RS [30]	75.38	70.95	82.94	73.05	47.7	46.44	51.93	43.23
RS-Focal [21]	76.45	72.05	83.42	74.52	51.14	48.9	54.79	48.30
RS+RW-Focal [27, 28]	71.96	63.14	81.09	71.73	49.07	47.80	52.09	46.20
RS+CB-Focal [5]	75.24	70.30	**83.53**	72.74	50.07	48.45	54.03	48.28
R-BCE-Focal	**77.39**	**72.44**	83.16	**76.77**	**52.75**	**50.2**	**56.52**	**50.02**

general improvement among classes and amend performance drop by regular re-sampling. Negative-tolerant regularization also benefits a large range of classes, and it leads to a remarkable improvement for tail classes as we expected, indicating an improved generalization ability after the suppression of negative samples is relaxed.

The Combination of Re-sampling and Various Re-weighting Methods. Re-sampling and traditional re-weighting methods based on the original distribution share a similar thought of drawing more importance to the rare classes, and they're usually performed at instance level. As a result, the combinations of them are at risk of redundancy: the head classes are over-ignored and the tail classes are over-emphasized. Our re-balanced weight is also calculated from the training distribution and it's designed to fine-tune and enhance re-sampling rather than doing repetitive jobs. So we combine the traditional re-weighting methods with re-sampling for comparison as performed in Table 2. Our superior performance shows the benefit of applying R-BCE to re-sampling.

4.5   Further Analysis

The Effect of Hyper-Parameters of Smoothing Function. The smoothing function Eq. 4 has three hyper-parameters, α applies an overall lift in weight, β and μ control the shape of the mapping function. We report the results of β with $\mu = 0.2$ fixed as shown in Fig. 4.

The Effect of λ in Negative-tolerant Regularization. To understand how λ and ν of Eq. 7 affect the results independently, we first fix $\nu = 0$ the same as in the main experiments and change λ in a large range, and then fix $\lambda = 2, 5$ for COCO-MLT and VOC-MLT, respectively, and change ν. The effect of ν is relatively small, and we would report the results in the supplementary material. Here in Fig. 7b, we observe that it performs the best at $\lambda = 5 - 10$ for VOC, with an improvement of about 2.5% for the tail classes, and 1% for head and medium classes. In COCO, head and medium classes are slightly affected when $\lambda < 3$ and tail classes have an improvement of about 1% at around $2 < \lambda < 3$. What's

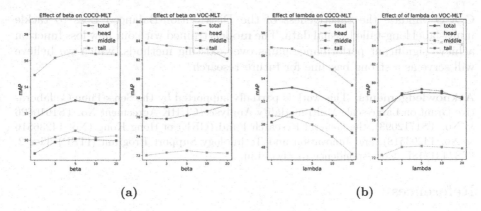

(a)   (b)

Fig. 7. (a). We show how mAP is effected by β independently. The total mAP has a peak at around $5 < \beta < 10$ for both datasets. And we observe an inverse tendency of results for different subsets when $\beta > 5$. (b). We show how mAP is effected by λ independently. We can observe a peak for both datasets away from $\lambda = 1$. COCO is more sensitive towards it and we finalize at choosing $\lambda = 2$ in the main experiment

worth noting is that, by adding the same form of regularization to positive logits, the results slightly drop as expected.

Group-Wise Analysis. Medium classes always have a better mAP on average than both the head and tail classes. The reason for this may be that, medium classes neither suffer from the over-fitting problem as tail classes do due to insufficient training samples, nor do they get hurt from the imbalance induced by re-sampling. Another explanation for this is that the average number of classes an instance has is gradually reduced from the head classes to the tail. This indicates a lower complexity and difficulty for the recognition task. For instance, quite a number of the training samples for the tail classes of VOC-MLT have only one ground-truth label. Phenomena led by this is that the defect of re-sampling is relieved and the mAP performance of the tail classes surprisingly outperforms head classes by a margin.

5   Conclusion

In this work, we propose a simple yet powerful loss function, Distribution-Balanced Loss, to tackle the multi-label long-tailed recognition problem. Multi-label long-tailed recognition problem has two intrinsic challenges, namely the co-occurrence of labels and the dominance of negative labels (when treated as multiple binary classification problems). To tackle these two obstacles, the Distribution-Balanced Loss consists of two key ingredients: 1) a new way to rebalance the weights that takes into account the impact caused by label co-occurrence, and 2) a negative tolerant regularization to mitigate the over-suppression of negative labels. Extensive experiments on both Pascal VOC and

COCO validate the effectiveness of the Distribution-Balanced Loss to tackle multi-label long-tailed visual data. The models trained with our new loss function achieve significant performance gains over existing methods, which we believe will serve as a strong baseline for future research.

Acknowledgements. This work is partially supported by the SenseTime Collaborative Grant on Large-scale Multi-modality Analysis (CUHK Agreement No. TS1610626 & No. TS1712093), the General Research Fund (GRF) of Hong Kong (No. 14236516 & No. 14203518), and Innovation and Technology Support Program (ITSP) Tier 2, ITS/431/18F. Correspondence to Ziwei Liu.

References

1. Buda, M., Maki, A., Mazurowski, M.A.: A systematic study of the class imbalance problem in convolutional neural networks. Neural Netw. **106**, 249–259 (2018)
2. Byrd, J., Lipton, Z.C.: What is the effect of importance weighting in deep learning? arXiv preprint arXiv:1812.03372 (2018)
3. Cao, K., Wei, C., Gaidon, A., Arechiga, N., Ma, T.: Learning imbalanced datasets with label-distribution-aware margin loss. In: Advances in Neural Information Processing Systems (NIPS), pp. 1565–1576 (2019)
4. Chen, Z.M., Wei, X.S., Wang, P., Guo, Y.: Multi-label image recognition with graph convolutional networks. In: Proceedings of the IEEE Conference on Computer Vision and Pattern Recognition (CVPR), pp. 5177–5186 (2019)
5. Cui, Y., Jia, M., Lin, T.Y., Song, Y., Belongie, S.: Class-balanced loss based on effective number of samples. In: Proceedings of the IEEE Conference on Computer Vision and Pattern Recognition (CVPR) (2019)
6. Deng, J., Dong, W., Socher, R., Li, L.J., Li, K., Fei-Fei, L.: ImageNet: a large-scale hierarchical image database. In: Proceedings of the IEEE Conference on Computer Vision and Pattern Recognition (CVPR), pp. 248–255. IEEE (2009)
7. Durand, T., Mehrasa, N., Mori, G.: Learning a deep ConvNet for multi-label classification with partial labels. In: Proceedings of the IEEE Conference on Computer Vision and Pattern Recognition (CVPR), pp. 647–657 (2019)
8. Everingham, M., Eslami, S.M.A., Van Gool, L., Williams, C.K.I., Winn, J., Zisserman, A.: The pascal visual object classes challenge: a retrospective. Int. J. Comput. Vis. (IJCV) **111**(1), 98–136 (2015)
9. Goyal, P., et al.: Accurate, large minibatch SGD: training ImageNet in 1 hour. arXiv preprint arXiv:1706.02677 (2017)
10. He, H., Garcia, E.A.: Learning from imbalanced data. IEEE Trans. Knowl. Data Eng. **21**(9), 1263–1284 (2009)
11. He, K., Zhang, X., Ren, S., Sun, J.: Deep residual learning for image recognition. In: Proceedings of the IEEE Conference on Computer Vision and Pattern Recognition (CVPR), pp. 770–778 (2016)
12. Horn, G.V., Perona, P.: The devil is in the tails: fine-grained classification in the wild. arXiv preprint arXiv:1709.01450 (2017)
13. Huang, C., Li, Y., Change Loy, C., Tang, X.: Learning deep representation for imbalanced classification. In: Proceedings of the IEEE Conference on Computer Vision and Pattern Recognition (CVPR), pp. 5375–5384 (2016)
14. Huang, Q., Liu, W., Lin, D.: Person search in videos with one portrait through visual and temporal links. In: Proceedings of the European Conference on Computer Vision (ECCV), pp. 425–441 (2018)

15. Huang, Q., Xiong, Y., Rao, A., Wang, J., Lin, D.: MovieNet: a holistic dataset for movie understanding. In: Proceedings of the European Conference on Computer Vision (ECCV) (2020)
16. Huang, Q., Yang, L., Huang, H., Wu, T., Lin, D.: Caption-supervised face recognition: training a state-of-the-art face model without manual annotation. In: Proceedings of the European Conference on Computer Vision (ECCV) (2020)
17. Japkowicz, N., Stephen, S.: The class imbalance problem: a systematic study. Intel. Data Anal. 6(5), 429–449 (2002)
18. Kang, B., et al.: Decoupling representation and classifier for long-tailed recognition. In: International Conference on Learning Representations (ICLR) (2020)
19. Khan, S., Hayat, M., Zamir, S.W., Shen, J., Shao, L.: Striking the right balance with uncertainty. In: Proceedings of the IEEE Conference on Computer Vision and Pattern Recognition (CVPR) (2019)
20. Lee, C.W., Fang, W., Yeh, C.K., Frank Wang, Y.C.: Multi-label zero-shot learning with structured knowledge graphs. In: Proceedings of the IEEE Conference on Computer Vision and Pattern Recognition (CVPR), pp. 1576–1585 (2018)
21. Lin, T.Y., Goyal, P., Girshick, R., He, K., Dollar, P.: Focal loss for dense object detection (2017)
22. Lin, T.Y., et al.: Microsoft COCO: common objects in context. In: Fleet, D., Pajdla, T., Schiele, B., Tuytelaars, T. (eds.) ECCV 2014. LNCS, vol. 8693, pp. 740–755. Springer, Cham (2014). https://doi.org/10.1007/978-3-319-10602-1_48
23. Liu, Z., Luo, P., Qiu, S., Wang, X., Tang, X.: DeepFashion: powering robust clothes recognition and retrieval with rich annotations. In: Proceedings of the IEEE Conference on Computer Vision and Pattern Recognition (CVPR) (2016)
24. Liu, Z., Luo, P., Wang, X., Tang, X.: Deep learning face attributes in the wild. In: Proceedings of the IEEE International Conference on Computer Vision (ICCV) (2015)
25. Liu, Z., et al.: Open compound domain adaptation. In: Proceedings of the IEEE Conference on Computer Vision and Pattern Recognition (CVPR) (2020)
26. Liu, Z., Miao, Z., Zhan, X., Wang, J., Gong, B., Yu, S.X.: Large-scale long-tailed recognition in an open world. In: Proceedings of the IEEE Conference on Computer Vision and Pattern Recognition (CVPR) (2019)
27. Mahajan, D., et al.: Exploring the limits of weakly supervised pretraining. In: Proceedings of the European Conference on Computer Vision (ECCV), pp. 181–196 (2018)
28. Mikolov, T., Sutskever, I., Chen, K., Corrado, G.S., Dean, J.: Distributed representations of words and phrases and their compositionality. In: Advances in Neural Information Processing Systems (NIPS), pp. 3111–3119 (2013)
29. Ren, M., Zeng, W., Yang, B., Urtasun, R.: Learning to reweight examples for robust deep learning. arXiv preprint arXiv:1803.09050 (2018)
30. Shen, L., Lin, Z., Huang, Q.: Relay backpropagation for effective learning of deep convolutional neural networks. In: Leibe, B., Matas, J., Sebe, N., Welling, M. (eds.) ECCV 2016. LNCS, vol. 9911, pp. 467–482. Springer, Cham (2016). https://doi.org/10.1007/978-3-319-46478-7_29
31. Tsoumakas, G., Katakis, I.: Multi-label classification: an overview. Int. J. Data Warehouse. Min. (IJDWM) 3(3), 1–13 (2007)
32. Wang, J., Yang, Y., Mao, J., Huang, Z., Huang, C., Xu, W.: CNN-RNN: a unified framework for multi-label image classification. In: Proceedings of the IEEE Conference on Computer Vision and Pattern Recognition (CVPR), pp. 2285–2294 (2016)

33. Wang, Y.X., Ramanan, D., Hebert, M.: Learning to model the tail. In: Advances in Neural Information Processing Systems (NIPS), pp. 7029–7039 (2017)
34. Wang, Z., Chen, T., Li, G., Xu, R., Lin, L.: Multi-label image recognition by recurrently discovering attentional regions (2017)
35. Xiong, Y., Huang, Q., Guo, L., Zhou, H., Zhou, B., Lin, D.: A graph-based framework to bridge movies and synopses. In: The IEEE International Conference on Computer Vision (ICCV), October 2019
36. Zhang, M.L., Zhou, Z.H.: ML-KNN: a lazy learning approach to multi-label learning. Pattern Recogn. **40**(7), 2038–2048 (2007)

Hamiltonian Dynamics for Real-World Shape Interpolation

Marvin Eisenberger$^{(\boxtimes)}$ and Daniel Cremers

Technical University of Munich, Garching, Germany
marvin.eisenberger@in.tum.de

Abstract. We revisit the classical problem of 3D shape interpolation and propose a novel, physically plausible approach based on Hamiltonian dynamics. While most prior work focuses on synthetic input shapes, our formulation is designed to be applicable to real-world scans with imperfect input correspondences and various types of noise. To that end, we use recent progress on dynamic thin shell simulation and divergence-free shape deformation and combine them to address the inverse problem of finding a plausible intermediate sequence for two input shapes. In comparison to prior work that mainly focuses on small distortion of consecutive frames, we explicitly model volume preservation and momentum conservation, as well as an anisotropic local distortion model. We argue that, in order to get a robust interpolation for imperfect inputs, we need to model the input noise explicitly which results in an alignment based formulation. Finally, we show a qualitative and quantitative improvement over prior work on a broad range of synthetic and scanned data. Besides being more robust to noisy inputs, our method yields exactly volume preserving intermediate shapes, avoids self-intersections and is scalable to high resolution scans.

Keywords: Shape interpolation · Registration · 3D computer vision

1 Introduction

Modeling realistic deformations of 3D shapes is at the heart of many computer vision applications. The central motivation in this context is to give meaning to sparse observations of a dynamically moving 3D object. Depending on the application, these measurements are given in the form of a point cloud, a triangle mesh, a voxel grid or a signed distance function (Fig. 1).

Electronic supplementary material The online version of this chapter (https://doi.org/10.1007/978-3-030-58548-8_11) contains supplementary material, which is available to authorized users.

© Springer Nature Switzerland AG 2020
A. Vedaldi et al. (Eds.): ECCV 2020, LNCS 12349, pp. 179–196, 2020.
https://doi.org/10.1007/978-3-030-58548-8_11

Fig. 1. An example interpolation (middle) on real scans from the FAUST dataset [7] and the final overlap (right). Here, the input correspondences were computed with Deep Functional Maps [36] (left).

In many cases, the sampling is not consistent over time and finding commonalities between observations is not a trivial task. While there are a lot of approaches that try to fuse scanned data for 3D reconstruction [45,53,54], relatively few work was dedicated to modeling the temporal transformation of the observed object directly. In this work, we revisit this classical challenge of 3D shape interpolation. Although there exists a multitude of elegant formulations, we will show that a lot of these approaches are mainly designed for synthetic shapes and therefore lack robustness to noisy real-world measurements.

The classical formulation is to define an interpolation as a sequence of shapes with minimal local distortion between consecutive frames [12,28,32]. While this is undoubtedly a reasonable assumption, it does not suffice in practice to account for the peculiarities of real-world data. For synthetic 3D objects, the ground-truth correspondences are typically known. For real scans, on the other hand, we need to first estimate them, e.g. by using a shape matching method. In practice, the resulting correspondences are not perfect and contain both outliers and fine-scale noise. This is problematic for an interpolation method that minimizes the local distortion between neighboring frames, because the noise from the faulty correspondences tends to distort the local geometry throughout the whole sequence. Moreover, most classical approaches do not model the global geometry of an object which can lead to artifacts like self-intersections.

Contribution. We propose a novel framework for real-world shape interpolation that is systematically derived from Hamiltonian dynamics. It resolves the above challenges by introducing additional, physically plausible modeling assumptions like volume preservation and momentum conservation. More specifically, we formulate shape interpolation as the inverse problems of a dynamic thin shell simulation. The Eulerian time-varying deformation fields are represented in a low rank manner which allows us to build volume preservation directly into our model. In qualitative and quantitative experiments, we demonstrate that our method gives rise to high-quality interpolations for real-world inputs.

2   Related Work

Shape interpolation has a long tradition in computer graphics. Originally it was developed for planar shapes [2,42,52] with [16] being a more recent formalism. A common approach for 3D surfaces is to define an interpolating trajectory as a geodesic in some higher dimensional shape space [12,26,27,60,60]. Most of these methods use some kind of deformation measure and then optimize for a sequence such that the local distortion between any two consecutive shapes is low. In [32] this is done with an as-Killing-as-possible energy and in [28] with a discrete shell energy motivated by [24]. Other popular examples of non-linear shape deformation are PriMo [9] and as-rigid-as-possible [56]. For a more thorough introduction to shape spaces, we refer the reader to the book of Younes [62].

An alternative approach to shape interpolation is to interpolate intrinsic quantities like dihedral angles before reconstructing the extrinsic geometry [2, 6,61]. One class of such intrinsic quantities are rotation-invariant or differential coordinates [1,34,35,51].

Sometimes shape deformation is stated as the time-dependent gradient flow wrt. some surface functional. Typically these functionals promote a smooth flow [14,15,18,58] but most of these methods focus on shape matching with less emphasis on the quality of the intermediate shapes.

Recently, more and more work was dedicated to processing collections of shapes in order to make interpolation more efficient. This can e.g. be achieved by constructing a low-dimensional subspace of admissible poses [3,21,29,57,64]. In practice, this greatly helps to reduce the computational cost of shape interpolation and even allows for interactive applications [47].

A common assumption of interactive shape deformation modeling is volume preservation. This can be obtained by defining a deformation as the flow of a divergence-free Eulerian vector field [4,22]. Recently, [19] extended this idea by constructing a divergence-free vector field basis that can be used to interpolate 3D objects. We will make use of this vector field representation and additionally formulate shape interpolation as the inverse problem of a dynamical thin shell simulation. The forward simulation corresponding to this is a well-known problem in computer graphics [44] with applications like cloth [23] or fluid [37] simulation. A recent formulation of this problem that is akin to our approach is projective dynamics [10,11]. Here, the Lagrangian gradient flow of a dynamical system is restated using the variational form of implicit Euler integration from [38] which leads to an efficient and extremely robust thin shell simulation.

3   Background

We briefly review important preliminary work on shape deformation and interpolation of non-rigidly deforming 3D objects. In this work, we focus on surface-based models like point clouds and 3D meshes. This allows for a compact representation and is in coherence with the output of real-world sensors. In particular, the set of observations $p = (p_1, \ldots, p_n)^\top \in \mathbb{R}^{n \times 3}$ consists of n points sampled from a two-dimensional Riemannian manifold $\mathcal{X}$. Depending on the application, these points are either part of a triangle mesh or embedded in a (knn-)graph.

3.1 Physical Assumptions for Shape Deformation

In order to find similarities between two non-rigid poses of an object, it helps to model geometric assumptions about the expected deformations directly. We review two common assumptions, namely small local distortions and volume preservation.

Local Distortion. A popular deformation energy to quantify the distortion between p and a deformed counterpart $p^* \in \mathbb{R}^{n \times 3}$ is the as-rigid-as-possible (arap) energy [56]:

$$\mathcal{W}_{\text{arap}}(p, p^*; (R_i)_{1 \leq i \leq n}) = \frac{1}{2} \sum_{i=1}^{n} \sum_{j \in \mathcal{N}(i)} \left\| R_i(p_j - p_i) - (p_j^* - p_i^*) \right\|_2^2. \quad (1)$$

The assumption behind this functional is that the local deformation of the geometry in the neighborhood $\mathcal{N}(i)$ of every vertex p_i is approximately rigid. I.e. one can find a rotation matrix $R_i \in SO(3)$ that approximately captures the transformation of the neighboring edges $p_j - p_i$. In turn, deviations of the deformation p^* from the approximate rigidity are penalized. The neighborhood $\mathcal{N}(i)$ for a given vertex i is defined as some set of adjacent vertices j to i.

There are multiple popular alternatives with the same flavor as $\mathcal{W}_{arap}$, including PriMo [9], discrete shells [24] and as-Killing-as-possible [55]. Most techniques penalize deformations of the local geometry and each one of them has certain advantages. In our formulation, we choose the arap energy because it is applicable directly for point clouds and because the optimization for p^* and R_i can be done efficiently in closed form.

Volume Preservation. Another common assumption for shape deformation is that the volume of the observed object is preserved over time [22]. This can be obtained by prescribing that the deformation is the flow induced by an underlying Eulerian deformation field $v : \mathbb{R}^3 \to \mathbb{R}^3$ which is divergence-free $\text{div}(v) = 0$. Recently, [19] proposed a formulation of a coarse-to-fine vector field basis that has the volume preservation built in as a hard constraint. A flow field v is then obtained as the linear combination of a finite subset of those divergence-free basis functions:

$$v(x; c) = \sum_{k=1}^{K} c_k \phi_k(x), \text{ where } \text{div}(\phi_k) = 0. \quad (2)$$

These deformation fields v are exactly volume preserving because the divergence is a linear operator, see [19] for more details. In practice, a relatively small number $K \approx 1000$ of coefficients $c = (c_1, \ldots, c_K)^\top \in \mathbb{R}^K$ suffices to represent arbitrary smooth, volume preserving vector fields v. We make use of this compact representation in this work. However, while [19] only considered stationary vector fields $v(x)$, in this work we consider time-dependent vector fields $v(t, x)$ in order to account for more complex shape variation.

3.2   Shape Interpolation

Computing an interpolation of two 3D objects $p = p^{(0)}$ and $q = p^{(T)}$ is a common problem in computer graphics and vision. In general, it is not a well-defined problem because there are typically infinitely many conceivable paths between $p^{(0)}$ and $p^{(T)}$. Therefore, we need to make additional assumptions about plausible sequences like small local distortions or volume preservation. The common way to do this is to define a deformation energy for the whole, time-discrete sequence $p^{(0)}, \ldots, p^{(T)}$ of intermediate shapes [12, 27, 28, 31, 32]:

$$E\big(p^{(1)}, \ldots, p^{(T-1)}\big) = \sum_{t=0}^{T-1} \mathcal{W}\big(p^{(t)}, p^{(t+1)}\big). \tag{3}$$

Here, $\mathcal{W}$ is some local distortion measure like $\mathcal{W}_{\mathrm{arap}}$ from Eq. (1). For symmetry reasons, the optimization is commonly done jointly for both the standard and the inverse sequence $p^{(T)}, \ldots, p^{(0)}$. W.l.o.g. we will consider the time interval $[0, t_{\max}] = [0, 1]$ which leads to a discrete step size $\tau = \frac{1}{T}$.

4   Interpolation of Real-World Objects

The implicit assumption behind most shape interpolation approaches is that the exact point-to-point correspondences between the two input surfaces p and q are known. While there is a lot of synthetic data where this is feasible, for scanned data the sampling of two given objects is typically not consistent, even if they approximate the same real-world surface $\mathcal{X}$. Not even the number of points of the two surfaces $p \in \mathbb{R}^{n \times 3}$ and $q \in \mathbb{R}^{m \times 3}$ is necessarily the same in the most general case. In order to compute an interpolation for this type of input data, we need to first estimate the surface correspondences between p and q.

Computing shape correspondences is a problem in itself and there is a variety of methods that focus on shape matching, either in the classical sense [20, 33, 41, 46, 48, 59] or using machine learning [8, 25, 36, 39, 43, 50]. The output of those methods is a point-to-point assignment of the surface $p \in \mathbb{R}^{n \times 3}$ to $q \in \mathbb{R}^{m \times 3}$ which can be represented with a matrix $\Pi \in \{0, 1\}^{n \times m}$. In principle, we can now transfer the points and neighborhood information from p to q and apply a classical interpolation method like [32] or [28] to $p \in \mathbb{R}^{n \times 3}$ and $\Pi q \in \mathbb{R}^{n \times 3}$. However, in practice the correspondences Π are not perfect and contain faulty or noisy matches. We found that most interpolation methods that assume perfect correspondences are not very robust to fine-scale noise, see Fig. 2.

[32]   [56]   [28]   [19]   Ours

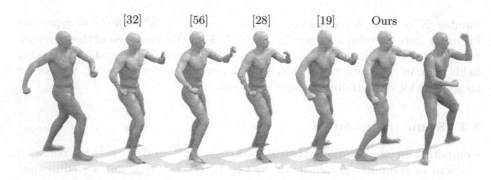

Fig. 2. A qualitative comparison of our approach with other popular shape interpolation methods. Here, we display the intermediate shapes at $t = 0.5$ for an example pair from SCAPE [5] with correspondences from BCICP [48]. Like us, [19] solves for an approximate alignment formulated as an IVP but the stationary vector field leads to slight distortions of the geometry (e.g. at the head and right arm). The other methods [28,32,56] solve a BVP and in certain areas the high frequency noise of the correspondences from BCICP leads to a severely degenerate geometry.

One possible way to make interpolation feasible for real scans is to acknowledge that the given matching Π is not perfect and to build this stochastic discrepancy directly into our model. In particular, we add Gaussian random noise η to the vertex position of the second shape Πq:

$$\tilde{q} := \Pi q + \eta, \text{ with } \eta \sim \mathcal{N}(0, \sigma). \tag{4}$$

Instead of finding intermediate shapes by solving a boundary value problem (BVP) as outlined in Eq. (3), we can then define an initial value problem (IVP) similar to [19]. In particular, we will optimize for a sequence $p^{(0)}, \ldots, p^{(T)}$ with $p = p^{(0)}$ and $p^{(T)} = \tilde{q} \approx \Pi q$.

5   From Hamiltonian Dynamics to Eulerian-Lagrangian Shape Interpolation

In this work, we model the motions of objects in an inertial frame of reference as a physical phenomenon that is governed by three aspects: internal forces, momentum conservation and volume preservation. Most existing interpolation techniques model internal forces in some way, yet they omit the momentum conservation and volume preservation. Without momentum conservation, the intermediate objects can be plausible but in many cases the motions lack temporal coherence. The volume preservation helps to constrain the optimization and prevents self-intersections, see Fig. 3. Our formulation combines the strengths of volume preserving fields [19] and projective dynamics [10] with those of classical interpolation methods [28,32].

5.1   Deformation Model

We systematically derive the evolution of a surface as a physical system from the Hamiltonian energy given by:

$$\mathcal{H}(p, v) = \frac{1}{2}\|v\|_2^2 + \mathcal{W}(p). \tag{5}$$

This energy consists of a kinetic energy term that models momentum conservation (with unit mass per point) and some potential energy component $\mathcal{W}$ that penalizes intrinsic distortions. The principles of Hamiltonian mechanics now prescribe how this system evolves over time:

$$\begin{cases} \dot{p} = \ \ \frac{\mathrm{d}\mathcal{H}}{\mathrm{d}v} = v. \\ \dot{v} = -\frac{\mathrm{d}\mathcal{H}}{\mathrm{d}p} = -\nabla\mathcal{W}(p). \end{cases} \tag{6}$$

We couple this with the volume preservation assumption by constraining v to the low rank vector field representation from Eq. (2). This allows us to model displacements of a shape $p(t) = \big(p_1(t), \ldots, p_n(t)\big)^{\mathsf{T}}$ at time t with only $K \ll n$ degrees of freedom:

$$\dot{p}_i(t) = v\big(p_i(t); c(t)\big) = \sum_{k=1}^{K} c_k(t)\phi_k\big(p_i(t)\big). \tag{7}$$

Besides providing a compact representation, this approach builds volume preservation directly into the deformation model, because $\mathrm{div}(v) = 0$. In [19], the authors model shape deformations in a similar way but with a stationary vector field $v\big(x; c(t)\big) = v\big(x; c\big)$. This leads to a well-constrained optimization problem with only K degrees of freedom $c_1, \ldots, c_K$ but it is also restrictive and lacks expressivity. Instead of using a constant vector field, following Eq. (6), we define a dynamic flow $v(t, x) = v\big(x; c(t)\big)$:

$$\begin{cases} \dot{v}\big(t, p(t)\big) = -\nabla\mathcal{W}\big(p(t)\big). \\ \mathrm{div}(v) = 0. \end{cases} \tag{8}$$

In our formulation, the internal forces are defined as the negative gradient of our anisotropic as as-rigid-as-possible potential $\mathcal{W}$ which we define in the next section.

5.2   Anisotropic As-rigid-As-Possible Deformation

For most 3D objects, not all parts are behaving similar in terms of local distortions. For example, regions near joints of a human body allow for more movement than most other parts of the surface. The classical as-rigid-as-possible potential that we reviewed in Eq. (1) penalizes distortions of the geometry uniformly in

[28]   [32]   Ours

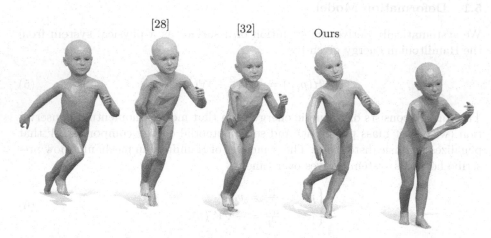

Fig. 3. A pair of synthetic shapes with ground-truth correspondences from the KIDS dataset [50] for which we show the intermediate shapes at $t = 0.5$. This example shows that many classical methods like [32] or [28] cannot detect self-intersections of different subparts. Here, the optimal path that minimizes a local distortion metric makes the right arm of the kid move through itself. Our method, on the other hand, avoids self-intersections by design: All deformations are expressed as a divergence-free Eulerian field, therefore the resulting flow has to be globally consistent in the sense that two close parts cannot have contradictory motions.

Fig. 4. An example from TOSCA where we color code the log-determinant of the covariance matrix $\log \det(\Sigma_i)$ for each point p_i. Red stands for a low value which corresponds to a high local distortion. This shows how our anisotropic as-rigid-as-possible energy (9) automatically adapts to objects consisting of inhomogeneous parts. Certain regions like joints allow for more local distortion throughout the sequence than others. Notice the difference between the hind legs, the head and the rest of the body. (Color figure online)

all directions and equal for all parts of the considered object. We generalize this idea and introduce an anisotropic as-rigid-as-possible energy:

$$\mathcal{W}\big(p(t); (R_i)_i, (\Sigma_i)_i\big) = \frac{1}{2} \sum_{i=1}^{n} \sum_{j \in \mathcal{N}(i)} \big\| (p_j(0) - p_i(0)) - R_i^\top (p_j(t) - p_i(t)) \big\|_{\Sigma_i}^2. \quad (9)$$

In this context, $\| \cdot \|_{\Sigma_i}$ denotes the standard Mahalanobis norm [40] with an unknown covariance matrix $\Sigma_i \in \mathbb{R}^{3 \times 3}$. This energy $\mathcal{W}$ allows our model to adapt the appropriate local behavior during the optimization, see Fig. 4 for an example. Moreover, the distortion is always computed in the reference frame of the first pose $p(0)$. This means that we only need to compute the distortion model of $p(0)$ and therefore we only need one local distortion matrix per vertex Σ_i for the whole sequence.

5.3   Time Discretization

In the time-discrete setting, we can approximate Eq. (7) and Eq. (8) using an implicit Euler intergration scheme:

$$\begin{cases} p^{(t+1)} = p^{(t)} + \tau v^{(t+1)}. & (10a) \\ v^{(t+1)} = v^{(t)} - \tau \nabla \mathcal{W}\big(p^{(t+1)}\big). & (10b) \\ \operatorname{div}\big(v^{(t+1)}\big) = 0. & (10c) \end{cases}$$

This is a Eulerian-Lagrangian scheme: The velocity field is represented on the surface $v^{(t)} \in \mathbb{R}^{n \times 3}$ but the divergence-free condition $\operatorname{div}\big(v^{(t+1)}\big) = 0$ is Eulerian. In order to make this interaction tractable, we will use the divergence-free vector field representation from Eq. (2) and combine it with the variational form of implicit Euler integration introduced in [38]. This allows us to restate this scheme as an optimization problem in terms of the vector field coefficients $c \in \mathbb{R}^K$:

$$\begin{cases} c^{(t+1)} = \underset{c,R}{\arg\min} \left\| \mathbf{v}\big(p^{(t)}; c\big) - \bar{v}^{(t)} \right\|_F^2 + \mathcal{W}\Big(p^{(t)} + \tau \mathbf{v}\big(p^{(t)}; c\big); R, \Sigma \Big). & (11a) \\[2mm] v_i^{(t+1)} = \mathbf{v}\big(p_i^{(t)}; c^{(t+1)}\big) = \sum_{k=1}^{K} c_k^{(t+1)} \phi_k\big(p_i^{(t)}\big). & (11b) \\[2mm] p^{(t+1)} = p^{(t)} + \tau v^{(t+1)}. & (11c) \\[2mm] \bar{v}^{(t+1)} = 2v^{(t+1)} - v^{(t)}. & (11d) \end{cases}$$

We refer the interested reader to [38] and [10] for more details on how this scheme is derived. The update of the coefficients c in (11a) can be computed using Gauss-Newton optimization. We use an additional extrapolation step (11d) to get a better prediction of the velocity $v^{(t+2)}$ which we justify in the following:

Theorem 1. *For continuously differentiable vector fields, the extrapolation step (11d) of Algorithm 11 yields an estimate $\bar{v}^{(t+1)}$ of $v^{(t+2)}$ with an error of order $\mathcal{O}(\tau^2)$. For the alternative scheme without step (11d) it is $\mathcal{O}(\tau)$.*

This result implies that (11d) leads to a qualitative improvement because a better estimate $\bar{v}^{(t+1)} \approx v^{(t+2)}$ provides a more faithful approximation in the next update step (11a) of c. See Appendix A for a proof of Thorem 1.

5.4 Interpolation Algorithm

We will now use the scheme (11) from last section to define an interpolation algorithm for two given shapes p and q. In each iteration, we initialize the scheme with $p^{(0)} := p$ and the unknown variables $c^{(0)} := \hat{c}$ and $(\hat{\Sigma}_i)_{1 \leq i \leq n}$. We then compute the deformed shapes $p^{(0)}, \ldots, p^{(T)}$ according to our scheme (11). Overall, this forward pass can be summarized as the differentiable solution operator $\mathcal{S}$:

$$\mathcal{S} : \begin{cases} \mathbb{R}^K \times \mathbb{R}^{n \times 3 \times 3} \to \mathbb{R}^{n \times 3 \times (T+1)} \\ (\hat{c}, \hat{\Sigma}) \mapsto (p^{(0)}, \ldots, p^{(T)}). \end{cases}$$ (12)

The goal is now to find the input parameters $\hat{c}$ and $\hat{\Sigma}$ that lead to a tight alignment of the deformed shape $p^{(T)}$ with q in accordance with Eq. (4). Together with our regularizer $\mathcal{W}$ from Eq. (9) this leads to the following energy:

$$E(p^{(0)}, \ldots, p^{(T)}; \Pi) := \frac{1}{2\sigma^2} \left\| p^{(T)} - \Pi q \right\|^2 + \sum_{t=0}^{T} \mathcal{W}(p^{(t)}).$$ (13)

Putting everything together, we can derive the following algorithm:

Algorithm 1. Volume preserving shape interpolation.

Require: $p \in \mathbb{R}^{n \times 3}$, $q \in \mathbb{R}^{m \times 3}$
  $\hat{c} \leftarrow 0 \in \mathbb{R}^K$
  $\hat{\Sigma}_i \leftarrow \mathrm{Id}_3 \in \mathbb{R}^{3 \times 3}$
  $\Pi \leftarrow \text{match\_shapes}(p, q) \in \{0, 1\}^{n \times m}$
  for $i = 1, \ldots, N_{\mathrm{it}}$ **do**
    $(\hat{c}, \hat{\Sigma}) \leftarrow (\hat{c}, \hat{\Sigma}) - \gamma \nabla E(\mathcal{S}(\hat{c}, \hat{\Sigma}); \Pi)$
  end for
  return $(p^{(0)}, \ldots, p^{(T)}) := \mathcal{S}(\hat{c}, \hat{\Sigma})$

In our implementation, we use a modern automatic differentiation toolbox to compute the gradient $\nabla E \circ \mathcal{S}$ wrt. $(\hat{c}, \hat{\Sigma})$ in Algorithm 1. The choice of algorithm to compute the input correspondences Π is not further specified here because it is more or less arbitrary. We show various different possibilities in our experiments.

Fig. 5. Two interpolated sequences for real scans of a puppet from SHREC'19 Isometry [17] and a human from FAUST [7] with input correspondences from Smooth Shells [20] and Deep Functional Maps [36] respectively.

6   Experiments

We verify the generality of our method on four different datasets with increasing complexity. The first two are the synthetic datasets TOSCA [13] and SCAPE [5] where we use the ground truth correspondences for the former and correspondences from BCICP [48] for the latter. The last two datasets SHREC'19 Isometry [17] and FAUST [7] contain reals scans of a puppet and different humans respectively, see Fig. 5. For those, we use correspondences from Smooth Shells [20] and FMNet [36]. Our experiments show that our formulation is applicable to a wide range of inputs with varying levels of noise. Figure 7 summarizes our quantitative evaluations on all datasets with comparisons to four other popular interpolation methods. The other methods are Geometric Modeling in Shape Space [32], Time-Discrete Geodesics in the Space of Shells [28], Divergence-Free Shape Correspondence by Deformation [19] and As-Rigid-As-Possible Surface Modeling [56]. Although the latter does not describe an interpolation algorithm explicitly, it is trivial to employ its shape deformation procedure in an interpolation pipeline by using Eq. (3). On the surface, our method is similar to [19] in the sense that both approaches compute divergence-free fields in a low rank basis. The decisive difference is that our method is based on a physically plausible formulation which, among other things, allows for time-dependent vector fields $v(t, x)$. This makes our method more expressive, see Fig. 6 for an example.

Error Metrics. In order to quantify the precision of a shape interpolation, we compute three different metrics for each pair of input shapes and plot the resulting cumulative curves in Fig. 7. In particular, we measure the conformal distortion [63, Eq. (6)] and volume change [30, Eq. (3)] of intermediate shapes and the

Ours   [19]

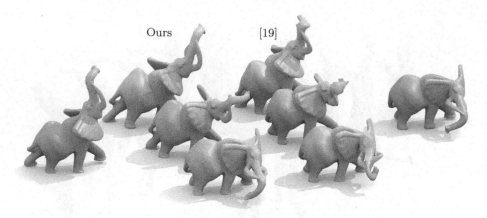

Fig. 6. A comparison of our method and divergence-free interpolation [19] on a pair of synthetic shapes (green). Both methods preserve the volume but for this large scale deformation the stationary vector field in [19] is too restrictive which leads to a distorted geometry for $t \geq 0.5$ (Color figure online)

Chamfer distance to the target shapes in % of the diameter for our method and the second alignment based method [19]. If we are strict, the notion of volume change is only meaningful for watertight meshes, which typically does not hold for real scans. Our argument regarding this is that in theory, a flow induced by a divergence-free deformation field is exactly volume preserving in terms of the underlying watertight real-world manifold $\mathcal{X}$. Remarkably, in this way we can even make sense of the notion of volume for a point cloud, assuming that it was sampled from a closed, continuous surface.

Implementation Details. The low rank vector field representation of divergence-free fields in our Scheme 11 is entirely decoupled of the input resolutions n and m. Moreover, the vector fields are represented in a spatially dense Eulerian basis which means that at any discrete time t, the resulting vector field $v(x; c^{(t)})$ can be computed for arbitrary points x in our domain, see [19] for more details. This allows us to efficiently perform the optimization in Algorithm 1 on a subsampled version of the input shapes p and q with a fixed resolution of $2k$ points. Afterwards, the computed vector field can be applied to the full resolution in a single forward pass without any skinning strategy or the like. For once, this makes our approach significantly faster but it also allows for an interpolation of very high resolution objects like those from FAUST ($\sim$200k vertices). Many other classical interpolation methods use some multiscale scheme to allow for higher resolutions [28,32], but there are still upper limits for them as to what is feasible in terms of computation cost. Our interpolation Algorithm 1 is directly applicable to point clouds, therefore we simply subsample both input shapes using Euclidean farthest point sampling. However, other subsampling strategies like remeshing are also possible if one wants to work directly with meshes. Finally, we use the same set of parameters for all experiments, see our implementation for details.

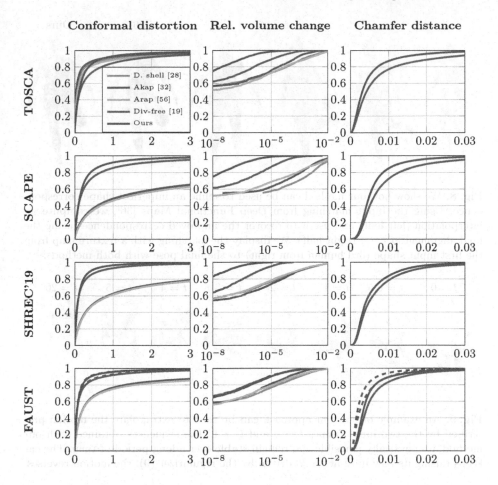

Fig. 7. Quantitative results and comparisons with other methods on four benchmarks. For FAUST, dashed lines correspond to the results on the high resolution scans. Those were only computed for our method and [19] because for the other methods the resolution of around 200k vertices is prohibitively high.

Additional Evaluations. As a proof of concept, we show that our physically plausible formulation allows for a broad range of applications beyond shape interpolation. For once, we can use our alignment at $t = 1$ to refine a shape matching which we show for a real scan of FAUST in Fig. 8. Furthermore, we can compute plausible shape extrapolations by simply simulating the forward integration for a longer period of time than $t = 1$. Remarkably, this can be done without any additional optimization, we simply compute an interpolation between p at $t = 0$ and q at $t = 1$ and then integrate our Scheme 11 until $t > 1$, see Fig. 9. Finally, we show that our method allows for input objects where only parts of the geometry are available, see Fig. 10. This is only feasible for an alignment based method, because the classical formulation as a BVP requires

[36]   Ours

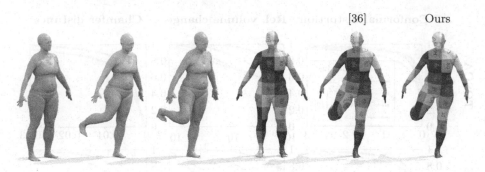

Fig. 8. We show how our method can be used to refine an imperfect shape correspondence. Using the input matching from Deep Functional Maps [36], we compute an interpolation (left half) and use it to recover the improved correspondences using the final alignment at $t = 1$ (right half). We display the matching with a texture map from the first input shape (3rd human from right) to the final pose with both methods.

$t = 0$ $t = 0.5$ $t = 1$ $t = 2$ $t = 3$ $t = 4$ $t = 5$ $t = 20$

Fig. 9. An example of how our approach can be used to extrapolate the motion prescribed by the two input frames $t = 0$ and $t = 1$. The sequences obtained with our method are physically plausible and remain stable over a long period of time. The cat keeps raising its paw until at $t = 2$, driven by the regularizer (9), the motion reverses.

Fig. 10. An example interpolation of a pair of partial shapes from the synthetic TOSCA cuts [49] dataset with our method.

that every vertex has a corresponding point on the other surface. Partial shape interpolation is an important preliminary result for many real world applications like scanning of dynamically moving 3D objects.

7 Conclusion

We presented a general and flexible approach to shape interpolation that is systematically derived from a formulation of Hamiltonian dynamics. For this, we employ recent advances in dynamic thin shell simulation to get a robust deformation model and solve its inverse problem by optimizing over the initial motion and anisotropic surface properties. We demonstrated that, in comparison to prior work, our approach is able to compute high quality, physically plausible interpolations of noisy real world inputs. In future work, we will apply our setup to a broader range of applications like 3D scanning of actions or mesh compression.

Acknowledgements. We would like to thank Zorah Lähner and Aysim Toker for useful discussions. We gratefully acknowledge the support of the ERC Consolidator Grant "3D Reloaded".

References

1. Alexa, M.: Differential coordinates for local mesh morphing and deformation. Vis. Comput. **19**(2–3), 105–114 (2003)
2. Alexa, M., Cohen-Or, D., Levin, D.: As-rigid-as-possible shape interpolation. In: Proceedings of the 27th Annual Conference on Computer Graphics and Interactive Techniques, SIGGRAPH 2000, pp. 157–164 (2000)
3. Allen, B., Curless, B., Popović, Z., Hertzmann, A.: Learning a correlated model of identity and pose-dependent body shape variation for real-time synthesis. In: Proceedings of the 2006 ACM SIGGRAPH/Eurographics Symposium on Computer Animation, pp. 147–156. Eurographics Association (2006)
4. Angelidis, A., Cani, M.P., Wyvill, G., King, S.: Swirling-sweepers: constant-volume modeling. Graph. Models **68**(4), 324–332 (2006)
5. Anguelov, D., Srinivasan, P., Koller, D., Thrun, S., Rodgers, J., Davis, J.: Scape: shape completion and animation of people. In: ACM Transactions on Graphics (TOG), vol. 24, pp. 408–416. ACM (2005)
6. Baek, S.Y., Lim, J., Lee, K.: Isometric shape interpolation. Comput. Graph. **46**, 257–263 (2015)
7. Bogo, F., Romero, J., Loper, M., Black, M.J.: FAUST: dataset and evaluation for 3D mesh registration. In: Proceedings IEEE Conference on Computer Vision and Pattern Recognition (CVPR). IEEE, Piscataway, June 2014
8. Boscaini, D., Masci, J., Rodolà, E., Bronstein, M.: Learning shape correspondence with anisotropic convolutional neural networks. In: Advances in Neural Information Processing Systems, pp. 3189–3197 (2016)
9. Botsch, M., Pauly, M., Gross, M.H., Kobbelt, L.: Primo: coupled prisms for intuitive surface modeling. In: Symposium on Geometry Processing, pp. 11–20. No. CONF (2006)
10. Bouaziz, S., Martin, S., Liu, T., Kavan, L., Pauly, M.: Projective dynamics: fusing constraint projections for fast simulation. ACM Trans. Graph. (TOG) **33**(4), 1–11 (2014)
11. Brandt, C., Eisemann, E., Hildebrandt, K.: Hyper-reduced projective dynamics. ACM Trans. Graph. (TOG) **37**(4), 1–13 (2018)

12. Brandt, C., von Tycowicz, C., Hildebrandt, K.: Geometric flows of curves in shape space for processing motion of deformable objects. In: Computer Graphics Forum, vol. 35, pp. 295–305. Wiley Online Library (2016)
13. Bronstein, A.M., Bronstein, M.M., Kimmel, R.: Numerical geometry of non-rigid shapes. Springer (2008). http://tosca.cs.technion.ac.il/book/resources_data.html
14. Charpiat, G., Faugeras, O., Keriven, R.: Approximations of shape metrics and application to shape warping and empirical shape statistics. Found. Comput. Math. **5**(1), 1–58 (2005)
15. Charpiat, G., Maurel, P., Pons, J.P., Keriven, R., Faugeras, O.: Generalized gradients: priors on minimization flows. Int. J. Comput. Vision **73**(3), 325–344 (2007)
16. Chen, R., Weber, O., Keren, D., Ben-Chen, M.: Planar shape interpolation with bounded distortion. ACM Trans. Graph. (TOG) **32**(4), 1–12 (2013)
17. Dyke, R., Stride, C., Lai, Y., Rosin, P.: SHREC-19: shape correspondence with isometric and non-isometric deformations (2019)
18. Eckstein, I., Pons, J.P., Tong, Y., Kuo, C.C., Desbrun, M.: Generalized surface flows for mesh processing. In: Proceedings of the fifth Eurographics Symposium on Geometry Processing, pp. 183–192. Eurographics Association (2007)
19. Eisenberger, M., Lähner, Z., Cremers, D.: Divergence-free shape correspondence by deformation. In: Computer Graphics Forum, vol. 38, pp. 1–12. Wiley Online Library (2019)
20. Eisenberger, M., Lähner, Z., Cremers, D.: Smooth shells: Multi-scale shape registration with functional maps. arXiv preprint arXiv:1905.12512 (2019)
21. Fletcher, P.T., Lu, C., Pizer, S.M., Joshi, S.: Principal geodesic analysis for the study of nonlinear statistics of shape. IEEE Trans. Med. Imaging **23**(8), 995–1005 (2004)
22. von Funck, W., Theisel, H., Seidel, H.P.: Vector field based shape deformations. In: ACM Transactions on Graphics (TOG), vol. 25, pp. 1118–1125. ACM (2006)
23. Goldenthal, R., Harmon, D., Fattal, R., Bercovier, M., Grinspun, E.: Efficient simulation of inextensible cloth. In: ACM SIGGRAPH 2007 Papers, pp. 49-es (2007)
24. Grinspun, E., Hirani, A.N., Desbrun, M., Schröder, P.: Discrete shells. In: Proceedings of the 2003 ACM SIGGRAPH/Eurographics Symposium on Computer Animation, pp. 62–67. Eurographics Association (2003)
25. Groueix, T., Fisher, M., Kim, V.G., Russell, B.C., Aubry, M.: 3D-CODED: 3D correspondences by deep deformation. In: Ferrari, V., Hebert, M., Sminchisescu, C., Weiss, Y. (eds.) ECCV 2018. LNCS, vol. 11206, pp. 235–251. Springer, Cham (2018). https://doi.org/10.1007/978-3-030-01216-8_15
26. Heeren, B., Rumpf, M., Schröder, P., Wardetzky, M., Wirth, B.: Exploring the geometry of the space of shells. In: Computer Graphics Forum, vol. 33, pp. 247–256. Wiley Online Library (2014)
27. Heeren, B., Rumpf, M., Schröder, P., Wardetzky, M., Wirth, B.: Splines in the space of shells. Comput. Graph. Forum **35**(5), 111–120 (2016)
28. Heeren, B., Rumpf, M., Wardetzky, M., Wirth, B.: Time-discrete geodesics in the space of shells. In: Computer Graphics Forum, vol. 31, pp. 1755–1764. Wiley Online Library (2012)
29. Heeren, B., Zhang, C., Rumpf, M., Smith, W.: Principal geodesic analysis in the space of discrete shells. In: Computer Graphics Forum, vol. 37, pp. 173–184. Wiley Online Library (2018)
30. Hormann, K., Greiner, G.: Mips: An efficient global parametrization method. Technical report, Erlangen-Nuernberg University (Germany) Computer Graphics Group (2000)

31. Huber, P., Perl, R., Rumpf, M.: Smooth interpolation of key frames in a riemannian shell space. Comput. Aided Geom. Des. **52**, 313–328 (2017)
32. Kilian, M., Mitra, N.J., Pottmann, H.: Geometric modeling in shape space. In: ACM Transactions on Graphics (TOG), vol. 26, p. 64. ACM (2007)
33. Kim, V.G., Lipman, Y., Funkhouser, T.A.: Blended intrinsic maps. Trans. Graph. (TOG) **30**, 4 (2011)
34. Lipman, Y., Sorkine, O., Cohen-Or, D., Levin, D., Rossi, C., Seidel, H.P.: Differential coordinates for interactive mesh editing. In: Proceedings Shape Modeling Applications, 2004, pp. 181–190. IEEE (2004)
35. Lipman, Y., Sorkine, O., Levin, D., Cohen-Or, D.: Linear rotation-invariant coordinates for meshes. ACM Trans. Graph. (TOG) **24**(3), 479–487 (2005)
36. Litany, O., Remez, T., Rodolà, E., Bronstein, A., Bronstein, M.: Deep functional maps: structured prediction for dense shape correspondence. In: Proceedings of the IEEE International Conference on Computer Vision, pp. 5659–5667 (2017)
37. Macklin, M., Müller, M.: Position based fluids. ACM Trans. Graph. (TOG) **32**(4), 1–12 (2013)
38. Martin, S., Thomaszewski, B., Grinspun, E., Gross, M.: Example-based elastic materials. In: ACM SIGGRAPH 2011 Papers, pp. 1–8 (2011)
39. Masci, J., Boscaini, D., Bronstein, M., Vandergheynst, P.: Geodesic convolutional neural networks on riemannian manifolds. In: Proceedings of the IEEE International Conference on Computer Vision Workshops, pp. 37–45 (2015)
40. McLachlan, G.J.: Mahalanobis distance. Resonance **4**(6), 20–26 (1999)
41. Melzi, S., Ren, J., Rodola, E., Ovsjanikov, M., Wonka, P.: Zoomout: spectral upsampling for efficient shape correspondence. arXiv preprint arXiv:1904.07865 (2019)
42. Michor, P.W., Mumford, D.: Riemannian geometries on spaces of plane curves. arXiv preprint math/0312384 (2003)
43. Monti, F., Boscaini, D., Masci, J., Rodola, E., Svoboda, J., Bronstein, M.M.: Geometric deep learning on graphs and manifolds using mixture model CNNs. In: Proceedings of the IEEE Conference on Computer Vision and Pattern Recognition, pp. 5115–5124 (2017)
44. Müller, M., Heidelberger, B., Hennix, M., Ratcliff, J.: Position based dynamics. J. Vis. Commun. Image Represent. **18**(2), 109–118 (2007)
45. Newcombe, R.A., Fox, D., Seitz, S.M.: Dynamicfusion: reconstruction and tracking of non-rigid scenes in real-time. In: Proceedings of the IEEE Conference on Computer Vision and Pattern Recognition, pp. 343–352 (2015)
46. Ovsjanikov, M., Ben-Chen, M., Solomon, J., Butscher, A., Guibas, L.: Functional maps: a flexible representation of maps between shapes. ACM Trans. Graph. (TOG) **31**(4), 30 (2012)
47. von Radziewsky, P., Eisemann, E., Seidel, H.P., Hildebrandt, K.: Optimized subspaces for deformation-based modeling and shape interpolation. Comput. Graph. **58**, 128–138 (2016)
48. Ren, J., Poulenard, A., Wonka, P., Ovsjanikov, M.: Continuous and orientation-preserving correspondences via functional maps. ACM Trans. Graph. **37**(6), 248:1–248:16 (2018)
49. Rodolà, E., Cosmo, L., Bronstein, M.M., Torsello, A., Cremers, D.: Partial functional correspondence. In: Computer Graphics Forum, vol. 36, pp. 222–236. Wiley Online Library (2017)
50. Rodolà, E., Rota Bulo, S., Windheuser, T., Vestner, M., Cremers, D.: Dense non-rigid shape correspondence using random forests. In: Proceedings of IEEE Conference on Computer Vision and Pattern Recognition (CVPR) (2014)

51. Sassen, J., Heeren, B., Hildebrandt, K., Rumpf, M.: Geometric optimization using nonlinear rotation-invariant coordinates. arXiv preprint arXiv:1908.11728 (2019)
52. Sederberg, T.W., Gao, P., Wang, G., Mu, H.: 2-D shape blending: an intrinsic solution to the vertex path problem. In: Proceedings of the 20th Annual Conference on Computer Graphics and Interactive Techniques, pp. 15–18 (1993)
53. Slavcheva, M., Baust, M., Cremers, D., Ilic, S.: Killingfusion: Non-rigid 3D reconstruction without correspondences. In: Proceedings of the IEEE Conference on Computer Vision and Pattern Recognition, pp. 1386–1395 (2017)
54. Slavcheva, M., Baust, M., Ilic, S.: Sobolevfusion: 3D reconstruction of scenes undergoing free non-rigid motion. In: Proceedings of the IEEE Conference on Computer Vision and Pattern Recognition, pp. 2646–2655 (2018)
55. Solomon, J., Ben-Chen, M., Butscher, A., Guibas, L.: As-killing-as-possible vector fields for planar deformation. In: Computer Graphics Forum, vol. 30, pp. 1543–1552. Wiley Online Library (2011)
56. Sorkine, O., Alexa, M.: As-rigid-as-possible surface modeling. In: Symposium on Geometry Processing, vol. 4, pp. 109–116 (2007)
57. Sumner, R.W., Zwicker, M., Gotsman, C., Popović, J.: Mesh-based inverse kinematics. ACM Trans. Graph. (TOG) 24(3), 488–495 (2005)
58. Sundaramoorthi, G., Yezzi, A., Mennucci, A.C.: Sobolev active contours. Int. J. Comput. Vision 73(3), 345–366 (2007)
59. Vestner, M., et al.: Efficient deformable shape correspondence via kernel matching. In: International Conference on 3D Vision (3DV), October 2017
60. Wirth, B., Bar, L., Rumpf, M., Sapiro, G.: A continuum mechanical approach to geodesics in shape space. Int. J. Comput. Vision 93(3), 293–318 (2011)
61. Xu, D., Zhang, H., Wang, Q., Bao, H.: Poisson shape interpolation. In: Proceedings of the 2005 ACM Symposium on Solid and Physical Modeling, pp. 267–274 (2005)
62. Younes, Laurent: Shapes and Diffeomorphisms. AMS, vol. 171. Springer, Heidelberg (2019). https://doi.org/10.1007/978-3-662-58496-5
63. Zhang, C., Chen, T.: Efficient feature extraction for 2D/3D objects in mesh representation. In: Proceedings 2001 International Conference on Image Processing (Cat. No. 01CH37205), vol. 3, pp. 935–938. IEEE (2001)
64. Zhang, C., Heeren, B., Rumpf, M., Smith, W.A.: Shell PCA: statistical shape modelling in shell space. In: Proceedings of the IEEE International Conference on Computer Vision, pp. 1671–1679 (2015)

Learning to Scale Multilingual Representations for Vision-Language Tasks

Andrea Burns[1]([✉]), Donghyun Kim[1]([✉]), Derry Wijaya[1]([✉]), Kate Saenko[1,2]([✉]),
and Bryan A. Plummer[1]([✉])

[1] Boston University, Boston, MA 02215, USA
{aburns4,donhk,wijaya,saenko,bplum}@bu.edu
[2] MIT-IBM Watson AI Lab, Cambridge, MA 02142, USA

Abstract. Current multilingual vision-language models either require a
large number of additional parameters for each supported language, or
suffer performance degradation as languages are added. In this paper,
we-9*6 propose a Scalable Multilingual Aligned Language Representation (SMALR) that supports many languages with few model parameters without sacrificing downstream task performance. SMALR learns
a fixed size language-agnostic representation for most words in a multilingual vocabulary, keeping language-specific features for just a few.
We use a masked cross-language modeling loss to align features with
context from other languages. Additionally, we propose a cross-lingual
consistency module that ensures predictions made for a query and its
machine translation are comparable. The effectiveness of SMALR is
demonstrated with ten diverse languages, over twice the number supported in vision-language tasks to date. We evaluate on multilingual
image-sentence retrieval and outperform prior work by 3–4% with less
than 1/5th the training parameters compared to other word embedding
methods.

Keywords: Scalable vision-language models · Multilingual word
embeddings · Image-sentence retrieval

1 Introduction

Learning a good language representation is a fundamental component of addressing a vision-language task, such as phrase grounding [22,34] or visual question

---

Project page: http://ai.bu.edu/smalr.

Electronic supplementary material The online version of this chapter (https://
doi.org/10.1007/978-3-030-58548-8_12) contains supplementary material, which is
available to authorized users.

© Springer Nature Switzerland AG 2020
A. Vedaldi et al. (Eds.): ECCV 2020, LNCS 12349, pp. 197–213, 2020.
https://doi.org/10.1007/978-3-030-58548-8_12

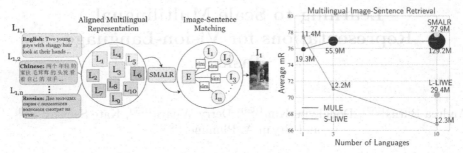

(a) Multilingual image-sentence retrieval   (b) MSCOCO multilingual retrieval

Fig. 1. (a) presents multilingual bidirectional retrieval. We embed sentences in ten languages with SMALR, which is used to compute the highest scoring image. (b) shows the effect of the number of training languages on performance for prior work MULE [23] and LIWE [41]. LIWE is the original model, hereafter referred to as S-LIWE. The plot contains two points: L-LIWE, [41] trained with a larger embedding (120-D vs. 24-D) for fair comparison, in orange, and SMALR, in yellow. The points are scaled to the number of parameters, P; specifically, their area is $(\frac{P}{10^6})^{\frac{3}{2}}$. SMALR is able to outperform all prior work with few parameters

answering [3,17]. Many recent methods have demonstrated that learning text representations aligned to images can boost performance across many vision-language tasks over traditional text-only trained representations [8,19,29,37,38]. This is often accomplished by using auxiliary vision-language tasks when learning the language representation (such as image-sentence retrieval, as shown in Fig. 1(a)). However, these methods often only support a single language. Although some work has addressed a multilingual scenario (e.g., [16,23,41]), these methods do not scale well to support many languages in terms of memory or performance (see Fig. 1(b)). As the number of languages grows, methods like LIWE [41] that use character-based recognition systems can save memory but suffer from performance degradation. In contrast, methods that learn to align word embeddings across languages can maintain (or even improve) performance as languages are added (e.g., [16,23]), but require additional parameters for the word embeddings that represent each new language's vocabulary. This becomes a challenge when scaling to support many languages, as an increasing majority of trainable parameters are required for representing each language (e.g. ~93% of parameters of [23] with ten languages). While pretrained word embeddings could be used without fine-tuning, e.g. Multilingual BERT [13] or MUSE [11], this comes at a significant cost in downstream task performance [8,23].

To address this trade-off between multilingual capacity and performance, we propose a *Scalable Multilingual Aligned Language Representation (SMALR)* model, which we demonstrate achieves strong task performance while also being highly compact compared to state-of-the-art word embedding methods [13,24,26]. As seen in Fig. 1, LIWE drops over 10% in performance going from supporting one to ten languages. MULE slightly increases performance with more languages, but requires 6x more parameters compared to its single language model. Our approach, SMALR, outperforms both with only 1/5th the parameters of MULE. We learn to efficiently represent each language by separating our language embedding into language-specific and language-agnostic token representations. As language follows a long-tailed distribution, only a few words occur often, with large portions of tokens occurring very rarely. For example, in the MSCOCO dataset [28] there are 25,126 unique tokens, but 61% of them occur less than 4 times. This suggests that having unique representations for every token in the vocabulary is unnecessary, as only a subset would affect downstream task performance significantly. Thus, we use a Hybrid Embedding Model (HEM) that contains language-specific embeddings for the common tokens, thereby providing a good representation for each language, and a compact language-agnostic representation for rare and uncommon words. This results in a model that needs far fewer unique embeddings than prior work without sacrificing performance.

We learn how to assign tokens to the language-agnostic representation in a pretraining step, which uses monolingual FastText embeddings [7] to map similar words to the same token, e.g. mapping "double-decker" in English and "impériale" in French to the same shared token. Once we obtain our language embeddings, our goal is to align them so that semantically similar words, even those from other languages, are embedded nearby. To accomplish this, we use a multilingual masked language model, where we randomly mask words and then predict them based on context. Unlike similar masking approaches used to train models such as BERT [13], we mask words of sentences from any two languages, say German and Chinese, which are semantically similar sentences referring to the same image, and use the context from each to predict both masked tokens. To further encourage cross-language alignment, we also use an adversarial language classifier and neighborhood constraints that have been used in prior work [23]. These universal language embeddings are provided as input to a multimodal model that learns to relate them to images. Finally, we use a cross-lingual consistency module that uses machine translations to reason about the image-sentence similarity across multiple languages, which we show significantly boosts performance. Figure 2 contains an overview of our model.

We use bidirectional image-sentence retrieval as the primary evaluation of our multilingual language representation. In this task, the goal is to retrieve a relevant sentence from a database given an image or to retrieve a relevant image from a database given a sentence. We augment current multilingual datasets Multi30K [6,14,15,43] and MSCOCO [27,28,31] using machine translations so that every image has at least five sentences across ten diverse languages: English (En), German (De), French (Fr), Czech (Cs), Chinese (Cn), Japanese (Ja), Arabic (Ar), Afrikaans (Af), Korean (Ko), and Russian (Ru). See the supplementary for details on our data augmentation procedure. This constitutes the highest number of languages used in multilingual learning for vision-language tasks to date, supporting more than double the number of visually-semantically aligned languages compared to prior work [5,11,16,23,36,41].

We list the contributions of our work below:

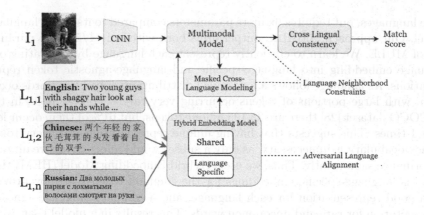

Fig. 2. The contributions of SMALR are in blue: a Hybrid Embedding Model (HEM), a Masked Cross-Language Model (MCLM), and a Cross-Lingual Consistency stage (CLC). HEM embeds input sentences as a mixture of language-specific and language-agnostic representations using a hard attention mechanism. The MCLM component provides an additional loss to enforce language alignment, while also augmenting the original dataset with masked sentences (Color figure online)

- SMALR, a scalable multilingual model for training visually-semantically aligned word embeddings that outperforms the state-of-the-art on multilingual image-sentence retrieval while also requiring few model parameters.
- A comparison to four types of vocabulary reduction methods that serve as baselines to complement our evaluation against prior work.
- A Masked Cross-Language Modeling (MCLM) procedure that further aligns the multilingual embedding, stabilizing variance in performance over all languages, and serves as an additional data augmentation technique.
- A Cross-Lingual Consistency (CLC) module, the first of its kind, that learns how to aggregate an ensemble of predictions across languages made with machine translations, which, combined with our SMALR architecture, results in a total improvement over the state-of-the-art by 3–4%.

2   Related Work

Transformer [39] based representation learning models have become increasingly popular since the release of BERT [13]. BERT transfers surprisingly well to other languages, despite having no multilingual training data or explicit multilingual loss [42]. However, [33] demonstrates that there is an unequal transfer between different language pairs, notably those with typological differences to English. Both BERT and M-BERT, its multilingual extension, have been shown to be dependent on the number of parameters in the model, which reaches 110M parameters for the smaller base model [21]. Thus, as also shown in [1], a large number of additional parameters is needed to counter the performance degradation caused by training with many languages. Using the better performing large

BERT model is impractical for many vision-language tasks as it contains 340M parameters, leaving little room in many GPUs memory for anything else.

Along with language-only BERT variants, a burst of multimodal BERT-like models have been designed specifically for vision-language tasks [26,29,37,38]. More traditional word embedding models have also been designed for multimodal tasks with the use of either visual-word co-occurrence frequencies [19], multi-task training [32], or both [8], and require significantly less training data to reach similar performance. While these efforts evaluate on many multimodal tasks such as Visual Question Answering [3], Visual Commonsense Reasoning [44], Phrase Grounding [34], and more, they only train and evaluate on a single language.

Recently, several multilingual methods have shown better performance on vision-language tasks than complicated transformer-based methods. LIWE [41] is a light-weight character embedding model that can represent many languages with few model parameters. LIWE uses a bidirectional gated recurrent unit (GRU) [9] to aggregate 24-D character embeddings for a text query that is encouraged to closely embed semantically similar images and sentences in other languages. Although LIWE represents a single language well, it suffers from significant performance loss when co-training on multiple languages as shown in Fig. 1(b). Gella *et al.* [16] learns how to relate an image to language-specific representations and also constrain semantically similar sentences across languages to embed nearby each other. MULE [23] learns a universal language embedding so that it can use a single language branch in the multimodal model, significantly reducing the number of parameters required to represent each language compared to Gella *et al.*. In addition, MULE combined the same cross-lingual constraints used in both Gella *et al.* and LIWE with an adversarial language classifier to further encourage alignment across languages. This results in a model that slightly improves performance as more languages are added as shown Fig. 1(b). However, MULE learns a word-level embedding that requires significantly more parameters than LIWE (approximately 8x more with ten languages), and thus capacity concerns remain when scaling to many languages.

3   Scalable Multilingual Aligned Language Representation

In this section we describe how we train our Scalable Multilingual Aligned Language Representation (SMALR) to bridge the gap between scalability and downstream vision-language task performance. To accomplish this, we assume we are provided with an image and sentences that describe it in multiple languages. The intuition behind our model is to first learn a universal language embedding which represents all languages, and then learn to relate it to corresponding images using a multimodal model. In our experiments our multimodal model uses a modified version [23] of the Embedding Network architecture [40], although our approach can be easily adapted to use other multimodal models. After obtaining image and sentence features, the Embedding Network uses two branches, one for each modality, and projects them into a joint semantic space where distances are meaningful. The image branch consists of two fully connected layers, while the language branch obtains a sentence representation by passing the final hidden state of a GRU through a fully connected layer.

Our approach is architecturally similar to MULE [23], but with notable distinctions. First, MULE learned a unique word embedding for every word in every language (*i.e.*, no shared tokens), whereas we learn an efficient universal embedding with our Hybrid Embedding Model (HEM) that consists of a mix of language-agnostic and language-specific word representations (Sect. 3.1). Then, we align our language representations using a novel Masked Cross-Language Model (MCLM) (Sect. 3.2) on both the input of the multimodal model and the final language representation of the multimodal model. This acts to supplement the neighborhood constraints, adversarial language classifier, and image-sentence matching losses used by MULE that we briefly review in Sect. 3.3. Finally, we also propose a Cross-Lingual Consistency (CLC) module that boosts model performance in downstream vision-language tasks using machine translation (Sect. 3.4). See Fig. 2 for an overview of our approach.

3.1 Efficient Multilingual Learning with a Hybrid Embedding Model

A significant challenge in multilingual representation learning is scaling to many languages, especially when there is a wide disparity in the available training data of different languages. This is more apparent for vision-language tasks where annotations are very expensive to collect, making it more difficult to learn a good visually-semantically aligned language representation than in monolingual settings [8, 26]. Inspired by work in low-resource neural machine translation [18], we propose a Hybrid Embedding Model (HEM) which projects low-frequency words across languages into a shared latent vocabulary, while allowing the top-K most frequent words in each language to maintain their own unique (language-specific) representation. The output of the HEM is the universal language embedding that is used as input to the multimodal model in Fig. 2 and is also used in the language alignment losses (Sect. 3.2 and Sect. 3.3). The value of K can be determined experimentally for any targeted downstream task; we use $K = 5000$.

The language-specific word embeddings used for common words roughly follow the implementation used in prior work [18,23]. We begin by using a monolingual pretrained FastText embedding [11] that has been reduced from 300-D to 50-D using Principal Component Analysis (PCA) [30]. These reduced features are used as input to a fully connected layer that projects them into a 512-D universal embedding space that we align across languages; the alignment is applied with the language-agnostic representations as well (see Sect. 3.2 and 3.3 for details on our language alignment procedures).

While our language-agnostic representation is similar to Gu *et al.* [18], it has some key differences. Specifically, Gu *et al.* project all words into a universal embedding space with learned language-specific mappings. A soft-attention module is used over the universal embedding features (as it assumes an aligned cross-lingual input) to obtain mixing weights; these weights are then used to combine the language-agnostic features. While this does enable feature sharing across languages, it does not reduce the number of trainable parameters in the network, as a language-specific representation is still necessary for all words in the vocabulary. Additionally, aggregating the features in the latent vocabulary using soft-attention weights per-word is costly, especially for large vocabularies. Instead, we perform a pretraining step where we learn both the initial representation of the latent vocabulary as well as how to assign the infrequent words to entries in it. We use a hard attention mechanism that is directly predicted from FastText features, in which each vocabulary word is mapped to only a single language-agnostic token, as opposed to an interpolation of many. This allows us to avoid both computing a language-specific representation for the uncommon words and aggregating the latent vocabulary features on a per-word basis.

To obtain our latent shared vocabulary in the pretraining step, we learn to embed semantically similar sentences near each other using a triplet loss. More formally, given a triplet of items (x, y^+, y^-) that can be decomposed into a positive pair (x, y^+) and a negative pair (x, y^-), a triplet loss is computed as:

$$L_{triplet}(x, y^+, y^-) = \max(0, m + d(x, y^+) - d(x, y^-)) \qquad (1)$$

where $d(x, y)$ is a distance function, and m is a scalar parameter. We use cosine distance for all triplet losses and set $m = 0.05$. Following the methodology of [23,40], we construct minibatches by providing semantically similar sentence pairs as input and consider any non-paired sentence as a negative example. These negatives are randomly sampled from each minibatch. We enumerate all triplets in the minibatch and compute the loss over the top-N most violated constraints, where $N = 10$ in our experiments. Note that these sentences may not come from the same language, so sentences referring to the same image in different languages are also used as positive pairs. To predict which latent embedding we map a source word to, we use sentence representations obtained by feeding FastText embeddings into a fully connected layer. With this mapping, we average the latent embeddings of each word for use in Eq. (1) during the pretraining step, which has been shown to be an efficient, high-performing representation [4,8].

Instead of deterministically mapping to the latent token which achieves the best score, we randomly choose from the top M scoring tokens with probability p, which we refer to as exploration parameters. This helps ensure that spurious mappings are not learned, typically resulting in a 2% performance improvement (see supplementary for a detailed comparison). While we freeze the latent token assignments when training the full model, we allow the features themselves to be fine-tuned. Our experiments use a latent vocabulary size of $40K$ tokens, with exploration parameters $p = 0.2$, $M = 20$. In practice not all latent tokens are used at the end of pretraining; these are dropped when training the full model.

3.2   Masked Cross-Language Modeling (MCLM)

Masked Language Modeling has proven to be useful in training language representations by masking some tokens of an input sentence and then trying to predict the missing tokens [13]. We present a generalization of this approach to a multilingual scenario to encourage stronger cross-language alignment. In MCLM, we assume we have paired sentences across different languages. These sentences need not be direct translations of each other, but, as our experiments will show, they simply need to be semantically related. This is important as vision-language datasets do not always have paired text queries that are direct translations, but are often independently generated instead (*e.g.* [15,27,31]).

Traditional Masked Language Modeling makes predictions about a single masked token using its surrounding words as context. The words immediately surrounding a token referring to the same entity between sentences in different languages may vary greatly due to differences in grammar. Thus, even using a dictionary between languages to identify word correspondences may not provide useful context. Instead, we use the intuition that semantically similar sentences should contain comparable information across languages, so a sentence in one language could be used as context to predict missing information in another. Conneau *et al.* [10] similarly use masking for improved language alignment. However, our approach does not require parallel data and may sample amongst any of the languages. Lastly, unlike [10] which computes its loss on the predicted word, our objective in Eq. (2) is computed on the fully reconstructed sentences.

More formally, for a pair of languages (i, j), we obtain sentences (S_i, S_j) such that both sentences describe the same image (*i.e.*, they are semantically similar to each other). Then, we randomly replace some portion of their words with a special MASK token to obtain masked representations (S_i^m, S_j^m). These are concatenated together and fed into a fully connected layer that is shared across language pairs to predict the missing information in both sentences $(S_i^{'}, S_j^{'})$. Our MCLM loss then compares this to the unmasked sentences, *i.e.*,

$$L_{mask} = ||\ell_2(S_i^m + S_i^{'}) - \ell_2(S_i)|| + ||\ell_2(S_j^m + S_j^{'}) - \ell_2(S_j)||, \qquad (2)$$

where ℓ_2 identifies vectors forced to have unit norm. Both average embedding and LSTM representations are used; details can be found in the supplementary. We compute the masking loss in Eq. (2) for all unique pairs of languages in our experiments, and found masking 20% of the words in the sentences worked best.

3.3   Multilingual Visual-Semantic Alignment

In this section we briefly review the visual-semantic alignment constraints used by MULE [23] that we also employ. First, we use neighborhood constraints [40] that we shall refer to as L_{nc} to encourage similar sentences to embed nearby each other using a triplet loss (*i.e.*, Eq. (1)). Just as with the MCLM module described in Sect. 3.2, these neighborhood constraints are applied to both the universal language embedding (*i.e.*, the output of the HEM module) as well as the final language representation from the multimodal model as shown in Fig. 2. The second component of the MULE alignment constraint consists of an adversarial language classifier. We shall refer to this classifier loss as L_{adv}, using the approach of [23], whose goal is to ensure that the representations of the different languages in the universal embedding have similar feature distributions.

The last component of the MULE constraint is used to train the multimodal model to embed the images and sentences near each other using a triplet loss. This uses a bidirectional triplet loss function, $i.e.$, for image I and paired sentences (Q^+, Q^-) representing a positive and negative sentence pair, respectively, and sentence Q and its paired images (I^+, I^-), this multimodal loss would be,

$$L_{mm} = L_{triplet}(I, Q^+, Q^-) + \lambda_1 L_{triplet}(Q, I^+, I^-) \quad (3)$$

where λ_1 is a scalar parameter, which we set to 1.5 in our experiments. In addition to using the unmasked sentence representations for the multimodal loss, we observe that most sentences retain their overall semantic meaning if you remove just a few words at random. Using this intuition, we also compute Eq. (3) using the masked sentences (S_i^m, S_j^m) from the MCLM module, which we found provides a small, but consistent improvement to performance. As a reminder, all triplet losses use the implementation details ($e.g.$ hyperparameter settings and hard-negative mining) as described in the first part of Sect. 3. Our total loss function to train SMALR is then,

$$L_{SMALR} = L_{mm} + \lambda_2 L_{mask} + \lambda_3 L_{adv} + \lambda_4 L_{nc} \quad (4)$$

where λ_{2-4} are scalar parameters that we set to (1e-4, 1e-6, 5e-2), respectively.

3.4 Cross-Lingual Consistency

Prior work on multilingual vision-language tasks has primarily focused on how to change training procedures or architectures to support multiple languages, and does not fully take advantage of this multilingual support at test time. In particular, we argue that semantically similar sentences in different languages may capture complementary information, and therefore, considering the predictions made in other languages may improve performance. We validate our intuition by obtaining machine translations of a query in the other languages supported by our model. More formally, suppose we have a set of languages L. Given a query q in language $l_i \in L$, we translate q to all other supported languages in $L \setminus \{l_i\}$ and use this as input to our Cross-Lingual Consistency (CLC) module.

We propose two variants of CLC: CLC-A and CLC-C. CLC-A simply averages matching scores over all languages, and does not require any additional parameters. CLC-C, on the other hand, uses a small Multilayer Perceptron (MLP) to aggregate the scores of each language, which enables us to consider the relative information present in each language's predictions. This MLP has two layers with input size $|L|$ and 32 hidden layer units ($i.e.$, it has 352 learnable parameters) and all parameters are initialized with uniform weight. We train the CLC-C module separately to SMALR using the validation set for 30 iterations. No minibatches are employed ($i.e.$, it is trained with all image-sentence pairs at once) and it is trained using the multimodal triplet loss described in Eq. (3).

4 Experimental Setup

Datasets. SMALR is evaluated on bidirectional retrieval with Multi30K [6,14, 15] and MSCOCO [27,28,31]. The Multi30K dataset is built off of Flickr30K [43], which originally contained 31,783 images and five English descriptions per image. [6,14,15] obtained annotations in German, French, and Czech, resulting in a four-language dataset. Multi30K contains five descriptions per image in English and German, but only one per image in French and Czech; the latter two were collected as human-generated translations of the English annotations. We use the 29K/1K/1K train/test/val splits from the original dataset [43].

MSCOCO is approximately four times the size of Multi30K, with 123,287 images. There are five human-generated captions per image in English, but significantly fewer in Chinese and Japanese. YJ Captions [31] introduced Japanese annotations for MSCOCO, but only provides five captions per image for a subset of about 26K images. [27] extended MSCOCO with a total of 22,218 Chinese captions for 20,341 images. We use train/test/validation splits as defined in [23].

We augment both datasets with machine translations so every image contains at least five sentences for ten languages: English, German, Czech, French, Chinese, Japanese, Arabic, Afrikaans, Korean, and Russian. All models we compare to are trained using this augmented training set. For languages with no human-generated sentences, we use machine translations at test time as well. We found using translations at test time did not affect the relative performance of different methods in our experiments. See the supplementary for details.

Visual Features. We use ResNet-152 [20] features trained on ImageNet [12] as input to the Embedding Network (EmbN) [40]. As done in [23], we average visual features over ten 448×448 image crops. This generates an image embedding of size 2048, which is then passed through a pair of fully connected layers. The resulting 512-D embedding can be used in the shared image-sentence embedding space. The learning rate was set to $1e^{-3}$ for the HEM and LA models; remaining hyperparameters are consistent with those in [23].

Note that all LIWE [41] experiments use bottom-up Faster R-CNN [35] visual features trained on Visual Genome [25]. This represents a significant increase in the annotation cost compared to our approach, which doesn't use these annotations. Visual Genome also contains MSCOCO [28] images, which means that there is train/test contamination, as LIWE's features are extracted using the pretrained, publicly available model from [2].

Metrics. For our results, we report the mean Recall (mR) across Recall@K, with $K \in [1, 5, 10]$, for both the image-sentence and sentence-image directions per language. All recall values can be found in the supplementary. We also provide an average mR across all languages to serve as a global performance metric: "A" in Tables 1 and 2. The human average, "HA," refers to the average mR over the languages which have human-generated annotations (*i.e.* English, Chinese, and Japanese for MSCOCO, and English, German, French, and Czech for Multi30K).

Comparative Evaluation. We compare the following methods:

- **Frequency Thresholding:** We drop words that occur fewer than t times in the training set. Results are reported in Fig. 3.
- **PCA Reduction:** We use PCA [30] to reduce the size of the initial 300-D FastText word embeddings. Results are reported in Fig. 3.
- **Dictionary Mapping:** We map words that occur fewer than t times in non-English languages to English using dictionaries [11]. By mapping rare words in other languages to English, some information may be lost, but the token will still exist indirectly in the vocabulary. Results are reported in Fig. 3.
- **Language-Agnostic (LA):** We compare to only using a latent vocabulary as described in Sect. 3.1 with 40K tokens, *i.e.*not using any language specific features. Results are in Tables 1 and 2.
- **HEM:** We evaluate our full hybrid embedding model (Sect. 3.1), which uses a mix of language-agnostic and language-specific representations. This baseline does not include MCLM nor CLC. Results are in Tables 1 and 2.
- **SMALR:** Our base SMALR is composed of the HEM (Sect. 3.1) and MCLM (Sect. 3.2) components of our model. We compare to our complete SMALR which makes use of CLC variants (CLC-A and CLC-C, described in Sect. 3.4). Results are in Tables 1 and 2.

Note that the first line of Tables 1 and 2, **Trans To En**, refers to using machine translation on non-English sentences, and then using an English-only trained Embedding Network [40], providing a strong baseline method to compare to.

5   Multilingual Image-Sentence Retrieval Results

We provide results for MSCOCO and Multi30K in Table 1 and Table 2, respectively, which contain comparisons to prior work on fewer languages (a), adaptations of prior work to our setting (b), and our model variants (c). SMALR obtains consistent performance gains when evaluating on ten languages over the state-of-the-art (S-LIWE, line 3(b)) while also being more efficient than high-performing models like MULE (line 5(b)). SMALR outperforms S-LIWE by 11 points on MSCOCO and 5.8 points on Multi30K (line 3(c) versus 3(b)). A parameter comparison is later shown in Fig. 3. SMALR's initial Language-Agnostic (LA) baseline alone is able to boost performance over previous scalable method LIWE by 2–7 points. The HEM, which combines language-agnostic and language-specific embeddings as described in Sect. 3.1, consistently improves upon the fully language-agnostic vocabulary, even though they share the same latent vocabulary size of 40K tokens. This points to the utility of our hybrid embedding space, which improves performance upon LA by 3.4 average mR on MSCOCO and 2.4 average mR on Multi30K while adding only a few parameters.

When MCLM losses are added, referred to as SMALR in Tables 1 and 2 (line 3(c)), mR improves for nearly all languages. This is significant, because we find more compact models like LIWE degrade with additional languages when using the same number of parameters (S-LIWE). The LA baseline is still able

Table 1. MSCOCO multilingual bidirectional retrieval results. (a) contains results from prior work, (b) contains reproductions of two state-of-the art methods evaluated for our scenario using their code, and (c) contains variants of our model

	Model	En	De[a]	Fr[a]	Cs[a]	Cn	Ja	Ar[a]	Af[a]	Ko[a]	Ru[a]	HA	A
(a)	Trans. to En [23]	75.6	–	–	–	72.2	66.1	–	–	–	–	71.3	–
	EmbN [40]	76.8	–	–	–	73.5	73.2	–	–	–	–	74.5	–
	PAR. EmbN [16]	78.3	–	–	–	73.5	76.0	–	–	–	–	75.9	–
	MULE [23]	79.5	–	–	–	74.8	76.3	–	–	–	–	76.9	–
(b)	(1) S-LIWE [41][b]	80.9	–	–	–	–	73.6	–	–	–	–	–	–
	(2) S-LIWE[b]	77.4	–	–	–	–	66.6	–	–	–	–	–	–
	(10) S-LIWE[b]	77.3	67.4	68.5	66.9	64.5	65.8	63.8	66.2	63.1	63.6	69.2	66.7
	(10) L-LIWE[b]	79.1	71.2	70.3	70.1	70.0	69.6	67.5	68.9	66.2	69.6	72.9	70.3
	MULE [23]	79.0	77.2	76.8	77.8	75.6	75.9	77.2	77.8	74.3	77.3	76.8	76.9
(c)	Language-Agnostic	75.0	74.3	74.1	73.4	72.3	72.1	74.4	74.7	71.6	72.7	73.1	73.5
	HEM	78.7	77.3	76.4	77.9	76.7	76.3	77.0	76.7	**75.5**	77.0	77.3	76.9
	SMALR	79.3	**78.4**	**77.8**	**78.6**	76.7	77.2	**77.9**	**78.2**	75.1	**78.0**	77.7	**77.7**
	SMALR-CLC-A	81.2	–	–	–	79.6	75.0	–	–	–	–	78.6	–
	SMALR-CLC-C	**81.5**	–	–	–	**80.1**	**77.5**	–	–	–	–	**79.7**	–

[a] uses translations from English for testing
[b] visual features trained using outside dataset that includes some test images

to outperform L-LIWE on MSCOCO and Multi30K, in which LIWE learns an embedding five fold larger to try to compensate for the increased number and diversity of languages (120-D instead of 24-D embedding). This suggests that the masking process may help regain some semantic information that is lost when tokens are mapped to the language-agnostic space.

We next evaluate two CLC variants that use machine translations at test time (described in Sect. 3.4) on top of SMALR: an average ensemble over all languages (CLC-A), and a weighted ensemble which makes use of a simple classifier (CLC-C). CLC-A uses no additional test-time parameters, and increases the human average performance by 1–3 points, with a larger gain on Multi30K. This may be because more languages can be leveraged on Multi30K (four versus three, compared to MSCOCO). Surprisingly, English performance improves the most amongst CLC-A metrics on Multi30K, demonstrating that certain image-sentence pairs can be better retrieved from the queries in other languages, which may better capture the visual semantics of the same image. CLC-C further improves the human average over CLC-A by 0.9 points on MSCOCO and 0.5 points on Multi30K, using negligible additional parameters.

Parameter Reduction Method Comparison. We present a comparison of baseline vocabulary reduction techniques, described in Sect. 4, against prior works LIWE and MULE, in addition to our method SMALR (consisting of only HEM and MCLM components in Fig. 3). The frequency thresholding and dictionary mapping labels represent the threshold with which we drop infrequent words or map them to English (*e.g.* the blue 50 data point represents dropping words that occur fewer than 50 times). PCA point labels represent the dimensionality we reduce our input vectors to (*e.g.* 300D → 50D, 100D, or 200D).

Table 2. Multi30K multilingual bidirectional retrieval results. (a) contains results from prior work, (b) contains reproductions of two state-of-the art methods evaluated for our scenario using their code, and (c) contains variants of our model

	Model	En	De	Fr	Cs	Cn[a]	Ja[a]	Ar[a]	Af[a]	Ko[a]	Ru[a]	HA	A
(a)	Trans. to En [23]	71.1	48.5	46.7	46.9	–	–	–	–	–	–	53.3	–
	EmbN [40]	72.0	60.3	54.8	46.3	–	–	–	–	–	–	58.4	–
	PAR. EmbN [16]	69.0	62.6	60.6	54.1	–	–	–	–	–	–	61.6	–
	MULE [23]	70.3	64.1	62.3	57.7	–	–	–	–	–	–	63.6	–
(b)	(1) S-LIWE [41][b]	**76.3**	**72.1**	–	–	–	–	–	–	–	–	–	–
	(2) S-LIWE[b]	75.6	66.1	–	–	–	–	–	–	–	–	–	–
	(10) S-LIWE[b]	75.2	65.2	51.8	50.0	54.1	56.2	62.7	62.8	54.5	63.1	60.6	59.6
	(10) L-LIWE[b]	75.9	66.7	53.3	51.3	56.9	56.3	65.0	63.7	57.1	65.4	61.9	61.2
	MULE [23]	70.7	63.6	63.4	59.4	**64.2**	**67.3**	65.8	67.3	63.6	65.4	64.3	65.1
(c)	Language-Agnostic	65.5	61.3	59.9	54.0	59.4	64.7	63.9	66.5	60.3	60.3	60.2	61.6
	HEM	69.2	62.8	63.3	60.0	62.4	66.3	64.5	66.8	62.3	62.6	63.8	64.0
	SMALR	69.6	64.7	64.5	61.1	64.0	66.7	**66.0**	**67.4**	**64.2**	**65.7**	65.0	**65.4**
	SMALR-CLC-A	74.1	68.9	65.2	64.5	–	–	–	–	–	–	68.2	–
	SMALR-CLC-C	74.5	69.8	**65.9**	**64.8**	–	–	–	–	–	–	**68.7**	–

[a]uses translations from English for testing
[b]visual features trained using outside dataset

In our comparison of vocabulary reduction methods, frequency thresholding with $t = 50$ and vanilla language-agnostic vocabularies (LA) obtain better performance than both LIWE variants on Multi30K, without adding significantly more parameters, as shown on the right of Fig. 3. While more model parameters are needed for MSCOCO, due to the increased vocabulary size, all baselines and prior work MULE significantly outperform LIWE. This demonstrates that more-complex character-based models do not necessarily obtain competitive performance with few parameters when addressing a larger multilingual scenario.

SMALR outperforms all baselines for MSCOCO, as seen on the left of Fig. 3, outperforming S-LIWE by over 10 points and using fewer parameters than L-LIWE. We also find that average mean recall performance on MSCOCO is more robust to vocabulary reduction, with a maximum range of about 1.5 average mR between the most extreme reduction and the least. We believe this may be due to the size discrepancy between the two datasets, as MSCOCO is approximately four times the size of Multi30K. PCA reduction appears to have a more linear effect as parameters increase on both datasets. Since Multi30K performance is more sensitive to the number of parameters, it is significant that our SMALR model, in green, (which does not yet make use of our cross-lingual consistency module in Fig. 3) outperforms all other models while having less than 20M parameters, 1/5th the parameter count of high performing MULE.

In addition to SMALR outperforming MULE on both datasets while using significantly fewer trainable parameters, we find MULE even fails to outperform simple baselines such as dictionary mapping on MSCOCO. This exposes that the large number of parameters used in MULE are unnecessary for performance gains. While SMALR uses more parameters during training than S-LIWE, we have far fewer test-time parameters. We reduce the computation needed for

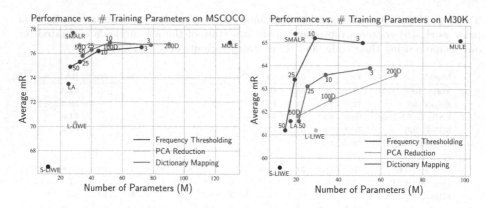

Fig. 3. We compare three types of vocabulary reduction: frequency thresholding, PCA dimensionality reduction, and mapping rare words to English with the use of dictionaries. The left-hand side evaluates on MSCOCO, the right on Multi30K. We have additional standalone points for the small LIWE (S-LIWE), large LIWE (L-LIWE), MULE, language agnostic vocabulary (LA), and our model, SMALR

evaluation by using precomputed language representations from training. This reduces the entire SMALR model to the image-sentence matching model with our CLC add-on, totaling only 7.1M parameters, now fewer than S-LIWE.

6   Conclusion

We have presented a Scalable Multilingual Aligned Representation (SMALR), which addresses the trade-off between multilingual model size and downstream vision-language task performance. Our approach is modular, and thus can be used as a drop-in language representation for any vision-language method/task. SMALR outperforms all prior work on the task of multilingual image-sentence retrieval on average across ten diverse languages, with the use of a hybrid embedding model, masked cross-language modeling loss, and cross-lingual consistency module. Our hybrid embedding model significantly reduces the input to a language model by mapping most tokens to a fixed size, shared vocabulary. The masking procedure aligns our diverse set of languages and uses the multimodal model to provide additional alignment with visual grounding. We find that both cross-lingual consistency modules better aggregates retrieved results, boosting performance with minimal additional parameters. This is all accomplished with less than 20M trainable parameters, significantly reducing oversized prior work by 1/5th, while improving performance over the state-of-the-art by 3–4%.

Acknowledgements. This work is funded in part by the NSF, DARPA LwLL, and DARPA XAI grants, including NSF grant 1838193.

References

1. Aharoni, R., Johnson, M., Firat, O.: Massively multilingual neural machine translation. In: Proceedings of the 2019 Conference of the North American Chapter of the Association for Computational Linguistics: Human Language Technologies, Volume 1 (Long and Short Papers), June 2019
2. Anderson, P., et al.: Bottom-up and top-down attention for image captioning and visual question answering. In: The IEEE Conference on Computer Vision and Pattern Recognition (CVPR) (2018)
3. Antol, S., et al.: VQA: visual question answering. In: The IEEE International Conference on Computer Vision (ICCV) (2015)
4. Arora, S., Liang, Y., Ma, T.: A simple but tough-to-beat baseline for sentence embeddings. In: International Conference on Learning Representations (ICLR) (2017)
5. Artetxe, M., Labaka, G., Agirre, E.: Learning principled bilingual mappings of word embeddings while preserving monolingual invariance. In: Empirical Methods in Natural Language Processing (EMNLP), pp. 2289–2294 (2016)
6. Barrault, L., Bougares, F., Specia, L., Lala, C., Elliott, D., Frank, S.: Findings of the third shared task on multimodal machine translation. In: Proceedings of the Third Conference on Machine Translation: Shared Task Papers, pp. 304–323 (2018)
7. Bojanowski, P., Grave, E., Joulin, A., Mikolov, T.: Enriching word vectors with subword information. Trans. Assoc. Comput. Linguist. (TACL) 5, 135–146 (2017)
8. Burns, A., Tan, R., Saenko, K., Sclaroff, S., Plummer, B.A.: Language features matter: effective language representations for vision-language tasks. In: The IEEE International Conference on Computer Vision (ICCV) (2019)
9. Cho, K., van Merrienboer, B., Gülçehre, Ç., Bougares, F., Schwenk, H., Bengio, Y.: Learning phrase representations using RNN encoder-decoder for statistical machine translation. In: Empirical Methods in Natural Language Processing (EMNLP) (2014)
10. Conneau, A., Lample, G.: Cross-lingual language model pretraining. In: Advances in Neural Information Processing Systems (NeurIPS) (2019)
11. Conneau, A., Lample, G., Ranzato, M., Denoyer, L., Jégou, H.: Word translation without parallel data. In: International Conference on Learning Representations (ICLR) (2018)
12. Deng, J., Dong, W., Socher, R., Li, L.J., Li, K., Fei-Fei, L.: ImageNet: a large-scale hierarchical image database. In: The IEEE Conference on Computer Vision and Pattern Recognition (CVPR) (2009)
13. Devlin, J., Chang, M.W., Lee, K., Toutanova, K.: Bert: pre-training of deep bidirectional transformers for language understanding. arXiv:1810.04805v1 (2018)
14. Elliott, D., Frank, S., Barrault, L., Bougares, F., Specia, L.: Findings of the second shared task on multimodal machine translation and multilingual image description. arXiv:1710.07177 (2017)
15. Elliott, D., Frank, S., Sima'an, K., Specia, L.: Multi30k: multilingual English-German image descriptions. arXiv:1605.00459 (2016)
16. Gella, S., Sennrich, R., Keller, F., Lapata, M.: Image pivoting for learning multilingual multimodal representations. In: Empirical Methods in Natural Language Processing (EMNLP) (2017)

17. Goyal, Y., Khot, T., Summers-Stay, D., Batra, D., Parikh, D.: Making the V in VQA matter: elevating the role of image understanding in Visual Question Answering. In: The IEEE Conference on Computer Vision and Pattern Recognition (CVPR) (2017)
18. Gu, J., Hassan, H., Devlin, J., Li, V.O.: Universal neural machine translation for extremely low resource languages. In: Proceedings of the 2018 Conference of the North American Chapter of the Association for Computational Linguistics: Human Language Technologies (ACL-HLT) (2018)
19. Gupta, T., Schwing, A., Hoiem, D.: Vico: word embeddings from visual co-occurrences. In: The IEEE International Conference on Computer Vision (ICCV) (2019)
20. He, K., Zhang, X., Ren, S., Sun, J.: Deep residual learning for image recognition. arXiv:1512.03385 (2015)
21. K, K., Wang, Z., Mayhew, S., Roth, D.: Cross-lingual ability of multilingual bert: an empirical study. arXiv:1912.07840 (2019)
22. Kazemzadeh, S., Ordonez, V., Matten, M., Berg, T.: ReferitGame: referring to objects in photographs of natural scenes. In: Empirical Methods in Natural Language Processing (EMNLP) (2014)
23. Kim, D., Saito, K., Saenko, K., Sclaroff, S., Plummer, B.A.: Mule: multimodal universal language embedding. In: AAAI Conference on Artificial Intelligence (2020)
24. Klein, B., Lev, G., Sadeh, G., Wolf, L.: Fisher vectors derived from hybrid Gaussian-Laplacian mixture models for image annotation. In: The IEEE Conference on Computer Vision and Pattern Recognition (CVPR) (2015)
25. Krishna, R., et al.: Visual genome: Connecting language and vision using crowd-sourced dense image annotations. Int. J. Comput. Vis. (IJCV) (2017)
26. Li, L.H., Yatskar, M., Yin, D., Hsieh, C.J., Chang, K.W.: VisualBERT: a simple and performant baseline for vision and language. arXiv:1908.03557 (2019)
27. Li, X., et al.: COCO-CN for cross-lingual image tagging, captioning and retrieval. IEEE Trans. Multimedia (2019)
28. Lin, T.-Y., et al.: Microsoft COCO: common objects in context. In: Fleet, D., Pajdla, T., Schiele, B., Tuytelaars, T. (eds.) ECCV 2014. LNCS, vol. 8693, pp. 740–755. Springer, Cham (2014). https://doi.org/10.1007/978-3-319-10602-1_48
29. Lu, J., Batra, D., Parikh, D., Lee, S.: ViLBERT: pretraining task-agnostic visiolinguistic representations for vision-and-language tasks. arXiv:1908.02265 (2019)
30. Maćkiewicz, A., Ratajczak, W.: Principal components analysis (PCA). Comput. Geosci. 19(3), 303–342 (1993)
31. Miyazaki, T., Shimizu, N.: Cross-lingual image caption generation. In: Conference of the Association for Computational Linguistics (ACL) (2016)
32. Nguyen, D.K., Okatani, T.: Multi-task learning of hierarchical vision-language representation. In: The IEEE Conference on Computer Vision and Pattern Recognition (CVPR) (2019)
33. Pires, T., Schlinger, E., Garrette, D.: How multilingual is multilingual bert? arXiv:1906.01502 (2019)
34. Plummer, B.A., Wang, L., Cervantes, C.M., Caicedo, J.C., Hockenmaier, J., Lazebnik, S.: Flickr30k entities: collecting region-to-phrase correspondences for richer image-to-sentence models. In: The IEEE International Conference on Computer Vision (ICCV) (2015)
35. Ren, S., He, K., Girshick, R., Sun, J.: Faster R-CNN: towards real-time object detection with region proposal networks. In: Advances in Neural Information Processing Systems (NeurIPS) (2015)

36. Smith, S.L., Turban, D.H.P., Hamblin, S., Hammerla, N.Y.: Offline bilingual word vectors, orthogonal transformations and the inverted softmax. arXiv:1702.03859 (2017)
37. Su, W., et al.: Vl-BERT: pre-training of generic visual-linguistic representations. arXiv:1908.08530 (2019)
38. Tan, H., Bansal, M.: Lxmert: learning cross-modality encoder representations from transformers. In: Proceedings of the 2019 Conference on Empirical Methods in Natural Language Processing (EMNLP) (2019)
39. Vaswani, A., et al.: Attention is all you need. In: Guyon, I., et al. (eds.) Advances in Neural Information Processing Systems (NeurIPS), pp. 5998–6008 (2017)
40. Wang, L., Li, Y., Huang, J., Lazebnik, S.: Learning two-branch neural networks for image-text matching tasks. IEEE Trans. Pattern Anal. Mach. Intell.(TPAMI) **41**(2), 394–407 (2018)
41. Wehrmann, J., Souza, D.M., Lopes, M.A., Barros, R.C.: Language-agnostic visual-semantic embeddings. In: The IEEE International Conference on Computer Vision (ICCV) (2019)
42. Wu, S., Dredze, M.: Beto, Bentz, Becas: the surprising cross-lingual effectiveness of Bert. arXiv:1904.09077 (2019)
43. Young, P., Lai, A., Hodosh, M., Hockenmaier, J.: From image descriptions to visual denotations: new similarity metrics for semantic inference over event descriptions. Trans. Assoc. Comput. Linguist. (TACL) **2**, 67–78 (2014)
44. Zellers, R., Bisk, Y., Farhadi, A., Choi, Y.: From recognition to cognition: visual commonsense reasoning. In: The IEEE Conference on Computer Vision and Pattern Recognition (CVPR) (2019)

Multi-modal Transformer
for Video Retrieval

Valentin Gabeur[1,2]([✉]), Chen Sun[2], Karteek Alahari[1], and Cordelia Schmid[2]

[1] Inria, Univ. Grenoble Alpes, Inria, CNRS, Grenoble INP, LJK,
38000 Grenoble, France
karteek.alahari@inria.fr
[2] Google Research, Meylan, France
{valgab,chensun,cordelias}@google.com

Abstract. The task of retrieving video content relevant to natural language queries plays a critical role in effectively handling internet-scale datasets. Most of the existing methods for this caption-to-video retrieval problem do not fully exploit cross-modal cues present in video. Furthermore, they aggregate per-frame visual features with limited or no temporal information. In this paper, we present a multi-modal transformer to jointly encode the different modalities in video, which allows each of them to attend to the others. The transformer architecture is also leveraged to encode and model the temporal information. On the natural language side, we investigate the best practices to jointly optimize the language embedding together with the multi-modal transformer. This novel framework allows us to establish state-of-the-art results for video retrieval on three datasets. More details are available at http://thoth.inrialpes.fr/research/MMT.

Keywords: Video · Language · Retrieval · Multi-modal · Cross-modal · Temporality · Transformer · Attention

1 Introduction

Video is one of the most popular forms of media due to its ability to capture dynamic events and its natural appeal to our visual and auditory senses. Online video platforms are playing a major role in promoting this form of media. However, the billions of hours of video available on such platforms are unusable if we cannot access them effectively, for example, by retrieving relevant content through queries.

In this paper, we tackle the tasks of caption-to-video and video-to-caption retrieval. In the first task of caption-to-video retrieval, we are given a query in the form of a caption (e.g., "How to build a house") and the goal is to retrieve

---

Electronic supplementary material The online version of this chapter (https://doi.org/10.1007/978-3-030-58548-8_13) contains supplementary material, which is available to authorized users.

A. Vedaldi et al. (Eds.): ECCV 2020, LNCS 12349, pp. 214–229, 2020.
https://doi.org/10.1007/978-3-030-58548-8_13

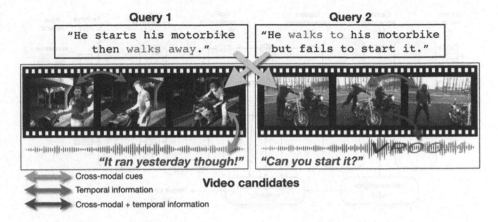

Fig. 1. When matching a text query with videos, the inherent cross-modal and temporal information in videos needs to be leveraged effectively, for example, with a video encoder that handles all the constituent modalities (appearance, audio, speech) jointly across the entire duration of the video. In this example, a video encoder will be able to distinguish between "someone walking *to*" and "someone walking *away*" only if it exploits the temporal information of events occurring in the video (red arrows). Also, in order to understand that a "motorbike failed to start", it needs to use cross-modal information (e.g., absence of noise after someone tried to start the engine, orange arrow). (Color figure online)

the videos best described by it (i.e., videos explaining how to build a house). In practice, given a test set of caption-video pairs, our aim is to provide, for each caption query, a ranking of all the video candidates such that the video associated with the caption query is ranked as high as possible. On the other hand, the task of video-to-caption retrieval focuses on finding among a collection of caption candidates the ones that best describe the query video.

A common approach for the retrieval problem is similarity learning [29], where we learn a function of two elements (a query and a candidate) that best describes their similarity. All the candidates can then be ranked according to their similarity with the query. In order to perform this ranking, the captions as well as the videos are represented in a common multi-dimensional embedding space, wherein similarities can be computed as a dot product of their corresponding representations. The critical question here is how to learn accurate representations of both caption and video to base our similarity estimation on.

The problem of learning representation of text has been extensively studied, leading to various methods [3,7,18,25,34], which can be used to encode captions. In contrast to these advances, learning effective video representation continues to be a challenge, and forms the focus of our work. This is in part due to the multimodal and temporal nature of video. Video data not only varies in terms of appearance, but also in possible motion, audio, overlaid text, speech, etc. Leveraging cross-modal relations thus forms a key to building effective video representations. As illustrated in Fig. 1, cues jointly extracted from all the constituent

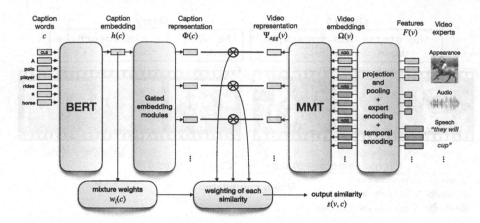

Fig. 2. Our cross-modal framework for similarity estimation. We use our Multi-modal Transformer (MMT, right) to encode video, and BERT (left) for text.

modalities are more informative than handling each modality independently. Hearing a motor sound right after seeing someone starting a bike tells us that the running bike is the visible one and not a background one. Another example is the case of a video of "a crowd listening to a talk", neither of the modalities "appearance" or "audio" can fully describe the scene, but when processed together, higher level semantics can be obtained.

Recent work on video retrieval does not fully exploit such cross-modal high-level semantics. They either ignore the multi-modal signal [15], treat modalities separately [16], or only use a gating mechanism to modulate certain modality dimensions [14]. Another challenge in representing video is its temporality. Due to the difficulty in handling variable duration of videos, current approaches [14, 16] discard long-term temporal information by aggregating descriptors extracted at different moments in the video. We argue that this temporal information can be important to the task of video retrieval. As shown in Fig. 1, a video of "someone walking *to* an object" and "someone walking *away* from an object" will have the same representation once pooled temporally, however, the movement of the person relative to the object is potentially important in the query.

We address the temporal and multi-modal challenges posed in video data by introducing our multi-modal transformer. It performs the task of processing features extracted from different modalities at different moments in video and aggregates them in a compact representation. Building on the transformer architecture [25], our multi-modal transformer exploits the self-attention mechanism to gather valuable cross-modal and temporal cues about events occurring in a video. We integrate our multi-modal transformer in a cross-modal framework, as illustrated in Fig. 2, which leverages both captions and videos, and estimates their similarity.

Contributions. In this work, we make the following three contributions: (i) First, we introduce a novel video encoder architecture for retrieval: Our multi-modal transformer processes effectively multiple modality features extracted at different times. (ii) We thoroughly investigate different architectures for language embedding, and show the superiority of the BERT model for the task of video retrieval. (iii) By leveraging our novel cross-modal framework we outperform prior state of the art for the task of video retrieval on MSRVTT [30], ActivityNet [12] and LSMDC [21] datasets. It is also the winning solution in the CVPR 2020 Video Pentathlon Challenge [4].

2 Related Work

We present previous work on language and video representation learning, as well as on visual-language retrieval.

Language Representations. Earlier work on language representations include bag of words [34] and Word2Vec [18]. A limitation of these representations is capturing the sequential properties in a sentence. LSTM [7] was one of the first successful deep learning models to handle this. More recently, the transformer [25] architecture has shown impressive results for text representation by implementing a self-attention mechanism where each word (or wordpiece [27]) of the sentence can attend to all the others. The transformer architecture, consisting of self-attention layers alternatively stacked with fully-connected layers, forms the base of the popular language modeling network BERT [3]. Burns et al. [1] perform an analysis of the different word embeddings and language models (Word2Vec [18], LSTM [7], BERT [3], etc.) used in vision-language tasks. They show that the pretrained and frozen BERT model [3] performs relatively poorly compared to a LSTM or even a simpler average embedding model. In this work, we show that for video retrieval, a pretrained BERT outperforms other language models, but it needs to be finetuned.

Video Representations. With a two-stream network, Simonyan et al. [22] have used complementary information from still frames and motion between frames to perform action recognition in videos. Carreira et al. [2] incorporated 3D convolutions in a two-stream network to better attend the temporal structure of the signal. S3D [28] is an alternative approach, which replaced the expensive 3D spatio-temporal convolutions by separable 2D and 1D convolutions. More recently, transformer-based methods, which leverage BERT pretraining [3], have been applied to S3D features in VideoBERT [24] and CBT [23]. While these works focus on visual signals, they have not studied how to encode the other multi-modal semantics, such as audio signals.

Visual-Language Retrieval. Harwath et al. [5] perform image and audio-caption retrieval by embedding audio segments and image regions in the same space and requiring high similarity between each audio segment and its corresponding image region. The method presented in [13] takes a similar approach for image-text retrieval by embedding images regions and words in a joint space. A high similarity is obtained for images that have matching words and image regions.

For videos, JSFusion [31] estimates video-caption similarity through dense pairwise comparisons between each word of the caption and each frame of the video. In this work, we instead estimate both a video embedding and a caption embedding and then compute the similarity between them. Zhang et al. [33] perform paragraph-to-video retrieval by assuming a hierarchical decomposition of the video and paragraph. Our method do not assume that the video can be decomposed into clips that align with sentences of the caption. A recent alternative is creating separate embedding spaces for different parts of speech (e.g., noun or verb) [26]. In contrast to this method, we do not pre-process the sentences but encode them directly through BERT.

Another work [17] leverages the large number of instructional videos in the HowTo100M dataset, but does not fully exploit the temporal relations. Our work instead relies on longer segments extracted from HowTo100M videos in order to learn temporal dependencies and address the problem of misalignment between speech and visual features. Mithun et al. [19,20] use three experts (Object, Activity and Place) to compute three corresponding text-video similarities. These experts however do not collaborate together as their respective similarities are simply summed together. A related approach [16] uses precomputed features from experts for text to video retrieval, where the overall similarity is obtained as a weighted sum of each expert's similarity. A recent extension [14] to this mixture of experts model uses a collaborative gating mechanism for modulating each expert feature according to the other experts. However, this collaborative gating mechanism only strengthens (or weakens) some dimensions of the input signal in a single step, and is therefore not able to capture high level inter-modality information. Our multi-modal transformer overcomes this limitation by attending to all available modalities over multiple self-attention layers.

3   Methodology

Our overall method relies on learning a function s to compute the similarity between two elements: text and video, as shown in Fig. 2. We then rank all the videos (or captions) in the dataset, according to their similarity with the query caption (or video) in the case of text-to-video (or video-to-text) retrieval. In other words, given a dataset of n video-caption pairs $\{(v_1, c_1), ..., (v_n, c_n)\}$, the goal of the learnt similarity function $s(v_i, c_j)$, between video v_i and caption c_j, is to provide a high value if $i = j$, and a low one if $i \neq j$. Estimating this similarity (described in Sect. 3.3) requires accurate representations for the video as well as the caption. Figure 2 shows the two parts focused on producing these representations (presented in Sects. 3.1 and 3.2 respectively) in our cross-modal framework.

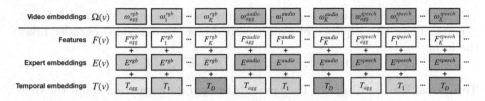

Fig. 3. Inputs to our multi-modal transformer. We combine feature semantics F, expert information E, and temporal cues T to form our video embeddings $\Omega(v)$, which are input to MMT.

3.1 Video Representation

The video-level representation is computed by our proposed multi-modal transformer (MMT). MMT follows the architecture of the transformer encoder presented in [25]. It consists of stacked self-attention layers and fully collected layers. MMT's input $\Omega(v)$ is a set of embeddings, all of the same dimension d_{model}. Each of them embeds the semantics of a feature, its modality, and the time in the video when the feature was extracted. This input is given by:

$$\Omega(v) = F(v) + E(v) + T(v), \qquad (1)$$

In the following, we describe those three components.

Features F. In order to learn an effective representation from different modalities inherent in video data, we begin with video feature extractors called "experts" [14,16,19,31]. In contrast to previous methods, we learn a joint representation leveraging both cross-modal and long-term temporal relationships among the experts. We use N pretrained experts $\{F^n\}_{n=1}^N$. Each expert is a model trained for a particular task that is then used to extract features from video. For a video v, each expert extracts a sequence $F^n(v) = [F_1^n, ..., F_K^n]$ of K features.

The features extracted by our experts encode the semantics of the video. Each expert F^n outputs features in $\mathbb{R}^{d_n}$. In order to project the different expert features into a common dimension d_{model}, we learn N linear layers (one per expert) to project all the features into $\mathbb{R}^{d_{model}}$.

A transformer encoder produces an embedding for each of its feature inputs, resulting in several embeddings for an expert. In order to obtain a unique embedding for each expert, we define an aggregated embedding F_{agg}^n that will collect and contextualize the expert's information. We initialize this embedding with a max pooling aggregation of all the corresponding expert's features as $F_{agg}^n = maxpool(\{F_k^n\}_{k=1}^K)$. The sequence of input features to our video encoder then takes the form:

$$F(v) = [F_{agg}^1, F_1^1, ..., F_K^1, ..., F_{agg}^N, F_1^N, ..., F_K^N]. \qquad (2)$$

Expert Embeddings E. In order to process cross-modality information, our MMT needs to identify which expert it is attending to. We learn N embeddings

$\{E_1, ..., E_N\}$ of dimension d_{model} to distinguish between embeddings of different experts. Thus, the sequence of expert embeddings to our video encoder takes the form:

$$E(v) = [E^1, E^1, ..., E^1, ..., E^N, E^N, ..., E^N].$$ (3)

Temporal Embeddings T. They provide temporal information about the time in the video where each feature was extracted to our multi-modal transformer. Considering videos of a maximum duration of t_{max} seconds, we learn $D = |t_{max}|$ embeddings $\{T_1, ..., T_D\}$ of dimension d_{model}. Each expert feature that has been extracted in the time range $[t, t+1)$ will be temporally embedded with T_{t+1}. For example, a feature extracted at 7.4s in the video will be temporally encoded with temporal embedding T_8. We learn two additional temporal embeddings T_{agg} and T_{unk}, which encode aggregated features and unknown temporal information features (for experts whose temporal information is unknown), respectively. The sequence of temporal embeddings of our video encoder then takes the form:

$$T(v) = [T_{agg}, T_1, ..., T_D, ..., T_{agg}, T_1, ..., T_D].$$ (4)

Multi-modal Transformer. The video embeddings $\Omega(v)$ defined as the sum of features, expert and temporal embeddings in (1), as shown in Fig. 3, are input to the transformer. They are given by: $\Omega(v) = F(v) + E(v) + T(v) = [\omega_{agg}^1, \omega_1^1, ..., \omega_K^1, ..., \omega_{agg}^N, \omega_1^N, ..., \omega_K^N]$. MMT contextualises its input $\Omega(v)$ and produces the video representation $\Psi_{agg}(v)$. As illustrated in Fig. 2, we only keep the aggregated embedding per expert. Thus, our video representation $\Psi_{agg}(v)$ consists of the output embeddings corresponding to the aggregated features, i.e.,

$$\Psi_{agg}(v) = MMT(\Omega(v)) = [\psi_{agg}^1, ..., \psi_{agg}^N].$$ (5)

The advantage of our MMT over the state-of-the-art collaborative gating mechanism [14] is two-fold: First, the input embeddings are not simply modulated in a single step but iteratively refined through several layers featuring multiple attention heads. Second, we do not limit our video encoder with a temporally aggregated feature for each expert, but provide all the extracted features instead, along with a temporal encoding describing at what moment of the video they were extracted from. Thanks to its self-attention modules, each layer of our multi-modal transformer is able to attend to all its input embeddings, thus extracting semantics of events occurring in the video over several modalities.

3.2   Caption Representation

We compute our caption representation $\Phi(c)$ in two stages: first, we obtain an embedding $h(c)$ of the caption, and then project it with a function g into N different spaces as $\Phi = g \circ h$. For the embedding function h, we use a pretrained BERT model [3]. Specifically, we extract our single caption embedding $h(c)$ from the [CLS] output of BERT. In order to match the size of this caption representation with that of video, we learn for function g as many gated embedding modules [16] as there are video experts. Our caption representation then consists of N embeddings, represented by $\Phi(c) = \{\phi^i\}_{i=1}^N$.

3.3   Similarity Estimation

We compute our final video-caption similarity s, as a weighted sum of each expert i's video-caption similarity $\langle \phi^i, \psi^i_{agg} \rangle$. It is given by:

$$s(v, c) = \sum_{i=1}^{N} w_i(c) \langle \phi^i, \psi^i_{agg} \rangle, \tag{6}$$

where $w_i(c)$ represents the weight for the ith expert. To obtain these mixture weights, we follow [16] and process our caption representation $h(c)$ through a linear layer and then perform a softmax operation, i.e.,

$$w_i(c) = \frac{e^{h(c)^\top a_i}}{\sum_{j=1}^{N} e^{h(c)^\top a_j}}, \tag{7}$$

where $(a_1, ..., a_N)$ are the weights of the linear layer. The intuition behind using a weighted sum is that a caption may not describe all the inherent modalities in video uniformly. For example, in the case of a video with a person in a red dress singing opera, the caption "a person in a red dress" provides no information relevant for audio. On the contrary, the caption "someone is singing" should focus on the audio modality for computing similarity. Note that $w_i(c)$, ϕ^i and ψ^i_{agg} can all be precomputed offline for each caption and for each video, and therefore the retrieval operation only involves dot product operations.

3.4   Training

We train our model with the bi-directional max-margin ranking loss [10]:

$$\mathcal{L} = \frac{1}{B} \sum_{i=1}^{B} \sum_{j \neq i} \Big[\max(0, s_{ij} - s_{ii} + m) + \max(0, s_{ji} - s_{ii} + m) \Big], \tag{8}$$

where B is the batch size, $s_{ij} = s(v_i, c_j)$, the similarity score between video v_i and caption c_j, and m is the margin. This loss enforces the similarity for true video-caption pairs s_{ii} to be higher than the similarity of negative samples s_{ij} or s_{ji}, for all $i \neq j$, by at least m.

4   Experiments

4.1   Datasets and Metrics

HowTo100M [17]. It is composed of more than 1 million YouTube instructional videos, along with automatically-extracted speech transcriptions, which form the captions. These captions are naturally noisy, and often do not describe the visual content accurately or are temporally misaligned with it. We use this dataset only for pre-training.

MSRVTT [30]. This dataset is composed of 10K YouTube videos, collected using 257 queries from a commercial video search engine. Each video is 10 to 30s long, and is paired with 20 natural sentences describing it, obtained from Amazon Mechanical Turk workers. We use this dataset for training from scratch and also for fine-tuning. We report results on the train/test splits introduced in [31] that uses 9000 videos for training and 1000 for test. We refer to this split as "1k-A". We also report results on the train/test split in [16] that we refer to as "1k-B". Unless otherwise specified, our MSRVTT results are with "1k-A".

ActivityNet Captions [12]. It consists of 20K YouTube videos temporally annotated with sentence descriptions. We follow the approach of [33], where all the descriptions of a video are concatenated to form a paragraph. The training set has 10009 videos. We evaluate our video-paragraph retrieval on the "val1" split (4917 videos). We use ActivityNet for training from scratch and fine-tuning.

LSMDC [21]. It contains 118,081 short video clips ($\sim$4–5 s) extracted from 202 movies. Each clip is annotated with a caption, extracted from either the movie script or the audio description. The test set is composed of 1000 videos, from movies not present in the training set. We use LSMDC for training from scratch and also fine-tuning.

Metrics. We evaluate the performance of our model with standard retrieval metrics: recall at rank N (R@N, higher is better), median rank (MdR, lower is better) and mean rank (MnR, lower is better). For each metric, we report the mean and the standard deviation over experiments with 3 random seeds. In the main paper, we only report recall@5, median and mean ranks, and refer the reader to the supplementary material for additional metrics.

4.2 Implementation Details

Pre-trained Experts. Recall that our video encoder uses pre-trained experts models for extracting features from each video modality. We use the following seven experts. **Motion** features are extracted from S3D [28] trained on the Kinetics action recognition dataset. **Audio** features are extracted using VGGish model [6] trained on YT8M. **Scene** embeddings are extracted from DenseNet-161 [9] trained for image classification on the Places365 dataset [35]. **OCR** features are obtained in three stages. Overlaid text is first detected using the pixel link text detection model. The detected boxes are then passed through a text recognition model trained on the Synth90K dataset. Finally, each character sequence is encoded with word2vec [18] embeddings. **Face** features are extracted in two stages. An SSD face detector is used to extract bounding boxes, which are then passed through a ResNet50 trained for face classification on the VGGFace2 dataset. **Speech** transcripts are extracted using the Google Cloud Speech to Text API, with the language set to English. The detected words are then encoded with word2vec. **Appearance** features are extracted from the final global average pooling layer of SENet-154 [8] trained for classification on ImageNet. For scene, OCR, face, speech and appearance, we use the features publicly released by [14], and compute the other features ourselves.

Training. For each dataset, we run a grid search on the corresponding validation set to estimate the hyperparameters. We use the Adam optimizer for all our experiments, and set the margin of the bidirectional max-margin ranking loss to 0.05. We also freeze our pre-trained expert models.

When pre-training on HowTo100M, we use a batch size of 64 video-caption pairs, an initial learning rate of 5e–5, which we decay by a multiplicative factor 0.98 every 10K optimisation steps, and train for 2 million steps. Given the long duration of most of the HowTo100M videos, we randomly sample 100 consecutive words in the caption, and keep 100 consecutive seconds of video data, closest in time to the selected words.

When training from scratch or finetuning on MSRVTT or LSMDC, we use a batch size of 32 video-caption pairs, an initial learning rate of 5e-5, which we decay by a multiplicative factor 0.95 every 1K optimisation steps. We train for 50K steps. We use the same settings when training from scratch or finetuning on ActivityNet, except for 0.90 as the multiplicative factor.

To compute our caption representation $h(c)$, we use the "BERT-base-cased" checkpoint of the BERT model and finetune it with a dropout probability of 10%. To compute our video representation $\Psi_{agg}(v)$, we use MMT with 4 layers and 4 attention heads, a dropout probability of 10%, a hidden size d_{model} of 512, and an intermediate size of 3072.

For datasets with short videos (MSRVTT and LSMDC), we use all the 7 experts and limit video input to 30 features per expert, and BERT input to the first 30 wordpieces. For datasets containing longer videos (HowTo100M and ActivityNet), we only use motion and audio experts, and limit our video input to 100 features per expert and our BERT input to the first 100 wordpieces. In cases where an expert is unavailable for a given video, e.g., no speech was detected, we set the aggregated feature F_{agg}^n to a zero vector. We refer the reader to the supplementary material for a study of the model complexity.

4.3   Ablation Studies and Comparisons

We will first show the advantage of pretraining our model on a large-scale, uncurated dataset. We will then perform ablations on the architecture used for our language and video encoders. Finally, we will present the relative importance of the pretrained experts used in this work, and compare with related methods.

Pretraining. Table 1 shows the advantage of pretraining on HowTo100M, before finetuning on the target dataset (MSRVTT in this case). We also evaluated the impact of pretraining on ActivityNet and LSMDC; see Table 5 and Table 6.

Language Encoder. We evaluated several architectures for caption representation, as shown in Table 2. Similar to the observation made in [1], we obtain poor results from a frozen, pretrained BERT. Using the [CLS] output from a pretrained and frozen BERT model is in fact the worst result. We suppose this is because the output was not trained for caption representation, but for a very different task: next sentence prediction. Finetuning BERT greatly improves performance; it is the best result. We also compare with GrOVLE [1] embeddings,

Table 1. Advantage of pretraining on HowTo100M then finetuning on MSRVTT. Impact of removing the stop words. Performance reported on MSRVTT.

Method	Caption	Text $\longrightarrow$ Video		
		R@5↑	MdR↓	MnR↓
Pretraining without finetuning (zero-shot setting)	All words	6.9	160.0	240.2
	w/o stop words	**14.4**	**66.0**	**148.1**
Training from scratch on MSRVTT	All words	**54.0**$_{\pm0.2}$	**4.0**$_{\pm0.0}$	**26.7**$_{\pm0.9}$
	w/o stop words	50.0$_{\pm0.6}$	5.3$_{\pm0.5}$	28.5$_{\pm0.9}$
Pretraining then finetuning on MSRVTT	All words	**57.1**$_{\pm1.0}$	**4.0**$_{\pm0.0}$	**24.0**$_{\pm0.8}$
	w/o stop words	55.0$_{\pm0.7}$	4.3$_{\pm0.5}$	24.3$_{\pm0.3}$

Table 2. Comparison of different architectures for caption embedding when training from scratch on MSRVTT.

Word embeddings		Aggregation	Text $\longrightarrow$ Video		
			R@5↑	MdR↓	MnR↓
GrOVLE	frozen	maxpool	31.8$_{\pm0.4}$	14.7$_{\pm0.5}$	63.1$_{\pm1.3}$
		LSTM	36.4$_{\pm0.8}$	10.3$_{\pm0.9}$	44.2$_{\pm0.1}$
	finetuned	maxpool	34.6$_{\pm0.1}$	12.0$_{\pm0.0}$	52.3$_{\pm0.8}$
		LSTM	40.3$_{\pm0.5}$	8.7$_{\pm0.5}$	38.1$_{\pm0.7}$
BERT	frozen	maxpool	39.4$_{\pm0.8}$	9.7$_{\pm0.5}$	46.5$_{\pm0.2}$
		LSTM	36.4$_{\pm1.8}$	10.7$_{\pm0.5}$	42.2$_{\pm0.6}$
	finetuned	maxpool	44.2$_{\pm1.2}$	7.3$_{\pm0.5}$	35.6$_{\pm0.4}$
		LSTM	40.1$_{\pm1.0}$	8.7$_{\pm0.5}$	37.4$_{\pm0.5}$
	Frozen	BERT-frozen	17.1$_{\pm0.2}$	34.7$_{\pm1.2}$	98.8$_{\pm0.8}$
	Finetuned	BERT-finetuned	**54.0**$_{\pm0.2}$	**4.0**$_{\pm0.0}$	**26.7**$_{\pm0.9}$

frozen or finetuned, aggregated with a max-pooling operation or a 1-layer LSTM and a fully-connected layer. We show that pretrained BERT embeddings aggregated by a max-pooling operation perform better than GrOVLE embeddings processed by a LSTM (best results from [1] for the text-to-clip task).

We also analysed the impact of removing stop words from the captions in Table 1. In a zero-shot setting, i.e., trained on HowTo100M, evaluated on MSRVTT without finetuning, removing the stop words helps generalize, by bridging the domain gap—HowTo100M speech is very different from MSRVTT captions. This approach was adopted in [15]. However, we observe that when finetuning, it is better to keep all the words as they contribute to the semantics of the caption.

Video Encoder. We evaluated the influence of different architectures for computing video embeddings on the MSRVTT 1k-A test split.

Table 3. Ablation studies on the video encoder of our framework with MSRVTT. **(a) Influence of the architecture and input.** With max-pooled features as input, we compare our transformer architecture (MMT) with the variant not using an encoder (NONE) and the one with Collaborative Gating [14] (COLL). We also show that MMT can attend to all extracted features, as detailed in the text. **(b) Importance of initializing F_{agg}^n features.** We compare zero-vector initialisation, mean pooling and max pooling of the expert features. **(c) Influence of the size of the multi-modal transformer.** We compare different values for number-of-layers × number-of-attention-heads.

(a) Encoder architecture and input

		Text $\longrightarrow$ Video		
Encoder	Input	R@5$\uparrow$	MdR$\downarrow$	MnR$\downarrow$
NONE	max pool	$50.9_{\pm 1.5}$	$5.3_{\pm 0.5}$	$28.6_{\pm 0.5}$
COLL	max pool	$51.3_{\pm 0.8}$	$5.0_{\pm 0.0}$	$29.5_{\pm 1.8}$
MMT	max pool	$52.5_{\pm 0.7}$	$5.0_{\pm 0.0}$	$27.2_{\pm 0.7}$
MMT	shuffled feats	$53.3_{\pm 0.2}$	$5.0_{\pm 0.0}$	$27.4_{\pm 0.7}$
MMT	ordered feats	$\mathbf{54.0_{\pm 0.2}}$	$\mathbf{4.0_{\pm 0.0}}$	$\mathbf{26.7_{\pm 0.9}}$

(b) F_{agg}^n initialisation

	Text $\longrightarrow$ Video		
F_{agg}^n init	R@5$\uparrow$	MdR$\downarrow$	MnR$\downarrow$
zero	$50.2_{\pm 0.9}$	$5.7_{\pm 0.5}$	$28.5_{\pm 1.3}$
mean pool	$\mathbf{54.2_{\pm 0.3}}$	$5.0_{\pm 0.0}$	$27.1_{\pm 0.9}$
max pool	$54.0_{\pm 0.2}$	$\mathbf{4.0_{\pm 0.0}}$	$\mathbf{26.7_{\pm 0.9}}$

(c) Model size

		Text $\longrightarrow$ Video		
Layers	Heads	R@5$\uparrow$	MdR$\downarrow$	MnR$\downarrow$
2	2	$53.2_{\pm 0.4}$	$5.0_{\pm 0.0}$	$\mathbf{26.7_{\pm 0.4}}$
4	4	$\mathbf{54.0_{\pm 0.2}}$	$\mathbf{4.0_{\pm 0.0}}$	$26.7_{\pm 0.9}$
8	8	$53.9_{\pm 0.3}$	$4.7_{\pm 0.5}$	$\mathbf{26.7_{\pm 0.7}}$

In Table 3a, we evaluate variants of our encoder architecture and its input. Similar to [16], we experiment with directly computing the caption-video similarities on each max-pooled expert features, i.e., no video encoder (NONE in the table). We compare this with the collaborative gating architecture (COLL) [14] and our MMT variant using only the aggregated features as input. For the first two variants without MMT, we adopt the approach of [16] to deal with missing modalities by re-weighting $w_i(c)$. We also show the superior performance of our multi-modal transformer in contextualising the different modality embeddings compared to the collaborative gating approach. We argue that our MMT is able to extract cross-modal information in a multi-stage architecture compared to collaborative gating, which is limited to modulating the input embeddings. Table 3a also highlights the advantage of providing MMT with **all** the extracted features, instead of only aggregated ones. Temporally aggregating each expert's features ignores information about multiple events occurring in a same video (see the last three rows). As shown by the influence of ordered and randomly shuffled features on the performance, MMT has the capacity to make sense of the relative ordering of events in a video.

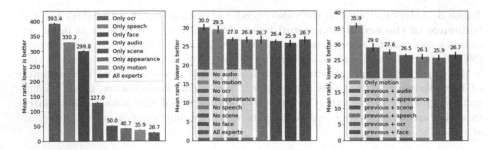

Fig. 4. MSRVTT performance (mean rank; lower is better) after training from scratch, when using only one expert (left), when using all experts but one (middle), when gradually adding experts by greedy search (right).

Table 3b shows the importance of initialising the expert aggregation feature F_{agg}^n. Since the output of our video encoder is extracted from the "agg" columns, it is important to initialise them with an appropriate representation of the experts' features. The transformer being a residual network architecture, initializing F_{agg}^n input embeddings with a zero vector leads to a low performance. Initializing with max pooling aggregation of each expert performs better than mean pooling. Finally, we analyze the impact of the size of our multi-modal transformer model in Table 3c. A model with 4 layers and 4 attention heads outperforms both a smaller model (2 layers and 2 attention heads) and a larger model (8 layers and 8 attention heads).

Comparison of the Different Experts. In Fig. 4, we show an ablation study when training our model on MSRVTT using only one expert (left), using all experts but one (middle), or gradually adding experts by greedy search (right). In the case of using only one expert, we note that the motion expert provides the best results. We attribute the poor performance of OCR, speech and face to the fact that they are absent from many videos, thus resulting in a zero vector input to our video encoder. While the scene expert shows a decent performance, if used alone, it does not contribute when used along other experts, perhaps due to the semantics it encodes being captured already by other experts like appearance or motion. On the contrary, the audio expert alone does not provide a good performance, but it contributes the most when used in conjunction with the others, most likely due to the complementary cues it provides, compared to the other experts.

Comparison to Prior State of the Art. We compare our method on three datasets: MSRVTT (Table 4), ActivityNet (Table 5) and LSMDC (Table 6). While MSRVTT and LSMDC contain short video-caption pairs (average video duration of 13s for MSRVTT, one-sentence captions), ActivityNet contains much longer videos (several minutes) and each video is captioned with multiple sentences. We consider the concatenation of all these sentences as the caption. We show that our method obtains state-of-the-art results on all the three datasets.

Table 4. Retrieval performance on the MSRVTT dataset. 1k-A and 1k-B denote test sets of 1000 randomly sampled caption-video pairs used in [31] and [16] resp.

Method	SplitSplit	Text $\longrightarrow$ Video			Video $\longrightarrow$ Text		
		R@5↑	MdR↓	MnR↓	R@5↑	MdR↓	MnR↓
Random baseline	1k-A	0.5	500.0	500.0	0.5	500.0	500.0
JSFusion [31]	1k-A	31.2	13	–	–	–	–
HT [17]	1k-A	35.0	12	–	–	–	–
CE [14]	1k-A	48.8$_{\pm0.6}$	6.0$_{\pm0.0}$	28.2$_{\pm0.8}$	50.3$_{\pm0.5}$	5.3$_{\pm0.6}$	25.1$_{\pm0.8}$
Ours	1k-A	**54.0**$_{\pm0.2}$	**4.0**$_{\pm0.0}$	**26.7**$_{\pm0.9}$	**56.0**$_{\pm0.9}$	**4.0**$_{\pm0.0}$	**23.6**$_{\pm1.0}$
HT-pretrained [17]	1k-A	40.2	9	–	–	–	–
Ours-pretrained	1k-A	**57.1**$_{\pm1.0}$	**4.0**$_{\pm0.0}$	**24.0**$_{\pm0.8}$	**57.5**$_{\pm0.6}$	**3.7**$_{\pm0.5}$	**21.3**$_{\pm0.6}$
Random baseline	1k-B	0.5	500.0	500.0	0.5	500.0	500.0
MEE [16]	1k-B	37.9	10.0	–	–	–	–
JPose [26]	1k-B	38.1	9	–	41.3	8.7	–
MEE-COCO [16]	1k-B	39.2	9.0	–	–	–	–
CE [14]	1k-B	46.0$_{\pm0.4}$	7.0$_{\pm0.0}$	35.3$_{\pm1.1}$	46.0$_{\pm0.5}$	6.5$_{\pm0.5}$	30.6$_{\pm1.2}$
Ours	1k-B	**49.1**$_{\pm0.4}$	**6.0**$_{\pm0.0}$	**29.5**$_{\pm1.6}$	**49.4**$_{\pm0.4}$	**6.0**$_{\pm0.0}$	**24.5**$_{\pm1.8}$

Table 5. Retrieval performance on the ActivityNet dataset.

Method	Text $\longrightarrow$ Video			Video $\longrightarrow$ Text		
	R@5↑	MdR↓	MnR↓	R@5↑	MdR↓	MnR↓
Random baseline	0.1	2458.5	2458.5	0.1	2458.5	2458.5
FSE [33]	44.8$_{\pm0.4}$	7	–	43.1$_{\pm1.1}$	7	–
CE [14]	47.7$_{\pm0.6}$	6.0$_{\pm0.0}$	23.1$_{\pm0.5}$	46.6$_{\pm0.7}$	6.0$_{\pm0.0}$	24.4$_{\pm0.5}$
HSE [33]	49.3	–	–	48.1	–	–
Ours	**54.2**$_{\pm1.0}$	**5.0**$_{\pm0.0}$	**20.8**$_{\pm0.4}$	**54.8**$_{\pm0.4}$	**4.3**$_{\pm0.5}$	**21.2**$_{\pm0.5}$
Ours-pretrained	**61.4**$_{\pm0.2}$	**3.3**$_{\pm0.5}$	**16.0**$_{\pm0.4}$	**61.1**$_{\pm0.2}$	**4.0**$_{\pm0.0}$	**17.1**$_{\pm0.5}$

Table 6. Retrieval performance on the LSMDC dataset.

Method	Text $\longrightarrow$ Video			Video $\longrightarrow$ Text		
	R@5↑	MdR↓	MnR↓	R@5↑	MdR↓	MnR↓
Random baseline	0.5	500.0	500.0	0.5	500.0	500.0
CT-SAN [32]	16.3	46	–	–	–	–
JSFusion [31]	21.2	36	–	–	–	–
CCA [11] (rep. by [16])	21.7	33	–	–	–	–
MEE [16]	25.1	27	–	–	–	–
MEE-COCO [16]	25.6	27	–	–	–	–
CE [14]	26.9$_{\pm1.1}$	25.3$_{\pm3.1}$	–	–	–	–
Ours	**29.2**$_{\pm0.8}$	21.0$_{\pm1.4}$	76.3$_{\pm1.9}$	29.3$_{\pm1.1}$	22.5$_{\pm0.4}$	77.1$_{\pm2.6}$
Ours-pretrained	**29.9**$_{\pm0.7}$	19.3$_{\pm0.2}$	75.0$_{\pm1.2}$	28.6$_{\pm0.3}$	20.0$_{\pm0.0}$	76.0$_{\pm0.8}$

The gains obtained through MMT's long term temporal encoding are particularly noticeable on the long videos of ActivityNet.

5 Summary

We introduced multi-modal transformer, a transformer-based architecture capable of attending multiple features extracted at different moments, and from different modalities in video. This leverages both temporal and cross-modal cues, which are crucial for accurate video representation. We incorporate this video encoder along with a caption encoder in a cross-modal framework to perform caption-video matching and obtain state-of-the-art results for video retrieval. As future work, we would like to improve temporal encoding for video and text.

Acknowledgments. We thank the authors of [14] for sharing their codebase and features, and Samuel Albanie, in particular, for his help with implementation details. This work was supported in part by the ANR project AVENUE.

References

1. Burns, A., Tan, R., Saenko, K., Sclaroff, S., Plummer, B.A.: Language features matter: effective language representations for vision-language tasks. In: ICCV (2019)
2. Carreira, J., Zisserman, A.: Quo vadis, action recognition? In: CVPR, A New Model and the Kinetics Dataset (2017)
3. Devlin, J., Chang, M.W., Lee, K., Toutanova, K.: BERT: pre-training of deep bidirectional transformers for language understanding. In: NAACL-HLT (2019)
4. Gabeur, V., Sun, C., Alahari, K., Schmid, C.: CVPR 2020 video pentathlon challenge: multi-modal transformer for video retrieval. In: CVPR Video Pentathlon Workshop (2020)
5. Harwath, D., Recasens, A., Surís, D., Chuang, G., Torralba, A., Glass, J.: Jointly discovering visual objects and spoken words from raw sensory input. In: Ferrari, V., Hebert, M., Sminchisescu, C., Weiss, Y. (eds.) ECCV 2018. LNCS, vol. 11210, pp. 659–677. Springer, Cham (2018). https://doi.org/10.1007/978-3-030-01231-1_40
6. Hershey, S., et al.: CNN architectures for large-scale audio classification. In: ICASSP (2017)
7. Hochreiter, S., Schmidhuber, J.: Long short-term memory. Neural Comput. **9**(8) (1997)
8. Hu, J., Shen, L., Albanie, S., Sun, G., Wu, E.: Squeeze-and-excitation networks. IEEE Trans. Pattern Anal. Mach. Intell. (2019)
9. Huang, G., Liu, Z., Weinberger, K.Q.: Densely connected convolutional networks. In: CVPR (2016)
10. Karpathy, A., Joulin, A., Fei-Fei, L.: Deep fragment embeddings for bidirectional image sentence mapping. In: NIPS (2014)
11. Klein, B., Lev, G., Sadeh, G., Wolf, L.: Associating neural word embeddings with deep image representations using fisher vectors. In: CVPR (2015)
12. Krishna, R., Hata, K., Ren, F., Fei-Fei, L., Niebles, J.C.: Dense-captioning events in videos. In: ICCV (2017)
13. Lee, K.-H., Chen, X., Hua, G., Hu, H., He, X.: Stacked cross attention for image-text matching. In: Ferrari, V., Hebert, M., Sminchisescu, C., Weiss, Y. (eds.) ECCV 2018. LNCS, vol. 11208, pp. 212–228. Springer, Cham (2018). https://doi.org/10.1007/978-3-030-01225-0_13
14. Liu, Y., Albanie, S., Nagrani, A., Zisserman, A.: Use what you have: video retrieval using representations from collaborative experts. arXiv abs/1907.13487 (2019)

15. Miech, A., Alayrac, J.B., Smaira, L., Laptev, I., Sivic, J., Zisserman, A.: End-to-end learning of visual representations from uncurated instructional videos. arXiv e-prints arXiv:1912.06430, December 2019

16. Miech, A., Laptev, I., Sivic, J.: Learning a text-video embedding from incomplete and heterogeneous data. arXiv abs/1804.02516 (2018)

17. Miech, A., Zhukov, D., Alayrac, J.B., Tapaswi, M., Laptev, I., Sivic, J.: HowTo100M: learning a text-video embedding by watching hundred million narrated video clips. In: ICCV (2019)

18. Mikolov, T., Chen, K., Corrado, G.S., Dean, J.: Efficient estimation of word representations in vector space. In: ICLR (2013)

19. Mithun, N.C., Li, J., Metze, F., Roy-Chowdhury, A.K.: Learning joint embedding with multimodal cues for cross-modal video-text retrieval. In: ICMR (2018)

20. Mithun, N.C., Li, J., Metze, F., Roy-Chowdhury, A.K.: Joint embeddings with multimodal cues for video-text retrieval. Int. J. Multimedia Inf. Retrieval 8(1), 3–18 (2019). https://doi.org/10.1007/s13735-018-00166-3

21. Rohrbach, A., Rohrbach, M., Tandon, N., Schiele, B.: A dataset for movie description. In: CVPR (2015)

22. Simonyan, K., Zisserman, A.: Two-stream convolutional networks for action recognition in videos. In: NIPS (2014)

23. Sun, C., Baradel, F., Murphy, K., Schmid, C.: Learning video representations using contrastive bidirectional transformer. arXiv:1906.05743 (2019)

24. Sun, C., Myers, A., Vondrick, C., Murphy, K., Schmid, C.: VideoBERT: a joint model for video and language representation learning. In: ICCV (2019)

25. Vaswani, A., et al.: Attention is all you need. In: NIPS (2017)

26. Wray, M., Larlus, D., Csurka, G., Damen, D.: Fine-grained action retrieval through multiple parts-of-speech embeddings. In: ICCV (2019)

27. Wu, Y., et al.: Google's neural machine translation system: bridging the gap between human and machine translation. arXiv:1609.08144 (2016)

28. Xie, S., Sun, C., Huang, J., Tu, Z., Murphy, K.: Rethinking spatiotemporal feature learning: speed-accuracy trade-offs in video classification. In: Ferrari, V., Hebert, M., Sminchisescu, C., Weiss, Y. (eds.) ECCV 2018. LNCS, vol. 11219, pp. 318–335. Springer, Cham (2018). https://doi.org/10.1007/978-3-030-01267-0_19

29. Xing, E.P., Ng, A.Y., Jordan, M.I., Russell, S.: Distance metric learning, with application to clustering with side-information. In: NIPS (2002)

30. Xu, J., Mei, T., Yao, T., Rui, Y.: MSR-VTT: a large video description dataset for bridging video and language. In: CVPR (2016)

31. Yu, Y., Kim, J., Kim, G.: A joint sequence fusion model for video question answering and retrieval. In: Ferrari, V., Hebert, M., Sminchisescu, C., Weiss, Y. (eds.) ECCV 2018. LNCS, vol. 11211, pp. 487–503. Springer, Cham (2018). https://doi.org/10.1007/978-3-030-01234-2_29

32. Yu, Y., Ko, H., Choi, J., Kim, G.: End-to-end concept word detection for video captioning, retrieval, and question answering. In: CVPR (2017)

33. Zhang, B., Hu, H., Sha, F.: Cross-modal and hierarchical modeling of video and text. In: Ferrari, V., Hebert, M., Sminchisescu, C., Weiss, Y. (eds.) ECCV 2018. LNCS, vol. 11217, pp. 385–401. Springer, Cham (2018). https://doi.org/10.1007/978-3-030-01261-8_23

34. Zhang, Y., Jin, R., Zhou, Z.H.: Understanding bag-of-words model: a statistical framework. Int. J. Mach. Learn. Cybernet. 1, 43–52 (2010)

35. Zhou, B., Lapedriza, À., Khosla, A., Oliva, A., Torralba, A.: Places: a 10 million image database for scene recognition. IEEE Trans. Pattern Anal. Mach. Intell. 40, 1452–1464 (2018)

Feature Representation Matters: End-to-End Learning for Reference-Based Image Super-Resolution

Yanchun Xie[1], Jimin Xiao[1(✉)], Mingjie Sun[1], Chao Yao[2], and Kaizhu Huang[1,3]

[1] School of Advanced Technology, Xi'an Jiaotong-Liverpool University, Suzhou, China
jimin.xiao@xjtlu.edu.cn
[2] University of Science and Technology Beijing, Beijing, China
[3] Alibaba-Zhejiang University Joint Institute of Frontier Technologies, Hangzhou, China

Abstract. In this paper, we are aiming for a general reference-based super-resolution setting: it does not require the low-resolution image and the high-resolution reference image to be well aligned or with a similar texture. Instead, we only intend to transfer the relevant textures from reference images to the output super-resolution image. To this end, we engaged neural texture transfer to swap texture features between the low-resolution image and the high-resolution reference image. We identified the importance of designing a super-resolution task-specific features rather than classification oriented features for neural texture transfer, making the feature extractor more compatible with the image synthesis task. We develop an end-to-end training framework for the reference-based super-resolution task, where the feature encoding network prior to matching and swapping is jointly trained with the image synthesis network. We also discovered that learning the high-frequency residual is an effective way for the reference-based super-resolution task. Without bells and whistles, the proposed method E2ENT^2 achieved better performance than state-of-the method (i.e., SRNTT with five loss functions) with only two basic loss functions. Extensive experimental results on several datasets demonstrate that the proposed method E2ENT^2 can achieve superior performance to existing best models both quantitatively and qualitatively.

Keywords: Super-resolution · Reference-based · Feature matching · Feature swapping · CUFED5 · Flickr1024

1 Introduction

Image super-resolution (SR) is an essential task in computer vision, aiming to transfer low-resolution (LR) images to their high-resolution (HR) counterparts. SR remains to be a long-standing and ill-posed problem due to the non-unique

© Springer Nature Switzerland AG 2020
A. Vedaldi et al. (Eds.): ECCV 2020, LNCS 12349, pp. 230–245, 2020.
https://doi.org/10.1007/978-3-030-58548-8_14

mapping between high and low-resolution samples. A single low resolution (LR) image could correspond to multiple high resolution (HR) images. A large number of deep SR models have been proposed to solve this problem in recent years [1,3,7,10,11,13]. However, in case of a large upsampling factor, recovering an HR image requires to provide sufficient information to fill the missing contents in the LR image.

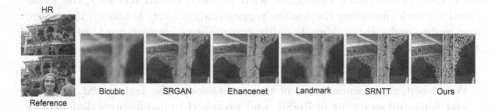

Fig. 1. Left: High resolution image (up) and reference (bottom). Right: zoomed results of different SR algorithms, including SRGAN [11], Ehancenet [13], Landmark [22], SRNTT [24], and ours. Our end-to-end learning method produces the best result.

Reference-based super-resolution (RefSR) is a new SR branch in recent years, which has been proven to be effective in recovering lost high-frequency details in the LR images [16,22,27,28]. These reference-based methods generally require reference images to have similar content with the LR image or with proper alignment. For example, prior work [28] focuses on RefSR for light field images where the LR image and the HR reference image are very similar as they have relatively small disparities. It estimates the transformation by optical flow and uses the multi-scale warping technique for feature alignment. For these RefSR methods, if the reference images do not possess relevant textures with the LR image, their performance would significantly degrade and even be worse than signal image SR methods.

In this paper, we are aiming for a more general RefSR setting: it does not require the LR image and the HR reference image to be well aligned or with a similar texture. Instead, we only intend to transfer the relevant texture from reference images to the output SR image. Ideally, a robust RefSR algorithm should outperform single image super-resolution (SISR) when a better reference image is provided, whilst achieving comparable performance when reference images do not possess relevant texture at all.

Based on this goal, SRNTT [24] proposes a neural texture transfer approach that breaks the limitation of reference images. In SRNTT, local texture matching is conducted in the feature space, and the matched textures are transferred to the synthesized high-resolution image through a deep neural network. However, there are three main issues for SRNTT: (1) the features used in this image synthesis task are extracted from a VGG net. Initially designed for image classification, VGG may not lead to the best features for SR. (2) WIth the fixed VGG net, SRNTT does not take advantage of the end-to-end learning in the SR task.

(3) VGG features in shallow layers involve a high computational and enormous memory cost, making it time-consuming to process images with large size.

In this paper, we argue that the matching feature does matter for neural texture transfer in RefSR. Thus, we analyze the feature extractor in the RefSR method and propose to use features designed for SR (i.e., SRGAN [11]) instead of features designed for classification (VGG). Such features, on the other hand, are more compatible with the image synthesis network where the adversarial loss is used [5]. Secondly, Distinctive with previous RefSR methods, the whole neural network, including the feature representation part, is able to be trained in an end-to-end manner. Visual quality comparisons between our approach and other state-of-the-art methods are shown in Fig. 1.

Our contributions are summarized as follows:

- We identified the importance of using a task-specific feature extractor for matching and swapping in RefSR, and proposed to use features designed for SR (i.e., SRGAN [11]) instead of features designed for classification (VGG), making the feature extractor more compatible with the image synthesis task.
- We designed an end-to-end training framework for the RefSR task, where the feature extraction network for matching and swapping is jointly trained with the image synthesis network. We also discovered that learning the high-frequency residual is an effective and efficient way for the reference-based super-resolution task. Without bells and whistles, we achieved better performance than the state-of-the method (i.e., SRNTT [24] with five loss functions) with only two basic loss functions.
- We evaluated our method in RefSR datasets, achieving the new quantitatively results (24.01dB for PSNR, 0.705 for SSIM) in the CUFED5 dataset. Qualitative results also demonstrate the superiority of our method.

2   Related Work

2.1   Image Super-Resolution

Deep learning based methods have been applied to image SR in recent years [3,9, 10,12,23], and significant progress have been obtained due to its powerful feature representation ability. These methods learn an end-to-end mapping from LR to HR directly with a mean squared loss function, treating the super-resolution as a regression problem. SRGAN [11] considers both perceptual similarity loss and adversarial loss for super-resolution. The perceptual similarity is obtained by computing the feature distance extracted from the VGG middle layer. The adversarial loss enables us to generate realistic visual results for humans by using a discriminator to distinguish between real HR images and super-resolved images generated from generators.

The super-resolution performance has been boosted with deep features and residual learning. For example, Dong et al. first introduced a three-layer convolutional network SRCNN [3] for image super-resolution. After that, Kim et al. reformed the problem based on residual learning and proposed VDSR [9] and

DRCN [10] with deeper layers. Lim et al. proposed two very deep multi-scale super-resolution networks EDSR and MDSR [12] by modifying residual units and further improve the performance. Zhang et al. [23] proposed a residual in residual structure to allows focusing on learning high-frequency information and a channel attention mechanism to rescale channel-wise features by considering inter-dependencies among channels adaptively.

2.2 Reference-Based Super-Resolution

Different from single image super-resolution with the only low-resolution image provided, RefSR methods utilize additional images that have more texture information to assist the recovery process. Generally, the reference images contain similar objects, scenes, or texture with the low-resolution image. The reference images can be obtained from different frames in a video sequence, different viewpoints in light field images or multiview videos, or by web retrieval. Many works study the reference-based super-resolution by extra examples or similar scenes from web [14,15,17]. Other works [21,26–28] use reference images from different viewpoints to enhance light field images. These works mostly build the mapping from LR to HR patches, and fuse the HR patches at the pixel level or using a shallow model. To overcome inter-patch misalignment and the grid effect, Cross-Net [28] uses optical flow to spatially align the reference feature map with the LR feature map and then aggregates them into SR images. SRNTT [24] further proposes a neural texture transfer approach to improve the matching and fusing ability. In their approach, VGG features with semantically relevant textures from reference images are transferred to the LR image.

Unlike the flow and wrapping based approach [28], our method could further handle the images with much larger disparities than that in light field data. Different from the existing neural texture transfer approach [24], our texture matching and swapping part is end-to-end trainable.

3 Our Method

In this section, our proposed method, namely End-to-End learning for Neural Texture Transfer (E2ENT^2), will be introduced in detail. We first present the network framework of our proposed E2ENT^2, as shown in Fig. 2, which consists of 3 key blocks, including (1) a feature encoding module which extracts features from the LR input and reference images; (2) a newly designed match and swap (MS) module which identifies similar LR-HR feature pairs and conducts feature swapping, where gradients can back-propagate through it to enable end-to-end learning; (3) an image synthesis module which fuses the LR image feature and swapped feature, and outputs the SR image.

3.1 Notations

The input of our network includes a LR input image I_{in}, an HR reference image I_{ref} and a corresponding LR reference image $I_{ref}^{\downarrow}$. $I_{ref}^{\downarrow}$ is the down-sampled

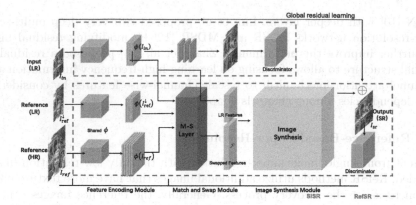

Fig. 2. The framework of our proposed network. The network consists of three main modules: feature encoding module, match and swap module, and image synthesis module. The network takes the LR image, HR reference image as input, and outputs the super-resolved image.

version of the HR reference image I_{ref}. I_{in} is with size $W_{in} \times H_{in}$; I_{ref} is with size $W_{ref} \times H_{ref}$, which does not need to be the same size as I_{in}, and $I_{ref}^{\downarrow}$ is with size $\frac{W_{ref}}{r} \times \frac{H_{ref}}{r}$, with r being the super-resolution ratio.

After the feature encoding module, we get feature maps $\phi(I_{in})$, $\phi(I_{ref})$ and $\phi(I_{ref}^{\downarrow})$ for I_{in}, I_{ref} and $I_{ref}^{\downarrow}$, respectively. The feature map size is $W_{in} \times H_{in}$ for $\phi(I_{in})$, $W_{ref} \times H_{ref}$ for $\phi(I_{ref})$, and $\frac{W_{ref}}{r} \times \frac{H_{ref}}{r}$ for $\phi(I_{ref})$. In other words, the feature map shares the same width and height with the image, so that could minimize the loss of details.

Feature maps $\phi(I_{in})$, $\phi(I_{ref}^{\downarrow})$ and $\phi(I_{ref})$ are fed to the match and swap module ψ, and a new swapped feature map $\mathcal{F}$ is obtained,

$$\mathcal{F} = \psi(\phi(I_{in}), \phi(I_{ref}^{\downarrow}), \phi(I_{ref})), \qquad (1)$$

where the size of $\mathcal{F}$ is $rW_{in} \times rH_{in}$.

Finally, the swapped feature $\mathcal{F}$ together with the LR feature $\phi(I_{in})$ are fed into the image synthesis module ζ to generate the super-resolution image I_{sr}, as

$$I_{sr} = \zeta(\mathcal{F}, \phi(I_{in})), \qquad (2)$$

where the size of I_{sr} is $rW_{in} \times rH_{in}$.

3.2   Feature Encoding Module

Single image super-resolution benefits a lot from skip-connections [9,10,12,23], and various deep learning models have achieved state-of-the-art performance. Thus, we propose to utilize the residual learning in the SR feature encoding module to improve the accuracy of feature representation for the reference-based super-resolution task.

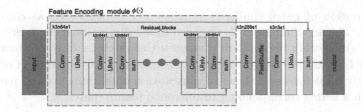

Fig. 3. The structure of our single-image super-resolution (SISR) branch with the residual connection. The network consists of several residual blocks for feature encoding. The feature encoding module is marked within the dashed line.

Our proposed RefSR network shares the same feature encoding module $\phi(\cdot)$ in the SISR branch to produce features for I_{in}, I_{ref} and $I_{ref}^{\downarrow}$. The SISR branch has a deep residual-based structure without the BN layer, as shown in Fig. 3. The SISR branch is composed of stacked residual blocks with 3×3 Conv kernels and followed by pixelshuffle layers for upsampling. The skip connections allow the network to focus on informative features rather than the LR features. After the feature encoding module, we can get $\phi(I_{in})$, $\phi(I_{ref})$ and $\phi(I_{ref}^{\downarrow})$.

In addition to being used in the RefSR branch, $\phi(I_{in})$ is also passed to the rest of the SISR branch to complete a SISR task, which ensures feature consistency between the two standalone SR tasks. Meanwhile, introducing a shared trainable feature encoding module in both SISR and RefSR can generate discriminative features for the match and swap module due to end-to-end learning.

To further enhance the subjective visual quality of the SR image, we also adopt a discriminator for adversarial learning in both SISR and RefSR branches.

3.3 Match and Swap Module

To transfer the semantically relevant texture from reference images to the output SR image, we adopt a patch-based feature match and swap module. As shown in Fig. 4, the match and swap module takes the feature maps obtained in the encoding stage as input, including $\phi(I_{in})$, $\phi(I_{ref})$ and $\phi(I_{ref}^{\downarrow})$. This module outputs a fused feature map $\mathcal{F}$.

Forward Pass. Our proposed matching process is conducted at patch level, which is a 3×3 feature block. Firstly, we crop $\phi(I_{in})$, $\phi(I_{ref}^{\downarrow})$ and $\phi(I_{ref})$ into 3×3, 3×3 and $3r \times 3r$ patches with stride 1, 1 and r, respectively. These patches are indexed based on the horizontal and vertical position. Matching similarity is computed between patches in $\phi(I_{in})$ and $\phi(I_{ref}^{\downarrow})$.

To recover the missing details as much as possible, in the feature matching process, for each LR feature patch in $\phi(I_{in})$, we need to search for the most similar feature patch in $\phi(I_{ref}^{\downarrow})$, and the corresponding feature patch in $\phi(I_{ref})$ will be used to replace the original patch.

Computation of patch similarity is efficiently implemented as convolution operations. The matching result is recorded in a 3-dimensional similarity map $\mathcal{S}$, with $\mathcal{S}_i(x, y)$ denoting the similarity between the patch centered at the location (x, y) in $\phi(I_{in})$ and the i-th reference patch in $\phi(I_{ref}^{\downarrow})$. Computation of $\mathcal{S}_i$ can be efficiently implemented as a set of convolution operations over all patches in $\phi(I_{in})$ with a kernel corresponding to reference feature patch i:

$$\mathcal{S}_i = \phi\left(I_{in}\right) * \frac{\mathcal{P}_i\left(\phi\left(I_{ref}^{\downarrow}\right)\right)}{\left\|\mathcal{P}_i\left(\phi\left(I_{ref}^{\downarrow}\right)\right)\right\|}, \tag{3}$$

where $\mathcal{P}_i(\cdot)$ denotes to sample the i-th patch from a feature map, $*$ is a 2D convolution operation, and $\|\cdot\|$ is used to get the feature length (L1). Note that $\mathcal{S}_i$ is a 2-dimensional map.

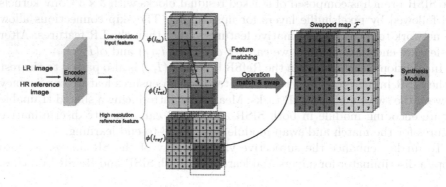

Fig. 4. Illustration of the forward pass in the match and swap module. Feature patch matching are conducted based on the feature similarity between $\phi(I_{in})$, $\phi(I_{ref}^{\downarrow})$. The corresponding matched HR reference feature patches replace the LR features, and finally a swapped feature $\mathcal{F}$ is produced.

After the feature matching, we can obtain a swapped feature map $\mathcal{F}$ based on the 3D similarity map $\mathcal{S}$. Each patch in $\mathcal{F}$ centered at (x, y) is defined as:

$$\mathcal{F}_{(x,y)}^p = \mathcal{P}_{i^*}\left(\phi\left(I_{ref}\right)\right), i^* = \arg\max_i \mathcal{S}_i(x, y), \tag{4}$$

where i^* is the patch index for the most similar one in the reference feature. $\mathcal{P}_{i^*}(\cdot)$ denotes to sample the i^*-th patch from a feature map. Note that the patch size of $\mathcal{P}_{i^*}\left(\phi\left(I_{ref}\right)\right)$ is r^2 times that of $\mathcal{P}_{i^*}\left(\phi\left(I_{ref}^{\downarrow}\right)\right)$. Therefore, after swapping, the feature size of $\mathcal{F}$ is r^2 times that of $\phi(I_{in})$.

In the forward pass, we use $\mathcal{K}_{(x,y)}$ to record the number of times that the reference patch centered at (x, y) in $\phi(I_{ref})$ is selected for swapping, and use $\mathcal{Q}_{(x,y)}$ to record a list of patch center coordinates for all the LR patches in

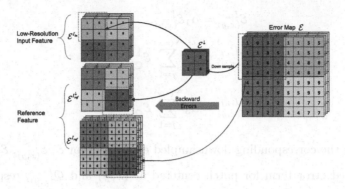

Fig. 5. Illustration of the error backward propagation in the match and swap module. Error gathered at $\mathcal{F}$ from the loss layer backward propagates through the match and swap module to the image encoding module. In this figure, we assume $\alpha 1 = \alpha 2 = \alpha 3 = 1$ for simplicity.

$\phi(I_{in})$ that matches with the reference patch centered at (x, y) in the matching process. $\mathcal{K}_{(x,y)}$ and $\mathcal{Q}_{(x,y)}$ will be used in the gradient backpropagation process.

We conduct the feature matching at low-resolution (using $\phi(I_{in})$ and $\phi(I_{ref}^{\downarrow})$) to boost the matching speed for fast training. Traditional feature matching methods [4,24] use a bicubic up-sampling strategy on the LR image to get an up-sampled image that shares the same spatial size as an HR image. However, such operation brings exponential computation in the feature matching process, especially when the image size is large.

Backward Pass. To have an end-to-end training, we design a mechanism to enable the gradients to back-propagate through the match and swap module, from the image synthesis module to the feature encoding module, as shown in Fig. 5.

The error term $\mathcal{E} = \partial \mathcal{J} / \partial \mathcal{F}$ for $\mathcal{F}$ can be calculated from the loss layer, with $\mathcal{J}$ being the loss function. $\mathcal{E}$ is with the same size as the swapped map $\mathcal{F}$. Notice that the argmax function in Eq. (4) is non-differentiable, a new mechanism to back-propagate $\mathcal{E}$ to the feature encoding module is needed.

As demonstrated in Fig. 4, features $\phi(I_{in})$, $\phi(I_{ref})$ and $\phi(I_{ref}^{\downarrow})$ all affect the swapped map $\mathcal{F}$. We define the error term for $\phi(I_{in})$, $\phi(I_{ref})$ and $\phi(I_{ref}^{\downarrow})$ are $\mathcal{E}^{I_{in}}$, $\mathcal{E}^{I_{ref}}$ and $\mathcal{E}^{I_{ref}^{\downarrow}}$, respectively. Since the feature matching location information, $\mathcal{K}_{(x,y)}$ and $\mathcal{Q}_{(x,y)}$, are recorded in the forward process, for each matching patch centered at (x, y), we have their error terms:

$$\mathcal{E}^{I_{in}}_{(x,y)} = \alpha_1 \mathcal{E}^{\downarrow}_{(x,y)},$$

$$\mathcal{E}^{I^{\downarrow}_{ref}}_{(x,y)} = \alpha_2 \sum_{j=1}^{\mathcal{K}_{(x,y)}} \mathcal{E}^{\downarrow}_{\mathcal{Q}^j_{(x,y)}},$$

$$\mathcal{E}^{I_{ref}}_{(x,y)} = \alpha_3 \sum_{j=1}^{\mathcal{K}_{(x,y)}} \mathcal{E}_{\mathcal{Q}^j_{(x,y)}}, \qquad (5)$$

where $\mathcal{E}^{\downarrow}$ is the corresponding downsampled error term for $\mathcal{E}$; $\mathcal{E}^{\downarrow}_{(x,y)}$, $\mathcal{E}^{\downarrow}_{\mathcal{Q}^j_{(x,y)}}$ are downsampled error term for patch centered at (x,y) and $\mathcal{Q}^j_{(x,y)}$, respectively; $\mathcal{E}_{\mathcal{Q}^j_{(x,y)}}$ is the error term for patch centered at $\mathcal{Q}^j_{(x,y)}$. α_1, α_2 and α_3 are different weighting factors. Considering that each reference feature patch could have multiple matches with patches in $\phi(I_{in})$, the corresponding error terms are accumulated multiple times for $\mathcal{E}^{I^{\downarrow}_{ref}}$ and $\mathcal{E}^{I_{ref}}$.

We construct the whole error map $\mathcal{E}^{I_{in}}$, $\mathcal{E}^{I^{\downarrow}_{ref}}$ and $\mathcal{E}^{I_{ref}}$ in the feature encoding module by accumulating error terms for all the patches along with their coordinates. For the overlapped regions covered by multiple patches, the average error value is used.

Finally, the error map $\mathcal{E}^{I_{in}}$, $\mathcal{E}^{I^{\downarrow}_{ref}}$ and $\mathcal{E}^{I_{ref}}$ are used for the parameter update in the convolution layers of the feature encoding module:

$$\frac{\partial \mathcal{J}(\mathcal{W})}{\partial \mathcal{W}} = \mathcal{E}^{I_{in}} \frac{\partial \phi(I_{in})}{\partial \mathcal{W}} + \mathcal{E}^{I^{\downarrow}_{ref}} \frac{\partial \phi(I^{\downarrow}_{ref})}{\partial \mathcal{W}} + \mathcal{E}^{I_{ref}} \frac{\partial \phi(I_{ref})}{\partial \mathcal{W}}, \qquad (6)$$

where $\mathcal{W}$ is the parameter set, and η is the update rate.

3.4   Image Synthesis Module

In the image synthesis module, the LR image I_{in}, its features $\phi(I_{in})$, and the swapped feature map $\mathcal{F}$ are used to fuse and synthesize the SR image with residual learning. The swapped feature $\mathcal{F}$ contains HR textures to recover the details.

Similar to the structure in our feature encoding module, we also utilize the stacked residual blocks to fuse the high-frequency features to the SR image. As shown in Fig. 6, the first set of residual blocks on the left mainly focuses on upsampling the LR features $\phi(I_{in})$ for the next stage, while the second set of residual blocks focuses on the information fusion between the two kinds of features. The features at the concatenation operation are with the same feature size and they are concatenated at the channel dimension.

The final output super-resolution image I_{sr} can be defined as:

$$I_{sr} = I^{\uparrow}_{in} + Res2([\mathcal{F} \oplus Res1(\phi(I_{in}))]), \qquad (7)$$

where $I^{\uparrow}_{in}$ is a bilinear interpolated upsampled input, $Res1$ and $Res2$ represent the left and right residual connection blocks, respectively, and $\oplus$ is the concatenation operation. The detailed structure of the image synthesis network is

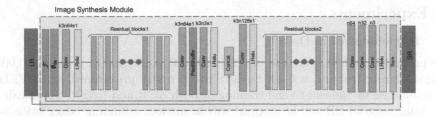

Fig. 6. The structure of the feature transfer and image synthesis network. The network consists of several residual blocks for feature decoding. The image synthesis module is marked with the dashed line.

shown in Fig. 6. Note that feature $\mathcal{F}$ and $\phi(I_{in})$ used for image synthesis are all obtained from our SR task, instead of coming from a classification model, *e.g.*, VGG [24].

The skip-connection between the LR image and the SR image could increase the image synthesis stability by making the network focus more on the high-frequency details during the training.

To further enhance the subjective visual quality of the SR image, we also adopt discriminators for adversarial learning in both SISR and RefSR branches.

3.5 Loss Function

Reconstruction Loss. Generally, the mean squared error (MSE) loss function is used in the SR task to achieve high PSNR. While in our work, we adopt the L1 norm to precisely measure the pixel difference. The L1 norm can sharpen the super-resolution image compared to that of MSE loss [25], though its PSNR is slightly lower than that of MSE loss.

$$\mathcal{L}_{rec} = \left\| I^{SR} - GT \right\|_1.\tag{8}$$

Adversarial Loss. We introduce adversarial learning in our RefSR method, the loss function is define as:

$$\mathcal{L}_D = -\mathbb{E}_{x_{real}}\left[\log\left(D\left(x_{real}, x_{fake}\right)\right)\right] - \mathbb{E}_{x_{fake}}\left[\log\left(1 - D\left(x_{fake}, x_{real}\right)\right)\right],\tag{9}$$

where D is an relativistic average discriminator [8]. Respectively, x_{real} and x_{fake} are the groundtruth and generated output of our network.

$$\mathcal{L}_G = -\mathbb{E}_{x_{real}}\left[\log\left(1 - D\left(x_{real}, x_{fake}\right)\right)\right] - \mathbb{E}_{x_{fake}}\left[\log\left(D\left(x_{fake}, x_{real}\right)\right)\right],\tag{10}$$

It is observed using this adversarial loss [8] can make our training faster and more stable compared to a standard GAN objective. We also empirically conclude that the generated results possess higher perceptual quality than that of a standard GAN objective.

4 Experiments

4.1 Implementation Details

The proposed method is trained on CUFED [20], consisting of around 100,000 images. During training, a GAN-based SISR is firstly pre-trained on CUFED. Then followed by the end-to-end training of both SISR and RefSR. Specifically, each image of CUFED will be cropped within random bounding boxes twice, to generate two different patches with similar content. The crops image pair (input and reference) will be used for end-to-end training. Adam optimizer is used with a learning rate of 1e-4 throughout the training. The weights for L_{rec}, L_{adv}, is 1e-2 and 1e-5, respectively. The number of residual blocks is 16 for both encoder and decoder. The network is trained with the CUFED dataset for 20 epochs with two basic losses. In all our designated experiments, no augmentation other than image translation is applied.

The proposed method is evaluated on the datasets CUFED5 [24], SUN hays [15] and Flickr1024 [19], containing 126, 80 and 112 image pairs respectively. Each image pair contains one input image and one reference image for the evaluation of reference-based SR methods. To evaluate single-image SR methods, all images in these datasets are viewed as individual images. Moreover, compared with CUFED5 and SUN hays datasets, Flickr1024 is a stereo image dataset with higher resolution and similarity, and we use its testset for evaluation. The evaluation relies on two common metrics, including PSNR and SSIM.

Table 1. A quantitative comparison of our approach with other SR methods on CUFED5 and SUN Hays dataset. The used super-resolution ratio is 4×4. PSNR and SSIM are used as the evaluation metrics.

Method	CUFED5 [24]		SUN Hays [15]	
	PSNR	SSIM	PSNR	SSIM
Bicubic	22.64	0.646	27.25	0.742
DRCN [10]	23.56	0.692	–	–
EnhanceNet [13]	22.58	0.651	25.46	0.669
SRGAN [11]	22.93	0.656	26.42	0.696
Ours-SISR	23.75	0.697	26.72	0.712
Landmark [22]	23.23	0.674	-	-
SRNTT [24]	23.64	0.684	26.79	0.727
E2ENT^2-MSE (ours)	**24.24**	**0.724**	**28.50**	**0.789**
E2ENT^2 (ours)	24.01	0.705	28.13	0.765

4.2 Evaluations

The proposed method is compared with some related methods, which are classified into two groups. Methods in the first group are designed for single-image

Table 2. A quantitative comparison of our approach with other SR methods on Flickr1024 dataset. The used super-resolution ratio is 4×4. PSNR and SSIM are used as the evaluation metrics.

Method	Flickr1024 Test Set [19]	
	PSNR	SSIM
SteroSR [6]	21.77	0.617
PASSRnet [18]	21.31	0.600
SRGAN [11]	21.67	0.567
SRNTT [24]	22.02	0.637
E2ENT^2 (ours)	**22.89**	**0.680**

SR, including Bicubic, DRCN [10], EnhanceNet [13] and SRGAN [11]. Methods in the second groups are designed for reference-based SR, including Landmark [22], SRNTT [24], SteroSR [6] and PASSRnet [18]. The quantitative results are summarized in Table 1 and Table 2.

For the evaluation of reference-based SR methods, the proposed method also outperforms other methods and boosts PSNR by 0.6 dB on CUFED5 and 1.71 dB on SUN Hays against the previous state-of-the-art method (SRNTT). The SSIM gain over SRNTT is also substantial, being 0.040 and 0.062 for CUFED5 and SUN Hays, respectively. E2ENT^2-MSE denotes that the MSE loss is used to replace the L1 reconstruction loss. When evaluated on a stereo dataset (Flickr1024), as shown in Table 2, where the reference images are highly relevant, the proposed method shows a great advantage over the SISR method (SRGAN) and other RefSR based methods, demonstrating its robustness under different similarity levels between LR input images and HR reference images.

Some visualization comparisons are reported in Fig. 7, including indoor objects, buildings, and natural scenes. For a clear illustration, some image patches are zoomed in to fully demonstrate the exquisite textures and details of the SR images generated by the proposed method. A user study is conducted, seven algorithms, including both single/reference-based image super-resolution results, are given to the respondents. The statistical results are shown in Fig. 8, compared with single image super-resolution methods, respondents favor the results of reference-based methods more.

4.3 Ablation Study

Impact of Feature Encoding Module. The first ablation study is about the impact of different feature encoding methods, with the comparisons reported in Table 3.

To do this, firstly, the SISR branch is pre-trained on the SR dataset, and the encoder of this pre-trained SISR will be utilized later. Then, SISR and RefSR are trained in an end-to-end way on the CUFED dataset, obtaining the feature encoding method E2ENT^2 in Table 3. Secondly, we replace the feature

242   Y. Xie et al.

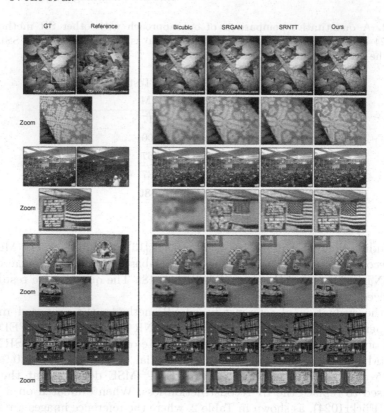

Fig. 7. Visualizations of generated images with different SR methods on CUFED5 (first 4 rows) and Flickr1024 datasets (last rows). Best viewed in color, and zoom-in mode. (Color figure online)

encoding module of E2ENT^2 with VGG (pre-trained on ImageNet [2]) and train the remaining network to obtain the model Feature-VGG. Similarly, by replacing the feature encoding module of E2ENT^2 with the encoder of the pre-trained SISR in the first step, we train the model Feature-preSISR. As can be observed from Table 3, E2ENT^2 obtains the highest PSNR and SSIM among all settings. The results demonstrate the effectiveness of the proposed trainable feature encoding module.

Besides, we calculate the feature distance (L1) between the swapped feature map $\mathcal{F}$ and that of the ground truth HR image without the match and swap module. The small feature distance of E2ENT^2 indicates that the feature of E2ENT^2 is closer to the feature of the HR ground truth image than others.

Impact of Gradient Allocation. The second ablation study is about the influence of gradient allocation, which is controlled through variable weights $(\alpha_1, \alpha_2, \alpha_3)$ in Eq. (5). As can be observed from Table 4, parameter set $(\alpha_1, \alpha_2, \alpha_3) = (0.25, 0.25, 0.50)$ outperforms $(1, 0, 0)$, $(0, 1, 0)$ and $(0, 0, 1)$, indicating

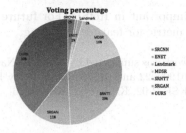

Fig. 8. The user study results. Our method is compared with different SR algorithms, more respondents favor our E2ENT results than that of SRNTT.

Table 3. A comparison study of three different feature coding methods. The used super-resolution ratio is 4×4. PSNR and SSIM are used as the evaluation metrics.

Feature type	PSNR	SSIM	Feature distance
Feature-VGG	22.85	0.647	106.77
Feature-preSISR	23.46	0.678	58.94
E2ENT^2(ours)	24.01	0.705	25.77

that only to consider the gradient for one feature in $\{\phi(I_{in}), \phi(I_{ref}^{\downarrow}), \phi(I_{ref})\}$ is not sufficient for the proposed method. We allocate slightly higher value to α_3 ($\alpha_3 = 0.5$), because the selected image patch in $\phi(I_{ref})$ will be finally used in $\mathcal{F}$. However, the similarity metric in the matching operation relies on both the LR features $\phi(I_{in})$ and the reference features $\phi(I_{ref}^{\downarrow})$, meaning that we can not neglect them during the gradient propagation process; thus, we set $\alpha_1 = \alpha_2 = 0.25$.

Table 4. A comparison of different settings for $(\alpha_1, \alpha_2, \alpha_3)$.

Weights	Different combinations			
$(\alpha_1, \alpha_2, \alpha_3)$	(1, 0, 0)	(0, 1, 0)	(0, 0, 1)	(0.25, 0.25, 0.5)
PSNR	23.75	23.67	23.83	24.01
SSIM	0.697	0.672	0.695	0.705

5   Conclusions

In this paper, we explored a generalized problem for image super-resolution by utilizing high-resolution reference images. We proposed a match and swap module to obtain similar texture and high-frequency information from reference images, where end-to-end learning is enabled by properly distributing the gradients to the prior feature encoding module. Experiment results indicating that the matching feature is important in RefSR. For future work, we are going to study a better similarity metric for feature matching.

Acknowledgment. The work was supported by National Natural Science Foundation of China under 61972323, 61902022 and 61876155, and Key Program Special Fund in XJTLU under KSF-T-02, KSF-P-02, KSF-A-01, KSF-E-26.

References

1. Ahn, N., Kang, B., Sohn, K.A.: Fast, accurate, and lightweight super-resolution with cascading residual network. In: Proceedings of the European Conference on Computer Vision, pp. 252–268 (2018)
2. Deng, J., Dong, W., Socher, R., Li, L.J., Li, K., Fei-Fei, L.: ImageNet: a large-scale hierarchical image database. In: 2009 IEEE Conference on Computer Vision and Pattern Recognition, pp. 248–255. IEEE (2009)
3. Dong, C., Loy, C.C., He, K., Tang, X.: Image super-resolution using deep convolutional networks. IEEE Trans. Pattern Anal. Mach. Intell. **38**(2), 295–307 (2015)
4. Freedman, G., Fattal, R.: Image and video upscaling from local self-examples. ACM Trans. Graph. **30**(2), 1–11 (2011)
5. Goodfellow, I., et al.: Generative adversarial nets. In: Advances in Neural Information Processing Systems, pp. 2672–2680 (2014)
6. Jeon, D.S., Baek, S.H., Choi, I., Kim, M.H.: Enhancing the spatial resolution of stereo images using a parallax prior. In: Proceedings of the IEEE Conference on Computer Vision and Pattern Recognition, pp. 1721–1730 (2018)
7. Johnson, J., Alahi, A., Fei-Fei, L.: Perceptual losses for real-time style transfer and super-resolution. In: Leibe, B., Matas, J., Sebe, N., Welling, M. (eds.) ECCV 2016. LNCS, vol. 9906, pp. 694–711. Springer, Cham (2016). https://doi.org/10.1007/978-3-319-46475-6_43
8. Jolicoeur-Martineau, A.: The relativistic discriminator: a key element missing from standard GAN. arXiv preprint arXiv:1807.00734 (2018)
9. Kim, J., Kwon Lee, J., Mu Lee, K.: Accurate image super-resolution using very deep convolutional networks. In: Proceedings of the IEEE Conference on Computer Vision and Pattern Recognition, pp. 1646–1654 (2016)
10. Kim, J., Kwon Lee, J., Mu Lee, K.: Deeply-recursive convolutional network for image super-resolution. In: Proceedings of the IEEE Conference on Computer Vision and Pattern Recognition, pp. 1637–1645 (2016)
11. Ledig, C., et al.: Photo-realistic single image super-resolution using a generative adversarial network. In: Proceedings of the IEEE Conference on Computer Vision and Pattern Recognition, pp. 4681–4690 (2017)
12. Lim, B., Son, S., Kim, H., Nah, S., Mu Lee, K.: Enhanced deep residual networks for single image super-resolution. In: Proceedings of the IEEE Conference on Computer Vision and Pattern Recognition Workshops, pp. 136–144 (2017)
13. Sajjadi, M.S., Scholkopf, B., Hirsch, M.: EnhanceNet: single image super-resolution through automated texture synthesis. In: Proceedings of the IEEE International Conference on Computer Vision, pp. 4491–4500 (2017)
14. Salvador, J.: Example-Based Super Resolution. Academic Press (2016)
15. Sun, L., Hays, J.: Super-resolution from internet-scale scene matching. In: 2012 IEEE International Conference on Computational Photography, pp. 1–12. IEEE (2012)

16. Timofte, R., De Smet, V., Van Gool, L.: Anchored neighborhood regression for fast example-based super-resolution. In: The IEEE International Conference on Computer Vision, December 2013
17. Timofte, R., De Smet, V., Van Gool, L.: Anchored neighborhood regression for fast example-based super-resolution. In: Proceedings of the IEEE International Conference on Computer Vision, pp. 1920–1927 (2013)
18. Wang, L., et al.: Learning parallax attention for stereo image super-resolution. In: Proceedings of the IEEE Conference on Computer Vision and Pattern Recognition, pp. 12250–12259 (2019)
19. Wang, Y., Wang, L., Yang, J., An, W., Guo, Y.: Flickr1024: a large-scale dataset for stereo image super-resolution. In: Proceedings of the IEEE International Conference on Computer Vision Workshops, p. 0 (2019)
20. Wang, Y., Lin, Z., Shen, X., Mech, R., Miller, G., Cottrell, G.W.: Event-specific image importance. In: The IEEE Conference on Computer Vision and Pattern Recognition (2016)
21. Wang, Y., Liu, Y., Heidrich, W., Dai, Q.: The light field attachment: Turning a DSLR into a light field camera using a low budget camera ring. IEEE Trans. Visual Comput. Graphics **23**(10), 2357–2364 (2016)
22. Yue, H., Sun, X., Yang, J., Wu, F.: Landmark image super-resolution by retrieving web images. IEEE Trans. Image Process. **22**(12), 4865–4878 (2013)
23. Zhang, Y., Li, K., Li, K., Wang, L., Zhong, B., Fu, Y.: Image super-resolution using very deep residual channel attention networks. In: Ferrari, V., Hebert, M., Sminchisescu, C., Weiss, Y. (eds.) ECCV 2018. LNCS, vol. 11211, pp. 294–310. Springer, Cham (2018). https://doi.org/10.1007/978-3-030-01234-2_18
24. Zhang, Z., Wang, Z., Lin, Z., Qi, H.: Image super-resolution by neural texture transfer. In: Proceedings of the IEEE Conference on Computer Vision and Pattern Recognition, pp. 7982–7991 (2019)
25. Zhao, H., Gallo, O., Frosio, I., Kautz, J.: Loss functions for image restoration with neural networks. IEEE Trans. Comput. Imaging **3**(1), 47–57 (2016)
26. Zheng, H., Guo, M., Wang, H., Liu, Y., Fang, L.: Combining exemplar-based approach and learning-based approach for light field super-resolution using a hybrid imaging system. In: Proceedings of the IEEE International Conference on Computer Vision Workshops, pp. 2481–2486 (2017)
27. Zheng, H., Ji, M., Han, L., Xu, Z., Wang, H., Liu, Y., Fang, L.: Learning cross-scale correspondence and patch-based synthesis for reference-based super-resolution. In: Proceedings of the British Machine Vision Conference (2017)
28. Ferrari, V., Hebert, M., Sminchisescu, C., Weiss, Y. (eds.): ECCV 2018. LNCS, vol. 11210. Springer, Cham (2018). https://doi.org/10.1007/978-3-030-01231-1

RobustFusion: Human Volumetric Capture with Data-Driven Visual Cues Using a RGBD Camera

Zhuo Su[1], Lan Xu[1,2], Zerong Zheng[1], Tao Yu[1], Yebin Liu[1], and Lu Fang[1(✉)]

[1] Tsinghua University, Beijing, China
fanglu@sz.tsinghua.edu.cn
[2] ShanghaiTech University, Shanghai, China

Abstract. High-quality and complete 4D reconstruction of human activities is critical for immersive VR/AR experience, but it suffers from inherent self-scanning constraint and consequent fragile tracking under the monocular setting. In this paper, inspired by the huge potential of learning-based human modeling, we propose RobustFusion, a robust human performance capture system combined with various data-driven visual cues using a single RGBD camera. To break the orchestrated self-scanning constraint, we propose a data-driven model completion scheme to generate a complete and fine-detailed initial model using only the front-view input. To enable robust tracking, we embrace both the initial model and the various visual cues into a novel performance capture scheme with hybrid motion optimization and semantic volumetric fusion, which can successfully capture challenging human motions under the monocular setting without pre-scanned detailed template and owns the reinitialization ability to recover from tracking failures and the disappear-reoccur scenarios. Extensive experiments demonstrate the robustness of our approach to achieve high-quality 4D reconstruction for challenging human motions, liberating the cumbersome self-scanning constraint.

Keywords: Dynamic reconstruction · Volumetric capture · Robust · RGBD camera

1 Introduction

With the recent popularity of virtual and augmented reality (VR and AR) to present information in an innovative and immersive way, the 4D (3D spatial plus 1D time) content generation evolves as a cutting-edge yet bottleneck technique.

Z. Su and L. Xu—Equal Contribution.

Electronic supplementary material The online version of this chapter (https://doi.org/10.1007/978-3-030-58548-8_15) contains supplementary material, which is available to authorized users.

© Springer Nature Switzerland AG 2020
A. Vedaldi et al. (Eds.): ECCV 2020, LNCS 12349, pp. 246–264, 2020.
https://doi.org/10.1007/978-3-030-58548-8_15

Reconstructing the 4D models of challenging human activities conveniently for better VR/AR experience has recently attracted substantive attention of both the computer vision and computer graphics communities.

Early solutions [27,28,34,52,53] requires pre-scanned templates or two to four orders of magnitude more time than is available for daily usages such as immersive tele-presence. Recent volumetric approaches have eliminated the reliance of a pre-scanned template model and led to a profound progress in terms of both effectiveness and efficiency, by leveraging the RGBD sensors and high-end GPUs. The high-end solutions [7,9,10,24,61] rely on multi-view studio setup to achieve high-fidelity reconstruction but are expensive and difficult to be deployed, leading to the high restriction of the wide applications for daily usage. Besides, a number of approaches [16,22,35,45,60,65–67] adopt the most common single RGBD camera setup with a temporal fusion pipeline to achieve complete reconstruction. However, these single-view approaches suffer from careful and orchestrated motions, especially when the performer needs to turn around carefully to obtain complete reconstruction. When the captured model is incomplete, the non-rigid tracking in those newly fused regions is fragile, leading to inferior results and impractical usage for VR/AR applications. On the other hand, the learning-based techniques have achieved significant progress recently for human attribute prediction using only the RGB input. This overcomes inherent constraint of existing monocular volumetric capture approaches, since such data-driven visual cues encode various prior information of human models such as motion [6,25,32] or geometry [2,42,68]. However, researchers did not explore these solutions to strengthen the volumetric performance capture.

In this paper, we attack the above challenges and propose *RobustFusion* – the first human volumetric capture system combined with various data-driven visual cues using only a single RGBD sensor, which does not require a pre-scanned template and outperforms existing state-of-the-art approaches significantly. Our novel pipeline not only eliminates the tedious self-scanning constraint but also captures challenging human motions robustly with the re-initialization ability to handle the severe tracking failures or disappear-reoccur scenarios, whilst still maintaining light-weight computation and monocular setup.

To maintain the fast running performance for the wide daily usages, we utilize those light-weight data-driven visual cues including implicit occupancy representation, human pose, shape and body part parsing. Combining such light-weight data-driven priors with the non-rigid fusion pipeline to achieve more robust and superior human volumetric capture is non-trivial. More specifically, to eliminate the inherent orchestrated self-scanning constraint of single-view capture, we first combine the data-driven implicit occupancy representation and the volumetric fusion within a completion optimization pipeline to generate an initial complete human model with fine geometric details. Such complete model is utilized to initialize both the performance capture parameters and the associated human priors. To enable robust tracking, based on both the initial complete model and the various visual cues, a novel performance capture scheme is proposed to combine the non-rigid tracking pipeline with human pose, shape and parsing priors, through a hybrid and flip-flop motion optimization and an effective semantic volumetric fusion strategy. Our hybrid optimization handles challenging fast human motions and recovers from tracking failures and the disappear-reoccur scenarios, while the volumetric fusion strategy estimates semantic motion tracking behavior to achieve robust and precise geometry update and avoid deteriorated fusion model caused by challenging fast motion and self-occlusion. To summarize, the main contributions of RobustFusion include:

- We propose a robust human volumetric capture method, which is the first to embrace various data-driven visual cues under the monocular setting without pre-scanned template, achieving significant superiority to state-of-the-arts.
- To eliminate the tedious self-scanning constraint, we propose a novel optimization pipeline to combine data-driven occupancy representation with volumetric fusion only using the front-view input.
- We propose an effective robust performance capture scheme with human pose, shape and parsing priors, which can handle challenging human motions with reinitialization ability.

2 Related Work

Human Performance Capture. Marker-based performance capture systems are widely used [55,56,59] but they are costly and quite intrusive to wear the marker suites. Thus, markerless performance capture [1,5,54] technologies have been widely investigated. The multi-view markerless approaches require studio-setup with a controlled imaging environment [7,11,15,23,24,29,48], while recent work [38,40,44] even demonstrates robust out-of-studio capture but synchronizing and calibrating multi-camera systems are still cumbersome. Some recent work only relies on a light-weight single-view setup [19,63,64] and even enables hand-held capture [20,37,57,58] or drone-based capture [60,62] for more practical application of performance capture. However, these methods require pre-scanned template model or can only reconstruct naked human model.

Recently, free-form dynamic reconstruction methods combine the volumetric fusion [8] and the nonrigid tracking [17,27,49,71]. The high-end solutions [9,10,61] rely on multi-view studio to achieve high-fidelity reconstruction but are difficult to be deployed for daily usage, while some work [18,22,35,45–47] adopt the most common single RGBD camera setting. Yu et al. [51,65,66] constrain the motion to be articulated to increase tracking robustness, while HybridFusion [67] utilizes extra IMU sensors for more reliable reconstruction. Xu et al. [60] further model the mutual gains between capture view selection and reconstruction. Besides, some recent work [31,36] combine the neural rendering techniques to provide more visually pleasant results. However, these methods still suffer from careful and orchestrated motions, especially for a tedious self-scanning process where the performer need to turn around carefully to obtain complete reconstruction. Comparably, our approach is more robust for capturing challenging motions with reinitialization ability, and eliminates the self-scanning constraint.

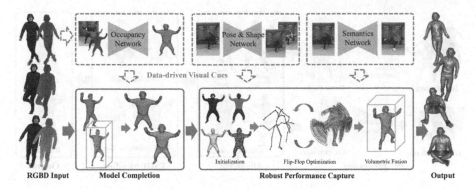

Fig. 1. The pipeline of RobustFusion. Assuming monocular RGBD input with various data-driven human visual priors, our approach consists of a model completion stage (Sect. 4) and a robust performance capture stage (Sect. 5) to generate live 4D results.

Data-Driven Human Modeling. Early human modeling techniques [12,43] are related to the discriminative approaches of performance capture, which take advantage of data driven machine learning strategies to convert the capture problem into a regression or pose classification problem. With the advent of deep neural networks, recent approaches obtain various attributes of human successfully from only the RGB input, which encodes rich prior information of human models. Some recent work [6,25,26,32,41] learns the skeletal pose and even human shape prior by using human parametric models [3,30]. Various approaches [2,39,50,68,69] propose to predict human geometry from a single RGB image by utilizing parametric human model as a basic estimation. Several work [21,33,42] further reveals the effectiveness of learning the implicit occupancy directly for detailed geometry modeling. However, such predicted geometry lacks fine details for face region and clothes wrinkle, which is important for immersive human modeling. Besides, researchers [13,14,70] propose to fetch the semantic information of human model. Several works [19,63,64] leverage learnable pose detections [6,32] to improve the accuracy of human motion capture, but these methods rely on pre-scanned template models. However, even though these visual attributes yield huge potential for human performance modeling, researchers pay less attention surprisingly to explicitly combine such various data-driven visual priors with the existing volumetric performance capture pipeline. In contrast, we explore to build a robust volumetric capture algorithm on top of these visual priors and achieve significant superiority to previous capture methods.

3   Overview

RobustFusion marries volumetric capture to various data-driven human visual cues, which not only eliminates the tedious self-scanning constraint but also

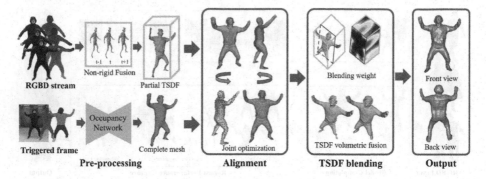

Fig. 2. Model completion pipeline. Assuming the front-view RGBD input, both a partial TSDF volume and a complete mesh are generated, followed by the alignment and blending operations to obtain a complete human model with fine geometry details.

captures challenging human motions robustly handling the severe tracking failures or disappear-reoccur scenarios. As illustrated in Fig. 1, our approach takes a RGBD video from Kinect v2 as input and generates 4D meshes, achieving considerably more robust results than previous methods. Similar to [61,66], we utilize TSDF [8] volume and ED model [49] for representation.

Model Completion. Only using the front-view input, we propose to combine the data-driven implicit occupancy network with the non-rigid fusion to eliminate the orchestrated self-scanning constraint of monocular capture. A novel completion optimization scheme is adopted to generate a high-quality watertight human model with fine geometric details.

Motion Initialization. Before the tracking stage, we further utilize the watertight mesh to initialize both the human motions and the visual priors. A hybrid motion representation based on the mesh is adopted, while various human pose and parsing priors based on the front-view input are associated to the mesh.

Robust Tracking. The core of our pipeline is to solve the hybrid motion parameters from the canonical frame to current camera view. We propose a robust tracking scheme which utilizes the reliable visual priors to optimize both the skeletal and non-rigid motions in an iterative flip-flop manner. Our scheme can handle challenging motions with the reinitialization ability.

Volumetric Fusion. After estimating the motions, we fuse the depth stream into a canonical TSDF volume to provide temporal coherent results. Based on various visual priors, our adaptive strategy models semantic tracking behavior to avoid deteriorated fusion caused by challenging motions. Finally, dynamic atlas [61] is adopted to obtain 4D textured reconstruction results.

4   Model Completion

To eliminate the orchestrated self-scanning constraint and the consequent fragile tracking of monocular capture, we propose a model completion scheme using only the front-view RGBD input. As illustrated in Fig. 2, our completion scheme combines the data-driven implicit representation and the non-rigid fusion to obtain a complete human model with fine geometry details.

Pre-processing. To generate high-fidelity geometry details, we utilize the traditional ED-based non-rigid alignment method [35,60] to fuse the depth stream into live partial TSDF volume. Once the average accumulated TSDF weight in the front-view voxels reaches a threshold (32 in our setting), a data-driven occupancy regression network is triggered to generate a watertight mesh from only the triggered RGBD frame. To this end, we pre-train the PIFu [42] network using 1820 scans from Twindom, which learns a pixel-aligned implicit function by combining an image encoder and an implicit occupancy regression. To improve the scale and pose consistency, both the depth image and the human parsing image are added to the input of the image encoder.

Alignment. Note that the unique human prior can serve as a reliable reference to eliminate the misalignment between the partial TSDF and the complete mesh caused by their different input modalities. Thus, we adopt the double-layer motion representation [61,66], which combines the ED model and the linear human body model SMPL [30]. For any 3D vertex $\mathbf{v}_c$, let $\tilde{\mathbf{v}}_c = ED(\mathbf{v}_c; G)$ denote the warped position after ED motion, where G is the non-rigid motion field. As for the SMPL model [30], the body model $\bar{\mathbf{T}}$ deforms into the morphed model $T(\boldsymbol{\beta}, \boldsymbol{\theta})$ with the shape parameters $\boldsymbol{\beta}$ and pose parameters $\boldsymbol{\theta}$. For any vertex $\bar{\mathbf{v}} \in \bar{\mathbf{T}}$, let $W(T(\bar{\mathbf{v}}; \boldsymbol{\beta}, \boldsymbol{\theta}); \boldsymbol{\beta}, \boldsymbol{\theta})$ denote the corresponding posed 3D position. Please refer to [61,66] for details about the motion representation.

To align the partial TSDF and the complete mesh, we jointly optimize the unique human shape $\boldsymbol{\beta}_0$ and skeleton pose $\boldsymbol{\theta}_0$, as well as the ED non-rigid motion field G_0 from the TSDF volume to the complete mesh as follows:

$$E_{\text{comp}}(G_0, \boldsymbol{\beta}_0, \boldsymbol{\theta}_0) = \lambda_{\text{vd}} E_{\text{vdata}} + \lambda_{\text{md}} E_{\text{mdata}} + \lambda_{\text{bind}} E_{\text{bind}} + \lambda_{\text{prior}} E_{\text{prior}}. \quad (1)$$

Here the volumetric data term E_{vdata} measures the misalignment error between the SMPL model and the reconstructed geometry in the partial TSDF volume:

$$E_{\text{vdata}}(\boldsymbol{\beta}_0, \boldsymbol{\theta}_0) = \sum_{\bar{\mathbf{v}} \in \bar{\mathbf{T}}} \psi(\mathbf{D}(W(T(\bar{\mathbf{v}}; \boldsymbol{\beta}_0, \boldsymbol{\theta}_0); \boldsymbol{\beta}_0, \boldsymbol{\theta}_0))), \quad (2)$$

where $\mathbf{D}(\cdot)$ takes a point in the canonical volume and returns the bilinear interpolated TSDF, and $\psi(\cdot)$ is the robust Geman-McClure penalty function.

The mutual data term E_{mdata} further measures the fitting from both the TSDF volume and the SMPL model to the complete mesh, which is formulated as the sum of point-to-plane distances:

$$E_{\text{mdata}} = \sum_{(\bar{\mathbf{v}}, \mathbf{u}) \in \mathcal{C}} \psi(\mathbf{n}_{\mathbf{u}}^T(W(T(\bar{\mathbf{v}}; \boldsymbol{\beta}_0, \boldsymbol{\theta}_0)) - \mathbf{u})) + \sum_{(\tilde{\mathbf{v}}_c, \mathbf{u}) \in \mathcal{P}} \psi(\mathbf{n}_{\mathbf{u}}^T(\tilde{\mathbf{v}}_c - \mathbf{u})), \quad (3)$$

where $\mathcal{C}$ and $\mathcal{P}$ are the correspondence pair sets found via closest searching; $\mathbf{u}$ is a corresponding 3D vertex on the complete mesh. Besides, the pose prior term E_{prior} from [4] penalizes the unnatural poses while the binding term E_{bind}

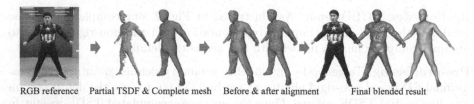

RGB reference   Partial TSDF & Complete mesh   Before & after alignment   Final blended result

Fig. 3. The results of our model completion pipeline.

from [66] constrains both the non-rigid and skeletal motions to be consistent. We solve the resulting energy E_{comp} under the Iterative Closest Point (ICP) framework, where the non-linear least squares problem is solved using Levenberg-Marquardt (LM) method with a custom designed Preconditioned Conjugate Gradient (PCG) solver on GPU [10,18].

TSDF Blending. After the alignment, we blend both the partial volume and the complete mesh seamlessly in the TSDF domain. For any 3D voxel $\mathbf{v}$, $\tilde{\mathbf{v}}$ denotes its warped position after applying the ED motion field; $\mathbf{N}(\mathbf{v})$ denotes the number of non-empty neighboring voxels of $\mathbf{v}$ in the partial volume which indicates the reliability of the fused geometry; $\mathbf{D}(\mathbf{v})$ and $\mathbf{W}(\mathbf{v})$ denote its TSDF value and accumulated weight, respectively. Then, to enable smooth blending, we calculate the corresponding projective SDF value $\mathbf{d}(\mathbf{v})$ and the updating weight $\mathbf{w}(\mathbf{v})$ as follows:

$$\mathbf{d}(\mathbf{v}) = (\mathbf{u} - \tilde{\mathbf{v}})\mathbf{sgn}(\mathbf{n}_\mathbf{u}^T(\mathbf{u} - \tilde{\mathbf{v}})), w(\mathbf{v}) = 1/(1 + \mathbf{N}(\mathbf{v})). \quad (4)$$

Here, recall that $\mathbf{u}$ is the corresponding 3D vertex of $\tilde{\mathbf{v}}$ on the complete mesh and $\mathbf{n}_\mathbf{u}$ is its normal; $\mathbf{sgn}(\cdot)$ is the sign function to distinguish positive and negative SDF. The voxel is further updated by the following blending operation:

$$\mathbf{D}(\mathbf{v}) \leftarrow \frac{\mathbf{D}(\mathbf{v})\mathbf{W}(\mathbf{v}) + \mathbf{d}(\mathbf{v})w(\mathbf{v})}{\mathbf{W}(\mathbf{v}) + w(\mathbf{v})}, \mathbf{W}(\mathbf{v}) \leftarrow \mathbf{W}(\mathbf{v}) + w(\mathbf{v}). \quad (5)$$

Finally, as illustrated in Fig. 3, marching cubes algorithm is adopted to obtain a complete and watertight human model with fine geometry details, which further enables robust motion initialization and tracking in Sect. 5.

5   Robust Performance Capture

As illustrated in Fig. 4, a novel performance capture scheme is proposed to track challenging human motions robustly with re-initialization ability with the aid of reliable data-driven visual cues.

Initialization. Note that the final complete model from Sect. 4 provides a reliable initialization for both the human motion and the utilized visual priors. To this end, before the tracking stage, we first re-sample the sparse ED nodes $\{\mathbf{x}_i\}$ on the mesh to form a non-rigid motion field, denoted as G. Besides, we

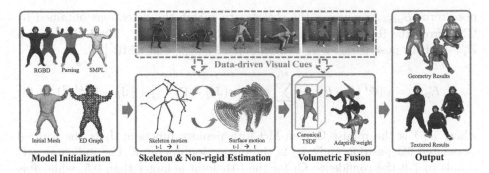

Fig. 4. The pipeline of our robust performance capture scheme. We first initialize both the motions and visual priors. Then, both skeletal and non-rigid motions are optimized with the associated visual priors. Finally, an adaptive volumetric fusion scheme is adopted to generated 4D textured results.

rig the mesh with the output pose parameters θ_0 from its embedded SMPL model in Sect. 4 and transfer the SMPL skinning weights to the ED nodes $\{\mathbf{x}_i\}$. Then, for any 3D point $\mathbf{v}_c$ in the capture volume, let $\tilde{\mathbf{v}}_c$ and $\hat{\mathbf{v}}_c$ denote the warped positions after the embedded deformation and skeletal motion, respectively. Note that the skinning weights of $\mathbf{v}_c$ for the skeletal motion are given by the weighted average of the skinning weights of its knn-nodes. Please refer to [30,61,66] for more detail about the motion formulation. To initialize the pose prior, we apply OpenPose [6] on the RGBD image to obtain the 2D and lifted 3D joint positions, denoted as $\mathbf{P}_l^{2D}$ and $\mathbf{P}_l^{3D}$, respectively, with a detection confidence $\mathbf{C}_l$. Then, we find the closest vertex from the watertight mesh to $\mathbf{P}_l^{3D}$, denoted as $\mathbf{J}_l$, which is the associated marker position for the l-th joint. To utilize the semantic visual prior, we apply the light-weight human parsing method [70] to the triggered RGB image to obtain a human parsing image L. Then, we project each ED node $\mathbf{x}_i$ into L to obtain its initial semantic label $\mathbf{l}_i$. After the initialization, inspired by [19,64], we propose to optimize the motion parameters G and θ in an iterative flip-flop manner, so as to fully utilize the rich motion prior information of the visual cues to capture challenging motions.

Skeletal Pose Estimation. During each ICP iteration, we first optimize the skeletal pose θ of the watertight mesh, which is formulated as follows:

$$E_{\text{smot}}(\theta) = \lambda_{\text{sd}} E_{\text{sdata}} + \lambda_{\text{pose}} E_{\text{pose}} + \lambda_{\text{prior}} E_{\text{prior}} + \lambda_{\text{temp}} E_{\text{temp}}. \qquad (6)$$

Here, the dense data term E_{sdata} measures the point-to-plane misalignment error between the warped geometry in the TSDF volume and the depth input:

$$E_{\text{sdata}} = \sum_{(\mathbf{v}_c,\mathbf{u})\in\mathcal{P}} \psi(\mathbf{n}_{\mathbf{u}}^T(\hat{\mathbf{v}}_c - \mathbf{u})), \qquad (7)$$

where $\mathcal{P}$ is the corresponding set found via a projective searching; $\mathbf{u}$ is a sampled point on the depth map while $\mathbf{v}_c$ is the closet vertex on the fused surface. The pose term E_{pose} encourages the skeleton to match the detections obtained by CNN from the RGB image, including the 2D position $\mathbf{P}_l^{2D}$, lifted 3D position $\mathbf{P}_l^{3D}$ and the pose parameters $\boldsymbol{\theta}_d$ from OpenPose [6] and HMR [25]:

$$E_{\text{pose}} = \psi(\Phi^T(\boldsymbol{\theta} - \boldsymbol{\theta}_d)) + \sum_{l=1}^{N_J} \phi(l)(\|\pi(\hat{\mathbf{J}}_l) - \mathbf{P}_l^{2D}\|_2^2 + \|\hat{\mathbf{J}}_l - \mathbf{P}_l^{3D}\|_2^2), \quad (8)$$

where $\psi(\cdot)$ is the robust Geman-McClure penalty function; $\hat{\mathbf{J}}_l$ is the warped associated 3D position and $\pi(\cdot)$ is the projection operator. The indicator $\phi(l)$ equals to 1 if the confidence $\mathbf{C}_l$ for the l-th joint is larger than 0.5, while Φ is the vectorized representation of $\{\phi(l)\}$. The prior term E_{prior} from [4] penalizes the unnatural poses, while the temporal term E_{temp} encourages coherent deformations by constraining the skeletal motion to be consistent to the previous ED motion:

$$E_{\text{temp}} = \sum_{\mathbf{x}_i} \|\hat{\mathbf{x}}_i - \tilde{\mathbf{x}}_i\|_2^2, \quad (9)$$

where $\tilde{\mathbf{x}}_i$ is the warped ED node using non-rigid motion from previous iteration.

Non-rigid Estimation. To capture realistic non-rigid deformation, on top of the pose estimation result, we solve the surface tracking energy as follows:

$$E_{\text{emot}}(G) = \lambda_{\text{ed}} E_{\text{edata}} + \lambda_{\text{reg}} E_{\text{reg}} + \lambda_{\text{temp}} E_{\text{temp}}. \quad (10)$$

Here the dense data term E_{edata} jointly measures the dense point-to-plane misalignment and the sparse landmark-based projected error:

$$E_{\text{edata}} = \sum_{(\mathbf{v}_c, \mathbf{u}) \in \mathcal{P}} \psi(\mathbf{n}_{\mathbf{u}}^T(\tilde{\mathbf{v}}_c - \mathbf{u})) + \sum_{l=1}^{N_J} \phi(l)\|\pi(\tilde{\mathbf{J}}_l) - \mathbf{P}_l^{2D}\|_2^2, \quad (11)$$

where $\tilde{\mathbf{J}}_l$ is the warped associated 3D joint of the l-th joint in the fused surface. The regularity term E_{reg} from [66] produces locally as-rigid-as-possible (ARAP) motions to prevent over-fitting to depth inputs. Besides, the $\hat{\mathbf{x}}_i$ after the skeletal motion in the temporal term E_{temp} is fixed during current optimization.

Both the pose and non-rigid optimizations in Eq. 6 and Eq. 10 are solved using LM method with the same PCG solver on GPU [10,18]. Once the confidence $\mathbf{C}_l$ reaches 0.9 and the projective error $\|\pi(\tilde{\mathbf{J}}_l) - \mathbf{P}_l^{2D}\|_2^2$ is larger than 5.0 for the l-th joint, the associated 3D position $\mathbf{J}_l$ on the fused surface is updated via the same closest searching strategy of the initialization stage. When there is no human detected in the image, our whole pipeline will be suspended until the number of detected joints reaches a threshold (10 in our setting).

Volumetric Fusion. To temporally update the geometric details, after above optimization, we fuse the depth into the TSDF volume and discard the voxels which are collided or warped into invalid input to achieve robust geometry update. To avoid deteriorated fusion caused by challenging motion or reinitialization, an effective adaptive fusion strategy is proposed to model semantic motion

Fig. 5. 4D reconstructed results of the proposed RobustFusion system.

tracking behavior. To this end, we apply the human parsing method [70] to current RGB image to obtain a human parsing image L. For each ED node $\mathbf{x}_i$, recall that $\mathbf{l}_i$ is its associated semantic label during initialization while $L(\pi(\tilde{\mathbf{x}}_i)$ is current corresponding projected label. Then, for any voxel $\mathbf{v}$, we formulate its updating weight $\mathbf{w}(\mathbf{v})$ as follows:

$$\mathbf{w}(\mathbf{v}) = \exp(\frac{-\|\Phi^T(\theta^* - \theta_d)\|_2^2}{2\pi}) \sum_{i \in \mathcal{N}(v_c)} \frac{\varphi(\mathbf{l}_i, L(\pi(\tilde{\mathbf{x}}_i)))}{|\mathcal{N}(v_c)|}, \qquad (12)$$

where θ^* is the optimized pose; $\mathcal{N}(v_c)$ is the collection of the knn-nodes of $\mathbf{v}$; $\varphi(\cdot, \cdot)$ denote an indicator which equals to 1 only if the two input labels are the same. Note that such robust weighting strategy measures the tracking performance based on human pose and semantic priors. Then, $\mathbf{w}(\mathbf{v})$ is set to be zero if it's less than a truncated threshold (0.2 in our setting), so as to control the minimal integration and further avoid deteriorated fusion of severe tracking failures. Finally, the voxel is updated using Eq. 5 and the dynamic atlas scheme [61] is adopted to obtain 4D textured reconstruction.

6   Experiment

In this section, we evaluate our RobustFusion system on a variety of challenging scenarios. We run our experiments on a PC with a NVIDIA GeForce GTX

Fig. 6. Qualitative comparison. Our results overlay better with the RGB images.

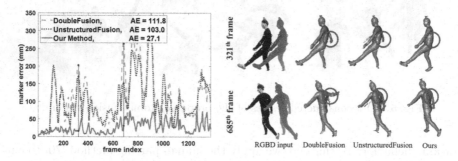

Fig. 7. Quantitative comparison against UnstructuredFusion [61] and DoubleFusion [66]. Left: the error curves. Right: the reconstruction results.

TITAN X GPU and an Intel Core i7-7700K CPU. Our optimized GPU code takes 15 s for the model completion. The following robust performance capture pipeline runs at 123 ms per frame, where the visual priors collecting takes 90 ms, the flip-flop optimization takes around 18 ms with 4 ICP iteration, and 15 ms for all the remaining computations. In all experiments, we use the following empirically determined parameters: $\lambda_{vd} = 1.0$, $\lambda_{md} = 2.0$, $\lambda_{bind} = 1.0$, $\lambda_{prior} = 0.01$, $\lambda_{sdata} = 4.0$, $\lambda_{pose} = 2.0$, $\lambda_{temp} = 1.0$, $\lambda_{edata} = 4.0$ and $\lambda_{reg} = 5.0$. Figure 5 demonstrates the results of RobustFusion, where both the challenging motion and the fine-detailed geometry are faithfully captured.

6.1 Comparison

We compare our RobustFusion against the state-of-the-art methods DoubleFusion [66] and UnstructuredFusion [61]. For fair comparison, we modify [61] into the monocular setting by removing their online calibration stage. As shown in Fig. 6, our approach achieves significantly better tracking results especially for

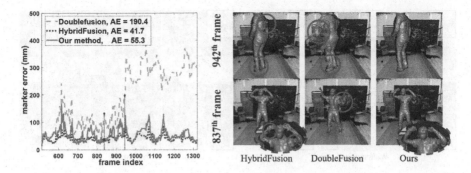

Fig. 8. Quantitative comparison against HybridFusion [67] and DoubleFusion [66]. Left: the error curves. Right: the reconstruction results.

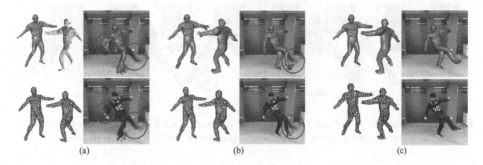

Fig. 9. Evaluation of model completion. (a–c) are the results without completion, with completion only using skeleton optimization and SMPL-based ED-graph, with our full model completion, respectively. Each triple includes the output mesh, corresponding ED-graph and the overlaid tracking geometry/texture results for a following frame.

challenging self-occluded and fast motions. Then, we utilize the sequence from [61] with available ground truth captured via the OptiTrack system to compute the per-frame mean error of all the markers as well as the average mean error of the whole sequence (AE). As illustrated in Fig. 7, our approach achieves the highest tracking accuracy with the aid of various visual priors.

We further compare against HybridFusion [67], which uses extra IMU sensors. We utilize the challenging sequence with ground truth from [67] and remove their orchestrated self-scanning process before the tracking stage (the first 514 frames). As shown in Fig. 8, our approach achieves significantly better result than DoubleFusion and even comparable performance than HybridFusion only using the RGBD input. Note that HybridFusion still relies on the self-scanning stage for sensor calibration and suffers from missing geometry caused by the body-worn IMUs, while our approach eliminates such tedious self-scanning and achieves more complete reconstruction.

Fig. 10. Evaluation of robust tracking. Our approach with various human visual priors achieves superior results for challenging motions and has reinitialization ability.

Fig. 11. Evaluation of robust tracking. Our approach with the flip-flop optimization strategy achieves more visually pleasant 4D geometry for challenging motions.

6.2   Evaluation

Model Completion. As shown in Fig. 9 (a), without model completion, only partial initial geometry with SMPL-based ED-graph leads to inferior tracking results. The result in Fig. 9 (b) is still imperfect because only the skeletal pose is optimized during completion optimization and only SMPL-based ED-graph is adopted for motion tracking. In contrast, our approach with model completion in Fig. 9 (c) successfully obtains a watertight and fine-detailed human mesh to enable both robust motion initialization and tracking.

Robust Tracking. We further evaluate our robust performance capture scheme. In Fig. 10, we compare to the results using traditional tracking pipeline [61,66] without data-driven visual priors in two scenarios where fast motion or disappear-reoccurred case happens. Note that our variation without visual priors suffers from severe accumulated error and even totally tracking lost, while our approach achieves superior tracking results for these challenging cases. Furthermore, we compare to the baseline which jointly optimizes skeletal and non-rigid motions without our flip-flop strategy. As shown in Fig. 11, our approach with flip-flop strategy makes full use of the visual priors, achieving more robust and visually pleasant reconstruction especially for the challenging human motions.

Adaptive Fusion. To evaluate our adaptive fusion scheme based on the pose and semantic cues, we compare to the variation of our pipeline using traditional

Fig. 12. Evaluation of the adaptive fusion. (a, d) Reference color images. (b, e) The results without adaptive fusion. (c, f) The results with adaptive fusion.

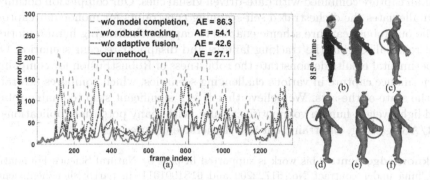

Fig. 13. Quantitative evaluation. (a) Numerical error curves. (b) RGBD input. (c)–(e) The results of three baselines without model completion, robust tracking and adaptive fusion, respectively. (f) The reconstruction results of our full pipeline.

volumetric fusion [35,61,67]. As shown in Fig. 12, the results without adaptive fusion suffer from severe accumulated error, especially for those regions with high-speed motions. In contrast, our adaptive fusion successfully models semantic tracking behavior and avoids deteriorated fusion.

For further analysis of the individual components of RobustFusion, we utilize the sequence with ground truth from [61]. We compute the per-frame mean error for the three variation of our approach without model completion, prior-based robust tracking and adaptive fusion, respectively. Figure 13 shows our full pipeline consistently outperforms the three baselines, yielding the lowest AE. This not only highlights the contribution of each algorithmic component but also illustrates that our approach can robustly capture human motion details.

7   Discussion

Limitation. First, we cannot handle surface splitting topology changes like clothes removal, which we plan to address by incorporating the key-volume update technique [10]. Our method is also restricted to human reconstruction, without modeling human-object interactions. This could be alleviated in the future by combining the static object reconstruction methods into current framework. As is common for learning methods, the utilized visual cue regressions fail for extreme poses not seen in training, such as severe and extensive (self-)occlusion. Fortunately, our approach is able to instantly recover robustly with our reinitialization ability as soon as the occluded parts become visible again. Our current pipeline turns to utilize the data-driven cues in an optimization framework. It's an promising direction to jointly model both visual cues and volumetric capture in an end-to-end learning-based framework.

Conclusions. We have presented a superior approach for robust volumetric human capture combined with data-driven visual cues. Our completion optimization alleviates the orchestrated self-scanning constraints for monocular capture, while our robust capture scheme enables to capture challenging human motions and reinitialize from the tracking failures and disappear-reoccur scenarios. Our experimental results demonstrate the robustness of RobustFusion for compelling performance capture in various challenging scenarios, which compares favorably to the state-of-the-arts. We believe that it is a significant step to enable robust and light-weight human volumetric capture, with many potential applications in VR/AR, gaming, entertainment and immersive telepresence.

Acknowledgement. This work is supported in part by Natural Science Foundation of China under contract No. 61722209 and 6181001011, in partbyShenzhenScience-andTechnologyResearchandDevelopmentFunds (JCYJ201805071 83706645).

References

1. de Aguiar, E., et al.: Performance capture from sparse multi-view video **27**(3), 98:1–10 (2008)
2. Alldieck, T., Pons-Moll, G., Theobalt, C., Magnor, M.: Tex2Shape: detailed full human body geometry from a single image. In: The IEEE International Conference on Computer Vision (ICCV), October 2019
3. Anguelov, D., Srinivasan, P., Koller, D., Thrun, S., Rodgers, J., Davis, J.: Scape: shape completion and animation of people. In: ACM SIGGRAPH 2005 Papers, SIGGRAPH 2005, pp. 408–416. Association for Computing Machinery, New York (2005). https://doi.org/10.1145/1186822.1073207
4. Bogo, F., Kanazawa, A., Lassner, C., Gehler, P., Romero, J., Black, M.J.: Keep It SMPL: automatic estimation of 3D human pose and shape from a single image. In: Leibe, B., Matas, J., Sebe, N., Welling, M. (eds.) ECCV 2016. LNCS, vol. 9909, pp. 561–578. Springer, Cham (2016). https://doi.org/10.1007/978-3-319-46454-1_34
5. Bregler, C., Malik, J.: Tracking people with twists and exponential maps. In: Computer Vision and Pattern Recognition (CVPR) (1998). https://doi.org/10.1109/CVPR.1998.698581
6. Cao, Z., Simon, T., Wei, S.E., Sheikh, Y.: Realtime multi-person 2d pose estimation using part affinity fields. In: Computer Vision and Pattern Recognition (CVPR) (2017)
7. Collet, A., et al.: High-quality streamable free-viewpoint video. ACM Trans. Graph. (TOG) **34**(4), 69 (2015)
8. Curless, B., Levoy, M.: A volumetric method for building complex models from range images. In: Proceedings of the 23rd Annual Conference on Computer Graphics and Interactive Techniques, SIGGRAPH 1996, pp. 303–312. ACM, New York (1996). https://doi.org/10.1145/237170.237269

9. Dou, M., et al.: Motion2Fusion: real-time volumetric performance capture. ACM Trans. Graph. **36**(6), 246:1–246:16 (2017)
10. Dou, M., et al.: Fusion4D: real-time performance capture of challenging scenes. In: ACM SIGGRAPH Conference on Computer Graphics and Interactive Techniques (2016)
11. Gall, J., Rosenhahn, B., Brox, T., Seidel, H.P.: Optimization and filtering for human motion capture. Int. J. Comput. Vis. (IJCV) **87**(1–2), 75–92 (2010)
12. Ganapathi, V., Plagemann, C., Koller, D., Thrun, S.: Real time motion capture using a single time-of-flight camera (2010)
13. Gong, K., Liang, X., Li, Y., Chen, Y., Yang, M., Lin, L.: Instance-level human parsing via part grouping network. In: Ferrari, V., Hebert, M., Sminchisescu, C., Weiss, Y. (eds.) ECCV 2018. LNCS, vol. 11208, pp. 805–822. Springer, Cham (2018). https://doi.org/10.1007/978-3-030-01225-0_47
14. Gong, K., Liang, X., Zhang, D., Shen, X., Lin, L.: Look into person: self-supervised structure-sensitive learning and a new benchmark for human parsing. In: The IEEE Conference on Computer Vision and Pattern Recognition (CVPR), July 2017
15. Guo, K., et al.: The relightables: volumetric performance capture of humans with realistic relighting. ACM Trans. Graph. **38**(6) (2019)
16. Guo, K., et al.: TwinFusion: high framerate non-rigid fusion through fast correspondence tracking. In: International Conference on 3D Vision (3DV), pp. 596–605 (2018)
17. Guo, K., Xu, F., Wang, Y., Liu, Y., Dai, Q.: Robust non-rigid motion tracking and surface reconstruction using L0 regularization. In: Proceedings of the IEEE International Conference on Computer Vision, pp. 3083–3091 (2015)
18. Guo, K., Xu, F., Yu, T., Liu, X., Dai, Q., Liu, Y.: Real-time geometry, albedo and motion reconstruction using a single RGBD camera. ACM Trans. Graph. (TOG) (2017)
19. Habermann, M., Xu, W., Zollhöfer, M., Pons-Moll, G., Theobalt, C.: LiveCap: real-time human performance capture from monocular video. ACM Trans. Graph. (TOG) **38**(2), 14:1–14:17 (2019)
20. Hasler, N., Rosenhahn, B., Thormahlen, T., Wand, M., Gall, J., Seidel, H.P.: Markerless motion capture with unsynchronized moving cameras. In: Computer Vision and Pattern Recognition (CVPR), pp. 224–231 (2009)
21. Huang, Z., et al.: Deep volumetric video from very sparse multi-view performance capture. In: Ferrari, V., Hebert, M., Sminchisescu, C., Weiss, Y. (eds.) ECCV 2018. LNCS, vol. 11220, pp. 351–369. Springer, Cham (2018). https://doi.org/10.1007/978-3-030-01270-0_21
22. Innmann, M., Zollhöfer, M., Nießner, M., Theobalt, C., Stamminger, M.: VolumeDeform: real-time volumetric non-rigid reconstruction, October 2016
23. Joo, H., et al.: Panoptic studio: a massively multiview system for social motion capture. In: Proceedings of the IEEE International Conference on Computer Vision, pp. 3334–3342 (2015)
24. Joo, H., Simon, T., Sheikh, Y.: Total capture: a 3D deformation model for tracking faces, hands, and bodies. In: The IEEE Conference on Computer Vision and Pattern Recognition (CVPR), June 2018
25. Kanazawa, A., Black, M.J., Jacobs, D.W., Malik, J.: End-to-end recovery of human shape and pose. In: Computer Vision and Pattern Recognition (CVPR) (2018)
26. Kovalenko, O., Golyanik, V., Malik, J., Elhayek, A., Stricker, D.: Structure from articulated motion: accurate and stable monocular 3D reconstruction without training data. Sensors **19**(20) (2019)

27. Li, H., Adams, B., Guibas, L.J., Pauly, M.: Robust single-view geometry and motion reconstruction **28**(5), 175 (2009)
28. Li, H., et al.: Temporally coherent completion of dynamic shapes. ACM Trans. Graph. **31** (2012)
29. Liu, Y., Gall, J., Stoll, C., Dai, Q., Seidel, H.P., Theobalt, C.: Markerless motion capture of multiple characters using multiview image segmentation. IEEE Trans. Pattern Anal. Mach. Intell. **35**(11), 2720–2735 (2013)
30. Loper, M., Mahmood, N., Romero, J., Pons-Moll, G., Black, M.J.: SMPL: a skinned multi-person linear model. ACM Trans. Graph. **34**(6), 248:1–248:16 (2015)
31. Martin-Brualla, R., et al.: Lookingood: enhancing performance capture with real-time neural re-rendering. ACM Trans. Graph. **37**(6) (2018)
32. Mehta, D., et al.: VNect: real-time 3D human pose estimation with a single RGB camera. ACM Trans. Graph. (TOG) **36**(4) (2017)
33. Mescheder, L., Oechsle, M., Niemeyer, M., Nowozin, S., Geiger, A.: Occupancy networks: learning 3D reconstruction in function space. In: The IEEE Conference on Computer Vision and Pattern Recognition (CVPR), June 2019
34. Mitra, N.J., Floery, S., Ovsjanikov, M., Gelfand, N., Guibas, L., Pottmann, H.: Dynamic geometry registration. In: Symposium on Geometry Processing (2007)
35. Newcombe, R.A., Fox, D., Seitz, S.M.: DynamicFusion: reconstruction and tracking of non-rigid scenes in real-time, June 2015
36. Pandey, R., et al.: Volumetric capture of humans with a single RGBD camera via semi-parametric learning. In: The IEEE Conference on Computer Vision and Pattern Recognition (CVPR), June 2019
37. Pavlakos, G., et al.: Expressive body capture: 3D hands, face, and body from a single image. In: Proceedings IEEE Conference on Computer Vision and Pattern Recognition (CVPR), pp. 10975–10985, June 2019. http://smpl-x.is.tue.mpg.de
38. Pavlakos, G., Zhou, X., Derpanis, K.G., Daniilidis, K.: Harvesting multiple views for marker-less 3D human pose annotations. In: Computer Vision and Pattern Recognition (CVPR) (2017)
39. Pumarola, A., Sanchez-Riera, J., Choi, G.P.T., Sanfeliu, A., Moreno-Noguer, F.: 3Dpeople: modeling the geometry of dressed humans. In: The IEEE International Conference on Computer Vision (ICCV), October 2019
40. Robertini, N., Casas, D., Rhodin, H., Seidel, H.P., Theobalt, C.: Model-based outdoor performance capture. In: International Conference on 3D Vision (3DV) (2016). http://gvv.mpi-inf.mpg.de/projects/OutdoorPerfcap/
41. Rogez, G., Schmid, C.: Mocap guided data augmentation for 3D pose estimation in the wild. In: Neural Information Processing Systems (NIPS) (2016)
42. Saito, S., Huang, Z., Natsume, R., Morishima, S., Kanazawa, A., Li, H.: PIFu: pixel-aligned implicit function for high-resolution clothed human digitization. In: The IEEE International Conference on Computer Vision (ICCV), October 2019
43. Shotton, J., et al.: Real-time human pose recognition in parts from single depth images (2011)
44. Simon, T., Joo, H., Matthews, I., Sheikh, Y.: Hand keypoint detection in single images using multiview bootstrapping. In: Computer Vision and Pattern Recognition (CVPR) (2017)
45. Slavcheva, M., Baust, M., Cremers, D., Ilic, S.: KillingFusion: non-rigid 3D Reconstruction without Correspondences. In: IEEE Conference on Computer Vision and Pattern Recognition (CVPR) (2017)
46. Slavcheva, M., Baust, M., Ilic, S.: SobolevFusion: 3D reconstruction of scenes undergoing free non-rigid motion. In: IEEE/CVF Conference on Computer Vision and Pattern Recognition (CVPR) (2018)

47. Slavcheva, M., Baust, M., Ilic, S.: Variational level set evolution for non-rigid 3D reconstruction from a single depth camera. IEEE Trans. Pattern Anal. Mach. Intell. (PAMI) (2020)

48. Stoll, C., Hasler, N., Gall, J., Seidel, H.P., Theobalt, C.: Fast articulated motion tracking using a sums of Gaussians body model. In: International Conference on Computer Vision (ICCV) (2011)

49. Sumner, R.W., Schmid, J., Pauly, M.: Embedded deformation for shape manipulation. ACM Trans. Graph. (TOG) **26**(3), 80 (2007)

50. Tang, S., Tan, F., Cheng, K., Li, Z., Zhu, S., Tan, P.: A neural network for detailed human depth estimation from a single image. In: The IEEE International Conference on Computer Vision (ICCV), October 2019

51. Yu, T., Zhao, J., Huang, Y., Li, Y., Liu, Y.: Towards robust and accurate single-view fast human motion capture. IEEE Access (2019)

52. Taylor, J., Shotton, J., Sharp, T., Fitzgibbon, A.: The vitruvian manifold: inferring dense correspondences for one-shot human pose estimation. In: 2012 IEEE Conference on Computer Vision and Pattern Recognition, pp. 103–110 (2012)

53. Tevs, A., et al.: Animation cartography-intrinsic reconstruction of shape and motion. ACM Trans. Graph. (TOG) (2012)

54. Theobalt, C., de Aguiar, E., Stoll, C., Seidel, H.P., Thrun, S.: Performance capture from multi-view video. In: Ronfard, R., Taubin, G. (eds.) Image and Geometry Processing for 3-D Cinematography, pp. 127–149. Springer, Heidelberg (2010). https://doi.org/10.1007/978-3-642-12392-4_6

55. Vicon Motion Systems (2019). https://www.vicon.com/

56. Vlasic, D., et al.: Practical motion capture in everyday surroundings **26**, 3 (2007)

57. Wu, C., Stoll, C., Valgaerts, L., Theobalt, C.: On-set performance capture of multiple actors with a stereo camera **32**, 6 (2013)

58. Xiang, D., Joo, H., Sheikh, Y.: Monocular total capture: posing face, body, and hands in the wild. In: The IEEE Conference on Computer Vision and Pattern Recognition (CVPR), June 2019

59. Xsens Technologies B.V. (2019) https://www.xsens.com/

60. Xu, L., Cheng, W., Guo, K., Han, L., Liu, Y., Fang, L.: FlyFusion: realtime dynamic scene reconstruction using a flying depth camera. IEEE Trans. Vis. Comput. Graph., 1 (2019)

61. Xu, L., Su, Z., Han, L., Yu, T., Liu, Y., FANG, L.: UnstructuredFusion: realtime 4D geometry and texture reconstruction using commercial RGBD cameras. IEEE Trans. Pattern Anal. Mach. Intell., 1 (2019)

62. Xu, L., et al.: FlyCap: markerless motion capture using multiple autonomous flying cameras. IEEE Trans. Visual Comput. Graphics **24**(8), 2284–2297 (2018)

63. Xu, L., Xu, W., Golyanik, V., Habermann, M., Fang, L., Theobalt, C.: EventCap: monocular 3D capture of high-speed human motions using an event camera. arXiv e-prints (2019)

64. Xu, W., et al.: MonoPerfCap: human performance capture from monocular video. ACM Trans. Graph. (TOG) **37**(2), 27:1–27:15 (2018)

65. Yu, T., et al.: BodyFusion: real-time capture of human motion and surface geometry using a single depth camera. In: The IEEE International Conference on Computer Vision (ICCV). ACM, October 2017

66. Yu, T., et al.: DoubleFusion: real-time capture of human performances with inner body shapes from a single depth sensor. Trans. Pattern Anal. Mach. Intell. (TPAMI) (2019)

67. Zheng, Z., et al.: HybridFusion: real-time performance capture using a single depth sensor and sparse IMUs. In: Ferrari, V., Hebert, M., Sminchisescu, C., Weiss, Y. (eds.) ECCV 2018. LNCS, vol. 11213, pp. 389–406. Springer, Cham (2018). https://doi.org/10.1007/978-3-030-01240-3_24

68. Zheng, Z., Yu, T., Wei, Y., Dai, Q., Liu, Y.: DeepHuman: 3D human reconstruction from a single image. In: The IEEE International Conference on Computer Vision (ICCV), October 2019

69. Zhu, H., Zuo, X., Wang, S., Cao, X., Yang, R.: Detailed human shape estimation from a single image by hierarchical mesh deformation. In: The IEEE Conference on Computer Vision and Pattern Recognition (CVPR), June 2019

70. Zhu, T., Oved, D.: Bodypix github repository (2019). https://github.com/tensorflow/tfjs-models/tree/master/body-pix

71. Zollhöfer, M., et al.: Real-time non-rigid reconstruction using an RGB-D camera. ACM Trans. Graph. (TOG) 33(4), 156 (2014)

Surface Normal Estimation of *Tilted* Images via Spatial Rectifier

Tien Do$^{(\boxtimes)}$, Khiem Vuong, Stergios I. Roumeliotis, and Hyun Soo Park

University of Minnesota, Minneapolis, USA
{doxxx104,vuong067,stergios,hspark}@umn.edu

Abstract. In this paper, we present a spatial rectifier to estimate surface normals of tilted images. Tilted images are of particular interest as more visual data are captured by arbitrarily oriented sensors such as body-/robot-mounted cameras. Existing approaches exhibit bounded performance on predicting surface normals because they were trained using gravity-aligned images. Our two main hypotheses are: (1) visual scene layout is indicative of the gravity direction; and (2) not all surfaces are equally represented by a learned estimator due to the structured distribution of the training data, thus, there exists a transformation for each tilted image that is more responsive to the learned estimator than others. We design a spatial rectifier that is learned to transform the surface normal distribution of a tilted image to the rectified one that matches the gravity-aligned training data distribution. Along with the spatial rectifier, we propose a novel truncated angular loss that offers a stronger gradient at smaller angular errors and robustness to outliers. The resulting estimator outperforms the state-of-the-art methods including data augmentation baselines not only on ScanNet and NYUv2 but also on a new dataset called Tilt-RGBD that includes considerable roll and pitch camera motion.

Keywords: Surface normal estimation · Spatial rectifier · Tilted images

1 Introduction

We live within structured environments where various scene assumptions, e.g., Manhattan world assumption [4], can be reasonably applied to model their 3D scene geometry. Nonetheless, the motion of daily-use cameras that observe these scenes is rather unrestricted, generating diverse visual data. For instance, *embodied sensor measurements*, e.g., body-mounted cameras and agile robot-mounted cameras that are poised to enter our daily spaces, often capture images

---

Electronic supplementary material The online version of this chapter (https://doi.org/10.1007/978-3-030-58548-8_16) contains supplementary material, which is available to authorized users.

© Springer Nature Switzerland AG 2020
A. Vedaldi et al. (Eds.): ECCV 2020, LNCS 12349, pp. 265–280, 2020.
https://doi.org/10.1007/978-3-030-58548-8_16

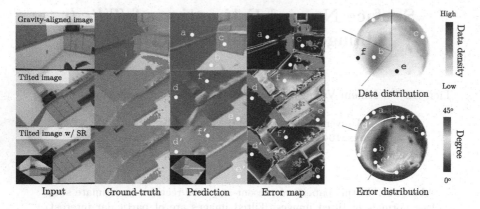

Fig. 1. We present a spatial rectifier that is designed to learn to warp a tilted image for surface normal estimation. While a state-of-the-art surface normal estimator [14] produces accurate prediction on an upright image (top row), it performs poorly on a tilted image (middle row) because of the bias in the surface normal distribution of the training data. Data distribution illustrates the density of surface normals in the ScanNet training dataset, which is highly correlated with the error distribution. We leverage the spatial rectifier to transform the surface normal in erroneous regions to nominal orientation regions, i.e., (d, e, f) → (d', e', f') in error distribution, allowing accurate surface normal prediction (bottom row).

from tilted camera placement (i.e., non-zero roll and pitch). Understanding scene geometry, e.g., surface normal, from such non-upright imagery can benefit numerous real-world applications such as augmented reality [1,14], 3D reconstruction [23,27,28,30,34,39], and 3D robot navigation.

Existing learning-based frameworks, however, exhibit limited capability to predict surface normals of images taken from cameras with arbitrary pointing directions because not all scene surfaces are equally represented in deep neural networks. Figure 1 shows the error distribution of surface normals predicted by a state-of-the-art network FrameNet [14] after being trained on the ScanNet dataset [5]. Albeit accurate on an upright image (regions a, b, and c in the top row), the prediction on a tilted image (regions d, e, and f in the middle row) is highly erroneous. This performance degradation is mainly caused by a domain gap between the training and testing data, i.e., the training data is predominantly composed of gravity-aligned images while the testing tilted images have large roll and pitch angles. This poses a major limitation for surface normal estimation on embodied sensor measurements using the state-of-the-art estimator.

We address this challenge by leveraging a *spatial rectifier*. Our main hypothesis is that it is possible to learn a global transformation that can transform a tilted image such that its surface normal distribution can be matched to that of gravity-aligned images. For instance, a homography warping induced by a pure 3D rotation can warp image features such that the surface normals in the underrepresented regions are transformed to the densely distributed regions (d, e, f) → (d', e', f') as shown in Fig. 1. We design the spatial rectifier parameterized by the gravity direction and a principle direction of the tilted image that maximizes a distribution match, i.e., minimizes the domain gap between the tilted image and the upright training data. The spatial rectifier is learned by synthesizing tilted images using a family of homographies induced by the gravity and principle directions in an end-to-end fashion. We demonstrate that the spatial rectifier is highly effective for surface normal estimation in embodied sensor data, outperforming the state-of-the-art methods including data augmentation baselines not only on ScanNet [5] and NYUv2 [25] but also a new dataset called Tilt-RGBD with a large roll and pitch camera motion. The bottom row of Fig. 1 illustrates surface normal estimation with the spatial rectifier applied to a tilted image where the estimation error is substantially lower than the one without the spatial rectifier (the middle row).

Further, we integrate two new design factors to facilitate embodied sensor measurements in practice. (1) Robustness: there exists a mismatch between the loss used in existing approaches and their evaluation metric. On the one hand, due to the noisy nature of ground truth surface normals [13], the evaluation metric relied on measures in the angle domain. On the other hand, L_2 loss or cosine similarity between 3D unit vectors is used to train the estimators for convergence and differentiability. This mismatch introduces sub-optimal performance of their normal estimators. We address this mismatch by introducing a new loss function called *truncated angular loss* that is robust to outliers and accurate in the region of small angular error below 7.5°. (2) Resolution vs. context: Learning surface normal requires accessing both global scene layout and fine-grained local features. Hierarchical representations, such as the feature pyramid network [11,17], have been used to balance the representational expressibility and computational efficiency while their receptive fields are bounded by the coarsest feature resolution. In contrast, we propose a new design that further increases the receptive fields by incorporating dilated convolutions in the decoding layers. The resulting network architecture significantly reduces the computation (70% FLOPS reduction as compared to the state-of-the-art DORN [11]) while maintaining its accuracy. This allows us to use a larger network (e.g., ResNeXt-101; 57% FLOPS reduction) that achieves higher accuracy and faster inference (45 fps).

To the best of our knowledge, this is the first paper that addresses the challenges in surface normal estimation of tilted images without external sensors. Our contributions include: (1) A novel spatial rectifier that achieves better performance by learning to warp a tilted image to match the surface normal distribution of the training data; (2) A robust angular loss for handling outliers in the ground truth data; (3) An efficient deep neural network design that accesses both global and local visual features, outperforming the state-of-the-art methods in computational efficiency and accuracy; and (4) A new dataset called Tilt-RGBD that includes a large variation of roll and pitch camera angles measured by body-mounted cameras.

Table 1. Surface normal estimation methods and performance comparison. The median measures the median of angular error and 11.25° denotes the percentage of pixels whose angular errors are less than 11.25° ($\downarrow$ / $\uparrow$: the lower/higher the better). Note that Zeng et al. [37] uses RGBD as an input. AL stands for angular loss ($\cos^{-1}(\cdot)$).

Related Work	Training data	Testing data	Loss	Backbone	Median (°) $\downarrow$	11.25° $\uparrow$
Li et al. [21]	NYUv2	NYUv2	L_2	AlexNet	27.8	19.6
Chen et al. [3]			L_2	AlexNet	15.78	39.17
Eigen and Fergus [7]			L_2	AlexNet/VGG	13.2	44.4
SURGE [31]			L_2	AlexNet/VGG×4	12.17	47.29
Bansal et al. [1]			L_2	VGG-16	12.0	47.9
GeoNet [27]			L_2	Deeplabv3	11.8	48.4
FrameNet [14]	ScanNet		L_2	DORN	11.0	50.7
VPLNet [32]			AL	UPerNet	9.83	54.3
FrameNet [14]	ScanNet	ScanNet	L_2	DORN	7.7	62.5
Zeng et al. [37]			L_1	VGG-16	7.47	65.65
VPLNet [32]			AL	UPerNet	6.68	66.3

2 Related Work

This paper includes three key innovations to enable accurate and efficient surface normal estimation for tilted images: Spatial rectification, robust loss, and network design. We briefly review most related work regarding datasets, accuracy, loss function, and network designs summarized in Table 1.

Gravity Estimation and Spatial Rectification. Estimating gravity is one the fundamental problems in visual scene understanding. Visual cues such as vanishing points in indoor scene images [20,24] can be leveraged to estimate the gravity from images without external sensors such as an IMU. In addition, learning-based methods have employed visual semantics to predict gravity [9,26,35]. This gravity estimate is, in turn, beneficial to recognize visual semantics, e.g., single view depth prediction [8,29]. In particular, a planar constraint enforced by the gravity direction and visual semantics (segmentation) is used to regularize depth prediction on KITTI [8]. Furthermore, a monocular SLAM method is used to correct in-plane rotation of images, allowing accurate depth prediction [29]. Unlike existing methods that rely on external sensors or offline gravity estimation, we present a spatial rectifier that can be trained in an end-to-end fashion in conjunction with the surface normal estimator. Our spatial rectifier, inspired by the spatial transformer networks [15], transforms an image by a 3D rotation parameterized by the gravity and the principle directions.

Robust Loss. There is a mismatch between the training loss and the evaluation metric for surface normal estimation. Specifically, while the cosine similarity or L_2 loss is used for training $[1,3,7,13,14,21,27,31,37\text{--}39]^1$, the evaluation metric

---

[1] Classification-based surface normal learning is not considered as its accuracy is bounded by the class resolution [19,33].

is, in contrast, based on the absolute angular error, e.g., L_1 or L_0. This mismatch causes not only slow convergence but also sub-optimality. Notably, Zeng et al. [37] and Liao et al. [22] proposed to use a L_1 measure on unit vectors and spherical regression loss, respectively, to overcome the limitations of L_2 loss. More recently, UprightNet [35] and VPLNet [32] employed an angular loss (AL), and showed its high effectiveness in both gravity and surface normal predictions, respectively. In this work, we propose a new angular loss called the truncated angular loss that increases robustness to outliers in the training data.

Efficient Network Design. Surface normal estimation requires to access both global and local features. Modern designs of surface normal estimators leverage a large capacity to achieve accurate predictions at high resolution. For instance, FrameNet [14] employs the DORN [11] architecture, a modification of DeepLabv3 [2] that removes multiple spatial reductions (2×2 max pool layers) to achieve high resolution surface normal estimation at the cost of processing a larger number of activations. Recently, new hierarchical designs (e.g., panoptic feature pyramid networks [17]) have shown comparable performance by accessing both global and local features while significantly reducing computational cost in instance segmentation tasks. Our design is inspired by such hierarchical designs through atrous convolution in decoding layers, which outperforms DORN on accuracy with lower memory footprint, achieving real-time performance.

3   Method

We present a new spatial rectifier to estimate the surface normals of tilted images. The spatial rectifier is jointly trained with the surface normal estimator by synthesizing randomly oriented images.

3.1   Spatial Rectifier

Given a tilted image $\mathcal{I}$, a surface normal estimator takes as input a pixel coordinate $\mathbf{x} \in \mathcal{R}(\mathcal{I})$ and outputs surface normal direction $\mathbf{n} \in \$^2$, $f_\phi(\mathbf{x}; \mathcal{I}) = \mathbf{n}$, where $\mathcal{R}(\mathcal{I})$ is the coordinate range of the image $\mathcal{I}$. The surface normal estimator is parameterized by ϕ.

Consider a spatial rectifier:

$$\overline{\mathcal{I}}(\mathbf{x}) = \mathcal{I}(W_{\mathbf{g},\mathbf{e}}(\mathbf{x})),$$

where $W_{\mathbf{g},\mathbf{e}}$ is the spatial rectifier that warps a tilted image $\mathcal{I}$ to the rectified image $\overline{\mathcal{I}}$, parameterized by the unit vectors $\mathbf{g}$ and $\mathbf{e}$ expressing the gravity and the principle directions, respectively, in the tilted image coordinate system (Fig. 2). We define the spatial rectifier using a homography induced by a pure rotation mapping from $\mathbf{g}$ to $\mathbf{e}$, i.e., $W_{\mathbf{g},\mathbf{e}} = \mathbf{K}\mathbf{R}\mathbf{K}^{-1}$

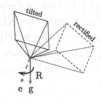

Fig. 2. Geometry.

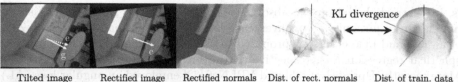

Tilted image   Rectified image   Rectified normals   Dist. of rect. normals   Dist. of train. data

Fig. 3. Spatial rectifier uses the tilted image to synthesize the rectified image by maximizing the surface normal distribution match. The rectified surface normal distribution is matched with that of the gravity aligned images.

where $\mathbf{K}$ is the matrix comprising the intrinsic parameters of the image $\mathcal{I}$. The rotation $\mathbf{R}$ can be written as:

$$\mathbf{R}(\mathbf{g}, \mathbf{e}) = \mathbf{I}_3 + 2\mathbf{e}\mathbf{g}^\mathsf{T} - \frac{1}{1 + \mathbf{e}^\mathsf{T}\mathbf{g}} \left(\mathbf{e} + \mathbf{g}\right)\left(\mathbf{e} + \mathbf{g}\right)^\mathsf{T}, \tag{1}$$

where $\mathbf{I}_3$ is the 3×3 identity matrix.

The key innovation of our spatial rectifier is a learnable $\mathbf{e}$ that transforms the surface normal distribution from an underrepresented region to a densely distributed region in the training data, i.e., (d, e, f) $\rightarrow$ (d', e', f') in Fig. 1. In general, the principle direction determines the minimum amount of rotation of the tilted image towards the rectified one whose surface normals follow the distribution of the training data. A special case is when $\mathbf{e}$ is aligned with one of the image's principle directions, e.g., the y-axis or $\mathbf{e} = \begin{bmatrix} 0 & 1 & 0 \end{bmatrix}^\mathsf{T}$, which produces a gravity-aligned image. We hypothesize that this principle direction and the gravity can be predicted by using the visual semantics of the tilted image:

$$h_\theta(\mathcal{I}) = \begin{bmatrix} \mathbf{g}^\mathsf{T} & \mathbf{e}^\mathsf{T} \end{bmatrix}, \tag{2}$$

where h_θ is a learnable estimator parameterized by $\boldsymbol{\theta}$.

We compute the ground truth of the principle direction $\mathbf{e}$ that maximizes the area of visible pixels in the rectified image while minimizing the KL divergence [18] of two surface normal distributions:

$$\mathbf{e}^* = \underset{\mathbf{e}}{\mathrm{argmin}} \, D_{KL}\left(P(\mathbf{e}) \| Q\right) + \lambda_\mathbf{e} V\left(\overline{\mathcal{I}}\left(W_{\mathbf{g},\mathbf{e}}\right)\right), \tag{3}$$

where Q is the discretized surface normal distribution from all training data and $P(\mathbf{e})$ is the surface normal distribution of the rectified image given $\mathbf{e}$ (Fig. 3). Minimizing the KL divergence allows us to find a transformation from the tilted image to the rectified image ($\mathbf{g} \rightarrow \mathbf{e}$), the surface normal distribution of which optimally matches the training data. V counts the number of invisible pixels in the rectified image:

$$V(\overline{\mathcal{I}}) = \int_{\mathcal{R}(\overline{\mathcal{I}})} \delta\left(\overline{\mathcal{I}}(\mathbf{x})\right) \mathrm{d}\mathbf{x}, \tag{4}$$

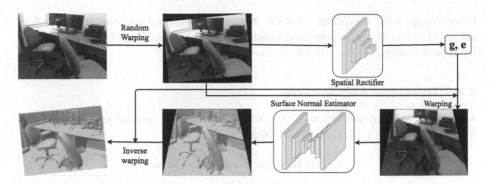

Fig. 4. We jointly train the spatial rectifier and surface normal estimator by randomly synthesizing tilted images. A training image is warped to produce a tilted image. The rectifier is learned to predict **g** and **e** in the tilted image coordinate system. This allows us to warp the tilted image to the rectified image whose surface normal distribution matches that of the training data. The surface normal estimator is used to predict the rectified surface normals and warp back to the tilted surface normals.

where $\delta(\cdot)$ is the Kronecker delta function that outputs one if the pixel at $\mathbf{x}$ is invisible, i.e., $\overline{\mathcal{I}}(\mathbf{x}) = 0$, and zero otherwise. This ensures a minimal transformation of the tilted image, facilitating learning surface normals from the rectified image. The scalar factor $\lambda_{\mathbf{e}}$ balances the relative importance of distribution matching and visibility. Figure 3 illustrates the optimal **e** that maximizes the surface normal distribution match between the rectified image and the gravity-aligned training images (e.g., ScanNet [5]).

3.2 Surface Normal Estimation by Synthesis

We learn the parameters of the neural networks (ϕ and θ) by minimizing the following loss:

$$\mathcal{L}(\phi, \theta) = \sum_{\mathcal{D}} \sum_{\mathbf{x} \in \mathcal{R}(\mathcal{I})} \mathcal{L}_{\mathrm{TAL}}(\mathbf{n}_{\mathbf{x};\mathcal{I}}, \widehat{\mathbf{n}}_{\mathbf{x};\mathcal{I}}) + \lambda \mathcal{L}_{\mathrm{TAL}}(h_\theta(\mathcal{I}), [\widehat{\mathbf{g}}^\mathsf{T} \, \widehat{\mathbf{e}}^\mathsf{T}]),$$

where $\mathcal{D}$ is the training dataset. For each training image, we synthesize a tilted image $\mathcal{I}$ and its corresponding ground truth surface normals $\widehat{\mathbf{n}}_{\mathbf{x};\mathcal{I}}$, gravity $\widehat{\mathbf{g}}$, and principle axis $\widehat{\mathbf{e}}$. λ controls the importance between the surface normal and the spatial rectification. $\mathcal{L}_{\mathrm{TAL}}$ is the truncated angular loss (Sect. 3.3). The predicted surface normals of the tilted image $\mathbf{n}_{\mathbf{x};\mathcal{I}}$ is obtained by transforming the rectified one $f_\phi(\mathbf{x}; \overline{\mathcal{I}})$ as follows:

$$\mathbf{n}_{\mathbf{x};\mathcal{I}} = \mathbf{R}_{\mathbf{g},\mathbf{e}}^\mathsf{T} f_\phi(\mathbf{W}_{\mathbf{g},\mathbf{e}}^{-1}(\mathbf{x}); \overline{\mathcal{I}}),$$

where $\mathbf{R}_{\mathbf{g},\mathbf{e}}$ and $\mathbf{W}_{\mathbf{g},\mathbf{e}}$ are defined in Sect. 3.1. Note that since f_ϕ takes rectified images as inputs, it does not require a large capacity network to memorize all possible tilted orientations. Figure 4 illustrates our approach that synthesizes a tilted image with a random rotation where its predicted parameters ($\mathbf{g}$ and $\mathbf{e}$) are used to rectify the synthesized image. We measure the loss of surface normals by transforming back to the tilted image coordinates.

3.3 Truncated Angular Loss

L_2 loss on 3D unit vectors was used to train a surface normal estimator in the literature [1,3,7,13,14,21,27,31,37–39]:

$$\mathcal{L}_{L_2}(\mathbf{n}, \widehat{\mathbf{n}}) = \|\mathbf{n} - \widehat{\mathbf{n}}\|_2^2 = 1 - \mathbf{n}^\mathsf{T}\widehat{\mathbf{n}}$$

where $\mathbf{n}$ and $\widehat{\mathbf{n}}$ are the estimated surface normals and its ground truth, respectively. This measures the Euclidean distance between two unit vectors, which is equivalent to the cosine similarity. A key limitation of this loss is the vanishing gradient around small angular error, i.e., $\partial\mathcal{L}_{L_2}/\partial\xi \approx 0$ if $\xi \approx 0$, where $\xi = \cos^{-1}(\mathbf{n}^\mathsf{T}\widehat{\mathbf{n}})$. As a result, the estimator is less sensitive to small errors.

On the other hand, the angular error or the cardinality of the pixel set with an angular threshold has been used for their evaluation metric to address considerable noise in the ground truth surface normals [13], i.e.,

$$E_{\text{angular}} = \cos^{-1}(\mathbf{n}^\mathsf{T}\widehat{\mathbf{n}}), \quad \text{or}$$
$$E_{\text{card}}(\xi) = |\{\cos^{-1}(\mathbf{n}^\mathsf{T}\widehat{\mathbf{n}}) < \xi\}|, \quad \xi = 11.25°, 22.5°, 30°$$

This mismatch between the evaluation metric and the training loss introduces sub-optimality in the trained estimator. To address this issue, we propose the truncated angular loss (TAL) derived from the angular loss:

$$\mathcal{L}_{\text{TAL}}(\mathbf{n}, \widehat{\mathbf{n}}) = \begin{cases} 0, & 1 - \epsilon \leq \mathbf{n}^\mathsf{T}\widehat{\mathbf{n}} \\ \cos^{-1}(\mathbf{n}^\mathsf{T}\widehat{\mathbf{n}}), & 0 \leq \mathbf{n}^\mathsf{T}\widehat{\mathbf{n}} < 1 - \epsilon \\ \frac{\pi}{2} - \mathbf{n}^\mathsf{T}\widehat{\mathbf{n}}, & \mathbf{n}^\mathsf{T}\widehat{\mathbf{n}} < 0 \end{cases}$$

where $\epsilon = 10^{-6}$. When $\mathbf{n}^\mathsf{T}\widehat{\mathbf{n}} \approx 1$, i.e., as the angular error is close to zero, the loss is clamped zero to avoid the infinite gradient of inverse cosine. More importantly, when $\mathbf{n}^\mathsf{T}\widehat{\mathbf{n}} < 0$, i.e., where the angular error is large (outlier), TAL is linear, i.e., it assigns the constant weight for each $\mathbf{n}^\mathsf{T}\widehat{\mathbf{n}}$ in this region, similar to L_2, whereas AL assigns larger weight for larger $\mathbf{n}^\mathsf{T}\widehat{\mathbf{n}}$, as shown in Fig. 5. This makes the training with TAL less sensitive to outliers than AL.

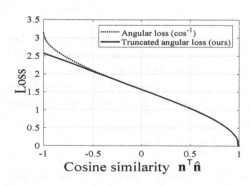

Fig. 5. Truncated angular loss.

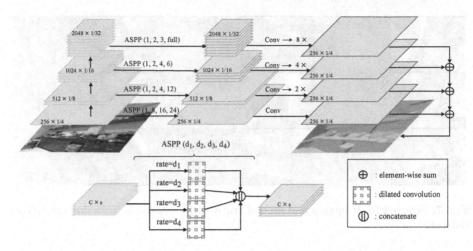

Fig. 6. Illustration of the surface normal estimation network architecture.

3.4 Surface Normal Estimator Design

For the surface normal estimator module, we design a new network architecture inspired by the Panoptic FPN [17] that has shown strong performance on high-resolution prediction tasks while maintaining its compact representation. We employ an asymmetric encoder-decoder structure with the following modifications: (i) We do not suppress the number of channels of each pyramid feature immediately to $C = 128$; and (ii) We make use of the atrous spatial pyramid pooling (ASPP) module similar to DeepLabv3 [2].

The network encodes an image with multi-scale feature maps ranging from 1/4 to 1/32 for spatial resolution and 256 to 2048 for channel dimension. For the decoder, we assemble these multi-scale features with ASPP to increase the receptive fields as shown in Fig. 6. The decoded feature maps are combined and upsampled to produce a quarter resolution feature maps and summed to predict surface normals.

4 Results

We evaluate the performance of our surface normal estimation quantitatively and qualitatively. All networks used in the evaluation, including ours, take as input 320×240 resolution images and output the same size surface normals. h_θ (Sect. 3.1) makes use of ResNet-18 backbone. The networks are trained with a batch size of 16 and optimized by the Adam [16] optimizer with a learning rate 10^{-4}. We use an NVIDIA Tesla V100 GPU (with 32 GB of memory) to train for 20 epochs and report the best epoch on the corresponding dataset's validation set. Our code, dataset and pretrained networks are available at https://github.com/MARSLab-UMN/TiltedImageSurfaceNormal.

Fig. 7. Selected RGB images and ground truth surface normals from the Tilt-RGBD dataset.

Evaluation Metrics. For the quantitative evaluation of surface normal prediction, we employ standard metrics used in [1,10]: (a) Mean absolute of the error (mean), (b) Median of absolute error (median), (c) Root mean square error (RMSE), and (d) Percentage of pixels with angular error below a threshold ξ with $\xi = 5°$, $7.5°$, $11.25°$, $22.5°$, $30.0°$.

4.1   Evaluation Dataset

We evaluate our method using ScanNet [5], NYUv2 [25], and a new dataset from body-mounted cameras called *Tilt-RGBD* that includes tilted images. All networks are trained and validated by ScanNet and tested on NYUv2 and Tilt-RGBD. No domain adaptation is applied.

ScanNet. ScanNet [5] is an indoor RGB-D video dataset that spans a large variation of scenes. We use the images from their standard training/validation/ testing splits by extracting every 10th frame: 189,916 images for training, 53,193 images for validation, and 20,942 images for testing. For the ground truth surface normals, we make use of the ground truth generated by FrameNet [14]. For comparison with FrameNet, we make use of the training (199,720)/testing (64,319) split that was originally used in the paper.

NYUv2. NYUv2 [25] was captured by MS Kinect, which includes 654 testing images. The ground truth surface normals are generated by Ladicky et al. [19]. Note that we evaluate only on the valid pixels following Zhang et al. [40].

Tilt-RGBD. We collect a new dataset with body-mounted cameras (Azure Kinect) that includes 24 different scenes. Each image instance is associated with the gravity measured by an IMU following [6]. Two types of scenes are collected. (i) *Gravity-aligned images*: We control the camera orientation to be aligned with gravity. 14 scenes and 2,403 images are used. (ii) *Tilted images*: We capture images without controlling the camera's pose; this process generates a large variation of tilted images that includes 14 scenes and 2,428 images. We follow

Table 2. Comparison between all baselines in term of generalization capacity on Tilt-RGBD on gravity-aligned and tilted images.

Gravity-aligned images	Mean	Median	RMSE	5°	7.5°	11.25°	22.5°	30.0°
FrameNet	11.57	6.59	18.19	37.68	55.69	71.23	86.85	90.93
DORN+TAL	9.54	4.96	16.60	50.33	67.56	79.74	90.24	92.96
DORN+TAL+SR	9.40	5.49	15.76	45.09	65.75	80.71	91.31	93.75
DFPN+TAL	9.74	5.40	16.30	46.05	65.45	78.90	90.40	93.14
DFPN+TAL+AUG	10.19	5.62	17.57	43.89	64.39	79.07	89.95	92.62
DFPN+TAL+IMU	8.62	**4.26**	15.57	**56.90**	**72.42**	82.03	91.17	93.68
DFPN+TAL+SR	**8.59**	4.42	**15.24**	55.80	72.37	**82.26**	**91.39**	**94.01**
Tilted images	Mean	Median	RMSE	5°	7.5°	11.25°	22.5°	30.0°
FrameNet	17.65	10.51	25.38	24.62	38.35	52.25	73.65	81.17
DORN+TAL	14.33	7.09	22.76	37.15	52.08	65.03	80.92	85.79
DORN+TAL+SR	11.77	6.59	18.95	37.00	56.05	72.35	87.29	90.72
DFPN+TAL	14.53	7.84	22.47	32.01	48.30	62.92	81.41	86.43
DFPN+TAL+AUG	12.89	6.86	21.28	34.81	54.28	71.20	85.75	89.10
DFPN+TAL+IMU	11.46	**5.22**	19.88	**48.30**	**62.92**	74.24	86.29	89.75
DFPN+TAL+SR	**11.22**	5.82	**18.88**	43.03	61.19	**75.02**	**87.53**	**90.80**

the approach of [19] to generate ground truth surface normals from the depth images. Figure 7 shows representative RGB images along with their ground truth surface normals.

4.2 Baseline

FrameNet. FrameNet [14] uses the DORN design that jointly predicts surface normal, tangent, and bitangent directions. We obtain the provided pre-trained network on ScanNet to perform our experiments;

VPLNet. VPLNet [32] incorporates lines and vanishing point directions to improve surface normals prediction, and is currently the state-of-the-art surface normal estimator on ScanNet and NYUv2. Note that we are only able to compare with VPLNet on ScanNet and NYUv2 using the reported results [32] since neither the pretrained network nor the implementation is provided. **DORN+TAL** DORN [11] is a high-capacity network based on the ResNet-101 [12] backbone. We use our truncated angular loss (TAL) to train the network; **DORN+TAL+SR** We train DORN with our proposed spatial rectifier (SR) described in Sect. 3.2; **DFPN+TAL** Dilated feature pyramid network (DFPN) is our proposed network described in Sect. 3.4 with our truncated angular loss (TAL). We use DFPN with ResNet-101 backbone; **DFPN+TAL+AUG** We train our DFPN with data augmentation (AUG) by synthesizing tilted images with random rotations. This network does not take

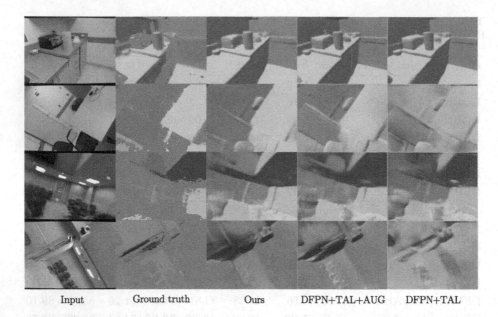

Input   Ground truth   Ours   DFPN+TAL+AUG   DFPN+TAL

Fig. 8. Qualitative results: We compare ours (DFPN+TAL+SR) with DFPN+ TAL+AUG and DFPN+TAL.

rectified images and therefore, is expected to memorize all the tilted representations; **DFPN+TAL+IMU** We train our DFPN with rectified images by assuming the gravity is given (from the ground plane normal direction for Scan-Net); **DFPN+TAL+SR (ours)** We train our DFPN with our spatial rectifier (SR) described in Sect. 3.2.

4.3   Surface Normal Estimation on Tilt-RGBD

We compare our method with the baselines on Tilt-RGBD on both gravity-aligned and tilted images summarized in Table 2. We observe that all networks trained on ScanNet perform excellent on unseen gravity-aligned frames indicating that the ScanNet dataset contains sufficient scene diversity. Nevertheless, the best performance in terms of median and tight thresholds ($\xi = 5°, 7.5°$) belongs to DFPN+TAL+IMU. This is due to the fact that gravity direction estimated from the IMU is highly accurate and is employed directly as the surface normal direction of the dominating part of the scene, e.g., floors, ceilings. Our method achieves on-par performance with DFPN+TAL+IMU since the gravity estimate from the network is less accurate.

For tilted images, there is a significant performance degradation in all metrics, e.g., the percentage drop in 11.25° for FrameNet (by 19%), DFPN+TAL (by 16%), DORN+TAL (by 14%) while DORN+TAL+SR, DFPN+TAL+AUG, DFPN+TAL+IMU, and DFPN+TAL+SR show less degradation due to either data augmentation or external sensor information. DFPN+TAL+SR performs

Table 4. Accuracy comparison between different network architectures.

Network	Backbone	ScanNet							NYUv2					
		Mean	Median	RMSE	11.25°	22.5°	30°		Mean	Median	RMSE	11.25°	22.5°	30°
P-FPN+TAL	ResNet-101	15.8	8.0	24.8	61.8	78.8	84.2		17.0	8.9	26.0	57.2	75.6	82.1
DORN+TAL	ResNet-101	15.1	**7.4**	24.1	**63.8**	79.6	84.9		16.6	8.5	25.7	58.5	76.4	82.6
DFPN+TAL	ResNet-101	15.4	7.5	24.6	63.3	79.4	84.7		16.9	8.6	26.2	58.2	76.1	82.2
DFPN+TAL	ResNeXt-101	**15.0**	7.5	**23.9**	**63.8**	**80.0**	**85.2**		**16.2**	**8.2**	**25.3**	**59.5**	**77.1**	**83.2**

significantly better than augmenting the data DFPN+TAL+AUG and on par with DFPN+TAL+IMU while it does not require an external sensor. In addition, DFPN+TAL+SR also outperforms DORN+TAL+SR although DORN has higher capacity than our DFPN (Sect. 4.4), which suggests that the SR is more compatible with our DFPN. Figure 8 shows our qualitative results where we compare our method with the baselines.

4.4 Network Efficiency

We compare our proposed DFPN+TAL with DORN+TAL in terms of number of floating operations (FLOPS), number of parameters, actual memory consumption, and inference time summarized in Table 3 as well as surface normal estimation accuracy in Table 4. With the ResNet-101 backbone, our DFPN +TAL shows improvement in terms of accuracy over the Panoptic FPN (P-FPN) [17] and performs comparably to DORN+TAL while highly efficient [less than 1/3 of DORN+TAL in terms of FLOPs, memory, and

Network	Backbone	# params	FLOPS	Memory	FPS
P-FPN+TAL	ResNet-101	53.5M	17.3 G	6.9 GB	68.5
DORN+TAL	ResNet-101	99.5M	97.0 G	13.2 GB	18.6
DFPN+TAL	ResNet-101	93.0M	29.2 G	7.8 GB	43.7
DFPN+TAL	ResNeXt-101	137.3M	42.6 G	11.5 GB	30.0

Table 3. Comparison between different network architectures in terms of FLOPS, number of parameters, memory consumption (in training), and inference time (FPS with batch size of 4).

inference time (FPS)]. With efficient DFPN+TAL, we further increase the network capacity to ResNeXt-101 [36], producing more accurate prediction. It outperforms DORN+TAL in terms of accuracy both on ScanNet and NYUv2 with smaller FLOPS (2.4× faster training time), memory consumption, and realtime inference (30 FPS on NVIDIA GTX 1660).

4.5 Surface Normal Training Loss

We compare three losses used for surface normal estimation: L_2, angular loss (AL), and truncated angular loss (TAL). As shown in Fig. 9 and Table 5 (Top), changing from L_2 to AL or TAL leads to faster convergence and significantly lower error metrics. Furthermore, the TAL loss performs on-par with AL at loose thresholds such as 22.5° and 30° and shows improvement in particular at

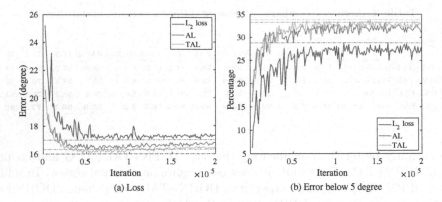

Fig. 9. (a) Loss values and (b) percentage of pixels with angular error less than 5° over training iterations evaluated on NYUv2. Truncated angular loss (TAL) outperforms L_2 and Angular loss (AL).

Table 5. (Top) Comparison between different loss functions. (Bottom) Comparison with the state-of-the-art methods.

Loss/Method	ScanNet								NYUv2							
	Mean	Median	RMSE	5°	7.5°	11.25°	22.5°	30°	Mean	Median	RMSE	5°	7.5°	11.25°	22.5°	30°
L_2	15.7	8.2	24.3	30.4	46.7	61.2	78.7	84.4	17.1	9.4	25.7	28.0	42.1	55.7	75.2	82.0
AL	15.1	7.5	24.0	33.9	50.1	63.8	79.9	85.2	16.7	8.6	25.6	30.9	45.3	58.3	76.6	82.8
TAL	15.0	7.5	23.9	34.2	50.2	63.8	80.0	85.2	16.2	8.2	25.3	33.4	47.0	59.5	77.1	83.2
FrameNet [14]	14.7	7.7	22.8	33.4	48.9	62.5	80.1	85.8	18.6	11.0	26.8	23.2	36.6	50.7	72.0	79.5
VPLNet [32]	13.8	6.8	–	–	–	66.3	81.8	87.0	18.0	9.8	–	–	–	54.3	73.8	80.7
DFPN+TAL	12.6	6.0	21.1	42.8	57.5	69.3	83.9	88.6	16.1	8.1	25.1	33.6	47.3	59.8	77.4	83.4

tight thresholds such as 5° and 7.5°, both on ScanNet and NYUv2. In addition, we also show in the Table 5 (Bottom) that our method outperforms FrameNet and VPLNet in all of the metrics both on ScanNet and NYUv2. Note that in order to ensure the fairness in comparison with FrameNet [14] and VPLNet [32], we train and evaluate our method on their modified ScanNet data split. Our results suggest that TAL outperforms the L_2 and AL losses, which allows us to achieve the state-of-the-art performance across different train/test splits while maintaining its generalization capability.

5   Summary

In this paper, we present a new spatial rectifier to estimate surface normals from a tilted image. The spatial rectifier is learned to warp the tilted image to the rectified image such that its surface normal distribution matches that of the training data. We train the spatial rectifier jointly with the surface normal estimator by synthesizing tilted images from the ScanNet dataset. To facilitate practical deployment of surface normal estimation, we design a new efficient network that produces the state-of-the-art accuracy while maintaining low computational burden. Further, we propose a truncated angular loss that overcomes a key limitation of the L_2 loss, resulting in accurate estimation, especially in the region of small angular error. Our method outperforms the state-of-the-art baselines not only on ScanNet and NYUv2 but also on a new Tilt-RGBD dataset that includes large roll and pitch camera motions.

Acknowledgements. This work is supported by NSF IIS-1328722 and NSF CAREER IIS-1846031.

References

1. Bansal, A., Russell, B., Gupta, A.: Marr revisited: 2D–3D alignment via surface normal prediction. In: CVPR (2016)
2. Chen, L.C., Papandreou, G., Kokkinos, I., Murphy, K., Yuille, A.L.: DeepLab: semantic image segmentation with deep convolutional nets, atrous convolution, and fully connected CRFs. TPAMI (2017)
3. Chen, W., Xiang, D., Deng, J.: Surface normals in the wild. In: CVPR (2017)
4. Coughlan, J.M., Yuille, A.L.: The Manhattan world assumption: Regularities in scene statistics which enable Bayesian inference. In: NeurIPS (2001)
5. Dai, A., Chang, A.X., Savva, M., Halber, M., Funkhouser, T., Nießner, M.: Scan-Net: richly-annotated 3D reconstructions of indoor scenes. In: CVPR (2017)
6. Do, T., Neira, L., Yang, Y., Roumeliotis, S.I.: Attitude tracking from a camera and an accelerometer on gyro-less devices. In: ISRR (2019)
7. Eigen, D., Fergus, R.: Predicting depth, surface normals and semantic labels with a common multi-scale convolutional architecture. In: CVPR (2015)
8. Fei, X., Wong, A., Soatto, S.: Geo-supervised visual depth prediction. RA-L (2019)
9. Fischer, P., Dosovitskiy, A., Brox, T.: Image orientation estimation with convolutional networks. In: GCPR (2015)
10. Fouhey, D.F., Gupta, A., Hebert, M.: Data-driven 3D primitives for single image understanding. In: ICCV (2013)
11. Fu, H., Gong, M., Wang, C., Batmanghelich, K., Tao, D.: Deep ordinal regression network for monocular depth estimation. In: CVPR (2018)
12. He, K., Zhang, X., Ren, S., Sun, J.: Deep residual learning for image recognition. In: CVPR (2016)
13. Hickson, S., Raveendran, K., Fathi, A., Murphy, K., Essa, I.: Floors are flat: leveraging semantics for real-time surface normal prediction. arXiv:1906.06792 (2019)
14. Huang, J., Zhou, Y., Funkhouser, T., Guibas, L.J.: FrameNet: learning local canonical frames of 3D surfaces from a single RGB image. In: ICCV (2019)
15. Jaderberg, M., Simonyan, K., Zisserman, A., Kavukcuoglu, K.: Spatial transformer networks. In: NeurIPS (2015)
16. Kingma, D.P., Ba, J.: Adam: a method for stochastic optimization. In: ICLR, vol. 1412 (2014)
17. Kirillov, A., Girshick, R., He, K., Dollár, P.: Panoptic feature pyramid networks. In: CVPR (2019)
18. Kullback, S., Leibler, R.: On information and sufficiency. Ann. Math. Stat. (1951)
19. Ladický, L., Zeisl, B., Pollefeys, M.: Discriminatively trained dense surface normal estimation. In: Fleet, D., Pajdla, T., Schiele, B., Tuytelaars, T. (eds.) ECCV 2014. LNCS, vol. 8693, pp. 468–484. Springer, Cham (2014). https://doi.org/10.1007/978-3-319-10602-1_31

20. Lee, J.K., Yoon, K.J.: Real-time joint estimation of camera orientation and vanishing points. In: CVPR (2015)
21. Li, B., Shen, C., Dai, Y., Van Den Hengel, A., He, M.: Depth and surface normal estimation from monocular images using regression on deep features and hierarchical CRFs. In: CVPR (2015)
22. Liao, S., Gavves, E., Snoek, C.G.: Spherical regression: learning viewpoints, surface normals and 3D rotations on N-spheres. In: CVPR (2019)
23. Liu, C., Kim, K., Gu, J., Furukawa, Y., Kautz, J.: PlaneRCNN: 3D plane detection and reconstruction from a single image. In: CVPR (2019)
24. Mirzaei, F.M., Roumeliotis, S.I.: Optimal estimation of vanishing points in a Manhattan world. In: CVPR (2011)
25. Silberman, N., Hoiem, D., Kohli, P., Fergus, R.: Indoor segmentation and support inference from RGBD images. In: Fitzgibbon, A., Lazebnik, S., Perona, P., Sato, Y., Schmid, C. (eds.) ECCV 2012. LNCS, vol. 7576, pp. 746–760. Springer, Heidelberg (2012). https://doi.org/10.1007/978-3-642-33715-4_54
26. Olmschenk, G., Tang, H., Zhu, Z.: Pitch and roll camera orientation from a single 2D image using convolutional neural networks. In: CRV (2017)
27. Qi, X., Liao, R., Liu, Z., Urtasun, R., Jia, J.: Geonet: Geometric neural network for joint depth and surface normal estimation. In: CVPR (2018)
28. Qiu, J., et al.: DeepLiDAR: deep surface normal guided depth prediction for outdoor scene from sparse lidar data and single color image. In: CVPR (2019)
29. Saito, Y., Hachiuma, R., Yamaguchi, M., Saito, H.: In-plane rotation-aware monocular depth estimation using SLAM. In: Ohyama, W., Jung, S.K. (eds.) IW-FCV 2020. CCIS, vol. 1212, pp. 305–317. Springer, Singapore (2020). https://doi.org/10.1007/978-981-15-4818-5_23
30. Tang, J., Folkesson, J., Jensfelt, P.: Sparse2Dense: from direct sparse odometry to dense 3-D reconstruction. RA-L (2019)
31. Wang, P., Shen, X., Russell, B., Cohen, S., Price, B., Yuille, A.L.: SURGE: surface regularized geometry estimation from a single image. In: NeurIPS (2016)
32. Wang, R., Geraghty, D., Matzen, K., Szeliski, R., Frahm, J.M.: VPLNET: deep single view normal estimation with vanishing points and lines. In: NeurIPS (2020)
33. Wang, X., Fouhey, D., Gupta, A.: Designing deep networks for surface normal estimation. In: CVPR (2015)
34. Weerasekera, C.S., Latif, Y., Garg, R., Reid, I.: Dense monocular reconstruction using surface normals. In: ICRA (2017)
35. Xian, W., Li, Z., Fisher, M., Eisenmann, J., Shechtman, E., Snavely, N.: UprightNet: geometry-aware camera orientation estimation from single images. In: CVPR (2019)
36. Xie, S., Girshick, R., Dollár, P., Tu, Z., He, K.: Aggregated residual transformations for deep neural networks. In: CVPR (2017)
37. Zeng, J., et al.: Deep surface normal estimation with hierarchical RGB-D fusion. In: CVPR (2019)
38. Zhan, H., Weerasekera, C.S., Garg, R., Reid, I.: Self-supervised learning for single view depth and surface normal estimation. arXiv:1903.00112 (2019)
39. Zhang, Y., Funkhouser, T.: Deep depth completion of a single RGB-D image. In: CVPR (2018)
40. Zhang, Y., et al.: Physically-based rendering for indoor scene understanding using convolutional neural networks. In: CVPR (2017)

Multimodal Shape Completion via Conditional Generative Adversarial Networks

Rundi Wu[1], Xuelin Chen[2], Yixin Zhuang[1], and Baoquan Chen[1](✉)

[1] Center on Frontiers of Computing Studies, Peking University, Beijing, China
baoquan@pku.edu.cn
[2] Shandong University, Qingdao, China

Abstract. Several deep learning methods have been proposed for completing partial data from shape acquisition setups, i.e., filling the regions that were missing in the shape. These methods, however, only complete the partial shape with a single output, ignoring the ambiguity when reasoning the missing geometry. Hence, we pose a *multi-modal* shape completion problem, in which we seek to complete the partial shape with multiple outputs by learning a one-to-many mapping. We develop the first multimodal shape completion method that completes the partial shape via conditional generative modeling, without requiring paired training data. Our approach distills the ambiguity by conditioning the completion on a learned multimodal distribution of possible results. We extensively evaluate the approach on several datasets that contain varying forms of shape incompleteness, and compare among several baseline methods and variants of our methods qualitatively and quantitatively, demonstrating the merit of our method in completing partial shapes with both diversity and quality.

Keywords: Shape completion · Multimodal mapping · Conditional generative adversarial network

1 Introduction

Shape completion, which seeks to reason the geometry of the missing regions in incomplete shapes, is a fundamental problem in the field of computer vision, computer graphics and robotics. A variety of solutions now exist for efficient shape acquisition. The acquired shapes, however, are often incomplete, e.g., incomplete

---

R. Wu and X. Chen—Equal contribution.

Electronic supplementary material The online version of this chapter (https://doi.org/10.1007/978-3-030-58548-8_17) contains supplementary material, which is available to authorized users.

© Springer Nature Switzerland AG 2020
A. Vedaldi et al. (Eds.): ECCV 2020, LNCS 12349, pp. 281–296, 2020.
https://doi.org/10.1007/978-3-030-58548-8_17

work in the user modeling interface, and incomplete scans resulted from occlusion. The power of shape completion enables the use of these incomplete data in downstream applications, e.g., virtual walk-through, intelligent shape modeling, path planning.

With the rapid progress made in deep learning, many data-driven methods have been proposed and demonstrated effective in shape completion [5,7,11,12, 18,24,26,28,30–33,36,37]. However, most approaches in this topic have focused on completing the partial shape with a *single* result, learning a one-to-one mapping for shape completion. In contrast, we model a distribution of potential completion results, as the shape completion problem is *multimodal* in nature, especially when the incompleteness causes significant ambiguity. For example, as shown in Fig. 1, a partial chair can be completed with different types of chairs. Hence, we pose the multimodal shape completion problem, which seeks to associate each incomplete shape with multiple different complete shapes.

Fig. 1. We present a point-based shape completion network that can complete the partial shape with multiple plausible results. Here we show a sampling of our results, where a partial chair is completed with different types of chairs.

In this work, we propose a first point-based multimodal shape completion method, in which the multimodality of potential outputs is distilled in a low-dimensional latent space, enabling random sampling of completion results at inference time. The challenge to achieve this is two-fold. First, modeling the multimodality in the high-dimensional shape space and mapping it into a low-dimensional latent space is challenging. A common problem is *mode collapse*, where only a subset of the modes are represented in the low-dimensional space. Second, the ground-truth supervision data, on which most learning-based methods rely for success, is extremely hard to acquire in our problem. Without the availability of particular supervision data (i.e., for each training incomplete shape, multiple corresponding complete shapes are required), it is challenging to learn the multimodal mapping for shape completion in an unsupervised manner *without* any paired data.

We address the challenge by completing the partial shape in a conditional generative modeling setting. We design a conditional generative adversarial network (cGAN) wherein a generator learns to map incomplete training data, combined with a latent vector sampled from a learned multimodal shape distribution, to a suitable latent representation such that a discriminator cannot differentiate between the mapped latent variables and the latent variables obtained from complete training data (i.e., complete shape models). An encoder is introduced to encode mode latent vectors from complete shapes, learning the multimodal distribution of all possible outputs. We further apply this encoder to the completion output to extract and recover the input latent vector, forcing the bijective mapping between the latent space and the output modes. The mode encoder is trained to encode the multimodality in an explicit manner (Sect. 3.3), alleviating the aforementioned mode collapse issue. By conditioning the generation of completion results on the learned multimodal shape distribution, we achieve multimodal shape completion.

We extensively evaluate our method on several datasets that contain varying forms of shape incompleteness. We compare our method against several baseline methods and variants of our method, and rate the different competing methods using a combination of established metrics. Our experiments demonstrate the superiority of our method compared to other alternatives, producing completion results with both high diversity and quality, all the while remaining faithful to the partial input shape.

2   Related Work

Shape Completion. With the advancement of deep learning in 3D domain, many deep learning methods have been proposed to address the shape completion challenge. Following the success of CNN-based 2D image completion networks, 3D convolutional neural networks applied on voxelized inputs have been widely adopted for 3D shape completion task [7,12,24,26,28,30–33]. To avoid geometric information loss resulted from quantizing shapes into voxel grids, several approaches [5,11,18,36,37] have been develop to work directly on point sets to reason the missing geometry. While all these methods resort to learning a parameterized model (i.e., neural networks) as a mapping from incomplete shapes to completed shapes, the learned mapping function remains injective. Consequently, these methods can only complete the partial shape with a single deterministic result, ignoring the ambiguity of the missing regions.

Generative Modeling. The core of generative modeling is parametric modeling of the data distribution. Several classical approaches exist for tackling the generative modeling problem, e.g., restricted Boltzmann machines [25], and autoencoders [13,29]. Since the introduction of Generative Adversarial Networks (GANs) [8], it has been widely adopted for a variety of generative tasks. In 2D image domain, researchers have utilized GANs in tasks ranging from image generation [2,3,19], image super-resolution [15], to image inpainting in 2D domain. In the context of 3D, a lot of effort has also been put on the task of content generation [1,6,9,21]. The idea of using GANs to generatively model the missing regions for shape completion has also been explored in the pioneering works [5,11]. In this work, we achieve multimodal shape completion by utilizing GANs to reason the missing geometry in a conditional generative modeling setting.

Deep Learning on Points. Our method is built upon recent success in deep neural networks for point cloud representation learning. Although many improvements to PointNet [22] have been proposed [16,17,23,27,38], the simplicity and effectiveness of PointNet and its extension PointNet++ make them popular in many other analysis tasks [10,34–36]. To achieve point cloud generation, [1] proposed to train GANs in the latent space produced by a PointNet-based autoencoder, reporting significant performance gain in point cloud generation. Similar to [5], we also leverage the power of such point cloud generation model to complete shapes, but conditioning the generation on a learned distribution of potential completion outputs for multimodal shape completion.

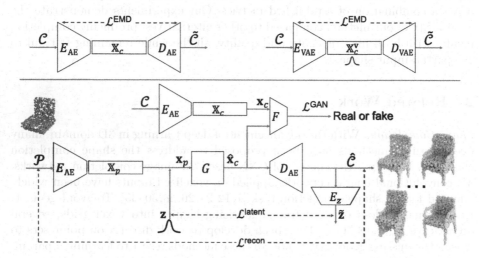

Fig. 2. The proposed network architecture for learning multimodal shape completion. We use the encoder E_{VAE} of a shape variational autoencoder as the mode encoder E_z to encode the multimodality from shapes explicitly.

3   Method

Given a partial shape domain $\mathcal{P} \subset \mathbb{R}^{K \times 3}$, we seek to learn a multimodal mapping from $\mathcal{P}$ to the complete shape domain $\mathcal{C} \subset \mathbb{R}^{N \times 3}$, achieving the goal of multimodal shape completion. Unlike existing methods, which are primarily limited to producing a single deterministic completion output $\hat{\mathbf{C}}$ from a partial shape $\mathbf{P} \in \mathcal{P}$, our method learns a mapping that could sample the completion $\hat{\mathbf{C}}$ from the conditional distribution of possible completions, producing diverse completion results. During training, our method only requires access to a set of partial point sets, and a set of complete point sets. It is important to note that there is no any paired completion instances in $\mathcal{C}$ for point sets in $\mathcal{P}$.

Following the spirit of [5], without paired training data, we address the multimodal shape completion problem via adversarial training on the latent space learned from point sets, while introducing a low-dimensional code $\mathbf{z} \in \mathbb{R}^Z$, of which the latent space is learned by modeling the multimodality presented in possible outputs, as a conditional input in addition to the partial shape. To enable stochastic sampling, we desire $\mathbf{z}$ to be drawn from a prior distribution $p(\mathbf{z})$; a standard Gaussian distribution $\mathcal{N}(0, \mathcal{I})$ is used in this work.

More specifically, we learn two class-specific point set manifolds, $\mathbb{X}_p$ for the partial shapes, and $\mathbb{X}_c$ for the complete shapes. Solving the multimodal shape completion problem then amounts to learning a mapping $\mathbb{X}_p \rightarrow \mathbb{X}_c$ in a conditional generative modeling setting between respective latent spaces. Hence, we train a generator $G \colon (\mathbb{X}_p, p(\mathbf{z})) \rightarrow \mathbb{X}_c$, to perform the multimodal mapping with the latent code $\mathbf{z}$ as a condition input. In absence of paired training data, we opt to use adversarial training to complete shapes in a generative modeling setting. Furthermore, to force the generator to use the latent code $\mathbf{z}$, we introduce a encoder E_z to recover $\mathbf{z}$ from the completion output, forcing the connection between the latent space and the shape space to be invertible. Figure 2 shows the setup of the proposed multimodal shape completion network. All network modules are detailed next.

3.1 Learning Latent Spaces for Point Sets

The latent space of a given set of point sets is obtained by training an autoencoder, which encodes the given input to a low-dimension latent feature and then decodes to reconstruct the original input.

For point sets coming from the complete point sets $\mathcal{C}$, we learn an encoder network E_{AE} that maps $\mathbf{C}$ from the original parameter space $\mathbb{R}^{N \times 3}$, defined by concatenating the coordinates of the N points, to a lower-dimensional latent space $\mathbb{X}_c$. A decoder network D_{AE} performs the inverse transformation back to $\mathbb{R}^{N \times 3}$ giving us a reconstructed point set $\tilde{\mathbf{C}}$ with also N points. The encoder-decoders are trained with reconstruction loss:

$$\mathcal{L}^{\mathrm{EMD}} = \mathbb{E}_{\mathbf{C} \sim p(\mathbf{C})} d^{\mathrm{EMD}}(\mathbf{C}, D_{\mathrm{AE}}(E_{\mathrm{AE}}(\mathbf{C}))), \qquad (1)$$

where $\mathbf{C} \sim p(\mathbf{C})$ denotes point set samples drawn from the set of complete point sets, $d^{\mathrm{EMD}}(X_1, X_2)$ is the Earth Mover's Distance (EMD) between point sets X_1, X_2. Once trained, the network weights are held fixed and the *complete* latent code $\mathbf{x}_c = E_{\mathrm{AE}}(\mathbf{C})$, $\mathbf{x}_c \in \mathbb{X}_c$ for a complete point set $\mathbf{C}$ provides a compact representation for subsequent training and implicitly captures the manifold of complete data. As for the point set coming from the partial point sets $\mathcal{P}$, instead of training another autoencoder for its latent parameterization, we directly feed the partial point sets to E_{AE} obtained above for producing *partial* latent space $\mathbb{X}_p$, which in [5] is proved to yield better performance in subsequent adversarial training. Note that, to obtain $\mathbf{x}_p \in \mathbb{X}_p$, we duplicate the partial point set of K points to align with the number of complete point set before feed it to E_{AE}.

3.2 Learning Multimodal Mapping for Shape Completion

Next, we setup a min-max game between the generator and the discriminator to perform the multimodal mapping between the latent spaces. The generator is trained to fool the discriminator such that the discriminator fails to reliably tell if the latent variable comes from original $\mathbb{X}_c$ or the remapped $\mathbb{X}_p$. The mode encoder is trained to model the multimodal distribution of possible complete point sets, and is further applied to the completion output to encode and recover the input latent vector, encouraging the use of conditional mode input.

Formally, the latent representation of the input partial shape $\mathbf{x}_p = E_{\mathrm{AE}}(\mathbf{P})$, along with a Gaussian-sampled condition $\mathbf{z}$, is mapped by the generator to $\hat{\mathbf{x}}_c = G(\mathbf{x}_p, \mathbf{z})$. Then, the task of the discriminator F is to distinguish between latent representations $\hat{\mathbf{x}}_c$ and $\mathbf{x}_c = E_{\mathrm{AE}}(\mathbf{C})$. The mode encoder E_z will encode the completion point set, which can be decoded from the completion latent code $\hat{\mathbf{C}} = D_{\mathrm{AE}}(\hat{\mathbf{x}}_c)$, to reconstruct the conditional input $\tilde{\mathbf{z}} = E_z(\hat{\mathbf{C}})$. We train the mapping function using a GAN. Given training examples of complete latent variables $\mathbf{x}_c$, remapped partial latent variables $\hat{\mathbf{x}}_c$, and Gaussian samples $\mathbf{z}$, we seek to optimize the following training losses over the generator G, the discriminator F, and the encoder E_z:

Adversarial Loss. We add the adversarial loss to train the generator and discriminator. In our implementation, we use least square GAN [19] for stabilizing the training. Hence, the adversarial losses minimized for the generator and the discriminator are defined as:

$$\mathcal{L}_F^{\mathrm{GAN}} = \mathbb{E}_{\mathbf{C}\sim p(\mathbf{C})}[F(E_{\mathrm{AE}}(\mathbf{C})) - 1]^2 + \mathbb{E}_{\mathbf{P}\sim p(\mathbf{P}),\mathbf{z}\sim p(\mathbf{z})}[F(G(E_{\mathrm{AE}}(\mathbf{P}),\mathbf{z}))]^2 \quad (2)$$

$$\mathcal{L}_G^{\mathrm{GAN}} = \mathbb{E}_{\mathbf{P}\sim p(\mathbf{P}),\mathbf{z}\sim p(\mathbf{z})}[F(G(E_{\mathrm{AE}}(\mathbf{P}),\mathbf{z})) - 1]^2, \quad (3)$$

where $\mathbf{C} \sim p(\mathbf{C})$, $\mathbf{P} \sim p(\mathbf{P})$ and $\mathbf{z} \sim p(\mathbf{z})$ denotes samples drawn from, respectively, the set of complete point sets, the set of partial point sets, and $\mathcal{N}(0, \mathcal{I})$.

Partial Reconstruction Loss. Similar to previous work [5], we add a reconstruction loss to encourage the generator to *partially* reconstruct the partial input, so that the completion output is faithful to the input partial:

$$\mathcal{L}_G^{\mathrm{recon}} = \mathbb{E}_{\mathbf{P}\sim p(\mathbf{P}),\mathbf{z}\sim p(\mathbf{z})}[d^{\mathrm{HL}}(\mathbf{P}, D_{\mathrm{AE}}(G(E_{\mathrm{AE}}(\mathbf{P}),\mathbf{z})))]), \quad (4)$$

where d^{HL} denotes the unidirectional Hausdorff distance from the partial point set to the completion point set.

Latent Space Reconstruction. A reconstruction loss on the $\mathbf{z}$ latent space is also added to encourage G to use the conditional mode vector $\mathbf{z}$:

$$\mathcal{L}_{G,E_z}^{\mathrm{latent}} = \mathbb{E}_{\mathbf{P}\sim p(\mathbf{P}),\mathbf{z}\sim p(\mathbf{z})}[\|\mathbf{z}, E_z(D_{\mathrm{AE}}(G(E_{\mathrm{AE}}(\mathbf{P}),\mathbf{z})))\|_1], \quad (5)$$

Hence, our full objective function for training the multimodal shape completion network is described as:

$$\underset{(G,E_z)}{\mathrm{argmin}}\ \underset{F}{\mathrm{argmax}}\ \mathcal{L}_F^{\mathrm{GAN}} + \mathcal{L}_G^{\mathrm{GAN}} + \alpha\mathcal{L}_G^{\mathrm{recon}} + \beta\mathcal{L}_{G,E_z}^{\mathrm{latent}}, \quad (6)$$

where α and β are importance weights for the partial reconstruction loss and the latent space reconstruction loss, respectively.

3.3   Explicitly-Encoded Multimodality

To model the multimodality of possible completion outputs, we resort to an explicit multimodality encoding strategy, in which E_z is trained as a part to explicitly reconstruct the complete shapes. More precisely, a variational autoencoder (E_{VAE}, D_{VAE}) is pre-trained to encode complete shapes into a standard Gaussian distribution $\mathcal{N}(0, \mathcal{I})$ and then decode to reconstruct the original shapes. Once trained, E_{VAE} can encode complete point sets as $\mathbf{x}_c^v \in \mathbb{X}_c^v$, and can be used to recover the conditional input $\mathbf{z}$ from the completion output. Hence, the mode encoder is set to $E_z = E_{VAE}$ and held fixed during the GAN training.

Another strategy is to implicitly encode the multimodality, in which the E_z is jointly trained to map complete shapes into a latent space without being trained as a part of explicit reconstruction. Although it has been shown effective to improve the diversity in the multimodal mapping learning [39,40], we demonstrate that using explicitly-encoded multimodality in our problem yields better performance. In Sect. 4.2, we present the comparison against variants of using implicit multimodality encoding.

3.4   Implementation Details

In our experiments, a partial shape is represented by $K = 1024$ points and a complete shape by $N = 2048$ points. The point set (variational) autoencoder follows [1,5]: using a PointNet[22] as the encoder and a 3-layer MLP as the decoder. The autoencoder encodes a point set into a latent vector of fixed dimension $|\mathbf{x}| = 128$. Similar to [1,5], we use a 3-layer MLP for both generator G and discriminator F. The E_z also uses the PointNet to map a point set into a latent vector $\mathbf{z}$, of which the length we set to $|\mathbf{z}| = 64$. Unless specified, the trade-off parameters α and β in Eq. 6 are set to 6 and 7.5, respectively, in our experiments. For training the point set (variational) autoencoder, we use the Adam optimizer[14] with an initial learning rate 0.0005, $\beta_1 = 0.9$ and train 2000 epochs with a batch size of 200. To train the GAN, we use the Adam optimizer with an initial learning rate 0.0005, $\beta_1 = 0.5$ and train for a maximum of 1000 epochs with a batch size of 50. More details about each network module are presented in the supplementary material.

4   Experiments

In this section, we present results produced from our method on multimodal shape completion, and both quantitative and qualitative comparisons against several baseline methods and variants of our method, along with a set of experiments for evaluating different aspects of our method.

Datasets. Three datasets are derived to evaluate our method under different forms of shape incompleteness: (A) PartNet dataset simulates part-level incompleteness in the user modeling interface. With the semantic segmentation provided in the original PartNet dataset [20], for each provided point set, we remove

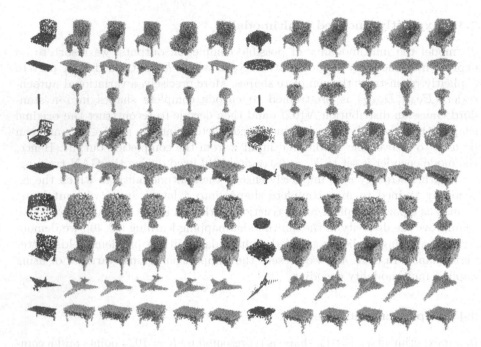

Fig. 3. Our multimodal shape completion results. We show result examples, where the input partial shape is colored in grey and is followed by five different completions in yellow. From top to bottom: `PartNet` (rows 1–3), `PartNet-Scan` (rows 4–6), and `3D-EPN` (rows 7–9).

Fig. 4. Shape completion guided by reference shapes. The completion result varies accordingly when the reference shape changes.

points of randomly selected parts to create a partial point set with at least one part. (B) `PartNet-Scan` dataset resembles the scenario where the partial scan suffers from part-level incompleteness. For each shape in [20], we randomly remove parts and virtually scan residual parts to obtain a partial scan with part-level incompleteness. (C) 3D-EPN dataset [7] is derived from ShapeNet [4] and provides simulated partial scans with arbitrary incompleteness. Scans are represented as Signed Distance Field but we only use the provided point cloud representations. Last, the complete point sets provided in PartNet [20] serve as the complete training data for the first two datasets, while for 3D-EPN dataset, we use the complete virtual scan of ShapeNet objects as the complete training data. We use Chair, Table and Lamp categories for PartNet and PartNet-scan, and use Chair, Airplane and Table categories for 3D-EPN. In all our experiments, we train separate networks for each category in each dataset. More details about data processing can be found in the supplementary material.

Evaluation Measures. For each partial shape in the test set, we generate $k = 10$ completion results and adopt the following measures for quantitative evaluation:

- Minimal Matching Distance (MMD) measures the *quality* of the completed shape. We calculates the Minimal Matching Distance (as described in [1]) between the set of completion shapes and the set of test shapes.
- Total Mutual Difference (TMD) measures the completion *diversity* for a partial input shape, by summing up all the difference among the k completion shapes of the same partial input. For each shape i in the k generated shapes, we calculate its average Chamfer distance d_i^{CD} to the other $k - 1$ shapes. The diversity is then calculated as the sum $\sum_{i=1}^{k} d_i^{CD}$.
- Unidirectional Hausdorff Distance (UHD) measures the completion *fidelity* to the input partial. We calculate the average Hausdorff distance from the input partial shape to each of the k completion results.

More in-depth description can be found in the supplementary material.

4.1 Multimodal Completion Results

We first present qualitative results of our method on multimodal shape completion, by using randomly sampled $\mathbf{z}$ from the standard Gaussian distribution. Figure 3 shows a collection of our multimodal shape completion results on the aforementioned datasets. More visual examples can be found in supplementary material.

To allow more explicit control over the modes in completion results, the mode condition $\mathbf{z}$ can also be encoded from a user-specified shape. As shown in Fig. 4, this enables us to complete the partial shape under the guidance of a given reference shape. The quantitative evaluation of our results, along with the comparison results, is presented next.

4.2 Comparison Results

We present both qualitative and quantitative comparisons against baseline methods and variants of our method:

Table 1. Quantitative comparison results on PartNet (top), PartNet-Scan (middle) and 3D-EPN (bottom). Top two methods on each measure are highlighted. Note that MMD (quality), TMD (diversity) and UHD (fidelity) presented in the tables are multiplied by 10^3, 10^2 and 10^2, respectively.

PartNet	MMD (lower is better)				TMD (higher is better)				UHD (lower is better)			
Method	Chair	Lamp	Table	Avg	Chair	Lamp	Table	Avg	Chair	Lamp	Table	Avg.
pcl2pcl	1.90	2.50	1.90	2.10	0.00	0.00	0.00	0.00	4.88	4.64	4.78	**4.77**
KNN-latent	1.39	1.72	1.30	**1.47**	2.28	4.18	2.36	2.94	8.58	8.47	7.61	8.22
Ours-im-l2z	1.74	2.36	1.68	1.93	3.74	2.68	3.59	**3.34**	8.41	6.37	7.21	7.33
Ours-im-pc2z	1.90	2.55	1.54	2.00	1.01	0.56	0.51	0.69	6.65	5.40	5.38	**5.81**
Ours	1.52	1.97	1.46	**1.65**	2.75	3.31	3.30	**3.12**	6.89	5.72	5.56	6.06

PartNet-Scan	MMD (lower is better)				TMD (higher is better)				UHD (lower is better)			
Method	Chair	Lamp	Table	Avg	Chair	Lamp	Table	Avg	Chair	Lamp	Table	Avg.
pcl2pcl	1.96	2.36	2.09	2.14	0.00	0.00	0.00	0.00	5.20	5.34	4.73	**5.09**
KNN-latent	1.40	1.80	1.39	**1.53**	3.09	4.47	2.85	3.47	8.79	8.41	7.50	8.23
Ours-im-l2z	1.79	2.58	1.92	2.10	3.85	3.18	4.75	**3.93**	7.88	6.39	7.40	7.22
Ours-im-pc2z	1.65	2.75	1.84	2.08	1.91	0.50	1.86	1.42	7.50	5.36	5.68	**6.18**
Ours	1.53	2.15	1.58	**1.75**	2.91	4.16	3.88	**3.65**	6.93	5.74	6.24	6.30

3D-EPN	MMD (lower is better)				TMD (higher is better)				UHD (lower is better)			
Method	Chair	Plane	Table	Avg	Chair	Plane	Table	Avg	Chair	Plane	Table	Avg.
pcl2pcl	1.81	1.01	3.12	1.98	0.00	0.00	0.00	0.00	5.31	9.71	9.03	**8.02**
KNN-latent	1.45	0.93	2.25	**1.54**	2.24	1.13	3.25	2.21	8.94	9.54	12.70	10.40
Ours-im-l2z	1.91	0.86	2.78	1.80	3.84	2.17	4.27	**3.43**	9.53	10.60	9.36	9.83
Ours-im-pc2z	1.61	0.91	3.19	1.90	1.51	0.82	1.67	1.33	8.18	9.55	8.50	**8.74**
Ours	1.61	0.82	2.57	**1.67**	2.56	2.03	4.49	**3.03**	8.33	9.59	9.03	8.98

- pcl2pcl [5], which also uses GANs to complete via generative modeling without paired training data. Without the conditional input, this method, however, cannot complete with diverse shapes for a single partial input.
- KNN-latent, which retrieves a desired number of best candidates from the latent space formed by our complete point set autoencoder, using k-nearest neighbor search algorithm.
- Ours-im-l2z, a variant of our method as described in Sect. 3.3, jointly trains the E_z to implicitly model the multimodality by mapping complete data into a low-dimensional space. The E_z can either take input as complete latent codes (denoted by Ours-im-l2z) or complete point clouds (denoted by Ours-im-pc2z) to map to z space.
- Ours-im-pc2z, in which, as stated above, the E_z takes complete point clouds as input to encode the multimodality in an implicit manner.

More details of the above methods can be found in the supplementary material.

We present quantitative comparison results in Table 1. We can see that, by rating these methods using the combination of the established metrics, our method outperforms other alternatives with high completion quality (low MMD) and diversity (high TMD), while remains faithful to the partial input. More specifically, pcl2pcl has the best fidelity (lowest UHD) to the input partial shape, the

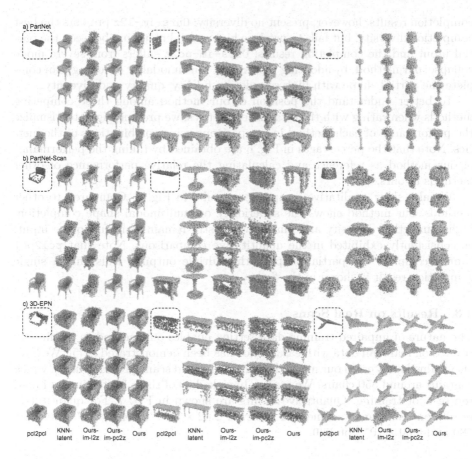

a) PartNet

b) PartNet-Scan

c) 3D-EPN

pcl2pcl KNN- Ours- Ours- Ours pcl2pcl KNN- Ours- Ours- Ours pcl2pcl KNN- Ours- Ours- Ours
        latent im-l2z im-pc2z   latent im-l2z im-pc2z   latent im-l2z im-pc2z

Fig. 5. Qualitative comparison on each three categories of PartNet (top), PartNet-Scan (middle) and 3D-EPN (bottom) dataset. Our method produces results that are both diverse and plausible.

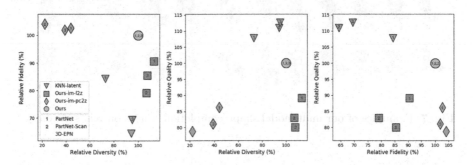

Fig. 6. Comparisons using metrics combinations. Our results present high diversity, quality and fidelity in comparisons using combinations of metrics. *Relative* performance is plotted, and pcl2plc is excluded as it fails to present completion diversity.

completion results, however, present no diversity; Ours-im-l2z presents the best completion diversity, but fails to produce high partial matching between the partial input and the completion results; Ours-im-pc2z suffers from severe mode collapse; our method, by adopting the explicit multimodality encoding, can complete the partial shape with high completion fidelity, quality and diversity.

To better understand the position of our method among those competing methods when rating with the established metrics, we present Fig. 6 to visualize the performance of each method in comparisons using combinations of the metrics. Note that the percentages in Fig. 6 are obtained by taking the performance of our method as reference and calculating the relative performance of other methods to ours.

We also show qualitative comparison results in Fig. 5. Compared to other methods, our method shows the superiority on multimodal shape completion: high completion diversity and quality while still remains faithful to the input, as consistently exhibited in the quantitative comparisons. Note that pcl2pcl cannot complete the partial shape with multiple outputs, thus only a single completion result is shown in the figure.

4.3 Results on Real Scans

The nature of unpaired training data setting enables our method to be directly trained on real scan data with ease, which has been demonstrated in [5]. We have also trained and tested our method on the real-world scans provided in [5], which contains around 550 chairs. We randomly picked 50 of them for testing and used the rest for training. Qualitative results are shown in Fig. 7. For quantitative results on real scans, our method achieved 2.42×10^{-3} on MMD, 3.17×10^{-2} on TMD and 8.60×10^{-2} on UHD.

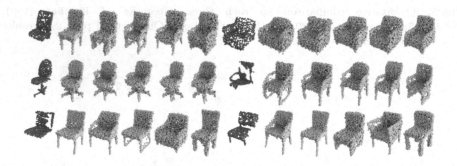

Fig. 7. Examples of our multimodal shape completion results on real scan data.

4.4 More Experiments

We also present more experiments conducted to evaluate different aspects of our method.

Effect of Trade-Off Parameters. Minimizing the $\mathcal{L}^{\text{GAN}}$, $\mathcal{L}^{\text{recon}}$ and $\mathcal{L}^{\text{latent}}$ in Eq. 6 corresponds to the improvement of completion quality, fidelity, and diversity, respectively. However, conflict exists when simultaneously optimizing these three terms, e.g., maximizing the completion fidelity to the partial input would potentially compromise the diversity presented in the completion results, as part of the completion point set is desired to be fixed. Hence, we conduct experiments and show in Fig. 8 to see how the diversity and fidelity vary with respect to the change of trade-off parameters α and β.

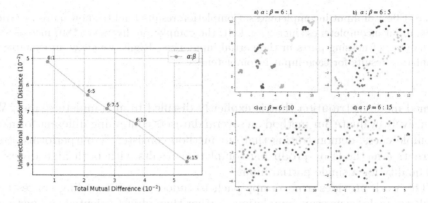

Fig. 8. Effect of trade-off parameters. Left: the completion fidelity (UHD) decreases and the completion diversity (TMD) increases as we set β to be larger. Right: t-SNE visualization of completion latent vectors under different parameter settings. Dots with identical color indicates completion latent vectors resulted from the same partial shape. a) setting a larger weight for $\mathcal{L}^{\text{recon}}$ to encourage completion fidelity leads to mode collapse within the completion results of each partial input; b)-d) setting larger weights for $\mathcal{L}^{\text{latent}}$ to encourage completion diversity results in more modes.

Effect of Input Incompleteness. The shape completion problem has more ambiguity when the input incompleteness increases. Thus, it is desired that the model can complete with more diverse results under increasing incompleteness of the input. To this end, we test our method on `PartNet` dataset, where the incompleteness can be controlled by the number of missing parts. Fig. 9 shows how diversity in our completion results changes with respect to the number of missing parts in the partial input.

5   Conclusion

We present a point-based shape completion framework that can produce multiple completion results for the incomplete shape. At the heart of our approach lies a generative network that completes the partial shape in a conditional generative modeling setting. The generation of the completions is conditioned on a

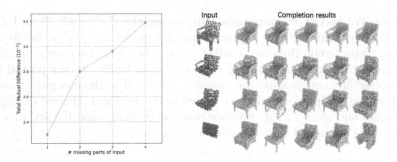

Fig. 9. Effect of input incompleteness. Completion results tend to show more variation as the input incompleteness increases. Left: the completion diversity (TMD) increases as the number of missing parts in the partial input rises. Right: we show an example of completion results for each input incompleteness.

learned mode distribution that is explicitly distilled from complete shapes. We extensively evaluate our method on several datasets containing different forms of incompleteness, demonstrating that our method consistently outperforms other alternative methods by producing completion results with both high diversity and quality for a single partial input.

The work provides a novel approach to indoor scene modeling suggestions and large-scale scene scan completions, rather than object-oriented completions. While the completion diversity in our method is already demonstrated, the explicit multimodality encoding module is, nonetheless, suboptimal and could be potentially improved. Our method shares the same limitations as many of the counterparts: not producing shapes with fine-scale details and requiring the input to be canonically oriented. A promising future direction would be to investigate the possibility of guiding the mode encoder network to concentrate on the missing regions for multimodality extraction, alleviating the compromise between the completion diversity and the faithfulness to the input.

Acknowledgements. We thank the anonymous reviewers for their valuable comments. This work was supported in part by National Key R&D Program of China (2018YFB1403901, 2019YFF0302902) and NSFC (61902007).

References

1. Achlioptas, P., Diamanti, O., Mitliagkas, I., Guibas, L.: Learning representations and generative models for 3D point clouds. In: International Conference on Machine Learning (ICML), pp. 40–49 (2018)
2. Alec, R., Luke, M., Soumith, C.: Unsupervised representation learning with deep convolutional generative adversarial networks. In: International Conference on Learning Representations (ICLR) (2016)
3. Arjovsky, M., Chintala, S., Bottou, L.: Wasserstein generative adversarial networks. In: International Conference on Machine Learning (ICML), pp. 214–223 (2017)

4. Chang, A.X., et al.: ShapeNet: an Information-Rich 3D Model Repository. Technical report, arXiv:1512.03012 [cs.GR], Stanford University – Princeton University – Toyota Technological Institute at Chicago (2015)
5. Chen, X., Chen, B., Mitra, N.J.: Unpaired point cloud completion on real scans using adversarial training. In: International Conference on Learning Representations (ICLR) (2020)
6. Chen, Z., Zhang, H.: Learning implicit fields for generative shape modeling. In: Conference on Computer Vision and Pattern Recognition (CVPR), pp. 5939–5948 (2019)
7. Dai, A., Ruizhongtai Qi, C., Nießner, M.: Shape completion using 3D-encoder-predictor CNNs and shape synthesis. In: Conference on Computer Vision and Pattern Recognition (CVPR), pp. 5868–5877 (2017)
8. Goodfellow, I., et al.: Generative adversarial nets. In: Advances in Neural Information Processing Systems (NeurIPS), pp. 2672–2680 (2014)
9. Groueix, T., Fisher, M., Kim, V.G., Russell, B., Aubry, M.: AtlasNet: a Papier-Mâché approach to learning 3D surface generation. In: Conference on Computer Vision and Pattern Recognition (CVPR) (2018)
10. Guerrero, P., Kleiman, Y., Ovsjanikov, M., Mitra, N.J.: PCPNET learning local shape properties from raw point clouds. Comput. Graph. Forum. **37**, 75–85 (2018)
11. Gurumurthy, S., Agrawal, S.: High fidelity semantic shape completion for point clouds using latent optimization, pp. 1099–1108. IEEE (2019)
12. Han, X., Li, Z., Huang, H., Kalogerakis, E., Yu, Y.: High-resolution shape completion using deep neural networks for global structure and local geometry inference. In: International Conference on Computer Vision (ICCV), pp. 85–93 (2017)
13. Hinton, G.E., Salakhutdinov, R.R.: Reducing the dimensionality of data with neural networks. Science **313**(5786), 504–507 (2006)
14. Kingma, D.P., Ba, J.: Adam: a method for stochastic optimization (2014)
15. Ledig, C., et al.: Photo-realistic single image super-resolution using a generative adversarial network. In: Conference on Computer Vision and Pattern Recognition (CVPR), pp. 4681–4690 (2017)
16. Li, J., Chen, B.M., Hee Lee, G.: SO-Net: self-organizing network for point cloud analysis. In: Conference on Computer Vision and Pattern Recognition (CVPR), pp. 9397–9406 (2018)
17. Li, Y., Bu, R., Sun, M., Wu, W., Di, X., Chen, B.: PointCNN: convolution on X-transformed points. In: Advances in Neural Information Processing Systems (NeurIPS), pp. 820–830 (2018)
18. Liu, M., Sheng, L., Yang, S., Shao, J., Hu, S.M.: Morphing and sampling network for dense point cloud completion. In: Association for the Advancement of Artificial Intelligence (AAAI) (2019)
19. Mao, X., Li, Q., Xie, H., Lau, R.Y., Wang, Z., Paul Smolley, S.: Least squares generative adversarial networks. In: International Conference on Computer Vision (ICCV), pp. 2794–2802 (2017)
20. Mo, K., et al.: PartNet: a large-scale benchmark for fine-grained and hierarchical part-level 3D object understanding. In: The IEEE Conference on Computer Vision and Pattern Recognition (CVPR), June 2019
21. Park, J.J., Florence, P., Straub, J., Newcombe, R., Lovegrove, S.: DeepSDF: learning continuous signed distance functions for shape representation. In: Conference on Computer Vision and Pattern Recognition (CVPR), pp. 165–174 (2019)
22. Qi, C.R., Su, H., Mo, K., Guibas, L.J.: PointNet: deep learning on point sets for 3D classification and segmentation. In: Conference on Computer Vision and Pattern Recognition (CVPR), pp. 652–660 (2017)

23. Qi, C.R., Yi, L., Su, H., Guibas, L.J.: Pointnet++: deep hierarchical feature learning on point sets in a metric space. In: Advances in Neural Information Processing Systems (NeurIPS), pp. 5099–5108 (2017)
24. Sharma, A., Grau, O., Fritz, M.: VConv-DAE: deep volumetric shape learning without object labels. In: Hua, G., Jégou, H. (eds.) ECCV 2016. LNCS, vol. 9915, pp. 236–250. Springer, Cham (2016). https://doi.org/10.1007/978-3-319-49409-8_20
25. Smolensky, P.: Information processing in dynamical systems: foundations of harmony theory. Technical report. Colorado University at Boulder Department of Computer Science (1986)
26. Stutz, D., Geiger, A.: Learning 3D shape completion under weak supervision. Int. J. Comput. Vis. (IJCV), 1–20 (2018)
27. Su, H., et al.: SplatNet: sparse lattice networks for point cloud processing. In: Conference on Computer Vision and Pattern Recognition (CVPR), pp. 2530–2539 (2018)
28. Thanh Nguyen, D., Hua, B.S., Tran, K., Pham, Q.H., Yeung, S.K.: A field model for repairing 3D shapes. In: Conference on Computer Vision and Pattern Recognition (CVPR), pp. 5676–5684 (2016)
29. Vincent, P., Larochelle, H., Bengio, Y., Manzagol, P.A.: Extracting and composing robust features with denoising autoencoders. In: International Conference on Machine Learning (ICML), pp. 1096–1103 (2008)
30. Wang, W., Huang, Q., You, S., Yang, C., Neumann, U.: Shape inpainting using 3D generative adversarial network and recurrent convolutional networks. In: International Conference on Computer Vision (ICCV), pp. 2298–2306 (2017)
31. Wu, R., Zhuang, Y., Xu, K., Zhang, H., Chen, B.: PQ-NET: a generative part Seq2Seq network for 3D shapes. arXiv preprint arXiv:1911.10949 (2019)
32. Wu, Z., et al.: 3D ShapeNets: a deep representation for volumetric shapes. In: Conference on Computer Vision and Pattern Recognition (CVPR), pp. 1912–1920 (2015)
33. Yang, B., Rosa, S., Markham, A., Trigoni, N., Wen, H.: 3D object dense reconstruction from a single depth view. arXiv preprint arXiv:1802.00411 1(2), 6 (2018)
34. Yin, K., Huang, H., Cohen-Or, D., Zhang, H.: P2P-NET: bidirectional point displacement net for shape transform. ACM Trans. Graph. (TOG) 37(4), 1–13 (2018)
35. Yu, L., Li, X., Fu, C.-W., Cohen-Or, D., Heng, P.-A.: EC-Net: an edge-aware point set consolidation network. In: Ferrari, V., Hebert, M., Sminchisescu, C., Weiss, Y. (eds.) ECCV 2018. LNCS, vol. 11211, pp. 398–414. Springer, Cham (2018). https://doi.org/10.1007/978-3-030-01234-2_24
36. Yu, L., Li, X., Fu, C.W., Cohen-Or, D., Heng, P.A.: PU-Net: point cloud upsampling network. In: Conference on Computer Vision and Pattern Recognition (CVPR), pp. 2790–2799 (2018)
37. Yuan, W., Khot, T., Held, D., Mertz, C., Hebert, M.: PCN: point completion network. In: International Conference on 3D Vision (3DV), pp. 728–737 (2018)
38. Zaheer, M., Kottur, S., Ravanbakhsh, S., Poczos, B., Salakhutdinov, R.R., Smola, A.J.: Deep sets. In: Advances in Neural Information Processing Systems (NeurIPS), pp. 3391–3401 (2017)
39. Zhu, J.Y., et al.: Toward multimodal image-to-image translation. In: Advances in Neural Information Processing Systems (NeurIPS), pp. 465–476 (2017)
40. Zhu, J.Y., et al.: Visual object networks: image generation with disentangled 3D representations. In: Advances in Neural Information Processing Systems (NeurIPS), pp. 118–129 (2018)

Generative Sparse Detection Networks for 3D Single-Shot Object Detection

JunYoung Gwak[1]([⊠]), Christopher Choy[2], and Silvio Savarese[1]

[1] Stanford University, Stanford, USA
{jgwak,ssilvio}@stanford.edu
[2] NVIDIA, Santa Clara, USA
cchoy@nvidia.com

Abstract. 3D object detection has been widely studied due to its potential applicability to many promising areas such as robotics and augmented reality. Yet, the sparse nature of the 3D data poses unique challenges to this task. Most notably, the observable surface of the 3D point clouds is disjoint from the center of the instance to ground the bounding box prediction on. To this end, we propose Generative Sparse Detection Network (GSDN), a fully-convolutional single-shot sparse detection network that efficiently generates the support for object proposals. The key component of our model is a generative sparse tensor decoder, which uses a series of transposed convolutions and pruning layers to expand the support of sparse tensors while discarding unlikely object centers to maintain minimal runtime and memory footprint. GSDN can process unprecedentedly large-scale inputs with a single fully-convolutional feed-forward pass, thus does not require the heuristic post-processing stage that stitches results from sliding windows as other previous methods have. We validate our approach on three 3D indoor datasets including the large-scale 3D indoor reconstruction dataset where our method outperforms the state-of-the-art methods by a relative improvement of 7.14% while being 3.78 times faster than the best prior work.

Keywords: Single shot detection · 3D object detection · Generative sparse network · Point cloud

1 Introduction

3D reconstructions have become more commonplace as a complete reconstruction pipeline become built into consumer devices, such as mobile phones or head-mounted displays, for applications in robotics and augmented reality. Among these applications, perceptions on 3D reconstructions is the first step allowing users to interact with a virtual world in 3D. For example, indoor navigation

Electronic supplementary material The online version of this chapter (https://doi.org/10.1007/978-3-030-58548-8_18) contains supplementary material, which is available to authorized users.

© Springer Nature Switzerland AG 2020
A. Vedaldi et al. (Eds.): ECCV 2020, LNCS 12349, pp. 297–313, 2020.
https://doi.org/10.1007/978-3-030-58548-8_18

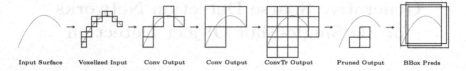

Fig. 1. The top-down view of the cross-section of our simplified 3D sparse anchor generation pipeline: a 3D scanner samples the surface of an object which we convert to a sparse tensor. Then, an encoder extracts hierarchical sparse tensor features with a series of convolutions. During the decoder stage, we apply a transposed convolution to upsample and expand the support of the sparse tensor. Finally, we prune out unnecessary supports that do not contain anchors and make bounding box anchor predictions.

applications can aid a user to localize objects, and mixed reality applications need to track objects to give users information relevant to the current status of their surroundings. Many of these virtual-reality and mixed-reality applications require identifying and detecting 3D objects in real-time.

However, unlike 2D images where the input is in a densely packed array, 3D data is scanned or reconstructed as a set of points or a triangular mesh. These data occupy a small portion of the 3D space and pose unique challenges for 3D object detection. First, the space of interest is three dimensional which requires cubic complexity to save or process data. Second, the data of interest is very sparse, and all information is sampled from the surface of objects.

Many previous 3D object detectors proposed various methods to process cubically growing sparse 3D data, and can be categorized into one of two branches: 3D object detection by converting sparse 3D data into a dense representation [1,13,15,19,28] or by directly feeding a set of points into multi-layer perceptrons [24,35]. First, dense 3D representation for indoor object detection [1,13,28] uses volumetric features which have memory and computational complexity of $O(N^3)$ where N is the resolution of the space. This representation requires large memory, which prevents the utilization of deep networks and requires cropping the scenes and stitching the results to process large or high-resolution scenes. Second, multi-layer perceptrons that process a scene as a set of points limit the number of points a network can process. Thus, as the size of the point cloud increases, the method suffers from either low-resolution input which makes it difficult to scale the method up for larger scenes (see Sect. 5.2) or apply sliding-window style cropping and stitching which prevents the network to see a larger context [35].

We instead propose to resolve the cubic complexity with our hierarchical sparse tensor encoder, adopting a sparse tensor network [8] to efficiently process a large scene fully-convolutionally. As we use a sparse representation, our network is fast and memory-efficient compared with a single-shot method that uses dense tensors [13]. It allows our network to adopt extremely deep architectures while requiring a fraction of the memory and computation. Also, compared with multi-layer perceptrons, our method scales to large scenes without sacrificing point

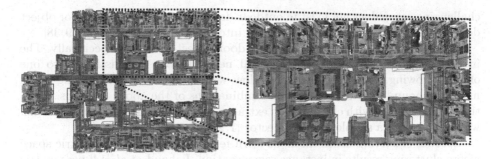

Fig. 2. Detection results on the entire S3DIS building 5: Our proposed method can process 78M points, 13984 m³, 53 room building as a whole in a *single fully-convolutional feed-forward pass*, only using 5G of GPU memory. Left: bird-eye-view of the entire building 5, Right: partial view of the same building.

density or the receptive field size of a network by cropping a scene into smaller windows [24,35].

Another key challenge of a 3D object detector is that the support of the input 3D scans and the support of the object bounding box anchors are disjoint. In other words, we have samples of 3D points on the surface of the objects, but not on the center of the object where a bounding box anchor is located. This is due to the fact that many objects are convex and we cannot directly observe the object center. For this, we propose a generative sparse tensor decoder that repeatedly upsamples the support of input to expand and cover the support of anchors while discarding unlikely object centers to maintain minimal runtime and memory footprint (Fig. 1).

To sum, we propose Generative Sparse Detector Network (GSDN), a deep *fully-convolutional* single-shot 3D object detection algorithm with a sparse tensor network. Our single-shot 3D object detection network consists of two components: an hierarchical sparse tensor encoder which efficiently extracts deep hierarchical features, and a generative sparse tensor decoder which expands the support of the sparse input to ground object proposals on. Experimentally, GSDN outperforms the state-of-the-art methods on two large-scale indoor datasets while being faster than the best prior work. We also analyze the speed and memory footprint of the model and demonstrate the extreme scalability of our method on orders of magnitudes larger 3D scenes (Fig. 2).

2   Related Work

3D Indoor Object Detection. In a 3D indoor setting or 3D indoor datasets [1,5], the distribution of object placement creates unique challenges: objects such as lamps and ceiling lights can be placed on a wall or a ceiling, or objects can be placed on top of another object such as a desk or a bed. However, such challenging setup does not exist in outdoor datasets and most 3D outdoor object detectors simply project the 3D problem into a 2D ground plane [15,19,38].

Thus, in this section, we cover 3D indoor object detection specifically. The indoor 3D object detection using neural networks can be classified into one of the following categories: sliding-window with classification, clustering-based methods, bounding-box proposal, or combinations of the above methods. First, the sliding window with classification extracts a 3D patch for object classification which is used as a simple object detector [1,28].

Second, clustering-based methods learn features or vectors in a metric space where clustering results in instance segmentation. Lahoud et al. [14] uses metric learning to train the feature space. Liu et al. [17], Yi et al. [36], Wang et al. [33], and Qi et al. [24] predict object centers per 3D point and cluster the center votes.

Third, the bounding box proposal methods adopt 2D rectangular bounding box proposal methods to 3D. Wang et al. [32] proposed Vote3D, which predicts 3D bounding boxes on a sparse grid for object detection. Yang et al. [35] directly predicts bounding boxes from MLP of global point cloud features. Hou et al.[13] makes a straight-forward 3D extension of region proposal networks on dense voxels. GSDN is a bounding box proposal method with a crucial difference in maintaining the sparsity of the input point cloud and target anchor space, enabling much faster inference on many orders of magnitude larger scene with better performance than state-of-the-art methods.

3D Generative Networks. Generating 3D shapes from a neural network can be classified into two broad categories: continuous 3D point representations [20,23,31,37] and discrete grid representations [2,4,6,7,30]. Specifically, within the discrete representations, some use sparse representations for 3D reconstruction which allow a high-resolution voxel or signed-distance-function (SDF) reconstruction [2,6,7,30]. Unlike previous works that focus on the shapes of objects, we use the generative process to predict the bounding box anchors. Also, compared with some sparse generative processes that subdivide voxels [6,30], our method extends the support with transposed convolutions to cover bounding box anchors which are located behind 3D surface observations.

Sparse Tensor Networks. A conventional neural network processes a dense tensor such as temporal data, images, or videos using a series of linear operations and non-linear operations. Most of the linear operations also use dense tensors for parametrization. In mobile and embedded systems, a sparse parametrization of neural networks [10,21,22] has been widely studied to compress a neural network. Graham et al. [9] instead proposes to take *spatially* sparse tensors as inputs and generate *spatially* sparse feature maps. Using a sparse tensor as an input has gained more popularity since its success on 3D data processing [2,3,8,9]. We adopt these spatially sparse networks, or sparse tensor networks to scale detection networks to an unprecedented depth and to handle extremely large scenes. Additionally, we propose to *dynamically* generate new coordinates to efficiently support bounding box center coordinates that are often missing in surface-scanned inputs.

3   Preliminaries

In this section, we briefly go over the basic 3D representation, a sparse tensor, and introduce basic operations that are critical for the generative sparse tensor network. Throughout the paper, we will use lowercase letters for variable scalars, t; uppercase letters for constants, N; lowercase bold letters for vectors, $\mathbf{v}$; uppercase bold letters for matrices, $\mathbf{R}$; Euler scripts for tensors, $\mathscr{T}$; and calligraphic symbols for sets, $\mathcal{C}$.

3.1   Sparse Tensor

A tensor is a multi-dimensional array that can represent high-dimensional data. A sparse tensor of order-D, $\mathscr{T} \in \mathbb{R}^{N_1 \times N_2 \times \dots \times N_D}$, is a D-dimensional array where majority of its elements are 0. Adopting the conventional sparse matrix representation, a sparse matrix can be represented as a set of non-zero coordinates $\mathcal{C} = \text{supp}(\mathscr{T})$ where supp is the set-theoretic support operator as in standard mathematical terminology, and corresponding features $\mathcal{F}$.

$$\mathscr{T}[x_i^1, x_i^2, \dots, x_i^D] = \begin{cases} \mathbf{f}_i & \text{if } (x_i^1, x_i^2, \dots, x_i^D) \in \mathcal{C} \\ 0 & \text{otherwise} \end{cases} \tag{1}$$

where x_d^i denotes d-th axis coordinate of the i-th non-zero element and $\mathbf{f}_i$ is the feature associated to the i-th non-zero element. These non-zero elements contain information that are equivalent to a sparse tensor $\mathcal{T} \Leftrightarrow (\mathcal{C}, \mathcal{F})$. These sets can also be converted to matrices of COOrdinate representation (COO) $\mathbf{C}, \mathbf{F}$ where each row is an element of the corresponding coordinate and feature sets $(\mathcal{C}, \mathcal{F})$.

3.2   Sparse Tensor for 3D Data Representation

The 3D data of interest in this work uses point clouds or meshes to represent 3D surfaces. We can represent a mesh or a point cloud as a sparse tensor by discretizing the coordinates of vertices or points. This process requires defining the discretization step size (voxel size) which is a hyperparameter that affects the performance of a neural network [2,3].

4   Generative Sparse Detection Networks

In this section, we propose the generative sparse detection networks for 3D object detection. Unlike the 2D object detection networks [16,27], we use a sparse tensor as the 3D representation throughout the network including the intermediate features. Thus, all layers such as convolution and batch normalization are well defined for sparse tensors [2,8]. Throughout the paper, we will implicitly refer to all tensors as sparse tensors and layers as sparse tensor counterparts.

The network consists mainly of two parts: a hierarchical sparse tensor encoder and a generative sparse tensor decoder. The first part of the network generates sparse tensor feature maps that can sufficiently capture geometry and identity of objects and the second part proposes new supports based on the feature maps.

4.1   Hierarchical Sparse Tensor Encoder

We use residual networks [12], specifically high-dimensional variants proposed in Choy *et al.* [2], as the backbone of our model. Note that the backbone network can be replaced with more modern and recent variants. The network consists of residual blocks and strided convolutions that reduce the resolution of the space and increase the receptive field size exponentially. First, the network takes a high-resolution sparse tensor as an input $\mathscr{T}_0$ and generate hierarchical feature maps $\mathscr{T}_l$ with a series of downsampling and residual blocks $f_l(\cdot; \mathbf{W}_l)$ for $l \in [1, ..., L]$. The encoder can be represented succinctly as (Fig. 3)

$$\mathscr{T}_l \leftarrow f_l(\mathscr{T}_{l-1}; \mathbf{W}_l) \text{ for } l \in [1, ..., L]$$

We cache all of the hierarchical sparse tensor feature maps $\mathscr{T}_l$ for $l \in [1, ..., L]$ which will be fed into the generative sparse tensor decoder.

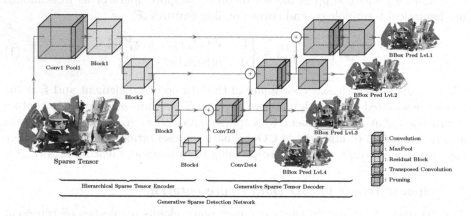

Fig. 3. Network overview: generative sparse detection networks process a sparse tensor input first with a series of strided convolutions followed by a few residual network blocks to generates hierarchical sparse tensor feature maps (Sect. 4.1). The second stage upsamples the sparse tensor feature maps using transposed convolution and pruning (Sect. 4.2). Note that all feature maps are sparse tensors and all layers process sparse tensors fully-convolutionally.

4.2   Generative Sparse Tensor Decoder

The second half of the network expands the support of the hierarchical sparse tensors feature maps $\mathscr{T}_l$ to cover the support for bounding box anchors. We approximate this process with transposed convolutions (also known as upconvolution, deconvolution). Given an input sparse tensor $\mathscr{T}$, we create an output sparse tensor $\mathscr{T}'$ that $\text{supp}(\mathscr{T}) \subset \text{supp}(\mathscr{T}')$. Yet, not all voxels generated from this process contain bounding box anchors and can be dynamically removed to save the memory and computation cost. Thus, we propose *sparsity pruning*, where we *dynamically* determine which coordinates to prune based on learned parameters. By applying a transposed convolution followed by sparsity pruning, we increase the resolution of the space while limiting the memory and computation cost. Without pruning, the number of coordinates grows cubically after every transposed convolution and our training pipeline fails from lack of memory. Additionally, we make skip connections between the hierarchical sparse tensor feature maps and the upsampled sparse tensors to recover the fine details of the input.

4.2.1 Transposed Convolution and Sparsity Pruning

We use transposed convolutions with the kernel size greater than 2 to not just upsample, but expand the support of a sparse tensor. This process affects the sparsity pattern of a sparse tensor and the support of the output sparse tensor is the stencil or outer-product of the convolution kernel shape on the input sparsity pattern $\mathrm{supp}(\mathscr{T}') = \mathcal{C} \otimes [-K, ..., K]^3$. Mathematically, a transposed convolution on a 3D sparse tensor $\mathscr{T}$ with $\mathrm{supp}(\mathscr{T}) = \mathcal{C}$ can be defined as follows:

$$\mathscr{T}'[x,y,z] = \sum_{i,j,k \in \mathcal{N}(x,y,z)} \mathbf{W}[x-i, y-j, z-k]\mathscr{T}[i,j,k] \text{ for } (x,y,z) \in \mathcal{C}' \quad (2)$$

where $\mathcal{C}' = \mathcal{C} \otimes [-K, ..., K]^3$, $\mathcal{N}(x,y,z) = \{(i,j,k) | \|x-i| \le K, |y-j| \le K, |z - k| < K, (i,j,k) \in \mathcal{C}\}$, W is the 3D convolution kernel weights and $2K + 1$ is the convolution kernel size. This results in denser sparsity pattern on the output tensor $\mathscr{T}'$ with $\mathrm{supp}(\mathscr{T}') = \mathcal{C} \otimes [-K, ..., K]^3$. Note that unlike the subdivision, the transposed convolution expands a sparse point into an arbitrarily large dense region and multiple regions could overlap with each other (Fig. 4).

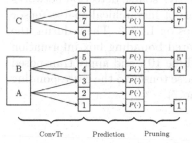

ConvTr   Prediction   Pruning

Fig. 4. Expansion and pruning: transposed convolution upsamples a low-resolution sparse tensor into a high-resolution sparse tensor. Then, we prune out some of the upsampled coordinates with sparsity predictions $P_s(\cdot)$.

After a transposed convolution, not all the newly created coordinates contain object bounding box anchors. Thus, we remove some of these voxels that have a small probability of containing bounding box anchors. We denote a function that returns the probability given features at each voxel as $P_s(\cdot)$ and remove all voxels $P_s(\cdot) < \tau$, where τ is the sparsity pruning confidence threshold.

$$\mathbf{p} = P_s(\mathscr{T}; \mathbf{W}_P) \quad (3)$$

$$\mathscr{T}' = \mathrm{SparsityPruning}(\mathscr{T}, \mathbf{p} < \tau) \quad (4)$$

4.2.2 Skip Connection and Sparse Tensor Addition

The upsampled sparse tensor feature maps from the generative process have gone through extreme spatial compression that allows neurons to see larger context, but have lost spatial resolution. To recover the fine details of the input, we create the skip connections to the cached feature map from the encoder [2,3]. Since both the upsampled feature map and the lower layer feature map are all sparse tensors, we use sparse tensor addition. This process also expands the support to be the union of the supports of both sparse tensors.

4.3   Multi-scale Bounding Box Anchor Prediction

Every voxel after the sparsity pruning potentially contains bounding box anchors. Therefore, we make a direct prediction of the bounding box parameters for every layer of the pruned sparse tensors. Specifically, for each k anchor box, the network predicts 1 object anchor likelihood score, 6 offsets relative to the anchor box, and c semantic class scores. This results in $(c + 7)k$ outputs per voxel.

To capture as many shape variations, we use bounding box anchors with different aspect ratios. Specifically, for each anchor ratio seed a_r, we use all unique permutations of $\left[\sqrt{a_r}, \sqrt{a_r}, \frac{1}{\sqrt{a_r}}\right]$ as the aspect ratios of an anchor. In total, we use $k = 13$ anchors with $a_r \in \{1, 2, 4, \frac{1}{2}, \frac{1}{4}\}$ including the identity ratio.

However, even with these various anchor ratios, it is difficult to capture the extreme scale variation among 3D objects. Thus, we predict anchors at various stages of the decoder to capture the scale variation of 3D objects similar to Liu *et al.* [18]. We construct the anchors at each level to double the size of the anchors at the previous level.

4.4   Summary of GSDN Feed Forward

We summarize the feed forward pass of the generative sparse detection networks in Algorithm 1. The algorithm generates L levels of hierarchical sparse tensor feature maps from the previous level feature maps on Line 3. Then, during the generative phase, we extract anchors and associated bounding box information (Line 8), predict sparsity and prune out voxels (Line 10), and apply transposed convolution (Line 12). We add the upsampled sparse tensor to the corresponding sparse tensor feature map from the encoder (Line 7).

4.5   Losses

The generative sparse detection network has to predict four types of outputs: sparsity prediction, anchor prediction, semantic class, and bounding box regression. First, the sparsity and anchor prediction are binary classification problems. However, the majority of the predictions are negative as many voxels does not contain positive anchors. Thus, we use balanced cross entropy loss:

$$L_{\mathrm{b}}(\hat{\mathbf{y}}, \mathbf{y}) = -\frac{1}{2|\mathcal{P}|} \sum_{i \in \mathcal{P}} \log(P(\hat{\mathbf{y}}_i)) - \frac{1}{2|\mathcal{N}|} \sum_{i \in \mathcal{N}} \log(1 - P(\hat{\mathbf{y}}_i))$$

where $\mathcal{P} = \{i | y_i = 1\}$ and $\mathcal{N} = \{i | y_i = 0\}$ are the set of indices with positive and negative labels respectively. We define an anchor to be positive if any of the

---

Algorithm 1: Generative Sparse Detection Networks

Input: $\mathscr{T}, f_l(\cdot; \mathbf{W}_l), f_l^{\mathrm{Tr}}(\cdot; \mathbf{W}_l^{\mathrm{Tr}}), g_l^b(\cdot; \mathbf{W}_l^b), P_s(\cdot; \mathbf{G}_l^s)$ for $l \in [1, ..., L], \tau_s$
Output: $\{\mathbf{B}_l\}_l$ for $l \in [1, ..., L]$

1 $\mathscr{T}_0 \leftarrow \mathscr{T}$
   /* Hierarchical Sparse Tensor Encoder § 4.1   */
2 **for** $l \leftarrow 1, ...L$ **do**
3     $\lfloor$ $\mathscr{T}_l \leftarrow f_l(\mathscr{T}_{l-1})$ // Hierarchical feature tensors
   /* Generative Sparse Tensor Decoder § 4.2   */
4 $\mathscr{T}_L^{\mathrm{Tr}} \leftarrow \mathscr{T}_L$
5 **for** $l \leftarrow L, ..., 1$ **do**
6   **if** $l < L$ **then**
7       $\lfloor$ $\mathscr{T}_l^{\mathrm{Tr}} \leftarrow \mathscr{T}_l^{\mathrm{Tr}} + \mathscr{T}_l$ // Skip connection §4.2.2
8     $\mathbf{B}_l \leftarrow g_l^b(\mathscr{T}_l^{\mathrm{Tr}})$ // Anchor predictions §4.3
9     $\mathbf{p}_l \leftarrow P_l^s(\mathscr{T}_l^{\mathrm{Tr}})$ // Sparsity predictions
10    $\mathscr{T}_l^{\mathrm{Tr}} \leftarrow \mathrm{SparsityPruning}(\mathscr{T}_l^{\mathrm{Tr}}, \mathbf{p}_l < \tau)$ // Pruning §4.2.1
11   **if** $l > 1$ **then**
12      $\lfloor$ $\mathscr{T}_{l|1}^{\mathrm{Tr}} \leftarrow f_l^{\mathrm{Tr}}(\mathscr{T}_l^{\mathrm{Tr}})$ // Transposed convolution §4.2.1
13 **return** $\{\mathbf{B}_l\}_l$

---

anchors in a voxel overlaps with any ground-truth bounding boxes for 3D IoU > 0.35 and negative if 3D IoU < 0.2. As the sparsity prediction must contain all anchors in subsequent levels, we define a sparsity to be positive if any of the subsequent positive anchor associated to the current voxel is positive. We do not enforce loss on anchors that have 0.2 $<$3D IoU < 0.35.

Finally, for positive anchors, we train semantic class prediction of the highest overlapping ground-truth bounding box class with the standard cross entropy, L_{class}, and bounding box center and size regression parameterized by difference of the center location relative to the size of the anchor and the log difference of the size of the bounding box with the Huber loss [27], L_{reg}. The final loss is the weighted sum of all losses:

$$L = \lambda_s L_s + \lambda_{\mathrm{anc}} L_{\mathrm{anc}} + \lambda_{\mathrm{class}} L_{\mathrm{class}} + \lambda_{\mathrm{reg}} L_{\mathrm{reg}}$$

where we use $\lambda_s = 1$, $\lambda_{\mathrm{anc}} = 1$, $\lambda_{\mathrm{class}} = 1$, $\lambda_{\mathrm{reg}} = 0.1$ for all of our experiments.

4.6 Prediction Post-processing

We train the network to overestimate the number of bounding box anchors as we label all anchors with 3D IoU >0.35 as positives. We filter out overlapping predictions with non-maximum suppression and merge them by computing score-weighted average of all removed bounding boxes to fine tune the final predictions similar to Redmon et al. [26].

5 Experiments

We evaluate our method on three 3D indoor datasets and compare with state-of-the-art object detection methods (Sect. 5.1). We also make a detailed analysis of the speed and memory footprint of our method (Sect. 5.2). Finally, we demonstrate the scalability of our proposed method on extremely large scenes (Sect. 5.3).

Table 1. Object detection mAP on the ScanNet v2 validation set. DSS, MRCNN 2D-3D, FPointNet are from [13]. GSPN from [24]. Our method, despite being single-shot, outperforms all previous state-of-the-art methods.

Method	Single Shot	mAP@0.25	mAP@0.5
DSS [13,28]	✗	15.2	6.8
MRCNN 2D-3D [11,13]	✗	17.3	10.5
F-PointNet [25]	✗	19.8	10.8
GSPN [24,36]	✗	30.6	17.7
3D-SIS [13]	✓	25.4	14.6
3D-SIS [13] + 5 views	✓	40.2	22.5
VoteNet [24]	✗	58.6	33.5
GSDN (Ours)	✓	**62.8**	**34.8**

Table 2. Class-wise mAP@0.25 object detection result on the ScanNet v2 validation set. Our method outperforms previous state-of-the-art on majority of the semantic classes.

	cab	bed	chair	sofa	tabl	door	wind	bkshf	pic	cntr	desk	curt	fridg	showr	toil	sink	bath	ofurn	mAP
Hou et al. [13]	12.75	63.14	65.98	46.33	26.91	7.95	2.79	2.30	0.00	6.92	33.34	2.47	10.42	12.17	74.51	22.87	58.66	7.05	25.36
Hou et al. [13] + 5 views	19.76	69.71	66.15	71.81	36.06	30.64	10.88	27.34	0.00	10.00	46.93	14.06	**53.76**	35.96	87.60	42.98	84.30	16.20	40.23
Qi et al. [24]	36.27	**87.92**	88.71	**89.62**	58.77	**47.32**	38.10	44.62	7.83	56.13	**71.69**	47.23	45.37	57.13	94.94	54.70	92.11	37.20	58.65
GSDN (Ours)	**41.58**	82.50	**92.14**	86.95	**61.05**	42.41	**40.66**	**51.14**	**10.23**	**64.18**	71.06	**54.92**	40.00	**70.54**	**99.97**	**75.50**	**93.23**	**53.07**	**62.84**

Datasets. We evaluate our method on the ScanNet dataset [5], annotated 3D reconstructions of 1500 indoor scenes with instance labels of 18 semantic classes. We follow the experiment protocol of Qi *et al.* [24] to define axis-aligned bounding boxes that encloses all points of an instance without any margin as the ground truth bounding boxes.

The second dataset is the Stanford Large-Scale 3D Indoor Spaces (S3DIS) dataset [1]. It contains 3D scans of 6 buildings with 272 rooms, each with instance and semantic labels of 7 structural elements such as floor and ceiling, and five furniture classes. We train and evaluate our method on the official furniture split and use the most-widely used *Area 5* for our test split. We follow the same procedure as above to generate ground-truth bounding boxes from instance labels.

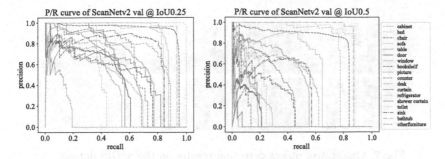

Fig. 5. Per-class precision/recall curve of ScanNetV2 validation object detection.

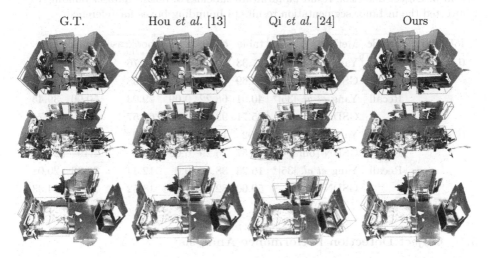

Fig. 6. Qualitative object detection results on the ScanNet dataset.

Finally, we demonstrate the scalability of GSDN on the Gibson environment [34] as it contains high-quality reconstructions of 575 multi-story buildings.

Metrics. We adopt the average precision (AP) and class-wise mean AP (mAP) to evaluate the performance of object detectors following the widely used convention of 2D object detection. We consider a detection as a positive match when a 3D intersection-over-union(IoU) between the prediction and the ground-truth bounding box is above a certain threshold.

Training Hyper-Parameters. We train our models using SGD optimizer with exponential decay of learning rate from 0.1 to 1e-3 for 120k iterations with the batch size 16. As our model can process an entire scene fully-convolutionally, we do not make smaller crops of a scene. We use high-dimensional ResNet34 [2,12] for the encoder. For all experiments, we use voxel size of 5cm, transpose kernel size of 3, with $L = 4$ scale hierarchy, sparsity pruning confidence $\tau = 0.3$, and 3D NMS threshold 0.2.

G.T.   Ours   G.T.   Ours

Fig. 7. Qualitative object detection results on the S3DIS dataset.

Table 3. Object detection result on furniture subclass of S3DIS dataset building 5. *: Converted the instance segmentation results to bounding boxes for reference

IoU Thres	Metric	Method	table	chair	sofa	bookcase	board	avg
0.25	AP	Yang et al. [35]*	27.33	53.41	9.09	14.76	29.17	26.75
		GSDN (ours)	73.69	98.11	20.78	33.38	12.91	47.77
	Recall	Yang et al. [35]*	40.91	68.22	9.09	29.03	50.00	39.45
		GSDN (ours)	85.71	98.84	36.36	61.57	26.19	61.74
0.5	AP	Yang et al. [35]*	4.02	17.36	0.0	2.60	13.57	7.51
		GSDN (ours)	36.57	75.29	6.06	6.46	1.19	25.11
	Recall	Yang et al. [35]*	16.23	38.37	0.0	12.44	33.33	20.08
		GSDN (ours)	50.00	82.56	18.18	18.52	2.38	34.33

5.1 Object Detection Performance Analysis

We compare the object detection performance of our proposed method with the previous state-of-the-art methods on Table 1 and Table 2. Our method, despite being a single-shot detector, outperforms all two-stage baselines with 4.2% mAP@0.25 and 1.3% mAP@0.5 performance gain and outperforms the state-of-the-art on the majority of semantic classes.

We also report the S3DIS detection results on Table 3. We compare the performance of our method against Yang et al. [35], a detection-based instance segmentation method, where we use the scene-level instance segmentation result as a proxy of the object detection the network learned at $1\,\mathrm{m} \times 1\,\mathrm{m}$ blocks. Our method in contrast to Yang et al. [35] takes the whole scene as an input, thus does not require slow pre-processing and post-processing, and is not limited by the cropped receptive field.

We plot class-wise precision-recall curves of ScanNet validation set on Fig. 5. We found that some of the PR curves drop sharply, which indicates that the simple aspect-ratio anchors have a low recall.

Finally, we visualize qualitative results of our method on Fig. 6 and Fig. 7. In general, we found that our method suffers from detecting thin structures such as bookcase and board, which may be resolved by adding more extreme-

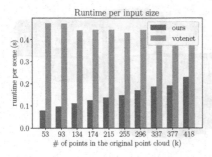

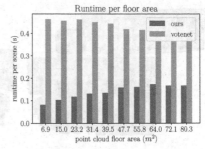

Fig. 8. Runtime comparison on ScanNet v2 validation set: Qi *et al.* [24] samples a constant number of points from a scene and their post-processing is inversely proportional to the density, whereas our method scales linearly to the number of points, and sublinearly to the floor area while being significantly faster.

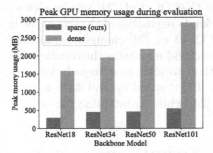

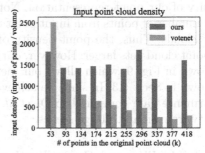

Fig. 9. Left: Memory usage comparison on ScanNet dataset evaluation: Our proposed sparse encoder and decoder maintains low memory usage compared to the dense counterparts. Right: Point cloud density on ScanNet dataset. Our model maintains constant input point cloud density compared to Qi *et al.* [24], which samples constant number of points regardless of the size of the input.

shaped anchors. Please refer to the supplementary materials for the class-wise breakdown of mAP@0.5 on the ScanNet dataset and class-wise precision-recall curves for the S3DIS dataset.

5.2 Speed and Memory Analysis

We analyze the memory footprint and runtime in Fig. 9 and Fig. 8. For the memory analysis, we compare our method with the dense object detector [13] and measured the peak memory usage on ScanNetV2 validation set. As expected, our proposed network maintains extremely low memory consumption regardless of the depth of the network while that of the dense counterparts grows noticeably.

For runtime analysis, we compare the network feed forward and post-processing time of our method with Qi *et al.* [24] in Fig. 8. On average, our method takes 0.12 s while Qi *et al.* [24] takes 0.45 s to process a scene of ScanNetV2 validation set. Moreover, the runtime of our method grows linearly to the

Fig. 10. Detection on a Gibson environment scene *Uvalda* [34]: GSDN can process a 17-room building with 1.4M points in a single *fully-convolutional* feed-forward.

number of points and sublinearly to the floor area of the point cloud, due to the sparsity of our point representation. Note that Qi *et al.* [24] subsamples a constant number of points from input point clouds regardless of the size of the input point clouds. Thus, the point density of Qi *et al.* [24] changes significantly as the point cloud gets larger. However, our method maintains the constant density as shown in Fig. 9, which allows our method to scale to extremely large scenes as shown in Sect. 5.3. In sum, we achieve 3.78× speed up and 4.2% mAP@0.25 performance gain compared to Qi *et al.* [24] while maintaining the same point density from small to large scenes.

5.3 Scalability and Generalization of GSDN on Extremely Large Inputs

We qualitatively demonstrate the scalability and generalization ability of our method on large scenes from the S3DIS dataset [1] and the Gibson environment [34]. First, we process the entire building 5 of S3DIS which consists of 78M points, $13984\,m^3$ volume, and 53 rooms. GSDN takes 20 s for a single feedforward of the entire scene including data pre-processing and post-processing. The model uses 5G GPU memory to detect 573 instances of 3D objects, which we visualized on Fig. 2.

Similarly, we train our network on ScanNet dataset [5] which only contain single-floor 3D scans. However, we tested the network on multi-story buildings. On Fig. 10, we visualize our detection results on the scene named *Uvalda* from Gibson, which is a 3-story building with $173\,m^2$ floor area. Note that our fully-convolutional network, which was only trained on single-story 3D scans, generalizes to multi-story buildings without any ad-hoc pre-processing or post-processing. GSDN takes 2.2 s to process the building from the raw point cloud and takes up 1.8G GPU memory to detect 129 instances of 3D objects.

6   Conclusion

In this work, we present the Generative Sparse Detection Network (GSDN) for single-shot fully-convolutional 3D object detection. GSDN maintains sparsity throughout the network by generating object centers using the proposed generative sparse tensor decoder. GSDN can efficiently process large-scale point clouds without cropping the scene into smaller windows to take advantage of the full receptive field. Thus, GSDN outperforms the previous state-of-the-art method by 4.2 mAP@0.25 while being 3.78× faster. In the follow-up work, we will examine and adopt various image detection techniques to boost the accuracy of GSDN.

References

1. Armeni, I., et al.: 3D semantic parsing of large-scale indoor spaces. In: Proceedings of the IEEE International Conference on Computer Vision and Pattern Recognition (2016)
2. Choy, C., Gwak, J., Savarese, S.: 4D spatio-temporal ConvNets: Minkowski convolutional neural networks. In: Proceedings of the IEEE Conference on Computer Vision and Pattern Recognition, pp. 3075–3084 (2019)
3. Choy, C., Park, J., Koltun, V.: Fully convolutional geometric features. In: ICCV (2019)
4. Choy, C.B., Xu, D., Gwak, J.Y., Chen, K., Savarese, S.: 3D-R2N2: a unified approach for single and multi-view 3D object reconstruction. In: Leibe, B., Matas, J., Sebe, N., Welling, M. (eds.) ECCV 2016. LNCS, vol. 9912, pp. 628–644. Springer, Cham (2016). https://doi.org/10.1007/978-3-319-46484-8_38
5. Dai, A., Chang, A.X., Savva, M., Halber, M., Funkhouser, T., Nießner, M.: ScanNet: richly-annotated 3D reconstructions of indoor scenes. In: Proceedings of the IEEE Conference on Computer Vision and Pattern Recognition, pp. 5828–5839 (2017)
6. Dai, A., Diller, C., Nießner, M.: SG-NN: sparse generative neural networks for self-supervised scene completion of RGB-D scans. arXiv preprint arXiv:1912.00036 (2019)
7. Dai, A., Ritchie, D., Bokeloh, M., Reed, S., Sturm, J., Nießner, M.: ScanComplete: large-scale scene completion and semantic segmentation for 3D scans. In: Proceedings of the Computer Vision and Pattern Recognition (CVPR). IEEE (2018)
8. Graham, B., Engelcke, M., van der Maaten, L.: 3D semantic segmentation with submanifold sparse convolutional networks. In: CVPR (2018)
9. Graham, B., van der Maaten, L.: Submanifold sparse convolutional networks. arXiv preprint arXiv:1706.01307 (2017)
10. Han, S., Mao, H., Dally, W.J.: Deep compression: compressing deep neural networks with pruning, trained quantization and Huffman coding. arXiv preprint arXiv:1510.00149 (2015)
11. He, K., Gkioxari, G., Dollár, P., Girshick, R.: Mask R-CNN. In: Proceedings of the IEEE International Conference on Computer Vision, pp. 2961–2969 (2017)
12. He, K., Zhang, X., Ren, S., Sun, J.: Deep residual learning for image recognition. In: Proceedings of the IEEE Conference on Computer Vision and Pattern Recognition, pp. 770–778 (2016)
13. Hou, J., Dai, A., Nießner, M.: 3D-SIS: 3D semantic instance segmentation of RGB-D scans. In: Proceedings of the IEEE Conference on Computer Vision and Pattern Recognition, pp. 4421–4430 (2019)
14. Lahoud, J., Ghanem, B., Pollefeys, M., Oswald, M.R.: 3D instance segmentation via multi-task metric learning. In: Proceedings of the IEEE International Conference on Computer Vision, pp. 9256–9266 (2019)

15. Li, B., Zhang, T., Xia, T.: Vehicle detection from 3D lidar using fully convolutional network. arXiv preprint arXiv:1608.07916 (2016)
16. Lin, T.Y., Dollár, P., Girshick, R., He, K., Hariharan, B., Belongie, S.: Feature pyramid networks for object detection. In: Proceedings of the IEEE Conference on Computer Vision and Pattern Recognition, pp. 2117–2125 (2017)
17. Liu, C., Furukawa, Y.: MASC: multi-scale affinity with sparse convolution for 3D instance segmentation. arXiv preprint arXiv:1902.04478 (2019)
18. Liu, W., et al.: SSD: single shot MultiBox detector. In: Leibe, B., Matas, J., Sebe, N., Welling, M. (eds.) ECCV 2016. LNCS, vol. 9905, pp. 21–37. Springer, Cham (2016). https://doi.org/10.1007/978-3-319-46448-0_2
19. Maturana, D., Scherer, S.: VoxNet: a 3D convolutional neural network for real-time object recognition. In: IROS (2015)
20. Mescheder, L., Oechsle, M., Niemeyer, M., Nowozin, S., Geiger, A.: Occupancy networks: learning 3D reconstruction in function space. In: Proceedings IEEE Conference on Computer Vision and Pattern Recognition (CVPR) (2019)
21. Narang, S., Elsen, E., Diamos, G., Sengupta, S.: Exploring sparsity in recurrent neural networks. arXiv preprint arXiv:1704.05119 (2017)
22. Parashar, A., et al.: SCNN: an accelerator for compressed-sparse convolutional neural networks. ACM SIGARCH Comput. Archit. News **45**(2), 27–40 (2017)
23. Park, J.J., Florence, P., Straub, J., Newcombe, R., Lovegrove, S.: DeepSDF: learning continuous signed distance functions for shape representation. In: The IEEE Conference on Computer Vision and Pattern Recognition (CVPR), June 2019
24. Qi, C.R., Litany, O., He, K., Guibas, L.J.: Deep Hough voting for 3D object detection in point clouds. In: Proceedings of the IEEE International Conference on Computer Vision, pp. 9277–9286 (2019)
25. Qi, C.R., Liu, W., Wu, C., Su, H., Guibas, L.J.: Frustum PointNets for 3D object detection from RGB-D data. In: Proceedings of the IEEE Conference on Computer Vision and Pattern Recognition, pp. 918–927 (2018)
26. Redmon, J., Farhadi, A.: YOLO9000: better, faster, stronger. In: Proceedings of the IEEE Conference on Computer Vision and Pattern Recognition, pp. 7263–7271 (2017)
27. Ren, S., He, K., Girshick, R., Sun, J.: Faster R-CNN: towards real-time object detection with region proposal networks. In: Advances in Neural Information Processing Systems, pp. 91–99 (2015)
28. Song, S., Xiao, J.: Deep sliding shapes for amodal 3D object detection in RGB-D images. In: CVPR (2016)
29. Tange, O., et al.: GNU parallel-the command-line power tool. USENIX Mag. **36**(1), 42–47 (2011)
30. Tatarchenko, M., Dosovitskiy, A., Brox, T.: Octree generating networks: efficient convolutional architectures for high-resolution 3D outputs. In: IEEE International Conference on Computer Vision (ICCV) (2017). http://lmb.informatik.uni-freiburg.de/Publications/2017/TDB17b
31. Tchapmi, L.P., Kosaraju, V., Rezatofighi, S.H., Reid, I., Savarese, S.: TopNet: structural point cloud decoder. In: The IEEE Conference on Computer Vision and Pattern Recognition (CVPR) (2019)
32. Wang, D.Z., Posner, I.: Voting for voting in online point cloud object detection. In: Robotics: Science and Systems, vol. 1, pp. 10–15607 (2015)
33. Wang, W., Yu, R., Huang, Q., Neumann, U.: SGPN: similarity group proposal network for 3D point cloud instance segmentation. In: Proceedings of the IEEE Conference on Computer Vision and Pattern Recognition, pp. 2569–2578 (2018)

34. Xia, F., Zamir, A.R., He, Z.Y., Sax, A., Malik, J., Savarese, S.: Gibson Env: real-world perception for embodied agents. In: 2018 IEEE Conference on Computer Vision and Pattern Recognition (CVPR). IEEE (2018)

35. Yang, B., et al.: Learning object bounding boxes for 3D instance segmentation on point clouds. In: Advances in Neural Information Processing Systems, pp. 6737–6746 (2019)

36. Yi, L., Zhao, W., Wang, H., Sung, M., Guibas, L.J.: GSPN: generative shape proposal network for 3D instance segmentation in point cloud. In: Proceedings of the IEEE Conference on Computer Vision and Pattern Recognition, pp. 3947–3956 (2019)

37. Yuan, W., Khot, T., Held, D., Mertz, C., Hebert, M.: PCN: point completion network. In: 2018 International Conference on 3D Vision (3DV) (2018)

38. Zhou, Y., Tuzel, O.: VoxelNet: end-to-end learning for point cloud based 3D object detection. In: The IEEE Conference on Computer Vision and Pattern Recognition (CVPR), June 2018

Grounded Situation Recognition

Sarah Pratt[1]([⊠]), Mark Yatskar[1]([⊠]), Luca Weihs[1]([⊠]), Ali Farhadi[2]([⊠]),
and Aniruddha Kembhavi[1,2]([⊠])

[1] Allen Institute for AI, Seattle, USA
{sarahp,marky,lucaw,anik}@allenai.org
[2] University of Washington, Seattle, USA
ali@cs.uw.edu

Abstract. We introduce Grounded Situation Recognition (GSR), a task that requires producing structured semantic summaries of images describing: the primary activity, entities engaged in the activity with their roles (e.g. agent, tool), and bounding-box groundings of entities. GSR presents important technical challenges: identifying semantic saliency, categorizing and localizing a large and diverse set of entities, overcoming semantic sparsity, and disambiguating roles. Moreover, unlike in captioning, GSR is straightforward to evaluate. To study this new task we create the Situations With Groundings (SWiG) dataset which adds 278,336 bounding-box groundings to the 11,538 entity classes in the *imSitu* dataset. We propose a Joint Situation Localizer and find that jointly predicting situations and groundings with end-to-end training handily outperforms independent training on the entire grounding metric suite with relative gains between 8% and 32%. Finally, we show initial findings on three exciting future directions enabled by our models: conditional querying, visual chaining, and grounded semantic aware image retrieval. Code and data available at https://prior.allenai.org/projects/gsr.

Keywords: Situation recognition · Scene understanding · Grounding

1 Introduction

Situation Recognition [60] is the task of recognizing the activity happening in an image, the actors and objects involved in this activity, and the roles they play. The structured image descriptions produced by situation recognition are drawn from FrameNet [5], a formal verb lexicon that pairs every verb with a frame of semantic roles, as shown in Fig. 1. These semantic roles describe how objects in the image participate in the activity described by the verb.

Electronic supplementary material The online version of this chapter (https://doi.org/10.1007/978-3-030-58548-8_19) contains supplementary material, which is available to authorized users.

© Springer Nature Switzerland AG 2020
A. Vedaldi et al. (Eds.): ECCV 2020, LNCS 12349, pp. 314–332, 2020.
https://doi.org/10.1007/978-3-030-58548-8_19

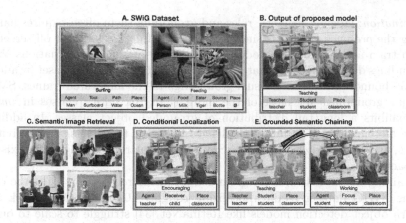

Fig. 1. A Two examples from our dataset: semantic frames describe primary activities and relevant entities. Groundings are bounding-boxes colored to match roles. **B** Output of our model (dev set image). **C** Top-4 nearest neighbors to B using model predictions. Beyond visual similarity, these images are clearly semantically similar. **D** Output of the conditional model: given a bounding-box (yellow-dashed), predicts a relevant frame. **E** Example of grounded semantic chaining: given query boxes we are able to chain situations together. E.g. the teacher teaches students so they may work on a project (Color figure online)

As such, situation recognition generalizes several computer vision tasks such as image classification, activity recognition, and human object interaction. It is related to the task of image captioning, which also typically describes the salient objects and activities in an image using natural language. However, in contrast to captioning, it has the advantages of always producing a structured and complete (with regards to semantic roles) output and it does not suffer from the well known challenges of evaluating natural language captions.

While situation recognition addresses *what* is happening in an image, *who* is playing a part in this and *what* their roles are, it does not address a critical aspect of visual understanding: ***where*** the involved entities lie in the image. We address this shortcoming and present Grounded Situation Recognition (GSR), a task that builds upon situation recognition and requires one to not just identify the situation observed in the image but also visually ground the identified roles within the corresponding image. GSR presents the following technical challenges. *Semantic saliency*: in contrast to recognizing all entities in the image, it requires identifying the key objects and actors in the context of the primary activity being presented. *Semantic sparsity*: grounded situation recognition suffers from the problem of semantic sparsity [59], with many combinations of roles and groundings rarely seen in training. This challenge requires models to learn from limited data. *Ambiguity*: grounding roles into images often requires disambiguating between multiple observed entities of the same category. *Scale*: the scales of the grounded entities vary vastly with some entities also being absent in the image (in which case models are responsible for detecting this absence).

Halucination: labeling semantic roles and grounding them often requires haluci-nating the presence of objects since they may be fully occluded or off screen.

To train and benchmark models on GSR, we present the Situations With Groundings dataset (SWiG) that builds upon the large *imSitu* dataset by adding 278,336 bounding-box-based visual groundings to the annotated frames. SWiG contains groundings for most of the more than 10k entity classes in *imSitu* and exhibits a long tail distribution of grounded object classes. In addition to the aforementioned technical challenges of GSR, the diversity of activities, images, and grounded classes, makes SWiG particularly challenging for existing approaches.

Training neural networks for grounded situation recognition using the chal-lenging SWiG dataset requires localizing roughly 10k categories; a task that modern object detection models like RetinaNet [34] struggle to scale to out of the box. We first propose modifications to RetinaNet that enables us to train large-class-cardinality object detectors. Using these modifications, we then cre-ate a strong baseline, the Independent Situation Localizer (ISL), that indepen-dently predicts the situation and groundings and uses late fusion to produce the desired outputs. Our proposed model, the Joint Situation Localizer (JSL), jointly predicts the situation and grounding conditioned on the context of the image. During training, JSL backpropagates gradients through the entire net-work. JSL demonstrates the effectiveness of joint structured semantic prediction and grounding by improving both semantic role prediction and grounding and obtaining huge relative gains of between 8% and 32% points over ISL on the entire suite of grounding metrics.

Grounded situation recognition opens up several exciting avenues for future research. First, it enables us to build a Conditional Situation Localizer (CSL); a model that outputs a grounded situation conditioned on an input image and a specified region of interest within the image. CSL allows us to query *what* is happening in an image in regards to a specified *query object or region*. This is particularly revealing when entities are involved in multiple situations within an image or when an image consists of a large number of visible entities. Second, we show that such pointed conditioning models enable us to tackle higher order semantic relations amongst activities in images via visual chaining. Third, we show that grounded situation recognition models can serve as effective image retrieval mechanisms that can condition on linguistic as well as visual inputs and are able to retrieve images with the desired semantics.

In summary our contributions include: (i) proposing Grounded Situation Recognition, a task to identify the observed salient situation and ground the corresponding roles within the image, (ii) presenting the SWiG dataset towards building and benchmarking models for this task, (iii) showing that joint struc-tured semantic prediction and grounding models improve both semantic role prediction and grounding by large margins, but also noting that there is still con-siderable ground for future improvements; (iv) revealing several exciting avenues for future research that exploit grounded situation recognition data to build mod-els for semantic querying, visual chaining, and image retrieval. Our new dataset, code, and trained model weights will be publicly released.

2   Related Work

Grounded Situation Recognition is related to several areas of research at the intersection of vision and language and we now present a review of these below.

Describing Activities in Images. While recognizing actions in videos has been a major focus area [21,25,47,48,50], describing activities from images has also received a lot of attention (see Gella *et al.* [15] for a more detailed overview).

Early works [10,13,19,23,29,57,58] framed this as a classification problem amongst a few verbs (running/walking/etc.) or few verb-object tuples (riding bike/riding horse/etc.). More recent work has focused on human object interactions [8,30,45,61] with more classes; but the classes are either arbitrarily chosen or obtained by starting with a set of images and then labeling them with actions. Also, the relationships include Subject-Verb-Object triples or subsets thereof. In contrast, the *imSitu* dataset for situation recognition uses linguistic resources to define a large and more comprehensive space of possible situations, ensuring a fairly balanced datasets despite the large number of verbs (roughly 500) and modeling a detailed set of semantic roles per verb obtained from FrameNet [5].

Image captioning is another popular setup to describe the salient actions taking place in an image with several datasets [1,9,46] and many recent neural models that perform well [3,24,53]. One serious drawback to image captioning is the well known challenge of evaluation which has led to a number of proposed metrics [2,6,32,38,52]; but these problems continue to persist. Situation recognition does not face this issue and has clearly established metrics for evaluation owing to its structured frame output.

Other relevant works include visual sense disambiguation [16], visual semantic role labelling [20], and scene graph generation [28] with the latter two described in more detail below.

Visual Grounding. In contrast to associating full images with actions or captions, past works have also associated regions to parts of captions. This includes *visual grounding* i.e. associating words in a caption to regions in an image and *referring expression generation* i.e. producing a caption to unambiguously describe a region of interest; and there are several interesting datasets here.

Flickr30k-Entities [40] is a large dataset for grounded captioning. v-COCO [20] is more focused on semantic role labeling for human interactions with human groundings, action labels and relevant object groundings. Compared to SWiG, the verbs (26 vs 504) and semantic roles per verb (up to 2 vs up to 6) are fewer. HICO-Det [7] has 117 actions, but they only involve 80 objects, compared to nearly 10,000 objects in SWiG. In addition to these human centric datasets, SWiG also contains actions by animals and objects.

Large referring expression datasets include RefClef [26], RefCOCO [37] and RefCOCO+ collected using a two person game, RefCOCOg collected by standard crowdsourcing and GuessWhat?! [54] that combines dialog and visual grounding.

An all encompassing vision and language dataset is Visual Genome (VG) [28] containing scene graphs: dense structured representations for images with objects, attributes, relations, groundings and QA. VG differs from SWiG in a few ways. Scene graphs are dense while situations capture salient activities. Also, relations in scene graphs are binary and tend to favor part and positional relations (the top 10 relations in VG are of this nature and cover 66% of the total) while SWiG contains more roles per verb, has 504 verbs drawn from language and has a good coverage of data per verb. Finally, dense annotations are notoriously hard to obtain; and it is well known that VG suffers from missing relations, rendering evaluation tricky.

Situation Recognition Models. Yatskar *et al.* [60] present a conditional random field model fed by CNN features and extend it with semantic sparsity augmentation [59]. Mallya *et al.* [36] improve the accuracy by using a specialized verb predictor and an RNN for noun prediction. Li *et al.* [31] use Graph Neural Nets to capture joint dependencies between roles. Most recently, Suhail *et al.* [51] achieved state of the art accuracy using attention graph neural nets. Our proposed grounded models build upon the RNN based approach of [36] owing to its simplicity and high accuracy; but our methods to combine situation recognition models with detectors can be applied to any of the aforementioned approaches.

Large-Class-Cardinality Object Detection. While most popular object detectors are built and evaluated on datasets [14,35] with few classes, some past works have addressed the problem of building detectors for thousands of classes. This includes YOLO-9000 [43], DLM-FA [56], R-FCN-3000 [49], and CS-R-FCN [18]. Our modifications to RetinaNet borrow some ideas from these works.

3   GSR and SWiG

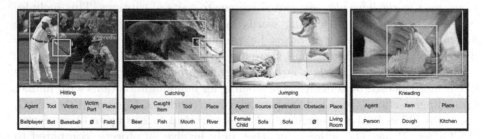

Fig. 2. Grounded situations from the SWiG dataset. This figure showcases the variability of images, situations and groundings across the dataset. Some challenges seen in this figure are absent roles (first image), animals as agents (second image) contrasting datasets that only focus on human interactions, ambiguity resolution (two female children in the third image), matching groundings for two roles (sofa in the third image) and partial occlusion (person only partially visible in the fourth image)

Task. Grounded Situation Recognition (GSR) builds upon situation recognition and requires one to identify the salient activity, the entities involved, the semantic roles they play and the locations of each entity in the image. The frame representation is drawn from the linguistic resource FrameNet and the visual groundings are akin to bounding boxes produced by object detectors. More formally, given an input image, the goal is to produce three outputs. (a) **Verb**: classifying the salient activity into one of 504 visually groundable verbs (one in which it is possible to view the action, for example, 'talking' is visible, but 'thinking' is not). (b) **Frame**: consists of 1 to 6 semantic role values i.e. nouns associated with the verb (each verb has its own pre-defined set of roles). For e.g., Fig. 2 shows that 'kneading' consists of 3 roles: 'Agent', 'Item', and 'Place'. Every image labeled with the verb 'kneading' will have the same roles but may have different nouns filled in at each role based on the contents of the image. A role value can also be $\varnothing$ indicating that a role does not exist in an image (Fig. 2c). (c) **Groundings**: each grounding is described with coordinates $[x_1, y_1, x_2, y_2]$ if the noun in grounded in the image. It is possible for a noun to be labeled in the frame but not grounded, for example in cases of occlusion.

Data. SWiG builds on top of *imSitu* [60]. SWiG retains the original images, frame annotations and splits from *imSitu* with a total of 126,102 images spanning 504 verbs. For each image, there are three frames by three different annotators, with a length between 1 and 6 roles and an average frame length of 3.55.

Bounding-box annotations were obtained using Amazon's Mechanical Turk framework with each role annotated by three workers and the resulting boxes combined by averaging their extents. In total, SWiG contains 451,916 noun slots across all images. Of these 435,566 are non-$\varnothing$. Of these 278,336 (63.9%) have bounding boxes. The missing bounding boxes correspond to objects that are not visible or to 'Place' which is never annotated with a bounding box as the location of an action is always the entire image.

SWiG exhibits a huge variability in the number of groundings per noun (see Fig. 3a). For instance 'man' appears over 100k times while others occur only once. Unlike other detection datasets such as MS-COCO [9], SWiG contains a long-tail distribution of grounded objects similar to the real world.

Figure 3b shows the frequency with which different roles are grounded in the image. Note that, like nouns, roles also have an uneven distribution. Almost all situations are centered around an 'Agent' but very few situations use a 'Firearm'. This plot shows how often each role is grounded invariant to its absolute frequency. Some roles are much more frequently salient, demonstrating the linguistic frame's ability to capture both concrete and abstract concepts related to situations. Objects filling roles like 'Firearm'/'Teacher' are visible nearly every time they are relevant to a situation. However, the noun taking on the role of the 'Event' cannot usually be described by a particular object in the image. Only one role ('Place') is never grounded in the image.

Figure 3c shows the distribution of grounding scale and aspect ratio for a sample of nouns. Many nouns exhibit high variability across the dataset (1st

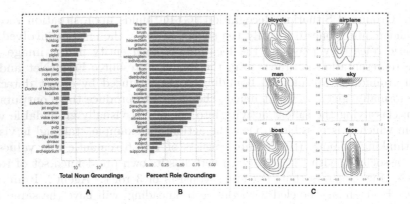

Fig. 3. Dataset visualizations. (A) Number of groundings per noun. Note the log scale and the fact that this only shows a small sample. (B) Frequency with which different roles are grounded in the image. (C) Distribution of grounding scale (y-axis) and aspect ratio (x-axis) conditioned on some nouns

column), but some nouns have strong priors that may be used by models (2nd column).

Figure 4 shows the variability of appearance of groundings across verbs. Figure 4a indicates the scale and aspect ratio of every occurrence of the noun 'Rope' for verbs where this noun occurs at least 60 times. Each point is an instance and the color represents the verb label for the image where that instance appears. A large scale indicates that at least one side of the bounding box is large in relation to the image. A large aspect ratio indicates that the height of the bounding box is much greater than the width. This plot shows that the verb associated with an image gives a strong prior towards the physical attributes of an object. In this case, knowing that a rope appears in a situation with the verb 'drag' or 'pull', indicates that it is likely to have a horizontal alignment. If the situation is 'hoisting' or 'climbing' then the rope is likely to have a vertical alignment.

Figure 4b shows the scale and aspect ratio of the role 'Agent', invariant to the noun, for a variety of verbs. The clustering of colors in the plot indicates that the verb gives a strong prior to the size and aspect ratio of the 'Agent'. However, this also demonstrates the non-triviality of the task. This is especially evident in the images depicting the agent for 'Mowing' compared to 'Harvesting'. While knowing the verb gives a strong indication as to the appearance of the 'Agent', it is not trivial to distinguish between the two verbs given just the agent. The correlation between object appearance and actions demonstrates the importance of combining situation understanding with groundings, but we must still maintain the entire context to complete the task.

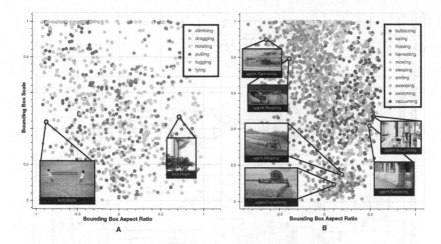

Fig. 4. Scale and aspect ratio distributions across nouns and roles. (A) Every occurrence of the noun 'Rope' for verbs - showing that verb gives a strong prior towards the physical attributes of an object. (B) The role 'Agent', invariant to the noun - shows priors but also the challenges of the task

4   Methods

Grounded situation recognition involves recognizing the salient situation and grounding the associated role values via bounding boxes; indicating that a model for this task must perform the roles of situation recognition and object detection. We present a novel method Joint Situation Localization (JSL) with a strong baseline, the Independent Situation Localization (ISL).

Situation Recognition Model. The proposed ISL and JSL models represent techniques to combine situation recognition and detection models and can be applied to all past situation recognition models. In this work, we select the RNN without fusion from [36] since: (i) it achieves a high accuracy while having a simple architecture (Table 1). (ii) We were able to *upgrade* it with a reimplementation, new backbone, label smoothing, and hyper-parameter tuning resulting in huge gains (Table 1) over the reported numbers, beating graph nets [31] and much closer to attention graph nets [51] (the current state-of-the-art on *imSitu*). (iii) Code and models for attention graph nets are not released, rendering reproducibility challenging, especially given the complexity of the method.

As in the top of Fig. 5, ResNet-50 embeddings are used for verb prediction and then an LSTM [22] sequentially predicts the noun for each role in the frame. The order of the roles is dictated by the dataset. The loss is a sum of cross entropy with label smoothing on the predicted nouns, and cross entropy on the verb.

Large-Class-Cardinality Object Detection. We use a modified version of RetinaNet [34] as our baseline detector within the proposed models. RetinaNet is

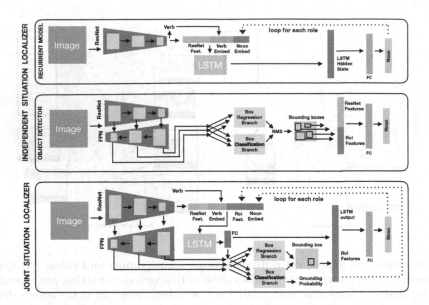

Fig. 5. Model schematics for the proposed ISL and JSL models

a single stage detector with a Feature Pyramidal Network (FPN) [33] for multi-scale features, multiple anchors to account for varied aspect ratios and two heads: a classification head that assigns each anchor to a class and a regression head that modifies the anchor to better localize any object in that location. RetinaNet does not scale well to the 10k classes in SWiG out of the box.

We make 3 modifications to RetinaNet to scale it to 10,000 classes, as seen in the middle of Fig. 5. (i) **Objectness**: instead of each anchor predicting a score for each class, each anchor now predicts an "objectness score". Non-Maximum Suppression (NMS) is performed on the boxes, the top 100 are chosen and featurized using RoI Align [44]. These local features combined with global ResNet features are classified into the ∼10,000 noun categories. The resulting memory savings are huge. In RetinaNet, the classification branch output tensor has dimensions $\sum_{i=1}^{n}(W_i \times H_i \times A \times K)$ where W_i, H_i indicate the spatial dimensional of the features for the ith output of the FPN, A indicates the number of anchor boxes, and K indicates the number of classes. This does not fit on a single TITAN RTX GPU for $K = 10,000$ for any reasonable batch size. In contrast, our modification reduces the tensor dimension to $\sum_{i=1}^{n}(W_i \times H_i \times A \times P)$ where P is the number of image regions we consider and is set to 100. With these modifications we are able to train with a batch size of 64 on 4 TITAN RTX GPUs with 24 GB of memory. (ii) **Drop fine scale**: we exclude the finest grain of features from FPN since anchors at this scale do not overlap with a significant portion of our data leading to computation savings. (iii) **Anchor selection**: anchor box aspect ratios are assigned using aspect ratio clustering on our training data, as in [43].

As in [34], we use a focal loss for classification and an L_1 loss for regression, with a binary cross entropy loss for noun prediction.

Independent Situation Localizer (ISL). The ISL independently runs the situation recognizer and detector and combines their results. The RNN model produces a prediction for each noun in the frame. The detector obtains a distribution over all possible object categories for each of the top 100 bounding boxes. Then for each noun in the frame, we assign the grounding with the highest score for that noun. This allows an object that is assigned to one class by the detector to eventually get assigned to another class as long as the score for the latter class is high enough. If all of the box scores for a noun are below a threshold or the role is 'Place', it is considered ungrounded.

Joint Situation Localizer (JSL). We propose JSL as a method to simultaneously classify a situation and locate objects in that situation. This allows for a role's noun and grounding to be conditioned on the nouns and groundings of previous roles and the verb. It also allows features to be shared potential patterns between nouns and positions (like in Fig. 4) to be exploited. We refer the reader to the supplementary material and our code for model details, but point out key differences between JSL and ISL here.

JSL (shown in the bottom of Fig. 5) uses similar backbones as ISL but with key differences: (i) rather than predicting localization for every object in the image at the same time (as is done in object detection), JSL predicts the location of the objects recurrently (as is done for predicting nouns in situation recognition). (ii) In contrast to the RNN model in ISL, the LSTM accepts as input the verb embedding, global ResNet features of the image, embedding of the noun predicted at the previous time step and local ResNet features of the bounding box predicted at the previous time step. (iii) In contrast to the detector in ISL, the localization is now conditioned on the situation and current role being predicted. In ISL, FPN features feed directly into the classification and regression branches, but in JSL the FPN features are combined with the LSTM hidden state and then fed in. (iv) The JSL also uses the classification branch to produce an explicit score indicating the likelihood that an object is grounded. (v) Only one noun needs to be localized at each time step, which means that only anchor boxes relevant to that one grounding will be marked as positive during training, given that the noun is visible and grounded in the training data.

The loss includes focal loss and L_1 loss from the detector and cross entropy loss for the grounding and verb. Additionally, we use cross entropy with label smoothing for the noun loss, and sum this over all three annotator predictions. This results in the following total loss: $\mathcal{L} = L_1(reg) + FL_{0.25,2}(class) + CE(verb) + CE(ground) + \sum_{i=1}^{3} CE_{0.2}(noun_i)$.

Similar to previous works [36,51], we found that using a separate ResNet backbone to predict the verb achieved a boost in accuracy. However, the JSL architecture with this additional ResNet backbone still maintains the same number of parameters and ResNet backbones as the ISL model.

5   Experiments

Implementation Details. ISL and JSL use two ResNet-50 backbones and maintain an equal number of parameters (~108 million). We train our models via gradient descent using the Adam Optimizer [27] with momentum parameters of $\beta = (0.9, 0.999)$. We use 4 24 GB TITAN RTX GPUs for approximately 20 h. For comprehensive training and model details, including learning rate schedules, batch sizes, and layer sizes, please see our supplementary material.

Metrics. We report five metrics, three standard ones from prior situation recognition work: (i) **verb** to measure verb prediction accuracy, (ii) **value** to measure accuracy when predicting a noun for a given role, (iii) **value-all** to measure the accuracy in correctly predicting all nouns in a frame simultaneously; and introduce two new grounding metrics: (iv) **grounded-value** to measure accuracy in predicting the correct noun and grounding for a given role. A grounding is considered correct if it has an IoU of at least 0.5 with the ground truth. (v) **grounded-value-all** to measures how frequently both the noun and the groundings are predicted correctly for the entire frame. Note that if a noun does not have a grounding, the model must also predict this correctly. All these metrics are calculated for each verb and then averaged across verbs so as to not unfairly bias this metric toward verbs with more annotations or longer semantic frames.

Since these metrics are highly dependent on verb accuracy, they have the potential to obfuscate model differences with regards to noun prediction and grounding. Hence we report them in 3 settings: **Ground-Truth-Verb**: the ground truth verb is assumed to be known. **Top-1-Verb**: *verb* reports the accuracy of the top 1 predicted verb and all noun and groundings are considered incorrect if the verb is incorrect. **Top-5-Verb**: *verb* corresponds to the top-5 accuracy of verb prediction. Noun and grounding predicitons are taken from the model conditioning on the correct verb having been predicted.

Results. The top section of Table 1 shows past *imSitu* models for the dev set while the lower section illustrates the efficacy of jointly training a model for grounding and situation recognition. The yellow rows indicate the base RNN model used in this work and the green row shows the large upgrades to this model across all metrics. ISL achieves reasonable results, especially for ground truth verbs. However, JSL improves over ISL across every metric while using an equal number of parameters. This includes substantial improvements on all grounding metrics (ranging from relative improvements of 8.6% for *Ground-Truth-Verb–ground-value* to 32.9% for *Top-1-Verb–grounded-value-all*).

The ability to improve across both the grounding and non-grounding scores demonstrate the value in combining grounding with the task of situation recognition. Not only can the context of the situation improve the models ability to locate objects, but locating these objects improves the models ability to understand them. This is further emphasized by the models ability to predict the correct noun under the *GroundTruthVerb* setting.

Table 1. Evaluation of models on the SWiG dev set. * indicates our implementation. Yellow rows indicate the base RNN model architecture with numbers from the paper. Green shows the upgraded version of this RNN model used in our proposed models

Method	top-1 predicted verb					top-5 predicted verbs					ground truth verbs			
	verb	value	value-all	grnd value	grnd value-all	verb	value	value-all	grnd value	grnd value-all	value	value-all	grnd value	grnd value-all
Prior Models for Situation Recognition														
CRF [60]	32.25	24.56	14.28	-	-	58.64	42.68	22.75	-	-	65.90	29.50	-	-
CRF+Aug [59]	34.20	25.39	15.61	-	-	62.21	46.72	25.66	-	-	70.80	34.82	-	-
RNN w/o Fusion[36]	35.35	26.80	15.77	-	-	61.42	44.84	24.31	-	-	68.44	32.98	-	-
RNN w/ Fusion[36]	36.11	27.74	16.60	-	-	63.11	47.09	26.48	-	-	70.48	35.56	-	-
GraphNet [31]	36.93	27.52	19.15	-	-	61.80	45.23	29.98	-	-	68.89	41.07	-	-
Kernel GraphNet[51]	43.21	35.18	19.46	-	-	68.55	56.32	30.56	-	-	73.14	41.48	-	-
RNN based models														
RNN w/o Fusion [36]	35.35	26.80	15.77	-	-	61.42	44.84	24.31	-	-	68.44	32.98	-	-
Updated RNN*	38.83	30.47	18.23	-	-	65.74	50.29	28.59	-	-	72.77	37.49	-	-
ISL*	38.83	30.47	18.23	22.47	7.64	65.74	50.29	28.59	36.90	11.66	72.77	37.49	52.92	15.00
JSL*	39.60	31.18	18.85	25.03	10.16	67.71	52.06	29.73	41.25	15.07	73.53	38.32	57.50	19.29

Table 2. Evaluation of models on the SWiG test set. * indicates our implementation

Method	Top-1 predicted verb					Top-5 predicted verbs					Ground truth verbs			
	verb	value	value all	grnd value	grnd value-all	verb	value	value-all	grnd value	grnd value-all	value	value-all	grnd value	grnd value-all
RNN based models														
Updated RNN*	39.36	30.09	18.62	–	–	65.51	50.16	28.47	–	–	72.42	37.10	–	–
ISL*	39.36	30.09	18.62	22.73	7.72	65.51	50.16	28.47	36.60	11.56	72.42	37.10	52.19	14.58
JSL*	39.94	31.44	18.87	24.86	9.66	67.60	51.88	29.39	40.60	14.72	73.21	37.82	56.57	18.45

Importantly, in spite of using the simpler RNN based backbone for situation recognition, JSL achieves state of the art numbers on the GroundTruthVerb-Value metric, beating the more complex Kernel GraphNet model demonstrating the benefits of joint prediction. This indicates that further improvements may be obtained by incorporating better backbones. Additionally, it is interesting to note that the model achieves this high *value* even though it does not achieve state of the art in *value-all*. This indicates another potential benefit of the model. While more total frames contain a mistake, JSL is still able to recover some partial information and recognize more total objects. One explanation is that grounding may contribute to the ability to have a *partial* understanding of more complicated images where other models fail completely. Finally, test set metrics are shown in Table 2 and qualitative results in Fig. 9.

6 Discussion

Grounded situation recognition and SWiG open up several exciting directions for future research. We present initial findings for some of these explorations.

Fig. 6. Qualitative results for semantic image retrieval. For the query figure of a surfer in action, ResNet and Object Detection based methods struggle to match the fine semantics. Grounded situation based retrieval leads to the correct semantics with matching viewpoints

Grounded Semantic Aware Image Retrieval. Over the past few years, large improvements have been obtained in content based image retrieval (CBIR) by employing visual representations from CNNs [4,17,41,42,55]. CNN features work well particularly when evaluated on datasets requiring instance retrieval [39] or category retrieval [11], but unsurprisingly do not do well when the intent of the query is finding a matching situation. We perform a small study for image retrieval using the dev set in SWiG. We partition this set into a query set and a retrieval set and perform retrieval using four representations: (i) ResNet-50 embeddings, (ii) bag of objects obtained from our modified RetinaNet object detector, (iii) situations obtained from our baseline RNN model, and (iv) grounded situations obtained from JSL. Details regarding the setup and distance functions for each are presented in the supplementary material.

Figure 6 shows a qualitative result. Resnet-50 retrieves images that look similar (all have water) but have the wrong situations. The same goes for object detection. Situation based retrieval gets the semantics correct (most of the retrieved images contain surfing). Grounded situations provide the additional detail of not just similar semantics but also similar arrangement of objects, since the locations of the entities are also taken into account. Furthermore, the proposed method also produces explainable outputs via the grounded situations in the retrieved images; arguably more useful than CBIR explanations via heatmaps [12]. This approach can also be extended to structured queries (obtained from text) and a mix of text and image based queries.

Conditional Grounded Situation Recognition. JSL accepts the entire image as an input and produces groundings for the salient situation. But images may contain entities in multiple situations (a person *sitting* and *discussing* and *drinking coffee*) or multiple entities. Conditioning on a localized object or region can enable us to *query* an image regarding the entity of interest. Note that a

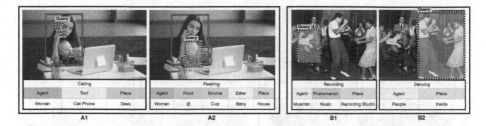

Fig. 7. Qualitative results using the Conditional Situation Localizer. A1 & A2: The woman is taking part in multiple situations with different entities in the scene. These situations are invoked via different queries. B1 & B2: Querying the person with a guitar vs querying the group of people also reveals their corresponding situations

query entity may be an actor (*what is this person doing?*) or an object (*What situation is this object involved in?*) or a location (*What is happening at this specific location?*) in the scene. A small modification to JSL results in a Conditional Situation Localizer (CSL) model (details in supplement), which enables this exploration. Figure 7a shows that a query box around the cellphone invokes *calling* while the baby invokes *feeding*. Figure 7b shows that a query box may have 1 or more entities within it.

Fig. 8. Grounded semantic chaining. When a person looks at this image, they may infer several things. A father is teaching his son to use the grill. They are barbecuing some meat with the intent of feeding friends and family who are sitting at a nearby table. Using the conditional localizer followed by spatial and semantic chaining produces situations and relationships-between-situations. These are shown via colored boxes, text and arrows. Conditional inputs are shown with dashed yellow boxes. Notice the similarity between the higher level semantics output by this chaining model and the inferences about the image that you may draw

Grounded Semantic Chaining. Pointed conditional models such as CSL, when invoked on a set of bounding boxes (obtained via object detection), enable us to chain together situations across multiple parts of image. While a situation addresses local semantics, chaining of situations enables us to address higher

328   S. Pratt et al.

Fig. 9. Qualitative results for the proposed JSL model. First two rows show examples with correctly classified situations and detected groundings; and demonstrates the diversity of situations in the data. Third row shows classification errors. Note that some of them are perfectly plausible answers. Fourth row shows incorrect groundings; some of which are only partially wrong but get counted as errors nonetheless

order semantics across an entire image. Visual chaining can be obtained using spatial and semantic proximity between groundings in different situations. While scene graphs are a formalism towards this, they only enable binary relations between entities; and this data in Visual Genome [28] has a large focus on part and spatial relations. Since SWiG contains a diverse set of verbs with comprehensive semantic roles, visual chains obtained by CSL tend to be very revealing and are an interesting direction to pursue in future work. Figure 8 shows an interesting example of querying multiple persons and then chaining the results using groundings, revealing: a man is helping his son while barbecuing meat and the people are dining on the hamburger that is being grilled by the man.

7   Conclusion

We introduce Grounded Situation Recognition (GSR) and the SWiG dataset. Our experiments reveal that simultaneously predicting the semantic frame and groundings results in huge gains over independent prediction. We also show exciting directions for future research.

References

1. Agrawal, H., et al.: nocaps: novel object captioning at scale. In: International Conference on Computer Vision abs/1812.08658 (2019)
2. Anderson, P., Fernando, B., Johnson, M., Gould, S.: SPICE: semantic propositional image caption evaluation. In: Leibe, B., Matas, J., Sebe, N., Welling, M. (eds.) ECCV 2016. LNCS, vol. 9909, pp. 382–398. Springer, Cham (2016). https://doi.org/10.1007/978-3-319-46454-1_24
3. Anderson, P., et al.: Bottom-up and top-down attention for image captioning and visual question answering. In: Proceedings of the IEEE Conference on Computer Vision and Pattern Recognition, pp. 6077–6086 (2018)
4. Babenko, A., Slesarev, A., Chigorin, A., Lempitsky, V.S.: Neural codes for image retrieval. arXiv abs/1404.1777 (2014)
5. Baker, C.F., Fillmore, C.J., Lowe, J.B.: The Berkeley FrameNet project. In: Proceedings of the 17th International Conference on Computational Linguistics - Volume 1, COLING 1998, pp. 86–90. Association for Computational Linguistics, Stroudsburg (1998). https://doi.org/10.3115/980451.980860
6. Banerjee, S., Lavie, A.: METEOR: an automatic metric for MT evaluation with improved correlation with human judgments. In: IEEvaluation@ACL (2005)
7. Chao, Y.W., Liu, Y., Liu, X., Zeng, H., Deng, J.: Learning to detect human-object interactions. In: 2018 IEEE Winter Conference on Applications of Computer Vision (WACV), pp. 381–389 (2017)
8. Chao, Y.W., Wang, Z., He, Y., Wang, J., Deng, J.: HICO: a benchmark for recognizing human-object interactions in images. In: Proceedings of the IEEE International Conference on Computer Vision, pp. 1017–1025 (2015)
9. Chen, X., et al.: Microsoft COCO captions: data collection and evaluation server. arXiv preprint arXiv:1504.00325 (2015)
10. Delaitre, V., Laptev, I., Sivic, J.: Recognizing human actions in still images: a study of bag-of-features and part-based representations. In: BMVC (2010)
11. Deng, J., Dong, W., Socher, R., Li, L.J., Li, K., Fei-Fei, L.: ImageNet: a large-scale hierarchical image database. In: CVPR (2009)
12. Dong, B., Collins, R., Hoogs, A.: Explainability for content-based image retrieval. In: The IEEE Conference on Computer Vision and Pattern Recognition (CVPR) Workshops, June 2019
13. Everingham, M., Eslami, S.A., Van Gool, L., Williams, C.K., Winn, J., Zisserman, A.: The pascal visual object classes challenge: a retrospective. Int. J. Comput. Vision 111(1), 98–136 (2015)
14. Everingham, M., Gool, L.V., Williams, C.K.I., Winn, J.M., Zisserman, A.: The pascal visual object classes (VOC) challenge. Int. J. Comput. Vision 88, 303–338 (2009)
15. Gella, S., Keller, F.: An analysis of action recognition datasets for language and vision tasks. arXiv abs/1704.07129 (2017)

16. Gella, S., Lapata, M., Keller, F.: Unsupervised visual sense disambiguation for verbs using multimodal embeddings. arXiv abs/1603.09188 (2016)
17. Gordo, A., Almazán, J., Revaud, J., Larlus, D.: Deep image retrieval: learning global representations for image search. In: Leibe, B., Matas, J., Sebe, N., Welling, M. (eds.) ECCV 2016. LNCS, vol. 9910, pp. 241–257. Springer, Cham (2016). https://doi.org/10.1007/978-3-319-46466-4_15
18. Guo, Y., Li, Y., Wang, S.: CS-R-FCN: cross-supervised learning for large-scale object detection. CoRR abs/1905.12863 (2020). http://arxiv.org/abs/1905.12863
19. Gupta, A., Kembhavi, A., Davis, L.S.: Observing human-object interactions: using spatial and functional compatibility for recognition. IEEE Trans. Pattern Anal. Mach. Intell. **31**, 1775–1789 (2009)
20. Gupta, S., Malik, J.: Visual semantic role labeling. arXiv preprint arXiv:1505.04474 (2015)
21. Heilbron, F.C., Escorcia, V., Ghanem, B., Niebles, J.C.: ActivityNet: a large-scale video benchmark for human activity understanding. In: 2015 IEEE Conference on Computer Vision and Pattern Recognition (CVPR), pp. 961–970 (2015)
22. Hochreiter, S., Schmidhuber, J.: Long short-term memory. Neural Comput. **9**, 1735–1780 (1997)
23. Ikizler, N., Cinbis, R.G., Pehlivan, S., Sahin, P.D.: Recognizing actions from still images. In: 2008 19th International Conference on Pattern Recognition, pp. 1–4 (2008)
24. Karpathy, A., Li, F.F.: Deep visual-semantic alignments for generating image descriptions. In: CVPR (2015)
25. Kay, W., et al.: The kinetics human action video dataset. arXiv abs/1705.06950 (2017)
26. Kazemzadeh, S., Ordonez, V., Matten, M., Berg, T.L.: ReferitGame: referring to objects in photographs of natural scenes. In: EMNLP (2014)
27. Kingma, D.P., Ba, J.: Adam: a method for stochastic optimization. In: Bengio, Y., LeCun, Y. (eds.) International Conference on Learning Representations (2015)
28. Krishna, R., et al.: Visual genome: connecting language and vision using crowd-sourced dense image annotations. Int. J. Comput. Vision **123**, 32–73 (2016). https://doi.org/10.1007/s11263-016-0981-7
29. Le, D.T., Bernardi, R., Uijlings, J.R.R.: Exploiting language models to recognize unseen actions. In: ICMR 2013 (2013)
30. Le, D.T., Uijlings, J.R.R., Bernardi, R.: TUHOI: Trento universal human object interaction dataset. In: VL@COLING (2014)
31. Li, R., Tapaswi, M., Liao, R., Jia, J., Urtasun, R., Fidler, S.: Situation recognition with graph neural networks. In: Proceedings of the IEEE International Conference on Computer Vision, pp. 4173–4182 (2017)
32. Lin, C.Y.: ROUGE: a package for automatic evaluation of summaries. In: ACL 2004 (2004)
33. Lin, T.Y., Dollár, P., Girshick, R., He, K., Hariharan, B., Belongie, S.: Feature pyramid networks for object detection (2016)
34. Lin, T.Y., Goyal, P., Girshick, R., He, K., Dollár, P.: Focal loss for dense object detection (2017)
35. Lin, T.-Y., et al.: Microsoft COCO: common objects in context. In: Fleet, D., Pajdla, T., Schiele, B., Tuytelaars, T. (eds.) ECCV 2014. LNCS, vol. 8693, pp. 740–755. Springer, Cham (2014). https://doi.org/10.1007/978-3-319-10602-1_48
36. Mallya, A., Lazebnik, S.: Recurrent models for situation recognition (2017)

37. Mao, J., Huang, J., Toshev, A., Camburu, O.M., Yuille, A.L., Murphy, K.: Generation and comprehension of unambiguous object descriptions. In: 2016 IEEE Conference on Computer Vision and Pattern Recognition (CVPR), pp. 11–20 (2015)
38. Papineni, K., Roukos, S., Ward, T., Zhu, W.J.: BLEU: a method for automatic evaluation of machine translation. In: ACL (2001)
39. Philbin, J., Chum, O., Isard, M., Sivic, J., Zisserman, A.: Object retrieval with large vocabularies and fast spatial matching. In: 2007 IEEE Conference on Computer Vision and Pattern Recognition, pp. 1–8 (2007)
40. Plummer, B.A., Wang, L., Cervantes, C.M., Caicedo, J.C., Hockenmaier, J., Lazebnik, S.: Flickr30k entities: collecting region-to-phrase correspondences for richer image-to-sentence models. In: ICCV (2015)
41. Radenović, F., Tolias, G., Chum, O.: CNN image retrieval learns from BoW: unsupervised fine-tuning with hard examples. In: Leibe, B., Matas, J., Sebe, N., Welling, M. (eds.) ECCV 2016. LNCS, vol. 9905, pp. 3–20. Springer, Cham (2016). https://doi.org/10.1007/978-3-319-46448-0_1
42. Razavian, A.S., Sullivan, J., Maki, A., Carlsson, S.: Visual instance retrieval with deep convolutional networks. CoRR abs/1412.6574 (2014)
43. Redmon, J., Farhadi, A.: YOLO9000: better, faster, stronger (2016)
44. Ren, S., He, K., Girshick, R., Sun, J.: Faster R-CNN: towards real-time object detection with region proposal networks. In: Cortes, C., Lawrence, N.D., Lee, D.D., Sugiyama, M., Garnett, R. (eds.) Advances in Neural Information Processing Systems 28, pp. 91–99. Curran Associates, Inc. (2015)
45. Ronchi, M.R., Perona, P.: Describing common human visual actions in images. arXiv preprint arXiv:1506.02203 (2015)
46. Sharma, P., Ding, N., Goodman, S., Soricut, R.: Conceptual captions: a cleaned, hypernymed, image alt-text dataset for automatic image captioning. In: Proceedings of the 56th Annual Meeting of the Association for Computational Linguistics (Volume 1: Long Papers), pp. 2556–2565 (2018)
47. Sigurdsson, G.A., Gupta, A., Schmid, C., Farhadi, A., Alahari, K.: Charades-Ego: a large-scale dataset of paired third and first person videos. arXiv abs/1804.09626 (2018)
48. Sigurdsson, G.A., Varol, G., Wang, X., Farhadi, A., Laptev, I., Gupta, A.: Hollywood in homes: crowdsourcing data collection for activity understanding. arXiv abs/1604.01753 (2016)
49. Singh, B., Li, H., Sharma, A., Davis, L.S.: R-FCN-3000 at 30fps: decoupling detection and classification. In: 2018 IEEE/CVF Conference on Computer Vision and Pattern Recognition, pp. 1081–1090 (2017)
50. Soomro, K., Zamir, A.R., Shah, M.: UCF101: a dataset of 101 human actions classes from videos in the wild. arXiv abs/1212.0402 (2012)
51. Suhail, M., Sigal, L.: Mixture-kernel graph attention network for situation recognition. In: Proceedings of the IEEE International Conference on Computer Vision, pp. 10363–10372 (2019)
52. Vedantam, R., Zitnick, C.L., Parikh, D.: CIDEr: consensus-based image description evaluation. In: 2015 IEEE Conference on Computer Vision and Pattern Recognition (CVPR), pp. 4566–4575 (2014)
53. Vinyals, O., Toshev, A., Bengio, S., Erhan, D.: Show and tell: a neural image caption generator. In: 2015 IEEE Conference on Computer Vision and Pattern Recognition (CVPR), pp. 3156–3164 (2014)

54. de Vries, H., Strub, F., Chandar, A.P.S., Pietquin, O., Larochelle, H., Courville, A.C.: Guesswhat?! Visual object discovery through multi-modal dialogue. In: 2017 IEEE Conference on Computer Vision and Pattern Recognition (CVPR), pp. 4466–4475 (2016)
55. Yang, F., Hinami, R., Matsui, Y., Ly, S., Satoh, S.: Efficient image retrieval via decoupling diffusion into online and offline processing. In: AAAI (2018)
56. Yang, H., Wu, H., Chen, H.: Detecting 11k classes: large scale object detection without fine-grained bounding boxes. In: 2019 IEEE/CVF International Conference on Computer Vision (ICCV), pp. 9804–9812 (2019)
57. Yao, B., Jiang, X., Khosla, A., Lin, A.L., Guibas, L.J., Li, F.F.: Human action recognition by learning bases of action attributes and parts. In: 2011 International Conference on Computer Vision, pp. 1331–1338 (2011)
58. Yao, B., Li, F.F.: Grouplet: a structured image representation for recognizing human and object interactions. In: 2010 IEEE Computer Society Conference on Computer Vision and Pattern Recognition, pp. 9–16 (2010)
59. Yatskar, M., Ordonez, V., Zettlemoyer, L., Farhadi, A.: Commonly uncommon: semantic sparsity in situation recognition (2016)
60. Yatskar, M., Zettlemoyer, L.S., Farhadi, A.: Situation recognition: visual semantic role labeling for image understanding. In: 2016 IEEE Conference on Computer Vision and Pattern Recognition (CVPR), pp. 5534–5542 (2016)
61. Zhuang, B., Wu, Q., Shen, C., Reid, I.D., van den Hengel, A.: HCVRD: a benchmark for large-scale human-centered visual relationship detection. In: AAAI (2018)

Learning Modality Interaction
for Temporal Sentence Localization
and Event Captioning in Videos

Shaoxiang Chen[1], Wenhao Jiang[2], Wei Liu[2], and Yu-Gang Jiang[1]([✉])

[1] Shanghai Key Lab of Intelligent Information Processing,
School of Computer Science, Fudan University, Shanghai, China
{sxchen13,ygj}@fudan.edu.cn
[2] Tencent AI Lab, Bellevue, USA
cswhjiang@gmail.com, wl2223@columbia.edu

Abstract. Automatically generating sentences to describe events and temporally localizing sentences in a video are two important tasks that bridge language and videos. Recent techniques leverage the multimodal nature of videos by using off-the-shelf features to represent videos, but interactions between modalities are rarely explored. Inspired by the fact that there exist cross-modal interactions in the human brain, we propose a novel method for learning pairwise modality interactions in order to better exploit complementary information for each pair of modalities in videos and thus improve performances on both tasks. We model modality interaction in both the sequence and channel levels in a pairwise fashion, and the pairwise interaction also provides some explainability for the predictions of target tasks. We demonstrate the effectiveness of our method and validate specific design choices through extensive ablation studies. Our method turns out to achieve state-of-the-art performances on four standard benchmark datasets: MSVD and MSR-VTT (event captioning task), and Charades-STA and ActivityNet Captions (temporal sentence localization task).

Keywords: Temporal sentence localization · Event captioning in videos · Modality interaction

1 Introduction

Neuroscience researches [3,5,15] have discovered that the early sensory processing chains in the human brain are not unimodal, information processing in one

---

S. Chen–Part of the work is done when the author was an intern at Tencent AI Lab.

Electronic supplementary material The online version of this chapter (https://doi.org/10.1007/978-3-030-58548-8_20) contains supplementary material, which is available to authorized users.

A. Vedaldi et al. (Eds.): ECCV 2020, LNCS 12349, pp. 333–351, 2020.
https://doi.org/10.1007/978-3-030-58548-8_20

modality (*e.g.*, auditory) can affect another (*e.g.*, visual), and there is a system in the brain for modulating cross-modal interactions. However, modality interactions are largely overlooked in the research of high-level video understanding tasks, such as event captioning [57,58,71] and temporal sentence localization [8,17,34]. Both tasks involve natural language descriptions and are substantially more challenging than recognition tasks. Thus, it is crucial to utilize information from each of the available modalities and capture inter-modality complementary information to better tackle these tasks.

Recent event captioning methods [9,26,37,47,60] mostly adopt an encoder-decoder structure, where the encoder aggregates video features and the decoder (usually LSTM [24] or GRU [13]) generates sentences based on the aggregation results. The video features stem mainly from the visual appearance modality, which are usually extracted with off-the-shelf CNNs (Convolutional Neural Networks) [21,46,50,51] that are pre-trained to recognize objects and can output high-level visual representations for still images. Using features from the visual modality solely can generally work well on video event captioning. Recent works [9,11,37,41,43,44,73] suggest that further improvements can be obtained by additionally leveraging motion and audio representations. However, the limitation of these works is that the features from multiple modalities are simply concatenated without considering their relative importances or the high-level interactions among them, so the great potential of multiple modalities has not been fully explored. There exist a few works [25,28,69,76] that learn to assign importance weights to individual modalities via cross-modal attention in the encoder, but modality interactions are still not explicitly handled. Temporal sentence localization in videos is a relatively new problem [17]. Although various approaches [10,49,74] have been proposed and significant progresses have been made, this problem has not been discussed in a multimodal setting. Most recently, Rahman *et al.* [41] emphasized the importance of jointly considering video and audio to tackle dense event captioning, in which sentence localization is a subtask. Apart from the visual, motion, and audio modalities, utilizing semantic attributes is gaining popularity in recent methods [1,10,36,64] for both event captioning and sentence localization.

In order to better exploit multimodal features for understanding video contents, we propose a novel and generic method for modeling modality interactions that can be leveraged to effectively improve performances on both the sentence localization and event captioning tasks. Our proposed Pairwise Modality Interaction (PMI) explicitly models sequence-level interactions between each pair of feature sequences by using a channel-gated bilinear model, and the outputs of each interacting pair are fused with importance weights. Such a modeling provides some explainability for the predictions.

Our main contributions are as follows:

- We propose a novel multimodal interaction method that uses a Channel-Gated Modality Interaction model to compute pairwise modality interactions (PMI), which better exploits intra- and inter-modality information in videos.

Utilizing PMI achieves significant improvements on both the video event captioning and temporal sentence localization tasks.
- Based on modality interaction within video and text, we further propose a novel sentence localization method that builds video-text local interaction for better predicting the position-wise video-text relevance. To the best of our knowledge, this is also the first work that addresses sentence localization in a multimodal setting.
- Extensive experiments on the MSVD, MSR-VTT, ActivityNet Captions, and Charades-STA datasets verify the superiority of our method compared against state-of-the-art methods on both tasks.

2 Related Works

Temporal Sentence Localization. Gao *et al.* [17] proposed the temporal sentence localization task recently, and it has attracted growing interests from both the computer vision and natural language processing communities. Approaches for this task can be roughly divided into two groups, *i.e.*, proposal-based methods and proposal-free methods. TALL [17] uses a multimodal processing module to fuse visual and textual features for sliding window proposals, and then predicts a ranking score and temporal boundaries for each proposal. NSGV [8] performs interaction between sequentially encoded sentence and video via an LSTM, and then predicts K proposals at each time step. Proposal-free methods usually regress the temporal boundaries. As the most representative one, ABLR [74] iteratively applies co-attention between visual and textual features to encourage interactions, and finally uses the interacted features to predict temporal boundaries.

Event Captioning. The S2VT [57] method is the first attempt at solving video captioning using an encoder-decoder network, in which two layers of LSTMs [24] first encode the CNN-extracted video features and then predict a sentence word-by-word. Later works are mostly based on the encoder-decoder structure, and improvements are made for either the encoder or decoder. Yao *et al.* [71] applied temporal attention to the video features, which enables the encoder to assign an importance weight to each video feature during decoding, and this method is also widely adopted by the following works. Some works [4,12,37,61,78] tried to improve the encoder by considering the temporal structures inside videos. Another group of works [9,33,70] are focused on exploiting spatial information in video frames by applying a dynamic attention mechanism to aggregate frame features spatially. Utilizing multimodal (appearance, motion, and audio) features is also common in recent works, but only a few works [25,36,69,76] tried to handle the relative importances among different modalities using cross-modal attention. Most recently, some works [1,36,75] have proven that incorporating object/semantic attributes into video captioning is effective. As for the decoder, LSTM has been commonly used as the decoder for video captioning, and some recent attempts have also been made to using non-recurrent decoders such as CNN [7] or the Transformer [77] structure.

Modality Interaction. There are some works trying to use self-attention to model modality interaction. Self-attention has been proven effective on both vision [65] and language [55] tasks. Its effectiveness in sequence modeling can be attributed to that it computes a response at one position by attending to all positions in a sequence, which better captures long-range dependencies. AutoInt [48] concatenates features from different modalities and then feeds them to a multi-head self-attention module for capturing interactions. For the referred image segmentation task, Ye *et al.* [72] introduced CMSA (Cross-Modal Self-Attention), which operates on the concatenation of visual features, word embeddings, and spatial coordinates to model long-range dependencies between words and spatial regions. DFAF [18] is a visual question answering (VQA) method, which applies self-attention for regional feature sequences and word embedding sequences to model inter-modality interactions, and also models intra-modality interactions for each sequence. We note that modality interaction is common in VQA methods, but they usually pool the multimodal feature sequences into a single vector using bilinear or multi-linear pooling [16,30,31,35]. And VQA methods are more focused on the interaction between visual and textual modalities, so they do not fully exploit the modality interactions within videos.

Compared to these existing methods, our proposed Pairwise Modality Interaction (PMI) has two distinctive features: (1) modality interactions are captured in a pairwise fashion, and information flow between each pair of modalities in videos is explicitly considered in both the sequence level and channel level; (2) the interaction does not pool the feature sequences (*i.e.*, temporal dimension is preserved), and the interaction results are fused by their importance weights to provide some explainability.

3   Proposed Approach

3.1   Overview

We first give an overview of our approach. As shown in Fig. 1, multimodal features are first extracted from a given video and then fed to a video modality interaction module, where a Channel-Gated Modality Interaction is performed for all pairs of modalities to exploit intra- and inter-modality information. The interaction results are tiled into a high-dimensional tensor and we then use a simple fully-connected network to efficiently compute the importance weights to transform this tensor into a feature sequence. This process to model pairwise modality interaction is abbreviated as PMI.

For sentence localization, the text features are also processed with modality interaction to exploit its intra-modality information. Then video and textual features are locally interacted in order to capture the complex association between these two modalities at each temporal location. Finally, a light-weight convolutional network is applied as the localization head to process the feature sequence and output the video-text relevance score and boundary prediction.

For video captioning, since the focus of this paper is to fully exploit multimodal information, we do not adopt a sophisticated decoder architecture and

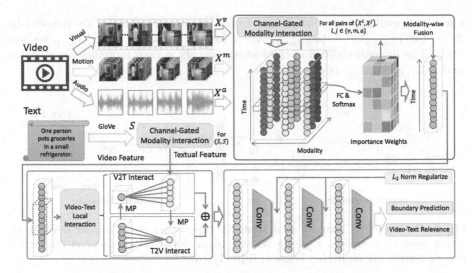

Fig. 1. The framework of our approach. The multimodal features from a video are processed with Channel-Gated Modality Interaction (see Fig. 2) for each pair of modalities, and a weighted modality-wise fusion is then executed to obtain an aggregated video feature (Blue box). **Note that this feature can also be used for video captioning, but the two tasks are not jointly trained**. For temporal sentence localization, the word embedding features also interact with themselves to exploit intra-sentence information, resulting in a textual feature. The video and textual features then interact locally at each temporal position (Green box), and the resulting feature is fed to a light-weight convolutional network with layer-wise norm regularization to produce predictions (Orange box). Each colored circle represents a feature vector. (Color figure online)

only use a two-layer LSTM with temporal attention on top of the video modality interaction. However, due to the superiority of PMI, state-of-the-art performances are still achieved. Note that video modality interaction can be used in either a sentence localization model or an event captioning model, but the models are trained separately.

3.2 Video Modality Interaction

Given an input video $V = \{f_i\}_{i=1}^{F}$, where f_i is the i-th frame, multimodal features can be extracted using off-the-shelf deep neural networks. In this paper, three apparent modalities in videos are adopted, which are visual modality, motion modality, and audio modality. Given features from these modalities, a sequence of features can be learned to represent the latent semantic modality[1].

---

[1] For fair comparison, we do not include this modality when comparing with state-of-the-art methods, but will demonstrate some qualitative results with the latent semantic modality. The corresponding learning method is placed in the Supplementary Material.

The corresponding feature sequences from the above modalities are denoted by $X^v = \{x_n^v\}_{n=1}^N$, $X^m = \{x_n^m\}_{n=1}^N$, $X^a = \{x_n^a\}_{n=1}^N$, and $X^l = \{x_n^l\}_{n=1}^N$, respectively. The dimensionalities of the feature vectors in each modality are denoted as d_v, d_m, d_a, and d_l, respectively.

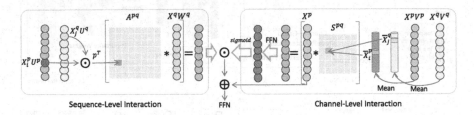

Fig. 2. Overview of Channel-Gated Modality Interaction. The Channel-Level Interaction results are used as a gating variable to modulate Sequence-Level Interaction results. Details are illustrated below in Eqs. (1)–(6).

We propose to explicitly model modality interaction between a pair of feature sequences, denoted by X^p and X^q, where $p \in \{a, m, v, l\}$ and $q \in \{a, m, v, l\}$. Note that p and q can be the same modality, and in that case, the interaction exploits intra-modality information. As shown in Fig. 2, the interaction can be formulated as

$$\text{INT}(X^p, X^q) = \text{FFN}\big(\text{BA}(X^p, X^q) \odot \text{CG}(X^p, X^q) \oplus X^p\big), \qquad (1)$$

where $\text{BA}(\cdot)$ is the bilinear attention model that performs sequence-level modality interaction, $\text{CG}(\cdot)$ is a channel gating mechanism based on the channel-level interaction and is used to modulate the sequence-level interaction output, a residual connection is introduced with $\oplus X^p$, and $\text{FFN}(\cdot)$ is a position-wise feed-forward network that projects its input into a lower dimension[2].

Sequence-Level Interaction. We use a low-rank bilinear model to consider the interaction between each pair of elements in feature sequences X^p and X^q:

$$A_{ij}^{pq} = p^T\big(\rho(X_i^p U^p) \odot \rho(X_j^q U^q)\big), \quad \mathcal{A}_{ij}^{pq} = \text{Softmax}_j(A_{ij}^{pq}), \qquad (2)$$

where X_i^p is the i-th element of X^p, X_j^q is the j-th element of X^q, and $U^p \in \mathbb{R}^{d_p \times d}$ and $U^q \in \mathbb{R}^{d_q \times d}$ are low-rank projection matrices ($d < min(d_p, d_q)$). $\odot$ denotes element-wise multiplication (Hadamard product), and ρ denotes ReLU non-linearity. $p \in \mathbb{R}^d$ projects the element interaction into a scalar, so that $A^{pq} \in \mathbb{R}^{N \times N}$ can be normalized into a bilinear attention map by applying column-wise softmax. Then the output of the bilinear model is

$$\text{BA}(X^p, X^q) = \mathcal{A}^{pq}(X^q W^q). \qquad (3)$$

---

[2] Details about this FFN can be found in the Supplementary Material.

In the matrix multiplication of $\boldsymbol{A}^{pq}$ and $\boldsymbol{X}^q\boldsymbol{W}^q$, a relative position embedding [42] is injected to make the sequence-level interaction to be position-aware.

Channel-Level Interaction. In order to modulate the sequence-level interaction result, we devise a gate function based on fine-grained channel-level interaction. We first obtain a channel representation of $\boldsymbol{X}^p$ and $\boldsymbol{X}^p$ as

$$\overline{\boldsymbol{X}}^p = \mathrm{Mean}_n(\boldsymbol{X}^p\boldsymbol{V}^p), \quad \overline{\boldsymbol{X}}^q = \mathrm{Mean}_n(\boldsymbol{X}^q\boldsymbol{V}^q), \tag{4}$$

where $\mathrm{Mean}(\cdot)$ is sequence-wise mean-pooling, and $\boldsymbol{V}^p, \boldsymbol{V}^q$ are used to project $\boldsymbol{X}^p$ and $\boldsymbol{X}^p$ to lower dimension for efficient processing. Similarly, we also compute a channel-to-channel attention map

$$S_{ij}^{pq} = f_{chn}(\overline{\boldsymbol{X}}_i^p, \overline{\boldsymbol{X}}_j^q), \quad \mathcal{S}_{ij}^{pq} = \mathrm{Softmax}_i(S_{ij}^{pq}), \tag{5}$$

where $f_{chn}(\cdot)$ is a function for computing channel-level interaction. Since each element in $\overline{\boldsymbol{X}}^p$ and $\overline{\boldsymbol{X}}^q$ is a scalar, we simply use $f_{chn}(a, b) = -(a - b)^2$. Then the output of the gate function is

$$\mathrm{CG}(\boldsymbol{X}^p, \boldsymbol{X}^q) = \sigma\big(\mathrm{FFN}(\boldsymbol{X}^p\mathcal{S}^{pq})\big), \tag{6}$$

where σ is the Sigmoid function, so the output has values in $[0, 1]$.

Modality-Wise Fusion. Given M modalities, there will be M^2 pairs of interacting modalities, and they are tiled as a high-dimensional tensor $\boldsymbol{X}^{MI} \in \mathbb{R}^{N \times M^2 \times d}$. The information in $\boldsymbol{X}^{MI}$ needs to be further aggregated before feeding it to target tasks. Simple concatenation or pooling can achieve this purpose. Here, we consider the importance of each interacting result by using a position-wise fully-connected layer to predict importance weights:

$$\boldsymbol{e}_n = \boldsymbol{X}_n^{MI}\boldsymbol{W}_n^a + \boldsymbol{b}_n^a, \quad \alpha_n = \mathrm{Softmax}_m(\boldsymbol{e}_n),$$
$$\widehat{\boldsymbol{X}}_n = \sum\nolimits_{m=1}^{M^2} \alpha_{nm}\boldsymbol{X}_{nm}^{MI}. \tag{7}$$

Finally, the fusion result $\widehat{\boldsymbol{X}} \in \mathbb{R}^{N \times d}$ is the modality-interacted representation of a video.

3.3 Sentence Localization

The sentence is represented as a sequence of word-embedding vectors $Y = \{\boldsymbol{w}_l\}_{l=1}^L$, which is also processed with the CGMI to exploit its intra-modality information, yielding a textual feature $\widehat{\boldsymbol{Y}}$. For sentence localization, it is crucial to capture the complex association between the video and textual modalities at each temporal location, and then predict each location's relevance to the sentence.

Video-Text Local Interaction. Based on the above intuition, we propose Video-Text Local Interaction. For each temporal location $t \in [1, N]$ of $\widehat{X}$, a local window $\widetilde{X} = \{\widehat{X}_n\}_{n=t-w}^{t+w}$ is extracted to interact with the textual feature $\widehat{Y}$. As shown in Fig. 1, the local video-to-text interaction is modeled as

$$Z_t^{xy} = \text{BA}(\text{Mean}(\widetilde{X}), \widehat{Y}), \quad \widehat{Z}_t^{xy} = \text{MM}(Z_t^{xy}, \text{Mean}(\widetilde{X})). \qquad (8)$$

Here instead of gating, we use a more efficient multimodal processing unit $\text{MM}(a, b) = W^T[a||b||a \odot b||a \oplus b]$ to encourage further interaction of both modalities. Likewise, text-to-video interaction $\widehat{Z}_t^{yx}$ is computed given $\widetilde{X}$ and $\text{Mean}(\widehat{Y})$, and then fused with the video-to-text interaction result

$$Z_t = \widehat{Z}_t^{xy} \oplus \widehat{Z}_t^{yx}. \qquad (9)$$

Localization Head. We apply a light-weight convolutional network upon the video-text interacted sequence Z to produce predictions. Each layer can be formulated as

$$C^k = Conv(C^{k-1}||\text{Mean}(\widehat{Y})), \qquad (10)$$

where $k = 1, .., K$, and $C_0 = Z$. We apply Instance Normalization [54] and LeakyReLU [66] activation to each layer's output. Since we are computing the video-text relevance in a layer-wise fashion, we impose an ℓ_2 norm regularization on each layer's output to obtain a more robust feature

$$\text{Loss}_{norm} = \sum_{n=1}^{N}(||C_n^k||_2 - \beta_k)^2, \qquad (11)$$

where $|| \cdot ||$ is the ℓ_2 norm of a vector. The K-th layer output C^K has 1 output channel, which is normalized using Softmax, representing the Video-Text Relevance $r \in [0, 1]^N$. Then a fully connected layer with two output units is applied to r to produce a boundary prediction $b \in \mathbb{R}^2$. The loss for the predictions is

$$\text{Loss}_{pred} = \text{Huber}(b - \hat{b}) - \lambda_r \frac{\sum_n \hat{r}_n \log(r_n)}{\sum_n \hat{r}_n}, \qquad (12)$$

where $\hat{b}$ is the ground-truth temporal boundary, $\text{Huber}(\cdot)$ is the Huber loss function, and $\hat{r}_n = 1$ if n is in the ground-truth temporal region, otherwise $\hat{r}_n = 0$. The overall loss is

$$\text{Loss}_{loc} = \text{Loss}_{pred} + \lambda_n \text{Loss}_{norm}, \qquad (13)$$

where λ_n, λ_r are constant weights used to balance the loss terms.

3.4   Event Captioning

After the video modality interaction result is obtained, we use a standard bi-directional LSTM for encoding and a two-layer LSTM network with temporal attention [71] to generate sentences as in previous works [9,59,69]. The sentence generation is done in a word-by-word fashion. At every time step, a set of temporal attention weights is computed based on the LSTM hidden states and video features, which is then used to weighted-sum the video features into a single vector. This dynamic feature vector is fed to the LSTM with the previously generated word to predict the next word[3]. We would like to emphasize again that video modality interaction can be used as a basic video feature encoding technique for either sentence localization or event captioning, but we do not perform multi-task training for these two.

4   Experiments

In this section, we provide experimental analysis of our model design and present comparisons with the state-of-the-art methods on both temporal sentence localization and video captioning.

4.1   Experimental Settings

MSVD Dataset [6]. MSVD is a well-known video captioning dataset with 1,970 videos. The average length is 9.6 s, and each video has around 40 sentence annotations on average. We adopt the same common dataset split as in prior works [4,69,71]. Thus, we have 1,200/100/670 videos for training, validation, and testing, respectively.

MSR-VTT Dataset [68]. MSR-VTT is a large-scale video captioning dataset with 10,000 videos. The standard split [68] for this dataset was provided. Hence, we use 6,513/497/2,990 videos for training, validation, and testing, respectively, in our experiments. In this dataset, each video is associated with 20 sentence annotations and is of length 14.9 s on average.

ActivityNet Captions Dataset [32] **(ANet-Cap)**. ANet-Cap is built on the ActivityNet dataset [22] with 19,994 untrimmed videos (153 s on average). The standard split is 10,009/4,917/5,068 videos for training, validation, and testing, respectively. There are 3.74 *temporally localized* sentences per video on average. Since the testing set is not publicly available, we evaluate our method on the validation set as previous works [62,67].

Charades-STA Dataset [17]. Charades-STA is built on 6,672 videos from the Charades [45] dataset. The average duration of the videos is 29.8 s. There are 16,128 *temporally localized* sentence annotations, which give 2.42 sentences per video. The training and testing sets contain 12,408 and 3,720 annotations, respectively.

---

[3] Due to the space limit and that caption decoder is not the focus of this work, we omit formal descriptions here. We also move some experiments and analysis below to the Supplementary Material.

Table 1. Performance comparison of video modality interaction strategies on MSVD.

#	Method	B@4	M	C
0	Concat w/o Interact (Baseline)	45.28	31.60	62.57
1	Concat + Interact	46.24	32.03	66.10
2	Pairwise Interact + Concat Fusion	47.86	33.73	75.30
3	Pairwise Interact + Sum Fusion	51.37	34.01	78.42
4	Pairwise Interact + Weighted Fusion (ours)	54.68	36.40	95.17
5	Intra-modality Interactions only	49.92	34.76	88.46
6	Inter-modality Interactions only	47.30	32.72	70.20
7	(Intra+Inter)-modality (ours)	54.68	36.40	95.17

Table 2. Performances (%) of different localizer settings on the Charades-STA dataset.

#	PMI	VTLI	ℓ_2-Norm	IoU $= 0.3$	IoU $= 0.5$	IoU $= 0.7$
0	✗	✗	✗	51.46	35.34	15.81
1	✓	✗	✗	53.22	37.05	17.36
2	✓	✓	✗	54.37	38.42	18.63
3	✓	✓	✓	55.48	39.73	19.27

We evaluate the captioning performance of our method on MSVD and MSR-VTT with commonly used metrics, *i.e.*, BLEU [38], METEOR [14], and CIDEr [56]. ANet-Cap and Charades-STA are used to evaluate sentence localization performance. We adopt the same evaluation metric used by previous works [17], which computes "Recall@1,IoU=m" (denoted by $r(m, s_i)$), meaning the percentage of the top-1 results having IoU larger than m with the annotated segment of a sentence s_i. The overall performance on a dataset of N sentences is the average score of all the sentences $\frac{1}{N} \sum_{i=1}^{N} r(m, s_i)$.

Implementation Details. The sentences in all datasets are converted to lowercase and then tokenized. For the captioning task, randomly-initialized word embedding vectors of dimension 512 are used, which are then jointly fine-tuned with the model. For the sentence localization task, we employ the GloVe [40] word embedding as previous works. We use Inception-ResNet v2 [50] and C3D [52] to extract visual and motion features. For the audio features, we employ the MFCC (Mel-Frequency Cepstral Coefficients) on the captioning task and SoundNet [2] on the sentence localization task. We temporally subsample the feature sequences to length 32 for event captioning, and 128 for sentence localization. The bilinear attention adopts 8 attention heads, and the loss weights λ_r and λ_n are set to 5 and 0.001, respectively. In all of our experiments, the batch size is set to 32 and the Adam optimizer with learning rate 0.0001 is used to train our model.

4.2   Ablation Studies

Firstly, we perform extensive experiments to validate the design choices in our approach. We study the effect of different modality interaction strategies on the MSVD dataset, and the effects of sentence localizer components on the Charades-STA dataset. All experiments use Inception-ResNet v2 and C3D features.

On the MSVD dataset, we design 8 different variants and their performances are summarized in Table 1. In variant 0, which is a baseline, multimodal features are concatenated and directly fed to the caption decoder. Variant 1 treats the concatenated features as one modality and performs intra-modality interaction. In variants 2–4, PMI is performed and different fusion strategies are adopted. In variants 5–7, we study the ablation of intra- and inter-modality interactions.

Why Pairwise? We perform modality interaction in a pairwise fashion in our model, and this is the main distinctive difference from existing methods [48, 72], which employ feature concatenation. As shown in Table 1, while concatenating all modalities into one and performing intra-modality interaction can gain performance improvements over the baseline (#1 vs. #0), concatenating after pairwise interaction has a more significant advantage (#2 vs. #1). We also compare the effects of different aggregation strategies after pairwise interaction (#2–4), and weighted fusion (in PMI) yields the best result with a clear margin, which also indicates that the interactions between different modality pairs produce unique information of different importances.

Effect of Inter-modality Complementarity. We then inspect the intra- and inter-modality interactions separately. Table 1 (#5–7) shows that intra-modality interaction can already effectively exploit information in each modality compared to the baseline. Inter-modality complementarity alone is not sufficient for captioning, but it can be combined with intra-modality information to obtain a further performance boost, which again validates our design of pairwise interaction.

Effect of Sentence Localizer Components. The PMI, video-text local interaction (VTLI), and ℓ_2-norm regularization are the key components of the sentence localization model. As can be observed from Table 2, incorporating each component consistently leads to a performance boost.

Table 3. Video captioning performances of our proposed PMI and other state-of-the-art multimodal fusion methods on the MSVD dataset. Meanings of features can be found in Table 4.

Method	Features	B@4	M	C
AF [25]	V+C	52.4	32.0	68.8
TDDF [76]	V+C	45.8	33.3	73.0
MA-LSTM [69]	G+C	52.3	33.6	70.4
MFATT [36]	R152+C	50.8	33.2	69.4
GRU-EVE [1]	IRV2+C	47.9	35.0	78.1
XGating [59]	IRV2+I3D	52.5	34.1	88.7
HOCA [28]	IRV2+I3D	52.9	35.5	86.1
PMI-CAP	V+C	49.74	33.59	77.11
PMI-CAP	G+C	51.55	34.64	74.51
PMI-CAP	R152+C	52.07	34.34	77.35
PMI-CAP	IRV2+C	54.68	36.40	95.17
PMI-CAP	IRV2+I3D	55.76	36.63	95.68

Table 4. Performances of our proposed model and other state-of-the-art methods on the MSVD and MSR-VTT datasets. R*, G, V, C, IV4, R3D, IRV2, Obj, and A mean ResNet, GoogLeNet, VGGNet, C3D, Inception-V4, 3D ResNeXt, Inception-ResNet v2, Object features, and audio features, respectively. Note that audio track is only available on MSR-VTT, and for fair comparison, we use the MFCC audio representation as [9, 11]. Please refer to the original papers for the detailed feature extraction settings.

Dataset	MSVD				MSR-VTT			
Method	Features	B@4	M	C	Features	B@4	M	C
STAT [53]	G+C+Obj	51.1	32.7	67.5	G+C+Obj	37.4	26.6	41.5
M³ [63]	V+C	51.78	32.49	–	V+C	38.13	26.58	–
DenseLSTM [79]	V+C	50.4	32.9	72.6	V+C	38.1	26.6	42.8
PickNet [12]	R152	52.3	33.3	76.5	R152	41.3	27.7	44.1
hLSTMat [47]	R152	53.0	33.6	73.8	R152	38.3	26.3	–
VRE [44]	R152	51.7	34.3	86.7	R152+A	43.2	28.0	48.3
MARN [39]	R101+R3D	48.6	35.1	92.2	R101+R3D	40.4	28.1	47.1
OA-BTG [75]	R200+Obj	56.9	36.2	90.6	R200+Obj	41.4	28.2	46.9
RecNet [60]	IV4	52.3	34.1	80.3	IV4	39.1	26.6	42.7
XGating [59]	IRV2+I3D	52.5	34.1	88.7	IRV2+I3D	42.0	28.1	49.0
MM-TGM [11]	IRV2+C	48.76	34.36	80.45	IRV2+C+A	44.33	29.37	49.26
GRU-EVE [1]	IRV2+C	47.9	35.0	78.1	IRV2+C	38.3	28.4	48.1
MGSA [9]	IRV2+C	53.4	35.0	86.7	IRV2+C+A	45.4	28.6	50.1
PMI-CAP	IRV2+C	54.68	36.40	95.17	IRV2+C	42.17	28.79	49.45
PMI-CAP	–	–	–	–	IRV2+C+A	43.96	29.56	50.66

4.3 Comparison with State-of-the-Art Methods

Results on the Video Event Captioning Task. We abbreviate our approach as PMI-CAP for video captioning. To demonstrate the superiority of our proposed pairwise modality interaction, we first compare our method with state-of-the-art methods that focus on the fusion of multimodal features for video captioning. For fair comparison, we use the same set of features as each compared method. As shown in Table 3, our PMI-CAP has outperformed all the compared methods when using the same features. The improvement in the CIDEr metric is especially significant, which is 10.8% on average. This shows that our pairwise modality interaction can really utilize multimodal features more effectively.

Table 5. Performances (%) of our proposed model and other state-of-the-art methods on the Charades-STA dataset. * means our implementation.

Method	IoU = 0.3	IoU = 0.5	IoU = 0.7
Random	14.16	6.05	1.59
VSA-RNN [29]	–	10.50	4.32
VSA-STV [29]	–	16.91	5.81
MCN [23]	32.59	11.67	2.63
ACRN [34]	38.06	20.26	7.64
ROLE [35]	37.68	21.74	7.82
SLTA [27]	38.96	22.81	8.25
CTRL [17]	–	23.63	8.89
VAL [49]	–	23.12	9.16
ACL [19]	–	30.48	12.20
SAP [10]	–	27.42	13.36
SM-RL [64]	–	24.36	11.17
QSPN [67]	54.7	35.6	15.8
ABLR* [74]	51.55	35.43	15.05
TripNet [20]	51.33	36.61	14.50
CBP [62]	–	36.80	18.87
PMI-LOC (C)	**55.48**	**39.73**	**19.27**
PMI-LOC (C+IRV2)	56.84	41.29	20.11
PMI-LOC (C+IRV2+A)	58.08	42.63	21.32

Table 4 shows the performance comparison on the MSVD and MSR-VTT datasets. We adopt the set of features commonly used by recent state-of-the-art methods [1,9,59], which are Inception-ResNet v2 and C3D for visual and motion modalities, respectively. Among the competitive methods, OA-BTG [75] utilizes object-level information from an external detector, and MARN [39] uses a more advanced 3D CNN to extract motion features. We do not exploit spatial

Table 6. Performances (%) of our proposed model and other state-of-the-art methods on the ActivityNet Captions dataset.

Method	IoU = 0.3	IoU = 0.5	IoU = 0.7
Random	12.46	6.37	2.23
QSPN [67]	45.3	27.7	13.6
TGN [8]	43.81	27.93	–
ABLR [74]	55.67	36.79	–
TripNet [20]	48.42	32.19	13.93
CBP [62]	54.30	35.76	17.80
PMI-LOC (C)	**59.69**	**38.28**	**17.83**
PMI-LOC (C+IRV2)	60.16	39.16	18.02
PMI-LOC (C+IRV2+A)	61.22	40.07	18.29

information like MGSA [9] and VRE [44], or use a sophisticated decoder as hLSTMat [47] and MM-TGM [11], while we emphasize that PMI may be used along with most of these methods. Overall, our PMI-CAP achieves state-of-the-art performances on both MSVD and MSR-VTT.

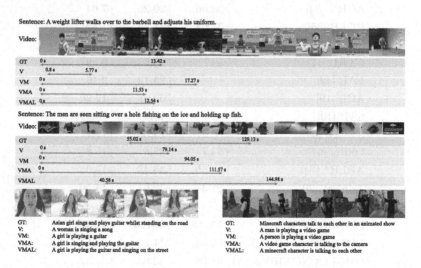

Fig. 3. Qualitative results of temporal sentence localization and event captioning. The results are generated using our model but with different combinations of modalities.

Results on the Sentence Localization Task. As previously introduced, current state-of-the-art methods for sentence localization haven't considered this problem in a multimodal setting and only use the C3D feature. Thus we present results with only C3D feature to fairly compare with these methods and also report performances under multimodal settings. Our approach is abbreviated as PMI-LOC for sentence localization. Table 5 shows results on the widely-used Charades-STA dataset. Our PMI-LOC outperforms all compared methods in all metrics. Further experiments with multimodal features show even higher localization accuracies, which verify the effectiveness of our modality interaction method. As shown in Table 6, on the large-scale ActivityNet Captions dataset, our method also achieves state-of-the-art performances.

4.4 Qualitative Results

We show some qualitative results in Figs. 3, 4, and 5 to demonstrate the effectiveness of our modality interaction method and how it provides expainability to the final prediction of the target tasks. Note that in addition to the visual (V), motion (M), and audio (A) modalities, we also utilize the previously mentioned latent semantics (L) modality to comprehensively explore the video content.

Figure 3 indicates that by utilizing more modalities, the model gets more complementary information through modality interaction and achieves better performance for both temporal sentence localization and event captioning. The event captioning examples in Fig. 4 show that each type of events has its modality interaction pattern. The sports video (top) has distinctive visual and motion patterns that are mainly captured by visual-motion modality interaction. The cooking video (middle) has unique visual cues and sounds made by kitchenware, so the important interactions are between the visual and audio modalities and within the audio modality. For the animated video (bottom), latent semantics modality is important when the other modalities are not sufficient to capture its contents. Similar observations can also be made on the sentence localization examples in Fig. 5.

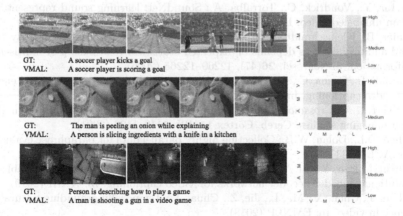

Fig. 4. Qualitative results of video event captioning with visualization of the modality importance weights.

Fig. 5. Qualitative results of temporal sentence localization with visualization of the modality importance weights.

5   Conclusions

In this paper, we proposed pairwise modality interaction (PMI) for tackling the temporal sentence localization and event captioning tasks, and performed fine-grained cross-modal interactions in both the sequence and channel levels to better understand video contents. The extensive experiments on four benchmark datasets on both tasks consistently verify the effectiveness of our proposed method. Our future work will extend the proposed modality interaction method to cope with other video understanding tasks.

Acknowledgement. Shaoxiang Chen is partially supported by the Tencent Elite Internship program.

References

1. Aafaq, N., Akhtar, N., Liu, W., Gilani, S.Z., Mian, A.: Spatio-temporal dynamics and semantic attribute enriched visual encoding for video captioning. In: CVPR (2019)
2. Aytar, Y., Vondrick, C., Torralba, A.: SoundNet: learning sound representations from unlabeled video. In: NIPS (2016)
3. Baier, B., Kleinschmidt, A., Müller, N.G.: Cross-modal processing in early visual and auditory cortices depends on expected statistical relationship of multisensory information. J. Neurosci. **26**(47), 12260–12265 (2006)
4. Baraldi, L., Grana, C., Cucchiara, R.: Hierarchical boundary-aware neural encoder for video captioning. In: CVPR (2017)
5. Calvert, G.A.: Crossmodal processing in the human brain: insights from functional neuroimaging studies. Cereb. Cortex **11**(12), 1110–1123 (2001)
6. Chen, D.L., Dolan, W.B.: Collecting highly parallel data for paraphrase evaluation. In: ACL (2011)
7. Chen, J., Pan, Y., Li, Y., Yao, T., Chao, H., Mei, T.: Temporal deformable convolutional encoder-decoder networks for video captioning. In: AAAI (2019)
8. Chen, J., Chen, X., Ma, L., Jie, Z., Chua, T.: Temporally grounding natural sentence in video. In: EMNLP (2018)
9. Chen, S., Jiang, Y.: Motion guided spatial attention for video captioning. In: AAAI (2019)
10. Chen, S., Jiang, Y.: Semantic proposal for activity localization in videos via sentence query. In: AAAI (2019)

11. Chen, S., Chen, J., Jin, Q., Hauptmann, A.G.: Video captioning with guidance of multimodal latent topics. In: ACM MM (2017)
12. Chen, Y., Wang, S., Zhang, W., Huang, Q.: Less is more: picking informative frames for video captioning. In: Ferrari, V., Hebert, M., Sminchisescu, C., Weiss, Y. (eds.) ECCV 2018. LNCS, vol. 11217, pp. 367–384. Springer, Cham (2018). https://doi.org/10.1007/978-3-030-01261-8_22
13. Chung, J., Gülçehre, Ç., Cho, K., Bengio, Y.: Empirical evaluation of gated recurrent neural networks on sequence modeling. arXiv preprint arXiv:1412.3555 (2014)
14. Denkowski, M.J., Lavie, A.: Meteor universal: language specific translation evaluation for any target language. In: WMT@ACL (2014)
15. Eckert, M.A., Kamdar, N.V., Chang, C.E., Beckmann, C.F., Greicius, M.D., Menon, V.: A cross-modal system linking primary auditory and visual cortices: evidence from intrinsic fMRI connectivity analysis. Hum. Brain Mapp. **29**(7), 848–857 (2008)
16. Fukui, A., Park, D.H., Yang, D., Rohrbach, A., Darrell, T., Rohrbach, M.: Multimodal compact bilinear pooling for visual question answering and visual grounding. In: EMNLP (2016)
17. Gao, J., Sun, C., Yang, Z., Nevatia, R.: TALL: temporal activity localization via language query. In: ICCV (2017)
18. Gao, P., et al.: Dynamic fusion with intra- and inter-modality attention flow for visual question answering. In: CVPR (2019)
19. Ge, R., Gao, J., Chen, K., Nevatia, R.: MAC: mining activity concepts for language-based temporal localization. In: WACV (2019)
20. Hahn, M., Kadav, A., Rehg, J.M., Graf, H.P.: Tripping through time: efficient localization of activities in videos. arXiv preprint arXiv:1904.09936 (2019)
21. He, K., Zhang, X., Ren, S., Sun, J.: Deep residual learning for image recognition. In: CVPR (2016)
22. Heilbron, F.C., Escorcia, V., Ghanem, B., Niebles, J.C.: ActivityNet: a large-scale video benchmark for human activity understanding. In: CVPR (2015)
23. Hendricks, L.A., Wang, O., Shechtman, E., Sivic, J., Darrell, T., Russell, B.C.: Localizing moments in video with natural language. In: ICCV (2017)
24. Hochreiter, S., Schmidhuber, J.: Long short-term memory. Neural Comput. **9**(8), 1735–1780 (1997)
25. Hori, C., et al.: Attention-based multimodal fusion for video description. In: ICCV (2017)
26. Hu, Y., Chen, Z., Zha, Z., Wu, F.: Hierarchical global-local temporal modeling for video captioning. In: ACM MM (2019)
27. Jiang, B., Huang, X., Yang, C., Yuan, J.: Cross-modal video moment retrieval with spatial and language-temporal attention. In: ICMR (2019)
28. Jin, T., Huang, S., Li, Y., Zhang, Z.: Low-rank HOCA: efficient high-order cross-modal attention for video captioning. In: EMNLP-IJCNLP (2019)
29. Karpathy, A., Li, F.: Deep visual-semantic alignments for generating image descriptions. In: CVPR (2015)
30. Kim, J., Jun, J., Zhang, B.: Bilinear attention networks. In: NeurIPS (2018)
31. Kim, J., On, K.W., Lim, W., Kim, J., Ha, J., Zhang, B.: Hadamard product for low-rank bilinear pooling. In: ICLR (2017)
32. Krishna, R., Hata, K., Ren, F., Fei-Fei, L., Niebles, J.C.: Dense-captioning events in videos. In: ICCV (2017)
33. Li, X., Zhao, B., Lu, X.: MAM-RNN: multi-level attention model based RNN for video captioning. In: IJCAI (2017)

34. Liu, M., Wang, X., Nie, L., He, X., Chen, B., Chua, T.: Attentive moment retrieval in videos. In: ACM SIGIR (2018)
35. Liu, Z., Shen, Y., Lakshminarasimhan, V.B., Liang, P.P., Zadeh, A., Morency, L.: Efficient low-rank multimodal fusion with modality-specific factors. In: ACL (2018)
36. Long, X., Gan, C., de Melo, G.: Video captioning with multi-faceted attention. TACL **6**, 173–184 (2018)
37. Pan, P., Xu, Z., Yang, Y., Wu, F., Zhuang, Y.: Hierarchical recurrent neural encoder for video representation with application to captioning. In: CVPR (2016)
38. Papineni, K., Roukos, S., Ward, T., Zhu, W.: BLEU: a method for automatic evaluation of machine translation. In: ACL (2002)
39. Pei, W., Zhang, J., Wang, X., Ke, L., Shen, X., Tai, Y.: Memory-attended recurrent network for video captioning. In: CVPR (2019)
40. Pennington, J., Socher, R., Manning, C.D.: GloVe: global vectors for word representation. In: EMNLP (2014)
41. Rahman, T., Xu, B., Sigal, L.: Watch, listen and tell: multi-modal weakly supervised dense event captioning. In: ICCV (2019)
42. Shaw, P., Uszkoreit, J., Vaswani, A.: Self-attention with relative position representations. In: NAACL-HLT (2018)
43. Shen, Z., et al.: Weakly supervised dense video captioning. In: CVPR (2017)
44. Shi, X., Cai, J., Joty, S.R., Gu, J.: Watch it twice: video captioning with a refocused video encoder. In: ACM MM (2019)
45. Sigurdsson, G.A., Varol, G., Wang, X., Farhadi, A., Laptev, I., Gupta, A.: Hollywood in homes: crowdsourcing data collection for activity understanding. In: Leibe, B., Matas, J., Sebe, N., Welling, M. (eds.) ECCV 2016. LNCS, vol. 9905, pp. 510–526. Springer, Cham (2016). https://doi.org/10.1007/978-3-319-46448-0_31
46. Simonyan, K., Zisserman, A.: Very deep convolutional networks for large-scale image recognition. In: ICLR (2015)
47. Song, J., Gao, L., Guo, Z., Liu, W., Zhang, D., Shen, H.T.: Hierarchical LSTM with adjusted temporal attention for video captioning. In: IJCAI (2017)
48. Song, W., et al.: AutoInt: automatic feature interaction learning via self-attentive neural networks. In: CIKM (2019)
49. Song, X., Han, Y.: VAL: visual-attention action localizer. In: Hong, R., Cheng, W.-H., Yamasaki, T., Wang, M., Ngo, C.-W. (eds.) PCM 2018. LNCS, vol. 11165, pp. 340–350. Springer, Cham (2018). https://doi.org/10.1007/978-3-030-00767-6_32
50. Szegedy, C., Ioffe, S., Vanhoucke, V., Alemi, A.A.: Inception-v4, inception-ResNet and the impact of residual connections on learning. In: AAAI (2017)
51. Szegedy, C., et al.: Going deeper with convolutions. In: CVPR (2015)
52. Tran, D., Bourdev, L.D., Fergus, R., Torresani, L., Paluri, M.: Learning spatiotemporal features with 3D convolutional networks. In: ICCV (2015)
53. Tu, Y., Zhang, X., Liu, B., Yan, C.: Video description with spatial-temporal attention. In: ACM MM (2017)
54. Ulyanov, D., Vedaldi, A., Lempitsky, V.S.: Instance normalization: the missing ingredient for fast stylization. arXiv preprint arXiv:1607.08022 (2016)
55. Vaswani, A., et al.: Attention is all you need. In: NIPS (2017)
56. Vedantam, R., Zitnick, C.L., Parikh, D.: CIDEr: consensus-based image description evaluation. In: CVPR (2015)
57. Venugopalan, S., Rohrbach, M., Donahue, J., Mooney, R.J., Darrell, T., Saenko, K.: Sequence to sequence - video to text. In: ICCV (2015)
58. Venugopalan, S., Xu, H., Donahue, J., Rohrbach, M., Mooney, R.J., Saenko, K.: Translating videos to natural language using deep recurrent neural networks. In: NAACL-HLT (2015)

59. Wang, B., Ma, L., Zhang, W., Jiang, W., Wang, J., Liu, W.: Controllable video captioning with POS sequence guidance based on gated fusion network. In: ICCV (2019)
60. Wang, B., Ma, L., Zhang, W., Liu, W.: Reconstruction network for video captioning. In: CVPR (2018)
61. Wang, J., Jiang, W., Ma, L., Liu, W., Xu, Y.: Bidirectional attentive fusion with context gating for dense video captioning. In: CVPR (2018)
62. Wang, J., Ma, L., Jiang, W.: Temporally grounding language queries in videos by contextual boundary-aware prediction. In: AAAI (2020)
63. Wang, J., Wang, W., Huang, Y., Wang, L., Tan, T.: M3: multimodal memory modelling for video captioning. In: CVPR (2018)
64. Wang, W., Huang, Y., Wang, L.: Language-driven temporal activity localization: a semantic matching reinforcement learning model. In: CVPR (2019)
65. Wang, X., Girshick, R.B., Gupta, A., He, K.: Non-local neural networks. In: CVPR (2018)
66. Xu, B., Wang, N., Chen, T., Li, M.: Empirical evaluation of rectified activations in convolutional network. arXiv preprint arXiv:1505.00853 (2015)
67. Xu, H., He, K., Plummer, B.A., Sigal, L., Sclaroff, S., Saenko, K.: Multilevel language and vision integration for text-to-clip retrieval. In: AAAI (2019)
68. Xu, J., Mei, T., Yao, T., Rui, Y.: MSR-VTT: a large video description dataset for bridging video and language. In: CVPR (2016)
69. Xu, J., Yao, T., Zhang, Y., Mei, T.: Learning multimodal attention LSTM networks for video captioning. In: ACM MM (2017)
70. Yang, Z., Han, Y., Wang, Z.: Catching the temporal regions-of-interest for video captioning. In: ACM MM (2017)
71. Yao, L., et al.: Describing videos by exploiting temporal structure. In: ICCV (2015)
72. Ye, L., Rochan, M., Liu, Z., Wang, Y.: Cross-modal self-attention network for referring image segmentation. In: CVPR (2019)
73. Yu, H., Wang, J., Huang, Z., Yang, Y., Xu, W.: Video paragraph captioning using hierarchical recurrent neural networks. In: CVPR (2016)
74. Yuan, Y., Mei, T., Zhu, W.: To find where you talk: temporal sentence localization in video with attention based location regression. In: AAAI (2019)
75. Zhang, J., Peng, Y.: Object-aware aggregation with bidirectional temporal graph for video captioning. In: CVPR (2019)
76. Zhang, X., Gao, K., Zhang, Y., Zhang, D., Li, J., Tian, Q.: Task-driven dynamic fusion: reducing ambiguity in video description. In: CVPR (2017)
77. Zhou, L., Zhou, Y., Corso, J.J., Socher, R., Xiong, C.: End-to-end dense video captioning with masked transformer. In: CVPR (2018)
78. Zhu, L., Xu, Z., Yang, Y.: Bidirectional multirate reconstruction for temporal modeling in videos. In: CVPR (2017)
79. Zhu, Y., Jiang, S.: Attention-based densely connected LSTM for video captioning. In: ACM MM (2019)

Unpaired Learning of Deep Image Denoising

Xiaohe Wu[1], Ming Liu[1], Yue Cao[1], Dongwei Ren[2], and Wangmeng Zuo[1,3]($\boxtimes$)

[1] Harbin Institute of Technology, Harbin, China
csxhwu@gmail.com, csmliu@outlook.com, cscaoyue@gmail.com, wmzuo@hit.edu.com
[2] University of Tianjin, Tianjin, China
rendongweihit@gmail.com
[3] Peng Cheng Lab, Shenzhen, China

Abstract. We investigate the task of learning blind image denoising networks from an unpaired set of clean and noisy images. Such problem setting generally is practical and valuable considering that it is feasible to collect unpaired noisy and clean images in most real-world applications. And we further assume that the noise can be signal dependent but is spatially uncorrelated. In order to facilitate unpaired learning of denoising network, this paper presents a two-stage scheme by incorporating self-supervised learning and knowledge distillation. For self-supervised learning, we suggest a dilated blind-spot network (D-BSN) to learn denoising solely from real noisy images. Due to the spatial independence of noise, we adopt a network by stacking 1×1 convolution layers to estimate the noise level map for each image. Both the D-BSN and image-specific noise model (CNN_{est}) can be jointly trained via maximizing the constrained log-likelihood. Given the output of D-BSN and estimated noise level map, improved denoising performance can be further obtained based on the Bayes' rule. As for knowledge distillation, we first apply the learned noise models to clean images to synthesize a paired set of training images, and use the real noisy images and the corresponding denoising results in the first stage to form another paired set. Then, the ultimate denoising model can be distilled by training an existing denoising network using these two paired sets. Experiments show that our unpaired learning method performs favorably on both synthetic noisy images and real-world noisy photographs in terms of quantitative and qualitative evaluation. Code is available at https://github.com/XHWXD/DBSN.

Keywords: Image denoising · Unpaired learning · Convolutional networks · Self-supervised learning

---

Electronic supplementary material The online version of this chapter (https://doi.org/10.1007/978-3-030-58548-8_21) contains supplementary material, which is available to authorized users.

A. Vedaldi et al. (Eds.): ECCV 2020, LNCS 12349, pp. 352–368, 2020.
https://doi.org/10.1007/978-3-030-58548-8_21

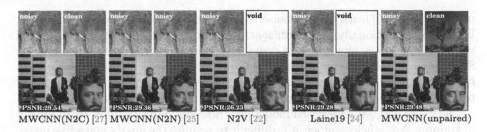

MWCNN(N2C) [27]   MWCNN(N2N) [25]   N2V [22]   Laine19 [24]   MWCNN(unpaired)

Fig. 1. Supervision settings for CNN denoisers, including Noise2Clean (MWCNN(N2C) [27]), Noise2Noise [25] (MWCNN(N2N)), Noise2Void (N2V [22]), Self-supervised learning (Laine19 [24]), and our unpaired learning scheme.

1   Introduction

Recent years have witnessed the unprecedented success of deep convolutional neural networks (CNNs) in image denoising. For additive white Gaussian noise (AWGN), numerous CNN denoisers, e.g., DnCNN [40], RED30 [30], MWCNN [27], N3Net [34], and NLRN [26], have been presented and achieved noteworthy improvement in denoising performance against traditional methods such as BM3D [9] and WNNM [14]. Subsequently, attempts have been made to apply CNN denoisers for handling more sophisticated types of image noise [18,35] as well as removing noise from real-world noisy photographs [2,6,8,15].

Albeit breakthrough performance has been achieved, the success of most existing CNN denoisers heavily depend on supervised learning with large amount of paired noisy-clean images [2,6,15,30,38,40,42]. On the one hand, given the form and parameters of noise model, one can synthesize noisy images from noise-less clean images to constitute a paired training set. However, real noise usually is complex, and the in-camera signal processing (ISP) pipeline in real-world photography further increases the complexity of noise, making it difficult to be fully characterized by basic parametric noise model. On the other hand, one can build the paired set by designing suitable approaches to acquire the nearly noise-free (or clean) image corresponding to a given real noisy image. For real-world photography, nearly noise-free images can be acquired by averaging multiple noisy images [2,31] or by aligning and post-processing low ISO images [33]. Unfortunately, the nearly noise-free images may suffer from over-smoothing issue and are cost-expensive to acquire. Moreover, such nearly noise-free image acquisition may not be applicable to other imaging mechanisms (e.g., microscopy or medical imaging), making it yet a challenging problem for acquiring noisy-clean image pairs for other imaging mechanisms.

Instead of supervised learning with paired training set, Lehtinen et al. [25] suggest a Noise2Noise (N2N) model to learn the mapping from pairs of noisy instances. However, it requires that the underlying clean images in each pair are exactly the same and the noises are independently drawn from the same distribution, thereby limiting its practicability. Recently, Krull et al. [22] introduce a practically more feasible Noise2Void (N2V) model which adopts a blind-spot network (BSN) to learn CNN denoisers solely from noisy images. Unfortunately, BSN is computationally very inefficient in training and fails to exploit the pixel value at blind spot, giving rise to degraded denoising performance (See Fig. 1). Subsequently, self-supervised model [24] and probabilistic N2V [23] have been further suggested to improve training efficiency via masked convolution [24] and to improve denoising performance via probabilistic inference [23,24].

N2V [22] and self-supervised model [24], however, fail to exploit clean images in training. Nonetheless, albeit it is difficult to acquire the nearly noise-free image corresponding to a given noisy image, it is practically feasible to collect a set of unpaired clean images. Moreover, specially designed BSN architecture generally is required to facilitate self-supervised learning, and cannot employ the progress in state-of-the-art networks [26,27,30,34,38,40–42] to improve denoising performance. Chen et al. [8] suggest an unpaired learning based blind denoising method GCBD based on the generative adversarial network (GAN) [17], but only achieve limited performance on real-world noisy photographs.

In this paper, we present a two-stage scheme, i.e., self-supervised learning and knowledge distillation, to learn blind image denoising network from an unpaired set of clean and noisy images. Instead of GAN-based unpaired learning [7,8], we first exploit only the noisy images to learn a BSN as well as an image-specific noise level estimation network CNN_{est} for image denoising and noise modeling. Then, the learned noise models are applied to clean images for synthesizing a paired set of training images, and we also use the real noisy images and the corresponding denoising results in the first stage to form another paired set. As for knowledge distillation, we simply train a state-of-the-art CNN denoiser, e.g., MWCNN [27], using the above two paired sets.

In particular, the clean image is assumed to be spatially correlated, making it feasible to exploit the BSN architecture for learning blind denoising network solely from noisy images. To improve the training efficiency, we present a novel dilated BSN (i.e., D-BSN) leveraging dilated convolution and fully convolutional network (FCN), allowing to predict the denoising result of all pixels with a single forward pass during training. We further assume that the noise is pixel-wise independent but can be signal dependent. Hence, the noise level of a pixel can be either a constant or only depends on the individual pixel value. Considering that the noise model and parameters may vary with different images, we suggest an image-specific CNN_{est} by stacking 1×1 convolution layers to meet the above requirements. Using unorganized collections of noisy images, both D-BSN and CNN_{est} can be jointly trained via maximizing the constrained log-likelihood. Given the outputs of D-BSN and CNN_{est}, we use the Bayes' rule to obtain the denoising result in the first stage. As for a given clean image in the second stage, an image-specific CNN_{est} is randomly selected to synthesize a noisy image.

Extensive experiments are conducted to evaluate our D-BSN and unpaired learning method, e.g., MWCNN(unpaired). On various types of synthetic noise (e.g., AWGN, heteroscedastic Gaussian, multivariate Gaussian), our D-BSN is efficient in training and is effective in image denoising and noise modeling. While our MWCNN(unpaired) performs better than the self-supervised model Laine19 [24], and on par with the fully-supervised counterpart (e.g., MWCNN [27]) (See Fig. 1). Experiments on real-world noisy photographs further validate the effectiveness of our MWCNN(unpaired). As for real-world noisy photographs, due to the effect of demosaicking, the noise violates the pixel-wise independent noise assumption, and we simply train our blind-spot network on pixel-shuffle down-sampled noisy images to circumvent this issue. The results show that our MWCNN(unpaired) also performs well and significantly surpasses GAN-based unpaired learning GCBD [8] on DND [33].

The contributions are summarized as follows:

1. A novel two-stage scheme by incorporating self-supervised learning and knowledge distillation is presented to learn blind image denoising network from an unpaired set of clean and noisy images. In particular, self-supervised learning is adopted for image denoising and noise modeling, consequently resulting in two complementary paired set to distill the ultimate denoising network.
2. A novel dilated blind-spot network (D-BSN) and an image-specific noise level estimation network CNN_{est} are elaborated to improve the training efficiency and to meet the assumed noise characteristics. Using unorganized collections of noisy images, D-BSN and CNN_{est} can be jointly trained via maximizing the constrained log-likelihood.
3. Experiments on various types of synthetic noise show that our unpaired learning method performs better than N2V [22] and Laine19 [24], and on par with its fully-supervised counterpart. MWCNN(unpaired) also performs well on real-world photographs and significantly surpasses GAN-based unpaired learning (GCBD) [8] on the DND [33] dataset.

2   Related Work

2.1   Deep Image Denoising

In the last few years, significant progress has been made in developing deep CNN denoisers. Zhang et al. [40] suggested DnCNN by incorporating residual learning and batch normalization, and achieved superior performance than most traditional methods for AWGN removal. Subsequently, numerous building modules have been introduced to deep denoising networks, such as dilated convolution [41], channel attention [4], memory block [38], and wavelet transform [27]. To fulfill the aim of image denoising, researchers also modified the representative network architectures, e.g., U-Net [27,30], Residual learning [40], and non-local network [26], and also introduced several new ones [19,34]. For handling AWGN with different noise levels and spatially variant noise, FFDNet was proposed by taking both noise level map and noisy image as network input. All these studies have consistently improved the denoising performance of deep networks [4,19,26,27,30,34,38,41]. However, CNN denoisers for AWGN usually generalize poorly to complex noise, especially real-world noisy images [33].

Besides AWGN, complex noise models, e.g., heteroscedastic Gaussian [33] and Gaussian-Poisson [11], have also been suggested, but are still not sufficient to approximate real sensor noise. As for real-world photography, the introduction of ISP pipeline makes the noise both signal-dependent and spatially correlated, further increasing the complexity of noise model [13,28]. Instead of noise modeling, several methods have been proposed to acquire nearly noise-free images by averaging multiple noisy images [2,31] or by aligning and post-processing low ISO images [33]. With the set of noisy-clean image pairs, several well-designed CNNs were developed to learn a direct mapping for removing noise from real-world noisy RAW and sRGB images [6,15,39]. Based upon the GLOW architecture [20], Abdelhamed et al. [1] introduced a deep compact noise model, i.e., Noise Flow, to characterize the real noise distribution from noisy-clean image pairs. In this work, we learn a FCN with 1×1 convolution from noisy images for modeling pixel-independent signal-dependent noise. To exploit the progress in CNN denoising, we further train state-of-the-art CNN denoiser using the synthetic noisy images generated with the learned noise model.

2.2   Learning CNN Denoisers Without Paired Noisy-Clean Images

Soltanayev and Chun [37] developed a Steins unbiased risk estimator (SURE) based method on noisy images. Zhussip et al. [44] further extended SURE to learn CNN denoisers from correlated pairs of noisy images. But the above methods only handle AWGN and require that noise level is known.

Lehtinen et al. [25] suggested to learn an N2N model from a training set of noisy image pairs, which avoids the acquisition of nearly noise-free images but remains limited in practice. Subsequently, N2V [22] (Noise2Self [5]) has been proposed to learn (calibrate) denoisers by requiring the output at a position does not depend on the input value at the same position. However, N2V [22] is inefficient in training and fails to exploit the pixel value at blind spot. To address these issues, Laine19 [24] and probabilistic N2V [23] were then suggested by introducing masked convolution [24] and probabilistic inference [23,24].

N2V and follow-up methods are solely based on noisy images without exploiting unpaired clean images. Chen et al. [8] presented a GAN-based model, i.e., GCBD, to learn CNN denoisers using unpaired noisy and clean images, but its performance on real-world noisy photographs still falls behind. In contrast to [8], we develop a non-GAN based method for unpaired learning. Our method involves two stage in training, i.e., self-supervised learning and knowledge distillation. As opposed to [22–24], our self-supervised learning method elaborately incorporates dilated convolution with FCN to design a D-BSN for improving training efficiency. For modeling pixel-independent signal-dependent noise, we adopt an image-specific FCN with 1×1 convolution. And constrained log-likelihood is then introduced to train D-BSN and image-specific CNN_{est}.

3   Proposed Method

In this section, we present our two-stage training scheme, i.e., self-supervised learning and knowledge distillation, for learning CNN denoisers from an unpaired set of noisy and clean images. After describing the problem setting and assumptions, we first introduce the main modules of our scheme and explain the knowledge distillation stage in details. Then, we turn to self-supervised learning by describing the dilated blind-spot network (D-BSN), image-specific noise model CNN_{est} and self-supervised loss. For handling real-world noisy photographs, we finally introduce a pixel-shuffle down-sampling strategy to apply our method.

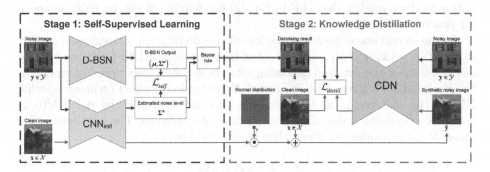

Fig. 2. Illustration of our two-stage training scheme involving self-supervised learning and knowledge distillation.

3.1   Two-Stage Training and Knowledge Distillation

This work tackles the task of learning CNN denoisers from an unpaired set of clean and noisy images. Thus, the training set can be given by two independent sets of clean images $\mathcal{X}$ and noisy images $\mathcal{Y}$. Here, $\mathbf{x}$ denotes a clean image from $\mathcal{X}$, and $\mathbf{y}$ a noisy image from $\mathcal{Y}$. With the unpaired setting, both the real noisy observation of $\mathbf{x}$ and the noise-free image of $\mathbf{y}$ are unavailable. Denote by $\tilde{\mathbf{x}}$ the underlying clean image of $\mathbf{y}$. The real noisy image $\mathbf{y}$ can be written as,

$$\mathbf{y} = \tilde{\mathbf{x}} + \mathbf{n}, \tag{1}$$

where $\mathbf{n}$ denotes the noise in $\mathbf{y}$. Following [22], we assume that the image $\mathbf{x}$ is spatially correlated and the noise $\mathbf{n}$ is pixel-independent and signal-dependent Gaussian. That is, the noise variance (or noise level) at pixel i is determined only by the underlying noise-free pixel value $\tilde{x}_i$ at pixel i,

$$\mathrm{var}(n_i) = g_{\tilde{\mathbf{x}}}(\tilde{x}_i). \tag{2}$$

Thus, $g_{\tilde{\mathbf{x}}}(\tilde{\mathbf{x}})$ can be regarded as a kind of noise level function (NLF) in multivariate heteroscedastic Gaussian model [11,33]. Instead of linear NLF in [11,33],

$g_{\tilde{\mathbf{x}}}(\tilde{\mathbf{x}})$ can be any nonlinear function, and thus is more expressive in noise modeling. We also note that the NLF may vary with images (e.g., AWGN with different variance), and thus image-specific $g_{\tilde{\mathbf{x}}}(\tilde{\mathbf{x}})$ is adopted for the flexibility issue.

With the above problem setting and assumptions, Fig. 2 illustrates our two-stage training scheme involving self-supervised learning and knowledge distillation. In the first stage, we elaborate a novel blind-spot network, i.e., D-BSN, and an image-specific noise model $\mathrm{CNN_{est}}$ (Refer to Sect. 3.2 for details). Then, self-supervised loss is introduced to jointly train D-BSN and $\mathrm{CNN_{est}}$ solely based on $\mathcal{Y}$ via maximizing the constrained log-likelihood (Refer to Sect. 3.3 for details). For a given real noisy image $\mathbf{y} \in \mathcal{Y}$, D-BSN and $\mathrm{CNN_{est}}$ collaborate to produce the first stage denoising result $\hat{\mathbf{x}}_{\mathbf{y}}$ and the estimated NLF $g_{\mathbf{y}}(\mathbf{y})$. It is worth noting that we modify the NLF in Eq. (2) by defining it on the noisy image $\mathbf{y}$ for practical feasibility.

In the second stage, we adopt the knowledge distillation strategy, and exploit $\mathcal{X}$, $\mathcal{Y}$, $\hat{\mathcal{X}}^{(1)} = \{\hat{\mathbf{x}}_{\mathbf{y}} | \mathbf{y} \in \mathcal{Y}\}$, and the set of image-specific NLFs $\{g_{\mathbf{y}}(\mathbf{y}) | \mathbf{y} \in \mathcal{Y}\}$ to distill a state-of-the-art deep denoising network in a fully-supervised manner. On the one hand, for a given clean image $\mathbf{x} \in \mathcal{X}$, we randomly select an image-specific NLF $g_{\mathbf{y}}(\mathbf{y})$, and use $g_{\mathbf{y}}(\mathbf{x})$ to generate a NLF for $\mathbf{x}$. Denote by $\mathbf{n}_0 \sim \mathcal{N}(0, 1)$ a random Gaussian noise of zero mean and one variance. The synthetic noisy image $\tilde{\mathbf{y}}$ corresponding to $\mathbf{x}$ can then be obtained by,

$$\tilde{\mathbf{y}} = \mathbf{x} + g_{\mathbf{y}}(\mathbf{x}) \cdot \mathbf{n}_0. \tag{3}$$

Consequently, we build the first set of paired noisy-clean images $\{(\mathbf{x}, \tilde{\mathbf{y}}) | \mathbf{x} \in \mathcal{X}\}$. On the other hand, given a real noisy image $\mathbf{y}$, we have its denoising result $\hat{\mathbf{x}}_{\mathbf{y}}$ in the first stage, thereby forming the second set of paired noisy-clean images $\{(\hat{\mathbf{x}}_{\mathbf{y}}, \mathbf{y}) | \mathbf{y} \in \mathcal{Y}\}$.

The above two paired sets are then used to distill a state-of-the-art convolutional denoising network (CDN) by minimizing the following loss,

$$\mathcal{L}_{distill} = \sum_{\mathbf{x} \in \mathcal{X}} \|\mathrm{CDN}(\tilde{\mathbf{y}}) - \mathbf{x}\|^2 + \lambda \sum_{\mathbf{y} \in \mathcal{Y}} \|\mathrm{CDN}(\mathbf{y}) - \hat{\mathbf{x}}_{\mathbf{y}}\|^2, \tag{4}$$

where $\lambda = 0.1$ is the tradeoff parameter. We note that the two paired sets are complementary, and both benefit the denoising performance. In particular, for $\{(\mathbf{x}, \tilde{\mathbf{y}}) | \mathbf{x} \in \mathcal{X}\}$, the synthetic noisy image $\tilde{\mathbf{y}}$ may not fully capture real noise complexity when the estimated image-specific noise model $\mathrm{CNN_{est}}$ is not accurate. Nonetheless, the clean image $\mathbf{x}$ are real, which is beneficial to learn denoising network with visually pleasing result and fine details. As for $\{(\hat{\mathbf{x}}_{\mathbf{y}}, \mathbf{y}) | \mathbf{y} \in \mathcal{Y}\}$, the noisy image $\mathbf{y}$ is real, which is helpful in compensating the estimation error in noise model. The denoising result $\hat{\mathbf{x}}_{\mathbf{y}}$ in the first stage may suffer from the over-smoothing effect, which, fortunately, can be mitigated by the real clean images in $\{(\mathbf{x}, \tilde{\mathbf{y}}) | \mathbf{x} \in \mathcal{X}\}$. In our two-stage training scheme, the convolutional denoising network CDN can be any existing CNN denoisers, and we consider MWCNN [27] as an example in our implementation.

Our two-stage training scheme offers a novel and effective approach to train CNN denoisers with unpaired learning. In contrast to GAN-based method [8], we present a self-supervised method for joint estimation of denoising result and image-specific noise model. Furthermore, knowledge distillation with two complementary paired sets is exploited to learn a deep denoising network in a fully-supervised manner. The network structures and loss function of our self-supervised learning method are also different with [22–24], which will be explained in the subsequent subsections.

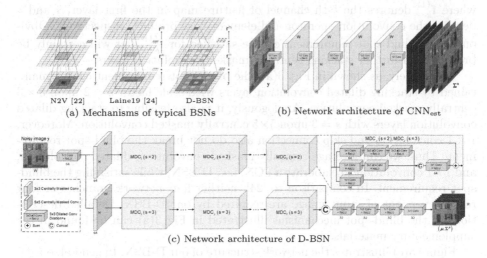

(a) Mechanisms of typical BSNs   (b) Network architecture of CNN$_{est}$

(c) Network architecture of D-BSN

Fig. 3. Mechanisms of BSNs, and network structures of D-BSN and CNN$_{est}$.

3.2 D-BSN and CNN$_{est}$ for Self-supervised Learning

Blind-spot network (BSN) generally is required for self-supervised learning of CNN denoisers. Among existing BSN solutions, N2V [22] is computationally very inefficient in training. Laine et al. [24] greatly multigate the efficiency issue, but still require four network branches or four rotated versions of each input image, thereby leaving some leeway to further improve training efficiency. To tackle this issue, we elaborately incorporate dilated convolution with FCN to design a D-BSN. Besides, Laine et al. [24] assume the form of noise distribution is known, e.g., AWGN and Poisson noise, and adopt a U-Net to estimate the parameters of noise model. In comparison, our model is based on a more general assumption that the noise $\mathbf{n}$ is pixel-independent, signal-dependent, and image-specific. To meet this assumption, we exploit a FCN with 1×1 convolution (i.e., CNN$_{est}$) to produce an image-specific NLF for noise modeling. In the following, we respectively introduce the structures of D-BSN and CNN$_{est}$.

D-BSN. For the output at a position, the core of BSN is to exclude the effect of the input value at the same position (i.e., blind-spot requirement). For the first convolution layer, we can easily meet this requirement using masked convolution [32]. Denote by $\mathbf{y}$ a real noisy image, and $\mathbf{w}_k$ the k-th 3×3 convolutional kernel. We introduce a 3×3 binary mask $\mathbf{m}$, and assign 0 to the central element of $\mathbf{m}$ and 1 to the others. The centrally masked convolution is then defined as,

$$\mathbf{f}_k^{(1)} = \mathbf{y} * (\mathbf{w}_k \circ \mathbf{m}), \tag{5}$$

where $\mathbf{f}_k^{(1)}$ denotes the k-th channel of feature map in the first layer, $*$ and $\circ$ denotes the convolution operator and element-wise product, respectively. Obviously, the blind-spot requirement can be satisfied for $\mathbf{f}_k^{(1)}$, but will certainly be broken when further stacking centrally masked convolution layers.

Fortunately, as shown in Fig. 3(a), the blind-spot requirement can be maintained by stacking dilated convolution layers with scale factor $s = 2$ upon 3×3 centrally masked convolution. Analogously, it is also feasible to stacking dilated convolution layers with $s = 3$ upon 5×5 centrally masked convolution. Moreover, 1×1 convolution and skip connection also do not break the blind-spot requirement. Thus, we can leverage centrally masked convolution, dilated convolution, and 1×1 convolution to elaborate a FCN, i.e., D-BSN, while satisfying the blind-spot requirement. In comparison to [24], neither four network branches or four rotated versions of input image are required by our D-BSN. Detailed description for the blind-spot mechanisms illustrated in Fig. 3(a) can be found in the supplementary materials.

Figure 3(c) illustrates the network structure of our D-BSN. In general, our D-BSN begins with a 1×1 convolution layer, and follows by two network branches. Each branch is composed of a 3×3 (5×5) centrally masked convolution layer following by six multiple dilated convolution (MDC) modules with $s = 2$ ($s = 3$). Then, the feature maps of the two branches are concatenated and three 1×1 convolution layers are further deployed to produce the network output. As shown in Fig. 3(c), the MDC module adopts a residual learning formulation and involves three sub-branches. In these sub-branches, zero, one, and two 3×3 convolution layers are respectively stacked upon a 1×1 convolution layer. The outputs of these sub-branches are then concatenated, followed by another 1×1 convolution layer, and added with the input of the MDC module. Then, the last 1×1 convolution layer is deployed to produce the output feature map. Finally, by concatenating the feature maps from the two network branches, we further apply three 1×1 convolution layers to produce the D-BSN output. Please refer to Fig. 3(c) for the detailed structure of D-BSN.

CNN$_{est}$. The noise is assumed to be conditionally pixel-wise independent given the underlying clean image. We assume that the noise is signal-dependent multivariate Gaussian with the NLF $g_{\tilde{\mathbf{x}}}(\tilde{\mathbf{x}})$, and further require that the NLF is image-specific to improve the model flexibility. Taking these requirements into account, we adopt a FCN architecture CNN$_{est}$ with 1×1 convolution to learn the noise model. Benefited from all 1×1 convolution layers, the noise level at a position can be guaranteed to only depends on the input value at the same position. Note that the input of $g_{\tilde{\mathbf{x}}}(\tilde{\mathbf{x}})$ is a clean image and we only have noisy images in self-supervised learning. Thus, $\mathrm{CNN_{est}}$ takes the noisy image as the input and learns the NLF $g_{\mathbf{y}}(\mathbf{y})$ to approximate $g_{\tilde{\mathbf{x}}}(\tilde{\mathbf{x}})$. For an input image of C channels ($C = 1$ for gray level image and 3 for color image), the output at a position i is a $C \times C$ covariance matrix $\Sigma_i^{\mathbf{n}}$, thereby making it feasible in modeling channel-correlated noise. Furthermore, we require each noisy image has its own network parameters in $\mathrm{CNN_{est}}$ to learn image specific NLF. From Fig. 3(b), our $\mathrm{CNN_{est}}$ consists of five 1×1 convolution layers of 16 channels. And the ReLU nonlinearity [21] is deployed for all convolution layers except the last one.

3.3   Self-supervised Loss and Bayes Denoising

In our unpaired learning setting, the underlying clean image $\tilde{\mathbf{x}}$ and ground-truth NLF of real noisy image $\mathbf{y}$ are unavailable. We thus resort to self-supervised learning to train D-BSN and $\mathrm{CNN_{est}}$. For a given position i, we have $y_i = \tilde{x}_i + n_i$ with $n_i \sim \mathcal{N}(\mathbf{0}, \Sigma_i^{\mathbf{n}})$. Here, y_i, $\tilde{x}_i$, n_i, and $\mathbf{0}$ all are $C \times 1$ vectors. Let $\boldsymbol{\mu}$ be the directly predicted clean image by D-BSN. We assume $\boldsymbol{\mu} = \tilde{\mathbf{x}} + \mathbf{n}^{\mu}$ with $n_i^{\mu} \sim \mathcal{N}(\mathbf{0}, \Sigma_i^{\mu})$, and further assume that n_i and μ_i are independent. It is noted that $\boldsymbol{\mu}$ is closer to $\tilde{\mathbf{x}}$ than $\mathbf{y}$, and usually we have $|\Sigma_i^{\mathbf{n}}| \gg |\Sigma_i^{\mu}| \approx 0$. Considering that $\tilde{x}_i$ is not available, a new variable $\epsilon_i = y_i - \mu_i$ is introduced and it has $\epsilon_i \sim \mathcal{N}(\mathbf{0}, \Sigma_i^{\mathbf{n}} + \Sigma_i^{\mu})$. The negative log-likelihood of $y_i - \mu_i$ can be written as,

$$\mathcal{L}_{\epsilon_i} = \frac{1}{2}(y_i - \mu_i)^\top (\Sigma_i^{\mathbf{n}} + \Sigma_i^{\mu})^{-1}(y_i - \mu_i) + \frac{1}{2}\log|\Sigma_i^{\mathbf{n}} + \Sigma_i^{\mu}|, \qquad (6)$$

where $|\cdot|$ denotes the determinant of a matrix.

However, the above loss ignores the constraint $|\Sigma_i^{\mathbf{n}}| \gg |\Sigma_i^{\mu}| \approx 0$. Actually, when taking this constraint into account, the term $\log|\Sigma_i^{\mathbf{n}} + \Sigma_i^{\mu}|$ in Eq. (6) can be well approximated by its first-order Taylor expansion at the point $\Sigma_i^{\mathbf{n}}$,

$$\log|\Sigma_i^{\mathbf{n}} + \Sigma_i^{\mu}| \approx \log|\Sigma_i^{\mathbf{n}}| + \mathrm{tr}\left((\Sigma_i^{\mathbf{n}})^{-1}\Sigma_i^{\mu}\right), \qquad (7)$$

where $\mathrm{tr}(\cdot)$ denotes the trace of a matrix. Note that $\Sigma_i^{\mathbf{n}}$ and Σ_i^{μ} are treated equally in the left term. While in the right term, smaller Σ_i^{μ} and larger $\Sigma_i^{\mathbf{n}}$ are favored based on $\mathrm{tr}\left((\Sigma_i^{\mathbf{n}})^{-1}\Sigma_i^{\mu}\right)$, which is consistent with $|\Sigma_i^{\mathbf{n}}| \gg |\Sigma_i^{\mu}| \approx 0$.

Actually, μ_i and Σ_i^{μ} can be estimated as the output of D-BSN at position i, i.e., $\hat{\mu}_i = (\text{D-BSN}_{\mu}(\mathbf{y}))_i$ and $\hat{\Sigma}_i^{\mu} = (\text{D-BSN}_{\Sigma}(\mathbf{y}))_i$. $\Sigma_i^{\mathbf{n}}$ can be estimated as the output of $\mathrm{CNN_{est}}$ at position i, i.e., $\hat{\Sigma}_i^{\mathbf{n}} = (\mathrm{CNN_{est}}(\mathbf{y}))_i$. By substituting Eq. (7) into Eq. (6), and replacing μ_i, Σ_i^{μ} and $\Sigma_i^{\mathbf{n}}$ with the network outputs, we adopt the constrained negative log-likelihood for learning D-BSN and $\mathrm{CNN_{est}}$,

$$\mathcal{L}_{self} = \sum_i \frac{1}{2}\left\{(y_i - \hat{\mu}_i)^\top(\hat{\Sigma}_i^{\mu} + \hat{\Sigma}_i^{\mathbf{n}})^{-1}(y_i - \hat{\mu}_i) + \log|\hat{\Sigma}_i^{\mathbf{n}}| + \mathrm{tr}\left((\hat{\Sigma}_i^{\mathbf{n}})^{-1}\hat{\Sigma}_i^{\mu}\right)\right\}. \quad (8)$$

After self-supervised learning, given the output $\text{D-BSN}_{\mu}(\mathbf{y})$, $\text{D-BSN}_{\Sigma}(\mathbf{y})$ and $\mathrm{CNN_{est}}(\mathbf{y})$, the denoising result in the first stage can be obtained using the Bayes' rule to each pixel,

$$\hat{x}_i = (\hat{\Sigma}_i^{\mu} + \hat{\Sigma}_i^{\mathbf{n}})^{-1}(\hat{\Sigma}_i^{\mu} y_i + \hat{\Sigma}_i^{\mathbf{n}} \hat{\mu}_i). \qquad (9)$$

3.4   Extension to Real-World Noisy Photographs

Due to the effect of demosaicking, the noise in real-world photographs is spatially correlated and violates the pixel-independent noise assumption, thereby restricting the direct application of our method. Nonetheless, such assumption is critical in separating signal (spatially correlated) and noise (pixel-independent). Fortunately, the noise is only correlated within a short range. Thus, we can break this dilemma by training D-BSN on the pixel-shuffle downsampled images with factor 4. Considering the noise distributions on sum-images are different, we assign the 16 sub-images to 4 groups according to the Bayer pattern. The results of 16 sub-images are then pixel-shuffle upsampled to form an image of the original size, and the guided filter [16] with radius of 1 and penalty value of 0.01 is applied to obtain the final denoising image. We note that denoising on pixel-shuffle sub-images slightly degrades the performance. Nonetheless, our method can still obtain visually satisfying results on real-world noisy photographs.

4   Experimental Results

In this section, we first describe the implementation details and conduct ablation study of our method. Then, extensive experiments are carried out to evaluate our method on synthetic and real-world noisy images. The evaluation is performed on a PC with Intel(R) Core (TM) i9-7940X CPU @ 3.1 GHz and an Nvidia Titan RTX GPU. The source code and pre-trained models will be publicly available.

4.1   Implementation Details

Our unpaired learning consists of two stages: (i) self-supervised training of D-BSN and CNN_{est} and (ii) knowledge distillation for training MWCNN [27], which are respectively described as follows.

Self-supervised Training. For synthetic noises, the clean images are from the validation set of ILSVRC2012 (ImageNet) [10] while excluding the images smaller than 256×256. Several basic noise models, e.g., AWGN, multivariate Gaussian, and heteroscedastic Gaussian, are adopted to synthesize the noisy images $\mathcal{Y}$. While for real noisy images, we simply use the testing dataset as $\mathcal{Y}$. During the training, we randomly crop $48,000$ patches with size 80×80 in each epoch and finish the training after 180 epochs. The Adam optimizer is employed to train D-BSN and CNN_{est}. The learning rate is initialized as 1×10^{-4}, and is decayed by factor 10 after every 30 epochs until reaching 1×10^{-7}.

Knowledge Distillation. For gray level images, we adopt the BSD [36] training set as clean image set $\mathcal{X}$. For color images, DIV2K [3], WED [29] and CBSD [36] training set are used to form $\mathcal{X}$. Then, we exploit both $\{(\hat{\mathbf{x}}_{\mathbf{y}}, \mathbf{y}) | \mathbf{y} \in \mathcal{Y}\}$ and $\{(\mathbf{x}, \tilde{\mathbf{y}}) | \mathbf{x} \in \mathcal{X}\}$ to train a state-of-the-art CNN denoiser from scratch. And MWCNN with original setting [27] on learning algorithm is adopted to train the CNN denoiser on our data.

Table 1. Average PNSR(dB) results of different methods on the BSD68 dataset with noise levels 15, 25 and 50, and heteroscedastic Gaussian (HG) noise with $\alpha = 40$, $\delta = 10$.

Noise	Para	BM3D [9]	DnCNN [40]	NLRN [26]	N3Net [34]	N2V [22]	Laine19 [24]	GCBD [8]	D-BSN (ours)	N2C [27]	N2N [25]	Ours (full)
AWGN	$\sigma = 15$	31.07	31.72	31.88	–	–	–	31.37	31.24	31.86	31.71	31.82
	$\sigma = 25$	28.57	29.23	29.41	29.30	27.71	29.27	28.83	28.88	29.41	29.33	29.38
	$\sigma = 50$	25.62	26.23	26.47	26.39	–	–	–	25.81	26.53	26.52	26.51
HG	$\alpha = 40$ $\delta = 10$	23.84	–	–	–	–	–	–	29.13	30.16	29.53	30.10

4.2 Comparison of Different Supervision Settings

CNN denoisers can be trained with different supervision settings, such as N2C, N2N [25], N2V [22], Laine19 [24], GCBD [8], our D-BSN and MWCNN(unpaired). For a fair comparison, we retrain two MWCNN models with the N2C and N2N [25] settings, respectively. The results of N2V [22], Laine19 [24] and GCBD [8] are from the original papers.

Results on Gray Level Images. We consider two basic noise models, i.e., AWGN with $\sigma = 15, 25$ and 50, and heteroscedastic Gaussian (HG) [33] $n_i \sim \mathcal{N}(0, \alpha^2 x_i + \delta^2)$ with $\alpha = 40$ and $\delta = 10$. From Table 1, on BSD68 [36] our D-BSN performs better than N2V [22] and on par with GCBD [8], but is inferior to Laine19 [24]. Laine19 [24] learns the denoisers solely from noisy images and does not exploit unpaired clean images in training, making it still poorer than MWCNN(N2C). By exploiting both noisy and clean images, our MWCNN(unpaired) (also Ours(full) in Table 1) outperforms both Laine19 [24] and MWCNN(N2N) in most cases, and is on par with MWCNN(N2C). In terms of training time, Laine19 takes about 14 h using four Tesla V100 GPUs on NVIDIA DGX-1 servers. In contrast, our D-BSN takes about 10 h on two 2080Ti GPUs and thus is more efficient. In terms of testing time, Our MWCNN(unpaired) takes 0.020s to process a 320 × 480 image, while N2V [22] needs 0.034s and Laine19 [24] spends 0.044 s.

Results on Color Images. Besides AWGN and HG, we further consider another noise model, i.e., multivariate Gaussian (MG) [42] $n \sim \mathcal{N}(0, \Sigma)$ with $\Sigma = 75^2 \cdot U \Lambda U^T$. Here, U is a random unitary matrix, Λ is a diagonal matrix of three random values in the range $(0, 1)$. GCBD [8] and N2V [22] did not report their results on color image denoising. On CBSD68 [36] our MWCNN(unpaired) is consistently better than Laine19 [24], and notably outperforms MWCNN(N2N) [25] for HG noise. We have noted that both MWCNN(unpaired) and D-BSN perform well in handling multivariate Gaussian with cross-channel correlation.

Table 2. Average PNSR(dB) results of different methods on the CBSD68 dataset with noise levels 15, 25 and 50, heteroscedastic Gaussian (HG) noise with $\alpha = 40$, $\delta = 10$, and multivariate Gaussian (MG) noise.

Noise	Para	CBM3D [9]	CDnCNN [40]	FFDNet [42]	CBDNet [15]	Laine19 [24]	D-BSN (ours)	N2C [27]	N2N [25]	Ours (full)
AWGN	$\sigma = 15$	33.52	33.89	33.87	–	–	33.56	34.08	33.76	34.02
	$\sigma = 25$	30.71	30.71	31.21	–	31.35	30.61	31.40	31.22	31.40
	$\sigma = 50$	27.38	27.92	27.96	–	–	27.41	28.26	27.79	28.25
HG	$\alpha = 40\ \delta = 10$	23.21	-	28.67	30.89	-	30.53	32.10	31.13	31.72
MG	$\Sigma = 75^2 \cdot U\Lambda U^T \lambda_c \in (0,1)$	24.07	-	26.78	21.50	-	26.48	26.89	26.59	26.81
	$U^T U = I$	24.07	-	26.78	21.50	-	26.48	26.89	26.59	26.81

Table 3. Average PNSR(dB) results of different methods on KODAK24 and McMaster datasets.

Dataset	Para	CBM3D [9]	CDnCNN [40]	FFDNet [42]	Laine19 [24]	D-BSN (ours)	Ours (full)
KODAK24	$\sigma = 15$	34.28	34.48	34.63	–	33.74	34.82
	$\sigma = 25$	31.68	32.03	32.13	32.33	31.64	32.35
	$\sigma = 50$	28.46	28.85	28.98	-	28.69	29.36
McMaster	$\sigma = 15$	34.06	33.44	34.66	–	33.85	34.87
	$\sigma = 25$	31.66	31.51	32.35	32.52	31.56	32.54
	$\sigma = 50$	28.51	28.61	29.18	–	28.87	29.58

Table 4. The quantitative results (PSNR/SSIM) of different methods on the CC15 and DND (sRGB images) datasets.

Method	BM3D [9]	DnCNN [40]	CBDNet [15]	VDN [39]	N2S [5]	N2V [22]	GCBD [8]	MWCNN (unpaired)
Supervised	–	Yes	Yes	Yes	Not	Not	Not	Not
CC15	35.19	33.86	36.47	–	35.38	35.27	–	35.90
	0.9063	0.8636	0.9392	–	0.9204	0.9158	–	0.9370
DND	34.51	32.43	38.05	39.38	–	–	35.58	37.93
	0.8507	0.7900	0.9421	0.9518	–	–	0.9217	0.9373

4.3 Experiments on Synthetic Noisy Images

In this subsection, we assess our MWCNN(unpaired) in handling different types of synthetic noise, and compare it with the state-of-the-art image denoising methods. The competing methods include BM3D [9], DnCNN [40], NLRN [26] and N3Net [34] for gray level image denoising on BSD68 [36], and CBM3D [9], CDnCNN [40], FFDNet [42] and CBDNet [15] for color image denoising on BSD68 [36], KODAK24 [12], McMaster [43].

AWGN. From Table 1, it can be seen that our MWCNN(unpaired) performs on par with NLRN [26] and better than BM3D [9], DnCNN [40] and N3Net [34]. Tables 2 and 3 list the results of color image denoising. Benefited from unpaired learning and the use of MWCNN [27], our MWCNN(unpaired) outperforms CBM3D [9], CDnCNN [40] and FFDNet [42] by a non-trivial margin for all noise levels and on the three datasets.

Heteroscedastic Gaussian. We further test our MWCNN(unpaired) in handling heteroscedastic Gaussian (HG) noise which is usually adopted for modeling RAW image noise. It can be seen from Table 2 that all the CNN denoisers outperform CBM3D [9] by a large margin on CBSD68. CBDNet [15] takes both HG noise and in-camera signal processing pipeline into account when training the deep denoising network, and thus is superior to FFDNet [42]. Our MWCNN(unpaired) can leverage the progress in CNN denoisers and well exploit the unpaired noisy and clean images, and achieves a PSNR gain of 0.8dB against CBDNet [15].

Multivariate Gaussian. Finally, we evaluate our MWCNN(unpaired) in handling multivariate Gaussian (MG) noise with cross-channel correlation. Some image processing operations, e.g., image demosaicking, may introduce cross-channel correlated noise. From Table 2, FFDNet [42] is flexible in handle HG noise, while our unpaired learning method, MWCNN(unpaired), performs on par with FFDNet [42] on CBSD68.

4.4   Experiments on Real-World Noisy Photographs

Finally, we conduct comparison experiments on two widely adopted datasets of real-world noisy photographs, i.e, DND [33] and CC15 [31]. Our methods is compared with both traditional denoising method, i.e., CBM3D [9], deep Gaussian denoiser, i.e., DnCNN [40], deep blind denoisers, i.e., CBDNet [15] and VDN [39], and unsupervised learning methods, i.e., GCBD [8], N2S [25] and N2V [22], in terms of both quantitative and qualitative results. The average PSNR and SSIM metrics are presented in Table 4. On CC15, our MWCNN(unpaired) outperforms the other unsupervised learning methods (i.e., N2S [25] and N2V [22]) by a large margin (0.5dB in PSNR). On DND, our MWCNN(unpaired) achieves a PSNR gain of 2.3dB against GCBD. The results clearly show the merits of our methods in exploiting unpaired noisy and clean images. Actually, our method is only inferior to deep models specified for real-world noisy photography, e.g., CBDNet [15] and VDN [39]. Such results should not be criticized considering that our MWCNN(unpaired) has no access to neither the details of ISP [15] and the paired noisy-clean images [39]. Moreover, we adopt the pixel-shuffle downsampling to decouple spatially correlated noise, which also gives rise to moderate performance degradation of our method. Nonetheless, it can be seen from Fig. 4 that our MWCNN(unpaired) achieves comparable or better denoising results in comparison to all the competing methods on DND and CC15.

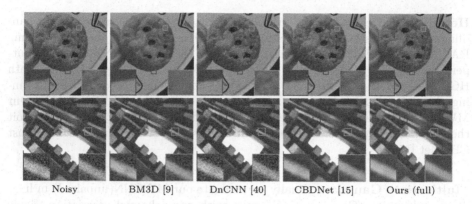

| Noisy | BM3D [9] | DnCNN [40] | CBDNet [15] | Ours (full) |

Fig. 4. Denoising results of different methods on real-world images from CC15(up) and DND(down) datasets.

5   Concluding Remarks

This paper presented a novel unpaired learning method by incorporating self-supervised learning and knowledge distillation for training CNN denoisers. In self-supervised learning, we proposed a dilated blind-spot network (D-BSN) and a FCN with 1×1 convolution, which can be efficiently trained via maximizing constrained log-likelihood from unorganized collections of noisy images. For knowledge distillation, the estimated denoising image and noise model are used to distill the state-of-the-art CNN denoisers, such as DnCNN and MWCNN. Experimental results showed that the proposed method is effective on both images with synthetic noise (e.g., AWGN, heteroscedastic Gaussian, multivariate Gaussian) and real-world noisy photographs.

Compared with [22,24] and GAN-based unpaired learning of CNN denoisers [8], our method has several merits in efficiently training blind-spot networks and exploiting unpaired noisy and clean images. However, there remain a number of challenges to be addressed: (i) our method is based on the assumption of pixel-independent heteroscedastic Gaussian noise, while real noise can be more complex and spatially correlated. (ii) In self-supervised learning, estimation error may be unavoidable for clean images and noise models, and it is interesting to develop robust and accurate distillation of CNN denoisers.

Acknowledgement. This work is partially supported by the National Natural Science Foundation of China (NSFC) under Grant No.s 61671182, U19A2073.

References

1. Abdelhamed, A., Brubaker, M.A., Brown, M.S.: Noise flow: noise modeling with conditional normalizing flows. In: ICCV, pp. 3165–3173 (2019)

2. Abdelhamed, A., Lin, S., Brown, M.S.: A high-quality denoising dataset for smart-phone cameras. In: CVPR, pp. 1692–1700 (2018)
3. Agustsson, E., Timofte, R.: NTIRE 2017 challenge on single image super-resolution: dataset and study. In: CVPR Workshops (2017)
4. Anwar, S., Barnes, N.: Real image denoising with feature attention. In: ICCV, pp. 3155–3164 (2019)
5. Batson, J., Royer, L.: Noise2Self: blind denoising by self-supervision, pp. 524–533 (2019)
6. Brooks, T., Mildenhall, B., Xue, T., Chen, J., Sharlet, D., Barron, J.T.: Unprocessing images for learned raw denoising. In: CVPR, pp. 11036–11045 (2019)
7. Bulat, A., Yang, J., Tzimiropoulos, G.: To learn image super-resolution, use a GAN to learn how to do image degradation first. In: Ferrari, V., Hebert, M., Sminchisescu, C., Weiss, Y. (eds.) ECCV 2018. LNCS, vol. 11210, pp. 187–202. Springer, Cham (2018). https://doi.org/10.1007/978-3-030-01231-1_12
8. Chen, J., Chen, J., Chao, H., Yang, M.: Image blind denoising with generative adversarial network based noise modeling. In: CVPR, pp. 3155–3164 (2018)
9. Dabov, K., Foi, A., Katkovnik, V., Egiazarian, K.: Image denoising by sparse 3-D transform-domain collaborative filtering. TIP $16(8)$, 2080–2095 (2007)
10. Deng, J., Dong, W., Socher, R., Li, L.J., Li, K., Fei-Fei, L.: ImageNet: a large-scale hierarchical image database. In: CVPR, pp. 248–255 (2009)
11. Foi, A., Trimeche, M., Katkovnik, V., Egiazarian, K.: Practical Poissonian-Gaussian noise modeling and fitting for single-image raw-data. TIP $17(10)$, 1737–1754 (2008)
12. Franzen, R.: Kodak lossless true color image suite, April 1999. http://r0k.us/graphics/kodak
13. Grossberg, M.D., Nayar, S.K.: Modeling the space of camera response functions. TPAMI $26(10)$, 1272–1282 (2004)
14. Gu, S., Zhang, L., Zuo, W., Feng, X.: Weighted nuclear norm minimization with application to image denoising. In: CVPR, pp. 2862–2869 (2014)
15. Guo, S., Yan, Z., Zhang, K., Zuo, W., Zhang, L.: Toward convolutional blind denoising of real photographs. In: CVPR, pp. 1712–1722 (2019)
16. He, K., Sun, J., Tang, X.: Guided image filtering. In: Daniilidis, K., Maragos, P., Paragios, N. (eds.) ECCV 2010. LNCS, vol. 6311, pp. 1–14. Springer, Heidelberg (2010). https://doi.org/10.1007/978-3-642-15549-9_1
17. Goodfellow, I.J., et al.: Generative adversarial networks. In: NIPS, p. 2672C2680 (2014)
18. Islam, M.T., Rahman, S.M., Ahmad, M.O., Swamy, M.: Mixed Gaussian-impulse noise reduction from images using convolutional neural network. Sig. Process. Image Commun. **68**, 26–41 (2018)
19. Jia, X., Liu, S., Feng, X., Zhang, L.: FOCNet: a fractional optimal control network for image denoising. In: CVPR, pp. 6054–6063 (2019)
20. Kingma, D.P., Dhariwal, P.: Glow: generative flow with invertible 1x1 convolutions. In: NIPS, pp. 10215–10224 (2018)
21. Krizhevsky, A., Sutskever, I., Hinton, G.E.: ImageNet classification with deep convolutional neural networks. In: NIPS, pp. 1097–1105 (2012)
22. Krull, A., Buchholz, T.O., Jug, F.: Noise2Void-learning denoising from single noisy images. In: CVPR, pp. 2129–2137 (2019)
23. Krull, A., Vicar, T., Jug, F.: Probabilistic Noise2Void: unsupervised content-aware denoising. arXiv preprint arXiv:1906.00651 (2019)
24. Laine, S., Karras, T., Lehtinen, J., Aila, T.: High-quality self-supervised deep image denoising. In: NIPS, pp. 6968–6978 (2019)

25. Lehtinen, J., et al.: Noise2Noise: learning image restoration without clean data. In: ICML, pp. 2965–2974 (2018)
26. Liu, D., Wen, B., Fan, Y., Loy, C.C., Huang, T.S.: Non-local recurrent network for image restoration. In: NIPS, pp. 1673–1682 (2018)
27. Liu, P., Zhang, H., Zhang, K., Lin, L., Zuo, W.: Multi-level wavelet-CNN for image restoration. In: CVPR Workshops, pp. 773–782 (2018)
28. Liu, X., Tanaka, M., Okutomi, M.: Practical signal-dependent noise parameter estimation from a single noisy image. TIP **23**(10), 4361–4371 (2014)
29. Ma, K., et al.: Waterloo exploration database: new challenges for image quality assessment models. TIP **26**(2), 1004–1016 (2016)
30. Mao, X., Shen, C., Yang, Y.B.: Image restoration using very deep convolutional encoder-decoder networks with symmetric skip connections. In: NIPS, pp. 2802–2810 (2016)
31. Nam, S., Hwang, Y., Matsushita, Y., Joo Kim, S.: A holistic approach to cross-channel image noise modeling and its application to image denoising. In: CVPR, pp. 1683–1691 (2016)
32. Oord, A.v.d., Kalchbrenner, N., Kavukcuoglu, K.: Pixel recurrent neural networks. In: ICML, p. 1747C1756 (2016)
33. Plotz, T., Roth, S.: Benchmarking denoising algorithms with real photographs. In: CVPR, pp. 1586–1595 (2017)
34. Plötz, T., Roth, S.: Neural nearest neighbors networks. In: NIPS, pp. 1087–1098 (2018)
35. Remez, T., Litany, O., Giryes, R., Bronstein, A.M.: Class-aware fully convolutional Gaussian and Poisson denoising. TIP **27**(11), 5707–5722 (2018)
36. Roth, S., Black, M.J.: Fields of experts. IJCV **82**(2), 205 (2009)
37. Soltanayev, S., Chun, S.Y.: Training deep learning based denoisers without ground truth data. In: NIPS, pp. 3257–3267 (2018)
38. Tai, Y., Yang, J., Liu, X.: Image super-resolution via deep recursive residual network. In: CVPR, pp. 3147–3155 (2017)
39. Yue, Z., Yong, H., Zhao, Q., Meng, D., Zhang, L.: Variational denoising network: toward blind noise modeling and removal. In: NIPS, pp. 1688–1699 (2019)
40. Zhang, K., Zuo, W., Chen, Y., Meng, D., Zhang, L.: Beyond a Gaussian denoiser: residual learning of deep CNN for image denoising. TIP **26**(7), 3142–3155 (2017)
41. Zhang, K., Zuo, W., Gu, S., Zhang, L.: Learning deep CNN denoiser prior for image restoration. In: Proceedings of the IEEE Conference on Computer Vision and Pattern Recognition, pp. 3929–3938 (2017)
42. Zhang, K., Zuo, W., Zhang, L.: FFDNet: toward a fast and flexible solution for CNN-based image denoising. TIP **27**(9), 4608–4622 (2018)
43. Zhang, L., Wu, X., Buades, A., Li, X.: Color demosaicking by local directional interpolation and nonlocal adaptive thresholding. J. Electron. Imaging **20**(2), 023016 (2011)
44. Zhussip, M., Soltanayev, S., Chun, S.Y.: Training deep learning based image denoisers from undersampled measurements without ground truth and without image prior. In: CVPR, pp. 10255–10264 (2019)

Self-supervising Fine-Grained Region Similarities for Large-Scale Image Localization

Yixiao Ge[1], Haibo Wang[3], Feng Zhu[2], Rui Zhao[2], and Hongsheng Li[1(✉)]

[1] The Chinese University of Hong Kong, Shatin, Hong Kong
yxge@link.cuhk.edu.hk, haibo@cumt.edu.cn
[2] SenseTime Research, Shanghai, China
{zhufeng,zhaorui}@sensetime.com
[3] China University of Mining and Technology, Xuzhou, China
haibo@cumt.edu.cn

Abstract. The task of large-scale retrieval-based image localization is to estimate the geographical location of a query image by recognizing its nearest reference images from a city-scale dataset. However, the general public benchmarks only provide noisy GPS labels associated with the training images, which act as weak supervisions for learning image-to-image similarities. Such label noise prevents deep neural networks from learning discriminative features for accurate localization. To tackle this challenge, we propose to self-supervise image-to-region similarities in order to fully explore the potential of difficult positive images alongside their sub-regions. The estimated image-to-region similarities can serve as extra training supervision for improving the network in generations, which could in turn gradually refine the fine-grained similarities to achieve optimal performance. Our proposed self-enhanced image-to-region similarity labels effectively deal with the training bottleneck in the state-of-the-art pipelines without any additional parameters or manual annotations in both training and inference. Our method outperforms state-of-the-arts on the standard localization benchmarks by noticeable margins and shows excellent generalization capability on multiple image retrieval datasets (Code of this work is available at https://github.com/yxgeee/SFRS.).

1 Introduction

Image-based localization (IBL) aims at estimating the location of a given image by identifying reference images captured at the same places from a geo-tagged database. The task of IBL has long been studied since the era of hand-crafted features [2,21,22,29,34,35] and has been attracting increasing attention with the advances of convolutional neural networks (CNN) [46], motivated by its wide

Electronic supplementary material The online version of this chapter (https://doi.org/10.1007/978-3-030-58548-8_22) contains supplementary material, which is available to authorized users.

© Springer Nature Switzerland AG 2020
A. Vedaldi et al. (Eds.): ECCV 2020, LNCS 12349, pp. 369–386, 2020.
https://doi.org/10.1007/978-3-030-58548-8_22

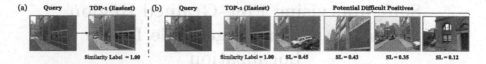

Fig. 1. (a) To mitigate the noise with weak GPS labels, existing works [1,28] only utilized the easiest top-1 image of the query for training. (b) We propose to adopt self-enhanced similarities as soft training supervisions to effectively explore the potential of difficult positives

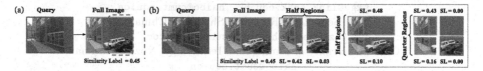

Fig. 2. (a) Even with true positive pairs, the image-level supervision provides misleading information for non-overlapping regions. (b) We further estimate fine-grained soft labels to refine the supervision by measuring query-to-region similarities

applications in SLAM [17,30] and virtual/augmented reality [5]. Previous works have been trying to tackle IBL as image retrieval [1,23,28], 2D-3D structure matching [27,38] or geographical position classification [19,43,44] tasks. In this paper, we treat the problem as an image retrieval task, given its effectiveness and feasibility in large-scale and long-term localization.

The fundamental challenge faced by image retrieval-based methods [1,23,28] is to learn image representations that are discriminative enough to tell apart repetitive and similar-looking locations in GPS-tagged datasets. It is cast as a weakly-supervised task because geographically close-by images may not depict the same scene when facing different directions.

To avoid being misled by noisy and weak GPS labels, most works [1,28] utilized on-the-fly first-ranking images of the queries as their positive training samples, *i.e.* to force the queries to be closer to their already nearest neighbors (Fig. 1(a)). Consequently, a paradox arises when only the most confident, or in other words, the *easiest* on-the-fly positives are utilized in the training process, but these images in turn result in a lack of robustness to varying conditions as the first-ranking images for queries might be too simple to provide enough supervisions for learning robust feature representations. To tackle the issue, we argue that difficult positives are needed for providing informative supervisions.

Identifying the true difficult positives, however, is challenging with only geographical tags provided from the image database. Therefore, the network is easy to collapse when naïvely trained with lower-ranking positive samples, which might not have overlapping regions with the queries at all. Such false positives might deteriorate the feature learning. Kim *et al.*[23] attempted to mine true positive images by verifying their geometric relations, but it is limited by the accuracy of off-the-shelf geometric techniques [9,18]. In addition, even if pairs of images are indeed positive pairs, they might still have non-corresponding regions.

The correct image-level labels might not necessarily be the correct region-level labels. Therefore, the ideal supervisions between paired positive images should provide region-level correspondences for more accurately supervising the learning of local features.

To tackle the above mentioned challenge, we propose to estimate informative *soft* image-to-region supervisions for training image features with noisy positive samples in a self-supervised manner. In the proposed system, an image retrieval network is trained in generations, which gradually improves itself with self-supervised image-to-region similarities. We train the network for the first generation following the existing pipelines [1,28], after which the network can be assumed to successfully capture most feature distribution of the training data. For each query image, however, its k-nearest gallery images according to the learned features might still contain noisy (false) positive samples. Directly utilizing them as difficult true positives in existing pipelines might worsen the learned features. We therefore propose to utilize their previous generation's query-gallery similarities to serve as the *soft* supervision for training the network in a new generation. The generated soft supervisions are gradually refined and sharpened as the network generation progresses. In this way, the system is no longer limited to learning only from the on-the-fly *easiest* samples, but can fully explore the potential of the difficult positives and mitigate their label noise with refined soft confidences (Fig. 1(b)).

However, utilizing only the image-level soft supervisions for paired positive images simply forces features from all their spatial regions to approach the same target scores (Fig. 2(a)). Such an operation would hurt the network's capability of learning discriminative local features. To mitigate this problem, we further decompose the matching gallery images into multiple sub-regions of different sizes. The query-to-region similarities estimated from the previous generation's network can serve as the refined soft supervisions for providing more fine-grained guidance for feature learning (Fig. 2(b)). The above process can iterate and train the network for multiple generations to progressively provide more accurate and fine-grained image-to-region similarities for improving the features.

Our contributions can be summarised as three-fold. (1) We propose to estimate and self-enhance the image similarities of top-ranking gallery images to fully explore the potential of difficult positive samples in image-based localization (IBL). The self-supervised similarities serve as refined training supervisions to improve the network in generations, which in turn, generates more accurate image similarities. (2) We further propose to estimate image-to-region similarities to provide region-level supervisions for enhancing the learning of local features. (3) The proposed system outperforms state-of-the-art methods on standard localization benchmarks by noticeable margins, and shows excellent generalization capability on both IBL and standard image retrieval datasets.

2 Related Work

Image-Based Localization (IBL). Existing works on image-based localization can be grouped into image retrieval-based [1,23,28], 2D-3D registration-based [27,38] and per-position classification-based [19,43,44] methods. Our work aims at solving the training bottleneck of the weakly-supervised image retrieval-based IBL problem. We therefore mainly discuss related solutions that cast IBL as an image retrieval task. Retrieval-based IBL is mostly related to the traditional place recognition task [3,6,24,39,41], which has long been studied since the era of hand-engineered image descriptors, e.g. SIFT [29], BoW [35], VLAD [2] and Fisher Vector [34]. Thanks to the development of deep learning [26], NetVLAD [1] successfully transformed dense CNN features for localization by proposing a learnable VLAD layer to effectively aggregate local descriptors with learnable semantic centers. Adopting the backbone of NetVLAD, later works further looked into multi-scale contextual information [23] or effective metric learning [28] to achieve better performance. Our work has the similar motivation with [23], i.e. to mine difficult positive samples for more effective local feature learning. However, [23] was only able to roughly mine positive images by adopting off-the-shelf geometric verification techniques [9,18] and required noticeable more parameters for both training and inference. In contrast, our method self-enhances fine-grained supervisions by gradually estimating and refining image-to-region soft labels without any additional parameters.

Self-supervised Label Estimation has been widely studied in self-supervised learning methods [8,16,25,32,33], where the network mostly learns to predict a collective label set by properly utilizing the capability of the network itself. Some works [4,11,12,14] proposed to create task-relative pseudo labels, e.g. [4] generated image-level pseudo labels for classification by clustering features on-the-fly. The others [13,15,31,48] attempted to optimize the network by dealing with pretext tasks whose labels are more easily to create, e.g. [31] solved jigsaw puzzles and [15] predicted the rotation of transformed images. Inspired by the self-supervised learning methods, we propose a self-supervised image-to-region label creation scheme for training the network in generations, which effectively mitigates the image-to-image label noise caused by weak geo-tagged labels.

Self-distillation. Training in generations via self-predicted soft labels has been investigated in self-distillation methods [10,45,47]. However, soft labels in these methods are not applicable to our task, as they focus on the classification problem with predefined classes. We successfully generalize self-distillation to the weakly-supervised IBL problem by proposing to generate soft labels for both query-to-gallery and query-to-region similarities in the retrieval task.

3 Method

We propose to self-supervise image-to-region similarities for tackling the problem of noisy pairwise image labels in image-based localization (IBL). In major public benchmarks, each training image is generally associated with a GPS location

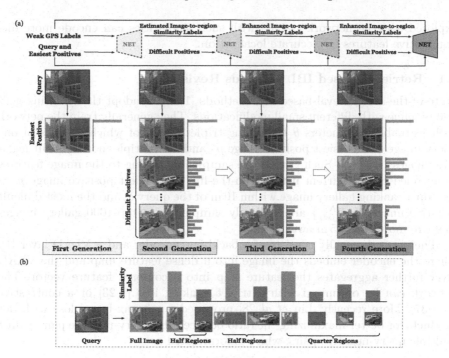

Fig. 3. Illustration of our proposed self-supervised image-to-region similarities for image localization, where the image-to-region similarities are gradually refined via training the network in generations as shown in (a). The sidebars for each difficult positive image demonstrate the soft similarity labels for the full image, left half, right half, top half, bottom half, top-left quarter, top-right quarter, bottom-left quarter and bottom-right quarter sub-regions respectively, detailed in (b), whereas the easiest positives only have one bar indicating the similarity label for the full image. Note that the most difficult negative samples utilized for joint training are not shown in the figure for saving space

tag. However, images geographically close to each other do not necessarily share overlapping regions. This is because the images might be captured from opposite viewing directions. The GPS tags could only help to geographically discover potential positive images with much label noise. Such a limitation causes bottleneck for effectively training the neural networks to recognize challenging positive pairs. There were previous methods [1,23,28] tackling this problem. However, they either ignored the potential of difficult but informative positive images, or required off-the-shelf time-consuming techniques with limited precision for filtering positives.

As illustrated in Fig. 3, the key of our proposed framework is to gradually enhance image-to-region similarities as the training proceeds, which can in turn act as soft and informative training supervisions for iteratively refining the network itself. In this way, the potential of difficult positive samples can be fully

374   Y. Ge et al.

explored and the network trained with such similarities can encode more discriminative features for accurate localization.

3.1 Retrieval-Based IBL Methods Revisit

State-of-the-art retrieval-based IBL methods [1,23,28] adopt the following general pipeline with different small modifications. They generally train the network with learnable parameters θ by feeding triplets, each of which consists of one query image q, its easiest positive image p^* and its multiple most difficult negative images $\{n_j|_{j=1}^{N}\}$. Such triplets are sampled according to the image features f_θ encoded by the current network on-the-fly. The easiest positive image p^* is the top-1 ranking gallery image within 10 m of the query q, and the most difficult negative images $\{n_j|_{j=1}^{N}\}$ are randomly sampled from top-1000 gallery images that are more than 25 m away from q.

The network usually consists of a backbone encoder and a VLAD layer [1], where the encoder embeds the image into a dense feature map and the VLAD layer further aggregates the feature map into a compact feature vector. The network can be optimized with a triplet ranking loss [1,23] or a contrastive loss [37]. More recently, Liu *et al.*[28] proposed a softmax-based loss with dot products for better maximizing the ratio between the query-positive pair against multiple query-negative pairs, which is formulated as

$$\mathcal{L}_{\text{hard}}(\theta) = -\sum_{j=1}^{N} \log \frac{\exp\langle f_\theta^q, f_\theta^{p^*}\rangle}{\exp\langle f_\theta^q, f_\theta^{p^*}\rangle + \exp\langle f_\theta^q, f_\theta^{n_j}\rangle}, \tag{1}$$

where θ is the network parameters, f_θ^q is the query image features encoded by the current network, and $\langle\cdot,\cdot\rangle$ denotes dot product between two vectors. Trained by this pipeline, the network can capture most feature distributions and generate acceptable localization results. Such a network acts as the **baseline** in our paper.

However, an obvious problem arises: the network trained with the easiest positive images alone cannot well adapt to challenging positive pairs, as there generally exists variations in viewpoints, camera poses and focal lengths in real-world IBL datasets. In addition, the image-level weak supervisions provide misleading information for learning local features. There are non-overlapping regions even in true positive pairs. The network trained with image-level supervisions might pull the non-overlapping regions to be close to each other in the feature space. Such incorrect image-level supervisions impede the feature learning process. To tackle such challenges, we propose to utilize self-enhanced fine-grained supervisions for supervising both image-to-image and image-to-region similarities. In this way, difficult positive images as well as their sub-regions can be fully exploited to improve features for localization.

3.2 Self-supervising Query-Gallery Similarities

A naïve way to utilize difficult positive images is to directly train the network with lower-ranking positive images (*e.g.* , top-k gallery images) as positive samples with "hard" (one-hot) labels in Eq. (1). However, this may make the training process deteriorate, because the network cannot effectively tell the true positives apart from the false ones when trained with the existing pipeline in Sect. 3.1. To mitigate the label noise and make full use of the difficult positive samples, we propose to self-supervise query-gallery similarities of top-ranking gallery images which also act as the refined soft supervisions, and to gradually improve the network with self-enhanced similarity labels in generations.

We use θ_ω to indicate the network parameters in different generations, denoted as $\{\theta_\omega|_{\omega=1}^{\Omega}\}$, where Ω is the total number of generations. In each generation, the network is initialized by the same ImageNet-pretrained [7] parameters and is trained *again* with new supervision set until convergence. We adopt the same training process for each generation, except for the first initial generation, for which we train the network parameter θ_1 following the general pipeline in Sect. 3.1. Before proceeding to training the 2nd-generation network, the query-gallery feature distances are estimated by the 1st-generation network. For each query image q, its k-reciprocal nearest neighbors $p_1, \cdots, p_k$ according to Euclidean distances are first obtained from the gallery image set. The query-gallery similarities can be measured by dot products and normalized by a softmax function with temperature τ_1 over the top-k images,

$$\mathcal{S}_{\theta_1}(q, p_1, \cdots, p_k; \tau_1) = \text{softmax}\left(\left[\langle f_{\theta_1}^q, f_{\theta_1}^{p_1}\rangle/\tau_1, \cdots, \langle f_{\theta_1}^q, f_{\theta_1}^{p_k}\rangle/\tau_1\right]^\top\right), \quad (2)$$

where $f_{\theta_1}^{p_i}$ is the encoded feature representations of the ith gallery image by the 1st-generation network, and τ_1 is temperature hyper-parameter that makes the similarity vector sharper or smoother.

Rather than naïvely treating all top-k ranking images as the true positive samples for training the next-generation network, we propose to utilize the relative similarity vector $\mathcal{S}_{\theta_1}$ as extra training supervisions. There are two advantages of adopting such supervisions: (1) We are able to use the top-ranking gallery images as candidate positive samples. Those plausible positive images are more difficult and thus more informative than the on-the-fly easiest positive samples for learning scene features. (2) To mitigate the inevitable label noise from the plausible positive samples, we propose to train the next-generation network to approach the relative similarity vectors $\mathcal{S}_{\theta_1}$, which "softly" measures the relative similarities between $(q, p_1), \cdots, (q, p_k)$ pairs. At earlier generations, we set the temperature parameter τ_ω to be large, as the relative similarities are less accurate. The large temperature makes the similarity vector $\mathcal{S}_{\theta_\omega}$ more equally distributed. At later generations, the network becomes more accurate. The relative similarity vector $\mathcal{S}_{\theta_\omega}$'s maximal response is more accurate to identify the true positive samples. Lower temperatures are used to make $\mathcal{S}_{\theta_\omega}$ concentrate on a small number of gallery images.

The relative similarity vectors $\mathcal{S}_{\theta_1}$ estimated by the 1st-generation network are used to supervise the 2nd-generation network via a "soft" cross-entropy loss,

$$\mathcal{L}_{\text{soft}}(\theta_2) = \ell_{ce}\left(\mathcal{S}_{\theta_2}(q, p_1, \cdots, p_k; 1), \mathcal{S}_{\theta_1}(q, p_1, \cdots, p_k; \tau_1)\right), \quad (3)$$

where $\ell_{ce}(y, \hat{y}) = -\sum_i \hat{y}(i) \log(y(i))$ denotes the cross-entropy loss. Note that only the learning target $\mathcal{S}_{\theta_1}$ adopts the temperature hyper-parameter to control its sharpness. In our experiments, the performance of the trained network generally increases with generations and saturates around the 4th generation, with temperatures set as 0.07, 0.06, 0.05, respectively. In each iteration, both the "soft" cross-entropy loss and the original hard loss (Eq. (1)) are jointly adopted as $\mathcal{L}(\theta) = \mathcal{L}_{\text{hard}}(\theta) + \lambda \mathcal{L}_{\text{soft}}(\theta)$ for supervising the feature learning.

By self-supervising query-gallery similarities, we could fully utilize the difficult positives obtained from the network of previous generations. The "soft" supervisions can effectively mitigate the label noise. However, even with the refined labels, paired positive images' local features from all their spatial regions are trained to approach the same similarity scores, which provide misleading supervision for those non-overlapping regions. The network's capability on learning discriminative local features is limited given only such image-level supervisions.

3.3 Self-supervising Fine-Grained Image-to-region Similarities

To tackle the above challenge, we propose to fine-grain the image-to-image similarities into image-to-region similarities to generate more detailed supervisions. Given the top-k ranking images $p_1, \cdots, p_k$ for each query q, instead of directly calculating image-level relative similarity vector $\mathcal{S}_{\theta_\omega}$, we can further decompose each plausible positive image's feature maps into 4 half regions (top, bottom, left, right halves) and 4 quarter regions (top-left, top-right, bottom-left, bottom-right quarters). Specifically, The gallery image p_i's feature maps m_i are first obtained by the network backbone before the VLAD layer. We split m_i into 4 half regions and 4 quarter regions, and then feed them into the aggregation VLAD respectively for obtaining 8 region feature vectors $\{f_\theta^{r_i^1}, f_\theta^{r_i^2}, \cdots, f_\theta^{r_i^8}\}$ corresponding to the 8 image sub-regions $\{r_i^1, r_i^2, \cdots, r_i^8\}$, each of which depicts the appearance of one sub-region of the gallery scene. Thus, the query-gallery supervisions are further fine-grained by measuring the relative query-to-region similarities,

$$\mathcal{S}_{\theta_\omega}^r(q, p_1, \cdots, p_k; \tau_\omega) = \text{softmax}\left([\langle f_{\theta_\omega}^q, f_{\theta_\omega}^{p_1} \rangle / \tau_\omega, \langle f_{\theta_\omega}^q, f_{\theta_\omega}^{r_1^1} \rangle / \tau_\omega, \cdots, \langle f_{\theta_\omega}^q, f_{\theta_\omega}^{r_1^8} \rangle / \tau_\omega, \right.$$
$$\left. \cdots, \langle f_{\theta_\omega}^q, f_{\theta_\omega}^{p_k} \rangle / \tau_\omega, \langle f_{\theta_\omega}^q, f_{\theta_\omega}^{r_k^1} \rangle / \tau_\omega, \cdots, \langle f_{\theta_\omega}^q, f_{\theta_\omega}^{r_k^8} \rangle / \tau_\omega] \right). \qquad (4)$$

The "soft" cross-entropy loss in Eq. (3) can be extended such that it can learn from the fine-grained similarities in Eq. (4) to encode more discriminative local features,

$$\mathcal{L}_{\text{soft}}^r(\theta_\omega) = \ell_{ce}\left(\mathcal{S}_{\theta_\omega}^r(q, p_1, \cdots, p_k; 1), \mathcal{S}_{\theta_{\omega-1}}^r(q, p_1, \cdots, p_k; \tau_{\omega-1}) \right). \qquad (5)$$

The image-to-region similarities can also be used for mining the most difficult negative gallery-image regions for each query q. The mined negative regions can be used in the hard loss in Eq. (1). For each query q, the accurate negative images could be easily identified by the geographical GPS labels. We propose to further mine the most difficult region n_j^* for each of the negative images n_j on-the-fly by measuring query-to-region similarities with the current network. The image-level softmax-based loss in Eq. (1) could be refined with the most difficult negative regions as

$$\mathcal{L}_{\text{hard}}^r(\theta_\omega) = -\sum_{j=1}^{N} \log \frac{\exp\langle f_{\theta_\omega}^q, f_{\theta_\omega}^{p^*}\rangle}{\exp\langle f_{\theta_\omega}^q, f_{\theta_\omega}^{p^*}\rangle + \exp\langle f_{\theta_\omega}^q, f_{\theta_\omega}^{n_j^*}\rangle}. \tag{6}$$

Note that the image-to-region similarities for selecting the most difficult negative regions are measured by the network on-the-fly, while those for supervising difficult positives in Eq. (5) are measured by the previous generation's network.

In our proposed framework, the network is trained with extended triplets, each of which consists of one query image q, the easiest positive image p^*, the difficult positive images $\{p_i|_{i=1}^k\}$, and the most difficult regions in the negative images $\{n_j^*|_{j=1}^N\}$. The overall objective function of generation ω is formulated as

$$\mathcal{L}(\theta_\omega) = \mathcal{L}_{\text{hard}}^r(\theta_\omega) + \lambda \mathcal{L}_{\text{soft}}^r(\theta_\omega), \tag{7}$$

where λ is the loss weighting factor. The multi-generation training can be performed similarly to that mentioned in Sect. 3.2 to self-supervise and to gradually refine the image-to-region similarities.

3.4 Discussions

Why Not Decompose Queries for Self-supervising Region-to-Region Similarities? We observe that region-to-region similarities might contain many superficially easy cases that are not informative enough to provide discriminative training supervisions for the network. For instance, the similarity between the sky regions of two overlapping images might be too easy to serve as effective training samples. Image-to-region similarities could well balance the robustness and granularity of the generated supervisions, as region-to-region similarities have more risk to be superficial.

Why Still Require $\mathcal{L}_{\text{hard}}^r$? We observe that the top-1 gallery images obtained from the k-reciprocal nearest neighbors can almost always act as true positives for training with hard supervisions in Eq. (6). The hard loss could stabilize the training process to avoid error amplification.

4 Experiments

4.1 Implementation Details

Datasets. We utilize the Pittsburgh benchmark dataset [42] for optimizing our image retrieval-based localization network following the experimental settings of state-of-the-art methods [1,28]. Pittsburgh consists of a large scale of panoramic images captured at different times and are associated with noisy GPS locations. Each panoramic image is projected to create 24 perspective images. For fair comparison, we use the subset Pitts30k for training and select the best model that achieves the optimal performance on the val-set of Pitts30k. The Pitts30k-train contains 7,416 queries and 10,000 gallery images, and the Pitts30k-val consists of 7,608 queries and 10,000 gallery images. We obtain the final retrieval results by ranking images in the large-scale Pitts250k-test, which contains 8,280 probes and 83,952 database images.

To verify the generalization ability of our method on different IBL tasks, we directly evaluate our models trained on Pitts30k-train on the Tokyo 24/7 [41] dataset, which is quite challenging since the queries were taken in varying conditions. In addition, we evaluate the model's generalization ability on standard image retrieval datasets, e.g. the Oxford 5k [35], Paris 6k [36], and Holidays [20].

Table 1. Comparison with state-of-the-arts on image-based localization benchmarks. Note that the network is only trained on Pitts30k-train and directly evaluated on both Tokyo 24/7 and Pitts250k-test datasets

Method	Tokyo 24/7 [41]			Pitts250k-test [42]		
	R@1	R@5	R@10	R@1	R@5	R@10
NetVLAD [1] (CVPR'16)	73.3	82.9	86.0	86.0	93.2	95.1
CRN [23] (CVPR'17)	75.2	83.8	87.3	85.5	93.5	95.5
SARE [28] (ICCV'19)	79.7	86.7	90.5	89.0	95.5	96.8
Ours	**85.4**	**91.1**	**93.3**	**90.7**	**96.4**	**97.6**

Evaluation. During inference, we perform PCA whitening whose parameters are learnt on the Pitts30k-train, reducing the feature dimension to 4,096. We follow the same evaluation metric of [1,28], where the top-k recall is measured on the localization datasets, Pitts250k-test [42] and Tokyo 24/7 [41]. The query image is determined to be successfully retrieved from top-k if at least one of the top k retrieved reference images locates within $d = 25$ m from the query image. As for the image-retrieval datasets, Oxford 5k [35], Paris 6k [36], and Holidays [20], the mean average precision (mAP) is adopted for evaluation.

Architecture. For fair comparison, we adopt the same architecture used in [1,28], which comprises of a VGG-16 [40] backbone and a VLAD layer [1] for encoding and aggregating feature representations. We use the ImageNet-pretrained [7] VGG-16 up to the last convolutional layer (i.e. conv5) before ReLU, as the backbone. Following [28], the whole backbone except the last convolutional block (i.e. conv5) is frozen when trained on image-based localization datasets.

Training Details. For data organization, each mini-batch contains 4 triplets, each of which consists of one query image, one easiest positive image, top-10

difficult positive images and 10 negative images. The negative images are sampled following the same strategy in [1,28]. Our model is trained by 4 generations, with 5 epochs in each generation. We empirically set the hyper-parameters $\lambda = 0.5, \tau_1 = 0.07, \tau_2 = 0.06, \tau_3 = 0.05$ in all experiments. The stochastic gradient descent (SGD) algorithm is utilized to optimize the loss function, with momentum 0.9, weight decay 0.001, and a constant learning rate $= 0.001$.

4.2 Comparison with State-of-the-arts

We compare with state-of-the-art image localization methods NetVLAD [1], CRN [23] and SARE [28] on localization datasets Pitts250k-test [42] and Tokyo 24/7 [41] in this experiment. Our model is only trained on Pitts30k-train without any Tokyo 24/7 images. CRN (Contextual Reweighting Network) improves NetVLAD by proposing a contextual feature reweighting module for selecting most discriminative local features for aggregation. While a Stochastic Attraction and Repulsion Embedding (SARE) loss function is proposed on top of VLAD-aggregated feature embeddings in [28]. None of the above methods have well handled the bottleneck of weakly-supervised training for image localization.

Experimental results are shown in Table 1. The proposed method achieves 90.7% rank-1 recall on Pitts250k-test, outperforming the second best 89.0% obtained by SARE, with an improvement of 1.7%. The feature embeddings learned by our method show very strong generalization capability on the challenging Tokyo 24/7 dataset where the rank-1 recall is significantly boosted to 85.4%, up to 5.7% performance improvement against SARE. The superior performances validate the effectiveness of our self-enhanced image-to-region similarities in learning discriminative features for image-based localization.

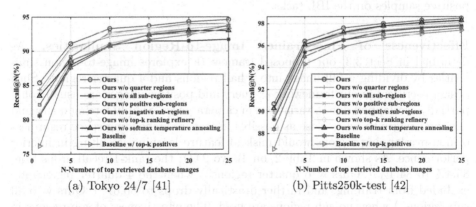

(a) Tokyo 24/7 [41]   (b) Pitts250k-test [42]

Fig. 4. Quantitative results of ablation studies on our proposed method in terms of top-k recall on Tokyo 24/7 and Pitts250k-test datasets. The models are only trained on Pitts30k-train set

380   Y. Ge et al.

4.3   Ablation Studies

We perform ablation studies on Tokyo 24/7 [41] and Pitts250k-test [42] to anal-
yse the effectiveness of the proposed method and shed light on the importance
of supervisions from self-enhanced fine-grained image-to-region similarities. We
illustrate quantitative results in Fig. 4, and show the detailed ablation experiments in Table 2. "Baseline" is our re-implementation of SARE [28], which is
trained with only the best-matching positive image using $\mathcal{L}_{hard}$ in Eq. (1).

**Directly Training with Noisy Difficult Positives in the Existing
Pipeline.** Existing image-based localization benchmarks only provide geo-
tagged images for learning image-to-image similarities. These GPS labels are
weak and noisy in finding true positive images for given query images. SARE
[28] and previous works only choose the easiest positive for training, *i.e.* the
most similar one to the query in the feature space. Our proposed approach
can effectively learn informative features from multiple difficult positives (top-k
positive images) by introducing self-enhanced image-region similarities as extra
supervisions, thus boost the rank-1 recall to 90.7% and 85.4% on Pitts250k-test
and Tokyo 24/7 respectively. To check whether such difficult (top-k) positive
images can be easily used without our proposed self-enhanced image-to-region
similarities to improve the final performance, we conduct an ablation experiment by training the "Baseline" with extra difficult positive images. This model
is denoted as "Baseline w/ top-k positives" in Table 2. It shows that naïvely
adding more difficult positive images for training causes drastic performance
drop, where the rank-1 recall decreases from 80.6% ("Baseline") to 76.2% on
Tokyo 24/7. The trend is similar on Pitts250k-test. This phenomenon validates
the effectiveness of our self-enhanced image-to-region similarities from difficult
positive samples on the IBL tasks.

Effectiveness of Fine-Grained Image-to-Region Similarities. As
described in Sect. 3.3, our proposed framework explores image-to-region similarities by dividing a full image into 4 half regions and 4 quarter regions. Such
design is critical in our framework as query and positive images are usually only
partially overlapped due to variations in camera poses. Simply forcing query and
positive images to be as close as possible in feature embedding space regardless
of non-overlapping regions would mislead feature learning, resulting in inferior
performance. As shown in Table 2, on Tokyo 24/7, the rank-1 recall drops from
85.4% to 84.4% ("Ours w/o quarter regions") when the 4 quarter regions are
excluded from training, and further drastically drops to 80.6% ("Ours w/o all
sub-regions") when no sub-regions are used. The effectiveness of sub-regions in
positive and negative images are compared by "Ours w/o positive sub-regions"
and "Ours w/o negative sub-regions" in Table 2. It shows that in both cases the
rank-1 recall is harmed, and the performance drop is even more significant when
positive sub-regions are ignored. The above observations indicate that difficult
positives are critical for feature learning in IBL tasks, which have not been well

investigated in previous works, while our fine-grained image-to-region similarities effectively help learn discriminative features from these difficult positives.

Table 2. Ablation studies for our proposed method on individual components. The models are only trained on Pitts30k-train set

Method	Tokyo 24/7 [41]		
	R@1	R@5	R@10
Baseline	80.6	87.6	90.8
Baseline w/top-k positives	76.2	88.6	90.8
Baseline w/regions	79.8	86.9	90.4
Ours w/o all sub-regions	80.6	88.0	90.9
Ours w/o positive sub-regions	80.8	88.6	91.1
Ours w/o negative sub-regions	82.2	89.2	92.7
Ours w/o quarter regions	84.4	90.5	92.1
Ours w/o top-k ranking refinery	83.8	90.2	92.4
Ours w/o softmax temperature annealing	83.5	88.6	90.8
Ours	**85.4**	**91.1**	**93.3**

Effectiveness of Self-enhanced Similarities as Soft Supervisions. When comparing "Ours w/o all sub-regions" and "Baseline w/ top-k positives", one may find that the only difference between these two methods is $\mathcal{L}_{\text{soft}}$ in Eq. (3). By adding the extra objective $\mathcal{L}_{\text{soft}}$ to "Baseline w/ top-k positives", the rank-1 recall is significantly improved from 76.1% to 80.6% (the same with "Baseline") on Tokyo 24/7. "Ours w/o all sub-regions" even outperforms "Baseline" on Pitts250k-test as shown in Fig. 4. The above comparisons demonstrate the effectiveness of using our self-enhanced similarities as soft supervisions for learning from difficult positive images even at the full image level.

More importantly, soft supervisions serve as a premise for image-to-region similarities to work. We evaluate the effects of soft supervisions at the region level, dubbed as "Baseline w/ regions", where the dataset is augmented with decomposed regions and only the noisy GPS labels are used. The result in Table 2 is even worse than "Baseline" since sub-regions with only GPS labels provide many too easy positives that are not informative enough for feature learning.

Benefit from Top-k Ranking Refinery. We propose to find k-reciprocal nearest neighbors of positive images for recovering more accurate difficult positives images for training in Sect. 3.2. Although our variant ("Ours w/o top-k ranking refinery") without using k-reciprocal nearest neighbors refining top-k images already outperforms "Baseline" by a large margin, ranking with k-reciprocal nearest neighbor further boosts the rank-1 recall by 1.6% on Tokyo 24/7. The

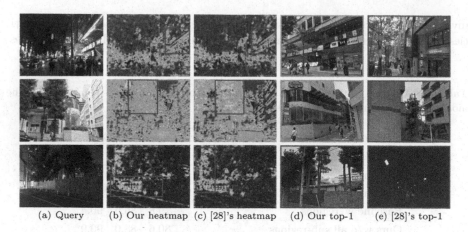

(a) Query   (b) Our heatmap   (c) [28]'s heatmap   (d) Our top-1   (e) [28]'s top-1

Fig. 5. Retrieved examples of our method and state-of-the-art SARE [28] on Tokyo 24/7 dataset [41]. The regions highlighted by red boxes illustrates the main differences.

superior performance demonstrates that k-reciprocal nearest neighbor ranking can more accurately identify true positive images.

Benefits from Softmax Temperature Annealing. We validate the effectiveness of the proposed temperature annealing strategy (Sect. 3.2) in this experiment by setting a constant temperature $\tau = 0.07$ in all generations. Comparisons between "Ours" and "Ours w/o softmax temperature annealing" in Fig. 4 and Table 2 show that temperature annealing is beneficial for learning informative features with our self-enhanced soft supervisions.

4.4   Qualitative Evaluation

To better understand the superior performance of our method on the IBL tasks, we visualize the learned feature maps before VLAD aggregation as heatmaps, shown in Fig. 5. In the first example, our method pays more attention on the discriminative shop signs than SARE, which provide valuable information for localization in city-scale street scenarios. In the second example, SARE incorrectly focuses on the trees, while our method learns to ignore such misleading regions by iteratively refining supervisions from fine-grained image-to-region similarities. Although both methods fail in the third example, our retrieved top-1 image is more reasonable with wall patterns similar to the query image.

4.5   Generalization on Image Retrieval Datasets

In this experiment, we evaluate the generalization capability of learned feature embeddings on standard image retrieval datasets by directly testing trained models without fine-tuning. The experimental results are listed in Table 3. "Full"

Table 3. Evaluation on standard image retrieval datasets in terms of mAP (%) to validate the generalization ability of the networks

Method	Oxford 5k [35]		Paris 6k [36]		Holidays [20]
	Full	Crop	Full	Crop	
NetVLAD [1] (CVPR'16)	69.1	71.6	78.5	79.7	**83.1**
CRN [23] (CVPR'17)	69.2	–	–	–	–
SARE [28] (ICCV'19)	71.7	75.5	82.0	81.1	80.7
Ours	**73.9**	**76.7**	**82.5**	**82.4**	80.5

means the feature embeddings are extracted from the whole image, while only cropped landmark regions are used in the "crop" setting. Our method shows good generalization ability on standard image retrieval tasks, and outperforms other competitors on most datasets and test settings. All compared methods do not perform well on Holidays due to the fact that Holidays contains lots of natural sceneries, which are difficult to find in our street-view training set.

5 Conclusion

This paper focuses on the image-based localization (IBL) task which aims to estimate the geographical location of a query image by recognizing its nearest reference images from a city-scale dataset. We propose to tackle the problem of weak and noisy supervisions by self-supervising image-to-region similarities. Our method outperforms state-of-the-arts on the standard localization benchmarks by noticeable margins.

Acknowledgements. This work is supported in part by SenseTime Group Limited, in part by the General Research Fund through the Research Grants Council of Hong Kong under Grants CUHK 14202217/14203118/14205615/14207814/14213616/14208417/14239816, in part by CUHK Direct Grant.

References

1. Arandjelovic, R., Gronat, P., Torii, A., Pajdla, T., Sivic, J.: NetVLAD: CNN architecture for weakly supervised place recognition. In: Proceedings of the IEEE Conference on Computer Vision and Pattern Recognition, pp. 5297–5307 (2016)
2. Arandjelovic, R., Zisserman, A.: All about VLAD. In: Proceedings of the IEEE Conference on Computer Vision and Pattern Recognition, pp. 1578–1585 (2013)
3. Arandjelović, R., Zisserman, A.: DisLocation: scalable descriptor distinctiveness for location recognition. In: Cremers, D., Reid, I., Saito, H., Yang, M.-H. (eds.) ACCV 2014. LNCS, vol. 9006, pp. 188–204. Springer, Cham (2015). https://doi.org/10.1007/978-3-319-16817-3_13
4. Caron, M., Bojanowski, P., Joulin, A., Douze, M.: Deep clustering for unsupervised learning of visual features. In: Ferrari, V., Hebert, M., Sminchisescu, C., Weiss, Y. (eds.) Computer Vision – ECCV 2018. LNCS, vol. 11218, pp. 139–156. Springer, Cham (2018). https://doi.org/10.1007/978-3-030-01264-9_9

5. Castle, R., Klein, G., Murray, D.W.: Video-rate localization in multiple maps for wearable augmented reality. In: 2008 12th IEEE International Symposium on Wearable Computers, pp. 15–22. IEEE (2008)

6. Chen, D.M., et al.: City-scale landmark identification on mobile devices. In: CVPR 2011, pp. 737–744. IEEE (2011)

7. Deng, J., Dong, W., Socher, R., Li, L.J., Li, K., Fei-Fei, L.: ImageNet: a large-scale hierarchical image database. In: 2009 IEEE Conference on Computer Vision and Pattern Recognition, pp. 248–255. IEEE (2009)

8. Doersch, C., Gupta, A., Efros, A.A.: Unsupervised visual representation learning by context prediction. In: Proceedings of the IEEE International Conference on Computer Vision, pp. 1422–1430 (2015)

9. Fischler, M.A., Bolles, R.C.: Random sample consensus: a paradigm for model fitting with applications to image analysis and automated cartography. Commun. ACM **24**(6), 381–395 (1981)

10. Furlanello, T., Lipton, Z.C., Tschannen, M., Itti, L., Anandkumar, A.: Born again neural networks. In: International Conference on Machine Learning (2018)

11. Ge, Y., Chen, D., Li, H.: Mutual mean-teaching: pseudo label refinery for unsupervised domain adaptation on person re-identification. In: International Conference on Learning Representations (2020)

12. Ge, Y., Chen, D., Zhu, F., Zhao, R., Li, H.: Self-paced contrastive learning with hybrid memory for domain adaptive object Re-ID (2020)

13. Ge, Y., et al.: FD-GAN: pose-guided feature distilling GAN for robust person re-identification. In: Advances in Neural Information Processing Systems, pp. 1229–1240 (2018)

14. Ge, Y., Zhu, F., Zhao, R., Li, H.: Structured domain adaptation with online relation regularization for unsupervised person Re-ID (2020)

15. Gidaris, S., Singh, P., Komodakis, N.: Unsupervised representation learning by predicting image rotations. arXiv preprint arXiv:1803.07728 (2018)

16. Goyal, P., Mahajan, D., Gupta, A., Misra, I.: Scaling and benchmarking self-supervised visual representation learning. In: Proceedings of the IEEE International Conference on Computer Vision, pp. 6391–6400 (2019)

17. Häne, C., et al.: 3D visual perception for self-driving cars using a multi-camera system: calibration, mapping, localization, and obstacle detection. Image Vis. Comput. **68**, 14–27 (2017)

18. Hartley, R., Zisserman, A.: Multiple View Geometry in Computer Vision. Cambridge University Press, Cambridge (2003)

19. Seo, P.H., Weyand, T., Sim, J., Han, B.: CPlaNet: enhancing image geolocalization by combinatorial partitioning of maps. In: Ferrari, V., Hebert, M., Sminchisescu, C., Weiss, Y. (eds.) ECCV 2018. LNCS, vol. 11214, pp. 544–560. Springer, Cham (2018). https://doi.org/10.1007/978-3-030-01249-6_33

20. Jegou, H., Douze, M., Schmid, C.: Hamming embedding and weak geometric consistency for large scale image search. In: Forsyth, D., Torr, P., Zisserman, A. (eds.) ECCV 2008. LNCS, vol. 5302, pp. 304–317. Springer, Heidelberg (2008). https://doi.org/10.1007/978-3-540-88682-2_24

21. Jégou, H., Douze, M., Schmid, C., Pérez, P.: Aggregating local descriptors into a compact image representation. In: 2010 IEEE Computer Society Conference on Computer Vision and Pattern Recognition, pp. 3304–3311. IEEE (2010)

22. Jegou, H., Perronnin, F., Douze, M., Sánchez, J., Perez, P., Schmid, C.: Aggregating local image descriptors into compact codes. IEEE Trans. Pattern Anal. Mach. Intell. **34**(9), 1704–1716 (2011)

23. Kim, H.J., Dunn, E., Frahm, J.M.: Learned contextual feature reweighting for image geo-localization. In: 2017 IEEE Conference on Computer Vision and Pattern Recognition, pp. 3251–3260. IEEE (2017)
24. Knopp, J., Sivic, J., Pajdla, T.: Avoiding confusing features in place recognition. In: Daniilidis, K., Maragos, P., Paragios, N. (eds.) ECCV 2010. LNCS, vol. 6311, pp. 748–761. Springer, Heidelberg (2010). https://doi.org/10.1007/978-3-642-15549-9_54
25. Kolesnikov, A., Zhai, X., Beyer, L.: Revisiting self-supervised visual representation learning. In: Proceedings of the IEEE Conference on Computer Vision and Pattern Recognition, pp. 1920–1929 (2019)
26. LeCun, Y., Bengio, Y., Hinton, G.: Deep learning. Nature **521**(7553), 436–444 (2015)
27. Liu, L., Li, H., Dai, Y.: Efficient global 2D–3D matching for camera localization in a large-scale 3D map. In: Proceedings of the IEEE International Conference on Computer Vision, pp. 2372–2381 (2017)
28. Liu, L., Li, H., Dai, Y.: Stochastic attraction-repulsion embedding for large scale image localization. In: Proceedings of the IEEE International Conference on Computer Vision, pp. 2570–2579 (2019)
29. Lowe, D.G.: Distinctive image features from scale-invariant keypoints. Int. J. Comput. Vision **60**(2), 91–110 (2004)
30. Mur-Artal, R., Montiel, J.M.M., Tardos, J.D.: ORB-SLAM: a versatile and accurate monocular SLAM system. IEEE Trans. Rob. **31**(5), 1147–1163 (2015)
31. Noroozi, M., Favaro, P.: Unsupervised learning of visual representations by solving jigsaw puzzles. In: Leibe, B., Matas, J., Sebe, N., Welling, M. (eds.) ECCV 2016. LNCS, vol. 9910, pp. 69–84. Springer, Cham (2016). https://doi.org/10.1007/978-3-319-46466-4_5
32. Noroozi, M., Pirsiavash, H., Favaro, P.: Representation learning by learning to count. In: Proceedings of the IEEE International Conference on Computer Vision, pp. 5898–5906 (2017)
33. Paulin, M., Douze, M., Harchaoui, Z., Mairal, J., Perronin, F., Schmid, C.: Local convolutional features with unsupervised training for image retrieval. In: Proceedings of the IEEE International Conference on Computer Vision, pp. 91–99 (2015)
34. Perronnin, F., Liu, Y., Sánchez, J., Poirier, H.: Large-scale image retrieval with compressed fisher vectors. In: 2010 IEEE Computer Society Conference on Computer Vision and Pattern Recognition, pp. 3384–3391. IEEE (2010)
35. Philbin, J., Chum, O., Isard, M., Sivic, J., Zisserman, A.: Object retrieval with large vocabularies and fast spatial matching. In: 2007 IEEE Conference on Computer Vision and Pattern Recognition, pp. 1–8. IEEE (2007)
36. Philbin, J., Chum, O., Isard, M., Sivic, J., Zisserman, A.: Lost in quantization: improving particular object retrieval in large scale image databases. In: IEEE Conference on Computer Vision and Pattern Recognition (2008)
37. Radenović, F., Tolias, G., Chum, O.: CNN image retrieval learns from BoW: unsupervised fine-tuning with hard examples. In: Leibe, B., Matas, J., Sebe, N., Welling, M. (eds.) ECCV 2016. LNCS, vol. 9905, pp. 3–20. Springer, Cham (2016). https://doi.org/10.1007/978-3-319-46448-0_1
38. Sattler, T., Leibe, B., Kobbelt, L.: Fast image-based localization using direct 2D-to-3D matching. In: 2011 International Conference on Computer Vision, pp. 667–674. IEEE (2011)
39. Schindler, G., Brown, M., Szeliski, R.: City-scale location recognition. In: 2007 IEEE Conference on Computer Vision and Pattern Recognition, pp. 1–7. IEEE (2007)

40. Simonyan, K., Zisserman, A.: Very deep convolutional networks for large-scale image recognition. arXiv preprint arXiv:1409.1556 (2014)
41. Torii, A., Arandjelovic, R., Sivic, J., Okutomi, M., Pajdla, T.: 24/7 place recognition by view synthesis. In: Proceedings of the IEEE Conference on Computer Vision and Pattern Recognition, pp. 1808–1817 (2015)
42. Torii, A., Sivic, J., Pajdla, T., Okutomi, M.: Visual place recognition with repetitive structures. In: Proceedings of the IEEE Conference on Computer Vision and Pattern Recognition, pp. 883–890 (2013)
43. Vo, N., Jacobs, N., Hays, J.: Revisiting IM2GPS in the deep learning era. In: Proceedings of the IEEE International Conference on Computer Vision, pp. 2621–2630 (2017)
44. Weyand, T., Kostrikov, I., Philbin, J.: PlaNet - photo geolocation with convolutional neural networks. In: Leibe, B., Matas, J., Sebe, N., Welling, M. (eds.) ECCV 2016. LNCS, vol. 9912, pp. 37–55. Springer, Cham (2016). https://doi.org/10.1007/978-3-319-46484-8_3
45. Xie, Q., Hovy, E., Luong, M.T., Le, Q.V.: Self-training with noisy student improves ImageNet classification. In: Proceedings of the IEEE Conference on Computer Vision and Pattern Recognition (2020)
46. Zeiler, M.D., Fergus, R.: Visualizing and understanding convolutional networks. In: Fleet, D., Pajdla, T., Schiele, B., Tuytelaars, T. (eds.) ECCV 2014. LNCS, vol. 8689, pp. 818–833. Springer, Cham (2014). https://doi.org/10.1007/978-3-319-10590-1_53
47. Zhang, M., Song, G., Zhou, H., Liu, Y.: Discriminability distillation in group representation learning. In: European Conference on Computer Vision (2020)
48. Zhou, H., Liu, J., Liu, Z., Liu, Y., Wang, X.: Rotate-and-render: unsupervised photorealistic face rotation from single-view images. In: Proceedings of the IEEE/CVF Conference on Computer Vision and Pattern Recognition, pp. 5911–5920 (2020)

Rotationally-Temporally Consistent Novel View Synthesis of Human Performance Video

Youngjoong Kwon[1,2]($\boxtimes$), Stefano Petrangeli[2], Dahun Kim[3], Haoliang Wang[2], Eunbyung Park[1], Viswanathan Swaminathan[2], and Henry Fuchs[1]

[1] University of North Carolina at Chapel Hill, Chapel Hill, USA
youngjoong@cs.unc.edu
[2] Adobe Research, San Jose, USA
[3] Korea Advanced Institute of Science and Technology, Daejeon, South Korea

Abstract. Novel view *video* synthesis aims to synthesize novel viewpoints videos given input captures of a human performance taken from multiple reference viewpoints and over consecutive time steps. Despite great advances in model-free novel view synthesis, existing methods present three limitations when applied to complex and time-varying human performance. First, these methods (and related datasets) mainly consider simple and symmetric objects. Second, they do not enforce explicit consistency across generated views. Third, they focus on static and non-moving objects. The fine-grained details of a human subject can therefore suffer from inconsistencies when synthesized across different viewpoints or time steps. To tackle these challenges, we introduce a human-specific framework that employs a learned 3D-aware representation. Specifically, we first introduce a novel siamese network that employs a gating layer for better reconstruction of the latent volumetric representation and, consequently, final visual results. Moreover, features from consecutive time steps are shared inside the network to improve temporal consistency. Second, we introduce a novel loss to explicitly enforce consistency across generated views both in *space* and in *time*. Third, we present the Multi-View Human Action (MVHA) dataset, consisting of near 1200 synthetic human performance captured from 54 viewpoints. Experiments on the MVHA, Pose-Varying Human Model and ShapeNet datasets show that our method outperforms the state-of-the-art baselines both in view generation quality and spatio-temporal consistency.

Keywords: Novel view video synthesis · Synthetic human dataset

---

Electronic supplementary material The online version of this chapter (https://doi.org/10.1007/978-3-030-58548-8_23) contains supplementary material, which is available to authorized users.

© Springer Nature Switzerland AG 2020
A. Vedaldi et al. (Eds.): ECCV 2020, LNCS 12349, pp. 387–402, 2020.
https://doi.org/10.1007/978-3-030-58548-8_23

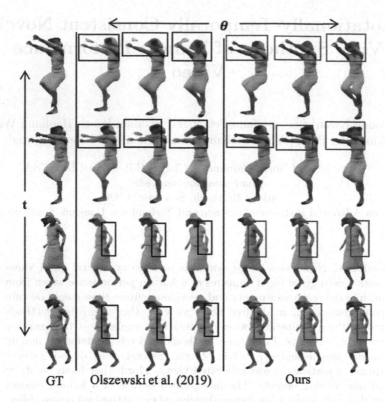

GT   Olszewski et al. (2019)   Ours

Fig. 1. Our method combines a siamese network architecture and rotational-temporal supervision for higher quality novel view video generation of human performance. Compared to Olszewski et al. [18], ours generate higher quality and more consistent results across views (left-right) and time steps (top-down). Model used for training with an unseen animation sequence, 4 input views.

1   Introduction

Novel View Synthesis (NVS) aims to synthesize new views of an object given different known viewpoints. Recently, a number of learning-based approaches have enabled the view synthesis by direct image or video generation without explicit 3D reconstruction or supervision [18,25]. Applying high-quality, accurate novel view synthesis to human action performance *videos* has a variety of applications in the area of AR/VR, telepresense, volumetric videos, and so on. Existing approaches present three shortcomings when applied to the human novel view video synthesis task. First, they focus on objects with simple shapes and strong symmetry, and perform quite poorly on deformable and asymmetric shapes like the human body. Second, current NVS methods do not enforce explicit consistency between different generated viewpoints, which does not guarantee consistency among generated views (Fig. 1, left). Third, current NVS methods focus on static objects, while human motion cannot be modeled by simple rigid transformations of existing (latent) volumetric representations. Dynamic articulations of body parts, like limbs and heads, can therefore suffer from significant inconsistencies when synthesizing novel views over time (Fig. 1, left). Moreover, existing NVS datasets are either not designed for the human NVS task [1], are too limited to support learning-based approaches [8], or are not publicly available [6,18]. In this paper, we therefore focus on synthesizing a temporally and spatially consistent novel view *video* of a human performance captured over time from fixed viewpoints. In particular, we present a novel end-to-end trainable video NVS network, combined with effective rotational and temporal consistency supervisions, and a synthetic dataset to further support research in this domain.

Contributions. Our model is based on recent NVS methods using latent volumetric representations that can be rendered by a learnable decoder [6,18,25]. It consists of a pair of siamese encoder-decoder networks, each receiving the input RGB video frames of the human performance captured from multiple viewpoints from two consecutive time steps. The temporal features of the inputs are shared between the two networks in order to enhance the novel view reconstruction video quality and temporal consistency. We also present a novel volume gating layer to improve the latent volumetric representation by adaptively attending on the valid volume points only when filling in missing parts during the novel view reconstruction. Moreover, we explicitly enforce rotational and temporal consistency across generated views to provide superior reconstruction performance, and demonstrate the effectiveness of this approach in capturing the complexity of human motion across different viewpoints and time steps. It is worth noting that the proposed rotational consistency supervision is applicable to image-level synthesis and non-human objects as well. Indeed, our approach evaluated on the ShapeNet [1] cars and chairs categories achieves state-of-the-art performance. Finally, we collect and publicly release the synthetic Multi-View Human Action (MVHA) dataset, composed of 30 different 3D human models animated with 40 different Mocap sequences, captured from 54 different viewpoints. Results on our MVHA, ShapeNet [1] and Pose-Varying Human Model (PVHM) [37] datasets confirm both quantitatively and qualitatively the superior performance of the proposed approach compared to start-of-the-art baselines for the NVS task.

The remainder of this paper is structured as follows. Section 2 presents related works, while Sect. 3 presents the network architecture, volume gating convolutions, temporal feature augmentation, and the MVHA dataset. Section 4 reports quantitative and qualitative results, while Sect. 5 concludes the paper.

2   Related Work

In this section, we review prior works in the areas of 2D-based novel view synthesis, 3D-based novel view synthesis, and existing datasets available for the human NVS task.

2D-Based Novel View Synthesis. Tatarchenko *et al.* and Yang *et al.* propose to synthesize novel views by regressing the pixel colors of the target view directly from the input image using a Convolutional Neural Network (CNN) [28,35]. Instead of starting from an empty state, pixel-flow based approaches leverage pixel-flow to generate high-quality, sharp results [19,27,36]. These works usually use flow prediction to directly sample input pixels to reconstruct the output view. Zhou *et al.* suggest moving pixels from an input to a target view leveraging bilinear sampling kernels [9,36]. The approach by Park *et al.* achieves high-quality results by moving only the pixels that can be seen in the novel view, and by hallucinating the empty parts using a completion network [19]. Park *et al.* also take advantage of the symmetry of objects from ShapeNet [1] by producing a symmetry-aware visibility map [19]. Our human novel view synthesis task cannot fully take advantage of this approach since humans can present highly asymmetric poses. Sun *et al.* further improve these results by aggregating an arbitrary number of input images [27]. Eslami *et al.* use a latent representation that can aggregate multiple input views, which shows good results on synthetic geometric scenes [3]. Unlike the previous NVS works that move or regress pixels, Shysheya *et al.* regress texture coordinates corresponding to a pre-defined texture map [24].

3D-Based Novel View Synthesis. Works embedding implicit spatial consistency in the NVS task using explicit or latent volumetric representations have shown promising reconstruction results. Several recent methods reconstruct an explicit occupancy volume from a single image, and render it using traditional rendering techniques [2,5,10,22,29,32–34]. Methods leveraging signed-distance-field-encoded volumes [20,23], or RGBα-encoded volumes [12] have achieved excellent quality while overcoming the memory limitations of voxel-based representations. Saito *et al.* predict the continuous inside/outside probability of a clothed human, and also infer an RGB value at given 3D positions of the surface geometry, resulting in a successful recovery of intricate details of garments [23]. We do not compare to these methods [20,23] as they require ground truth geometry for the supervision [20,23], and as Lombardi *et al.* does not support generalization to unseen subjects [12]. Rather than generating explicit occupancy volumes, several methods generate latent volumetric representations that can be rendered by a learnable decoder [10,15,16,18,22,25,26]. Sitzmann *et al.* introduce a persistent 3D feature embedding to address the inconsistency between views synthesized by generative networks, which can occur due to a lack of 3D understanding [25,26]. Olszewski *et al.* generate a latent volumetric representation that allows 3D transformations and a combination of different input view images [18]. Moreover, their network does not require any 3D supervision and produces state-of-the-art results for the NVS task on ShapeNet. The main difference between our work and those by Sitzmann *et al.* [25,26] is that we strengthen the spatial consistency by introducing explicit rotational consistency supervision, and by also introducing implicit and explicit temporal consistency to better cope with the human novel view video synthesis task, while Sitzmann *et al.* do not consider temporal aspect.

Datasets for Human Novel View Synthesis. The Human3.6M dataset provides 3.6 million human poses and corresponding images from 4 calibrated cameras [8]. Collecting these datasets requires complex setup with multiple cameras, which is expensive and time-consuming. To address this limitation, synthetic methods have been proposed. The SURREAL dataset provides a large number of images generated with SMPL body shapes and synthetic textures, together with body masks, optical flow and depth [13,31]. The Pose-Varying Human Model (PVHM) dataset provides RGB images, depth maps, and optical flows of 22 models [37]. The 3DPeople dataset contains 80 3D models of dressed humans performing 70 different motions, captured from 4 different camera viewpoint [21]. Many of the existing learning-based NVS works are built using the ShapeNet dataset [1], which provides 54 reference viewpoints. In contrast, most of the aforementioned human datasets only provide data from relatively few reference viewpoints, preventing existing NVS approaches to be directly applied to the human NVS tasks. The MVHA synthetic dataset we collect contains 30 3D models animated with 40 Mocap sequences each rendered from 54 viewpoints. Compared to the PVHM dataset, we provide much more diverse outfits (*e.g.*, short/long sleeve, pants, and skirts, different types of hats and glasses, etc.), complex motion sequences, and higher resolution images.

3 Proposed Method

3.1 Problem Definition

A human performance captured from view k, out of K available views, consists of T consecutive RGB frames $I_{1:T}^k := \{I_1^k, I_2^k, ..., I_T^k\}$. Given a set of input $I_{1:T}^k$ with $k = 1, \ldots, K$, our goal is to directly synthesize a novel view video $\hat{I}_{1:T}^q$ so that: 1) given time step t, views generated across different query viewpoints q are consistent among each other and 2) given query q, temporally consecutive frames are consistent among each other. We simplify the problem of optimizing $p(\hat{I}_{1:T}^q | I_{1:T}^k)$ by factorizing the conditional distribution to a product form:

$$p(\hat{I}_{1:T}^q | I_{1:T}^k) = \prod_{t=1}^{T} p(\hat{I}_t^q | I_t^k, I_{t-1}^k). \tag{1}$$

In our experiments, we sample two consecutive frames for each network feed-forward. During training, we augment the supervision signal using the symmetry between time steps t and $t-1$ by learning $p(\hat{I}_t^q, \hat{I}_{t-1}^q | I_t^k, I_{t-1}^k)$. The volume used in our paper is centered on the target object and its axis is aligned with the camera coordinate. Perspective effects caused by pinhole camera projection and camera intrinsic parameters are approximately learned by the encoder and decoder networks, rather than handled explicitly.

3.2 Network Architecture

Inspired by 3D structure-aware view synthesis pipelines [10,17,18,22,25], input pixels I_τ^k (with $\tau = t, t-1$) are transformed from 2D-to-3D and then 3D-to-2D throughout the layers to be mapped onto the target view pixels I_τ^q. An overview of the proposed architecture design is given in Fig. 2. Our model is a two-tower

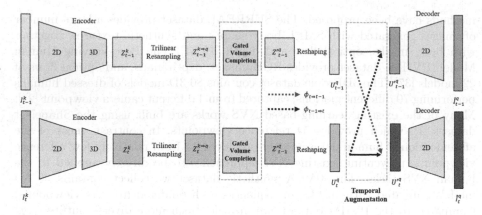

Fig. 2. Our network is a two-tower siamese network consisting of four blocks: encoder, volume completion, temporal augmentation, and decoder.

siamese network consisting of four blocks: encoder, volume completion, temporal augmentation, and decoder. For each tower, the encoder E takes I_τ^k as input, and generates a volumetric representation Z_τ^k of I_τ^k through a series of 2D-conv, reshape and 3D-conv layers. The 3D representation is then transformed to a target view $Z_\tau^{k \to q}$ via trilinear resampling. Multiple input views can be used by adding the representation over all k, $Z_\tau^q = \sum_k Z_\tau^{k \to q}$. These 3D features are fed into the proposed volume completion and are reshaped along the depth axis to become 2D features U_τ^q. U_τ^q then become $U_\tau^{\prime q}$ after going through the proposed temporal augmentation. The decoder D generates the output image I_τ^q starting from $U_\tau^{\prime q}$. The siamese network is coupled with two input-output pairs from consecutive time steps as $I_t^q, I_{t-1}^q = f(I_t^k, I_{t-1}^k)$, and the two towers are connected at the temporal fusion module.

Gated Volume Completion. The volumetric representation is initialized with the visible volume points generated from the source views I_τ^k (with $\tau = t, t-1$). The remaining *unseen* volume points should be hallucinated so as to generate plausible target views when rendered by the decoder. Our volume completion module consists of a series of 3D convolutions to inpaint the missing volume points. In vanilla 3D convolutional layers, all feature points are treated as the same valid ones, which is appropriate for tasks with complete inputs such as 3D object detection. In the presence of *empty* voxels in our completion problem, however, it is ambiguous whether current locations belong to the foreground voxels that should be hallucinated, or to the background that must remain unchanged. Vanilla 3D convolutions apply the same filters on all seen and unseen foreground, background and mixed voxels/features, leading to visual artifacts such as color discrepancy, blurriness and omission of shape details when synthesized to a target view by the decoder (Fig. 1).

To address this problem, we propose the volume gating convolutions to improve the latent volumetric representation computed by the network. The volume gating convolution learns a dynamic feature gating mechanism for each spatial location, *e.g.*, foreground or background voxels, seen or unseen voxels. Together, the gating operation can properly handle the uncertainty of voxel occupancy. Such explicit decoupling of the foreground object is also necessary to deal with real images with arbitrary background, which should remain unchanged in any possible target views. Specifically, we consider the formulation where the input 3D features are firstly used to compute gating values $g^{k \to q} \in \mathbb{R}^{H \times W \times D}$ (with H, W, and D being the width, height, and depth of the latent volumetric representation). By construction, however, the employed 3D volume is memory inefficient, *i.e.*, $O(n^3)$ w.r.t. the resolution, which is another limitation in creating the voxel-wise gating values [25]. Thus, we propose that full 3D gating may be conveniently approximated by decomposing it into 2D spatial gatings, *i.e.*, $O(n^2)$, along three canonical directions, *i.e.*, $g^{HW_{k \to q}} \in \mathbb{R}^{H \times W \times 1}$, $g^{HD_{k \to q}} \in \mathbb{R}^{H \times 1 \times D}$, and $g^{WD_{k \to q}} \in \mathbb{R}^{1 \times W \times D}$. It is worth noting that the gates depend only on input and query viewpoints k and q, and not on the timestep τ.

To obtain the 2D gating values, we first average-pool the input volume features $Z_\tau^{k \to q}$ along one spatial dimension (we will omit the superscript $k \to q$ for the remaining of this paragraph for ease of notation). For example, $Z_\tau^{HW} \in \mathbb{R}^{C \times H \times W \times 1}$ is obtained by average-pooling along the depth axis. We then apply average-pooling and max-pooling operations along the channel axis and concatenate them to generate an efficient feature descriptor $F_\tau^{HW} \in \mathbb{R}^{H \times W \times 2}$. The 2D gating is generated by applying a 5×5 convolution on the concatenated feature as $g^{HW} = \sigma(w(F_\tau^{HW})) \in \mathbb{R}^{H \times W \times 1}$. g^{HD} and g^{WD} are computed similarly and each of them achieves canonical-view volume carving. The full 3D gating is a geometrical mean of the three 2D gating values with shape broadcasting as $g = (g^{HW} \odot g^{HD} \odot g^{WD})^{1/3} \in \mathbb{R}^{H \times W \times D}$, which encodes *where* in HWD volume space to emphasize or suppress. The final output is a multiplication of the learned feature and gating value $Z_\tau'^{k \to q} = Z_\tau^{k \to q} \odot g^{k \to q}$, where g is copied along the channel axis. Our proposed volume gating is memory-efficient, easy to implement and performs significantly better at correcting color discrepancy and missing shape details in the generated views (see Sect. 4).

Temporal Feature Augmentation. Up to the completion of the 3D features, our siamese network treats human performance at each time step separately, producing $Z_{t-1}'^q$ and $Z_t'^q$ from each tower independently. These 3D features are then reshaped along the depth dimension to become frontal-view 2D features U_{t-1}^q and U_t^q with respect to the query viewpoint. At this point, we propose a temporal feature augmentation module to leverage complementary information between $t - 1$ and t. First, this approach allows us to improve the per-frame quality synthesis, since consecutive time steps might reveal more visible pixels of the same model and, therefore, leads to better occlusion handling. Second, temporal coherency is greatly improved as the generation of the final output is conditioned on both consecutive time steps.

Given the projected 2D features U_{t-1}^q and U_t^q, our temporal augmentation module learns the flow warping to align U_{t-1}^q onto U_t^q. The flow submodule receives an initial optical flow $\phi_{t \to t-1}^{init}$ computed by FlowNet2 [7], and refine it for more accurate deep feature flow. The warped feature $\hat{U}_{t-1}^q$ is then multiplied with a non-occlusion mask $M_{t\to t-1}$ calculated from the warping error as $e^{-\alpha\|I_t - W_{t\to t-1}(I_{t-1})\|_2^2}$, and then element-wise summed with the current features as $U_t^{\prime q} = U_t^q + M_{t\to t-1} \cdot \hat{U}_{t-1}^q$. The above operations are designed to be performed in a temporally bi-directional manner, *i.e.*, from $t-1$ to t and vice versa. Once $U_{t-1}^{\prime q}$ and $U_t^{\prime q}$ are obtained, they are fed into the decoders to generate the two consecutive frames of the query viewpoint $\hat{I}_{t-1}^q$ and $\hat{I}_t^q$.

3.3 Rotational and Temporal Supervision

Our goal is to generate novel view videos of a human performance that is consistent across query viewpoints and time steps, which cannot be guaranteed by simply providing multiple input viewpoints at each time step. To this end, we design a loss function that can explicitly enforce both rotational and temporal consistency in the generated views. It is worth noting that, during training, our network generates additional target views in addition to the query viewpoint q that are used to enforce rotational consistency. At testing time instead, only the query viewpoint is generated.

Query Loss. We first calculate the reconstruction loss on the individual query view (*e.g.*, between I_τ^q and $\hat{I}_\tau^q$) as follows:

$$\mathcal{L}_{query} = \lambda_R \mathcal{L}_R + \lambda_P \mathcal{L}_P + \lambda_S \mathcal{L}_S + \lambda_A \mathcal{L}_A \tag{2}$$

where $\mathcal{L}_R$ denotes the L_1 reconstruction loss, $\mathcal{L}_P$ is the L_2 loss in the feature space of the VGG-19 network, $\mathcal{L}_S$ is the SSIM loss, $\mathcal{L}_A$ is the adversarial loss calculated using the discriminator architecture from Tulyakov *et al.* [30].

Rotational Consistency Loss. To improve consistency, we let the network generate two additional target views $l \in L_a$ immediately adjacent to the query view. These additional views help the network to synthesize complex shapes, by providing additional information during the training process of the target view q on the actual shape of the human subject, and are generated from the same volumetric representation Z_τ^k. Similarly as for the query viewpoint, we compute the reconstruction loss $\mathcal{L}_{rot1}$ for each additional adjacent view (*e.g.*, between I_τ^l and $\hat{I}_\tau^l$) in the same way as in Eq. 2 (excluding term $\mathcal{L}_P$ in this case).

Next, to minimize inconsistencies between synthesized views, we consider the warping error between the query view and the adjacent views. We compute $W_{l\to q}$, the function warping an image according to the backward flow between the ground-truth images I_τ^q and I_τ^l computed by FlowNet2 [7]. Next, we compute $M_{l\to q}$, the binary occlusion mask between the I_τ^l and $W_{l\to q}(I_\tau^q)$ as $M_{l\to q} = e^{-\alpha\|I_\tau^l - W_{l\to q}(I_\tau^q)\|_2^2}$ [11]. We use a bi-linear sampling layer to warp images and set $\alpha = 50$ (with pixel range between $[0, 1]$) [9]. We apply this occlusion mask to both $\hat{I}_\tau^l$ (the additional generated view) and the warped generated query view $W_{l\to q}(\hat{I}_\tau^q)$, and compute the warping loss as follows:

$$\mathcal{L}_{rot2} = \left\| M_{l \to q} * (\hat{I}_r^l - W_{l \to q}(\hat{I}_r^q)) \right\|_1 \tag{3}$$

The final rotational loss is given by:

$$\mathcal{L}_{rot} = \frac{1}{|L_a|} \sum_{l \in L_a} \mathcal{L}_{rot1} + \lambda_R \mathcal{L}_{rot2} \tag{4}$$

Temporal Consistency Loss. The temporal loss is calculated as the sum of the bi-directional warping errors ($e.g.$, from $t - 1 \to t$ and $t \to t - 1$) between the generated query images $\hat{I}_{t-1}^q$ and $\hat{I}_t^q$. First, we calculate $W_{t \to t-1}$, the function warping the ground-truth image I_{t-1}^q towards I_t^q, computed by FlowNet2 [7]. Next, we calculate the binary occlusion mask $M_{t \to t-1}$, and apply it to the generated view at time t, $\hat{I}_t^q$, and the warped generated view at time $t - 1$, $W_{t \to t-1}(\hat{I}_{t-1}^q)$. We repeat the same calculation for the warping error $t \to t - 1$ and compute the final temporal loss as follows:

$$\mathcal{L}_{temp} = \left\| M_{t \to t-1} * (\hat{I}_t^q - W_{t \to t-1}(\hat{I}_{t-1}^q)) \right\|_1 + \left\| M_{t-1 \to t} * (\hat{I}_{t-1}^q - W_{t-1 \to t}(\hat{I}_t^q)) \right\|_1 \tag{5}$$

Overall Loss. The overall training loss $\mathcal{L}_{tot}$ is given by:

$$\mathcal{L}_{Total} = \frac{1}{|\{t, t-1\}|} \left(\sum_{\tau \in \{t, t-1\}} (\lambda_{query} \mathcal{L}_{query} + \lambda_{rot} \mathcal{L}_{rot}) + \lambda_{temp} \mathcal{L}_{temp} \right), \tag{6}$$

where $t - 1$ and t are the two consecutive time steps. The weights λ_{query}, λ_{rot}, λ_{temp} are set to 2, 1, 1, respectively.

3.4   Multi-View Human Action (MVHA) Dataset

In order to support the development of learning-based NVS solutions that are applicable to human performance, we introduce the Multi-View Human Action (MVHA) dataset. Compared with previous similar datasets that only provide 4 captured viewpoints [8], ours provide 54 different viewpoints for each unique model (18 azimuths and 3 elevations). Moreover, our dataset is not composed of static captures, but of synthetic human subjects moving in extremely diverse modality. The detailed description and samples of the MVHA dataset can be found in the supplementary material.

Body and Clothing Models. We generate fully textured meshes for 30 human characters using Adobe Fuse [4]. The distribution of the subjects' physical characteristics covers a broad spectrum of body shapes, skin tones, outfits and hair geometry. Each subject is dressed in a different outfit including a variety of garments, combining tight and loose clothes.

Table 1. Quantitative results (with 4 input views) on the MVHA dataset for the ablation study (1a) and comparison with TBN approach [18] under different testing scenarios (1b). Arrows indicate the direction of improvement for the metric under consideration.

Method	$L_1\downarrow$	SSIM↑	RL↓	TL↓
Base	.0825	.7364	.0433	.0280
Base+RC	.0736	.7541	.0390	.0163
Base+TC	.0743	.7506	.0414	.0149
Base+RC+TC	**.0722**	**.7596**	**.0370**	**.0146**

(a) Ablation study (seen model/unseen pose)

Testing Scenario	Method	$L1\downarrow$	SSIM↑	RL↓	TL↓
1. Seen model	TBN [18]	.1068	.7072	.0541	.0303
+ Unseen pose	Ours	**.0722**	**.7596**	**.0370**	**.0131**
2. Unseen model	TBN [18]	.0982	.7109	.0504	0.024
+ Seen pose	Ours	**.0806**	**.7438**	**.0458**	**.0146**
3. Unseen model	TBN [18]	.1130	.6867	.0564	.0282
+ Unseen pose	Ours	**.0863**	**.7279**	**.0480**	**.0150**

(b) Results and Generalization Capabilities

Mocap Sequences. We gather 40 realistic motion sequences from Adobe Mixamo [14]. These sequences include human movements with different complexity, from relatively static body motions (*e.g.*, standing) to very complex and dynamic motion patterns (*e.g.*, break-dance or punching).

Camera, Lights and Background. A 3D rendering software is used to apply different Mocap animation sequences to the 30 3D models. The illumination is composed of an ambient light plus a directional light source. We use a projective camera with 512×512 pixel resolution. The distance to the subject is fixed to ensure the whole body is in view at all times. Every sequence is rendered from 54 camera views, 18 azimuths at 20-degree intervals and 3 elevations at 10-degree intervals. For every rendered view and time step, we provide the final RGB image and associated binary segmentation mask. Custom background can be added by taking advantage of the released segmentation masks.

4   Experiments

In this section, we present the results in terms of view generation quality, and rotational and temporal consistency. We present quantitative and qualitative results on our MVHA dataset, as well as the PVHM and ShapeNet datasets [1,37]. To evaluate the view generation quality, we use the L_1 and SSIM scores between the generated and ground truth views. The consistencies between adjacent generated views and between consecutive time steps are evaluated using the Rotational Loss (RL), Eq. 4, and Temporal Loss (TL), Eq. 5, respectively. Unless otherwise stated, all results are reported for models not used during training. The details on the dataset splits, training process, additional results and video results are available in the supplementary material.

4.1   Results on the MVHA Dataset

In this section, we present the performance of the proposed approach for the MVHA dataset. Note that we used 18 azimuths with a fixed elevation throughout the experiments. We investigate the ablative impact of the proposed rotational

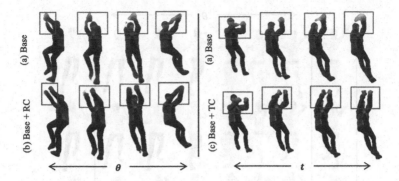

Fig. 3. Ablation study (left: fixed time step, varying query viewpoint; right: fixed query viewpoint, varying time step). Rotational and temporal supervision remarkably improve quality and consistency of the novel view synthesis, both across query viewpoints (left) and time steps (right). Model seen during training, unseen pose, 4 input views.

and temporal supervisions, the generalization ability of our method to unseen models and poses, and provide a visual analysis of the learned volume gating.

Ablation Study. The ablation study investigates the impact of the proposed Rotational Consistency (RC) and Temporal Consistency (TC) on the novel view synthesis quality, both quantitatively (Table 1a) and qualitatively (Fig. 3). Both methods improve the baseline performance and significantly reduce the rotational (Table 1a, rows 1–2) and temporal (Table 1a, rows 1–3) warping errors in the final results. Visual results confirm the gains brought by the proposed supervisions (Fig. 3). In addition, we find that rotational and temporal supervisions have a complementary contribution to the per-frame generation quality, each improving the baseline. With both RC and TC together, we achieve the best performances overall, across all metrics (Table 1a, row 4).

Visualization of Learned Volume Gating. As presented in Sect. 3.2, we allow the network to learn the volumetric mask g automatically. Memory-intensive 3D gating operations are decomposed into three 2D gatings along the depth, height and width axes of the latent volumetric representation. We visualize these learned gating values g_{HW}, g_{WD} and g_{HD} in Fig. 5. We observe the gating masks have different values at each spatial location, especially based on whether the current location is on the foreground or not. More interestingly, the three gating masks *attend to* the foreground shape captured at each corresponding canonical view. This also implies that our learned volumetric representation is indeed aware of 3D structure, and consequently, the learned volume gating layer can achieve soft volume carving.

4.2 Results on PVHM and ShapeNet Datasets

We compare our view generation quality with diverse well-known NVS methods [6,18,19,28,36,37]. We first compare the performance of these methods

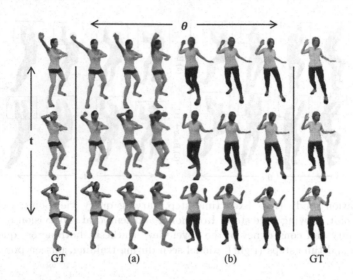

Fig. 4. Our method generates high-quality visual results in case of models unseen during training, both for seen poses (a) and unseen poses (b) (4 input views).

Table 2. Quantitative results on the PVHM dataset [37], for our method and several baseline methods.

Method	MSE↓	SSIM↑
Tatarchenko *et al.* [28]	96.83	.9488
Zhou *et al.* [36]	131.6	.9527
Park *et al.* [19]	85.35	.9519
Zhu *et al.* [37]	72.86	.9670
Huang *et al.* [6]	89.44	.9301
Olszewski *et al.* [18]	70.34	.9695
Ours	**61.68**	**.9807**

on the PVHM dataset [37], to show that our approach can produce superior results on a different dataset with similar characteristics as the one we collected (Table 2). We use the Mean Squared Error (MSE) and Structural Similarity Index (SSIM) to quantitatively compare the different approaches, as done by Zhu *et al.* [37]. Our method produces the best results overall, for both metrics (Table 2). We also show a qualitative comparison with some of the most recent human-specific methods [6,18,37] in Fig. 6. Zhu *et al.* [37] show artifacts on the face region, due to the failure of transferring pixels from visible parts. Huang *et al.* [6] cannot successfully recover the whole body shape. Olszewski *et al.* [18] show incomplete reconstruction of the arm and noticeable color alterations. Our method is able to reconstruct fine-grained details, including limbs and wrinkles, with higher fidelity with respect to the original colors. Both qualitative

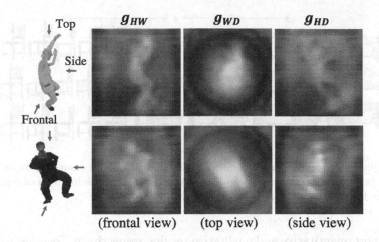

Fig. 5. Example visualization of the gate volumes, for three views and two models of the MVHA dataset.

Fig. 6. Qualitative results on the PVHM dataset, for our approach and different baselines [6,18,37]. Our approach results in overall superior performance.

Table 3. Quantitative results on the ShapeNet dataset [1], for our method and several baseline methods (4 input views), on both car and chair categories. We additionally report the rotational loss for our approach and Olszewski *et al.* [18].

Views	Methods	Car			Chair		
		$L_1 \downarrow$	SSIM↑	RL↓	$L_1 \downarrow$	SSIM↑	RL↓
4	Tatarchenko *et al.* [28]	.112	.890	–	.192	.900	–
	Zhou *et al.* [36]	.081	.924	–	.165	.891	–
	Sun *et al.* [27]	.062	.946	–	.111	.925	–
	Olszewski *et al.* [18]	.059	.946	.076	.107	.939	.073
	Ours	**.051**	**.960**	**.059**	**.087**	**.958**	**.061**

and quantitative results confirm that our approach results in better novel view synthesis reconstruction for human subjects (Fig. 4).

We finally demonstrate that the proposed method can generalize as well for static, inanimate objects such as 'cars' and 'chairs'. In this case, we compare our method to four baselines from recent literature [18,27,28,36]. Table 3 and

Fig. 7. Qualitative results on the ShapeNet dataset [1], for an example car and chair model (4 input views). Our approach outperforms TBN [18] both in terms of generation quality and consistency across generated views.

Fig. 7 report quantitative and qualitative results, respectively. Together with the L_1 and SSIM metric, we also report the rotational loss for ours and Olszewski et al. [18]. For all metrics, for both the car and chair categories, we outperform all baselines by a consistent margin (Table 3). Figure 7 visually confirms that the quality of the generated views is highly improved with our approach, and that we can keep a high degree of consistency across adjacent generated views.

5 Conclusion

In this paper, we propose a novel siamese network architecture employing volume gating convolutions and temporal feature augmentation to tackle the problem of novel view *video* synthesis of human performance. We also introduce explicit rotational and temporal supervision to guarantee high-quality reconstructions and consistency across the generated views. To support future research in this domain, we collect the Multi-View Human Action (MVHA) dataset, composed of near 1200 synthetic, animated human performance captured from 54 viewpoints. Quantitative and qualitative results on our MVHA, PVHM, and ShapeNet datasets confirm the gains brought by the proposed approach compared to state-of-the-art baselines.

Acknowledgments. Youngjoong Kwon was supported partly by Adobe Research and partly by the National Science Foundation grant 1816148. This work was done while Youngjoong Kwon and Dahun Kim were doing an internship at Adobe Research.

References

1. Chang, A.X., et al.: ShapeNet: an information-rich 3D model repository. arXiv preprint arXiv:1512.03012 (2015)
2. Choy, C.B., Xu, D., Gwak, J.Y., Chen, K., Savarese, S.: 3D-R2N2: a unified approach for single and multi-view 3D object reconstruction. In: Leibe, B., Matas, J., Sebe, N., Welling, M. (eds.) ECCV 2016. LNCS, vol. 9912, pp. 628–644. Springer, Cham (2016). https://doi.org/10.1007/978-3-319-46484-8_38

3. Eslami, S.A., et al.: Neural scene representation and rendering. Science **360**(6394), 1204–1210 (2018)
4. Fuse, A.: https://www.adobe.com/products/fuse.html
5. Girdhar, R., Fouhey, D.F., Rodriguez, M., Gupta, A.: Learning a predictable and generative vector representation for objects. In: Leibe, B., Matas, J., Sebe, N., Welling, M. (eds.) ECCV 2016. LNCS, vol. 9910, pp. 484–499. Springer, Cham (2016). https://doi.org/10.1007/978-3-319-46466-4_29
6. Huang, Z., et al.: Deep volumetric video from very sparse multi-view performance capture. In: Ferrari, V., Hebert, M., Sminchisescu, C., Weiss, Y. (eds.) ECCV 2018. LNCS, vol. 11220, pp. 351–369. Springer, Cham (2018). https://doi.org/10.1007/978-3-030-01270-0_21
7. Ilg, E., Mayer, N., Saikia, T., Keuper, M., Dosovitskiy, A., Brox, T.: FlowNet 2.0: evolution of optical flow estimation with deep networks. In: Proceedings of the IEEE Conference on Computer Vision and Pattern Recognition, pp. 2462–2470 (2017)
8. Ionescu, C., Papava, D., Olaru, V., Sminchisescu, C.: Human3.6M: large scale datasets and predictive methods for 3D human sensing in natural environments. IEEE Trans. Pattern Anal. Mach. Intell. **36**(7), 1325–1339 (2014)
9. Jaderberg, M., Simonyan, K., Zisserman, A., et al.: Spatial transformer networks. In: Advances in Neural Information Processing Systems, pp. 2017–2025 (2015)
10. Kar, A., Häne, C., Malik, J.: Learning a multi-view stereo machine. In: Advances in Neural Information Processing Systems, pp. 365–376 (2017)
11. Lai, W.-S., Huang, J.-B., Wang, O., Shechtman, E., Yumer, E., Yang, M.-H.: Learning blind video temporal consistency. In: Ferrari, V., Hebert, M., Sminchisescu, C., Weiss, Y. (eds.) ECCV 2018. LNCS, vol. 11219, pp. 179–195. Springer, Cham (2018). https://doi.org/10.1007/978-3-030-01267-0_11
12. Lombardi, S., Simon, T., Saragih, J., Schwartz, G., Lehrmann, A., Sheikh, Y.: Neural volumes: learning dynamic renderable volumes from images. ACM Trans. Graph. **38**(4) (2019). https://doi.org/10.1145/3306346.3323020
13. Loper, M., Mahmood, N., Romero, J., Pons-Moll, G., Black, M.J.: SMPL: a skinned multi-person linear model. ACM Trans. Graph. (TOG) **34**(6), 248 (2015)
14. Mixamo, A.: https://www.mixamo.com
15. Nguyen-Phuoc, T., Li, C., Theis, L., Richardt, C., Yang, Y.L.: HoloGAN: unsupervised learning of 3D representations from natural images. In: Proceedings of the IEEE International Conference on Computer Vision, pp. 7588–7597 (2019)
16. Nguyen-Phuoc, T., Richardt, C., Mai, L., Yang, Y.L., Mitra, N.: BlockGAN: learning 3D object-aware scene representations from unlabelled images. arXiv preprint arXiv:2002.08988 (2020)
17. Nguyen-Phuoc, T.H., Li, C., Balaban, S., Yang, Y.: RenderNet: a deep convolutional network for differentiable rendering from 3D shapes. In: Advances in Neural Information Processing Systems, pp. 7891–7901 (2018)
18. Olszewski, K., Tulyakov, S., Woodford, O., Li, H., Luo, L.: Transformable bottleneck networks. In: The IEEE International Conference on Computer Vision (ICCV), October 2019
19. Park, E., Yang, J., Yumer, E., Ceylan, D., Berg, A.C.: Transformation-grounded image generation network for novel 3D view synthesis. In: Proceedings of the IEEE Conference on Computer Vision and Pattern Recognition, pp. 3500–3509 (2017)
20. Park, J.J., Florence, P., Straub, J., Newcombe, R., Lovegrove, S.: DeepSDF: learning continuous signed distance functions for shape representation. arXiv preprint arXiv:1901.05103 (2019)

21. Pumarola, A., Sanchez, J., Choi, G., Sanfeliu, A., Moreno-Noguer, F.: 3DPeople: modeling the geometry of dressed humans. arXiv preprint arXiv:1904.04571 (2019)
22. Rezende, D.J., Eslami, S.A., Mohamed, S., Battaglia, P., Jaderberg, M., Heess, N.: Unsupervised learning of 3D structure from images. In: Advances in Neural Information Processing Systems, pp. 4996–5004 (2016)
23. Saito, S., Huang, Z., Natsume, R., Morishima, S., Kanazawa, A., Li, H.: PIFu: pixel-aligned implicit function for high-resolution clothed human digitization. arXiv preprint arXiv:1905.05172 (2019)
24. Shysheya, A., et al.: Textured neural avatars. In: Proceedings of the IEEE Conference on Computer Vision and Pattern Recognition, pp. 2387–2397 (2019)
25. Sitzmann, V., Thies, J., Heide, F., Nießner, M., Wetzstein, G., Zollhofer, M.: Deep-Voxels: learning persistent 3D feature embeddings. In: Proceedings of the IEEE Conference on Computer Vision and Pattern Recognition, pp. 2437–2446 (2019)
26. Sitzmann, V., Zollhöfer, M., Wetzstein, G.: Scene representation networks: continuous 3D-structure-aware neural scene representations. In: Advances in Neural Information Processing Systems, pp. 1121–1132 (2019)
27. Sun, S.-H., Huh, M., Liao, Y.-H., Zhang, N., Lim, J.J.: Multi-view to novel view: synthesizing novel views with self-learned confidence. In: Ferrari, V., Hebert, M., Sminchisescu, C., Weiss, Y. (eds.) ECCV 2018. LNCS, vol. 11207, pp. 162–178. Springer, Cham (2018). https://doi.org/10.1007/978-3-030-01219-9_10
28. Tatarchenko, M., Dosovitskiy, A., Brox, T.: Single-view to multi-view: reconstructing unseen views with a convolutional network. arXiv preprint arXiv:1511.06702 6 (2015)
29. Tulsiani, S., Zhou, T., Efros, A.A., Malik, J.: Multi-view supervision for single-view reconstruction via differentiable ray consistency. In: Proceedings of the IEEE Conference on Computer Vision and Pattern Recognition, pp. 2626–2634 (2017)
30. Tulyakov, S., Liu, M.Y., Yang, X., Kautz, J.: MoCoGAN: decomposing motion and content for video generation. In: Proceedings of the IEEE Conference on Computer Vision and Pattern Recognition, pp. 1526–1535 (2018)
31. Varol, G., et al.: Learning from synthetic humans. In: CVPR (2017)
32. Wu, J., Wang, Y., Xue, T., Sun, X., Freeman, B., Tenenbaum, J.: MarrNet: 3D shape reconstruction via 2.5 D sketches. In: Advances in Neural Information Processing Systems, pp. 540–550 (2017)
33. Wu, J., Zhang, C., Xue, T., Freeman, B., Tenenbaum, J.: Learning a probabilistic latent space of object shapes via 3D generative-adversarial modeling. In: Advances in Neural Information Processing Systems, pp. 82–90 (2016)
34. Yan, X., Yang, J., Yumer, E., Guo, Y., Lee, H.: Perspective transformer nets: learning single-view 3D object reconstruction without 3D supervision. In: Advances in Neural Information Processing Systems, pp. 1696–1704 (2016)
35. Yang, J., Reed, S.E., Yang, M.H., Lee, H.: Weakly-supervised disentangling with recurrent transformations for 3D view synthesis. In: Advances in Neural Information Processing Systems, pp. 1099–1107 (2015)
36. Zhou, T., Tulsiani, S., Sun, W., Malik, J., Efros, A.A.: View synthesis by appearance flow. In: Leibe, B., Matas, J., Sebe, N., Welling, M. (eds.) ECCV 2016. LNCS, vol. 9908, pp. 286–301. Springer, Cham (2016). https://doi.org/10.1007/978-3-319-46493-0_18
37. Zhu, H., Su, H., Wang, P., Cao, X., Yang, R.: View extrapolation of human body from a single image. In: Proceedings of the IEEE Conference on Computer Vision and Pattern Recognition, pp. 4450–4459 (2018)

Side-Aware Boundary Localization for More Precise Object Detection

Jiaqi Wang[1]([✉]), Wenwei Zhang[2], Yuhang Cao[1], Kai Chen[3], Jiangmiao Pang[4], Tao Gong[5], Jianping Shi[3], Chen Change Loy[2], and Dahua Lin[1]

[1] The Chinese University of Hong Kong, Shatin, Hong Kong
{wj017,dhlin}@ie.cuhk.edu.hk, yhcao6@gmail.com
[2] Nanyang Technological University, Singapore, Singapore
{wenwei001,ccloy}@ntu.edu.sg
[3] SenseTime Research, Beijing, China
chenkaidev@gmail.com, shijianping@sensetime.com
[4] Zhejiang University, Hangzhou, China
pangjiangmiao@gmail.com
[5] University of Science and Technology of China, Hefei, China
gongtao950513@gmail.com

Abstract. Current object detection frameworks mainly rely on bounding box regression to localize objects. Despite the remarkable progress in recent years, the precision of bounding box regression remains unsatisfactory, hence limiting performance in object detection. We observe that precise localization requires careful placement of each side of the bounding box. However, the mainstream approach, which focuses on predicting centers and sizes, is not the most effective way to accomplish this task, especially when there exists displacements with large variance between the anchors and the targets. In this paper, we propose an alternative approach, named as *Side-Aware Boundary Localization (SABL)*, where each side of the bounding box is respectively localized with a dedicated network branch. To tackle the difficulty of precise localization in the presence of displacements with large variance, we further propose a two-step localization scheme, which first predicts a range of movement through bucket prediction and then pinpoints the precise position within the predicted bucket. We test the proposed method on both two-stage and single-stage detection frameworks. Replacing the standard bounding box regression branch with the proposed design leads to significant improvements on Faster R-CNN, RetinaNet, and Cascade R-CNN, by 3.0%, 1.7%, and 0.9%, respectively. Code is available at https://github.com/open-mmlab/mmdetection.

Electronic supplementary material The online version of this chapter (https://doi.org/10.1007/978-3-030-58548-8_24) contains supplementary material, which is available to authorized users.

A. Vedaldi et al. (Eds.): ECCV 2020, LNCS 12349, pp. 403–419, 2020.
https://doi.org/10.1007/978-3-030-58548-8_24

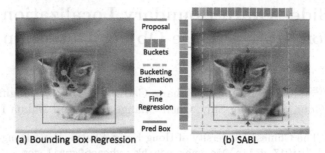

(a) Bounding Box Regression   (b) SABL

Fig. 1. The Illustration of Side-Aware Boundary Localization (SABL). (a) Common *Bounding box Regression* directly predicts displacements from proposals to ground-truth boxes. (b) SABL focuses on object boundaries and localizes them with a bucketing scheme comprising two steps: bucketing estimation and fine regression

1 Introduction

The development of new frameworks for object detection, *e.g.*, *Faster R-CNN* [28], *RetinaNet* [20], and *Cascade R-CNN* [1], has substantially pushed forward the state of the art. All these frameworks, despite the differences in their technical designs, have a common component, namely *bounding box regression*, for object localization.

Generally, bounding box regression is trained to align nearby proposals to target objects. In a common design, the bounding box regression branch predicts the offsets of the centers $(\delta x, \delta y)$ together with the relative scaling factors $(\delta w, \delta h)$ based on the features of RoI (Region of Interest). While this design has been shown to be quite effective in previous works, it remains very difficult to *precisely* predict the location of an object when there exists a displacement, with large variance, between the anchor and the target. This difficulty also limits the overall detection performance.

In recent years, various efforts have been devoted to improving the localization precision, such as cascading the localization process [1,9,16,32], and treating localization as a procedure to segment grid points [23]. Although being shown effective in boosting the accuracy of localization, adoption of these methods complicates the detection pipeline, resulting in considerable computational overhead.

In this work, we aim to explore a new approach to object localization that can effectively tackle *precise localization* with a lower overhead. Empirically, we observe that when we manually annotate a bounding box for an object, it is often much easier to align each side of the box to the object boundary than to move the box as a whole while tuning the size. Inspired by this observation, we propose a new design, named as **Side-Aware Boundary Localization (SABL)**, where each side of the bounding box is respectively positioned based on its surrounding context. As shown in Fig. 1, we devise a *bucketing* scheme to improve the localization precision. For each side of a bounding box, this scheme divides the target space into multiple *buckets*, then determines the bounding box via two steps. Specifically, it first searches for the *correct* bucket, *i.e.*, the one in which the boundary resides. Leveraging the centerline of the selected buckets as a coarse estimate, fine regression is then performed by predicting the offsets. This scheme allows very precise localization even in the presence of displacements with large variance. Moreover, to preserve precisely localized bounding boxes in the non-maximal suppression procedure, we also propose to adjust the classification score based on the bucketing confidences, which leads to further performance gains.

We evaluate the proposed SABL upon various detection frameworks, including two-stage [28], single-stage [20], and cascade [1] detectors. By replacing the existing bounding box regression branch with the proposed design, we achieve significant improvements on $COCO$ $test$-dev [21] without inflicting high computational cost, $i.e.$ 41.8% vs. 38.8% AP with only around 10% extra inference time on top of Faster R-CNN, 40.5% vs. 38.8% AP $without$ extra inference time on top of RetinaNet. Furthermore, we integrate SABL into Cascade R-CNN, where SABL achieves consistent performance gains on this strong baseline, $i.e.$, 43.3% vs. 42.4% AP.

2   Related Work

Object Detection. Object detection is one of the fundamental tasks for computer vision applications [15,34,36,38]. Recent years have witnessed a dramatic improvement in object detection [2,5,6,8,31,35,40,42]. The two-stage pipeline [10,28] has been the leading paradigm in this area. The first stage generates a set of region proposals, and then the second stage classifies and refines the coordinates of proposals by bounding box regression. This design is widely adopted in the later two-stage methods [7,13]. Compared to two-stage approaches, the single-stage pipeline [20,22,26,27] predicts bounding boxes directly. Despite omission of the proposal generation process, single-stage methods [20,22,39] require densely distributed anchors produced by sliding window. Recently, some works attempt to use anchor-free methods [17,18,30] for object detection. Intuitively, iteratively performing the classification and regression process could effectively improve the detection performance. Therefore, many attempts [1,9,16,25,32,37] apply cascade architecture to regress bounding boxes iteratively for progressive refinement.

Object Localization. Object localization is one of the crucial and fundamental modules for object detection. A common approach for object localization is to regress the center coordinate and the size of a bounding box [7,10,11,22,28]. This approach is widely adopted, yet the precision is unsatisfactory due to the large variance of regression target. Aiming for a more accurate localization, some methods [1,3,16,25,32,37] directly repeat the bounding box regression multiple times to further improve accuracy. However, such cascading pipeline expenses much more computational overhead. Some methods that try to reformat the object localization process. Grid R-CNN [23] adopts a grid localization mechanism to encode more clues for accurate object detection. It deals with localization as a procedure to segment grid points, which involves a heavy mask prediction

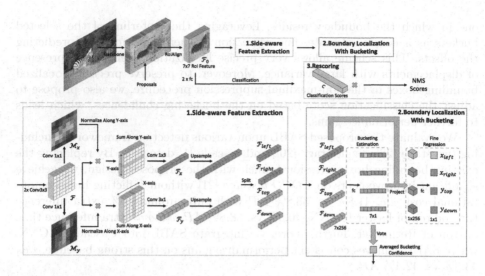

Fig. 2. Pipeline of **Side-Aware Boundary Localization (SABL)** for the two-stage detector (see above). First, RoI features are aggregated to produce side-aware features in the Side-Aware Feature Extraction module. Second, the Boundary Localization with Bucketing module is performed to localize the boundaries by a two-step *bucketing scheme*. Each boundary is first coarsely estimated into buckets and then finely regressed to more precise localization. Third, the confidences of buckets are adopted to assist the classification scores

process. CenterNet [41] combines the classification and regression to localize the object center. It predicts possible object centers on a keypoint heatmap and then adjusts the centers by regression. A similar idea is also adopted in 3D object detection [29]. However, they still fall into the tradition center localization and size estimation paradigm, and the localization precision is still unsatisfactory. LocNet [9] predicts probabilities for object borders or locations inside the object's bounding box. However, the resolution of RoI features limits the performance of LocNet because it needs to transfer the probability of pixels into the bounding box location. On the contrary, our method focuses on the boundaries of object bounding box and decomposes the localization process for each boundary with a bucketing scheme. We also leverage the bucketing estimation confidence to improve the classification results. Performing localization in one pass, SABL achieves substantial gains on both two-stage and single-stage pipelines while keeping their efficiency.

3   Side-Aware Boundary Localization

Accurate object localization is crucial for object detection. Most current methods directly regress the normalized displacements between proposals and ground-truth boxes. However, this paradigm may not provide satisfactory localization results in one pass. Some methods [1,16,32] attempt to improve localization performance with a cascading pipeline at the expense of considerable computational costs. A lightweight as well as effective approach thus becomes necessary.

We propose Side-Aware Boundary Localization (SABL) as an alternative for the conventional bounding box regression to locate the objects more accurately. As shown in Fig. 2, it first extracts horizontal and vertical features $\mathcal{F}_x$ and $\mathcal{F}_y$ by aggregating the RoI features $\mathcal{F}$ along X-axis and Y-axis, respectively, and then splits $\mathcal{F}_x$ and $\mathcal{F}_y$ into side-aware features $\mathcal{F}_{left}$, $\mathcal{F}_{right}$, $\mathcal{F}_{top}$ and $\mathcal{F}_{down}$ (Sect. 3.1). Then for each side of a bounding box, SABL first divides the target space into multiple buckets (as shown in Fig. 1) and searches for the one where the boundary resides via leveraging the side-aware features. It will refine the boundary location x_{left}, x_{right}, y_{top} and y_{down} by further predicting their offsets from the bucket's centerline (Sect. 3.2). Such a two-step *bucketing scheme* could reduce the regression variance and ease the difficulties of prediction. Furthermore, the confidence of estimated buckets could also help to adjust the classification scores and further improve the performance (Sect. 3.3). With minor modifications, SABL is also applicable for single-stage detectors (Sect. 3.4).

3.1 Side-Aware Feature Extraction

As shown in Fig. 2, we extract side-aware features $\mathcal{F}_{left}$, $\mathcal{F}_{right}$, $\mathcal{F}_{top}$, and $\mathcal{F}_{down}$ based on the $k \times k$ RoI features $\mathcal{F}$ ($k = 7$). Following typical conventions [10, 13,28], we adopt RoIAlign to obtain the RoI feature of each proposal. Then we utilize two 3×3 convolution layers to transform it to $\mathcal{F}$. To better capture direction-specific information of the RoI region, we employ the self-attention mechanism to enhance the RoI feature. Specifically, we predict two different attention maps from $\mathcal{F}$ with a 1×1 convolution, which are then normalized along the Y-axis and X-axis, respectively. Taking the attention maps $\mathcal{M}_x$ and $\mathcal{M}_y$, we aggregate $\mathcal{F}$ to obtain $\mathcal{F}_x$ and $\mathcal{F}_y$ as follows,

$$\mathcal{F}_x = \sum_y \mathcal{F}(y,:) * \mathcal{M}_x(y,:),$$
$$\mathcal{F}_y = \sum_x \mathcal{F}(:,x) * \mathcal{M}_y(:,x). \tag{1}$$

$\mathcal{F}_x$ and $\mathcal{F}_y$ are both a 1-D feature map of shape $1 \times k$ and $k \times 1$, respectively. They are further refined by a 1×3 or 3×1 convolution layer and upsampled by a factor of 2 through a deconvolution layer, resulting in $1 \times 2k$ and $2k \times 1$ features on the horizontal and vertical directions, respectively. Finally, the upsampled features are simply split into two halves, leading to the side-aware features $\mathcal{F}_{left}$, $\mathcal{F}_{right}$, $\mathcal{F}_{top}$ and $\mathcal{F}_{down}$.

3.2 Boundary Localization with Bucketing

As shown in the module 2 of Fig. 2, we decompose the localization process into a two-step *bucketing scheme*: bucketing estimation and fine regression. The candidate region of each object boundary is divided into buckets horizontally and

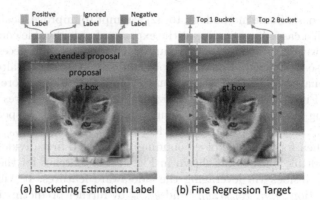

(a) Bucketing Estimation Label   (b) Fine Regression Target

Fig. 3. The localization target of **SABL** for bucketing estimation and fine regression on X-axis. The localization target for Y-axis can be calculated similarly

vertically. We first estimate in which bucket the boundary resides and then regress a more accurate boundary localization from this bucket.

Two-Step Bucketing Scheme. Given a proposal box, *i.e.*, $(B_{left}, B_{right}, B_{top}, B_{down})$, we relax the candidate region of boundaries by a scale factor of σ $(\sigma > 1)$, to cover the entire object. The candidate regions are divided into $2k$ buckets on both X-axis and Y-axis, with k buckets corresponding to each boundary. The width of each bucket on X-axis and Y-axis are therefore $l_x = (\sigma B_{right} - \sigma B_{left})/2k$ and $l_y = (\sigma B_{down} - \sigma B_{top})/2k$, respectively.

In the bucketing estimation step, we adopt a binary classifier to predict whether the boundary is located in or is the closest to the bucket on each side, based on the side-aware features. In the fine regression step, we apply a regressor to predict the offset from the centerline of the selected bucket to the ground-truth boundary.

Localization Targets. There are a bucketing estimation and a fine regression branch in the *bucketing scheme* to be trained. We follow the conventional methods [10,28] for label assigning and proposal sampling. The bucketing estimation determines the nearest buckets to the boundaries of a ground-truth bounding box by binary classification. As shown in Fig. 3, on each side, the bucket, whose centerline is the nearest to the ground-truth boundary, is labeled as 1 (positive sample), while the others are labeled as 0 (negative samples). To reduce the ambiguity in training, on each side, we ignore the bucket that is the second nearest to the ground-truth boundary because it is hard to be distinguished from the positive one. For each side, we ignore negative buckets when training the boundary regressor. To increase the robustness of the fine regression branch, we include both the nearest (labeled as "positive" in the bucketing estimation step) bucket and the second nearest (labeled as "ignore" in the bucketing estimation step) bucket to train the regressor. The regression target is the displacement between the bucket centerline and the corresponding ground-truth boundary. To ease the training difficulties of regressors, we normalize the target by l_x and l_y on the corresponding axes.

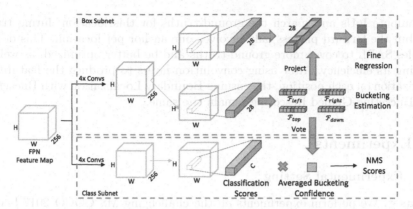

Fig. 4. Pipeline of **Side-Aware Boundary Localization (SABL)** for the single-stage detector. Since there is no RoI features, SABL adopts convolution layers to produce the feature for localization at each location. Then bucketing estimation and fine regression are performed based on this feature at each location. Furthermore, bucketing estimation confidence is leveraged to adjust the classification scores as well

3.3 Bucketing-Guided Rescoring

The bucketing scheme brings a natural benefit, *i.e.*, the bucketing estimation confidences can represent the reliability of predicted locations. With the aim at keeping the more accurately localized bounding boxes during non-maximal suppression (NMS), we utilize the localization reliability to guide the rescoring. Therefore, SABL averages the bucketing estimation confidence scores of four boundaries. The multi-category classification scores are multiplied by the averaged localization confidence, and then used for ranking candidates during NMS. The rescoring helps maintain the best box with both high classification confidence and accurate localization.

3.4 Application to Single-Stage Detectors

SABL can also be applied to single-stage detectors such as [20], with minor modifications. Since there is no proposal stage in single-stage detectors, Side-Aware Feature Extraction (SAFE) is not adopted and the feature extraction is performed following RetinaNet [20]. As shown in Fig. 4, on top of the FPN features, four convolution layers are adopted to classification and localization branches respectively. Following the state of the arts [17,30,41], Group Normalization (GN) [33] is adopted in these convolution layers. At each position of FPN feature maps, there is only one anchor used for detection following [32]. The size of this anchor is $\gamma * s$, where γ is a hyperparameter ($\gamma = 8$), and s is the stride of the current feature map. SABL learns to predict and classify one bounding box based on this anchor. The target assignment process follows the same setting as in [32]. Specifically, we utilize multiple (9 by default) anchors on each location to compute IoUs and match the ground-truths for this location during training, but the forward process only involves one anchor per location. This design enables SABL to cover more ground-truths and be better optimized, as well as keeping its efficiency. After using convolution layers to produce the feature for localization on each position, the ensuing Boundary Localization with Bucketing and Bucketing-Guided Rescoring remain the same.

4   Experiments

4.1   Experimental Setting

Dataset. We perform experiments on the challenging MS COCO 2017 benchmark [21]. We use the *train* split for training and report the performance on the *val* split for ablation study. Detection results for comparison with other methods are reported on the *test-dev* split if not further specified.

Implementation Details. During training, We follow the $1\times$ training scheduler [12] and use mmdetection [4] as the codebase. We train Faster R-CNN [28], Cascade R-CNN [1] and RetinaNet [20] with batch size of 16 for 12 epochs. We apply an initial learning rate of 0.02 for Faster R-CNN, Cascade R-CNN, and 0.01 for RetinaNet. ResNet-50 [14] with FPN [19] backbone is adopted if not further specified. The long edge and short edge of images are resized to 1333 and 800 respectively without changing the aspect ratio during training and inference if not otherwise specified. The scale factor σ is set as 1.7 and 3.0 for Faster R-CNN and RetinaNet, respectively. For Cascade R-CNN, we replace the original bbox head with the proposed SABL, and σ for three cascading stages are set as 1.7, 1.5 and 1.3, respectively. k is set to 7 for all experiments if not further specified. GN is adopted in RetinaNet and RetinaNet w/ SABL as in Sect. 3.4 but not in Faster R-CNN and Cascade R-CNN. Detection results are evaluated with the standard COCO metric. The runtime is measured on a single Tesla V100 GPU.

Training Details. The proposed framework is optimized in an end-to-end manner. For the two-stage pipeline, the RPN loss $\mathcal{L}_{rpn}$ and classification loss $\mathcal{L}_{cls}$ remain the same as Faster R-CNN [28]. We replace the bounding box regression loss by a bucketing estimation loss $\mathcal{L}_{bucketing}$ and a fine regression loss $\mathcal{L}_{reg}$. Specifically, $\mathcal{L}_{bucketing}$ adopts Binary Cross-Entropy Loss, $\mathcal{L}_{reg}$ applies Smooth L1 Loss. In summary, a general loss function can be written as follows: $\mathcal{L} = \lambda_1 \mathcal{L}_{rpn} + \mathcal{L}_{cls} + \lambda_2(\mathcal{L}_{bucketing} + \mathcal{L}_{reg})$, where $\lambda_1 = 1, \lambda_2 = 1$ for the two-stage pipeline, $\lambda_1 = 0, \lambda_2 = 1.5$ for the single stage pipeline.

4.2   Results

We show the effectiveness of SABL by applying SABL on RetinaNet, Faster R-CNN and Cascade R-CNN with ResNet-101 [14] with FPN [19] backbone. To be specific, we adopt SABL to Faster R-CNN and RetinaNet as described in

Table 1. Comparison to mainstream methods with ResNet-101 FPN backbone on COCO dataset. *m.s.* indicates multi-scale training. *Sch.* indicates training schedule. 50e indicates 50 epochs. *Data* indicates the results are evaluated on the corresponding data split of COCO dataset, *e.g.*, some two-stage detectors are evaluated on COCO val split

Method	Backbone	Sch.	AP	AP_{50}	AP_{75}	AP_S	AP_M	AP_L	FPS
RetinaNet [20]	ResNet-101	1×	38.8	60.0	41.7	21.9	42.1	48.6	13.0
FSAF [42] (m.s.)	ResNet-101	1.5×	40.9	61.5	44.0	24.0	44.2	51.3	12.4
FCOS [30] (m.s.)	ResNet-101	2×	41.5	60.7	45.0	24.4	44.8	51.6	13.5
GA-RetinaNet [32] (m.s.)	ResNet-101	2×	41.9	62.2	45.3	24.0	45.3	53.8	11.7
CenterNet [41] (m.s.)	Hourglass-104	50e	42.1	61.1	45.9	24.1	45.5	52.8	8.9
FoveaBox [17] (m.s.)	ResNet-101	2×	42.0	63.1	45.2	24.7	45.8	51.9	12.8
RepPoints [35] (m.s.)	ResNet-101	2×	42.6	63.5	46.2	25.4	46.2	53.3	12.2
RetinaNet w/ SABL	ResNet-101	1×	40.5	59.3	43.6	23.0	44.1	51.3	13.0
RetinaNet w/ SABL (m.s.)	ResNet-101	1.5×	42.7	61.4	46.0	25.3	46.8	53.5	13.0
RetinaNet w/ SABL (m.s.)	ResNet-101	2×	**43.2**	62.0	46.6	25.7	47.4	53.9	13.0
Method	Backbone	Data	AP	AP_{50}	AP_{75}	AP_S	AP_M	AP_L	FPS
Faster R-CNN [28]	ResNet-101	val	38.5	60.3	41.6	22.3	43.0	49.8	13.8
Faster R-CNN [28]	ResNet-101	test-dev	38.8	60.9	42.3	22.3	42.2	48.6	13.8
IoU-Net [16]	ResNet-101	val	40.6	59.0	–	–	–	–	–
GA-Faster R-CNN [32]	ResNet-101	test-dev	41.1	59.9	45.2	22.4	44.4	53.0	11.5
Grid R-CNN Plus [24]	ResNet-101	test-dev	41.4	60.1	44.9	23.4	44.8	52.3	11.1
Faster R-CNN w/ SABL	ResNet-101	val	41.6	59.5	45.0	23.5	46.5	54.6	12.4
Faster R-CNN w/ SABL	ResNet-101	test-dev	**41.8**	60.2	45.0	23.7	45.3	52.7	12.4
Cascade R-CNN [1]	ResNet-101	test-dev	42.4	61.1	46.1	23.6	45.4	54.1	11.2
Cascade R-CNN w/ SABL	ResNet-101	test-dev	**43.3**	60.9	46.2	23.8	46.5	55.7	8.8

Sect. 3. As shown in Table 1, SABL improves the *AP* of RetinaNet by 1.7% with no extra cost, and Faster R-CNN by 3.0% with only around 10% extra inference time. We further apply SABL to the powerful Cascade R-CNN. SABL improves the performance by 0.9% on this strong baseline.

The significant performance gains on various object detection architectures show that SABL is a generally efficient and effective bounding box localization method for object detection. We further compare SABL with other advanced detectors in Table 1. The reported performances here either come from the original papers or from released implementations and models. SABL exhibits the best performance among these methods and retains its efficiency. To make a fair comparison with other single stage-detectors, we employ multi-scale training, *i.e.*, randomly scaling the shorter edge of input images from 640 to 800 pixels, and the training schedule is extended to 2×. For two-stage detectors, the 1× training schedule is adopted. As shown in Table 1, Faster R-CNN w/ SABL outperforms recent two-stage detectors [16,24,32] that also aim at better localization precision. To be specific, IoU-Net [16] and GA-Faster RCNN [32] adopt iterative regression, and Grid R-CNN Plus [24] improves localization by leverag-

Table 2. The effects of each module in our design. *SAFE, BLB, BGR* denote Side-Aware Feature Extraction, Boundary Localization with Bucketing and Bucketing-Guided Rescoring, respectively

SAFE	BLB	BGR	AP	AP_{50}	AP_{75}	AP_{90}	AP_S	AP_M	AP_L
			36.4	58.4	39.3	8.3	21.6	40.0	47.1
✓			38.5	58.2	41.6	14.3	23.0	42.5	49.5
	✓		38.3	57.6	40.5	16.1	22.3	42.6	49.7
	✓	✓	39.0	57.5	41.9	17.1	22.6	43.2	50.9
✓	✓		39.0	57.9	41.4	17.8	22.7	43.4	49.9
✓	✓	✓	39.7	57.8	42.8	18.8	23.1	44.1	51.2

ing a keypoint prediction branch. The experimental results reveal the advantages of the proposed SABL among advanced localization pipelines.

4.3   Ablation Study

Model Design. We omit different components of SABL on two-stage pipeline to investigate the effectiveness of each component, including Side-Aware Feature Extraction (SAFE), Boundary Localization with Bucketing (BLB) and Bucketing-Guided Rescoring (BGR). The results are shown in Table 2. We use Faster R-CNN with ResNet-50 [14] w/ FPN [19] backbone as the baseline. Faster R-CNN adopts center offsets and scale factors of spatial sizes, *i.e.*, $(\delta x, \delta y, \delta w, \delta h)$ as regression targets and achieves 36.4% AP on COCO val set. SABL significantly improves the baseline by 3.3% AP, especially on high IoU thresholds, *e.g.*, SABL tremendously improves AP_{90} by 10.5%.

Side-Aware Feature Extraction (SAFE). In Table 2, we apply Side-Aware Feature Extraction (SAFE) as described in Sect. 3.1. In order to leverage the side-aware features, side-aware regression targets are required. We introduce *boundary regression* targets, *i.e.*, the offset of each boundary $(\delta x_1, \delta y_1, \delta x_2, \delta y_2)$. This simple modification improves the performance from 36.4% to 37.3%, demonstrating that localization by each boundary is more preferable than regressing the box as a whole. SAFE focuses on content of the corresponding side and further improves the performance from 37.3% to 38.5%. To verify that simply adding more parameters will not apparently improve the performance, we also train a Faster RCNN with *boundary regression* and 4conv1fc head. The 4conv1fc head contains four 3×3 convolution layers followed by one fully-connected layer. Although the 4conv1fc head is heavier than SAFE, it marginally improves the AP by 0.1%, *i.e.*, from 37.3% to 37.4%.

Boundary Localization with Bucketing (BLB). As described in Sect. 3.2, BLB divides the RoI into multiple buckets, it first determines which bucket the boundary resides and takes the centerline of the selected bucket as a coarse estimation of boundary. Then it performs fine regression to localize the boundary precisely.

Table 3. Number of convolution layers for Side-Aware Feature Extraction (SAFE) module. *2D Conv* indicates the number of 3×3 Convolution layers before $\mathcal{F}$. *1D Conv* indicates the number of 1×3 and 3×1 convolution layers before $\mathcal{F}_x$ and $\mathcal{F}_y$, respectively

2D Conv	1D Conv	AP	Param	FLOPS	2D Conv	1D Conv	AP	Param	FLOPS
0	1	38.3	40.8M	212G	2	0	39.5	41.6M	267G
0	2	38.3	41.2M	215G	2	1	39.7	42M	270G
1	1	39.3	41.4M	241G	2	2	39.6	42.4M	273G
1	2	39.4	41.8M	244G	3	1	39.7	42.6M	299G

Table 4. Comparison of different methods to aggregate the 2D RoI features into 1D features in SAFE module

Aggregating Method	AP
Max pooling	39.4
Average pooling	39.3
Attention mask	39.7

Table 5. Comparison of different settings of feature size in SAFE module

RoI	Upsample	AP	FLOPS
7	7	39.0	266G
7	14	39.7	270G
7	28	39.1	281G
14	14	39.7	443G

BLB achieves 38.3%, outperforming the popular bounding box regression by 1.9%. Combining BLB with SAFE further improves the AP to 39.0%.

Bucketing-Guided Rescoring (BGR). Bucketing-Guided Rescoring (BGR) is proposed to adjust the classification scores as in Sect. 3.3. The bucketing confidence can naturally be used to represent how confident the model believes that a boundary is precisely localized. We average the confidences of selected buckets for four boundaries and multiply it to classification scores before NMS. Applying the BGR further improves the performance by 0.7% AP.

Side-Aware Feature Extraction. Side-Aware Feature Extraction (SAFE) is used to aggregate the 2D RoI features to 1D features for X-axis and Y-axis, respectively. Here we perform a thorough ablation study for SAFE.

Parameters. In SAFE, after performing the RoI pooling, we apply two 3×3 convolution layers to obtain $\mathcal{F}$. We adopt one 1×3 and the other 3×1 convolution layers after aggregating the 1D features on horizontal and vertical directions to obtain $\mathcal{F}_x$ and $\mathcal{F}_y$, respectively. We investigate the influence of these convolution layers. As shown in Table 3, we list the performance as well as parameters and FLOPS under different settings. It's noteworthy that Faster R-CNN w/ SABL still achieves satisfactory performance with smaller computational cost. Thus the proposed method could be flexibly adjusted to fulfill different requirements of computational cost.

Feature Aggregating Method. As in Sect. 3.1, we apply a self-attention mechanism to aggregate 2D RoI features into 1D features. The max pooling and

Table 6. Influence of different localization pipelines. To crystallize the effectiveness of Boundary Localization of Bucketing (BLB), Side-Aware Feature Extraction (SAFE) and Bucketing Guided Rescoring (BGR) are not applied here

Localization Approach	AP	AP_{50}	AP_{75}	AP_{90}	AP_S	AP_M	AP_L
Bounding Box Regression	36.4	58.4	39.3	8.3	21.6	40.0	47.1
Boundary Regression	37.3	58.2	40.4	10.6	22.0	41.2	47.8
Bucketing	32.8	56.7	35.9	2.0	20.1	36.5	41.5
Iterative Bucketing	36.8	58.3	40.9	6.0	20.8	40.2	48.1
Center Localization with Bucketing (CLB)	36.9	57.6	39.5	11.4	20.8	41.2	47.7
Boundary Localization with Bucketing (BLB)	38.3	57.6	40.5	16.1	22.3	42.6	49.7

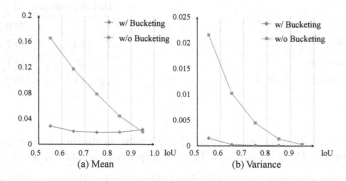

Fig. 5. Mean and variance of displacements from proposals to ground-truth boundaries *w.r.t.* the size of the ground-truth boxes with or without bucketing

average pooling are two alternative approaches in this procedure. In Table 4, experimental results reveal that the proposed attention mask is more effective than max or average pooling to aggregate RoI features.

Size of Side-Aware Features. In the Side-Aware Feature Extraction module, we first perform RoI-Pooling and get the RoI features with spatial size 7×7. The RoI features are aggregated into 1D features with size 1×7 and 7×1, and then upsampled to size of 1×14 and 14×1 by a deconvolution layer. We study RoI features size and upsampled features size. As shown in Table 5, our settings, *i.e.*, RoI size of 7 and upsampled size of 14, achieve the best trade-off between effectiveness and efficiency.

Boundary Localization with Bucketing. Here we discuss the effectiveness of different designs for localization. In our work, we propose Boundary Localization with Bucketing (BLB), that contains 3 key ideas, *i.e.*, localizing by boundaries, bucketing estimation and fine regression.

Table 7. Influence of different designs to generate regression targets. *Ignore* and *Top2-Reg* are described in *Target design*

Ignore	Top2-Reg	AP
		38.7
✓		39.1
	✓	39.4
✓	✓	39.7

Table 8. Influence of different hyper-parameters in RetinaNet w/ SABL, *i.e.*, scale factor σ, buckets number and localization loss weight λ_2. GN is not adopted in this table

σ	Bucket-Num	λ_2	AP	σ	Bucket-Num	λ_2	AP
2	7	1.5	37.3	3	9	1.5	37.2
3	**7**	**1.5**	**37.4**	3	7	1.0	36.8
4	7	1.5	36.9	3	7	1.25	37.2
3	5	1.5	36.9	3	7	1.75	37.2

As shown in Table 6, the proposed BLB achieves significantly higher performance than the widespread *Bounding Box Regression* (38.3% vs. 36.4%), that adopts center offsets and scale factors of spatial sizes, *i.e.*, $(\delta x, \delta y, \delta w, \delta h)$ as regression targets. Switching to *Boundary Regression* that regresses boundary offsets $(\delta x_1, \delta y_1, \delta x_2, \delta y_2)$, improves the *AP* by 0.9%. The result reveals that localizing the object boundaries is more preferable than localizing object centers. Moreover, to show the advantages of the proposed design to iterative *Bounding Box Regression*, we compare SABL with IoUNet [16], GA-Faster R-CNN [32], GA-RetinaNet [32] in Table 1.

Bucketing indicates adopting the centerline of the predicted bucket for each boundary as the final localization. It presents a much inferior performance. Due to the absence of fine regression, the localization quality is severely affected by the bucket width. Following LocNet [9], we design a heavy *Iterative Bucketing* where the bucketing step is performed iteratively. Although the performance is improved from 32.8% to 36.8%, it remains inferior to 38.3% of our method.

We also investigate a scheme to localize the object center with bucketing named *Center Localization with Bucketing (CLB)*. Bucketing estimation and fine regression are used to localize the object center, and width and height are then regressed as in the conventional *Bounding Box Regression*. CLB achieves 1.4% lower *AP* than BLB, which further validates the necessity of localizing object boundaries other than the center.

Figure 5 shows mean and variance of displacements from proposals to ground-truth boxes which are normalized by the size of ground-truth boxes. Without loss of generality, we choose the left boundary to calculate the statistic. The proposals are split into five groups according to their IoU with the ground-truth, *i.e.*, [0.5, 0.6), [0.6, 0.7), [0.7, 0.8), [0.8, 0.9), [0.9, 1.0). Regression with bucketing exhibits more stable distribution on displacements, easing the difficulties of regression and lead to more precise localization.

Target Design. We further study the training target designs for this module. 1) *Ignore*: During training bucketing estimation branch, we ignore the second nearest bucket to ease its ambiguity with the nearest bucket. 2) *Top2-Reg*: Dur-

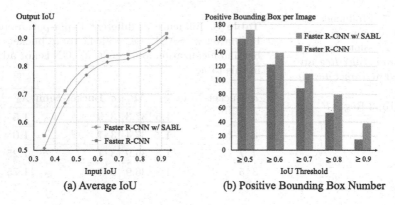

(a) Average IoU   (b) Positive Bounding Box Number

Fig. 6. Analysis of bounding boxes predicted by Faster R-CNN and Faster R-CNN w/ SABL without NMS. (a) Average IoU of proposals before and after localization branch. (b) Number of positive boxes per image with different IoU threshold after localization

ing training the fine regression branch, buckets with Top-2 displacements to the ground-truth boundaries are trained to reduce the influence of mis-classification in bucketing estimation. As shown in Table 7, two proposed designs bring substantial performance gains. In our study, the classification accuracy of Top-1, Top-2 and Top-3 buckets are 69.3%, 90.0% and 95.7%, respectively. We also try to train with Top-3 regression targets, however the performance remains 39.7%.

Scale Factor. We study the influence of different scale factors σ to enlarge proposals during generating localization targets. To be specific, when adopting σ of 1.1, 1.3, 1.5, **1.7**, 1.9, the performance are 39.2%, 39.4%, 39.6%, **39.7%**, 39.6%, respectively.

SABL for Single-Stage Detectors. For single-stage detectors, we take RetinaNet [20] as a baseline. Following conventions of recent single-stage methods [17,30,41], GN is adopted in the head of both RetinaNet and RetinaNet w/ SABL. GN improves RetinaNet from 35.6% to 36.6% and RetinaNet w/SABL from 37.4% to 38.5%. SABL shows consistent improvements over the baseline. Since the single-stage pipeline is different from the two-stage one, we study the hyper-parameters as shown in Table 8. Results reveal that the setting of $\sigma = 3, \lambda_2 = 1.5$ and a bucket number of 7 achieves the best performance.

Analysis of Localization Precision. To demonstrate the effectiveness of SABL on improving the localization quality, we perform quantitative analysis on Faster R-CNN and Faster R-CNN w/ SABL. We split the proposals into different bins ($[0.3, 0.4), [0.4, 0.5), \ldots, [0.9, 1)$) according to the IoUs with their nearest ground-truth object, and then compare the average IoU before and after the localization branch in each bin. As shown in Fig. 6(a), SABL achieves consistently higher IoU than the bounding box regression baseline in all bins, which reveals that both low and high quality proposals are more precisely localized.

Furthermore, in Fig. 6(b) we compare the IoU distribution of proposals after the localization branch. Specifically, we calculate the average number of positive boxes per image with different IoU threshold (*e.g.*, IoU $\geq$ 0.5). SABL results in more positive boxes under all thresholds, especially for high IoU thresholds, *e.g.*, $\geq$ 0.9. It contributes to the significant gains of Faster R-CNN w/ SABL on AP_{90} compared to Faster R-CNN. We also notice that although SABL achieves a higher overall AP and better localization precision across all IoU thresholds, AP_{50} is slightly lower. The situation of higher overall AP but lower AP_{50}, also occurs in a number of detectors [16,23,32] that aim at better localization. AP is affected by not only the localization quality but classification accuracy, and AP_{50} is more sensitive to classification since it does not require bounding boxes with high IoU. Localization and classification branches are jointly trained, and SABL is more optimized for the former. To improve AP_{50}, other efforts to obtain a higher classification accuracy are required, *e.g.*, reducing misclassified boxes, which is beyond the discussion and target of our method.

5   Conclusion

In this work, we propose **Side-Aware Boundary Localization (SABL)** to replace the conventional bounding box regression. We extract side-aware features which focus on the content of boundaries for localization. A lightweight two-step *bucketing scheme* is proposed to locate objects accurately based on the side-aware features. We also introduce a rescoring mechanism to leverage the bucketing confidence to keep high-quality bounding boxes. The proposed SABL exhibits consistent and significant performance gains on various object detection pipelines.

Acknowledgement. This work is partially supported by the SenseTime Collaborative Grant on Large-scale Multi-modality Analysis (CUHK Agreement No. TS1610626 & No. TS1712093), the General Research Fund (GRF) of Hong Kong (No. 14203518 & No. 14205719), SenseTime-NTU Collaboration Project and NTU NAP.

References

1. Cai, Z., Vasconcelos, N.: Cascade R-CNN: delving into high quality object detection. In: CVPR (2018)
2. Cao, Y., Chen, K., Loy, C.C., Lin, D.: Prime sample attention in object detection. In: CVPR (2020)
3. Chen, K., et al.: Hybrid task cascade for instance segmentation. In: CVPR (2019)
4. Chen, K., et al.: MMDetection: Open MMLab detection toolbox and benchmark. arXiv preprint arXiv:1906.07155 (2019)
5. Chen, K., et al.: Optimizing video object detection via a scale-time lattice. In: CVPR (2018)
6. Choi, J., Chun, D., Kim, H., Lee, H.J.: Gaussian YOLOv3: an accurate and fast object detector using localization uncertainty for autonomous driving. In: ICCV (2019)

7. Dai, J., Li, Y., He, K., Sun, J.: R-FCN: object detection via region-based fully convolutional networks. In: NIPS (2016)
8. Ghiasi, G., Lin, T., Pang, R., Le, Q.V.: NAS-FPN: learning scalable feature pyramid architecture for object detection. CoRR abs/1904.07392 (2019). http://arxiv.org/abs/1904.07392
9. Gidaris, S., Komodakis, N.: LocNet: improving localization accuracy for object detection. In: CVPR (2016)
10. Girshick, R.: Fast R-CNN. In: ICCV (2015)
11. Girshick, R., Donahue, J., Darrell, T., Malik, J.: Rich feature hierarchies for accurate object detection and semantic segmentation. In: CVPR (2014)
12. Girshick, R., Radosavovic, I., Gkioxari, G., Dollár, P., He, K.: Detectron (2018). https://github.com/facebookresearch/detectron
13. He, K., Gkioxari, G., Dollár, P., Girshick, R.B.: Mask R-CNN. In: ICCV (2017)
14. He, K., Zhang, X., Ren, S., Sun, J.: Deep residual learning for image recognition. In: CVPR (2016)
15. Huang, Q., Xiong, Y., Lin, D.: Unifying identification and context learning for person recognition. In: CVPR (2018)
16. Jiang, B., Luo, R., Mao, J., Xiao, T., Jiang, Y.: Acquisition of localization confidence for accurate object detection. In: Ferrari, V., Hebert, M., Sminchisescu, C., Weiss, Y. (eds.) Computer Vision – ECCV 2018. LNCS, vol. 11218, pp. 816–832. Springer, Cham (2018). https://doi.org/10.1007/978-3-030-01264-9_48
17. Kong, T., Sun, F., Liu, H., Jiang, Y., Shi, J.: FoveaBox: beyond anchor-based object detector. CoRR abs/1904.03797 (2019)
18. Law, H., Deng, J.: CornerNet: detecting objects as paired keypoints. In: Ferrari, V., Hebert, M., Sminchisescu, C., Weiss, Y. (eds.) Computer Vision – ECCV 2018. LNCS, vol. 11218, pp. 765–781. Springer, Cham (2018). https://doi.org/10.1007/978-3-030-01264-9_45
19. Lin, T., Dollár, P., Girshick, R.B., He, K., Hariharan, B., Belongie, S.J.: Feature pyramid networks for object detection. In: CVPR (2017)
20. Lin, T., Goyal, P., Girshick, R.B., He, K., Dollár, P.: Focal loss for dense object detection. In: ICCV (2017)
21. Lin, T.-Y., et al.: Microsoft COCO: common objects in context. In: Fleet, D., Pajdla, T., Schiele, B., Tuytelaars, T. (eds.) ECCV 2014. LNCS, vol. 8693, pp. 740–755. Springer, Cham (2014). https://doi.org/10.1007/978-3-319-10602-1_48
22. Liu, W., et al.: SSD: single shot MultiBox detector. In: Leibe, B., Matas, J., Sebe, N., Welling, M. (eds.) ECCV 2016. LNCS, vol. 9905, pp. 21–37. Springer, Cham (2016). https://doi.org/10.1007/978-3-319-46448-0_2
23. Lu, X., Li, B., Yue, Y., Li, Q., Yan, J.: Grid R-CNN. In: CVPR (2019)
24. Lu, X., Li, B., Yue, Y., Li, Q., Yan, J.: Grid R-CNN plus: faster and better. arXiv preprint arXiv:1906.05688 (2019)
25. Najibi, M., Rastegari, M., Davis, L.S.: G-CNN: an iterative grid based object detector. In: CVPR (2016)
26. Redmon, J., Divvala, S.K., Girshick, R.B., Farhadi, A.: You only look once: unified, real-time object detection. In: CVPR (2016)
27. Redmon, J., Farhadi, A.: YOLO9000: better, faster, stronger. In: CVPR (2017)
28. Ren, S., He, K., Girshick, R., Sun, J.: Faster R-CNN: towards real-time object detection with region proposal networks. In: NIPS (2015)
29. Shi, S., Wang, X., Li, H.: PointRCNN: 3D object proposal generation and detection from point cloud. In: CVPR (2019)
30. Tian, Z., Shen, C., Chen, H., He, T.: FCOS: fully convolutional one-stage object detection. CoRR abs/1904.01355 (2019)

31. Wang, J., Chen, K., Xu, R., Liu, Z., Loy, C.C., Lin, D.: CARAFE: content-aware reassembly of features. In: ICCV (2019)
32. Wang, J., Chen, K., Yang, S., Loy, C.C., Lin, D.: Region proposal by guided anchoring. In: CVPR (2019)
33. Wu, Y., He, K.: Group normalization. In: Ferrari, V., Hebert, M., Sminchisescu, C., Weiss, Y. (eds.) ECCV 2018. LNCS, vol. 11217, pp. 3–19. Springer, Cham (2018). https://doi.org/10.1007/978-3-030-01261-8_1
34. Xiong, Y., Huang, Q., Guo, L., Zhou, H., Zhou, B., Lin, D.: A graph-based framework to bridge movies and synopses. In: ICCV (2019)
35. Yang, Z., Liu, S., Hu, H., Wang, L., Lin, S.: RepPoints: point set representation for object detection. In: ICCV (2019)
36. Zhan, X., Pan, X., Dai, B., Liu, Z., Lin, D., Loy, C.C.: Self-supervised scene de-occlusion. In: CVPR (2020)
37. Zhang, S., Wen, L., Bian, X., Lei, Z., Li, S.Z.: Single-shot refinement neural network for object detection. In: CVPR (2018)
38. Zhang, W., Zhou, H., Sun, S., Wang, Z., Shi, J., Loy, C.C.: Robust multi-modality multi-object tracking. In: ICCV (2019)
39. Zhang, X., Wan, F., Liu, C., Ji, R., Ye, Q.: FreeAnchor: learning to match anchors for visual object detection. In: NIPS
40. Zhao, Q., et al.: M2Det: a single-shot object detector based on multi-level feature pyramid network. In: AAAI (2019)
41. Zhou, X., Wang, D., Krähenbühl, P.: Objects as points. arXiv preprint arXiv:1904.07850 (2019)
42. Zhu, C., He, Y., Savvides, M.: Feature selective anchor-free module for single-shot object detection. In: CVPR (2019)

SF-Net: Single-Frame Supervision
for Temporal Action Localization

Fan Ma[1][✉], Linchao Zhu[1], Yi Yang[1], Shengxin Zha[2], Gourab Kundu[2],
Matt Feiszli[2], and Zheng Shou[2]

[1] University of Technology Sydney, Ultimo, Australia
`fan.ma@student.uts.edu.au`
[2] Facebook AI, Menlo Park, USA

Abstract. In this paper, we study an intermediate form of supervision, i.e., **single-frame supervision**, for temporal action localization (TAL). To obtain the single-frame supervision, the annotators are asked to identify only a single frame *within* the temporal window of an action. This can significantly reduce the labor cost of obtaining full supervision which requires annotating the action *boundary*. Compared to the weak supervision that only annotates the video-level label, the single-frame supervision introduces extra temporal action signals while maintaining low annotation overhead. To make full use of such single-frame supervision, we propose a unified system called **SF-Net**. First, we propose to predict an actionness score for each video frame. Along with a typical category score, the actionness score can provide comprehensive information about the occurrence of a potential action and aid the temporal boundary refinement during inference. Second, we mine pseudo action and background frames based on the single-frame annotations. We identify pseudo action frames by adaptively expanding each annotated single frame to its nearby, contextual frames and we mine pseudo background frames from all the unannotated frames across multiple videos. Together with the ground-truth labeled frames, these pseudo-labeled frames are further used for training the classifier. In extensive experiments on THU-MOS14, GTEA, and BEOID, SF-Net significantly improves upon state-of-the-art weakly-supervised methods in terms of both segment localization and single-frame localization. Notably, SF-Net achieves comparable results to its fully-supervised counterpart which requires much more resource intensive annotations. The code is available at https://github.com/Flowerfan/SF-Net.

Keywords: Single-frame annotation · Action localization

Electronic supplementary material The online version of this chapter (https://doi.org/10.1007/978-3-030-58548-8_25) contains supplementary material, which is available to authorized users.

© Springer Nature Switzerland AG 2020
A. Vedaldi et al. (Eds.): ECCV 2020, LNCS 12349, pp. 420–437, 2020.
https://doi.org/10.1007/978-3-030-58548-8_25

1 Introduction

Recently, weakly-supervised Temporal Action Localization (TAL) has attracted substantial interest. Given a training set containing only video-level labels, we aim to detect and classify each action instance in long, untrimmed testing videos. In the fully-supervised annotation, the annotators usually need to rollback the video for repeated watching to give the precise temporal boundary of an action instance when they notice an action while watching the video [41]. For the weakly-supervised annotation, annotators just need to watch the video once to give labels. They can record the action class once they notice an unseen action. This significantly reduces annotation resources: video-level labels use fewer resources than annotating the start and end times in the fully-supervised setting (Fig. 1).

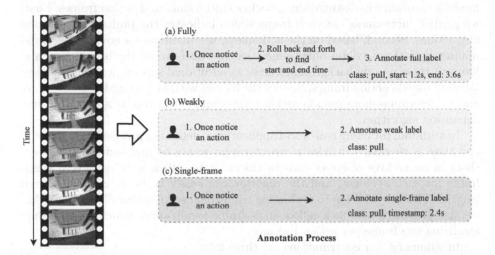

Fig. 1. Different ways of annotating actions while watching a video. (a) Annotating actions in the fully-supervised way. The start and end time of each action instance are required to be annotated. (b) Annotating actions in the weakly-supervised setting. Only action classes are required to be given. (c) Annotating actions in our single-frame supervision. Each action instance should have one timestamp. Note that the time is automatically generated by the annotation tool. Compared to the weakly-supervised annotation, the single-frame annotation requires only a few extra pauses to annotate repeated seen actions in one video.

Despite the promising results achieved by state-of-the-art weakly-supervised TAL work [25,27,29], their localization performance is still inferior to fully-supervised TAL work [6,18,28]. In order to bridge this gap, we are motivated to utilize single-frame supervision [23]: for each action instance, only one single positive frame is pointed out. The annotation process for single-frame supervision is almost the same as it in the weakly-supervised annotation. The annotators only watch the video once to record the action class and timestamp when they notice each action. It significantly reduces annotation resources compared to full supervision.

In the image domain, Bearman et al. [2] were the first to propose point supervision for image semantic segmentation. Annotating at point-level was extended by Mettes et al. [22] to video domain for spatio-temporal localization, where each action frame requires one spatial point annotation during training. Moltisanti et al. [23] further reduced the required resources by proposing single-frame supervision and developed a method that selects frames with a very high confidence score as pseudo action frames.

However, [23] was designed for whole video classification. In order to make full use of single-frame supervision for our TAL task, the unique challenge of **localizing temporal boundaries of actions** remains unresolved. To address this challenge in TAL, we make three innovations to improve the localization model's capability in distinguishing background frames and action frames. First, we predict "actionness" at each frame which indicates the probability of being any actions. Second, based on the actionness, we investigate a novel background mining algorithm to determine frames that are likely to be the background and leverage these pseudo background frames as additional supervision. Third, when labeling pseudo action frames, besides the frames with high confidence scores, we aim to determine more pseudo action frames and thus propose an action frame expansion algorithm.

In addition, for many real-world applications, detecting precise start time and end time is overkill. Consider a reporter who wants to find some car accident shots in an archive of street camera videos: it is sufficient to retrieve a single frame for each accident, and the reporter can easily truncate clips of desired lengths. Thus, in addition to evaluating traditional segment localization in TAL, we also propose a new task called single-frame localization, which requires only localizing one frame per action instance.

In summary, our contributions are three-fold:

(1) To our best knowledge, this is the first work to use single-frame supervision for the challenging problem of localizing temporal boundaries of actions. We show that the single-frame annotation significantly saves annotation time compared to fully-supervised annotation.
(2) We find that single-frame supervision can provide strong cue about the background. Thus, from frames that are not annotated, we propose two novel methods to mine likely background frames and action frames, respectively. These likely background and action timestamps are further used as pseudo ground truth for training.
(3) We conduct extensive experiments on three benchmarks, and the performances on both segment localization and single-frame localization tasks are largely boosted.

2   Related Work

Action Recognition. Action recognition has recently witnessed an increased focus on trimmed videos. Both temporal and spatial information is significant for classifying the video. Early works mainly employed hand-crafted features to solve this task. IDT [34] had been widely used across many video-related tasks. Recently, various deep neural networks were proposed to encode spatial-temporal video information. Two-stream network [31] adopted optical flow to learn temporal motion, which had been used in many latter works [4,31,36]. Many 3D convolutional networks [4,11,33] are also designed to learn action embeddings. Beyond fully-supervised action recognition, a few works focus on self-supervised video feature learning [43] and few-shot action recognition [44]. In this paper, we focus on single-frame supervision for temporal action localization.

Point Supervision. Bearman et al. [2] first utilized the point supervision for image semantic segmentation. Mettes et al. [22] extended it to spatio-temporal localization in video, where the action is pointed out by one spatial location in each action frame. We believe this is overkill for temporal localization, and demonstrate that single-frame supervision can achieve very promising results already. Recently, single-frame supervision has been used in [23] for video-level classification, but this work does not address identifying temporal boundaries. Note that Alwassel et al. [1] proposed to spot action in the video during inference time but targeted detecting one action instance per class in one video while our proposed single-frame localization task aims to detect every instance in one video.

Fully-Supervised Temporal Action Localization. Approaches of temporal action localization trained in full supervision have mainly followed a proposal-classification paradigm [6,8,12,18,28,30], where temporal proposals are generated first and then classified. Other categories of methods, including sequential decision-making [1] and single-shot detectors [17] have also been studied. Given full temporal boundary annotations, the proposal-classification methods usually filter out the background frames at the proposal stage via a binary actionness classifier. Activity completeness has also been studied in the temporal action localization task. Zhao et al. [42] used a structural temporal pyramid pooling followed by an explicit binary classifier to evaluate the completeness of an action instance. Yuan et al. [38] structured an action into three parts to model its temporal evolution. Chéron et al. [7] handled the spatio-temporal action localization with various supervisions. Long et al. [21] proposed a Gaussian kernel to dynamically optimize temporal scale of action proposals. However, these methods use fully temporal annotations, which are resource intensive.

Weakly-Supervised Temporal Action Localization. Multiple Instance Learning (MIL) has been widely used in weakly-supervised temporal action localization. Without temporal boundary annotations, temporal action score sequence has been widely used to generate action proposals [19,24,25,35]. Wang et al. [35] proposed UntrimmedNet composed of a classification module and a

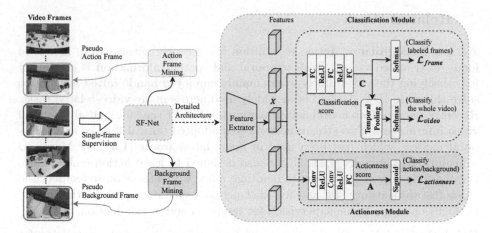

Fig. 2. Overall training framework of our proposed SF-Net. Given single-frame supervision, we employ two novel frame mining strategies to label pseudo action frames and background frames. The detailed architecture of SF-Net is shown on the right. SF-Net consists of a classification module to classify each labeled frame and the whole video, and an actionness module to predict the probability of each frame being action. The classification module and actionness module are trained jointly with three losses explained in Sect. 3.3.

selection module to reason about the temporal duration of action instances. Nguyen *et al.* [25] introduced a sparsity regularization for video-level classification. Shou *et al.* [29] and Liu [20] investigated score contrast in the temporal dimension. Hide-and-Seek [32] randomly removed frame sequences during training to force the network to respond to multiple relevant parts. Liu *et al.* [19] proposed a multi-branch network to model the completeness of actions. Narayan *et al.* [24] introduced three-loss forms to guide the learning discriminative action features with enhanced localization capabilities. Nguyen [26] used attention modules to detect foreground and background for detecting actions. Despite the improvements over time, the performances of weakly-supervised methods are still inferior to the fully-supervised method.

3   Method

In this section, we define our tasks, present architecture of our SF-Net, and finally discuss details of training and inference, respectively.

3.1   Problem Definition

A training video can contain multiple action classes and multiple action instances. Unlike the full supervision setting, which provides temporal boundary annotation of each action instance, in our single-frame supervision setting, each instance only has one frame pointed out by annotator with timestamp t and action class y. Note that $y \in \{1, \ldots, N_c\}$ where N_c is the total number of classes and we use index 0 to represent the background class.

Given a testing video, we perform two temporal localization tasks: (1) **Segment localization**. We detect the start time and end time for each action instance with its action class prediction. (2) **Single-frame localization**. We output the timestamp of each detected action instance with its action class prediction. The evaluation metrics for these two tasks are explained in Sect. 4.

3.2   Framework

Overview. Our overall framework is presented in Fig. 2. During training, learning from single-frame supervision, SF-Net mines pseudo action and background frames. Based on the labeled frames, we employ three losses to jointly train a classification module to classify each labeled frame and the whole video, and an actionness module to predict the probability of each frame being action. In the following, we outline the framework while details of frame mining strategies and different losses are described in Sect. 3.3.

Feature Extraction. For a training batch of N videos, the features of all frames are extracted and stored in a feature tensor $X \in R^{N \times T \times D}$, where D is the feature dimension, and T is the number of frames. As different videos vary in the temporal length, we simply pad zeros when the number of frames in a video is less than T.

Classification Module. The classification module outputs the score of being each action class for all frames in the input video. To classify each labeled frame, we feed X into three Fully-Connected (FC) layers to get the classification score $C \in \mathcal{R}^{N \times T \times N_c + 1}$. The classification score C is then used to compute frame classification loss $\mathcal{L}_{frame}$. We also pool C temporally as described in [24] to compute video-level classification loss $\mathcal{L}_{video}$.

Actionness Module. As shown in Fig. 2, our model has an actionness branch of identifying positive action frames. Different from the classification module, the actionness module only produces a scalar for each frame to denote the probability of being contained in an action segment. To predict an actionness score, we feed X into two temporal convolutional layers followed by one FC layer, resulting in an actionness score matrix $A \in \mathcal{R}^{N \times T}$. We apply sigmoid on A and then compute a binary classification loss $\mathcal{L}_{actionness}$.

3.3   Pseudo Label Mining and Training Objectives

Action Classification at Labeled Frames. We use cross entropy loss for the action frame classification. As there are NT frames in the input batch of videos and most of the frames are unlabeled, we first filter the labeled frames for classification. Suppose we have K labeled frames where $K \ll NT$. We can get classification activations of K labeled frames from C. These scores are fed to a Softmax layer to get classification probability $\mathbf{p}^l \in \mathcal{R}^{K \times N_c+1}$ for all labeled frames. The classification loss of annotated frames in the batch of videos is formulated as:

$$\mathcal{L}_{frame}^l = -\frac{1}{K} \sum_i^K \mathbf{y}_i \log \mathbf{p}_i^l, \qquad (1)$$

where the $\mathbf{p}_i^l$ denote the prediction for the ith labeled action frame.

Pseudo Labeling of Frames. With only a single label per action instance, the total number of positive examples is quite small and may be difficult to learn from. While we do not use full temporal annotation, it is clear that actions are longer events spanning consecutive frames. To increase the temporal information available to the model, we design an action frame mining and a background frame mining strategy to introduce more frames into the training process.

(a) Action frame mining: We treat each labeled action frame as an anchor frame for each action instance. We first set the expand radius r to limit the maximum expansion distance to the anchor frame at t. Then we expand the past from $t-1$ frame and the future from $t+1$ frame, separately. Suppose the action class of the anchor frame is represented by y_i. If the current expanding frame has the same predicted label with the anchor frame, and the classification score at y_i class is higher than that score of the anchor frame multiplying a predefined value ξ, we then annotate this frame with label y_i and put it into the training pool. Otherwise, we stop the expansion process for the current anchor frame.

(b) Background frame mining: The background frames are also important and widely used in localization methods [19,26] to boost the model performance. Since there is no background label under the single-frame supervision, our proposed model manages to localize background frames from all the unlabeled frames in the N videos. At the beginning, we do not have supervision about where the background frames are. But explicitly introducing a background class can avoid forcing classifying a frame into one of the action classes. Our proposed background frame mining algorithm can offer us the supervision needed for training such a background class so as to improve the discriminability of the classifier. Suppose we try to mine ηK background frames, we first gather the classification scores of all unlabeled frames from C. The η is the ratio of background frames to labeled frames. These scores are then sorted along background class to select the top ηK scores $\mathbf{p}^b \in \mathcal{R}^{\eta K}$ as the score vector of the background frames. The pseudo background classification loss is calculated on the top ηK frames by,

$$\mathcal{L}_{frame}^b = -\frac{1}{\eta K} \sum \log \mathbf{p}^b, \qquad (2)$$

The background frame classification loss assists the model with identifying irrelevant frames. Different from background mining in [19,26] which either require

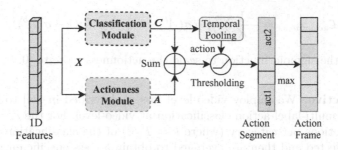

Fig. 3. The inference framework of SF-Net. The classification module outputs the classification score C of each frame for identifying possible target actions in the given video. The action module produces the actionness score determining the possibility of a frame containing target actions. The actionness score together with the classification score are used to generate action segment based on the threshold.

extra computation source to generate background frames or adopt a complicated loss for optimization, we mining background frames across multiple videos and use the classification loss for optimization. The selected pseudo background frames may have some noises in the initial training rounds. As the training evolves and the classifier's discriminability improves, we are able to reduce the noises and detect background frames more correctly. With the more correct background frames as supervision signals, the classifier's discriminability can be further boosted. In our experiments, we observed that this simple background mining strategy allows for better action localization results. We incorporate the background classification loss with the labeled frame classification loss to formulate the single-frame classification loss

$$\mathcal{L}_{frame} = \mathcal{L}^l_{frame} + \frac{1}{N_c}\mathcal{L}^b_{frame} \tag{3}$$

where N_c is the number of action classes to leverage the influence from background class.

Actionness Prediction at Labeled Frames. In fully-supervised TAL, various methods learn to generate action proposals that may contain the potential activities [6,18,37]. Motivated by this, we design the actionness module to make the model focus on frames relevant to target actions. Instead of producing the temporal segment in proposal methods, our actionness module produces the actionness score for each frame. The actionness module is in parallel with the classification module in our SF-Net. It offers extra information for temporal action localization. We first gather the actionness score $A^l \in \mathcal{R}^K$ of labeled frames in the training videos. The higher the value for a frame, the higher probability of that frame belongs to a target action. We also use the background frame mining strategy to get the actionness score $A^b \in \mathcal{R}^{\eta K}$. The actionness loss is calculated by,

$$\mathcal{L}_{actionness} = -\frac{1}{K}\sum \log \sigma(A^l) - \frac{1}{\eta K}\sum \log(1 - \sigma(A^b)), \qquad (4)$$

where σ is the sigmoid function to scale the actionness score to $[0, 1]$.

Full Objective. We employ video-level loss as described in [24] to tackle the problem of multi-label action classification at video-level. For the ith video, the top-k activations per category (where $k = T_i/8$) of the classification activation $C(i)$ are selected and then are averaged to obtain a class-specific encoding $r_i \in \mathcal{R}^{C+1}$ as in [24,27]. We average all the frame label predictions in the video v_i to get the video-level ground-truth $q_i \in \mathcal{R}^{N_c+1}$. The video-level loss is calculated by

$$\mathcal{L}_{video} = -\frac{1}{N}\sum_{i=1}^{N}\sum_{j=1}^{N_c} q_i(j) \log \frac{\exp(r_i(j))}{\sum_{N_c+1}\exp(r_i(k))}, \qquad (5)$$

where $q_i(j)$ is the jth value of q_i representing the probability mass of video v_i belong to j^{th} class.

Consequently, the total training objective for our proposed method is

$$\mathcal{L} = \mathcal{L}_{frame} + \alpha\mathcal{L}_{video} + \beta\mathcal{L}_{actionness}, \qquad (6)$$

where $\mathcal{L}_{frame}$, $\mathcal{L}_{video}$, and $\mathcal{L}_{actionness}$, denote the frame classification loss, video classification loss, and actionness loss, respectively. α and β are the hyperparameters leveraging different losses.

3.4   Inference

During the test stage, we need to give the temporal boundary for each detected action. We follow previous weakly-supervised work [25] to predict video-level labels by temporally pooling and thresholding on the classification score. As shown in Fig. 3, we first obtain the classification score C and actionness score A by feeding input features of a video to the classification module and actionness module. Towards segment localization, we follow the thresholding strategy in [25, 26] to keep the action frames above the threshold and consecutive action frames constitute an action segment. For each predicted video-level action class, we localize each action segment by detecting an interval that the sum of classification score and actionness score exceeds the preset threshold at every frame inside the interval. We simply set the confidence score of the detected segment to the sum of its highest frame classification score and the actionness score. Towards single frame localization, for the action instance, we choose the frame with the maximum activation score in the detected segment as the localized action frame.

4 Experiment

4.1 Datasets

THUMOS14. There are 1010 validation and 1574 test videos from 101 action categories in THUMOS14 [13]. Out of these, 20 categories have temporal annotations in 200 validation and 213 test videos. The dataset is challenging, as it contains an average of 15 activity instances per video. Similar to [14,16], we use the validation set for training and test set for evaluating our framework.

GTEA. There are 28 videos of 7 fine-grained types of daily activities in a kitchen contained in GTEA [15]. An activity is performed by four different subjects, and each video contains about 1800 RGB frames, showing a sequence of 7 actions, including the background action.

BEOID. There are 58 videos in BEOID [9]. There is an average of 12.5 action instances per video. The average length is about 60s, and there are 30 action classes in total. We randomly split the untrimmed videos in an 80–20% proportion for training and testing, as described in [23].

Table 1. Comparison between different methods for simulating single-frame supervision on THUMOS14. "Annotation" means that the model uses human annotated frame for training. "TS" denotes that the single-frame is sampled from action instances using a uniform distribution, while "TS in GT" is using a Gaussian distribution near the mid timestamp of each activity. The AVG for segment localization is the average mAP from IoU 0.1 to 0.7.

Position	mAP@hit	Segment mAP@IoU			
		0.3	0.5	0.7	AVG
Annotation	**60.2** ± 0.70	**53.3** ± 0.30	28.8 ± 0.57	9.7 ± 0.35	**40.6** ± 0.40
TS	57.6 ± 0.60	52.0 ± 0.35	**30.2** ± 0.48	**11.8** ± 0.35	40.5 ± 0.28
TS in GT	52.8 ± 0.85	47.4 ± 0.72	26.2 ± 0.64	9.1 ± 0.41	36.7 ± 0.52

4.2 Implementation Details

We use I3D network [4] trained on the Kinetics [5] to extract video features. For the RGB stream, we rescale the smallest dimension of a frame to 256 and perform the center crop of size 224×224. For the flow stream, we apply the TV-L1 optical flow algorithm [39]. We follow the two-stream fusion operation in [24] to integrate predictions from both appearance (RGB) and motion (Flow) branches. The inputs to the I3D models are stacks of 16 frames.

On all datasets, we set the learning rate to 10^{-3} for all experiments, and the model is trained with a batch size of 32 using the Adam [14]. Loss weight hyper-parameters α and β are set to 1. The model performance is not sensitive to

Table 2. Single-frame annotation differences between different annotators on three datasets. We show the number of action segments annotated by Annotator 1, Annotator 2, Annotator 3, and Annotator 4. In the last column, we report the total number of the ground-truth action segments for each dataset.

Datasets	Annotator 1	Annotator 2	Annotator 3	Annotator 4	# of total segments
GTEA	369	366	377	367	367
BEOID	604	602	589	599	594
THUMOS14	3014	2920	2980	2986	3007

these hyper-parameters. For the hyper-parameter η used in mining background frames, we set it to 5 on THUMOS14 and set it to 1 on the other two datasets. The number of iterations is set to 500, 2000 and 5000 for GTEA, BEOID and THUMOS14, respectively.

4.3 Evaluation Metrics

(1) **Segment localization**: We follow the standard protocol, provided with the three datasets, for evaluation. The evaluation protocol is based on mean Average Precision (mAP) for different intersection over union (IoU) values for the action localization task.

(2) **Single-frame localization**: We also use mAP to compare performances. Instead of measuring IoU, the predicted single-frame is regarded as correct when it lies in the temporal area of the ground-truth segment, and the class label is correct. We use mAP@hit to denote the mean average precision of selected action frame falling in the correct action segment.

4.4 Annotation Analysis

Single-Frame Supervision Simulation. First, to simulate the single-frame supervision based on ground-truth boundary annotations existed in the above three datasets, we explore the different strategies to sample a single-frame for each action instance. We follow the strategy in [23] to generate single-frame annotations with uniform and Gaussian distribution (**Denoted by TS and TS in GT**). We report the segment localization at different IoU thresholds and frame localization results on THUMOS14 in Table 1. The model with each single-frame annotation is trained five times. The mean and standard deviation of mAP is reported in the Table. Compared to models trained on sampled frames, the model trained on human annotated frames achieves the highest mAP@hit. As the action frame is the frame with the largest prediction score in the prediction segment, the model with higher mAP@hit can assist with localizing action timestamp more accurately when people need to retrieve the frame of target actions. When sampling frames are from near middle timestamps to the action segment (TS

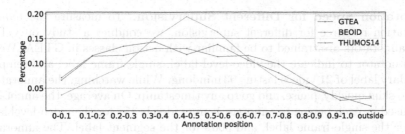

Fig. 4. Statistics of human annotated single-frame on three datasets. X-axis: single-frame falls in the relative portion of the whole action; Y-axis: percentage of annotated frames. We use different colors to denote annotation distribution on different datasets.

in GT), the model performs inferior to other models as these frames may not contain informative elements of complete actions. For the segment localization result, the model trained on truly single-frame annotations achieves higher mAP at small IoU thresholds, and the model trained on frames sampled uniformly from the action instance gets higher mAP at larger IoU thresholds. It may be originated by sampled frames of uniform distribution containing more boundary information for the given action instances.

Single-Frame Annotation. We also invite four annotators with different backgrounds to label a single frame for each action segment on three datasets. More details of annotation process can be found in the supplementary material. In Table 2, we have shown the action instances of different datasets annotated by different annotators. The ground-truth in the Table denotes the action instances annotated in the fully-supervised setting. From the Table, we obtain that the number of action instances by different annotators have a very low variance. The number of labeled frames is very close to the number of action segments in the fully-supervised setting. This indicates that annotators have common justification for the target actions and hardly miss the action instance despite that they only pause once to annotate single-frame of each action.

We also present the distribution of the relative position of single-frame annotation to the corresponding action segment. As shown in Fig. 4, there are rare frames outside of the temporal range of action instances from the ground-truth in the fully-supervised setting. As the number of annotated single frames is almost the same as the number of action segments, we can draw the inference that the single frame annotation includes all almost potential action instances. We obtain that annotators prefer to label frames near to the middle part of action instances. This indicates that humans can identify an action without watching the whole action segment. On the other hand, this will significantly reduce the annotation time compared with fully-supervised annotation as we can quickly skip the current action instance after single-frame annotation.

Annotation Speed for Different Supervision. To measure the required annotation resource for different supervision, we conduct a study on GTEA. Four annotators are trained to be familiar with action classes in GTEA. We ask the annotator to indicate the video-level label, single-frame label and temporal boundary label of 21 videos lasting 93 min long. While watching, the annotator is able to skim quickly, pause, and go to any timestamp. On average, the annotation time used by each person to annotate 1-min video is 45 s for the video-level label, 50 s for the single-frame label, and 300 s for the segment label. The annotation time for single-frame label is close to the annotation time for video-level label but much fewer than time for the fully-supervised annotation.

Table 3. Segment localization mAP results at different IoU thresholds on three datasets. Weak denotes that only video-level labels are used for training. All action frames are used in the full supervision approach. SF uses extra single frame supervision with frame level classification loss. SFB means that pseudo background frames are added into the training, while the SFBA adopts the actionness module, and the SFBAE indicates the action frame mining strategy added in the model. For models trained on single-frame annotations, we report mean and standard deviation results of five runs. AVG is the average mAP from IoU 0.1 to 0.7.

Dataset	Models	mAP@IoU				
		0.1	0.3	0.5	0.7	AVG
GTEA	Full	58.1	40.0	22.2	14.8	31.5
	Weak	14.0	9.7	4.0	3.4	7.0
	SF	50.0 ± 1.42	35.6 ± 2.61	**21.6 ± 1.67**	**17.7 ± 0.96**	30.5 ± 1.23
	SFB	52.9 ± 3.84	34.9 ± 4.72	17.2 ± 3.46	11.0 ± 2.52	28.0 ± 3.53
	SFBA	52.6 ± 5.32	32.7 ± 3.07	15.3 ± 3.63	8.5 ± 1.95	26.4 ± 3.61
	SFBAE	**58.0 ± 2.83**	**37.9 ± 3.18**	19.3 ± 1.03	11.9 ± 3.89	**31.0 ± 1.63**
BEOID	Full	65.1	38.6	22.9	7.9	33.6
	Weak	22.5	11.8	1.4	0.3	8.7
	SF	54.1 ± 2.48	24.1 ± 2.37	6.7 ± 1.72	1.5 ± 0.84	19.7 ± 1.25
	SFB	57.2 ± 3.21	26.8 ± 1.77	9.3 ± 1.94	1.7 ± 0.68	21.7 ± 1.43
	SFBA	**62.9 ± 1.68**	36.1 ± 3.17	12.2 ± 3.15	2.2 ± 2.07	27.1 ± 1.44
	SFBAE	**62.9 ± 1.39**	**40.6 ± 1.8**	**16.7 ± 3.56**	**3.5 ± 0.25**	**30.1 ± 1.22**
THUMOS14	Full	68.7	54.5	34.4	16.7	43.8
	Weak	55.3	40.4	20.4	7.3	30.8
	SF	58.6 ± 0.56	41.3 ± 0.62	20.4 ± 0.55	6.9 ± 0.33	31.7 ± 0.41
	SFB	60.8 ± 0.65	44.5 ± 0.37	22.9 ± 0.38	7.8 ± 0.46	33.9 ± 0.31
	SFBA	68.7 ± 0.33	52.3 ± 1.21	28.2 ± 0.42	**9.7 ± 0.51**	39.9 ± 0.43
	SFBAE	**70.0 ± 0.64**	**53.3 ± 0.3**	**28.8 ± 0.57**	**9.7 ± 0.35**	**40.6 ± 0.40**

Table 4. Segment localization results on THUMOS14 dataset. The mAP values at different IoU thresholds are reported, and the column AVG indicates the average mAP at IoU thresholds from 0.1 to 0.5. * denotes the single-frame labels are simulated based on the ground-truth annotations. # denotes single-frame labels are manually annotated by human annotators.

Supervision	Method	mAP @IoU							
		0.1	0.2	0.3	0.4	0.5	0.6	0.7	AVG
Full	S-CNN [30]	47.7	43.5	36.3	28.7	19.0	–	5.3	35.0
Full	CDC [28]	–	–	40.1	29.4	23.3	–	7.9	–
Full	R-C3D [37]	54.5	51.5	44.8	35.6	28.9	–	–	43.1
Full	SSN [42]	60.3	56.2	50.6	40.8	29.1	–	–	47.4
Full	Faster- [6]	59.8	57.1	53.2	48.5	42.8	**33.8**	**20.8**	52.3
Full	BMN [16]	–	–	56.0	47.4	38.8	29.7	20.5	–
Full	P-GCN [40]	**69.5**	**67.8**	**63.6**	**57.8**	**49.1**	–	–	**61.6**
Weak	Hide-and-Seek [32]	36.4	27.8	19.5	12.7	6.8	–	–	20.6
Weak	UntrimmedNet [35]	44.4	37.7	28.2	21.1	13.7	–	–	29.0
Weak	W-TALC [10]	49.0	42.8	32.0	26.0	18.8	–	6.2	33.7
Weak	AutoLoc [29]	–	–	35.8	29.0	21.2	13.4	5.8	–
Weak	STPN [25]	52.0	44.7	35.5	25.8	16.9	9.9	4.3	35.0
Weak	W-TALC [27]	55.2	49.6	40.1	31.1	22.8	–	7.6	39.7
Weak	Liu *et al.* [19]	57.4	50.8	41.2	32.1	23.1	15.0	7.0	40.9
Weak	Nguyen *et al.* [26]	**60.4**	**56.0**	**46.6**	**37.5**	**26.8**	**17.6**	**9.0**	**45.5**
Weak	3C-Net [24]	59.1	53.5	44.2	34.1	26.6	–	8.1	43.5
Single-frame simulation*	Moltisanti *et al.* [23]	24.3	19.9	15.9	12.5	9.0	–	–	16.3
Single-frame simulation*	SF-Net	68.3	62.3	52.8	**42.2**	**30.5**	**20.6**	**12.0**	51.2
Single-frame#	SF-Net	**71.0**	**63.4**	**53.2**	40.7	29.3	18.4	9.6	**51.5**

4.5 Analysis

Effectiveness of Each Module, Loss, and Supervision. To analyze the contribution of the classification module, actionness module, background frame mining strategy, and the action frame mining strategy, we perform a set of ablation studies on THUMOS14, GTEA and BEOID datasets. The segment localization mAP at different thresholds is presented in Table 3. We also compare the model with only weak supervision and the model with full supervision. The model with weak supervision is implemented based on [24].

We observe that the model with single-frame supervision outperforms the weakly-supervised model. And large performance gain is obtained on GTEA and BEOID datasets as the single video often contains multiple action classes, while action classes in one video are fewer in THUMOS14. Both background frame mining strategy and action frame mining strategy boost the performance on BEOID and THUMOS14 by putting more frames into the training, the performance on GTEA decreases mainly due to that GTEA contains almost no background frame. In this case, it is not helpful to employ background mining and the actionness module which aims for distinguishing background against action. The actionness module works well for the BEOID and THUMOS14 datasets, although the actionness module only produces one score for each frame.

Comparisons with State-of-the-art. Experimental results on THUMOS14 testing set are shown in Table 4. Our proposed single-frame action localization method is compared to existing methods for weakly-supervised temporal action localization, as well as several fully-supervised ones. Our model outperforms the previous weakly-supervised methods at all IoU thresholds regardless of the choice of feature extraction network. The gain is substantial even though only one single-frame for each action instance is provided. The model trained on human annotated frames achieves higher mAP at lower IoU compared to model trained on sampling frames uniformly from action segments. The differences come from the fact that the uniform sampling frames from ground-truth action segments contain more information about temporal boundaries for different actions. As there are many background frames in the THUMOS14 dataset, the single frame supervision assists the proposed model with localizing potential action frames among the whole video. Note that the supervised methods have the regression module to refine the action boundary, while we simply threshold on the score sequence and still achieve comparable results.

5   Conclusions

In this paper, we have investigated how to leverage single-frame supervision to train temporal action localization models for both segment localization and single-frame localization during inference. Our SF-Net makes full use of single-frame supervision by predicting actionness score, pseudo background frame mining and pseudo action frame mining. SF-Net significantly outperforms weakly-supervised methods in terms of both segment localization and single-frame localization on three standard benchmarks.

Acknowledgements. This research was partially supported by ARC DP200100938 and Facebook.

References

1. Alwassel, H., Caba Heilbron, F., Ghanem, B.: Action search: spotting actions in videos and its application to temporal action localization. In: Ferrari, V., Hebert, M., Sminchisescu, C., Weiss, Y. (eds.) ECCV 2018. LNCS, vol. 11213, pp. 253–269. Springer, Cham (2018). https://doi.org/10.1007/978-3-030-01240-3_16
2. Bearman, A., Russakovsky, O., Ferrari, V., Fei-Fei, L.: What's the point: semantic segmentation with point supervision. In: Leibe, B., Matas, J., Sebe, N., Welling, M. (eds.) ECCV 2016. LNCS, vol. 9911, pp. 549–565. Springer, Cham (2016). https://doi.org/10.1007/978-3-319-46478-7_34

3. Caba Heilbron, F., Escorcia, V., Ghanem, B., Carlos Niebles, J.: ActivityNet: a large-scale video benchmark for human activity understanding. In: Proceedings of the IEEE Conference on Computer Vision and Pattern Recognition, pp. 961–970 (2015)
4. Carreira, J., Zisserman, A.: Quo vadis, action recognition? A new model and the kinetics dataset. In: The IEEE Conference on Computer Vision and Pattern Recognition (CVPR), July 2017
5. Carreira, J., Zisserman, A.: Quo vadis, action recognition? A new model and the kinetics dataset. In: proceedings of the IEEE Conference on Computer Vision and Pattern Recognition, pp. 6299–6308 (2017)
6. Chao, Y.W., Vijayanarasimhan, S., Seybold, B., Ross, D.A., Deng, J., Sukthankar, R.: Rethinking the faster R-CNN architecture for temporal action localization. In: Proceedings of the IEEE Conference on Computer Vision and Pattern Recognition, pp. 1130–1139 (2018)
7. Chéron, G., Alayrac, J.B., Laptev, I., Schmid, C.: A flexible model for training action localization with varying levels of supervision. In: Advances in Neural Information Processing Systems, pp. 942–953 (2018)
8. Dai, X., Singh, B., Zhang, G., Davis, L.S., Qiu Chen, Y.: Temporal context network for activity localization in videos. In: Proceedings of the IEEE International Conference on Computer Vision, pp. 5793–5802 (2017)
9. Damen, D., Leelasawassuk, T., Haines, O., Calway, A., Mayol-Cuevas, W.W.: You-Do, I-Learn: discovering task relevant objects and their modes of interaction from multi-user egocentric video. In: BMVC, vol. 2, p. 3 (2014)
10. Ding, L., Xu, C.: Weakly-supervised action segmentation with iterative soft boundary assignment. In: Proceedings of the IEEE Conference on Computer Vision and Pattern Recognition, pp. 6508–6516 (2018)
11. Feichtenhofer, C., Fan, H., Malik, J., He, K.: SlowFast networks for video recognition. In: The IEEE International Conference on Computer Vision (ICCV), October 2019
12. Gao, J., Yang, Z., Nevatia, R.: Cascaded boundary regression for temporal action detection. arXiv preprint arXiv:1705.01180 (2017)
13. Idrees, H., et al.: The THUMOS challenge on action recognition for videos "in the wild". Comput. Vis. Image Underst. 155, 1–23 (2017)
14. Kingma, D.P., Ba, J.: Adam: a method for stochastic optimization. arXiv preprint arXiv:1412.6980 (2014)
15. Lei, P., Todorovic, S.: Temporal deformable residual networks for action segmentation in videos. In: Proceedings of the IEEE Conference on Computer Vision and Pattern Recognition, pp. 6742–6751 (2018)
16. Lin, T., Liu, X., Li, X., Ding, E., Wen, S.: BMN: boundary-matching network for temporal action proposal generation. In: The IEEE International Conference on Computer Vision (ICCV), October 2019
17. Lin, T., Zhao, X., Shou, Z.: Single shot temporal action detection. In: Proceedings of the 25th ACM International Conference on Multimedia, pp. 988–996. ACM (2017)
18. Lin, T., Zhao, X., Su, H., Wang, C., Yang, M.: BSN: boundary sensitive network for temporal action proposal generation. In: Ferrari, V., Hebert, M., Sminchisescu, C., Weiss, Y. (eds.) ECCV 2018. LNCS, vol. 11208, pp. 3–21. Springer, Cham (2018). https://doi.org/10.1007/978-3-030-01225-0_1
19. Liu, D., Jiang, T., Wang, Y.: Completeness modeling and context separation for weakly supervised temporal action localization. In: The IEEE Conference on Computer Vision and Pattern Recognition (CVPR), June 2019

20. Liu, Z., et al.: Weakly supervised temporal action localization through contrast based evaluation networks. In: Proceedings of the IEEE International Conference on Computer Vision, pp. 3899–3908 (2019)
21. Long, F., Yao, T., Qiu, Z., Tian, X., Luo, J., Mei, T.: Gaussian temporal awareness networks for action localization. In: Proceedings of the IEEE Conference on Computer Vision and Pattern Recognition, pp. 344–353 (2019)
22. Mettes, P., van Gemert, J.C., Snoek, C.G.M.: Spot on: action localization from pointly-supervised proposals. In: Leibe, B., Matas, J., Sebe, N., Welling, M. (eds.) ECCV 2016. LNCS, vol. 9909, pp. 437–453. Springer, Cham (2016). https://doi.org/10.1007/978-3-319-46454-1_27
23. Moltisanti, D., Fidler, S., Damen, D.: Action recognition from single timestamp supervision in untrimmed videos. In: Proceedings of the IEEE Conference on Computer Vision and Pattern Recognition, pp. 9915–9924 (2019)
24. Narayan, S., Cholakkal, H., Khan, F.S., Shao, L.: 3C-Net: category count and center loss for weakly-supervised action localization. In: The IEEE International Conference on Computer Vision (ICCV), October 2019
25. Nguyen, P., Liu, T., Prasad, G., Han, B.: Weakly supervised action localization by sparse temporal pooling network. In: Proceedings of the IEEE Conference on Computer Vision and Pattern Recognition, pp. 6752–6761 (2018)
26. Nguyen, P.X., Ramanan, D., Fowlkes, C.C.: Weakly-supervised action localization with background modeling. In: The IEEE International Conference on Computer Vision (ICCV), October 2019
27. Paul, S., Roy, S., Roy-Chowdhury, A.K.: W-TALC: weakly-supervised temporal activity localization and classification. In: Ferrari, V., Hebert, M., Sminchisescu, C., Weiss, Y. (eds.) ECCV 2018. LNCS, vol. 11208, pp. 588–607. Springer, Cham (2018). https://doi.org/10.1007/978-3-030-01225-0_35
28. Shou, Z., Chan, J., Zareian, A., Miyazawa, K., Chang, S.F.: CDC: convolutional-de-convolutional networks for precise temporal action localization in untrimmed videos. In: The IEEE Conference on Computer Vision and Pattern Recognition (CVPR), July 2017
29. Shou, Z., Gao, H., Zhang, L., Miyazawa, K., Chang, S.-F.: AutoLoc: weakly-supervised temporal action localization in untrimmed videos. In: Ferrari, V., Hebert, M., Sminchisescu, C., Weiss, Y. (eds.) ECCV 2018. LNCS, vol. 11220, pp. 162–179. Springer, Cham (2018). https://doi.org/10.1007/978-3-030-01270-0_10
30. Shou, Z., Wang, D., Chang, S.F.: Temporal action localization in untrimmed videos via multi-stage CNNs. In: The IEEE Conference on Computer Vision and Pattern Recognition (CVPR), June 2016
31. Simonyan, K., Zisserman, A.: Two-stream convolutional networks for action recognition in videos. In: Advances in Neural Information Processing Systems, pp. 568–576 (2014)
32. Singh, K.K., Lee, Y.J.: Hide-and-seek: forcing a network to be meticulous for weakly-supervised object and action localization. In: 2017 IEEE International Conference on Computer Vision (ICCV), pp. 3544–3553. IEEE (2017)
33. Tran, D., Bourdev, L., Fergus, R., Torresani, L., Paluri, M.: Learning spatiotemporal features with 3D convolutional networks. In: The IEEE International Conference on Computer Vision (ICCV), December 2015
34. Wang, H., Schmid, C.: Action recognition with improved trajectories. In: The IEEE International Conference on Computer Vision (ICCV), December 2013
35. Wang, L., Xiong, Y., Lin, D., Van Gool, L.: Untrimmednets for weakly supervised action recognition and detection. In: Proceedings of the IEEE Conference on Computer Vision and Pattern Recognition, pp. 4325–4334 (2017)

36. Wang, L., et al.: Temporal segment networks: towards good practices for deep action recognition. In: Leibe, B., Matas, J., Sebe, N., Welling, M. (eds.) ECCV 2016. LNCS, vol. 9912, pp. 20–36. Springer, Cham (2016). https://doi.org/10.1007/978-3-319-46484-8_2
37. Xu, H., Das, A., Saenko, K.: R-C3D: region convolutional 3D network for temporal activity detection. In: Proceedings of the IEEE International Conference on Computer Vision, pp. 5783–5792 (2017)
38. Yuan, Z., Stroud, J.C., Lu, T., Deng, J.: Temporal action localization by structured maximal sums. In: Proceedings of the IEEE Conference on Computer Vision and Pattern Recognition, pp. 3684–3692 (2017)
39. Zach, C., Pock, T., Bischof, H.: A duality based approach for realtime TV-L^1 optical flow. In: Hamprecht, F.A., Schnörr, C., Jähne, B. (eds.) DAGM 2007. LNCS, vol. 4713, pp. 214–223. Springer, Heidelberg (2007). https://doi.org/10.1007/978-3-540-74936-3_22
40. Zeng, R., et al.: Graph convolutional networks for temporal action localization. In: The IEEE International Conference on Computer Vision (ICCV), October 2019
41. Zhao, H., Torralba, A., Torresani, L., Yan, Z.: HACS: human action clips and segments dataset for recognition and temporal localization. In: Proceedings of the IEEE International Conference on Computer Vision, pp. 8668–8678 (2019)
42. Zhao, Y., Xiong, Y., Wang, L., Wu, Z., Tang, X., Lin, D.: Temporal action detection with structured segment networks. In: Proceedings of the IEEE International Conference on Computer Vision, pp. 2914–2923 (2017)
43. Zhu, L., Yang, Y.: ActBERT: learning global-local video-text representations. In: The IEEE Conference on Computer Vision and Pattern Recognition (CVPR) (2020)
44. Zhu, L., Yang, Y.: Label independent memory for semi-supervised few-shot video classification. IEEE Trans. Pattern Anal. Mach. Intell. (2020). https://doi.org/10.1109/TPAMI.2020.3007511

Negative Margin Matters: Understanding Margin in Few-Shot Classification

Bin Liu[1], Yue Cao[2(✉)], Yutong Lin[2,3], Qi Li[1], Zheng Zhang[2], Mingsheng Long[1], and Han Hu[2]

[1] Tsinghua University, Beijing, China
liubinthss@gmail.com, liqi17thu@gmail.com, mingsheng@tsinghua.edu.cn
[2] Microsoft Research Asia, Beijing, China
{yuecao,v-yutlin,zhez,hanhu}@microsoft.com
[3] Xi'an Jiaotong University, Xi'an, China

Abstract. This paper introduces a negative margin loss to metric learning based few-shot learning methods. The negative margin loss significantly outperforms regular softmax loss, and achieves state-of-the-art accuracy on three standard few-shot classification benchmarks with few bells and whistles. These results are contrary to the common practice in the metric learning field, that the margin is zero or positive. To understand why the negative margin loss performs well for the few-shot classification, we analyze the discriminability of learned features w.r.t different margins for training and novel classes, both empirically and theoretically. We find that although negative margin reduces the feature discriminability for training classes, it may also avoid falsely mapping samples of the same novel class to multiple peaks or clusters, and thus benefit the discrimination of novel classes. Code is available at https://github.com/bl0/negative-margin.few-shot.

Keywords: Few-shot classification · Metric learning · Large margin loss

1 Introduction

Recent success on visual recognition tasks [2,4,13,17,38,42] heavily relies on the massive-scale manually labeled training data, which is too expensive in many real scenarios. In contrast, humans are capable of learning new concepts with only a few examples, yet it still remains a challenge for modern machine learning systems. Hence, learning to generalize the knowledge in base classes (with sufficient

---

B. Liu and Y. Cao—Equal contribution.
Y. Lin—The work is done when Yutong Lin is an intern at MSRA.

Electronic supplementary material The online version of this chapter (https://doi.org/10.1007/978-3-030-58548-8_26) contains supplementary material, which is available to authorized users.

A. Vedaldi et al. (Eds.): ECCV 2020, LNCS 12349, pp. 438–455, 2020.
https://doi.org/10.1007/978-3-030-58548-8_26

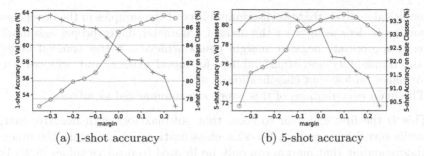

(a) 1-shot accuracy   (b) 5-shot accuracy

Fig. 1. The one-shot and five-shot accuracy on novel classes (in red) and base classes (in blue) w.r.t different margins in cosine softmax loss on mini-ImageNet. As we expect, applying larger margin to softmax loss can achieve better accuracy on base classes. But surprisingly, applying appropriate negative margin to softmax loss can achieve state-of-the-art few-shot accuracy on novel classes. (Color figure online)

annotated examples) to novel classes (with a few labeled examples), also known as few-shot learning, has attracted more and more attention [3, 7, 9, 10, 19, 25, 35–37, 43–45].

An important direction of few-shot classification is meta learning, which aims to learn a meta-learner on base classes and generalizes it to novel classes. Metric learning based methods [3, 7, 25], are an important series of the meta-learning methods, and perform metric learning in the base classes and then transfer the learned metrics to the novel classes. For example, [3] proved that simply using standard softmax loss or cosine softmax loss for learning metrics in base classes can achieve the state-of-the-art few-shot classification performance via learning a linear classifier on novel classes.

In the metric learning area, a common view is that the standard softmax loss is insufficient for discrimination on different training classes. Several previous approaches integrate the large and positive margin to the softmax loss [22] or the cosine softmax loss [6, 47] so as to enforce the score of ground truth class larger than that of other classes by at least a margin. This could help to learn highly-discriminative deep features and result in remarkable performance improvement on visual recognition tasks, especially on face recognition [6, 22, 47].

Consequently, it inspires us to adopt this large-margin softmax loss to learn better metrics for few-shot classification. As we expected, shown as the blue curves in Fig. 1, the metrics learned by large-margin softmax with positive margin are more discriminative on training classes, resulting in higher few-shot accuracy on the validation set of training classes. But in the standard open-set setting of few-shot classification, shown as red curves in Fig. 1, we surprisingly find out that adding the positive margin in softmax loss would hurt the performance.

From our perspective, the positive margin would make the learned metrics more discriminative to training classes. But for novel classes, positive margin would map the samples of the same class to multiple peaks or clusters in base classes (shown in Fig. 3 and Fig. 7) and hurt their discriminability. We then give a theoretical analysis that the discriminability of the samples in the novel classes is monotonic decreasing w.r.t the margin parameter under proper assumption. Instead, appropriate negative margin could achieve a better tradeoff between the discriminability and transferability for novel classes, and achieves better performance on few-shot classification.

The main contributions of this paper are summarized as follows:

1. This is the first endeavor to show that softmax loss with negative margin works surprisingly well on few-shot classification, which breaks the inherent understanding that margin can only be limited to positive values [6,22,47].
2. We provide insightful intuitive explanation and the theoretical analysis about why negative margin works well for few-shot classification.
3. The proposed approach with negative margin achieves state-of-the-art performance on three widely-used few-shot classification benchmarks.

2 Related Work

Few-Shot Classification. The existing representative few-shot learning methods can be broadly divided into three categories: *gradient-based* methods, *hallucination-based* methods, and *metric-based* methods.

Gradient-based methods tackle the few-shot classification by learning the task-agnostic knowledge. [9,27,29,31,39] focus on learning a suitable initialization of the model parameters which can quickly adapt to new tasks with a limited number of labeled data and a small number of gradient update steps. Another line of works aims at learning an optimizer, such as LSTM-based meta learner [37] and weight-update mechanism with an external memory [28], for replacing the stochastic gradient descent optimizer. However, it is challenging to solve the dual or bi-level optimization problem of these works, so their performance is not competitive on large datasets. Recently, [1,19] alleviate the optimization problem by closed-form model like SVM, and achieve better performance on few-shot classification benchmark of large dataset.

Hallucination-based methods attempt to address the limited data issue by learning an image generator from base classes, which is adopted to hallucinate new images in novel classes [12,48]. [12] presents a way of hallucinating additional examples for novel classes by transferring modes of variation from base classes. [48] learns to hallucinate examples that are useful for classification by the end-to-end optimization of both classifier and hallucinator. As hallucination-based methods can be considered as the supplement and are always adopted with other few-shot methods, we follow [3] to exclude these methods in our experimental comparison and leave it to future work.

Metric-based methods aim at learning a transferable distance metric. MatchingNet [45] computes cosine similarity between the embeddings of labeled images and unlabeled images, to classify the unlabeled images. ProtoNet [43] represents each class by the mean embedding of the examples inside this class, and the classification is performed based on the distance to the mean embedding of each class. RelationNet [44] replaces the non-parametric distance in ProtoNet to a parametric relation module. Recently, [3,7,25] reveal that the simple pre-training and fine-tuning pipeline (following the standard transfer learning paradigm) can achieve surprisingly competitive performance with the state-of-the-art few-shot classification methods.

Based on this simple paradigm, our work is the first endeavor towards explicitly integrating the margin parameter to the softmax loss, and mostly importantly breaks the inherent understanding that the margin can be only restricted as positive values, with both intuitive understanding and theoretical analysis. With an appropriate negative margin, our approach could achieve the state-of-the-art performance on three standard few-shot classification benchmarks.

Margin Based Metric Learning. Metric learning aims to learn a distance metric between examples, and plays a critical role in many tasks, such as classification [49], clustering [51], retrieval [20] and visualization [24].

In practice, the margin between data points and the decision boundary plays a significant role in achieving strong generalization performance. [16] develops a margin theory and shows that the margin loss leads to an informative generalization bound for classification task. In the past decades, the idea of margin-based metric learning has been widely explored in SVM [40], k-NN classification [49], multi-task learning [33], etc. In the deep learning era, many margin-based metric learning methods are proposed to enhance the discriminative power of the learned deep features, and show remarkable performance improvements in many tasks [19,20,30], especially in face verification [6,21,41,47]. For example, SphereFace [21], CosFace [47], and ArcFace [6] enforce the intra-class variance and inter-class diversity by adding the margin to cosine softmax loss.

However, as the tasks of previous works are based on close-set scenarios, they limit the margin parameter as positive values [6,21,47], where making the deep features more discriminative could be generalized to the validation set and improve the performance. For open-set scenarios, such as few-shot learning, increasing the margin would not enforce the inter-class diversity but unfortunately enlarge the intra-class variance for novel classes, as shown in Fig. 2, which would hurt the performance. In contrast, an appropriate negative margin would better tradeoff the discriminability and transferability of deep features in novel classes, and obtain better performance for few-shot classification.

3   Methodology

In a few-shot classification task, we are given two sets of data with different classes, formulated as $I^b = \{(\mathbf{x}_i, y_i)\}_{i=1}^{N^b}$ as the base training set with C^b base classes for the first training stage, and $I^n = \{(\mathbf{x}_i', y_i')\}_{i=1}^{N^n}$ as the novel training set with C^n novel classes for the second training stage. For the novel training set, each class has K samples, where $K = 1$ or 5, and $C^n = 5$ is the standard setting [3,7,19,25,35,36,44]. This is called C^n-way K-shot learning. Few-shot classification aims to learn both discriminative and transferable feature representations from the abundant labeled data in base classes, such that the features can be easily adapted for the novel classes with few labeled examples.

3.1 Negative-Margin Softmax Loss

In image classification, the softmax loss is built upon the feature representation of deep networks $\mathbf{z}_i = f_\theta(\mathbf{x}_i) \in \mathbb{R}^D$ ($f_\theta(\cdot)$ denotes the backbone network with the parameters θ), its corresponding label y_i and the linear transform matrix $\mathbf{W} = [W_1, W_2, ..., W_{C^b}] \in \mathbb{R}^{D \times C^b}$. Recently, introducing the **large and positive** margin parameter to the softmax loss is widely explored in metric learning [6, 22, 47]. Hence, we directly integrate the margin parameter to the softmax loss to learn the transferable metrics, aiming at benefiting the few-shot classification on novel classes. The general formulation of large-margin softmax loss is defined as

$$L = -\frac{1}{N} \sum_{i=1}^{N} \log \frac{e^{\beta \cdot \left(\mathbf{s}\left(z_i, W_{y_i}\right) - m \right)}}{e^{\beta \cdot \left(\mathbf{s}\left(z_i, W_{y_i}\right) - m \right)} + \sum_{j=1, j \neq y_i}^{C} e^{\beta \cdot \mathbf{s}(z_i, W_j)}}, \tag{1}$$

where m is the margin parameter, β denotes the temperature parameter which defines how much strength to enlarge the gap between the largest logit and other logits. And $\mathbf{s}(\cdot, \cdot)$ denotes the similarity function between two input vectors.

It's worth noting that all the previous works on large-margin softmax loss restrict the margin as positive values [6, 22, 47]. This is because that previous works focus on the close-set scenarios, the loss with larger margin leads to the smaller intra-class variance and the larger between-class variance, which will help to classify examples in the same classes. This is also validated in Fig. 1, that the softmax loss with larger margin could improve the classification accuracy on the validation set of training classes.

However, the situations are different in the open-set scenarios. Learned metrics which are too discriminative to training classes may hurt their transferability to the novel classes. So applying appropriate negative margin to softmax loss aims to tradeoff the discriminability on training classes and the transferability to novel classes of the learned metrics.

Here we formulate two instantiations of Eq. 1 with different similarity functions. By taking the inner-product similarity $\mathbf{s}(\mathbf{z}_i, W_j) = W_j^T \mathbf{z}_i$ into Eq. 1, the **negative-margin softmax loss** (abbreviated as Neg-Softmax) could be obtained. By taking the cosine similarity $\mathbf{s}(\mathbf{z}_i, W_j) = \frac{W_j^T \mathbf{z}_i}{\|\mathbf{z}_i\| \|W_j\|}$ into Eq. 1, we can formulate the **negative-margin cosine softmax loss** (abbreviated as Neg-Cosine). The detailed loss functions could be found at the Appendix. These two loss functions are adopted at the pre-training stage.

3.2 Discriminability Analysis of Deep Features w.r.t Different Margins

We analyze the discriminability of the deep features extracted by the deep model with different margins, to understand why negative margin works well on novel classes. For simplicity, we only analyze the cosine softmax loss, and it is direct to extend the analysis and conclusion to standard softmax loss.

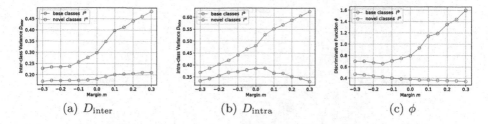

Fig. 2. Inter-class variance D_{inter}, intra-class variance D_{intra}, and discriminative function ϕ w.r.t margin m on both base and novel classes of mini-ImageNet. As the margin increases, the features of base classes is more discriminative, while that of novel classes is less discriminative.

We denote the pre-trained backbone network trained with margin parameter m as $f_{\theta(m)}$. For class j in base classes or novel classes, denote the set of examples labeled with class j as $I_j = \{(x_i, y_i)|y_i = j\}$. We compute the class center $\mu(I_j, m)$ for class j as the mean of the L2-normalized feature embeddings as

$$\mu(I_j, m) = \frac{1}{|I_j|} \sum_{(\mathbf{x}_i, y_i) \in I_j} \frac{f_{\theta(m)}(\mathbf{x}_i)}{\|f_{\theta(m)}(\mathbf{x}_i)\|_2}. \tag{2}$$

The dataset $I = I_1 \cup I_2 \cup \cdots \cup I_C$ with C classes could be base dataset I^b with a large number of base classes or novel dataset I^n with small number of novel classes (such as 5 for 5-way few shot learning). Then we define the inter-class variance $D_{\text{inter}}(I, m)$, and intra-class variance $D_{\text{intra}}(I, m)$ as

$$D_{\text{inter}}(I, m) = \frac{1}{C(C-1)} \sum_{j=1}^{C} \sum_{k=1, k \neq j}^{C} \|\mu(I_j, m) - \mu(I_k, m)\|_2^2,$$

$$D_{\text{intra}}(I, m) = \frac{1}{C} \sum_{j=1}^{C} (\frac{1}{|I_j|} \sum_{(\mathbf{x}_i, y_i) \in I_j} \left\| \frac{f_{\theta(m)}(x_i)}{\|f_{\theta(m)}(x_i)\|} - \mu(I_j, m) \right\|_2^2). \tag{3}$$

For every two classes, the inter-class variance is the squared L2 distance between their class centers. For each class, the intra-class variance is the squared L2 distance between every sample in this class and the class center.

If inter-class variance becomes larger or intra-class variance becomes smaller, the deep features would be more discriminative. So we follow [26] to define the discriminative function $\phi(I, m)$ as the inter-class variance divided by the intra-class variance:

$$\phi(I, m) = \frac{D_{\text{inter}}(I, m)}{D_{\text{intra}}(I, m)}. \tag{4}$$

To measure the discriminability of the deep features with different margins, we plot the inter-class variance D_{inter}, intra-class variance D_{intra}, and discriminative function ϕ w.r.t margin m on both the base and novel classes of mini-ImageNet, respectively. As shown in Fig. 2, for base classes (red curves), as the margin

444   B. Liu et al.

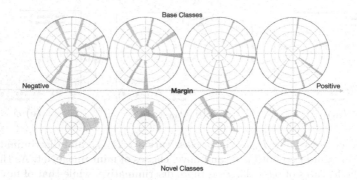

Fig. 3. The visualizations of the data distributions on angular space with different margins, on base classes (the first row) or novel classes (the second row) of MNIST. Plots from left to right denotes the margins from negative to positive. For each figure, we plot the histogram of the occurrence for each angle. Different colors denote the data points belonging to different classes. (Color figure online)

increases, the inter-class variance increases a lot, meanwhile the intra-class variance does not change much, so the features of base classes become more discriminative. This is widely observed in previous works [6,22,47], and motivates them to introduce large and also positive margin to softmax loss for close-set scenarios.

But for novel classes (blue curves), the situation is just on the contrary. As the margin increases, the inter-class variance does not change much, but the intra-class variance increases a lot, so the features of base classes become less discriminative. This indicates that larger margin may hurt the classification on the novel classes. This is also verified in the real few-shot classification task, shown as red curves in Fig. 1, larger and positive margin will achieve worse performance of few-shot classification on novel classes. Instead, the appropriate negative margin could achieve the best performance, which may lead to a better tradeoff on discriminability and transferability for novel classes.

3.3   Intuitive Explanation

To better understand how the margin works, we perform the visualization on the data distributions in the angular space trained on MNIST[1], as shown in Fig. 3. We choose seven classes as the base classes for pre-training, and adopt the other three classes as the novel classes. We first train this deep model with 2-dimensional output features using cosine softmax loss with different margins on the base classes. Then we normalize the 2-D features to obtain the direction of each data point, and visualize the count of each direction (also known as the data distributions in angular space) on both base (first row) and novel classes (second row) using the models trained with different margins.

---

[1] This technique is widely used to characterize the feature embedding under the softmax-related objectives [21,47,54].

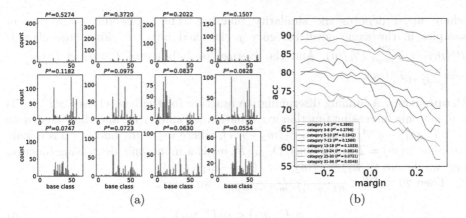

Fig. 4. We first sort the 36 novel classes according to the probability of sample pairs in the same novel class j classified into the same base class P_j^s (one of every 3 categories are plotted for clarity) on mini-ImageNet. For each novel class, (a) shows the histogram of samples in this class to be classified to 64 base classes. (b) shows the accuracy curves w.r.t different margins for novel classes with different averaged P^s.

As shown in the first row in Fig. 3, with larger and even positive margin (from left to right), the clusters for each training class are getting thinner and higher, and the angle differences between different class centers are getting larger. This matches our previous observation in Fig. 2, that enlarging the margin leads to the smaller intra-class variance and larger inter-class variance on the base classes.

However, with larger margin, less data points would lie in the space far from all centers, which to some extent makes the output space much narrower. As shown on the right side of the second row in Fig. 3, as novel classes are different to base classes, model with large margin may map the data points of the same class in novel classes to multiple peaks or clusters belonging to different base classes. Then the intra-class variance for novel classes would increase accordingly, making the classification of novel classes more difficult. Instead, as shown on the left side of second row in Fig. 3, the appropriate negative margin would not enforce the data points in novel classes too close to the training center, and may alleviate the multi-peak issue, which could benefit the classification on novel classes.

3.4 Theoretical Analysis

After giving the intuitive explanation that why negative margin works well on novel classes, we then prove this claim theoretically. Denote the parameter of the classifier joint pre-trained with backbone on base classes with margin m as $\mathbf{W}(m)$, the probability of a sample in the novel category j classified by pre-trained backbone $f_{\theta(m)}$ and classifier $W(m)$ as a base category k is

$$P_{jk}(m) = \frac{1}{|I_j|} \sum_{(x_i, y_i) \in I_j} \frac{\exp\left(\beta \mathbf{s}(f_{\theta(m)}(x), W_k(m))\right)}{\sum_{k'=1}^{C^b} \exp\left(\beta \mathbf{s}(f_{\theta(m)}(x), W_{k'}(m))\right)}, \tag{5}$$

where $\mathbf{s}(\cdot, \cdot)$ denotes the similarity function. The probability of a pair of samples in the same novel category j classified into the same base class is $P_j^s(m) = \sum_{k=1}^{C^b} P_{jk}^2(m)$. And the average probability of $P_j^s(m)$ is $P^s(m) = \frac{1}{|C^n|} \sum_{j=1}^{C^n} P_j^s(m)$.

Proposition. Assuming discriminative function for the base classes $\phi(I^b, m)$ is a monotonic increasing function w.r.t margin parameter m, and then we denote $\phi^{-1}(I^b, m_1) - \phi^{-1}(I^b, m_2) = r \cdot (m_2 - m_1)$, where $m_2 > m_1$ and $r > 0$ is a scale variable. $\psi(m) = D_{\text{inter}}(I^n, m)/D_{\text{inter}}(I^b, m)$ is a monotonic decreasing function and we denote $\psi(m_1) - \psi(m_2) = t \cdot (m_2 - m_1), t > 0$.

Then $\forall 0 < P^s < \frac{t}{t(1 - \phi^{-1}(I^b, m_1)) + r\psi(m_1)}$, we have[2]:

$$\phi(I^n, m_2) < \phi(I^n, m_1). \tag{6}$$

The above proposition proves that the discriminative function on the novel classes $\phi(I^n, m)$ is a monotonic decreasing function w.r.t m under proper assumption and a measurable condition about the similarity between base and novel classes using P^s. The proposition indicates that an appropriate value of "negative" margin could work well for discriminating the samples in novel classes.

Figure 4 shows the actual behavior of mini-ImageNet dataset. We first sort the 36 novel classes according to the probability of sample pairs in the same novel class j classified into the same base class P_j^s (one of every 3 categories are plotted for clarity) on mini-ImageNet. And the histograms of the samples in novel classes to be classified to 64 base classes is shown in Fig. 4(a). Figure 4(b) shows the accuracy curves w.r.t different margins for novel classes with different averaged P^s. With smaller P^s, the histograms of novel classes become more diverse (shown in Fig. 4(a)) and their accuracies become lower (shown in Fig. 4(b)). Importantly, most subsets of novel classes favor negative margins, implying the condition in the Proposition is not hard to reach.

3.5 Framework

Following the standard transfer learning paradigm [8,52], we adopt a two-stage training pipeline for few-shot classification, including pre-training stage to perform metric learning on the abundant labeled data in base classes, and fine-tuning stage to learn a classifier to recognize novel classes. This pipeline is widely adopted in recent few-shot learning methods [3,7,25].

In the pre-training stage, we aim at training the backbone network $f_\theta(\cdot)$ with abundant labeled data I^b in base classes, driven by metric learning loss, such as softmax loss in [3]. In our paper, we adopt the negative-margin softmax loss, which could learn more transferable representations for few-shot learning. In the fine-tuning stage, as there are only few labeled samples in I^n for training (e.g. 5-way 1-shot learning only contains 5 training samples), we follow [3] to fix the parameters of the backbone $f_\theta(\cdot)$, and only train a new classifier from scratch

---

[2] Proof is attached in the supplemental material.

by the softmax loss. Note that, the computation of similarity (such as inner-product similarity or cosine similarity) in softmax loss is the same as that in the pre-training stage.

4 Experiments

4.1 Setup

Datasets and Scenarios. Following [3], we address the few-shot classification problem under three different scenarios: (1) generic object recognition; (2) fine-grained image classification; and (3) cross-domain adaptation.

For the generic scenario, the widely-used few-shot classification benchmark: mini-ImageNet, is used to evaluate the effectiveness of the proposed Negative-Margin Softmax Loss. The mini-ImageNet dataset, firstly proposed by [45], consists of a subset of 100 classes from the ILSVRC-2012 [5], and contains 600 images for each classes. Following the commonly-used evaluation protocol of [37], we split the 100 classes into 64 base, 16 validation, and 20 novel classes for pre-training, validation, and testing. To validate the effectiveness of our model on the large dataset, we further conduct ablation study on the ImageNet-1K dataset following the setting in [12,48].

For the fine-grained image classification, we use CUB-200-2011 dataset [46] (hereinafter referred as CUB), which consists of 200 classes and 11,788 images in total. Followingit stg the standard setting of [14], we split the classes in the dataset into 100 base classes, 50 validation classes, and 50 novel classes.

For the cross-domain adaptation scenario, we use mini-ImageNet $\rightarrow$ CUB [3], in which the 100 classes in mini-ImageNet, the 50 validation and 50 novel classes in CUB are adopted as base, validation and novel classes respectively, to evaluate the performance of the proposed Negative-Margin Softmax Loss in the presence of domain shift.

Implementation Details. For fair comparison, we evaluate our model with four commonly used backbone networks, namely Conv-4 [45], ResNet-12 [32], ResNet-18 [3] and WRN-28-10 [25,53]. Besides the differences in network depth and architecture, the expected input size of Conv-4 and ResNet-12 is 84×84, and that of ResNet-18 is 224×224, while WRN-28-10 takes 80×80 images as input.

Our implementation is based on PyTorch [34]. In the training stage, the backbone network and classifier are trained from scratch, with a batch size of 256. The models are trained for 200, 400 and 400 epochs in the CUB, mini-ImageNet and mini-ImageNet $\rightarrow$ CUB, respectively. We adopt the Adam [15] optimizer with initial learning rate 3e-3 and cosine learning rate decay [23]. We apply the same data argumentation as [3], including random cropping, horizontal flipping and color jittering.

In the fine-tuning stage, each episode contains 5 classes and each class contains 1 or 5 support images to train a new classifier from scratch and 16 query images to test the accuracy. The final performance is reported as the mean classification accuracy over 600 random sampled episodes with the 95% confidence

448   B. Liu et al.

Table 1. Few-shot classification results on the mini-ImageNet dataset.[†] indicates the method using the combination of base and validation classes to train the meta-learner

Backbone	Method	1 shot	5 shot
Conv-4	MAML [9]	48.70 ± 1.84	63.11 ± 0.92
	ProtoNet [43]	49.42 ± 0.78	68.20 ± 0.66
	MatchingNet [45]	48.14 ± 0.78	63.48 ± 0.66
	RelationNet [44]	50.44 ± 0.82	65.32 ± 0.70
	MAML+Meta-dropout [18]	51.93 ± 0.67	67.42 ± 0.52
	R2D2 [1]	51.20 ± 0.60	68.80 ± 0.10
	Neg-Softmax (ours)	47.65 ± 0.78	67.27 ± 0.66
	Neg-Cosine (ours)	**52.84 ± 0.76**	**70.41 ± 0.66**
ResNet-12	SNAIL [27]	55.71 ± 0.99	68.88 ± 0.92
	TADAM [32]	58.50 ± 0.30	76.70 ± 0.30
	MetaOptNet-SVM [19]	62.64 ± 0.61	78.63 ± 0.46
	Neg-Softmax (ours)	62.58 ± 0.82	80.43 ± 0.56
	Neg-Cosine (ours)	**63.85 ± 0.81**	**81.57 ± 0.56**
ResNet-18	SNCA [50]	57.80 ± 0.80	72.80 ± 0.70
	Baseline [3]	51.75 ± 0.80	74.27 ± 0.63
	Baseline++ [3]	51.87 ± 0.77	75.68 ± 0.63
	Neg-Softmax (ours)	59.02 ± 0.81	78.80 ± 0.61
	Neg-Cosine (ours)	**62.33 ± 0.82**	**80.94 ± 0.59**
WRN-28-10	Activation to Parameter[†] [36]	59.60 ± 0.41	73.74 ± 0.19
	LEO[†] [39]	61.76 ± 0.08	77.59 ± 0.12
	Fine-tuning [7]	57.73 ± 0.62	78.17 ± 0.49
	Cosine + rotation [11]	**62.93 ± 0.45**	79.87 ± 0.33
	Neg-Softmax (ours)	60.04 ± 0.79	80.90 ± 0.60
	Neg-Cosine (ours)	61.72 ± 0.81	**81.79 ± 0.55**

interval. Note that all the hyper-parameters are determined by the performance on the validation classes.

4.2   Results

Results on Mini-ImageNet. For the generic object recognition scenario, we evaluate our methods on the widely-used mini-ImageNet dataset. For fair comparison with existing methods which uses different network architecture as backbone, we evaluate our methods with all four commonly used backbone networks. The 5-way 1-shot and 5-shot classification results on the novel classes of the mini-ImageNet dataset are listed in Table 1. We find that by simply adopting appropriate negative margin in standard softmax loss, our Neg-Softmax achieves competitive results with the existing state-of-the-art methods. It is worth noting

Table 2. The few-shot classification accuracy on the novel classes (also known as test classes) of the CUB dataset and cross-domain setting with ResNet-18 as the backbone

Method	CUB		mini-ImageNet→CUB
	1 shot	5 shot	5 shot
MAML [9]	69.96 ± 1.01	82.70 ± 0.65	51.34 ± 0.72
ProtoNet [43]	71.88 ± 0.91	87.42 ± 0.48	62.02 ± 0.70
MatchingNet [45]	72.36 ± 0.90	83.64 ± 0.60	53.07 ± 0.74
RelationNet [44]	67.59 ± 1.02	82.75 ± 0.58	57.71 ± 0.73
Baseline [3]	65.51 ± 0.87	82.85 ± 0.55	65.57 ± 0.70
Baseline++ [3]	67.02 ± 0.90	83.58 ± 0.54	62.04 ± 0.76
Neg-Softmax (ours)	71.48 ± 0.83	87.30 ± 0.48	**69.30 ± 0.73**
Neg-Cosine (ours)	**72.66 ± 0.85**	**89.40 ± 0.43**	67.03 ± 0.76

that our Neg-Cosine achieves the state-of-the-art performance for both 1-shot and 5-shot settings on almost all four backbones on mini-ImageNet.

Results on CUB. On the fine-grained dataset CUB, we compared the proposed method with several state-of-the-art methods with ResNet-18 as backbone. The results are showed in Table 2, in which the results of the comparison methods are directly borrowed from [3]. It shows that the proposed Neg-Cosine outperforms all the comparison methods on both 1-shot and 5-shot settings. Furthermore, Neg-Softmax also achieves highly competitive performance on both 1-shot and 5-shot settings.

Results on mini-ImageNet → CUB. In the real-world applications, there may be a signification domain shift between the base and novel classes. So we evaluate our methods on a cross domain scenario: mini-ImageNet → CUB, where we pre-train the backbone on a generic object recognition dataset, and transfer it to a fine-grained dataset. We follow [3] to report the 5-shot results with ResNet-18 backbone, as shown in Table 2. We can observe that both Neg-Softmax and Neg-Cosine are significantly better than all the comparison methods. Specifically, Neg-Softmax outperforms Baseline [3], the state-of-the-art method on the mini-ImageNet → CUB, by a large margin of 3.73%.

Results on ImageN1K dataset. To validate that negative margin works well on large dataset, we follow [12,48] to run an ablation study on the ImageNet-1K dataset. We train ResNet-10 with standard cosine softmax loss and proposed Neg-Cosine for 90 epochs on the base classes. The learning rate starts at 0.1 and is divided by 10 every 30 epochs. The weight decay is 0.0001 and the temperature factor is 15. In the fine-tuning stage, we train a new linear classifier using SGD for 10000 iterations. The top-5 accuracy is reported in Table 4, which shows that the accuracies of Neg-Cosine are consistently better than standard cosine softmax loss with margin = 0 and LogReg [48].

4.3   Analysis

This section presents a comprehensive analysis of the proposed approach. In the following experiments, we use Neg-Cosine with ResNet-18 backbone as default.

Effects of Negative Margin. Table 3 shows the 1-shot and 5-shot accuracy of the standard softmax, cosine softmax and our proposed Neg-Softmax, Neg-Cosine on the validation classes of mini-ImageNet, CUB and mini-ImageNet → CUB. By adopting appropriate negative margin, Neg-Softmax and Neg-Cosine yields significant performance gains over standard softmax loss and cosine softmax loss on all three benchmarks. Interestingly, Neg-Cosine outperforms Neg-Softmax in the in-domain setting, such as mini-ImageNet and CUB, while Neg-Softmax could achieve better performance than Neg-Cosine in the cross-domain setting. This is also observed in [3].

Accuracy w.r.t Different Margins. Figure 5 shows the 1-shot accuracy and 5-shot accuracy on validation classes of mini-ImageNet dataset w.r.t different margins in Neg-Cosine and Neg-Softmax. As we expect, as the margin gets negative and smaller, both the 1-shot accuracy and 5-shot accuracy of Neg-Cosine and Neg-Softmax first increase and then decrease, demonstrating a desirable bell-shaped curve. Hence, adopting appropriate negative margin yields significant performance gains over both standard softmax loss and cosine softmax loss on 1-shot and 5-shot classification of mini-ImageNet.

Table 3. The few-shot accuracy of standard softmax, cosine softmax and our proposed Neg-Softmax, Neg-Cosine on validation classes of three standard benchmarks

Method	mini-ImageNet		CUB		mini-ImageNet→CUB	
	1 shot	5 shot	1 shot	5 shot	1 shot	5 shot
Softmax	45.98 ± 0.79	75.25 ± 0.61	58.32 ± 0.87	80.21 ± 0.59	46.87 ± 0.78	67.68 ± 0.71
Neg-Softmax	56.95 ± 0.82	78.87 ± 0.57	59.54 ± 0.88	80.60 ± 0.57	47.74 ± 0.73	68.58 ± 0.70
Cosine	59.49 ± 0.90	79.58 ± 0.59	66.39 ± 0.93	82.17 ± 0.58	42.96 ± 0.76	61.99 ± 0.75
Neg-Cosine	63.68 ± 0.86	82.02 ± 0.57	69.17 ± 0.85	85.60 ± 0.56	44.51 ± 0.85	64.04 ± 0.75

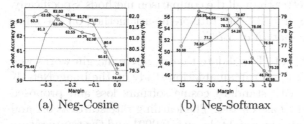

(a) Neg-Cosine   (b) Neg-Softmax

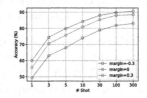

Fig. 5. The 1-shot (on red) and 5-shot (on blue) accuracy on validation classes of mini-ImageNet w.r.t different margins in Neg-Cosine and Neg-Softmax

Fig. 6. Accuracy w.r.t # shots of validation classes on the mini-ImageNet dataset for margin = −0.3, 0 and 0.3

Table 4. Top-5 accuracy on ImageN1K dataset with ResNet-10 as backbone.

Method	1 shot	5 shot
LogReg [48]	38.4	64.8
Cosine	42.1	64.0
Neg-Cosine	43.8	66.3

Table 5. Test accuracy on 5-way mini-ImageNet of various regularization techniques

Negative margin	Weight decay	Drop block	1 shot	5 shot
			54.51 ± 0.79	75.70 ± 0.62
✓			60.25 ± 0.81	80.07 ± 0.58
✓	✓		62.21 ± 0.83	80.81 ± 0.59
✓	✓	✓	62.33 ± 0.82	80.94 ± 0.59

Various Regularization Techniques. Table 5 shows the importance of regularizations on Neg-Cosine, which reveals that integrating various regularization techniques steadily improves the 1-shot and 5-shot test accuracy on mini-ImageNet benchmark. Firstly, by simply adopting negative margin, the test accuracy increased by 5.74% and 4.37% on the 1-shot and 5-shot settings, respectively. Based on our approach, weight decay and DropBlock could further improve the performance. After integrating all regularizations together, our method achieves state-of-the-art accuracy of 62.33% and 80.94% for the 1-shot and 5-shot settings respectively on novel classes of mini-ImageNet.

More Shots. We conduct an experiment by varying the number of shots from 1 (few shot) to 300 (many shot) and report the classification accuracy of the validation classes on the mini-ImageNet dataset in Figure 6. It shows that the test accuracy of margin $= -0.3$ is consistently higher than that of margin=0 from 1-shot to 300-shot settings, which prove that the negative margin could benefit the open-set scenarios with more shots.

T-SNE   Visualization.
Figure 7 shows the t-SNE [24] visualizations. As shown in the first row, compared with negative margin, the feature embedding of zero and positive margin exhibit more discriminative structures and achieve better 1-shot accuracy on the base classes. However, the second row shows that enlarging the margin parameter would break the cluster structure of the novel classes and make the classification of novel classes harder. Instead, the appropriate negative margin retain the better

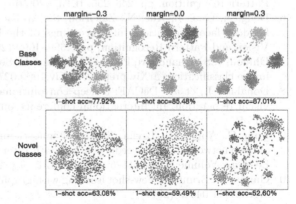

Fig. 7. The t-SNE visualizations of the feature embeddings and the corresponding 1-shot accuracy in the base and novel classes of mini-ImageNet dataset for the softmax loss with negative, zero and positive margin respectively

cluster structure for novel classes. Thus the few-shot classification accuracy of negative margin is better than that of zero and positive margin.

5 Conclusion

In this paper, we unconventionally propose to adopt appropriate negative-margin to softmax loss for few-shot classification, which surprisingly works well for the open-set scenarios of few-shot classification. We then provide the intuitive explanation and the theoretical proof to understand why negative margin works well for few-shot classification. This claim is also demonstrated via sufficient experiments. With the negative-margin softmax loss, our approach achieves the state-of-the-art performance on all three standard benchmarks of few-shot classification. In the future, the negative margin may be applied in more general open-set scenarios that do not restrict the number of samples in novel classes.

References

1. Bertinetto, L., Henriques, J.F., Torr, P., Vedaldi, A.: Meta-learning with differentiable closed-form solvers. In: International Conference on Learning Representations (2019). https://openreview.net/forum?id=HyxnZh0ct7
2. Carreira, J., Zisserman, A.: Quo vadis, action recognition? A new model and the kinetics dataset. In: proceedings of the IEEE Conference on Computer Vision and Pattern Recognition, pp. 6299–6308 (2017)
3. Chen, W.Y., Liu, Y.C., Kira, Z., Wang, Y.C., Huang, J.B.: A closer look at few-shot classification. In: International Conference on Learning Representations (2019)
4. Dai, J., et al.: Deformable convolutional networks. In: Proceedings of the IEEE International Conference on Computer Vision, pp. 764–773 (2017)
5. Deng, J., Dong, W., Socher, R., Li, L.J., Li, K., Fei-Fei, L.: ImageNet: a large-scale hierarchical image database. In: 2009 IEEE Conference on Computer Vision and Pattern Recognition, pp. 248–255. IEEE (2009)
6. Deng, J., Guo, J., Xue, N., Zafeiriou, S.: ArcFace: additive angular margin loss for deep face recognition. In: Proceedings of the IEEE Conference on Computer Vision and Pattern Recognition, pp. 4690–4699 (2019)
7. Dhillon, G.S., Chaudhari, P., Ravichandran, A., Soatto, S.: A baseline for few-shot image classification. arXiv preprint arXiv:1909.02729 (2019)
8. Donahue, J., et al.: DeCAF: a deep convolutional activation feature for generic visual recognition. In: International conference on machine learning, pp. 647–655 (2014)
9. Finn, C., Abbeel, P., Levine, S.: Model-agnostic meta-learning for fast adaptation of deep networks. In: Proceedings of the 34th International Conference on Machine Learning-Volume 70, pp. 1126–1135. JMLR. org (2017)
10. Garcia, V., Bruna, J.: Few-shot learning with graph neural networks. arXiv preprint arXiv:1711.04043 (2017)
11. Gidaris, S., Bursuc, A., Komodakis, N., Pérez, P., Cord, M.: Boosting few-shot visual learning with self-supervision. arXiv preprint arXiv:1906.05186 (2019)
12. Hariharan, B., Girshick, R.: Low-shot visual recognition by shrinking and hallucinating features. In: Proceedings of the IEEE International Conference on Computer Vision, pp. 3018–3027 (2017)
13. He, K., Zhang, X., Ren, S., Sun, J.: Deep residual learning for image recognition. arXiv preprint arXiv:1512.03385 (2015)

14. Hilliard, N., Phillips, L., Howland, S., Yankov, A., Corley, C.D., Hodas, N.O.: Few-shot learning with metric-agnostic conditional embeddings. arXiv preprint arXiv:1802.04376 (2018)
15. Kingma, D.P., Ba, J.: Adam: a method for stochastic optimization. arXiv preprint arXiv:1412.6980 (2014)
16. Koltchinskii, V., Panchenko, D., et al.: Empirical margin distributions and bounding the generalization error of combined classifiers. Ann. Stat. **30**(1), 1–50 (2002)
17. Krizhevsky, A., Sutskever, I., Hinton, G.E.: ImageNet classification with deep convolutional neural networks. In: Advances in Neural Information Processing Systems, pp. 1097–1105 (2012)
18. Lee, H.B., Nam, T., Yang, E., Hwang, S.J.: Meta dropout: learning to perturb latent features for generalization. In: International Conference on Learning Representations (2020). https://openreview.net/forum?id=BJgd81SYwr
19. Lee, K., Maji, S., Ravichandran, A., Soatto, S.: Meta-learning with differentiable convex optimization. In: Proceedings of the IEEE Conference on Computer Vision and Pattern Recognition, pp. 10657–10665 (2019)
20. Liu, B., Cao, Y., Long, M., Wang, J., Wang, J.: Deep triplet quantization. In: 2018 ACM Multimedia Conference on Multimedia Conference - MM 2018 (2018)
21. Liu, W., Wen, Y., Yu, Z., Li, M., Raj, B., Song, L.: SphereFace: deep hypersphere embedding for face recognition. In: Proceedings of the IEEE Conference on Computer Vision and Pattern Recognition, pp. 212–220 (2017)
22. Liu, W., Wen, Y., Yu, Z., Yang, M.: Large-margin softmax loss for convolutional neural networks. arXiv preprint arXiv:1612.02295 (2016)
23. Loshchilov, I., Hutter, F.: SGDR: stochastic gradient descent with warm restarts. arXiv preprint arXiv:1608.03983 (2016)
24. van der Maaten, L., Hinton, G.: Visualizing high-dimensional data using t-SNE. JMLR, November 2008
25. Mangla, P., Singh, M., Sinha, A., Kumari, N., Balasubramanian, V.N., Krishnamurthy, B.: Charting the right manifold: Manifold mixup for few-shot learning. arXiv preprint arXiv:1907.12087 (2019)
26. Mika, S., Ratsch, G., Weston, J., Scholkopf, B., Mullers, K.R.: Fisher discriminant analysis with kernels. In: Neural Networks for Signal Processing IX: Proceedings of the 1999 IEEE Signal Processing Society Workshop (cat. no. 98th8468), pp. 41–48. IEEE (1999)
27. Mishra, N., Rohaninejad, M., Chen, X., Abbeel, P.: A simple neural attentive meta-learner. In: International Conference on Learning Representations (2018). https://openreview.net/forum?id=B1DmUzWAW
28. Munkhdalai, T., Yu, H.: Meta networks. In: Proceedings of the 34th International Conference on Machine Learning-Volume 70, pp. 2554–2563. JMLR. org (2017)
29. Munkhdalai, T., Yuan, X., Mehri, S., Trischler, A.: Rapid adaptation with conditionally shifted neurons. In: ICML, pp. 3661–3670 (2018)
30. Narayanaswamy, V.S., Thiagarajan, J.J., Song, H., Spanias, A.: Designing an effective metric learning pipeline for speaker diarization. In: ICASSP 2019–2019 IEEE International Conference on Acoustics, Speech and Signal Processing (ICASSP), pp. 5806–5810. IEEE (2019)
31. Nichol, A., Achiam, J., Schulman, J.: On first-order meta-learning algorithms. arXiv preprint arXiv:1803.02999 (2018)
32. Oreshkin, B., López, P.R., Lacoste, A.: TADAM: task dependent adaptive metric for improved few-shot learning. In: Advances in Neural Information Processing Systems, pp. 721–731 (2018)

33. Parameswaran, S., Weinberger, K.Q.: Large margin multi-task metric learning. In: Advances in Neural Information Processing Systems, pp. 1867–1875 (2010)
34. Paszke, A., et al.: Automatic differentiation in PyTorch. In: NIPS Autodiff Workshop (2017)
35. Qi, H., Brown, M., Lowe, D.G.: Low-shot learning with imprinted weights. In: Proceedings of the IEEE Conference on Computer Vision and Pattern Recognition, pp. 5822–5830 (2018)
36. Qiao, S., Liu, C., Shen, W., Yuille, A.L.: Few-shot image recognition by predicting parameters from activations. In: Proceedings of the IEEE Conference on Computer Vision and Pattern Recognition, pp. 7229–7238 (2018)
37. Ravi, S., Larochelle, H.: Optimization as a model for few-shot learning (2016). https://openreview.net/forum?id=rJY0-Kcll
38. Ren, S., He, K., Girshick, R., Sun, J.: Faster R-CNN: towards real-time object detection with region proposal networks. In: Advances in Neural Information Processing Systems, pp. 91–99 (2015)
39. Rusu, A.A., et al.: Meta-learning with latent embedding optimization. In: International Conference on Learning Representations (2019). https://openreview.net/forum?id=BJgklhAcK7
40. Schölkopf, B., Smola, A.J., Bach, F., et al.: Learning with Kernels: Support Vector Machines, Regularization, Optimization, and Beyond. MIT Press, Cambridge (2002)
41. Schroff, F., Kalenichenko, D., Philbin, J.: FaceNet: a unified embedding for face recognition and clustering. In: Proceedings of the IEEE Conference on Computer Vision and Pattern Recognition, pp. 815–823 (2015)
42. Simonyan, K., Zisserman, A.: Very deep convolutional networks for large-scale image recognition. arXiv preprint arXiv:1409.1556 (2014)
43. Snell, J., Swersky, K., Zemel, R.: Prototypical networks for few-shot learning. In: Advances in Neural Information Processing Systems, pp. 4077–4087 (2017)
44. Sung, F., Yang, Y., Zhang, L., Xiang, T., Torr, P.H., Hospedales, T.M.: Learning to compare: relation network for few-shot learning. In: Proceedings of the IEEE Conference on Computer Vision and Pattern Recognition, pp. 1199–1208 (2018)
45. Vinyals, O., Blundell, C., Lillicrap, T., Wierstra, D., et al.: Matching networks for one shot learning. In: Advances in Neural Information Processing Systems, pp. 3630–3638 (2016)
46. Wah, C., Branson, S., Welinder, P., Perona, P., Belongie, S.: The Caltech-UCSD Birds-200-2011 dataset. Technical report CNS-TR-2011-001, California Institute of Technology (2011)
47. Wang, H., et al.: CosFace: large margin cosine loss for deep face recognition. In: Proceedings of the IEEE Conference on Computer Vision and Pattern Recognition, pp. 5265–5274 (2018)
48. Wang, Y.X., Girshick, R., Hebert, M., Hariharan, B.: Low-shot learning from imaginary data. In: Proceedings of the IEEE Conference on Computer Vision and Pattern Recognition, pp. 7278–7286 (2018)
49. Weinberger, K.Q., Saul, L.K.: Distance metric learning for large margin nearest neighbor classification. J. Mach. Learn. Res. **10**(1), 207–244 (2009)
50. Wu, Z., Efros, A.A., Yu, S.X.: Improving generalization via scalable neighborhood component analysis. In: Ferrari, V., Hebert, M., Sminchisescu, C., Weiss, Y. (eds.) ECCV 2018. LNCS, vol. 11211, pp. 712–728. Springer, Cham (2018). https://doi.org/10.1007/978-3-030-01234-2_42

51. Xing, E.P., Jordan, M.I., Russell, S.J., Ng, A.Y.: Distance metric learning with application to clustering with side-information. In: Advances in Neural Information Processing Systems, pp. 521–528 (2003)
52. Yosinski, J., Clune, J., Bengio, Y., Lipson, H.: How transferable are features in deep neural networks? In: Advances in Neural Information Processing Systems, pp. 3320–3328 (2014)
53. Zagoruyko, S., Komodakis, N.: Wide residual networks. In: 2016 Proceedings of the British Machine Vision Conference (2016). https://doi.org/10.5244/C.30.87
54. Zheng, Y., Pal, D.K., Savvides, M.: Ring loss: convex feature normalization for face recognition. In: Proceedings of the IEEE Conference on Computer Vision and Pattern Recognition, pp. 5089–5097 (2018)

Particularity Beyond Commonality: Unpaired Identity Transfer with Multiple References

Ruizheng Wu[1(✉)], Xin Tao[2], Yingcong Chen[3], Xiaoyong Shen[4],
and Jiaya Jia[1,4]

[1] The Chinese University of Hong Kong, Sha Tin, Hong Kong
{rzwu,leojia}@cse.cuhk.edu.hk
[2] Kuaishou Technology, Beijing, China
jiangsutx@gmail.com
[3] CSAIL, MIT, Cambridge, USA
ycchen@csail.mit.edu
[4] SmartMore, Shenzhen, China
xiaoyong@smartmore.com

Abstract. Unpaired image-to-image translation aims to translate images from the source class to target one by providing sufficient data for these classes. Current few-shot translation methods use multiple reference images to describe the target domain through extracting common features. In this paper, we focus on a more specific identity transfer problem and advocate that particular property in each individual image can also benefit generation. We accordingly propose a new multi-reference identity transfer framework by simultaneously making use of particularity and commonality of reference. It is achieved via a semantic pyramid alignment module to make proper use of geometric information for individual images, as well as an attention module to aggregate for the final transformation. Extensive experiments demonstrate the effectiveness of our framework given the promising results in a number of identity transfer applications.

1 Introduction

Reference images are usually used as a supplement in image translation and image editing tasks [7,11,25,57]. They provide guidance of content (e.g., poses, expression) [7,57], style (such as texture) [11] or category information (e.g., identities, expression labels) [3,25] for the final results. In this paper, we focus on identity transfer tasks, including clothes and face identity transfer.

In certain tasks, multiple reference images as guidance are also available. Image translation frameworks of [25,51] utilize these multiple inputs, usually

Electronic supplementary material The online version of this chapter (https://doi.org/10.1007/978-3-030-58548-8_27) contains supplementary material, which is available to authorized users.

unseen classes for generation, proved to be effective to achieve promising results. We term them as *few-shot-based* methods. There is still an enormous room to explore the appropriate way to use multiple references.

Few-Shot *vs.* Multi-reference. Multiple reference images contain variation in many dimensions while keeping one common attribute. For example, in face identity transfer, reference images vary from poses and expression while maintaining the same identity. Few-shot-based methods [25,51], contrarily, diminish variation in the unconcerned attributes and only focus on *commonality*.

This paper forms a new point of view that *particularity* inside each reference also provides useful clues for generation. Intuitively, in the task of identity transfer, if the reference image shares the similar poses/expression as the input content image, then the desired output can copy a lot of patches from the reference. In [12,36,46], the outputs are generated by warping from one reference image in image space. However, these methods are mostly applicable for images with similar poses (e.g. frontal faces), when the pose of a reference image differs greatly from the content image, the warping technique in these methods is very likely to fail. To obtain more robust results, we consider using multiple reference images, as more references can provide complementary for generation. For example, in Fig. 4(b), the 3rd reference image provides a frontal face, and the 2nd reference provides a contemptuous mouth for the final output.

With this new motivation, we propose an intriguing way of using multiple reference images and name it *multi-reference* method. In this method, we obtain *particularity* from individual reference with an alignment module, and we adaptively assign weights with an attention module to references for fusion as *commonality*. The effectiveness of this new line of approach can be well proved on the unpaired identity transfer task.

Semantic Alignment. To make full use of *particularity* inside each reference, we carefully design the alignment module for individual references. There exist prior image translation tasks involving alignment, which estimate pixel-wise correspondence in image space [12,44,46]. It is widely known that pixel-level alignment among images in different domains may result in unwanted distortion given domains of, for instance, cartoon and real faces. In this case, we believe semantic-level alignment is more important than the pixel-level one, which takes context information into consideration for deeper image understanding.

We thus introduce a semantic pyramid to represent different levels of image features, and a new module named semantic pyramid alignment to align images hierarchically. The module starts from the highest semantic level and progressively refine estimated correspondence in lower levels. Unlike previous multi-level feature matching [1,23] that mainly searches sparse or dense correspondence in a feature extraction network (e.g., pre-trained VGG), we instead accomplish results by semantic alignment in an end-to-end fashion. Our alignment module empirically outperforms several single-level alignment baselines.

Our contribution is as follows. 1) We propose a multi-reference framework, which takes advantage of priors in each individual reference image, and adaptively fuses and generates result images. 2) We propose a semantic pyramid alignment module, which aligns references semantically in multiple levels. 3) We design an attention module to adaptively assign weights for reference fusion, along with an effective category classification and comparison discriminator to enforce image generation in a specific domain. 4) We achieve promising results on several unpaired identity transfer tasks with only a few reference images.

2   Related Work

Unpaired Image Translation. Unpaired Image Translation is a task to translate source domain images to target domain ones without ground-truth. Methods of [18,50,57] use cycle consistency loss for image reconstruction, which greatly improves performance without paired data. To translate multiple target domains in one trained network, methods of [7,33,46] introduce class conditions as extra input for translation. In another stream, some works [6,24,26,30,40,45] address this task with the assumption that images in source and target domains share the same latent space. Moreover, methods of [17,21,47] disentangle the latent space into style, structure, and content partition, so that the generator can fuse features from different spaces.

To better utilize the feature from input images, some approaches [8,12,37,46, 52] propose to warp features [8,37] or pixels [12,46] from inputs to the translated results. Different from them, our alignment is accomplished on semantic pyramid for higher robustness and it is done on multiple references simultaneously. Similar to our approach, which fuses priors from multiple references, a view synthesis network [56] is also proposed to synthesize a novel view of a scene or an object, while we focus more on translation among different identities.

Recently, research [4,25,42,51] also involves few-shot unpaired image translation, which is more challenging since only one or several target/source domain images are provided for training or inference. Different from the most related work [25], our framework takes both the geometric priors from each reference and global context of all references into consideration, and thus achieves decent results. The method of [42] also tackles temporal alignment and video frame fusion, while we consider more on semantic alignment among different domains and adaptively fuse the clues from references.

Face Image Generation. To generate faces with different poses or expressions, research works [2,10,12,16,36,44,46] have been proposed to synthesize face or head images by warping a single or multiple reference images. The work of [2, 12] warp images with face landmark features, while in [10,36,44,46], the input images are warped with learned warping field. But there is a chance to produce distortion or visual artifacts when head rotation, large motion or occlusion exists.

Alternatively, deep convolutional networks were considered [3,9,22,31,33,34, 41,48,51,54,55]. In [9,33,34,41,48,51,54], faces are generated with the guidance of segmentation map, facial landmark, boundary map, or pose/expression parameters. Bao et al. [3] disentangle identity and facial attribute for face synthesis.

Natsume et al. [31] swap faces between two identities, while Zakharov et al. [51] synthesize talking heads by adopting few-shot adversarial learning strategy. They do not take full advantage of the geometric clues from multiple input faces. As for the work of Sungjoo et al. [14], they also consider multiple reference images for final generation. They apply adaptive weights for under-aligned feature blending and average the independently aligned feature, while we combine the two steps of aligning different references and adaptive aggregation without the help of extra landmark input.

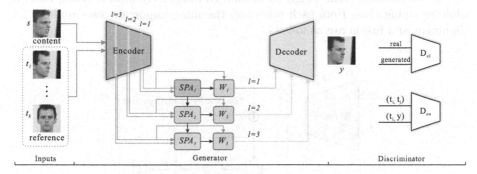

Fig. 1. The overall architecture of our multi-reference framework, where $\{SPA_i\}$ denotes the semantic pyramid alignment module, $\{W_i\}$ indicates the attention module for fusion. l refers to the feature level. All reference images are aligned with the content image in multi-level feature space, and references are adaptively fused with the attention module for the decoder to generate the final result.

3 Our Method

Our method is to translate an image from the source class to the target one under the condition that the target class is unseen in the training set and is only specified by one or a few reference images. Specifically, given one content image s from the source class $\mathbb{S}$ and k reference images $\{t_i\}_{i=1,2,...,k}$ from the target class $\mathbb{T}$, we generate output y in the target class $\mathbb{T}$ while preserving the content of pose, expressions, and shape from s.

To this end, we propose a multi-reference framework (Fig. 1), which makes appropriate use of the clues from references to generate the final result. In the generation stage, we firstly align each reference and the content image by applying the semantic pyramid alignment module, and then adaptively fuse all reference features with an attention sub-network. Finally, we decode the fused features to obtain the output hierarchically [35]. In the discrimination stage, both category classification and comparison are adopted in discriminator establishment.

3.1   Multi-reference Guided Generator

To utilize multiple reference images, current few-shot image generation frameworks [25,51] extract a spatially invariant embedding vector for each reference. The vectors are averaged as global context for decoding. Although these methods extract *commonality* from references, *particularity* in each individual reference is discarded, which however also provides vitally important clues for generation.

To address this issue, we consider *particularity* by aligning each individual reference with the content image for generation, without sacrificing *commonality* due to the important weighted fusion. To adaptively obtain global context while retaining clues from each reference, the most important two modules for alignment and fusion are as follows.

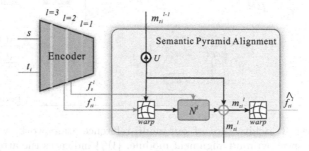

Fig. 2. Median-level structure in semantic pyramid alignment module.

In the alignment module, for each reference, we align it with the content image semantically. Rather than directly aligning on image space, we start from the high-level feature space to learn a coarse semantic correspondence, which is then propagated to lower feature space for refinement. It not only enables the semantic context aligned in high-level feature space, but also preserves textures from references in low-level one. In the fusion stage, we estimate the weight map for reference by attention module in each feature space. Noted that we do not map the image into a 1D embedding vector and instead produce a 2D feature map to preserve the spatial structure of each reference.

Semantic Pyramid Alignment. As mentioned above, we align the reference and content images in multi-level deep feature space, which is referred to as semantic pyramid. The pipeline is described in Algorithm 1, where l refers to different levels, and $l = 1, 2, 3$ indicates high-, median- and low-level feature space. Besides, i indexes reference images with number k in total.

Starting from the highest level, we estimate feature correspondence between current and each reference to obtain the coarsest optical flow map, which is then fed into the lower level for refinement. For subsequent median and low levels, we update the coarse optical flow by estimating a residual.

The procedure is to first up-sample the optical flow map $\{m_{ti}^{l-1}\}$ from the high level. It is used to warp the current reference feature f_{ti}^l to $w(f_{ti}^l)$ that is coarsely aligned with content f_s^l. We then estimate the dense correspondence between $w(f_{ti}^l)$ and f_s^l in a finer level with the alignment network N^l. The network output is a residual for the upsampled flow map from the last level. We sum them up to get the final flow map $\widehat{f_{ti}^l}$. The structure of median-level alignment module is shown in Fig. 2. With optical flow maps estimated on multiple levels, the reference image feature is warped for our deployment.

Feature Fusion with Attention. By aligning features from all reference images, we select useful regions for generation. Since output image preserves pose, expressions, and shape from the content image, we thus search similar patches with the content image from all references.

Algorithm 1: Semantic Pyramid Alignment

Input: Alignment network N^l, content image feature f_s^l, reference image
        feature f_{ti}^l, where $l = 1, 2, 3$ and $i = 1, 2, ..., k$.
Output: Flow maps $\{m_{ti}^l\}$, warped features $\{\widehat{f_{ti}^l}\}$.
for $l = 1; l \leq 3; l = l + 1$ do
    if $l == 1$ then
        for $i = 1; i \leq k; i = i + 1$ do
            $m_{ti}^l = N^l(f_s^l, f_{ti}^l)$;
        end
    else
        for $i = 1; i \leq k; i = i + 1$ do
            $U(m_{ti}^{l-1}) = Upsample(m_{ti}^{l-1}) \times 2.0$;
            $w(f_{ti}^l) = warp(f_{ti}^l, U(m_{ti}^{l-1}))$;
            $m_{ti}^l = N^l(f_s^l, w(f_{ti}^l)) + U(m_{ti}^{l-1})$;
        end
    end
    $\widehat{f_{ti}^l} = warp(f_{ti}^l, m_{ti}^l)$;
end
return $\{m_{ti}^l\}, \{\widehat{f_{ti}^l}\}$;

For each level in feature space, as shown in Fig. 3, we firstly flatten and transpose the content feature f_s^l and all reference ones $\widehat{f_t^l}$ into an appropriate shape, then we calculate their similarity at each pixel location by batch matrix multiplication. This operation results in attention weights for each reference, and then they are normalized with *softmax* on the dimension of reference number. Finally, we apply the normalized weights on reference features $\widehat{f_t^l}$ and obtain the output o^l by another matrix multiplication for fusion. o^l is fed to decoder for final generation. Noted that it is related but different from self-attention blocks [43,49,53]. In self-attention blocks, the similarity matrix is computed among features of pixels, while we compute the similarity among features of different references.

3.2 Discriminators

To distinguish the generation of the translated images from the real ones, we employ two discriminators, i.e. category classification discriminator D_{cl} and category comparison discriminator D_{co}.

Category Classification. Following the discriminator in [25], we build a multi-task adversarial discriminator to distinguish between the generated and real images with multiple categories. For $|\mathbb{S}|$ categories in the training set, the discriminator D_{cl} produces $|\mathbb{S}|$ output, and we treat each as binary classification.

The adversarial loss is applied to specific class output. When updating D_{cl} for a real image of source class s, we penalize D_{cl} if its sth output is negative. For fake images of class s, we penalize D_{cl} if its sth output is positive. As for generator, we only penalize generator if the sth output of D_{cl} is negative.

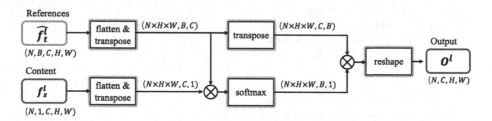

Fig. 3. The procedure of feature fusion in attention mechanism, where n, b, c, h and w refer to batch size, reference number, channel, height, and width respectively.

Category Comparison. Only category classification discriminator is not enough to enforce generation of unseen categories, since the classification discriminator is trained only with the known categories in the training set. When an image from an unseen category is required to be generated, the generator needs a similar category in the training dataset. Therefore, in the case that there is no similar category, we need an auxiliary network to strengthen the relation between reference inputs and output.

To this end, we design another discriminator D_{co} for comparing the category between two images. We treat two images as a positive sample if they belong to the same category. Otherwise, they are negative. Besides, the generated and real images in the same category are viewed as another negative sample. The discriminator helps preserve the identity in unseen classes.

3.3 Training

Random Warping for Reference Images. Directly training of our proposed framework can be easily trapped into generating one of the most similar reference images. To avoid this trivial solution, which copies one of the reference images, we randomly produce the warping parameters and apply them to distort the reference images as a perturbation.

Loss Functions. We adopt two adversarial losses $\mathcal{L}_{GAN}^{cl}$ and $\mathcal{L}_{GAN}^{co}$ for two kinds of discriminators of D_{cl} and D_{co}. D_{cl} is a $|\mathbb{S}|$ binary discriminator, and D_{co} is a conditional binary discriminator. The two loss functions are

$$\mathcal{L}_{GAN}^{cl}(G, D_{cl}) = \mathbb{E}_{\mathbf{s}}[-logD_{cl}(\mathbf{s})] + \mathbb{E}_{\mathbf{s},\{\mathbf{t_1},...,\mathbf{t_k}\}}[log(1 - D_{cl}(\mathbf{y}))],$$
$$\mathcal{L}_{GAN}^{co}(G, D_{co}) = \mathbb{E}_{\mathbf{t_i},\mathbf{t_j}}[-logD_{co}(\mathbf{t_i}, \mathbf{t_j})] + \mathbb{E}_{\mathbf{s},\{\mathbf{t_1},...,\mathbf{t_k}\}}[log(1 - D_{cl}(G(\mathbf{t_i},\mathbf{y})))],$$

Where s, $\{t_i\}$ and y indicate source class image, target image set, and translated image respectively.

We also adopt reconstruction loss by sampling content and reference images in the same category. In this case, the output image is the same as the content one. It is expressed as

$$\mathcal{L}_{REC}(G) = \mathbb{E}_{\mathbf{s},\{\mathbf{s_1},...,\mathbf{s_k}\}}\|\mathbf{s} - G(\mathbf{s}, \{\mathbf{s_1},...,\mathbf{s_k}\})\|_1, \qquad (1)$$

where $\{\mathbf{s_1},...,\mathbf{s_k}\}$ indicates a set of k random samples from category $\mathbb{S}$.

4 Experiments

4.1 Datasets and Implementation

Datasets. To verify the effectiveness of our framework, we conduct experiments on two kinds of datasets of faces and human body. Face/clothes identity transfer tasks are accomplished. For face identity transfer, we conduct experiments on RaFD [20], Multi-PIE [13] and CelebA [29], while we utilize DeepFashion [28] for clothes identity transfer.

Both RaFD and Multi-PIE contain face images with a clean background. RaFD contains 67 identities, and we use the first session of Multi-PIE with 249 identities. CelebA is a more challenging dataset with complicated background, which contains 10,177 identities, while different images of the same person may vary widely. As for Deepfashion dataset, we utilize its group of 'Blouses_Shirts', and we split them into 1,438 styles for training and 189 for test.

Implementation. We implement our method with PyTorch [32] on a TITAN Xp card. We train our framework with resolution 128×128, and set batch size as 6. Adam [19] optimizer with learning rate $1e-4$ is adopted for both generator and discriminators. We train our framework with 3 reference images (3-shot). Any number of references can be fed into our framework for testing. At the stage of inference, the class/identity of reference images is unseen in training.

4.2   Quantitative Evaluation Metrics

We set up quantitative evaluation metrics as follows.

Classification Accuracy (Acc). Similar to that of [7,25], we train a classifier for testset. We adopt a pre-trained Inception-v3 [39] as backbone and replace the fully connected layer with a new one that produce specific class number for corresponding dataset. We evaluate classification accuracy of generated images.

Distribution Discrepancy (mFID). To obtain distribution discrepancy, we firstly extract features with a deep face feature extractor VGGFace2 [5] for face dataset and VGG16 [38] for DeepFashion dataset, then we use FID [15] to measure feature distribution discrepancy between real and generated faces for each category and obtain the average as mFID.

Inception Score (IS). We utilize the fine-tuned Inception-v3 [39] in 'Classification Accuracy' to calculate the inception score between real and generated images, which measures the realism of generated images.

Perceptual Distance (Per). We also measure if our results preserve the content by calculating the $\mathcal{L}_2$ distance between the content image and output in feature space produced by the 'conv_5' layer in pre-trained VGG16 on ImageNet.

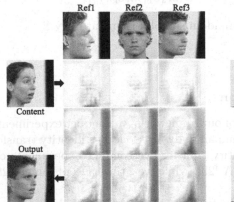 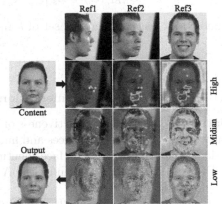

(a) Optical flow visualization among different level features, 'Ref*i*' indicates the *i*th reference. For each reference, we visualize the corresponding optical flow from high to low level feature space in 3 rows.

(b) Attention weight visualization (blue → low, red→ high) on different levels. Similar to (a), we show the visualization of images and low-, median- and high-level attention maps.

Fig. 4. Visualization of (a) pyramid alignment and (b) attention maps for fusion. (Color figure online)

4.3   Analysis of Different Components

Visualization of Pyramid Alignment Module. In our proposed framework, hierarchical features from the encoder are aligned by estimating correspondence between content and reference images. The dense correspondence on multiple levels is visualized in Fig. 4(a). We notice that optical flow (dense correspondence) is gradually refined from high to low levels.

Visualization of Attention Map for Fusion. The attention weight on each level for selection and fusion is visualized in Fig. 4(b). The attention module produces different weights on regions – regions that are similar between the reference and the content images gain large attention weights. For example, in the high level space, most face regions of 'Ref3' receive great weight as they share similar pose with the content image; the mouth of 'Ref2' is also important with the same reason.

Besides, from high to low levels, attention weights dilute for each reference image, since features are gradually aligned and constructed by our alignment module and decoder layers.

Semantic Pyramid *vs*. Single Level. In the alignment and attention module, we utilize multiple level features in a hierarchical mechanism. To verify the effectiveness, we compare with simplified versions with only one level. The versions include the highest level feature (i.e. the last layer of encoder) for alignment and fusion and the lowest level where alignment and fusion are directly done on RGB image space. The results shown in Fig. 5 indicate that the output is easily trapped due to model collapse with only the highest level feature, while distortion emerges when processing on RGB image level.

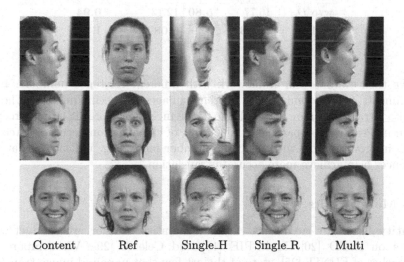

Content   Ref   Single_H   Single_R   Multi

Fig. 5. Results of multi-level and simplified single-level frameworks, where 'Single_H' and 'Single_R' means that alignment is conducted on only the highest feature and RGB-image levels respectively. 'Multi' indicates our complete semantic pyramid alignment.

Effectiveness of Discriminators. We utilize two discriminators for category classification and comparison. To study the roles of each discriminator, we conduct ablation study on discriminators. The quantitative results are shown in Table 1, where 'w/o D_{co}', 'w/o D_{cl}' and 'Full' indicate the proposed framework without category classification discriminator, framework without category comparison discriminator and our full system.

In Table 1, we observe that without the category classification module, translation accuracy drops significantly while the inception score gets better. It means that category comparison is beneficial for higher-quality image generation, while category classification helps specific category image generation. Note that the translation accuracy and mean FID reach the best values when we utilize both.

Reference Quality and Reference Number. Since we make use of clues from references for generation, the quality of references makes difference for final generation. We conduct experiments on RaFD dataset, and sample images with different poses and expressions as reference images. Results in Fig. 6 show that FUNIT [25] differs little even with the change of references, while our proposed framework generates better results with more suitable reference images. The more similar poses content and reference images are, the higher-quality images we translate. This attributes to the fact that similar poses provide realistic clues for final generation.

Table 1. Quantitative comparison among different discriminators on RaFD dataset.

	Acc(%) ↑	IS ↑	mFID(1e3) ↓	Per ↓
w/o D_{co}	68.84	2.41	7.70	1.41
w/o D_{cl}	16.77	**5.80**	12.13	**0.94**
Full	**73.03**	2.49	**6.08**	1.39

We then evaluate our framework with different reference numbers. In the experiments, we randomly select reference images to specific number. They are fed into a trained framework. In Table 2, the inception score (IS) and perceptual distance (Per) are comparable, while scores of classification accuracy (Acc) and mFID improve greatly with reference number increasing, indicating that more references help achieve more accurate translation.

4.4   More Results

Identity Transfer on Faces. For face identity transfer, we conduct experiments on RaFD [20], Multi-PIE [13] and CelebA [29]. We compare with approaches of FUNIT [25], state-of-the-art few-shot unpaired image translation framework; Tow [3], the method to synthesize faces with identity and attributes from two faces respectively, and FOMM [37], the latest motion transfer method.

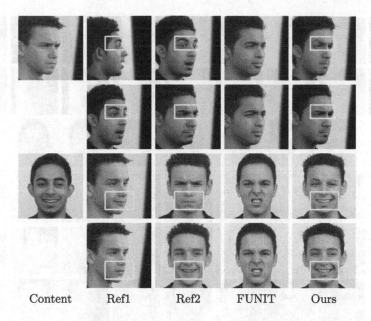

| Content | Ref1 | Ref2 | FUNIT | Ours |

Fig. 6. Effect of reference quality for final generation. The highlight regions show the main effect with different references. In the first case, similar eyebrow between 'Ref2' and content leads to better eyebrow in translation (2nd row). In the second example, smile expression in 'Ref2' helps the generation of smiling face (4th row).

Table 2. Quantitative comparison regarding reference numbers.

Ref	Acc(%) ↑	IS ↑	mFID(1e3) ↓	Per ↓	Runtime(ms) ↓
1	53.12	**3.31**	9.55	1.56	**51.1**
3	73.03	2.49	6.08	**1.39**	53.0
5	90.01	1.46	4.50	1.47	59.8
10	**90.21**	1.47	**4.39**	1.44	71.5

We also compare with StarGAN [7], state-of-the-art multi-class unpaired image translation method. Since StarGAN is a full-shot framework, we follow the 'fair' ('Star-F') and 'unfair' ('Star-U') setting of [25].

As for Star-F, we firstly train StarGAN with seen categories in the training dataset. During testing, we estimate class association vectors [25] for input reference images. Then the estimated vectors are used as the target condition for inference. For the 'unfair' setting of StarGAN ('Star-U'), we train the framework with images of unseen categories in the test set. We adopt 3-shot training in our experiments, i.e., 3 reference images are provided in training.

Visual Comparison. We show the visual comparison in Fig. 7. As for the results on RaFD and Multi-PIE dataset, Star-F produces results with correct pose and

| Content | Ref1 | Ref2 | FUNIT | Star-F | Star-U | FOMM | Tow | Ours |

Fig. 7. Visual comparison on different face datasets. The results are generated with 3 references, while only 2 references are shown here.

expression. Yet the identity is incorrect. Contrarily, Star-U achieves results with correct identities; but the output is blurry with more visual artifacts. They are mainly caused by a lack of sufficient training data with only 3 reference samples provided. FOMM generates satisfied results when reference and content image share similar poses, while it generates distorted results in the cases that large geometric transform is required. As for Tow, center faces are translated to another identity but the results are kind of blurry. FUNIT generates decent output with fewer artifacts, and the identity is similar to one person in the training set, while it fails to generate images with the satisfying categories specified by the reference images. Since our method makes appropriate use of the clues from

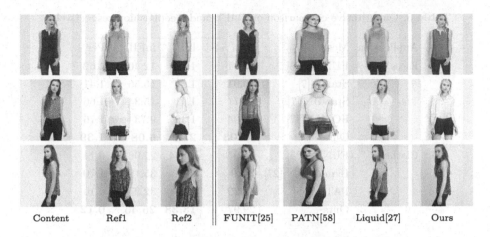

|Content|Ref1|Ref2|FUNIT[25]|PATN[58]|Liquid[27]|Ours|

Fig. 8. Visual comparison on clothes identity transfer. We resize PATN results from 256 × 176 to 128 × 128, which make them slightly misaligned with the input.

each reference, our results preserve identities from references best. Besides, the alignment module in semantic pyramid greatly improves the generation quality.

For CelebA dataset, reference images vary much though they belong to the same identity, and thus it would be difficult to define target identity with only these few references. As a result, Star-F and Star-U cannot generate decent output since there are too many identities but few samples provided for each in the training set. As for FUNIT, it can generate decent output, while our method can produce more promising results with few varied references.

Quantitative Comparison. We also make comparison quantitatively in Table 3. Our method achieves both the highest classification accuracy and IS score among all compared methods on face dataset. It indicates that our method consistently guarantees correct category and ensures high-quality results. Besides, we achieve the lowest mFID score and perceptual distance with content images, showing that our results also preserve the content well from the input.

Identity Transfer on Clothes. For the application of clothes translation on human body, we utilize DeepFashion dataset for method comparison. We evaluate different frameworks regarding the group of 'Blouses_Shirts'. Different styles of clothes are regarded as the categories in our framework.

We compare our method with LiquidGAN [27] and PATN [58], which are used for pose translation. Their pre-trained models on DeepFashion dataset are adopted for test. We also compare with FUNIT here. The results are shown in Fig. 8. Both LiquidGAN and our method translate the input images to the correct categories of clothes, while our generated images are more natural.

The quantitative comparison is listed in Table 3. Since LiquidGAN [27] and PATN [58] are single reference frameworks, we generate results with all 3 testing

Table 3. Quantitative comparison on RaFD and DeepFashion respectively.

Application	Methods	Acc(%) ↑	IS ↑	mFID ↓	Per ↓
Faces	FUNIT [25]	38.31	1.83	12.40	1.62
	Star-F [7]	16.91	1.82	15.56	1.41
	Star-UN [7]	31.71	1.41	15.34	1.66
	FOMM [37]	33.94	1.39	9.73	1.76
	Ours	**73.03**	**2.49**	**6.08**	**1.39**
Clothes	FUNIT [25]	0.69	10.74	25.96	0.14
	LiquidGAN [27]	**36.9**	2.04	35.19	1.70
	PATN [58]	12.29	2.19	32.24	0.16
	Ours	29.97	**13.38**	**25.45**	**0.12**

references and average their evaluation metrics scores. Noted that we can achieve the best scores on both IS and mFID, which indicates the high quality of our results. Besides, we also preserve poses from content images well with the shortest perceptual distance, while LiquidGAN and our method achieve comparable classification accuracy.

5   Conclusion

In this paper, we propose a multi-reference framework for unpaired identity transfer, which makes decent use of clues from each individual reference. A well-designed semantic pyramid alignment module is introduced to extract *particularity* from each reference. References are also adaptively fused as *commonality* for generation with the attention module. We conduct extensive experiments and achieve promising results on some unpaired identity transfer applications.

References

1. Aberman, K., Liao, J., Shi, M., Lischinski, D., Chen, B., Cohen-Or, D.: Neural best-buddies: sparse cross-domain correspondence. ACM Trans. Graph. **37**, 1–14 (2018)
2. Averbuch-Elor, H., Cohen-Or, D., Kopf, J., Cohen, M.F.: Bringing portraits to life. ACM Trans. Graph. **36**, 1–13 (2017)
3. Bao, J., Chen, D., Wen, F., Li, H., Hua, G.: Towards open-set identity preserving face synthesis. In: CVPR (2018)
4. Benaim, S., Wolf, L.: One-shot unsupervised cross domain translation. In: NeurIPS (2018)
5. Cao, Q., Shen, L., Xie, W., Parkhi, O.M., Zisserman, A.: VGGFace2: a dataset for recognising faces across pose and age. In: 2018 13th IEEE International Conference on Automatic Face & Gesture Recognition (FG 2018) (2018)
6. Chen, Y.C., Xu, X., Tian, Z., Jia, J.: Homomorphic latent space interpolation for unpaired image-to-image translation. In: CVPR (2019)

7. Choi, Y., Choi, M., Kim, M., Ha, J.W., Kim, S., Choo, J.: StarGAN: unified generative adversarial networks for multi-domain image-to-image translation. In: CVPR (2018)
8. Dong, H., Liang, X., Gong, K., Lai, H., Zhu, J., Yin, J.: Soft-gated warping-GAN for pose-guided person image synthesis. In: NeurIPS (2018)
9. Fu, C., Hu, Y., Wu, X., Wang, G., Zhang, Q., He, R.: High fidelity face manipulation with extreme pose and expression. arXiv preprint arXiv:1903.12003 (2019)
10. Ganin, Y., Kononenko, D., Sungatullina, D., Lempitsky, V.: DeepWarp: photorealistic image resynthesis for gaze manipulation. In: Leibe, B., Matas, J., Sebe, N., Welling, M. (eds.) ECCV 2016. LNCS, vol. 9906, pp. 311–326. Springer, Cham (2016). https://doi.org/10.1007/978-3-319-46475-6_20
11. Gatys, L.A., Ecker, A.S., Bethge, M.: Image style transfer using convolutional neural networks. In: CVPR (2016)
12. Geng, J., Shao, T., Zheng, Y., Weng, Y., Zhou, K.: Warp-guided GANs for single-photo facial animation. In: SIGGRAPH Asia 2018 Technical Papers (2018)
13. Gross, R., Matthews, I., Cohn, J., Kanade, T., Baker, S.: Multi-pie. Image Vis. Comput. **28**, 807–813 (2010)
14. Ha, S., Kersner, M., Kim, B., Seo, S., Kim, D.: MarioNETte: few-shot face reenactment preserving identity of unseen targets. In: AAAI (2020)
15. Heusel, M., Ramsauer, H., Unterthiner, T., Nessler, B., Klambauer, G., Hochreiter, S.: GANs trained by a two time-scale update rule converge to a nash equilibrium. arXiv preprint arXiv:1706.08500 (2017)
16. Huang, R., Zhang, S., Li, T., He, R.: Beyond face rotation: global and local perception GAN for photorealistic and identity preserving frontal view synthesis. In: ICCV (2017)
17. Huang, X., Liu, M.Y., Belongie, S., Kautz, J.: Multimodal unsupervised image-to-image translation. arXiv preprint arXiv:1804.04732 (2018)
18. Kim, T., Cha, M., Kim, H., Lee, J.K., Kim, J.: Learning to discover cross-domain relations with generative adversarial networks. In: ICML (2017)
19. Kingma, D.P., Ba, J.: Adam: a method for stochastic optimization. arXiv preprint arXiv:1412.6980 (2014)
20. Langner, O., Dotsch, R., Bijlstra, G., Wigboldus, D.H., Hawk, S.T., Van Knippenberg, A.: Presentation and validation of the radboud faces database. Cogn. Emot. **24**, 377–1388 (2010)
21. Lee, H.-Y., Tseng, H.-Y., Huang, J.-B., Singh, M., Yang, M.-H.: Diverse image-to-image translation via disentangled representations. In: Ferrari, V., Hebert, M., Sminchisescu, C., Weiss, Y. (eds.) ECCV 2018. LNCS, vol. 11205, pp. 36–52. Springer, Cham (2018). https://doi.org/10.1007/978-3-030-01246-5_3
22. Li, M., Zuo, W., Zhang, D.: Deep identity-aware transfer of facial attributes. arXiv preprint arXiv:1610.05586 (2016)
23. Liao, J., Yao, Y., Yuan, L., Hua, G., Kang, S.B.: Visual attribute transfer through deep image analogy. ACM Trans. Graph. (TOG) **36**(4), 1–15 (2017)
24. Liu, M.Y., Breuel, T., Kautz, J.: Unsupervised image-to-image translation networks. In: NeurIPS (2017)
25. Liu, M.Y., et al.: Few-shot unsupervised image-to-image translation. In: ICCV (2019)
26. Liu, M.Y., Tuzel, O.: Coupled generative adversarial networks. In: NeurIPS (2016)
27. Liu, W., Piao, Z., Min, J., Luo, W., Ma, L., Gao, S.: Liquid warping GAN: a unified framework for human motion imitation, appearance transfer and novel view synthesis. In: ICCV (2019)

28. Liu, Z., Luo, P., Qiu, S., Wang, X., Tang, X.: DeepFashion: powering robust clothes recognition and retrieval with rich annotations. In: CVPR (2016)
29. Liu, Z., Luo, P., Wang, X., Tang, X.: Deep learning face attributes in the wild. In: ICCV (2015)
30. Murez, Z., Kolouri, S., Kriegman, D., Ramamoorthi, R., Kim, K.: Image to image translation for domain adaptation. arXiv preprint arXiv:1712.00479 (2017)
31. Natsume, R., Yatagawa, T., Morishima, S.: FSNet: an identity-aware generative model for image-based face swapping. In: Jawahar, C.V., Li, H., Mori, G., Schindler, K. (eds.) ACCV 2018. LNCS, vol. 11366, pp. 117–132. Springer, Cham (2019). https://doi.org/10.1007/978-3-030-20876-9_8
32. Paszke, A., et al.: Automatic differentiation in PyTorch. In: NIPS-W (2017)
33. Pumarola, A., Agudo, A., Martinez, A.M., Sanfeliu, A., Moreno-Noguer, F.: GANimation: anatomically-aware facial animation from a single image. In: Ferrari, V., Hebert, M., Sminchisescu, C., Weiss, Y. (eds.) ECCV 2018. LNCS, vol. 11214, pp. 835–851. Springer, Cham (2018). https://doi.org/10.1007/978-3-030-01249-6_50
34. Qian, S., et al.: Make a face: towards arbitrary high fidelity face manipulation. In: ICCV (2019)
35. Ronneberger, O., Fischer, P., Brox, T.: U-Net: convolutional networks for biomedical image segmentation. In: Navab, N., Hornegger, J., Wells, W.M., Frangi, A.F. (eds.) MICCAI 2015. LNCS, vol. 9351, pp. 234–241. Springer, Cham (2015). https://doi.org/10.1007/978-3-319-24574-4_28
36. Shu, Z., Sahasrabudhe, M., Alp Güler, R., Samaras, D., Paragios, N., Kokkinos, I.: Deforming autoencoders: unsupervised disentangling of shape and appearance. In: Ferrari, V., Hebert, M., Sminchisescu, C., Weiss, Y. (eds.) ECCV 2018. LNCS, vol. 11214, pp. 664–680. Springer, Cham (2018). https://doi.org/10.1007/978-3-030-01249-6_40
37. Siarohin, A., Lathuilière, S., Tulyakov, S., Ricci, E., Sebe, N.: First order motion model for image animation. In: NeurIPS (2019)
38. Simonyan, K., Zisserman, A.: Very deep convolutional networks for large-scale image recognition. arXiv preprint arXiv:1409.1556 (2014)
39. Szegedy, C., Vanhoucke, V., Ioffe, S., Shlens, J., Wojna, Z.: Rethinking the inception architecture for computer vision. In: CVPR (2016)
40. Taigman, Y., Polyak, A., Wolf, L.: Unsupervised cross-domain image generation. arXiv preprint arXiv:1611.02200 (2016)
41. Tripathy, S., Kannala, J., Rahtu, E.: ICface: interpretable and controllable face reenactment using GANs. arXiv preprint arXiv:1904.01909 (2019)
42. Wang, T.C., Liu, M.Y., Tao, A., Liu, G., Kautz, J., Catanzaro, B.: Few-shot video-to-video synthesis. In: NeurIPS (2019)
43. Wang, X., Girshick, R., Gupta, A., He, K.: Non-local neural networks. In: CVPR (2018)
44. Wiles, O., Koepke, A.S., Zisserman, A.: X2Face: a network for controlling face generation using images, audio, and pose codes. In: Ferrari, V., Hebert, M., Sminchisescu, C., Weiss, Y. (eds.) ECCV 2018. LNCS, vol. 11217, pp. 690–706. Springer, Cham (2018). https://doi.org/10.1007/978-3-030-01261-8_41
45. Wolf, L., Taigman, Y., Polyak, A.: Unsupervised creation of parameterized avatars. In: ICCV (2017)
46. Wu, R., Tao, X., Gu, X., Shen, X., Jia, J.: Attribute-driven spontaneous motion in unpaired image translation. In: ICCV (2019)
47. Wu, W., Cao, K., Li, C., Qian, C., Loy, C.C.: TransGaGa: geometry-aware unsupervised image-to-image translation. In: CVPR (2019)

48. Wu, W., Zhang, Y., Li, C., Qian, C., Loy, C.C.: ReenactGAN: learning to reenact faces via boundary transfer. In: Ferrari, V., Hebert, M., Sminchisescu, C., Weiss, Y. (eds.) ECCV 2018. LNCS, vol. 11205, pp. 622–638. Springer, Cham (2018). https://doi.org/10.1007/978-3-030-01246-5_37
49. Yi, P., Wang, Z., Jiang, K., Jiang, J., Ma, J.: Progressive fusion video super-resolution network via exploiting non-local spatio-temporal correlations. In: ICCV (2019)
50. Yi, Z., Zhang, H.R., Tan, P., Gong, M.: DualGAN: unsupervised dual learning for image-to-image translation. In: ICCV (2017)
51. Zakharov, E., Shysheya, A., Burkov, E., Lempitsky, V.: Few-shot adversarial learning of realistic neural talking head models. arXiv preprint arXiv:1905.08233 (2019)
52. Zhan, F., Zhu, H., Lu, S.: Spatial fusion GAN for image synthesis. In: CVPR (2019)
53. Zhang, H., Goodfellow, I., Metaxas, D., Odena, A.: Self-attention generative adversarial networks. In: ICML (2019)
54. Zhang, Y., Zhang, S., He, Y., Li, C., Loy, C.C., Liu, Z.: One-shot face reenactment. In: BMVC (2019)
55. Zhou, H., Liu, Y., Liu, Z., Luo, P., Wang, X.: Talking face generation by adversarially disentangled audio-visual representation. In: AAAI (2019)
56. Zhou, T., Tulsiani, S., Sun, W., Malik, J., Efros, A.A.: View synthesis by appearance flow. In: Leibe, B., Matas, J., Sebe, N., Welling, M. (eds.) ECCV 2016. LNCS, vol. 9908, pp. 286–301. Springer, Cham (2016). https://doi.org/10.1007/978-3-319-46493-0_18
57. Zhu, J.Y., Park, T., Isola, P., Efros, A.A.: Unpaired image-to-image translation using cycle-consistent adversarial networks. In: ICCV (2017)
58. Zhu, Z., Huang, T., Shi, B., Yu, M., Wang, B., Bai, X.: Progressive pose attention transfer for person image generation. In: CVPR (2019)

Tracking Objects as Points

Xingyi Zhou[1]([✉]), Vladlen Koltun[2], and Philipp Krähenbühl[1]

[1] UT Austin, Austin, USA
zhouxy@cs.utexas.edu
[2] Intel Labs, Hillsboro, USA

Abstract. Tracking has traditionally been the art of following interest points through space and time. This changed with the rise of powerful deep networks. Nowadays, tracking is dominated by pipelines that perform object detection followed by temporal association, also known as tracking-by-detection. We present a simultaneous detection and tracking algorithm that is simpler, faster, and more accurate than the state of the art. Our tracker, CenterTrack, applies a detection model to a pair of images and detections from the prior frame. Given this minimal input, CenterTrack localizes objects and predicts their associations with the previous frame. That's it. CenterTrack is simple, online (no peeking into the future), and real-time. It achieves 67.8% MOTA on the MOT17 challenge at 22 FPS and 89.4% MOTA on the KITTI tracking benchmark at 15 FPS, setting a new state of the art on both datasets. CenterTrack is easily extended to monocular 3D tracking by regressing additional 3D attributes. Using monocular video input, it achieves 28.3% AMOTA@0.2 on the newly released nuScenes 3D tracking benchmark, substantially outperforming the monocular baseline on this benchmark while running at 28 FPS.

Keywords: Multi-object tracking · Conditioned detection · 3D object tracking

1 Introduction

In early computer vision, tracking was commonly phrased as following interest points through space and time [35,42]. Early trackers were simple, fast, and reasonably robust. However, they were liable to fail in the absence of strong low-level cues such as corners and intensity peaks. With the advent of high-performing object detection models [9,30], a powerful alternative emerged: tracking-by-detection (or more precisely, tracking-after-detection) [2,40,49]. These models rely on a given accurate recognition to identify objects and then link them up through time in a separate stage. Tracking-by-detection leverages the power of deep-learning-based object detectors and is currently the dominant tracking paradigm. Yet the best-performing object trackers are not without drawbacks. Many rely on slow and complex association strategies to link detected boxes through time [14,40,47,49]. Recent work on *simultaneous detection and tracking* [1,8] has made progress in alleviating some of this complexity. Here, we show how

---

Electronic supplementary material The online version of this chapter (https://doi.org/10.1007/978-3-030-58548-8_28) contains supplementary material, which is available to authorized users.

A. Vedaldi et al. (Eds.): ECCV 2020, LNCS 12349, pp. 474–490, 2020.
https://doi.org/10.1007/978-3-030-58548-8_28

Fig. 1. We track objects by tracking their centers. We learn a 2D offset between two adjacent frames and associate them based on center distance.

combining ideas from point-based tracking and simultaneous detection and tracking further simplifies tracking.

We present a point-based framework for joint detection and tracking, referred to as CenterTrack. Each object is represented by a single point at the center of its bounding box. This center point is then tracked through time (Fig. 1). Specifically, we adopt the recent CenterNet detector to localize object centers [55]. We condition the detector on two consecutive frames, as well as a heatmap of prior tracklets, represented as points. We train the detector to also output an offset vector from the current object center to its center in the previous frame. We learn this offset as an attribute of the center point at little additional computational cost. A greedy matching, based solely on the distance between this predicted offset and the detected center point in the previous frame, suffices for object association. The tracker is end-to-end trainable and differentiable.

Tracking objects as points simplifies two key components of the tracking pipeline. First, it simplifies tracking-conditioned detection. If each object in past frames is represented by a single point, a constellation of objects can be represented by a heatmap of points [4]. Our tracking-conditioned detector directly ingests this heatmap and reasons about all objects jointly when associating them across frames. Second, point-based tracking simplifies object association across time. A simple displacement prediction, akin to sparse optical flow, allows objects in different frames to be linked. This displacement prediction is conditioned on prior detections. It learns to jointly detect objects in the current frame and associate them to prior detections.

While the overall idea is simple, subtle details matter in making this work. Tracked objects in consecutive frames are highly correlated. With the previous-frame heatmap given as input, CenterTrack could easily learn to repeat the predictions from the preceding frame, and thus refuse to track without incurring a large training error. We prevent this through an aggressive data-augmentation scheme during training. In fact, our data augmentation is aggressive enough for the model to learn to track objects from *static images*. That is, CenterTrack can be successfully trained on static image datasets (with "hallucinated" motion), with no real video input.

CenterTrack is purely local. It only associates objects in adjacent frames, without reinitializing lost long-range tracks. It trades the ability to reconnect long-range tracks for simplicity, speed, and high accuracy in the local regime. Our experiments indicate that this trade-off is well worth it. CenterTrack outperforms complex tracking-by-detection strategies on the MOT [27] and KITTI [12] tracking benchmarks. We further apply the approach to monocular 3D object tracking on the nuScenes dataset [3]. Our monocular tracker achieves 28.3% AMOTA@0.2, outperforming the monocular baseline by a factor of 3, while running at 22 FPS. It can be trained on labelled video sequences, if available, or on static images with data augmentation. Code is available at https://github.com/xingyizhou/CenterTrack.

2 Related Work

Tracking-by-Detection. Most modern trackers [2,7,22,32,34,40,46,49,57] follow the tracking-by-detection paradigm. An off-the-shelf object detector [9,29,30,50] first finds all objects in each individual frame. Tracking is then a problem of bounding box association. SORT [2] tracks bounding boxes using a Kalman filter and associates each bounding box with its highest overlapping detection in the current frame using bipartite matching. DeepSORT [46] augments the overlap-based association cost in SORT with appearance features from a deep network. More recent approaches focus on increasing the robustness of object association. Tang et al. [40] leverage person-reidentification features and human pose features. Xu et al. [49] take advantage of the spatial locations over time. BeyondPixel [34] uses additional 3D shape information to track vehicles.

These methods have two drawbacks. First, the data association discards image appearance features [2] or requires a computationally expensive feature extractor [10, 34,40,49]. Second, detection is separated from tracking. In our approach, association is almost free. Association is *learned* jointly with detection. Also, our detector takes the previous tracking results as an input, and can learn to recover missing or occluded objects from this additional cue.

Joint Detection and Tracking. A recent trend in multi-object tracking is to convert existing detectors into trackers and combine both tasks in the same framework. Feichtenhofer et al. [8] use a siamese network with the current and past frame as input and predict inter-frame offsets between bounding boxes. Integrated detection [54] uses tracked bounding boxes as additional region proposals to enhance detection, followed by bipartite-matching-based bounding-box association. Tracktor [1] removes the box association by directly propagating identities of region proposals using bounding box regression. In video object detection, Kang et al. [16,17] feed stacked consecutive frames into the network and do detection for a whole video segment. And Zhu et al. [58] use flow to warp intermediate features from previous frames to accelerate inference.

Our method belongs to this category. The difference is that all of these works adopt the FasterRCNN framework [30], where the tracked boxes are used as region proposals. This assumes that bounding boxes have a large overlap between frames, which is not true in low-framerate regimes. As a consequence, Tracktor [1] requires a motion model [5,6] for low-framerate sequences. Our approach instead provides the tracked predictions as an additional point-based heatmap input to the network. The network is then able to reason about and match objects anywhere in its receptive field even if the boxes have no overlap at all.

Motion Prediction. Motion prediction is another important component in a tracking system. Early approaches [2,46] used Kalman filters to model object velocities. Held et al. [13] use a regression network to predict four scalars for bounding box offset between frames for single-object tracking. Xiao et al. [48] utilize an optical flow estimation network to update joint locations in human pose tracking. Voigtlaender et al. [44] learn a high-dimensional embedding vector for object identities for simultaneous object tracking and segmentation. Our center offset is analogous to sparse optical flow, but is learned together with the detection network and does not require dense supervision.

Heatmap-Conditioned Keypoint Estimation. Feeding the model predictions as an additional input to a model works across a wide range of vision tasks [43], especially for keypoint estimation [4,11,28]. Auto-context [43] feeds the mask prediction back into the network. Iterative-Error-Feedback (IEF) [4] takes another step by rendering predicted keypoint coordinates into heatmaps. PoseFix [28] generates heatmaps that simulate test errors for human pose refinement.

Our tracking-conditioned detection framework is inspired by these works. A rendered heatmap of prior keypoints [4,11,28,43] is especially appealing in tracking for two reasons. First, the information in the previous frame is freely available and does not slow down the detector. Second, conditional tracking can reason about occluded objects that may no longer be visible in the current frame. The tracker can simply learn to keep those detections from the prior frame around.

3D Object Detection and Tracking. 3D trackers replace the object detection component in standard tracking systems with 3D detection from monocular images [29] or 3D point clouds [36,56]. Tracking then uses an off-the-shelf identity association model. For example, 3DT [14] detects 2D bounding boxes, estimates 3D motion, and uses depth and order cues for matching. AB3D [45] achieves state-of-the-art performance by combining a Kalman filter with accurate 3D detections [36].

3   Preliminaries

Our method, CenterTrack, builds on the CenterNet detector [55]. CenterNet takes a single image $I \in \mathbb{R}^{W \times H \times 3}$ as input and produces a set of detections $\{(\mathbf{p}_i, \mathbf{s}_i)\}_{i=0}^{N-1}$ for each class $c \in \{0, \ldots, C-1\}$. CenterNet identifies each object through its center point $\mathbf{p} \in \mathbb{R}^2$ and then regresses to a height and width $\mathbf{s} \in \mathbb{R}^2$ of the object's bounding box. Specifically, it produces a low-resolution heatmap $\hat{Y} \in [0, 1]^{\frac{W}{R} \times \frac{H}{R} \times C}$ and a size map $\hat{S} \in \mathbb{R}^{\frac{W}{R} \times \frac{H}{R} \times 2}$ with a downsampling factor $R = 4$. Each local maximum $\hat{\mathbf{p}} \in \mathbb{R}^2$ (also called peak, whose response is the strongest in a 3×3 neighborhood) in the heatmap $\hat{Y}$ corresponds to a center of a detected object with confidence $\hat{w} = \hat{Y}_{\hat{\mathbf{p}}}$ and object size $\hat{\mathbf{s}} = \hat{S}_{\hat{\mathbf{p}}}$.

Given an image with a set of annotated objects $\{\mathbf{p}_0, \mathbf{p}_1, \ldots\}$, CenterNet uses a training objective based on the focal loss [21,24]:

$$L_k = \frac{1}{N} \sum_{xyc} \begin{cases} (1 - \hat{Y}_{xyc})^\alpha \log(\hat{Y}_{xyc}) & \text{if } Y_{xyc} = 1 \\ (1 - Y_{xyc})^\beta (\hat{Y}_{xyc})^\alpha \log(1 - \hat{Y}_{xyc}) & \text{otherwise} \end{cases}, \quad (1)$$

where $Y \in [0, 1]^{\frac{W}{R} \times \frac{H}{R} \times C}$ is a ground-truth heatmap corresponding to the annotated objects. N is the number of objects, and $\alpha = 2$ and $\beta = 4$ are hyperparameters of the

Image $I^{(t)}$ Image $I^{(t-1)}$ Tracks $T^{(t-1)}$ Detections $\hat{Y}^{(t)}$ Size $\hat{S}^{(t)}$ Offset $\hat{O}^{(t)}$

Fig. 2. Illustration of our framework. The network takes the current frame, the previous frame, and a heatmap rendered from tracked object centers as inputs, and produces a center detection heatmap for the current frame, the bounding box size map, and an offset map. At test time, object sizes and offsets are extracted from peaks in the heatmap.

focal loss. For each center $\mathbf{p}$ of class c, we render a Gaussian-shaped peak into $Y_{:,:,c}$ using a rendering function $Y = \mathcal{R}(\{\mathbf{p}_0, \mathbf{p}_1, \ldots\})$ [21]. Formally, the rendering function at position $\mathbf{q} \in \mathbb{R}^2$ is defined as

$$\mathcal{R}_{\mathbf{q}}(\{\mathbf{p}_0, \mathbf{p}_1, \ldots\}) = \max_i \exp\left(-\frac{(\mathbf{p}_i - \mathbf{q})^2}{2\sigma_i^2}\right).$$

The Gaussian kernel σ_i is a function of the object size [21].

The size prediction is only supervised at the center locations. Let $\mathbf{s}_i$ be the bounding box size of the i-th object at location $\mathbf{p}_i$. Size prediction is learned by regression

$$L_{size} = \frac{1}{N} \sum_{i=1}^{N} |\hat{S}_{\mathbf{p}_i} - \mathbf{s}_i|. \tag{2}$$

CenterNet further regresses to a refined center local location using an analogous L1 loss L_{loc}. The overall loss of CenterNet is a weighted sum of all three loss terms: focal loss, size, and local location regression.

4   Tracking Objects as Points

We approach tracking from a local perspective. When an object leaves the frame or is occluded and reappears, it is assigned a new identity. We thus treat tracking as the problem of propagating detection identities across *consecutive* frames, without re-establishing associations across temporal gaps.

At time t, we are given an image of the current frame $I^{(t)} \in \mathbb{R}^{W \times H \times 3}$ and the previous frame $I^{(t-1)} \in \mathbb{R}^{W \times H \times 3}$, as well as the tracked objects in the previous frame $T^{(t-1)} = \{b_0^{(t-1)}, b_1^{(t-1)}, \ldots\}_i$. Each object $b = (\mathbf{p}, \mathbf{s}, w, id)$ is described by its center location $\mathbf{p} \in \mathbb{R}^2$, size $\mathbf{s} \in \mathbb{R}^2$, detection confidence $w \in [0, 1]$, and unique identity $id \in \mathbb{I}$. Our aim is to detect and track objects $T^{(t)} = \{b_0^{(t)}, b_1^{(t)}, \ldots\}$ in the current frame t, and assign objects that appear in both frames a consistent id.

There are two main challenges here. The first is finding all objects in every frame – including occluded ones. The second challenge is associating these objects through time. We address both via a single deep network, trained end-to-end. Section 4.1 describes a tracking-conditioned detector that leverages tracked detections from the previous frame to improve detection in the current frame. Section 4.2 then presents a simple offset prediction scheme that is able to link detections through time. Finally, Sects. 4.3 and 4.4 show how to train this detector from video or static image data.

4.1 Tracking-Conditioned Detection

As an object detector, CenterNet already infers most of the required information for tracking: object locations $\hat{\mathbf{p}}$, their size $\hat{\mathbf{s}} = \hat{S}_{\hat{\mathbf{p}}}$, and a confidence measure $\hat{w} = \hat{Y}_{\hat{\mathbf{p}}}$. However, it is unable to find objects that are not directly visible in the current frame, and the detected objects may not be temporally coherent. One natural way to increase temporal coherence is to provide the detector with additional image inputs from past frames. In CenterTrack, we provide the detection network with two frames as input: the current frame $I^{(t)}$ and the prior frame $I^{(t-1)}$. This allows the network to estimate the change in the scene and potentially recover occluded objects at time t from visual evidence at time $t-1$.

CenterTrack also takes prior detections $\{\mathbf{p}_0^{(t-1)}, \mathbf{p}_1^{(t-1)}, \ldots\}$ as additional input. How should these detections be represented in a form that is easily provided to a network? The point-based nature of our tracklets is helpful here. Since each detected object is represented by a single point, we can conveniently render all detections in a class-agnostic single-channel heatmap $H^{(t-1)} = \mathcal{R}(\{\mathbf{p}_0^{(t-1)}, \mathbf{p}_1^{(t-1)}, \ldots\})$, using the same Gaussian render function as in the training of point-based detectors. To reduce the propagation of false positive detections, we only render objects with a confidence score greater than a threshold τ. The architecture of CenterTrack is essentially identical to CenterNet, with four additional input channels. (See Fig. 2.)

Tracking-conditioned detection provides a temporally coherent set of detected objects. However, it does not link these detections across time. In the next section, we show how to add one additional output to point-based detection to track objects through space and time.

4.2 Association Through Offsets

To associate detections through time, CenterTrack predicts a 2D displacement as two additional output channels $\hat{D}^{(t)} \in \mathbb{R}^{\frac{W}{R} \times \frac{H}{R} \times 2}$. For each detected object at location $\hat{\mathbf{p}}^{(t)}$, the displacement $\hat{\mathbf{d}}^{(t)} = \hat{D}_{\hat{\mathbf{p}}^{(t)}}^{(t)}$ captures the difference in location of the object in the current frame $\hat{\mathbf{p}}^{(t)}$ and the previous frame $\hat{\mathbf{p}}^{(t-1)}$: $\hat{\mathbf{d}}^{(t)} = \hat{\mathbf{p}}^{(t)} - \hat{\mathbf{p}}^{(t-1)}$. We learn this displacement using the same regression objective as size or location refinement:

$$L_{off} = \frac{1}{N} \sum_{i=1}^{N} \left| \hat{D}_{\mathbf{p}_i^{(t)}} - (\mathbf{p}_i^{(t-1)} - \mathbf{p}_i^{(t)}) \right|, \tag{3}$$

where $\mathbf{p}_i^{(t-1)}$ and $\mathbf{p}_i^{(t)}$ are tracked ground-truth objects. Figure 2 shows an example of this offset prediction.

With a sufficiently good offset prediction, a simple greedy matching algorithm can associate objects across time. For each detection at position $\hat{p}$, we greedily associate it with the closest unmatched prior detection at position $\hat{p} - \hat{D}_{\hat{p}}$, in descending order of confidence $\hat{w}$. If there is no unmatched prior detection within a radius κ, we spawn a new tracklet. We define κ as the geometric mean of the width and height of the predicted bounding box for each tracklet. A precise description of this greedy matching algorithm is provided in supplementary material. The simplicity of this greedy matching algorithm again highlights the advantages of tracking objects as points. A simple displacement prediction is sufficient to link objects across time. There is no need for a complicated distance metric or graph matching.

4.3 Training on Video Data

CenterTrack is first and foremost an object detector, and trained as such. The architectural changed from CenterNet to CenterTrack are minor: four additional input channels and two output channels. This allows us to fine-tune CenterTrack directly from a pretrained CenterNet detector [55]. We copy all weights related to the current detection pipeline. All weights corresponding to additional inputs or outputs are initialized randomly. We follow the CenterNet training protocol and train all predictions as multi-task learning. We use the same training objective with the addition of offset regression L_{off}.

The main challenge in training CenterTrack comes in producing a realistic tracklet heatmap $H^{(t-1)}$. At inference time, this tracklet heatmap can contain an arbitrary number of missing tracklets, wrongly localized objects, or even false positives. These errors are not present in ground-truth tracklets $\{\mathbf{p}_0^{(t-1)}, \mathbf{p}_1^{(t-1)}, \ldots\}$ provided during training. We instead simulate this test-time error during training. Specifically, we simulate three types of error. First, we locally jitter each tracklet $\mathbf{p}^{(t-1)}$ from the prior frame by adding Gaussian noise to each center. That is, we render $p_i' = (x_i + r \times \lambda_{jt} \times w_i, y_i + r \times \lambda_{jt} \times h_i)$, where r is sampled from a Gaussian distribution. We use $\lambda_{jt} = 0.05$ in all experiments. Second, we randomly add false positives near ground-truth object locations by rendering a spurious noisy peak p_i' with probability λ_{fp}. Third, we simulate false negatives by randomly removing detections with probability λ_{fn}. λ_{fp} and λ_{fn} are set according to the statistics of our baseline model. These three augmentations are sufficient to train a robust tracking-conditioned object detector.

In practice, $I^{(t-1)}$ does not need to be the immediately preceding frame from time $t - 1$. It can be a different frame from the same video sequence. In our experiments, we randomly sample frames near t to avoid overfitting to the framerate. Specifically, we sample from all frames k where $|k - t| < M_f$, where $M_f = 3$ is a hyperparameter.

4.4 Training on Static Image Data

Without labeled video data, CenterTrack does not have access to a prior frame $I^{(t-1)}$ or tracked detections $\{\mathbf{p}_0^{(t-1)}, \mathbf{p}_1^{(t-1)}, \ldots\}$. However, we can simulate tracking on standard detection benchmarks, given only single images $I^{(t)}$ and detections $\{\mathbf{p}_0^{(t)}, \mathbf{p}_1^{(t)}, \ldots\}$. The idea is simple: we simulate the previous frame by randomly scaling and translating the current frame. As our experiments will demonstrate, this is surprisingly effective.

4.5   End-to-End 3D Object Tracking

To perform monocular 3D tracking, we adopt the monocular 3D detection form of CenterNet [55]. Specifically, we train output heads to predict object depth, rotation (encoded
as an 8-dimensional vector [14]), and 3D extent. Since the projection of the center of the
3D bounding box may not align with the center of the object's 2D bounding box (due
to perspective projection), we also predict a 2D-to-3D center offset. Further details are
provided in the supplement.

5   Experiments

We evaluate 2D multi-object tracking on the MOT17 [27] and KITTI [12] tracking
benchmarks. We also evaluate monocular 3D tracking on the nuScenes dataset [3].
Experiments on MOT16 can be found in the supplement.

5.1   Datasets and Evaluation Metrics

MOT. MOT17 contains 7 training sequences and 7 test sequences [27], The videos
were captured by stationary cameras mounted in high-density scenes with heavy occlusion. Only pedestrians are annotated and evaluated. The video framerate is 25–30 FPS.
The MOT dataset does not provide an official validation split. For ablation experiments,
we split each training sequence into two halves, and use the first half frames for training
and the second for validation. Our main results are reported on the test set.

KITTI. The KITTI tracking benchmark consists of 21 training sequences and 29 test
sequences [12]. They are collected by a camera mounted on a car moving through
traffic. The dataset provides 2D bounding box annotations for cars, pedestrians, and
cyclists, but only cars are evaluated. Videos are captured at 10 FPS and contain large
inter-frame motions. KITTI does not provide detections, and all entries use private
detection. We again split all training sequences into halves for training and validation.

nuScenes. nuScenes is a newly released large-scale driving dataset with 7 object
classes annotated for tracking [3]. It contains 700 training sequences, 150 validation
sequences, and 150 test sequences. Each sequence contains roughly 40 frames at 2 FPS
with 6 slightly overlapping images in a panoramic 360° view, resulting in 168k training, 36k validation, and 36k test images. The videos are sampled at 12 FPS, but frames
are only annotated and evaluated at 2 FPS. All baselines and CenterTrack only use
keyframes for training and evaluation. Due to the low framerate, the inter-frame motion
is significant.

Evaluation Metrics. We use the official evaluation metrics in each dataset.
The common metric is multi-object tracking accuracy [23,39]: $MOTA = 1 - \frac{\sum_t (FP_t + FN_t + IDSW_t)}{\sum_t GT_t}$, where GT_t, FP_t, FN_t, and $IDSW_t$ are the number of ground-
truth bounding boxes, false positives, false negatives, and identity switches in frame t,

respectively. MOTA does not rank tracklets according to confidence and is sensitive to the task-dependent output threshold θ [45]. The thresholds we use are listed in Sect. 5.2. The interplay between output threshold and true positive criteria matters. For 2D tracking [12,27], >0.5 bounding box IoU is a the true positive. For 3D tracking [3], bounding box center distance <2 m on the ground plane is the criterion for a true positive. When objects are successfully detected, but not tracked, they are identified as an identity switch (IDSW). The IDF1 metric measures the minimal cost change from predicted ids to the correct ids. In our ablation studies, we report false positve rate (FP) $\frac{\sum_t FP_t}{\sum_t GT_t}$, false negative rate (FN) $\frac{\sum_t FN_t}{\sum_t GT_t}$, and identity switches (IDSW) $\frac{\sum_t IDSW_t}{\sum_t GT_t}$ separately. In comparisons with other methods, we report the absolute numbers following the dataset convention [12,27]. We also report the Most Tracked ratio (MT) for the ratio of most tracked (>80% time) objects and Most Lost ratio (ML) for most lost (<20% time) objects [39].

nuScenes adopts a more robust metric, AMOTA, which is a weighted average of MOTA across different output thresholds. Specifically,

$$AMOTA = \frac{1}{n-1} \sum_{r \in \{\frac{1}{n-1}, \frac{2}{n-1}, \cdots, 1\}} MOTA_r$$

$$MOTA_r = max(0, 1 - \alpha \frac{IDSW_r + FP_r + FN_r - (1-r) \times P}{r \times P})$$

where r is a fixed recall threshold, $P = \sum_t GT_t$ is the total number of annotated objects among all frames, and $FP_r = \sum_t FP_{r,t}$ is the total number of false positive samples only considering the top confident samples that achieve the recall threshold r. The hyperparameters $n = 40$ and $\alpha = 0.2$ (AMOTA@0.2), or $\alpha = 1$ (AMOTA@1) are set by the benchmark organizers. The overall AMOTA is the average AMOTA among all 7 categories.

5.2   Implementation Details

Our implementation is based on CenterNet [55]. We use DLA [52] as the network backbone, optimized with Adam [20] with learning rate $1.25e - 4$ and batchsize 32. Data augmentations include random horizontal flipping, random resized cropping, and color jittering. For all experiments, we train the networks for 70 epochs. The learning rate is dropped by a factor of 10 at the 60th epoch. We test the runtime on a machine with an Intel Core i7-8086K CPU and a Titan Xp GPU. The runtimes depend on the number of objects for rendering and the input resolution in each dataset.

The MOT dataset [27] annotates each pedestrian as an amodal bounding box. That is, the bounding box always covers the whole body even when part of the object is out of the frame. In contrast, CenterNet [55] requires the center of each inferred bounding box to be within the frame. To handle this, we separately predict the visible and amodal bounding boxes [41]. Further details on this can be found in the supplement. We follow prior works [31,38,40,51,54] to pretrain on external data. We train our network on the CrowdHuman [33] dataset, using the static image training described in Sect. 4.4. Details on the CrowdHuman dataset and ablations of pretraining are in the supplement.

Table 1. Evaluation on the MOT17 test sets (top: public detection; bottom: private detection). We compare to published entries on the leaderboard. The runtime is calculated from the HZ column on the leaderboard. +D means detection time, which is usually >100 ms [30].

	Time (ms)	MOTA ↑	IDF1 ↑	MT ↑	ML ↓	FP ↓	FN ↓	IDSW ↓
Tracktor17 [1]	666+D	53.5	52.3	19.5	36.6	12201	248047	2072
LSST17 [10]	666+D	54.7	**62.3**	20.4	40.1	26091	228434	**1243**
Tracktor v2 [1]	666+D	56.5	55.1	21.1	35.3	**8866**	235449	3763
GMOT	167+D	55.4	57.9	22.7	34.7	20608	229511	1403
Ours (Public)	**57+D**	61.5	59.6	**26.4**	**31.9**	14076	**200672**	2583
Ours (Private)	57	67.8	64.7	34.6	24.6	18498	160332	3039

The default input resolution for MOT images is 1920×1080. We resize and pad the images to 960×544. We use random false positive ratio $\lambda_{fp} = 0.1$ and random false negative ratio $\lambda_{fn} = 0.4$. We only output tracklets that have a confidence of $\theta = 0.4$ or higher, and set the heatmap rendering threshold to $\tau = 0.5$. A controlled study of these hyperparameters is in the supplement.

For KITTI [12], we keep the original input resolution 1280×384 in training and testing. The hyperparameters are set at $\lambda_{fp} = 0.1$ and $\lambda_{fn} = 0.2$, with output threshold $\theta = 0.4$ and rendering threshold $\tau = 0.4$. We fine-tune our KITTI model from a nuScenes tracking model.

For nuScenes [3], we use input resolution 800×448. We set $\lambda_{fp} = 0.1$ and $\lambda_{fn} = 0.4$, and use output threshold $\theta = 0.1$ and rendering threshold $\tau = 0.1$. We first train our nuScenes model for 140 epochs for just 3D detection [55] and then fine-tune for 70 epochs for 3D tracking. Note that nuScenes evaluation is done per 360 panorama, not per image. We naively fuse all outputs from the 6 cameras together, without handling duplicate detections at the intersection of views [37].

Track Rebirth. Following common practice [1,54], we keep unmatched tracks "inactive" until they remain undetected for K consecutive frames. Inactive tracks can be matched to detections and regain their ID, but not appear in the prior heatmap or output. The tracker stays online. Rebirth only matters for the MOT test set, where we use $K = 32$. For all other experiments, we found rebirth not to be required ($K = 0$).

5.3 Public Detection

The MOT17 challenge only supports public detection. That is, participants are asked to use the provided detections. Public detection is meant to test a tracker's ability to associate objects, irrespective of its ability to detect objects. Our method operates in the private detection mode by default. For the MOT challenge we created a public-detection version of CenterTrack that uses the externally provided (public) detections and is thus fairly compared to other participants in the challenge. This shows that the advantages of CenterTrack are not due to the accuracy of the detections but are due to the tracking framework itself.

Table 2. Evaluation on the KITTI test set. We compare to all published entries on the leaderboard. Runtimes are from the leaderboard. +D means detection time.

	Time (ms)	MOTA ↑	MOTP ↑	MT ↑	ML ↓	IDSW ↓	FRAG ↓
AB3D [45]	4+D	83.84	85.24	66.92	11.38	9	**224**
BeyondPixel [34]	300+D	84.24	85.73	73.23	2.77	468	944
3DT [14]	30+D	84.52	85.64	73.38	2.77	377	847
mmMOT [53]	10+D	84.77	85.21	73.23	2.77	284	753
MOTSFusion [26]	440+D	84.83	85.21	3.08	2.77	275	759
MASS [18]	10+D	85.04	**85.53**	74.31	2.77	301	744
Ours	82	**89.44**	85.05	**82.31**	**2.31**	116	334

Table 3. Evaluation on the nuScenes test set. We compare to the official monocular 3D tracking baseline, which applies a state-of-the-art 3D tracker [45]. We list the average AMOTA@0.2, AMOTA@1, and AMOTP over all 7 categories.

	Time(ms)	AMOTA@0.2 ↑	AMOTA@1 ↑	AMOTP ↓
Mapillary [37] +AB3D [45]	–	6.9	1.8	1.8
Ours	45	**27.8**	**4.6**	**1.5**

Note that refining and rescoring the given bounding boxes is allowed and is commonly used by participants in the challenge [1,19,25]. Following Tracktor [1], we keep the bounding boxes that are close to an existing bounding box in the previous frame. We only initialize a new trajectory if it is near a public detection. All bounding boxes in our results are either near a public detection in the current frame or near a tracked box in the previous frame. The algorithm's diagram of this public-detection configuration can be found in the supplement. We use this public-detection configuration of CenterTrack for MOT17 test set evaluation and use the private-detection setting in our ablation studies.

5.4   Main Results

All three datasets – MOT17 [27], KITTI [12], and nuScenes [3] – host test servers with hidden annotations and leaderboards. We compare to all published results on these leaderboards. The numbers were accessed on Mar. 5th, 2020. We retrain CenterTrack on the full training set with the same hyperparameters in the ablation experiments.

Table 1 lists the results on the MOT17 challenge. We use our public configuration in Sect. 5.3 and do not pretrain on CrowdHuman [33]. CenterTrack significantly outperforms the prior state of the art even when restricted to the public-detection configuration. For example CenterTrack improves MOTA by 5 points (an 8.6% relative improvement) over Tracktor v2 [1].

The public detection setting ensures that all methods build on the same underlying detector. Our gains come from two sources. Firstly, the heatmap input makes our tracker better preserve tracklets from the previous frame, which results in a much lower rate of false negatives. And second, our simple learned offset is effective. (See Sect. 5.6 for

Table 4. Ablation study on MOT17, KITTI, and nuScenes. All results are on validation sets (Sect. 5.1). For each dataset, we report the corresponding official metrics. ↑ indicates that higher is better, ↓ indicates that lower is better.

	MOT17				KITTI				nuScenes	
	MOTA↑	FP↓	FN↓	IDSW↓	MOTA↑	FP↓	FN↓	IDSW↓	AMOTA@0.2↑	AMOTA@1↑
Detection only	63.6	3.5%	30.3%	2.5 %	84.3	4.3%	9.8%	1.5%	18.1	3.4
w/o offset	65.8	**4.5%**	**28.4%**	1.3%	87.1	**5.4%**	**5.8%**	1.6%	17.8	3.6
w/o heatmap	63.9	3.5%	30.3%	2.3%	85.4	4.3%	9.8%	0.4%	26.5	5.9
Ours	**66.1**	**4.5%**	**28.4%**	**1.0%**	**88.7**	**5.4%**	**5.8%**	**0.1%**	**28.3**	**6.8**

more analysis.) For reference, we also included a private detection version, where CenterTrack simultaneously detects and tracks objects (Table 1, bottom). It further improves the MOTA to 67.3%, and runs at 17 FPS end-to-end (including detection).

For IDF1 and id-switch, our local model is not as strong as offline methods such as LSST17 [10], but is better than other online methods [1]. We believe that there is an exciting avenue for future work in combining local trackers (such as our work) with stronger offline long-range models (such as SORT [2], LMP [40], and other ReID-based trackers [49,51]).

On KITTI [12], we submitted our best-performing model with flip testing [55]. The model runs at 82 ms and yields 89.44% MOTA, outperforming all published work (Table 2). Note that our model without flip testing runs at 45 ms with 88.7% MOTA on the validation set (vs. 89.63% with flip testing on the validation set). We avoid submitting to the test server multiple times following their test policy. The results again indicate that CenterTrack performs competitively with more complex methods.

On nuScenes [3], our monocular tracking method achieves an AMOTA@0.2 of 28.3% and an AMOTA@1 of 4.6%, outperforming the monocular baseline [37,45] by a large margin. There are two main reasons. Firstly, we use a stronger and faster 3D detector [55] (see the 3D detector comparison in the supplementary). More importantly, as shown in Table 6, the Kalman-filter-based 3D tracking baseline relies on hand-crafted motion rules [45], which are less effective in low-framerate regimes. Our method learns object motion from data and is much more stable at low framerates.

5.5 Ablation Studies

We first ablate our two main technical contributions: tracking-conditioned detection (Sect. 4.1) and offset prediction (Sect. 4.2) on all three datasets. Specifically, we compare our full framework with three baselines.

Detection Only runs a CenterNet detector at each individual frame and associates their identity only based on 2D center distance. This model does not use video data, but still uses two input images.

Without Offset uses just tracking-conditioned prediction with a predicted offset of zero. Every object is again associated to its closest object in the previous frame.

Table 5. Additional experiments on the MOT17 validation set. From top to bottom: our model, our model trained without simulating heatmap noise, our model trained on static images only, our model with Hungarian matching, and our model with track rebirth.

	MOTA ↑	IDF1 ↑	MT ↑	ML ↓	FP ↓	FN ↓	IDSW ↓
Ours	66.1	64.2	41.3	21.2	4.5%	28.4%	1.0%
w.o. noisy hm	34.4	46.2	26.3	42.2	7.3%	57.4%	0.9%
Static image	66.1	65.4	41.6	19.2	5.4%	27.5%	1.0%
w. Hungarian	66.1	61.0	40.7	20.9	4.5%	28.3%	1.0%
w. rebirth	66.2	69.4	39.5	22.1	3.9%	29.5%	0.4%

Without Heatmap predicts the center offset between frames and uses the updated center distance as the association metric, but the prior heatmap is not provided. The offset-based greedy association is used.

Table 4 shows the results. On all datasets, our full CenterTrack model performs significantly better than the baselines. Tracking-conditioned detection yields ~2% MOTA improvement on MOT and ~ 3% MOTA improvement on KITTI, with or without offset prediction. It produces more false positives but fewer false negatives. This is because with the heatmap prior, the network tends to predict more objects around the previous peaks, which are sometimes misleading. The merits of the heatmap outweigh the limitations and improve MOTA overall. Using the prior heatmap also significantly reduces IDSW on both datasets, indicating that the heatmap stabilizes detection.

Tracking offset prediction gives a huge boost on nuScenes and reduces IDSW consistently in MOT and KITTI. The effectiveness of the tracking offset appears to be related to the video framerate. When the framerate is high, motion between frames is small, and a zero offset is often a reasonable starting point for association. When framerate is low, as in the nuScenes dataset, motion between frames is large and static object association is considerably less effective. Our offset prediction scheme helps deal with such large inter-frame motion. Next, we ablate other components on MOT17.

Training with Noisy Heatmap. The 2nd row in Table 5 shows the importance of injecting noise into heatmaps during training (Sect. 4.3). Without noise injection, the model fails to generalize and yields dramatically lower accuracy. In particular, this model has a large false negative rate. One reason is that in the first frame, the input heatmap is empty. This model had a hard time discovering new objects that were not indicated in the prior heatmap.

Training on Static Images. We train a version of our model on static images only, as described in Sect. 4.4. The results are shown in Table 5 (3rd row, 'Static image'). As reported in this table, training on static images gives the same performance as training on videos on the MOT dataset. Separately, we observed that training on static images is less effective on nuScenes, where framerate is low.

Table 6. Comparing different motion models on MOT17, KITTI, and nuScenes. All results are on validation sets (Sect. 5.1). All experiments on the same dataset are from the same model.

| | MOT17 | | | | KITTI | | | | nuScenes | |
	MOTA↑	FP↓	FN↓	IDSW↓	MOTA↑	FP↓	FN↓	IDSW↓	AMOTA@0.2↑	AMOTA@1↑
No motion	65.8	4.5%	28.4%	1.3%	87.1	5.4%	5.8%	1.6%	17.8	3.6
Kalman filter	66.1	4.5%	28.4%	1.0%	87.9	5.4%	5.8%	0.9%	18.3	3.8
Optical flow	66.1	4.5%	28.4%	1.0%	88.4	5.4%	5.8%	0.4%	26.6	6.2
Ours	66.1	4.5%	28.4%	1.0%	88.7	5.4%	5.8%	0.1%	28.3	6.8

Matching Algorithm. We use a simple greedy matching algorithm based on the detection score, while most other trackers use the Hungarian algorithm. We show the performance of CenterTrack with Hungarian matching in the 4th row of Table 5. It does not improve performance. We choose greedy matching for simplicity.

Track Rebirth. We show CenterTrack with track rebirth ($K = 32$) in the last row of Table 5. While the MOTA performance keeps similar, it significantly increases IDF1 and reduces ID switch. We use this setting for our MOT test set submission. For other datasets and evaluation metrics no rebirth was required ($K = 0$).

5.6 Comparison to Alternative Motion Models

Our offset prediction is able to estimate object motion, but also performs a simple association, as current objects are linked to prior detections, which CenterTrack receives as one of its inputs. To verify the effectiveness of our learned association, we replace our offset prediction with three alternative motion models:

No Motion. We set the offset to zeros. It is copied from Table 4 for reference only.

Kalman Filter. The Kalman filter predicts each object's future state through an explicit motion model estimated from its history. It is the most widely used motion model in traditional real-time trackers [2,45,46]. We use the popular public implementation from SORT [2].

Optical Flow. As an alternative motion model, we use FlowNet2 [15]. The model was trained to estimate dense pixel motion for all objects in a scene. We run the strongest officially released FlowNet2 model (~150 ms/image pair), and replace our learned offset with the predicted optical flow at each predicted object center (Fig. 3).

The results are shown in Table 6. All models use the exact same detector. On the high-framerate MOT17 dataset, any motion model suffices, and even no motion model at all performs competitively. On KITTI and nuScenes, where the intra-frame motions are non-trivial, the hand-crafted motion rule of the Kalman filter performs significantly worse, and even the performance of optical flow degrades. This emphasizes that our offset model does more than just motion estimation. CenterTrack is conditioned on prior detections and can learn to snap offset predictions to exactly those prior detections. Our training procedure strongly encourages this through heavy data augmentation.

Fig. 3. Qualitative results on MOT (1st row), KITTI (2nd row), and nuScenes (3rd and 4th rows). Each row shows three consecutive frames. We show the predicted tracking offset in arrow. Tracks are coded by color. Best viewed on the screen. (Color figure online)

6   Conclusion

We presented an end-to-end simultaneous object detection and tracking framework. Our method takes two frames and a prior heatmap as input, and produces detection and tracking offsets for the current frame. Our tracker is purely local and associates objects greedily through time. It runs online (no knowledge of future frames) and in real time, and sets a new state of the art on the challenging MOT17, KITTI, and nuScenes 3D tracking benchmarks.

Acknowledgements. This work has been supported in part by the National Science Foundation under grant IIS-1845485.

References

1. Bergmann, P., Meinhardt, T., Leal-Taixe, L.: Tracking without bells and whistles. In: ICCV (2019)
2. Bewley, A., Ge, Z., Ott, L., Ramos, F., Upcroft, B.: Simple online and realtime tracking. In: ICIP (2016)
3. Caesar, H., et al.: nuScenes: a multimodal dataset for autonomous driving. In: CVPR (2020)
4. Carreira, J., Agrawal, P., Fragkiadaki, K., Malik, J.: Human pose estimation with iterative error feedback. In: CVPR (2016)

5. Choi, W., Savarese, S.: Multiple target tracking in world coordinate with single, minimally calibrated camera. In: ECCV (2010)
6. Evangelidis, G.D., Psarakis, E.Z.: Parametric image alignment using enhanced correlation coefficient maximization. IEEE Trans. Pattern Anal. Mach. Intell. **30**(10), 1858–1865 (2008)
7. Fang, K., Xiang, Y., Li, X., Savarese, S.: Recurrent autoregressive networks for online multi-object tracking. In: WACV (2018)
8. Feichtenhofer, C., Pinz, A., Zisserman, A.: Detect to track and track to detect. In: ICCV (2017)
9. Felzenszwalb, P.F., Girshick, R.B., McAllester, D., Ramanan, D.: Object detection with discriminatively trained part-based models. In: TPAMI (2009)
10. Feng, W., Hu, Z., Wu, W., Yan, J., Ouyang, W.: Multi-object tracking with multiple cues and switcher-aware classification. arXiv:1901.06129 (2019)
11. Fieraru, M., Khoreva, A., Pishchulin, L., Schiele, B.: Learning to refine human pose estimation. In: CVPR Workshops (2018)
12. Geiger, A., Lenz, P., Urtasun, R.: Are we ready for autonomous driving? The KITTI vision benchmark suite. In: CVPR (2012)
13. Held, D., Thrun, S., Savarese, S.: Learning to track at 100 FPS with deep regression networks. In: ECCV (2016)
14. Hu, H.N., et al.: Joint monocular 3D detection and tracking. In: ICCV (2019)
15. Ilg, E., Mayer, N., Saikia, T., Keuper, M., Dosovitskiy, A., Brox, T.: FlowNet 2.0: evolution of optical flow estimation with deep networks. In: CVPR (2017)
16. Kang, K., et al.: Object detection in videos with tubelet proposal networks. In: CVPR (2017)
17. Kang, K., et al.: T-CNN: tubelets with convolutional neural networks for object detection from videos. Circuits Syst. Video Technol. **28**(10), 2896–2907 (2017)
18. Karunasekera, H., Wang, H., Zhang, H.: Multiple object tracking with attention to appearance, structure, motion and size. IEEE Access **7**, 104423–104434 (2019)
19. Keuper, M., Tang, S., Andres, B., Brox, T., Schiele, B.: Motion segmentation and multiple object tracking by correlation co-clustering. IEEE Trans. Pattern Anal. Mach. Intell. **42**(1), 140–153 (2018)
20. Kingma, D.P., Ba, J.: Adam: a method for stochastic optimization. In: ICLR (2015)
21. Law, H., Deng, J.: CornerNet: detecting objects as paired keypoints. In: ECCV (2018)
22. Leal-Taixé, L., Canton-Ferrer, C., Schindler, K.: Learning by tracking: Siamese CNN for robust target association. In: CVPR Workshops (2016)
23. Leal-Taixé, L., Milan, A., Schindler, K., Cremers, D., Reid, I., Roth, S.: Tracking the trackers: an analysis of the state of the art in multiple object tracking. arXiv:1704.02781 (2017)
24. Lin, T.Y., Goyal, P., Girshick, R., He, K., Dollár, P.: Focal loss for dense object detection. In: ICCV (2017)
25. Long, C., Haizhou, A., Zijie, Z., Chong, S.: Real-time multiple people tracking with deeply learned candidate selection and person re-identification. In: ICME (2018)
26. Luiten, J., Fischer, T., Leibe, B.: Track to reconstruct and reconstruct to track. arXiv:1910.00130 (2019)
27. Milan, A., Leal-Taixé, L., Reid, I., Roth, S., Schindler, K.: MOT16: a benchmark for multi-object tracking. arXiv:1603.00831 (2016)
28. Moon, G., Chang, J., Lee, K.M.: PoseFix: model-agnostic general human pose refinement network. In: CVPR (2019)
29. Ren, J., et al.: Accurate single stage detector using recurrent rolling convolution. In: CVPR (2017)
30. Ren, S., He, K., Girshick, R., Sun, J.: Faster R-CNN: towards real-time object detection with region proposal networks. In: NIPS (2015)
31. Sadeghian, A., Alahi, A., Savarese, S.: Tracking the untrackable: learning to track multiple cues with long-term dependencies. In: ICCV (2017)

32. Schulter, S., Vernaza, P., Choi, W., Chandraker, M.: Deep network flow for multi-object tracking. In: CVPR (2017)
33. Shao, S., et al.: CrowdHuman: a benchmark for detecting human in a crowd. arXiv:1805.00123 (2018)
34. Sharma, S., Ansari, J.A., Murthy, J.K., Krishna, K.M.: Beyond pixels: leveraging geometry and shape cues for online multi-object tracking. In: ICRA (2018)
35. Shi, J., Tomasi, C.: Good features to track. In: CVPR (1994)
36. Shi, S., Wang, X., Li, H.: PointRCNN: 3D object proposal generation and detection from point cloud. In: CVPR (2019)
37. Simonelli, A., Bulò, S.R.R., Porzi, L., López-Antequera, M., Kontschieder, P.: Disentangling monocular 3D object detection. In: ICCV (2019)
38. Son, J., Baek, M., Cho, M., Han, B.: Multi-object tracking with quadruplet convolutional neural networks. In: CVPR (2017)
39. Stiefelhagen, R., Bernardin, K., Bowers, R., Garofolo, J., Mostefa, D., Soundararajan, P.: The CLEAR 2006 evaluation. In: Stiefelhagen, R., Garofolo, J. (eds.) CLEAR 2006. LNCS, vol. 4122, pp. 1–44. Springer, Heidelberg (2007). https://doi.org/10.1007/978-3-540-69568-4_1
40. Tang, S., Andriluka, M., Andres, B., Schiele, B.: Multiple people tracking by lifted multicut and person re-identification. In: CVPR (2017)
41. Tian, Z., Shen, C., Chen, H., He, T.: FCOS: fully convolutional one-stage object detection. In: ICCV (2019)
42. Tomasi, C., Kanade, T.: Detection and tracking of point features. Technical report CMU-CS-91-132, Carnegie Mellon University (1991)
43. Tu, Z.: Auto-context and its application to high-level vision tasks. In: CVPR (2008)
44. Voigtlaender, P., et al.: MOTS: multi-object tracking and segmentation. In: CVPR (2019)
45. Weng, X., Kitani, K.: A baseline for 3D multi-object tracking. arXiv:1907.03961 (2019)
46. Wojke, N., Bewley, A., Paulus, D.: Simple online and realtime tracking with a deep association metric. In: ICIP (2017)
47. Xiang, Y., Alahi, A., Savarese, S.: Learning to track: online multi-object tracking by decision making. In: ICCV (2015)
48. Xiao, B., Wu, H., Wei, Y.: Simple baselines for human pose estimation and tracking. In: ECCV (2018)
49. Xu, J., Cao, Y., Zhang, Z., Hu, H.: Spatial-temporal relation networks for multi-object tracking. In: ICCV (2019)
50. Yang, F., Choi, W., Lin, Y.: Exploit all the layers: fast and accurate CNN object detector with scale dependent pooling and cascaded rejection classifiers. In: CVPR (2016)
51. Yu, F., Li, W., Li, Q., Liu, Y., Shi, X., Yan, J.: POI: multiple object tracking with high performance detection and appearance feature. In: ECCV Workshops (2016)
52. Yu, F., Wang, D., Shelhamer, E., Darrell, T.: Deep layer aggregation. In: CVPR (2018)
53. Zhang, W., Zhou, H., Sun, S., Wang, Z., Shi, J., Loy, C.C.: Robust multi-modality multi-object tracking. In: ICCV (2019)
54. Zhang, Z., Cheng, D., Zhu, X., Lin, S., Dai, J.: Integrated object detection and tracking with tracklet-conditioned detection. arXiv:1811.11167 (2018)
55. Zhou, X., Wang, D., Krähenbühl, P.: Objects as points. arXiv:1904.07850 (2019)
56. Zhu, B., Jiang, Z., Zhou, X., Li, Z., Yu, G.: Class-balanced grouping and sampling for point cloud 3D object detection. arXiv:1908.09492 (2019)
57. Zhu, J., Yang, H., Liu, N., Kim, M., Zhang, W., Yang, M.H.: Online multi-object tracking with dual matching attention networks. In: ECCV (2018)
58. Zhu, X., Wang, Y., Dai, J., Yuan, L., Wei, Y.: Flow-guided feature aggregation for video object detection. In: ICCV (2017)

CPGAN: Content-Parsing Generative Adversarial Networks for Text-to-Image Synthesis

Jiadong Liang[1], Wenjie Pei[2], and Feng Lu[1,3]([✉])

[1] State Key Laboratory of VR Technology and Systems, School of CSE,
Beihang University, Beijing, China
lufeng@buaa.edu.cn
[2] Harbin Institute of Technology, Shenzhen, China
[3] Peng Cheng Laboratory, Shenzhen, China

Abstract. Typical methods for text-to-image synthesis seek to design effective generative architecture to model the text-to-image mapping directly. It is fairly arduous due to the cross-modality translation. In this paper we circumvent this problem by focusing on parsing the content of both the input text and the synthesized image thoroughly to model the text-to-image consistency in the semantic level. Particularly, we design a memory structure to parse the textual content by exploring semantic correspondence between each word in the vocabulary to its various visual contexts across relevant images during text encoding. Meanwhile, the synthesized image is parsed to learn its semantics in an object-aware manner. Moreover, we customize a conditional discriminator to model the fine-grained correlations between words and image sub-regions to push for the text-image semantic alignment. Extensive experiments on COCO dataset manifest that our model advances the state-of-the-art performance significantly (from **35.69** to **52.73** in Inception Score).

Keywords: Text-to-image synthesis · Content-Parsing · Generative Adversarial Networks · Memory structure · Cross-modality

1 Introduction

Text-to-image synthesis aims to generate an image according to a textual description. The synthesized image is expected to be not only photo-realistic but also consistent with the description in the semantic level. It has various potential applications such as artistic creation and interactive entertainment. Text-to-image synthesis is more challenging than other tasks of conditional

---

J. Liang and W. Pei—Contributed equally.

Electronic supplementary material The online version of this chapter (https://doi.org/10.1007/978-3-030-58548-8_29) contains supplementary material, which is available to authorized users.

© Springer Nature Switzerland AG 2020
A. Vedaldi et al. (Eds.): ECCV 2020, LNCS 12349, pp. 491–508, 2020.
https://doi.org/10.1007/978-3-030-58548-8_29

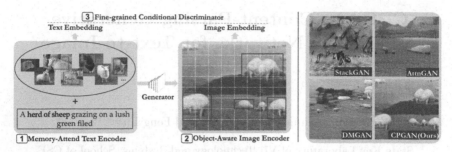

Fig. 1. Our model parses the input text by a customized memory-attended mechanism and parses the synthesized image in an object-aware manner. Besides, the proposed Fine-grained Conditional Discriminator is designed to push for the text-image alignment in the semantic level. Consequently, our CPGAN is able to generate more realistic and more consistent image than other methods.

image synthesis like label-conditioned synthesis [29] or image-to-image translation [13]. On one hand, the given text contains much more descriptive information than a label, which implies more conditional constraints for image synthesis. On the other hand, the task involves cross-modality translation which is more complicated than image-to-image translation. Most existing methods [4,9–11,17,32,34,39,43,44,49,50,52], for text-to-image synthesis are built upon the GANs [8], which has been validated its effectiveness in various tasks on image synthesis [2,25,48]. A pivotal example is StackGAN [49] which is proposed to synthesize images iteratively in a coarse-to-fine framework by employing stacked GANs. Subsequently, many follow-up works focus on refining this generative architecture either by introducing the attention mechanism [43,52] or modeling an intermediate representation to smoothly bridge the input text and generated image [10,11,17]. Whilst substantial progress has been made by these methods, one potential limitation is that these methods seek to model the text-to-image mapping directly during generative process which is fairly arduous for such cross-modality translation. Consider the example in Fig. 1, both StackGAN and AttnGAN can hardly correspond the word 'sheep' to an intact visual picture for a sheep correctly. It is feasible to model the text-to-image consistency more explicitly in the semantic level, which however requires thorough understanding for both text and image modalities. Nevertheless, little attention is paid by these methods to parsing content semantically for either the input text or the generated image. Recently this limitation is investigated by SD-GAN [44], which leverages the Siamese structure in the discriminator to learn semantic consistency between two textual descriptions. However, direct content-oriented parsing in the semantic level for both input text and the generated image is not performed in depth.

In this paper we focus on parsing the content of both the input text and the synthesized image thoroughly and thereby modeling the semantic correspondence between them. On the side of text modality, we design a memory mechanism to parse the textual content by capturing the various visual context information across relevant images in the training data for each word in the vocabulary. On the side of image modality, we propose to encode the generated image in an object-aware manner to extract the visual semantics. The obtained text embeddings and the image embeddings are then utilized to measure the text-image consistency in the semantic space. Besides, we also design a conditional discriminator to push for the semantic text-image alignment by modeling the fine-grained correlations locally between words and image sub-regions. Thus, a full-spectrum content parsing is performed by the resulting model, which we refer to as Content-Parsing Generative Adversarial Networks (CPGAN), to better align the input text and the generated image semantically and thereby improve the performance of text-to-image synthesis. Going back to the example in Fig. 1, our CPGAN successfully translates the textual description 'a herd of sheep grazing on a greed field' to a correct visual scene, which is more realistic than the generated results of other methods. We evaluate the performance of our CPGAN on COCO dataset both quantitatively and qualitatively, demonstrating that CPGAN pushes forward the state-of-the-art performance by a significant step. Moreover, the human evaluation performed on a randomly selected subset from COCO test set consistently shows that our model outperforms other two methods (StackGAN and AttnGAN). To conclude, the idea of our CPGAN to **parse the content on both the text side (by MATE, in** Sect. 3.2) **and the image side (by OAIE, in** Sect. 3.3) is novel, which tackles the cross-modality semantic alignment problem effectively and clearly distinguishes our CPGAN from existing methods. Along with a customized fine-grained conditional discriminator (FGCD, in Sect. 3.4), the CPGAN pushes forward the state-of-the-art performance significantly, from 35.69 to 52.73 in Inception Score.

2   Related Work

Text-to-Image Synthesis. Text-to-image synthesis was initially investigated based on pixelCNN [35,37], which suffers from highly computational cost during the inference phase. Meanwhile, the variational autoencoder (VAE) [23] was applied to text-to-image synthesis. A potential drawback of VAE-based synthesis methods is that the generated images by VAE tend to be blurry presumably. This limitation is largely mitigated by the GANs [8], which was promptly extended to various generative tasks in computer vision [2,3,13,19,20,25,28,48,51]. After Reed [34] made the first attempt to apply GAN to text-to-image synthesis, many follow-up works [10,11,17,32,43,49,50,52] focus on improving the generative architecture of GAN to refine the quality of generated images. A well-known example is StackGAN [49,50], which proposes to synthesize images in a coarse-to-fine framework. Following StackGAN, AttnGAN [43] introduces the attention mechanism which was widely used in computer vision tasks [21,45] to this framework. DMGAN [52] further refines the attention mechanism by utilizing a memory scheme. MirrorGAN [32] develops a text-to-image-to-text cycle framework to encourage text-image consistency. Another interesting line of research introduces an intermediate representation as a smooth bridge between the input text

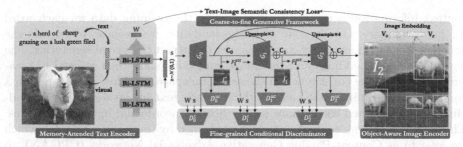

Fig. 2. Architecture of the proposed CPGAN. It follows the coarse-to-fine generative framework. We customize three components to perform content parsing: memory-attended text encoder for text, object-aware image encoder for image, and fine-graind conditional discriminator for the text-image alignment.

and the synthesized image [10,11,17,46]. To improve the semantic consistency between the generated image and the input text, ControlGAN [16] applies the matching scheme of DAMSM in AttnGAN [43] in all 3-level discriminators. In contrast, our Fine-Grained Conditional Discriminator (FGCD) proposes a novel discriminator structure to capture the local semantic correlations between each caption word and image regions. Whist these methods have brought about substantial progress, they seek to model the text-to-image mapping directly during generative process. Unlike these methods, we focus on content-oriented parsing of both text and image to obtain a thorough understanding of involved multimodal information. Recently Siamese network is leveraged to explore the semantic consistence either between two textual descriptions by SD-GAN [44] or two images by SEGAN [39]. LeicaGAN [31] adopts text-visual co-embeddings to replace input text with corresponding visual features. Lao et al. [15] parses the input text by learning two variables that are disentangled in the latent space. Text-SeGAN [5] focuses on devising a specific discriminator to regress the semantic relevance between text and image. CKD [47] parses the image content by a hierarchical semantic representation to enhance the semantic consistency and visual quality of synthesized images. However, deep content parsing in the semantic level for both text and image modalities is not performed.

Memory Mechanism. Memory networks were first proposed to tackle the limited memory of recurrent networks [14,38]. It was then extensively applied in tasks of natural language processing (NLP) [6,7,24,41] and computer vision (CV) [22,26,30]. Different from the initial motivation of memory networks that is to enlarge the modeling memory, we design a specific memory mechanism to build the semantic correspondence between a word to all its relevant visual features across training data during text parsing.

3   Content-Parsing Generative Adversarial Networks

The proposed Content-Parsing Generative Model for text-to-image synthesis focuses on parsing the involved multimodal information by three customized components. To be specific, the Memory-Attended Text Encoder employs the memory structure to explore the semantic correspondence between a word and its various visual contexts; the Object-Aware Image Encoder is designed to parse the generated image in the semantic level; the Fine-grained Conditional Discriminator is proposed to measure the consistency between the input text and the generated image for guiding optimization of the whole model. We will first present the overall architecture of the proposed CPGAN illustrated in Fig. 2, which follows the coarse-to-fine generative framework, then we will elaborate on the three aforementioned components specifically designed for content parsing.

3.1   Coarse-to-fine Generative Framework

Our proposed model synthesizes the output image from the given textual description in the classical coarse-to-fine framework, which has been extensively shown to be effective in generative tasks [17, 32, 43, 44, 49, 50]. As illustrated in Fig. 2, the input text is parsed by our Memory-Attended Text Encoder and the resulting text embedding is further fed into three cascaded generators to obtain coarse-to-fine synthesized images. Two different types of loss functions are employed to optimize the whole model jointly: 1) Generative Adversarial Losses to push the generated image to be realistic and meanwhile match the descriptive text by training adversarial discriminators and 2) Text-Image Semantic Consistency Loss to encourage the text-image alignment in the semantic level.

Formally, given a textual description X containing T words, the parsed text embeddings by the Memory-Attended Text Encoder (Sect. 3.2) are denoted as: $\mathbf{W}, \mathbf{s} = \text{TextEnc}(X)$. Herein $\mathbf{W} = \{\mathbf{w}_1, \mathbf{w}_2, \ldots \mathbf{w}_T\}$ consists of embeddings of T words in which $\mathbf{w}_t \in \mathbb{R}^d$ denotes the embedding for the t-th word. $\mathbf{s} \in \mathbb{R}^d$ is the global embedding for the whole sentence. Three cascaded generators $\{G_0, G_1, G_2\}$ are then employed to sequentially synthesize coarse-to-fine images $\{\tilde{I}_0, \tilde{I}_1, \tilde{I}_2\}$. We apply similar structure as Generative Network in AttnGAN [43]:

$$\tilde{I}_0, \mathbf{C_0} = G_0(\mathbf{z}, \mathbf{s}), \qquad \tilde{I}_\mathbf{i}, \mathbf{C_i} = G_i(\mathbf{C_{i-1}}, F_i^{att}(\mathbf{W}, \mathbf{C_{i-1}})), i = 1, 2, \qquad (1)$$

where $\mathbf{C}_i$ are the generated intermediate feature maps by G_i and F_i^{att} is an attention model designed to attend to the word embeddings $\mathbf{W}$ to each pixel of $\mathbf{C_{i-1}}$ in i-th generation stage. Note that the first-stage generator G_0 takes as input the noise vector $\mathbf{z}$ sampled from a standard Gaussian distribution to introduce the randomness. In practice, F_i^{att} and G_i are modeled as convolutional neural networks (CNNs), which are elaborated in the supplementary material. Different from AttnGAN, we introduce extra residual connection from $\mathbf{C_0}$ to $\mathbf{C_1}$ and $\mathbf{C_2}$ (via up-sampling) to ease the information propagation between generators.

To optimize the whole model, the generative adversarial losses are utilized by training generators and the corresponding discriminators alternately. In particular, we train two discriminators for each generative stage: 1) an unconditional discriminator D^{uc} to push the synthesized image to be realistic and 2) a conditional discriminator D^c to align the synthesized image and the input text. The generators are trained by minimizing following adversarial losses:

$$\mathcal{L}_G = \sum_{i=0}^{2} \mathcal{L}_{G_i}, \quad \mathcal{L}_{G_i} = -\frac{1}{2}\mathbb{E}_{\widetilde{I}_i \sim p_{G_i}} D_i^{uc}(\widetilde{I}_i) - \frac{1}{2}\mathbb{E}_{\widetilde{I}_i \sim p_{G_i}} D_i^c(\widetilde{I}_i, X). \quad (2)$$

Accordingly, the adversarial loss for the corresponding discriminators in the i-th generative stage is defined as:

$$\mathcal{L}_{D_i} = \frac{1}{2}\mathbb{E}_{I_i \sim p_{\text{data}_i}}[\max(0, 1 - D_i^{uc}(I_i))] + \frac{1}{3}\mathbb{E}_{\widetilde{I}_i \sim p_{G_i}}[\max(0, 1 + D_i^{uc}(\widetilde{I}_i))]$$

$$+ \frac{1}{2}\mathbb{E}_{I_i \sim p_{\text{data}_i}}[\max(0, 1 - D_i^c(I_i, X))] + \frac{1}{3}\mathbb{E}_{\widetilde{I}_i \sim p_{G_i}}[\max(0, 1 + D_i^c(\widetilde{I}_i, X))]$$

$$+ \frac{1}{3}\mathbb{E}_{I_i \sim p_{\text{data}_i}}[\max(0, 1 + D_i^c(I_i, \overline{X}))], \quad (3)$$

where X is the input descriptive text and I_i is the corresponding groudtruth image for the i-th generative stage. The negative pairs $(I_i, \overline{X})$ are also involved to improve the training robustness. Note that we formulate the adversarial losses in the form of Hinge loss rather than the negative log-likelihood due to the empirical superior performance of Hinge loss [25,48]. The modeling of unconditional discriminator D_i^{uc} is straightforward by CNNs (check supplementary material for details), it is however non-trivial to design an effective conditional discriminator D_i^c. For this reason, we propose the Fine-grained Conditional Discriminator in Sect. 3.4. While the adversarial losses in Eqs. 2, 3 push for the text-image consistency in an adversarial manner by the conditional discriminator, Text-Image Semantic Consistency Loss (TISCL) is proposed to optimize the semantic consistency directly. Specifically, the synthesized image and the input text are encoded respectively, then the obtained image embedding and the text embedding are projected to the same latent space to measure their consistency. We adopt DAMSM [43] to compute the non-matching loss between a textual description X and the corresponding image $\widetilde{I}$:

$$\mathcal{L}_{\text{TISCL}}(\widetilde{I}, X) = \mathcal{L}_{\text{DAMSM}}(\text{ImageEnc}(\widetilde{I}), \text{TextEnc}(X)). \quad (4)$$

The key difference between our TISCL and DAMSM lies in encoding mechanisms for both input text (TextEnc) and the synthesized image (ImageEnc). Our proposed Memory-Attended Text Encoder and Object-Aware Image Encoder focus on 1) distilling the underlying semantic information contained in text and image, and 2) capturing the semantic correspondence between them. We will discuss these two encoders in subsequent sections concretely.

3.2 Memory-Attended Text Encoder

The Memory-Attended Text Encoder is designed to parse the input text and learn meaningful text embeddings for downstream generators to synthesize realistic images. A potential challenge during text encoding is that a word may

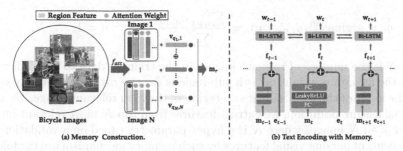

Fig. 3. (a) The memory $\mathbf{m}_r$ is constructed by considering salient regions from all relevant images across training data. (b) The learned memory and the word embedding are fused via LSTM structure to incorporate temporal information.

have multiple (similar but not identical) visual context information and correspond to more than one relevant images in training data. Typical text encoding methods which encode the text online during training can only focus on the text-image correspondence of the current training pair. Our Memory-Attended Text Encoder aims to capture full semantic correspondence between a word to various visual contexts from all its relevant images across training data. Thus, our model can achieve more comprehensive understanding for each word in the vocabulary and synthesize images of higher quality with more diversity.

Memory Construction. The memory is constructed as a mapping structure, wherein each item maps a word to its visual context representation. To learn the meaningful visual features from each relevant image for a given word, we detect salient regions in each image to the word and extract features from them. There are many ways to achieve this goal. We resort to existing models for image captioning, which is the sibling task of text-to-image synthesis, since we can readily leverage the capability of image-text modeling. In particular, we opt for the Bottom-Up and Top-Down (BUTD) Attention model [1] which extracts the salient visual features for each word in a caption at the level of objects.

Specifically, given an image-text pair $\langle I, X \rangle$, object detection is first performed on image I by pretrained Yolo-V3 [33] to select top-36 sub-regions (indicated by bounding boxes) w.r.t. the confidence score and the extracted features are denoted as $\mathbf{V} = \{\mathbf{v}_1, \mathbf{v}_2, \ldots, \mathbf{v}_{36}\}$. Note that we replace the Faster R-CNN with Yolo-V3 for object detection for computational efficiency. Then the pretrained BUTD Attention model is employed to measure the salience of each of 36 sub-regions for each word in the caption (text) X based on attention mechanism. In practice we only retain the visual feature of the most salient sub-region from each relevant image. Since a word may correspond to multiple relevant images, we extract salient visual features for each of the images the word is involved in. As shown in Fig. 3(a), the visual context features in the memory $\mathbf{m}_r$ for the r-th word in the vocabulary is modeled as the weighted average feature:

$$q_n = \mathrm{argmax}_{i=1}^{36} a_{i,n}, \quad \mathbf{m}_r = \frac{\sum_{n=1}^{N} a_{q_n,n} \mathbf{v}_{q_n,n}}{\sum_{n=1}^{N} a_{q_n,n}}, \quad n = 1, \ldots, N, \qquad (5)$$

where N is the number of relevant images in the training data to the r-th word; $a_{i,n}$ is the attention weight on i-th sub-regions for the n-th relevant image and q_n is the index of the most salient sub-region of the n-th relevant image. To avoid potential feature pollution, we extract features from top-K most relevant images instead of all N images where K is a hyper-parameter tuned on a validation set. The benefits of parsing visual features by such memory mechanism are twofold: 1) extract precise semantic features from the most salient region of relevant images for each word; 2) capture full semantic correspondence between a word to its various visual contexts. It is worth mentioning that both Yolo-V3 and BUTD Attention model are pretrained on MSCOCO dataset [18] which is also used for text-to-image synthesis, hence we do not utilize extra data in our method.

Text Encoding with Memory. Apart from the learned memory which parses the text from visual context information, we also encode the text by learning latent embedding directly for each word in the vocabulary to characterize the semantic distance among all words. To be specific, we aim to learn an embedding matrix $\mathbf{E} \in \mathbb{R}^{d \times K}$ consisting of d-dim embeddings for in total K words in the vocabulary. The learned word embedding $\mathbf{e}_i = \mathbf{E}[:, i]$ for the i-th word in the vocabulary is then fused with the learned memory $\mathbf{m}_i$ by concatenation: $\mathbf{f}_i = [\mathbf{e}_i; p(\mathbf{m}_i)]$, where $p(\mathbf{m}_i)$ is a nonlinear projection function to balance the feature dimensions between $\mathbf{m}_i$ and $\mathbf{e}_i$. In practice, we perform p by two fully-connected layers with a LeakReLU layer [42] in between, as illustrated in Fig. 3(b).

Given a textual description X containing T words, we employ a Bi-LSTM [12] structure to obtain final word embedding for each time step, which incorporates the temporal dependencies between words: $\mathbf{W}, \mathbf{s} = \text{Bi-LSTM}(\mathbf{f}_1, \mathbf{f}_2, ..., \mathbf{f}_T)$. Herein, $\mathbf{W} = \{\mathbf{w}_1, \mathbf{w}_2, ... \mathbf{w}_T\}$ consists of the embeddings of T words. We use $\mathbf{w}_T$ as the sentence embedding $\mathbf{s}$.

3.3 Object-Aware Image Encoder

The Object-Aware Image Encoder is proposed to parse the synthesized image by our generator in the semantic level. The obtained image-encoded features are prepared for the proposed TISCL (Eq. 4) to guide the optimization of the whole model by minimizing the semantic discrepancy between the input text and the synthesized image. Thus, the quality of the parsed image features are crucial to the performance of image synthesis by our model.

Besides learning global features of the whole image, typical way of attending to local image features is to extract features from equally-partitioned image sub-regions [43]. We propose to parse the image in object level to extract more physically-meaningful features. In particular, we employ Yolo-V3 (pretrained on MSCOCO) to detect salient bounding boxes with top confidence of object detection and learn features from them, which is exactly same as the corresponding

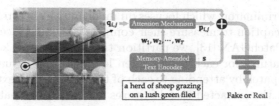

Fig. 4. The structure of Fine-grained Conditional Discriminator.

operations by Yolo-V3 in the section of memory construction 3.2. Formally, we extract visual features (1024-dim) of top 36 bounding boxes by Yolo-V3 for a given image I, denoted as $\mathbf{V}_o \in \mathbb{R}^{1024 \times 36}$. Another benefit of parsing images in object level is that it is consistent with our Memory-Attended Text Encoder, which parses text based on visual context information in object level.

The synthesized image in the early stage of training process cannot be sufficiently meaningful for performing object (salience) detection by Yolo-V3, which would adversely affects the image encoding quality. Hence, we also incorporate local features extracted from equally-partitioned sub-regions (8×8 in our implementation) like AttnGAN [43], which is denoted as $\mathbf{V}_e \in \mathbb{R}^{768 \times 64}$. This kind of two-pronged image encoding scheme is illustrated in Fig. 2.

Two kinds of extracted features $\mathbf{V}_o$ and $\mathbf{V}_e$ are then projected into latent spaces with the same dimension by linear transformation and concatenated together to derive the final image encoding features $\mathbf{V}_c$:

$$\mathbf{V}'_o = \mathbf{M}_o \mathbf{V}_o + \mathbf{b}_o, \quad \mathbf{V}'_e = \mathbf{M}_e \mathbf{V}_e + \mathbf{b}_e, \quad \mathbf{V}_c = [\mathbf{V}'_o; \mathbf{V}'_e], \qquad (6)$$

where $\mathbf{M}_o \in \mathbb{R}^{256 \times 1024}$ and $\mathbf{M}_e \in \mathbb{R}^{256 \times 768}$ are transformation matrices. The obtained image encoding feature $\mathbf{V}_c \in \mathbb{R}^{256 \times 100}$ is further fed into the DAMSM in Eq. 4 to compute the TISCL by measuring the maximal semantic consistency between each word of the input text and different sub-region of the image by attention mechanism[1].

3.4   Fine-Grained Conditional Discriminator

Conditional discriminator is utilized to distinguish whether a textual caption matches the image in a pair, thus to push the semantic alignment between the synthesized image and the input text by the corresponding adversarial loss. Typical way of designing conditional discriminator is to extract a feature embedding from the text and the image respectively, and then train a discriminator directly on the aggregated features. A potential limitation of such method is that only the global compatibility between the text and the image is considered whereas the local correlations between a word in the text and a sub-region of the image are not explored. Nevertheless, most salient correlations between an image and a caption are always reflected locally. To this end, we propose the Fine-grained

---

[1] Details are provided in the supplementary file.

Conditional Discriminator, which focuses on modeling local correlations between an image and a caption to measure their compatibility more accurately.

Inspired by PatchGAN [13], we partition the image into $N \times N$ patches and extract visual features for each patch. Then learn the contextual features from the text for each patch by attending to each of the word in the text. As illustrated in Fig. 4, suppose the extracted visual features for the (i, j)-th patch in the image are denoted as $\mathbf{q}_{i,j}, i, j \in 1, 2, \ldots, N$ and the word features in the text extracted by our text encoder are denoted as $\mathbf{W} = \{\mathbf{w}_1, \mathbf{w}_2, \cdots, \mathbf{w}_T\}$. We compute the contextual features for the (i, j)-th patch by attention mechanism:

$$a_n = \frac{\exp(\mathbf{q}_{i,j}^\top \mathbf{w}_n)}{\sum_{k=1}^T \exp(\mathbf{q}_{i,j}^\top \mathbf{w}_k)}, \quad \mathbf{p}_{i,j} = \sum_{k=1}^T a_k \mathbf{w}_k, \quad n = 1, 2, \ldots, T \qquad (7)$$

where a_n is the attention weight for n-th word in the text. The obtained contextual feature $\mathbf{p}_{i,j}$ is concatenated together with the visual feature $\mathbf{q}_{i,j}$ as well as the sentence embedding $\mathbf{s}$ for the discrimination to be real for fake. Note that the patch size (or the value of N) should be tuned to balance between capturing fine-grained local correlations and global text-image correlations.

4   Experiments

To evaluate the performance of CPGAN, we conduct experiments on COCO dataset [18] which is a widely used benchmark of text-to-image synthesis.

4.1   Experimental Setup

Dataset. Following the official 2014 data splits, COCO dataset contains 82,783 images for training and 40,504 images for validation. Each image has 5 corresponding textual descriptions by human annotation. Note that CUB [40] and Oxford-102 [27] are not selected since they are too easy (only one object is contained per image) to fully explore the potential of our model.

Evaluation Metrics. We adopt three metrics for quantitative evaluation: Inception score [36], R-precision [43] and SOA [10]. Inception score is extensively used to evaluate the quality of synthesized images taking into account both the authenticity and diversity of images. R-precision is used to measure the semantic consistency between the textual description and the synthesized image. SOA adopts a pre-trained object detection network to measure whether the objects specifically mentioned in the caption are recognizable in the generated images. Specifically, it includes two sub-metrics: SOA-C (average recall w.r.t. object category) and SOA-I (average recall w.r.t. image sample), which are defined as:

$$\text{SOA-C} = \frac{1}{|C|} \sum_{c \in C} \frac{1}{|I_c|} \sum_{i_c \in I_c} \text{Det}(i_c), \quad \text{SOA-I} = \frac{1}{\sum_{c \in C} |I_c|} \sum_{c \in C} \sum_{i_c \in I_c} \text{Det}(i_c), \quad (8)$$

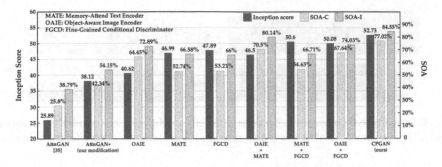

Fig. 5. Performance of ablation study both in Inception score and SOA.

where C and I_C refer to the set of categories and set of images in the category c respectively. $\mathrm{Det}(i_c) \in \{0, 1\}$ indicates whether the pre-trained detector successfully recognizes an object corresponding to class c in the image i_c.

Implementation Details. Our model is designed based on AttnGAN [43], hence AttnGAN is an important baseline to evaluate our model. We make several minor technical improvements over AttnGAN, which yield much performance gain. Specifically, we replace the binary cross-entropy function for adversarial loss with hinge-loss form. Besides, we adopt truncated Gaussian noise [2] as input noise for synthesis ($\mathbf{z}$ in Eq. 1). We observe that larger batch size in the training process can also lead to better performance. In our implementation, we use batch size of 72 samples instead of 14 samples in AttnGAN. Finally, the hyper-parameters in AttnGAN are carefully tuned. We call the resulting version based on these improvements as AttnGAN$^+$.

4.2   Ablation Study

We first conduct experiments to investigate the effectiveness of our proposed three modules respectively, i.e., Memory-Attended Text Encoder (MATE), Object-Aware Image Encoder (OAIE) and Fine-Grained Conditional Discriminator (FGCD). To this end, we perform ablation experiments which begins with AttnGAN$^+$, and then incrementally augments the text-to-image synthesis system with three modules. Figure 5 presents the performance measured by Inception score and SOA of all ablation experiments.

AttnGAN$^+$ Versus AttnGAN. It is shown in Fig. 5 that AttnGAN$^+$ performs much better than original AttnGAN, which benefits from the aforementioned technical improvements. We observe that increasing the batch size (from 14 to 72) during training brings about the largest performance gain (around 7 points in Inception score). Additionally, fine-tuning the hyper-parameters contributes another 4 points of improvement in Inception score to the performance. Besides,

Fig. 6. Qualitative comparison between different modules of our model for ablation study, the results of AttnGAN are also provided for reference. (Color figure online)

the substantial performance gains in SOA show that AttnGAN$^+$ could synthesis images containing more recognizable objects than AttnGAN.

Effect of Single Module. Equipping the system with each of three proposed modules individually boosts the performance substantially. Compared to AttnGAN$^+$, the performance is improved by 8.9 points, 2.5 points, and 9.8 points by MATE, OAIE and FGCD respectively in Inception score. SOA evaluation results also show large improvements by each of three modules. It is worth noting that OAIE performs best among three modules on SOA metrics emphasizing more on object-level semantics in synthesized images, which in turn validates that OAIE could effectively parse the image in object level. These improvements demonstrate the effectiveness of all three modules. Whilst sharing same generators with AttnGAN$^+$, all our three modules focus on parsing the content of the input text or the synthesized image. Therefore, it is implied that deeper semantic content parsing for the text by the memory-based mechanism helps the downstream generators to understand the input text more precisely. On the other hand, our OAIE encourages generators to generate more consistent images with the input text in object level under the guidance of our TISCL. Besides, FGCD steers the optimization of generators to achieve better alignment between the text and the image by the corresponding adversarial losses.

Effect of Combined Modules. We then combine every two of three modules together to further augment the text-to-image synthesis system. The experimental results in Fig. 5 indicate that the performances in Inception score are further enhanced compared to the results of single-module cases with the exception of MATE + OAIE. We surmise that this is because MATE performs similar operations as OAIE when learning the visual context information from images in the object level for each word in the vocabulary. Nevertheless, OAIE still advances the performances after being mounted over the single FGCD or MATE + FGCD. It can be observed that combined modules also perform much better than the corresponding single module on SOA metrics. Employing all three modules leads to our full CPGAN model and achieves the best performance in all metrics, which is better than all other single-module or double-module cases.

Qualitative Evaluation. To gain more insight into effectiveness of our three modules, we visualize the synthesized images for several examples by systems equipped with different modules and the baseline AttnGAN. Figure 6 presents the qualitative comparison. Compared to AttnGAN, the synthesized images by each of our three modules are more realistic and more consistent with the input text, which again reveals advantages of our proposed modules over AttnGAN. Benefiting from the content-oriented parsing mechanisms, our modules tend to generate more intact and realistic objects corresponding to the meaningful words in the input text, which are indicated with red or green arrows.

4.3 Comparison with State-of-the-arts

In this set of experiments, we compare our model with the state-of-the-art methods for text-to-image synthesis on COCO dataset.

Table 1. Performance of different text-to-image synthesis models on COCO dataset in terms of Inception score, R-precision SOA-C, SOA-I and model size.

Model	Inception score	R-precision	SOA-C	SOA-I	#Parameters
Reed [34]	7.88 ± 0.07	–	–	–	–
StackGAN [49]	8.45 ± 0.03	–	–	–	996M
StackGAN++ [50]	8.30 ± 0.03	–	–	–	466M
Lao [15]	8.94 ± 0.20	–	–	–	–
Infer [11]	11.46 ± 0.09	–	–	–	–
MirrorGAN [32]	26.47 ± 0.41	–	–	–	–
SEGAN [39]	27.86 ± 0.31	–	–	–	–
ControlGAN [16]	24.06 ± 0.60	–	–	–	–
SD-GAN [44]	35.69 ± 0.50	–	–	–	–
DMGAN [52]	30.49 ± 0.57	88.56%	33.44%	48.03%	223M
AttnGAN [43]	25.89 ± 0.47	82.98%	25.8%	38.79%	956M
objGAN [17]	30.29 ± 0.33	91.05%	27.14%	41.24%	–
OP-GAN [10]	28.57 ± 0.17	87.90%	33.11%	47.95%	1019M
AttnGAN⁺ (our modification)	38.12 ± 0.68	92.58%	42.34%	54.15%	956M
CPGAN (ours)	**52.73 ± 0.61**	**93.59%**	**77.02%**	**84.55%**	318M

Quantitative Evaluation. Table 1 reports the quantitative experimental results. Our model achieves the best performance in all four metrics and outperforms other methods significantly in terms of Inception score and SOA, which is owing to joint contributions from all three modules we proposed. Particularly, our CPGAN boosts the state-of-the-art performance by 47.74% in inception score, 130.32% in SOA-C and 78.12% in SOA-I. It proves that the synthesized images by our model not only have higher authenticity and diversity, but also are semantically consistent with the corresponding captions in object level. It is worth mentioning that our CPGAN contains much less parameters than StackGAN and AttnGAN, which also follow the coarse-to-fine generative framework.

The reduction of model size mainly benefits from two aspects: 1) a negligible amount of parameters are introduced by our proposed MATE and OAIE, 2) The parameter number of three-level discriminators are substantially reduced due to the adoption of Patch-based discriminating behavior in our proposed FGCD.

Human Evaluation. As a complement to the standard evaluation metrics, we also perform a human evaluation to compare our model with two classical models: StackGAN and AttnGAN. We randomly select 50 test samples and ask 100 human subjects to compare the quality of synthesized images by these three models and vote for the best for each sample. Note that three models' synthesized results are presented to human subjects randomly for each test sample. We calculate the rank-1 ratio for each model as the comparison metric, presented in Table 2. Averagely, our model achieves 63.73% of votes while AttnGAN wins on 28.33% votes and StackGAN performs worst. This human evaluation result is consistent with the quantitative results in terms of Inception score in Table 1.

Fig. 7. Qualitative comparison between our CPGAN with other classical models.

Table 2. Human evaluation results.

Fig. 8. Challenging examples.

Model	Rank-1 ratio
StackGAN [49]	7.94%
AttnGAN [43]	28.33%
CPGAN (ours)	**63.73%**

Qualitative Evaluation. To obtain a qualitative comparison, we visualize the synthesized images on randomly selected text samples by our models and other three classical models: StackGAN, AttnGAN and DMGAN, which is shown in

Fig. 7. It can be observed that our model is able to generate more realistic images than other two models, like 'sheep' , 'doughnuts' or 'sink'. Besides, the scenes in the generated image by our model are also more consistent with the given text than the other models, such as 'bench next to a patch of grass'.

Image synthesis from text is indeed a fairly challenging task that is far from solved. Take Fig. 8 as a challenging example, all models can hardly precisely interpret the the interaction ('ride') between 'man' and 'horse'. Nevertheless, our model still synthesizes more reasonable images than other two methods.

5   Conclusions

In this work, we have presented the Content-Parsing Generative Adversarial Networks (CPGAN) for text-to-image synthesis. The proposed CPGAN focuses on content-oriented parsing on both the input text and the synthesized image to learn the text-image consistency in the semantic level. Further, we also design a fine-grained conditional discriminator to model the local correlations between words and image sub-regions to push for the text-image alignment. Our model significantly improves the state-of-the-art performance on COCO dataset.

Acknowledgements. This work was supported by National Natural Science Foundation of China (NSFC) under Grant 61972012 and 61732016.

References

1. Anderson, P., et al.: Bottom-up and top down attention for image captioning and visual question answering. In: Proceedings of the IEEE Conference on Computer Vision and Pattern Recognition (CVPR), pp. 6077–6086 (2018)
2. Brock, A., Donahue, J., Simonyan, K.: Large scale GAN training for high fidelity natural image synthesis (2017)
3. Cao, C., Lu, F., Li, C., Lin, S., Shen, X.: Makeup removal via bidirectional tunable de-makeup network. IEEE Trans. Multimedia (TMM) **21**(11), 2750–2761 (2019)
4. Cha, M., Gwon, Y., Kung, H.: Adversarial nets with perceptual losses for text-to-image synthesis. In: 2017 IEEE 27th International Workshop on Machine Learning for Signal Processing (MLSP), pp. 1–6. IEEE (2017)
5. Cha, M., Gwon, Y.L., Kung, H.: Adversarial learning of semantic relevance in text to image synthesis. In: Proceedings of the AAAI Conference on Artificial Intelligence (AAAI), vol. 33, pp. 3272–3279 (2019)
6. Das, R., Zaheer, M., Reddy, S., Mccallum, A.: Question answering on knowledge bases and text using universal schema and memory networks. In: Proceedings of the 55th Annual Meeting of the Association for Computational Linguistics (ACL), pp. 358–365 (2017)
7. Feng, Y., Zhang, S., Zhang, A., Wang, D., Abel, A.: Memory-augmented neural machine translation. In: Proceedings of the 2017 Conference on Empirical Methods in Natural Language Processing (EMNLP), pp. 1390–1399 (2017)
8. Goodfellow, I., et al.: Generative adversarial nets. In: Advances in Neural Information Processing Systems (NIPS), pp. 2672–2680 (2014)

9. Hao, D., Yu, S., Chao, W., Guo, Y.: Semantic image synthesis via adversarial learning. In: Proceedings of the IEEE International Conference on Computer Vision (ICCV), pp. 5706–5714 (2017)
10. Hinz, T., Heinrich, S., Wermter, S.: Semantic object accuracy for generative text-to-image synthesis. arXiv:1910.13321 (2019)
11. Hong, S., Yang, D., Choi, J., Lee, H.: Inferring semantic layout for hierarchical text-to-image synthesis. In: Proceedings of the IEEE Conference on Computer Vision and Pattern Recognition (CVPR), pp. 7986–7994 (2018)
12. Huang, Z., Xu, W., Yu, K.: Bidirectional LSTM-CRF models for sequence tagging. arXiv preprint arXiv:1508.01991 (2015)
13. Isola, P., Zhu, J.Y., Zhou, T., Efros, A.A.: Image-to-image translation with conditional adversarial networks. In: Proceedings of the IEEE Conference on Computer Vision and Pattern Recognition (CVPR), pp. 1125–1134 (2017)
14. Weston, J., Chopra, S., Bordes, A.: Memory networks. In: International Conference on Learning Representations (ICLR) (2015)
15. Lao, Q., Havaei, M., Pesaranghader, A., Dutil, F., Jorio, L.D., Fevens, T.: Dual adversarial inference for text-to-image synthesis. In: Proceedings of the IEEE International Conference on Computer Vision (ICCV), pp. 7567–7576 (2019)
16. Li, B., Qi, X., Lukasiewicz, T., Torr, P.: Controllable text-to-image generation. In: Advances in Neural Information Processing Systems (NeurIPS), pp. 2063–2073 (2019)
17. Li, W., et al.: Object-driven text-to-image synthesis via adversarial training. In: Proceedings of the IEEE Conference on Computer Vision and Pattern Recognition (CVPR), pp. 12174–12182 (2019)
18. Lin, T.Y., et al.: Microsoft COCO: common objects in context. In: European Conference on Computer Vision (ECCV), pp. 740–755 (2014)
19. Liu, Y., Li, Y., You, S., Lu, F.: Unsupervised learning for intrinsic image decomposition from a single image. In: Proceedings of the IEEE/CVF Conference on Computer Vision and Pattern Recognition (CVPR), pp. 3248–3257 (2020)
20. Liu, Y., Lu, F.: Separate in latent space: unsupervised single image layer separation. In: Proceedings of the AAAI Conference on Artificial Intelligence (AAAI), pp. 11661–11668 (2020)
21. Lv, F., Lu, F.: Attention-guided low-light image enhancement. arXiv preprint arXiv:1908.00682 (2019)
22. Ma, C., Shen, C., Dick, A., Den Hengel, A.V.: Visual question answering with memory-augmented networks. In: Proceedings of the IEEE Conference on Computer Vision and Pattern Recognition (CVPR), pp. 6975–6984 (2018)
23. Mansimov, E., Parisotto, E., Ba, J., Salakhutdinov, R.: Generating images from captions with attention. In: International Conference on Learning Representations (ICLR) (2016)
24. Maruf, S., Haffari, G.: Document context neural machine translation with memory networks. In: Proceedings of the 56th Annual Meeting of the Association for Computational Linguistics (ACL), pp. 1275–1284 (2018)
25. Miyato, T., Kataoka, T., Koyama, M., Yoshida, Y.: Spectral normalization for generative adversarial networks. In: International Conference on Learning Representations (ICLR) (2018)
26. Mohtarami, M., Baly, R., Glass, J., Nakov, P., Màrquez, L., Moschitti, A.: Automatic stance detection using end-to-end memory networks. arXiv preprint arXiv:1804.07581 (2018)

27. Nilsback, M.E., Zisserman, A.: Automated flower classification over a large number of classes. In: 2008 Sixth Indian Conference on Computer Vision, Graphics and Image Processing, pp. 722–729. IEEE (2008)
28. Niu, Y., et al.: Pathological evidence exploration in deep retinal image diagnosis. In: Proceedings of the AAAI Conference on Artificial Intelligence (AAAI), vol. 33, pp. 1093–1101 (2019)
29. Odena, A., Olah, C., Shlens, J.: Conditional image synthesis with auxiliary classifier GANs. In: Proceedings of the 34rd International Conference on Machine Learning (ICML), pp. 2642–2651 (2017)
30. Pei, W., Zhang, J., Wang, X., Ke, L., Shen, X., Tai, Y.W.: Memory-attended recurrent network for video captioning. In: Proceedings of the IEEE Conference on Computer Vision and Pattern Recognition (CVPR), pp. 8347–8356 (2019)
31. Qiao, T., Zhang, J., Xu, D., Tao, D.: Learn, imagine and create: text-to-image generation from prior knowledge. In: Advances in Neural Information Processing Systems (NeurIPS), pp. 885–895 (2019)
32. Qiao, T., Zhang, J., Xu, D., Tao, D.: MirrorGAN: learning text-to-image generation by redescription. In: Proceedings of the IEEE Conference on Computer Vision and Pattern Recognition (CVPR), pp. 4321–4330 (2019)
33. Redmon, J., Farhadi, A.: YOLOv3: an incremental improvement. arXiv preprint arXiv:1804.02767 (2018)
34. Reed, S., Akata, Z., Yan, X., Logeswaran, L., Schiele, B., Lee, H.: Generative adversarial text to image synthesis. In: Proceedings of the 33rd International Conference on Machine Learning (ICML) (2016)
35. Reed, S., et al.: Parallel multiscale autoregressive density estimation. In: Proceedings of the 34rd International Conference on Machine Learning (ICML), pp. 2912–2921 (2017)
36. Salimans, T., Goodfellow, I., Zaremba, W., Cheung, V., Radford, A., Chen, X.: Improved techniques for training GANs. In: Advances in neural information processing systems (NIPS), pp. 2234–2242 (2016)
37. Reed, S., Van Den Oord, A., Kalchbrenner, N., Bapst, V., Botvinick, M., De Freitas, N.: Generating interpretable images with controllable structure. In: International Conference on Learning Representations (ICLR) (2017)
38. Sukhbaatar, S., Weston, J., Fergus, R., et al.: End-to-end memory networks. In: Advances in Neural Information Processing Systems (NIPS), pp. 2440–2448 (2015)
39. Tan, L., Li, Y., Zhang, Y.: Semantics-enhanced adversarial nets for text-to-image synthesis. In: Proceedings of the IEEE International Conference on Computer Vision (ICCV), pp. 10501–10510 (2019)
40. Wah, C., Branson, S., Welinder, P., Perona, P., Belongie, S.: The caltech-UCSD birds-200-2011 dataset (2011)
41. Wang, S., Mazumder, S., Liu, B., Zhou, M., Chang, Y.: Target-sensitive memory networks for aspect sentiment classification. In: Proceedings of the 56th Annual Meeting of the Association for Computational Linguistics (ACL), pp. 957–967 (2018)
42. Xu, B., Wang, N., Chen, T., Li, M.: Empirical evaluation of rectified activations in convolutional network. arXiv preprint arXiv:1505.00853 (2015)
43. Xu, T., et al.: AttnGAN: fine-grained text to image generation with attentional generative adversarial networks. In: Proceedings of the IEEE Conference on Computer Vision and Pattern Recognition (CVPR), pp. 1316–1324 (2018)
44. Yin, G., Liu, B., Sheng, L., Yu, N., Wang, X., Shao, J.: Semantics disentangling for text-to-image generation. In: Proceedings of the IEEE Conference on Computer Vision and Pattern Recognition (CVPR), pp. 2327–2336 (2019)

45. Yu, H., Cai, M., Liu, Y., Lu, F.: What I see is what you see: joint attention learning for first and third person video co-analysis. In: Proceedings of the 27th ACM International Conference on Multimedia (ACMMM), pp. 1358–1366 (2019)
46. Yuan, M., Peng, Y.: Bridge-GAN: interpretable representation learning for text-to-image synthesis. IEEE Trans. Circuits Syst. Video Technol. (TCSVT) (2019)
47. Yuan, M., Peng, Y.: CKD: cross-task knowledge distillation for text-to-image synthesis. IEEE Trans. Multimedia (TMM) (2019)
48. Zhang, H., Goodfellow, I., Metaxas, D., Odena, A.: Self-attention generative adversarial networks. In: Proceedings of the 36rd International Conference on Machine Learning (ICML) (2019)
49. Zhang, H., et al.: StackGAN: text to photo-realistic image synthesis with stacked generative adversarial networks. In: Proceedings of the IEEE International Conference on Computer Vision (ICCV), pp. 5907–5915 (2017)
50. Zhang, H., et al.: StackGAN++: realistic image synthesis with stacked generative adversarial networks. IEEE Trans. Pattern Anal. Mach. Intell. (TPAMI) **41**(8), 1947–1962 (2018)
51. Zhu, J.Y., Park, T., Isola, P., Efros, A.A.: Unpaired image-to-image translation using cycle-consistent adversarial networks. In: Proceedings of the IEEE International Conference on Computer Vision (ICCV), pp. 2223–2232 (2017)
52. Zhu, M., Pan, P., Chen, W., Yang, Y.: DM-GAN: dynamic memory generative adversarial networks for text-to-image synthesis, pp. 5802–5810 (2019)

Transporting Labels via Hierarchical Optimal Transport for Semi-Supervised Learning

Fariborz Taherkhani[✉], Ali Dabouei, Sobhan Soleymani, Jeremy Dawson, and Nasser M. Nasrabadi

West Virginia University, Morgantown, USA
{ft0009,ad0046,ssoleyma}@mix.wvu.edu,
{jeremy.dawson,nasser.nasrabadi}@mail.wvu.edu

Abstract. Semi-Supervised Learning (SSL) based on Convolutional Neural Networks (CNNs) have recently been proven as powerful tools for standard tasks such as image classification when there is not a sufficient amount of labeled data available during the training. In this work, we consider the general setting of the SSL problem for image classification, where the labeled and unlabeled data come from the same underlying distribution. We propose a new SSL method that adopts a hierarchical Optimal Transport (OT) technique to find a mapping from empirical unlabeled measures to corresponding labeled measures by leveraging the minimum amount of transportation cost in the label space. Based on this mapping, pseudo-labels for the unlabeled data are inferred, which are then used along with the labeled data for training the CNN. We evaluated and compared our method with state-of-the-art SSL approaches on standard datasets to demonstrate the superiority of our SSL method.

Keywords: Semi-Supervised Learning · Hierarchical optimal transport

1 Introduction

Training a CNN model relies on large annotated datasets, which are usually tedious and labor intensive to collect [30]. Two approaches are usually considered to address this problem: Transfer Learning (TL) and Semi-Supervised Learning (SSL). In TL [51], learning of a new task is improved by transferring knowledge from a related task which has already been learned. However, in SSL [41], learning of a new task is improved by using information from an input distribution that is provided by a large amount of unlabeled data. To make use of the unlabeled data, it is assumed that the underlying distribution of this data follows at least one of the following structural assumptions: continuity, clustering, or manifold [12]. In the continuity assumption [8,36,61], data points close to each other are more likely to belong to the same class. In the clustering assumption [13,25,61], data tend to form discrete clusters, and data in the same cluster are

© Springer Nature Switzerland AG 2020
A. Vedaldi et al. (Eds.): ECCV 2020, LNCS 12349, pp. 509–526, 2020.
https://doi.org/10.1007/978-3-030-58548-8_30

more likely to share the same label. In the manifold assumption [10,59], data lie approximately on a manifold of much lower dimension than the input space which can be classified by distances between probability measures on the manifold [55]. To quantify the difference between two probability measures on a manifold properly, modeling the geometrical structures of the manifold is required [4,5,53]. One of the methodologies used to model geometrical structures on the probability simplex (i.e., manifold of discrete probability measures) is grounded on the theory of Optimal Transport (OT) [46,53]. The Wasserstein distance, which arises from the idea of OT, exploits prior geometric knowledge of the base space in which random variables are valued [53]. Computing the Wasserstein distance between two random variables amounts to achieving a transportation plan which requires the minimal expected cost. The Wasserstein distance considers the metric properties of the base space in which a pattern is defined [5]. This characteristic of the Wasserstein distances has attracted a lot of attention for machine learning and computer vision tasks such as computing the barycenters [1,2] of multiple distributions [50], generating data [7], designing loss function [21], domain adaptation [15,18,27,32,48,57], and clustering [17,23,28,37].

Data are usually organized in a hierarchical structure, or taxonomy. For example, considering a set of data belonging to the same class in a dataset as a measure, we can think of all the data in the dataset as a measure of measures. Inspired by OT, which maps two measures with the minimum amount of transportation cost, we can think of using hierarchical OT to map two measure of measures such that the total transportation cost across the measures becomes minimum. In this paper, we propose an SSL method that leverages from hierarchical OT to map measures from an unlabeled set to measures in a labeled set with a minimum amount of the total transportation cost in the label space.

Our method stems from two basic premises: 1) Data in a given class in the labeled and unlabeled sets come from the same distribution. 2) Assume we are given three measures with roughly the same amount of data, where only two of these measures come from the same distribution. The OT cost between two measures from the same distribution is expected to be less than the OT cost between one of these measures and the measure from a different distribution. Following these premises, we thus expect that the hierarchical OT maps measures from the same distribution in the labeled and unlabeled sets such that the total transportation cost between two measure of measures becomes minimum. Based on this mapping, a pseudo-label for unlabeled data in each measure from the unlabeled set is inferred. These unlabeled data annotated by pseudo-labels are then used along with the labeled data to train a CNN. However, data in the unlabeled set are not labeled to allow us to identify the measures. Thus, following the clustering assumption in SSL and the role of OT in clustering [17,23,28,37], we can consider all the measures in the unlabeled set as a group of clusters which are identified by the Wasserstein barycenters of the unlabeled data.

2   Related Work

Pseudo-Labeling is one of the straightforward SSL techniques in which a model incorporates its own predictions on unlabeled data to achieve additional information during the training [20, 24, 33, 35, 44]. The main downside of these approaches is vulnerability to confirmation bias, i.e., they can not correct their own mistakes, when predictions of the model on unlabeled data are confident but incorrect. In such cases, the erroneous data can not contribute to the training and the error of the models is augmented during the training. This effect is intensified in cases where the distribution of the unlabeled data is different from that of labeled data. It has been shown that pseudo-labeling is practically similar to entropy regularization [42], in the sense that it forces the model to produce higher confident predictions for unlabeled data [33, 49]. However, in contrast to entropy regularization, it only forces these criteria onto data which have a low entropy prediction because of the threshold of confidence.

Consistency Regularization is considered as a way of using unlabeled data to explore a smooth manifold on which all of the data points are embedded [10]. This simple criterion has provided a set of methods , such as SWA [8], stochastic perturbations [45], π-model [31], Mean Teacher (MT) [52], and Virtual Adversarial Training (VAT) [38] that are currently considered as state-of-the-art for SSL. The original idea behind stochastic perturbations and π-model is pseudo-ensemble [9]. The pseudo-ensemble regularization techniques are usually designed such that the prediction of the model, $f_\theta(x)$, does not change significantly for realistic perturbed data ($x \rightarrow x'$). This goal is obtained by adding a loss term $d(f_\theta(x), f_\theta(x'))$ to the total loss of the model $f_\theta(x)$, where $d(.,.)$ is mean squared error or Kullback-Leibler divergence. The main downside of pseudo-ensemble methods, including π-model, is that they rely on a potentially unstable target prediction, which can immediately change during the training.

To address this issue, temporal ensembling [31] and MT [52], were proposed to obtain a more stable target output $f'_\theta(x)$. Temporal ensembling uses an exponentially accumulated average of outputs, $f_\theta(x)$, to make the target output smooth and consistent. Inspired by this method, MT uses a prediction function which is parametrized by an exponentially accumulated average of θ during the training. Similar to π-model, MT adds a mean squared error loss $d(f_\theta(x), f'_\theta(x))$ as a regularization term to the total loss function for training the network. It has been shown that MT outperforms temporal ensembling in practice [52]. In contrast to stochastic perturbation methods which rely on constructing $f_\theta(x)$ stochastically, VAT initially approximates a small perturbation r, and then adds it to x, which significantly changes the prediction of the model $f_\theta(x)$. Next, a consistency regularization technique is applied to minimize $d(f_\theta(x), f_\theta(x + r))$ with respect to θ which represents the parameters of the model.

3   Preliminaries

3.1   Discrete OT and Dual Form

For any $r \geq 1$, let the probability simplex be denoted by $\Delta_r = \{q \in \mathbb{R}^r : q_i \geq 0, \sum_{i=1}^{r} q_i = 1\}$, and also assume that $X = \{x_1, ..., x_n\}$ and $X' = \{x'_1, ..., x'_m\}$ are two sets of data points in $\mathbb{R}^d$ such that $\mathcal{X} = \sum_{i=1}^{n} a_i \delta_{x_i}$ and $\mathcal{X}' = \sum_{i=1}^{m} b_i \delta_{x'_i}$ in which δ_{x_i} is a Dirac unit mass located on point x_i, and a, b are the weighting vectors which belong to the probability simplex Δ_n and Δ_m, respectively. Then, the Wasserstein-p distance $W_p(\mathcal{X}, \mathcal{X}')$ between two discrete measures $\mathcal{X}$ and $\mathcal{X}'$ is the p-th root of the optimum of a network flow problem known as the transportation problem [11]. The transportation problem depends on two components: 1) matrix $M \in \mathbb{R}^{n \times m}$ which encodes the geometry of the data points by measuring the pairwise distance between elements in X and X' raised to the power p, 2) the transportation polytope $\pi(a, b) \in \mathbb{R}^{n \times m}$ which acts as a feasible set, characterized as a set of $n \times m$ non-negative matrices such that their row and column marginals are a and b, respectively. This means that the transportation plan should satisfy the marginal constraints. In other words, let $\mathbf{1}_m$ be an m-dimensional vector with all elements equal to one, then the transportation polytope is represented as follows: $\pi(a, b) = \{T \in \mathbb{R}^{n \times m} | T^\top \mathbf{1}_n = b, T\mathbf{1}_m = a\}$. Essentially, each element $T(i, j)$ indicates the amount of mass which is transported from i to j. Note that in the transportation problem, the matrix M is also considered as a cost parameter such that $M(i, j) = d(x_i, x'_j)^p$ where $d(.)$ is the Euclidean distance.

Let $\langle T, M \rangle$ denote the Frobenius dot-product between T and M matrices. Then, the discrete Wasserstein distance $W_p(\mathcal{X}, \mathcal{X}')$ is formulated by an optimum of a parametric linear program $\mathbf{p}(.)$ on a cost matrix M, and $n \times m$ number of variables parameterized by the marginals a and b as follows:

$$W_p(\mathcal{X}, \mathcal{X}') = \mathbf{p}(a, b, M) = \min_{T \in \pi(a,b)} \langle T, M \rangle. \tag{1}$$

The Wasserstein distance in (1) is a Linear Program (LP) and a subgradient of its solution can be calculated by Lagrange duality. The dual LP of (1) is:

$$\mathbf{d}(a, b, M) = \max_{(\alpha, \beta) \in C_M} \alpha^\top a + \beta^\top b, \tag{2}$$

where the polyhedron C_M of dual variables is as follows: [11]

$$C_M = \{(\alpha, \beta) \in \mathbb{R}^{m+n} | \alpha_i + \beta_j \leq M(i, j)\}. \tag{3}$$

Considering LP duality, the following equality is established: $\mathbf{d}(a, b, M) = \mathbf{p}(a, b, M)$ [11]. Computing the exact Wasserstein distance is time consuming. To alleviate this, in [16], Cuturi has introduced an interesting method that regularizes (1) using the entropy of the solution matrix, $H(T)$ (i.e., $\min \langle T, M \rangle + \gamma H(T)$, where γ is regularization strength). It has been shown that if T'_γ is the solution of the regularized version of (1) and α'_γ is its dual solution in (2), then

$\exists! u \in \mathbb{R}^n$, $v \in \mathbb{R}^m$ such that the solution matrix is $T'_\gamma = \text{diag}(u)K\text{diag}(v)$ and $\alpha'_\gamma = -\log(u)/\gamma + (\log(u)^\top 1_n)/(\gamma n))1_n$ where, $K = exp(-M/\gamma)$. The vectors u and v are updated iteratively between step 1 and 2 by using the well-known Sinkhorn algorithm as follows: step $1 : u = a/Kv$ and step $2 : v = b/K^\top u$ [16].

3.2 Hierarchical OT

Let θ be a Polish space and $S(\theta)$ be the space of Borel probability measures on θ. Since θ is a Polish space, $S(\theta)$ is also Polish space and can be metrized by the Wasserstein distance [40]. Considering the recursion of concepts, $S(S(\theta))$ is also a Polish space and is defined as a space of Borel probability measure on $S(\theta)$, which we can then define a Wasserstein distance on this space by using the Wasserstein metric in $S(\theta)$ (Section 3 in [40]). The concept of Wasserstein distance on the measure of measures, $S(S(\theta))$, which is also referred to as Hierarchical OT, is a practical and efficient solution to include structure in the regular OT distance [3,34,47,60]. Hierarchical OT is used to model the data which are organized in a hierarchical structure, and has been recently studied for tasks such as multimodal distribution alignment [34], document representation [60], multi-level clustering [23] and a similarity measure between two hidden Markov models [14].

Let $\mathcal{D} = \{\mathcal{X}_1, \mathcal{X}_2, ..., \mathcal{X}_n\}$ and $\mathcal{D}' = \{\mathcal{X}'_1, \mathcal{X}'_2, ..., \mathcal{X}'_m\}$ be two sets of measures such that $\mathcal{M} = \sum_{i=1}^n r_i \delta_{\mathcal{X}_i}$ and $\mathcal{M}' = \sum_{i=1}^m s_i \delta_{\mathcal{X}'_i}$ in which $\delta_{\mathcal{X}_i}$ is a Dirac mass located on the measure $\mathcal{X}_i$, and r and s denote the weighting vectors belonging to the probability simplex Δ_n and Δ_m, respectively. Then, the hierarchical OT distance between $\mathcal{M}$ and $\mathcal{M}'$ can be formulated by a linear program as follows:

$$W'_p(\mathcal{M}', \mathcal{M}) = \min_{T' \in \pi'(r,s)} \sum_{i=1}^n \sum_{j=1}^m T'(i,j) W_p(\mathcal{X}_i, \mathcal{X}'_j), \qquad (4)$$

where $\pi'(r,s) = \{T' \in \mathbb{R}^{m \times n} | T'^\top 1_m = r, T'^\top 1_n = s\}$, and $W_p(.,.)$ is the Wasserstein-p distance between two discrete measures $\mathcal{X}_i$ and $\mathcal{X}'_j$ which is obtained by Eq. (1). In Eq. (4), we have expanded Eq. (1) such that $T'(i,j)$ represents the amount of mass transported from $\delta_{\mathcal{X}_i}$ to $\delta_{\mathcal{X}'_j}$, and $W_p(.,.)$ is the ground metric which has been substituted by the Euclidean distance in Eq. (1) to represent hierarchical nature of the similarity metric between $\mathcal{M}'$ and $\mathcal{M}$.

3.3 Wasserstein Barycenters

Given $N >= 1$ probability measures with finite second moments $\{\mathcal{X}_1, \mathcal{X}_2, ..., \mathcal{X}_N\} \in S_2(\theta)$, their Wasserstein barycenters is a minimizer of F over $S_2(\theta)$ where [1]:

$$F(\tilde{\mathcal{X}}) = \inf_{\tilde{\mathcal{X}} \in S_2(\theta)} \frac{1}{N} \sum_{i=1}^N W_2^2(\tilde{\mathcal{X}}, \mathcal{X}_i). \qquad (5)$$

In the case where $\{\mathcal{X}_1, ..., \mathcal{X}_N\}$ are discrete measures with finite number of elements, each with size e_i, the problem of finding Wasserstein barycenters $\tilde{\mathcal{X}}$ on

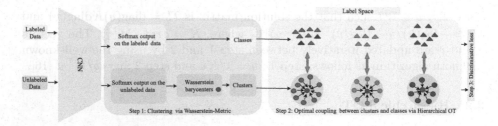

Fig. 1. At each epoch, a small amount of unlabeled data is processed through the current CNN and clustered into k groups. Then, the Wasserstein-2 distance is computed between theses groups and the ones formed by the labeled data. Next, a regularized OT is used to form an optimal coupling between the groups from the unlabeled data and the labeled ones, using the Wasserstein-2 distance as cost function (i.e., hierarchical OT). Finally, this coupling provides pseudo-labels for the selected unlabeled data to perform a gradient descent step on the CNN. Here, circles represent unlabeled data and triangles show the labeled data and their color indicate their labels. (Color figure online)

the space of $S_2(\theta)$ in (5) is recast to search only on a simpler space $\mathcal{O}_r(\theta)$, where $\mathcal{O}_r(\theta)$ is the set of probability measures with at most r support points in θ, and $r = \sum_{i=1}^{N} e_i - N + 1$ [6]. There are fast and efficient algorithms that find local solutions of the Wasserstein barycenters problem over $\mathcal{O}_r(\theta)$ for $r \geq 1$, which the use of these algorithms for clustering has also been studied in [17,58].

4   Method

Here, we describe our SSL model as is shown in Fig. 1. Here, data belonging to the same class is defined as a measure. Thus, all the initial labeled data hierarchically are considered as a measure of measures. Similarly, all the unlabeled data are also a measure of measures, each of which is constructed by data belonging to the same class. Following the basic premises mentioned earlier in the introduction, we use a hierarchical OT to predict pseudo-labels for the unlabeled measures to train a CNN. Our method is a three steps iterative algorithm. In the first step, we make a clustering assumption about the unlabeled data and consider all the unlabeled measures as a group of clusters which are identified by the Wasserstein barycenters of the unlabeled data. In the second step, we use the hierarchical OT to map each of the unlabeled measures to a corresponding labeled measure, based on which, a pseudo-label for the data within each of the clusters is predicted. Finally, unlabeled data annotated with pseudo-labels from the second step are used along with the initial labeled data to train the CNN.

4.1   Finding Unlabeled Measures via Wasserstein Metric

Given an image $z_i \in \mathbb{R}^{m \times n}$ from either the labeled or unlabeled dataset, CNN acts as a function $f(w, z_i) : \mathbb{R}^{m \times n} \to \mathbb{R}^c$ with the parameters w that maps

---

Algorithm 1 : Finding Unlabeled Measures via Wasserstein Metric

input: $Q \in \mathbb{R}^{c \times m}$.

1: **initialize:** $\mathcal{H} \in \mathbb{R}^{c \times k}$, $b = \mathbf{1}_m/m$, $t = 1$, $\eta = 0.5$.
2: **while** $\mathcal{H}$ and a have not converged **do**
3:    **Maximization Step:**
4:    set $\hat{a} = \tilde{a} = \mathbf{1}_m/m$.
5:    **while** not converged **do**
6:       $\beta = (t+1)/2$, $a \leftarrow (1 - \beta^{-1})\hat{a} + \beta^{-1}\tilde{a}$.
7:       $\alpha \leftarrow \alpha'$: dual optimal form of OT $\mathbf{d}(a, b, M_{\mathcal{HQ}})$.
8:       $\tilde{a} \leftarrow \tilde{a} \circ e^{-\beta\alpha}$; $\tilde{a} \leftarrow \tilde{a}/\tilde{a}^\top \mathbf{1}_n$.
9:       $\hat{a} \leftarrow (1 - \beta^{-1})\hat{a} + \beta^{-1}\tilde{a}$, $t \leftarrow t + 1$.
10:   **end while**
11:   $a \leftarrow \hat{a}$.
12:   **Expectation Step:**
13:   $T' \leftarrow$ optimal coupling for $\mathbf{p}(a, b, M_{\mathcal{HQ}})$.
14:   $\mathcal{H} \leftarrow (1 - \eta)\mathcal{H} + \eta(Q{T'}^\top)\text{diag}(a^{-1})$, $\eta \in [0, 1]$.
15: **end while**

---

z_i to a c-dimensional output, where c is number of the classes. Assume that $X = \{x_1, ..., x_m\}$ and $X' = \{x'_1, ..., x'_m\}$ are the sets of c-dimensional outputs represented by the CNN for the labeled and unlabeled images, respectively. Let $\mathcal{P}_i = 1/n_i \sum_{j=1}^{n_i} \delta_{x_j}$ denote a discrete measure constructed by labeled data in the i-th class, where δ_{x_j} is a Dirac unit mass on x_j and n_i is number of the data in the i-th class. Then, all of the labeled data form a measure of measures as follows: $\mathcal{M} = \sum_{i=1}^{c} \mu_i \delta_{\mathcal{P}_i}$, where $\mu_i = n_i/m$ represents amount of the mass in the measure $\mathcal{P}_i$ and $\delta_{\mathcal{P}_i}$ is a Dirac unit mass on the measure $\mathcal{P}_i$. Similarly, unlabeled data construct a measure of measures $\mathcal{M}' = \sum_{j=1}^{k} \nu_j \delta_{\mathcal{Q}_j}$, where each measure $\mathcal{Q}_i$, is created by unlabeled data belonging to the same class, $\nu_j = n'_j/m$ is amount of the mass in the measure $\mathcal{Q}_j$, and $\delta_{\mathcal{Q}_i}$ is a Dirac unit mass on $\mathcal{Q}_j$. However, data in the unlabeled set are not labeled to allow us to identify $\mathcal{Q}_j$.

One simple solution to find $\mathcal{Q}_j$, is to use the labels that are directly predicted by the CNN on the unlabeled data. In this case, there is no need to form unlabeled measures, since unlabeled data annotated by the CNN can be used directly for training the CNN. However, CNN as a classifier trained on a limited amount of the labeled data simply miss-classifies these unlabeled data. Thus, there is little option other than unsupervised methods, such as clustering to explore the unlabeled data belonging to the same class. This criterion stems from the structural assumption based on the clustering in SSL, where it is assumed that the data within the same cluster are more likely to share the same label. Inspired by the role of OT in clustering [17,23,28,37], we leverage the Wasserstein metric to explore these measures underlying the unlabeled data. Specifically, we use the k-means objective incorporated by a Wasserstein metric loss to find $\mathcal{Q}_j$.

Given m unlabeled data $x'_1, ..., x'_m \in \theta$, the k-means clustering as a vector quantization method [43] aims to find a set C containing at most k atoms $c_1, ..., c_k \in \theta$ such that the following objective is minimized:

$$J(C) = \inf_{c_1,\ldots,c_k} \frac{1}{m} \sum_{i=1}^{m} ||x_i' - c_j||^2. \tag{6}$$

Let $\mathcal{Q} = \frac{1}{m}\sum_{i=1}^{m} \delta_{x_i'}$ be the empirical measure of data $x_1', \ldots, x_m'$, where $\delta_{x_i'}$ is a Dirac unit mass on x_i'. Then (6) is equivalent to exploring a discrete probability measure $\mathcal{H}$ including finite number of support points which minimizes:

$$F(\mathcal{H}) = \inf_{\mathcal{H} \in \mathcal{O}_k(\theta)} W_2^2(\mathcal{H}, \mathcal{Q}). \tag{7}$$

When $N = 1$ in (5), then (7) can also be considered as a Wasserstein barycenters problem whose solution is studied in [1,17,58]. From this perspective as studied by [17], the algorithm for finding the Wasserstein barycenters introduces an alternative for the popular Loyd's algorithm to find local minimum of the k-means objective, where the maximization step (i.e., the assignment of the weight of each data point to its closest centroid) is equivalent to computing α' in dual form of OT (see Eq. (2)), while the expectation step (i.e., the re-centering step) is equivalent to updating $\mathcal{H}$ using the OT. Algorithm 1 presents clustering algorithm for exploring the unlabeled measures using the Wasserstein metric.

4.2   Mapping Measures via Hierarchical OT for Pseudo-Labeling

We design an OT cost function $f(.)$ to map the measures in $\mathcal{M}' = \sum_{j=1}^{k} \nu_j \delta_{\mathcal{Q}_j}$ to the measures in $\mathcal{M} = \sum_{i=1}^{c} \mu_i \delta_{\mathcal{P}_i}$ as follows:

$$f(\mu, \nu, G) = \min_{R \in \mathcal{T}(\mu,\nu)} \langle R, G \rangle - \omega H(R), \tag{8}$$

where R is the optimal coupling matrix in which $R(i,j)$ is amount of the mass that should be transported from $\mathcal{Q}_i$ to $\mathcal{P}_j$ to provide an OT plan between $\mathcal{M}'$ and $\mathcal{M}$. Thus, if the highest amount of the mass from $\mathcal{Q}_i$ is transported to $\mathcal{P}_r$ (i.e., $\mathcal{Q}_i$ is mapped to $\mathcal{P}_r$); the data belonging to the measure $\mathcal{Q}_i$ are annotated by r which is the label of the measure $\mathcal{P}_r$. Variable G is the pairwise similarity matrix between measures within $\mathcal{M}$ and $\mathcal{M}'$ in which $G(i,j) = W_p(\mathcal{Q}_i, \mathcal{P}_j)$ is the regularized Wasserstein distance between two clouds of data in $\mathcal{Q}_i$ and $\mathcal{P}_j$. Note that the ground metric used for computing $W_p(\mathcal{Q}_i, \mathcal{P}_j)$ is the Euclidean distance. Moreover, $\langle R, G \rangle$ is the Frobenius dot-product between R and G matrices, and $\mathcal{T}$ is transportation polytope defined as follows: $\mathcal{T}(\mu,\nu) = \{R \in \mathbb{R}^{c \times c} | R^\top \mathbf{1}_c = \nu, R\mathbf{1}_k = \mu\}$. Finally, $H(R)$ is entropy of the optimal coupling matrix R used for regularizing the OT, and ω is regularization strength in Eq. (8). The optimal solution for the regularized OT in (8) is obtained by Sinkhorn algorithm.

4.3   Training CNN in SSL Fashion

In the third step, we use the generic cross entropy as our discriminative loss function to train our CNN. Let $\{z_i\}_{i=1}^{b}$ be training batch annotated by true labels $\{y_i\}_{i=1}^{b}$, and $\{z_i'\}_{i=1}^{b}$ be training batch annotated by pseudo-labels $\{y_i'\}_{i=1}^{b}$,

---

Algorithm 2 : Transporting Labels via Hierarchical OT

---

input: LD: $Z_l = \{(z_l, y_l)\}_{l=1}^m$, UD: $Z_u = \{z_u'\}_{u=1}^n$, balancing coefficients: ξ, ω, γ, learning rate: r, batch size: b, number of clusters: k.

1: Train CNN parameters initially using the labeled data Z_l.
2: **repeat**
3:     Select $\{z_i'\}_{i=1}^m \subset Z_u$, where $m << n$.
4:     Compute $X = \{x_l\}_{l=1}^m$, $X' = \{x_u'\}_{u=1}^m$: softmax output on Z_l and $\{z_i'\}_{i=1}^m$, resp.
5:     $\{Q_1, ..., Q_k\} \leftarrow$ cluster on X' using Algorithm. 1.
6:     $\{\mathcal{P}_1, ..., \mathcal{P}_c\} \leftarrow$ group X to c classes.
7:     Compute μ_i, ν_i based on the amount of the mass in $\{Q_i\}_{i=1}^k$ and $\{\mathcal{P}_i\}_{i=1}^c$.
8:     **for** each Q_i and $\mathcal{P}_j$ **do**
9:         $G(i, j) \leftarrow W_2(Q_i, \mathcal{P}_j)$: using regularized OT.
10:     **end for**
11:     $R \leftarrow$ optimal coupling for $f(\mu, \nu, G)$ in Eq. (8): using regularized OT.
12:     $\{y_u'\}_{u=1}^m \leftarrow$ pseudo-label each cluster Q_i based on the highest amount of mass transport toward the labeled measure (i.e., argmax $R(i, :)$).
13:     **repeat**
14:         Select a mini-batch:$\{z_i, z_i'\}_{i=1}^b \subset (\{z_i'\}_{i=1}^m \cup Z_l)$.
15:         $w \leftarrow w - r\nabla_w[\mathcal{L}(w)]$, using Eq. (9).
16:     **until** for an epoch
17: **until** a fixed number of epochs

---

and c_i denotes barycenter of the cluster that sample z_i' belongs to it. Then, the total loss function $\mathcal{L}(.)$, used to train our CNN in an SSL fashion is as follows:

$$\mathcal{L}(w) = \sum_{i=1}^b \mathcal{L}_c(f(w, z_i), y_i) + \xi \left(\sum_{i=1}^b \mathcal{L}_c(f(w, z_i'), y_i') + \frac{1}{b} \sum_{i=1}^b ||f(w, z_i') - c_i||^2 \right), \quad (9)$$

where $f(w, z_i)$ is output of CNN for images z_i, and $\mathcal{L}_c(.)$ denotes cross entropy, and ξ is a balancing hyperparameter. Note that the third term in (9) is the center loss [56] which aims to reduce the distance between the unlabeled data and the barycenters of their corresponding cluster to perform a local consistency regularization [61]. For training, we initially train the CNN using the labeled data as a warm up step, and then use OT to provide pseudo-labels for the unlabeled data to train the CNN along with the labeled data for the next epochs. Specifically, after training the CNN using the labeled data, in each epoch, we randomly select the same amount of labeled data from the pool of unlabeled data to compute their pseudo-labels via hierarchical OT. Then, the CNN is trained in a mini-batch mode. Our overall SSL method is described in Algorithm 2.

Discussion on Time Complexity: Algorithm 2 has two main computational parts: 1) clustering the unlabeled data via Algorithm 1 (i.e. line 5), and 2) mapping the measures via HrOT (i.e. lines 8–11). For part (1): we used [17] and there is an analysis for its Time Complexity (TC) in [58] as follows: Let c, n, k, and i be the number of classes, unlabeled data, barycenters, and iterations in EM for Algorithm 1, resp. Based on the analysis provided in [58], in our case, $N = 1$ (# of distributions), $d = c$ (dimension) and $\tau = 1$ (adjusting the support points for barycenters every τ iterations). Thus, TC of part (1) is $\mathcal{O}(ink) + \mathcal{O}(inck) \approx \mathcal{O}(inck)$. For part (2): Since TC of computing the regularized OT distance between two sets of data each with size m is $\mathcal{O}(m^2)$ [22], then we need $\mathcal{O}((n/c)(n/k)(ck)) = \mathcal{O}(n^2)$ to compute matrix G in lines 8–10, and also we need $\mathcal{O}(ck)$ to find R in line 11. Thus, TC of part (2) is $\mathcal{O}(ck) + \mathcal{O}(n^2) \approx \mathcal{O}(n^2)$. By summing part(1) and (2), the total TC for inferring spudo-labels on n data is $\mathcal{O}(n^2 + inck)$. Note that i is not large due to smoothing [17].

5   Experiments and Setup

Setup: we conduct our experiments on SVHN [39], CIFAR-10/100 [29], and Mini-ImageNet [54] datasets. Mini-ImageNet [54] is subset of ImageNet [19] which consists of 100 classes and 600 images per class. For Mini-ImageNet, we follow the SSL setup in [24]. We use 500 and 100 images per class for the training and testing splits, res. Following the prior works in [8,24,31,52], we use a ResNet-18 network for the Mini-ImageNet, and a "13-layer" network for the SVHN and CIFAR-10/100 to evaluate our model. We use the typical configuration for SVHN and CIFAR-10/100 [52], and the same for the Mini-ImageNet, i.e., normalizing images using dataset mean and Standard Deviation (SD) together and then performing data augmentation by random horizontal flips and random 4 pixel translations [52]. For training, we use Adam optimizer [26] with a learning rate of 3×10^{-3} and the batch size is set to 128. The stopping criteria for the Sinkhorn algorithm is either maxIter = 10,000 or tolerance = 10^{-8}, where maxIter is the maximum number of iterations and tolerance is a threshold for the integrated stopping criterion based on the marginal differences. The barycenters in Algorithm. 1 are initially set to centroids obtained by k-means.

We follow suggested guidelines in [41] to evaluate our model. 1) We report performance of a fully-supervised baseline since the purpose of SSL is to significantly improve the fully-supervised baseline. 2) We vary the amount of labeled data when reporting the accuracy of our SSL method since a perfect SSL algorithm should remain effective even with small amount of labeled data. 3) We compare our method with the case where the unlabeled data are labeled by CNN using its own prediction during the training. 4) We study performance of the soft-pseudo-labels which can be generated by our model. 5) We compare the OT-based clustering method described in Algorithm. 1 vs k-means to show the effectiveness of Wasserstein metric for finding the unlabeled measures in our model. 7) We conduct ablation studies on the clustering resolution, and also we study effect of the center loss as a local consistency regularizer in our SSL model.

Hyperparameter Tuning: Following [8,41], we use a validation set of 5k images for CIFAR-10/100, and standard validation set of 7k images for SVHN to tune hyperparameters of our model. For CIFAR-10 and SVHN, we use 1k labeled data, and for CIFAR-100, we use 4k labeled data. The results shown in Fig. 2 are the mean and SD of error rate on validation set for CIFAR-10. Similarly,

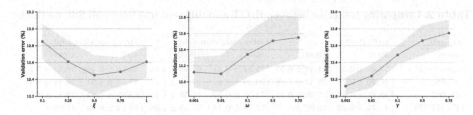

Fig. 2. Validation error for hyperparameter tuning on CIFAR-10.

Table 1. Comparing test error between HrOT and different baselines and SSL methods.

Dataset	CIFAR-10			SVHN		
Labels	1000	2000	4000	250	500	1000
Supervised	$45.89 \pm .97$	32.14 ± 0.84	21.79 ± 0.23	43.58 ± 1.98	23.78 ± 0.94	14.83 ± 0.79
TDCNN [49]	32.67 ± 1.93	22.99 ± 0.79	16.17 ± 0.37	22.90 ± 1.91	13.79 ± 1.24	8.77 ± 0.82
VAT [38]	-	-	11.36	-	-	5.42
π model [31]	-	-	12.36 ± 0.31	-	6.65 ± 0.53	4.82 ± 0.17
Temporal Ens [31]	-	-	12.16 ± 0.24	-	5.12 ± 0.13	4.42 ± 0.16
MT [52]	19.04 ± 0.51	14.35 ± 0.31	11.41 ± 0.25	4.35 ± 0.50	4.18 ± 0.27	3.95 ± 0.19
LP [24]	22.02 ± 0.88	15.66 ± 0.35	12.69 ± 0.29	-	-	-
LP+MT [24]	16.93 ± 0.70	13.22 ± 0.29	10.61 ± 0.28	-	-	-
SWA [8]	15.58	11.02	9.05	-	-	-
DOT	17.97 ± 0.47	14.46 ± 0.55	11.84 ± 0.20	5.14 ± 0.23	4.74 ± 0.35	4.11 ± 0.26
HrOT (k-means)	15.78 ± 0.65	13.16 ± 0.58	10.94 ± 0.32	4.89 ± 0.27	4.14 ± 0.30	3.86 ± 0.24
HrOT w/o CL	13.65 ± 0.36	10.44 ± 0.39	9.02 ± 0.44	4.82 ± 0.25	4.06 ± 0.24	3.61 ± 0.15
Soft HrOT	12.58 ± 0.34	9.56 ± 0.37	8.14 ± 0.49	4.32 ± 0.28	3.77 ± 0.21	3.55 ± 0.12
HrOT	$\mathbf{11.91 \pm 0.25}$	$\mathbf{8.87 + 0.32}$	$\mathbf{7.74 \pm 0.28}$	$\mathbf{4.19 \pm 0.16}$	$\mathbf{3.52 \pm 0.23}$	$\mathbf{3.06 \pm 0.09}$

for other datasets, we also tuned the hyperparameters. For SVHN, the values chosen for γ, ω and ξ are 0.01, 0.001, and 0.75, resp. For CIFAR-100, the values chosen for γ, ω and ξ are 0.001, 0.001, 0.5, res. For Mini-ImageNet, we used the same hyperparameters tuned for CIFAR-100.

5.1 Fully Supervised and Deep SSL Methods

We report the error rate of the "13-layer" CNN on CIFAR-10/100 and SVHN datasets and ResNet-18 on the Mini-ImageNet dataset for both cases where we only use the labeled data (i.e., Supervised in Table 1 and 2), and the case where we leverage the unlabeled data using the hierarchical OT technique during the training (i.e., HrOT in Table 1 and 2). All of the compared SSL methods in Table 1 and 2 use a common CNN architecture. Following the prior works [8,24,31,52], to compare HrOT with other SSL algorithms, we selected the same amount of data in the training set as the labeled data and the remaining as the unlabeled data for SVHN (73k), CIFAR-10/100 (50k) and Mini-ImageNet (50k) datasets. We run our SSL algorithm over 5 times with different random splits of labeled and unlabeled sets for each dataset, and we report the mean and SD of the test error rates. The results in Table 1 and 2 indicate the potential of our model for leveraging the unlabeled data in comparison to other SSL methods.

Table 2. Comparing test error between HrOT and different baselines and SSL methods.

Datasets	CIFAR-100		Mini-ImageNet-top1		Mini-ImageNet-top5	
Labels	4000	10000	4000	10000	4000	10000
Supervised	55.89 ± 0.26	41.07 ± 0.33	75.94 ± 0.41	61.59 ± 0.69	53.85 ± 0.46	38.59 ± 0.53
LP [24]	46.20 ± 0.76	38.43 ± 1.88	70.29 ± 0.81	57.58 ± 1.47	47.58 ± 0.94	36.14 ± 2.19
MT [24]	45.36 ± 0.49	36.08 ± 0.51	72.51 ± 0.22	57.55 ± 1.11	49.35 ± 0.22	32.51 ± 1.31
LP+MT [24]	43.73 ± 0.20	35.92 ± 0.47	72.78 ± 0.15	57.35 ± 1.66	50.52 ± 0.39	31.99 ± 0.55
SWA [8]	-	34.10 ± 0.31	-	-	-	-
DOT	44.28 ± 0.47	36.82 ± 0.33	73.84 ± 0.44	59.26 ± 0.52	48.22 ± 0.75	32.14 ± 0.48
HrOT (k-means)	42.06 ± 0.62	35.57 ± 0.64	72.04 ± 0.35	58.09 ± 0.43	46.47 ± 0.83	31.48 ± 0.33
HrOT w/o CL	40.66 ± 0.71	32.88 ± 0.36	68.94 ± 0.51	55.77 ± 0.83	44.97 ± 0.54	29.18 ± 0.26
Soft HrOT	40.02 ± 0.84	31.76 ± 0.31	68.49 ± 0.63	54.73 ± 0.70	44.16 ± 0.26	28.19 ± 0.24
HrOT	**38.98 ± 0.91**	**30.86 ± 0.56**	**67.66 ± 0.75**	**53.79 ± 0.46**	**43.38 ± 0.39**	**27.45 ± 0.59**

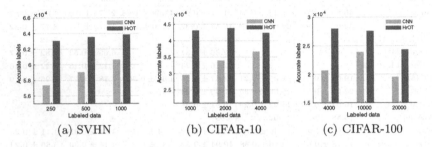

(a) SVHN   (b) CIFAR-10   (c) CIFAR-100

Fig. 3. Indicate the number of accurate predicted labels by HrOT and CNN.

5.2   Soft-Pseudo-Labels Based on Hierarchical OT

Other than particular manner in HrOT where we choose one particular pseudo-label based on the highest amount of mass transport from an unlabeled measure to a labeled measure to label the unlabeled measure, we also use "soft pseudo-labels" for training the CNN. In other words, instead of having one-hot target in the usual classification loss, we use the row of the transportation mass corresponding to the labeled measures as the target. The compared result in Table 1 and 2 show that using one-hot targets (HrOT) outperforms using soft pseudo-labels (Soft-HrOT). The reason can be supported by SSL methods based on the entropy minimization [42]. This set of methods forces the model to produce confident predictions (i.e., low entropy). Similarly, once we use one-hot targets, we essentially encourage the network to produce more confident predictions compared to using soft-pseudo labels.

5.3   Contribution of Hierarchical Optimal Transport to SSL

CNN trained on a limited amount of the labeled data simply miss-classifies the unlabeled data. Instead, we use the OT to cluster the unlabeled data and then map them with the labeled measures for pseudo-labeling. To compare these two criteria for pseudo-labeling, we report the number of accurate pseudo-labels obtained for the unlabeled training data using HrOT and the CNN by its own

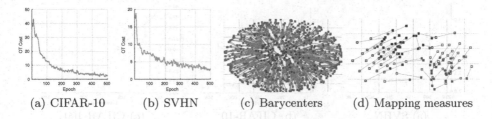

(a) CIFAR-10   (b) SVHN   (c) Barycenters   (d) Mapping measures

Fig. 4. (a, b) OT cost trend (c, d) mapping measures to barycenters and each other.

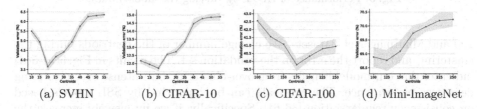

(a) SVHN   (b) CIFAR-10   (c) CIFAR-100   (d) Mini-ImageNet

Fig. 5. Validation error for different clustering resolution.

prediction (i.e., "13 layer" network). We experimentally show how HrOT has a greater positive influence on the training of CNN classifier. Essentially, this comparison allows us to know whether or not the CNN classifier can benefit from our method for producing pseudo-labels during the training, because, otherwise, the CNN can simply use its own predicted labels on the unlabeled training data during the training. To indicate the efficiency of HrOT, we change the number of labeled data in the training set and report the number of accurately predicted pseudo-labels by our CNN, and HrOT on the remaining unlabeled training data. Figure 3(a)-3(c) show that, for SVHN and CIFAR-10/100, the labels predicted by HrOT on the unlabeled training data are more accurate than the CNN, which means that the entire CNN can better benefit from HrOT than the case where it is trained solely by its own predicted labels. Moreover, we monitored the trend of transportation cost between the labeled and unlabeled measures obtained by Eq. (8) during the training. Experiments on SVHN and CIFAR-10 in Fig. 4(a) and Fig. 4(b) show that this cost reduces as the images fed into the CNN are represented by a better feature set during the training. In Fig. 4(c) and Fig. 4(d), we also visualized barycenters of the clusters, and mapping between measures of two classes in CIFAR-10 (i.e., bird and frog) using Sinkhorn algorithm. The Figures show that measures of different classes are separated properly after the training. Here, filled and unfilled squares represent unlabeled, and labeled data, respectively and the color of squares indicates their label.

5.4   Clustering Resolution

We study effect of the clustering resolution on the performance of our model. We use 1k labeled data for SVHN and CIFAR-10, and 4k labeled data for CIFAR-

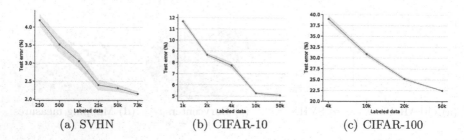

(a) SVHN	(b) CIFAR-10	(c) CIFAR-100

Fig. 6. Performance of HrOT by varying the labeled data.

100 and MiniImageNet datasets. We change number of the centriods during the clustering, and report the error on the validation set. The results in Fig. 5(a)-5(d) indicates that our model benefits from over-clustering but intense over-clustering decreases performance. The rationale can be supported by SSL models based on consistency regularization [36,61]. Specifically, if we intensively increase the number of the clusters, we only consider the global structure (or geometry) of the data in the label space, and then we ignore their local structure when transporting labels. This is not appropriate for SSL, since in such a case, we ignore the local consistency of data. However, if we cluster data in the label space via Wasserstein metric and then map them through HrOT, we exploit both local and global structure of the data in the label space during the transporting labels. To further validate this claim, we conducted following ablation study: in each batch, we solve an OT between labeled data $\{x_l\}_{u=1}^m$ and unlabeled data $\{x_u'\}_{u=1}^m$ directly and use the OT to assign pseudo-labels to the unlabeled data. We refer to this baseline as Direct-OT (DOT), our results in Table 1 and 2 indicate that mapping data directly using OT significantly reduces the performance of the CNN compared to our original model, HrOT which indicate importance of the clustering and considering hierarchical structure in OT for generating pseudo-labels in our SSL model.

In further study, instead of using OT to cluster, we use the regular k-means in our method. We refer to this baseline as HrOT (k-means). The compared results between HrOT and HrOT(k-means) in Table 1 and 2 shows the power of Wasserstein-metric in the k-means objective for finding unlabeled measures. Moreover, we ablated the center loss in Eq. (9) to see the effect of this term as a local consistency regularizer in our SSL model (i.e., HrOT w/o CL in in Tables 1 and 2). The compared results with HroT in Tables 1 and 2 show that this term can have relatively a positive influence on the performance of our model.

5.5   Varying Labeled Data

We evaluate how varying the amount of initial labeled data degrades the performance of HrOT in the very limited label regime. We gradually increase the number of labeled data during the training and report the performance of our SSL method on the testing set. Here, we run our SSL method over 5 times with different random splits of labeled and unlabeled sets for SVHN and CIFAR-10/100, and report the results in Fig. 6(a)-6(c). The results show that the performance of HrOT tends to level off as the number of labels increases.

6 Conclusion

We proposed a method which leverages optimal transport to train a CNN classifier in an SSL manner. We used the Wasserstein barycenters of the unlabeled data to identify the measures in the unlabeled set. Then, we used hierarchical optimal transport to map measures from the unlabeled set to measures in the labeled set with a minimum amount of the total transportation cost in the label space. Based on this mapping, pseudo-labels for the unlabeled data were inferred, which were then used along with the labeled data for training the CNN. Finally, we experimentally evaluated our SSL method to indicate its potential for leveraging the unlabeled data when labels are limited during the training.

References

1. Agueh, M., Carlier, G.: Barycenters in the wasserstein space. SIAM J. Math. Anal. **43**(2), 904–924 (2011)
2. Álvarez-Esteban, P.C., del Barrio, E., Cuesta-Albertos, J., Matrán, C.: A fixed-point approach to barycenters in wasserstein space. J. Math. Anal. Appl. **441**(2), 744–762 (2016)
3. Alvarez-Melis, D., Jaakkola, T., Jegelka, S.: Structured optimal transport. In: International Conference on Artificial Intelligence and Statistics, pp. 1771–1780 (2018)
4. Amari, S.: Information Geometry and Its Applications. AMS, vol. 194. Springer, Tokyo (2016). https://doi.org/10.1007/978-4-431-55978-8
5. Amari, S., Karakida, R., Oizumi, M.: Information geometry connecting wasserstein distance and kullback–leibler divergence via the entropy-relaxed transportation problem. Inform. Geom. **1**(1), 13–37 (2018). https://doi.org/10.1007/s41884-018-0002-8
6. Anderes, E., Borgwardt, S., Miller, J.: Discrete wasserstein barycenters: optimal transport for discrete data. Math. Methods Oper. Res. **84**(2), 389–409 (2016)
7. Arjovsky, M., Chintala, S., Bottou, L.: Wasserstein gan. arXiv preprint arXiv:1701.07875 (2017)
8. Athiwaratkun, B., Finzi, M., Izmailov, P., Wilson, A.G.: There are many consistent explanations of unlabeled data: why you should average. In: International Conference on Learning Representations (2019)
9. Bachman, P., Alsharif, O., Precup, D.: Learning with pseudo-ensembles. In: Advances in Neural Information Processing Systems, pp. 3365–3373 (2014)
10. Belkin, M., Niyogi, P., Sindhwani, V.: Manifold regularization: a geometric framework for learning from labeled and unlabeled examples. J. Mach. Learn. Res. **7**(Nov), 2399–2434 (2006)
11. Bertsimas, D., Tsitsiklis, J.N.: Introduction to Linear Optimization, vol. 6. Athena Scientific Belmont, MA (1997)
12. Chapelle, O., Scholkopf, B., Zien, A.: Semi-supervised learning (chapelle, o. et al., eds.; 2006) [book reviews]. IEEE Trans. Neural Netw. **20**(3), 542–542 (2009)

13. Chapelle, O., Weston, J., Schölkopf, B.: Cluster kernels for semi-supervised learning. In: Advances in Neural Information Processing Systems, pp. 601–608 (2003)
14. Chen, Y., Ye, J., Li, J.: Aggregated wasserstein distance and state registration for hidden markov models. IEEE Transactions on Pattern Analysis and Machine Intelligence (2019)
15. Courty, N., Flamary, R., Tuia, D., Rakotomamonjy, A.: Optimal transport for domain adaptation. IEEE Trans. Pattern Anal. Mach. Intell. 39(9), 1853–1865 (2017)
16. Cuturi, M.: Sinkhorn distances: lightspeed computation of optimal transport. In: Advances in Neural Information Processing Systems, pp. 2292–2300 (2013)
17. Cuturi, M., Doucet, A.: Fast computation of wasserstein barycenters. In: International Conference on Machine Learning, pp. 685–693 (2014)
18. Damodaran, B.B., Kellenberger, B., Flamary, R., Tuia, D., Courty, N.: Deepjdot: deep joint distribution optimal transport for unsupervised domain adaptation. In: European Conference on Computer Vision, pp. 467–483. Springer (2018)
19. Deng, J., Dong, W., Socher, R., Li, L.J., Li, K., Fei-Fei, L.: Imagenet: a large-scale hierarchical image database. In: 2009 IEEE Conference on Computer Vision and Pattern Recognition, pp. 248–255. IEEE (2009)
20. Dong-DongChen, W., WeiGao, Z.H.: Tri-net for semi-supervised deep learning. IJCAI (2018)
21. Frogner, C., Zhang, C., Mobahi, H., Araya, M., Poggio, T.A.: Learning with a wasserstein loss. In: Advances in Neural Information Processing Systems, pp. 2053–2061 (2015)
22. Genevay, A., Chizat, L., Bach, F., Cuturi, M., Peyré, G.: Sample complexity of sinkhorn divergences. In: The 22nd International Conference on Artificial Intelligence and Statistics, pp. 1574–1583 (2019)
23. Ho, N., Nguyen, X.L., Yurochkin, M., Bui, H.H., Huynh, V., Phung, D.: Multilevel clustering via wasserstein means. In: Proceedings of the 34th International Conference on Machine Learning, Vol. 70, pp. 1501–1509. JMLR. org (2017)
24. Iscen, A., Tolias, G., Avrithis, Y., Chum, O.: Label propagation for deep semi-supervised learning. In: Proceedings of the IEEE Conference on Computer Vision and Pattern Recognition, pp. 5070–5079 (2019)
25. Jia, Y., Kwong, S., Hou, J.: Semi-supervised spectral clustering with structured sparsity regularization. IEEE Signal Process. Lett. 25(3), 403–407 (2018)
26. Kingma, D.P., Ba, J.: Adam: a method for stochastic optimization. arXiv preprint arXiv:1412.6980 (2014)
27. Kolouri, S., Park, S.R., Thorpe, M., Slepcev, D., Rohde, G.K.: Optimal mass transport: signal processing and machine-learning applications. IEEE Signal Process. Mag. 34(4), 43–59 (2017)
28. Kolouri, S., Zou, Y., Rohde, G.K.: Sliced wasserstein kernels for probability distributions. In: Proceedings of the IEEE Conference on Computer Vision and Pattern Recognition, pp. 5258–5267 (2016)
29. Krizhevsky, A., Hinton, G.: Learning multiple layers of features from tiny images (2009)
30. Krizhevsky, A., Sutskever, I., Hinton, G.E.: Imagenet classification with deep convolutional neural networks. In: Advances in Neural Information Processing Systems, pp. 1097–1105 (2012)
31. Laine, S., Aila, T.: Temporal ensembling for semi-supervised learning. arXiv preprint arXiv:1610.02242 (2016)

32. Lee, C.Y., Batra, T., Baig, M.H., Ulbricht, D.: Sliced wasserstein discrepancy for unsupervised domain adaptation. In: Proceedings of the IEEE Conference on Computer Vision and Pattern Recognition, pp. 10285–10295 (2019)
33. Lee, D.H.: Pseudo-label: the simple and efficient semi-supervised learning method for deep neural networks. In: Workshop on Challenges in Representation Learning, vol. 3, p. 2. ICML (2013)
34. Lee, J., Dabagia, M., Dyer, E., Rozell, C.: Hierarchical optimal transport for multimodal distribution alignment. In: Advances in Neural Information Processing Systems, pp. 13453–13463 (2019)
35. Liu, X., Van De Weijer, J., Bagdanov, A.D.: Exploiting unlabeled data in cnns by self-supervised learning to rank. IEEE Trans. Pattern Anal. Mach. Intell. **41**(8), 1862–1878 (2019)
36. Luo, Y., Zhu, J., Li, M., Ren, Y., Zhang, B.: Smooth neighbors on teacher graphs for semi-supervised learning. In: Proceedings of the IEEE Conference on Computer Vision and Pattern Recognition, pp. 8896–8905 (2018)
37. Mi, L., Zhang, W., Gu, X., Wang, Y.: Variational wasserstein clustering. arXiv preprint arXiv:1806.09045 (2018)
38. Miyato, T., Maeda, S., Ishii, S., Koyama, M.: Virtual adversarial training: a regularization method for supervised and semi-supervised learning. IEEE Trans. Pattern Anal. Mach. Intell. **41**(8), 1979–1993 (2018)
39. Netzer, Y., Wang, T., Coates, A., Bissacco, A., Wu, B., Ng, A.Y.: Reading digits in natural images with unsupervised feature learning. In: NIPS Workshop on Deep Learning and Unsupervised Feature Learning, vol. 2011, p. 5 (2011)
40. Nguyen, X., et al.: Borrowing strength in hierarchical bayes: posterior concentration of the dirichlet base measure. Bernoulli **22**(3), 1535–1571 (2016)
41. Oliver, A., Odena, A., Raffel, C.A., Cubuk, E.D., Goodfellow, I.: Realistic evaluation of deep semi-supervised learning algorithms. In: Advances in Neural Information Processing Systems, pp. 3235–3246 (2018)
42. Pereyra, G., Tucker, G., Chorowski, J., Kaiser, Ł., Hinton, G.: Regularizing neural networks by penalizing confident output distributions. arXiv preprint arXiv:1701.06548 (2017)
43. Pollard, D.: Quantization and the method of k-means. IEEE Trans. Inform. Theory **28**(2), 199–205 (1982)
44. Rasmus, A., Berglund, M., Honkala, M., Valpola, H., Raiko, T.: Semi-supervised learning with ladder networks. In: Advances in Neural Information Processing Systems, pp. 3546–3554 (2015)
45. Sajjadi, M., Javanmardi, M., Tasdizen, T.: Regularization with stochastic transformations and perturbations for deep semi-supervised learning. In: Advances in Neural Information Processing Systems, pp. 1163–1171 (2016)
46. Santambrogio, F.: Optimal transport for applied mathematicians. Birkauser NY **55**, 58–63 (2015)
47. Schmitzer, B., Schnörr, C.: A hierarchical approach to optimal transport. In: Kuijper, A., Bredies, K., Pock, T., Bischof, H. (eds.) SSVM 2013. LNCS, vol. 7893, pp. 452–464. Springer, Heidelberg (2013). https://doi.org/10.1007/978-3-642-38267-3_38
48. Shen, J., Qu, Y., Zhang, W., Yu, Y.: Wasserstein distance guided representation learning for domain adaptation. In: Thirty-Second AAAI Conference on Artificial Intelligence (2018)
49. Shi, W., Gong, Y., Ding, C., MaXiaoyu Tao, Z., Zheng, N.: Transductive semi-supervised deep learning using min-max features. In: The European Conference on Computer Vision (ECCV), September 2018

50. Solomon, J., et al.: Convolutional wasserstein distances: efficient optimal transportation on geometric domains. ACM Trans. Graph. (TOG) **34**(4), 66 (2015)
51. Tan, C., Sun, F., Kong, T., Zhang, W., Yang, C., Liu, C.: A survey on deep transfer learning. In: Kůrková, V., Manolopoulos, Y., Hammer, B., Iliadis, L., Maglogiannis, I. (eds.) ICANN 2018. LNCS, vol. 11141, pp. 270–279. Springer, Cham (2018). https://doi.org/10.1007/978-3-030-01424-7_27
52. Tarvainen, A., Valpola, H.: Mean teachers are better role models: weight-averaged consistency targets improve semi-supervised deep learning results. In: Advances in Neural Information Processing Systems, pp. 1195–1204 (2017)
53. Villani, C.: Optimal transport: old and new, vol. 338. Springer Science & Business Media (2008)
54. Vinyals, O., Blundell, C., Lillicrap, T., Wierstra, D., et al.: Matching networks for one shot learning. In: Advances in Neural Information Processing Systems, pp. 3630–3638 (2016)
55. Vural, E., Guillemot, C.: A study of the classification of low-dimensional data with supervised manifold learning. J. Mach. Learn. Res. **18**, 1–157 (2017)
56. Wen, Y., Zhang, K., Li, Z., Qiao, Y.: A discriminative feature learning approach for deep face recognition. In: Leibe, B., Matas, J., Sebe, N., Welling, M. (eds.) ECCV 2016. LNCS, vol. 9911, pp. 499–515. Springer, Cham (2016). https://doi.org/10.1007/978-3-319-46478-7_31
57. Yan, Y., Li, W., Wu, H., Min, H., Tan, M., Wu, Q.: Semi-supervised optimal transport for heterogeneous domain adaptation. In: IJCAI, pp. 2969–2975 (2018)
58. Ye, J., Wu, P., Wang, J.Z., Li, J.: Fast discrete distribution clustering using wasserstein barycenter with sparse support. IEEE Trans. Signal Process. **65**(9), 2317–2332 (2017)
59. Yu, B., Wu, J., Ma, J., Zhu, Z.: Tangent-normal adversarial regularization for semi-supervised learning. In: Proceedings of the IEEE Conference on Computer Vision and Pattern Recognition, pp. 10676–10684 (2019)
60. Yurochkin, M., Claici, S., Chien, E., Mirzazadeh, F., Solomon, J.M.: Hierarchical optimal transport for document representation. In: Advances in Neural Information Processing Systems, pp. 1599–1609 (2019)
61. Zhou, D., Bousquet, O., Lal, T.N., Weston, J., Schölkopf, B.: Learning with local and global consistency. In: Advances in Neural Information Processing Systems, pp. 321–328 (2004)

MTI-Net: Multi-scale Task Interaction Networks for Multi-task Learning

Simon Vandenhende[1(✉)], Stamatios Georgoulis[2], and Luc Van Gool[1,2]

[1] ESAT-PSI, KU Leuven, Leuven, Belgium
simon.vandenhende@kuleuven.be
[2] CVL, ETH Zurich, Zurich, Switzerland

Abstract. In this paper, we argue about the importance of considering task interactions at multiple scales when distilling task information in a multi-task learning setup. In contrast to common belief, we show that tasks with high affinity at a certain scale are not guaranteed to retain this behaviour at other scales, and vice versa. We propose a novel architecture, namely MTI-Net, that builds upon this finding in three ways. First, it explicitly models task interactions at every scale via a multi-scale multi-modal distillation unit. Second, it propagates distilled task information from lower to higher scales via a feature propagation module. Third, it aggregates the refined task features from all scales via a feature aggregation unit to produce the final per-task predictions.

Extensive experiments on two multi-task dense labeling datasets show that, unlike prior work, our multi-task model delivers on the full potential of multi-task learning, that is, smaller memory footprint, reduced number of calculations, and better performance w.r.t. single-task learning. The code is made publicly available (https://github.com/SimonVandenhende/Multi-Task-Learning-PyTorch).

Keywords: Multi-task learning · Scene understanding

1 Introduction and Prior Work

The world around us is flooded with complex problems that require solving a multitude of tasks concurrently. An autonomous car should be able to detect all objects in the scene, localize them, understand what they are, estimate their distance and trajectory, etc., in order to safely navigate itself in its surroundings. In a similar vein, an intelligent advertisement system should be able to detect the presence of people in its viewpoint, understand their gender and age group, analyze their appearance, track where they are looking at, etc., in order to provide personalized content. The examples are countless. Understandably, this

Electronic supplementary material The online version of this chapter (https://doi.org/10.1007/978-3-030-58548-8_31) contains supplementary material, which is available to authorized users.

© Springer Nature Switzerland AG 2020
A. Vedaldi et al. (Eds.): ECCV 2020, LNCS 12349, pp. 527–543, 2020.
https://doi.org/10.1007/978-3-030-58548-8_31

calls for efficient computational models in which multiple learning tasks can be solved simultaneously.

Multi-task learning (MTL) [2,37] tackles this problem. Compared to the single-task case, where each individual task is solved separately by its own network, multi-task networks theoretically bring several advantages to the table. First, due to their layer sharing, the resulting memory footprint is substantially reduced. Second, as they explicitly avoid to repeatedly calculate the features in the shared layers, once for every task, they show increased inference speeds. Most importantly, they have the potential for improved performance if the associated tasks share complementary information, or act as a regularizer for one another. Evidence for the former has been provided in the literature for certain pairs of tasks, e.g. detection and classification [9,34], detection and segmentation [6,12], segmentation and depth estimation [7,47], while for the latter recent efforts point to that direction [42].

Motivated by these observations, researchers started designing architectures capable of learning shared representations from multi-task supervisory signals. Misra et al. [31] proposed to use "cross-stitch" units to combine features from multiple networks to learn a better combination of shared and task-specific representations. Kokkinos [19] introduced a multi-head architecture called UberNet that jointly handles as many as seven tasks in a unified framework, which can be trained end-to-end. Zamir et al. [49] modeled the structure of the visual tasks' space by finding transfer learning dependencies across a dictionary of twenty six tasks. Despite the progress reported by these or similar works [25,27,32,39,44], the joint learning of multiple tasks can lead to single-task performance degradation if information sharing happens between unrelated tasks. The latter is known as *negative transfer* [52], and has been well documented in [19], where an improvement in estimating normals leads to a decline in object detection, or in [12] where the multi-task version underperforms the single-task ones.

To remedy this situation, a group of methods carefully balance the losses of the individual tasks, in an attempt to find an equilibrium where no task declines significantly. For example, Kendall et al. [16] used the homoscedastic uncertainty of each individual task to re-weigh the losses. Gradient normalization [5] was proposed to balance the losses by adaptively normalizing the magnitude of each task's gradients. Similarly, Sinha et al. [41] tried to balance the losses by adapting the gradients magnitude, but differently, they employed adversarial training to this end. Dynamic task prioritization [10] proposed to dynamically sort the order of task learning, and prioritized 'difficult' tasks over 'easy' ones. Zhao et al. [52] introduced a modulation module to encourage feature sharing among 'relevant' tasks and disentangle the learning of 'irrelevant' tasks. Sener and Koltun [38] proposed to cast multi-task learning into a multi-objective optimization scheme, where the weighting of the different losses is adaptively changed such that a Pareto optimal solution is achieved.

In a different vein, Maninis et al. [29] followed a 'single-tasking' route. That is, in a multi-tasking framework they performed separate forward passes, one for each task, that activate shared responses among all tasks, plus some residual responses that are task-specific. Furthermore, to suppress the negative transfer issue they applied adversarial training on the gradients level that enforces them to be statistically indistinguishable across tasks during the update step.

Note that all aforementioned works so far follow a common pattern: they *directly* predict all task outputs from the same input in one processing cycle (i.e. all predictions are generated once, in parallel or sequentially, and are not refined afterwards). By doing so, they fail to capture commonalities and differences among tasks, that are likely fruitful for one another (e.g. depth discontinuities are usually aligned with semantic edges). Arguably, this might be the reason for the only moderate performance improvements achieved by this group of works (see [29]). To alleviate this issue, a few recent works first employed a multi-task network to make initial task predictions, and then leveraged features from these initial predictions in order to further improve each task output – in a one-off or recursive manner. In particular, Xu et al. [47] proposed to distil information from the initial predictions of other tasks, by means of spatial attention, before adding it as a residual to the task of interest. Zhang et al. [50] opted for sequentially predicting each task, with the intention to utilize information from the past predictions of one task to refine the features of another task at each iteration. In [51], they extended upon this idea. They used a recursive procedure to propagate similar cross-task and task-specific patterns found in the initial task predictions. To do so, they operated on the affinity matrices of the initial predictions, and not on the features themselves, as was the case before [47,50].

Although better performance improvements have been reported in these works, albeit for specific datasets (see [47]), they are all based on the principle that the interactions between tasks, which are essential in the distillation or propagation procedures described above, only happen at a fixed, local or global, scale[1]. For all we know, however, this is not always the case. In fact, two tasks with high pattern affinity at a certain scale are not guaranteed to retain this behaviour at other scales, and vice versa. Take for example the tasks of semantic segmentation and depth estimation, and consider the case where two cars at different distances are in front of our camera's viewpoint, with one partially occluding the other.

Looking at the local scale (i.e. patch level), the discontinuity in depth labels in the region in-between cars suggests that a similar pattern should be present in the semantic labels, i.e. there should be a change of semantic labels in the exact same region, despite the fact that this is incorrect. However, looking at the global scale this ambiguity can be resolved. An analogous observation can be made if we swapped the order of tasks, and went from global to local scale.

We conclude that pattern affinities should not be considered at the task level only, as existing works do [47,50,51], but be conditioned on the scale level too (for a more detailed discussion visit Sect. 2.2).

---

[1] With the exception of [50], where a first attempt for multi-scale processing happens at the decoding stage, in a strict sequential manner. Note that, their approach is only suitable for a pair of tasks, and can not be extended to multi-task learning.

In this paper, we go beyond these limitations and explicitly consider interactions at separate scales when propagating features across tasks. We propose a novel architecture, namely MTI-Net, that builds upon this idea. Starting from a multi-scale feature representation of the input image, generated from an off-the-shelf backbone network (e.g. [23,45]), we make an initial prediction for each task at each considered scale (four scales in our case). Next, for each task we distill information from other tasks by means of spatial attention to refine the features of the initial predictions. Note that this process happens at each scale separately in order to capture the unique task interactions that happen at each individual scale, as discussed above. To tackle the limited field-of-view at higher scales of the backbone network, which can hinder task predictions at these scales, we propose to propagate distilled task information from the lower scales. At the final stage, the distilled features of each task from all scales are aggregated to arrive at the final task predictions.

Our contributions are threefold: (1) we propose to explicitly consider multi-scale interactions when distilling information across tasks in multi-task networks; (2) we introduce an architecture that builds upon this idea with dedicated modules, i.e. multi-scale multi-modal distillation (Sect. 2.1), feature propagation across scales (Sect. 2.4), and feature aggregation (Sect. 2.5); (3) we overcome a common obstacle of performance degradation in multi-task networks, and observe that tasks can mutually benefit from each other, resulting in significant improvements w.r.t. their single-task counterparts.

2 Method

2.1 Multi-task Learning by Multi-modal Distillation

Visual tasks can be related. For example, they can share complementary information (surface normals and depth can directly be derived from each other), act as a regularizer for one another (using RGB-D images to predict scene semantics [11] improves the quality of the prediction due to the available depth information), and so on. Motivated by this observation, recent MTL methods [47,50,51] tried to explicitly distill information from other tasks, as a complementary signal to improve task performance. Typically, this is achieved by combining an existing backbone network, that makes initial task predictions, with a multi-step decoding process (see Fig. 1 (left)).

In more detail, the shared features of the backbone network are processed by a set of task-specific heads, that produce an initial prediction for every task. We further refer to the backbone and the task-specific heads as the *front-end* of the network. The task-specific heads produce a per-task feature representation of the scene that is more task-aware than the shared features of the backbone network. The information from these task-specific feature representations is then combined via a multi-modal distillation unit, before making the final task predictions. As shown in Fig. 1, it is possible that some tasks are only predicted in the front-end of the network. The latter are known as auxiliary tasks, since they serve as proxies in order to improve the performance on the final tasks.

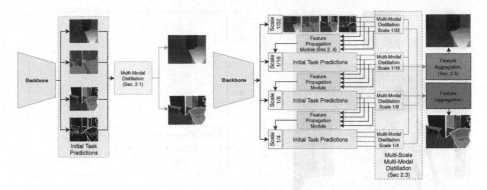

Fig. 1. An overview of different MTL architectures as described in Sect. 2. (**Left**) The architecture used in **PAD-Net** [47] and PAP-Net [51]. Features extracted from a backbone network are used to make initial task predictions. The task features are combined through a distillation unit before making the final task predictions. (**Right**) The architecture of the proposed **MTI-Net**. Starting from a backbone that extracts multi-scale features, initial task predictions are made at each scale. The task features are distilled separately at every scale, allowing our model to capture task interactions at multiple scales, i.e. receptive fields. After distillation, the distilled task features from all scales are aggregated to make the final task predictions. To boost performance, we extend our model with a feature propagation mechanism that passes distilled information from lower resolution task features to higher ones.

Prior works only differ in the way that the task-specific feature representations are combined. PAD-Net [47] distills information from other tasks by applying spatial attention to these features, before adding them as a residual. PAP-Net [51] recursively combines the pixel affinities from these features during the decoding step. Zhang et al. [50] sequentially predict one task in order to refine its features based on the features of the other task.

For brevity, we adopt the following notations. *Backbone features*: the shared features (at the last layer) of the backbone network; *Task features*: the task-specific feature representations (at the last layer) of each task-specific head; *Distilled task features*: the task features after multi-modal distillation; *Initial task predictions*: the per-task predictions at the front-end of the network; *Final task predictions*: the network outputs. Note that, backbone features, task features and distilled task features can be defined at a single scale or multiple scales.

2.2 Task Interactions at Different Scales

The approaches described in Sect. 2.1 follow a common pattern: they perform multi-modal distillation at a fixed scale, i.e. the backbone features. This rests on the assumption that all relevant task interactions can solely be modeled through a single filter operation with specific receptive field. For all we know, this is not always the case. In fact, tasks can influence each other differently for different receptive field sizes. Consider, for example, Fig. 2a. The local patches in the

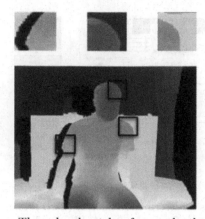

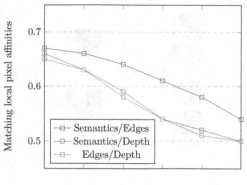

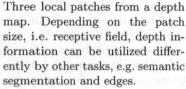

Three local patches from a depth map. Depending on the patch size, i.e. receptive field, depth information can be utilized differently by other tasks, e.g. semantic segmentation and edges.

To quantify task interactions w.r.t. scale, pixel affinities on the label space of each task, as defined in [51], are calculated in local patches. The correspondences in the affinity patterns between tasks are plotted as a function of the patch size, i.e. kernel dilation.

Fig. 2. Unlike the common belief, in this paper we question whether task interactions remain constant across all scales (see Sect. 2.2).

depth map provide little information about the semantics of the scene. However, when we enlarge the receptive field, the depth map reveals a person's shape, hinting at the scene's semantics. Note that the local patches can still provide valuable information, e.g. to improve the local alignment of edges between tasks.

To quantify the degree to which tasks share common local structures w.r.t. the size of the receptive field, we conduct the following experiment. We measure the pixel affinity in local patches on the label space of each task, using kernels of fixed size. The size of the receptive field can be selected by choosing the dilation for the kernel. We consider the tasks of semantic segmentation, depth estimation and edge detection on the NYUD-v2 dataset. A pair of semantic pixels is considered similar when both pixels belong to the same category. For the depth estimation task, we threshold the relative difference between pairs of pixels; pixels below the threshold are similar. Once the pixel affinities are calculated for every task, we measure how well similar and dissimilar pairs are matched across tasks. We repeat this experiment using different dilations for the kernel, effectively changing the receptive field. Figure 2b illustrates the result.

A first observation is that affinity patterns are matched well across tasks, with up to 65% of pair correspondence in some cases. This indicates that different tasks can share common structures in parts of the image. This is in agreement with a similar observation made earlier by [51]. A second observation is that the degree to which the affinity patterns are matched across tasks is dependent on the receptive field, which in turn, corresponds to the used dilation. This validates our initial assumption that the statistics of task interactions do not always remain constant, but rather depend on the scale, i.e. receptive field.

Based on these findings, in the next section we introduce a model that distills information from different tasks at multiple scales[2]. By doing so, we are able to capture the unique task interactions at each individual scale, overcoming the limitations of the models described in Sect. 2.1.

2.3   Multi-scale Multi-modal Distillation

We propose a multi-task architecture that explicitly takes into account task interactions at multiple scales. Our model is shown in Fig. 1 (right). First, an off-the-shelf backbone network extracts a multi-scale feature representation from the input image. Such multi-scale feature extractors have been used in semantic segmentation [17,35,45], object detection [23,45], pose estimation [33,43], etc. In Sect. 3 we verify our approach using two such backbones, i.e. HRNet [45] and FPN [23], but any multi-scale feature extractor can be used instead.

From the multi-scale feature representation (i.e. backbone features) we make initial task predictions at each scale. These initial task predictions at a particular scale are found by applying a set of task-specific heads to the backbone features extracted at that scale. The result is a per-task representation of the scene (i.e. task features) at a multitude of scales. Not only does this add deep supervision to our network, but the task features can now be distilled at each scale separately. This allows us to have multiple task interactions, each modeled for a specific receptive field size, as proposed in Sect. 2.2.

Next, we refine the task features by distilling information from the other tasks using a spatial attention mechanism [47]. Yet, our multi-modal distillation process is repeated at each scale, i.e. we apply multi-scale, multi-modal distillation. The distilled task features $F^o_{k,s}$ for task k at scale s are found as:

$$F^o_{k,s} = F^i_{k,s} + \sum_{l \neq k} \sigma \left(W_{k,l,s} F^i_{l,s} \right) \odot \left(W'_{k,l,s} F^i_{l,s} \right), \tag{1}$$

where $\sigma \left(W_{k,l,s} F^i_{l,s} \right)$ returns a per-scale spatial attention mask, that is applied to the task features $F^i_{l,s}$ from task l at scale s. Note that our approach is not necessarily limited to the use of spatial attention, but any type of feature distillation (e.g. squeeze-and-excitation [14]) can easily be plugged in. Through repetition, we calculate the distilled task features at every scale. As the bulk of filter operations is performed on low resolution feature maps, the computational overhead of our model is limited. We make a detailed resource analysis in Sect. 3.

2.4   Feature Propagation Across Scales

In Sect. 2.3 actions at each scale were performed in isolation. To sum up, we made initial task predictions at each scale, from which we refined the task features

---

[2] Cross-stitch nets [31] also exchange features at multiple scales, but in the encoder. A summary of differences with our approach is provided in the suppl. materials.

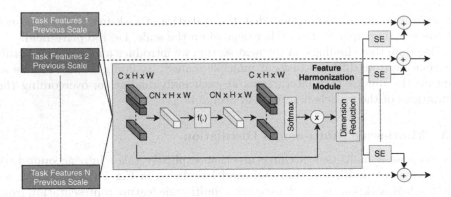

Fig. 3. Our Feature Propagation Module. First, task features from a lower scale are concatenated and mapped to a shared representation by the feature harmonization module. The task features are then refined by extracting information from the shared representation through squeeze-and-excitation (SE) [14], and are added as a residual to the original ones. Finally, these refined task features will be concatenated with the backbone features of the preceding higher scale.

through multi-modal distillation at each individual scale separately. However, as the higher resolution scales have a limited receptive field, the front-end of the network could have a hard time to make good initial task predictions at these scales, which in turn, would lead to low quality task features there. To remedy this situation we introduce a feature propagation mechanism, where the backbone features of a higher resolution scale are concatenated with the task features from the preceding lower resolution scale, before feeding them to the task-specific heads of the higher resolution scale to get the task features there.

A trivial implementation for our Feature Propagation Module (FPM) would be to just upsample the task features from the previous scale and pass them to the next scale. We opt for a different approach however, and design the FPM to behave similarly to the multi-modal distillation unit of Sect. 2.3, in order to model task interactions at this stage too. Figure 3 gives an overview of our FPM. We first use a *feature harmonization* block to combine the task features from the previous scale to a shared representation. We then use this shared representation to refine the task features from the previous scale, before passing them to the next scale. The refinement happens by selecting relevant information from the shared representation through a *squeeze-and-excitation* block [14]. Note that, since we refine the features from a single shared representation, instead of processing each task independently as done in the multi-modal distillation unit of Sect. 2.3, the computational cost is significantly smaller.

Feature Harmonization. The FPM receives as input the task features from N tasks of shape $C \times H \times W$. Our feature harmonization module combines the received task features into a shared representation. In particular, the set of N task features is first concatenated and processed by a learnable non-linear function f. The output is split into N chunks along the channel dimension, that

Table 1. Our multi-task learning benchmarks. We predict five tasks on PASCAL. On NYUD-v2 we only consider semantic segmentation and depth, but include edges and normals as auxiliary tasks. Distilled labels are marked with *.

Dataset	Edge	Seg	Parts	Normals	Saliency	Depth
PASCAL	✓	✓	✓	✓*	✓*	
NYUD-v2	✓	✓		✓		✓

match the original number of channels C. We then apply a softmax function along the task dimension to generate a task attention mask. The attended features are concatenated and further processed to reduce the number of channels from $N \cdot C$ to C. The output is a shared representation based on information from all tasks.

Refinement Through Squeeze-And-Excitation. The use of a shared representation can degrade performance when tasks are unrelated. We resolve this situation by applying a per-task channel gating function to the shared representation. This effectively allows each task to select the relevant features from the shared representation. The channel gating mechanism is implemented here as a squeeze-and-excitation (SE) block [14]. This is due to the fact that SE has shown great potential in MTL (e.g. [29]), yet other gating mechanisms could be used instead. After applying the SE module, the refined task features are added as a residual to the original task features.

2.5 Feature Aggregation

The multi-scale, multi-modal distillation described in Sect. 2.3 results in distilled task features at every scale. The latter are upsampled to the highest scale and concatenated, resulting in a final feature representation for every task. The final task predictions are found by decoding these final feature representations by task-specific heads again. All implementation details are discussed in Sect. 3. It is worth mentioning that our model has the possibility to add auxiliary tasks in the front-end of the network, similar to PAD-Net [47]. In our case however, the auxiliary tasks are predicted at multiple scales.

3 Experiments

3.1 Experimental Setup

Datasets. We perform our experimental evaluation on the PASCAL [8] and NYUD-v2 [40] datasets. Table 1 contains the tasks that we considered for each dataset. We use the original 795 train and 654 test images for the NYUD-v2 dataset. For PASCAL, we use the split from PASCAL-Context [4] which has annotations for semantic segmentation, human part segmentation and edge detection. We obtain the surface normals and saliency labels from [29], that distilled them from pre-trained state-of-the-art models [1,3].

Implementation Details. We build our approach on top of two different backbone networks, i.e. FPN [23] and HRNet [43]. We use the different output scales of the selected backbone networks to perform multi-scale operations. This translates to four scales (1/4, 1/8, 1/16, 1/32). The task-specific heads are implemented as two basic residual blocks [13]. All our experiments are conducted with pre-trained ImageNet weights.

We use the L1 loss for depth estimation and the cross-entropy loss for semantic segmentation on NYUD-v2. As in prior work [18,28,29], the edge detection task is trained with a positively weighted $w_{pos} = 0.95$ binary cross-entropy loss. We do not adopt a particular loss weighing strategy on NYUD-v2, but simply sum the losses together. On PASCAL, we reuse the training setup from [29] to facilitate a fair comparison. We reuse the loss weights from there. The initial task predictions in the front-end of the network use the same loss weighing as the final task predictions. In contrast to [47,50,51], we do not use a two-step training procedure where the front-end is pre-trained separately. Instead, we simply train the complete architecture end-to-end. We refer to the supplementary material for further implementation details.

Evaluation Metrics. We evaluate the performance of the backbone networks on the single tasks first. The optimal dataset F-measure (*odsF*) [30] is used to evaluate the edge detection task. The semantic segmentation, saliency estimation and human part segmentation tasks are evaluated using mean intersection over union (*mIoU*). We use the mean error (*mErr*) in the predicted angles to evaluate the surface normals. The depth estimation task is evaluated using the root mean square error (*rmse*). We measure the *multi-task learning performance* Δ_m as in [29], i.e. the multi-task performance of model m is defined as the average per-task drop in performance w.r.t. the single-task baseline b:

$$\Delta_m = \frac{1}{T} \sum_{i=1}^{T} (-1)^{l_i} \left(M_{m,i} - M_{b,i} \right) / M_{b,i}, \tag{2}$$

where $l_i = 1$ if a lower value means better for performance measure M_i of task i, and 0 otherwise. The single-task performance is measured for a fully-converged model that uses the same backbone network only for that task.

Baselines. On NYUD-v2, we compare MTI-Net against the state-of-the-art PAD-Net [47]. PAD-Net was originally designed for a single scale, but it is easy to plug-in a multi-scale backbone network and directly compare the two approaches. In contrast, a comparison with [50] is not possible, as this work was strictly designed for a pair of tasks, without any straightforward extension to the MTL setting. Finally, PAP-Net [51] adopts an architecture that is similar to PAD-Net, but the multi-modal distillation is performed recursively on the feature affinities. We chose to draw the comparison with the more generic PAD-Net, since it performs on par with PAP-Net (see Sect. 3.3).

Table 2. Ablation studies on (a) NYUD-v2 and (b) PASCAL using an HRNet-18 backbone network. Auxiliary tasks are indicated between brackets.

(a) Results on NYUD-v2.

Method	Seg ↑	Dep ↓	Δ_m% ↑
Single task	33.18	0.667	+0.00
MTL	32.09	0.668	−1.71
PAD-Net	32.80	0.660	−0.02
PAD-Net (N)	33.85	0.658	+1.65
PAD-Net (N+E)	32.92	0.655	+0.52
Ours (w/o FPM)	34.38	0.640	+3.85
Ours (w/o FPM) (N)	34.49	0.642	+3.84
Ours (w/o FPM) (N+E)	34.68	0.637	+4.48
Ours (w/ FPM)	35.12	0.620	+ 6.40
Ours (w/ FPM) (N)	36.22	**0.600**	+9.57
Ours (w/ FPM) (N+E)	**37.49**	0.607	**+10.91**

(b) Results on PASCAL.

Method	Seg ↑	Parts ↑	Sal ↑	Edge ↑	Norm ↓	Δ_m% ↑
Single task	60.07	60.74	67.18	69.70	14.59	+0.00
MTL (s)	54.53	59.54	65.60	-	-	−4.26
MTL (a)	53.60	58.45	65.13	70.60	15.08	−3.70
Ours (s)	64.06	62.39	68.09	-	-	+3.35
Ours (s)(E)	64.98	62.90	67.84	-	-	+3.98
Ours (s)(N)	63.74	61.75	67.90	-	-	+2.69
Ours (s)(E+N)	64.33	62.33	68.00	-	-	+3.36
Ours (a)	64.27	62.06	68.00	73.40	14.75	+2.74

On PASCAL, we compare our method against the state-of-the-art ASTMT [29]. Note that a direct comparison with ASTMT is also not straightforward, as this model is by design single-scale and heavily based on a DeepLab-v3+ (DLv3+) backbone network that contains dilated convolutions. Due to the latter, simply plugging the same DLv3+ backbone into MTI-Net would break the multi-scale features required to uniquely model the task interactions at a multitude of scales. Yet, we provide a fair comparison with ASTMT by combining it with a ResNet-50 FPN backbone, to show that it is not just using a multi-scale backbone that leads to improved results.

3.2 Ablation Studies

Network Components. In Table 2 we visualize the results of our ablation studies on NYUD-v2 and PASCAL with an HRNet18 backbone to verify how different components of our model contribute to the multi-task improvements. Additional results using different backbones are in the supplementary materials.

We focus on the smaller NYUD-v2 dataset first (see Table 2a), that contains arguably related tasks. These are semantic segmentation (Seg) and depth prediction (Dep) as main tasks, edge detection (E) and surface normals (N) as auxiliary tasks. The MTL baseline (i.e. a shared encoder with task-specific heads) has lower performance (−1.71%) than the single-task models. This is inline with prior work [29,44]. PAD-Net retains performance over the set of single-task models (−0.02%), and improves when adding the auxiliary tasks (+0.52%). Using our model without the FPM between scales further improves the results (w/o auxiliary tasks: +3.85%, w/auxiliary tasks: +4.48%). When including the FPM another significant boost in performance is achieved (+6.40%). Further adding the auxiliary tasks can help to improve the quality of our predictions (+10.91%).

Table 2b shows the ablation on PASCAL. We discriminate between a *small set (s)* and a *complete set (a)* of tasks. The small set contains the high-level (semantic and human parts segmentation) and mid-level (saliency) vision tasks. The complete set also adds the low-level (edges and normals) vision tasks. The MTL baseline leads to decreased performance, −4.26% and −3.70% on the small and complete set respectively. Instead, our model improves over the set of single-task

Table 3. Influence of using a different number of scales for the backbone network on NYUD-v2.

Method	Seg ↑	Dep ↓	Δ_m% ↑
ST	33.18	0.667	+0.00
1/4 (Pad-Net)	32.80	0.660	−0.02
1/4, 1/8	34.88	0.650	+3.80
1/4, 1/8, 1/16	35.01	0.630	+5.53
1/4, 1/8, 1/16, 1/32 (Ours)	35.12	0.620	+6.40

Table 4. Ablating the information flow within the proposed MTI-Net model on NYUD-v2.

Method	Seg ↑	Dep ↓	Δ_m% ↑
ST	33.18	0.667	+0.00
Front-end @ 1/32 scale	32.02	0.670	−1.87
Front-end @ 1/16 scale	33.02	0.660	+0.02
Front-end @ 1/8 scale	33.67	0.640	+2.72
Front-end @ 1/4 scale	34.05	0.633	+3.78
Final output		35.12	0.620	+6.40

Table 5. Comparison with the state-of-the-art on PASCAL.

Model	Backbone	Seg ↑		Parts ↑		Sal ↑		Edge ↑		Norm ↓		Δ_m ↑	Δ_m ↑
		ST	MT	ST	MT	ST	MT	ST	MT	ST	MT	(ST)	(R50-FPN)
ASTMT [29]	R26-DLv3+	64.9	64.6	57.1	57.3	64.2	64.7	71.3	71.0	14.9	15.0	−0.11	−3.42
	R50-DLv3+	68.3	68.0	60.70	61.1	65.4	65.7	72.7	72.4	14.6	14.7	−0.04	−0.08
	R50-FPN	67.7	66.8	61.8	61.1	67.2	66.1	71.1	70.9	14.8	14.7	−0.87	−0.87
PAD-Net [47]	HRNet-18	60.1	53.6	60.7	59.6	67.2	65.8	69.7	72.5	14.6	15.3	−3.08	−5.58
Ours	R18-FPN	64.5	65.7	57.4	61.6	66.4	66.8	68.2	73.9	14.8	14.6	+3.84	+0.29
	R50-FPN	67.7	66.6	61.8	63.3	67.2	66.6	71.1	74.9	14.8	14.6	+1.36	+1.36
	HRNet-18	60.1	64.3	60.7	62.1	67.2	68.0	69.7	73.4	14.6	14.8	+2.74	−0.02

models (+3.35%) on the small task set (s), where we obtain solid improvements on all tasks. We also report the influence of adding additional auxiliary tasks to the front-end of the network. Adding edges improves the multi-task performance to 3.98%, adding normals slightly decreases it to +2.69%, while adding both keeps it stable +3.36%. Finally, when learning all five tasks together, our model outperforms (+2.74%) the set of single-task models. In general, all tasks gain significantly, except for normals, where we observe a small decrease in performance. We argue that this is due to the inevitable negative transfer that characterizes all models with shared operations (also [29,47,51]). Yet, to the best of our knowledge, this is the first work to not only report overall improved multi-task performance, but also to maximize the gains over the single-task models, when jointly predicting an increasing and diverse set of tasks. We refer to Fig. 4 for qualitative results obtained with an HRNet-18 backbone.

Influence of Scales. So far, our experiments included all four scales of the backbone network (1/4, 1/8, 1/16, 1/32). Here, we study the influence of using a different number of scales for the backbone. Table 3 summarizes this ablation on NYUD-v2. Note that the use of a single scale (1/4) reduces our model to a PAD-Net like architecture. Using an increasing number of scales (1/4 vs + 1/8 vs + 1/16, ...) gradually improves performance. The results confirm our hypothesis from Sect. 2.2, i.e. task interactions should be modeled at multiple scales.

Table 6. Comparison with the state-of-the-art on NYUD-v2.

(a) Results on depth estimation.

Method	rmse	rel	δ_1	δ_2	δ_3
HCRF [20]	0.821	0.232	0.621	0.886	0.968
DCNF [24]	0.824	0.230	0.614	0.883	0.971
Wang [46]	0.745	0.220	0.605	0.890	0.970
NR forest [36]	0.774	0.187	-	-	-
Xu [48]	0.593	0.125	0.806	0.952	0.986
PAD-Net [47]	0.582	**0.120**	0.817	0.954	0.987
PAP-Net [51]	0.530	0.144	0.815	0.962	0.992
ST - HRNet48-V2	0.547	0.138	0.828	0.966	0.993
Ours - HRNet48-V2	**0.529**	0.138	**0.830**	**0.969**	**0.993**

(b) Results on semantic segmentation.

Method	pixel-acc	mean-acc	IoU
FCN [26]	60.0	49.2	29.2
Context [22]	70.0	53.6	40.6
Eigen [7]	65.6	45.1	34.1
B-SegNet [15]	68.0	45.8	32.4
RefineNet-101 [21]	72.8	57.8	44.9
PAD-Net [47]	75.2	62.3	50.2
TRL-ResNet50 [50]	76.2	56.3	46.4
PAP-Net [51]	**76.2**	62.5	**50.4**
ST - HRNet48-V2	73.4	58.1	45.7
Ours - HRNet48-V2	75.3	**62.9**	49.0

Table 7. Computational resource analysis (number of parameters and FLOPS).

(a) Results on NYUD-v2 (HRNet-18).

Method	Params (M)	FLOPS (G)	Δ_m%
Single Task	8.0	22.0	+0.00%
Multi-Task	−50%	−45%	−1.71%
PAD-Net	−15%	+204%	−0.02%
MTI-Net (Ours)	+57%	−13%	+6.40%

(b) Results on PASCAL (Res-50 FPN).

Method	Params (M)	FLOPS (G)	Δ_m%
Single Task	140	219	+0.00%
Multi-Task	−75.0%	−66%	−4.55%
ASTMT	−51.0%	−1.0%	−0.87%
Ours	−35.0%	−19.9%	+1.36%

Information Flow. To quantify the flow of information, we measure the performance of the initial task predictions at different locations in the front-end of the network. Table 4 illustrates the results on NYUD-v2. We observe that the performance gradually increases at the higher scales, due to the information that is being propagated from the lower scales via the FPM. The final prediction after aggregating the information from all scales is further improved substantially.

3.3 Comparison with the State-of-the-Art

Comparison on PASCAL. Table 5 visualizes the comparison of our model against ASTMT and PAD-Net on PASCAL. We report the multi-tasking performance both w.r.t. the single-task models using the same backbone (ST) and the single-task models based on the R50-FPN backbone. As explained, in the only possible fair comparison, i.e. when using the same R50-FPN backbone, our model achieves higher multi-tasking performance compared to ASTMT (+1.36% vs −0.87%). Yet, as ASTMT is by design single-scale and heavily based on DLv3+, we also report results using different backbones. Overall, MTI-Net achieves significantly higher gains over its single-task variants compared to ASTMT (see $\Delta_m \uparrow$ (ST)). Surprisingly, we find that our model with R18-FPN backbone even outperforms the deeper ASTMT R50-DLv3+ model in terms of multi-tasking performance (+0.29% vs −0.08%), despite the fact that the ASTMT single-task models perform better than ours, due to the use of the stronger DLv3+ backbone. Note that we are the first to report consistent multi-task improvements when solving such a diverse task dictionary. Finally, our model also outperforms PAD-Net in terms of multi-tasking performance (+2.74% vs −3.08%).

Comparison on NYUD-v2. Table 6 shows a comparison with the state-of-the-art approaches on NYUD-v2. We leave out methods that rely on extra input

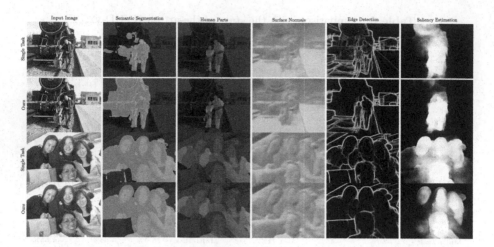

Fig. 4. Qualitative results on PASCAL. We compare the predictions made by a set of single-task models (first row for every image) against the predictions made by our MTI-Net (second row for every image). Differences can be seen for semantic segmentation, edge detection and saliency estimation.

modalities, or additional training data. As these methods are built on top of stronger single-scale backbones, we also use the multi-scale HRNet48-v2 backbone here. Again, our model improves w.r.t. the single-task models. Furthermore, we perform on par with the state-of-the-art on the depth estimation task, while performing slightly worse on the semantic segmentation task. We refer the reader to the supplementary materials for qualitative results.

Resource Analysis. We compare our model in terms of computational requirements against PAD-Net and ASTMT. The comparison with PAD-Net is performed on NYUD-v2 using the HRNet-18 backbone, while for the comparison with ASTMT on PASCAL we use a ResNet-50 FPN backbone. Table 7 reports the results relative to the single-tasking models. On NYUD-v2, MTI-Net reduces the number of FLOPS while improving the performance compared to the single-task models. The reason for the increased amount of parameters is the use of a shallow backbone, and the small number of tasks (i.e. 2). Furthermore, we significantly outperform PAD-Net in terms of FLOPS and multi-task performance. This is due to the fact that PAD-Net performs the multi-modal distillation at a single higher scale (1/4) with $4 \cdot C$ channels, C being the number of backbone channels at a single scale. Instead, we perform most of the computations at smaller scales (1/32, 1/16, 1/8), while operating on only C channels at the higher scale (1/4). On PASCAL, we significantly improve on all three metrics compared to the single-task models. We also outperform ASTMT in terms of FLOPS and multi-task performance, as the latter has to perform a separate forward pass per task.

4   Conclusion

We have shown the importance of modeling task interactions at multiple scales, enabling tasks to maximally benefit each other. We achieved this by introducing dedicated modules on top of an off-the-shelf multi-scale feature extractor, i.e. multi-scale multi-modal distillation, feature propagation across scales, and feature aggregation. Our multi-task model delivers on the full potential of multi-task learning, i.e. smaller memory footprint, reduced number of calculations and better performance. Our experiments show that our multi-task models consistently outperform their single-tasking counterparts by medium to large margins.

Acknowledgment. The authors acknowledge support by Toyota via the TRACE project and MACCHINA (KULeuven, C14/18/065).

References

1. Bansal, A., Chen, X., Russell, B., Gupta, A., Ramanan, D.: PixelNet: representation of the pixels, by the pixels, and for the pixels. arXiv preprint arXiv:1702.06506 (2017)
2. Caruana, R.: Multitask learning. Mach. Learn. **28**(1), 41–75 (1997). https://doi.org/10.1023/A:1007379606734
3. Chen, L.-C., Zhu, Y., Papandreou, G., Schroff, F., Adam, H.: Encoder-decoder with atrous separable convolution for semantic image segmentation. In: Ferrari, V., Hebert, M., Sminchisescu, C., Weiss, Y. (eds.) ECCV 2018. LNCS, vol. 11211, pp. 833–851. Springer, Cham (2018). https://doi.org/10.1007/978-3-030-01234-2_49
4. Chen, X., Mottaghi, R., Liu, X., Fidler, S., Urtasun, R., Yuille, A.: Detect what you can: detecting and representing objects using holistic models and body parts. In: CVPR, pp. 1971–1978 (2014)
5. Chen, Z., Badrinarayanan, V., Lee, C.Y., Rabinovich, A.: GradNorm: gradient normalization for adaptive loss balancing in deep multitask networks. In: ICML (2018)
6. Dvornik, N., Shmelkov, K., Mairal, J., Schmid, C.: BlitzNet: a real-time deep network for scene understanding. In: ICCV, pp. 4154–4162 (2017)
7. Eigen, D., Fergus, R.: Predicting depth, surface normals and semantic labels with a common multi-scale convolutional architecture. In: ICCV, pp. 2650–2658 (2015)
8. Everingham, M., Van Gool, L., Williams, C.K., Winn, J., Zisserman, A.: The PASCAL visual object classes (VOC) challenge. Int. J. Comput. Vision **88**(2), 303–338 (2010). https://doi.org/10.1007/s11263-009-0275-4
9. Girshick, R.: Fast R-CNN. In: ICCV, pp. 1440–1448 (2015)
10. Guo, M., Haque, A., Huang, D.-A., Yeung, S., Fei-Fei, L.: Dynamic task prioritization for multitask learning. In: Ferrari, V., Hebert, M., Sminchisescu, C., Weiss, Y. (eds.) ECCV 2018. LNCS, vol. 11220, pp. 282–299. Springer, Cham (2018). https://doi.org/10.1007/978-3-030-01270-0_17
11. Gupta, S., Girshick, R., Arbeláez, P., Malik, J.: Learning rich features from RGB-D images for object detection and segmentation. In: Fleet, D., Pajdla, T., Schiele, B., Tuytelaars, T. (eds.) ECCV 2014. LNCS, vol. 8695, pp. 345–360. Springer, Cham (2014). https://doi.org/10.1007/978-3-319-10584-0_23
12. He, K., Gkioxari, G., Dollár, P., Girshick, R.: Mask R-CNN. In: ICCV, pp. 2961–2969 (2017)

13. He, K., Zhang, X., Ren, S., Sun, J.: Deep residual learning for image recognition. In: CVPR, pp. 770–778 (2016)
14. Hu, J., Shen, L., Sun, G.: Squeeze-and-excitation networks. In: CVPR, pp. 7132–7141 (2018)
15. Kendall, A., Badrinarayanan, V., Cipolla, R.: Bayesian SegNet: model uncertainty in deep convolutional encoder-decoder architectures for scene understanding. arXiv preprint arXiv:1511.02680 (2015)
16. Kendall, A., Gal, Y., Cipolla, R.: Multi-task learning using uncertainty to weigh losses for scene geometry and semantics. In: CVPR (2018)
17. Kirillov, A., Girshick, R., He, K., Dollár, P.: Panoptic feature pyramid networks. In: CVPR, pp. 6399–6408 (2019)
18. Kokkinos, I.: Pushing the boundaries of boundary detection using deep learning. arXiv preprint arXiv:1511.07386 (2015)
19. Kokkinos, I.: UberNet: training a universal convolutional neural network for low-, mid-, and high-level vision using diverse datasets and limited memory. In: CVPR (2017)
20. Li, B., Shen, C., Dai, Y., Van Den Hengel, A., He, M.: Depth and surface normal estimation from monocular images using regression on deep features and hierarchical CRFs. In: CVPR, pp. 1119–1127 (2015)
21. Lin, G., Milan, A., Shen, C., Reid, I.: RefineNet: multi-path refinement networks for high-resolution semantic segmentation. In: CVPR, pp. 1925–1934 (2017)
22. Lin, G., Shen, C., Van Den Hengel, A., Reid, I.: Efficient piecewise training of deep structured models for semantic segmentation. In: CVPR, pp. 3194–3203 (2016)
23. Lin, T.Y., Dollár, P., Girshick, R., He, K., Hariharan, B., Belongie, S.: Feature pyramid networks for object detection. In: CVPR, pp. 2117–2125 (2017)
24. Liu, F., Shen, C., Lin, G., Reid, I.: Learning depth from single monocular images using deep convolutional neural fields. TPAMI 38(10), 2024–2039 (2015)
25. Liu, S., Johns, E., Davison, A.J.: End-to-end multi-task learning with attention. In: CVPR (2019)
26. Long, J., Shelhamer, E., Darrell, T.: Fully convolutional networks for semantic segmentation. In: CVPR, pp. 3431–3440 (2015)
27. Lu, Y., Kumar, A., Zhai, S., Cheng, Y., Javidi, T., Feris, R.: Fully-adaptive feature sharing in multi-task networks with applications in person attribute classification. In: CVPR (2017)
28. Maninis, K.K., Pont-Tuset, J., Arbeláez, P., Van Gool, L.: Convolutional oriented boundaries: from image segmentation to high-level tasks. TPAMI 40(4), 819–833 (2017)
29. Maninis, K.K., Radosavovic, I., Kokkinos, I.: Attentive single-tasking of multiple tasks. In: CVPR, pp. 1851–1860 (2019)
30. Martin, D.R., Fowlkes, C.C., Malik, J.: Learning to detect natural image boundaries using local brightness, color, and texture cues. TPAMI 5, 530–549 (2004)
31. Misra, I., Shrivastava, A., Gupta, A., Hebert, M.: Cross-stitch networks for multi-task learning. In: CVPR (2016)
32. Neven, D., De Brabandere, B., Georgoulis, S., Proesmans, M., Van Gool, L.: Fast scene understanding for autonomous driving. In: IV Workshops (2017)
33. Newell, A., Yang, K., Deng, J.: Stacked hourglass networks for human pose estimation. In: Leibe, B., Matas, J., Sebe, N., Welling, M. (eds.) ECCV 2016. LNCS, vol. 9912, pp. 483–499. Springer, Cham (2016). https://doi.org/10.1007/978-3-319-46484-8_29
34. Ren, S., He, K., Girshick, R., Sun, J.: Faster R-CNN: towards real-time object detection with region proposal networks. In: NIPS, pp. 91–99 (2015)

35. Ronneberger, O., Fischer, P., Brox, T.: U-Net: convolutional networks for biomedical image segmentation. In: Navab, N., Hornegger, J., Wells, W.M., Frangi, A.F. (eds.) MICCAI 2015. LNCS, vol. 9351, pp. 234–241. Springer, Cham (2015). https://doi.org/10.1007/978-3-319-24574-4_28
36. Roy, A., Todorovic, S.: Monocular depth estimation using neural regression forest. In: CVPR, pp. 5506–5514 (2016)
37. Ruder, S.: An overview of multi-task learning in deep neural networks. arXiv preprint arXiv:1706.05098 (2017)
38. Sener, O., Koltun, V.: Multi-task learning as multi-objective optimization. In: NIPS (2018)
39. Sermanet, P., Eigen, D., Zhang, X., Mathieu, M., Fergus, R., LeCun, Y.: OverFeat: integrated recognition, localization and detection using convolutional networks. arXiv preprint arXiv:1312.6229 (2013)
40. Silberman, N., Hoiem, D., Kohli, P., Fergus, R.: Indoor segmentation and support inference from RGBD images. In: Fitzgibbon, A., Lazebnik, S., Perona, P., Sato, Y., Schmid, C. (eds.) ECCV 2012. LNCS, vol. 7576, pp. 746–760. Springer, Heidelberg (2012). https://doi.org/10.1007/978-3-642-33715-4_54
41. Sinha, A., Chen, Z., Badrinarayanan, V., Rabinovich, A.: Gradient adversarial training of neural networks. arXiv preprint arXiv:1806.08028 (2018)
42. Standley, T., Zamir, A.R., Chen, D., Guibas, L., Malik, J., Savarese, S.: Which tasks should be learned together in multi-task learning? arXiv preprint arXiv:1905.07553 (2019)
43. Sun, K., Xiao, B., Liu, D., Wang, J.: Deep high-resolution representation learning for human pose estimation. In: CVPR, pp. 5693–5703 (2019)
44. Vandenhende, S., Georgoulis, S., De Brabandere, B., Van Gool, L.: Branched multi-task networks: deciding what layers to share. arXiv preprint arXiv:1904.02920 (2019)
45. Wang, J., et al.: Deep high-resolution representation learning for visual recognition. arXiv preprint arXiv:1908.07919 (2019)
46. Wang, P., Shen, X., Lin, Z., Cohen, S., Price, B., Yuille, A.L.: Towards unified depth and semantic prediction from a single image. In: CVPR, pp. 2800–2809 (2015)
47. Xu, D., Ouyang, W., Wang, X., Sebe, N.: Pad-Net: multi-tasks guided prediction-and-distillation network for simultaneous depth estimation and scene parsing. In: CVPR, pp. 675–684 (2018)
48. Xu, D., Wang, W., Tang, H., Liu, H., Sebe, N., Ricci, E.: Structured attention guided convolutional neural fields for monocular depth estimation. In: CVPR, pp. 3917–3925 (2018)
49. Zamir, A.R., Sax, A., Shen, W., Guibas, L.J., Malik, J., Savarese, S.: Taskonomy: disentangling task transfer learning. In: CVPR (2018)
50. Zhang, Z., Cui, Z., Xu, C., Jie, Z., Li, X., Yang, J.: Joint task-recursive learning for semantic segmentation and depth estimation. In: Ferrari, V., Hebert, M., Sminchisescu, C., Weiss, Y. (eds.) ECCV 2018. LNCS, vol. 11214, pp. 238–255. Springer, Cham (2018). https://doi.org/10.1007/978-3-030-01249-6_15
51. Zhang, Z., Cui, Z., Xu, C., Yan, Y., Sebe, N., Yang, J.: Pattern-affinitive propagation across depth, surface normal and semantic segmentation. In: CVPR, pp. 4106–4115 (2019)
52. Zhao, X., Li, H., Shen, X., Liang, X., Wu, Y.: A modulation module for multi-task learning with applications in image retrieval. In: Ferrari, V., Hebert, M., Sminchisescu, C., Weiss, Y. (eds.) ECCV 2018. LNCS, vol. 11205, pp. 415–432. Springer, Cham (2018). https://doi.org/10.1007/978-3-030-01246-5_25

Learning to Factorize and Relight a City

Andrew Liu[1]([✉]), Shiry Ginosar[2]([✉]), Tinghui Zhou[3]([✉]), Alexei A. Efros[2]([✉]),
and Noah Snavely[1]([✉])

[1] Google, Berkeley, USA
{ahliu,snavely}@google.com
[2] UC Berkeley, Berkeley, USA
{shiry,efros}@eecs.berkeley.edu
[3] Humen, Inc., San Francisco, USA

Abstract. We propose a learning-based framework for disentangling outdoor scenes into temporally-varying illumination and permanent scene factors. Inspired by the classic intrinsic image decomposition, our learning signal builds upon two insights: 1) combining the disentangled factors should reconstruct the original image, and 2) the permanent factors should stay constant across multiple temporal samples of the same scene. To facilitate training, we assemble a city-scale dataset of outdoor timelapse imagery from Google Street View, where the same locations are captured repeatedly through time. This data represents an unprecedented scale of spatio-temporal outdoor imagery. We show that our learned disentangled factors can be used to manipulate novel images in realistic ways, such as changing lighting effects and scene geometry. Please visit http://factorize-a-city.github.io/ for animated results.

1 Introduction

> *"The city of Sophronia is made up of two half-cities... One of the half-cities is permanent, the other is temporary."*
> —ITALO CALVINO, *Invisible Cities*

Imagine taking an image from every possible location on Earth at every possible time instant throughout history. Adelson and Bergen called this hypothetical construct the *plenoptic function* [2]. In practice, of course, it would be impossible to capture or store such a massive dataset. Yet, the data must also be highly redundant and compressible. There will be many images of the same view with slightly different illumination, many images capturing different places under the same conditions, etc. In other words, each image within this hypothetical dataset should have a low intrinsic dimensionality. Rather than store all pixels, we could instead store a small number of intrinsic, disentangled factors representing scene geometry, illumination conditions, etc.—if only we knew what those parameters were and how to reconstruct an image from them.

Electronic supplementary material The online version of this chapter (https://doi.org/10.1007/978-3-030-58548-8_32) contains supplementary material, which is available to authorized users.

A. Vedaldi et al. (Eds.): ECCV 2020, LNCS 12349, pp. 544–561, 2020.
https://doi.org/10.1007/978-3-030-58548-8_32

Input Image   Changing Sun Position   Changing Sky Illumination

Fig. 1. We learn to disentangle temporally-varying scene factors from permanent ones. We can manipulate the learned factors to relight scenes, e.g., by editing sun position and sky conditions. While we train our model on panoramas of NYC *(top)*, it generalizes at test time to images of other cities such as Paris *(bottom)*.

In this paper, we ask whether we can learn such a lower-dimensional representation from a sparse sampling of the plenoptic function on the scale of an entire city. Until recently, large-scale visual data that varies both in space and, separately, in time was difficult to obtain. Fortunately, there have been systematic efforts to capture the world through projects like Google Street View (GSV). While GSV is known for its worldwide coverage, it has also accumulated many samples of the world over time, powering features like Street View Time Machine (GSV-TM). However, GSV-TM still represents an extremely sparse sampling of the plenoptic function.

We use GSV to learn to factor a city's worth of outdoor panoramas into a single low-dimensional representation. In particular, we organize a large set of historical GSV panoramas of New York City into *assembled timelapses* at 100,000 fixed locations captured over time. These enable us to train an unsupervised model to disentangle two latent factors: illumination factors that vary over time, and geometric scene properties that are more permanent.

Once we learn a disentangled set of latent factors, we can synthesize missing data in our incomplete sampling of the plenoptic function by simply swapping or modifying the underlying factors. As illustrated in Fig. 1, our learned factorization can generate synthetic images of the same scene with completely novel illumination. Our disentangled factors are flexible enough to relight test scenes from a single panorama and can even be applied to entirely new cities like Paris.

2   Related Work

Intrinsic Images. Decomposing images into their underlying components is a well-studied problem [5]. For instance, the classic intrinsic images problem describes images as a combination of *reflectance* (i.e., scene albedo), and *shading* (effects induced by lighting) [1]. This problem is underconstrained as there are an infinite number of possible solutions for a single image. However, the regularities in natural scenes and lighting conditions allow for priors on the decomposition. While such priors can be manually crafted [4], many recent methods attempt to learn priors from data, using full supervision from synthetic data [23], sparse supervision from human annotations [6,39], or self-supervision from synthetic models [16]. Yet another kind of supervision comes from *timelapse videos* [24], which feature image sequences with constant reflectance but varying illumination. Such work hearkens back to classic work on deriving intrinsic images from image stacks [37], and is an inspiration for our work. However, while intrinsic image methods allow for editing reflectance or shading for a specific image, they use high-dimensional *pixel-level* descriptions of lighting that are not transferable across scenes. In our case, our model learns an illumination descriptor that can be meaningfully transferred from one image to another, e.g., to relight an image with an illumination from a completely different scene. Such "mix-and-match" capabilities are beyond the power of standard intrinsic images.

Inverse Graphics. An alternative way to factor visual appearance is via 3D reconstruction of the scene into underlying physical components like 3D shape, materials, and lighting. Such methods have been successful in several specific domains, including faces [33], single objects [18,41], or indoor scenes trained from synthetic data [25,32]. 3D reconstruction has also been used explicitly as a preprocess to aid in modeling visual appearance [19,26,27,30]. Most relevant to us are Martin-Brualla *et al.* [26], who organized millions of internet photos into a dense 3D and temporal reconstructions, and Meshry *et al.* [27], who employed a dense 3D reconstruction with a neural rendering pipeline to synthesize scene appearances. However, explicit 3D reconstruction methods require hundreds of images to create a 3D model and cannot generalize to novel test-time scenes. In contrast, we choose to handle geometry implicitly—allowing us to holistically learn to disentangle factors across many scenes composed of a few images each, and then generalize to novel settings, even single images.

Some recent inverse graphics methods learn to infer shape, appearance, and materials for new outdoor scenes, not just scenes observed during training. Yu and Smith train on multi-view stereo data using a physics-based inverse graphics model, and can infer explicit scene properties for novel test images, enabling relighting tasks [38]. Our work achieves a similar capability, but relies on a more implicit representation of geometry and illumination that can be learned solely from timelapse data, without requiring depth or surface normals during training.

Timelapse and Webcam Data. Timelapses are a popular source of data for capturing time-related effects. Applications include intrinsic images [24,37], scene-specific factorizations via physical shading models [34], illuminant transfer [21], analysis of worldwide temporal variations [15], motion denoising [31], learning temporal object transformations [40], and weather attribute manipulation [20]. However, prior work is limited by the variety and size of available data. The largest existing set of standard webcam data is the AMOS dataset of Jacobs *et al.* [15], which archived 29,445 webcams and 95 million images. BigTime [24] uses a much smaller set of 6,500 images from 195 timelapse sequences. Both

Fig. 2. *Left:* A Manhattan intersection. *Center:* Multiple Google Street View panoramic captures of this intersection forms an *assembled timelapse* stack. *Right:* The train and test split over the greater NYC area. Training stacks are drawn from the blue region, and test stacks from the yellow region. (Color figure online)

datasets sample *time* much more densely than *space*. In contrast, we leverage the vast amounts of data from Google Street View to create *assembled timelapses* of the same location captured at different times, across a large number of locations. This allows us to collect an order of magnitude more data than previously published [15]. We additionally note that data collection from Street View scales more easily than [15] which requires crawling the internet for webcam streams.

Learning from Street View. Google Street View (GSV), a large dataset of images sampling much of the world's streets, represents a compelling source of data for computer vision research. Researchers have utilized Google Street View images to learn about visual elements [9] or historical architectural styles [22] specific to certain cities like Paris, to predict non-visual city attributes [3,10,28], for localization [11], or to understand the relationship between satellite imagery and street-level views [35]. In our work we use historical GSV Time Machine imagery to observe how the world changes over time by assembling timelapses for a large number of locations. Such a large, comprehensive dataset is key to our unsupervised approach for learning to factor illumination from scene geometry.

3 Google Street View Time Machine Data

Google Street View (GSV) hosts an amazing quantity of panoramas capturing street scenes worldwide. Because GSV repeatedly captures many places over time, it can be treated as a sparse, imperfectly aligned, and irregularly-sampled collection of timelapse videos. These historical images are saved as part of the GSV-Time Machine (GSV-TM), which we mine to collect our dataset.

We focus on New York City, due to the richness of NYC scenes and the relative wealth of data. To assemble timelapses, we collect panoramas within NYC along with their timestamps and camera poses in a geographic coordinate system [8]. We greedily cluster nearby panoramas into sets of eight, which we refer to as *stacks*. The region we use and an example stack are shown in Fig. 2.

From the area shown in Fig. 2 (right) we collect ~100K assembled timelapse stacks for training (comprised of 800K individual panoramas stitched from 10 million captures) and 16K test stacks. We crop the sky and ground regions such

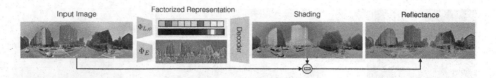

Fig. 3. Disentangling a single image. At test time, we *encode* a single image into disentangled time-varying and permanent factors. We train with the constraint that shading and reflectance images can be *decoded* from this learned factored representation.

that our final panoramas are 960×320. These sRGB panoramas can optionally be gamma-corrected before further processing.

4   Method

Our goal is to discover a low-dimensional representation of the world where temporally varying effects, such as different illumination conditions, are disentangled from permanent objects, such as buildings and roads.

One form of disentanglement is *intrinsic images*, a per-pixel decomposition into reflectance and shading images. However, such a disentangled representation is very low-level—a particular shading image cannot be used to relight a different scene. Instead, we seek to encode an image into higher-level latent factors capturing scene and illumination properties described above, as illustrated in Fig. 3. How can we find such a factorization? Our insight is that we should still be able to *decode* intrinsic images from our factored representation, as illustrated on the right side of Fig. 3. The decoded reflectance and shading images should recombine to form the original image, providing us with an autoencoder-style method for learning our high-level factorization [16]. However, such an image reconstruction framework alone would provide a very weak supervision signal. Our second insight is to learn from huge numbers of *timelapse stacks* mined from GSV-TM. Within such stacks, we assume the scene factors to be constant. This insight is inspired by the work of Li and Snavely, who learn intrinsic images from timelapse videos [24]. In our case we learn a high-level factorization that enables more powerful capabilities.

4.1   Encoder-Decoder Architecture

Figure 3 shows our encoder-decoder architecture with its learnt factored representation. Given an image, our encoders produce latent factors, capturing various temporal and permanent effects, that can be decoded to a log-shading intrinsic image. We use the intrinsic images equation ($\log(\text{Reflectance}) = \log(\text{Image}) - \log(\text{Shading})$) to compute a reflectance image by subtracting the temporally varying effects, represented by the shading image, from the original image.

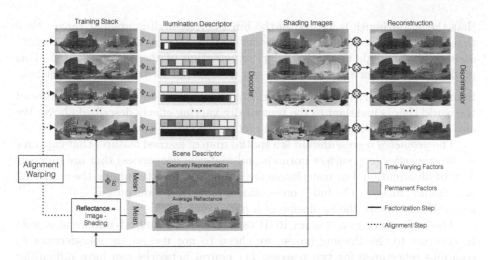

Fig. 4. Training with timelapses. We train encoders to disentangle an assembled timelapse stack into two factors: *illumination descriptors* that capture the time-varying aspects of each image, and a single *scene descriptor* that captures the permanent elements of the entire timelapse stack, such as the scene geometry. We train a generator to transform the disentangled factors into shading and reflectance images from which we can reconstruct the original images. As indicated by the dotted pathways, we also simultaneously solve for the alignment of the individual frames in the input timelapse.

Our model's latent factors are organized into two sets of descriptors, as shown in Fig. 4: an *illumination descriptor* represents temporally varying aspects of the scene and a *scene descriptor* represents the permanent aspects.

Illumination Descriptor: Our illumination descriptor captures the factors of the world that encode temporal variation like lighting. This descriptor is comprised of two disentangled sub-factors:

The *lighting context* $L \in \mathbb{R}^{32}$ is a global latent feature that captures the overall ambient illumination properties, such as atmospheric conditions and cloud cover. Our lighting context encoder Φ_L encodes an image to this embedding.

The *sun azimuth angle*, φ is an explicit factor representing the horizontal position of the sun in a given panorama. We model sun azimuth explicitly because, unlike illumination patterns, variations in sun azimuth have a simple geometric meaning, with a value in the range $[-\pi, \pi]$. Despite this simple parameterization, the effect of sun azimuth on a rendered scene is highly complex. Therefore an explicit azimuth factor allows our model to combine the factor's underlying mathematical simplicity with a network's ability to model complex behaviors.

Rather than regress to a scalar angle, we instead represent φ internally as a discretized distribution over sun angle (with $k = 40$ bins). Inspired by prior work on illumination estimation [12], our azimuth encoder Φ_φ is a horizontally fully-convolutional network that takes as input a panorama, and produces a 40-way softmax distribution φ, where each bin corresponds to the probability that the sun azimuth is located in the bin's corresponding angular range. Note that given this discrete distribution over angles, we can differentiably compute a single scalar angle as the (circular) expectation of the distribution, $\bar{\varphi}$. This predicted scalar sun angle is used by our decoder for normalizing sun position.

Scene Descriptor: Our scene descriptor captures the permanent structure of the world that is invariant to the temporally varying effects described above. We also divide this descriptor into two disentangled sub-factors:

The *geometry representation* is a spatial map of learned features that captures scene properties (e.g. surface normals and material properties) that are independent of illumination, but nonetheless are important to determining the rendering of a shading images. The fully convolutional encoder Φ_E outputs $E \in \mathbb{R}^{\frac{H}{8} \times \frac{W}{8} \times 16}$ where H and W are the resolution of a panorama.

The *reflectance image* is an RGB estimate of the underlying scene albedo. In contrast to the shading image, we chose to not use an encoder-decoder to compute reflectance for two reasons: (1) neural networks can have difficulties preserving high-frequency textures that are important for visual quality and (2) it suffices to predict only one intrinsic image component because its complement component has a closed form solution based on the intrinsic images equation.

Decoder: Given a set of learned factors (sun azimuth angle $\bar{\varphi}$, lighting context L, and geometry factor E), our decoder G is trained to generate an outdoor shading image. To facilitate training of G, one insight is that it is easier to learn to synthesize shading images with a fixed sun azimuth angle than with all possible angles. Further, we can normalize a panorama by its predicted sun azimuth angle by simply rotating it by the negative of that angle (i.e., circular horizontal translation). Hence, our decoder operates as follows: (1) use the predicted sun azimuth angle $\bar{\varphi}$ to rotate the geometry factor image E to a fixed sun angle, (2) decode the sun-normalized geometry image with lighting context L to a shading image, and (3) rotate the result back to the original coordinate frame.

We use the Spatial Adaptive Instance Normalization (SPADE) generator of Park *et al.* [29] to model the complex interactions between geometry and illumination in our decoder G. The SPADE generator takes the lighting context L as the network's noise input. We apply the insights from above and rotate the geometry representation E by $-\bar{\varphi}$ before using it as the SPADE conditioning.

While some prior works model shading with a grayscale image, such a model cannot capture real-world, colored illumination. Inspired by Sunkavalli *et al.* [34], we augment our decoder's gray-scale shading predictions with a bi-color assumption by additionally predicting two global color illuminants c_1 and c_2, corresponding to sunlight and skylight, and a per-pixel mixing weight M that models how much each pixel is illuminated by the sun or sky. For further details about the decoder architecture, please refer to the supplemental material.

4.2 Training

Learning to factor single images without *any* supervision is challenging—there is simply not enough information in a single image to disentangle scene factors

Fig. 5. Alignment results. We show stack averages, cropped for emphasis, before and after our alignment process. Aligning the estimated permanent reflectances rather than the input images results in good alignment and therefore crisp stack averages.

from illumination factors. However, a GSV-TM stack depicts the same underlying permanent scene under diverse temporally varying illuminations, providing a useful training signal. Our training procedure, shown in Fig. 4, learns to disentangle factors *within* a stack by separating the permanent geometry of the scene shared by all images in the stack from the varying lighting. The trained model can be applied to a single image at test time.

Given a timelapse stack, we run our encoder on individual frames to get a stack of encoded geometry representations and illumination descriptors. Because we assume the stack's geometry to be constant across time, we average the encoded geometry maps over the stack, resulting in a single shared geometry map, $\bar{E}$. From this shared geometry map, and the per-image illumination factors, our decoder produces a stack of shading and reflectance image pairs. As with geometry, we wish the scene's albedo to be constant across time. Accordingly, we impose a reflectance consistency loss $\mathcal{L}_{RC}$ that computes the L_1 distance between pairs of reflectance images from different frames. This loss encourages the encoder-decoder network to remove temporal variation from the encoded permanent factors such that the reflectances are constant across a stack.

As demonstrated in the right half of Fig. 4, we average the stack's reflectance images across frames to get the stack's shared reflectance. The shared reflectance is recomposited with the shading image of each frame in the stack to reconstruct the original pixels of each input frame. These reconstructions are used to drive the learning process via image synthesis losses.

4.3   Stack Alignment

Unlike traditional webcam data, our assembled GSV-TM timelapses do not come from stationary cameras. While each stack consists of nearby panoramas, they are not perfectly co-located and aligned. As shown in Fig. 5, the average of the stack reveals visible misalignment artifacts resulting from this parallax.

We could use 3D reconstruction methods as the basis for image alignment, but opted for a simpler 2D approach inspired by image congealing [13], and compute 2D warps that best align the images in each stack. Given a raw stack of imperfectly aligned images, we define Θ, an 8×32 grid of per-image control points initialized as the identity warp. The control points define a 2D spline used to differentiably warp each image within a stack to align with the rest.

To find the control points that best align images within a stack, we run gradient descent to minimize pixel alignment error. While one could use original image pixels to measure misalignment, we found that photometric differences across the stack due to varying lighting conditions led to poor alignments. Instead, we compute error on estimated *reflectance* images by reusing our previously defined reflectance consistency loss, $\mathcal{L}_{RC}$, to update alignment parameters. This approach is indicated by the dotted pathway in Fig. 4. By jointly minimizing alignment and intrinsic image decomposition, we create a positive feedback loop—as timelapse alignment improves, factorization becomes easier and vice versa.

4.4 Losses

Our losses are optimized over alignment parameters Θ, factorization encoders Φ_L Φ_φ, and Φ_E, and decoder G. We train a multi-scale patch discriminator [14,36] D to ensure that the stack reconstructions with shared reflectances look realistic.

Our primary loss for learning the disentanglement is the reflectance consistency loss $\mathcal{L}_{RC}$ described in Sect. 4.2. We include standard image generation losses on the reconstructed stack to ensure high quality synthesis results: a perceptual loss $\mathcal{L}_{VGG}$ [17], an adversarial loss $\mathcal{L}_{GAN}$ [7], and a feature matching loss $\mathcal{L}_{FM}$ [36]. Finally, because intrinsic images have a fundamental color ambiguity, we also include a white light penalty, $\mathcal{L}_{WL}$ that biases our encoder-decoder towards white-balanced reflectance outputs. Our overall objective function is:

$$\min_{\Theta} \max_{D} \min_{G,\Phi_L,\Phi_\varphi,\Phi_E} \mathcal{L}_{RC} + \mathcal{L}_{Gen} + \mathcal{L}_{GAN} \qquad (1)$$

where $\mathcal{L}_{Gen}$ is a weighted sum of $\mathcal{L}_{FM}, \mathcal{L}_{WL}, \mathcal{L}_{VGG}$ that measures the generative quality of the reconstructed images. We include additional descriptions, alignment results, insights, and analysis for reproducibility in the supplemental material.

5 Experiments

We evaluate our factorization method in two ways: 1) we compare to intrinsic image decomposition baselines in the single-scene setting, and 2) we apply our method to the task of transferring illumination descriptors across different scenes, a new capability enabled by our disentanglement. In both cases, we measure success by the quality of reconstructed images derived from swapping their disentangled factors with ones borrowed from other images as in [39].

Data. At test time, our network can take as input either an assembled timelapse stack or a single panorama. In order to align test-time stacks like those shown in Fig. 5, we estimate spline parameters by computing a gradient for alignment only, while keeping the weights of the factorization part of the network frozen. Below, we present results for stack as well as single-image inputs.

In particular, we show single-image test-time results on GSV imagery from cities never seen during training, such as Paris, as well as images from the Outdoor Laval HDR dataset [12]. This dataset contains HDR panoramas of outdoor scenes that are tonemapped to sRGB to match GSV. We use this data to compare to existing sRGB intrinsic image methods and to test generalization from GSV to a different domain of panoramas.

Baselines. Given the novelty of our problem, we perform model ablations to measure the individual benefits of various components. All ablated models are trained with the same losses and number of iterations as our full method. We report results on the following ablations:

– **Mono-color shading**: We ablate the bi-color shading by training our model with a mono-color assumption similar to that of Li and Snavely [24].
– **w/o alignment training**: Trained without the alignment feedback loop.
– **w/ unaligned test stacks**: Uses unaligned test stacks to measure the effect of ablating alignment at training (above) vs. at both training and test time.
– **w/o azimuth encoder**: Our model trained without an azimuth encoder nor normalizing for sun position.

Additionally, we consider the following baselines:

– **Pixel nearest neighbor**: Given a target image, we find the pixel-wise nearest neighbor in its aligned stack and report the error resulting from using that image as our synthesized result.
– **Weiss's MLE Intrinsics** [37]: use handcrafted priors on gradients extracted from image sequences.
– **Zhou et al.** [39]: learn to mimic human judgments of relative reflectance.
– **Li and Snavely's BigTime** [24] learn shading priors from image sequences.

5.1   Within-Scene Decomposition

Intrinsic image methods aim to decompose an image into shading and reflectance. The quality of a decomposition is measured by its ability to separate illumination effects, like cast shadows, from permanent properties such as albedo. In Fig. 6, we show reflectance and shading computed from a single image using our method and the two deep learning baselines. Both BigTime and Zhou et al. fail to remove cast shadows, as seen by residual shadows encoded in their reflectance. Unlike Zhou et al., our method produces shading images that are piecewise smooth, as expected for planar surfaces like building facades. BigTime struggles in outdoor settings because their single global illuminant cannot predict multiple illumination colors. Finally, both baselines incorrectly encode blue sky pixels as reflectance despite the fact that sky color is a temporal property. To further illustrate the advantages of our method over these baselines, Fig. 7 shows the

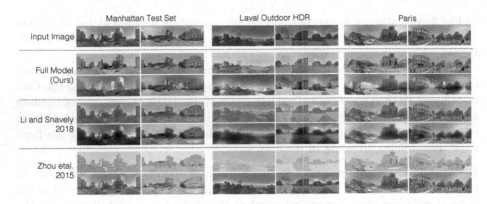

Fig. 6. Qualitative results on an intrinsic image decomposition task. We compare single-image decompositions of our method with Li and Snavely [24] and Zhou *et al.* [39]. Compared to the baselines, our reflectance images do not have residual shadows. Our method, trained on NYC, generalizes at test-time to Laval Outdoor HDR Panoramas [12] as well as to GSV imagery from Paris.

results of relighting pairs of images of the same scene by swapping reflectances within the pair. Unlike the baselines, our clean reflectance image allows us to relight the scene successfully.

Scene Consistency Verification. Since MLE Intrinsics [37] only works on timelapse stacks of single scenes, we devise a way to quantitatively compare to their method. We split our aligned test stacks to two smaller substacks of 4 images each. For each substack, each method predicts a single reflectance image and four shading images. Since both substacks capture the same underlying scene, the predicted reflectances should be consistent across the two. As in the case of single images (Fig. 6), we can test the consistency of the predicted reflectance for the depicted scene by swapping the predicted reflectance images between the two substacks and reconstructing the four input images in each substack from their shading and *swapped* reflectance images. We refer to this experiment as *scene consistency verification* because the reconstruction error is minimized when the predicted reflectances are identical for the two substacks.

We report the mean squared reconstruction error (MSE) between the input stack and the swap reconstructions in Table 1. Our method outperforms the three baselines at image reconstruction in this setting. We speculate that prior methods are hindered by their reliance on hand-defined shading priors and limited training data. In contrast, our massive dataset provides enough supervision for learning a good decomposition without shading priors. Interestingly, ablating the azimuth encoder does not degrade performance on this task, suggesting that a simpler setup is sufficient for within-scene illumination transfer.

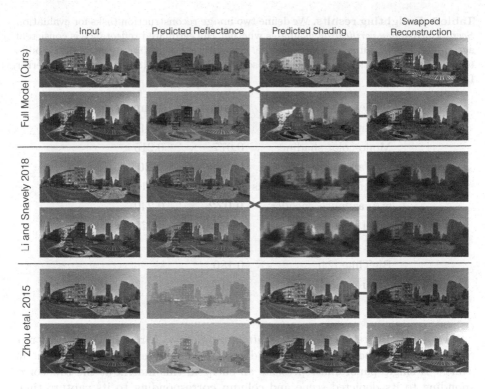

Fig. 7. Transferring illumination within a scene. Given a pair of images of the same scene under different illuminations (*left*), we disentangle the permanent and varying factors and decode their reflectance and shading (*middle*). To test the permanency of the estimated reflectance for the depicted scene, we swap reflectances within the pair and combine them with the estimated shading to reconstruct the original images (*right*). Red and blue paths connect the components used to reconstruct each image. Our method produces a reflectance, clean of any lighting, which can be safely swapped between captures of the same scene and still result in good reconstructions. (Color figure online)

5.2   Cross-Scene Factorization

Unlike intrinsic images methods, our factorization allows us to transfer illumination descriptors *across* scenes. Using our disentangled factors, we can synthesize a given scene under completely new lighting conditions, borrowed from a *different* location. For the purpose of evaluating the success of this cross-scene relighting process, we devise a way to compare the novel synthesis to ground truth. Namely, because illumination changes relatively slowly, we assume that images captured within 5 min across the city have the same illumination descriptor. Hence, we can relight a given scene, A, captured at time T_1 using illumination descriptors transferred from a different location, B, captured at time T_2. We then compare the resulting synthetic image of scene A at time T_2 to ground truth captures of scene A captured at a time close to T_2.

Table 1. Relighting results. We define two image reconstruction tasks for evaluation. *Scene consistency verification* evaluates whether the estimated reflectance is consistent across multiple captures of a single scene. *Space-time completion* evaluates the ability to transfer illumination across different scenes. We report MSE reconstruction error. Lower is better.

Model	Consistency	Completion
Full model (ours)	**0.071**	**0.196**
Mono-color shading	0.077	0.215
w/o alignment training	0.082	**0.201**
w/ unaligned test stacks	0.090	0.210
w/o azimuth encoder	**0.072**	0.240
Pixel nearest neighbors	0.274	0.278
MLE Intrinsic [37]	0.114	—
BigTime [24]	0.180	—
Zhou *et al.* [39]	0.217	—

We name this task *space-time matrix-completion*. A row in the matrix represents a unique point in "space" and a column represents a unique point in "time". A single panorama represents an entry in this matrix at the row corresponding to its depicted scene and column corresponding to its capture time. We can withhold entries in the matrix and reconstruct them by combining a scene descriptor derived from images in the same row, with an illumination descriptor extracted from a different scene from the same column. Table 1 shows the reconstruction MSE for each ablation between held-out and reconstructed views. Our full model and the *w/o alignment training* ablation show significant improvements over other ablations.

While alignment training does not significantly affect the performance of our model on this task (*w/o alignment training*), its performance degrades significantly on unaligned stacks (*w/ unaligned test stacks*). This indicates that alignment may be optional during training but is crucial for reconstruction. Additionally, unlike with the substack swap task, explicitly representating sun azimuth improves transferability of lighting descriptors across scenes.

6   Applications

We now present applications where we synthetically modify a panorama. These applications are uniquely enabled by our intrinsic factorization that disentangles time-varying effects from the permanent scene properties.

Changing Sun Position. Our model disentangles sun azimuth angle from scene and lighting context factors. Once a scene is factorized, we can visualize what a scene looks like when the sun angle is changed. Figure 8 shows examples of test scenes synthesized with new sun azimuth angles. Note that cast shadows and illumination on building faces change realistically with the rotation.

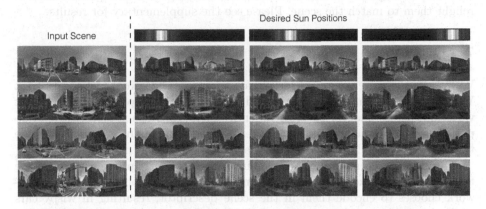

Fig. 8. Manipulating sun position. We can specify the sun position for an input scene and relight it realistically. Please see the supplemental video for full animations.

Fig. 9. Changing sky illumination. We can relight *novel* scenes by transferring the disentangled time-varying factors from one scene to another. Here we swap the illumination descriptors of a pair of input scenes to visualize what each scene might look like under a new illumination. The red and blue paths indicate the components used to reconstruct each relit scene. (Color figure online)

Relighting a *Novel* Scene. Our lighting context encodes the stylistic quality of illumination. As shown in Fig. 9, we can transfer the whole illumination descriptor, including sun azimuth, from one panorama to another with a new scene geometry. Results for transferring *only* lighting context can be found in the supplemental material. The supplemental material also demonstrates relighting a spatial sequence of panoramas from different times to a fixed illumination, thus producing a virtual drive through Manhattan.

Editing Scene Geometry. While shading and azimuth capture the essence of time, the scene descriptor encodes structures. By copy-pasting regions of the scene descriptors, we can transplant the buildings into new panoramas and relight them to match the scene. Please see the supplementary for results.

7 Discussion

We proposed a novel source of large-scale timelapse data from historical Street View data, and a learning-based method for factorizing temporal and permanent variations across imagery covering an entire city. Our learned factorization out-performs state-of-the-art intrinsic images methods, and enables cross-scene style transfer via manipulating our learned factors.

Our method has a few limitations. First, the scene descriptor learns to encode transient objects like cars. While moving objects are temporal effects, the net-work chooses to encode them in the scene descriptor, resulting in wispy cars appearing in the generator output. Second, high-frequency details such as cast shadows from tree branches are difficult to synthesize. Third, when the alignment module fails, the shared reflectance of a stack will appear blurry. Please see the supplemental material for examples of failure cases. Finally, when our permanence assumptions fail to hold—for instance when buildings are repainted or rebuilt—our assumption that the scene descriptor is constant across time is violated.

Despite these limitations, our work points towards a new approach to modeling and synthesizing the space of outdoor scenes, wherein we can learn to separate factors that persist at different time scales. An intriguing direction for future work is to expand to a richer range of timescales, for instance modeling transient effects (moving people, cars, etc.), effects with annual cycles (e.g., seasons), long-term changes like weathering, etc.

Acknowledgements. We would like to thank Richard Tucker, Richard Bowen, Ameesh Makadia, and Vincent Sitzmann for insightful discussions. We would also like to thank Angjoo Kanazawa and Tim Brooks for their help with preparing the manuscript. This work is supported, in part, by NSF grant IIS-1633310.

References

1. Adelson, E.H., Pentland, A.P.: The perception of shading and reflectance. In: Knill, D.C., Richards, W. (eds.) Perception as Bayesian Inference, pp. 409–423. Cambridge University Press, New York (1996)
2. Adelson, E.H., Bergen, J.R.: The plenoptic function and the elements of early vision. In: Landy, M., Movshon, J.A. (eds.) Computational Models of Visual Processing, pp. 3–20. MIT Press, Cambridge (1991)
3. Arietta, S.M., Efros, A.A., Ramamoorthi, R., Agrawala, M.: City forensics: using visual elements to predict non-visual city attributes. IEEE Trans. Visual Comput. Graphics **20**(12), 2624–2633 (2014). https://doi.org/10.1109/TVCG.2014.2346446

4. Barron, J.T., Malik, J.: Shape, illumination, and reflectance from shading. Trans. Pattern Anal. Mach. Intell. **37**(8), 1670–1687 (2015)
5. Barrow, H.G., Tenenbaum, J.M.: Recovering intrinsic scene characteristics from images. Comput. Vis. Syst. **2**(3–26), 2 (1978)
6. Bell, S., Bala, K., Snavely, N.: Intrinsic images in the wild. ACM Trans. Graphics (SIGGRAPH) **33**(4), 159:1–159:12 (2014). https://doi.org/10.1145/2601097. 2601206
7. Brock, A., Donahue, J., Simonyan, K.: Large scale GAN training for high fidelity natural image synthesis. In: International Conference on Learning Representations (2019)
8. Klingner, B., Martin, D., Roseborough, J.: Street view motion-from-structure-from-motion. In: Proceedings of the International Conference on Computer Vision (ICCV) (2013)
9. Doersch, C., Singh, S., Gupta, A., Sivic, J., Efros, A.A.: What makes Paris look like Paris? ACM Trans. Graphics (SIGGRAPH) **31**(4), 101:1–101:9 (2012)
10. Gebru, T., et al.: Using deep learning and google street view to estimate the demographic makeup of neighborhoods across the United States. Proc. Natl. Acad. Sci. **114**(50), 13108–13113 (2017). https://doi.org/10.1073/pnas.1700035114
11. Gronat, P., Obozinski, G., Sivic, J., Pajdla, T.: Learning and calibrating per-location classifiers for visual place recognition. In: Proceedings of the IEEE Conference on Computer Vision and Pattern Recognition (CVPR), June 2013
12. Hold-Geoffroy, Y., Sunkavalli, K., Hadap, S., Gambaretto, E., Lalonde, J.F.: Deep outdoor illumination estimation. In: Proceedings of the Computer Vision and Pattern Recognition (CVPR), July 2017
13. Huang, G.B., Jain, V., Learned-Miller, E.: Unsupervised joint alignment of complex images. In: Proceedings of the International Conference on Computer Vision (ICCV) (2007)
14. Isola, P., Zhu, J.Y., Zhou, T., Efros, A.A.: Image-to-image translation with conditional adversarial networks. In: Proceedings of the Computer Vision and Pattern Recognition (CVPR) (2016)
15. Jacobs, N., Roman, N., Pless, R.: Consistent temporal variations in many outdoor scenes. In: Proceedings of the Computer Vision and Pattern Recognition (CVPR), pp. 1–6, June 2007. https://doi.org/10.1109/CVPR.2007.383258
16. Janner, M., Wu, J., Kulkarni, T.D., Yildirim, I., Tenenbaum, J.: Self-supervised intrinsic image decomposition. In: Neural Information Processing Systems, pp. 5936–5946. Curran Associates, Inc. (2017)
17. Johnson, J., Alahi, A., Fei-Fei, L.: Perceptual losses for real-time style transfer and super-resolution. In: Leibe, B., Matas, J., Sebe, N., Welling, M. (eds.) ECCV 2016. LNCS, vol. 9906, pp. 694–711. Springer, Cham (2016). https://doi.org/10.1007/978-3-319-46475-6_43
18. Kanazawa, A., Tulsiani, S., Efros, A.A., Malik, J.: Learning category-specific mesh reconstruction from image collections. In: Ferrari, V., Hebert, M., Sminchisescu, C., Weiss, Y. (eds.) ECCV 2018. LNCS, vol. 11219, pp. 386–402. Springer, Cham (2018). https://doi.org/10.1007/978-3-030-01267-0_23
19. Laffont, P.Y., Bazin, J.C.: Intrinsic decomposition of image sequences from local temporal variations. In: Proceedings of the International Conference on Computer Vision (ICCV), December 2015
20. Laffont, P.Y., Ren, Z., Tao, X., Qian, C., Hays, J.: Transient attributes for high-level understanding and editing of outdoor scenes. ACM Trans. Graphics (SIGGRAPH) **33**(4), 1–11 (2014)

21. Lalonde, J.F., Efros, A.A., Narasimhan, S.G.: Webcam clip art: appearance and illuminant transfer from time-lapse sequences. ACM Trans. Graphics (SIGGRAPH) **28**(5), 1–10 (2009)
22. Lee, S., Maisonneuve, N., Crandall, D., Efros, A.A., Sivic, J.: Linking past to present: discovering style in two centuries of architecture. In: IEEE International Conference on Computational Photography (ICCP) (2015)
23. Li, Z., Snavely, N.: CGIntrinsics: better intrinsic image decomposition through physically-based rendering. In: Ferrari, V., Hebert, M., Sminchisescu, C., Weiss, Y. (eds.) ECCV 2018. LNCS, vol. 11207, pp. 381–399. Springer, Cham (2018). https://doi.org/10.1007/978-3-030-01219-9_23
24. Li, Z., Snavely, N.: Learning intrinsic image decomposition from watching the world. In: Proceedings of the Computer Vision and Pattern Recognition (CVPR) (2018)
25. Li, Z., Shafiei, M., Ramamoorthi, R., Sunkavalli, K., Chandraker, M.: Inverse rendering for complex indoor scenes: shape, spatially-varying lighting and SVBRDF from a single image. In: Proceedings of the Computer Vision and Pattern Recognition (CVPR) (2020)
26. Martin-Brualla, R., Gallup, D., Seitz, S.M.: Time-lapse mining from internet photos. ACM Trans. Graphics (SIGGRAPH) **34**(4), 62:1–62:8 (2015). https://doi.org/10.1145/2766903
27. Meshry, M., et al.: Neural rerendering in the wild. In: Proceedings of the Computer Vision and Pattern Recognition (CVPR) (2019)
28. Naik, N., Philipoom, J., Raskar, R., Hidalgo, C.: Streetscore - predicting the perceived safety of one million streetscapes. In: 2014 IEEE Conference on Computer Vision and Pattern Recognition Workshops, pp. 793–799, June 2014. https://doi.org/10.1109/CVPRW.2014.121
29. Park, T., Liu, M.Y., Wang, T.C., Zhu, J.Y.: Semantic image synthesis with spatially-adaptive normalization. In: Proceedings of the Computer Vision and Pattern Recognition (CVPR) (2019)
30. Philip, J., Gharbi, M., Zhou, T., Efros, A.A., Drettakis, G.: Multi-view relighting using a geometry-aware network. ACM Trans. Graphics (SIGGRAPH) **38**(4) (2019). http://www-sop.inria.fr/reves/Basilic/2019/PGZED19
31. Rubinstein, M., Liu, C., Sand, P., Durand, F., Freeman, W.T.: Motion denoising with application to time-lapse photography. In: Proceedings of the Computer Vision and Pattern Recognition (CVPR), pp. 313–320, June 2011
32. Sengupta, S., Gu, J., Kim, K., Liu, G., Jacobs, D.W., Kautz, J.: Neural inverse rendering of an indoor scene from a single image. In: Proceedings of the International Conference on Computer Vision (ICCV) (2019)
33. Sengupta, S., Kanazawa, A., Castillo, C.D., Jacobs, D.W.: SfSNet: learning shape, reflectance and illuminance of faces in the wild. In: Proceedings of the Computer Vision and Pattern Recognition (CVPR) (2018)
34. Sunkavalli, K., Matusik, W., Pfister, H., Rusinkiewicz, S.: Factored time-lapse video. ACM Trans. Graphics (SIGGRAPH) (2007). SIGGRAPH 2007. ACM, New York. https://doi.org/10.1145/1275808.1276504
35. Vo, N.N., Hays, J.: Localizing and orienting street views using overhead imagery. In: Leibe, B., Matas, J., Sebe, N., Welling, M. (eds.) ECCV 2016. LNCS, vol. 9905, pp. 494–509. Springer, Cham (2016). https://doi.org/10.1007/978-3-319-46448-0_30
36. Wang, T.C., Liu, M.Y., Zhu, J.Y., Tao, A., Kautz, J., Catanzaro, B.: High-resolution image synthesis and semantic manipulation with conditional GANs. In: Proceedings of the Computer Vision and Pattern Recognition (CVPR) (2018)
37. Weiss, Y.: Deriving intrinsic images from image sequences. In: Proceedings of the International Conference on Computer Vision (ICCV) (2001)

38. Yu, Y., Smith, W.A.: InverseRenderNet: learning single image inverse rendering. In: Proceedings of the Computer Vision and Pattern Recognition (CVPR) (2019)
39. Zhou, T., Krähenbähl, P., Efros, A.A.: Learning data-driven reflectance priors for intrinsic image decomposition. In: Proceedings of the International Conference on Computer Vision (ICCV) (2015)
40. Zhou, Y., Berg, T.L.: Learning temporal transformations from time-lapse videos. In: Leibe, B., Matas, J., Sebe, N., Welling, M. (eds.) ECCV 2016. LNCS, vol. 9912, pp. 262–277. Springer, Cham (2016). https://doi.org/10.1007/978-3-319-46484-8_16
41. Zhu, J.Y., et al.: Visual object networks: image generation with disentangled 3D representations. In: Neural Information Processing Systems (2018)

Region Graph Embedding Network for Zero-Shot Learning

Guo-Sen Xie[1](✉), Li Liu[1], Fan Zhu[1], Fang Zhao[1], Zheng Zhang[2,3](✉),
Yazhou Yao[5], Jie Qin[1], and Ling Shao[1,4]

[1] Inception Institute of Artificial Intelligence, Abu Dhabi, UAE
gsxiehm@gmail.com
[2] Harbin Institute of Technology, Shenzhen, China
darrenzz219@gmail.com
[3] Peng Cheng Laboratory, Shenzhen, China
[4] Mohamed bin Zayed University of Artificial Intelligence, Abu Dhabi, UAE
[5] Nanjing University of Science and Technology, Nanjing, China

Abstract. Most of the existing Zero-Shot Learning (ZSL) approaches
learn direct embeddings from global features or image parts (regions) to
the semantic space, which, however, fail to capture the appearance relationships between different local regions within a single image. In this
paper, to model the relations among local image regions, we incorporate the region-based relation reasoning into ZSL. Our method, termed
as Region Graph Embedding Network (RGEN), is trained end-to-end
from raw image data. Specifically, RGEN consists of two branches: the
Constrained Part Attention (CPA) branch and the Parts Relation Reasoning (PRR) branch. CPA branch is built upon attention and produces
the image regions. To exploit the progressive interactions among these
regions, we represent them as a *region graph*, on which the parts relation reasoning is performed with graph convolutions, thus leading to
our PRR branch. To train our model, we introduce both a *transfer* loss
and a *balance* loss to contrast class similarities and pursue the maximum response consistency among seen and unseen outputs, respectively.
Extensive experiments on four datasets well validate the effectiveness of
the proposed method under both ZSL and generalized ZSL settings.

Keywords: Zero-shot learning · Parts relation reasoning · Balance loss

1 Introduction

Humans can efficiently recognize instances from unseen categories, by simply
exploiting their past knowledge on seen class images as well as descriptions

Electronic supplementary material The online version of this chapter (https://
doi.org/10.1007/978-3-030-58548-8_33) contains supplementary material, which is
available to authorized users.

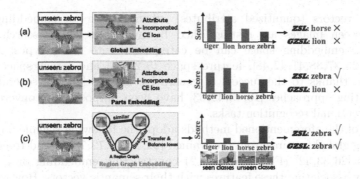

Fig. 1. Three Types of End-to-end ZSL and GZSL Models. (a) Global Embedding, which misclassifies unseen "zebra" to "horse" and "lion" under ZSL and GZSL, respectively. This happens because 1) the global features are not discriminative enough to distinguish these two confused classes, i.e., "zebra" and "horse", and 2) the domain bias under GZSL makes the prediction scores on seen classes ("tiger" and "lion") much higher than those on unseen classes ("horse" and "zebra"). **(b) Parts Embedding**, which correctly classifies "zebra" under ZSL. However, the domain bias still exists, which again results in a misclassification of "zebra" to "lion" under GZSL. **(c) Our RGEN** can distinguish "zebra" under both ZSL and GZSL with a better prediction confidence, which benefits from parts relation reasoning on the region graph and the new training losses (*transfer* and *balance* losses).

of both seen and unseen classes. This capability of perceiving unseen concepts is dubbed Zero-Shot Learning (ZSL) [24,35]. However, most of the available deep learning approaches [17,47,54,60] lack such a ZSL-like ability, e.g., the CNN models [46,68,70,75] usually suffer from insufficient (or no) training data. Moreover, annotating large amounts of data is both time consuming and costly [61,66,69,74,76], and novel (unseen) categories are constantly emerging in practical scenarios [58]. As such, ZSL [2,50] has become an important research topic for its potential to alleviate data annotation costs and handle unseen class recognition.

ZSL typically trains a model by merely leveraging seen class images and then apply it to unseen images, where the label sets of the seen and unseen classes are disjoint. In the ZSL paradigm, the testing label set is unrealistically constrained to only unseen class, hindering its application in the real world. Extending the label set to include both seen and unseen classes during testing leads to Generalized ZSL (GZSL) setting [8,50,65,67]. The semantic descriptions [12] collected for each seen and unseen category ensures efficient knowledge transfer between the two disjoint class sets, making both the ZSL and GZSL tasks feasible.

Semantic descriptions (such as attributes [12], sentences [38] and word vectors [41]) are shared information among seen and unseen categories, through which semantic knowledge is transferred from seen to unseen categories. Attributes are most commonly used and are thus adopted in this paper as well. Seminal works [2,4,14,33,39] on ZSL rely on seen images and their semantic vectors (quantized attributes) when finding an embedding space, where unseen images are distinguished by nearest neighbor search. Specifically, the embedding space can be categorized into three types: semantic space [9,23,37,38,41,42,48], feature space [5,29,40] and latent space [27,59]. Moreover, thanks to the success of generative models [16,71], several feature hallucinating approaches [10,43,49,63] have been proposed for converting ZSL into conventional recognition tasks.

Most of the aforementioned methods adopt the following scheme for ZSL: 1) extracting global features from pre-trained [22,29,62,72,73] or end-to-end trainable nets [26,33,42] (Fig. 1(a)) and 2) constructing embedding or generative models by associating these features with their semantic vectors. However, these approaches cannot efficiently capture the subtle differences between seen and unseen images [52], thus leading to undesirable semantic transfer. Very recently, attention based end-to-end models [52,78,80] have paved the way for discovering more discriminative part (region)[1] features by using semantic vectors as a guidance, showing remarkable improvements under ZSL but not GZSL. However, all these methods focus on direct parts embedding (Fig. 1(b)) of these part features and fail to capture appearance relationships among them. Additionally, issues with domain bias [8,15] still exist, meaning that the learned models merely rely on the seen categories, while ignoring the available unseen attributes.

To tackle the above challenges, in this paper, we first apply the attention method in [52] to generate the attended object regions on each input image. Then, we propose to perform the region-based relation modeling by Graph Convolutional Network (GCN) [21] (Sect. 3.3). Specifically, we represent each input image as a **Region Graph** (Fig. 1(c)) with each node in the graph representing an attended region in the image. The edges of these region nodes are their pairwise appearance similarities. As such, the updated features after the GCN reasoning can capture the appearance relationships among these local parts, which is a complementary cue for improving the ZSL performance. Furthermore, **Embedding** to the semantic space are conducted for both the original attended region features and the updated ones. On the other hand, to train our model, we first propose a *transfer* loss (detailed in Sect. 3.4, Eq. (9)) by transferring the class similarities from seen to unseen classes. The *transfer* loss is designed by extending the seen attributes guided compatibility loss [80] with the collaborative guidance of the contrastive similarity score between seen and unseen attributes. Moreover, to address the domain bias issue (Fig. 1(a)–(b)) in the end-to-end GZSL models [52,79,80], we propose a *balance* loss by minimizing the maximum response consistency between seen and unseen predictions. To this end, the end-to-end trainable **Network** architecture in Fig. 2 is termed as **Region Graph Embedding Network** (RGEN). Detailly, RGEN consists of the Constrained Part Attention (CPA) branch and the Parts Relation Reasoning (PRR) branch.

---

[1] In this paper, part and region are alternatively used.

To sum up, our contributions are: **(1)** We present a region graph embedding network which incorporates region-based relation reasoning into embedding learning. To the best of our knowledge, this is the first attempt to do this in ZSL domain. **(2)** We propose a novel *region graph* representation capturing relationships between attended parts in a single image; GCN-based parts relation reasoning on this graph is then performed. This leads to the complementary Parts Relation Reasoning (PRR) sub-branch. **(3)** We propose the *transfer* loss and *balance* loss to guide the end-to-end RGEN training. Especially, the novel *balance* loss is capable of tackling the severe domain bias problem in end-to-end GZSL models.

2   Related Works

(Generalized) Zero-Shot Learning. Early works [18,24] on ZSL rely on learning attribute classifiers, based on which the class posterior of a test image is deduced. However, associations among these attributes are not well exploited. More recently, a number of embedding based methods [50] have been proposed, which are usually accompanied by a compatibility loss and can effectively address the association issue. Among them, ALE [2] leverages a compatibility hinge loss for learning the association between images and attributes. LATEM [48] is a piecewise extension of ALE. A compatibility based ridge regression is utilized in ESZSL [39]. DEM [66], CMT [41], SJE [4], and DEVICE [14] are also competitive embedding based models. However, these methods usually achieve relatively inferior results, since they adopt global features and/or exploit shallow models. Currently, end-to-end CNN models, such as SCoRe [33], LDF [26], QSFL [42] and LFGAA [28], obtain the best performances. These methods extend the compatibility loss by adding the seen class attributes, and advocate learning more discriminative features. Nevertheless, they struggle to focus on the discriminative parts which are intrinsically accounting for better semantic transfer [11]. Methods designed for ZSL are applicable for GZSL [50], which is more appropriate for real-life applications as it searches the full label space during testing.

Part-Based ZSL. Initial works [1,11,64] utilized part annotations to discover discriminative part features for tackling fine-grained ZSL. However, part annotations are costly and labor-dependent. More recently, by pursing automatic part discovery [53], attention mechanisms [25,55–57] have been applied into ZSL and GZSL [30,52,78,80] for capturing multiple semantic regions, which can facilitate desirable knowledge transfer. These methods achieve remarkable improvements on ZSL, but the performance gains on GZSL are not satisfactory, indicating that they fail at solving the domain bias issue.

In this paper, to solve the realistic inductive ZSL and GZSL tasks (unseen images are inaccessible [19,27]), we propose a Region Graph Embedding Network (RGEN) with the *transfer* and *balance* losses as supervision. Specifically, the PRR branch is based on GCN [21,31] for relation reasoning. Although, GCN has been used in ZSL [20,45] for outputting the visual classifier for each object

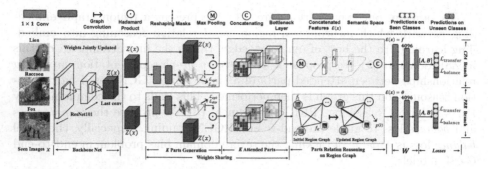

Fig. 2. Architecture. kRGEN consists of CPA and PRR branches. For CPA, the input x is first passed through the Backbone Net and K Parts Generation module (constrained by $\mathcal{L}_{\text{cpt}}$ and $\mathcal{L}_{\text{div}}$), thus producing K attended parts: $\{T_i\}_{i=1}^{K}$. Then, max pooling, concatenation ($\mathcal{E}(x)$), bottleneck layer embedding and semantic space embedding are carried out. For PRR, part features $\{f_i\}_{i=1}^{K}$ are taken as input node features of GCN to acquire updated node features $F^{(2)}$. Then, the same operations as CPA is conducted. Finally, $\mathcal{L}_{\text{transfer}}$ and $\mathcal{L}_{\text{balance}}$ (Sect. 3.4) are leveraged for training.

class, by feeding the word embedding for every object class as inputs; however, we are **the first to explicitly leverage GCN for reasoning about the parts relations within each single image for ZSL**, e.g., the "leg" image is dissimilar with the "head" image in Fig. 1(c). As such, our intuition of using GCN is completely different from [20,45]. To this end, our RGEN is related yet greatly different from current part- and GCN-based ZSL and GZSL methods.

3   Methodology

Task Definitions. We have N^s training samples from C^s seen classes which are defined as $\mathcal{S} = \{(x_i^s, y_i^s)\}_{i=1}^{N^s}$. $\mathcal{X}^{\mathcal{S}} = \{x_i^s\}_{i=1}^{N^s}$ and $\mathcal{Y}^{\mathcal{S}}$ are the training data set and its label set, respectively. The seen class label of the ith sample x_i^s is $y_i^s \in \mathcal{Y}^{\mathcal{S}}$. $\mathcal{A}^s = \{a_i^s\}_{i=1}^{C^s}$ represents the semantic vector set of seen classes. For ZSL, given an unseen testing set $\mathcal{U} = \{(x_i^u, y_i^u)\}_{i=1}^{N^u}$ with N^u samples, we want to predict the label $y_i^u \in \mathcal{Y}^{\mathcal{U}}$ for each x_i^u. More knowledge for $\mathcal{U}$ is provided by the semantic vector set $\mathcal{A}^u = \{a_i^u\}_{i=1}^{C^u}$ for the C^u unseen classes. The label sets of seen and unseen classes are disjoint, i.e., $\mathcal{Y}^{\mathcal{S}} \cap \mathcal{Y}^{\mathcal{U}} = \emptyset$. Meanwhile, for GZSL, the searched label space is expanded to $\mathcal{Y} = \mathcal{Y}^{\mathcal{S}} \cup \mathcal{Y}^{\mathcal{U}}$ by taking samples from both seen and unseen classes as the testing data. We further denote $a_i^s / a_i^u \in \mathbb{R}^Q$.

3.1   Overview

The Region Graph Embedding Network (RGEN) (Fig. 2) consists of two subbranches: the Constrained Part Attention (CPA) branch and the Parts Relation Reasoning (PRR) branch. Both branches are jointly trained by the proposed *transfer* and *balance* losses (Sect. 3.4). The CPA is capable of automatically discovering more discriminative regions, which applies [52] to generate attended object regions and is different from [52] as follows: 1) unlike [52] without any regularizations on attention masks, compactness and diversity are introduced for learning desirable parts; 2) *transfer* and *balance* losses are leveraged comparing to [52] which uses attribute incorporated cross-entropy loss (Fig. 1(b)). Moreover, PRR aims at capturing appearance relationships among the discovered parts by GCN-based [21] graph reasoning. The outputs of such GCNs are updated node features (with each node representing an attended region), which are further used to learn embedding to the semantic space.

We further add a bottleneck layer between the feature space and the low-dimensional semantic space to alleviate the loss of information caused by the extreme reduction in dimensions, e.g., 20,480D→85D on AWA2.

3.2 Constrained Part Attention Branch

Attention Parts Generation. We leverage the soft spatial attention [52] to map image x into a set of K part features. Specifically, suppose the last convolutional feature map w.r.t. x is $Z(x) \in \mathbb{R}^{H \times W \times C}$, with H,W,C being its height, width, and channel number, respectively. Then, K attention masks $\{M_i(x)\}_{i=1}^{K}$ are obtained by a 1×1 convolution G on $Z(x)$ and a Sigmoid thresholding:

$$\mathcal{M} = \text{Sigmoid}(G(Z(x))) \in \mathbb{R}^{H \times W \times K}, \quad M_i(x) = \mathcal{M}[:,:,i], \qquad (1)$$

where $M_i(x) \in \mathbb{R}^{H \times W}$ is the ith attention mask of input x. Based on these masks, we obtain K corresponding attentive feature maps $\{T_i(x)\}_{i=1}^{K}$ w.r.t. $Z(x)$:

$$T_i(x) = Z(x) \odot R(M_i(x)), \qquad (2)$$

where R reshapes the input to be the same shape as $Z(x)$, $\odot$ is an element-wise multiplication and $T_i(x) \in \mathbb{R}^{H \times W \times C}$. Finally, we apply global max-pooling to each $T_i(x)$, and thus get K part features $\{f_i(x)\}_{i=1}^{K}$ with $f_i(x) \in \mathbb{R}^{C}$.

$\{f_i(x)\}_{i=1}^{K}$ have two functions: 1) They are concatenated as a vector $f \in \mathbb{R}^{KC}$ (Fig. 2), which is connected to the bottleneck layer and then the semantic space. Finally the semantic layer output is supervised by the *transfer* and *balance* losses (Sect. 3.4). 2) They are taken as nodes and used to construct region graph, which is fed to GCNs [21] in the PRR branch for parts relation reasoning (Sect. 3.3).

Constrained Attention Masks. To discover more compact and divergent parts, we follow [78,80], which constrain the attention masks from the channel clustering. Here, we constrain masks from spatial attention. Specifically, the compact loss and divergent loss for K masks $\{M_i(x)\}_{i=1}^{K}$ (we drop x for ease of reading) on n_b batch data are:

$$\mathcal{L}_{\text{cpt}} = \frac{1}{K \times n_b} \sum_{j=1}^{n_b} \sum_{i=1}^{K} \sum_{h,w} \| M_i^{h,w} - \hat{M}_i^{h,w} \|_2^2,$$

$$\mathcal{L}_{\text{div}} = \frac{1}{K \times n_b} \sum_{j=1}^{n_b} \sum_{i=1}^{K} \sum_{h,w} M_i^{h,w} \tilde{M}_i^{h,w}, \tag{3}$$

where $\hat{M}_i$ is an ideal peaked attention map for the ith part; $\tilde{M}_i^{h,w} = \max_{j \neq i} M_j^{h,w}$ is the maximum activation of other masks at coordinate (h, w).

3.3   Parts Relation Reasoning Branch

Each of these acquired K part features $\{f_i(x)\}_{i=1}^{K}$ represents one attended region. When humans see these image regions (Fig. 1(c)), they can easily tell the appearance relationships among them. To imitate such human behavior in linking image regions, we employ GCN [21] to perform region-based relation modeling, which leads to the PRR branch (together with the afterward operations in bottom stream of Fig. 2). As validated in the experiments of Sect. 4.3 and Sect. 4.4, parts relation reasoning can help RGEN to achieve an improved performance.

We now construct a region graph $\Gamma \in \mathbb{R}^{K \times K}$ (with K part features as its K nodes) for each input image. In Γ, we have a high confidence edge between similar regions ("head"-"head") and a low confidence edge between dissimilar regions ("head"-"leg") (Fig.1(c)). Specifically, we first conduct l_2-normalization on each $f_i(x)$. Then, the dot-product is leveraged to calculate the pairwise similarity:

$$\Gamma_{ij} = \langle f_i(x), f_j(x) \rangle . \tag{4}$$

In this case, the dot-product calculation is equal to the cosine similarity metric and the graph has self-connections as well. We further calculate the degree matrix D of Γ with $D_{ii} = \sum_{j=1}^{K} \Gamma_{ij}$.

Given input as the region graph, we leverage GCN to perform reasoning on this graph. Specifically, we use a two-layer GCN propagation that is defined as:

$$F^{(l+1)} = \sigma(D^{-1} \Gamma F^{(l)} W^{(l)}), l = 0, 1, \tag{5}$$

where $F^{(0)} \in \mathbb{R}^{K \times C}$ are the stacked K part features, C is their dimension, $W^{(l)}, l = 0, 1$ are learnable parameters, and σ is the Relu activation function.

Finally, as CPA branch, the updated features $F^{(2)} \in \mathbb{R}^{K \times C}$ by GCNs further undergo a concatenation, a bottleneck layer and an embedding to the semantic space. In this case, the guidance losses are again the *transfer* and *balance* losses.

Here, GCN [21] is desirable due to: 1) It transfers original part features into new ones ($F^{(2)}$) by modeling parts relations automatically. 2) The parameters $W^{(l)}$ are jointly learned with the guidance of attributes. 3) It is entirely different from GCN with word embeddings as inputs [20,45], which learns visual classifier for each class for ZSL. To the best of our knowledge, this represents the first time GCN-based parts relation reasoning is used to tackle ZSL and GZSL tasks.

3.4 The Transfer and Balance Losses

To make ZSL and GZSL feasible, the achieved features ($\mathcal{E}(x)$ in Fig. 2) should be further embedded into a certain subspace. In this paper, we utilize semantic space as the embedding space. As such, given the ith seen image x_i^s and its ground-truth semantic vector $a_*^s \in \mathcal{A}^s$, suppose its embedded feature is collectively denoted as $\mathcal{E}(x_i^s)$, which equals to the concatenated rows of $F^{(2)}$ in Eq. (5) ($\boldsymbol{\theta}$ in Fig. 2) or the concatenated K part features ($\boldsymbol{f}$ in Fig. 2).

Revisit the ACE Loss. To associate image x_i^s with its true attribute information, the compatibility score τ_i^* is formulated as [2,26,33,42,52,80]:

$$\tau_i^* = \mathcal{E}(x_i^s) \boldsymbol{W} a_*^s, \tag{6}$$

where $\boldsymbol{W}$ are the embedding weights that need to be learned jointly, which is a two-layer MLP in our implementation (Fig. 2). Considering τ_i^* as a classification score in the cross-entropy (CE) loss, for seen data from a batch, the Attribute incorporated CE loss (ACE) becomes:

$$\mathcal{L}_{\text{ACE}} = -\frac{1}{n_b} \sum_{i=1}^{n_b} \log \frac{\exp(\tau_i^*)}{\sum_{a_j^s \in \mathcal{A}^s} \exp(\tau_i^j)}, \tag{7}$$

where $\tau_i^j = \mathcal{E}(x_i^s) \boldsymbol{W} a_j^s$, $j = 1, \cdots, C^s$ are the scores on C^s seen semantic vectors.

The Transfer Loss. Equation (7) introduces $\mathcal{A}^s$ for end-to-end training; however, there are two drawbacks: 1) The learned models are still biased towards seen classes, which is a common issue in ZSL and GZSL; and 2) The performances of these deep models are inferior on GZSL [52,80]. To alleviate these problems further, we incorporate unseen attributes $\mathcal{A}^u$ into RGEN.

In particular, we first define the l_2-normalized attribute matrix w.r.t. these C^s seen classes and C^u unseen classes as $A \in \mathbb{R}^{Q \times C^s}$ and $B \in \mathbb{R}^{Q \times C^u}$, respectively. Then, we leverage least square regression (LSR) to obtain the reconstruction coefficients $V \in \mathbb{R}^{C^u \times C^s}$ of each seen class attribute w.r.t. all unseen class attributes: $V = (B^\mathsf{T} B + \beta I)^{-1} B^\mathsf{T} A$, which is obtained by solving $\min_V \|A - BV\|_F^2 + \beta \|V\|_F^2$. The ith column of V represents the contrasting class similarity of a_i^s w.r.t. B. To this end, during RGEN training, besides Eq. (7), we propose the following loss:

$$\mathcal{L}_{\text{contra}} = -\frac{1}{n_b} \sum_{i=1}^{n_b} \sum_{j=1}^{C^u} v_{jy_i} \log \widetilde{\zeta_{ij}} + (1 - v_{jy_i}) \log(1 - \widetilde{\zeta_{ij}}), \tag{8}$$

where $\zeta_{ij} = \mathcal{E}(x_i^s) \boldsymbol{W} a_j^u$, $j = 1, \cdots, C^u$ are the scores w.r.t. C^u unseen semantic vectors for x_i^s, $\widetilde{\zeta_{ij}}$ is the softmax-layer normalization of ζ_{ij} and y_i is the column location in V w.r.t. ground-truth semantic vector of x_i^s. We formally call the loss combining $\mathcal{L}_{\text{contra}}$ and $\mathcal{L}_{\text{ACE}}$ the *transfer* loss:

$$\mathcal{L}_{\text{transfer}} = \mathcal{L}_{\text{ACE}} + \lambda_1 \mathcal{L}_{\text{contra}}. \tag{9}$$

The second term in Eq. (9) is related to [19] but differs from it as follows: 1) Our calculation of the prediction score (ζ_{ij}) is based on an end-to-end trained deep net and the compatibility score. 2) We calculate the contrasting class similarity using LSR regression while in [19] they use sparse coding.

Notably, we implement the *transfer* loss as a fully-connected layer by freezing the weights as $[A, B] \in \mathbb{R}^{Q \times (C^u + C^s)}$ during the training phase (Fig. 2). In this way, the seen and unseen attributes can guide the discovery of attention parts, and the relation reasoning among them.

The Balance Loss. To tackle the challenge of extreme domain bias in GZSL, especially encountered in end-to-end models [52, 80], we propose a *balance* loss by pursuing the maximum response consistency, among seen and unseen outputs.

Specifically, given the input seen sample x_i^s, we can get its prediction scores on seen class and unseen class attributes as $P_i^s = A^\mathsf{T} W^\mathsf{T} \mathcal{E}(x_i^s)^\mathsf{T} \in \mathbb{R}^{C^s \times 1}$ and $P_i^u = B^\mathsf{T} W^\mathsf{T} \mathcal{E}(x_i^s)^\mathsf{T} \in \mathbb{R}^{C^u \times 1}$, respectively. To balance these scores from the two sides (seen and unseen), the *balance* loss is proposed for batch data, as follows:

$$\mathcal{L}_{\text{balance}} = \frac{1}{n_b} \sum_{i=1}^{n_b} \| \max P_i^s - \max P_i^u \|_2^2, \tag{10}$$

where $\max P$ outputs the maximum value of the input vector P. The *balance* loss is only utilized for GZSL, and not ZSL, since balancing is not required when only unseen test images are available.

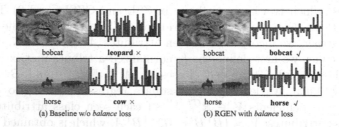

(a) Baseline w/o *balance* loss   (b) RGEN with *balance* loss

Fig. 3. Cyan and magenta bars are the predicted scores (before the softmax-layer in real-world model) on seen and unseen classes, respectively. Domain bias in (a) has been well addressed by adding our *balance* loss (see (b)). (Colo figure online)

The intuitions of balancing the predictions between seen (cyan bars in Fig. 3(a)) and unseen (magenta bars in Fig. 3(a)) outputs are two-folds: 1) From the perspective of prediction scaling for end-to-end networks, since no available training data for producing responses on these unseen locations (extreme training data imbalance), we want to balance the numerical scales between seen and unseen predictions. 2) Since some unseen test samples are correctly classified if we only consider their prediction scores on unseen locations (e.g., "zebra" under GZSL in Fig. 1(b)), we want to rescue these misclassified samples. Figure 3 is a real-world example, where we feed unseen images from AWA2 to the RGEN GZSL model (trained with *balance* loss) and its baseline w/o *balance* loss to observe the changing of the predicted scores.

3.5   Training Objective

As we have two branches (the CPA and the PRR), both are guided by our proposed *transfer* and *balance* losses during end-to-end training. However, we have only one stream of data as the input of our net, i.e., the backbone is shared. As such, the final loss for our proposed RGEN is as follows:

$$\mathcal{L}_{\text{RGEN}} = \eta_1 \mathcal{L}_{\text{CPA}} + \eta_2 \mathcal{L}_{\text{PRR}} + \eta_3 \mathcal{L}_{\text{cpt}} + \eta_4 \mathcal{L}_{\text{div}}. \tag{11}$$

The formulations of $\mathcal{L}_{\text{CPA}}$ and $\mathcal{L}_{\text{PRR}}$ are the same as follows:

$$\underbrace{\mathcal{L}_{\text{ACE}} + \lambda_1 \mathcal{L}_{\text{contra}}}_{\mathcal{L}_{\text{transfer}}} + \lambda_2 \mathcal{L}_{\text{balance}}, \tag{12}$$

where λ_1 and λ_2 again takes the same values for the two branches. The difference between $\mathcal{L}_{\text{CPA}}$ and $\mathcal{L}_{\text{PRR}}$ lies in that the concatenated embedding features f and θ (Fig. 2) come from $\{f_i(x)\}_{i=1}^K$ and $F^{(2)}$ (Eq. (5)) for them, respectively. Note that, we take $\eta_1 = 0.9$, $\eta_2 = 0.1$, $\eta_3 = 1.0$, and $\eta_4 = 1e$–4 for all datasets. The selections of λ_1 and λ_2 are further detailed in Sect. 4.2.

3.6   Zero-Shot Prediction

In RGEN framework, the unseen test image x^u is predicted in a fused manner. After obtaining the embedding features of x^u in the semantic space w.r.t. CPA and PRR branches, denoted as, $\psi_{\text{CPA}}(x^u)$ and $\psi_{\text{PRR}}(x^u)$, we calculate their fused result by the same combination coefficients (η_1, η_2) as the training phase, and then predict its label by:

$$y^{u*} = \arg \max_{c \in \mathcal{Y}^u/\mathcal{Y}} (\eta_1 \psi_{\text{CPA}}(x^u) + \eta_2 \psi_{\text{PRR}}(x^u))^\mathsf{T} a_c^u. \tag{13}$$

In Eq. (13), $\mathcal{Y}^u/\mathcal{Y}$ corresponds to ZSL/GZSL respectively. In our ablations, we show the performances when setting different combinations of η_1 and η_2 (Fig. 6).

4   Experiments

4.1   Datasets and Settings

We use four standard ZSL and GZSL datasets, i.e., SUN [36], CUB [44], AWA2 [50], and APY [12] to evaluate our RGEN. We use the Proposed Split (PS) [50] for evaluation, as this setting is more strict and does not contain any class overlapping with ImageNet classes [50]. Since images of AWA are not accessible, AWA2 is used instead. The details of these datasets can be found in [50].

RGEN is an end-to-end trainable embedding method. As such, it is fair to compare it with the same types of end-to-end models ([26, 28, 33, 42, 52, 80]). However, to comprehensively review the performance gains of RGEN over other methods, we further compare it with other non end-to-end methods (including the two-stage feature generation methods which are parallel solutions for tackling ZSL and GZSL), and these methods are based on the same ResNet101 features.

4.2   Implementation and Parameters

Almost all compared methods use the 2,048D ResNet101 features. As such, we use ResNet101 in Fig. 2 as our backbone net [17]. The size of input images for the four datasets is 224×224, which makes the size of the last convolutional feature map as $2048 \times 7 \times 7$. For all datasets, the RGEN is trained for a maximum of 40 epochs, with an initial learning rate of 0.001. The architecture for GCN is 2048D-Relu(1024D)-2048D for all datasets. Except for CUB (which has a higher 312D attributes and no bottleneck layer), the datasets all leverage a 4096D bottleneck layer before projecting to semantic space.

The parameters $\eta_1, \eta_2, \eta_3, \eta_4$ are fixed, as stated in Sect. 3.5 for all four datasets. The number of parts K is fixed as 10 and β is fixed as 5, for all used datasets. λ_1 is selected from $\{0.001, 0.01, 0.05, 0.07, 0.1\}$ and λ_2 from $\{0.01, 0.05, 0.07, 0.1\}$.

4.3   Zero-Shot Recognition

Mean Class Accuracy (MCA) is adopted as the evaluation metric for ZSL [50]. Table 1 shows the results. As can be seen, i) RGEN consistently outperforms most state-of-the-arts by a clear margin and performs best on four datasets among end-to-end models. For instance, RGEN achieves a MCA of 76.1% on CUB, which sets a new state-of-the-art on this dataset by a large margin than the counterparts. However, the performance gains on CUB/AWA2 are better than SUN/APY, this is because: 1) image number per class for SUN is about 20, which limits the RGEN training; 2) the prepared images for APY usually have an extreme aspect ratio, thus hindering the RGEN training. ii) The parts relation reasoning branch contributes to the performance improvements, e.g., the performance of RGEN w/o PRR is 75.0% and 72.5% on CUB and AWA2, respectively. This indicates that the parts relation reasoning must discover some underlying information that assists in semantic transfer, though the PRR branch alone does not achieve such a high MCA (Component Analysis in Table 1).

4.4   Generalized Zero-Shot Recognition

For GZSL, the searched label space includes both seen and unseen classes [50]. The evaluation for GZSL is different from ZSL as follows: The MCAs of the unseen/seen test samples are denoted as ts/tr, respectively. Then their Harmonic mean H is defined as $H = 2 \times tr \times ts/(tr + ts)$. The H score is the key evaluation

Table 1. ZSL and GZSL results (%) on used datasets. Results with underlines are obtained by further using extra word embeddings, however, our methods only utilize attributes. If there are no results on AWA2, results on AWA are shown. The best, the second best, and the third best results are marked in red, blue, and bold, respectively.

Methods	CUB ZSL MCA	CUB GZSL ts	tr	H	SUN ZSL MCA	SUN GZSL ts	tr	H	AWA2 ZSL MCA	AWA2 GZSL ts	tr	H	APY ZSL MCA	APY GZSL ts	tr	H
Non End-to-End																
CONSE(NeurIPS'14) [34]	34.3	1.6	72.2	3.1	38.8	6.8	39.9	11.6	44.5	0.5	90.6	1.0	26.9	0.0	91.2	0.0
CMT(NeurIPS'13) [41]	34.6	7.2	49.8	12.6	39.9	8.1	21.8	11.8	37.9	0.5	90.0	1.0	28.0	1.4	85.2	2.8
SSE(ICCV'15) [72]	43.9	8.5	46.9	14.4	51.5	2.1	36.4	4.0	61.0	8.1	82.5	14.8	34.0	0.2	78.9	0.4
LATEM(CVPR'16) [48]	49.3	15.2	57.3	24.0	55.3	14.7	28.8	19.5	55.8	11.5	77.3	20.0	35.2	0.1	73.0	0.2
ALE(TPAMI'13) [3]	54.9	23.7	62.8	34.4	58.1	21.8	33.1	26.3	62.5	14.0	81.8	23.9	39.7	4.6	73.7	8.7
DEVISE(NeurIPS'13) [14]	52.0	23.8	53.0	32.8	56.5	16.9	27.4	20.9	59.7	17.1	74.7	27.8	39.8	4.9	76.9	9.2
SJE(CVPR'15) [4]	53.9	23.5	59.2	33.6	53.7	14.7	30.5	19.8	61.9	8.0	73.9	14.4	32.9	3.7	55.7	6.9
ESZSL(ICML'15) [39]	53.9	12.6	63.8	21.0	54.5	11.0	27.9	15.8	58.6	5.9	77.8	11.0	38.3	2.4	70.1	4.6
SYNC(CVPR'16) [7]	55.6	11.5	70.9	19.8	56.3	7.9	43.3	13.4	46.6	10.0	90.5	18.0	23.9	7.4	66.3	13.3
SAE(CVPR'17) [23]	33.3	7.8	54.0	13.6	40.3	8.8	18.0	11.8	54.1	1.1	82.2	2.2	8.3	0.4	80.9	0.9
PSR(CVPR'18) [5]	56.0	24.6	54.3	33.9	61.4	20.8	37.2	26.7	63.8	20.7	73.8	32.3	38.4	13.5	51.4	21.4
DEM(CVPR'17) [66]	51.7	19.6	57.9	29.2	40.3	20.5	34.3	25.6	67.1	30.5	86.4	45.1	35.0	11.1	75.1	19.4
RN(CVPR'18) [58]	55.6	38.1	61.4	47.0	–	–	–	–	64.2	30.0	93.4	45.3	–	–	–	–
SP-AEN(CVPR'18) [9]	55.4	34.7	70.6	46.6	59.2	24.9	38.6	30.3	58.5	23.3	90.9	37.1	24.1	13.7	63.4	22.6
IIR(ICCV'19) [6]	63.8	30.4	65.8	41.2	63.5	22.0	34.1	26.7	67.9	17.6	87.0	28.9	–	–	–	–
TCN(ICCV'19) [19]	59.5	52.6	52.0	52.3	61.5	31.2	37.3	34.0	71.2	61.2	65.8	63.4	38.9	24.1	64.0	35.1
Feature generation methods																
f-CLSWGAN(CVPR'18) [49]	57.3	43.7	57.7	49.7	60.8	42.6	36.6	39.4	68.2	57.9	61.4	59.6	–	–	–	–
cycle-CLSWGAN(ECCV'18) [13]	58.4	45.7	61.0	52.3	60.0	49.4	33.6	40.0	66.3	56.9	64.0	60.2	–	–	–	–
f-VAEGAN-D2 w/o ft(CVPR'19) [51]	61.0	48.4	60.1	53.6	64.7	45.1	38.0	41.3	71.1	57.6	70.6	63.5	–	–	–	–
f-VAEGAN-D2 w ft(CVPR'19) [51]	72.9	63.2	75.6	68.9	65.6	50.1	37.8	43.1	70.3	57.1	76.1	65.2	–	–	–	–
End-to-End																
SCoRe(CVPR'17) [33]	62.7	–	–	–	–	–	–	–	61.6	–	–	–	–	–	–	–
QFSL(CVPR'18) [42]	58.8	33.3	48.1	39.4	56.2	30.9	18.5	23.1	63.5	52.1	72.8	60.7	–	–	–	–
LDF(CVPR'18) [26]	67.5	26.4	81.6	39.9	–	–	–	–	65.5	9.8	87.4	17.6	–	–	–	–
SGMA(NeurIPS'19) [80]	71.0	36.7	71.3	48.5	–	–	–	–	68.8	37.6	87.1	52.5	–	–	–	–
AREN(CVPR'19) [52]	71.8	38.9	78.7	52.1	60.6	19.0	38.8	25.5	67.9	15.6	92.9	26.7	39.2	9.2	76.9	16.4
LFGAA(ICCV'19) [28]	67.6	36.2	80.9	50.0	61.5	18.5	40.0	25.3	68.1	27.0	93.4	41.9	–	–	–	–
RGEN w/o PRR (Ours)	75.0	61.4	68.5	64.7	63.4	42.7	31.5	36.2	72.5	64.1	76.4	69.7	43.9	20.2	48.0	36.3
RGEN (Ours)	76.1	60.0	73.5	66.1	63.8	44.0	31.7	36.8	73.6	67.1	76.5	71.5	44.4	30.4	48.1	37.2

criterion for GZSL [50], since we want to correctly classify both seen/unseen test images as many as possible (i.e., a higher H score) in the real-world application.

We conclude from Table 1: **i)** Our RGEN achieves the best Hs compared to its end-to-end counterparts, e.g, we achieve a 71.5 H on AWA2, which represents the current best result. **ii)** The PRR branch can also effectively boost the performances of RGEN under GZSL. **iii)** Our *balance* loss contributes most to the performance improvements (Component Analysis in Table 4). **iv)** Compared with the two-stage feature generation method f-VAEGAN-D2 with finetuning [51], which belongs to a parallel solution for ZSL, performances of RGEN (w/o feature generating) are still on par, i.e., on four datasets, we achieve 3/4 better ZSL results, and 2/4 better H under GZSL. Notably, our H score on CUB is worse than [51], however, the latter further uses extra word embeddings [51].

4.5 Ablations

Effects of η_1 and η_2. For RGEN training (Eq. (11)) and testing (Eq. (13)), η_1 and η_2 have the same values in order to keep training and testing consistent. By taking their values from $\{0.0, 0.1, 0.2, \cdots, 0.9, 1.0\}$ and constraining $\eta_1+\eta_2=1.0$, we observe the MCA of RGEN w.r.t. different values of (η_1,η_2) for ZSL (Fig. 6).

Table 2. ZSL and GZSL results (%) with different GCN structures on CUB.

GCN layers	ZSL	GZSL			Best GCN structures	
	MCA	tr	ts	H	ZSL	GZSL
One-layer	75.2	69.7	61.5	65.4	2048-256	2048-128
Two-layer	76.1	60.0	73.5	66.1	2048-1024-2048	2048-1024-2048
Three-layer	75.1	68.5	59.3	63.6	2048-1024-1024-2048	2048-256-256-2048

We find that a small η_2 is better for assisting the RGEN model and, as such, we set $(\eta_1, \eta_2) = (0.9, 0.1)$ for all datasets.

Parts Number. K is fixed to 10 in all our experiments but we vary it from $\{1, 2, 5, 10, 15\}$ to observe the performances of our full RGEN model under both ZSL/GZSL. The results in Fig. 4 show that MCA (of ZSL/tr/ts) and H are stable with a small K, and $K=10$ is suitable for achieving satisfactory results.

Transfer Loss Coefficient. We show the results of MCA (of ZSL/tr/ts), and H when varying λ_1 over $\{0.0001, 0.001, 0.01, 0.05, 0.07, 0.1\}$ under ZSL/GZSL for RGEN. The results (Fig. 5) are stable for small values from $[0.001, 0.05]$.

Balance Loss Coefficient. The Balance loss is only used under GZSL training; therefore, we vary the value of λ_2 from $\{0.0, 0.01, 0.05, 0.07, 0.1\}$ and observe the MCA (of tr/ts) and H under these values. Figure 7 shows that a smaller coefficient always achieves better H/ts (with little sacrifice on tr) than the model w/o *balance* regularization ($\lambda_2=0.0$) and the overall changing tendency is stable.

GCN Architecture. We fix GCN in PRR branch as a two-layer one (2048-Relu(1024)-2048). We further investigate the influence of one- and three-layer GCN on ZSL/GZSL. Specifically, for one- and three-layer GCN, we vary the node dimension (of the output/middle layer) from $\{128, 256, 512, 1024, 2048\}$ to determine their best results, for fair comparisons with ours. Their best searched architectures are also shown in Table 2, which indicates that a two-layer GCN can better model the parts relation collaboratively with other parameters.

Component Analysis. The full RGEN consists of 1) Compact and Divergent (CD) regularizations; 2) Transfer loss; 3) Balance loss (for GZSL); and 4) PRR branch. We assume the PRR branch is trained by 1) and 2), and conduct a component analysis for 1), 2) and 4) under our best ZSL RGEN model (Table 3). As 2) is important for knowledge transfer in ZSL/GZSL, we take this as the indispensable loss for each GZSL model; as such, component analysis for GZSL includes 1), 3), and 4) (Table 4).

Table 3. Component analysis (MCA) of the best RGEN ZSL model.

ACE loss	✔	✔		✔				
CD regularization		✔			✔	✔		✔
Transfer loss			✔	✔			✔	✔
PRR branch	✔					✔	✔	
CUB	69.6	71.3	72.0	74.2	75.0	73.3	75.6	76.1
AWA2	70.1	69.4	69.9	72.3	72.5	70.4	73.0	73.6
SUN	59.8	62.0	62.4	63.1	63.4	63.1	63.3	63.8
APY	37.5	39.5	39.9	43.7	43.9	39.7	43.8	44.4

Table 4. Component analysis (H) of the best RGEN GZSL model.

Transfer loss	✔	✔	✔	✔	✔	✔	✔	✔
CD regularization	✔	✔				✔	✔	
PRR branch			✔	✔		✔	✔	
Balance loss					✔	✔	✔	✔
CUB	37.2	38.6	39.6	38.3	64.6	64.7	66.1	65.1
AWA2	12.5	14.1	14.7	14.9	69.9	69.7	71.5	71.3
SUN	21.2	23.7	24.3	23.6	35.9	36.4	36.8	36.7
APY	15.2	15.6	16.4	15.8	36.8	36.3	37.2	37.7

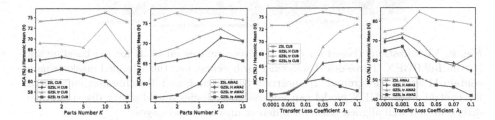

Fig. 4. (MCA,H,tr,ts)-K curves.   **Fig. 5.** (MCA,H,tr,ts)-λ_1 curves.

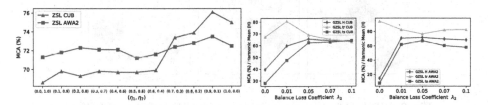

Fig. 6. MCA-(η_1, η_2) curves of RGEN.   **Fig. 7.** (H,tr,ts)-λ_2 curves.

4.6 Qualitative Analysis

We use unseen images from CUB under ZSL to visualize the attended parts (Fig. 8). Compared with the baseline (which achieves 71.3% MCA when trained by only the ACE loss, Table 3), RGEN can 1) discover more divergent parts w.r.t. objects; 2) suppress background and redundant foreground regions (maximum mask values in parts #1–4, 9–10 are all small and no similar masks exist among foreground parts #5–8); 3) automatically align the order relationships of different parts (parts #5–8 are consistent w.r.t. different unseen class images). However, the baseline has three drawbacks, as shown in Fig. 8. t-SNEs of the unseen test images on AWA2 under GZSL for RGEN and its variants are shown in Fig. 9.

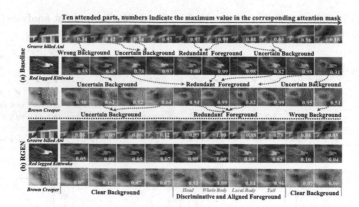

Fig. 8. Visualizations (based on [77]) of unseen images on CUB for (a) baseline and (b) our full RGEN. Three drawbacks exist for baseline, i.e., the attended masks usually contain 1) wrong background; 2) uncertain background; and 3) redundant foreground. By contrary, our RGEN model can address these issues and discover discriminative, divergent and well aligned parts for different unseen class images.

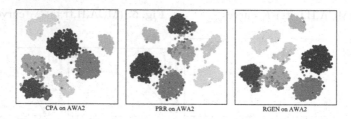

Fig. 9. t-SNE [32] of unseen test images for CPA, PRR, and RGEN.

5   Conclusions

The Region Graph Embedding Network (RGEN) is proposed for tackling ZSL and GZSL tasks. RGEN contains the constrained part attention and the parts relation reasoning branches. To guide RGEN training, the *transfer* and *balance* losses are integrated into the framework. The *balance* loss is especially valuable for alleviating the extreme domain bias in the deep GZSL models, providing intrinsic insights for solving GZSL. RGEN sets some new state-of-the-arts for both ZSL and GZSL, on several commonly used benchmarks.

Acknowledgments. This work was supported by the National Natural Science Foundation of China (Nos. 61702163 and 61976116), the Fundamental Research Funds for the Central Universities (Nos. 30920021135), and the Key Project of Shenzhen Municipal Technology Research (Nos. JSGG20200103103401723).

References

1. Akata, Z., Malinowski, M., Fritz, M., Schiele, B.: Multi-cue zero-shot learning with strong supervision. In: CVPR (2016)
2. Akata, Z., Perronnin, F., Harchaoui, Z., Schmid, C.: Label-embedding for attribute-based classification. In: CVPR (2013)
3. Akata, Z., Perronnin, F., Harchaoui, Z., Schmid, C.: Label-embedding for image classification. In: TPAMI (2016)
4. Akata, Z., Reed, S., Walter, D., Lee, H., Schiele, B.: Evaluation of output embeddings for fine-grained image classification. In: CVPR (2015)
5. Annadani, Y., Biswas, S.: Preserving semantic relations for zero-shot learning. In: CVPR (2018)
6. Cacheux, Y., Borgne, H., Crucianu, M.: Modeling inter and intra-class relations in the triplet loss for zero-shot learning. In: ICCV (2019)
7. Changpinyo, S., Chao, W.L., Gong, B., Sha, F.: Synthesized classifiers for zero-shot learning. In: CVPR (2016)
8. Chao, W.-L., Changpinyo, S., Gong, B., Sha, F.: An empirical study and analysis of generalized zero-shot learning for object recognition in the wild. In: Leibe, B., Matas, J., Sebe, N., Welling, M. (eds.) ECCV 2016. LNCS, vol. 9906, pp. 52–68. Springer, Cham (2016). https://doi.org/10.1007/978-3-319-46475-6_4
9. Chen, L., Zhang, H., Xiao, J., Liu, W., Chang, S.F.: Zero-shot visual recognition using semantics-preserving adversarial embedding network. In: CVPR (2018)
10. Elhoseiny, M., Elfeki, M.: Creativity inspired zero-shot learning. In: ICCV (2019)
11. Elhoseiny, M., Zhu, Y., Zhang, H., Elgammal, A.M.: Link the head to the "beak": zero shot learning from noisy text description at part precision. In: CVPR (2017)
12. Farhadi, A., Endres, I., Hoiem, D., Forsyth, D.: Describing objects by their attributes. In: CVPR (2009)
13. Felix, R., Kumar, V.B., Reid, I., Carneiro, G.: Multi-modal cycle-consistent generalized zero-shot learning. In: ECCV (2008)
14. Frome, A., Corrado, G.S., Shlens, J., Bengio, S., Dean, J., T. Mikolov, E.A.: DeViSE: a deep visual-semantic embedding model. In: NeurIPS (2013)
15. Fu, Y., Hospedales, T.M., Xiang, T., Gong, S.: Transductive multi-view zero-shot learning. In: TPAMI (2015)
16. Goodfellow, I., et al.: Generative adversarial nets. In: NeurIPS (2014)
17. He, K., Zhang, X., Ren, S., Sun, J.: Deep residual learning for image recognition. In: CVPR (2016)
18. Jayaraman, D., Grauman, K.: Zero-shot recognition with unreliable attributes. In: NeurIPS (2014)
19. Jiang, H., Wang, R., Shan, S., Chen, X.: Transferable contrastive network for generalized zero-shot learning. In: ICCV (2019)
20. Kampffmeyer, M., Chen, Y., Liang, X., Wang, H., Zhang, Y., Xing, E.: Rethinking knowledge graph propagation for zero-shot learning. In: CVPR (2019)
21. Kipf, T., Welling, M.: Semi-supervised classification with graph convolutional networks. arXiv:1609.02907 (2016)
22. Kodirov, E., Xiang, T., Fu, Z., Gong, S.: Unsupervised domain adaptation for zero-shot learning. In: ICCV (2015)
23. Kodirov, E., Xiang, T., Gong, S.: Semantic autoencoder for zero-shot learning. In: CVPR (2017)
24. Lampert, C.H., Nickisch, H., Harmeling, S.: Learning to detect unseen object classes by between-class attribute transfer. In: CVPR (2009)

25. Li, X., Yang, F., Cheng, H., Liu, W., Shen, D.: Contour knowledge transfer for salient object detection. In: ECCV (2018)
26. Li, Y., Zhang, J., Zhang, J., Huang, K.: Discriminative learning of latent features for zero-shot recognition. In: CVPR (2018)
27. Liu, S., Long, M., Wang, J., Jordan, M.: Generalized zero-shot learning with deep calibration network. In: NeurIPS (2018)
28. Liu, Y., Guo, J., Cai, D., He, X.: Attribute attention for semantic disambiguation in zero-shot learning. In: ICCV (2019)
29. Long, Y., Liu, L., Shen, F., Shao, L., Li, X.: Zero-shot learning using synthesised unseen visual data with diffusion regularisation. In: TPAMI (2017)
30. Lu, X., Wang, W., Ma, C., Shen, J., Shao, L., Porikli, F.: See more, know more: unsupervised video object segmentation with co-attention siamese networks. In: CVPR (2019)
31. Lu, X., Wang, W., Martin, D., Zhou, T., Shen, J., Luc, V.G.: Video object segmentation with episodic graph memory networks. In: Proceedings of the European Conference on Computer Vision (ECCV) (2020)
32. Maaten, L.V.D., Hinton, G.: Visualizing data using t-SNE. JMLR **9**, 2579–2605 (2008)
33. Morgado, P., Vasconcelos, N.: Semantically consistent regularization for zero-shot recognition. In: CVPR (2017)
34. Norouzi, M., et al.: Zero-shot learning by convex combination of semantic embeddings. In: NeurIPS (2014)
35. Palatucci, M., Pomerleau, D., Hinton, G.E., Mitchell, T.M.: Zero-shot learning with semantic output codes. In: NeurIPS (2009)
36. Patterson, G., Hays, J.: Sun attribute database: discovering, annotating, and recognizing scene attributes. In: CVPR (2012)
37. Qiao, R., Liu, L., Shen, C., van den Hengel, A.: Less is more: zero-shot learning from online textual documents with noise suppression. In: CVPR (2016)
38. Reed, S., Akata, Z., Lee, H., Schiele, B.: Learning deep representations of fine-grained visual descriptions. In: CVPR (2016)
39. Romera-Paredes, B., Torr, P.: An embarrassingly simple approach to zero-shot learning. In: ICML (2015)
40. Shen, Y., Qin, J., Huang, L., Liu, L., Zhu, F., Shao, L.: Invertible zero-shot recognition flows. In: Proceedings of the European Conference on Computer Vision (ECCV) (2020)
41. Socher, R., Ganjoo, M., Manning, C.D., Ng, A.: Zero-shot learning through cross-modal transfer. In: NeurIPS (2013)
42. Song, J., Shen, C., Yang, Y., Liu, Y., Song, M.: Transductive unbiased embedding for zero-shot learning. In: CVPR (2018)
43. Verma, V.K., Arora, G., Mishra, A., Rai, P.: Generalized zero-shot learning via synthesized examples. In: CVPR (2018)
44. Wah, C., Branson, S., Welinder, P., Perona, P., Belongie, S.: The caltech-UCSD birds-200-2011 dataset. In: Technical report (2011)
45. Wang, X., Ye, Y., Gupta, A.: Zero-shot recognition via semantic embeddings and knowledge graphs. In: CVPR (2018)
46. Wu, B., et al.: Tencent ml-images: a large-scale multi-label image database for visual representation learning. IEEE Access **7**, 172683–172693 (2019)
47. Wu, B., Jia, F., Liu, W., Ghanem, B., Lyu, S.: Multi-label learning with missing labels using mixed dependency graphs. Int. J. Comput. Vis. **126**, 875–896 (2018)
48. Xian, Y., Akata, Z., Sharma, G., Nguyen, Q., Hein, M., Schiele, B.: Latent embeddings for zero-shot classification. In: CVPR (2016)

49. Xian, Y., Lorenz, T., Schiele, B., Akata, Z.: Feature generating networks for zero-shot learning. In: CVPR (2018)
50. Xian, Y., Schiele, B., Akata, Z.: Zero-shot learning-the good, the bad and the ugly. In: CVPR (2017)
51. Xian, Y., Sharma, S., Saurabh, S., Akata, Z.: f-VAEGAN-D2: a feature generating framework for any-shot learning. In: CVPR (2019)
52. Xie, G.S., et al.: Attentive region embedding network for zero-shot learning. In: CVPR (2019)
53. Xie, G.S., Zhang, X.Y., Yang, W., Xu, M., Yan, S., Liu, C.L.: LG-CNN: from local parts to global discrimination for fine-grained recognition. Pattern Recogn. **71**, 118–131 (2017)
54. Xie, G.S., et al.: SRSC: selective, robust, and supervised constrained feature representation for image classification. IEEE Trans. Neural Netw. Learn. Syst. **31**, 4290–4302 (2019)
55. Xu, H., Saenko, K.: Ask, attend and answer: exploring question-guided spatial attention for visual question answering. In: Leibe, B., Matas, J., Sebe, N., Welling, M. (eds.) ECCV 2016. LNCS, vol. 9911, pp. 451–466. Springer, Cham (2016). https://doi.org/10.1007/978-3-319-46478-7_28
56. Xu, J., Zhao, R., Zhu, F., Wang, H., Ouyang, W.: Attention-aware compositional network for person re-identification. arXiv:1805.03344 (2018)
57. Xu, K., et al.: Show, attend and tell: neural image caption generation with visual attention. In: ICML (2015)
58. Yang, F.S.Y., Zhang, L., Xiang, T., Torr, P.H., Hospedales, T.M.: Learning to compare: Relation network for few-shot learning. In: CVPR (2018)
59. Yang, G., Liu, J., Xu, J., Li, X.: Dissimilarity representation learning for generalized zero-shot recognition. In: MM (2018)
60. Yao, Y., et al.: Exploiting web images for multi-output classification: from category to subcategories. IEEE Trans. Neural Netw. Learn. Syst. **31**, 2348–2360 (2020)
61. Yao, Y., Zhang, J., Shen, F., Hua, X., Xu, J., Tang, Z.: Exploiting web images for dataset construction: a domain robust approach. IEEE Trans. Multimedia **19**, 1771–1784 (2017)
62. Ye, M., Guo, Y.: Zero-shot classification with discriminative semantic representation learning. In: CVPR (2017)
63. Yu, H., Lee, B.: Zero-shot learning via simultaneous generating and learning. In: NeurIPS (2019)
64. Yu, Y., Ji, Z., Fu, Y., Guo, J., Pang, Y., Zhang, Z.: Stacked semantics-guided attention model for fine-grained zero-shot learning. In: NeurIPS (2018)
65. Yu, Y., Ji, Z., Han, J., Zhang, Z.: Episode-based prototype generating network for zero-shot learning. In: CVPR (2020)
66. Zhang, L., Xiang, T., Gong, S., et al.: Learning a deep embedding model for zero-shot learning. In: CVPR (2017)
67. Zhang, L., et al.: Towards effective deep embedding for zero-shot learning. IEEE Trans. Circ. Syst. Video Technol. **30**, 2843–2852 (2020)
68. Zhang, L., et al.: Adaptive importance learning for improving lightweight image super-resolution network. Int. J. Comput. Vis. **128**, 479–499 (2020)
69. Zhang, L., et al.: Unsupervised domain adaptation using robust class-wise matching. IEEE Trans. Circ. Syst. Video Technol. **29**, 1339–1349 (2018)
70. Zhang, L., Wei, W., Bai, C., Gao, Y., Zhang, Y.: Exploiting clustering manifold structure for hyperspectral imagery super-resolution. IEEE Trans. Image Process. **27**, 5969–5982 (2018)

71. Zhang, L., Wei, W., Zhang, Y., Shen, C., Van Den Hengel, A., Shi, Q.: Cluster sparsity field: an internal hyperspectral imagery prior for reconstruction. Int. J. Comput. Vis. **126**, 797–821 (2018)
72. Zhang, Z., Saligrama, V.: Zero-shot learning via semantic similarity embedding. In: ICCV (2015)
73. Zhang, Z., Saligrama, V.: Zero-shot learning via joint latent similarity embedding. In: CVPR (2016)
74. Zhang, Z., Liu, L., Shen, F., Shen, H.T., Shao, L.: Binary multi-view clustering. IEEE Trans. Pattern Anal. Mach. Intell. **41**, 1774–1782 (2018)
75. Zhao, F., Liao, S., Xie, G.S., Zhao, J., Zhang, K., Shao, L.: Unsupervised domain adaptation with noise resistible mutual-training for person re-identification. In: ECCV (2020)
76. Zhao, F., Zhao, J., Yan, S., Feng, J.: Dynamic conditional networks for few-shot learning. In: ECCV (2018)
77. Zhou, B., Khosla, A.A.L., Oliva, A., Torralba, A.: Learning deep features for discriminative localization. In: CVPR (2016)
78. Zhu, P., Wang, H., Saligrama, V.: Generalized zero-shot recognition based on visually semantic embedding. In: CVPR (2019)
79. Zhu, Y., Elhoseiny, M., Liu, B., Peng, X., Elgammal, A.: A generative adversarial approach for zero-shot learning from noisy texts. In: CVPR (2018)
80. Zhu, Y., Xie, J., Tang, Z., Peng, X., Elgammal, A.: Learning where to look: semantic-guided multi-attention localization for zero-shot learning. In: NeurIPS (2019)

GRAB: A Dataset of Whole-Body Human Grasping of Objects

Omid Taheri[✉], Nima Ghorbani, Michael J. Black, and Dimitrios Tzionas

Max Planck Institute for Intelligent Systems, Tübingen, Stuttgart, Germany
{otaheri,nghorbani,black,dtzionas}@tuebingen.mpg.de

Abstract. Training computers to understand, model, and synthesize human grasping requires a rich dataset containing complex 3D object shapes, detailed contact information, hand pose and shape, and the 3D body motion over time. While "grasping" is commonly thought of as a single hand stably lifting an object, we capture the motion of the entire body and adopt the generalized notion of "whole-body grasps". Thus, we collect a new dataset, called *GRAB* (GRasping Actions with Bodies), of whole-body grasps, containing full 3D shape and pose sequences of 10 subjects interacting with 51 everyday objects of varying shape and size. Given MoCap markers, we fit the full 3D body shape and pose, including the articulated face and hands, as well as the 3D object pose. This gives detailed 3D meshes over time, from which we compute contact between the body and object. This is a unique dataset, that goes well beyond existing ones for modeling and understanding how humans grasp and manipulate objects, how their full body is involved, and how interaction varies with the task. We illustrate the practical value of GRAB with an example application; we train GrabNet, a conditional generative network, to predict 3D hand grasps for unseen 3D object shapes. The dataset and code are available for research purposes at https://grab.is.tue.mpg.de.

1   Introduction

A key goal of computer vision is to estimate human-object interactions from video to help understand human behavior. Doing so requires a strong model of such interactions and learning this model requires data. However, capturing such data is not simple. Grasping involves both gross and subtle motions, as humans involve their whole body and dexterous finger motion to manipulate objects. Therefore, objects contact multiple body parts and not just the hands. This is difficult to capture with images because the regions of contact are occluded. Pressure sensors or other physical instrumentation, however, are also not a full solution as they can impair natural human-object interaction and do not capture full-body motion. Consequently, there are no existing datasets of complex

Electronic supplementary material The online version of this chapter (https://doi.org/10.1007/978-3-030-58548-8_34) contains supplementary material, which is available to authorized users.

A. Vedaldi et al. (Eds.): ECCV 2020, LNCS 12349, pp. 581–600, 2020.
https://doi.org/10.1007/978-3-030-58548-8_34

human-object interaction that contain full-body motion, 3D body shape, and detailed body-object contact. To fill this gap, we capture a novel dataset of full-body 3D humans dynamically interacting with 3D objects as illustrated in Fig. 1. By accurately tracking 3D body and object shape, we reason about contact resulting in a dataset with detail and richness beyond existing grasping datasets.

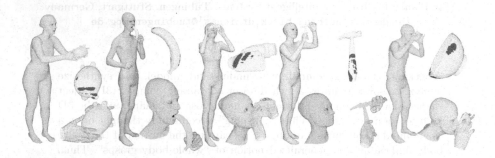

Fig. 1. Example "whole-body grasps" from the GRAB dataset. A "grasp" is usually thought of as a single hand interacting with an object. Using objects, however may involve more than just a single hand. From left to right: (i) passing a piggy bank, (ii) eating a banana, (iii) looking through binoculars, (iv) using a hammer, (v) drinking from a bowl. Contact between the object and the body is shown in red on the object; here contact areas are spatially extended to aid visualization. See the video on our website for a wide range of sequences with various objects and intents. (Color figure online)

Most previous work focuses on prehensile "grasps" [43]; i.e. a single human hand stably lifting or using an object. The hands, however, are only part of the story. For example, as infants, our earliest grasps involve bringing objects to the mouth [58]. Consider the example of drinking from a bowl in Fig. 1 (right). To do so, we must pose our body so that we can reach it, we orient our head to see it, we move our arm and hand to stably lift it, and then we bring it to our mouth, making contact with the lips, and finally we tilt the head to drink. As this and other examples in the figure illustrate, human grasping and using of everyday objects involves the *whole body*. Such interactions are fundamentally *three-dimensional*, and contact occurs between objects and multiple body parts.

Dataset. Such whole-body grasping [25] has received much less attention [4,37] than single hand-object grasping [3,10,14,27,43]. To model such grasping we need a dataset of humans interacting with varied objects, capturing the full 3D surface of both the body and objects. To solve this problem we adapt recent motion capture techniques, to construct a new rich dataset called **GRAB** for *"GRasping Actions with Bodies."* Specifically, we adapt MoSh++ [38] in two ways. First, MoSh++ estimates the 3D shape and motion of the body and hands from MoCap markers; here we extend this to include facial motion. For increased accuracy we first capture a 3D scan of each subject and fit the SMPL-X body model [46] to it. Then MoSh++ is used to recover the pose of the body, hands and face. Note that the face is important because it is involved in many interactions; see in Fig. 1 (second from left) how the mouth opens to eat a banana. Second, we also accurately capture the motion of 3D objects as they are manipulated by the subjects. To this end, we use small hemispherical markers on the objects and show that these do not impact grasping behavior. As a result, we obtain detailed 3D meshes for both the object and the human (with a full body, articulated fingers and face) moving over time while in interaction, as shown in Fig. 1. Using these meshes we then infer the body-object contact (red regions in Fig. 1). Unlike [5] this gives both the contact and the full body/hand pose over time. Interaction is dynamic, including in-hand manipulation and re-grasping. GRAB captures 10 different people (5 male and 5 female) interacting with 51 objects from [5]. Interaction takes place in 4 different contexts: lifting, handing over, passing from one hand to the other, and using, depending on the affordances and functionality of the object.

Applications. GRAB supports multiple uses of interest to the community. First, we show how GRAB can be used to gain insights into hand-object contact in everyday scenarios. Second, there is a significant interest in training models to grasp 3D objects [59]. Thus, we use GRAB to train a conditional variational autoencoder (cVAE) to generate plausible grasps for unseen 3D objects. Given a randomly posed 3D object, we predict plausible hand parameters (wrist pose and finger articulation) appropriate for grasping the object. To encode arbitrary 3D object shapes, we employ the recent basis point set (BPS) representation [52], whose fixed size is appropriate for neural networks. Then, by conditioning on a new 3D object shape, we sample from the learned latent space, and generate hand grasps for this object. We evaluate both quantitatively and qualitatively the resulting grasps and show that they look natural.

In summary, this work makes the following contributions: (1) we introduce a unique dataset capturing real "whole-body grasps" of 3D objects, including full-body human motion, object motion, in-hand manipulation and re-grasps; (2) to capture this, we adapt MoSh++ to solve for the body, face and hands of SMPL-X to obtain detailed moving 3D meshes; (3) using these meshes and tracked 3D objects we compute plausible contact on the object and the human and provide an analysis of observed patterns; (4) we show the value of our dataset for machine learning, by training a novel conditional neural network to generate 3D hand grasps for unseen 3D objects. The dataset, models, and code are available for research purposes at https://grab.is.tue.mpg.de.

2   Related Work

Hand Grasps: Hands are crucial for grasping and manipulating objects. For this reason, many studies focus on understanding grasps and defining taxonomies [3,10,14,27,43,51]. These works have explored the object shape and purpose of grasps [10], contact areas on the hand captured by sinking objects in ink [27], pose and contact areas [3] captured with an integrated data-glove [11] and tactile-glove [51], or number of fingers in contact with the object and thumb position [14]. A key element for these studies is capturing accurate hand poses, relative hand-object configurations and contact areas.

Whole-Body Grasps: Often people use more than a single hand to interact with objects. However, there are not many works in the literature on this topic [4,25]. Borras et al. [4] use MoCap data [39] of people interacting with a scene with multi-contact, and present a body pose taxonomy for such whole-body grasps. Hsiao et al. [25] focus on imitation learning with a database of whole-body grasp demonstrations with a human teleoperating a simulated robot. Although these works go in the right direction, they use unrealistic humanoid models and simple objects [4,25] or synthetic ones [25]. Instead, we use the SMPL-X model [46] to capture "whole-body", face and dexterous in-hand interactions.

Capturing Interactions with MoCap: MoCap is often used to capture, synthesize or evaluate humans interacting with scenes. Lee et al. [36] capture a 3D body skeleton interacting with a 3D scene and show how to synthesize new motions in new scenes. Wang et al. [73] capture a 3D body skeleton interacting with a large geometric objects. Han et al. [20] present a method for automatic labeling of hand markers, to speed up hand tracking for VR. Le et al. [35] capture a hand interacting with a phone to study the "comfortable areas", while Feit et al. [13] capture two hands interacting with a keyboard to study typing patterns. Other works [34,49] focus on graphics applications. Kry et al. [34] capture a hand interacting with a 3D shape primitive, instrumented with a force sensor. Pollard et al. [49] capture the motion of a hand to learn a controller for physically based grasping. Mandery et al. [39] sit between the above works, capturing humans interacting with both big and handheld objects, but without articulated faces and fingers. None of the previous work captures full 3D bodies, hands and faces together with 3D object manipulation and contact.

Capturing Contact: Capturing human-object contact is hard, because the human and object heavily occlude each other. One approach is instrumentation with touch and pressure sensors, but this might bias natural grasps. Pham et al. [47] predefine contact points on objects to place force transducers. More recent advances in tactile sensors allow accurate recognition of tactile patterns and handheld objects [65]. Some approaches [3] use a data glove [11] with an embedded tactile glove [51,67] but this combination is complicated and the two modalities can be hard to synchronize. A microscopic-domain tactile sensor [16] is introduced in [26], but is not easy to use on human hands. Mascaro et al. [40] attach a minimally invasive camera to detect changes in the coloration of fingernails. Brahmbhatt et al. [5] use a thermal camera to directly observe the "thermal print" of a hand on the grasped object. However, for this they only capture static grasps that last long enough for heat transfer. Consequently, even recent datasets that capture realistic hand-object [15,19] or body-scene [23,61] interaction avoid directly measuring contact.

3D Interaction Models: Learning a model of human-object interactions is useful for graphics and robotics to help avatars [12,63] or robots [21] interact with their surroundings, and for vision [18,42] to help reconstruct interactions from ambiguous data. However, this is a chicken-and-egg problem; to capture or synthesize data to learn a model, one needs such a model in the first place. For this reason, the community has long used hand-crafted approaches that exploit contact and physics, for body-scene [6,22,23,57,76], body-object [30,32, 37], or hand-object [24,45,54,55,62,68,69,72,77,78] scenarios. These approaches compute contact approximately; this may be rough for humans modeled as 3D skeletons [30,37] or shape primitives [6,32,62], or relatively accurate when using 3D meshes, whether generic [54,68], personalized [22,57,69,72], or based on 3D statistical models [23,24].

To collect training data, several works [54,55] use synthetic Poser [50] hand models, manually articulated to grasp 3D shape primitives. Contact points and forces are also annotated [54] through proximity and inter-penetration of 3D meshes. In contrast, Hasson et al. [24] use the robotics method GraspIt [41] to automatically generate 3D MANO [56] grasps for ShapeNet [7] objects and render synthetic images of the hand-object interaction. However, GraspIt optimizes for hand-crafted grasp metrics that do not necessarily reflect the distribution of human grasps (see Sup. Mat. Sect. C.2 of [24], and [17]). Alternatively, Garcia-Hernando et al. [15] use magnetic sensors to reconstruct a 3D hand skeleton and rigid object poses; they capture 6 subjects interacting with 4 objects. This dataset is used by [33,66] to learn to estimate 3D hand and object poses, but suffers from noisy poses and significant inter-penetrations (see Sect. 5.2 of [24]).

For bodies, Kim et al. [30] use synthetic data to learn to detect contact points on a 3D object, and then fit an interacting 3D body skeleton to them. Savva et al. [61] use RGB-D to capture 3D body skeletons of 5 subjects interacting in 30 3D scenes, to learn to synthesize interactions [61], affordance detection [60], or to reconstruct interaction from videos [42]. Mandery et al. [39] use optical MoCap to capture 43 subjects interacting with 41 tracked objects, both large and small. This is similar to our effort but they do not capture fingers or 3D body shape, so cannot reason about contact. Corona et al. [9] use this dataset to learn context-aware body motion prediction. Starke et al. [63] use Xsens IMU sensors [75] to capture the main body of a subject interacting with large objects, and learn to synthesize avatar motion in virtual worlds. Hassan et al. [23] use RGB-D and 3D scene constraints to capture 20 humans as SMPL-X [46] meshes interacting with 12 static 3D scenes, but do not capture object manipulation. Zhang et al. [79] use this data to learn to generate 3Dscene-aware humans.

We see that only parts of our problem have been studied. We draw inspiration from prior work, in particular [5,23,25,39]. We go beyond these by introducing a new dataset of real "whole-body" grasps, as described in the next section.

3 Dataset

To manipulate an object, the human needs to approach its 3D surface, and bring their skin to come in *physical contact* to apply forces. Realistically capturing such human-object interactions, especially with "whole-body grasps", is a challenging problem. First, the object may occlude the body and vice-versa, resulting in *ambiguous* observations. Second, for physical interactions it is crucial to reconstruct an accurate and detailed 3D *surface* for both the human and the object. Additionally, the capture has to work across multiple scales (body, fingers and face) and for objects of varying complexity. We address these challenges with a unique combination of state-of-the-art solutions that we adapt to the problem.

There is a fundamental trade-off with current technology; one has to choose between (a) accurate motion with instrumentation and without natural RGB images, or (b) less accurate motion but with RGB images. Here we take the former approach; for an extensive discussion we refer the reader to Sup. Mat.

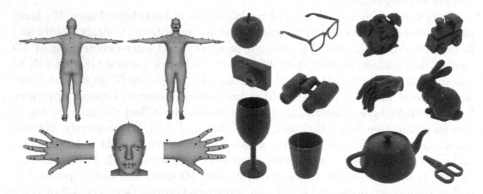

Fig. 2. MoCap markers used to capture humans and objects. **Left:** We attach 99 reflective markers per subject; 49 for the body, 14 for the face and 36 for the fingers. We use spherical 4.5 mm radius markers for the body and hemi-spherical 1.5 mm radius ones for the hands and face. **Right:** Example 3D printed objects from [5]. We glue 1.5 mm radius hemi-spherical markers (the gray dots) on the objects. These makers are small enough to be unobtrusive. The 6 objects on the right are mostly used by one or more hands, while the 6 on the left involve "whole-body grasps". (Color figure online)

3.1 Motion Capture (MoCap)

We use a Vicon system with 54 infrared "Vantage 16" [71] cameras that capture 16 MP at 120 fps. The large number of cameras minimized occlusions and the high frame rate captures temporal details of contact. The high resolution allows using small (1.5 mm radius) hemi-spherical markers. This minimizes their influence on finger and face motion and does not alter how people grasp objects. Details of the marker setup are shown in Fig. 2. Even with many cameras, motion

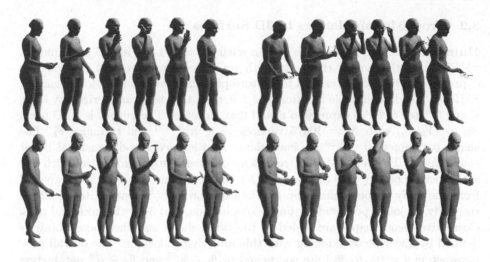

Fig. 3. We capture humans interacting with objects over time and reconstruct sequences of 3D meshes for both, as described in Sect. 3.1 and Sect. 3.2. Note the realistic and plausible placement of objects in the hands, and the "whole-body" involvement. The video on our website shows more examples.

capture of the body, face, and hands, together with objects, is uncommon because it is so challenging. MoCap markers become occluded, labels are swapped, and ghost makers appear. MoCap cleaning was done by four trained technicians using Vicon's Shōgun-Post software.

Capturing Human MoCap: To capture human motion, we use the marker set of Fig. 2 (left). The body markers are attached on a tight body suit with a velcro-based mounting at a distance of roughly $d_b = 9.5$ mm from the body surface. The hand and face markers are attached directly to the skin with special removable glue, therefore the distance to it is roughly $d_h = d_f \approx 0$ mm. Importantly, no hand glove is used and hand markers are placed only on the dorsal side, leaving the palmar side completely uninstrumented, for natural interactions.

Capturing Objects: To reconstruct interactions accurately, it is important to know the precise 3D object surface geometry. We therefore use the CAD object models of [5], and 3D print them with a Stratasys Fortus 360mc [64] printer; see Fig. 2 (right). Each object o is then represented by a known 3D mesh with vertices V_o. To capture object motion, we attach on the 1.5 mmhemi-spherical markers with strong glue directly to the object surface. We use at least 8 markers per object, empirically distributing them on the object so that at least 3 of them are always observed. The size and placement of the markers makes them unobtrusive. In Sup. Mat. we show empirical evidence that makers have minimal influence on grasping.

3.2   From MoCap Markers to 3D Surfaces

Human Model: We model the human with the SMPL-X [46] 3D body model. SMPL-X jointly models the body with an articulated face and fingers; this expressive body model is critical to capture physical interactions. More formally, SMPL-X is a differentiable function $M_b(\beta, \theta, \psi, \gamma)$ that is parameterized by body shape β, pose θ, facial expression ψ and translation γ. The output is a 3D mesh $M_b = (V_b, F_b)$ with $N_b = 10475$ vertices $V_b \in \mathbb{R}^{(N_b \times 3)}$ and triangles F_b. The shape parameters $\beta \in \mathbb{R}^{10}$ are coefficients in a learned low-dimensional linear shape space. This lets SMPL-X represent different subject identities with the same mesh topology. The 3D joints, $J(\beta)$, of a kinematic skeleton are regressed from the body shaped defined by β. The skeleton has 55 joints in total; 22 for the body, 15 joints per hand for finger articulation, and 3 for the neck and eyes. Corrective blend shapes are added to the body shape and then posed body is defined by linear blend skinning with this underlying skeleton. The overall pose parameters $\theta = (\theta_b, \theta_f, \theta_h)$ are comprised of $\theta_b \in \mathbb{R}^{66}$ and $\theta_f \in \mathbb{R}^9$ parameters in axis-angle representation for the main body and face joints correspondingly, with 3 degrees of freedom (DoF) per joint, and $\theta_h \in \mathbb{R}^{60}$ parameters in a lower-dimensional pose space for both hands, i.e. 30 DoF per hand following [24]. For more details, please see [46].

Model-Marker Correspondences: For the human body we define, a priori, the rough marker placement on the body as shown in Fig. 2 (left). Exact marker locations on individual subjects are then computed automatically using MoSh++ [38]. In contrast to the body, the objects have different shapes and mesh topologies. Markers are placed according to the object shape, affordances and expected occlusions during interaction; Fig. 2 (right), Therefore, we annotate object-specific vertex-marker correspondences, and do this once per object.

Human and Object Tracking: To ensure accurate human shape, we capture a 3D scan of each subject and fit SMPL-X to it following [56]. We fit these personalized SMPL-X models to our cleaned 3D marker observations using MoSh++ [38]. Specifically we optimize over pose, θ, expressions, ψ, and translation, γ, while keeping the known shape, β, fixed. The weights of MoSh++ for the finger and face data terms are tuned on a synthetic dataset, as described in Sup. Mat. An analysis of MoSh++ fitting accuracy is also provided in Sup. Mat.

Objects are simpler because they are rigid and we know their 3D shape. Given three or more detected markers, we solve for the rigid object pose $\theta_o \in \mathbb{R}^6$. Here we track the human and object separately and on a per-frame basis. Figure 3 shows that our approach captures realistic interactions and reconstructs detailed 3D meshes for both the human and the object, over time. The video on our website shows a wide range of reconstructed sequences.

3.3   Contact Annotation

Since contact cannot be directly observed, we estimate it using 3D proximity between the 3D human and object meshes. In theory, they come in contact when the distance between them is zero. In practice, however, we relax this and define contact when the distance, $d \leq \epsilon_{contact}$, for a threshold $\epsilon_{contact}$. This helps address: (1) measurement and fitting errors, (2) limited mesh resolution, (3) the fact that human soft tissue deforms when grasping an object, while the SMPL-X model cannot model this.

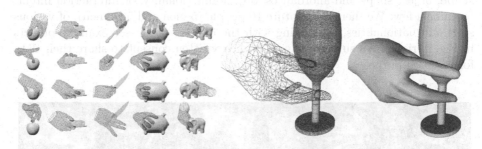

Fig. 4. Left: Accurate tracking lets us compute realistic contact areas (red) for each frame (Sect. 3.3). For illustration, we render only the hand of SMPL-X and spatially extend the red contact areas for visibility. **Right:** Detection of "intersection ring" triangles during contact annotation (Sect. 3.3). (Color figure online)

Given these issues, accurately estimating contact is challenging. Consider the hand grasping a wine glass in Fig. 4 (right), where the color rings indicate inter-sections. Ideally, the glass should be in contact with the thumb, index and middle fingers. "Contact under-shooting" results in fingers hovering close to the object surface, but not on it, like the thumb. "Contact over-shooting", results in fingers penetrating the object surface around the contact area, like the index (purple intersections) and middle finger (red intersections). The latter case is especially problematic for thin objects where a penetrating finger can pass through the object, intersecting it on two sides. In this example, we want to annotate con-tact only with the outer surface of the object and not the inner one.

We account for "contact over-shooting" cases with an efficient heuristic. We use a fast method [29,46] to detect intersections, cluster them in connected "intersection rings", $\mathcal{R}_b \subsetneq V_b$ and $\mathcal{R}_o \subsetneq V_o$, and label them with the intersecting body part, seen as purple and red rings in Fig. 4 (right). The "intersection ring", $\mathcal{R}_b$, segments the body mesh M_b to give the "penetrating sub-mesh" $\mathcal{M}_b \subsetneq M_b$. (1) When a body part gives only one intersection, we annotate the points $V_o^{\mathcal{C}} \subset V_o$ on the object all vertices enclosed by the ring $\mathcal{R}_o$ as being in contact. We then annotate as contact points, $V_b^{\mathcal{C}} \subset V_b$, on the body all vertices that lie close to $V_o^{\mathcal{C}}$ with a distance $d_{o \to b} \leq \epsilon_{contact}$. (2) In case of multiple intersections i we take into account only the ring $\mathcal{R}_b^i$ corresponding to the largest intersection subset, $\mathcal{M}_b^i$.

For body parts that are not found in contact above, there is the possibility of "contact under-shooting". To address this, we compute the distance from each object vertex V_o, to each non-intersecting body vertex V_b. We then annotate as contact vertices, V_o^C and V_b^C, the ones with $d_{o \to b} \leq \epsilon_{contact}$. We empirically find that $\epsilon_{contact} = 4.5$ mm works well for our purposes.

3.4   Dataset Protocol

Human-object interaction depends on various factors including the human body shape, object shape and affordances, object functionality, or interaction intent, to name a few. We therefore capture 10 people (5 men and 5 women), of various sizes and nationalities, interacting with the objects of [5]; see example objects in Fig. 2 (right). All subjects gave informed written consent to share their data for research purposes.

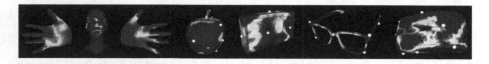

Fig. 5. Contact "heatmaps". **Left:** For the body we focus on "use" sequences to show "whole-body grasps". **Right:** For objects we include all intents. Object markers (light gray) are unobtrusive and can lie on "hot" (red) contact areas. (Color figure online)

Table 1. Size of the GRAB dataset. GRAB is sufficiently large to enable training of data-driven models of grasping as shown in Sect. 4.

Intent	"Use"	"Pass"	"Lift"	"Off-hand"	Total
# Sequences	579	414	274	67	1334
# Frames	605.796	335.733	603.381	77.549	1.622.459

For each object we capture interactions with 4 different intents, namely "use" and "pass" (to someone), borrowed from protocol of [5], as well as "lift" and "off-hand pass" (from one hand to the other). Figure 3 shows some example 3D capture sequences for the "use" intent. For each sequence we: (i) we randomize initial object placement to increase motion variance, (ii) we instruct the subject to follow an intent, (iii) the subject starts from a T-pose and approaches the object, (iv) they perform the instructed task, and (v) they leave the object and returns to a T-pose. The video on our website shows the richness of our protocol with a wide range of captured sequences.

3.5   Analysis

Dataset Analysis. The dataset contains 1334 sequences and over 1.6M frames of MoCap; Table 1 provides a detailed breakdown. Here we analyze those frames where we have detected contact between the human and the object. We assume that the object is static on a table and can move only due to grasping. Consequently, consider contact frames to be those in which the object's position deviates in the vertical direction by at least 5 mm from its initial position and in which at least 50 body vertices are in contact with the object. This results in 952, 514 contact frames that we analyze below. The exact thresholds of these contact heuristics have little influence on our analysis, see Sup. Mat.

By uniquely capturing the whole body, and not just the hand, interesting interaction patterns arise. By focusing on "use" sequences that highlight the object functionality, we observe that 92% of contact frames involve the right hand, 39% the left hand, 31% both hands, and 8% involve the head. For the first category the per-finger contact likelihood, from thumb to pinky, is 100%, 96%, 92%, 79%, 39% and for the palm 24%. For more results see Sup. Mat.

To visualize the fine-grained contact information, we integrate over time the binary per-frame contact maps, and generate "heatmaps" encoding the contact likelihood of contact across the whole body surface. Figure 5 (left) shows such "heatmaps" for "use" sequences. "Hot" areas (red) denote high likelihood of contact, while "cold" areas (blue) denote low likelihood. We see that both the hands and face are important for using everyday objects, highlighting the importance of capturing the whole interacting body. For the face, the "hot" areas are the lips, the nose, the temporal head area, and the ear. For hands, the fingers are more frequently in contact than the palm, with more contact on the right than left. The palm seems more important for right-hand grasps than for left-hand ones, possibly because all our subjects are right-handed. Contact patterns are also influenced by the size of the object and the size of the hand; see Sup. Mat. for a visualization.

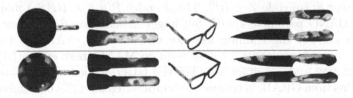

Fig. 6. Effect of interaction intent on contact during grasping. We show the "use" (top) and "pass" (bottom) intents for 4 different objects. (Color figure online)

Figure 6 shows the effect of the intent. Contact for "use" sequences complies with the functionality of the object; e.g. people do not touch the knife blade or the hot area of the pan, but they do contact the on/off button of the flashlight. For "pass" sequences subjects tend to contact one side of the object irrespective of affordances, leaving the other one free to be grasped by the receiving person.

For natural interactions it is important to have a minimally intrusive setup. While our MoCap markers are small and unobtrusive (Figure 2 (right)), we ask whether subjects may be biased in their grasps by these markers? Figure 5 (right) shows contact "heatmaps" for some objects across all intents. These clearly show that markers are often located in "hot" areas, suggesting that subjects do not avoid grasping these locations. Further analysis based on K-means clustering of grasps can be found in Sup. Mat.

4   GrabNet: Learning to Grab an Object

We show the value of the GRAB dataset with a challenging example application; we use it to train a model that generates plausible 3D MANO [56] grasps for a previously unseen 3D object. Our model, called GrabNet, is comprised of two main modules. First, we employ a conditional variational autoencoder (cVAE) [31], called CoarseNet, that generates an initial grasp. For this it learns a grasping embedding space Z conditioned on the object shape, that is encoded using the Basis Point Set (BPS) [52] representation as a set of distances from the basis points to the nearest object points. CoarseNet's grasps are reasonable, but realism can improve by refining contacts based on the distances D between the hand and the object. We do this with a second network, called RefineNet. The architecture of GrabNet is shown in Fig. 7, for more details see Sup. Mat.

Pre-processing. For training, we gather all frames with right-hand grasps that involve some minimal contact, for details see Sup. Mat. We then center each training sample, i.e. hand-object grasp, at the centroid of the object and compute the $BPS_o \in R^{4096}$ representation for the object, used for conditioning.

CoarseNet. We pass the object shape BPS_o along with initial MANO wrist rotation θ_{wrist} and translation γ to the encoder $Q(Z|\theta_{wrist}, \gamma, BPS_o)$ that produces a latent grasp code $Z \in R^{16}$. The decoder $P(\bar{\theta}, \bar{\gamma}|Z, BPS_o)$ maps Z and BPS_o to MANO parameters with full finger articulation $\bar{\theta}$, to generate a 3D grasping hand. For the training loss, we use standard cVAE loss terms (KL divergence, weight regularizer), a data term on MANO mesh edges (L1), as well as a penetration and a contact loss. For the latter, we learn candidate contact point weights from GRAB, in contrast to handcrafted ones [23] or weights learned from artificial data [24]. At inference time, given an unseen object shape BPS_o, we sample the latent space Z and decode our sample to generate a MANO grasp.

RefineNet. The grasps estimated by CoarseNet are plausible, but can be refined for improved contacts. For this, RefineNet takes as input the initial grasp $(\bar{\theta}, \bar{\gamma})$ and the distances D from MANO vertices to the object mesh. The distances are weighted according to the vertex contact likelihood learned from GRAB. Then, RefineNet estimates refined MANO parameters $(\hat{\theta}, \hat{\gamma})$ in 3 iterative steps as in [28], to give the final grasp. To train RefineNet, we generate a synthetic dataset; we sample CoarseNet grasps as ground truth and we perturb their hand pose parameters to simulate noisy input estimates. We use the same training losses as for CoarseNet.

GrabNet. Given an unseen 3D object, we first obtain an initial grasp estimate with CoarseNet, and pass this to RefineNet to obtain the final grasp estimate. For simplicity, the two networks are trained separately, but we expect end-to-end refinement to be beneficial, as in [24]. Figure 8 (right) shows some generated examples; our generations look realistic, as explained later in the evaluation section. For more qualitative results, see the video on our website and images in Sup. Mat.

Contact. As a free by-product of our 3D grasp predictions, we can compute contact between the 3D hand and object meshes, following Sect. 3.3. Contacts for GrabNet estimates are shown with red in Fig. 8 (right). Other methods for contact prediction, like [5], are pure bottom-up approaches that label a vertex as in contact or not, without explicit reasoning about the hand structure. In contrast, we follow a top-down approach; we first generate a 3D grasping hand, and then compute contact with explicit anthropomorphic reasoning.

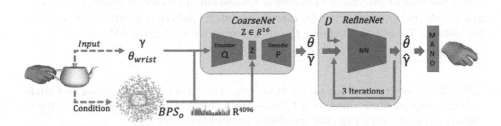

Fig. 7. GrabNet architecture. GrabNet generates MANO [56] grasps for unseen object shapes, encoded with a BPS [52] representation. It is comprised of two main modules. First, with CoarseNet we predict an initial plausible grasp. Second, we refine this with RefineNet to produce better contacts with the object.

Evaluation - CoarseNet/RefineNet. We first quantitatively evaluate the two main components, by computing the reconstruction vertex-to-vertex error. For CoarseNet the errors are 12.1 mm, 14.1 mm and 18.4 mm for the training, validation and test set respectively. For RefineNet the errors are 3.7 mm, 4.1 mm and 4.4 mm. The results show that the components, that are trained separately, work reasonably well before plugging them together.

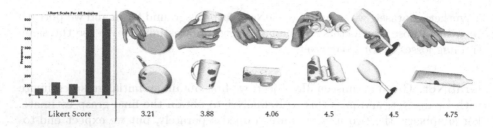

| Likert Score | 3.21 | 3.88 | 4.06 | 4.5 | 4.5 | 4.75 |

Fig. 8. Grasps generated by GrabNet for unseen objects; grasps look natural. As free by-product of 3D mesh generation, we get the red contact areas. For each grasp we show the average Likert score from all annotators. On the left the average Likert score is shown for all generated grasps. (Color figure online)

Evaluation - GrabNet. To evaluate GrabNet generated grasps, we perform a user study through AMT [1]. We take 6 test objects from the dataset and, for each object, we generate 20 grasps, mix them with 20 ground-truth grasps, and show them with a rotating 3D viewpoint to subjects. Then we ask participants how they agree with the statement "Humans can grasp this object as the video shows" on a 5-level Likert scale (5 is "strongly agree" and 1 is "strongly disagree"). To filter out the noisy subjects, namely the ones who do not understand the task or give random answers, we use catch trials that show implausible grasps. We remove subjects who rate these catch trials as realistic; see Sup. Mat. for details. Table 2 (left) shows the user scores for both ground-truth and generated grasps.

Table 2. GrabNet evaluation for 6 test objects. The "AMT" column shows user study results; grasp quality is rated from 1 (worst) to 5 (best). The "vertices" and "contact" columns evaluate grasps against the closest ground-truth one.

	AMT				Vertices	Contact
	Generation		Ground truth		Mean (cm)	%
Test object	Mean	Std	Mean	Std	N = 100	N = 20
Binoculars	4.09	0.93	4.27	0.80	2.56	4.00
Camera	4.40	0.79	4.34	0.76	2.90	3.75
Frying pan	3.19	1.30	4.49	0.67	3.58	4.16
Mug	4.13	1.00	4.36	0.78	1.96	3.25
Toothpaste	4.56	0.67	4.42	0.77	1.78	5.39
Wineglass	4.32	0.88	4.43	0.79	1.92	4.56
Average	4.12	1.04	4.38	0.77	2.45	4.18

Fig. 9. Comparison of GrabNet (right) to ContactDB [5] (left). For each estimation we render two views, following the presentation style of [5].

Evaluation - Contact. Figure 9 shows examples of contact areas (red) generated by [5] (left) and our approach (right). The method of [5] gives only 10 predictions per object, some with zero contact. Also, a hand is supposed to touch the whole red area; this is often not anthropomorphically plausible. Our contact is a product of MANO-based inference and is, by construction, anthropomorphically valid. Also, one can draw infinite samples from our learned grasping latent space. For further evaluation, we follow a protocol similar to [5] for our data. For every unseen test object we generate 20 grasps, and for each one we find both the closest ground-truth contact map and the closest ground-truth hand vertices, for comparison. Table 2 (right) reports the average error over all 20 predictions, in % for the former and cm for the latter case.

5 Discussion

We provide a new dataset to the community that goes beyond previous motion capture or grasping datasets. We believe that GRAB will be useful for a wide range of problems. Here we show that it provides enough data and variability to train a novel network to predict object grasping, as we demonstrate with GrabNet. But there is much more that can be done. Importantly, GRAB includes the full body motion, enabling a much richer modeling than GrabNet.

Limitations: By focusing on accurate MoCap, we do not have synced image data. However, GRAB can support image-based inference [8,28] by enabling rendering of synthetic human-object interaction [24,53,70] or learning priors to regularizing ill-posed inference of human-object interaction from 2D images [42].

Future Work: GRAB can support learning human-object interaction models [42,61], robotic grasping from imitation [74], learning mappings from MoCap markers to meshes [20], rendering synthetic images [24,53,70], inferring object shape/pose from interaction [2,44], or analysis of temporal patterns [48].

Acknowledgements. We thank S. Polikovsky, M. Höschle (MH) and M. Landry (ML) for the MoCap facility. We thank F. Mattioni, D. Hieber, and A. Valis for MoCap cleaning. We thank ML and T. Alexiadis for trial coordination, MH and F. Grimminger for 3D printing, V. Callaghan for voice recordings and J. Tesch for renderings. **Disclosure:**

In the last five years, MJB has received research gift funds from Intel, Nvidia, Facebook, and Amazon. He is a co-founder and investor in Meshcapade GmbH, which commercializes 3D body shape technology. While MJB is a part-time employee of Amazon, his research was performed solely at, and funded solely by, MPI.

References

1. Amazon Mechanical Turk. https://www.mturk.com
2. Behbahani, F.M.P., Singla–Buxarrais, G., Faisal, A.A.: Haptic SLAM: an ideal observer model for bayesian inference of object shape and hand pose from contact dynamics. In: Bello, F., Kajimoto, H., Visell, Y. (eds.) EuroHaptics 2016. LNCS, vol. 9774, pp. 146–157. Springer, Cham (2016). https://doi.org/10.1007/978-3-319-42321-0_14
3. Bernardin, K., Ogawara, K., Ikeuchi, K., Dillmann, R.: A sensor fusion approach for recognizing continuous human grasping sequences using hidden Markov models. IEEE Trans. Rob. (T-RO) **21**(1), 47–57 (2005)
4. Borras, J., Asfour, T.: A whole-body pose taxonomy for loco-manipulation tasks. In: IEEE/RSJ International Conference on Intelligent Robots and Systems (IROS), pp. 1578–1585 (2015)
5. Brahmbhatt, S., Ham, C., Kemp, C.C., Hays, J.: ContactDB: analyzing and predicting grasp contact via thermal imaging. In: IEEE/CVF Conference on Computer Vision and Pattern Recognition (CVPR) (2019)
6. Brubaker, M.A., Fleet, D.J., Hertzmann, A.: Physics-based person tracking using the anthropomorphic walker. Int. J. Comput. Vis. (IJCV) **87**(1), 140 (2009)
7. Chang, A.X., et al.: ShapeNet: An information-rich 3D model repository. arXiv:1512.03012 (2015)
8. Choutas, V., Pavlakos, G., Bolkart, T., Tzionas, D., Black, M.J.: Monocular expressive body regression through body-driven attention. In: Proceedings of the European Conference on Computer Vision (ECCV) (2020)
9. Corona, E., Pumarola, A., Alenyà, G., Moreno-Noguer, F.: Context-aware human motion prediction. In: IEEE/CVF Conference on Computer Vision and Pattern Recognition (CVPR) (2020)
10. Cutkosky, M.R.: On grasp choice, grasp models, and the design of hands for manufacturing tasks. IEEE Trans. Rob. Autom. **5**(3), 269–279 (1989)
11. Cyberglove III data glove. http://www.cyberglovesystems.com/cyberglove-iii
12. ElKoura, G., Singh, K.: Handrix: animating the human hand. In: Proceedings of the ACM SIGGRAPH/Eurographics Symposium on Computer Animation (2003)
13. Feit, A.M., Weir, D., Oulasvirta, A.: How we type: movement strategies and performance in everyday typing. In: Proceedings of the CHI Conference on Human Factors in Computing Systems (2016)
14. Feix, T., Romero, J., Schmiedmayer, H.B., Dollar, A.M., Kragic, D.: The GRASP taxonomy of human grasp types. IEEE Trans. Hum.-Mach. Syst. **46**(1), 66–77 (2016)
15. Garcia-Hernando, G., Yuan, S., Baek, S., Kim, T.K.: First-person hand action benchmark with rgb-d videos and 3d hand pose annotations. In: IEEE/CVF Conference on Computer Vision and Pattern Recognition (CVPR) (2018)
16. GelSight tactile sensor. http://www.gelsight.com
17. Goldfeder, C., Ciocarlie, M.T., Dang, H., Allen, P.K.: The Columbia grasp database. In: IEEE International Conference on Robotics and Automation (ICRA) (2009)

18. Hamer, H., Gall, J., Weise, T., Van Gool, L.: An object-dependent hand pose prior from sparse training data. In: IEEE/CVF Conference on Computer Vision and Pattern Recognition (CVPR) (2010)
19. Hampali, S., Oberweger, M., Rad, M., Lepetit, V.: HO-3D: a multi-user, multi-object dataset for joint 3D hand-object pose estimation. In: IEEE/CVF Conference on Computer Vision and Pattern Recognition (CVPR) (2019)
20. Han, S., Liu, B., Wang, R., Ye, Y., Twigg, C.D., Kin, K.: Online optical marker-based hand tracking with deep labels. ACM Trans. Graph. (TOG) **37**(4), 166:1–166:10 (2018)
21. Handa, A., et al.: DexPilot: vision based teleoperation of dexterous robotic hand-arm system. In: IEEE International Conference on Robotics and Automation (ICRA) (2019)
22. Hasler, N., Rosenhahn, B., Thormahlen, T., Wand, M., Gall, J., Seidel, H.: Markerless motion capture with unsynchronized moving cameras. In: IEEE/CVF Conference on Computer Vision and Pattern Recognition (CVPR) (2009)
23. Hassan, M., Choutas, V., Tzionas, D., Black, M.J.: Resolving 3D human pose ambiguities with 3D scene constrains. In: Proceedings of the IEEE/CVF International Conference on Computer Vision (ICCV) (2019)
24. Hasson, Y., et al.: Learning joint reconstruction of hands and manipulated objects. In: IEEE/CVF Conference on Computer Vision and Pattern Recognition (CVPR) (2019)
25. Hsiao, K., Lozano-Perez, T.: Imitation learning of whole-body grasps. In: IEEE/RSJ International Conference on Intelligent Robots and Systems (2006)
26. Johnson, M.K., Cole, F., Raj, A., Adelson, E.H.: Microgeometry capture using an elastomeric sensor. ACM Trans. Graph. (TOG) **30**(4), 46:1–46:8 (2011)
27. Kamakura, N., Matsuo, M., Ishii, H., Mitsuboshi, F., Miura, Y.: Patterns of static prehension in normal hands. Am. J. Occup. Therapy **34**(7), 437–445 (1980)
28. Kanazawa, A., Black, M.J., Jacobs, D.W., Malik, J.: End-to-end recovery of human shape and pose. In: IEEE/CVF Conference on Computer Vision and Pattern Recognition (CVPR) (2018)
29. Karras, T.: Maximizing parallelism in the construction of BVHs, octrees, and k-d trees. In: Proceedings of the ACM SIGGRAPH/Eurographics Conference on High-Performance Graphics (2012)
30. Kim, V.G., Chaudhuri, S., Guibas, L., Funkhouser, T.: Shape2pose: human-centric shape analysis. ACM Trans. Graph. (TOG) **33**(4), 120:1–120:12 (2014)
31. Kingma, D.P., Welling, M.: Auto-encoding variational bayes. In: International Conference on Learning Representations (ICLR) (2014)
32. Kjellstrom, H., Kragic, D., Black, M.J.: Tracking people interacting with objects. In: IEEE/CVF Conference on Computer Vision and Pattern Recognition (CVPR) (2010)
33. Kokic, M., Kragic, D., Bohg, J.: Learning task-oriented grasping from human activity datasets. IEEE Rob. Autom. Lett. (RA-L) **5**(2), 3352–3359 (2020)
34. Kry, P.G., Pai, D.K.: Interaction capture and synthesis. ACM Trans. Graph. (TOG) **25**(3), 872–880 (2006)
35. Le, H.V., Mayer, S., Bader, P., Henze, N.: Fingers' range and comfortable area for one-handed smartphone interaction beyond the touchscreen. In: Proceedings of the CHI Conference on Human Factors in Computing Systems (2018)
36. Lee, K.H., Choi, M.G., Lee, J.: Motion patches: building blocks for virtual environments annotated with motion data. ACM Trans. Graph. (TOG) **25**(3), 898–906 (2006)

37. Li, Z., Sedlar, J., Carpentier, J., Laptev, I., Mansard, N., Sivic, J.: Estimating 3D motion and forces of person-object interactions from monocular video. In: IEEE/CVF Conference on Computer Vision and Pattern Recognition (CVPR) (2019)
38. Mahmood, N., Ghorbani, N., Troje, N.F., Pons-Moll, G., Black, M.J.: AMASS: archive of motion capture as surface shapes. In: Proceedings of the IEEE/CVF International Conference on Computer Vision (ICCV) (2019)
39. Mandery, C., Terlemez, Ö., Do, M., Vahrenkamp, N., Asfour, T.: The KIT whole-body human motion database. In: International Conference on Advanced Robotics (ICAR) (2015)
40. Mascaro, S.A., Asada, H.H.: Photoplethysmograph fingernail sensors for measuring finger forces without haptic obstruction. IEEE Trans. Rob. Autom. (TRA) 17(5), 698–708 (2001)
41. Miller, A.T., Allen, P.K.: Graspit! a versatile simulator for robotic grasping. IEEE Rob. Autom. Mag. (RAM) 11(4), 110–122 (2004)
42. Monszpart, A., Guerrero, P., Ceylan, D., Yumer, E., Mitra, N.J.: iMapper: interaction-guided scene mapping from monocular videos. ACM Trans. Graph. (TOG) 38(4), 92:1–92:15 (2019)
43. Napier, J.R.: The prehensile movements of the human hand. J. Bone Joint Surg. 38(4), 902–913 (1956)
44. Oberweger, M., Wohlhart, P., Lepetit, V.: Generalized feedback loop for joint hand-object pose estimation. IEEE Trans. Pattern Anal. Mach. Intell. (TPAMI) 42(8), 1898–1912 (2020)
45. Oikonomidis, I., Kyriazis, N., Argyros, A.A.: Full dof tracking of a hand interacting with an object by modeling occlusions and physical constraints. In: Proceedings of the IEEE/CVF International Conference on Computer Vision (ICCV) (2011)
46. Pavlakos, G., et al.: Expressive body capture: 3D hands, face, and body from a single image. In: IEEE/CVF Conference on Computer Vision and Pattern Recognition (CVPR) (2019)
47. Pham, T., Kyriazis, N., Argyros, A.A., Kheddar, A.: Hand-object contact force estimation from markerless visual tracking. IEEE Trans. Pattern Anal. Mach. Intell. (TPAMI) 40(12), 2883–2896 (2018)
48. Pirk, S., et al.: Understanding and exploiting object interaction landscapes. ACM Trans. Graph. (TOG) 36(3), 31:1–31:14 (2017)
49. Pollard, N.S., Zordan, V.B.: Physically based grasping control from example. In: Proceedings of the ACM SIGGRAPH/Eurographics Symposium on Computer Animation (2005)
50. POSER: 3D rendering and animation software. https://www.posersoftware.com
51. Pressure Profile Systems Inc. (PPS). https://pressureprofile.com
52. Prokudin, S., Lassner, C., Romero, J.: Efficient learning on point clouds with basis point sets. In: IEEE/CVF Conference on Computer Vision and Pattern Recognition (CVPR) (2019)
53. Ranjan, A., Hoffmann, D.T., Tzionas, D., Tang, S., Romero, J., Black, M.J.: Learning multi-human optical flow. Int. J. Comput. Vis. (IJCV) 128, 873–890 (2020)
54. Rogez, G., Supančič III, J.S., Ramanan, D.: Understanding everyday hands in action from RGB-D images. In: Proceedings of the IEEE/CVF International Conference on Computer Vision (ICCV) (2015)
55. Romero, J., Kjellström, H., Kragic, D.: Hands in action: real-time 3D reconstruction of hands in interaction with objects. In: IEEE International Conference on Robotics and Automation (ICRA) (2010)

56. Romero, J., Tzionas, D., Black, M.J.: Embodied hands: modeling and capturing hands and bodies together. ACM Trans. Graph. (TOG) **36**(6), 245:1–245:17 (2017)
57. Rosenhahn, B., Schmaltz, C., Brox, T., Weickert, J., Cremers, D., Seidel, H.: Markerless motion capture of man-machine interaction. In: IEEE/CVF Conference on Computer Vision and Pattern Recognition (CVPR) (2008)
58. Ruff, H.A.: Infants' manipulative exploration of objects: effects of age and object characteristics. Dev. Psychol. **20**(1), 9 (1984)
59. Sahbani, A., El-Khoury, S., Bidaud, P.: An overview of 3D object grasp synthesis algorithms. Rob. Auton. Syst. (RAS) **60**(3), 326–336 (2012)
60. Savva, M., Chang, A.X., Hanrahan, P., Fisher, M., Nießner, M.: SceneGrok: inferring action maps in 3D environments. ACM Trans. Graph. (TOG) **33**(6), 212:1–212:10 (2014)
61. Savva, M., Chang, A.X., Hanrahan, P., Fisher, M., Nießner, M.: PiGraphs: learning interaction snapshots from observations. ACM Trans. Graph. (TOG) **35**(4), 1391–13912 (2016)
62. Sridhar, S., Mueller, F., Zollhoefer, M., Casas, D., Oulasvirta, A., Theobalt, C.: Real-time joint tracking of a hand manipulating an object from RGB-D input. In: Proceedings of the European Conference on Computer Vision (ECCV) (2016)
63. Starke, S., Zhang, H., Komura, T., Saito, J.: Neural state machine for character-scene interactions. ACM Trans. Graph. (TOG) **38**(6), 209:1 209:14 (2019)
64. Stratasys Fortus 360mc: 3D printing. https://www.stratasys.com/resources/search/white-papers/fortus-360mc-400mc
65. Sundaram, S., Kellnhofer, P., Li, Y., Zhu, J.Y., Torralba, A., Matusik, W.: Learning the signatures of the human grasp using a scalable tactile glove. Nature **569**(7758), 698–702 (2019)
66. Tekin, B., Bogo, F., Pollefeys, M.: H+O: unified egocentric recognition of 3D hand-object poses and interactions. In: IEEE/CVF Conference on Computer Vision and Pattern Recognition (CVPR) (2019)
67. Tekscan grip system: Tactile grip force and pressure sensing. https://www.tekscan.com/products-solutions/systems/grip-system
68. Tsoli, A., Argyros, A.A.: Joint 3D tracking of a deformable object in interaction with a hand. In: Proceedings of the European Conference on Computer Vision (ECCV) (2018)
69. Tzionas, D., Ballan, L., Srikantha, A., Aponte, P., Pollefeys, M., Gall, J.: Capturing hands in action using discriminative salient points and physics simulation. Int. J. Comput. Vis. (IJCV) **118**(2), 172–193 (2016)
70. Varol, G., et al.: Learning from synthetic humans. In: IEEE/CVF Conference on Computer Vision and Pattern Recognition (CVPR) (2017)
71. Vicon Vantage: Cutting edge, flagship camera with intelligent feedback and resolution. https://www.vicon.com/hardware/cameras/vantage
72. Wang, Y., et al.: Video-based hand manipulation capture through composite motion control. ACM Trans. Graph. (TOG) **32**(4), 43:1–43:14 (2013)
73. Wang, Z., Chen, L., Rathore, S., Shin, D., Fowlkes, C.: Geometric pose affordance: 3D human pose with scene constraints. arXiv:1905.07718 (2019)
74. Welschehold, T., Dornhege, C., Burgard, W.: Learning manipulation actions from human demonstrations. In: IEEE/RSJ International Conference on Intelligent Robots and Systems (IROS) (2016)
75. XSENS: Inertial motion capture. https://www.xsens.com/motion-capture
76. Yamamoto, M., Yagishita, K.: Scene constraints-aided tracking of human body. In: IEEE/CVF Conference on Computer Vision and Pattern Recognition (CVPR) (2000)

77. Ye, Y., Liu, C.K.: Synthesis of detailed hand manipulations using contact sampling. ACM Trans. Graph. (TOG) **31**(4), 41:1–41:10 (2012)
78. Zhang, H., Bo, Z.H., Yong, J.H., Xu, F.: InteractionFusion: real-time reconstruction of hand poses and deformable objects in hand-object interactions. ACM Trans. Graph. (TOG) **38**(4), 48:1–48:11 (2019)
79. Zhang, Y., Hassan, M., Neumann, H., Black, M.J., Tang, S.: Generating 3D people in scenes without people. In: IEEE/CVF Conference on Computer Vision and Pattern Recognition (CVPR) (2020)

DEMEA: Deep Mesh Autoencoders for Non-rigidly Deforming Objects

Edgar Tretschk[1], Ayush Tewari[1], Michael Zollhöfer[2], Vladislav Golyanik[1(✉)], and Christian Theobalt[1]

[1] Max Planck Institute for Informatics,
Saarland Informatics Campus, Saarbrücken, Germany
golyanik@mpi-inf.mpg.de
[2] Stanford University, Stanford, USA

Abstract. Mesh autoencoders are commonly used for dimensionality reduction, sampling and mesh modeling. We propose a general-purpose DEep MEsh Autoencoder (DEMEA) which adds a novel embedded deformation layer to a graph-convolutional mesh autoencoder. The embedded deformation layer (EDL) is a differentiable deformable geometric proxy which explicitly models point displacements of non-rigid deformations in a lower dimensional space and serves as a local rigidity regularizer. DEMEA decouples the parameterization of the deformation from the final mesh resolution since the deformation is defined over a lower dimensional embedded deformation graph. We perform a large-scale study on four different datasets of deformable objects. Reasoning about the local rigidity of meshes using EDL allows us to achieve higher-quality results for highly deformable objects, compared to directly regressing vertex positions. We demonstrate multiple applications of DEMEA, including non-rigid 3D reconstruction from depth and shading cues, non-rigid surface tracking, as well as the transfer of deformations over different meshes.

Keywords: Auto-encoding · Embedded deformation · Non-rigid tracking

1 Introduction

With the increasing volume of datasets of deforming objects enabled by modern 3D acquisition technology, the demand for compact data representations and compression grows. Dimensionality reduction of mesh data has multiple applications in computer graphics and vision, including shape retrieval, generation, interpolation, and completion. Recently, deep convolutional autoencoder networks were shown to produce compact mesh representations [2,6,31,37].

Electronic supplementary material The online version of this chapter (https://doi.org/10.1007/978-3-030-58548-8_35) contains supplementary material, which is available to authorized users.

© Springer Nature Switzerland AG 2020
A. Vedaldi et al. (Eds.): ECCV 2020, LNCS 12349, pp. 601–617, 2020.
https://doi.org/10.1007/978-3-030-58548-8_35

Dynamic real-world objects do not deform arbitrarily. While deforming, they preserve topology, and nearby points are more likely to deform similarly compared to more distant points. Current convolutional mesh autoencoders exploit this coherence by learning the deformation properties of objects directly from data and are already suitable for mesh compression and representation learning. On the other hand, they do not explicitly reason about the deformation field in terms of local rotations and translations. We show that explicitly reasoning about the local rigidity of meshes enables higher-quality results for highly deformable objects, compared to directly regressing vertex positions.

At the other end of the spectrum, mesh manipulation techniques such as As-Rigid-As-Possible Deformation [34] and Embedded Deformation [35] only require a single mesh and enforce deformation properties, such as smoothness and local rigidity, based on a set of hand-crafted priors. These hand-crafted priors are effective and work surprisingly well, but since they do not model the real-world deformation behavior of the physical object, they often lead to unrealistic deformations and artifacts in the reconstructions.

In this paper, we propose a general-purpose mesh autoencoder with a model-based deformation layer, combining the best of both worlds, *i.e.*, supervised learning with deformable meshes and a novel *differentiable embedded deformation* layer that models the deformable meshes using lower-dimensional deformation graphs with physically interpretable deformation parameters. While the core of our DEep MEsh Autoencoder (DEMEA) learns the deformation model of objects from data using the state-of-the-art convolutional mesh autoencoder (CoMA) [31], the novel embedded deformation layer decouples the parameterization of object motion from the mesh resolution and introduces local spatial coherence via vertex skinning. DEMEA is trained on mesh datasets of moderate sizes that have recently become available [3,4,24,26]. DEMEA is a general mesh autoencoding approach that can be trained for any deformable object class. We evaluate our approach on datasets of three objects with large deformations like articulated deformations (body, hand) and large non-linear deformations (cloth), and one object with small localized deformations (face). Quantitatively, DEMEA outperforms standard convolutional mesh autoencoder architectures in terms of vertex-to-vertex distance error. Qualitatively, we show that DEMEA produces visually higher fidelity results due to the physically based embedded deformation layer. We show several applications of DEMEA in computer vision and graphics. Once trained, the decoder of our autoencoders can be used for shape compression, high-quality depth-to-mesh reconstruction of human bodies and hands, and even poorly textured RGB-image-to-mesh reconstruction for deforming cloth. The low-dimensional latent space learned by our approach is meaningful and well-behaved, which we demonstrate by different applications of latent space arithmetic. Thus, DEMEA provides us a well-behaved general-purpose category-specific generative model of highly deformable objects.

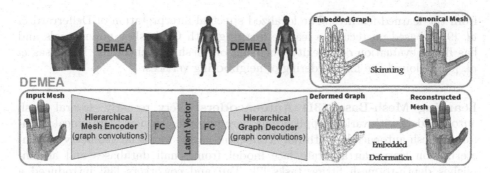

Fig. 1. Pipeline: DEMEA encodes a mesh using graph convolutions on a mesh hierarchy. The graph decoder first maps the latent vector to node features of the coarsest graph level. A number of upsampling and graph convolution modules infer the node translations and rotations of the embedded graph. An embedded deformation layer applies the node translations to a template graph, against which a template mesh is skinned. With the node rotations and the skinning, this deformed graph allows reconstructing a deformed mesh.

2 Related Work

Mesh Manipulation and Tracking. Our embedded deformation layer is inspired by as-rigid-as-possible modelling [34] and the method of Sumner *et al.* [35] for mesh editing and manipulation. While these methods have been shown to be very useful for mesh manipulation in computer graphics, to the best of our knowledge, this is the first time a model-based regularizer is used in a mesh autoencoder. Using a template for non-rigid object tracking from depth maps was extensively studied in the model-based setting [22,41]. Recently, Litany *et al.* [23] demonstrated a neural network-based approach for the completion of human body shapes from a single depth map.

Graph Convolutions. The encoder-decoder approach to dimensionality reduction with neural networks (NNs) for images was introduced in [17]. Deep convolutional neural networks (CNNs) allow to effectively capture contextual information of input data modalities and can be trained for various tasks. Lately, convolutions operating on regular grids have been generalized to more general topologically connected structures such as meshes and two-dimensional manifolds [7,29], enabling learning of correspondences between shapes, shape retrieval [5,27,28], and segmentation [40]. Masci *et al.* [27] proposed geodesic CNNs operating on Riemannian manifolds for shape description, retrieval, and correspondence estimation. Boscani *et al.* [5] introduced spatial weighting functions based on simulated heat propagation and projected anisotropic convolutions. Monti *et al.* [28] extended graph convolutions to variable patches through Gaussian mixture model CNNs. In FeaSTNet [38], the correspondences between filter weights and graph neighborhoods with arbitrary connectivities are established dynamically from the learned features. The localized spectral interpretation of Defferrard *et al.* [9] is based on recursive feature learning with Chebyshev polynomials and has linear evaluation complexity. Focusing on mesh autoencoding, Bouritsas *et al.* [6] exploited the fixed ordering of neighboring vertices.

Learning Mesh-Based 3D Autoencoders. Very recently, several mesh autoencoders with various applications were proposed. A new hierarchical variational mesh autoencoder with fully connected layers for facial geometry parameterization learns an accurate face model from small databases and accomplishes depth-to-mesh fitting tasks [2]. Tan and coworkers [36] introduced a mesh autoencoder with a rotation-invariant mesh representation as a generative model. Their network can generate new meshes by sampling in the latent space and perform mesh interpolation. To cope with meshes of arbitrary connectivity, they used fully-connected layers and did not explicitly encode neighbor relations. Tan *et al.* [37] trained a network with graph convolutions to extract sparse localized deformation components from meshes. Their method is suitable for large-scale deformations and meshes with irregular connectivity. Gao *et al.* [12] transferred mesh deformations by training a generative adversarial network with a cycle consistency loss to map shapes in the latent space, while a variational mesh autoencoder encodes deformations. The Convolutional facial Mesh Autoencoder (CoMA) of Ranjan *et al.* [31] allows to model and sample stronger deformations compared to previous methods and supports asymmetric facial expressions. The Neural 3DMM of Bouritsas *et al.* [6] improves quantitatively over CoMA due to better training parameters and task-specific graph convolutions. Similar to CoMA [31], our DEMEA uses spectral graph convolutions but additionally employs the embedded deformation layer as a model-based regularizer.

Learning 3D Reconstruction. Several supervised methods reconstruct rigid objects in 3D. Given a depth image, the network of Sinha *et al.* [33] reconstructs the observed surface of non-rigid objects. In its 3D reconstruction mode, their method reconstructs rigid objects from single images. Similarly, Groueix *et al.* [15] reconstructed object surfaces from a point cloud or single monocular image with an atlas parameterization. The approaches of Kurenkov *et al.* [21] and Jack *et al.* [18] deform a predefined object-class template to match the observed object appearance in an image. Similarly, Kanazawa *et al.* [19] deformed a template to match the object appearance but additionally support object texture. The Pixel2Mesh approach of Wang *et al.* [39] reconstructs an accurate mesh of an object in a segmented image. Initializing the 3D reconstruction with an ellipsoid, their method gradually deforms it until the appearance matches the observation. The template-based approaches [18,19,21], as well as Pixel2Mesh [39], produce complete 3D meshes.

Learning Monocular Non-rigid Surface Regression. Only a few supervised learning approaches for 3D reconstruction from monocular images tackle the deformable nature of non-rigid objects. Several methods [14,30,32] train networks for deformation models with synthetic thin plates datasets. These approaches can infer non-rigid states of the observed surfaces such as paper sheets or membranes. Still, their accuracy and robustness on real images are limited. Bednařík et al. [3] proposed an encoder-decoder network for texture-less surfaces relying on shading cues. They trained on a real dataset and showed an enhanced reconstruction accuracy on real images, but support only trained object classes. Fuentes-Jimenez et al. [11] trained a network to deform an object template for depth map recovery. They achieved impressive results on real image sequences but require an accurate 3D model of every object in the scene, which restricts the method's practicality. One of the applications of DEMEA is the recovery of texture-less surfaces from RGB images. Since a depth map as a data modality is closer to images with shaded surfaces, we train DEMEA in the depth-to-mesh mode on images instead of depth maps. As a result, we can regress surface geometry from shading cue.

3 Approach

In this section, we describe the architecture of the proposed DEMEA. We employ an embedded deformation layer to decouple the complexity of the learned deformation field from the actual mesh resolution. The deformation is represented relative to a canonical mesh $\mathcal{M} = (\mathbf{V}, \mathbf{E})$ with N_v vertices $\mathbf{V} = \{\mathbf{v}_i\}_{i=1}^{N_v}$, and edges $\mathbf{E}$. To this end, we define the encoder-decoder on a coarse deformation graph and use the embedded deformation layer to drive the deformation of the final high-resolution mesh, see Fig. 1. Our architecture is based on graph convolutions that are defined on a multi-resolution mesh hierarchy. In the following, we describe all components in more detail. We describe the employed spiral graph convolutions [6] in the supplemental document.

3.1 Mesh Hierarchy

The up- and downsampling in the convolutional mesh autoencoder is defined over a multi-resolution mesh hierarchy, similar to the CoMA [31] architecture. We compute the mesh hierarchy fully automatically based on quadric edge collapses [13], i.e., each hierarchy level is a simplified version of the input mesh. We employ a hierarchy with five resolution levels, where the finest level is the mesh. Given the multi-resolution mesh hierarchy, we define up- and downsampling operations [31] for feature maps defined on the graph. To this end, during downsampling, we enforce the nodes of the coarser level to be a subset of the nodes of the next finer level. We transfer a feature map to the next coarser level by a similar subsampling operation. The inverse operation, i.e., feature map upsampling, is implemented based on a barycentric interpolation of close features. During edge collapse, we project each collapsed node onto the closest triangle of the coarser level. We use the barycentric coordinates of this closest point with respect to the triangle's vertices to define the interpolation weights.

3.2   Embedded Deformation Layer (EDL)

Given a canonical mesh, we have to pick a corresponding coarse embedded deformation graph. We employ MeshLab's [8] more sophisticated implementation of quadric edge collapse to fully automatically generate the graph. See the supplemental document for details (Fig. 2).

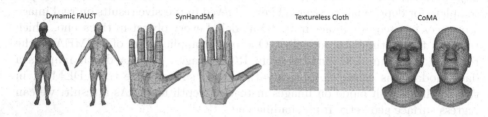

Fig. 2. Template mesh and the corresponding embedded deformation graph pairs automatically generated using [8].

The deformation graph is used as one of the two levels immediately below the mesh in the mesh hierarchy (depending on the resolution of the graph) of the autoencoder. When generating the mesh hierarchy, we need to enforce the subset relationship between levels. However, the quadric edge collapse algorithm of [31] might delete nodes of the embedded graph when computing intermediate levels between the mesh and the embedded graph. We ensure that those nodes are not removed by setting the cost of removing them from levels that are at least as fine as the embedded graph to infinity.

Our embedded deformation layer models a space deformation that maps the vertices of the canonical template mesh $\mathbf{V}$ to a deformed version $\hat{\mathbf{V}}$. Suppose $\mathcal{G} = (\mathbf{N}, \mathbf{E})$ is the embedded deformation graph [35] with L canonical nodes $\mathbf{N} = \{g_l\}_{i=1}^{L}$ and K edges $\mathbf{E}$, with $g_l \in \mathbb{R}^3$. The global space deformation is defined by a set of local, rigid, per-graph node transformations. Each local rigid space transformation is defined by a tuple $T_l = (R_l, t_l)$, with $R_l \in \mathbf{SO}(3)$ being a rotation matrix and $t_l \in \mathbb{R}^3$ being a translation vector. We enforce that $R_l^{\mathsf{T}} = R_l^{-1}$ and $\det(R_l) = 1$ by parameterizing the rotation matrices based on three Euler angles. Each T_l is anchored at the canonical node position g_l and maps every point $p \in \mathbb{R}^3$ to a new position in the following manner [35]:

$$T_l(p) = R_l[p - g_l] + g_l + t_l. \tag{1}$$

To obtain the final global space deformation $\mathbf{G}$, the local per-node transformations are linearly combined:

$$\mathbf{G}(p) = \sum_{l \in \mathcal{N}_p} w_l(p) \cdot T_l(p) . \tag{2}$$

Here, $\mathcal{N}_p$ is the set of approximate closest deformation nodes. The linear blending weights $w_l(p)$ for each position are based on the distance to the respective deformation node [35]. Please refer to the supplemental for more details.

The deformed mesh $\hat{\mathbf{V}} = \mathbf{G}(\mathbf{V})$ is obtained by applying the global space deformation to the canonical template mesh $\mathbf{V}$. The free parameters are the local per-node rotations $\boldsymbol{R}_l$ and translations $\boldsymbol{t}_l$, i.e., $6L$ parameters with L being the number of nodes in the graph. These parameters are input to our deformation layer and are regressed by the graph convolutional decoder.

3.3 Differentiable Space Deformation

Our novel EDL is fully differentiable and can be used during network training to decouple the parameterization of the space deformation from the resolution of the final high-resolution output mesh. This enables us to define the reconstruction loss on the final high-resolution output mesh and backpropagate the errors via the skinning transform to the coarse parameterization of the space deformation. Thus, our approach enables finding the best space deformation by only supervising the final output mesh.

3.4 Training

We train our approach end-to-end in Tensorflow [1] using Adam [20]. As loss we employ a dense geometric per-vertex ℓ_1-loss with respect to the ground-truth mesh. For all experiments, we use a learning rate of 10^{-4} and default parameters $\beta_1 = 0.9$, $\beta_2 = 0.999$, $\epsilon = 10^{-8}$ for Adam. We train for 50 epochs for Dynamic Faust, 30 epochs for SynHand5M, 50 epochs for the CoMA dataset and 300 epochs for the Cloth dataset. We employ a batch size of 8.

3.5 Reconstructing Meshes from Images/Depth

The image/depth-to-mesh network consists of an image encoder and a mesh decoder, see Fig. 5a. The mesh decoder is initialized from the corresponding mesh auto-encoder, the image/depth encoder is based on a ResNet-50 [16] architecture, and the latent code is shared between the encoder and decoder. We initialize the ResNet-50 component using pre-trained weights from ImageNet [10]. To obtain training data, we render synthetic depth maps from the meshes. We train with the same settings as for mesh auto-encoding.

3.6 Network Architecture Details

In the following, we provide more details of our encoder-decoder architectures.

Encoding Meshes. Input to the first layer of our mesh encoder is an $N_v \times 3$ tensor that stacks the coordinates of all N_v vertices. We apply four *downsampling modules*. Each module applies a graph convolution and is followed by a down-sampling to the next coarser level of the mesh hierarchy. We use spiral graph convolutions [6] and similarly apply an ELU non-linearity after each convolution. Finally, we take the output of the final module and apply a fully connected layer followed by an ELU non-linearity to obtain a latent space embedding.

Encoding Images/Depth. To encode images/depth, we employ a 2D convo-lutional network to map color/depth input to a latent space embedding. Input to our encoder are images of resolution 256×256 pixels. We modified the ResNet-50 [16] architecture to take single or three-channel input image. We furthermore added two additional convolution layers at the end, which are followed by global average pooling. Finally, a fully connected layer with a subsequent ELU non-linearity maps the activations to the latent space.

Decoding Graphs. The task of the graph decoder is to map from the latent space back to the embedded deformation graph. First, we employ a fully con-nected layer in combination with reshaping to obtain the input to the graph convolutional *upsampling modules*. We apply a sequence of three or four upsam-pling modules until the resolution level of the embedded graph is reached. Each upsampling module first up-samples the features to the next finer graph resolu-tion and then performs a graph convolution, which is then followed by an ELU non-linearity. Then, we apply two graph convolutions with ELUs for refinement and a final convolution without an activation function. The resulting tensor is passed to our embedded deformation layer.

4   Experiments

We evaluate DEMEA quantitatively and qualitatively on several challenging datasets and demonstrate state-of-the-art results for mesh auto-encoding. In Sect. 5, we show reconstruction from RGB images and depth maps and that the learned latent space enables well-behaved latent arithmetic. We use Tensorflow 1.5.0 [1] on Debian with an NVIDIA Tesla V100 GPU.

Datasets. We demonstrate DEMEA's generality on experiments with body (Dynamic Faust, DFaust [4]), hand (SynHand5M [26]), textureless cloth (Cloth [3]), and face (CoMA [31]) datasets. Table 1 gives the number of graph nodes used on each level of our hierarchical architecture. All meshes live in metric space.

Table 1. Number of vertices on each level of the mesh hierarchy. Bold levels denote the embedded graph. Note that except for Cloth these values were computed automatically based on [31].

	Mesh	1st	2nd	3rd	4th
DFaust [4]	6890	1723	**431**	108	27
CoMA [31]	5023	**1256**	314	79	20
SynHand5M [26]	1193	**299**	75	19	5
Cloth [3]	961	**256**	100	36	16

Table 2. Average per-vertex errors on the test sets of DFaust (cm), SynHand5M (mm), textureless cloth (mm) and CoMA (mm) for 8 and 32 latent dimensions.

	DFaust		SynHand5M		Cloth		CoMA	
	8	32	8	32	8	32	8	32
CA	6.35	**2.07**	8.12	2.60	**11.21**	6.50	**1.17**	0.72
MCA	**6.21**	2.13	**8.11**	2.67	11.64	6.59	1.20	0.71
Ours	6.69	2.23	8.12	**2.51**	11.28	6.40	1.23	0.81
FCA	6.51	2.17	15.10	2.95	15.63	5.99	1.77	**0.67**
FCED	6.26	2.14	14.61	2.75	15.87	**5.94**	1.81	0.73

DFaust [4]. The training set consists of 28,294 meshes. For the tests, we split off two identities (female 50004, male 50002) and two dynamic performances, *i.e.*, *one-leg jump* and *chicken wings*. Overall, this results in a test set with 12, 926 elements. For the depth-to-mesh results, we found the synthetic depth maps from the DFaust training set to be insufficient for generalization, *i.e.*, the test error was high. Thus, we add more pose variety to DFaust for the depth-to-mesh experiments. Specifically, we add $28k$ randomly sampled poses from the CMU Mocap[1] dataset to the training data, where the identities are randomly sampled from the SMPL [25] model ($14k$ female, $14k$ male). We also add $12k$ such samples to the test set ($6k$ female, $6k$ male).

Textureless Cloth [3]. For evaluating our approach on general non-rigidly deforming surfaces, we use the *textureless cloth* data set of Bednařík et al. [3]. It contains real depth maps and images of a white deformable sheet—observed in different states and differently shaded—as well as ground-truth meshes. In total, we select 3,861 meshes with consistent edge lengths. 3,167 meshes are used for training and 700 meshes are reserved for evaluation. Since the canonical mesh is a perfectly flat sheet, it lacks geometric features, which causes downsampling methods like [13,31] and [8] to introduce severe artifacts. Hence, we generate the entire mesh hierarchy for this dataset, see the supplemental. This hierarchy is also used to train the other methods for the performed comparisons.

SynHand5M [26]. For the experiments with hands, we take $100k$ random meshes from the synthetic *SynHand5M* dataset of Malik et al. [26]. We render the corresponding depth maps. The training set is comprised of $90k$ meshes, and the remaining $10k$ meshes are used for evaluation.

CoMA [31]. The training set contains 17,794 meshes of the human face in various expressions [31]. For tests, we select two challenging expressions, *i.e.*, *high smile* and *mouth extreme*. Thus, our test set contains 2,671 meshes in total.

---

[1] http://mocap.cs.cmu.edu/.

4.1 Baseline Architectures

We compare DEMEA to a number of strong baselines.

Convolutional Baseline. We consider a version of our proposed architecture, *convolutional ablation (CA)*, where the ED layer is replaced by learned upsampling modules that upsample to the mesh resolution. In this case, the extra refinement convolutions occur on the level of the embedded graph. We also consider *modified CA (MCA)*, an architecture where the refinement convolutions are moved to the end of the network, such that they operate on mesh resolution.

Fully-Connected Baseline. We also consider an almost-linear baseline, *FC ablation (FCA)*. The input is given to a fully-connected layer, after which an ELU is applied. The resulting latent vector is decoded using another FC layer that maps to the output space. Finally, we also consider an *FCED* network where the fully-connected decoder maps to the deformation graph, which the embedded deformation layer (EDL) in turn maps to the full-resolution mesh.

4.2 Evaluation Settings

Table 3. Average per-vertex errors on the test sets of DFaust (*cm*), SynHand5M (*mm*), textureless cloth (*mm*) and CoMA (*mm*) for 8 and 32 latent dimensions.

	DFaust		SynHand5M		Cloth		CoMA	
	8	32	8	32	8	32	8	32
w/GL	8.92	2.75	9.02	2.95	11.26	6.45	1.38	0.99
w/LP	7.71	**2.22**	8.00	2.52	11.46	7.96	1.25	**0.79**
Ours	**6.69**	2.23	8.12	**2.51**	11.28	**6.40**	1.23	0.81

Table 4. Average per-vertex errors on the test sets of DFaust (in *cm*), SynHand5M (in *mm*), textureless cloth (in *mm*) and CoMA (in *mm*) for 8 and 32 latent dimensions, compared with Neural 3DMM [6].

	DFaust		SynHand5M		Cloth		CoMA	
	8	32	8	32	8	32	8	32
N. 3DMM	7.09	1.99	8.50	2.58	12.64	6.49	1.34	**0.71**
Ours	**6.69**	2.23	**8.12**	**2.51**	**11.28**	**6.40**	**1.23**	0.81

We first determine how to integrate the EDL into the training. Our proposed architecture regresses node positions and rotations and then uses the EDL to obtain the deformed mesh, on which the reconstruction loss is applied.

As an alternative, we consider the *graph loss (GL)* with the ℓ_1 reconstruction loss directly on the graph node positions (where the vertex positions of the input mesh that correspond to the graph nodes are used as ground-truth). The GL setting uses the EDL only at test time to map to the full mesh, but not for training. Although the trained network predicts graph node positions t_l at test time, it does not regress graph node rotations R_l which are necessary for the EDL. We compute the missing rotation for each graph node l as follows: assuming that each node's neighborhood transforms roughly rigidly, we solve a small Procrustes problem that computes the rigid rotation between the 1-ring neighborhoods of l in the template graph and in the regressed network output. We directly use this rotation as R_l.

We also consider the alternative of estimating the local Procrustes rotation *inside the network during training (LP)*. We add a reconstruction loss on the deformed mesh as computed by the EDL. Here, we do not back-propagate through the rotation computation to avoid training instabilities.

Table 3 shows the quantitative results using the average per-vertex Euclidean error. Using the EDL during training leads to better quantitative results, as the network is aware of the skinning function and can move the graph nodes accordingly. In addition to being an order of magnitude faster than LP, regressing rotations either gives the best results or is close to the best. We use the EDL with regressed rotation parameters during training in all further experiments.

We use spiral graph convolutions [6], but show in the supplemental document that spectral graph convolutions [9] also give similar results.

4.3   Evaluations of the Autoencoder

Qualitative Evaluations. Our architecture significantly outperforms the baselines qualitatively on the DFaust and SynHand5M datasets, as seen in Figs. 3 and 4. Convolutional architectures without an embedded graph produce strong artifacts in the hand, feet and face regions in the presence of large deformations. Since EDL explicitly models deformations, we preserve fine details of the template under strong non-linear deformations and articulations of extremities.

Quantitative Evaluations. We compare the proposed DEMEA to the baselines on the autoencoding task, see Table 2.

While the fully-connected baselines are competitive for larger dimensions of the latent space, their memory demand increases drastically. On the other hand, they perform significantly worse for low dimensions on all datasets, except for DFaust. In this work, we are interested in low latent dimensions, *e.g.* less than 32, as we want to learn mesh representations that are as compact as possible. We also observe that adding EDL to the fully-connected baselines maintains their performance. Furthermore, the lower test errors of FCED on Cloth indicate that network capacity (and not EDL) limits the quantitative results.

On SynHand5M, Cloth and CoMA, the convolutional baselines perform on par with DEMEA. On DFaust, our performance is slightly worse, perhaps because other architectures can also fit to the high-frequency details and noise. EDL regularizes deformations to avoid artifacts, which also prevents fitting to high-frequency or small details. Thus, explicitly modelling deformations via the EDL and thereby avoiding artifacts has no negative impact on the quantitative performance. Since CoMA mainly contains small and local deformations, DEMEA does not lead to any quantitative improvement. This is more evident in the case of latent dimension 32, as the baselines can better reproduce noise and other high-frequency deformations.

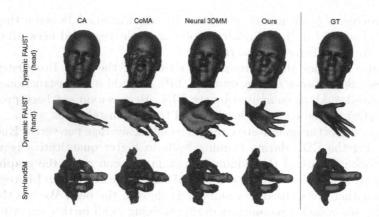

Fig. 3. In contrast to graph-convolutional networks that directly regress vertex positions, our embedded graph layer does not show artifacts. These results use a latent dimension of 32.

Comparisons. In extensive comparisons with several competitive baselines, we have demonstrated the usefulness of our approach for autoencoding strong non-linear deformations and articulated motion. Next, we compare DEMEA to the existing state-of-the-art CoMA approach [31]. We train their architecture on all mentioned datasets with a latent dimension of 8, which is also used in [31]. We outperform their method quantitatively on DFaust (6.7 cm *vs.* 8.4 cm), on SynHand5M (8.12 mm *vs.* 8.93 mm), on Cloth (1.13 cm *vs.* 1.44 cm), and even on CoMA (1.23 mm *vs.* 1.42 mm), where the deformations are not large. We also compare to Neural 3DMM [6] on latent dimensions 8 and 32, similarly to [31] on their proposed hierarchy. See Table 4 for the results. DEMEA performs better than Neural 3DMM in almost all cases. In Fig. 3, we show that DEMEA avoids many of the artifacts present in the case of [6,31] and other baselines.

5   Applications

5.1   RGB to Mesh

On the Cloth [3] dataset, we show that DEMEA can reconstruct meshes from RGB images. See Fig. 5b for qualitative examples with a latent dimension of 32.

On our test set, our proposed architecture achieves RGB-to-mesh reconstruction errors of 16.1 mm and 14.5 mm for latent dimensions 8 and 32, respectively. Bednařík *et al.* [3], who use a different split than us, report an error of 21.48 mm. The authors of IsMo-GAN [32] report results on their own split for IsMo-GAN and the Hybrid Deformation Model Network (HDM-net) [14]. On their split, HDM-Net achieves an error of 17.65 mm after training for 100 epochs using a batch size of 4. IsMo-GAN obtains an error of 15.79 mm. Under the same settings as HDM-Net, we re-train our approach without pre-training the mesh decoder.

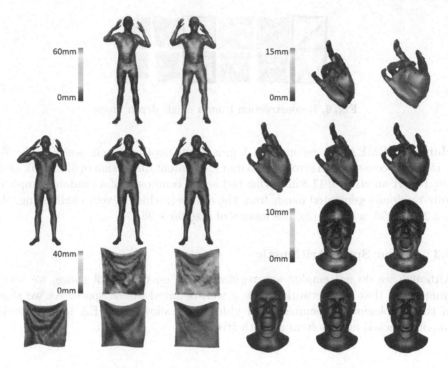

Fig. 4. Auto-encoding results on all four datasets. From left to right: ground-truth, ours with latent dimension 32, ours with latent dimension 8.

Fig. 5. (left) Image/depth-to-mesh pipeline: To train an image/depth-to-mesh reconstruction network, we employ a convolutional image encoder and initialize the decoder to a pre-trained graph decoder. (right) RGB-to-mesh results on our test set. From left to right: real RGB image, our reconstruction, ground-truth.

Our approach achieves test errors of 16.6 mm and 13.8 mm using latent dimensions of 8 and 32, respectively.

5.2   Depth to Mesh

Bodies. We train a network with a latent space dimension of 32. Quantitatively, we obtain an error of 2.3 cm on un-augmented synthetic data. Besides, we also apply our approach to real data, see Fig. 6a. To this end, we found it necessary to augment the depth images with artificial noise to lessen the domain gap. Video results are included in the supplementary.

Fig. 6. Reconstruction from a single depth image.

Hands. DEMEA can reconstruct hands from depth as well, see Fig. 6b. We achieve a reconstruction error of 6.73 mm for a latent dimension of 32. Malik *et al.* [26] report an error of 11.8 mm. Our test set is composed of a random sample of fully randomly generated hands from the dataset, which is very challenging. We use 256×256, whereas [26] use images of size 96×96.

5.3 Latent Space Arithmetic

Although we do not employ any regularization on the latent space, we found empirically that the network learns a well-behaved latent space. As we show in the supplemental document and video, this allows DEMEA to temporally smooth tracked meshes from a depth stream.

Latent Interpolation. We can linearly interpolate the latent vectors $\mathcal{S}$ and $\mathcal{T}$ of a source and a target mesh: $\mathcal{I}(\alpha) = (1 - \alpha)\mathcal{S} + \alpha\mathcal{T}$. Even for highly different poses and identities, these $\mathcal{I}(\alpha)$ yield plausible in-between meshes, see Fig. 7a.

Fig. 7. Latent space arithmetic.

Deformation Transfer. The learned latent space allows to transfer poses between different identities on DFaust. Let a sequence of source meshes $\mathbf{S} = \{\mathbf{M}_i\}_i$ of person A and a target mesh $\mathbf{M}_0'$ of person B be given, where w.l.o.g. $\mathbf{M}_0$ and $\mathbf{M}_0'$ correspond to the same pose. We now seek a sequence of target meshes $\mathbf{S}' = \{\mathbf{M}_i'\}_i$ of person B performing the same poses as person A in $\mathbf{S}$. We encode $\mathbf{S}$ and M_0' into the latent space of the mesh auto-encoder, yielding the corresponding latent vectors $\{\mathcal{M}_i\}_i$ and $\mathcal{M}_0'$. We define the identity difference $d = \mathcal{M}_0' - \mathcal{M}_0$ and set $\mathcal{M}_i' = \mathcal{M}_i + d$ for $i > 0$. Decoding $\{\mathcal{M}_i'\}_i$ using the mesh decoder than yields $\mathbf{S}'$. See Fig. 7b and the supplement for qualitative results.

6   Limitations

While the embedded deformation graph excels on highly articulated, non-rigid motions, it has difficulties accounting for very subtle actions. Since the faces in the CoMA [31] dataset do not undergo large deformations, our EDL-based architecture does not offer a significant advantage. Similar to all other 3D deep learning techniques, our approach also requires reasonably sized mesh datasets for supervised training, which might be difficult to capture or model. We train our network in an object-specific manner. Generalizing our approach across different object categories is an interesting direction for future work.

7   Conclusion

We proposed DEMEA—the first deep mesh autoencoder for highly deformable and articulated scenes, such as human bodies, hands, and deformable surfaces, that builds on a new differentiable embedded deformation layer. The deformation layer reasons about local rigidity of the mesh and allows us to achieve higher quality autoencoding results compared to several baselines and existing approaches. We have shown multiple applications of our architecture including non-rigid reconstruction from real depth maps and 3D reconstruction of texture-less surfaces from images.

Acknowledgments. This work was supported by the ERC Consolidator Grant 4DReply (770784), the Max Planck Center for Visual Computing and Communications (MPC-VCC), and an Oculus research grant.

References

1. Abadi, M., et al.: TensorFlow: Large-scale machine learning on heterogeneous systems (2015). https://www.tensorflow.org/
2. Bagautdinov, T., Wu, C., Saragih, J., Sheikh, Y., Fua, P.: Modeling facial geometry using compositional vaes (2018)
3. Bednařík, J., Fua, P., Salzmann, M.: Learning to reconstruct texture-less deformable surfaces. In: International Conference on 3D Vision (3DV) (2018)
4. Bogo, F., Romero, J., Pons-Moll, G., Black, M.J.: Dynamic FAUST: registering human bodies in motion. In: Computer Vision and Pattern Recognition (CVPR) (2017)
5. Boscaini, D., Masci, J., Rodolà, E., Bronstein, M.: Learning shape correspondence with anisotropic convolutional neural networks. In: International Conference on Neural Information Processing Systems (NIPS) (2016)
6. Bouritsas, G., Bokhnyak, S., Ploumpis, S., Bronstein, M., Zafeiriou, S.: Neural 3D morphable models: spiral convolutional networks for 3D shape representation learning and generation. In: International Conference on Computer Vision (ICCV) (2019)
7. Bruna, J., Zaremba, W., Szlam, A., LeCun, Y.: Spectral networks and locally connected networks on graphs. CoRR abs/1312.6203 (2013)

8. Cignoni, P., Callieri, M., Corsini, M., Dellepiane, M., Ganovelli, F., Ranzuglia, G.: MeshLab: an open-source mesh processing tool. In: Scarano, V., Chiara, R.D., Erra, U. (eds.) Eurographics Italian Chapter Conference. The Eurographics Association (2008). https://doi.org/10.2312/LocalChapterEvents/ItalChap/ItalianChapConf2008/129-136

9. Defferrard, M., Bresson, X., Vandergheynst, P.: Convolutional neural networks on graphs with fast localized spectral filtering. In: International Conference on Neural Information Processing Systems (NIPS) (2016)

10. Deng, J., Dong, W., Socher, R., Li, L.J., Li, K., Fei-Fei, L.: ImageNet: a large-scale hierarchical image database. In: Computer Vision and Pattern Recognition (CVPR) (2009)

11. Fuentes-Jimenez, D., Casillas-Perez, D., Pizarro, D., Collins, T., Bartoli, A.: Deep Shape-from-Template: Wide-Baseline, Dense and Fast Registration and Deformable Reconstruction from a Single Image. arXiv e-prints (2018)

12. Gao, L., Yang, J., Qiao, Y.L., Lai, Y.K., Rosin, P.L., Xu, W., Xia, S.: Automatic unpaired shape deformation transfer. ACM Trans. Graph. (TOG) **37**(6), 1–15 (2018)

13. Garland, M., Heckbert, P.S.: Surface simplification using quadric error metrics. In: ACM SIGGRAPH (1997)

14. Golyanik, V., Shimada, S., Varanasi, K., Stricker, D.: Hdm-net: monocular non-rigid 3D reconstruction with learned deformation model. In: International Conference on Virtual Reality and Augmented Reality (EuroVR) (2018)

15. Groueix, T., Fisher, M., Kim, V.G., Russell, B., Aubry, M.: AtlasNet: a Papier-Mâché approach to learning 3D surface generation. In: Computer Vision and Pattern Recognition (CVPR) (2018)

16. He, K., Zhang, X., Ren, S., Sun, J.: Deep residual learning for image recognition. In: Computer Vision and Pattern Recognition (CVPR) (2016)

17. Hinton, G.E., Salakhutdinov, R.R.: Reducing the dimensionality of data with neural networks. Science **313**, 504–507 (2006)

18. Jack, D., et al.: Learning free-form deformations for 3D object reconstruction. In: Jawahar, C.V., Li, H., Mori, G., Schindler, K. (eds.) ACCV 2018. LNCS, vol. 11362, pp. 317–333. Springer, Cham (2019). https://doi.org/10.1007/978-3-030-20890-5_21

19. Kanazawa, A., Tulsiani, S., Efros, A.A., Malik, J.: Learning category-specific mesh reconstruction from image collections. In: European Conference on Computer Vision (ECCV) (2018)

20. Kingma, D.P., Ba, J.: Adam: a method for stochastic optimization. In: International Conference on Learning Representations (ICLR) (2015)

21. Kurenkov, A., et al.: Deformnet: Free-form deformation network for 3D shape reconstruction from a single image. In: Winter Conference on Applications of Computer Vision (WACV) (2018)

22. Li, H., Adams, B., Guibas, L.J., Pauly, M.: Robust single-view geometry and motion reconstruction. In: ACM SIGGRAPH Asia (2009)

23. Litany, O., Bronstein, A., Bronstein, M., Makadia, A.: Deformable shape completion with graph convolutional autoencoders. In: Computer Vision and Pattern Recognition (CVPR) (2018)

24. Loper, M., Mahmood, N., Black, M.J.: Mosh: motion and shape capture from sparse markers. ACM Trans. Graph. (TOG) **33**, 1–13 (2014)

25. Loper, M., Mahmood, N., Romero, J., Pons-Moll, G., Black, M.J.: Smpl: a skinned multi-person linear model. ACM Trans. Graph. (TOG) **34**, 1–16 (2015)

26. Malik, J., et al.: Deephps: end-to-end estimation of 3D hand pose and shape by learning from synthetic depth. In: International Conference on 3D Vision (3DV) (2018)
27. Masci, J., Boscaini, D., Bronstein, M.M., Vandergheynst, P.: Geodesic convolutional neural networks on riemannian manifolds. In: International Conference on Computer Vision Workshop (ICCVW) (2015)
28. Monti, F., Boscaini, D., Masci, J., Rodola, E., Svoboda, J., Bronstein, M.M.: Geometric deep learning on graphs and manifolds using mixture model CNNs. In: Computer Vision and Pattern Recognition (CVPR) (2017)
29. Niepert, M., Ahmed, M., Kutzkov, K.: Learning convolutional neural networks for graphs. In: International Conference on Machine Learning (ICML) (2016)
30. Pumarola, A., Agudo, A., Porzi, L., Sanfeliu, A., Lepetit, V., Moreno-Noguer, F.: Geometry-aware network for non-rigid shape prediction from a single view. In: Computer Vision and Pattern Recognition (CVPR) (2018)
31. Ranjan, A., Bolkart, T., Sanyal, S., Black, M.J.: Generating 3D faces using convolutional mesh autoencoders. In: European Conference on Computer Vision (ECCV) (2018)
32. Shimada, S., Golyanik, V., Theobalt, C., Stricker, D.: IsMo-GAN: adversarial learning for monocular non-rigid 3D reconstruction. In: Computer Vision and Pattern Recognition Workshops (CVPRW) (2019)
33. Sinha, A., Unmesh, A., Huang, Q., Ramani, K.: Surfnet: generating 3D shape surfaces using deep residual networks. In: Computer Vision and Pattern Recognition (CVPR) (2017)
34. Sorkine, O., Alexa, M.: As-rigid-as-possible surface modeling. In: Eurographics Symposium on Geometry Processing (SGP) (2007)
35. Sumner, R.W., Schmid, J., Pauly, M.: Embedded deformation for shape manipulation. In: ACM SIGGRAPH (2007)
36. Tan, Q., Gao, L., Lai, Y.K., Xia, S.: Variational autoencoders for deforming 3D mesh models. In: Computer Vision and Pattern Recognition (CVPR) (2018)
37. Tan, Q., Gao, L., Lai, Y.K., Yang, J., Xia, S.: Mesh-based autoencoders for localized deformation component analysis. In: AAAI Conference on Artificial Intelligence (AAAI) (2018)
38. Verma, N., Boyer, E., Verbeek, J.: FeaStNet: Feature-steered graph convolutions for 3D shape analysis. In: Computer Vision and Pattern Recognition (CVPR) (2018)
39. Wang, N., Zhang, Y., Li, Z., Fu, Y., Liu, W., Jiang, Y.G.: Pixel2mesh: generating 3D mesh models from single RGB images. In: European Conference on Computer Vision (ECCV) (2018)
40. Yi, L., Su, H., Guo, X., Guibas, L.: Syncspeccnn: synchronized spectral CNN for 3D shape segmentation. In: Computer Vision and Pattern Recognition (CVPR) (2017)
41. Zollhöfer, M., et al.: Real-time non-rigid reconstruction using an RGB-D camera. ACM Trans. Graph. (TOG) 33, 1–12 (2014)

RANSAC-Flow: Generic Two-Stage Image Alignment

Xi Shen[1]([✉]), François Darmon[1,2], Alexei A. Efros[3], and Mathieu Aubry[1]

[1] LIGM (UMR 8049) - Ecole des Ponts, UPE, Marne-la-Vallée, France
xi.shen@enpc.fr
[2] Thales Land and Air Systems, Belfast, UK
[3] UC Berkeley, Berkeley, USA

Abstract. This paper considers the generic problem of dense alignment between two images, whether they be two frames of a video, two widely different views of a scene, two paintings depicting similar content, etc. Whereas each such task is typically addressed with a domain-specific solution, we show that a simple unsupervised approach performs surprisingly well across a range of tasks. Our main insight is that parametric and non-parametric alignment methods have complementary strengths. We propose a two-stage process: first, a feature-based parametric coarse alignment using one or more homographies, followed by non-parametric fine pixel-wise alignment. Coarse alignment is performed using RANSAC on off-the-shelf deep features. Fine alignment is learned in an unsupervised way by a deep network which optimizes a standard structural similarity metric (SSIM) between the two images, plus cycle-consistency. Despite its simplicity, our method shows competitive results on a range of tasks and datasets, including unsupervised optical flow on KITTI, dense correspondences on HPATCHES, two-view geometry estimation on YFCC100M, localization on AACHEN DAY-NIGHT, and, for the first time, fine alignment of artworks on the BRUGHEL DATASET. Our code and data are available at http://imagine.enpc.fr/~shenx/RANSAC-Flow/.

Keywords: Unsupervised dense image alignment · Applications to art

1 Introduction

Dense image alignment (also known as image registration) is one of the fundamental vision problems underlying many standard tasks from panorama stitching to optical flow. Classic work on image alignment can be broadly placed into two camps: parametric and non-parametric. Parametric methods assume that the two images are related by a global parametric transformation (e.g. affine, homography, etc), and use robust approaches, like RANSAC, to estimate this transformation. Non-parametric methods do not make any assumptions on

---

Electronic supplementary material The online version of this chapter (https://doi.org/10.1007/978-3-030-58548-8_36) contains supplementary material, which is available to authorized users.

A. Vedaldi et al. (Eds.): ECCV 2020, LNCS 12349, pp. 618–637, 2020.
https://doi.org/10.1007/978-3-030-58548-8_36

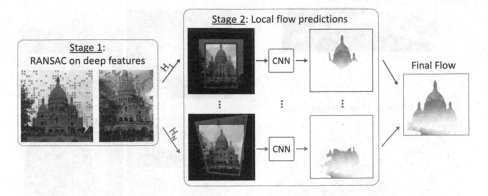

Fig. 1. Overview of RANSAC-Flow. Stage 1: given a pair of images, we compute sparse correspondences (using off-the-shelf deep features), use RANSAC to estimate a homography, and warp second image using it. Stage 2: given two coarsely aligned images, our self-supervised fine flow network generates flow predictions in the matchable region. To compute further homographies, we can remove matched correspondences, and iterate the process.

the type of transformation, and attempt to directly optimize some pixel agreement metric (e.g. brightness constancy constraint in optical flow and stereo). However, both approaches have flaws: parametric methods fail (albeit gracefully) if the parametric model is only an approximation for the true transform, while non-parametric methods have trouble dealing with large displacements and large appearance changes (e.g. two photos taken at different times from different views). It is natural, therefore, to consider a hybrid approach, combining the benefits of parametric and non-parametric methods together.

In this paper, we propose RANSAC-flow, a two-stage approach integrating parametric and non-parametric methods for generic dense image alignment. Figure 1 shows an overview. In the first stage, a classic geometry-verification method (RANSAC) is applied to a set of feature correspondences to obtain one or more candidate coarse alignments. Our method is agnostic to the particular choice of transformation(s) and features, but we've found that using multiple homographies and off-the-shelf self-supervised deep features works quite well. In the second non-parametric stage, we refine the alignment by predicting a dense flow field for each of the candidate coarse transformations. This is achieved by self-supervised training of a deep network to optimize a standard structural similarity metric (SSIM) [84] between the pixels of the warped and the original images, plus a cycle-consistency loss [92].

Despite its simplicity, the proposed approach turns out to be surprisingly effective. The coarse alignment stage takes care of large-scale viewpoint and appearance variations and, thanks to multiple homographies, is able to capture a piecewise-planar approximation of the scene structure. The learned local flow estimation stage is able to refine the alignment to the pixel level without relying on the brightness constancy assumption. As a result, our method produces

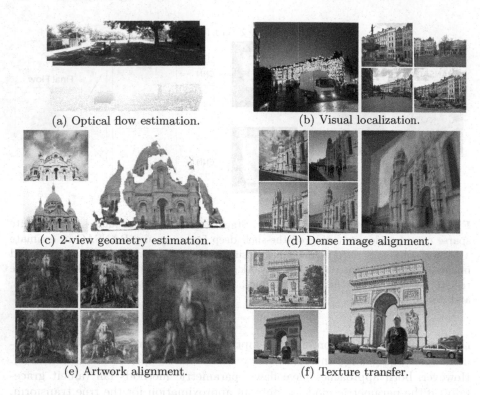

(a) Optical flow estimation.

(b) Visual localization.

(c) 2-view geometry estimation.

(d) Dense image alignment.

(e) Artwork alignment.

(f) Texture transfer.

Fig. 2. RANSAC-Flow provides competitive results on a wide variety of tasks and enables new challenging applications.

competitive results across a wide range of different image alignment tasks, as shown in Fig. 2: (a) unsupervised optical flow estimation on KITTI [48] and HPATCHES [5], (b) visual localization on AACHEN DAY-NIGHT [68], (c) 2-view geometry estimation on YFCC100M [78], (d) dense image alignment, and applications to (e) detail alignment in artwork and (f) texture transfer. Our code and data are available at http://imagine.enpc.fr/~shenx/RANSAC-Flow/.

2   Related Work

Feature-Based Image Alignment. The classic approach to align images with very different appearances is to use sparse local image features, such as SIFT [40], which are designed to deal with large viewpoint and illumination differences as well as clutter and occlusion. These features have to be used together with a geometric regularization step to discard false matches. This is typically done using RANSAC [6,8,18,58] to fit a simple geometric transformation (e.g. affine or homography) [77]. Recently, many works proposed to learn better local features [14,42,43,49,63,79]. Differentiable and trainable version of RANSAC have also been developed [54,56,59,87].

Using mid-level features [28–30,75] instead of local keypoints, proved to be beneficial for matching visual content across modalities, e.g. 3D models and paintings [3]. Recently, [72] learned deep mid-level features for matching across different visual media (drawings, oil paintings, frescoes, sketches, etc), and used them together with spatial verification to discover copied details in a dataset of thousands of artworks. [65] used deep feature map correlations as input to a regression network on synthetic image deformations to predict the parameters of an affine or thin-plate spline deformation. Finally, transformer networks [26] can also learn parametric alignment typically as a by-product of optimizing a classification task.

Direct Image Alignment. Direct, or pixel-based, alignment has its roots in classic optical flow methods, such as Lucas-Kanade [41], who solve for a dense flow field between a pair of images under a brightness constancy assumption. The main drawback is these methods tend to work only for very small displacements. This has been partially addressed with hierarchical flow estimation [77], as well as using local features in addition to pixels to increase robustness [4,9,22,61]. However, all such methods are still limited to aligning very similar images, where the brightness constancy assumption mostly holds. SIFT-Flow [38] was an early method that aimed at expanding optical flow-style approaches for matching pairs of images across physically distinct, and visually different scenes (and later generalized to joint image set alignment using cycle consistency [91]). Some approaches such as SCV [11] and MODS [50], were proposed to grow matches around initial warping. In the deep era, [39] showed that ConvNet activation features can be used for correspondence, achieving similar performance to SIFT-Flow. [12] proposed to learn matches with a Correspondence Contrastive loss, producing semi-dense matches. [66] introduced the idea of using 4D convolutions on the feature correlations to learn to filter neighbour consensus. Note that these latter works target semantic correspondences, whereas we focus on the case when all images depict the same physical scene.

Deep Flow Methods. Deep networks can be trained to predict optical flow and to be robust to drastic appearance changes, but require adapted loss and architectures. Flows can be learned in a completely supervised way using synthetic data, e.g. in [15,23], but transfer to real data remains a difficult problem. Unsupervised training through reconstruction has been proposed in several works, targeting brightness consistency [2,83], gradient consistency [60] or high SSIM [27,86]. This idea of learning correspondences through reconstruction has been applied to video, reconstructing colors [81], predicting weights for frame reconstruction [32,34], or directly optimizing feature consistency in the warped images [82]. Several papers have introduced cycle consistency as an additional supervisory signal for image alignment [82,92]. Recently, feature correlation became a key part of several architectures [23,76] aiming at predicting dense flows. Particularly relevant to us is the approach of [47] which includes a feature correlation layer in a U-Net [67] architecture to improve flow resolution.

A similar approach has been used in [36] which predicts dense correspondences. Recently, Glu-Net [55] learns dense correspondences by investigating the combined use of global and local correlation layers.

Hybrid Parametric/Non-parametric Image Alignment. Classic "plane + parallax" approaches [25,33,70,85] aimed to combine parametric and non-parametric alignment by first estimating a homography (plane) and then considering the violations from that homography (parallax). Similar ideas also appeared in stereo, e.g. model-based stereo [13]. Recently, [10,86] proposed to learn optical flow by jointly optimizing with depth and ego-motion for stereo videos. Our RANSAC-Flow is also related to the methods designed for geometric multi-model fitting, such as RPA [45], T-linkage [46] and Progressive-X [7].

3   Method

Our two-stage RANSAC-Flow method is illustrated in Fig. 1. In this section, we describe the coarse alignment stage, the fine alignment stage, and how they can be iterated to use multiple homographies.

3.1   Coarse Alignment by Feature-Based RANSAC

Our coarse parametric alignment is performed using RANSAC to fit a homography on a set of candidate sparse correspondences between the source and target images. We use off-the-shelf deep features (conv4 layer of a ResNet-50 network) to obtain these correspondences. We experimented with both pre-trained ImageNet features as well as features learned via MoCo self-supervision [20], and obtained similar results. We found it was crucial to perform feature matching at different scales. We fixed the aspect ratio of each image and extracted features at seven scales: 0.5, 0.6, 0.88, 1, 1.33, 1.66 and 2. Matches that were not symmetrically consistent were discarded. The estimated homography is applied to the source image and the result is given together with the target image as input to our fine alignment. We report coarse-only baselines in Experiments section for both features as "*ImageNet* [21]+*H*" and "*MoCo* [20]+*H*".

3.2   Fine Alignment by Local Flow Prediction

Given a source image I_s and a target image I_t which have already been coarsely aligned, we want to predict a fine flow $F_{s \to t}$ between them. We write $\mathbf{F}_{s \to t}$ as the mapping function associated to the flow $F_{s \to t}$. Since we only expect the fine alignment to work in image regions where the homography is a good approximation of the deformation, we predict a matchability mask $M_{s \to t}$, indicating which correspondences are valid. In the following, we first present our objective function, then how and why we optimize it using a self-supervised deep network.

Objective Function. Our goal is to find a flow that warps the source into an image similar to the target. We formalize this by writing an objective function composed of three parts: a reconstruction loss $\mathcal{L}_{rec}$, a matchability loss $\mathcal{L}_m$ and a cycle-consistency loss $\mathcal{L}_c$. Given the pair of images(I_s, I_t) the total loss is:

$$\mathcal{L}(I_s, I_t) = \mathcal{L}_{rec}(I_s, I_t) + \lambda \mathcal{L}_m(I_s, I_t) + \mu \mathcal{L}_c(I_s, I_t) \tag{1}$$

with λ and μ hyper-parameters weighting the contribution of the matchability and cycle loss. We detail these three components in the following paragraphs. Each loss is defined pixel-wise.

Matchability Loss. Our matchability mask can be seen as pixel-wise weights for the reconstruction and cycle-consistency losses. These losses will thus encourage the matchability to be zero. To counteract this effect, the matchability loss encourages the matchability mask to be close to one. Since the matchabiliy should be consistent between images, we define the cycle-consistent matchability at position (x,y) in I_t, (x',y') in I_s with $(x, y) = \mathbf{F}_{s \to t}(x', y')$ as:

$$M_t^{cycle}(x, y) = M_{t \to s}(x, y) M_{s \to t}(x', y') \tag{2}$$

where $M_{s \to t}$ is the matchability predicted from source to target and $M_{t \to s}$ the one predicted from target to source. M_t^{cycle} will be high only if both the matchability of the corresponding pixels in the source and target are high. The matchability loss encourages this cycle-consistent matchability to be close to 1:

$$\mathcal{L}_m(I_s, I_t) = \sum_{(x,y) \in I_t} |M_t^{cycle}(x, y) - 1| \tag{3}$$

Note that directly encouraging the matchability to be 1 leads to similar quantitative results, but using the cycle consistent matchability helps to identify regions that are not matchable in the qualitative results.

Reconstruction Loss. Reconstruction is the main term of our objective and is based on the idea that the source image warped with the predicted flow $F_{s \to t}$ should be aligned to the target image I_t. We use the structural similarity (SSIM) [84] as a robust similarity measure:

$$\mathcal{L}_{rec}^{SSIM}(I_s, I_t) = \sum_{(x,y) \in I_t} M_t^{cycle}(x, y) \left(1 - SSIM\left(I_s(x', y'), I_t(x, y)\right)\right) \tag{4}$$

Cycle Consistency Loss. We enforce cycle consistency of the flow for 2-cycles:

$$\mathcal{L}_c(I_s, I_t) = \sum_{(x,y) \in I_t} M_t^{cycle}(x, y) \|(x', y'), \mathbf{F}_{t \to s}(x, y)\|_2 \tag{5}$$

Optimization with Self-supervised Network. Optimizing objective functions similar to the one described above is common to most optical flow approaches. However, this is known to be an extremely difficult task because of the highly non-convex nature of the objective which typically has many bad local minima. Recent works on the priors implicit within deep neural network architectures [73,80] suggest that optimizing the flow as the output of a neural network might overcome these problems. Unfortunately, our objective is still too complex to obtain good result from optimization on just a single image pair. We thus built a larger database of image pairs on which we optimize the neural network parameters in a self-supervised way (i.e. without need for any annotations). The network could then be fine-tuned on the test image pair itself, but we have found that this single-pair optimization leads to unstable results. However, if several pairs similar to the test pair are available (i.e. we have access to the entire test set), the network can be fine-tuned on this test set which leads to some improvement, as can be seen in our experiments where we systematically report our results with and without fine-tuning.

To collect image pairs for the network training, we simply sample pairs of images representing the same scene and applied our coarse matching procedure. If it led to enough inliers, we added the pair to our training image set, if not we discarded it. For all the experiments, we sampled image pairs from the MegaDepth [37] scenes, using 20,000 image pairs from 100 scenes for training and 500 pairs from 30 different scenes for validation.

3.3 Multiple Homographies

The overall procedure described so far provides good results on image pairs where a single homography serves as a good (if not perfect) approximation of the overall transformation (e.g. planar scenes). This is, however, not the case for many image pairs with strong 3D effects or large objects displacements. To address this, we iterate our alignment algorithm to let it discover more homography candidates. At each iteration, we remove feature correspondences that were inliers for the previous homographies as well as from locations inside the previously predicted matchability masks, and recompute RANSAC again. We stop the procedure when not enough candidate correspondences remain. The full resulting flow is obtained by simply aggregating the estimated flows from each iteration together. The number of homographies considered depends on the input image pairs. For example, the average number of homographies we obtain from pairs for two-view geometry estimation in the YFCC100M [78] dataset is about five. While more complex combinations could be considered, this simple approach provides surprisingly robust results. In our experiments, we quantitatively validate the benefits of using these multiple homographies (*"multi-H"*).

3.4 Architecture and Implementation Details

In our fine-alignment network, the input source and target images (I_s, I_t) are first processed separately by a fully-convolutional *feature extractor* which outputs

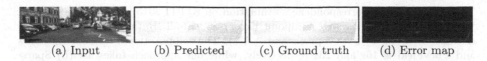

(a) Input   (b) Predicted   (c) Ground truth   (d) Error map

Fig. 3. Visual results on KITTI [48]. We show the predicted flow, ground-truth flow and the error map in (b), (c) and (d) respectively.

two feature maps (f_s, f_t). Each feature from the source image is then compared to features in a $(2K + 1) \times (2K + 1)$ square neighbourhood in the target image using cosine similarity, similar to [15,23]. This results in a $W \times H \times (2K + 1)^2$ similarity tensor s defined by:

$$s(i, j, (m + K + 1)(n + K + 1)) = \frac{f_s(i,j).f_t(i - m, j - n)}{\|f_s(i,j)\|\|f_t(i - m, j - n)\|} \qquad (6)$$

where $m, n \in [-K, ..., K]$ and "." denotes dot product. In all our experiments, we used $K = 3$. This similarity tensor is taken as input by two fully-convolutional *prediction networks* which predict flow and matchability.

Our *feature extractor* is similar to the *Conv3* feature extractor in ResNet-18 [21] but with minor modifications: the first 7×7 convolutional kernel of the network is replaced by a 3×3 kernel without stride and all the max-poolings and strided-convolution are replaced by their anti-aliasing versions proposed in [88]. These changes aim at reducing the loss of spatial resolution in the network, the output feature map being 1/8th of the resolution of the input images. The flow and matchability *prediction networks* are fully convolutional networks composed of three Conv+Relu+BN blocks (Convolution, Relu activation and Batch Normalization [24]) with 512, 256, 128 filters respectively and a final convolutional layer. The output flows and matchability are bilinearly upsampled to the resolution of the input images. Note we tried using up-convolutions, but this slightly decreased the performance while increasing the memory footprint.

We use Kornia [64] for homography warping. All images were resized so that their minimum dimension is 480 pixels. The hyper-parameters of our objective are set to $\lambda = 0.01$, $\mu = 1$. We provide a study of λ and μ in the supplementary material. The entire fine alignment model is learned from random initialization using the Adam optimizer [31] with a learning rate of 2e–4 and momentum terms β_1, β_2 set to 0.5, 0.999. We trained only with $\mathcal{L}_{rec}$ for the first 150 epochs then added $\mathcal{L}_c$ for another 50 epochs and finally trained with all the losses (Eq. 1) for the final 50 epochs. We use a mini-batch size of 16 for all the experiments. The whole training converged in approximately 30 h using a single GPU Geforce GTX 1080 Ti for the 20k image pairs from the MegaDepth. For fine-tuning on the target dataset, we used a learning rate of 2e–4 for 10K iterations.

4   Experiments

In this section, we evaluate our approach in terms of resulting correspondences (Sect. 4.1), downstream tasks (Sect. 4.2), as well as applications to texture

Table 1. (a) Dense correspondences evaluation on KITTI 2015 [48] and Hpatches [5]. We report the AEE (Average Endpoint Error) and Fl-all (Ratio of pixels where flow estimate is wrong by both 3 pixels and $\geq 5\%$). The computational time for EpicFlow and FlowField is 16s and 23s respectively, while our approach takes 4s. (b) Sparse correspondences evaluation on RobotCar [35,44] and MegaDepth [37]. We report the accuracy over all annotated alignments for pixel error smaller than d pixels. All the images are resized to have minimum dimension 480 pixels.

Method	KITTI 2015 [48]				Hpatches [5]				
	Train (AEE ↓)		Test (Fl-all ↓)		Viewpoint (AEE ↓)				
	noc	all	noc	all	1	2	3	4	5
Supervised approaches									
FlowNet2 [23,47,86]	4.93	10.06	6.94	10.41	5.99	15.55	17.09	22.13	30.68
PWC-Net [47,76]	–	10.35	**6.12**	**9.60**	4.43	11.44	15.47	20.17	28.30
Rocco [47,65]	–	–	–	–	9.59	18.55	21.15	27.83	35.19
DGC-Net [47]	–	–	–	–	1.55	5.53	8.98	11.66	16.70
DGC-Nc-Net [36]	–	–	–	–	1.24	4.25	8.21	9.71	13.35
Glu-Net [55]	6.86	9.79	–	–	0.59	4.05	7.64	9.82	14.89
Weakly supervised approaches									
ImageNet [21]+H	13.49	17.26	–	–	1.33	3.34	3.71	6.04	10.07
Cao et al. [10]	4.19	5.13	–	–	–	–	–	–	–
Unsupervised approaches									
Moco [20]+H	13.86	17.60	–	–	1.47	2.96	3.43	7.73	10.53
DeepMatching [47,62]	–	–	–	–	5.84	4.63	12.43	12.17	22.55
DSTFlow [60]	6.96	16.79	–	39	–	–	–	–	–
GeoNet [86]	6.77	10.81	–	–	–	–	–	–	–
EpicFlow [61,86]	4.45	**9.57**	16.69	26.29	–	–	–	–	–
FlowField [4]	–	–	10.98	19.80	–	–	–	–	–
Moco feature									
Ours	4.15	12.63	14.60	26.16	0.52	2.13	4.83	5.13	6.36
w/o fine-tuning	4.67	13.51	–	–	0.53	**2.04**	**2.32**	6.54	6.79
w/o Multi-H	7.04	14.02	–	–	–	–	–	–	–
ImageNet feature									
Ours	**3.87**	12.48	14.12	25.76	**0.51**	2.36	2.91	**4.41**	**5.12**
w/o fine-tuning	4.55	13.51	–	–	**0.51**	2.37	2.64	4.49	5.16
w/o Multi-H	6.74	13.77	–	–	–	–	–	–	–

(a) Dense correspondences evaluation on KITTI 2015 [48] and Hpatches [5].

Method	RobotCar [35,44]			MegaDepth [37]		
	Acc(≤ d pixels ↑)			Acc(≤ d pixels ↑)		
	1	3	5	1	3	5
ImageNet [21]+H	1.03	8.12	19.21	3.49	23.48	43.94
Moco [20]+H	1.08	8.77	20.05	3.70	25.12	45.45
SIFT-Flow [38]	1.12	8.13	16.45	8.70	12.19	13.30
NcNet [66]+H	0.81	7.13	16.93	1.98	14.47	32.80
DGC-Net [47]	1.19	9.35	20.17	3.55	20.33	34.28
Glu-Net [55]	**2.16**	**16.77**	**33.38**	25.2	51.0	56.8
Moco Feature						
Ours	2.10	16.07	31.66	**53.47**	**83.45**	**86.81**
w/o Multi-H	2.06	15.77	31.05	50.65	78.34	81.59
w/o Fine-tuning	2.09	15.94	31.61	52.60	**83.46**	86.80
ImageNet Feature						
Ours	2.10	16.09	31.80	53.15	83.34	86.74
w/o Multi-H	2.06	15.84	31.90	50.08	77.84	81.08
w/o Fine-tuning	2.09	16.00	31.90	52.80	83.31	86.64

(b) Sparse correspondences evaluation on the RobotCar [35,44] and MegaDepth [37].

transfer and artwork analysis (Sect. 4.3). We provide more visual results at http://imagine.enpc.fr/~shenx/RANSAC-Flow/.

4.1   Direct Correspondences Evaluation

Optical Flow. We evaluate the quality of our dense flow on the KITTI 2015 flow [48] and Hpatches [5] datasets and report the results in Table 1.

On KITTI [48], we evaluated both on the training and the test set since other approaches report results on one or the other. Note we could not perform an ablation study on the test set since the number of submissions to the online server is strictly limited. We report results both on non-occluded (noc) and all regions. Our results are on par with state of the art unsupervised and weakly supervised results on non-occluded regions, outperforming for example the recent approach [10,55]. Unsurprisingly, our method is much weaker on occluded regions since our algorithm is not designed specifically for optical flow performances and has no reason to handle occluded regions in a good way. We find that the largest errors are actually in occluded regions and image boundaries (Fig. 3). Interestingly, our ablations show that the multiple homographies is critical to our results even if the input images appear quite similar.

For completeness, we also present results on the Hpatches [5]. Note that Hpatches dataset is synthetically created by applying homographies to a set of real images, which would suggest that our coarse alignment alone should be

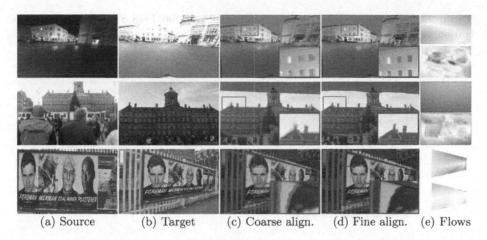

(a) Source   (b) Target   (c) Coarse align.   (d) Fine align.   (e) Flows

Fig. 4. Visual results on RobotCar [44] (1st row), Megadepth [37] (2nd row) and Hpatches [5] (3rd row) using one homography. We show the source and target in (a), (b). The overlapped images after coarse and fine alignment are in (c) and (d) with zoomed details. The coarse (top) and fine (bottom) flows are in (e).

enough. However, in practice, we have found that, due to the lack of feature correspondences, adding the fine flow network significantly boosts the results compared to using only our coarse approach.

While these results show that our approach is reasonable, these datasets only contain very similar and almost aligned pairs while the main goal of our approach is to be able to handle challenging cases with strong viewpoint and appearance variations.

Sparse Correspondences. Dense correspondence annotations are typically not available for extreme viewpoint and imaging condition variations. We thus evaluated our results on sparse correspondences available on the Robot-Car [35,44] and MegaDepth [37] datasets. In Robotcar, we evaluated on the correspondences provided by [35], which leads to approximately 340M correspondences. The task is especially challenging since the images correspond to different and challenging conditions (dawn, dusk, night, etc.) and most of the correspondences are on texture-less region such as roads where the reconstruction objective provides very little information. However, viewpoints in RobotCar are still very similar. To test our method on pairs of images with very different viewpoints, we used pairs of images from scenes of the MegaDepth [37] dataset that we didn't use for training and validation. Note that no real ground truth is available and we use as reference the result of SfM reconstructions. More precisely, we take 3D points as correspondences and randomly sample 1 600 pairs of images that shared more than 30 points, which results in 367K correspondences.

On both datasets, we evaluated several baselines which provide dense correspondences and were designed to handle large viewpoint changes, inluding

Table 2. (a) Two-view geometric estimation on YFCC100M [78,87]. (b) Visual Localization on Aachen night-time [68,69].

Method	mAP @5°	mAP@10°	mAP@20°
SIFT [40]	46.83	68.03	80.58
Contextdesc [42]	47.68	69.55	84.30
Superpoint [14]	30.50	50.83	67.85
PointCN [52,87]	47.98	–	–
PointNet++ [57,87]	46.23	–	–
N^3Net [54,87]	49.13	–	–
DFE [59,87]	49.45	–	–
OANet [87]	52.18	–	–
Moco feature			
Ours	64.88	73.31	81.56
w/o multi-H	61.10	70.50	79.24
w/o fine-tuning	63.48	72.93	81.59
ImageNet feature			
Ous	62.45	70.84	78.99
w/o multi-H	59.90	68.8	77.31
w/o fine-tuning	62.10	70.78	79.07

(a) Two-view geometry, YFCC100M [78]

Method	0.5m,2°	1m,5°	5m,10°
Upright RootSIFT [40]	36.7	54.1	72.5
DenseSfM [68]	39.8	60.2	84.7
HAN + HN++ [49,51]	39.8	61.2	77.6
Superpoint [14]	42.8	57.1	75.5
DELF [53]	39.8	61.2	85.7
D2-net [16]	44.9	66.3	88.8
R2D2 [63]	45.9	66.3	88.8
Moco feature			
Ours	44.9	68.4	88.8
w/o Multi-H	42.9	68.4	88.8
w/o Fine-tuning	41.8	68.4	88.8
ImageNet feature			
Ous	44.9	68.4	88.8
w/o Multi-H	43.9	66.3	88.8
w/o Fine-tuning	44.9	68.4	88.8

(b) Localization, Aachen night-time [68,69]

(a) Source   (b) Target   (c) Texture transfer

Fig. 5. Texture transfer: (a) source, (b) target and (c) texture transferred result.

SIFT-Flow [38], variants of NcNet [66], DGC-Net [47] and the very recent, concurrently developed Glu-Net [55]. In the results provided in Table 1, we can see that our approach is comparable to Glu-Net on RobotCar [35,44] but largely improves performances on MegeDepth [37]. We believe this is because by the large viewpoint variations on MegeDepth is better handled by our method. This qualitative difference between the datasets can be seen in the visual results in Fig. 4. Note that we can clearly see the effect of fine flows on the zoomed details.

4.2   Evaluation for Downstream Tasks

Given the limitations of the correspondence benchmarks discussed in the previous paragraph, and to demonstrate the practical interest of our results, we now evaluate our correspondences on two standard geometry estimation benchmarks where many results from competing approaches exist. Note that competing approaches typically use only sparse matches for these tasks, and being able to perform them using dense correspondences is a demonstration of the strength and originality of our method.

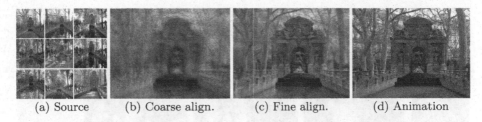

(a) Source   (b) Coarse align.   (c) Fine align.   (d) Animation

Fig. 6. Aligning a group of Internet images from the Medici Fountain, similar to [74]. We show the source images (a), the average image after coarse (b) and fine alignment (c). The animation (view with Acrobat Reader) is in (d) (see Supplementary material).

Two-View Geometry Estimation. Given a pair of views of the same scene, two-view geometry estimation aims at recovering their relative pose. To validate our approach, we follow the standard setup of [87] evaluating on 4×1000 image pairs for 4 scenes from YFCC100M [78] dataset and reporting mAP for different thresholds on the angular differences between ground truth and predicted vectors for both rotation and translation as the error metric. For each image pair, we use the flow we predict in regions with high matchability (> 0.95) to estimate an essential matrix with RANSAC and the 5-point algorithm [19]. To avoid correspondences in the sky, we used the pre-trained the segmentation network provided in [89] to remove them. While this require some supervision, this is reasonable since most of the baselines we compare to have been trained in a supervised way. As can be seen in Table 2, our method outperforms all the baselines by a large margin including the recent OANet [87] method which is trained with ground truth calibration of cameras. Also note that using multiple homographies consistently boosts the performance of our method.

Once the relative pose of the cameras has been estimated, our correspondences can be used to perform stereo reconstruction from the image pair as illustrated in Fig. 2(c) and in the project webpage. Note that contrary to many stereo reconstruction methods, we can use two very different input images.

Day-Night Visual Localization. Another task we performed is visual localization. We evaluate on the local feature challenge of the Visual Localization benchmark [68,69]. For each of the 98 night-time images contained in the dataset, up to 20 relevant day-time images with known camera poses are given. We followed evaluation protocol from [68] and first compute image matching for a list of image pairs and then give them as input to COLMAP [71] that provides a localisation estimation for the queries. To limit the number of correspondences we use only correspondences on a sparse set of keypoints using the Superpoint [14]. Our results are reported in Table 2(b) and are on par with state of the art results.

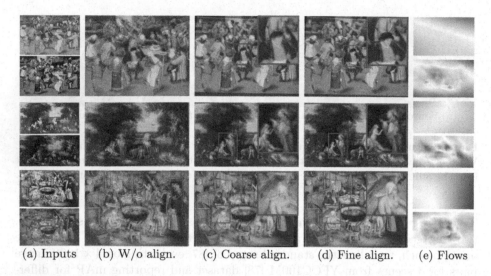

(a) Inputs   (b) W/o align.   (c) Coarse align.   (d) Fine align.   (e) Flows

Fig. 7. Aligning pairs of similar artworks from the Brueghel [1]: We show the pairs in (a). The average images without alignment, after coarse and fine alignment are in (b), (c) and (d). The coarse (top) and fine (bottom) flows are in (e).

4.3   Applications

One of the most exciting aspect of our approach is that it enables new applications based on the fine alignment of historical, internet or artistic images.

Texture Transfer. Our approach can be used to transfer texture between images. In Fig. 5 and 2(f) we show results using historical and modern images from the LTLL dataset [17]. We use the pre-trained segmentation network of [90], and transfer the texture from the source to the target building regions.

Internet Images Alignment. As visualized in Figs. 2(d) and 6, we can align sets of internet images, similar to [74]. Even if our image set is not precisely the same, much more details can be seen in the average of our fine-aligned images.

Artwork Analysis. Finding and matching near-duplicate patterns is an important problem for art historians. Computationally, it is difficult because the duplicate appearance can be very different [72]. In Fig. 7, we show visual results of aligning different versions of artworks from the Brueghel dataset [72] with our coarse and fine alignment. We can clearly see that a simple homography is not sufficient and that the fine alignment improves results by identifying complex displacements. The fine flow can thus be used to provide insights on Brueghel's copy process. Indeed, we found that some artworks were copied in a spatially consistent way, while in others, different parts of the picture were not aligned with each other. This can be clearly seen in the flows in Fig. 9, which are either

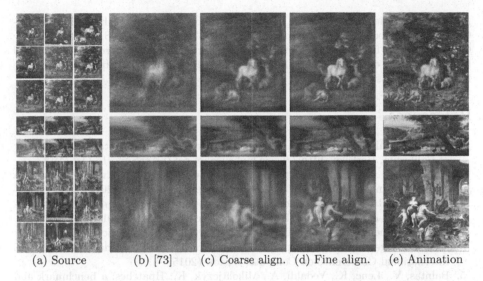

   (a) Source   (b) [73]   (c) Coarse align. (d) Fine align.   (e) Animation

Fig. 8. Aligning details discovered by [72]: (a) sources; average from [72] (b), with coarse (c) and fine (d) alignment; (e) animation (view with Acrobat Reader).

Fig. 9. Analyzing copy process from flow. The flow is smooth from the middle to the right one, while it is irregular from the middle to the left one (see Supplementary material).

very regular or very discontinuous. The same process can be applied to more than a single pair of images, as illustrated in Fig. 2(e) and 8 where we align together many similar details identified by [72]. Visualizing the succession of the finely aligned images allows to identify their differences.

5   Conclusion

We have introduced a new unsupervised method for generic dense image alignment which performs well on a wide range of tasks. Our main insight is to combine the advantages of parametric and non-parametric methods in a two-stage approach and to use multiple homography estimations as initializations for fine flow prediction. We also demonstrated it allows new applications for artwork analysis.

632   X. Shen et al.

Acknowledgements. This work was supported by ANR project EnHerit ANR-17-CE23-0008, project Rapid Tabasco, NSF IIS-1633310, grants from SAP and Berkeley CLTC, and gifts from Adobe. We thank Shiry Ginosar, Thibault Groueix and Michal Irani for helpful discussions, and Elizabeth Alice Honig for her help in building the Brueghel dataset.

References

1. Brueghel family: Jan brueghel the elder. The brueghel family database. University of California, Berkeley. http://www.janbrueghel.net/. Accessed 16 Oct 2018
2. Ahmadi, A., Patras, I.: Unsupervised convolutional neural networks for motion estimation. In: International Conference on Image Processing (2016)
3. Aubry, M., Russell, B.C., Sivic, J.: Painting-to-3D model alignment via discriminative visual elements. ACM Trans. Graph. (ToG) **33**, 1–14 (2014)
4. Bailer, C., Taetz, B., Stricker, D.: Flow fields: dense correspondence fields for highly accurate large displacement optical flow estimation. In: Proceedings of the IEEE International Conference on Computer Vision (2015)
5. Balntas, V., Lenc, K., Vedaldi, A., Mikolajczyk, K.: Hpatches: a benchmark and evaluation of handcrafted and learned local descriptors. In: Proceedings of the IEEE Conference on Computer Vision and Pattern Recognition (2017)
6. Barath, D., Matas, J.: Graph-cut ransac. In: Proceedings of the IEEE Conference on Computer Vision and Pattern Recognition (2018)
7. Barath, D., Matas, J.: Progressive-x: efficient, anytime, multi-model fitting algorithm. In: Proceedings of the IEEE International Conference on Computer Vision (2019)
8. Barath, D., Matas, J., Noskova, J.: Magsac: marginalizing sample consensus. In: Proceedings of the IEEE Conference on Computer Vision and Pattern Recognition (2019)
9. Brox, T., Bregler, C., Malik, J.: Large displacement optical flow. In: Proceedings of the IEEE Conference on Computer Vision and Pattern Recognition (2009)
10. Cao, Z., Kar, A., Hane, C., Malik, J.: Learning independent object motion from unlabelled stereoscopic videos. In: Proceedings of the IEEE Conference on Computer Vision and Pattern Recognition (2019)
11. Cech, J., Matas, J., Perdoch, M.: Efficient sequential correspondence selection by cosegmentation. IEEE Trans. Pattern Anal. Mach. Intell. **32**, 1568–1581 (2010)
12. Choy, C.B., Gwak, J., Savarese, S., Chandraker, M.: Universal correspondence network. In: Advances in Neural Information Processing Systems (2016)
13. Debevec, P.E., Taylor, C.J., Malik, J.: Modeling and rendering architecture from photographs: A hybrid geometry-and image-based approach. In: Proceedings of the 23rd Annual Conference on Computer Graphics and Interactive Techniques (1996)
14. DeTone, D., Malisiewicz, T., Rabinovich, A.: Superpoint: self-supervised interest point detection and description. In: Proceedings of the IEEE Conference on Computer Vision and Pattern Recognition Workshops (2018)
15. Dosovitskiy, A., et al.: Flownet: learning optical flow with convolutional networks. In: Proceedings of the IEEE International Conference on Computer Vision (2015)
16. Dusmanu, M., et al.: D2-net: a trainable CNN for joint description and detection of local features. In: Proceedings of the IEEE Conference on Computer Vision and Pattern Recognition (2019)
17. Fernando, B., Tommasi, T., Tuytelaars, T.: Location recognition over large time lags. Comput. Vis. Image Underst. **139**, 21–28 (2015)

18. Fischler, M.A., Bolles, R.C.: Random sample consensus: a paradigm for model fitting with applications to image analysis and automated cartography. Commun. ACM **24**, 381–395 (1981)
19. Hartley, R., Zisserman, A.: Multiple View Geometry in Computer Vision. Cambridge University Press, Cambridge (2003)
20. He, K., Fan, H., Wu, Y., Xie, S., Girshick, R.: Momentum contrast for unsupervised visual representation learning. In: Proceedings of the IEEE Conference on Computer Vision and Pattern Recognition (2020)
21. He, K., Zhang, X., Ren, S., Sun, J.: Deep residual learning for image recognition. In: Proceedings of the IEEE conference on Computer Vision and Pattern Recognition (2016)
22. Hu, Y., Song, R., Li, Y.: Efficient coarse-to-fine patchmatch for large displacement optical flow. In: Proceedings of the IEEE Conference on Computer Vision and Pattern Recognition (2016)
23. Ilg, E., Mayer, N., Saikia, T., Keuper, M., Dosovitskiy, A., Brox, T.: Flownet 2.0: evolution of optical flow estimation with deep networks. In: Proceedings of the IEEE Conference on Computer Vision and Pattern Recognition (2017)
24. Ioffe, S., Szegedy, C.: Batch normalization: accelerating deep network training by reducing internal covariate shift. In: International Conference on Machine Learning (2015)
25. Irani, M., Anandan, P., Cohen, M.: Direct recovery of planar-parallax from multiple frames. IEEE Trans. Pattern Anal. Mach. Intell. **24**, 1528–1534 (2002)
26. Jaderberg, M., Simonyan, K., Zisserman, A., et al.: Spatial transformer networks. In: Advances in Neural Information Processing Systems (2015)
27. Jason, J.Y., Harley, A.W., Derpanis, K.G.: Back to basics: unsupervised learning of optical flow via brightness constancy and motion smoothness. In: Proceedings of the European Conference on Computer Vision (2016)
28. Kim, S., Lin, S., Jeon, S.R., Min, D., Sohn, K.: Recurrent transformer networks for semantic correspondence. In: Advances in Neural Information Processing Systems (2018)
29. Kim, S., Min, D., Ham, B., Jeon, S., Lin, S., Sohn, K.: Fcss: fully convolutional self-similarity for dense semantic correspondence. In: Proceedings of the IEEE Conference on Computer Vision and Pattern Recognition (2017)
30. Kim, S., Min, D., Jeong, S., Kim, S., Jeon, S., Sohn, K.: Semantic attribute matching networks. In: Proceedings of the IEEE Conference on Computer Vision and Pattern Recognition (2019)
31. Kingma, D.P., Ba, J.: Adam: a method for stochastic optimization. In: International Conference for Learning Representations (2014)
32. Kong, S., Fowlkes, C.: Multigrid predictive filter flow for unsupervised learning on videos. arXiv preprint arXiv:1904.01693 (2019)
33. Kumar, R., Anandan, P., Hanna, K.: Direct recovery of shape from multiple views: a parallax based approach. In: Proceedings of 12th International Conference on Pattern Recognition (1994)
34. Lai, Z., Xie, W.: Self-supervised learning for video correspondence flow. In: BMVC (2019)
35. Larsson, M., Stenborg, E., Hammarstrand, L., Pollefeys, M., Sattler, T., Kahl, F.: A cross-season correspondence dataset for robust semantic segmentation. In: Proceedings of the IEEE Conference on Computer Vision and Pattern Recognition (2019)

36. Laskar, Z., Melekhov, I., Tavakoli, H.R., Ylioinas, J., Kannala, J.: Geometric image correspondence verification by dense pixel matching. In: Winter Conference on Applications of Computer Vision (2020)
37. Li, Z., Snavely, N.: Megadepth: learning single-view depth prediction from internet photos. In: Proceedings of the IEEE Conference on Computer Vision and Pattern Recognition (2018)
38. Liu, C., Yuen, J., Torralba, A.: Sift flow: dense correspondence across scenes and its applications. IEEE Trans. Pattern Anal. Mach. Intell. **33**, 978–994 (2010)
39. Long, J.L., Zhang, N., Darrell, T.: Do convnets learn correspondence? In: Advances in Neural Information Processing Systems (2014)
40. Lowe, D.G.: Distinctive image features from scale-invariant keypoints. Int. J. Comput. Vis. **60**, 91–110 (2004)
41. Lucas, B.D., Kanade, T., et al.: An iterative image registration technique with an application to stereo vision (1981)
42. Luo, Z., et al.: Contextdesc: local descriptor augmentation with cross-modality context. Proceedings of the IEEE Conference on Computer Vision and Pattern Recognition (2019)
43. Luo, Z., et al.: Geodesc: learning local descriptors by integrating geometry constraints. In: Proceedings of the European Conference on Computer Vision (2018)
44. Maddern, W., Pascoe, G., Linegar, C., Newman, P.: 1 year, 1000 km: the oxford robotcar dataset. Int. J. Rob. Res. **36**, 3–15 (2017)
45. Magri, L., Fusiello, A.: T-linkage: a continuous relaxation of j-linkage for multi-model fitting. In: Proceedings of the IEEE Conference on Computer Vision and Pattern Recognition (2014)
46. Magri, L., Fusiello, A.: Multiple structure recovery via robust preference analysis. Image Vis. Comp. **67**, 1–15 (2017)
47. Melekhov, I., Tiulpin, A., Sattler, T., Pollefeys, M., Rahtu, E., Kannala, J.: Dgcnet: dense geometric correspondence network. In: Winter Conference on Applications of Computer Vision (2019)
48. Menze, M., Geiger, A.: Object scene flow for autonomous vehicles. In: Proceedings of the IEEE Conference on Computer Vision and Pattern Recognition (2015)
49. Mishchuk, A., Mishkin, D., Radenovic, F., Matas, J.: Working hard to know your neighbor's margins: local descriptor learning loss. In: Advances in Neural Information Processing Systems (2017)
50. Mishkin, D., Matas, J., Perdoch, M.: Mods: fast and robust method for two-view matching. Comput. Vis. Image Underst. **141**, 81–93 (2015)
51. Mishkin, D., Radenovic, F., Matas, J.: Repeatability is not enough: learning affine regions via discriminability. In: Proceedings of the European Conference on Computer Vision (2018)
52. Moo Yi, K., Trulls, E., Ono, Y., Lepetit, V., Salzmann, M., Fua, P.: Learning to find good correspondences. In: Proceedings of the IEEE Conference on Computer Vision and Pattern Recognition (2018)
53. Noh, H., Araujo, A., Sim, J., Weyand, T., Han, B.: Large-scale image retrieval with attentive deep local features. In: Proceedings of the IEEE International Conference on Computer Vision (2017)
54. Plötz, T., Roth, S.: Neural nearest neighbors networks. In: Advances in Neural Information Processing Systems (2018)
55. Prune, T., Martin, D., Radu, T.: GLU-Net: global-local universal network for dense flow and correspondences. In: Proceedings of the IEEE Conference on Computer Vision and Pattern Recognition (2020)

56. Qi, C.R., Su, H., Mo, K., Guibas, L.J.: Pointnet: deep learning on point sets for 3D classification and segmentation. In: Proceedings of the IEEE Conference on Computer Vision and Pattern Recognition (2017)
57. Qi, C.R., Yi, L., Su, H., Guibas, L.J.: Pointnet++: deep hierarchical feature learning on point sets in a metric space. In: Advances in Neural Information Processing Systems (2017)
58. Raguram, R., Chum, O., Pollefeys, M., Matas, J., Frahm, J.M.: Usac: a universal framework for random sample consensus. IEEE Trans. Pattern Anal. Mach. Intell. **35**, 2022–2038 (2012)
59. Ranftl, R., Koltun, V.: Deep fundamental matrix estimation. In: Proceedings of the European Conference on Computer Vision (2018)
60. Ren, Z., Yan, J., Ni, B., Liu, B., Yang, X., Zha, H.: Unsupervised deep learning for optical flow estimation. In: Thirty-First AAAI Conference on Artificial Intelligence (2017)
61. Revaud, J., Weinzaepfel, P., Harchaoui, Z., Schmid, C.: Epicflow: edge-preserving interpolation of correspondences for optical flow. In: Proceedings of the IEEE Conference on Computer Vision and Pattern Recognition (2015)
62. Revaud, J., Weinzaepfel, P., Harchaoui, Z., Schmid, C.: Deepmatching: hierarchical deformable dense matching. Int. J. Comput. Vis. **120**, 300–323 (2016)
63. Revaud, J., Weinzaepfel, P., de Souza, C.R., Humenberger, M.: R2D2: repeatable and reliable detector and descriptor. In: Advances in Neural Information Processing Systems (2019)
64. Riba, E., Mishkin, D., Ponsa, D., Rublee, E., Bradski, G.: Kornia: an open source differentiable computer vision library for pytorch. In: Winter Conference on Applications of Computer Vision (2020)
65. Rocco, I., Arandjelovic, R., Sivic, J.: Convolutional neural network architecture for geometric matching. In: Proceedings of the IEEE Conference on Computer Vision and Pattern Recognition (2017)
66. Rocco, I., Cimpoi, M., Arandjelović, R., Torii, A., Pajdla, T., Sivic, J.: Neighbourhood consensus networks. In: Advances in Neural Information Processing Systems (2018)
67. Ronneberger, O., Fischer, P., Brox, T.: U-net: convolutional networks for biomedical image segmentation. In: International Conference on Medical image computing and computer-assisted intervention (2015)
68. Sattler, T., et al.: Benchmarking 6dof outdoor visual localization in changing conditions. In: Proceedings of the IEEE Conference on Computer Vision and Pattern Recognition (2018)
69. Sattler, T., Weyand, T., Leibe, B., Kobbelt, L.: Image retrieval for image-based localization revisited. In: BMVC (2012)
70. Sawhney, H.S.: 3D geometry from planar parallax. In: Proceedings of the IEEE Conference on Computer Vision and Pattern Recognition (1994)
71. Schonberger, J.L., Frahm, J.M.: Structure-from-motion revisited. In: Proceedings of the IEEE Conference on Computer Vision and Pattern Recognition (2016)
72. Shen, X., Efros, A.A., Aubry, M.: Discovering visual patterns in art collections with spatially-consistent feature learning. In: Proceedings IEEE Conference on Computer Vision and Pattern Recognition (2019)
73. Shocher, A., Cohen, N., Irani, M.: "zero-shot" super-resolution using deep internal learning. In: Proceedings of the IEEE Conference on Computer Vision and Pattern Recognition (2018)

74. Shrivastava, A., Malisiewicz, T., Gupta, A., Efros, A.A.: Data-driven visual similarity for cross-domain image matching. In: Proceedings of the 2011 SIGGRAPH Asia Conference (2011)
75. Singh, S., Gupta, A., Efros, A.A.: Unsupervised discovery of mid-level discriminative patches. In: Proceedings of the European Conference on Computer Vision (2012)
76. Sun, D., Yang, X., Liu, M.Y., Kautz, J.: Pwc-net: CNNs for optical flow using pyramid, warping, and cost volume. In: Proceedings of the IEEE Conference on Computer Vision and Pattern Recognition (2018)
77. Szeliski, R.: Image alignment and stitching: a tutorial. Found. Trends. Comput. Graph. Vis. **2**, 1–104 (2006)
78. Thomee, B., et al.: Yfcc100m: the new data in multimedia research. Commun. ACM **59**, 64–73 (2016)
79. Tian, Y., Fan, B., Wu, F.: L2-net: deep learning of discriminative patch descriptor in Euclidean space. In: Proceedings of the IEEE Conference on Computer Vision and Pattern Recognition (2017)
80. Ulyanov, D., Vedaldi, A., Lempitsky, V.: Deep image prior. In: Proceedings of the IEEE Conference on Computer Vision and Pattern Recognition (2018)
81. Vondrick, C., Shrivastava, A., Fathi, A., Guadarrama, S., Murphy, K.: Tracking emerges by colorizing videos. In: Proceedings of the European Conference on Computer Vision (2018)
82. Wang, X., Jabri, A., Efros, A.A.: Learning correspondence from the cycle-consistency of time. In: The IEEE Conference on Computer Vision and Pattern Recognition (2019)
83. Wang, Y., Yang, Y., Yang, Z., Zhao, L., Wang, P., Xu, W.: Occlusion aware unsupervised learning of optical flow. In: Proceedings of the IEEE Conference on Computer Vision and Pattern Recognition (2018)
84. Wang, Z., Bovik, A.C., Sheikh, H.R., Simoncelli, E.P.: Image quality assessment: from error visibility to structural similarity. IEEE Trans. Image Process. **13**, 600–612 (2004)
85. Wulff, J., Sevilla-Lara, L., Black, M.J.: Optical flow in mostly rigid scenes. In: Proceedings of the IEEE Conference on Computer Vision and Pattern Recognition (2017)
86. Yin, Z., Shi, J.: Geonet: unsupervised learning of dense depth, optical flow and camera pose. In: Proceedings of the IEEE Conference on Computer Vision and Pattern Recognition (2018)
87. Zhang, J., et al.: Learning two-view correspondences and geometry using order-aware network. Proceedings of the IEEE International Conference on Computer Vision (2019)
88. Zhang, R.: Making convolutional networks shift-invariant again. In: International Conference on Machine Learning (2019)
89. Zhou, B., Zhao, H., Puig, X., Fidler, S., Barriuso, A., Torralba, A.: Scene parsing through ade20k dataset. In: Proceedings of the IEEE Conference on Computer Vision and Pattern Recognition (2017)
90. Zhou, B., et al.: Semantic understanding of scenes through the ade20k dataset. Int. J. Comput. Vis. **127**, 302–321 (2018)

91. Zhou, T., Jae Lee, Y., Yu, S.X., Efros, A.A.: Flowweb: joint image set alignment by weaving consistent, pixel-wise correspondences. In: Proceedings of the IEEE Conference on Computer Vision and Pattern Recognition (2015)
92. Zhou, T., Krahenbuhl, P., Aubry, M., Huang, Q., Efros, A.A.: Learning dense correspondence via 3D-guided cycle consistency. In: Proceedings of the IEEE Conference on Computer Vision and Pattern Recognition (2016)

Semantic Object Prediction and Spatial Sound Super-Resolution with Binaural Sounds

Arun Balajee Vasudevan[1]([⊠]), Dengxin Dai[1], and Luc Van Gool[1,2]

[1] CVL, ETH Zurich, Zurich, Switzerland
{arunv,dai,vangool}@vision.ee.ethz.ch
[2] PSI, KU Leuven, Leuven, Belgium

Abstract. Humans can robustly recognize and localize objects by integrating visual and auditory cues. While machines are able to do the same now with images, less work has been done with sounds. This work develops an approach for dense semantic labelling of sound-making objects, purely based on binaural sounds. We propose a novel sensor setup and record a new audio-visual dataset of street scenes with eight professional binaural microphones and a 360° camera. The co-existence of visual and audio cues is leveraged for supervision transfer. In particular, we employ a cross-modal distillation framework that consists of a vision 'teacher' method and a sound 'student' method – the student method is trained to generate the same results as the teacher method. This way, the auditory system can be trained without using human annotations. We also propose two auxiliary tasks namely, a) a novel task on Spatial Sound Super-resolution to increase the spatial resolution of sounds, and b) dense depth prediction of the scene. We then formulate the three tasks into one end-to-end trainable multi-tasking network aiming to boost the overall performance. Experimental results on the dataset show that 1) our method achieves good results for all the three tasks; and 2) the three tasks are mutually beneficial – training them together achieves the best performance and 3) the number and the orientations of microphones are both important. The data and code will be released on the project page.

1 Introduction

Autonomous vehicles and other intelligent robots will have a substantial impact on people's daily life. While great progress has been made in the past years with visual perception systems [14,27,51], we argue that auditory perception and sound processing also play a crucial role in this context [8,15,34]. As known, animals such as bats, dolphins, and some birds have specialized on "hearing" their environment. To some extent, humans are able to do the same – to "hear"

Electronic supplementary material The online version of this chapter (https://doi.org/10.1007/978-3-030-58548-8_37) contains supplementary material, which is available to authorized users.

A. Vedaldi et al. (Eds.): ECCV 2020, LNCS 12349, pp. 638–655, 2020.
https://doi.org/10.1007/978-3-030-58548-8_37

(a) binaural sounds (b) super-resolved ones (c) semantic labels   (d) depth values

Fig. 1. An illustration of our three tasks: (a) input binaural sounds, (b) super-resolved binaural sounds – from azimuth angle 0° to 90°, (c) auditory semantic perception for three sound-making object classes, and (d) auditory depth perception.

the shape, distance, and density of objects around us [23,42]. In fact, humans surely need to use this capability for many daily activities such as for driving – certain alerting stimuli, such as horns of cars and sirens of ambulances, police cars and fire trucks, are meant to be heard, i.e. are primarily acoustic [34]. Auditory perception can be used to localize common objects like a running car, which is especially useful when visual perception fails due to adverse visual conditions or occlusions. Future intelligent robots are expected to have the same perception capability to be robust and to be able to interact with humans naturally.

Numerous interesting tasks have been defined at the intersection of visual and auditory sensing like sound source localization [56], scene-aware audio generation [33], geometry estimation for rooms using sound echos [3], sound source separation using videos [20], and scene [7] and object [19] recognition using audio cues. There are also works to learn the correlation of visual objects and sounds [55,56]. While great achievements have been made, previous methods mostly focus on specific objects, e.g. musical instruments or noise-free rooms, or on individual tasks only, e.g. sound localization or geometry estimation. This work aims to learn auditory perception for general, unconstrained environments.

This work primarily focuses on semantic object prediction based on binaural sounds aiming to replicate human auditory capabilities. To enhance the semantic prediction, this work proposes two auxiliary tasks: depth prediction from binaural sounds and spatial sound super-resolution (S³R). Studies [36,52] have shown that depth estimation and semantic prediction are correlated and are mutually beneficial. S³R is a novel task aiming to increase the directional resolution of audio signals, e.g. from Stereo Audio to Surround Audio. S³R, as an auxiliary task, is motivated from the studies [35,48] showing that humans are able to better localize the sounding sources by changing their head orientations. S³R is also a standalone contribution and has its own applications. For instance, spatially resolved sounds improve the spatial hearing effects in AR/VR applications. It offers better environmental perception for users and reduces the ambiguities of sound source localization [28]. The three tasks are demonstrated in Fig. 1.

In particular, we propose a sensor setup containing eight professional binaural microphones and a 360° camera. We used it to record the new 'Omni Auditory Perception Dataset' on public streets. The semantic labels of the sound-making objects and the depth of the scene are inferred from the video frames using well-established vision systems. The co-existence of visual and audio cues is leveraged for supervision transfer to train the auditory perception systems without using human annotations. In particular, we employ a cross-modal framework that consists of vision teacher methods and sound student methods to let the students imitate the performance of the teachers. For evaluation, we have manually annotated a test set. The task of S^3R has accurate ground truth to train, thanks to our multi-microphone rig. Finally, we formulate the semantic prediction task and the two auxiliary tasks into a multi-tasking network which can be trained in an end-to-end fashion.

We evaluate our method on our new Omni Auditory Perception Dataset. Extensive experimental results reveal that 1) our method achieves good results for auditory semantic perception, auditory depth prediction and spatial sound super-resolution; 2) the three tasks are mutually beneficial – training them together achieves the best results; and 3) both the number and orientations of microphones are important for auditory perception.

This work makes multiple contributions: 1) a novel approach to dense semantic label prediction for sound-making objects of multiple classes and an approach for dense depth prediction; 2) a method for spatial sound super-resolution, which is novel both as a standalone task and as an auxiliary task for auditory semantic prediction; and 3) a new Omni Auditory Perception Dataset with four pairs of binaural sounds (360° coverage) accompanied by synchronized 360° videos.

2   Related Works

Auditory Scene Analysis. Sound segregation is a well-established research field aiming to organize sound into perceptually meaningful elements [1]. Notable applications include background sounds suppression and speech recognition. Recent research has found that the motion from videos can be useful for the task [20,55]. Sound localization has been well studied with applications such as localizing sniper fire on the battle field, cataloging wildlife in rural areas, and localizing noise pollution sources in an urban environment. It also enriches human–robot interaction by complementing the robot's perceptual capabilities [6,41]. The task is often tackled by the beamforming technique with a microphone array [50], with a notable exception that relies on a single microphone only [45]. Another classical technique uses kalman filter that processes auto-correlations from multiple channels [32]. The recent advance of deep learning enables acoustic camera systems for real-time reconstruction of acoustic camera spherical maps [47].

Auditory scene analysis has been widely applied to automotive applications as well. For instance, auditory cues are used to determine the occurrence of abnormal events in driving scenarios [34]. An acoustic safety emergency system has also been proposed [17]. Salamon et al. have presented a taxonomy of urban sounds and a new dataset, UrbanSound, for automatic urban sound classification [44]. The closest to our work is the very recent work of car detection with stereo sounds [19] in which a 2D bounding-box is proposed for the sound-making object in an image frame. While being similar in spirit, our semantic prediction is dense and for multiple classes instead of a 2D bounding box prediction for a single class. Our method also includes dense depth prediction and spatial sound super-resolution. Binaural sounds are different from general stereo sounds and our method works with panoramic images to have omni-view perception. Another similar work to ours is [29]. We differ significantly in: 1) having multiple semantic classes, 2) working in unconstrained real outdoor environment, and 3) a multi-task learning setup.

Audio-Visual Learning. There is a large body of work that localize the sources of sounds in visual scenes [4,5,9,20,22,46,49,55,56]. The localization is mostly done by analyzing the consistency between visual motion cues and audio motion cues over a large collection of video data. These methods generally learn to locate image regions which produce sounds and separate the input sounds into a set of components that represent the sound for each pixel. Our work uses binaural sounds rather than monaural sounds. Our work localizes and recognizes sounds at the pixel level in image frames, but performs the task with sounds as the only inputs. Audio has also been used for estimating the geometric layout of indoor scenes [3,16,30,54]. The general idea is that the temporal relationships between the arrival times of echoes allows us to estimate the geometry of the environment. There are also works to add scene-aware spatial audio to 360° videos in typical indoor scenes [33,40] by using a conventional mono-channel microphone and a speaker and by analyzing the structure of the scene. The notable work by Owens et al. [38] have shown that the sounds of striking an object can be learned based on the visual appearance of the objects. By training a neural network to predict whether video frames and audio are temporally aligned [37], audio representations can be learned. Our S³R can be used as another self-learning method for audio representations.

Cross-Domain Distillation. The interplay among senses is basic to the sensory organization in human brains [18] and is the key to understand the complex interaction of the physical world. Fortunately, most videos like those available in the Internet contain both visual information and audio information, which provide a bridge linking those two domains and enable many interesting learning approaches. Aytar et al. propose an audio-based scene recognition method by cross-domain distillation to transfer supervision from the visual domain to the audio domain [7]. A similar system was proposed for emotion recognition [2]. Ambient sounds can provide supervision for visual learning as well [39].

3 Approach

3.1 Omni Auditory Perception Dataset

Training our method requires a large collection of omni-directional binaural audios and accompanying 360° videos. Since no public video dataset fulfills this requirement, we collect a new dataset with a custom rig. As shown in Fig. 2, we assembled a rig consisting of a 3Dio Omni Binaural Microphone, a 360°

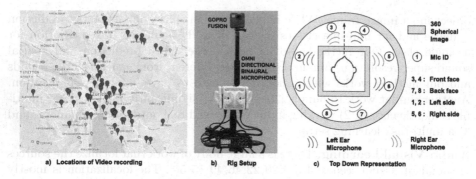

a) Locations of Video recording   b)   Rig Setup   c)   Top Down Representation

Fig. 2. Sensor and dataset: a) data capture locations, b) our custom rig and c) abstract depiction of our recording setup with sensor orientations and microphone ids.

GoPro Fusion camera and a Zoom F8 MultiTrack Field Recorder, all attached to a tripod. We mounted the GoPro camera on top of the 3Dio Omni Binaural Microphone by a telescopic pole to capture all of the sights with minimum occlusion from the devices underneath. This custom rig enables omni seeing and omni binaural hearing with 4 pairs of human ears. The 8 microphones are connected to 8 tracks of the MultiTrack Recorder. The recorder has 8 high-quality, super low-noise mic preamps to amplify the sounds, provides accurate control of microphone sensitivity and allows for accurate synchronization among all microphones. The GoPro Fusion camera captures 4K videos using a pair of 180° cameras housed on the front and back faces. The two cameras perform synchronized image capture and are later fused together to render 360° videos using the GoPro Fusion App. We further use the clap sound to synchronize the camera and 3Dio microphones. After the sensors are turned on, we do hand clapping near them. This clapping sounds recorded by both the binaural mics and the built-in mic of the camera are used to synchronize the binaural sounds and the video. The video and audio signals recorded by the camera are synchronized by default. Videos are recorded at 30 fps. Audios are recorded at 96 kHz.

We recorded videos on the streets of Zurich covering 165 locations within an area of 5 km × 5 km as shown in Fig. 2(a). We choose the locations next to road junctions, where we kept the rig stationary for the data recording of the traffic scenes. For each location, we recorded data for around 5–7 min. Our dataset consists of 165 city traffic videos and audios with an average length of 6.49 min, totalling 15 h. We post-process the raw video-audio data into 2 s segments, resulting in 64, 250 video clips. The videos contain numerous sound-making objects such as cars, trams, motorcycles, human, buses and trucks.

It is worth noticing that the professional 3Dio binaural mics simulate how human ears receive sound, which is different from the general stereo mics or monaural mics. Humans localize sound sources by using three primary cues [41]: interaural time difference (ITD), interaural level difference (ILD), and head-related transfer function (HRTF). ITD is caused by the difference between the times sounds reach the two ears. ILD is caused by the difference in sound pressure

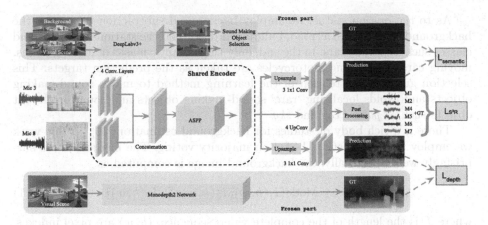

Fig. 3. The diagram of our method for the three considered tasks. The encoder is shared by all tasks and each task has its own decoder.

level reaching the two ears due to the acoustic shadow casted by the listener's head. HRTF is caused because the pinna and head affect the intensities of sound frequencies. All these cues are missing in monaural audio, thus in this work, we focus on learning to localize semantic objects with binaural audios. This difference makes our dataset more suitable for cognitive applications.

3.2 Auditory Semantic Prediction

Since it is costly to create a large collection of human annotations, we follow a teacher-student learning strategy to transfer knowledge from vision to audio [7, 19]. Thus, our auditory semantic object prediction system is composed of two key components: a teacher vision network and a student audio network. The difference to previous methods is that we learn to transfer precise semantic segmentation results instead of scene labels [7] or bounding boxes [19].

Vision Network. We employ the DeepLabv3+ model [12]. We pick the middle frame of our 2-s video clip as the target frame and feed it to the teacher network to generate the semantic map. During training, each target frame is fed into a Cityscapes [13] pre-trained DeepLabv3+ to assign a semantic label to each pixel. Since objects in many classes such as *sky*, *road* and *parked cars* are not sound-making, it is very challenging to predict their semantic masks. Therefore, an object selection policy needs to be designed in order to collect the semantic masks of major sound-making objects.

Sound-Making Object Collection. Our dataset contains numerous sound making objects such as cars, trams, motorcycles, pedestrians, buses and bicycles. The target objects must be constrained to make sure that the task is challenging but still achievable by current sensing systems and learning methods. In this work, we focus on *car*, *tram*, *motorcycle* due to their high occurrences in the datasets and because they produce sufficient noise when they move.

As to the motion status, we employ background subtraction to remove the background classes, such as road, building and sky, and the stationary foreground classes such as parked cars. In the end, only the semantic masks of moving trams, moving cars and moving motorcycles are taken as the prediction targets. This selection guides the attention of the learning method to major sound-making objects and avoids localizing 'rare' sound-making objects and sound-irrelevant objects such as *parked car* and *sky*.

There is a rich body of studies for background estimation [10], in this work we employ a simple method based on majority voting. The method works surprisingly well. Specifically, the background image is computed as

$$I_{bg}(h, w) = \text{Mode}\{I_1(h, w), I_2(h, w), ..., I_T(h, w)\}, \tag{1}$$

where T is the length of the complete video sequence, (h, w) are pixel indexes, and Mode{.} computes the number which appears most often in a set of numbers. Since the complete video sequence is quite long (about 5–7 min), the background estimation is accurate and reliable.

The sound-making objects are detected by the following procedure: given an video frame I_t and its corresponding background image I_{bg}, we use DeepLabv3+ to get their semantic segmentation results Y_t, and Y_{bg}. Figure 3 gives an illustration. The detection is done as

$$S(h, w) = \begin{cases} 1 \text{ if } Y_t(h, w) \in \{car, train, motorcycle\} \\ \quad \text{and } Y_t(h, w) \neq Y_{bg}(h, w), \\ 0 \text{ otherwise}, \end{cases} \tag{2}$$

where 1 indicates pixel locations of sound-making objects and 0 otherwise. Figure 5 shows examples of the detected background and the detected sound-making target objects (i.e. ground truth).

Audio Network. We treat auditory semantic prediction from the binaural sounds as a dense label prediction task. We take the semantic labels produced by the teacher vision network and filter by Eq. 2 as pseudo-labels, and then train a student audio network (BinauralSemanticNet) to predict the pseudo semantic labels directly from the audio signals. A cross-entropy loss is used. The description of the network architecture can be found in Sect. 3.5 and in the supple. material.

3.3   Auditory Depth Perception

Similar to auditory semantic object localization, our auditory depth prediction method is composed of a teacher vision network and a student audio network. It provides auxiliary supervision for semantic perception task.

Vision Network. We employ the MonoDepth2 model [25] given its good performance. We again pick the middle frame of our 2-s video clip as the target frame and feed it to the teacher network to generate the depth map. The model is pre-trained on KITTI [24]. We estimate the depth for the whole scene, similar to previous methods for holistic room layout estimation with sounds [3,30].

Audio Network. We treat depth prediction from the binaural sounds as a dense regression task. We take the depth values produced by the teacher vision network as pseudo-labels, and then train a student audio network (BinauralDepthNet) to regress the pseudo depth labels directly from the audio signals. The L2 loss is used. The description of the network architecture can be found in Sect. 3.5.

3.4 Spatial Sound Super-Resolution (S³R)

We leverage our omni-directional binaural microphones to design a novel task of spatial sound super-resolution (S³R), to provide auxiliary supervision for semantic perception task. The S³R task is motivated by the well-established studies [35,53] about the effects of head movement in improving the accuracy of sound localization. Previous studies also found that rotational movements of the head occur most frequently during source localization [48]. Inspired by these findings, we study the effect of head rotation on auditory semantic and depth perception. Our omni-directional binaural microphone set is an ideal device to simulate head rotations at four discrete angles, i.e. to an azimuth angle of 0° 90°, 180° and 270°, respectively.

Specifically, the S³R task is to train a neural network to predict the binaural audio signals at other azimuth angles given the signals at the azimuth angle of 0°. We denote the received signal by the left and right ears at azimuth 0° by $x^{L_0}(t)$ and $x^{R_0}(t)$, respectively. We then feed those two signals into a deep network to predict the binaural audio signals $x^{L_\alpha}(t)$ and $x^{R_\alpha}(t)$ at azimuth $\alpha°$. Inspired by [21], we predict the difference of the target signals to the input signals, instead of directly predicting the absolute values of the targets. This way, the network is forced to learn the subtle difference. Specifically, we predict the difference signals:

$$
\begin{aligned}
x^{DL_\alpha}(t) &= x^{L_0}(t) - x^{L_\alpha}(t) \\
x^{DR_\alpha}(t) &= x^{R_0}(t) - x^{R_\alpha}(t),
\end{aligned}
\tag{3}
$$

where $\alpha \in \{90°, 180°, 270°\}$. In order to leverage the image processing power of convolutional neural network, we follow the literature and choose to work with the spectrogram representation. Following [21], real and imaginary components of complex masks are predicted. The masks are multiplied with input spectrograms to get the spectrograms of the difference signals; the raw waveforms of the difference signals are then produced by applying Inverse Short-time Fourier Transform (ISTFT) [26]; and finally the target signals are reconstructed by adding back the reference raw waveform.

3.5 Network Architecture

Here, we present our multi-tasking audio network for all the three tasks. The network is composed of one shared encoder and three task-specific decoders. The pipeline of the method is shown in Fig. 3. As to the encoder, we convert the two channels of binaural sounds to log-spectrogram representations. Each spectrogram is passed through 4 strided convolutional (conv) layers with shared weights before they are concatenated. Each conv layer performs a 4×4 convolution with a stride of 2. Each conv layer is followed by a BN layer and a ReLU activation. The concatenated feature map is further passed to a Atrous Spatial Pyramid Pooling (ASPP) module [11]. ASPP has one 1×1 convolution and three 3×3 convolutions with dilation rates of 6, 12, and 18. Each of the convolutions has 64 filters and a BN layer. ASPP concatenates all the features and passes them through a 1×1 conv layer to generate binaural sound features. This feature map is taken as the input to our decoders.

Below, we present the three task-specific decoders. For the semantic prediction task, we employ a decoder to predict the dense semantic labels from the above feature map given by the shared encoder. The decoder comprises of an upsampling layer and three 1×1 conv layers. For the first two conv layers, each is followed by a BN and a ReLU activation; for the last one, it is followed by a softmax activation. We use the same decoder architecture for the depth prediction task, except that we use ReLU activation for the final conv layer of the decoder. For the S^3R task, we perform a series of 5 up-convolutions for the binaural feature map, each convolution layer is followed by a BN and a ReLU activation. The last layer is followed by a sigmoid layer which predicts a complex valued mask. We perform a few post processing steps to convert this mask to binaural sounds at other azimuth angles as mentioned in Sect. 3.4.

Loss Function. We train the complete model shown in Fig. 3 in an end-to-end fashion. We use a) cross-entropy loss for the semantic prediction task which is formulated as dense pixel labelling to 3 classes, b) L2 loss for the depth prediction task to minimize the distance between the predicted depth values and the ground-truth depth values, and c) L2 loss for the S^3R task to minimize the distance between the predicted complex spectrogram and the ground truths. Hence, the total loss L for our multi-tasking learning is

$$L = L_{semantic} + \lambda_1 L_{depth} + \lambda_2 L_{s^3r} \qquad (4)$$

where λ_1 and λ_2 are weights to balance the losses. The detailed network architecture will be provided in the supple. material.

4   Experiments

Data Preparation. Our dataset comprises of $64,250$ video segments, each of 2 s long. We split the samples into three parts: $51,400$ for training, $6,208$ for validation and $6,492$ for testing. We use 2-s segments following [19], which shows that performances are stable for ≥ 1 s segments. For each scene, a background image is also precomputed according to Eqn. 1. For the middle frame of each segment, we generate the ground truth for semantic segmentation task by using the Deeplabv3+ [12] pretrained on Cityscapes dataset [13] and the depth map by using the Monodepth2 [25] pretrained on KITTI dataset [24]. We use *AuditoryTestPseudo* to refer the test set generated this way. In order to more reliably evaluate the method, we manually annotate the middle frame of 80 test video segments for the three considered classes, namely car, train and motorcycle. We carefully select the video segments such that they cover diverse scenarios such as daylight, night, foggy and rainy. We use LabelMeToolbox [43] for the annotation and follow the annotation procedure of Cityscapes [13]. We call this test set *AuditoryTestManual*.

For all the experiments, a training or a testing sample consists of a 2-s video segment and eight 2-s audio channels. We preprocess audio samples following techniques from [21,56]. We keep the audio samples at 96 kHz and their amplitude is normalized to a desired RMS level, which we set to 0.1 for all the audio channels. For normalization, we compute mean RMS values of amplitude over the entire dataset separately for each channel. An STFT is applied to the normalized waveform, with a window size of 512 (5.3 ms), hop length of 160 (1.6 ms) resulting a Time-Frequency representation of size of 257×601 pixels. Video frames are resized to 960×1920 pixels to fit to the GPU.

Implementation Details. We train our complete model using Adam solver [31] with a learning rate of 0.00001 and we set a batch size of 2. We train our models on GeForce GTX 1080 Ti GPUs for 20 epochs. For joint training of all the three tasks, we keep $\lambda_1 = 0.2$ and $\lambda_2 = 0.2$ in Eq. 4.

Evaluation Metrics. We use the standard mean IoU for the semantic prediction task. For audio super resolution, we use MSE error for the spectrograms and the envelope error for the waveforms as used in [21]. For depth prediction, we employ RMSE, MSE, Abs Rel and Sq Rel by following [25].

Table 1. Results of auditory semantic prediction. The results of DeepLabv3+ on the background image (BG) and on the target middle frame (Visual) are reported for reference purpose. mIoU (%) is used. MC denotes Motorcycle.

Methods	Microphone		Auxiliary tasks		AuditoryTestPseudo				AuditoryTestManual			
	Mono	Binaural	S³R	Depth	Car	MC	Train	All	Car	MC	Train	All
BG					8.79	4.17	24.33	12.61	–	–	–	–
Visual					–	–	–	–	79.01	39.07	77.34	65.35
Mono	✓				33.53	7.86	24.99	22.12	30.13	9.21	24.1	21.14
Ours(B)		✓			35.80	19.51	40.71	32.01	35.30	13.28	35.48	28.02
Ours(B:D)		✓		✓	33.53	28.01	55.32	38.95	32.42	25.8	50.12	36.11
Ours(B:S)		✓	✓		35.62	36.81	**56.49**	42.64	**38.12**	26.5	49.02	37.80
Ours(B:SD)		✓	✓	✓	**35.81**	**38.14**	56.25	**43.40**	35.51	**28.51**	**50.32**	**38.01**

4.1 Auditory Semantic Prediction

We compare the performance of different methods and report the results in Table 1. The table shows our method learning with sounds can generate promising results for dense semantic object prediction. We also find that using binaural sounds *Ours(B)* generates significant better results than using *Mono* sound. This is mainly because the major cues for sound localization such as ILD, ITD, and HRTF are missing in *Mono* sound. Another very exciting observation is the joint training with the depth prediction task and the S³R task are beneficial to the semantic prediction task. We speculate that this is because all the three tasks benefit from a same common goal – reconstructing 3D surround sounds from Binaural sounds. Below, we present our ablation studies.

S³R and Depth Prediction Both Give a Boost. As can be seen in Table 1, by adding S³R or depth prediction as an auxiliary task, indicated by *Ours(B:S)* and *Ours(B:D)* respectively, improves the performance of our baseline *Ours(B)* clearly. We also observe that using both of the auxiliary tasks together, indicated by *Ours(B:SD)*, yields the best performance. This shows that both S³R and depth helps. This is because the three tasks share the same goal – extracting spatial information from binaural sounds.

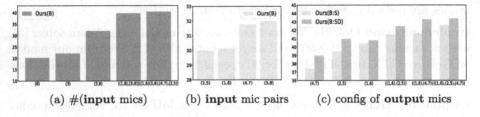

(a) #(**input** mics)   (b) **input** mic pairs   (c) config of **output** mics

Fig. 4. Semantic prediction ablation study results (mIoU) with different set of microphone used as inputs under *Ours(B)* in (a) and (b) and ablation on output microphones for S³R in *Ours(B:S)* and *Ours(B:SD)* in (c).

Adding Input Channels Increases the Performance. We compare the auditory semantic prediction (without using auxiliary tasks) accuracies in Fig. 4(a) for different set of input microphones. Here, we experiment with a) Mono sound from 3 (front) or 8 (back) microphone, b) binaural sounds from the pair (3,8), c) 2 pairs of binaural sound channels ((1,6),(3,8)) which faces in four orthogonal directions, and d) 4 pairs of binaural sound channels. We see that semantic prediction accuracy increases from 22.12% when using *Mono* sound to 40.32% when using all 8 channels. This shows that semantic prediction improves with the access to more sound channels.

Orientation of Microphones Matter. Figure 4(b) shows the auditory semantic prediction results (without using auxiliary tasks) from different orientations of the input binaural pairs for the same scene. The pair (3,8) is aligned in parallel with the front facing direction of the camera as can be seen in Fig. 2. We define this as orientation of 0°. Then, we have other pairs (1,6), (4,7) and (2,5) orientating at azimuth angles of 90°, 180° and 270° respectively. We observe that (3,8) outperforms all other pairs. Intuitively left-right pair is better for auditory perception, but for semantic prediction, especially for long-range objects, the mic facing straight to the objects may have the advantage of detecting weak audio signals. Also, at road intersections, many sounding objects move in the left-right direction, so the front-rear mic pair is effectively under the left-right configuration relative to those sounding objects. The optimal mic orientations for different tasks have not been well studied. Our dataset provides a great opportunity to study this problem.

Visual Scene   Detected Background   Semantic prediction   Semantic GT

Fig. 5. Qualitative results of auditory semantic prediction by our approach. First column shows the visual scene, the second for the computed background image, the third for the object masks predicted by our approach, and the fourth for the ground truth. The object classes are depicted in the following colours: Car, Train and Motorcycle.

Removing Output Channels Degrades the Performance. We vary the number of output microphone pairs for S³R under the two multi-tasking models *Ours(B:S)* and *Ours(B:SD)*. We fix the input to pair (3,8) and experiment with different number of output pairs, ranging from 1 to 3. The results are presented in Fig. 4(c) under these settings. We see that spatial sound resolution to 3 binaural pairs performs better than to 1 or 2 binaural pairs. The more output channels we have, the better the semantic prediction results are.

ASPP is a Powerful Audio Encoder. We have found in our experiments that ASPP is a powerful encoder for audio as well. We compare our audio encoder with and without the ASPP module. For instance, Mono sound with ASPP clearly outperforms itself without ASPP – adding ASPP improves the performance from 13.21 to 22.12 for mean IoU. The same trend is observed for other cases.

Table 2. Depth prediction results with Mono and binaural sounds under different multi-task settings. For all the metrics, lower score is better.

Microphone		Joint tasks		Metrics			
Mono	Binaural	Semantic	S³R	Abs Rel	Sq Rel	RMSE	MSE
✓				118.88	622.58	5.413	0.365
	✓			108.59	459.69	5.263	0.331
	✓		✓	90.43	400.53	5.193	0.318
	✓	✓		87.96	290.06	5.136	0.315
	✓	✓	✓	**84.24**	**222.41**	**5.117**	**0.310**

4.2 Auditory Depth Prediction

We report the depth prediction results in Table 2. The first row represents the depth prediction from Mono sounds alone while the second row represents our baseline method where the depth map is predicted from binaural sounds. As a simple baseline, we also compute the mean depth over the training dataset and evaluate over the test set. The RMSE and MSE scores are 15.024 and 0.864, respectively, which are 2.5 times worse than our binaural sound baseline. The multi-task learning with S³R and semantic prediction under shared audio encoder also provides pronounced improvements for depth prediction. Jointly training the three tasks also yields the best performance for depth prediction.

Table 3. a) S³R results. MSE1 and MSE2 represent mean squared error while ENV1 and ENV represent envelope error for the 2 output channels of binaural sounds. For all metrics, lower score is better. b) Subjective assessment of the generated binaural sounds.

Joint Tasks		Mic Ids		Metrics			
Semantic	Depth	In	Out	MSE-1	ENV-1	MSE-2	ENV-2
		(3,8)	(1,6)	0.1228	0.0298	0.1591	0.0324
	✓	(3,8)	(1,6)	0.0984	0.0221	0.1044	0.0267
✓		(3,8)	(1,6)	**0.0956**	**0.0214**	**0.1001**	**0.0243**
✓	✓	(3,8)	(1,6)	0.0978	0.0218	0.1040	0.0264

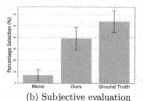

(a) S³R results in multi-tasking   (b) Subjective evaluation

4.3 Spatial Sound Super-Resolution

Table 3 shows the results of S³R as a stand-alone task (first row) and under the multi-task setting. To keep it simple, we estimate the sound signals of microphone pair $(1, 6)$ alone from the microphone pair $(3, 8)$. We can see from Fig. 2 that these two pairs are perpendicular in orientation, so the prediction is quite challenging. We can observe from Table 3 that the multi-task learning with semantic prediction task and depth prediction task outperforms the accuracy of the stand-alone S³R model. Hence, the multi-task learning also helps S³R – the same trend as for semantic perception and depth perception. We also conduct a user study for the subjective assessment of the generated binaural sounds. The participants listen to ground-truth binaural sounds, binaural sounds generated from *Mono* approach (Table 1) and from our approach. We present two (out of three) randomly picked sounds and ask the user to select a preferred one in temrs of binaural sound quality. Table 3(b) shows the percentage of times each method is chosen as the preferred one. We see that *Ours* is close to the ground truth selection implying that our predicted binaural sounds are of high quality.

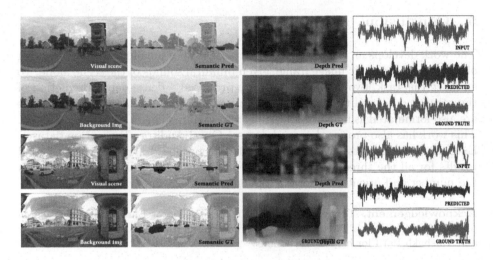

Fig. 6. Qualitative results of all three tasks. The object masks are depicted as highlighted colours in Car, Train and Motorcycle. Better view in color. (Color figure online)

4.4   Qualitative Results

We show qualitative results in Fig. 5 for the task of auditory semantic prediction. We also show the detected background image and the ground truth segmentation mask. The last three rows are devoted to the results in rainy, foggy and night conditions respectively. We observe that our model remains robust to adverse visual condition. Of course, if the rain is too big, it will become an adverse auditory condition. In Fig. 6, we show two results by the multi-task setting. We show the predictions and the ground truths for all the three tasks. It can be seen that the major sound-making objects can be properly detected. We see that the depth results reflect the general layout of the scene, though they are still coarser than the results of a vision system. This is valuable given the fact that binaural sounds are of very low-resolution – two channels in total.

4.5   Limitations and Future Work

Our teacher vision models are pretrained on perspective images and we apply them to panoramic images. This is due to the limited amount of datasets and annotations for panoramic images. We would like to note that most of the interesting objects appear in the equatorial region. For that region, the distortion of the panoramic images is less severe, and hence the results are less affected. In future, we plan to incorporate a 3D LiDAR to our sensor setup to get accurate ground-truth for depth prediction. We work with 3 object classes for semantic prediction task. This is because some classes are very rare in the middle-sized dataset. Due to this challenge, the recent works [19,29] only deals with one class in real world environment.

5   Conclusion

The work develops an approach to predict the semantic labels of sound-making objects in a panoramic image frame, given binaural sounds of the scene alone. To enhance this task, two auxiliary tasks are proposed – dense depth prediction of the scene and a novel task of spatial sound super-resolution. All the three tasks are also formulated as multi-task learning and is trained in an end-to-end fashion. This work has also proposed a novel dataset Omni Auditory Perception dataset. Extensive experiments have shown that 1) the proposed method achieves good results for all the three tasks; 2) the three tasks are mutually beneficial and 3) the number and orientations of microphones are both important.

Acknowledgement. This work is funded by Toyota Motor Europe via the research project TRACE-Zurich. We would like to thank Danda Pani Paudel, Suryansh Kumar and Vaishakh Patil for helpful discussions.

References

1. Computational auditory scene analysis: Comput. Speech Lang. **8**(4), 297–336 (1994)
2. Albanie, S., Nagrani, A., Vedaldi, A., Zisserman, A.: Emotion recognition in speech using cross-modal transfer in the wild. In: ACM Multimedia (2018)
3. Antonacci, F., et al.: Inference of room geometry from acoustic impulse responses. IEEE Trans. Audio Speech Lang Process. **20**(10), 2683–2695 (2012)
4. Arandjelovic, R., Zisserman, A.: Look, listen and learn. In: The IEEE International Conference on Computer Vision (ICCV) (2017)
5. Arandjelović, R., Zisserman, A.: Objects that sound. In: Proceedings of the European conference on computer vision (ECCV) (2018)
6. Argentieri, S., Danès, P., Souères, P.: A survey on sound source localization in robotics: from binaural to array processing methods. Comput. Speech Lang. **34**(1), 87–112 (2015)
7. Aytar, Y., Vondrick, C., Torralba, A.: SoundNet: learning sound representations from unlabeled video. In: Advances in Neural Information Processing Systems (NIPS) (2016)
8. Balajee Vasudevan, A., Dai, D., Van Gool, L.: Object referring in visual scene with spoken language. In: Winter Conference on Applications of Computer Vision (WACV) (2018)

9. Barzelay, Z., Schechner, Y.Y.: Harmony in motion. In: IEEE Conference on Computer Vision and Pattern Recognition (CVPR) (2007)

10. Brutzer, S., Höferlin, B., Heidemann, G.: Evaluation of background subtraction techniques for video surveillance. In: The IEEE Conference on Computer Vision and Pattern Recognition (CVPR) (2011)

11. Chen, L.C., Papandreou, G., Kokkinos, I., Murphy, K., Yuille, A.L.: DeepLab: semantic image segmentation with deep convolutional nets, atrous convolution, and fully connected CRFs. IEEE Trans. Pattern Anal. Mach. Intell. (TPAMI) **40**(4), 834–848 (2017)

12. Chen, L.-C., Zhu, Y., Papandreou, G., Schroff, F., Adam, H.: Encoder-decoder with atrous separable convolution for semantic image segmentation. In: Ferrari, V., Hebert, M., Sminchisescu, C., Weiss, Y. (eds.) ECCV 2018. LNCS, vol. 11211, pp. 833–851. Springer, Cham (2018). https://doi.org/10.1007/978-3-030-01234-2_49

13. Cordts, M., et al.: The cityscapes dataset for semantic urban scene understanding. In: Proceedings of the IEEE Conference on Computer Vision and Pattern Recognition (CVPR) (2016)

14. Delmerico, J., et al.: The current state and future outlook of rescue robotics. J. Field Robot. **36**(7), 1171–1191 (2019)

15. Deruyttere, T., Vandenhende, S., Grujicic, D., Van Gool, L., Moens, M.F.: Talk2Car: taking control of your self-driving car. In: EMNLP-IJCNLP (2019)

16. Dokmanic, I., Parhizkar, R., Walther, A., Lu, Y.M., Vetterli, M.: Acoustic echoes reveal room shape. Proc. Nat. Acad. Sci. **110**(30), 12186–12191 (2013)

17. Fazenda, B., Atmoko, H., Gu, F., Guan, L., Ball, A.: Acoustic based safety emergency vehicle detection for intelligent transport systems. In: ICCAS-SICE (2009)

18. Fendrich, R.: The merging of the senses. J. Cogn. Neurosci. **5**(3), 373–374 (1993)

19. Gan, C., Zhao, H., Chen, P., Cox, D., Torralba, A.: Self-supervised moving vehicle tracking with stereo sound. In: The IEEE International Conference on Computer Vision (ICCV) (2019)

20. Gao, R., Feris, R., Grauman, K.: Learning to separate object sounds by watching unlabeled video. In: Ferrari, V., Hebert, M., Sminchisescu, C., Weiss, Y. (eds.) ECCV 2018. LNCS, vol. 11207, pp. 36–54. Springer, Cham (2018). https://doi.org/10.1007/978-3-030-01219-9_3

21. Gao, R., Grauman, K.: 2.5 D visual sound. In: Proceedings of the IEEE Conference on Computer Vision and Pattern Recognition (CVPR), pp. 324–333 (2019)

22. Gao, R., Grauman, K.: Co-separating sounds of visual objects. In: The IEEE International Conference on Computer Vision (ICCV), October 2019

23. Gaver, W.W.: What in the world do we hear?: an ecological approach to auditory event perception. Ecol. Psychol. **5**(1), 1–29 (1993)

24. Geiger, A., Lenz, P., Stiller, C., Urtasun, R.: Vision meets robotics: the kitti dataset. Int. J. Robot. Res. (IJRR) **32**(11), 1231–1237 (2013)

25. Godard, C., Aodha, O.M., Firman, M., Brostow, G.J.: Digging into self-supervised monocular depth estimation. In: Proceedings of the IEEE International Conference on Computer Vision (CVPR), pp. 3828–3838 (2019)

26. Griffin, D., Lim, J.: Signal estimation from modified short-time Fourier transform. In: IEEE International Conference on Acoustics, Speech, and Signal Processing (ICASSP), vol. 8, pp. 804–807 (1983)

27. Hecker, S., Dai, D., Van Gool, L.: End-to-end learning of driving models with surround-view cameras and route planners. In: Ferrari, V., Hebert, M., Sminchisescu, C., Weiss, Y. (eds.) ECCV 2018. LNCS, vol. 11211, pp. 449–468. Springer, Cham (2018). https://doi.org/10.1007/978-3-030-01234-2_27

28. Huang, W., Alem, L., Livingston, M.A.: Human factors in augmented reality environments. Springer, New York (2012). https://doi.org/10.1007/978-1-4614-4205-9

29. Irie, G., et al.: Seeing through sounds: predicting visual semantic segmentation results from multichannel audio signals. In: IEEE International Conference on Acoustics, Speech and Signal Processing (ICASSP), pp. 3961–3964 (2019)

30. Kim, H., Remaggi, L., Jackson, P.J., Fazi, F.M., Hilton, A.: 3D room geometry reconstruction using audio-visual sensors. In: International Conference on 3D Vision (3DV), pp. 621–629 (2017)

31. Kingma, D.P., Ba, J.: Adam: a method for stochastic optimization. arXiv preprint arXiv:1412.6980 (2014)

32. Klee, U., Gehrig, T., McDonough, J.: Kalman filters for time delay of arrival-based source localization. EURASIP J. Adv. Signal Process. **2006**(1), 012378 (2006)

33. Li, D., Langlois, T.R., Zheng, C.: Scene-aware audio for 360° videos. ACM Trans. Graph **37**(4), 12 (2018)

34. Marchegiani, L., Posner, I.: Leveraging the urban soundscape: auditory perception for smart vehicles. In: IEEE International Conference on Robotics and Automation (ICRA) (2017)

35. McAnally, K.I., Martin, R.L.: Sound localization with head movement: implications for 3-d audio displays. Front. Neurosci. **8**, 210 (2014)

36. Mousavian, A., Pirsiavash, H., Košecká, J.: Joint semantic segmentation and depth estimation with deep convolutional networks. In: International Conference on 3D Vision (3DV), pp. 611–619 (2016)

37. Owens, A., Efros, A.A.: Audio-visual scene analysis with self-supervised multisensory features. In: Ferrari, V., Hebert, M., Sminchisescu, C., Weiss, Y. (eds.) ECCV 2018. LNCS, vol. 11210, pp. 639–658. Springer, Cham (2018). https://doi.org/10.1007/978-3-030-01231-1_39

38. Owens, A., Isola, P., McDermott, J., Torralba, A., Adelson, E.H., Freeman, W.T.: Visually indicated sounds. In: The IEEE Conference on Computer Vision and Pattern Recognition (CVPR) (2016)

39. Owens, A., Wu, J., McDermott, J.H., Freeman, W.T., Torralba, A.: Ambient sound provides supervision for visual learning. In: Leibe, B., Matas, J., Sebe, N., Welling, M. (eds.) ECCV 2016. LNCS, vol. 9905, pp. 801–816. Springer, Cham (2016). https://doi.org/10.1007/978-3-319-46448-0_48

40. Morgado, P., Vasconcelos, N., Langlois, T., Wang, O.: Self-supervised generation of spatial audio for 360 deg video. In: Neural Information Processing Systems (NIPS) (2018)

41. Rascon, C., Meza, I.: Localization of sound sources in robotics: a review. Robot. Auton. Syst. **96**, 184–210 (2017)

42. Rosenblum, L.D., Gordon, M.S., Jarquin, L.: Echolocating distance by moving and stationary listeners. Ecol. Psychol. **12**(3), 181–206 (2000)

43. Russell, B.C., Torralba, A., Murphy, K.P., Freeman, W.T.: LabelMe: a database and web-based tool for image annotation. Int. J. Comput. Vis. (IJCV) **77**(1–3), 157–173 (2008)

44. Salamon, J., Jacoby, C., Bello, J.P.: A dataset and taxonomy for urban sound research. In: ACM Multimedia (2014)

45. Saxena, A., Ng, A.Y.: Learning sound location from a single microphone. In: IEEE International Conference on Robotics and Automation (ICRA) (2009)

46. Senocak, A., Oh, T.H., Kim, J., Yang, M.H., So Kweon, I.: Learning to localize sound source in visual scenes. In: IEEE Conference on Computer Vision and Pattern Recognition (CVPR) (2018)

47. Simeoni, M.M.J.A., Kashani, S., Hurley, P., Vetterli, M.: DeepWave: a recurrent neural-network for real-time acoustic imaging. In: Neural Information Processing Systems (NIPS), p. 38 (2019)
48. Thurlow, W.R., Mangels, J.W., Runge, P.S.: Head movements during sound localizationtd. J. Acoust. Soc. Am. **42**(2), 489–493 (1967)
49. Tian, Y., Shi, J., Li, B., Duan, Z., Xu, C.: Audio-visual event localization in unconstrained videos. In: Ferrari, V., Hebert, M., Sminchisescu, C., Weiss, Y. (eds.) ECCV 2018. LNCS, vol. 11206, pp. 252–268. Springer, Cham (2018). https://doi.org/10.1007/978-3-030-01216-8_16
50. Tiete, J., Domínguez, F., da Silva, B., Segers, L., Steenhaut, K., Touhafi, A.: SoundCompass: a distributed MEMS microphone array-based sensor for sound source localization. Sensors **14**(2), 1918–1949 (2014)
51. Urmson, C., et al.: Autonomous driving in urban environments: boss and the urban challenge. J. Field Robot. **25**(8), 425–466 (2008). Special Issue on the 2007 DARPA Urban Challenge, Part I
52. Vandenhende, S., Georgoulis, S., Proesmans, M., Dai, D., Van Gool, L.: Revisiting multi-task learning in the deep learning era. arXiv (2020)
53. Wallach, H.: The role of head movements and vestibular and visual cues in sound localization. J. Exp. Psychol. **27**(4), 339 (1940)
54. Ye, M., Zhang, Y., Yang, R., Manocha, D.: 3D reconstruction in the presence of glasses by acoustic and stereo fusion. In: IEEE Conference on Computer Vision and Pattern Recognition (CVPR) (2015)
55. Zhao, H., Gan, C., Ma, W.C., Torralba, A.: The sound of motions. In: The IEEE International Conference on Computer Vision (ICCV) (2019)
56. Zhao, H., Gan, C., Rouditchenko, A., Vondrick, C., McDermott, J., Torralba, A.: The sound of pixels. In: Ferrari, V., Hebert, M., Sminchisescu, C., Weiss, Y. (eds.) ECCV 2018. LNCS, vol. 11205, pp. 587–604. Springer, Cham (2018). https://doi.org/10.1007/978-3-030-01246-5_35

Neural Object Learning for 6D Pose Estimation Using a Few Cluttered Images

Kiru Park$^{(\boxtimes)}$ ⓘ, Timothy Patten ⓘ, and Markus Vincze ⓘ

Vision for Robotics Group, Automation and Control Institute,
TU Wien, Vienna, Austria
{park,patten,vincze}@acin.tuwien.ac.at

Abstract. Recent methods for 6D pose estimation of objects assume either textured 3D models or real images that cover the entire range of target poses. However, it is difficult to obtain textured 3D models and annotate the poses of objects in real scenarios. This paper proposes a method, Neural Object Learning (NOL), that creates synthetic images of objects in arbitrary poses by combining only a few observations from cluttered images. A novel refinement step is proposed to align inaccurate poses of objects in source images, which results in better quality images. Evaluations performed on two public datasets show that the rendered images created by NOL lead to state-of-the-art performance in comparison to methods that use 13 times the number of real images. Evaluations on our new dataset show multiple objects can be trained and recognized simultaneously using a sequence of a fixed scene.

Keywords: 6D pose estimation · Object learning · Object model · Object modeling · Differentiable rendering · Object recognition

1 Introduction

The pose of an object is important information as it enables augmentation reality applications by displaying contents in correct locations and robots to grasp and place an object precisely. Recently, learned features from color images using Convolutional Neural Networks (CNN) have increased performance of object recognition tasks [9,10] including pose estimation [21,24,27,31,34,35,45]. These methods have achieved the best performance on benchmarks for pose estimation using household objects [13,19,43].

However, methods using CNNs require a large number of training images to cover potential view points of target objects in test environments. There have been two approaches to create training images for pose estimation methods: rendering synthetic images using textured 3D models [21,34] or cropping real

---

Electronic supplementary material The online version of this chapter (https:// doi.org/10.1007/978-3-030-58548-8_38) contains supplementary material, which is available to authorized users.

A. Vedaldi et al. (Eds.): ECCV 2020, LNCS 12349, pp. 656–673, 2020.
https://doi.org/10.1007/978-3-030-58548-8_38

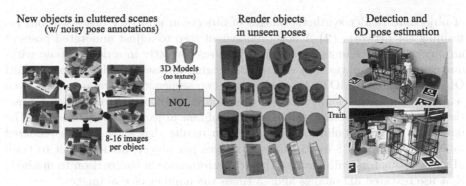

New objects in cluttered scenes
(w/ noisy pose annotations)

Render objects
in unseen poses

Detection and
6D pose estimation

Fig. 1. The proposed method, NOL, uses a few cluttered scenes that consist of new target objects of which to render images in arbitrary poses. Rendered images are used to train pipelines for 6D pose estimation and 2D detection of objects

images and pasting them on random background images [2,27,31]. Recently, state-of-the-art performance has been accomplished by using both synthetic and real images [23,24,28]. Unfortunately, both textured 3D models and large numbers of real images are difficult to obtain in the real world. Textured 3D models included in pose benchmarks are created with special scanning devices, such as the BigBIRD Object Scanning Rig [3] or a commercial 3D scanner [19]. During this scanning operation, objects are usually placed alone with a simple background and consistent lighting condition. Precise camera poses are obtained using visible markers or multiple cameras with known extrinsic parameters. In the real environment, however, target objects are placed in a cluttered scene, are often occluded by other objects and the camera pose is imprecise. Furthermore, the manual annotation of 6D poses is difficult and time consuming because it requires the association between 3D points and 2D image pixels to be known. Thus, it is beneficial to minimize the number of images with pose annotations that are required to train 6D pose estimators. This motivates us to develop a new approach to create images of objects from arbitrary view points using a small number of cluttered images for the purpose of training object detectors and pose estimators (Fig. 1).

In this paper, we propose Neural Object Learning (NOL), a method to synthesize images of an object in arbitrary poses using a few cluttered images with pose annotations and a non-textured 3D model of the object. For new objects, NOL requires 3D models and cluttered color images (less than 16 images in our evaluations) with pose annotations to map color information to vertices. To overcome pose annotation errors of source images, a novel refinement step is proposed to adjust poses of objects in the source images. Evaluation results show that images created by NOL are sufficient to train CNN-based pose estimation methods and achieves state-of-the-art performance.

In summary, this paper provides the following contributions: **(1)** Neural Object Learning that uses non-textured 3D models and a few cluttered images of objects to render synthetic images of objects in arbitrary poses without re-training the network. **(2)** A novel refinement step to adjust annotated poses of an object in source images to project features correctly in a desired pose without using depth images. **(3)** A new challenging dataset, Single sequence-Multi Objects Training (SMOT), that consists of two sequences for training and eleven sequences for evaluation, collected by a mobile robot, which represents a practical scenario of collecting training images of new objects in the real world. The dataset is available online[1]. **(4)** Evaluation results that show images rendered by NOL, which uses 8 to 16 cluttered images per object, are sufficient to train 6D pose estimators with state-of-the-art performance in comparison to methods that use textured 3D models and 13 times the number of real images.

The remainder of the paper is organized as follows. Related work is briefly introduced in Sect. 2. The detailed description of NOL and the rendering process are described in Sect. 3. We report results of evaluations in Sect. 4 and analyze effects of each component in Sect. 5. Lastly, we conclude the paper in Sect. 6.

2   Related Work

In this section, we briefly review the types of training images used to train 6D pose estimation methods. The approaches that create a 3D model and map a texture of a novel object are discussed. The recent achievements of differentiable rendering pipelines are also introduced.

Training Samples for 6D Pose Estimation. Previous work using CNNs requires a large number of training images of an object that covers a range of poses in test scenes sufficiently. Since it is difficult to annotate 6D poses of objects manually, synthetic training images are created using a textured 3D model of an object [21]. However, it is difficult to obtain a 3D model with high quality texture from the real world without a special device, such as the BigBIRD Object Scanning Rig [3], since precise pose information is required to correctly align texture images from different views. Using synthetic data introduces the domain gap between synthetic and real images, which should be specially treated with domain adaptation techniques [45]. It is possible to use only approximately 200 real images and apply various augmentation methods to successfully train pose estimation pipelines [2,27,31]. However, the performance highly depends on the range of poses in the real training samples. As discussed in [27], the limited coverage of poses in training images causes inaccurate results for novel poses. To overcome this limitation, both real images and synthetic images are used for training [23,24,28,45], which currently achieves state-of-the-art performance. The advantage of using both sources is that synthetic images supplement images for novel poses that are not observed in real images while real images regularize the network from over-fitting to synthetic images. However, both textured 3D models and more than 200 real images with pose annotations are difficult to obtain from the real world. Furthermore, textured 3D models in public datasets

---

[1] https://www.acin.tuwien.ac.at/en/vision-for-robotics/software-tools/smot/.

are captured separately from constrained environments [3,13,15,19] such as single objects with a simple background and precise camera pose localization tools. However, this well-constrained setup is difficult to replicate in real scenarios, e.g., a target object on a table is often occluded by other objects, camera poses are noisy without manual adjustments, and lighting conditions are not consistent. Thus, it is challenging to derive training images of new objects from cluttered scenes.

3D Object Modeling and Multi-view Texturing. RGB-D images have been used to build 3D models by presenting an object in front of a fixed camera while rotating a turn table [30], manipulating the object using a robot end-effector [22] or human hands [41]. Alternatively, a mobile robot is used to actively move a camera to build a model of a fixed object in [7]. Even though these methods produce good 3D models in terms of geometry, textures are not optimized or even explicitly considered. Depth images are also required to align different views. On the other hand, it is possible to map multiple images from different views to 3D mesh models using camera pose information [6,39]. These approaches produce 3D models with high-quality textures since their optimization tries to assign continuous source images to neighboring pixels. However, these methods requires depth images for correcting pose errors, which causes misalignment of color values and disconnected boundaries when different source images are not correctly aligned. Image based rendering (IBR) has been used to complete a large scene by in-painting occluded area using multiple images from different view points [29,33,37,42]. These methods re-project source images to a target image using the relative poses of view points. Then, projected images from different views are integrated with a weighted summation or an optimization based on different objective functions. However, IBR methods are designed to complete large-scale scenes and suffer from noisy estimation of the camera and object poses, which causes blurry or misaligned images.

Differentiable Rendering. The recent development of differentiable rendering pipelines enables the rendering process to be included during network training [12,20,36]. Therefore, the relationship between the 3D locations of each vertex and UV coordinates for textures are directly associated with pixel values of 2D rendered images, which have been used to create 3D meshes from a single 2D image [12,20]. Furthermore, it is possible to render trainable features of projected vertices, which can be trained to minimize loss functions defined in a 2D image space.

The purpose of NOL is to generate synthetic images to train CNN-based pose estimators for new objects while minimizing the effort for obtaining training data from real applications. The knowledge of 3D representation and a few observations of objects are usually sufficient for humans to recognize objects in a new environment. Likewise, NOL composes appearances of objects in arbitrary poses using 3D models and a few cluttered and unconstrained images without using depth images, which is sufficient to train a pose estimator and achieve state-of-the-art performance.

3   Neural Object Learning

The objective of NOL is to create an image X^D of an object in a desired pose T^D using K source images, $\{I^1, I^k...I^K\}$, with pose annotations, $\{T^1, T^k...T^K\}$, and object masks that indicate whether each pixel belongs to the object or not in a source image, $\{M^1, M^k...M^K\}$. We do not assume T^k or M^k to be accurate, which is common if the source images are collected without strong supervision such as marker-based localization or human annotation.

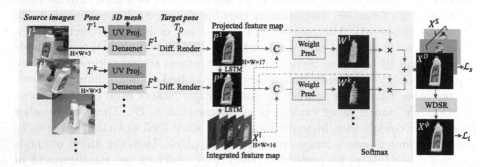

Fig. 2. An overview of the NOL architecture. X^D is used as a rendered output while X^ϕ is used to compute the image loss for training

3.1   Network Architecture

Fig. 2 depicts an overview of the network architecture. Firstly, source images are encoded and projected to compute an integrated feature map in a target pose. Secondly, the weighted sum of projected feature maps are computed by predicting weight maps. A decoder block produces a decoded image that is used to compute the image loss.

Integrated Feature Maps. Each source image, $I^k \in H \times W \times 3$, is encoded with a backbone network, Densenet-121 [17], to build feature pyramids using the outputs of the first four blocks. Each feature map from the pyramids is processed with a convolutional layer with 3×3 kernels to reduce the number of an output channel of each block to 4, 3, 3, and 3. Each feature map is resized to the size of the original input using bi-linear interpolation. In addition to the feature map with 13 channels, original color images (3 channels) and face angles with respect to camera views (one channel) are concatenated. As a result, the encoded feature map F^k of each input image I^k has 17 channels. The UV coordinates of each vertex are computed using 2D projected locations of each visible vertex in an input pose T^k. These UV coordinates are then projected to the target pose T^D using a differentiable renderer proposed in [12]. Feature values of each pixel in a projected feature map P^k are computed using bi-linear interpolation of surrounding feature values obtained from corresponding pixels from the encoded feature map F^k, which is similar to rendering an object with a separate texture image. The projected feature maps $P^{k \in K}$ are compiled by convolutional Long-Term and Short-Term memory (LSTM) layers to compute the integrated feature map X^I with 16 output channels at the same resolution of the projected feature maps. This LSTM layer enables the network to learn how to extract valuable pixels from different source images while ignoring outlier pixels caused by pose errors. It is also possible to use a different number of source images without changing the network architecture.

Weight Prediction Block. The integrated feature map X^I is concatenated with each projected feature map P^k to compute a corresponding weight map W^k in the weight prediction block, which implicitly encodes distances between P^k and X^I per pixel. The resulting weight maps $W^{k \in K}$ are normalized with the *Softmax* activation over K projected images. Therefore, the summation of the weighted maps over $W^{k \in K}$ is normalized for each pixel while keeping strong weights on pixels that have remarkably higher weights than others. The weighted sum of projected feature maps using the predicted weight maps produces the weighted feature map X^S. Since the first three channels of P^k represent projected color values from source images, the first three channels of X^S are a color image, which is referred to as the weighted rendering X^D and the output image of NOL during inference.

Decoder Block. Since X^D is obtained by the weighted summation of projected images, color values of pixels X^D are limited to the color range of projected pixels. However, when training the network, the color levels of the source images can be biased by applying randomized color augmentations while maintaining the original colors for the target image. This causes the weight prediction block to be over-penalized even though color levels of X^D are well balanced for given source images. This motivates us to add a module to compensate for these biased errors implicitly during training. An architecture used for the image super-resolution task, WDSR [44], is employed as a decoder block to predict the decoded rendering X^ϕ in order to compute the losses during training. A detailed analysis regarding the role of the decoder is presented in Sect. 5.

3.2 Training

The objective of training consists of two components. The first component, the *image loss*, renders a correct image $\mathcal{L}_i$ in a target pose. The second component, the *smooth loss*, minimizes the high frequency noise of the resulting images $\mathcal{L}_s$.

Image Loss. The image loss $\mathcal{L}_i$ computes the difference between a target image X^{GT} in a target pose and the decoded output X^ϕ. In addition to the standard L1 distance of each color channel, the feature reconstruction loss [18] is applied to guide the predicted images to be perceptually similar to the target image as formulated by

$$\mathcal{L}_i = \frac{1}{M^D} \sum_{p \in M^D} \lambda_i |X_p^\phi - X_p^{GT}|_1 + \lambda_f |\psi(X_p^\phi) - \psi(X_p^{GT})|_1, \tag{1}$$

where M^D is a binary mask that indicates whether each pixel has at least a valid projected value from any input $I^{k \in K}$, and $\psi(\cdot)$ denotes outputs of a backbone network with respect to the image. The outputs of the first two blocks of DenseNet [40] are used for the feature reconstruction loss. The parameters λ_i and λ_f are used to balance the losses.

Smooth Loss. Even if the objective function in Eq. 1 guides the network to reconstruct the image accurately, the penalty is not strong when the computed image has high frequency noise. This is the reason why IBR and image inpainting methods [29,37,42] usually employ a smooth term. This minimizes the gradient changes of neighboring pixels even if pixel values are obtained from different source images. Similarly, we add a loss function to ensure smooth transitions for neighboring pixels in terms of color values as well as encoded feature values. This is formulated as

$$\mathcal{L}_s = \frac{\lambda_s}{M^D} \sum_{p \in M^D} \nabla^2 X_p^S. \tag{2}$$

The loss function creates a penalty when the gradients of color and feature values of each pixel are inconsistent with those of neighboring pixels. In contrast to the image loss, the weighted feature map X^S is used directly instead of using the decoded output X^ϕ. Thus, the weight prediction block is strongly penalized when producing high frequency changes in the predicted weight maps $W^{k \in K}$ and the weighted feature map X^S.

Training using Synthetic Images with Pose Errors. Synthetic images are created to train the NOL network. 3D models from the YCB-Video dataset [43] are used while ignoring original textures and instead applying randomly sampled images from MS-COCO [26] as textures. After sampling a 3D model and a texture image, 10 images are rendered as a batch set in different poses with random background images. During training, one image from a batch set is chosen as a target image X^{GT} and its pose is set to a desired pose T^D, and the other K images are assigned as input images $I^{k \in K}$. To simulate different lighting conditions, color augmentations are applied to the input images while no augmentation is applied to the target image. Pose errors are also simulated by applying random perturbations to the actual poses $T^{k \in K}$ of the input images during training. As a result of the perturbations, vertices are projected to wrong 2D locations, which produces wrong UV coordinates per vertex and outlier pixels in projected feature maps at a desired pose. This pose augmentation forces the network to be robust to pose errors while attempting to predict an accurate image in the target pose. A total of 1,000 training sets, consisting of 10 images per set, are rendered for training. The same weights are used to render objects in all evaluations in the paper after training for 35 epochs. Detailed parameters used for data augmentation are listed in the supplementary material.

3.3 Gradient Based Pose Refinement and Rendering

The error in an input pose T^k causes crucial outlier pixels in the projected feature map P^k at the desired pose. Figure 3 shows an example of wrong pixels in the projected feature map obtained from the ground plane (blue) in the source image due to the error of the initial pose $T^k_{t=0}$. As discussed in Sec. 2, the differentiable renderer enables derivatives of 3D vertices of a 3D model to be computed with respect to the error defined in 2D space. Since 3D locations of vertices and UV coordinates in the desired pose are derived by matrix multiplications, which is differentiable, the gradient of each input pose T^k can be derived to specify a direction that decreases the difference between a desired feature map and each projected feature map P^k. The first prediction of NOL, $X^S_{t_0}$, without refinement is used as an initial desired target. The goal of the refinement step is to minimize the error, E^k, between the initial target and each projected feature map P^k. In every iteration, the partial derivative of the projection error E^k with respect to each input pose $T^k_{t=t_i}$ is computed by

$$\Delta T^k_{t_i} = \frac{\partial E^k}{\partial T^k_{t_i}} = \frac{\partial |X^S_{t_0} - P^k_{t_i}|}{\partial T^k_{t_i}}, \tag{3}$$

and the input pose at the next iteration $T^k_{t_i+1}$ is updated with a learning step δ, i.e. $T^k_{t_i+1} = T^k_{t_i} - \delta \Delta T^k_{t_i}$. In our implementation, translation components in T^k are directly updated using $\Delta T^k_{t_i}$. On the other hand, updated values for rotation components, $\mathbb{R}^{3\times3}$, do not satisfy constraints for the special orthogonal group, $SO(3)$. Thus, the rotation component of $\Delta T^k_{t_i}$ is updated in the Euler representation and converted back to the rotation matrix. As depicted in Fig. 3, the iterations of the refinement step correctly remove the pose error so that the projected image no longer contains pixels from the background, which decreases blur and mismatched boundaries in the final renderings. After refining every input pose $T^k_{t=t_0}$ until the error does not decrease or the number of iterations exceeds 50, the final output X^D is predicted using the refined poses $T^k_{t=t_f}$.

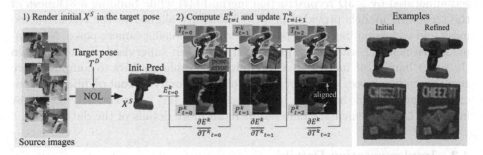

Fig. 3. An overview of the proposed pose refinement process and example results. The partial derivative of the projection error E^k is used to update each input pose T^k

4   Evaluation

This section presents the evaluation of the proposed NOL approach in relation to the task of 6D object pose estimation. We introduce datasets used in the evaluation and provide implementation details of NOL. The evaluation results show that the quality of NOL images are sufficient for pose estimation and leads to outperforming other methods trained with synthetic images using textured 3D models or real images.

4.1   Datasets

Three datasets are used for evaluation: LineMOD [13], LineMOD-Occlusion (LineMOD-Occ) [1], and a novel dataset created to reflect challenges of real environments. LineMOD and LineMOD-Occ have been used as standard benchmarks for 6D pose estimation of objects. LineMOD provides textured 3D models and 13 test sequences that have an annotated object per image. The 3D models are created by placing each object alone on a plane and performing a voxel-based 3D reconstruction [13]. LineMOD-Occ is created by additionally annotating eight objects presented in a test sequence in LineMOD. Previous works reporting results on these dataset have used either synthetic images using given 3D models [21,34] or 15% real images obtained from test sequences (183 images per object) [2,27,31,35] for training. In contrast to previous work, images created by NOL are used to train both a pose estimator and a 2D detector.

SMOT. A new dataset, Single sequence-Multi Objects Training, is created to reflect real noise when training images are collected from a real scenario, i.e. a mobile robot with a RGB-D camera collects a sequence of frames while driving around a table to learning multiple objects and tries to recognize objects in different locations. The dataset consists of two training sequences and eleven test sequences using eight target objects sampled from the YCB-Video [43] dataset. Two training sequences, that include four target objects per sequence, are collected by following trajectories around a small table. Camera poses of frames are self-annotated by a 3D reconstruction method [47] while building a 3D mesh of the scene. 3D models provided in YCB-Video are aligned to the reconstructed mesh and corresponding object poses are computed using camera poses. No manual adjustment is performed to preserve errors of self-supervised annotations. On the other hand, test images are collected with visible markers to compute more accurate camera poses while moving the robot manually in front of different types of tables and a bookshelf. As a result, each object has approximately 2,100 test images. The supplementary material includes more details of the dataset.

4.2   Implementation Details

For the NOL network, the resolution of input and target images are set to 256×256. A number of source images, K, is set to 8 for training and 6 for inference.

The loss weights are set to, $\lambda_i = 5$, $\lambda_f = 10$, and $\lambda_s = 1$. Detail parameters are reported in the supplementary material.

Sampling of Source Images. To render NOL images for training a pose estimator, source images are sampled from the training sequences of a dataset. For LineMOD and LineMOD-Occ, a maximum of 16 images per object are sampled from the same training splits of real images used in previous work [2,27,31,35]. Since objects are fully visible in the training set, images are simply sampled using pose annotations. In each sampling iteration, an image is randomly sampled and images that have similar poses (less than 300 mm translation and 45° rotation), are removed. The sampling is terminated when no more images remain. In contrast to LineMOD and LineMOD-Occ, the visibility of each object varies in the training set of SMOT. In order to minimize the number of source images, a frame with the highest value is selected at each sampling iteration by counting the number of visible vertices that have not been observed in the previously sampled frames. The sampling iteration is terminated when no frame adds additional observed vertices.

Rendering NOL Images. Each target object is rendered using NOL in uniformly sampled poses defined over an upper-hemisphere for every 5° for both azimuth and elevation. For each target pose, 6 images are chosen from sampled images using the same image sampling procedure while limiting the target vertices to visible vertices in the pose. As a result, 1296 images are rendered. The in-plane rotation is applied to rendered images by rotating them from −45° to 45° for every 15°. For LineMOD and LineMOD-Occ, synthetic images are also rendered in the same sampled poses using given 3D models to train a pose estimator for comparison. Rendering takes approximately 3.6 s per image, which consists of 19 ms for the initial prediction, 3597 ms for the pose refinement with maximum 25 iterations, and 19 ms for the final prediction.

Training Recognizers. To show whether NOL images are sufficient to estimate poses of objects in arbitrary poses using a recent RGB-based pose estimation method, an official implementation of state of the art for 6D pose estimation using color images, Pix2Pose [27], is used. This method is one of the most recent methods [24,27,45] that predict objects' coordinate values per pixel. To increase the training speed and decrease the number of training parameters, the discriminator and the GAN loss are removed. All other aspects are kept the same except for the number of training iterations, which is set to approximately 14K because of the decreased number of trainable parameters. Resnet-50 [10] is used as a backbone for the encoder and weights are initialized with pre-trained weights on ImageNet [4]. Since the online augmentation of Pix2Pose uses cropped patches of objects on a black background, no further image is created for training. On the other hand, images that contain multiple objects are created to train a 2D detector by pasting masked objects onto random background images. A total of 150,000 training images are created and used to train Retinanet [25] with Resnet-50 backbone.

4.3 Metrics

The AD{D|I} score [13] is a common metric used to evaluate pose estimation. This metric computes the average distance of vertices between a ground truth pose and a predicted pose (ADD). For symmetric objects, distances to the nearest vertices are used (ADI). The predicted pose is regarded as correct if the average distance is less than 10% of the diameter of each object.

The recent challenge for pose benchmark [14] proposes a new metric that consists of three different pose errors (VSD, MSSD, and MSPD) [16] and their average recall values over different thresholds. The metric is used to evaluate the performance on LineMOD-Occ because it is more suitable and also allows comparison with the results of the recent benchmark.

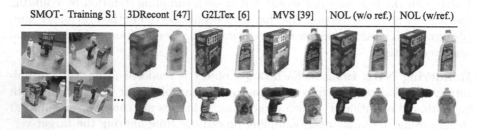

Fig. 4. Rendered results of SMOT objects using a training sequence. NOL successfully removes pixels from the background and other objects due to pose refinement

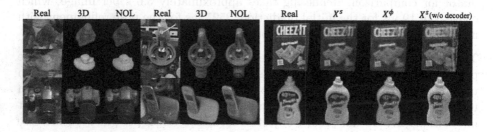

Fig. 5. Left: examples of rendered images of objects in LineMOD using NOL. Right: outputs of weighted renderings, decoded renderings, and weighted renderings after training without the decoder block

4.4 Quality of Rendered Images

Figure 4 shows the rendered images using a training sequence of SMOT. Renderings in *3DRecont* show object models extracted directly from a reconstructed 3D mesh of the training scene [47]. Both MVS [39] and G2LTex [6] use the same images sampled for rendering NOL images with the same pose annotations. Multi-view texturing methods create less blurry textures than NOL for planar surfaces since they try to map an image to a large area without combining pixels from other images. However, this induces misaligned results when the input poses are inaccurate even if depth images are used to optimize poses as in [6], e.g. doubled letters on the cheeze-it box. On the other hand, results of NOL after pose refinement (last column) removes these doubled textures by correcting pose errors using color images only, which is robust to depth registration errors. Furthermore, NOL successfully rejects outlier pixels from other objects and the background.

4.5 Pose Estimation: LineMOD

The left side of Table 1 shows the results when RGB images are used for pose estimation. Since no real image is directly cropped and used to train the pose estimator, the results are mainly compared against methods that use synthetic images only for training. The method trained with NOL images outperforms the same method trained with synthetic images using the given 3D models. This verifies that the quality of NOL images are more similar to appearances of real objects. The results of objects with metallic or shiny surfaces, e.g., *Camera*, *Phone*, and *Can*, show significant improvements against other results obtained with synthetic training images without any real observation. As depicted in

Table 1. Evaluation results on LineMOD. The ADD score is used except for *Eggbox* and *Glue* that use the ADI score

Types	Required data for training										
Texture		✓	✓	✓	✓	✓	–	✓	✓	–	✓
No. real images	14	–	–	–	–	183	14	–	–	183	183
GT 6D pose	✓	–	–	–	–	–	✓	–	–	✓	✓
Depth image	–	–	–	–	–	✓	–	–	–	✓	✓
Test type	RGB w/o refinement						RGB + ICP			RGB-D	
Training on	**NOL**	synthetic images				real+syn	**NOL**	syn		real+syn	
Method	[27]	[27]	[21]	[34]	[45]	[32]	[27]	[27]	[34]	[40]	[11]
Ape	35.4	10.0	2.6	4.0	**37.2**	19.8	**95.2**	92.7	20.6	92.3	**97.3**
Benchvise	55.6	13.4	15.1	20.9	66.8	**69.0**	**99.0**	90.4	64.3	93.2	**99.7**
Camera	37.5	4.4	6.1	30.5	24.2	**37.6**	**96.6**	77.9	63.2	94.4	**99.6**
Can	**65.5**	26.4	27.3	35.9	52.6	42.3	**97.6**	85.8	76.1	93.1	**99.5**
Cat	**38.1**	24.8	9.3	17.9	32.4	35.4	**98.6**	90.1	72.0	96.5	**99.8**
Driller	52.2	9.1	12.0	24.0	**66.6**	54.7	**98.0**	66.1	41.6	87.0	**99.3**
Duck	14.7	3.7	1.3	4.9	26.1	**29.4**	89.1	82.3	32.4	92.3	**98.2**
Eggbox	**93.7**	34.6	2.8	81.0	73.4	85.2	99.2	88.7	98.6	99.8	99.8
Glue	63.1	35.1	3.4	45.5	75.0	**77.8**	96.5	92.2	96.4	100	100
H.puncher	34.4	3.7	3.1	17.6	24.5	**36.0**	93.2	46.3	49.9	92.1	**99.9**
Iron	57.9	30.4	14.6	32.0	**85.0**	63.1	99.3	93.5	63.1	97.0	**99.7**
Lamp	54.2	6.7	11.4	**60.5**	57.3	75.1	96.6	39.3	91.7	95.3	**99.8**
Phone	41.8	13.8	9.7	33.8	29.1	**44.8**	92.8	79.1	71.0	92.8	**99.5**
Average	49.5	16.6	9.1	28.7	50.0	**51.6**	96.3	78.8	64.7	94.3	**99.4**

Fig. 5, NOL realizes the details of shiny and metallic materials by optimizing colors of each view separately. The performance is competitive to the best method that uses real color and depth images of objects for domain adaptation.

NOL images tend to contain noisy boundaries especially around the lower parts of objects where NOL mistakenly extracts pixels from the background table (see the bottom of *Phone* in Fig. 5). This limits the translation precision of predictions along the principle camera axis (z-axis). To decrease the translation errors, ICP refinement is applied to refine poses using depth images as reported on the right side of Table 1. The method trained with NOL images outperforms state-of-the-art trained with synthetic images. The result is competitive to state-of-the-art results in the last two columns even though the methods [11,40] use more than 13 times the number of real images for training.

4.6   Pose Estimation: LineMOD-Occ

The same models used in the LineMOD evaluation are used to test on LineMOD-Occ as reported in Table 2. Similar to the LineMOD evaluation, methods trained by synthetic images are mainly compared. The evaluation protocol used in the recent pose challenge [14] is applied with the same test target images. The result of [27] using synthetic images is obtained by re-training the network with Resnet-50 backbone, which performs better than the official result in the challenge [14].

The performance of this method is significantly improved by using images created by NOL for training with RGB inputs and with the inclusion of ICP refinement using depth images. Furthermore, using NOL images leads to the method outperforming state of the art using color images [24] and the best performing method on this dataset [38].

Table 2. Evaluation results on LineMOD-Occ. The results of other methods are cited from the last 6D pose challenge [14]

Type	RGB w/o refinement					RGB+ICP3D			Depth
Train source	**NOL**	Syn. using 3D models				**NOL**	Syn		3D model
Method	[27]	[27]	[34]	[24]	[45]	[27]	[27]	[34]	[38]
BOP score	**37.7**	20.0	14.6	37.4	16.9	**61.3**	45.3	23.7	58.2

Table 3. Evaluation results on the SMOT dataset

Type	RGB				RGB-D (ICP3D)			
3D Model	Precise			Recont	Precise			Recont
Train source	Real	G2Ltex	**NOL**	**NOL**	Real	G2Ltex	**NOL**	**NOL**
AD{D|I} score	25.0	25.7	**35.5**	22.5	86.5	82.0	**90.0**	80.6
mAP$_{IoU=50}$	88.3	90.2	**90.7**	73.9	–	–	–	–

4.7 Pose Estimation: SMOT

The pose estimator [27] and the 2D detection method [25] are trained using crops of entire real images where each object is visible more than 50%. This is an average of 364 images per objects. For G2Ltex and NOL, up to 16 images per object are sampled as explained in Sect. 4.2 to render training images.

Table 3 shows pose estimation and 2D detection results in terms of the AD{D|I} score and the mean Average Precision (mAP) [5]. The results using NOL images outperforms other methods using real images and models textured by G2Ltex for both RGB and RGB-D inputs. This is because real images do not fully cover target poses and objects are often occluded by other objects in training images. The comparison with G2LTex provides a quantitative verification regarding the better quality of renderings created by NOL using the same source images. The results denoted with *Recont* are obtained using reconstructed 3D models instead of precise 3D models for rendering. The performance drops significantly since the NOL images are noisier and blurrier due to geometrical errors of models. This indicates that precise 3D models are important for NOL to generate high-quality images.

5   Ablation Study

This section analyzes factors that influence the quality of NOL images. The perceptual similarity [46] is used to measure quality of generated images in comparison to the real images. We sample 10 test images per object in SMOT (80 images), render the objects at GT poses, and compare them with real images.

Components. Table 4 shows the most significant improvement comes from the decoder. The right side of Fig. 5 shows qualitative results of weighted renderings X^D, decoded renderings X^ϕ, and results after training the network without the decoder. As discussed in Sect. 3, the network trained with the decoder converges to produce X^D in a neutral color level as a reference image while the decoder absorbs over-penalized errors caused by randomly biased colors. The results denoted as *w/o LSTM* are derived by replacing the LSTM module with an simple average over projected features P^k. In this case, the results drops significantly since the LSTM module highlights valuable pixels among projected pixels. The refinement step consistently improves the image quality for all configurations.

Table 4. Perceptual similarity (smaller is better) of rendered images with different configurations

Setup	Components			Loss functions			
	All	w/o Decoder	w/o LSTM	$\mathcal{L}_1$ (RGB)	$\mathcal{L}_i$	$\mathcal{L}_i + \mathcal{L}_s$	$+\mathcal{L}_{GAN}$
w/o Ref	**0.181**	0.289	0.247	0.194	0.184	**0.181**	0.188
w/ Ref	**0.173**	0.279	0.241	0.184	0.177	**0.173**	0.183

Loss Functions. The best results are made with all proposed losses $\mathcal{L}_i+\mathcal{L}_s$. The perceptual loss in addition to the standard L1 loss significantly improves the performance by guiding the network to preserve perceptual details, like edges, with less blurry images while the smooth loss $\mathcal{L}_s$ additionally reduces the high-frequency noise. As the adversarial loss [8] provides better performance for image reconstruction tasks, the adversarial loss $\mathcal{L}_{GAN}$ is added to our loss function, which does not improve the result in our implementation.

6   Conclusion

This paper proposed a novel method that creates training images for pose estimators using a small number of cluttered images. To the best of our knowledge, this is the first attempt to learn multiple objects from a cluttered scene for 6D pose estimation, which minimizes the effort for recognizing a new object. Our code[2] and dataset are publicly available to motivate future research in this direction. The method can be further extended to optimize 3D models for reducing geometrical errors, which accomplishes the fully self-supervised learning of objects from cluttered scenes in real environments.

Acknowledgment. The research leading to these results has partially funded by the Austrian Science Fund (FWF) under grant agreement No. I3969-N30 (InDex) and the Austrian Research Promotion Agency (FFG) under grant agreement No. 879878 (K4R).

References

1. Brachmann, E., Krull, A., Michel, F., Gumhold, S., Shotton, J., Rother, C.: Learning 6D object pose estimation using 3D object coordinates. In: Fleet, D., Pajdla, T., Schiele, B., Tuytelaars, T. (eds.) ECCV 2014. LNCS, vol. 8690, pp. 536–551. Springer, Cham (2014). https://doi.org/10.1007/978-3-319-10605-2_35
2. Brachmann, E., Michel, F., Krull, A., Ying Yang, M., Gumhold, S., Rother, C.: Uncertainty-driven 6d pose estimation of objects and scenes from a single RGB image. In: Proceedings of the IEEE Conference on Computer Vision and Pattern Recognition (CVPR), pp. 3364–3372 (2016)
3. Calli, B., Walsman, A., Singh, A., Srinivasa, S.S., Abbeel, P., Dollar, A.M.: Benchmarking in manipulation research: using the Yale-CMU-Berkeley object and model set. IEEE Robot. Autom. Mag. (RAM) **22**, 36–52 (2015)
4. Deng, J., Dong, W., Socher, R., Li, L.J., Li, K., Fei-Fei, L.: ImageNet: a large-scale hierarchical image database. In: Proceedings of the IEEE Conference on Computer Vision and Pattern Recognition (CVPR), pp. 248–255 (2009)
5. Everingham, M., Van Gool, L., Williams, C.K., Winn, J., Zisserman, A.: The pascal visual object classes (VOC) challenge. Int. J. Comput. Vision (IJCV) **88**, 303–338 (2010). https://doi.org/10.1007/s11263-009-0275-4
6. Fu, Y., Yan, Q., Yang, L., Liao, J., Xiao, C.: Texture mapping for 3D reconstruction with RGB-D sensor. In: Proceedings of the IEEE Conference on Computer Vision and Pattern Recognition (CVPR), pp. 4645–4653 (2018)

[2] https://github.com/kirumang/NOL

7. Fäulhammer, T., et al.: Autonomous learning of object models on a mobile robot. IEEE Robot. Autom. Lett. (RA-L) **2**(1), 26–33 (2017). https://doi.org/10.1109/LRA.2016.2522086

8. Goodfellow, I., et al.: Generative adversarial nets. In: Advances in Neural Information Processing Systems 27 (NeurIPS), pp. 2672–2680. Curran Associates, Inc. (2014)

9. He, K., Gkioxari, G., Dollár, P., Girshick, R.: Mask R-CNN. In: Proceedings of the IEEE International Conference on Computer Vision (ICCV), pp. 2961–2969 (2017)

10. He, K., Zhang, X., Ren, S., Sun, J.: Deep residual learning for image recognition. In: Proceedings of the IEEE Conference on Computer Vision and Pattern Recognition (CVPR), pp. 770–778 (2016)

11. He, Y., Sun, W., Huang, H., Liu, J., Fan, H., Sun, J.: PVN3D: a deep pointwise 3D keypoints voting network for 6dof pose estimation. In: Proceedings of the IEEE Conference on Computer Vision and Pattern Recognition (CVPR), pp. 11632–11641 (2020)

12. Henderson, P., Ferrari, V.: Learning single-image 3D reconstruction by generative modelling of shape, pose and shading. Int. J. Comput. Vision **128**(4), 835–854 (2019). https://doi.org/10.1007/s11263-019-01219-8

13. Hinterstoisser, S., et al.: Model based training, detection and pose estimation of texture-less 3D objects in heavily cluttered scenes. In: Lee, K.M., Matsushita, Y., Rehg, J.M., Hu, Z. (eds.) ACCV 2012. LNCS, vol. 7724, pp. 548–562. Springer, Heidelberg (2013). https://doi.org/10.1007/978-3-642-37331-2_42

14. Hodaň, T., et al.: BOP: benchmark for 6D object pose estimation (2019). https://bop.felk.cvut.cz. Visited on 21 February 2020

15. Hodaň, T., Haluza, P., Obdržálek, Š., Matas, J., Lourakis, M., Zabulis, X.: T-LESS: an RGB-D dataset for 6D pose estimation of texture-less objects. In: IEEE Winter Conference on Applications of Computer Vision (WACV), pp. 880–888 (2017)

16. Hodaň, T., Matas, J., Obdržálek, Š.: On evaluation of 6D object pose estimation. In: Hua, G., Jégou, H. (eds.) ECCV 2016. LNCS, vol. 9915, pp. 606–619. Springer, Cham (2016). https://doi.org/10.1007/978-3-319-49409-8_52

17. Huang, G., Liu, Z., van der Maaten, L., Weinberger, K.Q.: Densely connected convolutional networks. In: Proceedings of the IEEE Conference on Computer Vision and Pattern Recognition (CVPR), pp. 4700–4708 (2017)

18. Johnson, J., Alahi, A., Fei-Fei, L.: Perceptual losses for real-time style transfer and super-resolution. In: Leibe, B., Matas, J., Sebe, N., Welling, M. (eds.) ECCV 2016. LNCS, vol. 9906, pp. 694–711. Springer, Cham (2016). https://doi.org/10.1007/978-3-319-46475-6_43

19. Kaskman, R., Zakharov, S., Shugurov, I., Ilic, S.: HomebrewedDB: RGB-D dataset for 6d pose estimation of 3d objects. In: The IEEE International Conference on Computer Vision Workshops (ICCVW) (2019)

20. Kato, H., Ushiku, Y., Harada, T.: Neural 3D mesh renderer. In: Proceedings of the IEEE Conference on Computer Vision and Pattern Recognition (CVPR), pp. 3907–3916 (2018)

21. Kehl, W., Manhardt, F., Tombari, F., Ilic, S., Navab, N.: SSD-6D: making RGB-based 3D detection and 6d pose estimation great again. In: Proceedings of the IEEE International Conference on Computer Vision (ICCV), pp. 1521–1529 (2017)

22. Krainin, M., Henry, P., Ren, X., Fox, D.: Manipulator and object tracking for in-hand 3D object modeling. Int. J. Robot. Res. (IJRR) **30**(11), 1311–1327 (2011)

23. Li, Y., Wang, G., Ji, X., Xiang, Y., Fox, D.: DeepIM: deep iterative matching for 6D pose estimation. Int. J. Comput. Vis. **128**(3), 657–678 (2019). https://doi.org/10.1007/s11263-019-01250-9
24. Li, Z., Wang, G., Ji, X.: CDPN: coordinates-based disentangled pose network for real-time RGB-based 6-DoF object pose estimation. In: Proceedings of the IEEE International Conference on Computer Vision (ICCV), pp. 7678–7687 (2019)
25. Lin, T.Y., Goyal, P., Girshick, R., He, K., Dollar, P.: Focal loss for dense object detection. In: Proceedings of the IEEE International Conference on Computer Vision (ICCV), pp. 2999–3007 (2017)
26. Lin, T.-Y., et al.: Microsoft COCO: common objects in context. In: Fleet, D., Pajdla, T., Schiele, B., Tuytelaars, T. (eds.) ECCV 2014. LNCS, vol. 8693, pp. 740–755. Springer, Cham (2014). https://doi.org/10.1007/978-3-319-10602-1_48
27. Park, K., Patten, T., Vincze, M.: Pix2Pose: pixel-wise coordinate regression of objects for 6D pose estimation. In: Proceedings of the IEEE International Conference on Computer Vision (ICCV), pp. 7668–7677 (2019)
28. Peng, S., Liu, Y., Huang, Q., Zhou, X., Bao, H.: PVNet: pixel-wise voting network for 6DofF pose estimation. In: Proceedings of the IEEE Conference on Computer Vision and Pattern Recognition (CVPR), pp. 4556–4565, June 2019
29. Philip, J., Drettakis, G.: Plane-based multi-view inpainting for image-based rendering in large scenes. In: Proceedings of the ACM SIGGRAPH Symposium on Interactive 3D Graphics and Games, pp. 1–11 (2018)
30. Prankl, J., Aldoma, A., Svejda, A., Vincze, M.: RGB-D object modelling for object recognition and tracking. In: IEEE/RSJ International Conference on Intelligent Robots and Systems (IROS), pp. 96–103 (2015)
31. Rad, M., Lepetit, V.: BB8: a scalable, accurate, robust to partial occlusion method for predicting the 3D poses of challenging objects without using depth. In: Proceedings of the IEEE International Conference on Computer Vision (ICCV), pp. 3828–3836 (2017)
32. Rad, M., Oberweger, M., Lepetit, V.: Domain transfer for 3D pose estimation from color images without manual annotations. In: Jawahar, C.V., Li, H., Mori, G., Schindler, K. (eds.) ACCV 2018. LNCS, vol. 11365, pp. 69–84. Springer, Cham (2019). https://doi.org/10.1007/978-3-030-20873-8_5
33. Shum, H., Kang, S.B.: Review of image-based rendering techniques. In: Visual Communications and Image Processing 2000, vol. 4067, pp. 2–13. International Society for Optics and Photonics (2000)
34. Sundermeyer, M., Marton, Z.-C., Durner, M., Brucker, M., Triebel, R.: Implicit 3D orientation learning for 6D object detection from RGB images. In: Ferrari, V., Hebert, M., Sminchisescu, C., Weiss, Y. (eds.) ECCV 2018. LNCS, vol. 11210, pp. 712–729. Springer, Cham (2018). https://doi.org/10.1007/978-3-030-01231-1_43
35. Tekin, B., Sinha, S.N., Fua, P.: Real-time seamless single shot 6D object pose prediction. In: Proceedings of the IEEE Conference on Computer Vision and Pattern Recognition (CVPR), pp. 292–301 (2018)
36. Thies, J., Zollhöfer, M., Nieundefinedner, M.: Deferred neural rendering: Image synthesis using neural textures. ACM Trans. Graph. **38**(4) (2019). https://doi.org/10.1145/3306346.3323035
37. Thonat, T., Shechtman, E., Paris, S., Drettakis, G.: Multi-view inpainting for image-based scene editing and rendering. In: International Conference on 3D Vision (3DV), pp. 351–359. IEEE (2016)
38. Vidal, J., Lin, C.Y., Lladó, X., Martí, R.: A method for 6D pose estimation of free-form rigid objects using point pair features on range data. Sensors **18**(8), 2678 (2018)

39. Waechter, M., Moehrle, N., Goesele, M.: Let there be color! large-scale texturing of 3D reconstructions. In: Fleet, D., Pajdla, T., Schiele, B., Tuytelaars, T. (eds.) ECCV 2014. LNCS, vol. 8693, pp. 836–850. Springer, Cham (2014). https://doi.org/10.1007/978-3-319-10602-1_54

40. Wang, C., et al.: DenseFusion: 6D object pose estimation by iterative dense fusion. In: Proceedings of the IEEE Conference on Computer Vision and Pattern Recognition (CVPR), pp. 3343–3352 (2019)

41. Wang, F., Hauser, K.: In-hand object scanning via RGB-D video segmentation. In: Proceedings of the IEEE International Conference on Robotics and Automation (ICRA), pp. 3296–3302 (2019)

42. Whyte, O., Sivic, J., Zisserman, A.: Get out of my picture! internet-based inpainting. In: British Machine Vision Conference (BMVC) (2009)

43. Xiang, Y., Schmidt, T., Narayanan, V., Fox, D.: PoseCNN: a convolutional neural network for 6D object pose estimation in cluttered scenes. Robotics: Science and Systems (RSS) (2018)

44. Yu, J., Fan, Y., Yang, J., Xu, N., Wang, Z., Wang, X., Huang, T.: Wide activation for efficient and accurate image super-resolution. arXiv preprint arXiv:1808.08718 (2018)

45. Zakharov, S., Shugurov, I., Ilic, S.: DPOD: 6D pose object detector and refiner. In: Proceedings of the IEEE International Conference on Computer Vision (ICCV), pp. 1941–1950 (2019)

46. Zhang, R., Isola, P., Efros, A.A., Shechtman, E., Wang, O.: The unreasonable effectiveness of deep features as a perceptual metric. In: Proceedings of the IEEE Conference on Computer Vision and Pattern Recognition (CVPR), pp. 586–595 (2018)

47. Zhou, Q.Y., Park, J., Koltun, V.: Open3D: a modern library for 3D data processing. arXiv:1801.09847 (2018)

Dense Hybrid Recurrent Multi-view Stereo Net with Dynamic Consistency Checking

Jianfeng Yan[1], Zizhuang Wei[1], Hongwei Yi[1(✉)], Mingyu Ding[2], Runze Zhang[3],
Yisong Chen[1], Guoping Wang[1], and Yu-Wing Tai[4]

[1] Peking University, Beijing, China
{haibao637,weizizhuang,hongweiyi,chenyisong,wgp}@pku.edu.cn
[2] HKU, Pokfulam, Hong Kong
myding@cs.hku.hk
[3] Tencent, Shenzhen, China
ryanrzzhang@tencent.com
[4] Kwai Inc., Beijing, China
yuwing@gmail.com

Abstract. In this paper, we propose an efficient and effective dense hybrid recurrent multi-view stereo net with dynamic consistency checking, namely D^2HC-RMVSNet, for accurate dense point cloud reconstruction. Our novel hybrid recurrent multi-view stereo net consists of two core modules: 1) a light DRENet (Dense Reception Expanded) module to extract dense feature maps of original size with multi-scale context information, 2) a HU-LSTM (Hybrid U-LSTM) to regularize 3D matching volume into predicted depth map, which efficiently aggregates different scale information by coupling LSTM and U-Net architecture. To further improve the accuracy and completeness of reconstructed point clouds, we leverage a dynamic consistency checking strategy instead of prefixed parameters and strategies widely adopted in existing methods for dense point cloud reconstruction. In doing so, we dynamically aggregate geometric consistency matching error among all the views. Our method ranks 1st on the complex outdoor *Tanks and Temples* benchmark over all the methods. Extensive experiments on the in-door DTU dataset show our method exhibits competitive performance to the state-of-the-art method while dramatically reduces memory consumption, which costs only 19.4% of R-MVSNet memory consumption. The codebase is available at https://github.com/yhw-yhw/D2HC-RMVSNet.

Keywords: Multi-view stereo · Deep learning · Dense hybrid recurrent-MVSNet · Dynamic consistency checking

J. Yan, Z. Wei and H. Yi—Equal Contribution.

Electronic supplementary material The online version of this chapter (https://doi.org/10.1007/978-3-030-58548-8_39) contains supplementary material, which is available to authorized users.

© Springer Nature Switzerland AG 2020
A. Vedaldi et al. (Eds.): ECCV 2020, LNCS 12349, pp. 674–689, 2020.
https://doi.org/10.1007/978-3-030-58548-8_39

1   Introduction

Dense point cloud reconstruction from multi-view stereo (MVS) information is a classic and important Computer Vision problem for decades, where stereo correspondences of more than two calibrated images are used to recover dense 3D representation [20,24,28,29,32]. While traditional MVS methods have achieved promising results, the recent advance in deep learning [6,10,14–16,21,33,34] allows the exploration of implicit representations of multi-view stereo, hence resulting in superior completeness and accuracy in MVS benchmarks [5,19] compared with traditional alternatives without learning.

However, those deep learning based MVS methods still have the following problems. First, due to the memory limitation, some methods like MVSNet [33] cannot deal with images with large resolutions. Then, RMVSNet [34] are proposed to solve this problem, while the completeness and accuracy of reconstruction are compromised. Second, heavy backbones with downsampling module have to be used to extract features in [6,15,21,33,34], which rely on large memory and lose information in the downsampling process. At last, those deep learning based MVS methods have to fuse the depth maps obtained by different images. The fusion criteria are set in a heuristic pre-defined manner for all data-sets, which lead to low complete results.

To tackle the above problems, we propose a novel deep learning based MVS method called D^2HC-RMVSNet with a network architecture and a dynamic algorithm to fuse depth maps in the postprocessing. The network architecture consists of 1) a newly designed lightweight backbone to extract features for the dense depth map regression, 2) a hybrid module coupling LSTM and U-Net to regularize 3D matching volume into predicted depth maps with different level information into LSTM. The dynamic algorithm to fuse depth maps attempts to aggregate the matching consistency among all the neighbor views to dynamically remain accurate and more reliable dense points in the final results.

Our main contributions are listed below:

- We propose a new lightweight DRENet to extract dense feature map for dense point cloud reconstruction.
- We design a hybrid architecture DHU-LSTM which absorbs both the merits of LSTM and U-Net to reduce the memory cost while maintains the reconstruction accuracy.
- We design a non-trivial dynamic consistency checking algorithm for filtering to remain more reliable and accurate depth values and obtain more complete dense point clouds.
- Our method ranks 1st on **Tanks and Temples** [19] among all methods and exhibits competitive performance to the state-of-the-art method on **DTU** [5] while dramatically reduces memory consumption.

2   Related Work

Deep neural network has made tremendous progress in many vision task [8, 9,17,36], including several attempts on multi-view stereo. The deep learning

676   J. Yan et al.

based MVS methods [6,10,14–16,21,33,34,37] generally first use the backbones
with some downsampling modules to extract features and the final layer of the
backbones with the most downsampled feature maps are output to the following
module. Hence, those methods cannot directly output depth maps with the same
resolution as the input images and may lose some information in those higher
resolutions, which may influence the accuracy of reconstructed results.

Then, plane-sweep volumes are pre-wraped from images as the input to those
networks [10,14,16]. The plane-sweep volumes are memory-consuming and those
methods cannot be trained end-to-end. To train the neural network in an end-
end fashion, MVSNet [33] and DPSNet [15] implicitly encodes multi-view camera
geometries into the network to build the 3D cost volumes by introducing the differential homography warping. P-MVSNet [21] utilizes a patch-wise matching
module to learn the isotropic matching confidence inside the cost volume. Point-
MVSNet [6] proposes a two-stage coarse-to-fine method to generate high resolution depth maps, where a coarse depth map is first yielded by the lower-resolution
version MVSNet [33] and depth errors are iteratively refined in the point cloud
format. However, this method is time-consuming and complicated to employ in
real applications since it consists of two different network architectures. In addition, the memory-consuming 3D-CNN modules adopted in those methods limit
their application for scalable 3D reconstruction from high resolution images. To
reduce memory consumption during the inference phase, R-MVSNet [34] leverages the recurrent gated recurrent unit (GRU) instead of 3D-CNN, whereas
compromises completeness and accuracy on 3D reconstruction.

All of the above methods have to fuse the depth maps from different reference
images to obtain the final reconstructed dense point clouds by following the post-
processing in the non-learning based MVS method COLMAP [25]. In the post-
processing, consistency is checked in a pre-defined manner, which is not robust
for different scenes and may miss some good points viewed by few images.

To improve the deep learning based MVS methods based on above analysis,
we propose a light DRENet specifically designed for the dense depth reconstruction, which outputs the same feature map size as input images with large receptive fields. Then a HU-LSTM (Hybrid U-LSTM) module is designed to reduce
the memory consumption while maintains the 3D reconstruction accuracy. At
last, we design a dynamic consistency checking algorithm for filtering to obtain
more accurate and complete dense point clouds.

3 Reconstruction Pipeline

Given a set of multi-view images and corresponding calibrated camera parameters calculated from Structure-from-Motion [26], our goal is to estimate the depth
map of each reference image and reconstruct dense 3D point cloud. First, each
input image is regarded as the reference image and fed to the effective Dense
Hybrid Recurrent MVSNet (DH-RMVSNet) with several neighbor images to
regress the corresponding dense depth map. Then, we use a dynamic consistency checking algorithm to filter all the estimated depth maps of multi-view

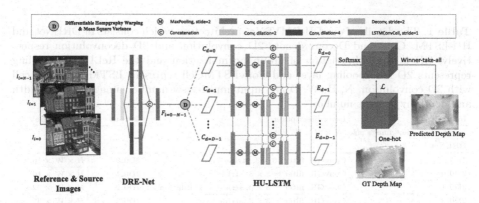

Fig. 1. The network architecture of DH-RMVSNet. 2D feature maps extracted from multi-view images by DRENet go through differentiable homography warping and mean square variance to generate 3D cost volumes. Our HU-LSTM processes 3D cost volume sequentially in depth direction for further training or depth prediction.

images to obtain more accurate and reliable depth values, by leveraging geometric consistency through all neighbor views. After achieving dense filtered reliable depth maps, we directly re-project and fuse all pixels with reliable depth values into 3D space to generate corresponding dense 3D point clouds.

In the following sections, we first introduce our efficient DH-RMVSNet and novel dynamic consistency checking. Then we evaluate our method on DTU [5] and *Tanks and Temples* [19] to prove the efficacy of our method. To evaluate the practicality and generalization on the large-scale dataset with wide range for wide real application, we extend our method on aerial photos in the *Blend-MVS* [35] dataset to reconstruct a large-scale scene.

4   Dense Hybrid Recurrent MVSNet

This section describes the details of our proposed network DH-RMVSNet as visualized in Fig. 1. We design a novel hybrid recurrent multi-view stereo network which absorbs both advantages of 3DCNN in MVSNet [33] and recurrent unit in R-MVSNet [34]. Specifically, our DH-RMVSNet leverages well both the accuracy of 3DCNN processing 3D dimension data and the efficiency of recurrent unit by sequentially processing. Therefore, our network can generate dense accurate depth maps and corresponding dense 3D reconstruction point clouds on the large-scale datasets. We first introduce our lightweight efficient image feature extractor DRENet in Sect. 4.1. Then we present HU-LSTM sub-network to sequentially regularize feature matching volumes in the depth hypothesis direction into 3D probability volume in Sect. 4.2. At last we introduce our training loss in Sect. 4.3.

Table 1. The details of our DH-RMVSNet architecture which consists of DRENet and HU-LSTM. Conv and Deconv denote 2D convolution and 2D deconvolution respectively, GR is the abbreviation of group normalization and the ReLU. MaxPooling represents 2D max-pooling layer and ConvLSTMCell represent LSTM recurrent cell with 2D convolution. N, H, W, D are input multi-view number, image height, width and depth hypothesis number.

Input	Layer description	Output	Output size
Input multi-view image size: $N \times H \times W \times 3$			
DRENet			
$I_{i=0\cdots N-1}$	ConvGR, filter=3 × 3, stride=1	2D0_0	$H \times W \times 16$
2D0_0	ConvGR, filter = 3 × 3, stride = 1	2D0_1	$H \times W \times 16$
2D0_1	ConvGR, filter = 3 × 3, stride = 1, dilation=2	2D0_2	$H \times W \times 32$
2D0_2	ConvGR, filter = 3 × 3, stride = 1	2D0_3	$H \times W \times 32$
2D0_2	ConvGR, filter = 3 × 3, stride = 1, dilation=3	2D1_1	$H \times W \times 32$
2D1_1	ConvGR, filter = 3 × 3, stride = 1	2D1_2	$H \times W \times 32$
2D0_2	ConvGR, filter = 3 × 3, stride = 1, dilation=4	2D2_1	$H \times W \times 32$
2D2_1	ConvGR, filter = 3 × 3, stride = 1	2D2_2	$H \times W \times 32$
[2D0_3, 2D1_2, 2D2_2]	ConvGR, filter=3 × 3, stride=1	$F_{i=0\cdots N-1}$	$H \times W \times 32$
HU-LSTM			
$\mathcal{C}(i)$	ConvLSTMCell, filter = 3 × 3	$\mathcal{C}_0(i)$	$H \times W \times 32$
$\mathcal{C}_0(i)$	MaxPooling, stride = 2	$\mathcal{C}'_0(i)$	$\frac{1}{2}H \times \frac{1}{2}W \times 32$
$\mathcal{C}'_0(i)\,\&\,\mathcal{C}_2(i-1)$	ConvLSTMCell, filter = 3 × 3	$\mathcal{C}_1(i)$	$H \times W \times 32$
$\mathcal{C}_1(i)$	MaxPooling, stride = 2	$\mathcal{C}'_1(i)$	$\frac{1}{4}H \times \frac{1}{4}W \times 32$
$\mathcal{C}'_1(i)\,\&\,\mathcal{C}_2(i-1)$	ConvLSTMCell, filter = 3 × 3	$\mathcal{C}_2(i)$	$H \times W \times 32$
$\mathcal{C}_2(i)$	DeConv, filter = 3 × 3, stride = 2	$\hat{\mathcal{C}}_2(i)$	$\frac{1}{2}H \times \frac{1}{2}W \times 32$
$[\mathcal{C}_1(i), \hat{\mathcal{C}}_2(i)\,[\,\&\,\mathcal{C}_3(i-1)$	ConvLSTMCell, filter = 3 × 3	$\mathcal{C}_3(i)$	$H \times W \times 32$
$\mathcal{C}_3(i)$	DeConv, filter = 3 × 3, stride = 2	$\hat{\mathcal{C}}_3(i)$	$\frac{1}{2}H \times \frac{1}{2}W \times 32$
$[\mathcal{C}_1(i), \hat{\mathcal{C}}_3(i)]\,\&\,\mathcal{C}_4(i-1)$	ConvLSTMCell, filter = 3 × 3	$\mathcal{C}_4(i)$	$H \times W \times 32$
$\mathcal{C}_4(i)$	Conv, filter = 3 × 3, stride = 1	$\mathcal{C}_H(i)$	$H \times W \times 1$

4.1 Image Feature Extractor

We design a Dense Receptive Expansion sub-network by concatenating feature maps from different dilated convolutional [38] layers to aggregate multi-scale contextual information without losing resolution. We term it DRENet whose weights are shared by multi-view images $\mathbf{I}_{i=0\cdots N-1}$. Most of previous multi-view stereo network, such as [21,33,34], usually use 2D convolutional layers with stride larger or equal than 2 to enlarge the receptive field and reduce the resolution at the same time for satisfying memory limitation. We introduce different dilated convolutional layers to generate multi-scale context information and preserve the resolution which leads to the possibility of dense depth map estimation. The details of DRENet are presented in Table 1.

Given N-view images, let $\mathbf{I}_{i=0}$ and $\mathbf{I}_{i=1\cdots N-1}$ denote the reference image and the neighbor source images respectively. We first use two usual convolutional layers to sum up local-wise pixel information, then we utilize three dilated convolutional layers with different dilated ratio $2, 3, 4$ to extract multi-scale context information without scarifying resolution. Thus, after concatenation, DRENet can extract the dense feature map $F_i \in \mathbb{R}^{C \times H \times W}$ efficiently, where C denotes the feature channel and H, W represent the height and width of the input image.

Following common practices [6,15,21,33,34], to build a 3D feature volume $\{V_i\}_{i=0}^{N-1}$, we utilize the differentiable homography to warp the extracted feature map between different views. And we adopt the same mean square variance to aggregate them into one cost volume C.

4.2 Hybrid Recurrent Regularization

There exists two different ways to regularize the cost volume C into one probability map $\mathcal{P}$. One is to utilize the 3DCNN U-Net in MVSNet [33] which can well leverage local wise information and multi-scale context information, but it can not directly be used to regress the original dense depth map estimation due to limited GPU memory especially for large resolution images. The other is to use stacked convolutional GRU in R-MVSNet [34] which is quiet efficient by sequentially processing the 3D volume through the depth direction but loss the aggregation of multi-scale context information.

Therefore, we absorb the merits in both two methods to propose a hybrid recurrent regularization network with more powerful recurrent convolutional cell than GRU, namely LSTMConvCell [31]. We construct a hybrid U-LSTM which is a novel 2D U-net architecture where each layer is LSTMConvCell, which can be processed sequentially. We term this module HU-LSTM. Our HU-LSTM can well aggregate multi-scale context information and easily process dense original size cost volumes with high efficiency at the same time. It costs 19.4% GPU memory of the previous recurrent method R-MVSNet [34]. The detailed architecture of HU-LSTM is demonstrated in Table 1.

Cost volume C can be viewed as D number 2D cost matching map $\{C(i)\}_{i=0}^{D-1}$ which are concatenated in the depth hypothesis direction. We denote the output of regularized cost matching map as $\{C_H(i)\}_{i=0}^{D-1}$ at i^{th} step during sequential processing. Therefore, $C_H(i)$ relies on the both current input cost matching map $C(i)$ and all previous states $C_H(0, \cdots, i-1)$. Different from GRU in R-MVSNet [34], we introduce more powerful recurrent unit named ConvLSTMCell which has three gates map to control the information flow and can well aggregate different scale context information.

Let $\mathbb{I}(i)$, $\mathbb{F}(i)$ and $\mathbb{O}(i)$ denote the input gate map, forget gate map and output gate map respectively. In the following part, $\odot$, '[]' and '*' represent the element-wise multiplication, the concatenation and the matrix multiplication respectively in convolutional layer.

The input gate map is used to select valid information from current input $\hat{C}(i)$ into the current state cell $C(i)$:

$$\mathbb{I}(i) = \sigma(\mathbb{W}_{\mathbb{I}} * [C(i), C_H(i-1)] + \mathbb{B}_{\mathbb{I}}), \tag{1}$$

$$\hat{C}(i) = tanh(\mathbb{W}_{\mathbb{C}} * [C(i), C_H(i-1)] + \mathbb{B}_{\mathbb{C}}), \tag{2}$$

while the forget gate map $\mathbb{F}(i)$ decides to filter useless information from previous state cell $\mathcal{C}(i-1)$ and combines the input information from the input gate map $\mathbb{I}(i)$ to generate current new state cell $\mathcal{C}(i)$:

$$\mathbb{F}(i) = \sigma(\mathbb{W}_{\mathbb{F}} * [\mathcal{C}(i), \mathcal{C}_H(i-1)] + \mathbb{B}_{\mathbb{F}}), \tag{3}$$

$$\mathcal{C}(i) = \mathbb{F}_i \odot \mathcal{C}_H(i-1) + \mathbb{I}_i \odot \hat{\mathcal{C}}(i), \tag{4}$$

Finally, the output gate map controls how much information from new current state cell $\mathcal{C}(i)$ will output, which is $\mathcal{C}_H(i)$:

$$\mathbb{O}(i) = \sigma(\mathbb{W}_{\mathbb{O}} * [\mathcal{C}(i), \mathcal{C}_H(i-1)] + \mathbb{B}_{\mathbb{O}}), \tag{5}$$

$$\mathcal{C}_H(i) = \mathbb{O}(i) \odot tanh(\mathcal{C}(i)), \tag{6}$$

where σ and $tanh$ represent $sigmoid$ and $tanh$ non-linear activation function respectively, $\mathbb{W}$ and $\mathbb{B}$ are learnable parameters in LSTM convolutional filter.

In our proposed HU-LSTM, by aggregating different scale context information to improve the robustness and accuracy of depth estimation, we adopt three LSTMConvCells to propragate different scale input feature maps with downsampling scale 0.5 and two LSTMConvCell to aggregate multi-scale context information as denoted in Table 1. Specifically, we input the 32-channel input cost map $\mathcal{C}_i$ to the first LSTMConvCell, and the output of each LSTMConvCell will be fed into next LSTMConvCell. Then the regularized cost matching volume $\{\mathcal{C}_H(i)\}_{i=0}^{D-1}$ goes through by a $softmax$ layer to generate corresponding the probability volume $\mathcal{P}$ for further calculating training loss.

4.3 Training Loss

Following MVSNet [33], we treat the depth regression task as multiple classification task and use the same cross entropy loss function $\mathcal{L}$ between the probability volumes $\mathcal{P}$ and ground truth depth map $\mathcal{G}$:

$$\mathcal{L} = \sum_{x \in x_{valid}} \sum_{i=0}^{D-1} -G(i,x) * log(P(i,x)), \tag{7}$$

where x_{valid} is the set of valid pixels in the ground truth, $G(i,x)$ represents the one-hot vector generated by the depth value of the ground truth $\mathcal{G}$ at pixel x and $P(i,x)$ is the corresponding depth estimated probability. During test phase, we do not need to save the whole probability map. To further improve the efficiency, the depth map is processed sequentially and the winner-take-all selection is used to generate the estimated depth map from regularized cost matching volume.

5   Dynamic Consistency Checking

The above DH-RMVSNet generates dense pixel-wise depth map for each input multi-view images. Before fusing all the estimated multi-view depth maps, it is necessary to filter out mismatched errors and store correct and reliable depths. All previous methods [6,21,33,34] just follow [27] to apply the geometric constraint to measure the depth estimation consistency among multiple views. However, those methods only use prefixed constant parameters. Specifically, the reliable depth value should satisfy both conditions: the pixel reprojection error less than τ_1 and the depth reprojection error less than τ_2 in at least three views, where $\tau_1 = 1$ and $\tau_2 = 0.01$ are pre-defined. These parameters are defined intuitively and not robust for different scenes, though they have large influence on the quality of reconstruct point cloud. For example, those depth values with much high reliable consistency in two views are filtered and a fixed number valid views also lose information in the views with slightly worse errors. Beside, using the fixed τ_1 and τ_2 may not filter enough mismatched pixels in different scenes.

In general, a estimated depth value is accurate and reliable when it has a very low reprojection error in few views, or a lower error in majority views. Therefore, we propose a novel dynamic consistency checking algorithm to select valid depth values, which is related to both the reprojection error and view numbers. By considering dynamic geometric matching cost as consistency among all neighbor views, it leads to more robust and complete dense 3D point clouds. We denote the estimated depth value $D_i(\boldsymbol{p})$ of a pixel $\boldsymbol{p}$ on reference image $\mathbf{I}_i$ through our DH-RMVSNET. The camera parameter is represented by $P_i = [M_i|t_i]$ in [13]. First we back-project the pixel p into 3D space to generate the corresponding 3D point X by:

$$X = M_i^{-1}(D_i(\boldsymbol{p}) \cdot \boldsymbol{p} - t_i), \tag{8}$$

Then we project the 3D point X to generate the projected pixel $\boldsymbol{q}$ on the neighbor view $\mathbf{I}_j$:

$$q = \frac{1}{d}P_j \cdot X, \tag{9}$$

where P_j is the camera parameter of neighbor view $\mathbf{I}_j$ and d is the depth from projection. In turn, we back-project the projected pixel $\boldsymbol{q}$ with estimated depth $D_j(\boldsymbol{q})$ on the neighbor view into 3D space and reproject back to the reference image denoted as $\boldsymbol{p}'$:

$$p' = \frac{1}{d'}P_j \cdot (M_j^{-1}(D_j(\boldsymbol{q}) \cdot \boldsymbol{q} - t_j)), \tag{10}$$

where d' is the depth value of the reprojected pixel $\boldsymbol{p}'$ on the reference image. Based on the above mentioned operation, the reprojection errors are calculated by:

$$\begin{aligned} \xi_p &= ||\boldsymbol{p} - \boldsymbol{p}'||_2, \\ \xi_d &= ||D_i(\boldsymbol{p}) - d'||_1 / D_i(\boldsymbol{p}). \end{aligned} \tag{11}$$

In order to quantify the depth matching consistency between two different views, we propose the dynamic matching consistency by considering dynamic matching consistency among all views. The dynamic matching consistency in different views is defined as:

$$c_{ij}(\boldsymbol{p}) = e^{-(\xi_p + \lambda \cdot \xi_d)}, \tag{12}$$

where λ is used to leverage the reprojection error in two different metrics. By aggregating the matching consistency from all the neighbor views to obtain the global dynamic multi-view geometric consistency $C_{geo}(\boldsymbol{p})$ as:

$$C_{geo}(\boldsymbol{p}) = \sum_{j=1}^{N} c_{ij}. \tag{13}$$

We calculate the dynamic geometric consistency for every pixel and filter out the outliers with $C_{geo}(\boldsymbol{p}) < \tau$. Benefiting from our proposed dynamic consistency checking algorithm, the filtered depth map is able to store more accurate and complete depth values compared with the previous intuitive fixed-threshold method. It improves the robustness, completeness and accuracy of 3D reconstructed point clouds.

6 Experiments

6.1 Implementation Details

Training. We train DH-RMVSNet on the DTU dataset [5], which contains 124 different indoor scenes which is split to three parts, namely *training*, *validation* and *evaluation*. Following common practices [6,7,14,16,33,34,37], we train our network on the training dataset and evaluate on the evaluation dataset. While the dataset only provides ground truth point clouds generated by scanners, to generate the ground truth depth map, we use the same rendering method as MVSNet [33]. Different from [33,34] which generate the depth map with $\frac{1}{4}$ size of original input image, our method generates the depth map with the same size as input image. Since MVSNet [33] only provides $\frac{1}{4}$ size depth map, thus we resize the training input image to $W \times H = 160 \times 128$ as same as the corresponding groundtruth depth map. We set the number of input images $N = 3$ and the depth hypotheses are sampled from 425 mm to 745 mm with depth plane number $D = 128$ in MVSNet [33]. We implement our network on **PyTorch** [23] and train the network end-to-end for 6 epochs using *Adam* [18] with an initial learning rate 0.001 which is decayed by 0.9 every epoch. Batch size is set to 6 on 2 NVIDIA TITAN RTX graphics cards.

Testing. For testing, we use the $N = 7$ views as input, and set $D = 256$ for depth plane hypothesis in an inverse depth manner in [34]. To evaluate *Tanks and Temples* dataset, the camera parameters are computed by OpenMVG [22] following MVSNet [33] and the input image resolution is set to 1920×1056. We test the BlendedMVS [35] dataset using original images of 768×576 resolution.

Filter and Fusion. After we generate estimated depth maps from DH-RMVSNet, we filter and fuse them to generate corresponding 3D dense point cloud. First, the depths with probability lower than $\phi = 0.4$ will be discarded. Then, we use our proposed dynamic global geometric consistency checking algorithm as further multi-view depth map filter with $\lambda = 200$ and $\tau = 1.8$. At last, we fuse all reliable depths into 3D space to generate 3D point cloud.

Table 2. Quantitative results on the DTU evaluation dataset [5] (lower is better). Our method D^2HC-RMVSNet exhibits a competitive reconstruction performance compared with state-of-the-art methods in terms of completeness and overall quality.

Method	Mean distance (mm)		
	Acc	Comp	*overall*
Tola [30]	0.342	1.190	0.766
Gipuma [11]	**0.283**	0.873	0.578
Colmap [25]	0.400	0.664	0.532
SurfaceNet [16]	0.450	1.040	0.745
MVSNet [33]	0.396	0.527	0.462
R-MVSNet [34]	0.385	0.459	0.422
P-MVSNet [21]	0.406	0.434	0.420
PointMVSNet [6]	0.361	0.421	0.391
PointMVSNet-HiRes [6]	0.342	0.411	**0.376**
D^2HC-RMVSNet	0.395	**0.378**	0.386

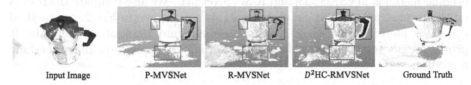

Input Image   P-MVSNet   R-MVSNet D^2HC-RMVSNet   Ground Truth

Fig. 2. Comparison on the reconstructed point clouds for the *Scan* 77 from the DTU [5] dataset with other methods [21,34]. Our method generates more complete and denser point cloud than other methods.

6.2 Datasets and Results

We first demonstrate the state-of-the-art performance of our proposed D^2HC-RMVSNet on the DTU [5] and *Tanks and Temples* [19], which outperforms its original methods, namely MVSNet [33] and R-MVSNet [34] with a significant margin. Specifically, our method ranks 1st in the complex large-scale outdoor *Tanks and Temples* benchmark over all existing methods. To investigate the practicality and scalability of our method, we extend our method on the aerial photos in *BlendedMVS* [35] to reconstruct a larger scale scenes.

DTU Dataset. We evaluate our proposed method on the DTU [5] *evaluation* dataset. We set $D = 256$ within the depth range [425 mm, 905 mm] for all scans and use the common evaluation metric in other methods [33,34]. Quantitative results are shown in Table 2. While Gipuma [11] performs the best regarding to accuracy, our method achieves the best completeness and the competitive *overall* quality of reconstruction results. Our proposed D^2HC-RMVSNet can both improve the accuracy and the completeness significantly compared with its

Table 3. Quantitative results on the *Tanks and Temples* benchmark [19]. The evaluation metric is *f-score* which higher is better. (L.H. and P.G. are the abbreviations of *Lighthouse* and *Playground* dataset respectively.)

Method	Rank	Mean	Family	Francis	Horse	L.H	M60	Panther	P.G	Train
COLMAP [26,27]	55.62	42.14	50.41	22.25	25.63	56.43	44.83	46.97	48.53	42.04
Pix4D [4]	53.38	43.24	64.45	31.91	26.43	54.41	50.58	35.37	47.78	34.96
MVSNet [33]	52.75	43.48	55.99	28.55	25.07	50.79	53.96	50.86	47.90	34.69
Point-MVSNet [6]	40.25	48.27	61.79	41.15	34.20	50.79	51.97	50.85	52.38	43.06
Dense R-MVSNet [34]	37.50	50.55	73.01	54.46	43.42	43.88	46.80	46.69	50.87	45.25
OpenMVS [3]	17.88	55.11	71.69	51.12	42.76	58.98	54.72	56.17	59.77	45.69
P-MVSNet [21]	17.00	55.62	70.04	44.64	40.22	**65.20**	55.08	55.17	60.37	54.29
CasMVSNet [12]	14.00	56.84	76.37	58.45	46.26	55.81	56.11	54.06	58.18	49.51
ACMM [32]	12.62	57.27	69.24	51.45	46.97	63.20	55.07	57.64	**60.08**	**54.48**
Altizure-HKUST-2019 [2]	9.12	59.03	**77.19**	**61.52**	42.09	63.50	59.36	58.20	57.05	53.30
DH-RMVSNet	10.62	57.55	73.62	53.17	46.24	58.68	59.38	58.31	58.26	52.77
D^2**HC-RMVSNet**	**5.62**	**59.20**	74.69	56.04	**49.42**	60.08	**59.81**	**59.61**	60.04	53.92

original methods MVSNet [33] and R-MVSNet [34]. We also compare the results on the reconstructed point clouds with [21,34]. As shown in Fig. 2, our method generates more complete and accurate point cloud than other methods. It proves the efficacy of our novel DH-RMVSNet and dynamic consistency checking algorithm.

Tanks and Temples Benchmark. Tanks and Temples Benchmark [19] is a large-scale outdoor dataset which consists of more complex environment, and it is quite typical for real captured situation, compared with DTU dataset which is taken under well-controlled environment with fixed camera trajectory.

We evaluate our method **without any fine-tuning** on the *Tanks and Temples* as denoted in Table 3. Our proposed D^2HC-RMVSNet ranks 1st over all existing methods. Specifically, our method outperforms all deep-learning based multi-view stereo methods with a big margin. It shows the stronger generalization compared with Point-MVSNet [6] while Point-MVSNet [6] is the state-of-the-art method on the DTU [5]. The mean *f-score* increases significantly from 50.55 to 59.20 (larger is better, date: Mar. 5, 2020) compared with Dense R-MVSNet [34], which demonstrates the efficacy and robustness of D^2HC-RMVSNet on the variant scenes. The reconstructed point clouds are shown in Fig. 3, it shows that our method generates accurate, delicate and complete point cloud. And we compare the Precision/Recall of the model *Family* with its original methods [33,34] at different error threshold in Fig. 4. It demonstrates our method achieves a significant improvement on the precision while maintains better recall than R-MVSNet [34], which leads to the best performance on the *Tanks and Temples*.

BlendedMVS. BlendedMVS is a new large-scale MVS dataset which is synthesized from 3D reconstructed models from Altizure [2]. The dataset contains over 113 different scenes with a variety of different camera trajectories. And each

(a) Family   (b) Francis

(c) Train   (d) M60

Fig. 3. Point cloud results on the *Tanks and Temples* [19] benchmark, our method generates accurate, delicate and complete reconstructed point clouds which show the strong generalization of our method on complex outdoor scenes.

scenes consists of 20 to 1000 input images including architectures, sculptures and small objects. To further evaluate the practicality and scalability of our propose D^2HC-RMVSNet, we directly test our method and R-MVSNet [34] on the provided *validation* dataset. For fair comparison, both methods are trained on the DTU [5] **without any fine-tuning** and we upsample the $\frac{1}{4}$ depth map from R-MVSNet to the same size of the depth map from our method, which is the original size of input images. As shown in Fig. 5, our method can well reconstruct the whole large scale scene and small cars in it, while R-MVSNet [34] fails on it. Our method can estimate the dense delicate accurate depth map with original size of input image in an inverse depth setting as in [34], because of our novel DRENet and HU-LSTM, which has more accuracy and stronger scalability

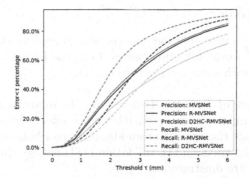

Fig. 4. Comparison on the Precision/Recall (in%) at different thresholds (within 6 mm) on the *Family* provided by [1] with MVSNet [33] and R-MVSNet [34].

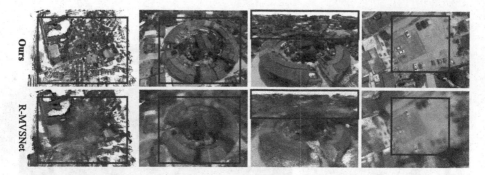

Fig. 5. Comparison of the reconstruction point cloud results on the *validation* set of *BlendedMVS* [35]. Our method can both well reconstruct a large-scale scene and small cars in it, while R-MVSNet [34] failed on it. *(Both methods are without fine-tuning.)*

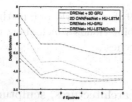

Fig. 6. Validation results of the mean average depth error with different network architectures during training.

Table 4. Comparison of the running time and memory consumption between our proposed D^2HC-RMVSNet and R-MVSNet [34] on the DTU [5].

Method	Input size	Output size	Mem. (GB)	Time (s)
R-MVSNet	1600 × 1196	400 × 296	6.7	**2.1**
Ours	400 × 296	400 × 296	**1.3**	2.6
Ours	800 × 600	800 × 592	2.4	8.0
Ours	1600 × 1200	1600 × 1196	6.6	29.15

of 3D point cloud reconstruction by aggregating multi-scale context information on the large-scale dataset. Due to our dynamical consistency checking algorithm, we can directly use our algorithm to remain dense reliable point cloud without any specific adjustment.

6.3 Ablation Study

In this section, we provide ablation experiments to analyze the strengths of the key components of our architecture. we perform the following studies on DTU *validation* dataset with same setting in Sect. 6.1.

Variant Components of Network Architecture. To quantitatively analyze how different network architecture in DH-RMVSNet affect the depth map reconstruction, we evaluate the average mean absolute error between estimated depth maps and the ground truth on the *validation* DTU dataset during training. The comparison results are illustrated in Fig. 6.

We replace our "DRENet" and "HU-LSTM" with "2DCNNFeatNet" and "3D GRU" in R-MVSNet [34] respectively to analysis the influence of our proposed feature encoder and cost volume regularization module. Compared with

"2DCNNFeatNet+HU-LSTM" in Fig. 6, our proposed "DRENet" can improve the accuracy slightly but with less inference time and memory consumption due to the light architecture. "HU-LSTM" achieves significant improvement with a big margin compared with "DRENet+3D GRU", which absorbs both the merits of efficacy in [33] and efficiency in [34] by aggregating multi-scale context information in sequential process. To further evaluate the difference in two different powerful recurrent gate units, LSTM and GRU, we replace the LSTM cell in our "HU-LSTM" with "GRU", denoted as "HU-GRU". The comparison between "HU-LSTM" and "HU-GRU" shows "LSTM" is more accurate and robust than "GRU" because of more gate map to control the information flow and have a better performance on the learning matching patterns.

Benefit from Dynamic Consistency Checking. To further study the influence and generalization of our dynamic consistency checking, we evaluate our DH-RMVSNet with common filtering algorithm as in previous methods [6,33,34] as shown in Table 3. Our proposed dynamic consistency checking algorithm significantly boosts the reconstruction results in all scenes on the *Tanks and Temples* benchmark, which shows the strong generalization and dynamic adaptation on the different scenes. It improves the *f-score* from 57.55 by DH-RMVSNet to 59.20, which leads to more accurate and complete reconstruction point clouds.

7   Discussion

Running Time and Memory Utility. For fair comparison on the running time and memory utility with R-MVSNet [34], we test our method with same depth sample number $D = 256$ on the GTX 1080Ti GPU. As shown in Table 4, our method inputs multi-images of only 400×296 resolution to generate the depth map of the same size as R-MVSNet [34], with only 19.4% memory consumption of R-MVSNet. Moreover, our method runs with 2.6 s per view, with a little extra inference time than R-MVSNet, which needs an extra 6.2 s refinement to enhance the performance. Our D^2HC-RMVSNet achieves significant improvement over R-MVSNet [34] both on the DTU and the *Tanks and Temples* benchmark, while our novel dynamic consistency checking takes negligible running time. Our method can generate dense depth maps with the same size of the input image with efficient memory consumption. It takes only 6.6 GB to process multi-view images with 1600×1200 resolution, which leads to a wide practicality for dense point cloud reconstruction.

Scalability and Generalization. Due to our light DRENet and HU-LSTM, our method shows more powerful general scalability than R-MVSNet [34] on the dense reconstruction with wide range. Our method can easily extend to aerial photos for the reconstruction of the big scene architectures in Fig. 5 and generate denser, more accurate and complete 3D point cloud reconstruction due to the original size depth map estimation from our D^2HC-RMVSNet.

8 Conclusions

We have presented a novel dense hybrid recurrent multi-view stereo network with dynamic consistency checking, denoted as D^2HC-RMVSNet, for dense accurate point cloud reconstruction. Our DH-RMVSNet well absorbs both the merits of the accuracy of 3DCNN and the efficiency of Recurrent unit, to design a new lightweight feature extractor DRENet and hybrid recurrent regularization module HU-LSTM. To further improve the robustness and completeness of 3D point cloud reconstruction, we propose a no-trivial dynamic consistency checking algorithm to dynamically aggregate geometric matching error among all views rather than use prefixed strategy and parameters. Experimental results show that our method ranks 1st on the complex outdoor *Tanks and Temples* and exhibits the competitive results on the *DTU* dataset, while dramatically reduces memory consumption, which costs only 19.4% of R-MVSNet memory consumption.

Acknowledgements. This project was supported by the National Key R&D Program of China (No. 2017YFB1002705, No. 2017YFB1002601) and NSFC of China (No. 61632003, No. 61661146002, No. 61872398).

References

1. https://www.tanksandtemples.org/
2. Altizure. https://www.altizure.com/
3. Openmvs. https://github.com/cdcseacave/openMVS
4. Pix4d. https://pix4d.com/
5. Aanæs, H., Jensen, R.R., Vogiatzis, G., Tola, E., Dahl, A.B.: Large-scale data for multiple-view stereopsis. IJCV **120**(2), 153–168 (2016)
6. Chen, R., Han, S., Xu, J., Su, H.: Point-based multi-view stereo network. arXiv preprint arXiv:1908.04422 (2019)
7. Chen, R., Han, S., Xu, J., Su, H.: Point-based multi-view stereo network. In: ICCV (2019)
8. Ding, M., et al.: Learning depth-guided convolutions for monocular 3D object detection. In: CVPR (2020)
9. Dosovitskiy, A., et al.: FlowNet: learning optical flow with convolutional networks. In: ICCV (2015)
10. Flynn, J., Neulander, I., Philbin, J., Snavely, N.: DeepStereo: learning to predict new views from the world's imagery. In: CVPR (2016)
11. Galliani, S., Lasinger, K., Schindler, K.: Massively parallel multiview stereopsis by surface normal diffusion. In: ICCV (2015)
12. Gu, X., Fan, Z., Zhu, S., Dai, Z., Tan, F., Tan, P.: Cascade cost volume for high-resolution multi-view stereo and stereo matching. arXiv preprint arXiv:1912.06378 (2019)
13. Hartley, R., Zisserman, A.: Multiple View Geometry in Computer Vision. Cambridge University Press, New York (2003)
14. Huang, P.H., Matzen, K., Kopf, J., Ahuja, N., Huang, J.B.: DeepMVS: learning multi-view stereopsis. In: CVPR (2018)
15. Im, S., Jeon, H.G., Lin, S., Kweon, I.S.: DpsNET: end-to-end deep plane sweep stereo. arXiv preprint arXiv:1905.00538 (2019)

16. Ji, M., Gall, J., Zheng, H., Liu, Y., Fang, L.: SurfaceNet: an end-to-end 3D neural network for multiview stereopsis. In: ICCV (2017)
17. Kendall, A., et al.: End-to-end learning of geometry and context for deep stereo regression. In: ICCV (2017)
18. Kingma, D.P., Ba, J.: Adam: a method for stochastic optimization. arXiv preprint arXiv:1412.6980 (2014)
19. Knapitsch, A., Park, J., Zhou, Q.Y., Koltun, V.: Tanks and temples: benchmarking large-scale scene reconstruction. TOG **36**(4), 78 (2017)
20. Lhuillier, M., Quan, L.: A quasi-dense approach to surface reconstruction from uncalibrated images. PAMI **27**(3), 418–433 (2005)
21. Luo, K., Guan, T., Ju, L., Huang, H., Luo, Y.: P-MVSNet: learning patch-wise matching confidence aggregation for multi-view stereo. In: ICCV (2019)
22. Moulon, P., Monasse, P., Marlet, R., et al.: OpenMVG. an open multiple view geometry library (2014)
23. Paszke, A., et al.: Automatic differentiation in PyTorch. In: NeurIPS Autodiff Workshop (2017)
24. Schonberger, J.L., Frahm, J.M.: Structure-from-motion revisited. In: CVPR (2016)
25. Schönberger, J.L., Zheng, E., Frahm, J.M., Pollefeys, M.: Pixelwise view selection for unstructured multi-view stereo. In: ECCV (2016)
26. Schönberger, J.L., Frahm, J.M.: Structure-from-motion revisited. In: CVPR (2016)
27. Schönberger, J.L., Zheng, E., Pollefeys, M., Frahm, J.M.: Pixelwise view selection for unstructured multi-view stereo. In: ECCV (2016)
28. Seitz, S.M., Curless, B., Diebel, J., Scharstein, D., Szeliski, R.: A comparison and evaluation of multi-view stereo reconstruction algorithms. In: CVPR (2006)
29. Strecha, C., Von Hansen, W., Van Gool, L., Fua, P., Thoennessen, U.: On benchmarking camera calibration and multi-view stereo for high resolution imagery. In: CVPR (2008)
30. Tola, E., Strecha, C., Fua, P.: Efficient large-scale multi-view stereo for ultra high-resolution image sets. Mach. Vis. Appl. **23**(5), 903–920 (2012)
31. Xingjian, S., Chen, Z., Wang, H., Yeung, D.Y., Wong, W.K., Woo, W.C.: Convolutional LSTM network: a machine learning approach for precipitation nowcasting. In: NeurIPS, pp. 802–810 (2015)
32. Xu, Q., Tao, W.: Multi-scale geometric consistency guided multi-view stereo. In: CVPR (2019)
33. Yao, Y., Luo, Z., Li, S., Fang, T., Quan, L.: MVSnet: depth inference for unstructured multi-view stereo. In: ECCV (2018)
34. Yao, Y., Luo, Z., Li, S., Fang, T., Quan, L.: MVSnet: depth inference for unstructured multi-view stereo. In: ECCV (2018)
35. Yao, Y., et al.: BlendedMVS: a large-scale dataset for generalized multi-view stereo networks. arXiv preprint arXiv:1911.10127 (2019)
36. Yi, H., et al.: MMFace: a multi-metric regression network for unconstrained face reconstruction. In: CVPR (2019)
37. Yi, H., et al.: Pyramid multi-view stereo net with self-adaptive view aggregation. arXiv preprint arXiv:1912.03001 (2019)
38. Yu, F., Koltun, V.: Multi-scale context aggregation by dilated convolutions. arXiv preprint arXiv:1511.07122 (2015)

Pixel-Pair Occlusion Relationship Map (P2ORM): Formulation, Inference and Application

Xuchong Qiu[1], Yang Xiao[1], Chaohui Wang[1(✉)], and Renaud Marlet[1,2]

[1] LIGM, Ecole des Ponts, Univ Gustave Eiffel, CNRS, ESIEE Paris,
Champs-sur-Marne, France
chaohui.wang@univ-eiffel.fr
[2] valeo.ai, Paris, France

Abstract. We formalize concepts around geometric occlusion in 2D images (i.e., ignoring semantics), and propose a novel unified formulation of both occlusion boundaries and occlusion orientations via a pixel-pair occlusion relation. The former provides a way to generate large-scale accurate occlusion datasets while, based on the latter, we propose a novel method for task-independent pixel-level occlusion relationship estimation from single images. Experiments on a variety of datasets demonstrate that our method outperforms existing ones on this task. To further illustrate the value of our formulation, we also propose a new depth map refinement method that consistently improve the performance of state-of-the-art monocular depth estimation methods.

Keywords: Occlusion relation · Occlusion boundary · Depth refinement

1 Introduction

Occlusions are ubiquitous in 2D images (cf. Fig. 1(a)) and constitute a major obstacle to address scene understanding rigorously and efficiently. Besides the joint treatment of occlusion when developing techniques for specific tasks [18, 19, 35–37, 40, 54], task-independent occlusion reasoning [24, 30, 42, 49, 51, 53] offers valuable occlusion-related features for high-level scene understanding tasks.

In this work, we are interested in one most valuable but challenging scenario of task-independent occlusion reasoning where the input is a single image and the output is the corresponding pixel-level occlusion relationship in the whole image domain (cf. Fig. 1(b)); the goal is to capture both the localization and orientation of the occlusion boundaries, similar to previous work such as [30, 49, 51, 53]. In this context, informative cues are missing compared to other usual scenarios of

Electronic supplementary material The online version of this chapter (https://doi.org/10.1007/978-3-030-58548-8_40) contains supplementary material, which is available to authorized users.

© Springer Nature Switzerland AG 2020
A. Vedaldi et al. (Eds.): ECCV 2020, LNCS 12349, pp. 690–708, 2020.
https://doi.org/10.1007/978-3-030-58548-8_40

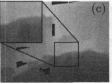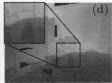

Fig. 1. Illustration of the proposed methods: (a) input image, (b) estimated horizontal occlusion relationship (a part of P2ORM) where red (resp. blue) pixels occlude (resp. are occluded by) their right-hand pixel, (c) depth estimation obtained by a state-of-the-art method [40], (d) our depth refinement based on occlusion relationships. (Color figure online)

occlusion reasoning, in particular semantics [38], stereo geometry [62] and interframe motion [10]. Moreover, the additional estimation of orientation further increases the difficulty compared to usual occlusion boundary estimation [2,10, 14]. Despite of recent progress achieved via deep learning [30,51,53], the study on pixel-level occlusion relationship in monocular images is still relatively limited and the state-of-the-art performance is still lagging.

Here, we formalize concepts around geometric occlusion in 2D images (i.e., ignoring semantics), and propose a unified formulation, called *Pixel-Pair Occlusion Relationship Map (P2ORM)*, that captures both localization and orientation information of occlusion boundaries. Our representation simplifies the development of estimation methods, compared to previous works [30,49,51,53]: a common ResNet-based [13] U-Net [45] outperforms carefully-crafted state-of-the-art architectures on both indoor and outdoor datasets, with either low-quality or high-quality ground truth. Besides, thanks to the modularity regarding pixel-level classification methods, better classifiers can be adopted to further improve the performance of our method. In addition, P2ORM can be easily used in scene understanding tasks to increase their performance. As an illustration, we develop a depth map refinement module based on P2ORM for monocular depth estimation (Fig. 1(c–d)). Experiments demonstrate that it significantly and consistently sharpens the edges of depth maps generated by a wide range of methods [8,9,20,22,25,27,28,40,58], including method targeted at sharp edges [40].

Moreover, our representation derives from a 3D geometry study that involves a first-order approximation of the observed 3D scene, offering a way to create high-quality occlusion annotations from a depth map with given or estimated surface normals. This allows the automated generation of large-scale, accurate datasets from synthetic data [26] (possibly with domain adaptation [61] for more realistic images) or from laser scanners [21]. Compared to manually annotated dataset that is commonly used [42], we generate a high-quality synthetic dataset of that is two orders of magnitude larger.

Our contributions are: (1) a formalization of geometric occlusion in 2D images; (2) a new formulation capturing occlusion relationship at pixel-pair level, from which usual boundaries and orientations can be computed; (3) an occlusion estimation method that outperforms the state-of-the-art on several datasets; (4) the illustration of the relevance of this formulation with an application to depth map refinement that consistently improves the performance of state-of-the-art monocular depth estimation methods. We will release our code and datasets.

Related Work

Task-Independent Occlusion Relationship in Monocular Images has long been studied due to the importance of occlusion reasoning in scene understanding. Early work often estimates occlusion relationship between simplified 2D models of the underlying 3D scene, such as blocks world [44], line drawings [5,48] and 2.1-D sketches [34]. Likewise, [17] estimates figure/ground labels using an estimated 3D scene layout. Another approach combines contour/junction structure and local shapes using a Conditional Random Field (CRF) to represent and estimate figure/ground assignment [42]. Likewise, [49] learns border ownership cues and impose a border ownership structure with structured random forests. Specific devices, e.g., with multi-flash imaging [41], have also been developed.

Recently, an important representation was used in several deep models to estimate occlusion relationship [30,51,53]: a pixel-level binary map encoding the localization of the occlusion boundary and an angle representing the oriented occlusion direction, indicating where the foreground lies w.r.t. the pixel.

This theme is also closely related to *occlusion boundary detection*, which ignores orientation. Existing methods often estimate occlusion boundaries from images sequences. To name a few, [2] detects T-junctions in space-time as a strong cue to estimate occlusion boundaries; [46] adds relative motion cues to detect occlusion boundaries based on an initial edge detector [31]; [10] further exploits both spatial and temporal contextual information in video sequences. Also, [1,29,59,60] detect object boundaries between specific semantic classes.

Monocular depth estimation is extremely valuable for geometric scene understanding, but very challenging due to its high ill-posedness. Yet significant progress has been made with the development of deep learning and large labeled datasets. Multi-scale networks better explore the global image context [7,8,22]. Depth estimation also is converted into an ordinal regression task to increase accuracy [9,23]. Other approaches propose a better regression loss [20] or the inclusion of geometric constraints from stereo image pairs [11,15].

Depth map refinement is often treated as a post-processing step, using CRFs [16,43,52,57]: an initial depth prediction is regularized based on pixel-wise and pairwise energy terms depending on various guidance signals. These methods now underperform state-of-the-art deep-learning-based methods without refinement [20,58] while being more computationally expensive. Recently, [39] predicts image displacement fields to sharpen initial depth predictions.

2   Formalizing and Representing Geometric Occlusion

In this section, we provide formal definitions and representations of occlusion in single images based on scene geometry information. It enables the generation of accurate datasets and the development of an efficient inference method.

We consider a camera located at C observing the surface $\mathcal{S}$ of a 3D scene. Without loss of generality, we assume $C = \mathbf{0}$. We note L a ray from C, and L_X the ray from C through 3D point X. For any surface patch S on $\mathcal{S}$ intersecting L, we note $L \cap S$ the closest intersection point to C, and $\|L \cap S\|$ it distance to C.

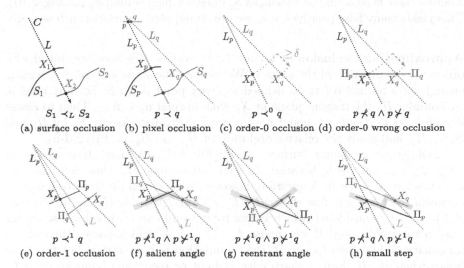

$$S_1 \prec_L S_2 \qquad p \prec q \qquad p \prec^0 q \qquad p \not\prec q \wedge p \not\succ q$$

(a) surface occlusion (b) pixel occlusion (c) order-0 occlusion (d) order-0 wrong occlusion

$$p \prec^1 q \qquad p \not\prec^1 q \wedge p \not\succ^1 q \qquad p \not\prec^1 q \wedge p \not\succ^1 q \qquad p \not\prec^1 q \wedge p \not\succ^1 q$$

(e) order-1 occlusion (f) salient angle (g) reentrant angle (h) small step

Fig. 2. Occlusion configurations (solid lines represent real or tangent surfaces, dotted lines are imaginary lines): (a) S_1 occludes S_2 along L; (b) p occludes q as S_p occludes S_q along L_p; (c) p occludes q at order 0 as $\|X_q\| - \|X_p\| \geq \delta > 0$, cf. Eq. (1); (d) no occlusion despite order-0 occlusion as Π_p, Π_q do not occlude one another; (e) p occludes q at order 1 as tangent plane Π_p occludes tangent plane Π_q in the $[L_p, L_q]$ cone, cf. Eq. (2); no occlusion for a (f) salient or (g) reentrant angle between tangent planes Π_p, Π_q, cf. Eq. (2); (h) tangent plane occlusion superseded by order-0 non-occlusion, cf. Eq. (2).

Approximating Occlusion at Order 0. Given two surface patches S_1, S_2 on $\mathcal{S}$ and a ray L (cf. Fig. 2(a)), we say that S_1 *occludes* S_2 *along* L, noted $S_1 \prec_L S_2$ (meaning S_1 comes before S_2 along L), iff L intersects both S_1 and S_2, and the intersection $X_1 = L \cap S_1$ is closer to C than $X_2 = L \cap S_2$, i.e., $\|X_1\| < \|X_2\|$.

Now given neighbor pixels $p, q \in \mathcal{P}$, that are also 3D points in the image plane, we say that p *occludes* q, noted $p \prec q$, iff there are surface patches S_p, S_q on $\mathcal{S}$ containing respectively X_p, X_q such that S_p occludes S_q along L_p, (cf. Fig. 2(b)). Assuming $L_p \cap S_q$ exists and $\|L_p \cap S_q\|$ can be approximated by $\|L_q \cap S_q\| =$

$\|X_q\|$, it leads to a common definition that we qualify as "order-0". We say that p *occludes* q *at order 0*, noted $p \prec^0 q$ iff X_q is deeper than X_p (cf. Fig. 2(c)):

$$p \prec^0 q \quad \text{iff} \quad \|X_p\| < \|X_q\|. \tag{1}$$

The depth here is w.r.t. the camera center ($d_p = \|X_p\|$), not to the image plane. This definition is constructive (can be tested) and the relation is antisymmetric. The case of a minimum margin $\|X_q\| - \|X_p\| \geq \delta > 0$ is considered below.

However, when looking at the same continuous surface patch $S_p = S_q$, the incidence angles of L_p, L_q on S_p, S_q may be such that order-0 occlusion is satisfied whereas there is no actual occlusion, as S_q does not pass behind S_p (cf. Fig. 2(d)). This yields many false positives, e.g., we observing planar surfaces such as walls.

Approximating Occlusion at Order 1. To address this issue, we consider an order-1 approximation of the surface. We assume the scene surface S is regular enough for a normal $\mathbf{n}_X$ to be defined at every point X on S. For any pixel p, we consider Π_p the tangent plane at X_p with normal $\mathbf{n}_p = \mathbf{n}_{X_p}$. Then to assess if p *occludes* q *at order 1*, noted $p \prec^1 q$, we approximate locally S_p by Π_p and S_q by Π_q, and study the relative occlusion of Π_p and Π_q, cf. Fig. 2(d-h).

Looking at a planar surface as in Fig. 2(d), we now have $p \prec^0 q$ as $\|X_p\| < \|X_q\|$, but $p \not\prec^1 q$ because Π_p does not occlude Π_q, thus defeating the false positive at order 0. A question, however, is on which ray L to test surface occlusion, cf. Fig. 2(a). If we choose $L = L_p$, cf. Fig. 2(b), only Π_q (approximating S_q) is actually considered, which is less robust and can lead to inconsistencies due to the asymmetry. If we choose $L = L_{(p+q)/2}$, which passes through an imaginary middle pixel $(p + q)/2$, the formulation is symmetrical but there are issues when Π_p, Π_q form a sharp edge (salient or reentrant) lying between L_p and L_q, cf. Fig. 2 (f–g), which is a common situation in man-made environments. Indeed, the occlusion status then depends on the edge shape and location w.r.t. $L_{(p+q)/2}$, which is little satisfactory. Besides, such declared occlusions are false positives.

To solve this problem, we define order-1 occlusion $p \prec^1 q$ as a situation where Π_p occludes Π_q along all rays L between L_p and L_q, which can simply be tested as $\|X_p\| < \|\Pi_q \cap L_p\|$ and $\|X_q\| > \|\Pi_p \cap L_q\|$. However, it raises yet another issue: there are cases where $\|X_p\| < \|X_q\|$, thus $p \prec^0 q$, and yet $\|\Pi_p \cap L\| > \|\Pi_q \cap L\|$ for all L between L_p and L_q, implying the inverse occlusion $p \succ^1 q$, cf. Fig. 2(h). This small-step configuration exists ubiquitously (e.g., book on a table, frame on a wall) but does not correspond to an actual occlusion. To prevent this paradoxical situation and also to introduce some robustness, as normals can be wrong due to estimation errors, we actually define order-1 occlusion so that it also implies order-0 occlusion. In the end, we say that p *occlude* q *at order 1* iff (i) p occludes q at order 0, (ii) Π_p occludes Π_q along all rays L between L_p and L_q, i.e.,

$$p \prec^1 q \quad \text{iff} \quad \|X_p\| < \|X_q\| \ \wedge \ \|X_p\| < \|\Pi_q \cap L_p\| \ \wedge \ \|X_q\| > \|\Pi_p \cap L_q\|. \tag{2}$$

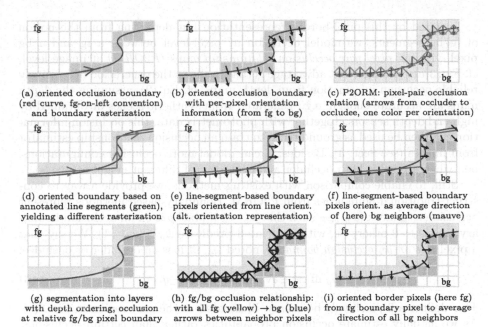

(a) oriented occlusion boundary (red curve, fg-on-left convention) and boundary rasterization

(b) oriented occlusion boundary with per-pixel orientation information (from fg to bg)

(c) P2ORM: pixel-pair occlusion relation (arrows from occluder to occludee, one color per orientation)

(d) oriented boundary based on annotated line segments (green), yielding a different rasterization

(e) line-segment-based boundary pixels oriented from line orient. (alt. orientation representation)

(f) line-segment-based boundary pixels orient. as average direction of (here) bg neighbors (mauve)

(g) segmentation into layers with depth ordering, occlusion at relative fg/bg pixel boundary

(h) fg/bg occlusion relationship: with all fg (yellow) → bg (blue) arrows between neighbor pixels

(i) oriented border pixels (here fg) from fg boundary pixel to average direction of all bg neighbors

Fig. 3. Some representations of occlusion and oriented occlusion.

Discretized Occlusion. In practice, we resort to a discrete formulation where p, q are neighboring pixels in image $\mathcal{P}$ and L_p passes through the center of p. We note $\mathcal{N}_p$ the immediate neighbors of p, considering either only the 4 horizontal and vertical neighbors $\mathcal{N}_p^4$, or including also in $\mathcal{N}_p^8$ the 4 diagonal pixels.

As distances (depths) $d_p = \|X_p\|$ can only be measured approximately, we require a minimum discontinuity threshold $\delta > 0$ to test any depth difference. A condition $d_p < d_q$ thus translates as $d_q - d_p \geq \delta$. However, to treat equally all pairs of neighboring pixels p, q, the margin δ has to be relative to the pixel distance $\|p - q\|$, which can be 1 or $\sqrt{2}$ due to the diagonal neighbors. Extending the first-order approximation, the relation $d_p < d_q$ is thus actually tested as $d_{pq} > \delta$ where $d_{pq} \overset{\text{def}}{=} (d_q - d_p)/\|q - p\|$, making δ a pixel-wise depth increasing rate.

Occlusion Relationship and Occlusion Boundary. Most of the literature on occlusion in images focuses on *occlusion boundaries*, that are imaginary lines separating locally a foreground (fg) from a background (bg). A problem is that they are often materialized as rasterized, 1-pixel-wide contours, that are not well defined, cf. Fig. 3(a). The fact is that vectorized occlusion delineations are not generally available in existing datasets, except for handmade annotations, that are coarse as they are made with line segments, with endpoints at discrete positions, only approximating the actual, ideal curve, cf. Fig. 3(d). An alternative representation [17,42,53] considers occlusion boundaries at the border pixels of two relative fg/bg segments (regions) rather than on a separating line (Fig. 3(g)).

Inspired by this pixel-border representation but departing from the notion of fg/bg segments, we model occlusion at pixel-level between a fg and a bg pixel, yielding *pixel-pair occclusion relationship maps (P2ORM)* at image level, cf. Fig. 3(c). An important advantage is that it allows the generation of relatively reliable occlusion information from depth maps, cf. Eq. (2), assuming the depth maps are accurate enough, e.g., generated from synthetic scenes or obtained by high-end depth sensors. Together with photometric data, this occlusion information can then be used as ground truth to train an occlusion relationship estimator from images (see Sect. 3). Besides, it can model more occlusion configurations, i.e., when a pixel is both occluder and occludee (of different neighbor pixels).

Still, to enable comparison with existing methods, we provide a way to construct traditional boundaries from P2ORM. Boundary-based methods represent occlusion as a mask $(\omega_p)_{p \in \mathcal{P}}$ such that $\omega_p = 1$ if pixel p is on an occlusion boundary, and $\omega_p = 0$ otherwise, with associated predicate $\dot{\omega}_p \overset{\text{def}}{=} (\omega_p = 1)$. We say that a pixel p *is on an occlusion boundary*, noted $\dot{\omega}_p$, iff it is an occluder or occludee:

$$\dot{\omega}_p \text{ iff } \exists q \in \mathcal{N}_p, \; p \prec q \lor p \succ q. \tag{3}$$

This defines a 2-pixel-wide boundary, illustrated as the grey region in Fig. 3(c). As we actually estimate occlusion probabilities rather than certain occlusions, this width may be thinned by thresholding or non-maximum suppression (NMS).

Occlusion Relationship and Oriented Occlusion Boundary. Related to the notions of segment-level occlusion relationship, figure/ground representation and boundary ownership [42,53], occlusion boundaries may be oriented to indicate which side is fg vs bg, cf. Fig. 3(b). It is generally modeled as the direction of the tangent to the boundary, conventionally oriented [17] (fg on the left, Fig. 3(a)). In practice, the boundary is modeled with line segments (Fig. 3(d)), whose orientation θ is transferred to their rasterized pixels [53] (Fig. 3(e)). Inaccuracies matter little here as the angle is only used to identify a boundary side.

The occlusion border formulation, based on fg/bg pixels (Fig. 3(g)), implicitly captures orientation information: from each fg pixel to each neighbor bg pixel (Fig. 3(h)). So does our modeling (Fig. 3(c)). To compare with boundary-based approaches, we define a notion of pixel occlusion orientation (that could apply to occlusion borders too (Fig. 3(i)), or even boundaries (Fig. 3(f))). We say that a pixel p is oriented as the sum v_p of the unitary directions of occluded or occluding neighboring pixels q, with angle $\theta_p = \text{atan2}(u_p^y, u_p^x) - \frac{\pi}{2}$ where $u_p = v_p / \|v_p\|$ and

$$v_p = \sum_{q \in \mathcal{N}_p} (\mathbb{1}(p \prec q) - \mathbb{1}(p \succ q)) \frac{q - p}{\|q - p\|}. \tag{4}$$

3   Pixel-Pair Occlusion Relationship Estimation

Modeling the Pixel-Pair Occlusion Relation. The occlusion relation is a binary property that is antisymmetric: $p \prec q \Rightarrow q \not\prec p$. Hence, to model

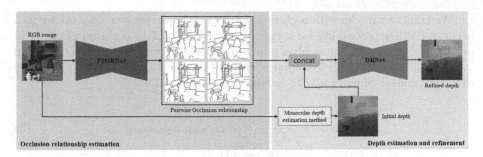

Fig. 4. Overview of our method. Left: a encoder-decoder structure followed by softmax takes an RGB image as input and outputs 4 classification maps (ω_p^i) where each pixel p in a map for inclination i actually represents a pair of pixels pq with $q = p + i$. The map $\omega_{pq}^i = \omega_p^i = r$ classifies p as occluded ($r = -1$), not involved in occlusion ($r = 0$) or occluding ($r = 1$), with probability ω_{pqr}^i. (If $\mathcal{N} = \mathcal{N}^4$, only 2 inclination maps are generated.) Colors blue, white and red represent respectively $r = -1$, 0 or 1. The top two images presents occlusion relationships along inclinations horizontal ($i = \mathsf{h}$) and vertical ($i = \mathsf{v}$); the bottom two, along inclinations diagonal ($i = \mathsf{d}$) and antidiagonal ($i = \mathsf{a}$). Right: A direct use of the occlusion relationship for depth map refinement. (Color figure online)

the occlusion relationship of neighbor pair pq, we use a random variable $\omega_{p,q}$ with only three possible values $r \in \{-1, 0, 1\}$ representing respectively: $p \succ q$ (p is occluded by q), $p \not\succ q \wedge p \not\prec q$ (no occlusion between p and q), and $p \prec q$ (p occludes q).

Since $\omega_{p,q} = -\omega_{q,p}$, a single variable per pair is enough. We assume a fixed total ordering $<$ on pixels (e.g., lexicographic order on image coordinates) and note $\omega_{pq} = \omega_{qp} =$ if $p < q$ then $\omega_{p,q}$ else $\omega_{q,p}$. We also define $\omega_{pqr} = \mathbb{P}(\omega_{pq} = r)$.

Concretely, we consider 4 inclinations, horizontal, vertical, diagonal, antidiagonal, with canonical displacements $\mathsf{h} = (1, 0)$, $\mathsf{v} = (0, 1)$, $\mathsf{d} = (1, 1)$, $\mathsf{a} = (1, -1)$, and we call $\mathcal{Q}_i = \{pq \mid p, q \in \mathcal{P}, q = p + i\}$ the set of pixel pairs with inclination $i \in \mathcal{I}^4 = \{\mathsf{h}, \mathsf{v}, \mathsf{d}, \mathsf{a}\}$. For the the 4-connectivity, we only consider $i \in \mathcal{I}^2 = \{\mathsf{h}, \mathsf{v}\}$.

Estimating the Occlusion Relation. For occlusion relationship estimation, we adopt a segmentation approach: we classify each valid pixel pair pq by scoring its 3 possible statuses $r \in \{-1, 0, 1\}$, from which we extract estimated probabilities $\hat{\omega}_{pqr}$. The final classification map is obtained as $\hat{\omega}_{pq} = \text{argmax}_r \, \hat{\omega}_{pqr}$.

Our architecture is sketched on Fig. 4 (left). The P2ORM estimator (named *P2ORNet*) takes an RGB image as input, and outputs its pixel-pair occlusion relationship map for the different inclinations. We use a ResNet-based [13] U-Net-like auto-encoder with skip-connections [45], cf. supplementary material (SM). It must be noted that this architecture is strikingly simple compared to more complex problem-specific architectures that have been proposed in the past [30,51,53]. Besides, our approach is not specifically bound to U-Net or ResNet; in the future, we may benefit from improvements in general segmentation methods.

We train our model with a *class-balanced cross-entropy loss* [56], taking into account the low probability for a pair pq to be labeled 1 (p occludes q) or -1 (q occludes p), given that most pixel pairs do not feature any occlusion. Our global loss $\mathcal{L}_{\text{occrel}}$ is a sum of $|\mathcal{I}|$ losses for each kind of pair inclination $i \in \mathcal{I}$, averaged over the number of pairs $|\mathcal{Q}_i|$ to balance each task $i \in \mathcal{I}$:

$$\mathcal{L}_{\text{occrel}} = \sum_{i \in \mathcal{I}} \frac{1}{|\mathcal{Q}_i|} \sum_{\substack{pq \in \mathcal{Q}_i \\ r \in \{-1,0,1\}}} -\alpha_r\, \omega_{pqr} \log(\hat{\omega}_{pqr}). \tag{5}$$

where $\hat{\omega}_{pqr}$ is the estimated probability that pair pq has occlusion status r, $\omega_{pqr} = \mathbb{1}(\omega_{pq} = r)$ where ω_{pq} is the ground truth (GT) occlusion status of pair pq, $\alpha_r = \mathbb{1}(r = 0) + \alpha \mathbb{1}(r \neq 0)$ and α accounts for the disparity in label frequency.

From Probabilistic Occlusion Relations to Occlusion Boundaries. As discussed with Eq. (3), occlusion boundaries can be generated from an occlusion relation. In case the relation is available with probabilities, as for an estimated $\hat{\omega}_{pqr}$, we define a probabilistic variant $\omega_p \in [0, 1]$: $\hat{\omega}_p = \frac{1}{|\mathcal{N}_p|} \sum_{q \in \mathcal{N}_p} (\hat{\omega}_{pq,-1} + \hat{\omega}_{pq,1})$.

As proposed in [6] and performed in many other methods, we operate a non-maximum suppression to get thinner boundaries. The final occlusion boundary map is given by thresholding $\text{NMS}((\omega_p)_{p \in \mathcal{P}})$ with a probability, e.g., 0.5.

Boundary orientations can then be generated as defined in Eq. (4). Given our representation, it has the following simpler formulation: $v_p = \sum_{q \in \mathcal{N}_p} \hat{\omega}_{pq} \frac{q-p}{\|q-p\|}$.

4   Application to Depth Map Refinement

Given an image, a depth map $(\tilde{d}_p)_{p \in \mathcal{P}}$ estimated by some method, and an occlusion relationship $(\hat{\omega}_{p,p+i})_{p \in \mathcal{P}, i \in \mathcal{I}}$ as estimated in Sect. 3, we produce a refined, more accurate depth map $(d_p)_{p \in \mathcal{P}}$ with sharper edges. To this end, we propose a U-Net architecture [45] (Fig. 4 (right)), named DRNet, where $(\tilde{d}_p)_{p \in \mathcal{P}}$ and the 8 maps $((\hat{\omega}_{p,p+i})_{p \in \mathcal{P}})_{i \in \mathcal{I} \cup (-\mathcal{I})}$ are stacked as a multi-channel input of the network.

As a pre-processing, we first use the GT depth map $(d_p^{\text{gt}})_{p \in \mathcal{P}}$ and normals $(\mathbf{n}_p^{\text{gt}})_{p \in \mathcal{P}}$ to compute the ground-truth occlusion relationship $(p \prec_{\text{gt}} q)_{p \in \mathcal{P}, q \in \mathcal{N}_p}$. We then train the network via the following loss:

$$\mathcal{L}_{\text{refine}} = \mathcal{L}_{\text{occonsist}} + \lambda \mathcal{L}_{\text{regul}} \tag{6}$$

$$\mathcal{L}_{\text{occonsist}} = \frac{1}{N} \sum_{p \in \mathcal{P}} \sum_{q \in \mathcal{N}_p^8} \begin{cases} \mathcal{B}(\log \delta, \log d_{pq}) & \text{if } p \prec_{\text{gt}} q \text{ and } d_{pq} < \delta \\ \mathcal{B}(\log \delta, \log D_{pq}) & \text{if } p \not\prec_{\text{gt}} q \text{ and } D_{pq} \geq \delta \\ 0 & \text{otherwise} \end{cases} \tag{7}$$

$$\mathcal{L}_{\text{regul}} = \frac{1}{|\mathcal{P}|} \sum_{p \in \mathcal{P}} \left(\mathcal{B}(\log \tilde{d}_p, \log d_p) + \|\nabla \log \tilde{d}_p - \nabla \log d_p\|^2 \right) \tag{8}$$

where $\mathcal{B}$ is the berHu loss [22], δ is the depth discontinuity threshold introduced in Sect. 2, N is the number of pixels p having a non-zero contribution to $\mathcal{L}_{\text{occonsist}}$, and D_{pq} is the order-1 depth difference at mid-pixel $(p+q)/2$, i.e., $D_{pq} = \min(d_{pq}, m_{pq})$ where $m_{pq} = (\|\Pi_q \cap L_{(p+q)/2}\| - \|\Pi_p \cap L_{(p+q)/2}\|)/\|q - p\|$ is the signed distance between tangent planes Π_p, Π_q along $L_{(p+q)/2}$.

$\mathcal{L}_{\text{occonsist}}$ penalizes refined depths d_p that are inconsistent with GT occlusion relationship $\prec_{\text{gt}}$, i.e., when p occludes q in the GT but not in the refinement, or when p does not occlude q in the GT but does it in the refinement. $\mathcal{L}_{\text{regul}}$ penalizes differences between the rough input depth and the refined output depth, which makes refined depths conditioned on input depths. The total loss $\mathcal{L}_{\text{refine}}$ tends to change depths only close to occlusion boundaries, preventing excessive drifts.

To provide occlusion information that has the same size as the depth map, as pixel-pair information is not perfectly aligned on the pixel grid, we turn pixel-pair data $(\omega_{p,p+i})_{p\in\mathcal{P},i\in\mathcal{I},p+i\in\mathcal{P}}$ into a pixelwise information: for a given inclination $i \in \mathcal{I}$, we define $\omega_p^i = \omega_{p,p+i}$. Thus, e.g., if $p \prec p+i$, then $\omega_p^i = 1$ and $\omega_{p+i}^i = -1$.

At test time, given the estimated occlusion relationships, we use NMS to sharpen depth edges. For this, we first generate pixelwise occlusion boundaries from the estimated P2ORM $(\hat{\omega}_{p,p+i})_{p\in\mathcal{P},i\in\mathcal{I}}$, pass them through NMS [6] and do thresholding to get a binary occlusion boundary map $(\omega_p)_{p\in\mathcal{P}}$ where $\omega_p \in \{0,1\}$. We then thin the estimated directional maps $(\omega_p^i)_{p\in\mathcal{P}}$ by setting $\omega_p^i \leftarrow 0$ if $\omega_p = 0$.

5   Experiments

Oriented Occlusion Boundary Estimation. Because of the originality of our approach, there is no other method to directly compare with. Yet to demonstrate its significance in task-independent occlusion reasoning, we translate our relation maps into oriented occlusion boundaries (cf. Sect. 3) to compare with SRF-OCC [49], DOC-DMLFOV [53], DOC-HED [53], DOOBNet [51][1], OFNet [30][1].

To disentangle the respective contributions of the P2ORM formulation and the network architecture, we also evaluate a "baseline" variant of our architecture, that relies on the usual paradigm of estimating separately boundaries and orientations [30,51,53]: we replace the last layer of our pixel-pair classifier by two separate heads, one for classifying the boundary and the other one for regressing the orientation, and we use the same loss as [30,51].

We evaluate on 3 datasets: BSDS ownership [42], NYUv2-OR, iBims-1-OR (cf. Table 1). We keep the original training and testing data of BSDS. NYUv2-OR is tested on a subset of NYUv2 [33] with occlusion boundaries from [39] and our labeled orientation. iBims-1-OR is tested on iBims-1 [21] augmented with

---

[1] As DOOBNet and OFNet are coded in Caffe, in order to have an unified platform for experimenting them on new datasets, we carefully re-implemented them in PyTorch (following the Caffe code). We could not reproduce exactly the same quantitative values provided in the original papers (ODS and OIS metrics are a bit less while AP is a bit better), probably due to some intrinsic differences between frameworks Caffe and PyTorch, however, the difference is very small (less than 0.03, cf. Table 2).

Table 1. Used and created occlusion datasets. (a) We only use 500 scenes and 20 images per scene (not all 500M images). (b) Training on NYUv2-OR uses all InteriorNet-OR images adapted using [61] with the 795 training images of NYUv2 as target domain. (c) Training on iBims-1-OR uses all InteriorNet-OR images w/o domain adaptation.

Dataset	InteriorNet-OR	BSDS ownership	NYUv2-OR	iBim-1-OR
Origin	[26]	[42]	[33]	[21]
Type	Synthetic	Real	Real	Real
Scene	Indoor	Outdoor	Indoor	Indoor
Resolution	640 × 480	481 × 321	640 × 480	640 × 480
Depth	Synthetic	N/A	Kinect v1	Laser scanner
Normals	Synthetic	N/A	N/A	Computed [4]
Relation annot	Ours from depth and normals	Ours from manual fig./ground [42]	Ours from boundaries and depth	Ours from depth and normals
Boundary annot	From relation	Manual [42]	Manual [39]	From relation
Orient. annot	From relation	Manual [53]	Manual (ours)	From relation
Annot. quality	High	Low	Medium	High
# train img. (orig.)	10,000$^{(a)}$	100	795$^{(b)}$	0$^{(c)}$
# train images	10,000$^{(a)}$	100	10,000$^{(b)}$	10,000$^{(c)}$
# testing images	0	100	654	100
α in $\mathcal{L}_{occrel}$	N/A	50	10	10

occlusion ground truth we generated automatically (cf. Sect. 2 and SM). As illustrated on Fig. 5, this new accurate ground truth is much more complete than the "distinct depth transitions" offered by iBims-1 [21], that are first detected on depth maps with [6], then manually selected. For training, a subset of InteriorNet [26] is used for NYUv2-OR and iBims-1-OR. For NYUv2-OR, because of the domain gap between sharp InteriorNet images and blurry NYUv2 images, the InteriorNet images are furthermore adapted with [61] using NYUv2 training images (see SM for the ablation study related to domain adaption).

We use the same protocol as [30,51] to compute 3 standard evaluation metrics, based on the Occlusion-Precision-Recall graph (OPR): F-measure with best fixed occlusion probability threshold over the all dataset (ODS), F-measure with best occlusion probability threshold for each image (OIS), and average precision over all occlusion probability thresholds (AP). Recall (R) is the proportion of

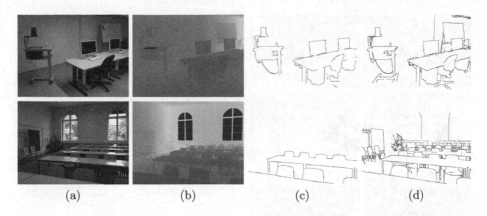

    (a)   (b)   (c)   (d)

Fig. 5. iBims-1-OR: (a) RGB images, (b) GT depth (invalid is black), (c) provided "distinct depth transitions" [21], (d) our finer and more complete occlusion boundaries.

Table 2. Oriented occlusion boundary estimation. *Our re-implementation.

Method	BSDS ownership			NYUv2-OR			iBims-1-OR		
Metric	ODS	OIS	AP	ODS	OIS	AP	ODS	OIS	AP
SRF-OCC [49]	.419	.448	.337	–	–	–	–	–	–
DOC-DMLFOV [53]	.463	.491	.369	–	–	–	–	–	–
DOC-HED [53]	.522	.545	.428	–	–	–	–	–	–
DOOBNet [51]	.555	.570	.440	–	–	–	–	–	–
OFNet [30]	.583	.607	.501	–	–	–	–	–	–
DOOBNet*	.529	.543	.433	.343	.370	.263	.421	.440	312
OFNet*	.553	.577	.520	.402	.431	.342	.488	.513	.432
Baseline	.571	.605	.524	.396	.428	.343	.482	.507	.431
Ours (4-connectivity)	.590	.612	.512	.500	.522	.477	.575	.599	.508
Ours (8-connectivity)	**.607**	**.632**	**.598**	**.520**	**.540**	**.497**	**.581**	**.603**	**.525**

correct boundary detections, while Precision (P) is the proportion of pixels with correct occlusion orientation w.r.t. all pixels detected as occlusion boundary.

Qualitative results are shown in Fig. 6, while Table 2 summarizes quantitative results. Our baseline is on par with the state-of-the-art on the standard BSDS ownership benchmark as well as on the two new datasets, hinting that complex specific architectures maybe buy little as a common ResNet-based U-Net is at least as efficient. More importantly, our method with 8-connectivity outperforms existing methods on all metrics by a large margin (up to 15 points), demonstrating the significance of our formulation on higher-quality annotations, as opposed to BSDS whose lower quality levels up performances. It could also be an illustration that classification is often superior to regression [32] as it does not average ambiguities. Lastly, the 4-connectivity variant shows that the ablation

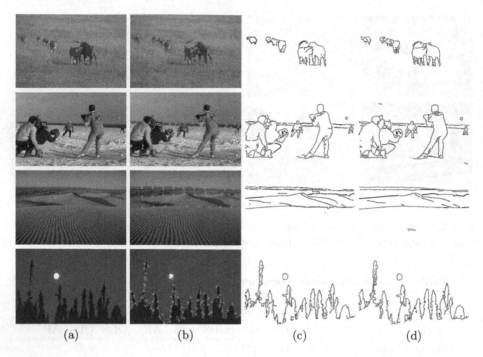

(a)   (b)   (c)   (d)

Fig. 6. Occlusion estimation on BSDS ownership dataset: (a) input RGB image, (b) ground-truth occlusion orientation, (c) OFNet estimation [30], (d) our estimation. green: correct boundary and orientation; red: correct boundary, incorrect orientation; **blue**: missed boundaries; gray: incorrect boundaries. (Color figure online)

of diagonal neighbors decreases the performance, thus assessing the relevance of 8-connectivity. (See SM for more results and ablation studies.)

Depth Map Refinement. To assess our refinement approach, we compare with [39], which is the current state-of-the-art for depth refinement on boundaries.

We evaluate based on depth maps estimated by methods that offer results on depth-edge metrics: [8,9,20,22,40,58] on NYUv2, and [8,22,25,27,28,40] on iBims-1. We train our network on InteriorNet-OR for ground truth, with input depth maps to refine estimated by SharpNet [40]. For a fair comparison, we follow the evaluation protocol of [39]. To assess general depth accuracy, we measure: mean absolute relative error (rel), mean $\log_{10}$ error ($\log_{10}$), Root Mean Squared linear Error (RMSE(lin)), Root Mean Squared log Error (RMSE(log)), and accuracy under threshold $(\sigma_i < 1.25^i)_{i=1,2,3}$. For depth-edge, following [21], we measure the accuracy ϵ_{acc} and completion ϵ_{comp} of predicted boundaries.

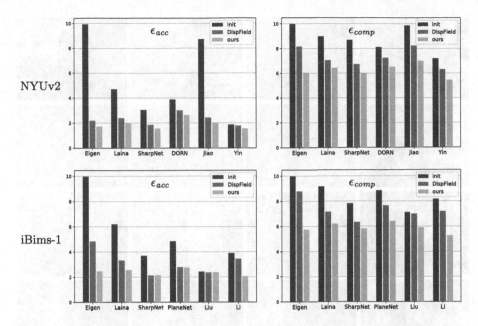

Fig. 7. Gain in edge quality after depth refinement for metrics ϵ_{acc} (left) and ϵ_{comp} (right) on NYUv2 (top) for respectively [8,9,20,22,40,58] and on iBimis-1 (bottom) for [8,22,25,27,28,40]: metric on input depth maps (blue), after refining with [39] (orange), and after our refinement (green). Lower metric value is better. (Color figure online)

Figure 7 summarizes quantitative results. We significantly improve edge metrics $\epsilon_{acc}, \epsilon_{comp}$ on NYUv2 and iBims-1, systematically outperforming [39] and showing consistency across the two different datasets. Not shown on the figure (see SM), the differences on general metrics after refinement are negligible ($< 1\%$), i.e., we improve sharpness without degrading the overall depth. Figure 8 illustrates the refinement on depth maps estimated by SharpNet [40]. We also outperform many methods based on image intensity [3,12,47,50,55] (see SM), showing the superiority of P2ORM for depth refinement w.r.t. image intensity.

In an extensive variant study (see SM), we experiment with possible alternatives: adding as input in the architecture (1) the original image, (2) the normal map, (3) the binary edges; (4) adding an extra loss term $\mathcal{L}_{\text{gtdepth}}$ that regularizes on the ground truth depths rather than on the estimated depth, or substituting $\mathcal{L}_{\text{gtdepth}}$ (5) for $\mathcal{L}_{\text{occonsist}}$ or (6) for $\mathcal{L}_{\text{regul}}$; using (7) d_{pq} only, or (8) D_{pq} only in Eq. (8). The alternative proposed here performs the best.

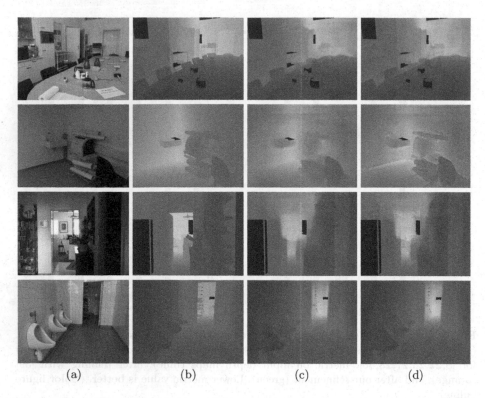

(a)   (b)   (c)   (d)

Fig. 8. Depth refinement: (a) input RGB image from iBims-1, (b) ground truth depth, (c) SharpNet depth prediction [40], (d) our refined depth.

6   Conclusion

In this paper, we propose a new representation of occlusion relationship based on pixel pairs and design a simple network architecture to estimate it. Translating our results into standard occlusion boundaries for comparison, we significantly outperform the state-of-the-art for both occlusion boundary and oriented occlusion boundary estimation. To illustrate the potential of our representation, we also propose a depth map refinement model that exploits our estimated occlusion relationships. It also consistently outperforms the state-of-the-art regarding depth edge sharpness, without degrading accuracy in the rest of the depth image. These results are made possible thanks to a new method to automatically generate accurate occlusion relationship labels from depth maps, at a large scale.

Acknowledgements.. We thank Yuming Du and Michael Ramamonjisoa for helpful discussions and for offering their GT annotations of occlusion boundaries for a large part of NYUv2, which we completed (NYUv2-OC++) [39]. This work was partly funded by the I-Site FUTURE initiative, through the DiXite project.

References

1. Acuna, D., Kar, A., Fidler, S.: Devil is in the edges: learning semantic boundaries from noisy annotations. In: Conference on Computer Vision and Pattern Recognition (CVPR), pp. 11075–11083 (2019)
2. Apostoloff, N., Fitzgibbon, A.: Learning spatiotemporal t-junctions for occlusion detection. In: Conference on Computer Vision and Pattern Recognition (CVPR), vol. 2, pp. 553–559. IEEE (2005)
3. Barron, J.T., Poole, B.: The fast bilateral solver. In: Leibe, B., Matas, J., Sebe, N., Welling, M. (eds.) ECCV 2016. LNCS, vol. 9907, pp. 617–632. Springer, Cham (2016). https://doi.org/10.1007/978-3-319-46487-9_38
4. Boulch, A., Marlet, R.: Fast and robust normal estimation for point clouds with sharp features. Comput. Graph. Forum (CGF) 31(5), 1765–1774 (2012)
5. Cooper, M.C.: Interpreting line drawings of curved objects with tangential edges and surfaces. Image Vis. Comput. 15(4), 263–276 (1997)
6. Dollár, P., Zitnick, C.L.: Fast edge detection using structured forests. IEEE Trans. Pattern Anal. Mach. Intell. (PAMI) 37(8), 1558–1570 (2014)
7. Eigen, D., Fergus, R.: Predicting depth, surface normals and semantic labels with a common multi-scale convolutional architecture. In: Conference on Computer Vision and Pattern Recognition (CVPR), pp. 2650–2658 (2015)
8. Eigen, D., Puhrsch, C., Fergus, R.: Depth map prediction from a single image using a multi-scale deep network. In: Ghahramani, Z., Welling, M., Cortes, C., Lawrence, N.D., Weinberger, K.Q. (eds.) Advances in Neural Information Processing Systems (NeurIPS), pp. 2366–2374. Curran Associates, Inc. (2014)
9. Fu, H., Gong, M., Wang, C., Batmanghelich, K., Tao, D.: Deep ordinal regression network for monocular depth estimation. In: Conference on Computer Vision and Pattern Recognition (CVPR), pp. 2002–2011 (2018)
10. Fu, H., Wang, C., Tao, D., Black, M.J.: Occlusion boundary detection via deep exploration of context. In: Conference on Computer Vision and Pattern Recognition (CVPR), pp. 241–250 (2016)
11. Godard, C., Mac Aodha, O., Brostow, G.J.: Unsupervised monocular depth estimation with left-right consistency. In: Conference on Computer Vision and Pattern Recognition (CVPR), pp. 270–279 (2017)
12. He, K., Sun, J., Tang, X.: Guided image filtering. In: Daniilidis, K., Maragos, P., Paragios, N. (eds.) ECCV 2010. LNCS, vol. 6311, pp. 1–14. Springer, Heidelberg (2010). https://doi.org/10.1007/978-3-642-15549-9_1
13. He, K., Zhang, X., Ren, S., Sun, J.: Deep residual learning for image recognition. In: Conference on Computer Vision and Pattern Recognition (CVPR), pp. 770–778 (2016)
14. He, X., Yuille, A.: Occlusion boundary detection using pseudo-depth. In: Daniilidis, K., Maragos, P., Paragios, N. (eds.) ECCV 2010. LNCS, vol. 6314, pp. 539–552. Springer, Heidelberg (2010). https://doi.org/10.1007/978-3-642-15561-1_39
15. Heise, P., Klose, S., Jensen, B., Knoll, A.: PM-Huber: patchmatch with Huber regularization for stereo matching. In: International Conference on Computer Vision (ICCV), pp. 2360–2367 (2013)
16. Heo, M., Lee, J., Kim, K.-R., Kim, H.-U., Kim, C.-S.: Monocular depth estimation using whole strip masking and reliability-based refinement. In: Ferrari, V., Hebert, M., Sminchisescu, C., Weiss, Y. (eds.) ECCV 2018. LNCS, vol. 11208, pp. 39–55. Springer, Cham (2018). https://doi.org/10.1007/978-3-030-01225-0_3

17. Hoiem, D., Efros, A.A., Hebert, M.: Recovering occlusion boundaries from an image. Int. J. Comput. Vis. (IJCV) **91**, 328–346 (2010)
18. Hong, Z., Chen, Z., Wang, C., Mei, X., Prokhorov, D., Tao, D.: Multi-store tracker (muster): a cognitive psychology inspired approach to object tracking. In: Conference on Computer Vision and Pattern Recognition (CVPR), pp. 749–758 (2015)
19. Ilg, E., Saikia, T., Keuper, M., Brox, T.: Occlusions, motion and depth boundaries with a generic network for disparity, optical flow or scene flow estimation. In: Ferrari, V., Hebert, M., Sminchisescu, C., Weiss, Y. (eds.) ECCV 2018. LNCS, vol. 11216, pp. 626–643. Springer, Cham (2018). https://doi.org/10.1007/978-3-030-01258-8_38
20. Jiao, J., Cao, Y., Song, Y., Lau, R.: Look deeper into depth: monocular depth estimation with semantic booster and attention-driven loss. In: Ferrari, V., Hebert, M., Sminchisescu, C., Weiss, Y. (eds.) ECCV 2018. LNCS, vol. 11219, pp. 55–71. Springer, Cham (2018). https://doi.org/10.1007/978-3-030-01267-0_4
21. Koch, T., Liebel, L., Fraundorfer, F., Körner, M.: Evaluation of CNN-based single-image depth estimation methods. In: Leal-Taixé, L., Roth, S. (eds.) ECCV 2018. LNCS, vol. 11131, pp. 331–348. Springer, Cham (2019). https://doi.org/10.1007/978-3-030-11015-4_25
22. Laina, I., Rupprecht, C., Belagiannis, V., Tombari, F., Navab, N.: Deeper depth prediction with fully convolutional residual networks. In: International Conference on 3D Vision (3DV), pp. 239–248. IEEE (2016)
23. Lee, J.H., Kim, C.S.: Monocular depth estimation using relative depth maps. In: Conference on Computer Vision and Pattern Recognition (CVPR), pp. 9729–9738 (2019)
24. Leichter, I., Lindenbaum, M.: Boundary ownership by lifting to 2.1-D. In: International Conference on Computer Vision (ICCV), pp. 9–16. IEEE (2008)
25. Li, J.Y., Klein, R., Yao, A.: A two-streamed network for estimating fine-scaled depth maps from single RGB images. In: International Conference on Computer Vision (ICCV), pp. 3392–3400 (2016)
26. Li, W., et al.: InteriorNet: mega-scale multi-sensor photo-realistic indoor scenes dataset. In: British Machine Vision Conference (BMVC) (2018)
27. Liu, C., Yang, J., Ceylan, D., Yumer, E., Furukawa, Y.: PlaneNet: piece-wise planar reconstruction from a single RGB image. In: Conference on Computer Vision and Pattern Recognition (CVPR), pp. 2579–2588 (2018)
28. Liu, F., Shen, C., Lin, G.: Deep convolutional neural fields for depth estimation from a single image. In: Conference on Computer Vision and Pattern Recognition (CVPR) (2015)
29. Liu, Y., Cheng, M.M., Fan, D.P., Zhang, L., Bian, J., Tao, D.: Semantic edge detection with diverse deep supervision. arXiv preprint arXiv:1804.02864 (2018)
30. Lu, R., Xue, F., Zhou, M., Ming, A., Zhou, Y.: Occlusion-shared and feature-separated network for occlusion relationship reasoning. In: International Conference on Computer Vision (ICCV) (2019)
31. Martin, D.R., Fowlkes, C.C., Malik, J.: Learning to detect natural image boundaries using local brightness, color, and texture cues. IEEE Trans. Pattern Anal. Mach. Intell. (PAMI) **26**(5), 530–549 (2004)
32. Massa, F., Marlet, R., Aubry, M.: Crafting a multi-task CNN for viewpoint estimation. In: British Machine Vision Conference (BMVC) (2016)
33. Silberman, N., Hoiem, D., Kohli, P., Fergus, R.: Indoor segmentation and support inference from RGBD images. In: Fitzgibbon, A., Lazebnik, S., Perona, P., Sato, Y., Schmid, C. (eds.) ECCV 2012. LNCS, vol. 7576, pp. 746–760. Springer, Heidelberg (2012). https://doi.org/10.1007/978-3-642-33715-4_54

34. Nitzberg, M., Mumford, D.B.: The 2.1-D Sketch. IEEE Computer Society Press (1990)
35. Oberweger, M., Rad, M., Lepetit, V.: Making deep heatmaps robust to partial occlusions for 3D object pose estimation. In: Ferrari, V., Hebert, M., Sminchisescu, C., Weiss, Y. (eds.) ECCV 2018. LNCS, vol. 11219, pp. 125–141. Springer, Cham (2018). https://doi.org/10.1007/978-3-030-01267-0_8
36. Peng, S., Liu, Y., Huang, Q., Zhou, X., Bao, H.: PVNet: pixel-wise voting network for 6DoF pose estimation. In: Conference on Computer Vision and Pattern Recognition (CVPR), pp. 4561–4570 (2019)
37. Rad, M., Lepetit, V.: BB8: a scalable, accurate, robust to partial occlusion method for predicting the 3D poses of challenging objects without using depth. In: International Conference on Computer Vision (ICCV), pp. 3828–3836 (2017)
38. Rafi, U., Gall, J., Leibe, B.: A semantic occlusion model for human pose estimation from a single depth image. In: Conference on Computer Vision and Pattern Recognition Workshops (CVPR Workshops), pp. 67–74 (2015)
39. Ramamonjisoa, M., Du, Y., Lepetit, V.: Predicting sharp and accurate occlusion boundaries in monocular depth estimation using displacement fields. In: Conference on Computer Vision and Pattern Recognition (CVPR), pp. 14648–14657 (2020)
40. Ramamonjisoa, M., Lepetit, V.: SharpNet: fast and accurate recovery of occluding contours in monocular depth estimation. In: International Conference on Computer Vision Workshops (ICCV Workshops) (2019)
41. Raskar, R., Tan, K.H., Feris, R., Yu, J., Turk, M.: Non-photorealistic camera: depth edge detection and stylized rendering using multi-flash imaging. ACM Trans. Graph. (TOG) 23(3), 679–688 (2004)
42. Ren, X., Fowlkes, C.C., Malik, J.: Figure/ground assignment in natural images. In: Leonardis, A., Bischof, H., Pinz, A. (eds.) ECCV 2006. LNCS, vol. 3952, pp. 614–627. Springer, Heidelberg (2006). https://doi.org/10.1007/11744047_47
43. Ricci, E., Ouyang, W., Wang, X., Sebe, N., et al.: Monocular depth estimation using multi-scale continuous CRFs as sequential deep networks. IEEE Trans. Pattern Anal. Mach. Intell.(PAMI) 41(6), 1426–1440 (2018)
44. Roberts, L.G.: Machine perception of three-dimensional solids. Ph.D. thesis, Massachusetts Institute of Technology (1963)
45. Ronneberger, O., Fischer, P., Brox, T.: U-Net: convolutional networks for biomedical image segmentation. In: International Conference on Medical Image Computing & Computer Assisted Intervention (MICCAI) (2015)
46. Stein, A.N., Hebert, M.: Occlusion boundaries from motion: low-level detection and mid-level reasoning. Int. J. Comput. Vis. (IJCV) 82, 325–357 (2008)
47. Su, H., Jampani, V., Sun, D., Gallo, O., Learned-Miller, E., Kautz, J.: Pixel-adaptive convolutional neural networks. In: Conference on Computer Vision and Pattern Recognition (CVPR), pp. 11166–11175 (2019)
48. Sugihara, K.: Machine Interpretation of Line Drawings, vol. 1. MIT press Cambridge (1986)
49. Teo, C., Fermuller, C., Aloimonos, Y.: Fast 2D border ownership assignment. In: Conference on Computer Vision and Pattern Recognition (CVPR) pp. 5117–5125 (2015)
50. Tomasi, C., Manduchi, R.: Bilateral filtering for gray and color images. In: International Conference on Computer Vision (ICCV), pp. 839–846 (1998)
51. Wang, G., Liang, X., Li, F.W.B.: DOOBNet: deep object occlusion boundary detection from an image. In: Asian Conference on Computer Vision (ACCV) (2018)

52. Wang, P., Shen, X., Russell, B., Cohen, S., Price, B., Yuille, A.L.: Surge: surface regularized geometry estimation from a single image. In: Advances in Neural Information Processing Systems (NeurIPS), pp. 172–180 (2016)
53. Wang, P., Yuille, A.: DOC: deep OCclusion estimation from a single image. In: Leibe, B., Matas, J., Sebe, N., Welling, M. (eds.) ECCV 2016. LNCS, vol. 9905, pp. 545–561. Springer, Cham (2016). https://doi.org/10.1007/978-3-319-46448-0_33
54. Wang, Y., Yang, Y., Yang, Z., Zhao, L., Wang, P., Xu, W.: Occlusion aware unsupervised learning of optical flow. In: Conference on Computer Vision and Pattern Recognition (CVPR), pp. 4884–4893 (2018)
55. Wu, H., Zheng, S., Zhang, J., Huang, K.: Fast end-to-end trainable guided filter. In: Conference on Computer Vision and Pattern Recognition (CVPR), pp. 1838–1847 (2018)
56. Xie, S., Tu, Z.: Holistically-nested edge detection. In: International Conference on Computer Vision (ICCV), pp. 1395–1403 (2015)
57. Xu, D., Ricci, E., Ouyang, W., Wang, X., Sebe, N.: Multi-scale continuous CRFs as sequential deep networks for monocular depth estimation. In: Conference on Computer Vision and Pattern Recognition (CVPR), pp. 5354–5362 (2017)
58. Yin, W., Liu, Y., Shen, C., Yan, Y.: Enforcing geometric constraints of virtual normal for depth prediction. In: International Conference on Computer Vision (ICCV) (2019)
59. Yu, Z., Feng, C., Liu, M.Y., Ramalingam, S.: CASENet: deep category-aware semantic edge detection. In: Conference on Computer Vision and Pattern Recognition (CVPR), pp. 5964–5973 (2017)
60. Yu, Z., et al.: Simultaneous edge alignment and learning. In: Ferrari, V., Hebert, M., Sminchisescu, C., Weiss, Y. (eds.) ECCV 2018. LNCS, vol. 11207, pp. 400–417. Springer, Cham (2018). https://doi.org/10.1007/978-3-030-01219-9_24
61. Zheng, C., Cham, T.-J., Cai, J.: T^2Net: synthetic-to-realistic translation for solving single-image depth estimation tasks. In: Ferrari, V., Hebert, M., Sminchisescu, C., Weiss, Y. (eds.) ECCV 2018. LNCS, vol. 11211, pp. 798–814. Springer, Cham (2018). https://doi.org/10.1007/978-3-030-01234-2_47
62. Zitnick, C.L., Kanade, T.: A cooperative algorithm for stereo matching and occlusion detection. IEEE Trans. Pattern Anal. Mach. Intell.(PAMI) **22**(7), 675–684 (2000)

MovieNet: A Holistic Dataset for Movie Understanding

Qingqiu Huang[✉], Yu Xiong, Anyi Rao, Jiaze Wang, and Dahua Lin

CUHK-SenseTime Joint Lab, The Chinese University of Hong Kong,
Shatin, Hong Kong
{hq016,xy017,ra018,dhlin}@ie.cuhk.edu.hk,
jzwang@link.cuhk.edu.hk

Abstract. Recent years have seen remarkable advances in visual understanding. However, how to understand a story-based long video with
artistic styles, *e.g.* movie, remains challenging. In this paper, we introduce MovieNet – a holistic dataset for movie understanding. MovieNet
contains $1,100$ movies with a large amount of multi-modal data,
e.g. trailers, photos, plot descriptions, *etc.*. Besides, different aspects of
manual annotations are provided in MovieNet, including 1.1 M characters with bounding boxes and identities, 42 K scene boundaries, 2.5 K
aligned description sentences, 65 K tags of place and action, and 92 K
tags of cinematic style. To the best of our knowledge, MovieNet is the
largest dataset with richest annotations for comprehensive movie understanding. Based on MovieNet, we set up several benchmarks for movie
understanding from different angles. Extensive experiments are executed
on these benchmarks to show the immeasurable value of MovieNet and
the gap of current approaches towards comprehensive movie understanding. We believe that such a holistic dataset would promote the researches
on story-based long video understanding and beyond. MovieNet will be
published in compliance with regulations at https://movienet.github.io.

1 Introduction

"You jump, I jump, right?" When Rose gives up the lifeboat and exclaims to
Jack, we are all deeply touched by the beautiful moving love story told by the
movie *Titanic*. As the saying goes, "Movies dazzle us, entertain us, educate
us, and delight us". Movie, where characters would face various situations and
perform various behaviors in various scenarios, is a reflection of our real world.
It teaches us a lot such as the stories took place in the past, the culture and
custom of a country or a place, the reaction and interaction of humans in different
situations, *etc.*. Therefore, to understand movies is to understand our world.

Q. Huang and Y. Xiong—Equal contribution.

Electronic supplementary material The online version of this chapter (https://
doi.org/10.1007/978-3-030-58548-8_41) contains supplementary material, which is
available to authorized users.

© Springer Nature Switzerland AG 2020
A. Vedaldi et al. (Eds.): ECCV 2020, LNCS 12349, pp. 709–727, 2020.
https://doi.org/10.1007/978-3-030-58548-8_41

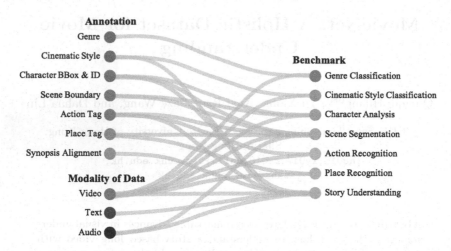

Fig. 1. The data, annotation, benchmark and their relations in MovieNet, which together build a holistic dataset for comprehensive movie understanding.

It goes not only for human, but also for an artificial intelligence system. We believe that movie understanding is a good arena for high-level machine intelligence, considering its high complexity and close relation to the real world. What's more, compared to web images [15] and short videos [7], the hundreds of thousands of movies in history containing rich content and multi-modal information become better nutrition for the data-hungry deep models.

Motivated by the insight above, we build a holistic dataset for movie understanding named *MovieNet* in this paper. As shown in Fig. 1, MovieNet comprises three important aspects, namely *data*, *annotation*, and *benchmark*.

First of all, MovieNet contains a large volume of data in multiple modalities, including movies, trailers, photos, subtitles, scripts and meta information like genres, cast, director, rating *etc.*. There are totally 3 K hour-long videos, 3.9 M photos, 10 M sentences of text and 7 M items of meta information in MovieNet.

From the annotation aspect, MovieNet contains massive labels to support different research topics of movie understanding. Based on the belief that middle-level entities, *e.g.* character, place, are important for high-level story understanding, various kinds of annotations on semantic elements are provided in MovieNet, including character bounding box and identity, scene boundary, action/place tag and aligned description in natural language. In addition, since movie is an art of filming, the cinematic styles, e.g., view scale, camera motion, lighting, *etc.*, are also beneficial for comprehensive video analysis. Thus we also annotate the view scale and camera motion for more than 46 K shots. Specifically, the annotations in MovieNet include: (1) 1.1 M characters with bounding boxes and identities; (2) 40 K scene boundaries; (3) 65 K tags of action and place; (4) 12 K description sentences aligned to movie segments; (5) 92 K tags of cinematic styles.

Based on the data and annotations in MovieNet, we exploit some research topics that cover different aspects of movie understanding, *i.e.* genre analysis,

Fig. 2. MovieNet is a holistic dataset for movie understanding, which contains massive data from different modalities and high-quality annotations in different aspects. Here we show some data (in blue) and annotations (in green) of *Titanic* in MovieNet. (Color figure online)

cinematic style prediction, character analysis, scene understanding, and movie segment retrieval. For each topic, we set up one or several challenging benchmarks. Then extensive experiments are executed to present the performances of different methods. By further analysis on the experimental results, we will also show the gap of current approaches towards comprehensive movie understanding, as well as the advantages of holistic annotations for throughout video analytics.

To the best of our knowledge, MovieNet is the first holistic dataset for movie understanding that contains a large amount of data from different modalities and high-quality annotations in different aspects. We hope that it would promote the researches on video editing, human-centric situation understanding, story-based video analytics and beyond.

2 Related Datasets

Existing Works. Most of the datasets of movie understanding focus on a specific element of movies, *e.g.* genre [49,66], character [1,3,19,26,29,39,51], action [5,6,18,32,37], scene [11,14,25,40,41,43] and description [47]. Also their scale is quite small and the annotation quantities are limited. For example, [3,19,51] take several episodes from TV series for character identification, [32] uses clips from twelve movies for action recognition, and [40] exploits scene segmentation with only three movies. Although these datasets focus on some important aspects of movie understanding, their scale is not enough for the data-hungry learning paradigm. Furthermore, the deep comprehension should go from middle-level elements to high-level story while each existing dataset can only support a single task, causing trouble for comprehensive movie understanding.

MovieQA. MovieQA [54] consists of 15 K questions designed for 408 movies. As for sources of information, it contains video clips, plots, subtitles, scripts, and DVS (Descriptive Video Service). To evaluate story understanding by QA is a good idea, but there are two problems. (1) Middle-level annotations, e.g., character identities, are missing. Therefore it is hard to develop an effective approach towards high-level understanding. (2) The questions in MovieQA come from the wiki plot. Thus it is more like a textual QA problem rather than story-based video understanding. A strong evidence is that the approaches based on textual plot can get a much higher accuracy than those based on "video+subtitle".

LSMDC. LSMDC [45] consists of 200 movies with audio description (AD) providing linguistic descriptions of movies for visually impaired people. AD is quite different from the natural descriptions of most audiences, limiting the usage of the models trained on such datasets. And it is also hard to get a large number of ADs. Different from previous work [45,54], we provide multiple sources of textual information and different annotations of middle-level entities in MovieNet, leading to a better source for story-based video understanding.

AVA. Recently, AVA dataset [24], an action recognition dataset with 430 15-min movie clips annotated with 80 spatial-temporal atomic visual actions, is proposed. AVA dataset aims at facilitating the task of recognizing atomic visual actions. However, regarding the goal of story understanding, the AVA dataset is not applicable since (1) The dataset is dominated by labels like *stand* and *sit*, making it extremely unbalanced. (2) Actions like *stand, talk, watch* are less informative in the perspective of story analytics. Hence, we propose to annotate semantic level actions for both action recognition and story understanding tasks.

MovieGraphs. MovieGraphs [55] is the most related one that provides graph-based annotations of social situations depicted in clips of 51 movies. The annotations consist of characters, interactions, attributes, *etc.*. Although sharing the same idea of multi-level annotations, MovieNet is different from MovieGraphs in three aspects: (1) MovieNet contains not only movie clips and annotations, but also photos, subtitles, scripts, trailers, *etc.*, which can provide richer data for various research topics. (2) MovieNet can support and exploit different aspects of movie understanding while MovieGraphs focuses on situation recognition only. (3) The scale of MovieNet is much larger than MovieGraphs.

Table 1. Comparison between MovieNet and related datasets in terms of data.

	# movie	Trailer	Photo	Meta	Script	Synop.	Subtitle	Plot	AD
MovieQA [54]	140						✓	✓	
LSMDC [45]	200				✓				✓
MovieGraphs [55]	51								
AVA [24]	430								
MovieNet	1,100	✓	✓	✓	✓	✓	✓	✓	

Table 2. Comparison between MovieNet and related datasets in terms of annotation.

	# character	# scene	# cine. tag	# aligned sent.	# action/place tag
MovieQA [54]	-	-	-	15 K	-
LSMDC [45]	-	-	-	128 K	-
MovieGraphs [55]	22 K	-	-	21 K	23 K
AVA [24]	116 K	-	-	-	360 K
MovieNet	1.1 M	42 K	92 K	25 K	65 K

3 Visit MovieNet: Data and Annotation

MovieNet contains various kinds of data from multiple modalities and high-quality annotations on different aspects for movie understanding. Figure 2 shows the data and annotations of the movie *Titanic* in MovieNet. Comparisons between MovieNet and other datasets for movie understanding are shown in Tables 1 and 2. All these demonstrate the tremendous advantage of MovieNet on both quality, scale and richness.

3.1 Data in MovieNet

Movie. We carefully selected and purchased the copies of 1, 100 movies, the criteria of which are (1) colored; (2) longer than 1 h; (3) cover a wide range of genres, years and countries.

Metadata. We get the meta information of the movies from IMDb and TMDb[1], including title, release date, country, genres, rating, runtime, director, cast, storyline, *etc.*. Here we briefly introduce some of the key elements, please refer

---

[1] IMDb: https://www.imdb.com; TMDb: https://www.themoviedb.org.

to supplementary material for detail: (1) Genre is one of the most important attributes of a movie. There are total 805 K genre tags from 28 unique genres in MovieNet. (2) For cast, we get both their names, IMDb IDs and the character names in the movie. (3) We also provide IMDb ID, TMDb ID and Douban ID of each movie, with which the researchers can get additional meta information from these websites conveniently. The total number of meta information in MovieNet is 375 K. Please note that each kind of data itself, even without the movie, can support some research topics [31]. So we try to get each kind of data as much as we can. Therefore the number here is larger than 1, 100. So as other kinds of data we would introduce below.

Subtitle. The subtitles are obtained in two ways. Some of them are extracted from the embedded subtitle stream in the movies. For movies without original English subtitle, we crawl the subtitles from YIFY[2]. All the subtitles are manually checked to ensure that they are aligned to the movies.

Trailer. We download the trailers from YouTube according to their links from IMDb and TMDb. We found that this scheme is better than previous work [10], which use the titles to search trailers from YouTube, since the links of the trailers in IMDb and TMDb have been manually checked by the organizers and audiences. Totally, we collect 60 K trailers belonging to 33 K unique movies.

Script. Script, where the movement, actions, expression and dialogs of the characters are narrated, is a valuable textual source for research topics of movie-language association. We collect around 2 K scripts from IMSDb and Daily Script[3]. The scripts are aligned to the movies by matching the dialog with subtitles.

Synopsis. A synopsis is a description of the story in a movie written by audiences. We collect 11 K high-quality synopses from IMDb, all of which contain more than 50 sentences. Synopses are also manually aligned to the movie, which would be introduced in Sect. 3.2.

Photo. We collect 3.9 M photos of the movies from IMDb and TMDb, including poster, still frame, publicity, production art, product, behind the scene and event.

3.2   Annotation in MovieNet

To provide a high-quality dataset supporting different research topics on movie understanding, we make great effort to clean the data and manually annotate various labels on different aspects, including character, scene, event and cinematic style. Here we just demonstrate the *content* and the *amount* of annotations due to the space limit. Please refer to supplementary material for details.

Cinematic Styles. Cinematic style, such as view scale, camera movement, lighting and color, is an important aspect of comprehensive movie understanding since

---

[2] https://www.yifysubtitles.com/.

[3] IMSDb: https://www.imsdb.com/; DailyScript: https://www.dailyscript.com/.

it influences how the story is telling in a movie. In MovieNet, we choose two kinds of cinematic tags for study, namely view scale and camera movement. Specifically, the view scale include five categories, *i.e. long shot, full shot, medium shot, close-up shot* and *extreme close-up shot*, while the camera movement is divided into four classes, *i.e. static shot, pans and tilts shot, zoom in* and *zoom out*. The original definitions of these categories come from [22] and we simplify them for research convenience. We totally annotate 47 K shots from movies and trailers, each with one tag of view scale and one tag of camera movement.

Character Bounding Box and Identity. Person plays an important role in human-centric videos like movies. Thus to detect and identify characters is a foundational work towards movie understanding. The annotation process of character bounding box and identity contains 4 steps: (1) Some key frames, the number of which is 758 K, from different movies are selected for bounding box annotation. (2) A detector is trained with the annotations in step-1. (3) We use the trained detector to detect more characters in the movies and manually clean the detected bounding boxes. (4) We then manually annotate the identities of all the characters. To make the cost affordable, we only keep the top 10 cast in credits order according to IMDb, which can cover the main characters for most movies. Characters not belong to credited cast were labeled as "others". In total, we got 1.1 M instances of 3,087 unique credited cast and 364 K "others".

Scene Boundary. In terms of temporal structure, a movie contains two hierarchical levels – shot, and scene. Shot is the minimal visual unit of a movie while scene is a sequence of continued shots that are semantically related. To capture the hierarchical structure of a movie is important for movie understanding. Shot boundary detection has been well solved by [48], while scene boundary detection, also named scene segmentation, remains an open question. In MovieNet, we manually annotate the scene boundaries to support the researches on scene segmentation, resulting in 42 K scenes.

Action/Place Tags. To understand the event(s) happened within a scene, action and place tags are required. Hence, we first split each movie into clips according to the scene boundaries and then manually annotated place and action tags for each segment. For place annotation, each clip is annotated with multiple place tags, e.g., {deck, cabin}. While for action annotation, we first detect sub-clips that contain characters and actions, then we assign multiple action tags to each sub-clip. We have made the following efforts to keep tags diverse and informative: (1) We encourage the annotators to create new tags. (2) Tags that convey little information for story understanding, e.g., *stand* and *talk*, are excluded. Finally, we merge the tags and filtered out 80 actions and 90 places with a minimum frequency of 25 as the final annotations. In total, there are 42 K segments with 19.6 K place tags and 45 K action tags.

Description Alignment. Since the event is more complex than character and scene, a proper way to represent an event is to describe it with natural language. Previous works have already aligned script [37], Descriptive Video Service (DVS) [45], book [67] or wiki plot [52–54] to movies. However, books cannot be well aligned since most of the movies would be quite different from their books. DVS transcripts are quite hard to obtain, limiting the scale of the datasets based on them [45]. Wiki plot is usually a short summary that cannot cover all the important events of the movie. Considering the issues above, we choose synopses as the story descriptions in MovieNet. The associations between the movie segments and the synopsis paragraphs are manually annotated by three different annotators with a coarse-to-fine procedure. Finally, we obtained 4,208 highly consistent paragraph-segment pairs.

4   Play with MovieNet: Benchmark and Analysis

With a large amount of data and holistic annotations, MovieNet can support various research topics. In this section, we try to analyze movies from five aspects, namely *genre, cinematic style, character, scene* and *story*. For each topic, we would set up one or several benchmarks based on MovieNet. Baselines with currently popular techniques and analysis on experimental results are also provided to show the potential impact of MovieNet in various tasks. The topics of the tasks have covered different perspectives of comprehensive movie understanding. But due to the space limit, here we can only touched the tip of the iceberg. More detailed analysis are provided in the supplementary material and more interesting topics to be exploited are introduced in Sect. 5.

Table 3. (a) Comparison between MovieNet and other benchmarks for genre analysis. (b) Results of some baselines for genre classification in MovieNet

(a)

	Genre	Movie	Trailer	Photo
MGCD [66]	4	-	1.2K	-
LMTD [49]	4	-	3.5K	-
MScope [10]	13	-	5.0K	5.0K
MovieNet	**21**	**1.1K**	**68K**	**1.6M**

(b)

Data	Model	r@0.5	p@0.5	mAP
Photo	VGG16 [50]	27.32	66.28	32.12
	ResNet50 [27]	**34.58**	**72.28**	**46.88**
Trailer	TSN-r50 [57]	17.95	**78.31**	43.70
	I3D-r50 [9]	16.54	69.58	35.79
	TRN-r50 [65]	**21.74**	77.63	**45.23**

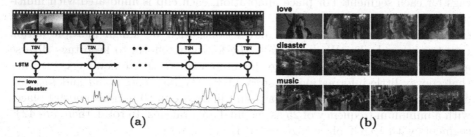

(a)   (b)

Fig. 3. (a) Framework of genre analysis in movies. (b) Some samples of genre-guided trailer generation for movie *Titanic*.

4.1 Genre Analysis

Genre is a key attribute for any media with artistic elements. To classify the genres of movies has been widely studied by previous works [10,49,66]. But there are two drawbacks for these works. (1) The scale of existing datasets is quite small. (2) All these works focus on image or trailer classification while ignore a more important problem, *i.e.* how to analyze the genres of a long video.

MovieNet provides a large-scale benchmark for genre analysis, which contains 1.1 K movies, 68 K trailers and 1.6 M photos. The comparison between different datasets are shown in Table 3a, from which we can see that MovieNet is much larger than previous datasets.

Based on MovieNet, we first provide baselines for both image-based and video-based genre classification, the results are shown Table 3b. Comparing the result of genre classification in small datasets [10,49] to ours in MovieNet, we find that the performance drops a lot when the scale of the dataset become larger. The newly proposed MovieNet brings two challenges to previous methods. (1) Genre classification in MovieNet becomes a long-tail recognition problem where the label distribution is extremely unbalanced. For example, the number of "Drama" is 40 times larger than that of "Sport" in MovieNet. (2) Genre is a high-level semantic tag depending on action, clothing and facial expression of the characters, and even BGM. Current methods are good at visual representation. When facing a problem that need to consider higher-level semantics, they would all fail. We hope MovieNet would promote researches on these challenging topics.

Another new issue to address is how to analyze the genres of a movie. Since movie is extremely long and not all segments are related to its genres, this problem is much more challenging. Following the idea of learning from trailers and applying to movies [30], we adopt the visual model trained with trailers as shot-level feature extractor. Then the features are fed to a temporal model to capture the temporal structure of the movie. The overall framework is shown in Fig. 3a. With this approach, we can get the genre response curve of a movie. Specifically, we can predict which part of the movie is more relevant to a specific genre. What's more, the prediction can also be used for genre-guided trailer generation, as shown in Fig. 3b. From the analysis above, we can see that MovieNet would promote the development of this challenging and valuable research topic.

Table 4. (a) Comparison between MovieNet and other benchmarks for cinematic style prediction. (b) Results of some baselines for cinematic style prediction in MovieNet

(a)

	Shot	Video	Scale	Move.
Lie 2014 [4]	327	327		✓
Sports 2007 [62]	1,364	8	✓	
Context 2011 [61]	3,206	4	✓	
Taxon 2009 [56]	5,054	7		✓
MovieNet	46,857	7,858	✓	✓

(b)

Method	Scale acc.	Move. acc.
I3D [9]	76.79	78.45
TSN [57]	84.08	70.46
TSN+R^{3}Net [16]	**87.50**	**80.65**

Table 5. Datasets for person analysis.

	ID	Instance	Source
COCO [35]	-	262 K	Web image
CalTech [17]	-	350 K	Surveillance
Market [64]	1,501	32 K	Surveillance
CUHK03 [33]	1,467	28 K	Surveillance
AVA [24]	-	426 K	Movie
CSM [28]	1,218	127 K	Movie
MovieNet	3,087	1.1 M	Movie

Fig. 4. Persons in different data sources

Table 6. Results of (a) Character detection and (b) Character identification

(a)

Train data	Method	mAP
COCO [35]	FasterRCNN	81.50
Caltech [17]	FasterRCNN	5.67
CSM [28]	FasterRCNN	89.91
MovieNet	FasterRCNN	92.13
	RetinaNet	91.55
	CascadeRCNN	**95.17**

(b)

Train Data	Cues	Method	mAP
Market [64]	body	r50-softmax	4.62
CUHK03 [33]	body	r50-softmax	5.33
CSM [28]	body	r50-softmax	26.21
MovieNet	body	r50-softmax	32.81
	body+face	two-step[36]	63.95
	body+face	PPCC[28]	**75.95**

4.2 Cinematic Style Analysis

As we mentioned before, cinematic style is about how to present the story to audience in the perspective of filming art. For example, a *zoom in* shot is usually used to attract the attention of audience to a specific object. In fact, cinematic style is crucial for both video understanding and editing. But there are few works focusing on this topic and no large-scale datasets for this research topic too.

Based on the tags of cinematic style we annotated in MovieNet, we set up a benchmark for cinematic style prediction. Specifically, we would like to recognize the view scale and camera motion of each shot. Comparing to existing datasets, MovieNet is the first dataset that covers both view scale and camera motion, and it is also much larger, as shown in Table 4a. Several models for video clip classification such as TSN [57] and I3D [9] are applied to tackle this problem, the results are shown in Table 4b. Since the view scale depends on the portion of the subject in the shot frame, to detect the subject is important for cinematic style prediction. Here we adopt the approach from saliency detection [16] to get the subject maps of each shot, with which better performances are achieved, as shown in Table 4b. Although utilizing subject points out a direction for this task, there is still a long way to go. We hope that MovieNet can promote the development of this important but ignored topic for video understanding.

4.3 Character Recognition

It has been shown by existing works [36,55,58] that movie is a human-centric video where characters play an important role. Therefore, to detect and

Table 7. Dataset for scene analysis.

	Scene	Action	Place
OVSD [46]	300	-	-
BBC [2]	670	-	-
Hollywood2 [37]	-	1.7 K	1.2 K
MovieGraph [55]	-	23.4 K	7.6 K
AVA [24]	-	360 K	-
MovieNet	**42 K**	**45.0 K**	**19.6 K**

Table 8. Datasets for story understanding in movies in terms of (1) number of sentences per movie; (2) duration (second) per segment.

Dataset	Sent./mov	Dur./seg
MovieQA [54]	35.2	202.7
MovieGraphs [55]	408.8	44.3
MovieNet	83.4	428.0

Table 9. Results of scene segmentation

Dataset	Method	AP(↑)	M_{iou}(↑)
OVSD [46]	MS-LSTM	0.313	0.387
BBC [2]	MS-LSTM	0.334	0.379
MovieNet	Grouping [46]	0.336	0.372
	Siamese [2]	0.358	0.396
	MS-LSTM	**0.465**	**0.462**

Table 10. Results of scene tagging

Tags	Method	mAP
Action	TSN [57]	14.17
	I3D [9]	20.69
	SlowFast [20]	**23.52**
Place	I3D [9]	7.66
	TSN [57]	**8.33**

identify characters is crucial for movie understanding. Although person/character recognition is not a new task, all previous works either focus on other data sources [33,35,64] or small-scale benchmarks [3,26,51], leading to the results lack of convincingness for character recognition in movies.

We proposed two benchmarks for character analysis in movies, namely, character detection and character identification. We provide more than 1.1 M instances from 3,087 identities to support these benchmarks. As shown in Table 5, MovieNet contains much more instances and identities comparing to some popular datasets about person analysis. The following sections will show the analysis on character detection and identification respectively.

Character Detection. Images from different data sources would have large domain gap, as shown in Fig. 4. Therefore, a character detector trained on general object detection dataset, e.g. COCO [35], or pedestrian dataset, e.g. CalTech [17], is not good enough for detecting characters in movies. This can be supported by the results shown in Table 6a. To get a better detector for character detection, we train different popular models [8,34,44] with MovieNet using toolboxes from [12,13]. We can see that with the diverse character instances in MovieNet, a Cascade R-CNN trained with MovieNet can achieve extremely high performance, i.e. 95.17% in mAP. That is to say, character detection can be well solved by a large-scale movie dataset with current SOTA detection models. This powerful detector would then benefit research on character analysis in movies.

Character Identification. To identify the characters in movies is a more challenging problem, which can be observed by the diverse samples shown in Fig. 4.

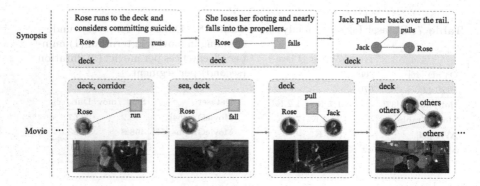

Fig. 5. Example of synopses paragraph and movie segment in MovieNet-MSR. It demonstrate the spatial-temporal structures of stories in movies and synopses. We can also see that character, action and place are the key element for story understanding.

We conduct different experiments based on MovieNet, the results are shown in Table 6b. From these results, we can see that: (1) models trained on ReID datasets are inefficient for character recognition due to domain gap; (2) to aggregate different visual cues of an instance is important for character recognition in movies; (3) the current state-of-the-art can achieve 75.95% mAP, which demonstrates that it is a challenging problem which need to be further exploited.

4.4   Scene Analysis

As mentioned before, scene is the basic semantic unit of a movie. Therefore, it is important to analyze the scenes in movies. The key problems in scene understanding is probably *where is the scene boundary* and *what is the content in a scene*. As shown in Table 7, MovieNet, which contains more than 43K scene boundaries and 65K action/place tags, is the only one that can support both *scene segmentation* and *scene tagging*. What's more, the scale of MovieNet is also larger than all previous works.

Scene Segmentation. We first test some baselines [2,46] for scene segmentation. In addition, we also propose a sequential model, named Multi-Semtantic LSTM (MS-LSTM) based on Bi-LSTMs [23,42] to study the gain brought by using multi-modality and multiple semantic elements, including audio, character, action and scene. From the results shown in Table 9, we can see that (1) Benefited from large scale and high diversity, models trained on MovieNet can achieve better performance. (2) Multi-modality and multiple semantic elements are important for scene segmentation, which highly raise the performance.

Action/Place Tagging. To further understand the stories within a movie, it is essential to perform analytics on the key elements of storytelling, *i.e.*, place and action. We would introduce two benchmarks in this section. Firstly, for action analysis, the task is multi-label action recognition that aims to recognize all the human actions or interactions in a given video clip. We implement three standard action recognition models, *i.e.*, TSN [57], I3D [9] and SlowFast Network [20] modified from [63] in experiments. Results are shown in Table 10. For place analysis, we propose another benchmark for multi-label place classification. We adopt I3D [9] and TSN [57] as our baseline models and the results are shown in Table 10. From the results, we can see that action and place tagging is an extremely challenging problem due to the high diversity of different instances.

4.5   Story Understanding

Web videos are broadly adopted in previous works [7,60] as the source of video understanding. Compared to web videos, the most distinguishing feature of movies is the story. Movies are created to tell stories and the most explicit way to demonstrate a story is to describe it using natural language, *e.g.* synopsis. Inspired by the above observations, we choose the task of movie segment retrieval with natural language to analyze the stories in movies. Based on the aligned synopses in MovieNet, we set up a benchmark for movie segment retrieval. Specifically, given a synopsis paragraph, we aim to find the most relevant movie segment that covers the story in the paragraph. It is a very challenging task due to the rich content in movie and high-level semantic descriptions in synopses. Table 8 shows the comparison of our benchmark dataset with other related datasets. We can see that our dataset is more complex in terms of descriptions compared with MovieQA [54] while the segments are longer and contain more information than those of MovieGraphs [55].

Generally speaking, a story can be summarized as *"somebody do something in some time at some place"*. As shown in Fig. 5, both stories represented by language and video can be composed as sequences of {character, action, place} graphs. That being said, to understand a story is to (1) recognize the key elements of story-telling, namely, character, action, place *etc.*; (2) analyze the spatial-temporal structures of both movie and synopsis. Hence, our method first leverage middle-level entities (*e.g.* character, scene), as well as multi-modality (*e.g.* subtitle) to assist retrieval. Then we explore the spatial-temporal structure from both movies and synopses by formulating middle-level entities into graph structures. Please refer to supplementary material for details.

Using Middle-Level Entities and Multi-modality. We adopt VSE [21] as our baseline model where the vision and language features are embedded into a joint space. Specifically, the feature of the paragraph is obtained by taking the average of Word2Vec [38] feature of each sentence while the visual feature is obtained by taking the average of the appearance feature extracted from ResNet [27] on each shot. We add subtitle feature to enhance visual feature. Then different semantic elements including character, action and cinematic style are aggregated in our framework. We are able to obtain action features and character features thanks to the models trained on other benchmarks on MovieNet, e.g., action recognition and character detection. Furthermore, we observe that the focused elements vary under different cinematic styles. For example, we should focus more on actions in a full shot while more on character and dialog in a close-up shot. Motivated by this observation, we propose a cinematic-style-guided

Table 11. Results of movie segment retrieval. Here, G stands for global appearance feature, S for subtitle feature, A for action, P for character and C for cinematic style.

Method	Recall@1	Recall@5	Recall@10	MedR
Random	0.11	0.54	1.09	460
G	3.16	11.43	18.72	66
G+S	3.37	13.17	22.74	56
G+S+A	5.22	13.28	20.35	52
G+S+A+P	18.50	43.96	55.50	7
G+S+A+P+C	18.72	44.94	56.37	7
MovieSynAssociation [59]	**21.98**	**51.03**	**63.00**	**5**

attention module that predicts the weights over each element (e.g., action, character) within a shot, which would be used to enhance the visual features. The experimental results are shown in Table 11. Experiments show that by considering different elements of the movies, the performance improves a lot. We can see that a holistic dataset which contains holistic annotations to support middle-level entity analyses is important for movie understanding.

Explore Spatial-Temporal Graph Structure in Movies and Synopses. Simply adding different middle-level entities improves the result. Moreover, as shown in Fig. 5, we observe that stories in movies and synopses persist two important structure: (1) the temporal structure in movies and synopses is that the story can be composed as a sequence of events following a certain temporal order. (2) the spatial relation of different middle-level elements, e.g., character co-existence and their interactions, can be formulated as graphs. We implement the method in [59] to formulate the above structures as two graph matching problems. The result are shown in Table 11. Leveraging the graph formulation for the internal structures of stories in movies and synopses, the retrieval performance can be further boosted, which in turn, show that the challenging MovieNet would provide a better source to story-based movie understanding.

5   Discussion and Future Work

In this paper, we introduce MovieNet, a holistic dataset containing different aspects of annotations to support comprehensive movie understanding. We introduce several challenging benchmarks on different aspects of movie understanding, *i.e.* discovering filming art, recognizing middle-level entities and understanding high-level semantics like stories. Furthermore, the results of movie segment retrieval demonstrate that integrating filming art and middle-level entities according to the internal structure of movies would be helpful for story understanding. These in turn, show the effectiveness of holistic annotations.

In the future, our work would go on in two aspects. (1) **Extending the Annotation.** In the future, we would further extend the dataset to include more movies and annotations. (2) **Exploring more Approaches and Topics.** To tackle the challenging tasks proposed above, we would explore more effective approaches. Besides, there are more meaningful and practical topics that can be addressed with MovieNet, such as movie deoldify, trailer generation, *etc.*

Acknowledgment. This work is partially supported by the SenseTime Collaborative Grant on Large-scale Multi-modality Analysis (CUHK Agreement No. TS1610626 & No. TS1712093), the General Research Fund (GRF) of Hong Kong (No. 14203518 & No. 14205719), and Innovation and Technology Support Program (ITSP) Tier 2, ITS/431/18F.

References

1. Arandjelovic, O., Zisserman, A.: Automatic face recognition for film character retrieval in feature-length films. In: Proceedings of the IEEE Conference on Computer Vision and Pattern Recognition. IEEE (2005)
2. Baraldi, L., Grana, C., Cucchiara, R.: A deep siamese network for scene detection in broadcast videos. In: 23rd ACM International Conference on Multimedia, pp. 1199–1202. ACM (2015)
3. Bauml, M., Tapaswi, M., Stiefelhagen, R.: Semi-supervised learning with constraints for person identification in multimedia data. In: Proceedings of the IEEE Conference on Computer Vision and Pattern Recognition (2013)
4. Bhattacharya, S., Mehran, R., Sukthankar, R., Shah, M.: Classification of cinematographic shots using lie algebra and its application to complex event recognition. IEEE Trans. Multimed. **16**(3), 686–696 (2014)
5. Bojanowski, P., Bach, F., Laptev, I., Ponce, J., Schmid, C., Sivic, J.: Finding actors and actions in movies. In: Proceedings of the IEEE International Conference on Computer Vision (2013)
6. Bojanowski, P., et al.: Weakly supervised action labeling in videos under ordering constraints. In: Fleet, D., Pajdla, T., Schiele, B., Tuytelaars, T. (eds.) ECCV 2014. LNCS, vol. 8693, pp. 628–643. Springer, Cham (2014). https://doi.org/10.1007/978-3-319-10602-1_41
7. Caba Heilbron, F., Escorcia, V., Ghanem, B., Carlos Niebles, J.: ActivityNet: a large-scale video benchmark for human activity understanding. In: Proceedings of the IEEE Conference on Computer Vision and Pattern Recognition, pp. 961–970 (2015)
8. Cai, Z., Vasconcelos, N.: Cascade R-CNN: delving into high quality object detection. In: Proceedings of the IEEE Conference on Computer Vision and Pattern Recognition, pp. 6154–6162 (2018)
9. Carreira, J., Zisserman, A.: Quo vadis, action recognition? a new model and the kinetics dataset. In: Proceedings of the IEEE Conference on Computer Vision and Pattern Recognition, pp. 6299–6308 (2017)
10. Cascante-Bonilla, P., Sitaraman, K., Luo, M., Ordonez, V.: Moviescope: Large-scale analysis of movies using multiple modalities. arXiv preprint arXiv:1908.03180 (2019)
11. Chasanis, V.T., Likas, A.C., Galatsanos, N.P.: Scene detection in videos using shot clustering and sequence alignment. IEEE Trans. Multimed. **11**, 89–100 (2008)
12. Chen, K., et al.: Hybrid task cascade for instance segmentation. In: The IEEE Conference on Computer Vision and Pattern Recognition (CVPR), June 2019

13. Chen, K., et al.: MMDetection: open MMLab detection toolbox and benchmark. arXiv preprint arXiv:1906.07155 (2019)
14. Del Fabro, M., Böszörmenyi, L.: State-of-the-art and future challenges in video scene detection: a survey. Multimed. Syst. **19**, 427–454 (2013). https://doi.org/10.1007/s00530-013-0306-4
15. Deng, J., Dong, W., Socher, R., Li, L.J., Li, K., Fei-Fei, L.: ImageNet: a large-scale hierarchical image database. In: 2009 IEEE Conference on Computer Vision and Pattern Recognition. IEEE (2009)
16. Deng, Z., et al.: R3net: recurrent residual refinement network for saliency detection. In: Proceedings of the 27th International Joint Conference on Artificial Intelligence, pp. 684–690. AAAI Press (2018)
17. Dollar, P., Wojek, C., Schiele, B., Perona, P.: Pedestrian detection: an evaluation of the state of the art. IEEE Trans. Pattern Anal. Mach. Intell. **34**(4), 743–761 (2011)
18. Duchenne, O., Laptev, I., Sivic, J., Bach, F.R., Ponce, J.: Automatic annotation of human actions in video. In: Proceedings of the IEEE International Conference on Computer Vision (2009)
19. Everingham, M., Sivic, J., Zisserman, A.: Hello my name is... buffy - automatic naming of characters in TV video. In: BMVC (2006)
20. Feichtenhofer, C., Fan, H., Malik, J., He, K.: Slowfast networks for video recognition. In: Proceedings of the IEEE International Conference on Computer Vision, pp. 6202–6211 (2019)
21. Frome, A., et al.: Devise: a deep visual-semantic embedding model. In: Advances in Neural Information Processing Systems, pp. 2121–2129 (2013)
22. Giannetti, L.D., Leach, J.: Understanding Movies, vol. 1. Prentice Hall Upper Saddle River, New Jersey (1999)
23. Graves, A., Schmidhuber, J.: Framewise phoneme classification with bidirectional LSTM and other neural network architectures. Neural Networks **18**(5–6), 602–610 (2005)
24. Gu, C., et al.: Ava: a video dataset of spatio-temporally localized atomic visual actions. In: Proceedings of the IEEE Conference on Computer Vision and Pattern Recognition, pp. 6047–6056 (2018)
25. Han, B., Wu, W.: Video scene segmentation using a novel boundary evaluation criterion and dynamic programming. In: IEEE International Conference on Multimedia and Expo. IEEE (2011)
26. Haurilet, M.L., Tapaswi, M., Al-Halah, Z., Stiefelhagen, R.: Naming TV characters by watching and analyzing dialogs. In: 2016 IEEE Winter Conference on Applications of Computer Vision (WACV). IEEE (2016)
27. He, K., Zhang, X., Ren, S., Sun, J.: Deep residual learning for image recognition. In: Proceedings of the IEEE Conference on Computer Vision and Pattern Recognition (2016)
28. Huang, Q., Liu, W., Lin, D.: Person search in videos with one portrait through visual and temporal links. In: Ferrari, V., Hebert, M., Sminchisescu, C., Weiss, Y. (eds.) ECCV 2018. LNCS, vol. 11217, pp. 437–454. Springer, Cham (2018). https://doi.org/10.1007/978-3-030-01261-8_26
29. Huang, Q., Xiong, Y., Lin, D.: Unifying identification and context learning for person recognition. In: The IEEE Conference on Computer Vision and Pattern Recognition (CVPR), June 2018
30. Huang, Q., Xiong, Y., Xiong, Y., Zhang, Y., Lin, D.: From trailers to storylines: an efficient way to learn from movies. arXiv preprint arXiv:1806.05341 (2018)

31. Huang, Q., Yang, L., Huang, H., Wu, T., Lin, D.: Caption-supervised face recognition: Training a state-of-the-art face model without manual annotation. In: Proceedings of the European Conference on Computer Vision (ECCV) (2020)
32. Laptev, I., Marszałek, M., Schmid, C., Rozenfeld, B.: Learning realistic human actions from movies. In: Proceedings of the IEEE Conference on Computer Vision and Pattern Recognition. IEEE Computer Society (2008)
33. Li, W., Zhao, R., Xiao, T., Wang, X.: Deepreid: deep filter pairing neural network for person re-identification. In: Proceedings of the IEEE Conference on Computer Vision and Pattern Recognition, pp. 152–159 (2014)
34. Lin, T.Y., Goyal, P., Girshick, R., He, K., Dollár, P.: Focal loss for dense object detection. In: Proceedings of the IEEE International Conference on Computer Vision, pp. 2980–2988 (2017)
35. Lin, T.-Y., et al.: Microsoft COCO: common objects in context. In: Fleet, D., Pajdla, T., Schiele, B., Tuytelaars, T. (eds.) ECCV 2014. LNCS, vol. 8693, pp. 740–755. Springer, Cham (2014). https://doi.org/10.1007/978-3-319-10602-1_48
36. Loy, C.C., et al.: Wider face and pedestrian challenge 2018: Methods and results. arXiv preprint arXiv:1902.06854 (2019)
37. Marszałek, M., Laptev, I., Schmid, C.: Actions in context. In: Proceedings of the IEEE Conference on Computer Vision and Pattern Recognition. IEEE Computer Society (2009)
38. Mikolov, T., Chen, K., Corrado, G., Dean, J.: Efficient estimation of word representations in vector space. arXiv preprint arXiv:1301.3781 (2013)
39. Nagrani, A., Zisserman, A.: From benedict cumberbatch to sherlock holmes: character identification in TV series without a script. BMVC (2017)
40. Park, S.B., Kim, H.N., Kim, H., Jo, G.S.: Exploiting script-subtitles alignment to scene boundary detection in movie. In: IEEE International Symposium on Multimedia. IEEE (2010)
41. Rao, A., et al.: A unified framework for shot type classification based on subject centric lens. In: Proceedings of the European Conference on Computer Vision (ECCV) (2020)
42. Rao, A., et al.: A local-to-global approach to multi-modal movie scene segmentation. In: Proceedings of the IEEE/CVF Conference on Computer Vision and Pattern Recognition, pp. 10146–10155 (2020)
43. Rasheed, Z., Shah, M.: Detection and representation of scenes in videos. IEEE Trans. Multimed. **7**, 1097–1105 (2005)
44. Ren, S., He, K., Girshick, R., Sun, J.: Faster R-CNN: towards real-time object detection with region proposal networks. In: Advances in Neural Information Processing Systems, pp. 91–99 (2015)
45. Rohrbach, A., Rohrbach, M., Tandon, N., Schiele, B.: A dataset for movie description. In: Proceedings of the IEEE Conference on Computer Vision and Pattern Recognition, pp. 3202–3212 (2015)
46. Rotman, D., Porat, D., Ashour, G.: Optimal sequential grouping for robust video scene detection using multiple modalities. Int. J. Semant. Comput. **11**(02), 193–208 (2017)
47. Shao, D., Xiong, Y., Zhao, Y., Huang, Q., Qiao, Y., Lin, D.: Find and focus: retrieve and localize video events with natural language queries. In: Ferrari, V., Hebert, M., Sminchisescu, C., Weiss, Y. (eds.) ECCV 2018. LNCS, vol. 11213, pp. 202–218. Springer, Cham (2018). https://doi.org/10.1007/978-3-030-01240-3_13
48. Sidiropoulos, P., Mezaris, V., Kompatsiaris, I., Meinedo, H., Bugalho, M., Trancoso, I.: Temporal video segmentation to scenes using high-level audiovisual features. IEEE Trans. Circuits Syst. Video Technol. **21**(8), 1163–1177 (2011)

49. Simões, G.S., Wehrmann, J., Barros, R.C., Ruiz, D.D.: Movie genre classification with convolutional neural networks. In: 2016 International Joint Conference on Neural Networks (IJCNN). IEEE (2016)
50. Simonyan, K., Zisserman, A.: Very deep convolutional networks for large-scale image recognition. arXiv preprint arXiv:1409.1556 (2014)
51. Tapaswi, M., Bäuml, M., Stiefelhagen, R.: knock! knock! who is it? probabilistic person identification in TV-series. In: IEEE Conference on Computer Vision and Pattern Recognition. IEEE (2012)
52. Tapaswi, M., Bäuml, M., Stiefelhagen, R.: Story-based video retrieval in TV series using plot synopses. In: Proceedings of International Conference on Multimedia Retrieval, p. 137. ACM (2014)
53. Tapaswi, M., Bäuml, M., Stiefelhagen, R.: Aligning plot synopses to videos for story-based retrieval. Int. J. Multimed. Inf. Retrieval 4(1), 3–16 (2015)
54. Tapaswi, M., Zhu, Y., Stiefelhagen, R., Torralba, A., Urtasun, R., Fidler, S.: MovieQA: understanding stories in movies through question-answering. In: Proceedings of the IEEE Conference on Computer Vision and Pattern Recognition (2016)
55. Vicol, P., Tapaswi, M., Castrejon, L., Fidler, S.: MovieGraphs: towards understanding human-centric situations from videos. In: Proceedings of the IEEE Conference on Computer Vision and Pattern Recognition (2018)
56. Wang, H.L., Cheong, L.F.: Taxonomy of directing semantics for film shot classification. IEEE Trans. Circuits Syst. Video Technol. 19(10), 1529–1542 (2009)
57. Wang, L., et al.: Temporal segment networks: towards good practices for deep action recognition. In: Leibe, B., Matas, J., Sebe, N., Welling, M. (eds.) ECCV 2016. LNCS, vol. 9912, pp. 20–36. Springer, Cham (2016). https://doi.org/10.1007/978-3-319-46484-8_2
58. Xia, J., Rao, A., Huang, Q., Xu, L., Wen, J., Lin, D.: Online multi-modal person search in videos. In: Vedaldi, A., Bischof, H., Brox, T., Frahm, J.-M. (eds.) ECCV 2020. LNCS, vol. 12357, pp. 174–190. Springer, Cham (2020). https://doi.org/10.1007/978-3-030-58610-2_11
59. Xiong, Y., Huang, Q., Guo, L., Zhou, H., Zhou, B., Lin, D.: A graph-based framework to bridge movies and synopses. In: Proceedings of the IEEE International Conference on Computer Vision, pp. 4592–4601 (2019)
60. Xu, J., Mei, T., Yao, T., Rui, Y.: MSR-VTT: a large video description dataset for bridging video and language. In: Proceedings of the IEEE Conference on Computer Vision and Pattern Recognition, pp. 5288–5296 (2016)
61. Xu, M., et al.: Using context saliency for movie shot classification. In: 2011 18th IEEE International Conference on Image Processing, pp. 3653–3656. IEEE (2011)
62. Yang, Y., Lin, S., Zhang, Y., Tang, S.: Statistical framework for shot segmentation and classification in sports video. In: Yagi, Y., Kang, S.B., Kweon, I.S., Zha, H. (eds.) ACCV 2007. LNCS, vol. 4844, pp. 106–115. Springer, Heidelberg (2007). https://doi.org/10.1007/978-3-540-76390-1_11
63. Zhao, Y., Xiong, Y., Lin, D.: Mmaction (2019). https://github.com/open-mmlab/mmaction
64. Zheng, L., Shen, L., Tian, L., Wang, S., Wang, J., Tian, Q.: Scalable person re-identification: a benchmark. In: Proceedings of the IEEE International Conference on Computer Vision, pp. 1116–1124 (2015)
65. Zhou, B., Andonian, A., Oliva, A., Torralba, A.: Temporal relational reasoning in videos. In: Ferrari, V., Hebert, M., Sminchisescu, C., Weiss, Y. (eds.) ECCV 2018. LNCS, vol. 11205, pp. 831–846. Springer, Cham (2018). https://doi.org/10.1007/978-3-030-01246-5_49

66. Zhou, H., Hermans, T., Karandikar, A.V., Rehg, J.M.: Movie genre classification via scene categorization. In: Proceedings of the 18th ACM International Conference on Multimedia, pp. 747–750. ACM (2010)
67. Zhu, Y., et al.: Aligning books and movies: towards story-like visual explanations by watching movies and reading books. In: Proceedings of the IEEE International Conference on Computer Vision, pp. 19–27 (2015)

Short-Term and Long-Term Context Aggregation Network for Video Inpainting

Ang Li[1]([✉]), Shanshan Zhao[2], Xingjun Ma[3], Mingming Gong[4], Jianzhong Qi[1], Rui Zhang[1], Dacheng Tao[2], and Ramamohanarao Kotagiri[1]

[1] School of Computing and Information Systems, The University of Melbourne, Melbourne, Australia
`angl4@student.unimelb.edu.au`,
`{jianzhong.qi,rui.zhang,kotagiri}@unimelb.edu.au`
[2] UBTECH Sydney AI Centre, School of Computer Science, Faculty of Engineering, The University of Sydney, Darlington 2008, Australia
`szha4333@uni.sydney.edu.au`, `dacheng.tao@sydney.edu.au`
[3] School of Information Technology, Deakin University, Geelong, Australia
`daniel.ma@deakin.edu.au`
[4] School of Mathematics and Statistics, The University of Melbourne, Melbourne, Australia
`mingming.gong@unimelb.edu.au`

Abstract. Video inpainting aims to restore missing regions of a video and has many applications such as video editing and object removal. However, existing methods either suffer from inaccurate short-term context aggregation or rarely explore long-term frame information. In this work, we present a novel context aggregation network to effectively exploit both short-term and long-term frame information for video inpainting. In the encoding stage, we propose **boundary-aware short-term context aggregation**, which aligns and aggregates, from neighbor frames, local regions that are closely related to the boundary context of missing regions into the target frame (The target frame refers to the current input frame under inpainting.). Furthermore, we propose **dynamic long-term context aggregation** to globally refine the feature map generated in the encoding stage using long-term frame features, which are dynamically updated throughout the inpainting process. Experiments show that it outperforms state-of-the-art methods with better inpainting results and fast inpainting speed.

Keywords: Video inpainting · Context aggregation

Electronic supplementary material The online version of this chapter (https://doi.org/10.1007/978-3-030-58548-8_42) contains supplementary material, which is available to authorized users.

© Springer Nature Switzerland AG 2020
A. Vedaldi et al. (Eds.): ECCV 2020, LNCS 12349, pp. 728–743, 2020.
https://doi.org/10.1007/978-3-030-58548-8_42

Fig. 1. Comparison with state-of-the-art CPNet [13] and FGNet [27]. The green areas in the input frames are the missing regions. Best viewed at zoom level 400%. (Color figure online)

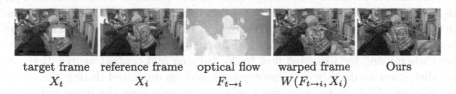

target frame	reference frame	optical flow	warped frame	Ours
X_t	X_i	$F_{t \to i}$	$W(F_{t \to i}, X_i)$	

Fig. 2. An example of missing regions' negative effects on flow-warping-based aggregation. Given a target frame X_t and a reference frame X_i, we get the optical flow $F_{t \to i}$ between them using a pretrained flow estimator. Then we warp X_i onto X_t and get the warped frame $W(F_{t \to i}, X_i)$. Heavy distortions can be found in $F_{t \to i}$ and $W(F_{t \to i}, X_i)$ within the missing regions. Our network alleviates this problem, producing more accurate aggregation.

1   Introduction

Video inpainting aims to restore missing regions in a video with plausible contents that are both spatially and temporally coherent [7,15]. It can benefit a wide range of practical video applications such as video editing, damage restoration, and undesired object removal. Whilst significant progress has been made in image inpainting [8,14,17,25,28,30,31], it is challenging to extend image inpainting methods to solve the video inpainting problem. Directly applying image inpainting methods on individual video frames may lose the inter-frame motion continuity and content dependency, which causes temporal inconsistencies and unexpected flickering artifacts.

Traditional video inpainting methods [7,15,24] utilize patch-based optimization strategies to fill missing regions with sampled patches from known regions. These methods often suffer from limited effectiveness and vulnerability to complex motions. Recently, deep learning-based video inpainting methods [11,13,16,21] have improved the inpainting performance by a large margin.

Most of them use encoder-decoder structures following a frame-by-frame inpainting pipeline to borrow information from reference frames[1] and perform different types of context aggregation to restore the target frame.

In spite of the encouraging results, deep learning-based methods still need to overcome the following limitations. First, they fail to make effective usage of short-term and long-term reference information in the input video. Studies [11,21] restrict the range of reference frames to nearby (short-term) frames of the target frame so as to maintain temporal consistency. When dealing with diverse motion patterns in videos (eg., slowly moving views or objects), short-term frames alone cannot provide sufficient information to restore the target frame. Other methods [13,16] often sample a set of fixed frames from the input video (eg., every 5-th frame) as the reference frames. Although this can exploit some long-term information, it tends to include irrelevant contexts, reduce temporal consistency, and increase the computation time. Second, how to achieve accurate context aggregation remains challenging. Since missing regions contain no visual information, it is difficult to find the most related local regions in reference frames for accurate context aggregation. For example, a recent method [11] uses estimated optical flows to warp reference frames onto the target frame and further aggregate them together. As shown in Fig. 2, the flow information within the missing region is inaccurate and is distorted. This will cause unexpected artifacts when using the distorted flow to do warping and context aggregation, and the distortion artifacts will be propagated and accumulated during the encoding process. While using more fixed reference frames from the input video may help mitigate the negative effects of missing regions, it inevitably brings more irrelevant or even noisy information into the target frame, which inadvertently does more harm than good.

In this paper, we aim to address the challenges above in a principled manner from the following three aspects: (1) We propose a novel framework for video inpainting that integrates the advantages of both short-term and long-term reference frames. Different from existing methods that only conduct context aggregation at the decoding stage, we propose to start context aggregation at the encoding stage with short-term reference frames. This can help provide more temporally consistent local contexts. Then, at the decoding stage, we refine the encoding-generated feature map with a further step of context aggregation on long-term reference frames. This refinement can deal with more complex motion patterns. (2) To better exploit short-term information, we propose *boundary-aware short-term context aggregation* at the encoding stage. Different from existing methods, here, we pay more attention to the boundary context of missing regions. Our intuition is that, in the target frame, the boundary area around the missing regions is more related to the missing regions than other areas of the frame. Considering the spatial and motion continuity of videos, if we can accurately locate and align the boundary context of missing regions with the corresponding regions in the reference frames, it would improve both the spatial and temporal consistency of the generated contents. This strategy can also

---

[1] The reference frames refer to other frames from the same video.

alleviate the impact of missing regions at context aggregation. (3) To better exploit long-term information, we propose a *dynamic long-term context aggregation* at the decoding stage. Since different videos have different motion patterns (eg., slow moving or back-and-forth moving), they have different contextual dependency between frames. Therefore, it is necessary to eliminate frames that are largely irrelevant with the target frame. Our dynamic strategy aims for the effective usage of long-term frame information. Specifically, instead of simply using fixed reference frames, we propose to dynamically update the long-term reference frames used for inpainting, according to similarities of other frames to the current target frame.

In summary, our main contributions are:

- We propose a novel framework for video inpainting that effectively integrates context information from both short-term and long-term reference frames.
- We propose a *boundary-aware short-term context aggregation* to better exploit the context information from short-term reference frames, by using the boundary information of the missing regions in the target frames.
- We propose a *dynamic long-term context aggregation* as a refinement operation to better exploit the context information from dynamically updated long-term reference frames.
- We empirically show that our proposed network outperforms the state-of-the-art methods with better inpainting results and fast inpainting speed.

2   Related Work

2.1   Image Inpainting

Traditional image inpainting methods [1,2] mostly perform inpainting by finding pixels or patches outside missing regions or from the entire image database. These methods often suffer from low generation quality, especially when dealing with complicated scenes or large missing regions [8,17].

Deep neural networks have been used to improve inpainting results [8,14, 17,19,25,28–31]. Pathak *et al.* [17] introduce the *Context Encoder* (CE) model where a convolutional encoder-decoder network is trained with the combination of an adversarial loss [6] and a reconstruction loss. Iizuka *et al.* [8] propose to utilize global and local discriminators to ensure consistency on both entirety and details. Yu *et al.* [30] propose the contextual attention module to restore missing regions with similar patches from undamaged regions in deep feature space.

2.2   Video Inpainting

Apart from spatial consistency in every restored frame, video inpainting also needs to solve a more challenging problem: how to make use of information in the whole video frame sequence and maintain temporal consistency between the restored frames. Traditional video inpainting methods adopt patch-based strategies. Wexler *et al.* [24] regard video inpainting as a global optimization problem by alternating between patch search and reconstruction steps. Newson et al. [15] extend this and improve the search algorithm by developing a 3D version of PatchMatch [1]. Huang et al. [7] introduce the optical flow optimization in spatial patches to enforce temporal consistency. These methods require heavy computations, which limit their efficiency and practical use.

Recently, several deep learning-based methods have been proposed [3,11,13, 16,21,27,32]. These works can be divided into two groups. The first group mainly relies on short-term reference information when inpainting the target frame. For example, VINet [11] uses a recurrent encoder-decoder network to collect information from adjacent frames via flow-warping-based context aggregation. Xu et al. [27] propose a multi-stage framework for video inpainting: they first use a deep flow completion network to restore the flow sequence, then perform forward and backward pixel propagation using the restored flow sequence, and finally use a pretrained image inpainting model to refine the results. The second group uses a fixed set of frames from the entire video as reference information. Wang et al. [21] propose a two-stage model with a combination of 3D and 2D CNNs. CPNet [13] conducts context aggregation by predicting affine matrices and applying affine transformation on fixedly sampled reference frames.

Although these video inpainting methods have shown promising results, they still suffer from ineffective usage of short-term and long-term frame reference information, and inaccurate context aggregation as discussed in Sect. 1.

3   Short-Term and Long-Term Context Aggregation Network

Given a sequence of continuous frames from a video $X := \{X_1, X_2, ..., X_T\}$ annotated with binary masks $M := \{M_1, M_2, ..., M_T\}$, a video inpainting network outputs the restored video $\hat{Y} := \{\hat{Y}_1, \hat{Y}_2, ..., \hat{Y}_T\}$. The goal is that $\hat{Y}$ should be spatially and temporally consistent with the ground truth video $Y := \{Y_1, Y_2, ..., Y_T\}$.

3.1   Network Overview

Our network is built upon a recurrent encoder-decoder architecture and processes the input video frame by frame in its temporal order. An overview of our proposed network is illustrated in Fig. 3. Different from existing methods, we start inpainting (context aggregation) at the encoding stage. Given current target frame X_t, we choose a group of neighboring frames $\{X_i\}$ with $i \in \{t-6, t-3, t+3, t+6\})$ as the short-term reference frames for X_t. During encoding, we have two sub-encoders: encoder a for the stream of target frame and encoder b for the other four streams of reference frames. The encoding process contains three feature spatial scales $\{\frac{1}{2}, \frac{1}{4}, \frac{1}{8}\}$. At each encoding scale, we perform **Boundary-aware Short-term Context Aggregation (BSCA)** between feature maps of the target frame and those of the short-term reference frames, to fill the missing regions in the *target feature map*. This module can

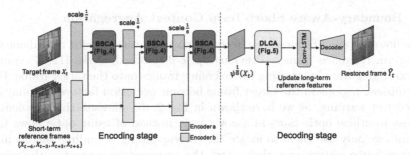

Fig. 3. Overview of our proposed network. In the encoding stage, we conduct Boundary-aware Short-term Context Aggregation (BSCA) (Sect. 3.2) using short-term frame information from neighbor frames, which is beneficial to context aggregation and generating temporally consistent contents. In the decoding stage, we propose the Dynamic Long-term Context Aggregation (DLCA) (Sect. 3.3), which utilizes dynamically updated long-term frame information to refine the encoding-generated feature map.

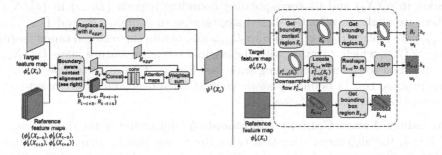

Fig. 4. Left: Boundary-aware Short-term Context Aggregation (BSCA) module. **Right:** The boundary-aware context alignment operation in BSCA. Here, $l \in \{\frac{1}{2}, \frac{1}{4}, \frac{1}{8}\}$ refers to the encoding scale.

accurately locate and aggregate relevant bounding regions in short-term reference frames and at the same time avoid distractions caused by missing regions in the target frame. At the decoding stage, our **Dynamic Long-term Context Aggregation (DLCA)** module refines the encoding-generated feature map using dynamically updated long-term features. This module stores long-term frame features selected from previously *restored* frames, and updates them according to their contextual correlation to the current target frame. Intuitively, it only keeps those long-term frame features that are more contextual relevant to the current target frame. We also adopt a convolutional LSTM (Conv-LSTM) layer to increase temporal consistency as suggested by Lai *et al.* [12]. Finally, the decoder takes the refined latent feature to generate the restored frame $\hat{Y}_t$. Since the missing regions are now filled with contents, we replace the target frame X_t by the restored frame $\hat{Y}_t$, which provides more accurate information for the following iterations.

3.2 Boundary-Aware Short-Term Context Aggregation

Optic flows between frames have been shown to be essential for alignment with short-term reference frames. Previous optic-flow-based works [11,27] conduct context aggregation by warping the reference frames onto the target frame. However, missing regions in the target frame become occlusion factors and may lead to incorrect warping, as we have shown in Fig. 2. To alleviate this problem, we propose to utilize optic flows in a novel way: instead of using optic flows to do warping, we only use them to locate the corresponding bounding regions in the reference frame feature map that match the surrounding context of the missing regions in the target frame feature map. Here, we define the surrounding context region as the non-missing pixels that are within a Euclidean distance d ($d = 8$ in our experiments) to the nearest pixels from the missing regions.

The structure of BSCA is illustrated in the left subfigure of Fig. 4. At a certain encoding scale $l \in \{\frac{1}{2}, \frac{1}{4}, \frac{1}{8}\}$, we have the target feature map $\phi_a^l(X_t)$ from encoder a and the reference feature maps $\{\phi_b^l(X_i)\}$ from encoder b as input for boundary-aware context aggregation. We first obtain the bounding region B_t of the missing region in $\phi_a^l(X_t)$ and its corresponding bounding regions $\{B_{t \to i}\}$ in $\{\phi_b^l(X_i)\}$. Then, we apply an attention-based aggregation to combine B_t and $\{B_{t \to i}\}$ as B_{aggr}. We replace B_t in $\phi_a^l(X_t)$ with B_{aggr} and obtain the restored target feature map $\psi^l(X_t)$, which, together with the original reference feature maps $\{\phi_b^l(X_i)\}$, is passed on to the next encoding scale (see Fig. 3). Two essential operations in this process, i.e., 1) boundary-aware context alignment and 2) attention-based aggregation, are detailed below.

Boundary-Aware Context Alignment. As illustrated in the right subfigure of Fig. 4, the alignment operation takes the target feature map $\phi_a^l(X_t)$ and a reference feature map $\phi_b^l(X_i)$ as inputs. In $\phi_a^l(X_t)$, we denote the missing region with white color. Then, surrounding region E_t in $\phi_a^l(X_t)$ is obtained (the yellow elliptical ring in $\phi_a^l(X_t)$). We further obtain the bounding box region of E_t and denote it by B_t. We use a pretrained FlowNet2 [9] to extract the flow information $F_{t \to i}$ between X_t and X_i, and then we downsample $F_{t \to i}$ to $F_{t \to i}^l$ for current encoding scale l. In $F_{t \to i}^l$, the corresponding flow information of E_t is denoted as $F_{t \to i}^l(E_t)$, which has the same position with E_t. With E_t and $F_{t \to i}^l(E_t)$, we can locate the corresponding region of E_t in $\phi_b^l(X_i)$, which is $E_{t \to i}$ (the yellow elliptical ring in $\phi_b^l(X_i)$). We also obtain the bounding box region of $E_{t \to i}$ as $B_{t \to i}$, and reshape it to the shape of B_t. To ensure the context coherence, we further refine the reshaped $B_{t \to i}$ using Atrous Spatial Pyramid Pooling (ASPP) [4]. With the aligned bounding regions B_t and $B_{t \to i}$, we can alleviate the impact of missing regions and achieve more accurate context aggregation.

Attention-Based Aggregation. Attention-based aggregation can help find the most relevant features from the reference feature maps, and eliminate irrelevant contents, eg., newly appeared backgrounds. We first append B_t into the set $\{B_{t \to i}\}$ to get a new set $\{B_{t \to i}, B_t\}$, denoted as $\{B_j'\}_{j=1}^5$. Then, we concatenate the elements in the new set along the channel dimension, and apply convolutional and softmax operations across different channels to obtain the attention maps $\{A_j\}_{j=1}^5$. Finally, the attention-based aggregation is performed as follows.

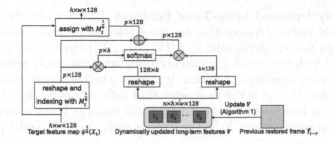

Fig. 5. The Dynamic Long-term Context Aggregation (DLCA) module.

$$B_{aggr} = \sum_{j=1}^{5} A_j B_j', \tag{1}$$

We replace B_t with the aggregated bounding region B_{aggr} in the target feature map $\phi_a^l(X_t)$. The replaced target feature map is processed by an ASPP module to get the $\psi^l(X_t)$.

3.3 Dynamic Long-Term Context Aggregation

Fixed sampling long-term reference frames [13,16] fail to consider the motion diversity of videos. Thus, they may inevitably bring more irrelevant or even noisy information. Since different videos have different motion patterns (eg., slow moving or back-and-forth moving), it results in different contextual dependency between frames. Therefore, it is necessary that the selected long-term reference information is contextually relevant to the current target frame. We use a dynamic strategy for the effective use of long-term reference information. The structure of this decoding-stage context aggregation module is illustrated in Fig. 5. It refines the feature map generated in the above encoding stage with 1) dynamically updated long-term features and 2) non-local-based aggregation.

Algorithm 1. Update Long-Term Features

Input: previous restored frame $\hat{Y}_{t-r}$, current target frame X_t, long-term features V
Output: updated V
1: distance = [ ]
2: $U_{X_t} = \text{Encoder\_b}(X_t)$
3: $U_{\hat{Y}_{t-r}} = \text{Encoder\_b}(\hat{Y}_{t-r})$
4: $d_{\hat{Y}_{t-r}} = \|U_{X_t} - U_{\hat{Y}_{t-r}}\|_1$
5: **for** V_r in V **do**
6:     $d_r = \|U_{X_t} - V_r\|_1$
7:     distance.append(d_r)
8: **end for**
9: $d_{max}, max = \text{get\_max\_and\_index}(\text{distance})$
10: **if** $d_{\hat{Y}_{t-r}} < d_{max}$ **then**
11:     $V.\text{remove}(V_{max})$
12:     $V.\text{append}(U_{\hat{Y}_{t-r}})$
13: **end if**

Dynamically Updated Long-Term Features. DLCA stores the features of the previously *restored* frames that are most relevant (in feature space) to the current target frame. Specifically, $V := \{V_1, V_2, ..., V_q\}$ stores a set of long-term feature maps with the length q, which are updated dynamically following Algorithm 1. At each inpainting iteration, it checks whether the feature map of a long-term frame $\hat{Y}_{t-r}$ (r is the parameter that defines how far from the current target frame to look back) can be incorporated into the V set according to its L_1 distance to target frame X_t in the feature space. Let U_{X_t} and $U_{\hat{Y}_{t-r}}$ be the feature maps of X_t and $\hat{Y}_{t-r}$ respectively, if the L_1 distance between $U_{\hat{Y}_{t-r}}$ and U_{X_t} is smaller than the maximum distance between a feature map in the current V set to U_{X_t}, then $U_{\hat{Y}_{t-r}}$ will replace the corresponding feature map (that has the maximum distance) in the V set. Note that these feature maps can be obtained using our encoder b. We suggest $r \geq 7$ to exploit long-term information (as short-term information from $\hat{Y}_{t-6/t-3}$ has already been considered by our BSCA module). At the beginning when $t < r$, we simply set $r = |t - r|$ to use all restored frames so far. With this dynamic updating policy, DLCA can automatically adjust the stored long-term frame features and remove irrelevant ones, regarding each target frame.

Non-local-Based Aggregation. Based on the long-term feature set V, we follow a typical approach [23] to perform non-local-based context aggregation between the target feature map $\psi^{\frac{1}{8}}(X_t)$ and feature maps stored in V, as shown in Fig. 5. Softmax is applied to obtain the normalized soft attention map over feature maps in V. The attention map is then utilized as weights to compute an aggregated feature map from V via weighted summation. Finally, the aggregated feature map replaces the feature map of missing regions.

3.4   Loss Function

The loss function used for training is:

$$\mathcal{L}_{total} = \mathcal{L}_{rec} + \lambda_{mre}\mathcal{L}_{mre} + \lambda_{per}\mathcal{L}_{per} + \lambda_{style}\mathcal{L}_{style}, \tag{2}$$

Here, $\mathcal{L}_{rec}$, $\mathcal{L}_{mre}$, $\mathcal{L}_{per}$, and $\mathcal{L}_{style}$ denote reconstruction loss, reconstruction loss of mask region, perceptual loss, and style loss respectively. The balancing weights λ_{mre}, λ_{per} and λ_{style} are empirically set to 2, 0.01, and 1, respectively.

The reconstruction loss and the reconstruction loss of the mask region are defined on pixels:

$$\mathcal{L}_{rec} = \sum_{t}^{T} \|\hat{Y}_t - Y_t\|_1. \tag{3}$$

$$\mathcal{L}_{mre} = \sum_{t}^{T} \|(1 - M_t) \odot (\hat{Y}_t - Y_t)\|_1, \tag{4}$$

where $\odot$ is the element-wise multiplication. To further enhance inpainting quality, we include two additional loss functions: perceptual loss [10] and style loss,

$$\mathcal{L}_{\mathrm{per}} = \sum_t^T \sum_s^S \frac{\|\sigma_s(\hat{Y}_t) - \sigma_s(Y_t)\|_1}{S},\tag{5}$$

$$\mathcal{L}_{\mathrm{style}} = \sum_t^T \sum_s^S \frac{\|G_s^\sigma(\hat{Y}_t) - G_s^\sigma(Y_t)\|_1}{S},\tag{6}$$

where σ_s is the s-th layer output of an ImageNet-pretrained VGG-16 [20] network, S is the number of chosen layers (i.e., $relu_{2\_2}$, $relu_{3\_3}$ and $relu_{4\_3}$), and G denotes the gram matrix multiplication [5].

4 Experiments

We evaluate and compare our model with state-of-the-art models qualitatively and quantitatively. We also conduct a comprehensive ablation study on our proposed model.

Table 1. Quantitative comparisons on YouTube-VOS and DAVIS datasets under three mask settings regarding 3 performance metrics: PSNR (higher is better), SSIM (higher is better) and VFID (lower is better). The rightmost column shows the average execution time to inpaint one video. The best results are in **bold**.

YouTube-VOS

Model	Square mask			Irregular mask			Object mask			Time (sec.)
	PSNR	SSIM	VFID	PSNR	SSIM	VFID	PSNR	SSIM	VFID	
VINet [11]	26.92	0.843	0.103	27.33	0.848	0.082	26.61	0.838	0.118	**33.6**
CPNet [13]	27.24	0.847	0.087	27.50	0.852	0.051	27.02	0.845	0.087	48.5
FGNet [27]	27.71	0.856	0.082	27.91	0.859	0.056	27.32	0.849	0.083	276.3
Ours	**27.76**	**0.858**	**0.076**	**28.12**	**0.866**	**0.047**	**27.45**	**0.853**	**0.075**	35.4

DAVIS

Model	Square mask			Irregular mask			Object mask			Time (sec.)
	PSNR	SSIM	VFID	PSNR	SSIM	VFID	PSNR	SSIM	VFID	
VINet [11]	27.88	0.863	0.060	28.67	0.874	0.043	27.02	0.850	0.068	**19.7**
CPNet [13]	27.92	0.862	0.049	28.81	0.876	0.031	27.48	0.855	0.049	28.2
FGNet [27]	28.32	0.870	0.045	29.37	0.880	0.033	**28.18**	0.864	0.046	194.8
Ours	**28.50**	**0.872**	**0.038**	**29.56**	**0.883**	**0.027**	28.13	**0.867**	**0.042**	21.5

Datasets. Following previous works [11,13], we train and evaluate our model on YouTube-VOS [26] and DAVIS [18] datasets. For YouTube-VOS, we use the 3471 training videos for training, and the 508 test videos for testing. For DAVIS, we use the 60 unlabeled videos to fine tune a pretrained model on YouTube-VOS, and the 90 videos with object mask annotations for testing. All video frames are resized to 424 × 240, and no pre-processing or post-processing is applied.

Mask Settings. To simulate the diverse and ever-changing real-world scenarios, we consider the following three mask settings for training and testing.

- Square mask: The same square region for all frames in a video, but has a random location and a random size ranging from 40 × 40 to 160 × 160 for different videos.
- Irregular mask: We use the irregular mask dataset [14] that consists of masks with arbitrary shapes and random locations.
- Object mask: Following [11,27], we use the foreground object masks in DAVIS [18] which has continuous motion and realistic appearance. Note that when quantitatively testing object masks on DAVIS dataset, we shuffle its video-mask pairs for more reasonable result, as it is a dataset for object removal and the ground-truth background (after removal) is unknown (without shuffling, the original objects will become the ground truth).

Baseline Models. We compare our model with three state-of-the-art video inpainting models: 1) VINet [11], a recurrent encoder-decoder network with flow-warping-based context aggregation; 2) CPNet [13], which conducts context aggregation by predicting affine matrices and applying affine transformation on fixedly sampled reference frames; and 3) FGNet [27], which consists of three stages: first restores flow between frames, then performs forward and backward warping with the restored flow, and finally utilizes an image inpainting model for post-processing.

4.1   Quantitative Results

We consider three metrics for evaluation: 1) *Peak Signal-to-Noise Ratio* (PSNR, measures image distortion), 2) *Structural Similarity* (SSIM, measures structure similarity) and 3) the video-based Fréchet Inception Distance (VFID, a video perceptual measure known to match well with human perception) [22]. As shown in Table 1, our model outperforms all baseline models according to all three metrics across all three mask settings on both datasets, a clear advantage of using both short-term and long-term information. In terms of execution time, our model is comparable to VINet, which has the least average execution time but worse performance than all other three models. Overall, our model achieves the best trade-off between performance and execution time on the two test sets.

4.2   Qualitative Results

To further inspect the visual quality of the inpainted videos, we show, in Fig. 6, three examples of the inpainted frames by our model and the compared baselines. As can be observed, frames inpainted by our models are generally of much higher

Fig. 6. Qualitative comparison of our proposed model with baseline models on DAVIS dataset. Better viewed at zoom level 400%. More video results can be found in supplementary material.

quality than those by VINet or CPNet, and also perceptibly better than the state-of-the-art model FGNet. For example, in the third example (right two columns), the car structures generated by VINet are highly distorted. This is mainly caused by the occlusion effect of mask regions in the target frame, and its limited exploration of long-term information. CPNet was able to restore the rough structures of the car with more information from its fixedly sampled long-term reference frames. However, blurriness or overlapping can still be found since those fixed-term reference frames also bring in a significant amount of irrelevant contexts. FGNet in general achieves sharper results than VINet or CPNet. However, it also generates artifacts in this example. This can be ascribed to inaccurate flow inpainting in the first stage of FGNet. In contrast, our model can generate more plausible contents with high spatial and temporal consistency.

4.3   User Study

We also conduct a user study to verify the effectiveness of our proposed network. We recruited 50 volunteers for this user study. We randomly select 20 videos from DAVIS test set. For each video, the original video with object mask and the anonymized results from VINet, CPNet, FGNet and our model are presented to the volunteers. The volunteers are then asked to rank the four models with 1, 2, 3 and 4 (1 is the best and 4 is the worst) based on the perceptual quality of the inpainted videos. The result in terms of the percentage of rank scores received by different models is shown in Fig. 7. Our model receives significantly more votes for rank 1 (the best) than the other three models, which verifies that our model can indeed generate more plausible results than existing models.

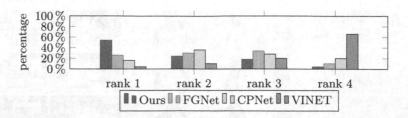

Fig. 7. User study results. For each rank (1 is the best), we collected 1000 votes (20 videos * 50 volunteers) in total. The y-axis indicates, within each rank, the percentage (out of 1000 notes) of the votes received by different models.

(a) Input   (b) Ours (w/o BSCA)   (c) Ours (scale $\frac{1}{8}$)   (d) Ours

Fig. 8. Ablation study on BSCA. Better viewed at zoom level 400%.

4.4   Ablation Study

We investigate the effectiveness of the two components of our network: Boundary-aware Short-term Context Aggregation (BSCA) and Dynamic Long-term Context Aggregation (DLCA). In Table 2, we report all the quantitative results under different ablation settings on the DAVIS test set with shuffled object masks.

Effectiveness of BSCA. As we described in Sect. 3.2, the purpose of BSCA is to alleviate the negative effects of missing regions in the target frame. Table 2 compares our full model with its two variants: 1) "w/o BSCA" (the first row in Table 2), which removes the BSCA module and directly uses the flows to warp reference feature maps onto the target feature map as [11]. Then it concatenates the warped reference features with the target feature map as the input for attention-based aggregation; 2) "scale $\frac{1}{8}$" (the second row in Table 2), which performs the BSCA module only at the $\frac{1}{8}$ encoding scale. The performance drop of these two variants justifies the effectiveness of the BSCA module and the multi-scale design at the encoding stage. As we further show in Fig. 8, the model without BSCA suffers from inaccurate feature alignment due to the occlusion effect of missing regions in the flows, thus producing distorted contents. Using BSCA only at the $\frac{1}{8}$ encoding scale apparently improves the results, but is still affected by the distortions from the previous scales. In contrast, using BSCA at multiple encoding scales lead to better results with temporally consistent details.

Table 2. Comparisons of different settings on BSCA and DLCA.

BSCA (scale $\frac{1}{8}$)	BSCA	DLCA (fixed)	DLCA	PSNR	SSIM	VFID
			✓	27.17	0.847	0.063
✓			✓	27.72	0.859	0.052
	✓			27.58	0.858	0.056
	✓	✓		27.85	0.862	0.048
	✓		✓	**28.13**	**0.867**	**0.042**

Effectiveness of DLCA. We test two other variants of our full model regarding the DLCA module: 1) "w/o DLCA" (the third row in Table 2), which directly removes the DLCA module; and 2) "fixed" (the fourth row in Table 2), which keeps the DLCA module but uses fixedly sampled reference features (rather than dynamic updated ones) that takes one frame for every five frames out of the entire input video sequence. Both variants exhibit performance degradation. Although fixedly sampled reference features can help, it is still less effective than using our dynamically updated long-term features. As shown in Fig. 9, the model without DLCA module (w/o DLCA) fails to recover the background after object removal due to the lack of long-term frame information. Although the model with fixedly sampled reference features successfully restores the background, blurriness and artifacts can be still be found. Figure 9 (e) and Fig. 9 (f) further illustrate the dynamic characteristic of our dynamic updating rule, which can effectively avoid irrelevant reference frames (eg., $\hat{Y}_{t-13}$ is not in our long-term feature set V) for the current target frame.

Dynamically Updated Long-Term Features. We investigate the impact of different lengths of dynamically updated long-term features V on the inpainting results. Small lengths of q is insufficient to capture long-term frame information, resulting in inferior performance. On the contrary, large lengths will include more irrelevant reference frames, which also leads to degraded performance. The best result is achieved at length $q = 10$. For the parameter r (long-term range), we empirically find that $r = 9$ works well across different settings.

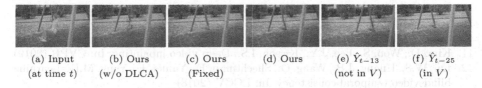

(a) Input	(b) Ours	(c) Ours	(d) Ours	(e) $\hat{Y}_{t-13}$	(f) $\hat{Y}_{t-25}$
(at time t)	(w/o DLCA)	(Fixed)		(not in V)	(in V)

Fig. 9. Ablation study on DLCA. Better viewed at zoom level 400%.

5 Conclusion

We studied the problem of video inpainting and addressed three limitations of existing methods: 1) ineffective usage of short-term or long-term reference frames; 2) inaccurate short-term context aggregation caused by missing regions in the target frame; and 3) fixed sampling of long-term contextual information. We therefore proposed a Short-term, and Long-term Context Aggregation Network with two complementary modules for the effective exploitation of both short-term and long-term information. We have empirically demonstrated the effectiveness of our proposed approach on benchmark datasets and provided a comprehensive understanding of each module of our model.

Acknowledgement. This research was supported by Australian Research Council Projects FL-170100117, IH-180100002, IC-190100031, LE-200100049.

References

1. Barnes, C., Shechtman, E., Finkelstein, A., Goldman, D.B.: PatchMatch: a randomized correspondence algorithm for structural image editing. ACM Trans. Graph. **28**(3), 24 (2009)
2. Bertalmio, M., Vese, L., Sapiro, G., Osher, S.: Simultaneous structure and texture image inpainting. IEEE Trans. Image Process. **12**(8), 880–889 (2003)
3. Chang, Y.L., Liu, Z.Y., Hsu, W.: Free-form video inpainting with 3D gated convolution and temporal patchgan. In: ICCV (2019)
4. Chen, L.C., Papandreou, G., Kokkinos, I., Murphy, K., Yuille, A.L.: DeepLab: semantic image segmentation with deep convolutional nets, atrous convolution, and fully connected CRFs. IEEE Trans. Pattern Anal. Mach. Intell. **40**(4), 834–848 (2017)
5. Gatys, L.A., Ecker, A.S., Bethge, M.: Image style transfer using convolutional neural networks. In: CVPR (2016)
6. Goodfellow, I.J., et al.: Generative adversarial nets. In: NIPS (2014)
7. Huang, J.B., Kang, S.B., Ahuja, N., Kopf, J.: Temporally coherent completion of dynamic video. ACM Trans. Graph. (TOG) **35**(6), 196 (2016)
8. Iizuka, S., Simo-Serra, E., Ishikawa, H.: Globally and locally consistent image completion. ACM Trans. Graph. (TOG) **36**(4), 1–14 (2017)
9. Ilg, E., Mayer, N., Saikia, T., Keuper, M., Dosovitskiy, A., Brox, T.: FlowNet 2.0: evolution of optical flow estimation with deep networks. In: CVPR (2017)
10. Johnson, J., Alahi, A., Fei-Fei, L.: Perceptual losses for real-time style transfer and super-resolution. In: ECCV (2016)
11. Kim, D., Woo, S., Lee, J.Y., Kweon, I.S.: Deep video inpainting. In: CVPR (2019)
12. Lai, W.S., Huang, J.B., Wang, O., Shechtman, E., Yumer, E., Yang, M.H.: Learning blind video temporal consistency. In: ECCV (2018)
13. Lee, S., Oh, S.W., Won, D., Kim, S.J.: Copy-and-paste networks for deep video inpainting. In: ICCV (2019)
14. Liu, G., Reda, F.A., Shih, K.J., Wang, T.C., Tao, A., Catanzaro, B.: Image inpainting for irregular holes using partial convolutions. In: ECCV (2018)
15. Newson, A., Almansa, A., Fradet, M., Gousseau, Y., Pérez, P.: Video inpainting of complex scenes. SIAM J. Imaging Sci. **7**(4), 1993–2019 (2014)

16. Oh, S.W., Lee, S., Lee, J.Y., Kim, S.J.: Onion-peel networks for deep video completion. In: ICCV (2019)
17. Pathak, D., Krahenbuhl, P., Donahue, J., Darrell, T., Efros, A.A.: Context encoders: feature learning by inpainting. In: CVPR (2016)
18. Perazzi, F., Pont-Tuset, J., McWilliams, B., Van Gool, L., Gross, M., Sorkine-Hornung, A.: A benchmark dataset and evaluation methodology for video object segmentation. In: CVPR (2016)
19. Sagong, M.C., Shin, Y.G., Kim, S.W., Park, S., Ko, S.J.: PEPSI: fast image inpainting with parallel decoding network. In: CVPR (2019)
20. Simonyan, K., Zisserman, A.: Very deep convolutional networks for large-scale image recognition. arXiv preprint arXiv:1409.1556 (2014)
21. Wang, C., Huang, H., Han, X., Wang, J.: Video inpainting by jointly learning temporal structure and spatial details. In: AAAI (2019)
22. Wang, T.C., et al.: Video-to-video synthesis. arXiv preprint arXiv:1808.06601 (2018)
23. Wang, X., Girshick, R., Gupta, A., He, K.: Non-local neural networks. In: CVPR (2018)
24. Wexler, Y., Shechtman, E., Irani, M.: Space-time video completion. In: CVPR (2004)
25. Xiong, W., et al.: Foreground-aware image inpainting. In: CVPR (2019)
26. Xu, N., et al.: YouTube-VOS: sequence-to-sequence video object segmentation. In: ECCV (2018)
27. Xu, R., Li, X., Zhou, B., Loy, C.C.: Deep flow-guided video inpainting. In: CVPR (2019)
28. Yang, C., Lu, X., Lin, Z., Shechtman, E., Wang, O., Li, H.: High-resolution image inpainting using multi-scale neural patch synthesis. In: CVPR (2017)
29. Yeh, R.A., Chen, C., Lim, T.Y., Schwing, A.G., Hasegawa-Johnson, M., Do, M.N.: Semantic image inpainting with deep generative models. In: CVPR (2017)
30. Yu, J., Lin, Z., Yang, J., Shen, X., Lu, X., Huang, T.S.: Generative image inpainting with contextual attention. In: CVPR (2018)
31. Zeng, Y., Fu, J., Chao, H., Guo, B.: Learning pyramid-context encoder network for high-quality image inpainting. In: CVPR (2019)
32. Zhang, H., Mai, L., Xu, N., Wang, Z., Collomosse, J., Jin, H.: An internal learning approach to video inpainting. In: CVPR (2019)

DH3D: Deep Hierarchical 3D Descriptors for Robust Large-Scale 6DoF Relocalization

Juan Du[1]([✉]), Rui Wang[1,2], and Daniel Cremers[1,2]

[1] Technical University of Munich, Garching, Germany
{duj,wangr,cremers}@in.tum.de
[2] Artisense, Garching, Germany

Abstract. For relocalization in large-scale point clouds, we propose the first approach that unifies global place recognition and local 6DoF pose refinement. To this end, we design a Siamese network that jointly learns 3D local feature detection and description directly from raw 3D points. It integrates FlexConv and Squeeze-and-Excitation (SE) to assure that the learned local descriptor captures multi-level geometric information and channel-wise relations. For detecting 3D keypoints we predict the discriminativeness of the local descriptors in an unsupervised manner. We generate the global descriptor by directly aggregating the learned local descriptors with an effective attention mechanism. In this way, local and global 3D descriptors are inferred in one single forward pass. Experiments on various benchmarks demonstrate that our method achieves competitive results for both global point cloud retrieval and local point cloud registration in comparison to state-of-the-art approaches. To validate the generalizability and robustness of our 3D keypoints, we demonstrate that our method also performs favorably without fine-tuning on the registration of point clouds that were generated by a visual SLAM system. Code and related materials are available at https://vision.in.tum.de/research/vslam/dh3d.

Keywords: Point clouds · 3D deep learning · Relocalization

1 Introduction

Relocalization within an existing 3D map is a critical functionality for numerous applications in robotics [3] and autonomous driving [35,53]. A common strategy is to split the problem into two subtasks, namely global place recognition and local 6DoF pose refinement. A lot of effort has been focused on tackling the

---

J. Du and R. Wang—Contributed equally.

Electronic supplementary material The online version of this chapter (https://doi.org/10.1007/978-3-030-58548-8_43) contains supplementary material, which is available to authorized users.

© Springer Nature Switzerland AG 2020
A. Vedaldi et al. (Eds.): ECCV 2020, LNCS 12349, pp. 744–762, 2020.
https://doi.org/10.1007/978-3-030-58548-8_43

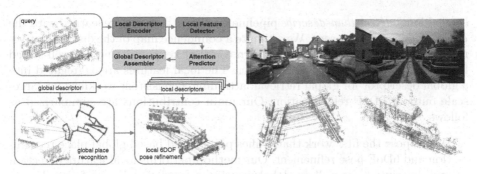

Fig. 1. Left: We propose a hierarchical network for large-scale point cloud based relocalization. The network consumes raw 3D points and performs local feature detection, description and global descriptor extraction in one forward pass. The global descriptor is used to retrieve similar scenes from the database. Accurate 6DoF pose is then obtained by matching the local features. Right: Our local descriptors trained on LiDAR points work favorably on the sparse point clouds generated by a visual SLAM method without fine-tuning. The point clouds are generated from sequences with different weathers, lighting conditions and scene layouts, thus *have significantly different distributions.*

problem using 2D images [6,44,46,51], where the 3D maps are usually defined as image feature points reconstructed in 3D using Structure from Motion (SfM). The coarse global place recognition is achieved by image retrieval, whereas accurate local 6DoF pose refinement is addressed separately by feature matching and PnP. With the progress of deep learning for image descriptor extraction [2,19] and 2D keypoints detection/description [14,15,40,60], image based methods have significantly gained in robustness to variations in viewpoint and illumination.

Alternatively one can tackle these variations by working on 3D point clouds since these are inherently invariant to such issues. Moreover, there exist numerous SLAM pipelines that generate accurate large-scale point clouds using sensory input from LiDAR [13,63,64] or camera [17,55]. While there is great potential to rely on such data, research on point cloud based relocalization is significantly less matured compared to the image-based counterpart [15,40]. Especially, deep learning on 3D descriptors emerged only roughly 3 years ago. With most of the early attempts focusing on small-scale tasks like object classification, detection and segmentation [28,37,39,57,68], only a limited number of networks have been proposed for large-scale localization [30,31,58]. Moreover, among these few attempts, global place recognition [1,65] and local 6DoF pose refinement [8,18,23] have been addressed isolatedly, despite the fact that both tasks depend on the same low level geometric clues.

In this paper, we propose a hierarchical deep network for large-scale point clouds based relocalization – see Fig. 1. The network directly consumes unordered 3D points and performs keypoint detection and description, as well as global point cloud descriptor extraction in a unified manner. In contrast to the conventional *detect-then-describe* pipeline, our local features are learned with the *detect-and-describe* concept. We estimate a confidence map of the discriminativeness of local features explicitly and learn to select keypoints that are well-suited for matching in an unsupervised manner. The local features are aggregated into a global descriptor for global retrieval, attaining a consistent workflow for large-scale outdoor 6DoF relocalization. Our main contributions are summarized as follows:

- We propose the first work that unifies point cloud based global place recognition and 6DoF pose refinement. Our method performs local feature detection and description, as well as global descriptor extraction in one forward pass, running significantly faster than previous methods.
- We propose to use FlexConv and SE block to integrate multi-level context information and channel-wise relations into the local features, thus achieve much stronger performance on feature matching and also boost the global descriptor.
- We introduce a *describe-and-detect* approach to explicitly learn a 3D keypoint detector in an unsupervised manner.
- Both our local and global descriptors achieve state-of-the-art performances on point cloud registration and retrieval across multiple benchmarks.
- Furthermore, our local descriptors trained on LiDAR data show competitive generalization capability when applied to the point clouds generated by a visual SLAM method, even though LiDAR and visual SLAM point clouds exhibit very different patterns and distributions.

2   Related Work

Handcrafted Local Descriptors encode local structural information as histograms over geometric properties e.g., surface normals and curvatures. Spin image (SI) [24] projects 3D points within a cylinder onto a 2D spin image. Unique Shape Context (USC) [52] deploys a unique local reference frame to improve the accuracy of the well-know 3D shape context descriptor. Point Feature Histogram (PFH) [43] and Fast PFH (FPFH) [42] describe the relationships between a point and its neighbors by calculating the angular features and normals. While these handcrafted methods have made great progress, they generalize poorly to large-scale scenarios and struggle to handle noisy real-world data.

Learned Local Descriptors. To cope with the inherent irregularity of point cloud data, researchers have suggested to convert 3D points to regular representations such as voxels and multi-view 3D images [38,47,49,50,54]. As a pioneering work, PointNet [37] proposed to apply deep networks directly to raw 3D points. Since then, new models [28,39,66] and flexible operators on irregular data [21,56,59] have been emerging. Accompanied by these progresses, learning-based 3D local descriptors such as 3DMatch [62], PPFNet [11] and PPF-FoldNet [10,12] have been proposed for segment matching, yet they are designed for RGB-D based indoor applications. Recently, Fully Convolutional

Geometric Features (FCGF) [8] was proposed to extract geometric features from 3D points. Yet, all these methods do not tackle feature detection and get their features rather by sampling. Another class of methods utilizes deep learning to reduce the dimensions of the handcrafted descriptors, such as Compact Geometric Features (CGF) [25] and LORAX [16]. In the realm of large-scale outdoor relocalization, 3DFeatNet [23] and L³-Net [31] extract local feature embedding using PointNet, whereas 3DSmoothNet [18] and DeepVCP [30] rely on 3D CNNs. In contrary to registration based on feature matching, DeepVCP and Deep Closest Point [58] learn to locate the correspondences in the target point clouds.

3D Keypoint Detectors. There are three representative hand-crafted 3D detectors. Intrinsic Shape Signatures (ISS) [67] selects salient points with large variations along the principal axes. SIFT-3D [29] constructs a scale-space of the curvature with the DoG operator. Harris-3D [48] calculates the Harris response of each 3D vertex based on first order derivatives along two orthogonal directions on the 3D surface. Despite the increasing number of learning-based 3D descriptors, only a few methods have been proposed to learn to detect 3D keypoints. 3DFeatNet [23] and DeepVCP [30] use an attention layer to learn to weigh the local descriptors in their loss functions. Recently, USIP [27] has been proposed to specifically detect keypoints with high repeatability and accurate localization. It establishes the current state of the art for 3D keypoint detectors.

Handcrafted Global Descriptors. Most 3D global descriptors describe places with handcrafted statistical information. Rohling et al. [41] propose to describe places by histograms of points elevation. Cop et al. [9] leverage LiDAR intensities and present DELIGHT. Cao et al. [5] transform a point cloud to a bearing-angle image and extract ORB features for bag-of-words aggregation.

Learned Global Descriptors. Granström et al. [20] describe point clouds with rotation invariant features and input them to a learning-based classifier for matching. LocNet [61] inputs range histogram features to 2D CNNs to learn a descriptor. In these methods, deep learning essentially plays the role of post-processing the handcrafted descriptors. Kim et al. [26] transform point clouds into scan context images and feed them into CNNs for place recognition. Point-NetVLAD [1] first tackles place recognition in an end-to-end way. The global descriptor is computed by a NetVLAD [2] layer on top of the feature map extracted using PointNet [37]. Following this, PCAN [65] learns attentions for points to produce more discriminative descriptors. These two methods extract local features using PointNet, which projects each point independently into a higher dimension and thus does not explicitly use contextual information.

3   Hierarchical 3D Descriptors Learning

For large-scale relocalization, an intuitive approach is to tackle the problem hierarchically in a coarse-to-fine manner: local descriptors are extracted, aggregated into global descriptors for coarse place recognition, and then re-used for accurate 6DoF pose refinement. While being widely adopted in image the

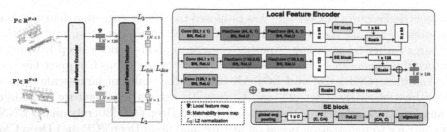

Fig. 2. Left: the flow of local feature descriptor and detector learning. Right: the architecture of our Local Feature Encoder. In Conv(D, $K \times K$) and FlexConv(D, k, d), D: output dimension, K: filter size, k: neighborhood size, d: dilation rate.

domain [33, 44, 45], this idea has not been addressed by the deep learning community for 3D. As a result, seeking for 6DoF relocalization for point clouds, one has to perform local feature detection, description, global descriptor extraction separately, possibly running an independent network for each. To address this problem, we design a hierarchical network operating directly on a point cloud, delivering local descriptors, a keypoint score map and a global descriptor in a single forward pass. Point cloud based relocalization thus can be performed hierarchically: a coarse search using the global descriptor retrieves 3D submap candidates, which are subsequently verified by local 3D feature matching to estimate the 6DoF poses. An overview of our system is provided in Fig. 1.

3.1   3D Local Feature Encoder and Detector

3DFeatNet [23] is a seminal work that learns both 3D local feature detection and description. Nevertheless, the following two points potentially limit its discriminative power: (1) Its detector is an attention map learned directly from the input points. During inference descriptors are only extracted for the keypoints defined by the attention map. Such classical *detect-then-describe* approach, as discussed in [15, 40], typically focuses on low-level structures of the raw input data, and cannot utilize the high level information encoded in the descriptors. (2) Its feature description is PointNet-based, the symmetric function of which tends to provide only limited structural information of local clusters. To resolve these limitations, we propose to use Flex Convolution (FlexConv) [21] and Squeeze-and-Excitation (SE) block [22] to respectively fuse multi-level spatial contextual information and channel-wise feature correlations into the local descriptors. The *describe-and-detect* pipeline [15, 40] is adopted to postpone our detection stage to employ higher-level information in the learned descriptors.

FlexConv Layer. Considering a 3D point $\boldsymbol{p}_l$ with its k neighbors $N_k(\boldsymbol{p}_l) = \{\boldsymbol{p}_{l_1}, \cdots, \boldsymbol{p}_{l_k}\}$ as a graph $\mathcal{G} = (\mathcal{V}, \mathcal{E})$, where $\mathcal{V} = \{0, .., k\}$ are the vertices and $\mathcal{E} \subseteq \mathcal{V} \times \mathcal{V}$ the edges. A 3D operator on an edge of this graph can be formulated as $\boldsymbol{e}_{ll_k} = f_\Theta(\boldsymbol{p}_l, \boldsymbol{p}_{l_k})$, with Θ the set of learnable parameters. As pointed out by [59], PointNet is a special case of $f_\Theta(\boldsymbol{p}_l, \boldsymbol{p}_{l_k}) = f_\Theta(\boldsymbol{p}_l)$, thus encodes only global shape information and ignores the local neighborhood structure. In contrary, FlexConv can be abstracted as $f_\Theta(\boldsymbol{p}_l, \boldsymbol{p}_{l_k}) = f_\Theta(\boldsymbol{p}_l - \boldsymbol{p}_{l_k}, \boldsymbol{p}_{l_k})$, therefore can effectively encode local information, which we believe is crucial for learning discriminative local descriptors. Formally, FlexConv is a generalization of the conventional grid-based convolution and is defined as:

$$f_{FlexConv}(\boldsymbol{p}_l) = \sum_{\boldsymbol{p}_{l_i} \in N_k(\boldsymbol{p}_l)} \omega(\boldsymbol{p}_{l_i}, \boldsymbol{p}_l) \cdot h(\boldsymbol{p}_{l_i}), \qquad (1)$$

where $h(\boldsymbol{p}_{l_i}) \in \mathbb{R}^C$ is a point-wise encoding function projecting a point to the high-dimensional feature space. It is convolved with a filter-kernel $\omega : \mathbb{R}^3 \times \mathbb{R}^3 \to \mathbb{R}^C$ that is computed by the standard scalar product in the Euclidean space, with learnable parameters $\theta \in \mathbb{R}^{C \times 3}$, $\theta_b \in \mathbb{R}^C : w(\boldsymbol{p}_l, \boldsymbol{p}_{l_i} \mid \theta, \theta_b) = \langle \theta, \boldsymbol{p}_l - \boldsymbol{p}_{l_i} \rangle + \theta_b$. It can be considered as a linear approximation of the traditional filter-kernel which uses the location information explicitly. In addition, ω is everywhere well-defined and allows us to perform back-propagation easily.

Squeeze-and-Excitation (SE) Block. While FlexConv models spatial connectivity patterns, SE blocks [22] are further used to explicitly model the channel-wise inter-dependencies of the output features from the FlexConv layers. Let $\boldsymbol{U} = \{u_1, \cdots, u_C\} \in \mathbb{R}^{N \times C}$ denote the input feature map to the SE block, where $u_c \in \mathbb{R}^N$ represents the c-th channel vector of the output from the last FlexConv layer. The squeeze operation first "squeezes" $\boldsymbol{U}$ into a channel-wise descriptor $z \in \mathbb{R}^C$ as $f_{sq} : \mathbb{R}^{N \times C} \to \mathbb{R}^C$, $z = f_{sq}(\boldsymbol{U})$, where f_{sq} is implemented as a global average pooling to aggregate across spatial dimensions. $z \in \mathbb{R}^C$ is an embedding containing global information which is then processed by the excitation operation $f_{ex} : \mathbb{R}^C \to \mathbb{R}^C$, $s = f_{ex}(z)$, where $s \in \mathbb{R}^C$ is implemented as two fully connected layers with ReLU to fully capture channel-wise dependencies and to learn a nonlinear relationship between the channels. In the end, the learned channel activations are used to recalibrate the input across channels achieving the attention selection of different channel-wise features $\tilde{u}_c = f_{scale}(u_c, s_c) = s_c \cdot u_c$, where $\tilde{U} = \{\tilde{u}_1, \cdots, \tilde{u}_C\}$ refers to the output of the SE block.

Encoder Architecture. The architecture of the encoder module is illustrated in Fig. 2. In comparison to 3DFeatNet [23] which relies on PointNet and only operates on one level of spatial granularity, our encoder extracts structural information from two spatial resolutions. At each resolution, the following operations are conducted: one 1×1 convolution, two consecutive FlexConv layers and a SE block. Taking a point cloud $\boldsymbol{P} = \{\boldsymbol{p}_1, \cdots, \boldsymbol{p}_N\} \in \mathbb{R}^{N \times 3}$ as input, the encoder fuses multi-level contextual information by adding the outputs from the two resolutions and produces the feature map $\boldsymbol{\Psi}$. It is then L2-normalized to give us the final local descriptor map $\boldsymbol{\mathcal{X}} = \{\boldsymbol{x}_1, \cdots, \boldsymbol{x}_N\} \in \mathbb{R}^{N \times D}$. Benefited from the better policies for integrating contextual information, when compared to 3DFeatNet, our local features are much more robust to points densities and distributions, thus generalize significantly better to point clouds generated by different sensors (more details in Sect. 4.4).

Description Loss. In feature space, descriptors of positive pairs are expected to be close and those of negative pairs should keep enough separability. Instead of using the simple Triplet Loss as in 3DFeatNet, we adopt the N-tuple loss [11] to learn to differentiate as many patches as possible. Formally, given two point clouds P, P', the two following matrices can be computed: a feature space distance matrix $D \in \mathbb{R}^{N \times N}$ with $D(i,j) = \|x_i - x_j\|$, a correspondence matrix $M \in \mathbb{R}^{N \times N}$ with $M_{i,j} \in \{0,1\}$ indicating whether two point patches $p_i \in P$ and $p_j \in P'$ form a positive pair, i.e., if their distance is within a pre-defined threshold. The N-tuple loss is formulated as:

$$L_{desc} = \sum{}^* \left(\frac{M \circ D}{\|M\|_F^2} + \eta \frac{max(\mu - (1-M) \circ D, 0)}{N^2 - \|M\|_F^2} \right), \qquad (2)$$

where $\sum^*(\cdot)$ is element-wise sum, $\circ$ element-wise multiplication, $\|\cdot\|_F$ the Frobenius norm, η a hyper-parameter balancing matching and non-matching pairs. The loss is divided by the number of true/false matches to remove the bias introduced by the larger number of negatives.

3D Local Feature Detection. Contrary to the classical *detect-then-describe* approaches, we postpone the detection to a later stage. To this end, we produce a keypoint saliency map $S \in \mathbb{R}^{N \times 1}$ from the extracted point-wise descriptors instead of from the raw input points. Our saliency map is thus estimated based on the learned local structure encoding and thus is less fragile to low level artifacts in the raw data, and provides significantly better generalization. Another benefit of the *describe-and-detect* pipeline is that feature description and detection can be performed in one forward pass, unlike the *detect-then-describe* approaches that usually need two stages. Our detector consumes the local feature map Ψ by a series of four 1×1 convolution layers, terminated by the sigmoid activation function (more details in the supplementary document).

As there is no standard definition of a keypoint's discriminativeness for outdoor point clouds, keypoint detection cannot be addressed by supervised learning. In our case, since the learned descriptors are to be used for point cloud registration, we propose to optimize keypoint confidences by leveraging the quality of descriptor matching. Local descriptor matching essentially boils down to nearest neighbor search in the feature space. Assuming the descriptor is informative enough and the existence of correspondence is guaranteed, a reliable keypoint, i.e., a keypoint with a high score s, is expected to find the correct match with high probability. We therefore can measure the quality of the learned detector using $\eta_i = (1 - s_i) \cdot (1 - M_{i,j}) + s_i \cdot M_{i,j}$, where $s_i \in [0,1]$ is an element of S and j refers to the nearest neighbor in Ψ'. A simple loss function can be formulated as $L_{det} = \frac{1}{N} \sum_{i=1}^{N} 1 - \eta_i$. However, we find that only using the nearest neighbor to define η is too strict on the learned feature quality and the training can be unstable. We thus propose a new metric called average successful rate (AR): given a point $p_i \in P$ and its feature $\psi_i \in \Psi$, we find the k nearest neighbors in Ψ'. The AR of p_i is computed as: $AR_i = \frac{1}{k} \sum_{j=1}^{k} c_{ij}$, where $c_{ij} = 1$ if at least one correct correspondence can be found in the first j candidates, otherwise is

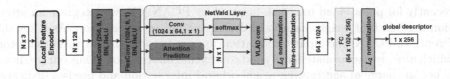

Fig. 3. The architecture of the global descriptor assembler.

$0.^1$ Now we can measure η with AR which is a real number in the range $[0, 1]$ instead of a binary number and the loss above can be rewritten as:

$$L_{det} = \frac{1}{N} \sum_{i=1}^{N} 1 - [\kappa(1 - s_i) + s_i \cdot \mathrm{AR}_i], \qquad (3)$$

where $\kappa \in [0, 1]$ is a hyperparameter indicating the minimum expected AR per keypoint. To minimize the new loss, the network should predict s_i to be close to 0 if $\mathrm{AR}_i < \kappa$ and to be near 1 conversely.

3.2 Global Descriptor Learning

As a key concept of this work, we propose to re-use the local descriptors for global retrieval. This early sharing of the computation is natural as both local and global descriptors are based on the same low-level geometric clues. The upcoming question is, how can the local descriptors be aggregated to a global one? While there exist many ways to do so, e.g., pooling, PointNet++ [39], FlexConv, PointCNN [28], Dynamic Graph CNN [59], we claim that the PCAN [65] (Point-NetVLAD extended by adding attention) gives the best performance among the many and provide an ablation study in the supplementary material.

Our global aggregation network is depicted in Fig. 3. Before the NetVLAD module, two FlexConv layers are added to project the local features to a higher dimension for a more retrieval relevant encoding. The attention predictor takes these features and outputs a per-point attention map, which is followed by a NetVLAD layer to generate a compact global representation. As the output of a NetVLAD layer usually has very high dimension which indicates expensive nearest neighbor search, a FC layer is used to compress it into a lower dimension. The global descriptor assembler is trained using the same lazy quadruplet loss as used in [1,65], to better verify our idea of using the learned local descriptor for global descriptor aggregation.

Attention Map Prediction. As it has been observed for image retrieval, visual cues relevant to place recognition are generally not uniformly distributed across an image. Therefore focusing on important regions is the key to improve the performance [7,34]. However, such attention mechanism has only been explored

---

[1] E.g., if the first correct correspondence appears as the 3rd nearest neighbor, then AR in the case of $k = 5$ is $(0 + 0 + 1 + 1 + 1)/5 = 0.6$.

recently for point cloud retrieval. Inspired by PCAN [65], we integrate an attention module that weighs each local descriptor before aggregation. As a key difference to PCAN, the input to our attention predictor is the learned descriptors which already encapsulate fairly good contextual information. Thus our predictor is not in charge of aggregating neighborhood information and needs a dedicated design to reflect such benefit. We thus construct our attention predictor by only chaining up three 1×1 Conv layers followed by softmax to ensure the sum of the attention weights is 1. We will show that, although our attention predictor has a much simpler structure than PCAN, yet it is effective. When combined with our descriptor, it still offers better global retrieval performance. More details on the network structure are provided in the supplementary document.

4   Experiments

The LiDAR point clouds from the Oxford RobotCar dataset [32] are used to train our network. Additionally, the ETH dataset [36] and the point clouds generated by Stereo DSO [55], a direct visual SLAM method are used to test the generalization ability of the evaluated methods. The margin used in L_{desc} is set to $\mu = 0.5$, the minimum expected AR in Eq. 3 $\kappa = 0.6$ with $k = 5$, N in Eq. 2 is set to 512. We use a weighted sum of L_{desc} and L_{det} as the loss function to train our network $L = L_{desc} + \lambda L_{det}$.

To train the local part of our network, we use Oxford RobotCar and follow the data processing procedures in [23]. We use 35 traversals and for each create 3D submaps using the provided GPS/INS poses with a 20 m trajectory and a 10 m interval. The resulting submaps are downsampled using a voxel grid with grid size of 0.2 m. In total 20,731 point clouds are collected for training. As the provided ground truth poses are not accurate enough to obtain cross-sequence point-to-point correspondences, we generate training samples with synthetic transformations: for a given point cloud we create another one by applying an arbitrary rotation around the upright axis and then adding Gaussian noise $\mathcal{N}(0, \sigma_{noise})$ with $\sigma_{noise} = 0.02$ m. Note that as a point cloud is centered wrt. its centroid before entering the network, no translation is added to the synthetic transformations. For the global part, we use the dataset proposed in PointNetVLAD [1]. Specifically, for each of the 23 full traversals out of the 44 selected sequences from Oxford RobotCar, a testing reference map is generated consisting of the submaps extracted in the testing part of the trajectory at 20 m intervals. More details on preparing the training data are left to the supplementary document.

Runtime. For a point cloud with 8192 points, our local (including feature description and keypoint detection) and global descriptors can be extracted in one forward pass in 80 ms. As comparison, 3DFeatNet takes 400 ms (detection) + 510 ms (NMS) + 18 ms (512 local descriptors); 3DSmoothNet needs 270 ms (preprocessing) + 144 ms (512 local descriptors).

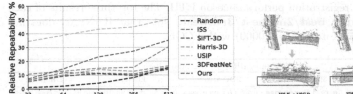

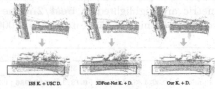

Fig. 4. Relative repeatability when different number of keypoints are detected.

Fig. 5. Qualitative point cloud registration on Oxford RobotCar. Green lines show the inliers of RANSAC. (Color figure online)

Table 1. Point cloud registration performance on Oxford RobotCar. Each row in the table corresponds to a keypoint detector and each column refers to a local 3D descriptor. In each cell, we show Relative Translational Error (RTE), Relative Rotation Error (RRE), the registration success rate and the average number of RANSAC iterations. The methods are evaluated on the testing set provided by [23]. The top three results of each metric are highlighted in **best**/2nd best/3rd best.

	RTE (m) / RRE(°) / Succ. (%) / Iter.				
	FPFH	3DSmoothNet	3DFeatNet	FCGF	DH3D
Random	0.44/1.84/89.8/7135	0.34/1.39/96.2/7274	0.43/1.62/90.5/9898	0.61/ 2.01/39.87/7737	0.33/1.31/92.1/6873
ISS	0.39/1.60/92.3/7171	0.32/1.21/96.8/6301	0.31/1.08/97.7/7127	0.56/1.89/43.99/7799	**0.30**/1 04/97.9/4986
Harris-3D	0.54/2.31/47.5/9997	0.31/1.19/97.4/5236	0.35/1.33/95.0/9214	0.57/1.99/46.82/7636	0.34/1.20/96.4/5985
3DFeatNet	0.43/2.01/73.7/9603	0.34/1.34/95.1/7280	**0.30**/1.07/98.1/2940	0.55/1.89/43.35/5958	0.32/1.24/95.4/2489
USIP	0.36/1.55/84.3/5663	**0.28/0.93**/98.0/**584**	**0.28/0.81/99.1/523**	0.41/1.73/53.42/3678	**0.30**/1.21/96.5/**1537**
DH3D	0.75/1.85/55.6/8697	0.32/1.22/96.0/3904	**0.28**/1.04/**98.2**/2908	0.38/1.48/49.47/4069	**0.23/0.95/98.5**/1972

4.1 3D Keypoint Repeatability

We use *relative repeatability* to quantify the performance of our keypoint detector. Given two point clouds $\{P, P'\}$ related by a transformation T, a keypoint detector detects keypoints $K = [K_1, K_2, \cdots, K_m]$ and $K' = [K'_1, K'_2, \cdots, K'_m]$ from them. $K_i \in K$ is repeatable if the distance between $T(K_i)$ and its nearest neighbor $K'_j \in K'$ is less than 0.5 m. *Relative repeatability* is then defined as $|K_{rep}|/|K|$ with K_{rep} the repeatable keypoints. We use the Oxford RobotCar testing set provided by 3DFeatNet [23], which contains 3426 point cloud pairs constructed from 794 point clouds. We compare to three handcrafted 3D detectors, ISS [67], SIFT-3D [29] and Harris-3D [48] and two learned ones 3DFeatNet [23] and USIP [27]. The results are presented in Fig. 4. As the most recently proposed learning based 3D detector that is dedicatedly designed for feature repeatability, USIP apparently dominates this benchmark. It is worth noting that the keypoints detected by USIP are highly clustered, which is partially in favor of achieving a high repeatability. Moreover, USIP is a pure detector, while 3DFeatNet and ours learn detection and description at the same time. Our detector outperforms all the other competitors by a large margin when detecting more than 64 keypoints. In the case of 256 keypoints, our repeatability is roughly 1.75x than the best follower 3DFeatNet. This clearly demonstrates that, when

Table 2. Point cloud registration performance on ETH. The top three results of each metric are highlighted in best/2nd best/3rd best. Note that RANSAC does not converge within the max. iterations (10000) with FCGF.

	RTE (m) / RRE(°) / Succ. (%) / Iter.				
	SI	3DSmoothNet	3DFeatNet	FCGF	DH3D
Random	0.36/4.36/95.2/7535	0.18/2.73/100/986	0.30/4.06/95.2/6898	0.69/52.87/17.46/10000	0.25/3.47/100/5685
ISS	0.37/5.07/93.7/7706	0.15/2.40/100/986	0.31/3.86/90.5/6518	0.65/24.78/6.35/10000	0.19/2.80/93.8/3635
Harris-3D	0.35/4.83/90.5/8122	0.15/2.41/100/788	0.27/3.96/88.9/6472	0.43/55.70/6.35/10000	0.22/3.47/93.4/4524
3DFeatNet	0.35/5.77/87.3/7424	0.17/2.73/100/1795	0.33/4.50/95.2/6058	0.52/47.02/3.17/10000	0.27/3.58/93.7/6462
USIP	0.32/4.06/92.1/6900/	0.18/2.61/100/1604	0.31/3.49/82.5/7060	0.54/27.62/15.87/10000	0.29/3.29/95.2/4312
DH3D	0.42/4.65/81.3/7922	0.38/3.49/100/5108	0.36/2.38/95.5/3421	0.56/48.01/15.87/10000	0.3/2.02/95.7/3107

learning detector and descriptors together, *describe-and-detect* is superior than *detect-then-describe*. It is yet interesting to see how the key ideas of USIP [27] can be merged into this concept.

4.2   Point Cloud Registration

Geometric registration is used to evaluate 3D feature matching. A SE3 transformation is estimated based on the matched keypoints using RANSAC. We compute Relative Translational Error (RTE) and Relative Rotation Error (RRE) and consider a registration is successful when RTE and RRE are below 2 m and $5°$, respectively. We compare to two handcrafted (ISS [67] and Harris-3D [48]) and two learned (3DFeatNet [23] and USIP [27]) detectors, and three handcrafted (SI [24], USC [52] and FPFH [42]) and three learned (3DSmoothNet [18], 3DFeatNet [23] and FCGF [8]) descriptors. The average RTE, RRE, the success rate and the average number of RANSAC iterations on Oxford RobotCar of each detector-descriptor combination are shown in Table 1. Note that due to space limitation, only the best performing handcrafted descriptor is shown (the same applies to Table 2 and 4). Although USIP shows significantly better performance on repeatability, our method now delivers competitive or even better results when applied for registration, where both keypoint detector and local feature encoder are needed. This on one hand demonstrates the strong discriminative power of our local descriptors, on the other hand also supports our idea of learning detector and descriptors in the *describe-and-detect* manner. Another thing to point out is that FCGF was trained on KITTI, which might explain its relatively bad results in this evaluation. Some qualitative results can be found in Fig. 5.

Unlike Oxford RobotCar, the ETH dataset [36] contains largely unstructured vegetations and much denser points, therefore is used to test the generalizability. The same detectors and descriptors as above are tested and the results are shown in Table 2. We notice that 3DSmoothNet shows the best performances on ETH. One important reason is that 3DSmoothNet adopts a voxel grid of size 0.02 m to downsample the point clouds, while our DH3D and other methods use size 0.1 m. Thus 3DSmoothNet has finer resolution and is more likely to produce smaller errors. Apart from that, our detector performs fairly well (last row) and when combine with our descriptor, it achieves the smallest rotation error. With all the detectors, FCGF descriptors cannot make RANSAC converge within the maximum number of iterations. The bad performance of FCGF is also noticed by the authors of [4] and was discussed on their GitHub page[2].

4.3 Point Cloud Retrieval

Table 3. Average recall (%) at top 1% and top 1 for Oxford RobotCar.

Method	@1%	@1
PN_MAX	73.44	58.46
PN_VLAD	81.01	62.18
PCAN	83.81	69.76
Ours-4096	**84.26**	**73.28**
Ours-8192	**85.30**	**74.16**

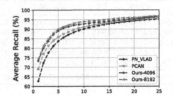

Fig. 6. Average recall of the top 25 retrievals on Oxford RobotCar.

We compare our method against the two state-of-the-art approaches, PCAN [65] and PointNetVLAD (PN_VLAD) [1]. We also report the results of PN_MAX presented in PN_VLAD, which consists of the original PointNet architecture with the maxpool layer and a fully connected layer to produce a global descriptor. Note that both PN_VLAD and PCAN take submaps of size 4096, whereas ours takes 8192 to favor local feature extraction. We thus add a downsampling layer before the final NetVLAD layer to make sure the same size of 4096 points enter the final aggregation procedure. For demonstration, we also report the results of using the default setting. We first evaluate the average recall at top 1% and top1 and show the results in Table 3. Our method with both settings outperforms all the other methods. We further show the recall curves of the top25 retrieval matches in Fig. 6, where both of our networks consistently outperform the other two state-of-the-art approaches. The evaluation results prove that our network can effectively take advantage of the informative local features and produce more discriminative global descriptors. Also note that even with the number of points halved before entering the NetVLAD layer, the performance drop of our method is very small, which shows that our local features integrate sufficient contextual information for global retrieval. Some qualitative retrieval results are provided in the supplementary document.

---

[2] https://github.com/XuyangBai/D3Feat/issues/1.

4.4   Application to Visual SLAM

Table 4. Generalization of point cloud registration for visual SLAM. In this experiment, point clouds are generated by running Stereo DSO [55] on Oxford. For learning based methods, models trained on LiDAR points are used without fine-tuning. The top three results of each metric are highlighted in **best**/**2nd best**/**3rd best**.

	RTE (m) / RRE(°) / Succ. (%) / Iter.				
	FPFH	3DSmoothNet	3DFeatNet	FCGF	DH3D
Random	0.56/2.82/53.13/9030	0.70/2.19/73.1/6109	0.72/2.37/69.0/9661	0.51/2.65/74.93/5613	0.70/2.23/71.9/7565
ISS	0.56/3.03/43.58/9210	0.67/2.15/79.1/6446	0.58/2.41/71.9/9776	0.51/2.57/71.94/6015	0.48/**1.72**/**90.2**/6312
Harris-3D	0.49/2.67/45.67/9130	0.48/2.07/74.9/6251	0.66/2.26/64.5/9528	0.48/2.63/74.03/5482	0.39/2.27/68.1/7860
3DFeatNet	0.62/3.05/35.52/7704	**0.38**/2.22/66.6/5235	0.92/1.97/84.1/8071	0.54/2.64/60.90/**4409**	0.74/2.38/80.9/7124
USIP	0.54/2.98/48.96/7248	0.39/2.27/77.3/5593	0.85/2.24/69.9/8389	0.51/2.65/67.46/**3846**	0.65/2.45/68.1/6824
DH3D	0.60/2.92/48.96/8914	**0.36**/2.01/77.9/5764	0.41/**1.84**/**89.3**/7818	0.48/2.43/69.55/**5002**	**0.36**/**1.58**/**90.6**/7071

Fig. 7. Registration on point clouds generated by Stereo DSO [55]. The first two columns display frames from the reference and the query sequences. The last two columns show the matches found by RANSAC and the point clouds after alignment.

In this section, we demonstrate the generalization capability of our method to a different sensor modality by evaluating its performance on the point clouds generated by Stereo DSO [55]. As a direct method, Stereo DSO has the advantage of delivering relatively dense 3D reconstructions, which can provide stable geometric structures that are less affected by image appearance changes. We therefore believe that it is worth exploring extracting 3D descriptors from such reconstructions which can be helpful for loop-closure and relocalization for visual SLAM. To justify this idea, Stereo DSO is used to generate point clouds of eight traversals from Oxford RobotCar, which covers a wide range of day-time and weather conditions. This gives us 318 point cloud pairs with manually annotated relative poses. Each point cloud is cropped with a radius of 30 m and randomly rotated around the vertical axis. We use the same parameters as in Sect. 4.2 without fine-tuning our network and evaluate the geometric registration performance against other methods in Table 4. As shown in the table, our approach achieves the best rotation error (1.58°) and success rate (90.6%) and second best translation error (0.36 m) among all the evaluated methods. It can also be noticed that most evaluated methods show significant inferior performances compared to the results in Table 1, e.g., the successful rates of 3DFeatNet+3DFeatNet, USIP+3DSmoothNet and USIP+3DFeatNet drop from 98.1%, 98.0% and 99.1% to 84.1%, 77.3% and 69.9%, respectively. This is largely because the point clouds extracted from LiDAR scannings have quite different distributions as those from Stereo DSO. Our model is still able to achieve a successful rate of 90.6%, showing the least degree of degeneracy. This further demonstrates the good generalization ability of the proposed method. Some qualitative results are shown in Fig. 7.

4.5   Ablation Study

Effectiveness of Different Components. We carry out three experiments to explore the contributions of different components of our method: (1) We remove the detector module and only L_{desc} is used to train the local feature encoder; (2) The weak supervision at the submap level proposed by 3DFeatNet [23] is used (details in the supplementary material); (3) We remove the SE blocks. As shown in Table 5, the largest performance decrease comes with (2), which verifies our idea of generating a supervision signal by synthesizing transformations. Results of (1) indicate that learning an effective confidence map S can improve the quality of the learned local descriptors for matching. The results of (3) show that SE blocks contribute to learning more informative local descriptors and therefore are helpful to 3D feature matching.

Robustness Test. We assess the robustness of our model for both point cloud retrieval and registration against three factors, i.e., noise, rotation and downsampling: We add Gaussian noise $\mathcal{N}(0, \sigma_{noise})$ to the point clouds; The range of the rotation test is set between 0 and 90°; Point clouds are downsampled using a set of factors α. For the local part, as shown in Fig. 8, our descriptors has shown excellent rotation invariance. When noise is added, our method can still achieves >90% success rate for $\sigma_{noise} < 0.15$ m. The performance significantly drops for $\sigma_{noise} > 0.2$ m, possibly due to the fact that training samples are filtered by a voxel grid with size 0.2 m, thus strong noise tends to heavily change the underlying point distribution. A similar explanation applies to the case of downsampling for a factor $\alpha > 2$. Nevertheless, our model can still guarantee 90% success rate for $\alpha \leqslant 1.5$. We conduct the same robustness tests for our global descriptors. Figure 9(a) demonstrates that our global descriptors possess good robustness to noise. Contrary to the local descriptors, the global descriptor seems to be less robust against rotations, which needs further investigation. Similar to the local descriptor, the quality of global feature is not affected too much for $\alpha \leqslant 1.5$.

Table 5. Effects of different components for
point cloud registration on Oxford RobotCar.

Method	RTE (m)	RRE (°)	Succ.	Iter.
w/o L_{det}	0.43	1.52	93.72	3713
Weak Sup.	0.48	1.78	90.82	3922
w/o SE	0.39	1.24	95.18	3628
Default	**0.23**	**0.95**	**98.49**	**1972**

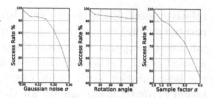

Fig. 8. Local detector and descriptor robustness test evaluated by the success rate of point cloud registration.

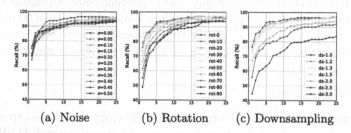

(a) Noise   (b) Rotation   (c) Downsampling

Fig. 9. Global descriptor robustness against random noise, rotation and downsampling. The x axes show the number of top retrieved matches.

5 Conclusion

We introduced a hierarchical network for the task of large-scale point cloud based relocalization. Rather than pursuing the traditional strategy of detect-then-describe or separately computing local and global descriptors, our network performs local feature detection, local feature description and global descriptor extraction in a single forward pass. Experimental results demonstrate the state-of-the-art performance of both our local and global descriptors across multiple benchmarks. Our model trained on LiDAR points also shows favorable generalization ability when applied to point clouds generated by visual SLAM methods. Future work is focused on further exploring the robustness of the learned descriptors to various perturbations.

References

1. Angelina Uy, M., Hee Lee, G.: PointNetVLAD: deep point cloud based retrieval for large-scale place recognition. In: Proceedings of the IEEE Conference on Computer Vision and Pattern Recognition, pp. 4470–4479 (2018)
2. Arandjelovic, R., Gronat, P., Torii, A., Pajdla, T., Sivic, J.: NetVLAD: CNN architecture for weakly supervised place recognition. In: Proceedings of the IEEE Conference on Computer Vision and Pattern Recognition, pp. 5297–5307 (2016)
3. Axelrod, B., Kaelbling, L.P., Lozano-Pérez, T.: Provably safe robot navigation with obstacle uncertainty. Int. J. Rob. Res. **37**(13–14), 1760–1774 (2018)

4. Bai, X., Luo, Z., Zhou, L., Fu, H., Quan, L., Tai, C.L.: D3Feat: joint learning of dense detection and description of 3D local features. arXiv:2003.03164 [cs.CV] (2020)
5. Cao, F., Zhuang, Y., Zhang, H., Wang, W.: Robust place recognition and loop closing in laser-based SLAM for UGVs in urban environments. IEEE Sens. J. **18**(10), 4242–4252 (2018)
6. Chen, Z., et al.: Deep learning features at scale for visual place recognition. In: 2017 IEEE International Conference on Robotics and Automation (ICRA), pp. 3223–3230. IEEE (2017)
7. Chen, Z., Liu, L., Sa, I., Ge, Z., Chli, M.: Learning context flexible attention model for long-term visual place recognition. IEEE Robot. Autom. Lett. **3**(4), 4015–4022 (2018)
8. Choy, C., Park, J., Koltun, V.: Fully convolutional geometric features. In: Proceedings of the IEEE International Conference on Computer Vision, pp. 8958–8966 (2019)
9. Cop, K.P., Borges, P.V., Dubé, R.: DELIGHT: an efficient descriptor for global localisation using LiDAR intensities. In: 2018 IEEE International Conference on Robotics and Automation (ICRA), pp. 3653–3660. IEEE (2018)
10. Deng, H., Birdal, T., Ilic, S.: PPF-FoldNet: unsupervised learning of rotation invariant 3D local descriptors. In: Ferrari, V., Hebert, M., Sminchisescu, C., Weiss, Y. (eds.) ECCV 2018. LNCS, vol. 11209, pp. 620–638. Springer, Cham (2018). https://doi.org/10.1007/978-3-030-01228-1_37
11. Deng, H., Birdal, T., Ilic, S.: PPFNet: global context aware local features for robust 3D point matching. In: Proceedings of the IEEE Conference on Computer Vision and Pattern Recognition, pp. 195–205 (2018)
12. Deng, H., Birdal, T., Ilic, S.: 3D local features for direct pairwise registration. In: Proceedings of the IEEE Conference on Computer Vision and Pattern Recognition, pp. 3244–3253 (2019)
13. Deschaud, J.E.: IMLS-SLAM: scan-to-model matching based on 3D data. In: 2018 IEEE International Conference on Robotics and Automation (ICRA), pp. 2480–2485. IEEE (2018)
14. DeTone, D., Malisiewicz, T., Rabinovich, A.: SuperPoint: self-supervised interest point detection and description. In: Proceedings of the IEEE Conference on Computer Vision and Pattern Recognition Workshops, pp. 224–236 (2018)
15. Dusmanu, M., et al.: D2-Net: a trainable CNN for joint description and detection of local features. In: Proceedings of the IEEE Conference on Computer Vision and Pattern Recognition, pp. 8092–8101 (2019)
16. Elbaz, G., Avraham, T., Fischer, A.: 3D point cloud registration for localization using a deep neural network auto-encoder. In: Proceedings of the IEEE Conference on Computer Vision and Pattern Recognition, pp. 4631–4640 (2017)
17. Engel, J., Stückler, J., Cremers, D.: Large-scale direct SLAM with stereo cameras. In: 2015 IEEE/RSJ International Conference on Intelligent Robots and Systems (IROS), pp. 1935–1942. IEEE (2015)
18. Gojcic, Z., Zhou, C., Wegner, J.D., Wieser, A.: The perfect match: 3D point cloud matching with smoothed densities. In: Proceedings of the IEEE Conference on Computer Vision and Pattern Recognition, pp. 5545–5554 (2019)
19. Gordo, A., Almazán, J., Revaud, J., Larlus, D.: Deep image retrieval: learning global representations for image search. In: Leibe, B., Matas, J., Sebe, N., Welling, M. (eds.) ECCV 2016. LNCS, vol. 9910, pp. 241–257. Springer, Cham (2016). https://doi.org/10.1007/978-3-319-46466-4_15

20. Granström, K., Schön, T.B., Nieto, J.I., Ramos, F.T.: Learning to close loops from range data. Int. J. Rob. Res. **30**(14), 1728–1754 (2011)
21. Groh, F., Wieschollek, P., Lensch, H.P.A.: Flex-convolution. In: Jawahar, C.V., Li, H., Mori, G., Schindler, K. (eds.) ACCV 2018. LNCS, vol. 11361, pp. 105–122. Springer, Cham (2019). https://doi.org/10.1007/978-3-030-20887-5_7
22. Hu, J., Shen, L., Sun, G.: Squeeze-and-excitation networks. In: Proceedings of the IEEE Conference on Computer Vision and Pattern Recognition, pp. 7132–7141 (2018)
23. Yew, Z.J., Lee, G.H.: 3DFeat-Net: weakly supervised local 3D features for point cloud registration. In: Ferrari, V., Hebert, M., Sminchisescu, C., Weiss, Y. (eds.) ECCV 2018. LNCS, vol. 11219, pp. 630–646. Springer, Cham (2018). https://doi.org/10.1007/978-3-030-01267-0_37
24. Johnson, A.E.: Spin-Images: A Representation for 3D Surface Matching. Carnegie Mellon University, Pittsburgh (1997)
25. Khoury, M., Zhou, Q.Y., Koltun, V.: Learning compact geometric features. In: Proceedings of the IEEE International Conference on Computer Vision, pp. 153–161 (2017)
26. Kim, G., Kim, A.: Scan context: egocentric spatial descriptor for place recognition within 3D point cloud map. In: 2018 IEEE/RSJ International Conference on Intelligent Robots and Systems (IROS), pp. 4802–4809. IEEE (2018)
27. Li, J., Lee, G.H.: USIP: unsupervised stable interest point detection from 3D point clouds. In: Proceedings of the IEEE International Conference on Computer Vision, pp. 361–370 (2019)
28. Li, Y., Bu, R., Sun, M., Wu, W., Di, X., Chen, B.: PointCNN: convolution on x-transformed points. In: Advances in Neural Information Processing Systems, pp. 820–830 (2018)
29. Lowe, D.G.: Local feature view clustering for 3D object recognition. In: Proceedings of the 2001 IEEE Computer Society Conference on Computer Vision and Pattern Recognition, vol. 1, p. I. IEEE (2001)
30. Lu, W., Wan, G., Zhou, Y., Fu, X., Yuan, P., Song, S.: DeepVCP: an end-to-end deep neural network for point cloud registration. In: Proceedings of the IEEE International Conference on Computer Vision, pp. 12–21 (2019)
31. Lu, W., Zhou, Y., Wan, G., Hou, S., Song, S.: L3-Net: towards learning based lidar localization for autonomous driving. In: Proceedings of the IEEE Conference on Computer Vision and Pattern Recognition, pp. 6389–6398 (2019)
32. Maddern, W., Pascoe, G., Linegar, C., Newman, P.: 1 year, 1000 km: the Oxford RobotCar dataset. Int. J. Rob. Res. **36**(1), 3–15 (2017)
33. Mur-Artal, R., Montiel, J.M.M., Tardos, J.D.: ORB-SLAM: a versatile and accurate monocular SLAM system. IEEE Trans. Rob. **31**(5), 1147–1163 (2015)
34. Noh, H., Araujo, A., Sim, J., Weyand, T., Han, B.: Large-scale image retrieval with attentive deep local features. In: Proceedings of the IEEE International Conference on Computer Vision, pp. 3456–3465 (2017)
35. Ort, T., Paull, L., Rus, D.: Autonomous vehicle navigation in rural environments without detailed prior maps. In: 2018 IEEE International Conference on Robotics and Automation (ICRA), pp. 2040–2047. IEEE (2018)
36. Pomerleau, F., Liu, M., Colas, F., Siegwart, R.: Challenging data sets for point cloud registration algorithms. Int. J. Rob. Res. **31**(14), 1705–1711 (2012)
37. Qi, C.R., Su, H., Mo, K., Guibas, L.J.: PointNet: deep learning on point sets for 3D classification and segmentation. In: Proceedings of the IEEE Conference on Computer Vision and Pattern Recognition, pp. 652–660 (2017)

38. Qi, C.R., Su, H., Nießner, M., Dai, A., Yan, M., Guibas, L.J.: Volumetric and multi-view CNNs for object classification on 3D data. In: Proceedings of the IEEE conference on Computer Vision and Pattern Recognition, pp. 5648–5656 (2016)
39. Qi, C.R., Yi, L., Su, H., Guibas, L.J.: PointNet++: deep hierarchical feature learning on point sets in a metric space. In: Advances in Neural Information Processing Systems, pp. 5099–5108 (2017)
40. Revaud, J., Weinzaepfel, P., de Souza, C.R., Humenberger, M.: R2D2: repeatable and reliable detector and descriptor. In: NeurIPS (2019)
41. Röhling, T., Mack, J., Schulz, D.: A fast histogram-based similarity measure for detecting loop closures in 3-D LiDAR data. In: 2015 IEEE/RSJ International Conference on Intelligent Robots and Systems (IROS), pp. 736–741. IEEE (2015)
42. Rusu, R.B., Bradski, G., Thibaux, R., Hsu, J.: Fast 3D recognition and pose using the viewpoint feature histogram. In: 2010 IEEE/RSJ International Conference on Intelligent Robots and Systems, pp. 2155–2162. IEEE (2010)
43. Rusu, R.B., Marton, Z.C., Blodow, N., Beetz, M.: Persistent point feature histograms for 3D point clouds. In: Proceedings of the 10th International Conference on Intelligent Autonomous Systems (IAS-10), Baden-Baden, Germany, pp. 119–128 (2008)
44. Sarlin, P.E., Cadena, C., Siegwart, R., Dymczyk, M.: From coarse to fine: robust hierarchical localization at large scale. In: Proceedings of the IEEE Conference on Computer Vision and Pattern Recognition, pp. 12716–12725 (2019)
45. Sarlin, P.E., Debraine, F., Dymczyk, M., Siegwart, R., Cadena, C.: Leveraging deep visual descriptors for hierarchical efficient localization. arXiv preprint arXiv:1809.01019 (2018)
46. Sattler, T., et al.: Are large-scale 3D models really necessary for accurate visual localization? In: Proceedings of the IEEE Conference on Computer Vision and Pattern Recognition, pp. 1637–1646 (2017)
47. Simonovsky, M., Komodakis, N.: Dynamic edge-conditioned filters in convolutional neural networks on graphs. In: Proceedings of the IEEE Conference on Computer Vision and Pattern Recognition, pp. 3693–3702 (2017)
48. Sipiran, I., Bustos, B.: Harris 3D: a robust extension of the Harris operator for interest point detection on 3D meshes. Vis. Comput. 27(11), 963 (2011). https://doi.org/10.1007/s00371-011-0610-y
49. Su, H., et al.: SplatNet: sparse lattice networks for point cloud processing. In: Proceedings of the IEEE Conference on Computer Vision and Pattern Recognition, pp. 2530–2539 (2018)
50. Su, H., Maji, S., Kalogerakis, E., Learned-Miller, E.: Multi-view convolutional neural networks for 3D shape recognition. In: Proceedings of the IEEE International Conference on Computer Vision, pp. 945–953 (2015)
51. Taira, H., et al.: InLoc: indoor visual localization with dense matching and view synthesis. In: Proceedings of the IEEE Conference on Computer Vision and Pattern Recognition, pp. 7199–7209 (2018)
52. Tombari, F., Salti, S., Di Stefano, L.: Unique shape context for 3D data description. In: Proceedings of the ACM Workshop on 3D Object Retrieval, pp. 57–62. ACM (2010)
53. Wang, P., Yang, R., Cao, B., Xu, W., Lin, Y.: Dels-3D: deep localization and segmentation with a 3D semantic map. In: Proceedings of the IEEE Conference on Computer Vision and Pattern Recognition, pp. 5860–5869 (2018)
54. Wang, P.S., Liu, Y., Guo, Y.X., Sun, C.Y., Tong, X.: O-CNN: octree-based convolutional neural networks for 3D shape analysis. ACM Trans. Graph. (TOG) 36(4), 72 (2017)

55. Wang, R., Schworer, M., Cremers, D.: Stereo DSO: large-scale direct sparse visual odometry with stereo cameras. In: Proceedings of the IEEE International Conference on Computer Vision, pp. 3903–3911 (2017)
56. Wang, S., Suo, S., Ma, W.C., Pokrovsky, A., Urtasun, R.: Deep parametric continuous convolutional neural networks. In: Proceedings of the IEEE Conference on Computer Vision and Pattern Recognition, pp. 2589–2597 (2018)
57. Wang, W., Yu, R., Huang, Q., Neumann, U.: SGPN: similarity group proposal network for 3D point cloud instance segmentation. In: Proceedings of the IEEE Conference on Computer Vision and Pattern Recognition, pp. 2569–2578 (2018)
58. Wang, Y., Solomon, J.M.: Deep closest point: learning representations for point cloud registration. In: Proceedings of the IEEE International Conference on Computer Vision, pp. 3523–3532 (2019)
59. Wang, Y., Sun, Y., Liu, Z., Sarma, S.E., Bronstein, M.M., Solomon, J.M.: Dynamic graph CNN for learning on point clouds. ACM Trans. Graph. (TOG) **38**(5), 1–12 (2019)
60. Yi, K.M., Trulls, E., Lepetit, V., Fua, P.: LIFT: learned invariant feature transform. In: Leibe, B., Matas, J., Sebe, N., Welling, M. (eds.) ECCV 2016. LNCS, vol. 9910, pp. 467–483. Springer, Cham (2016). https://doi.org/10.1007/978-3-319-46466-4_28
61. Yin, H., Wang, Y., Tang, L., Ding, X., Xiong, R.: LocNet: global localization in 3D point clouds for mobile robots. In: Proceedings of the 2018 IEEE Intelligent Vehicles Symposium (IV), Changshu, China, pp. 26–30 (2018)
62. Zeng, A., Song, S., Nießner, M., Fisher, M., Xiao, J., Funkhouser, T.: 3DMatch: learning local geometric descriptors from RGB-D reconstructions. In: Proceedings of the IEEE Conference on Computer Vision and Pattern Recognition, pp. 1802–1811 (2017)
63. Zhang, J., Singh, S.: LOAM: lidar odometry and mapping in real- time. In: Robotics: Science and Systems Conference (RSS), Berkeley, CA, July 2014
64. Zhang, J., Singh, S.: Visual-lidar odometry and mapping: low-drift, robust, and fast. In: 2015 IEEE International Conference on Robotics and Automation (ICRA), pp. 2174–2181. IEEE (2015)
65. Zhang, W., Xiao, C.: PCAN: 3D attention map learning using contextual information for point cloud based retrieval. In: Proceedings of the IEEE Conference on Computer Vision and Pattern Recognition, pp. 12436–12445 (2019)
66. Zhao, Y., Birdal, T., Deng, H., Tombari, F.: 3D point capsule networks. In: Proceedings of the IEEE Conference on Computer Vision and Pattern Recognition, pp. 1009–1018 (2019)
67. Zhong, Y.: Intrinsic shape signatures: a shape descriptor for 3D object recognition. In: 2009 IEEE 12th International Conference on Computer Vision Workshops, pp. 689–696. IEEE (2009)
68. Zhou, Y., Tuzel, O.: VoxelNet: end-to-end learning for point cloud based 3D object detection. In: Proceedings of the IEEE Conference on Computer Vision and Pattern Recognition, pp. 4490–4499 (2018)

Face Super-Resolution Guided by 3D Facial Priors

Xiaobin Hu[1,2], Wenqi Ren[2(✉)], John LaMaster[1], Xiaochun Cao[2,6],
Xiaoming Li[3], Zechao Li[4], Bjoern Menze[1], and Wei Liu[5]

[1] Informatics, Technische Universität München, Munich, Germany
[2] SKLOIS, IIE, CAS, Beijing, China
rwq.renwenqi@gmail.com
[3] Harbin Institute of Technology, Harbin, China
[4] NJUST, Nanjing, China
[5] Tencent AI Lab, Bellevue, USA
[6] Peng Cheng Laboratory, Cyberspace Security Research Center, Shenzhen, China

Abstract. State-of-the-art face super-resolution methods employ deep convolutional neural networks to learn a mapping between low- and high-resolution facial patterns by exploring local appearance knowledge. However, most of these methods do not well exploit facial structures and identity information, and struggle to deal with facial images that exhibit large pose variations. In this paper, we propose a novel face super-resolution method that explicitly incorporates 3D facial priors which grasp the sharp facial structures. Our work is the first to explore 3D morphable knowledge based on the fusion of parametric descriptions of face attributes (e.g., identity, facial expression, texture, illumination, and face pose). Furthermore, the priors can easily be incorporated into any network and are extremely efficient in improving the performance and accelerating the convergence speed. Firstly, a 3D face rendering branch is set up to obtain 3D priors of salient facial structures and identity knowledge. Secondly, the Spatial Attention Module is used to better exploit this hierarchical information (i.e., intensity similarity, 3D facial structure, and identity content) for the super-resolution problem. Extensive experiments demonstrate that the proposed 3D priors achieve superior face super-resolution results over the state-of-the-arts.

Keywords: Face super-resolution · 3D facial priors · Facial structures and identity knowledge

B. Menze, W. Liu—Contributed equally to this work.

Electronic supplementary material The online version of this chapter (https://doi.org/10.1007/978-3-030-58548-8_44) contains supplementary material, which is available to authorized users.

© Springer Nature Switzerland AG 2020
A. Vedaldi et al. (Eds.): ECCV 2020, LNCS 12349, pp. 763–780, 2020.
https://doi.org/10.1007/978-3-030-58548-8_44

1   Introduction

Face images provide crucial clues for human observation as well as computer analysis [20,45]. However, the performance of most face image tasks, such as face recognition and facial emotion detection [11,32], degrades dramatically when the resolution of a facial image is relatively low. Consequently, face super-resolution, also known as face hallucination, was coined to restore a high-resolution face image from its low-resolution counterpart.

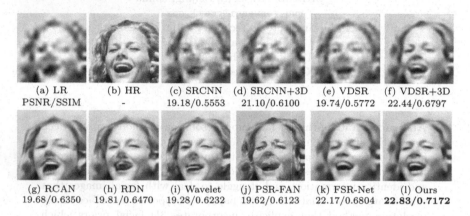

(a) LR | (b) HR | (c) SRCNN | (d) SRCNN+3D | (e) VDSR | (f) VDSR+3D
PSNR/SSIM | - | 19.18/0.5553 | 21.10/0.6100 | 19.74/0.5772 | 22.44/0.6797

(g) RCAN | (h) RDN | (i) Wavelet | (j) PSR-FAN | (k) FSR-Net | (l) Ours
19.68/0.6350 | 19.81/0.6470 | 19.28/0.6232 | 19.62/0.6123 | 22.17/0.6804 | **22.83/0.7172**

Fig. 1. Visual comparison with state-of-the-art face hallucination methods (×8). (a) 16 × 16 LR input. (b) 128 × 128 HR ground-truth. (c) Super-Resolution Convolutional Neural Network (SRCNN) [7]. (d) SRCNN incorporating our 3D facial priors. (e) Very Deep Super-Resolution Network (VDSR) [17]. (f) VDSR incorporating our 3D facial priors. (g) Very Deep Residual Channel Attention Network (RCAN) [42]. (h) Residual Dense Network (RDN) [43]. (i) Wavelet-based CNN for Multi-scale Face Super-Resolution (Wavelet-SRNet) [14]. (j) Progressive Face Super-Resolution using the facial landmark (PSR-FAN) [16]. (k) End-to-End Learning Face Super-Resolution with Facial Priors (FSRNet) [4]. (l) Our proposed method by embedding the 3D facial priors into the Spatial Attention Module (SAM3D).

Although a great influx of deep learning methods [3,5,9,24,36–39,44,46,47] have been successfully applied in face Super-Resolution (SR) problems, super-resolving arbitrary facial images, especially at high magnification factors, is still an open and challenging problem due to the ill-posed nature of the SR problem and the difficulty in learning and integrating strong priors into a face hallucination model. Some researches [4,10,16,28,35,41] on exploiting the face priors to assist neural networks in capturing more facial details have been proposed. A face hallucination model incorporating identity priors was presented in [10]. However, the identity prior was extracted only from the multi-scale up-sampling results in the training procedure and therefore cannot provide extra priors to guide the network. Yu et al. [35] employed facial component heatmaps to encourage the upsampling stream to generate super-resolved faces with higher-quality details, especially for large pose variations. Kim et al. [16] proposed a face alignment network (FAN) for landmark heatmap extraction to boost the performance of face SR. Chen et al. [4] utilized the heatmaps and parsing maps for face SR problems. Although these 2D priors provide global component regions, these methods cannot learn the 3D reconstruction of detailed edges, illumination, and expression priors. In addition, all of these aforementioned face SR approaches ignore facial structure and identity recovery.

In contrast to the aforementioned approaches, we propose a novel face super resolution method by exploiting 3D facial priors to grasp sharp face structures and identity knowledge. Firstly, a deep 3D face reconstruction branch is set up to explicitly obtain 3D face render priors which facilitate the face super-resolution branch. Specifically, the 3D facial priors contain rich hierarchical features, such as low-level (e.g., sharp edge and illumination) and perception level (e.g., identity) information. Then, a spatial attention module is employed to adaptively integrate the 3D facial prior into the network, in which we employ a spatial feature transform (SFT) [34] to generate affine transformation parameters for spatial feature modulation. Afterwards, it encourages the network to learn the spatial inter-dependencies of features between 3D facial priors and input images after adding the attention module into the network. As shown in Fig. 1, by embedding the 3D rendered face priors, our algorithm generates clearer and sharper facial structures without any ghosting artifacts compared with other 2D prior-based methods.

The main contributions of this paper are:

- A novel face SR model is proposed by explicitly exploiting facial structure in the form of facial prior estimation. The estimated 3D facial prior provides not only spatial information of facial components but also their 3D visibility information, which is ignored by the pixel-level content and 2D priors (e.g., landmark heatmaps and parsing maps).
- To well adapt to the 3D reconstruction of low-resolution face images, we present a new skin-aware loss function projecting the constructed 3D coefficients onto the rendered images. In addition, we use a feature fusion-based network to better extract and integrate the face rendered priors by employing a spatial attention module.
- Our proposed 3D facial prior has a high flexibility because its modular structure allows for easy plug-in of any SR methods (e.g., SRCNN and VDSR). We qualitatively and quantitatively evaluate the proposed algorithm on multi-scale face super-resolution, especially at very low input resolutions. The proposed network achieves better SR criteria and superior visual quality compared to state-of-the-art face SR methods.

2   Related Work

Face hallucination relates closely to the natural image super-resolution problem. In this section, we discuss recent research on super-resolution and face hallucination to illustrate the necessary context for our work.

Super-Resolution Neural Networks. Recently, neural networks have demonstrated a remarkable capability to improve SR results. Since the pioneering network [7] demonstrates the effectiveness of CNN to learn the mapping between LR and HR pairs, a lot of CNN architectures have been proposed for SR [8,12,18,19,30,31]. Most of the existing high-performance SR networks have residual blocks [17] to go deeper in the network architecture, and achieve better performance. EDSR [22] improved the performance by removing unnecessary batch normalization layers in residual blocks. A residual dense network (RDN) [43] was proposed to exploit the hierarchical features from all the convolutional layers. Zhang et al. [42] proposed the very deep residual channel attention networks (RCAN) to discard abundant low-frequency information which hinders the representational ability of CNNs. Wang et al.[34] used a spatial feature transform layer to introduce the semantic prior as an additional input of the SR network. Huang et al. [14] presented a wavelet-based CNN approach that can ultra-resolve a very low-resolution face image in a unified framework. Lian et al. [21] proposed a Feature-Guided Super-Resolution Generative Adversarial Network (FG-SRGAN) for unpaired image super-resolution. However, these networks require a lot of time to train the massive parameters to obtain good results. In our work, we largely decrease the training parameters, but still achieve superior performance in the SR criteria (SSIM and PSNR) and visible quality.

Facial Prior Knowledge. Exploiting facial priors in face hallucination, such as spatial configuration of facial components [29], is the key factor that differentiates it from generic super-resolution tasks. There are some face SR methods that use facial prior knowledge to super-resolve LR faces. Wang and Tang [33] learned subspaces from LR and HR face images, and then reconstructed an HR output from the PCA coefficients of the LR input. Liu et al. [23] set up a Markov Random Field (MRF) to reduce ghosting artifacts because of the misalignments in LR images. However, these methods are prone to generating severe artifacts, especially with large pose variations and misalignments in LR images. Yu and Porikli [38] interweaved multiple spatial transformer networks [15] with the deconvolutional layers to handle unaligned LR faces. Dahl et al. [5] leveraged the framework of PixelCNN [26] to super-resolve very low-resolution faces. Zhu et al. [47] presented a cascade bi-network, dubbed CBN, to localize LR facial components first and then upsample the facial components; however, CBN may produce ghosting faces when localization errors occur. Recently, Yu et al. [35] used a multi-task convolutional neural network (CNN) to incorporate structural information of faces. Grm et al. [10] built a face recognition model that acts as identity priors for the super-resolution network during training. Yu et al. [4] constructed an end-to-end SR network to incorporate the facial landmark heatmaps and parsing maps. Kim et al. [16] proposed a compressed version of the face alignment network (FAN) to obtain landmark heatmaps for the SR network in a progressive method. However, existing face SR algorithms only employ 2D priors without considering high-dimensional information (3D). In this paper, we exploit the 3D face reconstruction branch to extract the 3D facial structure, detailed edges, illumination, and identity priors to guide face image super-resolution.

3D Face Reconstruction. The 3D shapes of facial images can be restored from unconstrained 2D images by the 3D face reconstruction. In this paper, we employ the 3D Morphable Model (3DMM) [1,2,6] based on the fusion of parametric descriptions of face attributes (e.g., gender, identity, and distinctiveness) to reconstruct the 3D facial priors. The 3D reconstructed face will inherit the facial features and present the clear and sharp facial components.

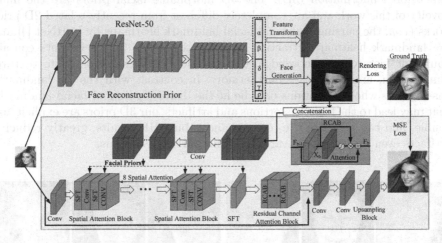

Fig. 2. The proposed face super-resolution architecture. Our model consists of two branches: the top block is a ResNet-50 Network to extract the 3D facial coefficients and restore a sharp face rendered structure. The bottom block is dedicated to face super-resolution guided by the facial coefficients and rendered sharp face structures which are concatenated by the Spatial Feature Transform (SFT) layer.

Closest to ours is the work of Ren et al. [28] which utilizes the 3D priors in the task of face video deblurring. Our method differs in several important ways. First, instead of simple priors concatenation, we employ the Spatial Feature Transform Block to incorporate the 3D priors in the intermediate layer by adaptively adjusting the modulation parameter pair. Specifically, the outputs of the SFT layer are adaptively controlled by the modulation parameter pair by applying an affine transformation spatially to each intermediate feature map. Second, the attention mechanism is embedded into the network as a guide to bias the allocation of most informative components and the interdependency between the 3D priors and input.

3   The Proposed Method

The proposed face super-resolution framework presented in Fig. 2 consists of two branches: the 3D rendering network to extract the facial prior and the spatial attention module aiming to exploit the prior for the face super-resolution problem. Given a low-resolution face image, we first use the 3D rendering branch to extract the 3D face coefficients. Then a high-resolution rendered image is generated using the 3D coefficients and regarded as the high-resolution facial prior which facilitates the face super-resolving process in the spatial attention module.

3.1   Motivations and Advantages of 3D Facial Priors

Existing face SR algorithms only employ 2D priors without considering high dimensional information (3D). The 3D morphable facial priors are the main novelty of this work and are completely different from recently related 2D prior works (*e.g.*, the parsing maps and facial landmark heatmaps by FSRNet [4] and the landmark heatmap extraction by FAN [16]). The 3D coefficients contain abundant hierarchical knowledge, such as identity, facial expression, texture, illumination, and face pose. Furthermore, in contrast with the 2D landmark-based priors whose attentions only lie at the distinct points of facial landmarks that may lead to the facial distortions and artifacts, our 3D priors are explicit and visible, and can generate the realistic and robust HR results, greatly reducing artifacts even for large pose variations and partial occlusions.

(a) LR inputs (b) Rendered priors (c) Ground truth   (d) LR inputs   (e) Rendered priors (f) Ground truth

Fig. 3. The rendered priors from our method. (a) and (d) low-resolution inputs. (b) and (e) our rendered face structures. (c) and (f) ground-truths. As shown, the reconstructed facial structures provide clear spatial locations and sharp visualization of facial components even for large pose variations (e.g., left and right facial pose positions) and partial occlusions.

Given low-resolution face images, the generated 3D rendered reconstructions are shown in Fig. 3. The rendered face predictions contain the clear spatial knowledge and sharp visual quality of facial components which are close to the ground-truth, even in images containing large pose variations as shown in the second row of Fig. 3. Therefore, we concatenate the reconstructed face image as an additional feature in the super-resolution network. The face expression, identity, texture, the element-concatenation of illumination, and face pose are transformed into four feature maps and fed into the spatial feature transform block of the super-resolution network.

For real-world applications of the 3D face morphable model, there are typical problems to overcome, including large pose variations and partial occlusions. As shown in the supplementary material, the morphable model can generate realistic reconstructions of large pose variations, which contain faithful visual quality of facial components. The 3D model is also robust and accurately restores the rendered faces partially occluded by glasses, hair, etc. In comparison with other SR algorithms which are blind to unknown degradation types, our 3D model can robustly generate the 3D morphable priors to guide the SR branch to grasp the clear spatial knowledge and facial components, even for complicated real-world applications. Furthermore, our 3D priors can be plugged into any network and largely improve the performance of existing SR networks (e.g., SRCNN and VDSR demonstrated in Sect. 5).

3.2 Formulation of 3D Facial Priors

It is still a challenge for state-of-the-art edge prediction methods to acquire very sharp facial structures from low-resolution images. Therefore, a 3DMM-based model is proposed to localize the precise facial structure by generating the 3D facial images which are constructed by the 3D coefficient vector. In addition, there exist large face pose variations, such as in-plane and out-of-plane rotations. A large amount of data is needed to learn the representative features varying with the facial poses. To address this problem, an inspiration came from the idea that the 3DMM coefficients can analytically model the pose variations with a simple mathematical derivation [2,6] and do not require a large training set. As such, we utilize a face rendering network based on ResNet-50 to regress a face coefficient vector. The output of the ResNet-50 is the representative feature vector of $x = (\alpha, \beta, \delta, \gamma, \rho) \in \mathbb{R}^{239}$, where $\alpha \in \mathbb{R}^{80}, \beta \in \mathbb{R}^{64}, \delta \in \mathbb{R}^{80}, \gamma \in \mathbb{R}^9$, and $\rho \in \mathbb{R}^6$ represent the identity, facial expression, texture, illumination, and face pose [6], respectively.

According to the Morphable model [1], we transform the face coefficients to a 3D shape $\mathbf{S}$ and texture $\mathbf{T}$ of the face image as

$$\mathbf{S} = \mathbf{S}(\alpha, \beta) = \overline{\mathbf{S}} + \mathbf{B}_{id}\alpha + \mathbf{B}_{exp}\beta, \tag{1}$$

and

$$\mathbf{T} = \mathbf{T}(\delta) = \overline{\mathbf{T}} + \mathbf{B}_t\delta, \tag{2}$$

where $\overline{\mathbf{S}}$ and $\overline{\mathbf{T}}$ are the average values of face shape and texture, respectively. $\mathbf{B}_t$, $\mathbf{B}_{id}$, and $\mathbf{B}_{exp}$ denote the base vectors of texture, identity, and expression calculated by the PCA method. We set up the illumination model by assuming a Lambertian surface for faces, and estimate the scene illumination with Spherical Harmonics (SH) [27] to derive the illumination coefficient $\gamma \in \mathbb{R}^9$. The 3D face pose $\rho \in \mathbb{R}^6$ is represented by rotation $\mathbf{R} \in SO(3)$ and translation $\mathbf{t} \in \mathbb{R}^3$.

To stabilize the rendered faces, a modified L_2 loss function for the 3D face reconstruction is presented based on a paired training set

$$\ell_r = \frac{1}{L} \sum_{j=1}^{L} \frac{\sum_{i \in M} A^i \left\| I_j^i - R_j^i(B(\boldsymbol{x})) \right\|_2}{\sum_{i \in M} A^i}, \quad (3)$$

where j is the paired image index, L is the total number of training pairs, i and M denote the pixel index and face region, respectively, I represents the sharp image, and A is a skin color based attention mask obtained by training a Bayes classifier with Gaussian Mixture Models [6]. In addition, x represents the LR (input) images, $B(x)$ denotes the regressed coefficients obtained by the ResNet-50 with input x as input, and finally R denotes the image rendered with the 3D coefficients $B(x)$. Rendering is the process to project the constructed 3D face onto the 2D image plane with the regressed pose and illumination. We use a ResNet-50 network to regress these coefficients by modifying the last fully-connected layer to 239 neurons (the same number of the coefficient parameters).

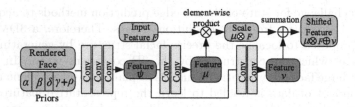

Fig. 4. The structure of the SFT layer. The rendered faces and feature vectors are regarded as the guidance for face super-resolution.

Coefficient Feature Transformation. Our 3D face priors consist of two parts: one directly from the rendered face region (*i.e.,* the RGB input), and the other from the feature transformation of the coefficient parameters. The coefficient parameters $\alpha, \beta, \delta, \gamma, \rho$ represent the identity, facial expression, texture, illumination, and face pose priors, respectively. The coefficient feature transformation procedure is described as follows: firstly, the coefficients of identity, expression, texture, and the element-concatenation of illumination and face pose $(\gamma + \rho)$ are reshaped to four matrices by setting extra elements to zeros. Afterwards, these four matrices are expanded to the same size as the LR images (16×16 or 32×32) by zero-padding, and then scaled to the interval [0,1]. Finally, the coefficient features are concatenated with the priors of the rendered face images.

3.3   Spatial Attention Module

To exploit the 3D face rendered priors, we propose a Spatial Attention Module (SAM) to grasp the precise locations of face components and the facial identity. The proposed SAM consists of three parts: a spatial feature transform block, a residual channel attention block, and an upscale block.

Spatial Feature Transform Block. The 3D face priors (rendered faces and coefficient features) are imported into the spatial attention transform block [34] after a convolutional layer. The structure of the spatial feature transform layer is shown in Fig. 4. The SFT layer learns a mapping function Θ that provides a modulation parameter pair (μ, ν) according to the priors ψ, such as segmentation probability. Here, the 3D face priors are taken as the input. The outputs of the SFT layer are adaptively controlled by the modulation parameter pair by applying an affine transformation spatially to each intermediate feature map. Specifically, the intermediate transformation parameters (μ, ν) are derived from the priors ψ by the mapping function:

$$(\mu, \nu) = \Theta(\psi), \qquad (4)$$

The intermediate feature maps are modified by scaling and shifting feature maps according to the transformation parameters:

$$SFT(F|\mu, \nu) = \mu \otimes F + \nu, \qquad (5)$$

where F denotes the feature maps, and $\otimes$ indicates element-wise multiplication. At this step, the SFT layer implements the spatial-wise transformation.

Residual Channel Attention Block. An attention mechanism can be viewed as a guide to bias the allocation of available processing resources towards the most informative components of the input [13]. Consequently, the channel mechanism is presented to explore the most informative components and the interdependency between the channels. Inspired by the residual channel network [42], the attention mechanism is composed of a series of residual channel attention blocks (RCAB) shown in Fig. 2. For the b-th block, the output F_b of RCAB is obtained by:

$$F_b = F_{b-1} + C_b(X_b) \cdot X_b, \qquad (6)$$

where C_b denotes the channel attention function. F_{b-1} is the block's input, and X_b is calculated by two stacked convolutional layers. The upscale block is progressive deconvolutional layers (also known as transposed convolution).

4   Experimental Results

To evaluate the performances of the proposed face super-resolution network, we qualitatively and quantitatively compare our algorithm against nine start-of-the-art super-resolution and face hallucination methods including: the Very Deep Super Resolution Network (VDSR) [17], the Very Deep Residual Channel Attention Network (RCAN) [42], the Residual Dense Network (RDN) [43], the Super-Resolution Convolutional Neural Network (SRCNN) [7], the Transformative Discriminative Autoencoder (TDAE) [38], the Wavelet-based CNN for Multi-scale Face Super Resolution (Wavelet-SRNet) [14], the deep end-to-end trainable face SR network (FSRNet) [4], face SR generative adversarial network (FSRGAN) [4] incorporating the 2D facial landmark heatmaps and parsing maps, and the progressive face Super Resolution network via face alignment network (PSR-FAN) [16] using 2D landmark heatmap priors. We use the open-source implementations from the authors and train all the networks on the same dataset for a fair comparison. For simplicity, we refer to the proposed network as Spatial Attention Module guided by 3D priors, or SAM3D. In addition, to demonstrate the plug-in characteristic of the proposed 3D facial priors, we propose two models of SRCNN+3D and VDSR+3D by embedding the 3D facial prior as an extra input channel to the basic backbone of SRCNN [7] and VDSR [17]. The implementation code will be made available to the public. More analyses and results can be found in the supplementary material.

4.1 Datasets and Implementation Details

CelebA [25] and Menpo [40] datasets are used to verify the performance of the algorithm. The training phase uses 162,080 images from the CelebA dataset. In the testing phase, 40,519 images from the CelebA test set are used along with the large-pose-variation test set from the Menpo dataset. The every facial pose test set of Menpo (left, right and semi-frontal) contains 1000 images, respectively. We follow the protocols of existing face SR methods (e.g., [4,16,35,36]) to generate the LR input by the bicubic downsampling method. The HR ground-truth images are obtained by center-cropping the facial images and then resizing them to the 128×128 pixels. The LR face images are generated by downsampling HR ground-truths to 32×32 pixels ($\times 4$ scale) and 16×16 pixels ($\times 8$ scale). In our network, the ADAM optimizer is used with a batch size of 64 for training, and input images are center-cropped as RGB channels. The initial learning rate is 0.0002 and is divided by 2 every 50 epochs. The whole training process takes 2 days with an NVIDIA Titan X GPU.

4.2 Quantitative Results

Quantitative evaluation of the network using PSNR and the structural similarity (SSIM) scores for the CelebA test set is listed in Table 1. Furthermore, to analyze the performance and stability of the proposed method with respect to large face pose variations, three cases corresponding to different face poses (left, right, and semifrontal) of the Menpo test data are listed in Table 2.

CelebA Test: As shown in Table 1, VDSR+3D (the basic VDSR model [17] guided by the proposed 3D facial priors) achieves significantly better results (1 dB higher than the remaining best method and 2 dB higher than the basic VDSR method in $\times 8$ SR) even for the large-scale parameter methods, such as RDN and RCAN. It is worth noting that VDSR+3D still performs slightly worse than the proposed algorithm of SAM3D. These results demonstrate that the proposed 3D priors make a significant contribution to the performance improvement (average 1.6 dB improvement) of face super-resolution. In comparison with 2D

Fig. 5. Comparison of state-of-the-art methods: magnification factors ×4 and the input resolution 32 × 32. Our algorithm is able to exploit the regularity present in face regions rather than other methods. Best viewed by zooming in on the screen.

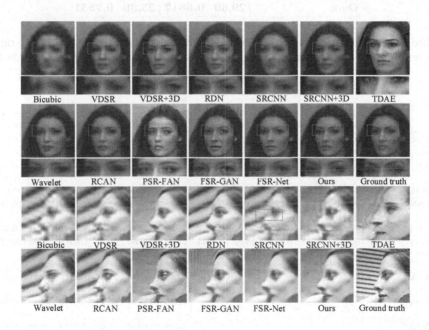

Fig. 6. Comparison with state-of-the-art methods: magnification factors ×8 and the input resolution 16 × 16. Best viewed by zooming in on the screen.

Table 1. Quantitative results on the CelebA test dataset. The best results are highlighted in bold.

–	CelebA			
Scale	×4		×8	
	PSNR	SSIM	PSNR	SSIM
Bicubic	27.16	0.8197	21.90	0.6213
VDSR [17]	28.13	0.8554	22.76	0.6618
RCAN [42]	29.04	0.8643	23.26	0.7362
RDN [43]	29.06	0.8650	23.69	0.7484
SRCNN [7]	27.57	0.8452	22.51	0.6659
TDAE [38]	–	–	20.10	0.5802
Wavelet-SRNet [14]	28.42	0.8698	23.08	0.7147
FSRGAN [4]	–	–	22.27	0.6010
FSRNet [4]	–	–	22.62	0.6410
PSR-FAN [16]	–	–	22.66	0.6850
VDSR+3D	29.29	0.8727	24.66	0.7127
Ours	**29.69**	**0.8817**	**25.39**	**0.7551**

Table 2. Quantitative results of different large facial pose variations (e.g., left, right, and semifrontal) on the Menpo test dataset. The best results are highlighted in bold.

–	Menpo											
Scale	×4						×8					
Pose	Left		Right		Semi-frontal		Left		Right		Semi-frontal	
	PSNR	SSIM	PSNR	SSIM	PSNR	SSIM	PSNR	SSIM	PSNR	SSIM	PSNR	SSIM
Bicubic	26.36	0.7923	26.19	0.7791	24.92	0.7608	22.09	0.6423	21.99	0.6251	20.68	0.5770
VDSR [17]	26.99	0.8024	26.85	0.7908	25.63	0.7794	22.28	0.6315	22.20	0.6163	20.98	0.5752
RCAN [42]	27.47	0.8259	27.27	0.8145	26.11	0.8080	21.94	0.6543	21.87	0.6381	20.60	0.5938
RDN [43]	27.39	0.8263	27.21	0.8150	26.06	0.8088	22.30	0.6706	22.24	0.6552	21.02	0.6160
SRCNN [7]	26.92	0.8038	26.74	0.7913	25.50	0.7782	22.38	0.6408	22.32	0.6272	21.08	0.5857
TDAE [38]	–	–	–	–	–	–	21.22	0.5678	20.22	0.5620	19.88	0.5521
Wavelet-SRNet [14]	26.97	0.8122	26.81	0.8001	25.72	0.7945	21.86	0.6360	21.72	0.6166	20.57	0.5779
FSRGAN [4]	–	–	–	–	–	–	23.00	0.6326	22.84	0.6173	22.00	0.5938
FSRNet [4]	–	–	–	–	–	–	23.56	0.6896	23.43	0.6712	22.03	0.6382
PSR-FAN [16]	–	–	–	–	–	–	22.04	0.6239	21.89	0.6114	20.88	0.5711
VDSR+3D	28.62	0.8439	28.89	0.8326	26.99	0.8236	23.45	0.6845	23.25	0.6653	21.83	0.6239
Ours	**28.98**	**0.8510**	**29.29**	**0.8408**	**27.29**	**0.8332**	**23.80**	**0.7071**	**23.57**	**0.6881**	**22.15**	**0.6501**

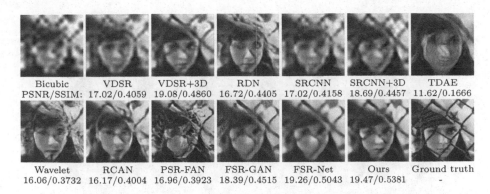

Fig. 7. Visual comparison with state-of-the-art methods (×8). The results by the proposed method have fewer artifacts on face components (e.g., eyes, mouth, and nose).

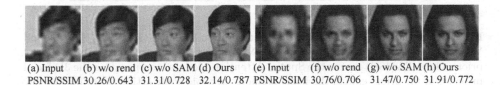

Fig. 8. Ablation study results: Comparisons between our proposed model with different configurations, with PSNR and SSIM relative to the ground truth. (a) and (e) are the inputs. (b) and (f) are the SR results without using the rendered priors. (c) and (g) are the SR results without the Spatial Attention Module. (d) and (h) are our SR results.

priors based methods (*e.g.*, FSRNet and PSR-FAN), our algorithm performs much better (2.73 dB higher than PSR-FAN and 2.78 dB higher than FSRNet).

Menpo Test: To verify the effectiveness and stability of the proposed network towards face pose variations, the quantitative results on the dataset with large pose variations are reported in Table 2. While ours (SAM3D) is the best method superior than the others, VDSR+3D also achieves 1.8 dB improvement compared with the basic VDSR method in the ×4 magnification factor. Our 3D facial priors based method is still the most effective approach to boost the SR performance compared with 2D heatmaps and parsing maps priors.

4.3 Qualitative Evaluation

The qualitative results of our methods at different magnifications (×4 and ×8) are shown respectively in Figs. 5 and 6. It can be observed that our proposed method recovers clearer faces with finer component details (e.g., noses, eyes, and mouths). The outputs of most methods (*e.g.*, PSR-FAN, RCAN, RDN, and Wavelet-SRNet) contain some artifacts around facial components such as eyes

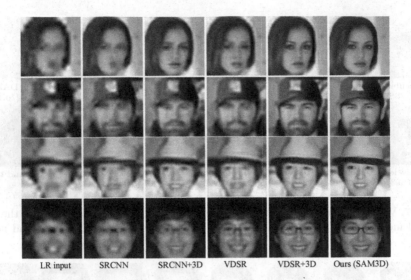

LR input   SRCNN   SRCNN+3D   VDSR   VDSR+3D   Ours (SAM3D)

Fig. 9. Qualitative evaluation with different ablation configurations: SRCNN+3D and VDSR+3D denote the basic method (SRCNN and VDSR) incorporating the 3D facial priors; Ours (SAM3D) means the Spatial Attention Module incorporating the 3D facial priors. Our 3D priors enable the basic methods to avoid some artifacts around the key facial components and to generate sharper edges.

and nose, as shown in Figs. 1 and 7, especially when facial images are partially occluded. After adding the rendered face priors, our results show clearer and sharper facial structures without any ghosting artifacts, which illustrates that the proposed 3D priors help the network understand the spatial location and the entire face structure and largely avoid the artifacts and significant distortions in facial attributes which are common in facial landmark priors, because the attention is applied merely to the distinct points of facial landmarks.

5   Analyses and Discussions

Ablation Study: In this section, we conduct an ablation study to demonstrate the effectiveness of each module. We compare the proposed network with and without using the rendered 3D face priors and the Spatial Attention Module (SAM) in terms of PSNR and SSIM on the ×8 scale test data. As shown in Fig. 8(b) and (f), the baseline method without using the rendered faces and SAM tends to generate blurry faces that cannot capture sharp structures. Figure 8(c) and (g) show clearer and sharper facial structures after adding the 3D rendered priors. By using both SAM and 3D priors, the visual quality is further improved in Fig. 8(d) and (h). The quantitative comparisons between (VDSR, our VDSR+3D, and our SAM3D) in Tables 1 and 2 also illustrate the effectiveness of the proposed rendered priors and the spatial attention module.

To verify the advantage of 3D facial structure priors in terms of the convergence and accuracy, three different configurations are designed: basic methods (*i.e.,* SRCNN [7] and VDSR [17]); basic methods incorporating 3D facial priors (*i.e.,* SRCNN+3D and VDSR+3D); the proposed method using the Spatial Attention Module and 3D priors (SAM3D). The validation accuracy curve of each configuration along the epochs is plotted to show the effectiveness of each block. The priors are easy to insert into any network. They only marginally increase the number of parameters, but significantly improve the accuracy and convergence of the algorithms as shown in Supplementary Fig. 3. The basic methods of SRCNN and VDSR incorporating the facial rendered priors tend to avoid some artifacts around key facial components and generate sharper edges compared to the baseline methods without the facial priors. By adding the Spatial Attention Module, it helps the network better exploit the priors and easily enables to generate sharper facial structures as shown in Fig. 9.

Results on Real-World Images: For real-world LR images, we provide the quantitative and qualitative analysis on 500 LR faces from the WiderFace (x4) dataset in Supplementary Table 1 and Fig. 1.

Model Size and Running Time: We evaluate the proposed method and STOA SR methods on the same server with an Intel Xeon W-2123 CPU and an NVIDIA TITAN X GPU. Our proposed SAM3D, embedded with 3D priors, are more lightweight and less time-consuming, shown in Supplementary Fig. 2.

6 Conclusions

In this paper, we proposed a face super-resolution network that incorporates the novel 3D facial priors of rendered faces and multi-dimensional knowledge. In the 3D rendered branch, we presented a face rendering loss to encourage a high-quality guided image providing clear spatial locations of facial components and other hierarchical information (*i.e.,* expression, illumination, and face pose). Compared with the existing 2D facial priors whose attentions are focused on the distinct points of landmarks which may result in face distortions, our 3D priors are explicit, visible and highly realistic, and can largely decrease the occurrence of face artifacts. To well exploit 3D priors and consider the channel correlation between priors and inputs, we employed the Spatial Feature Transform and Attention Block. The comprehensive experimental results have demonstrated that the proposed method achieves superior performance and largely decreases artifacts in contrast with the SOTA methods.

Acknowledgement. This work is supported by the National Key R&D Program of China under Grant 2018AAA0102503, Zhejiang Lab (NO.2019NB0AB01), Beijing Education Committee Cooperation Beijing Natural Science Foundation (No. KZ201910005007), National Natural Science Foundation of China (No. U1736219) and Peng Cheng Laboratory Project of Guangdong Province PCL2018KP004.

References

1. Blanz, V., Vetter, T.: A morphable model for the synthesis of 3D faces. In ACM SIGGRAPH (1999)
2. Booth, J., Roussos, A., Zafeiriou, S., Ponniah, A., Dunaway, D.: A 3D morphable model learnt from 10,000 faces. In: CVPR (2016)
3. Cao, Q., Lin, L., Shi, Y., Liang, X., Li, G.: Attention-aware face hallucination via deep reinforcement learning. In: CVPR (2017)
4. Chen, Y., Tai, Y., Liu, X., Shen, C., Yang, J.: FSRNet: end-to-end learning face super-resolution with facial priors. In: CVPR (2018)
5. Dahl, R., Norouzi, M., Shlens, J.: Pixel recursive super resolution. In: ICCV (2017)
6. Deng, Y., Yang, J., Xu, S., Chen, D., Jia, Y., Tong, X.: Accurate 3D face reconstruction with weakly-supervised learning: from single image to image set. In: CVPRW (2019)
7. Dong, C., Loy, C., He, K., Tang, X.: Image super-resolution using deep convolutional networks. TPAMI **38**(2), 295–307 (2016)
8. Dong, C., Loy, C.C., Tang, X.: Accelerating the super-resolution convolutional neural network. In: Leibe, B., Matas, J., Sebe, N., Welling, M. (eds.) ECCV 2016. LNCS, vol. 9906, pp. 391–407. Springer, Cham (2016). https://doi.org/10.1007/978-3-319-46475-6_25
9. Fritsche, M., Gu, S., Timofte, R.: Frequency separation for real-world super-resolution. In: CVPRW (2019)
10. Grm, K., Scheirer, W., Štruc, V.: Face hallucination using cascaded super-resolution and identity priors. TIP **29**, 2150–2165 (2019)
11. Han, C., Shan, S., Kan, M., Wu, S., Chen, X.: Face recognition with contrastive convolution. In: Ferrari, V., Hebert, M., Sminchisescu, C., Weiss, Y. (eds.) ECCV 2018. LNCS, vol. 11213, pp. 120–135. Springer, Cham (2018). https://doi.org/10.1007/978-3-030-01240-3_8
12. Haris, M., Shakhnarovich, G., Ukita, N.: Deep back projection networks for super-resolution. In: CVPR (2018)
13. Hu, J., Shen, L., Sun, G.: Squeeze-and-excitation networks. In: CVPR (2018)
14. Huang, H., He, R., Sun, Z., Tan, T.: Wavelet-SRNet: a wavelet-based CNN for multi-scale face super resolution. In: ICCV (2017)
15. Jaderberg, M., Simonyan, K., Zisserman, A.: Spatial transformer networks. In: NIPS (2015)
16. Kim, D., Kim, M., Kwon, G., Kim, D.: Progressive face super-resolution via attention to facial landmark. In: BMVC (2019)
17. Kim, J., Lee, J., Lee, K.: Accurate image super-resolution using very deep convolutional networks. In: CVPR (2016)
18. Kim, J., Lee, J., Lee, K.: Deeply recursive convolutional network for image super-resolution. In: CVPR (2016)
19. Lai, W., Huang, J., Ahuja, N., Yang, M.: Deep Laplacian pyramid networks for fast and accurate super-resolution. In: CVPR (2017)
20. Li, Z., Tang, J., Zhang, L., Yang, J.: Weakly-supervised semantic guided hashing for social image retrieval. Int. J. Comput. Vision **128**(8), 2265–2278 (2020). https://doi.org/10.1007/s11263-020-01331-0
21. Lian, S., Zhou, H., Sun, Y.: A feature-guided super-resolution generative adversarial network for unpaired image super-resolution. In: NIPS (2019)
22. Lim, B., Son, S., Kim, H., Nah, S., Lee, K.: Enhanced deep residual networks for single image super-resolution. In: CVPRW, pp. 1646–1654 (2017)

23. Liu, C., Shum, H., Freeman, W.: Face hallucination: theory and practice. Int. J. Comput. Vision **75**(1), 115–134 (2007). https://doi.org/10.1007/s11263-006-0029-5

24. Liu, W., Lin, D., Tang, X.: Hallucinating faces: TensorPatch super-resolution and coupled residue compensation. In: CVPR (2005)

25. Liu, Z., Luo, P., Wang, X., Tang, X.: Deep learning face attributes in the wild. In: ICCV (2015)

26. Oord, A., Kalchbrenner, N., Kavukcuoglu, K.: Pixel recurrent neural networks. In: ICML (2016)

27. Ramamoorthi, R., Hanrahan, P.: An efficient representation for irradiance environment maps. In: SIGGRAPH Annual Conference on Computer Graphics and Interactive Techniques, pp. 497–500 (2001)

28. Ren, W., Yang, J., Deng, S., Wipf, D., Cao, X., Tong, X.: Face video deblurring via 3D facial priors. In: ICCV (2019)

29. Shen, Z., Lai, W., Xu, T., Kautz, J., Yang, M.: Deep semantic face deblurring. In: CVPR (2018)

30. Shi, W., et al.: Real-time single image and video super-resolution using an efficient sub-pixel convolutional neural network. In: CVPR (2016)

31. Tai, Y., Yang, J., Liu, X.: Image super-resolution via deep recursive residual network. In: CVPR (2017)

32. Thies, J., Zollhofer, M., Stamminger, M., Theobalt, C., Nießner, M.: Face2Face: real-time face capture and reenactment of RGB videos. In: CVPR (2016)

33. Wang, X., Tang, X.: Hallucinating face by eigen transformation. Trans. Syst. Man Cybern. C **35**(3), 425–434 (2005)

34. Wang, X., Yu, K., Dong, C., Loy, C.: Recovering realistic texture in image super-resolution by deep spatial feature transform. In: CVPR (2018)

35. Yu, X., Fernando, B., Ghanem, B., Porikli, F., Hartley, R.: Face super-resolution guided by facial component heatmaps. In: Ferrari, V., Hebert, M., Sminchisescu, C., Weiss, Y. (eds.) ECCV 2018. LNCS, vol. 11213, pp. 219–235. Springer, Cham (2018). https://doi.org/10.1007/978-3-030-01240-3_14

36. Yu, X., Fernando, B., Hartley, R., Porikli, F.: Super-resolving very low-resolution face images with supplementary attributes. In: CVPR (2018)

37. Yu, X., Porikli, F.: Ultra-resolving face images by discriminative generative networks. In: Leibe, B., Matas, J., Sebe, N., Welling, M. (eds.) ECCV 2016. LNCS, vol. 9909, pp. 318–333. Springer, Cham (2016). https://doi.org/10.1007/978-3-319-46454-1_20

38. Yu, X., Porikli, F.: Hallucinating very low-resolution unaligned and noisy face images by transformative discriminative autoencoders. In: CVPR (2017)

39. Yu, X., Porikli, F.: Imagining the unimaginable faces by deconvolutional networks. TIP **27**(6), 2747–2761 (2018)

40. Zafeiriou, S., Trigeorgis, G., Chrysos, G., Deng, J., Shen, J.: The menpo facial landmark localisation challenge: a step towards the solution. In: CVPRW (2017)

41. Zhang, K., et al.: Super-identity convolutional neural network for face hallucination. In: Ferrari, V., Hebert, M., Sminchisescu, C., Weiss, Y. (eds.) ECCV 2018. LNCS, vol. 11215, pp. 196–211. Springer, Cham (2018). https://doi.org/10.1007/978-3-030-01252-6_12

42. Zhang, Y., Li, K., Li, K., Wang, L., Zhong, B., Fu, Y.: Image super-resolution using very deep residual channel attention networks. In: Ferrari, V., Hebert, M., Sminchisescu, C., Weiss, Y. (eds.) ECCV 2018. LNCS, vol. 11211, pp. 294–310. Springer, Cham (2018). https://doi.org/10.1007/978-3-030-01234-2_18

43. Zhang, Y., Tian, Y., Kong, Y., Zhong, B., Fu, Y.: Residual dense network for image super-resolution. In: CVPR (2018)
44. Zhao, J., Xiong, L., Li, J., Xing, J., Yan, S., Feng, J.: 3D-aided dual-agent GANs for unconstrained face recognition. TPAMI **41**, 2380–2394 (2019)
45. Zhao, W., Chellappa, R., Phillips, P.J., Rosenfeld, A.: Face recognition: a literature survey. ACM Comput. Surv. (CSUR) **35**(4), 399–458 (2003)
46. Zhou, E., Fan, H.: Learning face hallucination in the wild. In: AAAI (2015)
47. Zhu, S., Liu, S., Loy, C.C., Tang, X.: Deep cascaded bi-network for face hallucination. In: Leibe, B., Matas, J., Sebe, N., Welling, M. (eds.) ECCV 2016. LNCS, vol. 9909, pp. 614–630. Springer, Cham (2016). https://doi.org/10.1007/978-3-319-46454-1_37

Label Propagation with Augmented Anchors: A Simple Semi-supervised Learning Baseline for Unsupervised Domain Adaptation

Yabin Zhang[1,2,3], Bin Deng[1,2], Kui Jia[1,2](✉), and Lei Zhang[3,4]

[1] South China University of Technology, Guangzhou, China
zhang.yabin@mail.scut.edu.cn, bindeng.scut@gmail.com, kuijia@scut.edu.cn
[2] Pazhou Lab, Guangzhou, China
[3] DAMO Academy, Alibaba Group, Hangzhou, China
[4] Department of Computing, The Hong Kong Polytechnic University,
Hong Kong, Hong Kong
cslzhang@comp.polyu.edu.hk

Abstract. Motivated by the problem relatedness between unsupervised domain adaptation (UDA) and semi-supervised learning (SSL), many state-of-the-art UDA methods adopt SSL principles (e.g., the cluster assumption) as their learning ingredients. However, they tend to overlook the very domain-shift nature of UDA. In this work, we take a step further to study the proper extensions of SSL techniques for UDA. Taking the algorithm of label propagation (LP) as an example, we analyze the challenges of adopting LP to UDA and theoretically analyze the conditions of affinity graph/matrix construction in order to achieve better propagation of true labels to unlabeled instances. Our analysis suggests a new algorithm of Label Propagation with Augmented Anchors (A^2LP), which could potentially improve LP via generation of unlabeled virtual instances (i.e., the augmented anchors) with high-confidence label predictions. To make the proposed A^2LP useful for UDA, we propose empirical schemes to generate such virtual instances. The proposed schemes also tackle the domain-shift challenge of UDA by alternating between pseudo labeling via A^2LP and domain-invariant feature learning. Experiments show that such a simple SSL extension improves over representative UDA methods of domain-invariant feature learning, and could empower two state-of-the-art methods on benchmark UDA datasets. Our results show the value of further investigation on SSL techniques for UDA problems.

Keywords: Domain adaptation · Semi-supervised learning · Label propagation

Electronic supplementary material The online version of this chapter (https://doi.org/10.1007/978-3-030-58548-8_45) contains supplementary material, which is available to authorized users.

© Springer Nature Switzerland AG 2020
A. Vedaldi et al. (Eds.): ECCV 2020, LNCS 12349, pp. 781–797, 2020.
https://doi.org/10.1007/978-3-030-58548-8_45

1 Introduction

As a specific setting of transfer learning [32], unsupervised domain adaptation (UDA) is to predict labels of given instances on a target domain, by learning classification models assisted with labeled data on a source domain that has a different distribution from the target one. Impressive results have been achieved by learning domain-invariant features [27,43,45], especially the recent ones based on adversarial training of deep networks [12,36,42,47]. These methods are primarily motivated by the classical UDA theories [3,4,30,46] that specify the success conditions of domain adaptation, where domain divergences induced by hypothesis space of classifiers are typically involved.

While a main focus of these methods is on designing algorithms to learn domain-invariant features, they largely overlook a UDA nature that shares the same property with the related problem of semi-supervised learning (SSL)—both UDA and SSL argue for a principle that the (unlabeled) instances of interest satisfy basic assumptions (e.g., the cluster assumption [6]), although in SSL, the unlabeled instances follow the same distribution as that of the labeled ones. Given the advantages of SSL methods over models trained with labeled data only [5], it is natural to apply the SSL techniques to domain-invariant features learned by seminal UDA methods [12,27] so as to boost the performance further. We note that ideal domain alignment can hardly be achieved in practice. Although state-of-the-art results have already been achieved by the combination of vanilla SSL techniques and domain-invariant feature learning [9,17,18,25,29,47], they typically neglect the issue that SSL methods are designed for data of the same domain, and their direct use for data with shifted distributions (e.g., in UDA tasks) could result in deteriorated performance.

To this end, we investigate how to extend SSL techniques for UDA problems. Take the SSL method of label propagation (LP) [48] as an example. When there exists such a shift of distributions, edges of an LP graph constructed by affinity relations of data instances could be of low reliability, thus preventing its direct use in UDA problems. To tackle the issue, we analyze in this paper the conditions of the affinity graph (and the corresponding affinity matrix) for better propagation of true labels to unlabeled instances. Our analysis suggests a new algorithm of Label Propagation with Augmented Anchors (A^2LP), which could potentially improve LP via generation of unlabeled virtual instances (i.e., the augmented anchors) with high-confidence label predictions. To make the proposed A^2LP particularly useful for UDA, we generate such virtual instances via a weighted combination of unlabeled target instances, using weights computed by the entropy of their propagated soft cluster assignments, considering that instances of low entropy are more confident in terms of their predicted labels. We iteratively do the steps of (1) using A^2LP to get pseudo labels of target instances, and (2) learning domain-invariant features with the obtained pseudo-labeled target instances and labeled source ones, where the second step, in turn, improves the quality of pseudo labels of target instances. Experiments on benchmark UDA datasets show that our proposed A^2LP significantly improves over the LP algorithm, and alternating steps of A^2LP and domain-invariant feature learning give state-of-the-art results. We finally summarize our main contributions as follows.

- Motivated by the relatedness between SSL and UDA, we study in this paper the technical challenge that prevents the direct use of graph-based SSL methods in UDA problems. We analyze the conditions of the affinity graph/matrix construction for better propagation of true labels to unlabeled instances, which suggests a new algorithm of A^2LP. A^2LP could potentially improve LP via generation of unlabeled virtual instances (i.e., the augmented anchors) with high-confidence label predictions.
- To make the proposed A^2LP useful for UDA, we generate virtual instances as augmented anchors via a weighted combination of unlabeled target instances, where weights are computed based on the entropy of propagated soft cluster assignments of target instances. Our A^2LP based UDA method alternates in obtaining pseudo labels of target instances via A^2LP, and using the obtained pseudo-labeled target instances, together with the labeled source ones, to learn domain-invariant features. The second step is expected to enhance the quality of pseudo labels of target instances.
- We conduct careful ablation studies to investigate the influence of graph structure on the results of A^2LP. Empirical evidences on benchmark UDA datasets show that our proposed A^2LP significantly improves over the original LP, and the alternating steps of A^2LP and domain-invariant feature learning give state-of-the-art results, confirming the value of further investigating the SSL techniques for UDA problems. The codes are available at https://github.com/YBZh/Label-Propagation-with-Augmented-Anchors.

2   Related Works

In this section, we briefly review the UDA methods, especially these [9,11,17,23, 25,29,35,38,47] involving SSL principles as their learning ingredients, and the recent works [2,7,21,39,48,51] on the LP technique.

Unsupervised Domain Adaptation. Motivated by the theoretical bound proposed in [3,4,46], the dominant UDA methods target at minimizing the discrepancy between the two domains, which is measured by various statistic distances, such as Maximum Mean Discrepancy (MMD) [27], Jensen-Shannon divergence [12] and Wasserstein distance [37]. They assume that once the domain discrepancy is minimized, the classifier trained on source data only can also perform well on the target ones. Given the advantages of SSL methods over models trained with labeled data only [5], it is natural to apply SSL techniques on domain-invariant features to boost the results further. Recently, state-of-the-art results are achieved by involving the SSL principles in UDA, although they may not have emphasized this point explicitly. Based on the cluster assumption, entropy regularization [13] is adopted in UDA methods [23,29,38,47] to encourage low density separation of category decision boundaries, which is typically used in conjunction with the virtual adversarial training [31] to incorporate the locally-Lipschitz constraint. The vanilla LP method [52] is adopted in [9,17,25] together with the learning of domain-invariant features. Based on the mean teacher model of [41], a self-ensembling (SE) algorithm [11] is proposed to penalize the prediction differences between student and teacher networks for the same input target instance. Inspired by the tri-training [50], three task classifiers are asymmetrically used in [35]. However, taking the comparable LP-based methods [9,17,25] as an example, they adopt the vanilla LP algorithm directly with no consideration of the UDA nature of domain shift. By contrast, we analyze the challenges of adopting LP in UDA, theoretically characterize the conditions of potential improvement (cf. **Proposition** 1), and accordingly propose the algorithmic extension of LP for UDA. Such a simple algorithmic extension improves the results dramatically on benchmark UDA datasets.

Label Propagation. The LP algorithm is based on a graph whose nodes are data instances (labeled and unlabeled), and edges indicate the similarities between instances. The labels of labeled data can propagate through the edges in order to label all nodes. Following the above principle, a series of LP algorithms [39,48,51] and the graph regularization methods [2,7,21] have been proposed for the SSL problems. Recently, Iscen *et al.* [19] revisit the LP algorithm for SSL problems with the iterative strategy of pseudo labeling and network retraining. Liu *et al.* [26] study the LP algorithm for few-shot learning. Unlike them, we investigate the LP algorithm for UDA problems and alleviate the performance deterioration brought by domain shift via the introduction of virtual instances as well as the domain-invariant feature learning.

3 Semi-supervised Learning and Unsupervised Domain Adaptation

Given data sets $X_L = \{x_1, ..., x_l\}$ and $X_U = \{x_{l+1}, ..., x_n\}$ with each $x_i \in \mathcal{X}$, the first l instances have labels $Y_L = \{y_1, ..., y_l\}$ with each $y_i \in \mathcal{Y} = \{1, ..., K\}$ and the remaining $n - l$ instances are unlabeled. We also write them collectively as $X = \{X_L, X_U\}$. The goal of both SSL and UDA is to predict the labels of the unlabeled instances in X_U[1]. In UDA, the labeled data in X_L and unlabeled data in X_U are drawn from two different distributions of the source one $\mathcal{D}_s$ and the target one $\mathcal{D}_t$. Differently, in SSL, the source and target distributions are assumed to be the same, i.e., $\mathcal{D}_s = \mathcal{D}_t$.

3.1 Semi-supervised Learning Preliminaries

We denote $\psi : \mathcal{X} \to \mathbb{R}^K$ as the mapping function parameterized by $\theta = \{\theta_e, \theta_c\}$, where θ_e indicates the parameters of a feature extractor $\phi : \mathcal{X} \to \mathbb{R}^d$ and θ_c indicates the parameters of a classifier $f : \mathbb{R}^d \to \mathbb{R}^K$. Let $\mathcal{P}$ denote the set of $n \times K$ probability matrices. A matrix $P = [p_1^T; ...; p_n^T] \in \mathcal{P}$ corresponds to a

---

[1] We formulate in this paper both the SSL and UDA under the transductive learning setting [5].

classification on the dataset $X = \{X_L, X_U\}$ by labeling each instance $\boldsymbol{x}_i$ as a label $\hat{y}_i = \arg\max_j \boldsymbol{p}_{ij}$. Each $\boldsymbol{p}_i \in [0,1]^K$ indicates classification probabilities of the instance $\boldsymbol{x}_i$ to K classes.

The general goal of SSL can be stated as finding $\boldsymbol{P}$ by minimizing the following meta objective:

$$\mathcal{Q}(\boldsymbol{P}) = \mathcal{L}(X_L, Y_L; \boldsymbol{P}) + \lambda \mathcal{R}(X; \boldsymbol{P}), \tag{1}$$

where $\mathcal{L}$ represents the supervised loss term that applies only to the labeled data, and $\mathcal{R}$ is the regularizer with λ as a trade-off parameter. The purpose of regularizer $\mathcal{R}$ is to make the learning decision to satisfy the underlying assumptions of SSL, including the smoothness, cluster, and manifold assumptions [5].

For SSL methods based on cluster assumption (e.g., low density separation [6]), their regularizers are concerned with unlabeled data only. As such, they are more amenable to be used in UDA problems, since the domain shift is not an issue to be taken into account. A prominent example of SSL regularizer is the entropy minimization (EM) [13], whose use in UDA can be instantiated as:

$$\mathcal{Q}(\boldsymbol{P}) = \underbrace{\sum_{i=1}^{l} \ell(\boldsymbol{p}_i, y_i)}_{\mathcal{L}(X_L, Y_L; \boldsymbol{P})} + \lambda \underbrace{\sum_{i=l+1}^{n} \sum_{j=1}^{K} -\boldsymbol{p}_{ij} \log \boldsymbol{p}_{ij}}_{\mathcal{R}(X; \boldsymbol{P})}, \tag{2}$$

where ℓ represents a typical loss function (e.g., cross-entropy loss). Objectives similar to (2) are widely used in UDA methods, together with other useful ingredients such as adversarial learning of aligned features [15,29,38,47].

3.2 From Graph-Based Semi-supervised Learning to Unsupervised Domain Adaptation

Different from the above EM like methods, the graph-based SSL methods that are based on local (and global) smoothness rely on the geometry of the data, and thus their regularizers are concerned with both labeled and unlabeled instances. The key of graph-based methods is to build a graph whose nodes are data instances (labeled and unlabeled), and edges represent similarities between instances. Such a graph is represented by the affinity matrix $\boldsymbol{A} \in \mathbb{R}_+^{n \times n}$, whose elements $\boldsymbol{a}_{ij}$ are non-negative pairwise similarities between instances $\boldsymbol{x}_i$ and $\boldsymbol{x}_j$. Here, we choose the LP algorithm [48] as an instance for exploiting the advantages of graph-based SSL methods. Denote $\boldsymbol{Y} = [\boldsymbol{y}_1^T; ...; \boldsymbol{y}_n^T] \in \mathcal{P}$ as the label matrix with $\boldsymbol{y}_{ij} = 1$ if $\boldsymbol{x}_i$ is labeled as $y_i = j$ and $\boldsymbol{y}_{ij} = 0$ otherwise. The goal of LP is to find a $\boldsymbol{F} = [\boldsymbol{f}_1^T; ...; \boldsymbol{f}_n^T] \in \mathbb{R}_+^{n \times K}$ by minimizing

$$\mathcal{Q}(\boldsymbol{F}) = \underbrace{\sum_{i=1}^{n} \|\boldsymbol{f}_i - \boldsymbol{y}_i\|^2}_{\mathcal{L}(X_L, Y_L; \boldsymbol{F})} + \lambda \underbrace{\sum_{i,j}^{n} \boldsymbol{a}_{ij} \|\frac{\boldsymbol{f}_i}{\sqrt{d_{ii}}} - \frac{\boldsymbol{f}_j}{\sqrt{d_{jj}}}\|^2}_{\mathcal{R}(X; \boldsymbol{F})}, \tag{3}$$

and then the resulting probability matrix P is given by $p_{ij} = f_{ij}/\sum_j f_{ij}$, where $D \in \mathbb{R}^{n \times n}_+$ is a diagonal matrix with its (i, i)-element d_{ii} equal to the sum of the i-th row of A. From the above optimization objective, we can easily see that a good affinity matrix is the key success factor of the LP algorithm. So, the straightforward question is that what is the good affinity matrix? As an analysis, we assume the true label of each data instance x_i is y_i, then the regularizer of the objective (3) can be decomposed as:

$$\mathcal{R}(X; F) = \sum_{y_i = y_j} a_{ij} \| \frac{f_i}{\sqrt{d_{ii}}} - \frac{f_j}{\sqrt{d_{jj}}} \|^2$$
$$+ \sum_{y_i \neq y_j} a_{ij} \| \frac{f_i}{\sqrt{d_{ii}}} - \frac{f_j}{\sqrt{d_{jj}}} \|^2. \tag{4}$$

Obviously, a good affinity matrix should make its element a_{ij} as large as possible if instances x_i and x_j are in the same class, and at the same time make those a_{ij} as small as possible otherwise. Therefore, it is rather easy to construct such a good affinity matrix in the SSL setting where the all data are drawn from the same underlying distribution. However, in UDA, due to the domain shift between labeled and unlabeled data, those values of elements a_{ij} of the same class pairs between labeled and unlabeled instances would be significantly reduced, which would prevent its use in the UDA problems as illustrated in Fig. 1.

(a) SSL (93.5%)   (b) UDA (64.8%)   (c) UDA with Anchors (79.5%)

Fig. 1. Visualization of sub-affinity matrices for the settings of (a) SSL, (b) UDA, and (c) UDA with augmented anchors, and their corresponding classification results via the LP. The row-wise and column-wise elements are the unlabeled and labeled instances, respectively. For illustration purposes, we keep elements connecting instances of the same class unchanged, set the others to zero, and sort all instances in the category order using the ground truth category of all data. As we can see, the augmented anchors present better connections with unlabeled target instances compared to the labeled source instances in UDA.

4   Label Propagation with Augmented Anchors

In this section, we first analyze conditions of the corresponding affinity matrix for better propagation of true labels to unlabeled instances, which motivate our

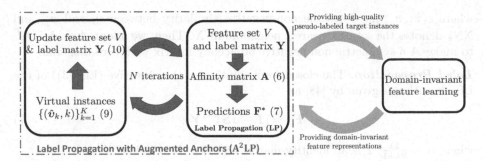

Fig. 2. An illustration of the overall framework of alternating steps of pseudo labeling via A^2LP and domain-invariant feature learning. The dashed line rectangle illustrates the algorithm of A^2LP, where we iteratively do the steps of (1) augmenting the feature set V and label matrix Y with the generated virtual instances and (2) generating virtual instances by the LP algorithm based on the updated feature set V and label matrix Y.

proposed A^2LP algorithm. Let Acc be the classification accuracy in X_U by the solution of the LP (Eq. (3)), i.e.,

$$Acc := \frac{|\{\boldsymbol{x}_i \in X_U : \hat{y}_i = y_i\}|}{|X_U|}, \qquad (5)$$

where $\hat{y}_i = \arg\max_j \boldsymbol{f}_{ij}^*$ with $\boldsymbol{F}^* = [\boldsymbol{f}_1^{*T}; ...; \boldsymbol{f}_n^{*T}]$ the solution of Eq. (3).

Proposition 1. *Assume the data satisfy the ideal cluster assumption, i.e., $a_{ij} = 0$ for all $y_i \neq y_j$. Enhancing one zero-valued element $\boldsymbol{a}_{mn}$ between a data instance $\boldsymbol{x}_m$ (labeled or unlabeled) and a labeled instance $\boldsymbol{x}_n \in X_L$ to a positive number, where $y_m = y_n$, the Acc (5) non-decreases, and increases under the condition when originally $\hat{y}_m \neq y_m$.*

The proof of Proposition 1 can be found in the appendices.

Remark 1. Under the assumption of Proposition 1, if we can augment the labeled set X_L with one virtual instance with the true label, whose neighbors are exactly the instances with the same label, then based on Proposition 1, the LP algorithm can get increasing (non-decreasing) Acc (Eq. (5)) in X_U.

4.1 The Proposed Algorithms

Based on the above analysis, we propose the algorithm of Label Propagation with Augmented Anchors (A^2LP), as illustrated in Fig. 2. We detail the A^2LP method as follows.

Nearest Neighbor Graph. We construct the feature set $V = \{\boldsymbol{v}_1, \cdots, \boldsymbol{v}_l, \boldsymbol{v}_{l+1}, \cdots, \boldsymbol{v}_n\}$, where $\boldsymbol{v}_i := \phi_{\theta_e}(\boldsymbol{x}_i)$. The affinity matrix $\boldsymbol{A} \in \mathbb{R}_+^{n \times n}$ is constructed with elements:

$$\boldsymbol{a}_{ij} := \begin{cases} \varepsilon(\boldsymbol{v}_i, \boldsymbol{v}_j), & \text{if } i \neq j \wedge \boldsymbol{v}_i \in \text{NN}_k(\boldsymbol{v}_j) \\ 0, & otherwise \end{cases} \qquad (6)$$

where $\varepsilon(\boldsymbol{v}_i, \boldsymbol{v}_j)$ measures the non-negative similarity between $\boldsymbol{v}_i$ and $\boldsymbol{v}_j$, and NN_k denotes the set of k nearest neighbors in X. Then, we adopt $\boldsymbol{A} = \boldsymbol{A} + \boldsymbol{A}^T$ to make $\boldsymbol{A}$ a symmetric non-negative adjacency matrix with zero diagonal.

Label Propagation. The closed-form solution of the objective (Eq. (3)) of the LP algorithm is given by [48] as

$$\boldsymbol{F}^* = (\boldsymbol{I} - \alpha \boldsymbol{S})^{-1} \boldsymbol{Y}, \tag{7}$$

where $\alpha = \frac{2\lambda}{2\lambda+1}$, $\boldsymbol{I}$ is an identity matrix and $\boldsymbol{S} = \boldsymbol{D}^{-1/2} \boldsymbol{A} \boldsymbol{D}^{-1/2}$.

LP with Augmented Anchors. Suggested by the Remark 1, we generate virtual instances via a weighted combination of unlabeled target instances, using weights computed by the entropy of their propagated soft cluster assignments, considering that instances of low entropy are more confident in terms of their predicted labels. In particular, we first obtain the pseudo labels of unlabeled target instances by solving Eq. (7), and then we assign the weight w_i to each unlabeled instance $\boldsymbol{x}_i$ by

$$w_i := 1 - \frac{H(\boldsymbol{p}_i^*)}{\log(K)}, \tag{8}$$

where $H(\cdot)$ is the entropy function and $\boldsymbol{p}_{ij}^* = \boldsymbol{f}_{ij}^* / \sum_j \boldsymbol{f}_{ij}^*$. We have $w_i \in [0,1]$ since $0 \le H(\boldsymbol{p}_i^*) \le \log K$. The virtual instances $\{(\hat{\boldsymbol{v}}_{n+k}, k)\}_{k=1}^K$ can then be calculated as:

$$(\hat{\boldsymbol{v}}_{n+k}, k) = \left(\sum_{\boldsymbol{x}_i \in X_U} \frac{\mathbb{1}(k = \hat{y}_i) w_i \phi_{\theta_e}(\boldsymbol{x}_i)}{\sum_{\boldsymbol{x}_j \in X_U} \mathbb{1}(k = \hat{y}_j) w_j}, k \right), \tag{9}$$

where $\mathbb{1}(\cdot)$ is the indicator function. The virtual instances generated by Eq. (9) are relatively robust to the label noise and their neighbors are probably the instances of the same label due to the underlying cluster assumption.

Then, we iteratively do the steps of (1) augmenting the feature set V and label matrix $\boldsymbol{Y}$ with the generated virtual instances and (2) generating virtual instances by the LP algorithm based on updated feature set V and label matrix $\boldsymbol{Y}$. The updating strategies of feature set V and label matrix $\boldsymbol{Y}$ are as follows:

$$V = V \cup \{\hat{\boldsymbol{v}}_{n+1}, \cdots, \hat{\boldsymbol{v}}_{n+K}\}, \boldsymbol{Y} = \begin{bmatrix} \boldsymbol{Y} \\ \boldsymbol{I} \end{bmatrix}, n = n + K. \tag{10}$$

The iterative steps empirically converge in less than 10 iterations, as illustrated in Sect. 5.1. The implementation of our A²LP is summarized in Algorithm 1 (line 2 to 10).

Alternating Steps of Pseudo Labeling and Domain-Invariant Feature Learning. Although our proposed A²LP can largely alleviate the performance degradation of applying LP to UDA tasks via the introduction of virtual instances, learning domain-invariant features across labeled source data and unlabeled target data is fundamentally important, especially when the domain

---

Algorithm 1 Alternating steps of pseudo labeling via A²LP and domain-invariant feature learning.

Input:
Labeled data: $\{X_L, Y_L\} = \{(\boldsymbol{x}_1, y_1), ..., (\boldsymbol{x}_l, y_l)\}$
Unlabeled data: $X_U = \{\boldsymbol{x}_{l+1}, ..., \boldsymbol{x}_n\}$
Model parameters: $\theta = [\theta_e, \theta_c]$
Procedure:
  1: **while** Not Converge **do**
  2:     Construct feature set V and label matrix $\boldsymbol{Y}$;
  3:     **for** iter $= 1$ to N **do**   ▷ Pseudo labeling via A²LP
  4:        Compute affinity matrix $\boldsymbol{A}$ by Eq. (6);
  5:        $\boldsymbol{A} \leftarrow \boldsymbol{A} + \boldsymbol{A}^T$;
  6:        $\boldsymbol{S} \leftarrow \boldsymbol{D}^{-1/2}\boldsymbol{A}\boldsymbol{D}^{-1/2}$;
  7:        Get predictions $\boldsymbol{F}^*$ by Eq. (7);
  8:        Calculate the virtual instances by Eq. (9);
  9:        Update V and $\boldsymbol{Y}$ with virtual instances by Eq. (10);
10:    **end for**
11:    Remove added virtual instances, and $n = n - NK$;
12:    **for** iter $= 1$ to M **do**   ▷ Domain-invariant feature learning
13:        Update parameters θ by domain-invariant feature learning (e.g., [44,22]);
14:    **end for**
15: **end while**

---

shift is unexpectedly large. To illustrate the advantage of our proposed A²LP on generating high-quality pseudo labels of unlabeled data, and to justify the efficacy of the alternating steps of pseudo labeling via SSL methods and domain-invariant feature learning, we empower state-of-the-art UDA methods [22,44] by replacing their pseudo label generators with our A²LP, and keep other settings unchanged. Empirical results in Sect. 5.2 testify the efficacy of our A²LP.

Time Complexity of A²LP. Computation of our proposed algorithm is dominated by constructing the affinity matrix (6) via k-nearest neighbor graph and solving the closed-form solution (7). Brute-force implementations of them are computationally expensive for datasets with large numbers of instances. Fortunately, the $O(n^2)$ complexity of full affinity matrix construction of the k-nearest neighbor graph can be largely improved via NN-Descent [10], giving rise to an almost linear empirical complexity of $O(n^{1.1})$. Given that the matrix $(\boldsymbol{I} - \alpha\boldsymbol{S})$ is positive-definite, the label predictions $\boldsymbol{F}^*$ (7) can be achieved by solving the following linear system with the conjugate gradient (CG) [16,53]:

$$(\boldsymbol{I} - \alpha\boldsymbol{S})\boldsymbol{F}^* = \boldsymbol{Y}, \tag{11}$$

which is known to be faster than the closed-form solution (7). Empirical results in the appendices show that such accelerating strategies significantly reduce the time consumption and hardly suffer performance penalties.

5   Experiments

Office-31 [34] is a standard UDA dataset including three diverse domains: Amazon (**A**) from Amazon website, Webcam (**W**) by web camera, and DSLR (**D**) by digital SLR camera. There are $4,110$ images of 31 categories shared across three domains. **ImageCLEF-DA** [1] is a balanced dataset containing three domains: Caltech-256 (**C**), ImageNet ILSVRC 2012 (**I**), and Pascal VOC 2012 (**P**). There are 12 categories and 600 images in each domain. **VisDA-2017** [33] is a dataset with large domain shift from the synthetic data (**Syn.**) to real images (**Real**). There are about 280K images across 12 categories.

We implement our A²LP based on PyTorch. We adopt the ResNet [14] pretrained on the ImageNet dataset [8] excluding the last fully connected (FC) layer as the feature extractor ϕ. In the alternating training step, we fine-tune the feature extractor ϕ and train a classifier f of one FC layer from scratch. We update all parameters by stochastic gradient descent with momentum of 0.9, and the learning rate of the classifier f is 10 times that of the feature extractor ϕ. We employ the annealing strategy of learning rate [12] following $\eta_p = \frac{\eta_0}{(1+\mu p)^\beta}$, where p is the process of training iterations linearly changing from 0 to 1, $\eta_0 = 0.01$ and $\mu = 10$. Following [22], we set $\beta = 0.75$ for datasets of Office-31 [34] and ImageCLEF-DA [1], while for VisDA-2017 dataset, $\beta = 2.25$. We adopt the cosine similarity, i.e., $\varepsilon(v_i, v_j) = \frac{<v_i, v_j>}{\|v_i\|\|v_j\|}$, to construct the affinity matrix (6) and compare it with other two alternatives in Sect. 5.1. We empirically set α as 0.75 and 0.5 for the VisDA-2017 dataset and datasets of Office-31 and ImageCLEF-DA, respectively. We use all labeled source data and all unlabeled target data in the training process, following the standard protocols for UDA [12,27]. For each adaptation task, we report the average classification accuracy and the standard error on three random experiments.

5.1   Analysis

Various Similarity Metrics. In this section, we conduct ablative experiments on the **C** → **P** task of the ImageCLEF-DA dataset to analyze the influences of graph structures to results of A²LP. To study the impact of similarity measurements, we construct the affinity matrix with other two alternative similarity measurements, namely the Euclidean distance-based similarity $\varepsilon(v_i, v_j) = \exp(-\|v_i - v_j\|^2/2)$ introduced in [49] and the scalar product-based similarity $\varepsilon(v_i, v_j) = \max(v_i^T v_j, 0)^3$ adopted in [20]. We also set k to different values to investigate its influence. Results are illustrated in Fig. 3. We empirically observe that results with cosine similarity are consistently better than those with the other two alternatives. We attribute the advantages of cosine similarity to the adopted FC layer-based classifier, where the cosine similarities between features and weights of the classifier dominate category predictions. The results of A²LP with affinity matrix constructed by the cosine similarity are stable under a wide range of k (i.e., $5 \sim 30$). Results with the full affinity matrix (i.e., $k = n$) are generally lower than that with the nearest neighbor graph. We empirically

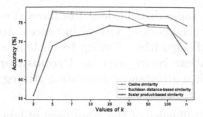

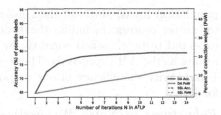

Fig. 3. An illustration of the accuracy (%) of A²LP with affinity matrix constructed with different similarity measurements and different values of k. Results are reported on the $\mathbf{C} \rightarrow \mathbf{P}$ task of the ImageCLEF-DA dataset based on a 50-layer ResNet. When $k = n$, we construct the full affinity matrix as in [49]. Please refer to Sect. 5.1 for the definitions of similarities.

Fig. 4. An illustration of the accuracy (%) of pseudo labels of unlabeled instances (left y-axis) and the percent (%) of connection weight (PoW) of the same class pairs between labeled and unlabeled data (right y-axis) of our proposed A²LP on the tasks of SSL and UDA. The A²LP degenerates to LP [48] when the number of iteration N is set to 1. Please refer to Sect. 5.1 for the detailed settings.

set $k = 20$ in the experiments for the Office-31 and ImageCLEF-DA datasets, and set $k = 100$ for the VisDA-2017 dataset, where the number of instances is considerably large.

A²LP on UDA and SSL. In this section, we observe the behaviors of the A²LP on UDA and SSL tasks. The goal of the experiment is to observe the results with augmented virtual instances in LP. For the labeled data, we randomly sample 1000 instances per class in the synthetic image set of the VisDA-2017 dataset. For the SSL task, we randomly sample another 1000 instances per class in the synthetic image set of the VisDA-2017 dataset as the unlabeled data, whereas 1000 instances are sampled randomly in each class of the real image set to construct unlabeled data in the UDA task. We denote the constructed UDA task as VisDA-2017-Small for ease of use. The mean prediction accuracy of all unlabeled instances is reported. To give insights of the different results, we illustrate the percent of connection weight (PoW) of the same class pairs between labeled and unlabeled data in the constructed k-nearest neighbor graph using ground-truth labels of unlabeled data, which is calculated as: $PoW = \frac{W_{lu}}{W_{all}}$, where W_{lu} is the similarities sum of connections of the same class pairs between labeled and unlabeled data in the affinity matrix $\mathbf{A}$ and $W_{all} = \sum_{i,j} a_{ij}$.

The results are illustrated in Fig. 4. In the UDA task, the initial PoW (i.e., $N = 1$) is too low to enable the labels of labeled data propagate to all the unlabeled target data. As the A²LP proceeds, the labeled data are augmented with virtual instances with true labels, whose neighbors involve unlabeled instances sharing the same label. Thus the PoW increases, leading to more accurate predictions of unlabeled data. In the SSL task, cluster centers of labeled and unlabeled data are positioned to be close and (statistically) in relatively dense population areas, since all data follow the same distribution. Instances close to cluster centers, including the k nearest neighbors of virtual instances, are expected to be classified correctly by the LP algorithm, leading to unchanged results as the A^{2}LP proceeds. These observations corroborate the **Proposition 1** and verify the efficacy of our proposed virtual instances generation strategy for UDA.

Robustness to Noise. We investigate the influence of the noise level of label predictions on the results of A^{2}LP. As illustrated in Table 1, the A^{2}LP is robust to the label noise. Specifically, as the noise level increases, results of A2LP degrade and are worse than that of the vanilla LP when the noise level is larger than $\sim$60%.

Table 1. Results of A^{2}LP with different noise levels of initial label predictions on the **P** $\to$ **C** task of the ImageCLEF-DA dataset. We replace the initial label predictions from the LP (i.e., the Line 7 of the Algorithm 1) with a manually defined setting, where the noise level of L% indicates that the virtual instances (i.e., Eq. (8)) are calculated with unlabeled target data, L% of which are assigned with random and wrong pseudo labels. Note that we set N = 2 (cf. Line 3 of Algorithm 1) here.

Noise level (%)	0	10	30	50	60	70	80	100	Vanilla LP
Acc. (%) of A^{2}LP	92.8	92.8	92.3	91.8	91.0	90.7	90.3	90.0	91.2

A^{2}LP Variant. We propose a degenerated variant of A^{2}LP by representing the entire labeled source data with several representative surrogate instances in the A^{2}LP process, which can largely alleviate the computation cost of the LP algorithm. More specifically, we replace the features of source data $\{v_1, \cdots, v_l\}$ with K source category centers $\{v_k = \sum_{i=1}^{l} \frac{\mathbb{1}(k=y_i)v_i}{\sum_{j=1}^{l} \mathbb{1}(k=y_j)}\}_{k=1}^{K}$ with category labels $\{1, \cdots, K\}$ (only the Line 2 of the Algorithm 1 is updated accordingly). As illustrated in Table 2, the result of A^{2}LP variant is slightly lower than that of the A^{2}LP on the VisDA-2017-Small task. Note that we only adopt the A^{2}LP variant in tasks involving the entire VisDA-2017 dataset unless otherwise specified.

Table 2. Comparison between the A^{2}LP and its degenerated variant on the VisDA-2017-Small task based on a 50-layer ResNet.

Methods	A^{2}LP	A^{2}LP variant
Acc. (%)	79.3	77.9

Table 3. Illustration of effects of the entropy-based instance weights (9) in A^{2}LP based on a 50-layer ResNet.

Methods	A $\to$ W	W $\to$ A
A^{2}LP	87.7	75.9
A^{2}LP ($w_i = 1, \forall i$)	87.4	75.4

Effects of Instance Weighting in A^{2}LP. We investigate the effects of entropy-based instance weights in reliable virtual instances generation (9) of

Table 4. Results on the Office31 dataset [34] (ResNet-50).

Methods	A → W	D → W	W → D	A → D	D → A	W → A	Avg.
Source Only	68.4 ± 0.2	96.7 ± 0.1	99.3 ± 0.1	68.9 ± 0.2	62.5 ± 0.3	60.7 ± 0.3	76.1
DAN [27]	80.5 ± 0.4	97.1 ± 0.2	99.6 ± 0.1	78.6 ± 0.2	63.6 ± 0.3	62.8 ± 0.2	80.4
DANN [12]	82.0 ± 0.4	96.9 ± 0.2	99.1 ± 0.1	79.7 ± 0.4	68.2 ± 0.4	67.4 ± 0.5	82.2
CDAN+E [28]	94.1 ± 0.1	98.6 ± 0.1	**100.0 ± .0**	92.9 ± 0.2	71.0 ± 0.3	69.3 ± 0.3	87.7
SymNets [47]	90.8 ± 0.1	98.8 ± 0.3	**100.0 ± .0**	93.9 ± 0.5	74.6 ± 0.6	72.5 ± 0.5	88.4
DADA [40]	92.3 ± 0.1	99.2 ± 0.1	100.0 ± 0.0	93.9 ± 0.2	74.4 ± 0.1	74.2 ± 0.1	89.0
CAN [22]	**94.5 ± 0.3**	**99.1 ± 0.2**	99.8 ± 0.2	95.0 ± 0.3	78.0 ± 0.3	77.0 ± 0.3	90.6
LP	81.1	96.8	99.0	82.3	71.6	73.1	84.0
A^{2}LP (ours)	87.7	98.1	99.0	87.8	75.8	75.9	87.4
MSTN (reproduced)	92.7 ± 0.5	98.5 ± 0.2	99.8 ± 0.2	89.9 ± 0.3	74.6 ± 0.3	75.2 ± 0.5	88.5
empowered by A^{2}LP	93.1 ± 0.2	98.5 ± 0.1	99.8 ± 0.2	94.0 ± 0.2	76.5 ± 0.3	76.7 ± 0.3	89.8
CAN (reproduced)	94.0 ± 0.5	98.5 ± 0.1	99.7 ± 0.1	94.8 ± 0.4	**78.1 ± 0.2**	76.7 ± 0.3	90.3
empowered by A^{2}LP	93.4 ± 0.3	98.8 ± 0.1	**100.0 ± .0**	**96.1 ± 0.1**	**78.1 ± 0.1**	**77.6 ± 0.1**	**90.7**

Table 5. Results on the ImageCLEF-DA dataset [1] (ResNet-50).

Methods	I → P	P → I	I → C	C → I	C → P	P → C	Avg.
Source Only	74.8 ± 0.3	83.9 ± 0.1	91.5 ± 0.3	78.0 ± 0.2	65.5 ± 0.3	91.2 ± 0.3	80.7
DAN [27]	74.5 ± 0.4	82.2 ± 0.2	92.8 ± 0.2	86.3 ± 0.4	69.2 ± 0.4	89.8 ± 0.4	82.5
DANN [12]	75.0 ± 0.6	86.0 ± 0.3	96.2 ± 0.4	87.0 ± 0.5	74.3 ± 0.5	91.5 ± 0.6	85.0
CDAN+E [28]	77.7 ± 0.3	90.7 ± 0.2	**97.7 ± 0.3**	91.3 ± 0.3	74.2 ± 0.2	94.3 ± 0.3	87.7
SymNets [47]	**80.2 ± 0.3**	93.6 ± 0.2	97.0 ± 0.3	**93.4 ± 0.3**	78.7 ± 0.3	96.4 ± 0.1	89.9
LP	77.1	89.2	93.0	87.5	69.8	91.2	84.6
A^{2}LP (ours)	79.3	91.8	96.3	91.7	78.1	96.0	88.9
MSTN (reproduced)	78.3 ± 0.2	92.5 ± 0.3	96.5 ± 0.2	91.1 ± 0.1	76.3 ± 0.3	94.6 ± 0.4	88.2
empowered by A^{2}LP	79.6 ± 0.3	92.7 ± 0.3	96.7 ± 0.1	92.5 ± 0.2	78.9 ± 0.2	96.0 ± 0.1	89.4
CAN (reproduced)	78.5 ± 0.3	93.0 ± 0.3	97.3 ± 0.2	91.0 ± 0.3	77.2 ± 0.2	**97.0 ± 0.2**	89.0
empowered by A^{2}LP	79.8 ± 0.2	**94.3 ± 0.3**	**97.7 ± 0.2**	93.0 ± 0.3	**79.9 ± 0.1**	96.9 ± 0.2	**90.3**

A^{2}LP in this section. As illustrated in Table 3, A^{2}LP improves over A^{2}LP ($w_i = 1, \forall i$), where all unlabeled instances are weighted equally, supporting that instances of low entropy are more confident in terms of their predicted labels.

5.2 Results

We report the classification results on the Office-31 [34], ImageCLEF-DA [1], and VisDA-2017 [33] datasets in Table 4, Table 5, and Table 6, respectively. Results of other methods are either directly reported from their original papers if available or quoted from [24,28]. Compared to classical methods [12,27] aiming at domain-invariant feature learning, the vanilla LP generally achieves better results via the graph-based SSL principle, certifying the efficacy of the SSL principles in UDA tasks. Our A^{2}LP improves over the LP significantly on all three UDA benchmarks, justifying the efficacy of the introduction of virtual instances for UDA. Additionally, we reproduce the state-of-the-art UDA methods of Moving

Table 6. Results on the VisDA-2017 dataset. The A^2LP reported is the degenerated variant detailed in Sect. 5.1. Full results are presented in the appendices.

Methods	Acc. based on a ResNet50	Acc. based on a ResNet101
Source Only	45.6	50.8
DAN [27]	53.0	61.1
DANN [12]	55.0	57.4
MCD [36]	–	71.9
CDAN+E [28]	70.0	–
LPJT [25]	–	74.0
DADA [40]	–	79.8
Lee *et al.* [24]	76.2	81.5
CAN [22]	–	87.2
LP	69.8	73.9
A^2LP (ours)	78.7	82.7
MSTN (reproduced)	71.9	75.2
empowered by A^2LP	81.5	83.7
CAN (reproduced)	85.6	87.2
empowered by A^2LP	**86.5**	**87.6**

Semantic Transfer Network (MSTN) [44] and Contrastive Adaptation Network (CAN) [22] with the released codes[2]; by replacing the pseudo label generators of MSTN and CAN with our A^2LP, we improve their results noticeably and achieve the new state of the art, testifying the effectiveness of the combination of A^2LP and domain-invariant feature learning.

6   Conclusion

Motivated by the relatedness of problem definitions between UDA and SSL, we study the use of SSL principles in UDA, especially the graph-based LP algorithm. We analyze the conditions of affinity graph/matrix to achieve better propagation of true labels to unlabeled instances, and accordingly propose a new algorithm of A^2LP, which potentially improves LP via generation of unlabeled virtual instances. An empirical scheme of virtual instance generation is particularly proposed for UDA via a weighted combination of unlabeled target instances. By iteratively using A^2LP to get high-quality pseudo labels of target instances and learning domain-invariant features involving the obtained pseudo-labeled target instances, new state of the art is achieved on three datasets, confirming the value of further investigating SSL techniques for UDA problems.

Acknowledgment. This work is supported in part by the Guangdong R&D key project of China (Grant No.: 2019B010155001), the National Natural Science Foundation of China (Grant No.: 61771201), and the Program for Guangdong Introducing Innovative and Enterpreneurial Teams (Grant No.: 2017ZT07X183).

---

[2] https://github.com/Mid-Push/Moving-Semantic-Transfer-Network.
https://github.com/kgl-prml.

References

1. Imageclef-da dataset. http://imageclef.org/2014/adaptation/
2. Belkin, M., Matveeva, I., Niyogi, P.: Regularization and semi-supervised learning on large graphs. In: Shawe-Taylor, J., Singer, Y. (eds.) COLT 2004. LNCS (LNAI), vol. 3120, pp. 624–638. Springer, Heidelberg (2004). https://doi.org/10.1007/978-3-540-27819-1_43
3. Ben-David, S., Blitzer, J., Crammer, K., Kulesza, A., Pereira, F., Vaughan, J.W.: A theory of learning from different domains. Mach. Learn. **79**(1–2), 151–175 (2010)
4. Ben-David, S., Blitzer, J., Crammer, K., Pereira, F.: Analysis of representations for domain adaptation. In: Advances in Neural Information Processing Systems, pp. 137–144 (2007)
5. Chapelle, O., Schölkopf, B., Zien, A. (eds.): Semi-Supervised Learning. The MIT Press (2006). https://doi.org/10.7551/mitpress/9780262033589.001.0001
6. Chapelle, O., Zien, A.: Semi-supervised classification by low density separation. In: AISTATS, vol. 2005, pp. 57–64. Citeseer (2005)
7. Delalleau, O., Bengio, Y., Roux, N.L.: Efficient non-parametric function induction in semi-supervised learning. In: Proceedings of the Tenth International Workshop on Artificial Intelligence and Statistics (2005). http://www.gatsby.ucl.ac.uk/aistats/fullpapers/204.pdf
8. Deng, J., Dong, W., Socher, R., Li, L.J., Li, K., Fei-Fei, L.: Imagenet: a large-scale hierarchical image database. In: 2009 IEEE Conference on Computer Vision and Pattern Recognition, pp. 248–255. IEEE (2009)
9. Ding, Z., Li, S., Shao, M., Fu, Y.: Graph adaptive knowledge transfer for unsupervised domain adaptation. In: Ferrari, V., Hebert, M., Sminchisescu, C., Weiss, Y. (eds.) ECCV 2018. LNCS, vol. 11206, pp. 36–52. Springer, Cham (2018). https://doi.org/10.1007/978-3-030-01216-8_3
10. Dong, W., Moses, C., Li, K.: Efficient k-nearest neighbor graph construction for generic similarity measures. In: Proceedings of the 20th International Conference on World Wide Web, pp. 577–586 (2011)
11. French, G., Mackiewicz, M., Fisher, M.: Self-ensembling for visual domain adaptation. In: International Conference on Learning Representations (2018)
12. Ganin, Y., et al.: Domain-adversarial training of neural networks. J. Mach. Learn. Res. **17**(1), 1–35 (2016)
13. Grandvalet, Y., Bengio, Y.: Semi-supervised learning by entropy minimization. In: Advances in Neural Information Processing Systems, vol. 17, pp. 529–536 (2005)
14. He, K., Zhang, X., Ren, S., Sun, J.: Deep residual learning for image recognition. In: Proceedings of the IEEE Conference on Computer Vision and Pattern Recognition, pp. 770–778 (2016)
15. He, R., Lee, W.S., Ng, H.T., Dahlmeier, D.: Adaptive semi-supervised learning for cross-domain sentiment classification. In: Proceedings of the 2018 Conference on Empirical Methods in Natural Language Processing, pp. 3467–3476 (2018)
16. Hestenes, M.R., Stiefel, E., et al.: Methods of conjugate gradients for solving linear systems. J. Res. Nat. Bur. Stand. **49**(6), 409–436 (1952)
17. Hou, C.A., Tsai, Y.H.H., Yeh, Y.R., Wang, Y.C.F.: Unsupervised domain adaptation with label and structural consistency. IEEE Trans. Image Process. **25**(12), 5552–5562 (2016)
18. Hui Tang, K.C., Jia, K.: Unsupervised domain adaptation via structurally regularized deep clustering. In: IEEE Conference on Computer Vision and Pattern Recognition (CVPR) (2020)

19. Iscen, A., Tolias, G., Avrithis, Y., Chum, O.: Label propagation for deep semi-supervised learning. In: Proceedings of the IEEE Conference on Computer Vision and Pattern Recognition, pp. 5070–5079 (2019)
20. Iscen, A., Tolias, G., Avrithis, Y., Furon, T., Chum, O.: Efficient diffusion on region manifolds: recovering small objects with compact CNN representations. In: Proceedings of the IEEE Conference on Computer Vision and Pattern Recognition, pp. 2077–2086 (2017)
21. Joachims, T.: Transductive learning via spectral graph partitioning. In: Proceedings of the 20th International Conference on Machine Learning (ICML-03), pp. 290–297 (2003)
22. Kang, G., Jiang, L., Yang, Y., Hauptmann, A.G.: Contrastive adaptation network for unsupervised domain adaptation. In: Proceedings of the IEEE Conference on Computer Vision and Pattern Recognition, pp. 4893–4902 (2019)
23. Kumar, A., et al.: Co-regularized alignment for unsupervised domain adaptation. In: In: Advances in Neural Information Processing Systems, pp. 9345–9356 (2018)
24. Lee, S., Kim, D., Kim, N., Jeong, S.G.: Drop to adapt: Learning discriminative features for unsupervised domain adaptation. In: Proceedings of the IEEE International Conference on Computer Vision, pp. 91–100 (2019)
25. Li, J., Jing, M., Lu, K., Zhu, L., Shen, H.T.: Locality preserving joint transfer for domain adaptation. IEEE Trans. Image Process. **28**(12), 6103–6115 (2019)
26. Liu, Y., et al.: Learning to propagate labels: transductive propagation network for few-shot learning. arXiv preprint arXiv:1805.10002 (2018)
27. Long, M., Cao, Y., Wang, J., Jordan, M.I.: Learning transferable features with deep adaptation networks. In: Proceedings of the 32Nd International Conference on International Conference on Machine Learning - Volume 37, ICML 2015, pp. 97–105. JMLR.org (2015). http://dl.acm.org/citation.cfm?id=3045118.3045130
28. Long, M., Cao, Z., Wang, J., Jordan, M.I.: Conditional adversarial domain adaptation. In: Advances in Neural Information Processing Systems, pp. 1640–1650 (2018)
29. Long, M., Zhu, H., Wang, J., Jordan, M.I.: Unsupervised domain adaptation with residual transfer networks. In: Advances in Neural Information Processing Systems, pp. 136–144 (2016)
30. Mansour, Y., Mohri, M., Rostamizadeh, A.: Domain adaptation: Learning bounds and algorithms. In: 22nd Conference on Learning Theory, COLT 2009 (2009)
31. Miyato, T., Maeda, S.i., Koyama, M., Ishii, S.: Virtual adversarial training: a regularization method for supervised and semi-supervised learning. IEEE Trans. Pattern Anal. Mach. Intell. **41**(8), 1979–1993 (2018)
32. Pan, S.J., Yang, Q., et al.: A survey on transfer learning. IEEE Trans. Knowl. Data Eng. **22**(10), 1345–1359 (2010)
33. Peng, X., Usman, B., Kaushik, N., Hoffman, J., Wang, D., Saenko, K.: VisDA: the visual domain adaptation challenge. arXiv preprint arXiv:1710.06924 (2017)
34. Saenko, K., Kulis, B., Fritz, M., Darrell, T.: Adapting visual category models to new domains. In: Daniilidis, K., Maragos, P., Paragios, N. (eds.) ECCV 2010. LNCS, vol. 6314, pp. 213–226. Springer, Heidelberg (2010). https://doi.org/10.1007/978-3-642-15561-1_16
35. Saito, K., Ushiku, Y., Harada, T.: Asymmetric tri-training for unsupervised domain adaptation. arXiv preprint arXiv:1702.08400 (2017)
36. Saito, K., Watanabe, K., Ushiku, Y., Harada, T.: Maximum classifier discrepancy for unsupervised domain adaptation. In: Proceedings of the IEEE Conference on Computer Vision and Pattern Recognition, pp. 3723–3732 (2018)

37. Shen, J., Qu, Y., Zhang, W., Yu, Y.: Wasserstein distance guided representation learning for domain adaptation. In: AAAI, pp. 4058–4065 (2018)
38. Shu, R., Bui, H.H., Narui, H., Ermon, S.: A DIRT-T approach to unsupervised domain adaptation. arXiv preprint arXiv:1802.08735 (2018)
39. Szummer, M., Jaakkola, T.: Partially labeled classification with Markov random walks. In: Advances in Neural Information Processing Systems, pp. 945–952 (2002)
40. Tang, H., Jia, K.: Discriminative adversarial domain adaptation. In: The Thirty-Fourth AAAI Conference on Artificial Intelligence, AAAI 2020, pp. 5940–5947. AAAI Press (2020)
41. Tarvainen, A., Valpola, H.: Mean teachers are better role models: weight-averaged consistency targets improve semi-supervised deep learning results. In: Advances in Neural Information Processing Systems, pp. 1195–1204 (2017)
42. Tzeng, E., Hoffman, J., Darrell, T., Saenko, K.: Simultaneous deep transfer across domains and tasks. In: Proceedings of the IEEE International Conference on Computer Vision, pp. 4068–4076 (2015)
43. Tzeng, E., Hoffman, J., Zhang, N., Saenko, K., Darrell, T.: Deep domain confusion: maximizing for domain invariance. arXiv preprint arXiv:1412.3474 (2014)
44. Xie, S., Zheng, Z., Chen, L., Chen, C.: Learning semantic representations for unsupervised domain adaptation. In: International Conference on Machine Learning, pp. 5419–5428 (2018)
45. Yan, H., Ding, Y., Li, P., Wang, Q., Xu, Y., Zuo, W.: Mind the class weight bias: Weighted maximum mean discrepancy for unsupervised domain adaptation. In: The IEEE Conference on Computer Vision and Pattern Recognition (CVPR), vol. 3 (2017)
46. Zhang, Y., Deng, B., Tang, H., Zhang, L., Jia, K.: Unsupervised multi-class domain adaptation: theory, algorithms, and practice. CoRR abs/2002.08681 (2020)
47. Zhang, Y., Tang, H., Jia, K., Tan, M.: Domain-symmetric networks for adversarial domain adaptation. In: Proceedings of the IEEE Conference on Computer Vision and Pattern Recognition, pp. 5031–5040 (2019)
48. Zhou, D., Bousquet, O., Lal, T.N., Weston, J., Schölkopf, B.: Learning with local and global consistency. In: Advances in Neural Information Processing Systems, pp. 321–328 (2004)
49. Zhou, D., Bousquet, O., Lal, T.N., Weston, J., Schölkopf, B.: Learning with local and global consistency. In: Thrun, S., Saul, L.K., Schölkopf, B. (eds.) Advances in Neural Information Processing Systems, vol. 16, pp. 321–328 (2004)
50. Zhou, Z.H., Li, M.: Tri-training: exploiting unlabeled data using three classifiers. IEEE Trans. Knowl. Data Eng. 17(11), 1529–1541 (2005)
51. Zhu, X., Ghahramani, Z.: Learning from labeled and unlabeled data with label propagation. Technical report, School of Computer Science, Carnegie Mellon University (2002)
52. Zhu, X., Ghahramani, Z., Lafferty, J.D.: Semi-supervised learning using gaussian fields and harmonic functions. In: Proceedings of the 20th International Conference on Machine Learning (ICML-03), pp. 912–919 (2003)
53. Zhu, X., Lafferty, J., Rosenfeld, R.: Semi-supervised learning with graphs. Ph.D. thesis, Carnegie Mellon University, Language Technologies Institute, School of . . . (2005)

Are Labels Necessary for Neural Architecture Search?

Chenxi Liu[1]($\boxtimes$), Piotr Dollár[2], Kaiming He[2], Ross Girshick[2], Alan Yuille[1], and Saining Xie[2]

[1] Johns Hopkins University, Baltimore, USA
cxliu@jhu.edu
[2] Facebook AI Research, Menlo Park, USA

Abstract. Existing neural network architectures in computer vision—whether designed by humans or by machines—were typically found using both images and their associated labels. In this paper, we ask the question: can we find high-quality neural architectures using only images, but no human-annotated labels? To answer this question, we first define a new setup called Unsupervised Neural Architecture Search (UnNAS). We then conduct two sets of experiments. In sample-based experiments, we train a large number (500) of diverse architectures with either supervised or unsupervised objectives, and find that the architecture rankings produced with and without labels are highly correlated. In search-based experiments, we run a well-established NAS algorithm (DARTS) using various unsupervised objectives, and report that the architectures searched without labels can be competitive to their counterparts searched with labels. Together, these results reveal the potentially surprising finding that labels are not necessary, and the image statistics alone may be sufficient to identify good neural architectures.

Keywords: Neural architecture search · Unsupervised learning

1 Introduction

Neural architecture search (NAS) has emerged as a research problem of searching for architectures that perform well on target data and tasks. A key mystery surrounding NAS is what factors contribute to the success of the search. Intuitively, using the target data and tasks during the search will result in the least domain gap, and this is indeed the strategy adopted in early NAS attempts [26,35]. Later, researchers [36] started to utilize the transferability of architectures, which enabled the search to be performed on different data and labels (*e.g.*, CIFAR-10) than the target (*e.g.*, ImageNet). However, what has not changed is that *both the images and the (semantic) labels* provided in the dataset need to be used

Electronic supplementary material The online version of this chapter (https://doi.org/10.1007/978-3-030-58548-8_46) contains supplementary material, which is available to authorized users.

© Springer Nature Switzerland AG 2020
A. Vedaldi et al. (Eds.): ECCV 2020, LNCS 12349, pp. 798–813, 2020.
https://doi.org/10.1007/978-3-030-58548-8_46

in order to search for an architecture. In other words, existing NAS approaches perform search in the *supervised* learning regime.

In this paper, we take a step towards understanding what role supervision plays in the success of NAS. We ask the question: How indispensable are labels in neural architecture search? Is it possible to find high-quality architectures using images only? This corresponds to the important yet underexplored *unsupervised setup* of neural architecture search, which we formalize in Sect. 3.

With the absence of labels, the quality of the architecture needs to be estimated in an unsupervised fashion during the search phase. In the present work, we conduct two sets of experiments using three unsupervised training methods [11,21,34] from the recent self-supervised learning literature.[1] These two sets of experiments approach the question from complementary perspectives. In *sample-based experiments*, we randomly sample 500 architectures from a search space, train and evaluate them using supervised *vs.* self-supervised objectives, and then examine the rank correlation (when sorting models by accuracy) between the two training methodologies. In *search-based experiments*, we take a well-established NAS algorithm, replace the supervised search objective with a self-supervised one, and examine the quality of the searched architecture on tasks such as ImageNet classification and Cityscapes semantic segmentation. Our findings include:

- The architecture rankings produced by supervised and self-supervised pretext tasks are *highly correlated*. This finding is consistent across two datasets, two search spaces, and three pretext tasks.
- The architectures searched without human annotations are *comparable in performance* to their supervised counterparts. This result is consistent across three pretext tasks, three pretext datasets, and two target tasks. There are even cases where unsupervised search outperforms supervised search.
- Existing NAS approaches typically use *labeled* images from a *smaller* dataset to learn transferable architectures. We present evidence that using *unlabeled* images from a *large* dataset may be a more promising approach.

We conclude that labels are not necessary for neural architecture search, and the deciding factor for architecture quality may hide within the image pixels.

2 Related Work

Neural Architecture Search. Research on the NAS problem involves designing the search space [33,36] and the search algorithm [25,35]. There are special focuses on reducing the overall time cost of the search process [18,19,23], or on extending to a larger variety of tasks [4,10,17,27]. Existing works on NAS all use human-annotated labels during the search phase. Our work is orthogonal to existing NAS research, in that we explore the unsupervised setup.

---

[1] Self-supervised learning is a form of unsupervised learning. The term "unsupervised learning" in general emphasizes "without *human*-annotated labels", while the term "self-supervised learning" emphasizes "producing labels from data".

Architecture Transferability. In early NAS attempts [26,35], the search phase and the evaluation phase typically operate on the same dataset and task. Later, researchers realized that it is possible to relax this constraint. In these situations, the dataset and task used in the search phase are typically referred as the *proxy* to the target dataset and task, reflecting a notion of architecture transferability. [36] demonstrated that CIFAR-10 classification is a good proxy for ImageNet classification. [18] measured the rank correlation between these two tasks using a small number of architectures. [15] studied the transferability of 16 architectures (together with trained weights) between more *supervised* tasks. [14] studied the role of architecture in several self-supervised tasks, but with a small number of architectures and a different evaluation method (*i.e.* linear probe). Part of our work studies architecture transferability at a larger scale, *across* supervised and unsupervised tasks.

Unsupervised Learning. There is a large literature on unsupervised learning, *e.g.*, [8,9,11,12,21,22,29–31,34]. In the existing literature, methods are generally developed to learn the *weights* (parameters) of a fixed architecture without using labels, and these weights are evaluated by transferring to a target *supervised* task. In our study, we explore the possibility of using such methods to learn the *architecture* without using labels, rather than the weights. Therefore, our subject of study is simultaneously the unsupervised generalization of NAS, and the architecture level generalization of unsupervised learning.

3   Unsupervised Neural Architecture Search

The goal of this paper is to provide an answer to the question asked in the title: are labels necessary for neural architecture search? To formalize this question, in this section, we define a new setup called Unsupervised Neural Architecture Search (UnNAS). Note that UnNAS represents a general problem setup instead of any specific algorithm for solving this problem. We instantiate UnNAS with specific algorithms and experiments to explore the importance of labels in neural architecture search.

3.1   Search Phase

The traditional NAS problem includes a search phase: given a pre-defined search space, the search algorithm explores this space and estimates the performance (*e.g.* accuracy) of the architectures sampled from the space. The accuracy estimation can involve full or partial training of an architecture. Estimating the accuracy requires access to the labels of the dataset. So the traditional NAS problem is essentially a *supervised learning* problem.

We define UnNAS as the counterpart *unsupervised learning* problem. It follows the definition of the NAS problem, only except that there are no human-annotated labels provided for estimating the performance of the architectures.

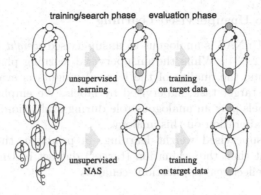

Fig. 1. Unsupervised neural architecture search, or UnNAS, is a new problem setup that helps answer the question: are labels necessary for neural architecture search? In traditional unsupervised learning (top panel), the *training phase* learns the weights of a fixed architecture; then the *evaluation phase* measures the quality of the weights by training a classifier (either by fine-tuning the weights or using them as a fixed feature extractor) using supervision from the target dataset. Analogously, in UnNAS (bottom panel), the *search phase* searches for an architecture without using labels; and the *evaluation phase* measures the quality of the architecture found by an UnNAS algorithm by training the architecture's weights using supervision from the target dataset.

An algorithm for the UnNAS problem still explores a pre-defined search space,[2] but it requires other criteria to estimate how good a sampled architecture is.

3.2   Evaluation Phase

Generally speaking, the goal of the NAS problem is to find an architecture. The weights of the found architecture are *not* necessarily the output of a NAS algorithm. Instead, the weights are optimized after the search phase in an evaluation phase of the NAS problem: it includes training the found architecture on a target dataset's training split, and validating the accuracy on the target dataset's validation split. We note that *"training* the architecture weights" is part of the NAS *evaluation* phase—the labels in the target dataset, both training and validation splits, play a role of evaluating an architecture.

Based on this context, we define *the evaluation phase of UnNAS* in the same way: training the weights of the architecture (found by an UnNAS algorithm) on a target dataset's training split, and validating the accuracy on the target dataset's validation split, both *using* the labels of the target dataset. We remark that using labels during the evaluation phase does *not* conflict with the definition of UnNAS: the search phase is unsupervised, while the evaluation phase requires labels to examine how good the architecture is.

---

[2] Admittedly, the search space design is typically heavily influenced by years of manual search, usually with supervision.

3.3   Analogy to Unsupervised Learning

Our definition of UnNAS is analogous to unsupervised *weight* learning in existing literature [11,21,34]. While the unsupervised learning phase has no labels for training weights, the quality of the learned weights is evaluated by transferring them to a target task, supervised by labels. We emphasize that in the UnNAS setup, labels play an analogous role during evaluation. See Fig. 1 for an illustration and elaboration on this analogy.

Similar to unsupervised weight learning, in principle, the search dataset should be different than the evaluation (target) dataset in UnNAS in order to more accurately reflect real application scenarios.

4   Experiments Overview

As Sect. 3 describes, an architecture discovered in an *unsupervised* fashion will be evaluated by its performance in a *supervised* setting. Therefore, we are essentially looking for some type of architecture level *correlation* that can reach across the unsupervised *vs.* supervised boundary, so that the unsupervisedly discovered architecture could be reliably *transferred* to the supervised target task. We investigate whether several existing self-supervised pretext tasks (described in Sect. 4.1) can serve this purpose, through two sets of experiments of complementary nature: *sample-based* (Sect. 5) and *search-based* (Sect. 6). In *sample-based*, each network is trained and evaluated individually, but the downside is that we can only consider a small, random subset of the search space. In *search-based*, the focus is to find a top architecture from the entire search space, but the downside is that the training dynamics during the search phase does not exactly match that of the evaluation phase.

4.1   Pretext Tasks

We explore three unsupervised training methods (typically referred to as *pretext tasks* in self-supervised learning literature): rotation prediction, colorization, and solving jigsaw puzzles. We briefly describe them for completeness.

- *Rotation prediction* [11] (Rot): the input image undergoes one of four preset rotations (0, 90, 180, and 270 degrees), and the pretext task is formulated as a 4-way classification problem that predicts the rotation.
- *Colorization* [34] (Color): the input is a grayscale image, and the pretext task is formulated as a pixel-wise classification problem with a set of pre-defined color classes (313 in [34]).
- *Solving jigsaw puzzles* [21] (Jigsaw): the input image is divided into patches and randomly shuffled. The pretext task is formulated as an image-wise classification problem that chooses from one out of K preset permutations.[3]

---

[3] On ImageNet, each image is divided into $3 \times 3 = 9$ patches and K is 1000 selected from $9! = 362,880$ permutations [21]; on CIFAR-10 in which images are smaller, we use $2 \times 2 = 4$ patches and K is $4! = 24$ permutations.

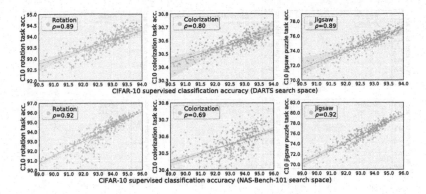

Fig. 2. Correlation between supervised classification accuracy *vs.* pretext task accuracy on CIFAR-10 ("C10"). Top panel: DARTS search space. Bottom panel: NAS-Bench-101 search space. The straight lines are fit with robust linear regression [13] (same for Fig. 3 and Fig. 4).

All three pretext tasks are image or pixel-wise classification problems. Therefore, we can compute the classification accuracy of the pretext task (on a validation set). Based on this pretext task accuracy, we can analyze its correlation with the supervised classification accuracy (also on a validation set), as in our sample-based experiments (Sect. 5). Also, since these pretext tasks all use cross entropy loss, it is also straightforward to use them as the training objective in standard NAS algorithms, as done in our search-based experiments (Sect. 6).

5 Sample-Based Experiments

5.1 Experimental Design

In *sample-based experiments*, we first randomly sample 500 architectures from a certain search space. We train each architecture from scratch on the pretext task and get its pretext task accuracy (*e.g.*, the 4-way classification accuracy in the rotation task), and also train the same architecture from scratch on the supervised classification task (*e.g.*, 1000-way classification on ImageNet).[4]

With these data collected, we perform two types of analysis. For the *rank correlation* analysis, we empirically study the statistical rank correlations between the pretext task accuracy and the supervised accuracy, by measuring the Spearman's Rank Correlation [28], denoted as ρ. For the *random experiment* analysis, we follow the setup proposed in [2] and recently adopted in [24]. Specifically, for each experiment size m, we sample m architectures from our pool of $n = 500$ architectures. For each pretext task, we select the architecture with the highest pretext task accuracy among the m. This process is repeated $\lceil n/m \rceil$ times,

---

[4] These architectures only have small, necessary differences when trained towards these different tasks (*i.e.*, having 4 neurons or 1000 neurons at the output layer).

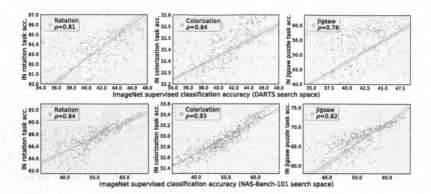

Fig. 3. Correlation between supervised classification accuracy *vs.* pretext task accuracy on ImageNet ("IN"). Top panel: DARTS search space. Bottom panel: NAS-Bench-101 search space.

to compute the mean and error bands (± 2 standard deviations) of the top-1 accuracy of these $\lceil n/m \rceil$ architectures *on the target dataset/task*.

These two studies provide complementary views. The *rank correlation* analysis aims to provide a global picture for *all* architectures sampled from a search space, while the *random experiment* analysis focuses on the *top* architectures in a random experiment of varying size.

We study two search spaces: the DARTS search space [19], and the NAS-Bench-101 search space [33]. The latter search space was built for benchmarking NAS algorithms, so we expect it to be less biased towards the search space for a certain algorithm. The experiments are conducted on two commonly used datasets: CIFAR-10 [16] and ImageNet [7].

5.2   Implementation Details

Details of sampling the 500 architectures are described in Appendix. In the DARTS search space, regardless of the task, each network is trained for 5 epochs with width 32 and depth 22 on ImageNet; 100 epochs with width 16 and depth 20 on CIFAR-10. In the NAS-Bench-101 search space, regardless of the task, each network is trained for 10 epochs with width 128 and depth 12 on ImageNet; 100 epochs with width 128 and depth 9 on CIFAR-10. Please refer to [19,33] for the respective definitions of *width* and *depth*. The performance of each network on CIFAR-10 is the average of 3 independent runs to reduce variance.

We remark that the small-scale of CIFAR-10 dataset and the short training epochs on ImageNet are the compromises we make to allow for more diverse architectures (*i.e.* 500).

5.3   Results

High Rank Correlation Between Supervised Accuracy and Pretext Accuracy on the <u>Same</u> Dataset. In Fig. 2 and Fig. 3, we show the scatter

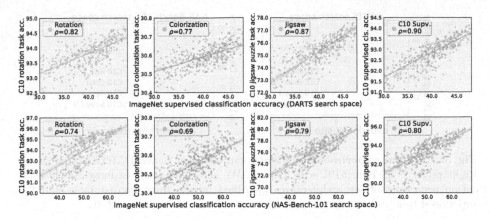

Fig. 4. Correlation between ImageNet supervised classification accuracy *vs.* CIFAR-10 ("C10") pretext task accuracy. Rankings of architectures are highly correlated between supervised classification and three unsupervised tasks, as measured by Spearman's rank correlation (ρ). We also show rank correlation using CIFAR-10 supervised proxy in the rightmost panel. Top panel: DARTS search space. Bottom panel: NAS-Bench-101 search space.

plots of the 500 architectures' supervised classification accuracy (horizontal axis) and pretext task accuracy (vertical axis) on CIFAR-10 and ImageNet, respectively. We see that this rank correlation is typically higher than 0.8, regardless of the dataset, the search space, and the pretext task. This type of consistency and robustness indicates that this phenomenon is general, as opposed to dataset/search space specific.

The same experiment is performed on both the DARTS and the NAS-Bench-101 search spaces. The rank correlations on the NAS-Bench-101 search space are generally higher than those on the DARTS search space. A possible explanation is that the architectures in NAS-Bench-101 are more diverse, and consequently their accuracies have larger gaps, *i.e.*, are less affected by training noise.

Interestingly, we observe that among the three pretext tasks, colorization consistently has the lowest correlation on CIFAR-10, but the highest correlation on ImageNet. We suspect this is because the small images in CIFAR-10 make the learning of per-pixel colorization difficult, and consequently the performance after training is noisy.

High Rank Correlation Between Supervised Accuracy and Pretext Accuracy <u>Across</u> Datasets. In Fig. 4 we show the *across dataset* rank correlation analysis, where the pretext task accuracy is measured on CIFAR-10, but the supervised classification accuracy is measured on ImageNet.

On the DARTS search space (top panel), despite the image distribution shift brought by different datasets, for each of the three pretext tasks (left three plots), the correlation remains consistently high ($\sim$0.8). This shows that across the entire search space, the relative ranking of an architecture is likely to be

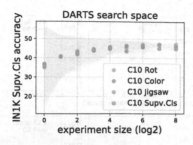
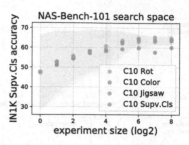

Fig. 5. Random experiment efficiency curves. Left panel: DARTS search space. Right panel: NAS-Bench-101 search space. We show the range of ImageNet classification accuracies of top architectures identified by the three pretext tasks and the supervised task under various experiment sizes. See text for more details.

similar, whether under the unsupervised pretext task accuracy or under the supervised target task accuracy. In the rightmost panel, we compare with a scenario where instead of using an unsupervised pretext task, we use a proxy task of CIFAR-10 supervised classification. The correlation is $\rho = 0.90$ in this case. As the CIFAR-10 supervised classification is a commonly used proxy task in existing NAS literature [36], it gives a reference on ρ's value.

In the bottom panel we show more analysis of this kind by replacing the search space with NAS-Bench-101. Although the architectures are quite different, the observations are similar. In all cases, the self-supervised pretext task accuracy is highly correlated to the supervised classification accuracy.

Better Pretext Accuracy Translates to Better Supervised Accuracy. In addition to the rank correlation analysis, we also perform the random experiment analysis. Figure 5 shows the *random experiment efficiency* curve for DARTS and NAS-Bench-101 search spaces. Again, the pretext accuracies are obtained on CIFAR-10, and the target accuracies are from ImageNet. By design of this experiment, as the experiment size m increases, the *pretext* accuracies of the $\lceil n/m \rceil$ architectures should increase. Figure 5 shows that the *target* accuracies of these $\lceil n/m \rceil$ architectures also increase with m. In addition, at each experiment size, most unsupervised pretext objectives perform similarly compared to the commonly used supervised CIFAR-10 proxy. The overall trends are also comparable. This shows that the architecture rankings produced with and without labels are not only correlated across the entire search space, but also towards the top of the search space, which is closer to the goal of UnNAS.

6 Search-Based Experiments

6.1 Experimental Design

In *search-based experiments*, the idea is to run a well-established NAS algorithm, except that we make the minimal modification of replacing its supervised search objective with an unsupervised one. Following the UnNAS setup, we then examine (by training from scratch) how well these unsupervisedly discovered architectures perform on supervised target tasks. Since all other variables are controlled to be the same, search-based experiments can easily compare between the supervised and unsupervised counterparts of a NAS algorithm, which can help reveal the importance of labels.

The NAS algorithm we adopt is DARTS (short for "differentiable architecture search") [19] for its simplicity. DARTS formulates the activation tensor selection and operation selection as a categorical choice, implemented as a softmax function on a set of continuous parameters (named α). These parameters are trained in a similar fashion as the architecture weights, by backpropagation from a loss function. After this training, the softmax outputs are discretized and produce an architecture.

The self-supervised objectives that we consider are, still, those described in Sect. 4.1: rotation prediction (Rot), colorization (Color), and solving jigsaw puzzles (Jigsaw). For comparison, we also perform NAS search with supervised objectives, e.g. classification (Supv.Cls) or semantic segmentation (Supv.Seg). To help distinguish, we name the method NAS-DARTS if the search objective is supervised, and UnNAS-DARTS if the search objective is unsupervised.

6.2   Implementation Details

Search Phase. We use three different datasets for architecture search: ImageNet-1K (IN1K) [7], ImageNet-22K (IN22K) [7], and Cityscapes [6]. IN1K is the standard ImageNet benchmark dataset with 1.2M images 1K categories. IN22K is the full ImageNet dataset that has ~14M images 22K categories. Cityscapes is a dataset of street scenes that has drastically different image statistics. Note that UnNAS-DARTS will only access the images provided in the dataset, while NAS-DARTS will additionally access the (semantic) labels provided in the respective dataset. The search phase will operate only within the training split, without accessing the true validation or test split.

We report in Appendix the hyper-parameters we used. One major difference between our experiments and DARTS [19] is that the images in the search datasets that we consider are much larger in size. We use 224×224 random crops for search on IN1K/IN22K, and 312×312 for search on Cityscapes following [17]. To enable DARTS training with large input images, we use 3 stride-2 convolution layers at the beginning of the network to reduce spatial resolution. This design, together with appropriately chosen number of search epochs (see Appendix), allows UnNAS search to be efficient (~2 GPU days on IN1K/Cityscapes, ~10 GPU days on IN22K, regardless of task) despite running on larger images.

Evaluation Phase. We use two distinct datasets and tasks for UnNAS evaluation: (1) ImageNet-1K (IN1K) for image classification. The performance metric is top-1 accuracy on the IN1K validation set. (2) Cityscapes for semantic segmentation. We use the train_fine set (2975 images) for training. The performance metric is mean Intersection-over-Union (mIoU) evaluated on the val set (500 images).

Table 1. ImageNet-1K classification results of the architectures searched by NAS and UnNAS algorithms. Rows in gray correspond to invalid UnNAS configurations where the search and evaluation datasets are the same. † is our training result of the DARTS architecture released in [19].

Method	Search dataset & Task	Top-1 acc.	FLOPs (M)	Params (M)
NAS-DARTS [19]	CIFAR-10 Supv.Cls	73.3	574	4.7
NAS-P-DARTS [5]	CIFAR-10 Supv.Cls	75.6	557	4.9
NAS-PC-DARTS [32]	CIFAR-10 Supv.Cls	74.9	586	5.3
NAS-PC-DARTS [32]	IN1K Supv.Cls	75.8	597	5.3
NAS-DARTS†	CIFAR-10 Supv.Cls	$74.9_{\pm0.08}$	538	4.7
NAS-DARTS	IN1K Supv.Cls	$76.3_{\pm0.06}$	590	5.3
UnNAS-DARTS	IN1K Rot	$75.8_{\pm0.18}$	558	5.1
UnNAS-DARTS	IN1K Color	$75.7_{\pm0.12}$	547	4.9
UnNAS-DARTS	IN1K Jigsaw	$75.9_{\pm0.15}$	567	5.2
NAS-DARTS	IN22K Supv.Cls	$75.9_{\pm0.09}$	585	5.2
UnNAS-DARTS	IN22K Rot	$75.7_{\pm0.23}$	549	5.0
UnNAS-DARTS	IN22K Color	$75.9_{\pm0.21}$	547	5.0
UnNAS-DARTS	IN22K Jigsaw	$75.9_{\pm0.31}$	559	5.1
NAS-DARTS	Cityscapes Supv.Seg	$75.8_{\pm0.13}$	566	5.1
UnNAS-DARTS	Cityscapes Rot	$75.9_{\pm0.19}$	554	5.1
UnNAS-DARTS	Cityscapes Color	$75.2_{\pm0.15}$	594	5.1
UnNAS-DARTS	Cityscapes Jigsaw	$75.5_{\pm0.06}$	566	5.0

For IN1K evaluation, we fix depth to 14 and adjust the width to have #FLOPs $\in [500,600]$M. Models are trained for 250 epochs with an auxiliary loss weighted by 0.4, batch size 1024 across 8 GPUs, cosine learning rate schedule [20] with initial value 0.5, and 5 epochs of warmup. For Cityscapes evaluation, we fix depth to 12 and adjust the width to have #Params $\in [9.5,10.5]$M. We train the network for 2700 epochs, with batch size 64 across 8 GPUs, cosine learning rate schedule with initial value 0.1. For both ImageNet and Cityscapes evaluations, we report the mean and standard deviation of 3 independent trainings of the same architecture. More implementation details are described in Appendix.

We note that under our definition of the UnNAS setup, the same dataset should not be used for both search and evaluation (because this scenario is unrealistic); We provide the IN1K→IN1K and Cityscapes→Cityscapes results purely as a reference. Those settings are analogous to the linear classifier probe for IN1K in conventional unsupervised learning research.

6.3 Results

In search-based experiments, the architectures are evaluated on both ImageNet classification, summarized in Table 1, and Cityscapes semantic segmentation,

Table 2. Cityscapes semantic segmentation results of the architectures searched by NAS and UnNAS algorithms. These are trained from scratch: there is no fine-tuning from ImageNet checkpoint. Rows in gray correspond to an illegitimate setup where the search dataset is the same as the evaluation dataset. † is our training result of the DARTS architecture released in [19].

Method	Search dataset & Task	mIoU	FLOPs (B)	Params (M)
NAS-DARTS†	CIFAR-10 Supv.Cls	$72.6_{\pm 0.55}$	121	9.6
NAS-DARTS	IN1K Supv.Cls	$73.6_{\pm 0.31}$	127	10.2
UnNAS-DARTS	IN1K Rot	$73.6_{\pm 0.29}$	129	10.4
UnNAS-DARTS	IN1K Color	$72.2_{\pm 0.56}$	122	9.7
UnNAS-DARTS	IN1K Jigsaw	$73.1_{\pm 0.17}$	129	10.4
NAS-DARTS	IN22K Supv.Cls	$72.4_{\pm 0.29}$	126	10.1
UnNAS-DARTS	IN22K Rot	$72.9_{\pm 0.23}$	128	10.3
UnNAS-DARTS	IN22K Color	$73.6_{\pm 0.41}$	128	10.3
UnNAS-DARTS	IN22K Jigsaw	$73.1_{\pm 0.59}$	129	10.4
NAS-DARTS	Cityscapes Supv.Seg	$72.4_{\pm 0.15}$	128	10.3
UnNAS-DARTS	Cityscapes Rot	$73.0_{\pm 0.25}$	128	10.3
UnNAS-DARTS	Cityscapes Color	$72.5_{\pm 0.31}$	122	9.5
UnNAS-DARTS	Cityscapes Jigsaw	$74.1_{\pm 0.39}$	128	10.2

summarized in Table 2. We provide visualization of all NAS-DARTS and UnNAS-DARTS cell architectures in Appendix.

UnNAS Architectures Perform Competitively to Supervised Counterparts. We begin by comparing NAS-DARTS and UnNAS-DARTS when they are performed on the same search dataset. This would correspond to every four consecutive rows in Table 1 and Table 2, grouped together by horizontal lines.

As discussed earlier, strictly speaking the IN1K→IN1K experiment is not valid under our definition of UnNAS. For reference, we gray out these results in Table 1. NAS-DARTS on IN1K dataset has the highest performance among our experiments, achieving a top-1 accuracy of 76.3%. However, the UnNAS algorithm variants with Rot, Color, Jigsaw objectives all perform very well (achieving 75.8%, 75.7% and 75.9% top-1 accuracy, respectively), closely approaching the results obtained by the supervised counterpart. This suggests it might be desirable to perform architecture search on the target dataset directly, as also observed in other work [3].

Two valid UnNAS settings include IN22K→IN1K and Cityscapes→IN1K for architecture search and evaluation. For IN22K→IN1K experiments, NAS and UnNAS results across the board are comparable. For Cityscapes→IN1K experiments, among the UnNAS architectures, Rot and Jigsaw perform well, once again achieving results comparable to the supervised search. However, there is a drop for UnNAS-DARTS search with Color objective (with a 75.2% top-1 accuracy). We hypothesize that this might be owing to the fact that the color distribution in Cityscapes images is not as diverse: the majority of the pixels are from *road* and *ground* categories of gray colors.

In general, the variances are higher for Cityscapes semantic segmentation (Table 2), but overall UnNAS-DARTS architectures still perform competitively to NAS-DARTS architectures, measured by mIoU. For the Cityscapes→Cityscapes experiment, we observe that searching with segmentation objective directly leads to inferior result (mean 72.4% mIoU), compared to the architectures searched for ImageNet classification tasks. This is different from what has been observed in Table 1. However, under this setting, our UnNAS algorithm, in particular the one with the Jigsaw objective shows very promising results (mean 74.1% mIoU). In fact, when the search dataset is IN22K or Cityscapes, all UnNAS-DARTS architectures perform *better* than the NAS-DARTS architecture. This is the opposite of what was observed in IN1K→IN1K. Results for Cityscapes→Cityscapes are grayed out for the same reason as before (invalid under UnNAS definition).

NAS and UnNAS Results Are Robust Across a Large Variety of Datasets and Tasks. The three search datasets that we consider are of different nature. For example, IN22K is 10 times larger than IN1K, and Cityscapes images have a markedly different distribution than those in ImageNet. In our experiments, NAS-DARTS/UnNAS-DARTS architectures searched on IN22K do not significantly outperform those searched on IN1K, meaning that they do not seem to be able to enjoy the benefit of having more abundant images. This reveals new opportunities in designing better algorithms to exploit bigger datasets for neural architecture search. For Cityscapes→IN1K experiments, it is interesting to see that after switching to a dataset with markedly distinct search images (urban street scenes), we are still able to observe decent performance. The same goes for the reverse direction IN1K/IN22K→Cityscapes, which implies that the search does not severely overfit to the images from the dataset.

In addition to this robustness to the *search dataset* distribution, NAS and UnNAS also exhibit robustness to *target dataset and task*. Classification on ImageNet and segmentation on Cityscapes are different in many ways, but among different combinations of search dataset and task (whether supervised or unsupervised), we do not observe a case where the same architecture performs well on one but poorly on the other.

UnNAS Outperforms Previous Methods. Finally, we compare our UnNAS-DARTS results against existing works. We first note that we are able to achieve a better baseline number with the NAS-DARTS architecture (searched with the CIFAR-10 proxy) compared to what was reported in [19]. This is mainly due to better hyper-parameter tuning we adopt from [5] for the *evaluation phase* model training. This baseline sets up a fair ground for all the UnNAS experiments; we use the same evaluation phase hyper-parameters across different settings.

On ImageNet, our UnNAS-DARTS architectures can comfortably outperform this baseline by up to 1% classification accuracy. In fact, the extremely competitive UnNAS-DARTS results also outperform the previous best result (75.8%) on this search space, achieved with a more sophisticated NAS algorithm [32].

On Cityscapes, there have not been many works that use DARTS variants as the backbone. The closest is Auto-DeepLab [17], but we use a lighter architecture (in that we do not have the Decoder) and shorter training iterations, so the results are not directly comparable. Nonetheless, according to our evaluation, the UnNAS architectures perform favorably against the DARTS architecture released in [19] (discovered with the CIFAR-10 proxy). The best UnNAS-DARTS variant (Cityscapes Jigsaw) achieves 74.1% mIoU, which outperforms this baseline by 1.5% on average. Overall, our experiments demonstrate that exploring neural architecture search with unsupervised/self-supervised objectives to improve target task performance might be a fruitful direction.

Outperforming previous methods was far from the original goal of our study. Nonetheless, the promising results of UnNAS suggest that in addition to developing new *algorithms* and finding new *tasks*, the role of *data* (in our case, more/larger images) and *paradigm* (in our case, no human annotations) is also worth attention in future work on neural architecture search.

7   Discussion

In this paper, we challenge the common practice in neural architecture search and ask the question: do we really need labels to successfully perform NAS? We approach this question with two sets of experiments. In sample-based experiments, we discover the phenomenon that the architecture rankings produced with and without labels are highly correlated. In search-based experiments, we show that the architectures learned without accessing labels perform competitively, not only relative to their supervised counterpart, but also in terms of absolute performance. In both experiments, the observations are consistent and robust across various datasets, tasks, and/or search spaces. Overall, the findings in this paper indicate that labels are *not* necessary for neural architecture search.

How to learn and transfer useful representations to subsequent tasks in an unsupervised fashion has been a research topic of extensive interest, but the discovery of neural network architectures has been driven solely by supervised tasks. As a result, current NAS products or AutoML APIs typically have the strict prerequisite for users to *"put together a training dataset of labeled images"* [1]. An immediate implication of our study is that the job of the user could potentially be made easier by dropping the labeling effort. In this sense, UnNAS could be especially beneficial to the many applications where data constantly comes in at large volume but labeling is costly.

At the same time, we should still ask: if not labels, then what factors are needed to reveal a good architecture? A meaningful unsupervised task seems to be important, though the several pretext tasks considered in our paper do not exhibit significant difference in either of the two experiments. In the future we plan to investigate even more and even simpler unsupervised tasks. Another possibility is that the architecture quality is mainly decided by the image statistics, and since the datasets that we consider are all natural images, the correlations are high and the results are comparable. This hypothesis would also suggest an interesting, alternative direction: that instead of performing NAS again and again for every specific labeled task, it may be more sensible to perform NAS once on large amounts of unlabeled images that capture the image distribution.

References

1. Google AutoML Vision API Tutorial (2019). https://cloud.google.com/vision/automl/docs/tutorial. Accessed 14 Nov 2019
2. Bergstra, J., Bengio, Y.: Random search for hyper-parameter optimization. J. Mach. Learn. Res. **13**(1), 281–305 (2012)
3. Cai, H., Zhu, L., Han, S.: ProxylessNAS: direct neural architecture search on target task and hardware. In: ICLR (2019)
4. Chen, L.C., et al.: Searching for efficient multi-scale architectures for dense image prediction. In: NeurIPS (2018)
5. Chen, X., Xie, L., Wu, J., Tian, Q.: Progressive differentiable architecture search: bridging the depth gap between search and evaluation. In: ICCV (2019)
6. Cordts, M., et al.: The Cityscapes dataset for semantic urban scene understanding. In: CVPR (2016)
7. Deng, J., Dong, W., Socher, R., Li, L.J., Li, K., Fei-Fei, L.: ImageNet: a large-scale hierarchical image database. In: CVPR (2009)
8. Doersch, C., Gupta, A., Efros, A.A.: Unsupervised visual representation learning by context prediction. In: ICCV (2015)
9. Dosovitskiy, A., Springenberg, J.T., Riedmiller, M., Brox, T.: Discriminative unsupervised feature learning with convolutional neural networks. In: NeurIPS (2014)
10. Ghiasi, G., Lin, T.Y., Le, Q.V.: NAS-FPN: learning scalable feature pyramid architecture for object detection. In: CVPR (2019)
11. Gidaris, S., Singh, P., Komodakis, N.: Unsupervised representation learning by predicting image rotations. In: ICLR (2018)
12. He, K., Fan, H., Wu, Y., Xie, S., Girshick, R.: Momentum contrast for unsupervised visual representation learning. In: CVPR (2020)
13. Huber, P.J.: Robust statistics. In: Lovric, M. (ed.) International Encyclopedia of Statistical Science, pp. 1248–1251. Springer, Heidelberg (2011). https://doi.org/10.1007/978-3-642-04898-2_594
14. Kolesnikov, A., Zhai, X., Beyer, L.: Revisiting self-supervised visual representation learning. In: CVPR (2019)
15. Kornblith, S., Shlens, J., Le, Q.V.: Do better ImageNet models transfer better? In: CVPR (2019)
16. Krizhevsky, A.: Learning multiple layers of features from tiny images. Technical report, Citeseer (2009)
17. Liu, C., et al.: Auto-DeepLab: hierarchical neural architecture search for semantic image segmentation. In: CVPR (2019)
18. Liu, C., et al.: Progressive neural architecture search. In: Ferrari, V., Hebert, M., Sminchisescu, C., Weiss, Y. (eds.) ECCV 2018. LNCS, vol. 11205, pp. 19–35. Springer, Cham (2018). https://doi.org/10.1007/978-3-030-01246-5_2

19. Liu, H., Simonyan, K., Yang, Y.: Darts: Differentiable architecture search. In: ICLR (2019)
20. Loshchilov, I., Hutter, F.: SGDR: stochastic gradient descent with warm restarts. In: ICLR (2017)
21. Noroozi, M., Favaro, P.: Unsupervised learning of visual representations by solving jigsaw puzzles. In: Leibe, B., Matas, J., Sebe, N., Welling, M. (eds.) ECCV 2016. LNCS, vol. 9910, pp. 69–84. Springer, Cham (2016). https://doi.org/10.1007/978-3-319-46466-4_5
22. van den Oord, A., Li, Y., Vinyals, O.: Representation learning with contrastive predictive coding. arXiv:1807.03748 (2018)
23. Pham, H., Guan, M.Y., Zoph, B., Le, Q.V., Dean, J.: Efficient neural architecture search via parameter sharing. In: ICML (2018)
24. Radosavovic, I., Johnson, J., Xie, S., Lo, W.Y., Dollár, P.: On network design spaces for visual recognition. In: ICCV (2019)
25. Real, E., Aggarwal, A., Huang, Y., Le, Q.V.: Regularized evolution for image classifier architecture search. In: AAAI (2019)
26. Real, E., et al.: Large-scale evolution of image classifiers. In: ICML (2017)
27. So, D.R., Liang, C., Le, Q.V.: The evolved transformer. In: ICML (2019)
28. Spearman, C.: The proof and measurement of association between two things. Am. J. Psychol. (1904)
29. Tian, Y., Krishnan, D., Isola, P.: Contrastive multiview coding. arXiv:1906.05849 (2019)
30. Wang, X., Gupta, A.: Unsupervised learning of visual representations using videos. In: ICCV (2015)
31. Wu, Z., Xiong, Y., Yu, S., Lin, D.: Unsupervised feature learning via non-parametric instance discrimination. In: CVPR (2018)
32. Xu, Y., et al..: Pc-darts: Partial channel connections for memory-efficient differentiable architecture search. In: ICLR (2020)
33. Ying, C., Klein, A., Christiansen, E., Real, E., Murphy, K., Hutter, F.: NAS-Bench-101: Towards reproducible neural architecture search. In: ICML (2019)
34. Zhang, R., Isola, P., Efros, A.A.: Colorful image colorization. In: Leibe, B., Matas, J., Sebe, N., Welling, M. (eds.) ECCV 2016. LNCS, vol. 9907, pp. 649–666. Springer, Cham (2016). https://doi.org/10.1007/978-3-319-46487-9_40
35. Zoph, B., Le, Q.V.: Neural architecture search with reinforcement learning. In: ICLR (2017)
36. Zoph, B., Vasudevan, V., Shlens, J., Le, Q.V.: Learning transferable architectures for scalable image recognition. In: CVPR (2018)

Author Index

Printed in the United States
By Bookmasters